Transportation Engineering

Theory, Practice, and Modeling

Dušan Teodorović
University of Belgrade

Milan Janić
Delft University of Technology

AMSTERDAM • BOSTON • HEIDELBERG • LONDON
NEW YORK • OXFORD • PARIS • SAN DIEGO
SAN FRANCISCO • SINGAPORE • SYDNEY • TOKYO
Butterworth-Heinemann is an imprint of Elsevier

Butterworth-Heinemann is an imprint of Elsevier
The Boulevard, Langford Lane, Kidlington, Oxford OX5 1GB, United Kingdom
50 Hampshire Street, 5th Floor, Cambridge, MA 02139, United States

Library of Congress Cataloging-in-Publication Data
A catalog record for this book is available from the Library of Congress

British Library Cataloguing-in-Publication Data
A catalogue record for this book is available from the British Library

ISBN: 978-0-12-803818-5

For information on all Butterworth Heinemann publications visit our website at https://www.elsevier.com/

Publisher: Joe Hayton
Acquisition Editor: Ken McCombs
Editorial Project Manager: Peter Jardim
Production Project Manager: Vijayaraj Purushothaman
Cover Designer: Victoria Pearson

Typeset by SPi Global, India

To my wife, children, and grandchildren
Dušan Teodorović
To my wife and son
Milan Janić

Contents

About the Authors xvii
Foreword xix
Preface xxi
Acknowledgments xxiii

CHAPTER 1 Introduction **1**
References 3

CHAPTER 2 Transportation Systems **5**
2.1 Background 5
2.2 History of Transportation 6
2.3 Transportation Sector and Transportation Modes 15
2.3.1 Components of Transportation Modes 16
2.3.2 Structure of Transportation Modes 17
2.3.3 Technologies of Transport Modes 17
2.3.4 Relationships Between Transport Modes 20
2.4 Characteristics of Transport Modes and Their Systems 22
2.4.1 Introduction 22
2.4.2 Urban and Sub/Urban Road and Rail-Based Transit Systems for Passengers 22
2.4.3 Urban and Sub/Urban Transport Systems for Freight Shipments 29
2.4.4 Interurban Road Transport Systems 30
2.4.5 Interurban Rail Transport Systems 34
2.4.6 Inland Waterways and Sea Shipping Systems for Cargo Shipments 44
2.4.7 Air Transport System 51
2.5 Transportation Systems Topics: Planning, Control, Congestion, Safety, and Environment Protection 56
2.6 Problems 59
References 59
Websites 62

CHAPTER 3 Traffic and Transportation Analysis Techniques **63**
3.1 Object Motion and Time-Space Diagrams 63
3.2 Transportation Networks Basics 69
3.3 Optimal Paths in Transportation Networks 72
3.3.1 Finding Shortest Path in a Transportation Network 72
3.3.2 Dijkstra's Algorithm 73
3.3.3 Shortest Paths Between All Pairs of Nodes 77

3.4 Mathematical Programming Applications in Traffic and Transportation ... 81
3.4.1 Linear Programming in Traffic and Transportation ... 82
3.4.2 Integer Programming ... 88
3.4.3 Dimensionality of the Traffic and Transportation Engineering Problems ... 91
3.4.4 Complexity of Algorithms ... 92
3.5 Probability Theory and Traffic Phenomena ... 93
3.5.1 Probability Theory Basics ... 94
3.5.2 Random Variables and Probability Distributions ... 96
3.6 Queueing in Transportation Systems ... 106
3.6.1 Elements of Queueing Systems ... 107
3.6.2 *D*/*D*/1 Queueing ... 109
3.6.3 Little's Law ... 109
3.6.4 *M*/*M*/1 Queueing ... 114
3.6.5 *M*/*M*/s Queueing ... 115
3.6.6 Queueing Theory and Investments in Transportation Facilities Expansion ... 119
3.7 Simulation ... 121
3.7.1 The Monte Carlo Simulation Method ... 121
3.8 MultiAttribute Decision Making Methods ... 124
3.8.1 Attribute Weights ... 125
3.8.2 Minimax Method ... 126
3.8.3 Maximax Method ... 127
3.8.4 Simple Additive Weighting Method ... 127
3.8.5 TOPSIS ... 128
3.9 Data Envelopment Analysis (DEA) ... 130
3.9.1 Ratios ... 130
3.9.2 DEA Basics ... 132
3.10 Computational Intelligence Techniques ... 135
3.10.1 The Concept of Fuzzy Sets ... 136
3.10.2 The Fuzzy Sets Basics ... 138
3.10.3 Basic Elements of Fuzzy Systems ... 140
3.10.4 Artificial Neural Networks ... 149
3.11 Problems ... 155
References ... 160

CHAPTER 4 Traffic Flow Theory ... 163
4.1 Traffic Flow Phenomenon ... 163
4.2 Measurements of the Basic Flow Variables ... 164
4.3 Vehicle Headways and Flow ... 165

4.4 Poisson Distribution of the Number of Arrivals and the Exponential Distribution of Headways 166
4.5 Normal Distribution and Pearson Type III Distribution of Headway 168
4.5.1 Speeds 169
4.6 Speed-Density Relationship 171
4.7 Flow-Density Relationship 173
4.8 Speed-Flow Relationship 175
4.9 Fundamental Diagram of Traffic Flow 182
4.10 Shock Waves 182
4.11 Micro-Simulation Traffic Models 186
4.12 Car Following Models 187
4.12.1 The Car-Following Model Based on Fuzzy Inference Rules 188
4.13 Network Flow Diagram 189
4.13.1 Link-Based Measurements 190
4.13.2 Generalized Traffic Flow Variables 190
4.13.3 Trajectory-Based Measurements 192
4.14 Problems 193
References 195

CHAPTER 5 Capacity and Level of Service 197
5.1 Introduction 197
5.2 Highway Capacity and Level of Service 198
5.2.1 Highway Capacity and Traffic Demand Variations 199
5.2.2 Freeways 201
5.2.3 Methodology for the Capacity Analysis, LOS, and the Lane Requirements 205
5.2.4 The Number of Lanes Required to Deliver the Target LOS 209
5.3 "Ultimate" and "Practical" Capacity of Bus Stations 212
5.4 Rail Inter-Urban Transport Systems 215
5.4.1 General 215
5.5 Inland Waterway Freight/Cargo Transportation System 236
5.5.1 General 236
5.5.2 Classification 237
5.5.3 Infrastructure Network 239
5.5.4 Transport Service Network 250
5.6 Maritime Freight/Cargo Transport System 253
5.6.1 General 253
5.6.2 Ports 253
5.6.3 Shipping Lines 263

5.7 Air Transport System ... 269
5.7.1 General ... 269
5.7.2 Airports ... 269
5.7.3 Air Traffic Control ... 282
5.8 Problems ... 287
References ... 290
Websites ... 292

CHAPTER 6 Traffic Control ... 293
6.1 Introduction ... 293
6.2 Traffic Control at Signalized Intersections ... 294
6.2.1 Fixed-Time Control at the Isolated Intersection ... 296
6.2.2 Vehicle Delays at Signalized Intersections ... 300
6.2.3 The Determination of Timing for Fixed-Time Signals ... 306
6.2.4 Signal Phasing Selection ... 307
6.2.5 Volume Adjustment (Calculation of Equivalent Straight-Through Passenger Cars) ... 308
6.2.6 Critical Lane Volumes Selection ... 310
6.2.7 Change Interval Calculation ... 311
6.2.8 Cycle Length Calculation ... 312
6.2.9 Green Time Allocation ... 314
6.2.10 Pedestrian Crossing Time Check ... 314
6.2.11 Actuated Signal Control ... 315
6.3 Traffic Control for Arterial Streets ... 317
6.3.1 Adaptive Control Strategies ... 319
6.4 Area-Wide Traffic Control Systems ... 320
6.5 Traffic Control Signal Needs Studies ... 324
6.6 Intelligent Transportation Systems ... 325
6.6.1 ITS Architecture ... 326
6.6.2 ITS User Services ... 327
6.6.3 Autonomous Vehicles ... 329
6.6.4 Autonomous Intersection Management ... 330
6.7 Freeway Traffic Control ... 332
6.7.1 Freeway Traffic Control Measures ... 333
6.7.2 Ramp Metering ... 333
6.7.3 Driver Information and Guidance Systems ... 336
6.8 Transportation Demand Management ... 337
6.8.1 Ridesharing (Carpooling) ... 338
6.8.2 Remote Parking and Park and Ride ... 339
6.8.3 Improved Walkability ... 339
6.8.4 Telework ... 339

6.8.5 Congestion Pricing 339
6.8.6 Congestion Charges 341
6.9 HOV Facilities 344
6.10 Highway Space Inventory Control System 344
6.11 Auctions 345
6.12 Rail Traffic Control 345
6.12.1 Background 345
6.12.2 Infrastructure 346
6.12.3 Supportive Facilities and Equipment 347
6.12.4 The Workload and Capacity of Train Dispatcher(s) 359
6.13 Air Traffic Control 362
6.13.1 Background 362
6.13.2 Infrastructure 363
6.13.3 Supportive Facilities and Equipment 369
6.13.4 The ATC Separation Rules and Procedures 370
6.13.5 The ATC Staff—Controller and Pilots 371
6.13.6 Automation 373
6.13.7 The Workload and Capacity of ATC Controller(s) 376
6.14 Problems 379
References 380
Websites 384

CHAPTER 7 Public Transportation Systems 387
7.1 Introduction 387
7.2 Number of Transported Passengers Versus Number of Served Vehicles 388
7.3 Urban Public Transit 390
7.3.1 Road-Based Urban Transit Systems 393
7.3.2 Rail-Based Urban Transit Systems 395
7.3.3 Complementarity of the Systems 397
7.4 Infrastructure of Urban Transit Systems 398
7.4.1 Stops/Stations in Urban Transit Systems 399
7.4.2 Urban Transit Systems Links and Indicators of Network Size 401
7.5 Public Transportation Availability 405
7.6 Passenger Flows in Public Transportation 407
7.7 Passenger Flows Along a Transit Line 408
7.8 Service Frequency and Headways 410
7.8.1 The Maximum Service Frequency 412
7.8.2 Passenger Waiting Time 413
7.8.3 Headway Determination by "Square Root Formula" 413
7.8.4 Headway Determination by Maximum Load Method 415

7.9 Timetable ... 416
7.10 Transit Line Capacity ... 418
7.10.1 Transit Line Capacity Utilization ... 420
7.11 The Performances of the Urban Transit Network ... 423
7.12 Public Transit Network Types ... 429
7.13 The Public Transit Network Design ... 431
7.13.1 Simple Greedy Algorithm for Public Transit Network Design ... 434
7.14 Service Frequencies Determination in Transit Network ... 438
7.15 Vehicle Scheduling in Public Transit ... 441
7.16 Crew Scheduling in Public Transit ... 444
7.17 Disruption Management in Public Transit ... 445
7.18 Public Transit Planning Process ... 447
7.19 Demand-Responsive Transportation Systems ... 448
7.19.1 Type of Routing and Scheduling in DRT ... 448
7.19.2 Dial-a-Ride ... 450
7.20 Interurban Road Transport Systems ... 453
7.20.1 Introduction ... 453
7.20.2 Service Networks in an Interurban Road Transportation ... 453
7.21 Air Transportation ... 456
7.21.1 Air Transportation Demand ... 457
7.21.2 Airline Supply and Airline Capacity ... 459
7.22 Air Transportation Networks ... 463
7.23 Flight Frequencies ... 469
7.23.1 Flight Frequency Satisfying Demand ... 470
7.23.2 Flight Frequency Gaining Market Share ... 470
7.23.3 Flight Frequency Minimizing the Total Route Cost ... 472
7.24 Airline Transport Work and Productivity ... 475
7.25 Fleet Size ... 478
7.26 Level of Service ... 479
7.27 Airline Scheduling ... 481
7.28 Airline Schedule Planning Process ... 483
7.29 Airline Revenue Management ... 486
7.30 Problems ... 489
References ... 490

CHAPTER 8 Transportation Demand Analysis ... 495
8.1 Introduction ... 495
8.2 Transportation Demand and Transportation Supply ... 496
8.3 Transportation Demand Modeling ... 497

8.4 Transportation Demand Forecasting Techniques 499
8.4.1 Time Series Models 501
8.4.2 Trend Projection 504
8.5 Four-Step Planning Procedure 509
8.5.1 Trip Generation 511
8.5.2 Trip Distribution 513
8.5.3 Gravity Model 514
8.5.4 Modal Split 518
8.5.5 Route Choice and Traffic Assignment 521
8.6 User Equilibrium and System Optimum 523
8.6.1 Formulation of the User Equilibrium Problem 527
8.7 Heuristic Algorithms for Finding User-Equilibrium Flow Pattern 534
8.7.1 Capacity Restraint Algorithm 534
8.7.2 FHWA Algorithm 536
8.7.3 Incremental Assignment Algorithm 537
8.8 System Optimal Route Choice 540
8.9 Price of Anarchy 542
8.10 Braess Paradox and Transportation Capacity Expansions 543
8.11 Dynamic Traffic Assignment 546
8.12 Transportation Demand Analysis Based on Discrete Choice Models 547
8.13 Logit Model 549
8.13.1 Independence of Irrelevant Alternatives Property 553
8.13.2 Logit Model Estimation 554
8.14 Application of the Computational Intelligence Techniques for the Prediction of Travel Demand 557
8.15 Activity-Based Travel Demand Models 559
8.15.1 Basic Characteristics of the Activity-Based Travel Models 560
8.16 Problems 563
References 565

CHAPTER 9 Freight Transportation and Logistics 569
9.1 Logistics Systems Basics 569
9.1.1 Reverse Logistics 572
9.2 Road Freight Transport Infrastructure 573
9.2.1 "Ultimate" and "Practical" Capacity and Service Level of Road Truck Roads 574
9.2.2 "Ultimate" and "Practical" Capacity of Road Freight Terminals and Their Level-of-Service 575
9.3 Service Networks of the Road Freight Transport Operators 577
9.3.1 Capacities and Service Level of the Road Freight Transport Service Networks 577

9.4 City Logistics ... 581
9.4.1 Urban Freight Transport Basics ... 582
9.4.2 Urban Freight Distribution Systems ... 584
9.5 Basics of Location Theory ... 595
9.5.1 Location Problems Classification ... 596
9.5.2 Measuring Distances Between Facilities and Demand-Generating Nodes ... 597
9.5.3 The Location Set Covering Problem ... 598
9.5.4 Maximal Covering Location Problem ... 603
9.5.5 Medians ... 604
9.5.6 Hub Location ... 610
9.6 Vehicle Routing and Scheduling ... 611
9.6.1 VRPs Types ... 612
9.6.2 Vehicle Routing and Scheduling Problems Complexity ... 614
9.6.3 Traveling Salesman Problem ... 614
9.6.4 Vehicle Routing Problem ... 616
9.6.5 Clark-Wright's "Savings" Algorithm for the VRP ... 618
9.6.6 Sweep Algorithm for the VRP ... 624
9.7 Problems ... 630
References ... 632
Website ... 634

CHAPTER 10 Transport Economics ... 635
10.1 Introduction ... 635
10.2 Definition of the Main Terms ... 637
10.2.1 Transport Sector/Industry ... 637
10.2.2 Fixed and Variable Costs ... 637
10.2.3 Economies of Scale and Economies of Scope ... 638
10.2.4 The Cost Function and Revenues ... 639
10.2.5 Relationship Between Demand and Supply ... 642
10.3 Transportation Projects Evaluation ... 643
10.4 Cost-Benefit Analysis ... 644
10.5 Infrastructure Cost ... 647
10.5.1 Urban Mass Transit Systems ... 647
10.5.2 Road ... 653
10.5.3 Rail ... 654
10.5.4 Inland Waterways ... 655
10.5.5 Ports ... 657
10.5.6 Airports ... 659

10.6 Operating Costs and Revenues 660
10.6.1 Individual Cars 660
10.6.2 Urban Mass Transit Systems 663
10.6.3 Interurban Mass Transit Systems 671
10.6.4 Inland Waterways Cargo Transport 681
10.6.5 Maritime Cargo Transport 684
10.6.6 Air 695
10.6.7 Intermodal—Rail/Road Freight Transport 706
References 714
Websites 717

CHAPTER 11 Transportation, Environment, and Society 719
11.1 Introduction 719
11.2 Categorization and Modeling Impacts 720
11.2.1 Congestion 720
11.2.2 Noise 721
11.2.3 Traffic Incidents/Accidents (Safety) 723
11.2.4 Energy/Fuel Consumption and Emissions of GHG 724
11.2.5 Land Use 732
11.2.6 Waste 733
11.3 Road-Based Systems 733
11.3.1 Congestion 733
11.3.2 Noise 737
11.3.3 Traffic Accidents/Incidents (Safety) 744
11.3.4 Energy/Fuel Consumption and Emissions of GHG 748
11.3.5 Land Use 760
11.4 Rail-Based Systems 762
11.4.1 Congestion 762
11.4.2 Noise 763
11.4.3 Traffic Accidents/Incidents (Safety) 770
11.4.4 Energy/Fuel Consumption and Emissions of GHG 780
11.4.5 Land Use 789
11.5 Water-Based Systems 791
11.5.1 Congestion 791
11.5.2 Noise 792
11.5.3 Traffic Accidents/Incidents (Safety) 793
11.5.4 Energy/Fuel Consumption and Emissions of GHG 797
11.5.5 Land Use 812
11.5.6 Waste 814

11.6 Air-Based Systems ... 816
11.6.1 Congestion ... 816
11.6.2 Noise ... 820
11.6.3 Traffic Accidents and Incidents (Safety) ... 826
11.6.4 Energy/Fuel Consumption and Emissions of GHG ... 829
11.6.5 Land Use ... 837
11.7 Costs of Impacts—Externalities ... 840
11.7.1 Definition ... 840
11.7.2 Some Modeling ... 842
11.7.3 Some Estimation/Quantification ... 848
References ... 851
Website ... 857

Index ... 859

About the Authors

Dr. Dušan Teodorović is Professor at the Faculty of Transport and Traffic Engineering, University of Belgrade, Serbia. He has been elected member of the *Serbian Academy of Sciences and Arts and the European Academy of Sciences and Arts*. Dr. Teodorović has been Honorable Senator of the University of Ljubljana, Slovenia since 2011. He was Vice Rector of the University of Belgrade. He has worldwide academic experience with formal appointments at Virginia Tech, United States. Professor Teodorović was visiting professor and visiting scholar at many universities in Europe, the United States, and Asia. Since 2006, he has been Professor Emeritus of Virginia Tech, United States. Dr. Teodorović has authored or coauthored numerous research publications in peer-reviewed international journals, book chapters, and conference proceedings. He has published the books *Airline Operations Research* and *Traffic Control and Transportation Planning*; *A Fuzzy Sets and Neural Networks Approach*. Professor Teodorović is the Editor of the *Routledge Handbook of Transportation*. His primary research interests are in operations research and computational intelligence applications in transportation engineering.

Dr. Milan Janić is a Senior Researcher at the Faculty of Civil Engineering and Geosciences (Department of Transport & Planning) of Delft University of Technology, The Netherlands. He is also Research Professor at the University of Belgrade, Serbia. Previously, he was a Leader of the Research Program and Senior Researcher at Manchester Metropolitan University, Loughborough University, and the Institute of Transport of the Slovenian Railways. Dr. Janić has been involved in many research and planning projects on both a national and international scale for almost 30 years. He has also published numerous papers in peer-reviewed journals. In addition to contributing to many edited books, he has published the following books: *Advanced Transport Systems*: *Analysis*, *Modelling*, *and Evaluation of Performances*, *Greening Airports*: *Advanced Technology and Operations*, *Airport Analysis*, *Planning and Design*: *Demand*, *Capacity and Congestion*, *The Sustainability of Air Transportation*: *A Quantitative Analysis and Assessment*, and *Air Transport System Analysis and Modelling*: *Capacity*, *Quality of Services and Economics*.

Foreword

It is with great pleasure that I am writing this foreword to *Transportation Engineering: Multi-Modal Transportation Networks, Theory, Practice and Modeling* by Dušan Teodorović and Milan Janić. I have known the two authors for a long time—close to 40 years by now. Both have enjoyed highly productive and distinguished careers at major universities, not only in their native lands of Serbia and Slovenia, but also in the United States, the United Kingdom, and the Netherlands. Their experience as educators and expertise as researchers are reflected in this valuable book. They have witnessed first-hand and have contributed themselves to the development and evolution of the transportation field during their professional lifetimes.

Since 2001, I have been co-teaching with various colleagues at MIT a course called "Transportation Systems Analysis: Supply and Performance," which introduces fundamental quantitative models and their application to first-year graduate students in our Transportation Program. The course attempts to provide an overview that spans all modes of transport and cuts across the disciplines of traffic engineering, transportation science, transportation economics, and operations research. While developing the course and in the years that have followed, I have been struck by the absence of a textbook that covers this material in an integrated fashion and at an adequate and consistent mathematical level. The available books tend to be focused primarily on a single mode and, usually, on only specific aspects of the mode, such as urban traffic, highway traffic, public transit, or air traffic management. Moreover, their mathematical level is all too often either elementary—so that the reader cannot appreciate the power of the existing analytical tools—or uneven, alternating between very basic and too advanced. As a result, my colleagues and I have relied on our own course notes, supplemented by selected readings of landmark papers or book chapters.

This new book therefore constitutes a most welcome addition to the transportation literature, as it addresses the central issue I identified in the previous paragraph—lack of integration across transportation modes, disciplines, and methodological approaches. Taking advantage of their complementary areas of expertise, the authors have managed to write a textbook that truly spans all modes. Having devoted a large part of my own research to air transportation, I am particularly pleased to see that airports and air traffic have been given their "fair share" of attention. Air transport has become the dominant mode of long-haul passenger transportation on a global scale and it is essential that transportation professionals and students become familiar with some of the most important models that describe air traffic movement and processes. An added benefit, in this respect, is that air traffic models typically treat vehicles (the airplanes) as discrete objects. Thus, these models expose the reader to a set of methodologies, such as integer programming, that focus on individual, "atomistic" entities. Air traffic models therefore offer a perspective different from the continuous flow models that are generally used to study road and highway traffic.

Another welcome feature of the book is that it does not draw a priori any dividing lines between modes of transport, but is structured around a set of major common themes, such as "traffic flow theory," "capacity and level of service," "traffic control," and "transportation planning" and "environmental impacts." This makes it possible to integrate the material better and facilitates the highlighting of the similarities and differences among the methodologies and modeling approaches used for each mode.

The major common themes are themselves arranged in a logical sequence, so that the reader can progress from an understanding of the "physics" of the *individual elements* of transportation networks—e.g., flows in a road segment, traffic light settings at an isolated intersection, capacity of a runway—to studying the performance of the *networks* as a whole. In this respect, the authors take great pains to emphasize the importance of considering how the pieces fit together into *systems*. The book is definitely "network-centric" and discusses the design of transportation networks, including multi-modal ones, their optimization at the planning level and the operations level, their control in the long run ("demand management") and in real time, and the economics and environmental impacts of the different modes.

The centerpiece of the book lies, I believe, in the presentation of fundamental *models*, i.e., mathematical abstractions and constructs that are used to represent essential features of actual transportation systems. In the transportation field, we are fortunate at this point to have an abundance of such models. As the authors make clear, the appropriate model to use depends on the questions that one wishes to answer. Any given transportation system can be modeled in several different ways, according to the question at hand. One can have models that are: deterministic or probabilistic/stochastic, depending on the extent to which one wishes to account for uncertainty; static or dynamic over time; macroscopic or microscopic depending on the level of detail one wishes to capture; and analytically based or simulation-based depending on the methodology used—with numerous additional subcategories when it comes to analytical models. The reader will find all these types of models in different parts of this book.

During the second half of the 20th century, studies in transportation engineering and transportation science made huge strides toward developing a knowledge base of methodologies and models for studying, quantifying, and predicting the behavior of transportation systems. As a result, we now have a much better understanding of how to plan, design, manage, and operate transportation networks. The first two decades of the 21st century have added to this arsenal of tools the ability to collect and process, often in real time, enormous amounts of data about the state of transportation networks. However, to take full advantage of this newfound capability, one must be familiar with this knowledge base of methodologies and models. This book, I believe, makes an important contribution toward providing this crucial background for students and professionals alike, including a much-needed historical perspective. I am quite certain that the readers will agree with this assessment.

Amedeo R. Odoni

T. Wilson Chair Professor Emeritus of Aeronautics and Astronautics

Professor Emeritus of Civil and Environmental Engineering

Massachusetts Institute of Technology

March 2016

When once you have tasted flight, you will forever walk the earth with your eyes turned skyward, for there you have been, and there you will always long to return.
Leonardo da Vinci

Preface

Joint writing a book is not a simple task. It is similar to playing the piano with four hands. We met, for the first time, in 1974, at the University of Belgrade, Serbia, when the first author was a teaching assistant, and another undergraduate student. We had a different professional careers in the countries in which we lived. During occasional meetings, we have been strengthened in our belief that we have similar views on important transportation engineering issues. During our professional careers, we were professors, and visiting scholars at the universities in Europe, the United States, and Asia.

For many years, we have been studying fascinating traffic phenomena. We thought, at one point, that it was time to write our book about transportation engineering fundamentals. We strongly believe that each new book in a certain area opens up new views to the reader. We have tried in this book to touch on urban and road transportation, air transportation, railways, inland water transportation, and logistics. This book begins with the story about the earliest discovered paths, made by animals, and adapted by humans, found near *Jericho*, and arrives at the issues related to the autonomous car (self-driving car, driverless car, robotic car) that are now around us. The following is a brief description of the book chapters.

Chapter 2 introduces the reader to the field of transportation engineering. The chapter covers the history of transportation, offers basic definitions and classification of the transportation systems, and describes the most important transportation systems issues: planning, control, congestion, safety, and environment protection.

Chapter 3 deals with traffic and transportation analysis techniques. This chapter covers object motion and time-space diagrams, transportation networks basics, mathematical programming applications in traffic and transportation, the relationship between the probability theory and traffic phenomena, queueing theory, simulation techniques, and computational intelligence techniques.

Chapter 4 covers traffic flow theory basics. The chapter describes measurements of the basic flow variables, speed-density relationship, flow-density relationship, speed-flow relationship, fundamental diagram of traffic flow, micro-simulation traffic models, car following models, and network flow diagram.

Chapter 5 involves capacity and level of service of different transportation modes (highways, urban transit systems, urban freight transport systems, rail interurban transport systems, inland waterway freight/cargo transport systems, maritime freight/cargo transport systems, air transport systems, and air traffic control systems).

Chapter 6 describes traffic control techniques related to the road, rail, and air traffic control systems. The chapter covers a variety of traffic control measures, methods, and strategies that should be implemented in order to use the existing transportation infrastructure optimally.

Chapter 7 deals with public transportation systems. The chapter describes public transportation basics, public transit network types, public transit network design, vehicle scheduling in public transit, public transit planning process, demand-responsive transportation systems, and air transportation basics.

Chapter 8 covers transportation planning methods and techniques. The chapter describes transportation demand modeling, the four step planning procedure (trip generation, trip distribution, modal split, and route choice), Wardrop's principles and traffic networks equilibrium conditions, the Braess paradox and transportation capacity expansions, dynamic traffic assignment problems, transportation demand analysis based on discrete choice models, and activity-based travel demand models.

Chapter 9 describes relationship between logistic systems and transportation. The chapter elaborates principles and techniques of city logistics (distribution of goods from warehouses to shops and supermarkets, emergency services, waste collection, street cleaning and sweeping in one city). This chapter covers basics of location theory, and vehicle routing and scheduling techniques.

Chapter 10 involves basic transportation economics concepts (fixed and variable costs in transportation, economies of scale, relationship between demand and supply, infrastructure costs, etc.). Transportation economics concepts and methods are described for every transportation mode.

Chapter 11 deals with the impacts of transportation systems on society and the environment. There is continuous construction, expansion, and maintenance of the transportation systems in the world. Different transportation systems have enormous impacts on energy use, air, water and soil quality, noise level, land use, and nature conservation. Chapter 11 analyzes the direct impacts of transportation systems on the society and environment, and their costs/externalities. The major impacts on the society taken into account include congestion, noise, and traffic incidents/accidents (ie, safety). The main impacts on the environment considered involve the energy/fuel consumption and related emissions of GHG (Green House Gases), land use, and waste. Considering the importance that transportation has on human society and environment, Chapter 11 contains material that exceeds the boundaries of university textbooks. The material given in Chapter 11 is related to the ideas of changing regulation and standards in the transportation sector, behavioral change of the participants in transportation (more pedestrian and bicycle transportation), as well as appearance of the new vehicle types and new fuel technologies.

The book is planned for use by students at the senior undergraduate level, and at the graduate level, interested in transportation engineering, civil engineering, city and regional planning, urban geography, economics, public administration, and management science. We believe that the book could also be useful for self-study, as well as for professionals working in the area of transportation.

The journey of writing this book was more than exciting. We tried, all the time, to find a balance between the explanation of the traffic phenomena, mathematical rigor, and real-world examples. We believe that the new generations of traffic engineers and planners will be more capable of reducing energy consumption and emissions from the transportation systems. We strongly believe that the future of transportation is the "green transportation." It is, in a way, the basic message of our book. The main motivation for writing this book is the desire to send this message to many young people in the world who are beginning to deal with complex transportation engineering problems.

Dušan Teodorović
Milan Janić

Acknowledgments

Few people have contributed to the development of this book. Our sincere gratitude goes to our young colleagues Dr. Milica Šelmić, and Dr. Miloš Nikolić of the Faculty of Traffic and Transport Engineering of the University of Belgrade (Belgrade, Serbia), who read parts of the manuscript and helped us in the technical preparation of the manuscript for the press. We would also like to thank Ph.D. student Aleksandar Jovanović for his technical support. We would also like to express our gratitude to Professor Bart van Arem, Chief of Department Transport & Planning of the Faculty of Civil Engineering and Geosciences of the Delft University of Technology (Delft, The Netherlands) for providing conditions for efficient work on this book. Last but not least, the encouragement of Kenneth P. McCombs, Senior Acquisitions Editor of Elsevier, is also greatly appreciated.

At the time of Julius Caesar, in the first century BC, in order to solve traffic congestion problems, a law was passed prohibiting the use of private vehicles in Rome during the first ten hours of daylight.

INTRODUCTION 1

The development of human civilization has been characterized by constant migration of population. Various ethnic groups, in search of a better life, traveled, explored, and inhabited new lands. Throughout history, humans continually migrated to new regions, establishing new settlements and creating states. The Hawaiian islands, Easter Island, Tuvalu, Samoa, Tahiti, Cook Islands, and numerous other islands in the Pacific Ocean were inhabited 2000–3000 years ago. The Vikings also developed an ocean-going tradition. They established colonies in Iceland and Greenland, and explored the coast of North America and interior of Russia more than 1000 years ago. Migration has led to changes in racial, ethnical, linguistic, economic, and cultural characteristics of the population. Europe in the past was characterized by significant migration of Germans and Slavs. The Americas, Australia, and New Zealand were inhabited predominantly by European migrants from the late 16th century through to the 20th century. The greatest intercontinental migration in human history happened between the American Civil War and the World War I, when millions of people from Europe crossed the Atlantic

Transportation Engineering. http://dx.doi.org/10.1016/B978-0-12-803818-5.00001-9

Ocean and came to the United States of America. Advances in transportation technologies partially enabled such a huge migration (Matthews, 1960; Fitzpatrick and Callaghan, 2008; Nichols Busch, 2008; Harvey, 2010; Clark, 2015).

The necessity for transportation evolves from a variety of man's activities. People travel for business, professional, cultural, or private reasons. Thousands of vehicles and passengers that travel from one place to another create flows of cars on highways, bicycles on streets, peoples in bus and metro stations, pedestrians on crossings, and aircraft on airport taxiways and runways. Passenger and freight flows are the consequences of spatial interaction among various regions (Vickrey, 1969; Gazis, 2002; Janić, 2014).

In ancient times as well as modern, people developed different transportation systems (whether based on animal-drawn wheeled vehicles or on Boeing-747 aircraft). The analysis and design of transportation systems are the essence of transportation engineering. Traffic engineers and planners are constantly faced with the question of how to plan, design, and maintain high-quality transportation system and livable human communities. Transportation engineering deals with planning, design, operations, control, management, maintenance, and rehabilitation of transportation systems, services, and components. The subareas that are parts of transportation engineering are transportation planning, traveler behavior, design and analysis of transportation networks, traffic flows analysis, analysis and control of traffic operations, queueing analysis, vehicle routing and scheduling, logistics and supply chain management, etc. Within transportation engineering, technology, mathematics, physics, computer science, social sciences, and cultural heritage converge. Transportation engineering methods and techniques have a high impact on transportation system performances (level of service, capacity, safety, reliability, resource consumption, environment, economics, etc.).

The basic elements of transportation modes are vehicles (aircraft, balloon, bicycle, boat, bus, cable car, car, electric vehicle, helicopter, locomotive, motorcycle, sailboat, ship, submarine, tractor, train, tram, tricycle, trolleybus, truck, unmanned aerial vehicle, van, wagon, etc.), guideways (street, highway, airway, railroad track, canal, etc.), transportation terminals (bus terminal, container terminal, marine terminal, airport terminal, railway terminal, freight terminal, etc.), and control policies (visual flight rules, instrument flight rules, railway signaling, fixed-time control, actuated signal control, adaptive control, etc.).

A range of control systems in transportation were originally created, mainly to improve traffic safety. Later on, engineers started with the development of control systems, intending to reduce traffic congestion. This congestion is an outcome of many decisions that different users make. Traffic and transportation systems are composed of decentralized individuals (pedestrians, drivers, passengers, dispatchers, operators, air traffic controllers, vehicles, vessels, aircraft, etc.) and each individual acts together with other individuals in accordance with localized knowledge. Occasionally, individuals collaborate, and at other times they are in conflict. They interact, simultaneously, with transportation infrastructure and the environment. Through the aggregation of the individual interactions, the global picture of the transportation system emerges.

Transportation and traffic systems are, in essence, different from other technical systems. Their performances depend a great deal on the users' behavior. A good understanding of the human decision-making mechanism is one of the key factors in the transportation planning process, as well as in developing appropriate real-time traffic control. There are continuous construction and expansion of traffic networks. Before the construction of a new bridge, road expansion, or development of a toll road, it is necessary to study how the potential users of the facility will react. For example, the following research questions should be properly answered in the case of route choice between a toll and nontoll road: how do the characteristics of competitive routes influence route choice when there is a toll and a nontoll road? How do travelers' characteristics influence route choice? Research to date has provided some answers to these questions. A proper understanding of the human decision-making mechanism is of high importance, since it has been shown that building additional roads, in some cases, does not automatically produce a reduction in total travel time in the transportation network.

Some existing transportation systems are characterized by outstanding performances. For example, the annual average delay of *Shinkansen* trains in Japan is only 0.9 min per operational train (including delays due to unmanageable causes, such as natural disasters). At the same time, no accidents resulting in fatalities or injuries to passengers on board have happened since operations commenced in 1964. The maximum speed of *Shinkansen* trains is about 300 km/h. On average, there are 342 daily departures, offering more than 1300 seats per train.

On the other hand, various transportation systems in many countries in the world are inefficient, and not reliable enough. They are also great consumers of energy and great polluters. Day after day, a number of the scheduled flights are canceled, traffic incidents happen on highways, some links in a city traffic network are fully congested, etc. Traffic engineers and operators must be also capable to obtain practical, "good" solutions for the complex transportation problems caused by random events.

Some transportation networks are exceptionally big. They are characterized by complex relationships between specific nodes and links, and they are repeatedly congested. Consequently, it is not simple to observe and study them, and to find suitable solutions for traffic problems. Most frequently, it is not possible to control the entire network in a centralized way.

Transportation science and transportation engineering offer various techniques related to transportation modeling, transportation planning, and traffic control. These techniques should be used for predicting travel and freight demand, planning new transportation networks, and developing traffic control strategies. The range of engineering concepts and methods should be used to make future transportation systems safer, more cost-effective, and "greener."

REFERENCES

Clark, J., 2015. Ships, clocks, and stars: the quest for longitude: National Maritime Museum Greenwich, London. J. Transp. Hist. 36, 124–126.

Fitzpatrick, S.M., Callaghan, R., 2008. Magellan's crossing of the Pacific: using computer simulations to examine oceanographic effects on one of the world's greatest voyages. J. Pac. Hist. 43 (2), 145–165.

Gazis, D., 2002. The origins of traffic theory. Oper. Res. 50 (1), 69–77. 50th Anniversary Issue (Jan.–Feb.).

Harvey, E., 2010. Pavage grants and urban street paving in medieval England, 1249–1462. J. Transp. Hist. 31, 151–163.

Janić, M., 2014. Advanced Transport Systems: Analysis, Modelling and Evaluation of Performances. Springer-Verlag, London.

Matthews Jr., K., 1960. The embattled driver in ancient Rome. Expedition 2, 22–27.

Nichols Busch, T., 2008. Connecting an empire: eighteenth-century Russian roads, from Peter to Catherine. J. Transp. Hist. 29, 240–257.

Vickrey, W., 1969. Congestion theory and transport investment. Am. Econ. Rev. Pap. Proc. 59, 251–261.

Where were located the earliest stone surfaced roads, and what are the Amber Routes? Who were Vigiles in ancient Rome? What are transportation systems? When the London Underground, the world's first underground railway, was opened? Who were Wilbur and Orville Wright? What is the length of the major part of the U.S. road network? What is the number of handled containers at Singapore? When the first autonomous cars appeared? What are radial, diametrical, tangential, circumferential transit networks? What are hub-and-spoke networks in air transportation?

TRANSPORTATION SYSTEMS 2

2.1 BACKGROUND

Passengers and goods travel over the land, under the land, over the oceans, and over the sky. Thanks to the development of technology and transportation systems, the world has become a "global village," and mankind has achieved economic and cultural development. Developed transportation systems have facilitated the development of political, economic, cultural, touristic, and sporting relations among people in the modern world. Sustainable and efficient transportation is one of the most important factors for the survival and progress of modern civilization.

Large-scale rail and sea transportation were created in the 19th century. Road and air transportation were established in the 20th century. Guideway and vehicle construction were the most important problems faced by the engineers who dealt with traffic and transportation problems at that time (ship building, road building, etc.).

Over time, there was a need to accommodate traffic efficiently and safely. Traffic control systems grew up step by step. At the very beginning, red and green traffic lights at the intersections were used with the aim of making the intersections safer. Very quickly, these lights became a powerful traffic control tool. At the present time, in all branches of transportation there are local, government, and international organizations that establish and maintain safety standards and various operating procedures and rules. These organizations also take care of licensing of pilots, drivers, dispatchers, etc.

Modern, large-scale transportation networks, their complexity, high level of congestion, and high transportation costs call for comprehensive approach to transportation planning and traffic control

Transportation Engineering. http://dx.doi.org/10.1016/B978-0-12-803818-5.00002-0

problems. Transportation engineering methods and techniques are quantitative. By these methods and techniques traffic engineers try to optimize traffic operations. They are quantitative, since the majority of traffic and transportation engineering problems are very complex. At the same time, wrong decisions in the transportation arena are environmentally damaging and highly costly.

Henry Ford started the age of automobile transportation. Traffic problems were present in engineering and economics even before his time. Because of their importance, traffic network equilibrium problems have been studied in the 19th and 20th centuries, dating back to Kohl (1841), Pigou (1920), and Knight (1924). With more and more cars on the streets and roads, by the middle of the 20th century, traffic problems demanded much more consideration. The roots of transportation science (Gazis, 2002; Boyce et al., 2005) could be traced back to seminal works of Wardrop (1952), Beckmann et al. (1956), and Prigogine and Herman (1971). In his influential paper, Braess (1968) proved that adding extra capacity to a traffic network can, in some cases, reduce the network's overall performance.

During the last six decades, a great number of models, methods, and techniques that deal with different transportation planning and traffic control issues have been developed. Methods and techniques developed within the framework of one transportation mode can easily be applied, with certain modifications, for solving similar problems in another branch of transportation. We try in this book to cover all branches of traffic and transportation. The following is the fundamental problem that we study in this book: How to plan, design, and maintain high-quality transportation systems and livable human communities?

2.2 HISTORY OF TRANSPORTATION

The word "*way*" has its roots in the Middle English *wey*, the Latin *veho* ("I carry"), and the Sanskrit *vah* ("carry," "go," or "move"). According to *Encyclopedia Britannica*, the earliest discovered paths, made by animals, and adapted by humans, are found near *Jericho*. These paths date from about 6000 BC. Domesticated animals were early transportation modes. Over the last few thousand years, horses have been used as riding animals in Central Asia.

Archeological sites indicate that the cities of the Indus civilization in *Sindh*, *Balochistān*, and the *Punjab* paved their major streets with burned bricks (3250–2750 BC), cemented with bitumen.

The invention of the wheel has led to the development of wheeled vehicles. The earliest stone surfaced road, built around 2000 BC, was 50-km long road from Gortyna on the south coast to Knossos on the north coast of the island Crete.

The Persian Royal Road connected *Susa*, the ancient capital of Persia, with the *Aegean Sea*. The Royal Road was rebuilt by the Persian king *Darius the Great* in the 5th century BC. The road, which was about 2400-km long, passed through Anatolia. Persian Royal messengers were able to traverse the whole road in 9 days, thanks to a system of relays.

The *Grand Canal* (*Beijing-Hangzhou Canal*) represents a series of waterways in eastern and northern China. The length of the Grand Canal is 1800 km. The Beijing–Hangzhou Canal is the world's longest man-made waterway. The oldest parts of the Canal were built in the 4th century BCE. The development of China's imperial road system started in about 220 BC. By AD 700, the road system has length of about 40,000 km. The desire to expand the silk trade to areas in the Central Asia populated mainly nomads, has led to the creation of the "Silk Road." The development of the Silk Road started about 300 BC (Lei Luo et al., 2014). Silk Road, in fact represented a network of roads that linked China with Central Asia, northern India, and the Roman Empire. Chinese silk, jade, and spices for centuries were transported to the Roman Empire. Caravans had a rest stop at a distance of 30–40 km which allowed 10-h travel during the day. The Silk Road alignment mostly followed the existing terrain, always trying to find a path that creates the least resistance to the journey. According to *Encyclopedia Britannica*, the Silk Road was connected with roads in Roman Empire and was the longest road system on Earth in AD 200.

Many paths and ways in eastern and central Europe created, by 1500 BC, trading network known as the Amber Routes. Etruscan and Greek traders used these routes, and transported amber and tin from northern Europe to the Mediterranean and Adriatic seas.

The ancient Romans were remarkable builders. They planned, designed, built, and maintained the road system with a total length 85,000 km. The most famous Roman road was the Via Appia (Berechman, 2003). The construction of this road started in 312 BC. Gradually, Rome expanded from a city-state to an empire. In order to occupy and control the new territories Romans built all-weather roads. The Romans had also road classification. The most important roads were public roads (*viae publicae*). The Romans had, 29 public roads radiating from Rome. About 300 BC Romans started to use a range of forms of cement-like materials. This invention significantly helped the Roman roads construction. The Romans adopted pavement structure methods and surveying techniques from the Greeks, Carthaginians, Phoenicians, and Egyptians.

The streets were very narrow in ancient Rome. Each chariot driver would send a runner to the other end of the street (Matthews, 1960). The task of a runner was to keep the traffic from approaching from the opposite direction. With the help of runners, streets were becoming one-way, but the allowed direction of movement was constantly changing. Traffic officers were called *Vigiles*, and frequently were resolving disputes between chariot drivers in relation to the right-of-way. Romans calculated that every day has 12 h of daylight. During *Julius Caesar*, in order to solve traffic congestion problems, a law was passed prohibiting the use of private vehicles during the first 10 h of daylight. Business deliveries were made during the night. The traffic in some residential streets was forbidden both by day and night.

The Inca built the road system that was more than 40,000-km long. The road system was spread from the Quito, Ecuador to the Mendoza, Argentina. The Inca used runners for relaying messages throughout the Inca's empire. The goods were transported by llamas.

The pavage tolls existed in medieval England (Harvey, 2010). The tolls were charged by town authorities on freights passing into and out of towns for sale. The collected money was used for paving and repairing town's streets. In the 14th century, all English towns had the men with the title Scavenger, who had been elected to clean and repair the streets. The archeological confirmations from many English towns show that town's streets were paved primarily with flat stones. In order to earn the right to assess this toll, towns were obliged to apply to the Crown for a pavage grant. The earliest pavage grant recorded in the Patent Rolls was presented to the town of Beverley (Yorkshire) in Feb. 1249, in the rule of Henry III. Approximately one hundred towns received pavage grants. Through the pavage grants, and pavage tools activities, both English towns and the King, tried to maintain and improve the transport system, to reinforce and increase trade inside and between English towns, and to contribute to the prosperity and economic success of the towns.

The magnetic compass was invented in both China and Europe in about AD 1200. Over the years, numerous technical improvements have been made in the magnetic compass. Its discovery has, to a large extent, improved navigation on the seas and oceans, and greatly contributed to the discovery of new territories.

Christopher Columbus (in Italian *Cristoforo Colombo*, in Spanish *Cristóbal Colón*), Italian, born in Genoa, sailed in the late 15th century, in the service of the Spanish crown, four times over the Atlantic Ocean (1492–93, 1493–96, 1498–1500, and 1502–04) and discovered the Americas. Columbus was maritime explorer. The Columbus's fleet departed from Palos de la Frontera on Aug. 3, 1492. They spent almost a month in the Canaries, and left San Sebastián de la Gomera on Sep. 6, 1492. The fleet was composed of the *Santa María*, *Pinta*, and *Niña*. After a 5-week voyage across the ocean, on Oct. 12, 1492, San Salvador, an island in the Bahamas, was sighted (Fig. 2.1).

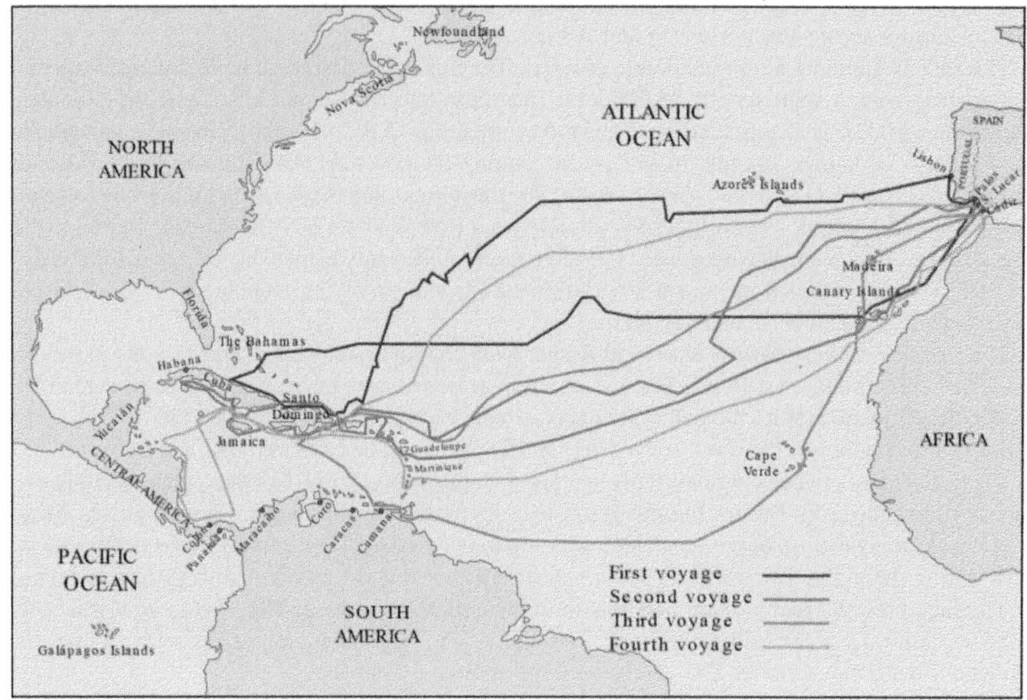

FIG. 2.1

Columbus's voyages.

"Viajes de colon en" by Viajes_de_colon.svg: Phirosiberiaderivative work: Phirosiberia (talk)—Viajes_de_colon.svg. Licensed under CC BY-SA 3.0 via Wikimedia Commons—http://commons.wikimedia.org/wiki/File:Viajes_de_colon_en.svg#mediaviewer/File:Viajes_de_colon_en.svg.

Christopher Columbus, the master navigator, sailor, admiral, and maritime explorer, discovered Bahamas, Cuba, Caribbean coast of Venezuela and Central America. Columbus's voyages made possible further expeditions and colonization of the Americas. At the same time, Columbus voyages gave a great motivation to other maritime explorers to investigate the unknown seas, oceans, and territories.

Vasco da Gama was the first European explorer who arrived in India by sea. Da Gama sailed from Lisbon on Jul. 8, 1497. His fleet had four vessels (*São Gabriel*, *São Rafael*, *Berrio*, and a 200-ton storeship). Vasco da Gama's fleet arrived in *Calicut* on May 20, 1498.

Amerigo Vespucci was a merchant and explorer-navigator. Vespucci made voyages over the Atlantic Ocean (1499–1500, 1501–02). The name for the Americas is developed from his given name.

Leonardo da Vinci (1452–1519) was a painter, sculptor, architect, musician, mathematician, engineer, inventor, anatomist, geologist, botanist, writer, and philosopher. He also had ideas about town planning and organization of traffic in the ideal city. Leonardo believed that the spread of the epidemic, to a significant extent, is caused by population density in cities. *Codex Atlanticus*, that represents a collection of Leonardo's drawing, writings, inventions, and philosophical meditations, contains also the study of the *Ideal City*, which spanned across a river. Leonardo proposed to divert the river in several branches upstream of the city. All these branches were parallel to the mainstream and they are once again flowed into the river downstream. Ideal City had a three-level of networks. The lowest level was a network of canals. This network was designed for the freight transportations, as well as for the transportation of the city's waste. The network of roads for passengers and common people was designed at the intermediate level of the city. The upper level are, according to Leonardo's proposal, contained palaces, gardens, and walkways for the gentry.

The Portuguese explorer *Ferdinand Magellan* (in Portuguese, *Fernão de Magalhães*) was the leader of the first expedition to circumnavigate the globe (Fig. 2.2).

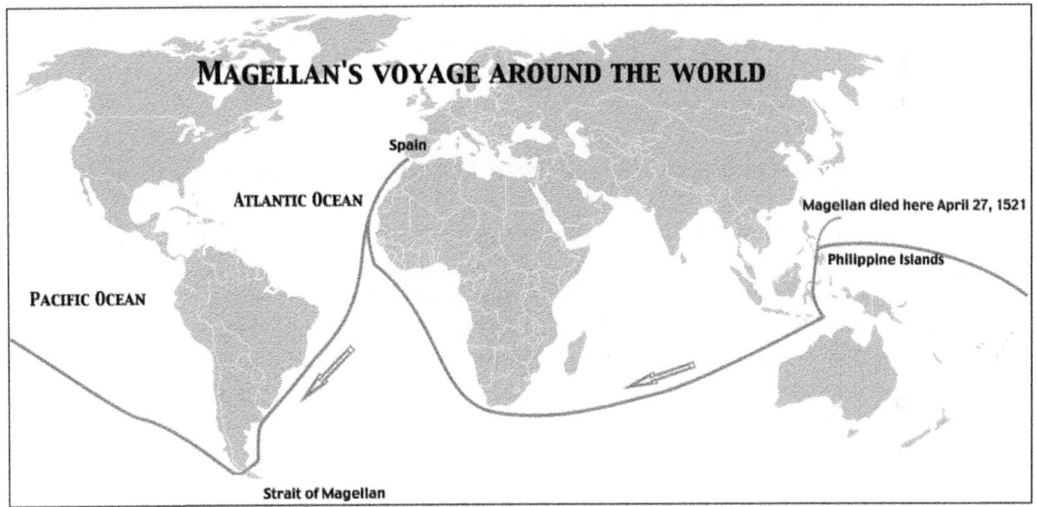

FIG. 2.2

Magellan's voyage around the world.

On Sep. 20, 1519, Magellan departed from Spain, and led the expedition with five ships and 237 crew members (Fitzpatrick and Callaghan, 2008). They sailed across the Atlantic Ocean, along the coast of South America, and found a passage (Strait of Magellan) that connects the Atlantic and the Pacific Ocean. They sailed across Pacific Ocean to the island of Guam and finally arrived to the Philippines, where, in fighting with the natives, Magellan was killed. The Magellan's expedition was terminated by the *Juan Sebastián del Cano*. The expedition continued westward to Spain, and made the first circumnavigation of the Earth. The *Victoria*, the only remaining ship from the expedition, returned to Spain on Sep. 8, 1522 with only 18 crew members surviving.

James Cook was a British explorer, navigator, cartographer, and captain in the Royal Navy. He explored the seaways and coasts of Canada (1759, 1763–67). In 1775, he made first chart of Newfoundland. Cook made three voyages to the Pacific Ocean. He was first European who arrived to Hawaii, and

Australia. He was also first explorer who circumnavigate New Zealand. James Cook was killed in Hawaii, during his third voyage in the Pacific in 1779. James Cook transformed the map of the world more than any other sailor and explorer in history.

Safe travel over the seas and oceans, as well as the improvement of marine exploration and marine trade required a precise measurement of latitude and longitude. Among others, the difficult and complex longitude problem was studied by *Galileo Galilei* and *Isaac Newton*. The British Government passed the Longitude Act in 1714. The Act provided for a reward of £20,000 for anyone who could find a method to determine accurate longitude that could be "*tried and found practicable and useful at sea*." Sir Isaac Newton became one of the Commissioners of Longitude. The "longitude problem" was solved by John Harrison, a self-educated carpenter (Clark, 2015). He developed an accurate *marine chronometer*, and was granted the Copley Medal by the Royal Society in 1749.

Steam vehicles were vehicles powered by a steam engine for use on land, rails, or for agricultural work. The base for the use of steam power is related to the experimental work of the French physicist *Denis Papin*. The first steam road vehicle appeared in 1769, the first successful steamboat demo happened in 1786, while the first successful railroad steam locomotive demo occurred in 1804. The development of internal combustion engine (ICE) technology, in the beginning of the 20th century, caused the termination of the steam engine.

It is now more than 200 years since the inauguration of the first commercial steamboat service in Europe (Williams and Armstrong, 2014). The *Comet* first ran on the Clyde, between Glasgow and Helensburgh in 1812. Over the next 10 years, the total of 142 steamboats were constructed. Between 1812 and 1822, many routes served by steamboats appeared (Glasgow-Helensburgh, Bristol-Bath, London-Rotterdam, Liverpool-Dublin, Glasgow-Belfast, Dover-Calais, London-Calais, etc). Steamboats were capable to operate to a schedule. In this way, they offered to passengers reliable services. Early steamboats provided local ferry, river, coastal, and short sea international services. Steamboat services highly influenced daily activities of the people. Scheduled steamboat services with exact departure and arrival times created a regularity to people's activities, and highly increased the mobility of the population. Goods were dispatched and received more quickly and news and information rapidly distributed. The impact of steamboats on the economy and society was so high that the journal *The Kaleidoscope* observed in 1822 that steamboats were "one of the noblest inventions of the age."

The father of the modern roads was Scottish engineer *John Loudon McAdam* (1756–1836). McAdam started to build roads with a smooth hard surface (Blondé, 2010). The Lancaster Turnpike in the United States connected Philadelphia and Lancaster in Pennsylvania. The Lancaster Turnpike had the length of 62 miles. This road was the first planned road in the United States and was built between 1793 and 1795.

The *corvée* was the road maintenance system, which Russia's rulers copied from the French in the 17th century. This maintenance concept required farmers to take care of the roads near their homes (Nichols Busch, 2008). *Peter the Great* applied this practice by charging landowners for negligent farmers. The first paved roads in Russia were road from St. Petersburg to Tsarskoe Selo and road running from Moscow to Tver. These paved roads involved a well-planned foundation topped with smooth stone.

The construction and good maintenance of paved roads considerably decreased the total transportation costs, as well as travel times. There are examples that the trip from Brussels to Terhulpen in Belgium, in a carriage drawn by four horses, lasted whole day. After the construction of the paved roads, the travel time on the same route in a carriage drawn by three horses was ~3 h. Road improvements that happened in many countries in the 18th century significantly also improved carriage schedule reliability, since the influence of weather on carriage travel times was much lower.

The *Locomotion No. 1*, was built by George and Robert Stephenson in 1825. The Locomotion No. 1, was the first steam locomotive to transport passengers on a public rail line (between the Stockton and Darlington, United Kingdom). The First Transcontinental Railroad (Pacific Railroad) was opened for traffic on May 10, 1869. This railroad line connected Pacific coast (San Francisco bay area) with the eastern US rail network. The opening of Pacific Railroad led to a very rapid settlement and rapid economic development of the American West. This line connected two oceans and enabled considerably faster, more efficient and more economical transport of goods and passengers.

The *Erie Canal* that connects Great Lakes with New York City was opened on Oct. 26, 1825. The Canal, which contains 36 locks, ran from Albany, on the Hudson River to Buffalo at Lake Erie. The Erie Canal, 584 km long, connected Great Lakes with the Atlantic Ocean. The construction of the Erie Canal took 8 years. Boats, capable to carry 30 tons, were pulled by mules and horses. Transportation costs were significantly decreased by opening Erie Canal, while the New York City became a major commercial center. In 1918, the New York State Barge Canal took the place of the Erie Canal.

Clipper ships were fast sailing ships in 1850s. These vessels appeared as a result of high demand for tea from China. They sailed all over the world. The clipper ships were narrow, and had three masts and a square rig.

The *Suez Canal* was opened in Nov. 17, 1869. Ferdinand de Lesseps received in 1854 an Act of Concession to construct the Canal, and construction began in 1859. There were 486 ship transits through the Canal in 1870. The Suez Canal, that separates Africa from Asia, connects the Mediterranean and the Red seas. The Suez Canal is the shortest maritime route between Europe and the Indian Ocean. The length of the Suez Canal is equal to 163 km (between Port Said and Suez). Crude petroleum, petroleum products, wood, metals, fabricated metals, and cement are the main cargoes transported.

The *Metropolitan Railway*, which is part of the London Underground, was the world's first underground railway, opened in Jan. 1863. The line was connecting Paddington and Farringdon. In total, 38,000 passengers were transported on the opening day. The first nickname *Tube* is today used by general public for the whole Underground system in London. Today, the London Underground has 270 stations, and 402 km of track.

The *Panama Canal* was opened in Aug. 15, 1914. The canal connects Atlantic Ocean and Pacific Ocean. The length of the Panama Canal from shoreline to shoreline is around 65 km. The total length from the deep water in the Atlantic Ocean to deep water in the Pacific Ocean is about 82 km. There were only 807 ship transits in the year of canal opening. Through the years canal traffic increased all the time. The total number of recorded ship transits in 1970 was 15,523. More than 200 million of metric tons of cargo were transported through the Panama Canal in 2013. From the very beginning of its opening, Panama Canal brought enormous benefits to the world's trade, economy, and transportation. For example, ships that sail from the east to the west coasts of the United States shorten their trip by about 15,000 km by sailing through the Panama Canal. Voyages between Europe and East Asia, and Australia, the east coast of the United States and East Asia, and Europe and the west coast of North America are also significantly shortened. Petroleum products, grains, coal, and motors vehicles are the main commodities that are transported through the Panama Canal (Fig. 2.3).

FIG. 2.3

Panama canal expansion (http://micanaldepanama.com/expansion/photos/#prettyPhoto/37).

The Panama Canal Expansion project began on Sep. 2007, at a total cost of US$5.2 billion. This project is the biggest project at the Panama Canal from the time when the Canal was open. The project is nearly finished. The main achievement of the project is the new lane of traffic along the Canal. The new set of locks is also constructed. In this way, the waterway's capacity is doubled. The current set of locks allows the passage of vessels that can carry up to 5000 TEUs. Subsequent to the extension the vessels will be able to travel, all the way through the Canal, with up to 13,000/14,000 TEUs.

Karl Friedrich Benz (1844–1929) was a German engine designer and car engineer. He is usually recognized as the inventor of the modern car. He was granted the patent for its creation in 1886. *Henry Ford* (1863–1947), an American industrialist revolutionized transportation in the United States by mass production of automobiles. The Federal Aid Road Act of 1916 was the first federal highway funding law. It significantly helped in expanding and improving the US road system. The Federal-Aid Highway Act of 1956 enabled the construction of 66,000 km of the Interstate Highway System in the United States.

Wilbur and Orville Wright, bicycle builders, made the first airplane flight in the morning on Dec. 17, 1903. It happened at Kill Devil Hills, Kitty Hawk, North Carolina. The Wright brothers made, for the first time in history, a heavier-than-air machine, capable of flying. In his first flight, Orville Wright flew about 36 m and stayed in the air for about 12 s. On the fourth flight on that day, Wilbur Wright flew about 260 m during 59 s (Fig. 2.4).

The first commercial air service started on Jan. 1, 1914 from St. Petersburg, Florida to Tampa, Florida. American *Charles Lindbergh* made the first nonstop flight over the Atlantic Ocean. He departed on May 20, 1927 from the Roosevelt Field New York's Long Island and arrived after 33 h and 30 min to Le Bourget Field in Paris, France. The total flown distance was 5800 km. Charles Lindbergh was flying in the single-seat, single-engine aircraft. American *Amelia Earhart*, was the first woman to fly solo across the Atlantic Ocean (May 20–21, 1932). The first jumbo jet (360 seats Boeing 747) arrived from

FIG. 2.4

Beginning of the Orville Wright's first flight.

New York to London's Heathrow airport in 1970. The jumbo jets significantly increased passenger capacity and lowered operating costs, enabling more people to use long-distance air transportation services.

Soviet cosmonaut *Yuri Gagarin* became the first human in space on Apr. 12, 1961. Gagarin's spacecraft *Vostok* accomplished an orbit of the Earth on Apr. 12, 1961. The spaceflight that landed the first humans on the Moon was *Apollo 11*. It happened on Jul. 20, 1969. The crew members of Apollo 11 were *Neil A. Armstrong*, Commander, *Michael Collins*, Command Module Pilot, and *Buzz Aldrin*, Lunar Module Pilot. NASA started in 1981 space shuttle program. Space shuttle (*Columbia*, *Challenger*, *Discovery*, *Atlantis* and *Endeavour*) became the first reusable space vehicles. The *International Space Station* operated by the Canadian Space Agency (CSA), European Space Agency (ESA), Japan Aerospace Exploration Agency (JAXA), National Aeronautics and Space Administration (NASA), and the Russian Federal Space Agency (Roscosmos), has been constantly in use since Nov. 2000. From that time more than two hundreds researchers from 15 countries visited and worked in this station. The successful research results have been achieved in biology and biotechnology, earth and space sciences, human research, physical sciences, and technology.

Japan was the first country in the world that constructed railway lines reserved exclusively for high speed travel. This has been an amazing achievement, since in the second part of the 19th century wheeled transport vehicles in Japan were accessible just for the Imperial family and the highest aristocrats. (These vehicles called *goshoguruma* were ox drawn carts. The drivers of these vehicles usually walked along the oxen.) The railway high-speed line, between Tokyo and Osaka, was opened on Oct. 1, 1964. A decade later this high-speed line reached one billion yearly transported passengers. The first *Shinkansen* trains ran at speeds of ~200 km/h. Nowadays, *Shinkansen* trains run at speeds of up to 300 km/h. These trains have been known for punctuality, comfort and, especially safety, since they have not had any fatal accidents in its history. Together with the French TGV, Spanish AVE, and German ICE trains, the *Shinkansen* trains are among the fastest trains in the world (Fig. 2.5).

FIG. 2.5

Toyota Prius modified with Google's experimental driverless technology.

Autonomous car (driverless car, self-driving car, robotic car) are capable of sensing its surroundings and navigating, and traveling with no human input. The first autonomous cars appeared in the 1980s (Carnegie Mellon University, Mercedes-Benz and Bundeswehr University Munich). A *Toyota Prius* modified with Google's experimental driverless technology was licensed by the Nevada Department of Motor Vehicles (DMV) in May 2012. This was the first license issue in the United States for a self-driven car.

Throughout the history of human civilization, there have been very strong relationships between economic growth, life style, and transport development. Roads improvements and the development of new transportation technologies have undoubtedly contributed that many countries were transformed from a conglomeration of separated villages and towns toward united countries. In this way, due to its political and economic importance, transportation became less of a local and more of a national priority.

Until the end of the 19th century, most of the residents of cities have walked or used horse-drawn vehicles. Horse drawn vehicles had low speed and capacity. On the other hand, traffic congestion in the modern meaning of the word, did not exist, as well as air pollution, high level of noise, and a high level of traffic accidents.

During the first half of the 20th century, public transit in cities started to be based on electric tramways, buses, and trolleybuses. New transport technology has enabled a large number of citizens to increase the distance between the residence and place of work. Cities have been expanding, simultaneously changing its urban structure. High increase of car ownership is one of the basic characteristics of developed countries over the last 80 years. This has led to the creation of cities (primarily in the United States and Canada) characterized by existing freeway network in urban areas. Although dense freeway networks were built with the idea to reduce traffic congestion, many cities in the world suffer from high levels of traffic congestion (Newman and Kenworthy, 1989). A large part of the population nowadays uses private cars to go to work, traveling long distances. In this way, traffic congestion level is constantly increasing. Congestion on existing freeways could not be eliminated by building more freeways. Building more roads is the main supply-side strategy. The expanding transportation capacities are not always the most excellent solution for traffic congestion problems. The transportation strategies that try to decrease the demand for existing transportation systems represent an alternative to strategies that promote building and expanding transportation infrastructure.

One of the articles included in the *Lex Julia Municipalis* is related to the legislation about the driving of carts, animals and people in and out of Rome (45 BC). Since ancient times, people were constantly trying to shorten the travel times, to reduce travel costs, to make traffic safer, and to reduce level of traffic congestion. The development of transportation technology, traffic control, and transportation planning techniques should primarily contribute to a better quality of life of people in the future.

2.3 TRANSPORTATION SECTOR AND TRANSPORTATION MODES

Transportation sector performs transport services in order to satisfy demand for mobility of people and transport of freight shipments. A range of socioeconomic activities in a society, as well as land uses induce transportation demand. The demand is represented by the number of passengers and/or volumes of cargo to be transported between given origins and destinations during a given (specified) period of time. The supply component in the transportation sector consists of transport services provided by different transport modes and their particular systems. Airports, highways, streets, and ports should be able to meet transportation demand and offer acceptable level-of-service to the users. The transport supply component contributes to the economy of a region, country, and continent it serves. The demand and supply component are in permanent interaction.

The transportation modes constituting the supply component are generally classified according to the way of performing their operations of transporting people and freight shipments. In general, the basic land-based transport modes include road, rail, and pipeline. The water-based mode includes inland waterways and sea shipping. The air transport is the air-based mode. The specific mode is intermodal transport consisting of combinations of particular basic modes and their systems. In addition, the mode not carrying out physical entities but just information is telecommunications. Fig. 2.6 shows a simplified scheme of the structure of transport sector, its modes and their systems.

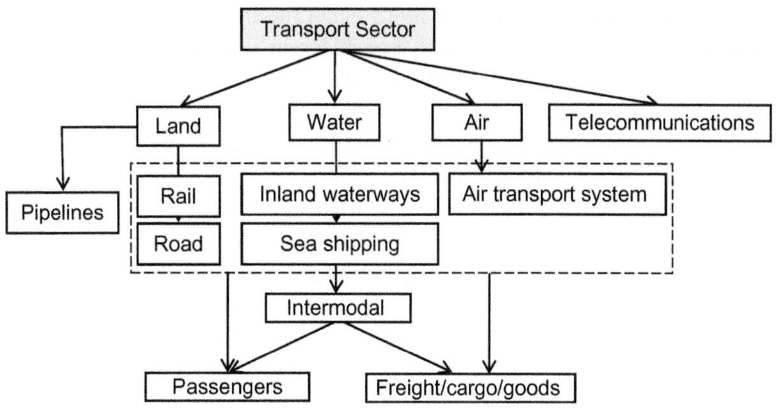

FIG. 2.6

Structure of transportation sector.

All transportation modes, except pipelines and telecommunications, are considered in this book.

Regarding the spatial scale, the road and rail transport mode can operate in urban, suburban, and interurban scale. In this way, they serve the urban areas, regions, and a country. The inland waterways mode operates at the regional, country, and continental scale. The sea shipping (maritime) transport mode usually operates at the country, continent, and intercontinental scale. The air transport mode operates at the country, continent, and intercontinental scale. The intermodal transport, depending on the modal combination, operates at the country, continental, and intercontinental scale. At any spatial scale, transportation modes serve passenger and air freight demand during the specified period of time.

The transportation modes are generally characterized by: components, technologies, and performances.

2.3.1 COMPONENTS OF TRANSPORTATION MODES

Each transport mode and its particular systems consist of transport infrastructure, vehicles, supporting facilities, equipment, and staff. These elements are used for setting up transport networks for serving passenger and freight demand, according to the specified rules and procedures. The latest are intended to provide efficiency and effectiveness of carried out transport services on the one hand, and safe operations of the system on the other.

The infrastructure of road transport mode generally includes roads, bridges, tunnels, and passenger and freight/cargo terminals/stations. The infrastructure of rail transport mode mainly consists of the rail lines, shunting yards, passenger stations, and freight/cargo terminals. The inland waterways infrastructure includes rivers and channels as lines, and corresponding ports with passenger and freight/cargo/goods terminals. The infrastructure of sea shipping (maritime) transport mode is land-based road and rail, and water-based access channels and sea-ports with passenger and freight/cargo/goods terminals. The air transport infrastructure consists of airports and their land-based ground access system operating mainly at the suburban/regional spatial scale. The intermodal transport mode integrates parts of infrastructure of different modes. Typical are intermodal or multimodal passenger and freight terminals facilitating vehicles of different modes.

The vehicles operated by the road transport mode generally include, on the one hand, diesel/petrol powered passenger cars and buses, and trucks, all of different size and other technical-technological characteristics. On the other hand, there are the electricity-powered trolleybuses and BEV (battery electric vehicles) (cars). The rail transport mode operates passenger and freight trains. Specifically, in urban and suburban areas, usually the electric powered "passenger trains" are streetcars (tramways), LRT (light rail transit), and subway (metro) systems. In interurban areas these are conventional and HS (high speed) passenger and freight/cargo/goods electricity and/or diesel-powered trains. They both consist of a certain number of corresponding cars/wagons pulled by electric- or diesel-powered locomotive(s). The inland waterways operate vessels/barges of different size/capacity, which can be diesel-powered self-propelled and/or pushed by diesel-powered towboats. The sea shipping transport mode operates exclusively diesel-powered passenger and cargo sea vessels of different size/capacity. The former (ie, passenger ones) are usually cruisers. The latter can be dedicated for type of freight and the level of its consolidation such as bulk, oil and gas tanker, and container vessels. The air transport mode operates as vehicles jet-fuel/kerosene powered passenger and cargo aircraft of different size/capacity. The intermodal transport mode operates vehicles of different modes synchronized at both passenger and freight/cargo intermodal or multimodal terminals.

The main supporting facilities and equipment of all these modes and their systems are traffic signaling, control, management systems, power-supply systems, and facilities/equipment for facilitating with the customers including administration.

The infrastructure, vehicles, and facilities and equipment of the transportation modes and their systems are operated and maintained by the qualified and dedicated workers.

2.3.2 STRUCTURE OF TRANSPORTATION MODES

Transport modes generally consist of two types of systems: the first is intended for serving passenger and the other for serving freight/cargo/goods demand. Each of these consists of subsystems, which will be called systems.

The classification of systems within each mode is carried out at three levels: (i) type of the system (passengers, freight), spatial scale of operation (urban/suburban/regional, interurban), and carrier type (individual, group) (Vuchic, 2007). The simplified schemes of classification of particular modes are shown in Fig. 2.7A–D. Only the systems enabling just transportation of passengers and freight shipments are shown. The schemes do not show the infrastructure and supporting facilities and equipment.

Fig. 2.7A shows classification of systems of the road transport mode. It should be mentioned that individual walking and cycling in urban and sub/urban areas in the scope of passenger systems are not particularly considered.

Fig. 2.7B shows the systems within the rail transport mode serving both passengers and freight/cargo/goods shipments in urban/suburban and interurban areas.

Fig. 2.7C shows the systems at inland waterways and sea shipping transport mode for serving passengers and freight/cargo/goods shipments. It should be mentioned that shipping lines for passengers (general and cruise) are not considered.

Fig. 2.7D shows that the air transport mode consists of two airline systems: that carrying out passengers (conventional and LCC (low cost carriers)) and that carrying out freight cargo shipments.

2.3.3 TECHNOLOGIES OF TRANSPORT MODES

Technologies of particular transport modes mainly relate to the mechanical characteristics of their vehicles and ways such as: support, guidance, propulsion, and control (Vuchic, 2007).

2.3.3.1 Support

Support is the vertical contact between a vehicle and its riding surface. At the road transport mode these are tires and the surface of road usually made of asphalt and/or concrete. In the case of the rail transport mode and all its systems support are the steel wheels on the steel tracks. For inland waterways and sea shipping mode and systems, these are the vehicles' bodies floating on the water. At air transport mode, the support is again the aircraft tires and the asphalt or concrete runway surface while maneuvering at airports, and just the vehicle/aircraft body and surrounding air while flying between these airports. The specific support has been magnets creating magnetic fields enabling both levitation and propulsion for vehicles of the MAGLEV (MAGnetic LEVitation) system moving along the dedicated guideway (Janić, 2014).

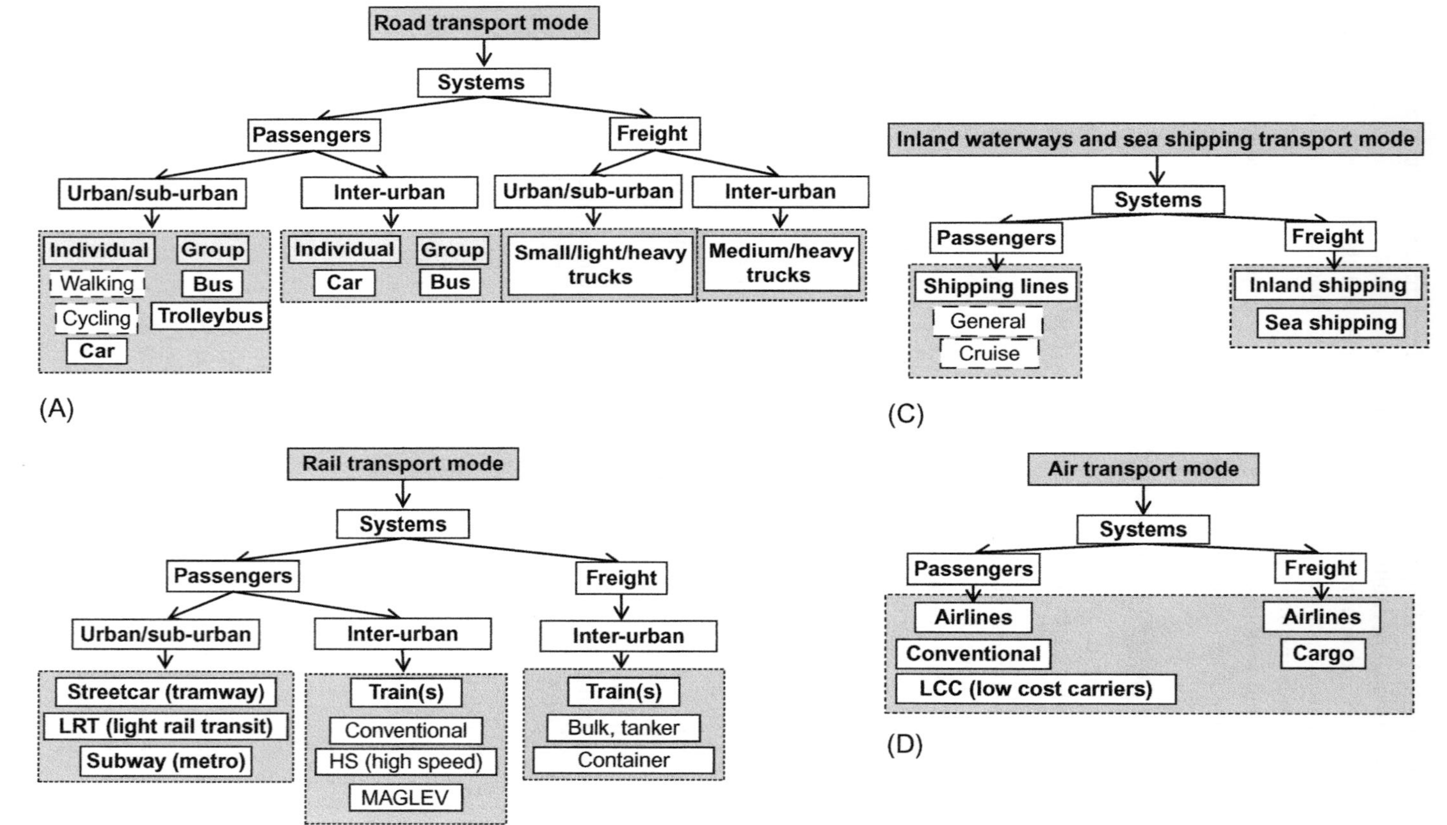

FIG. 2.7

Schemes of the structure of particular transport modes. (A) Road transport mode, (B) rail transport mode, (C) inland waterways and sea (maritime) shipping transport mode, and (D) air transport mode.

2.3.3.2 Guidance

Guidance implies the lateral guidance of vehicles operated by particular transport modes and their systems (Vuchic, 2007). For examples, the vehicles of road transport mode such as cars, buses, trolleybuses, and trucks are steered (by the driver(s)) while their lateral stability is enabled by adhesion between the wheels and the road (asphalt/concrete) surface. The forthcoming driverless cars are supposed to be steered automatically while being continuously monitored by drivers. The vehicles of rail transport mode are guided by the conical form of wheels and their adhesion with the steel tracks. The inland waterways and sea vessels/ships are primarily guided by the course keeping auto-pilot. More recently, the additional more sophisticated systems have been gradually implemented such as track-keeping, ie, guiding them along a prespecified path, station-keeping, ie, keeping the vessel's/ship's position relative to another ship constant while this other vessel/ship is moving, and evasion, ie, minimizing potential collision between more ships. In addition, many vessels are equipped with the active stability systems, that consist of the stabilizer fins on the side of vessel or tanks in which fluid is pumped around to counteract the motion of the vessel/ship (Zuidweg, 1970). It should also be pointed out that adhesion between vessel's hull and water is very low. At the air transport mode the aircraft are steered (by pilots) while maneuvering at airports thanks to adhesion between their tires and the runway asphalt surface, and by a range of stabilizers while flying in the air. These stabilizers are in the form of an aerodynamic surface including one or more movable control surfaces, which provide horizontal-longitudinal (pitch) and/or vertical-directional (yaw) stability and control of aircraft (https://en.wikipedia.org/wiki/Stabilizer_%28aeronautics%29).

2.3.3.3 Propulsion

Propulsion refers to the vehicles' propulsion units and the method of transferring acceleration/deceleration forces.

Propulsion units

The common propulsion units at the vehicles operated by road transport mode and its systems such as cars, buses, and trucks are ICEs. They are usually powered by gasoline, diesel, and/or LNG (liquid natural gas) fuel. The forthcoming BEVs are supposed to be propelled by electromotors using the electric energy stored in batteries on-board the vehicle. The batteries are recharged from the power grid (at home or at street/shop charging stations). In addition, HVs (hydrogen vehicles) and HFCVs (hydrogen fuel cell vehicles) are supposed to be powered by hydrogen fuel (Janić, 2014). The vehicles (locomotives/train sets) operated by rail transport mode and its systems are mainly propelled by diesel ICE (diesel locomotives) and electromotors (electric locomotives). The former are powered by diesel fuel and the latter by electricity from the power grid above the tracks. The steam-engines are worth to mention despite they are becoming an increase rarity rather for museums than for the commercial use by the rail transport mode. The inland waterways and sea vessels/ships are mainly propelled by diesel engines consuming diesel fuel. The forthcoming are ICEs powered by biofuels, LNG, and LH_2 (liquid hydrogen). The commercial aircraft operated by air transport mode are propelled by piston ICEs powered by aviation gasoline and turbojet and turbofan (jet) engines powered by jet fuel-kerosene. Both fuels are derivative of crude oil.

Methods of transferring tractive force

The predominant methods of transferring force are friction/adhesion at the vehicles operated by road and rail transport mode, and generally propeller at the vehicles operated by inland waterways, sea, and air transport mode. Specifically, at MAGLEV system, it is magnetic force.

2.3.4 RELATIONSHIPS BETWEEN TRANSPORT MODES

The relationships between particular transportation modes and their systems operating in the same transport markets can generally be competition and cooperation. Competition implies an intention to attract as much as possible volumes of passengers and/or freight/cargo/goods shipments during the specified period of time, ie, to gain as high as possible market share under given conditions. Cooperation implies providing integrated transport services between users'—passengers and freight/cargo goods shipments'—ultimate and/or final origins and destinations by the systems operated by different transport modes.

A typical example of modal competition is between the road and rail transit systems operating in urban areas such as buses and/or trolleybuses, streetcars (tramways), and subways (metros). They can compete between themselves and/or each of them individually or all together can compete with individual cars. The main competitive tools of these systems are the quality and cost (price) of services. In addition, road and rail public transit systems, operating in urban areas, can cooperate with each other by providing the integrated door-to-door transport services for passengers. This could be achieved by coordinating schedules at the intermodal transfer locations/points, as well as by common service charging. Some examples of the intermodal transfer locations/points are common bus/streetcar (tramway) stops or bus/trolleybus/streetcar (tramway) stops quite close to the subway (metro) stations. In addition, examples of the regional cooperation are integrated bus/LRT services and all possible combinations of integrated services provided by the urban and suburban/regional transit systems operated by different transport modes. In all these cases, the integrated services need again to be of the sufficient quality and reasonable cost (price) in order to be competitive to the individual car use.

Typical example of competition and cooperation of transport modes serving passengers is that between HSR (high speed rail) and air passenger transport (Fig. 2.8). The evidence so far has indicated that HSR has managed through competition to take over from air transport the certain volumes of passenger demand on the short- to medium-haul routes (350/400–800/900 km and travel time up to 4 h).

The location of HSR stations are in city centers. This enables more convenient/easier boarding/deboarding procedures, which all made the generalized "door-to-door" travel time and related costs quite comparable, if not superior, to that of air transportation. Nevertheless, this superiority of the generalized travel time/costs has decreased with an increasing length of route, ie, travel time, causing generally decreasing of the HSR market share as shown in Fig. 2.9.

The relative market share of HSR (that of air passenger transport is complement to 100%) has decreased linearly (Europe, Japan) and more than linearly (China) with increasing of the line/route travel time within the given range.

Cooperation between HSR and air passenger transportation has been taking different forms: by offering integrated air-rail services such as those by Lufthansa at Frankfurt (Germany) and Air France-KLM at Paris Charles de Gaulle airport; by replacing air transport flights by HSR services, such as Frankfurt-Stuttgart and Frankfurt-Cologne (Germany), and Paris (France)-Brussels (Belgium)

FIG. 2.8

High speed train.

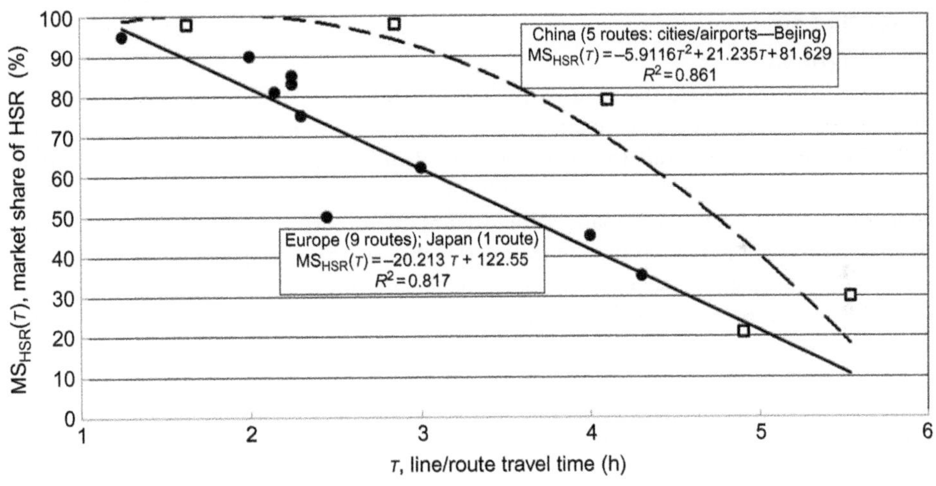

FIG. 2.9

Relationship between the market shares of HSR and APT, and the line/route travel time (Janić, 2016).

where rail tickets have been offered by Emirates, American Airlines, and United Airlines, and by deploying HSR services as one leg of the hub-and-spoke operations.

The typical competition between transportation modes serving freight/cargo/goods shipments have traditionally taken place between inland transport modes—rail and road. The evidence so far has indicated that the road has traditionally had higher market share.

Cooperation between different transport modes serving freight demand has also been taking place in terms of offering integrated door-to-door transport services to particular shippers and receivers. The most common has been that between inland rail and road freight transport modes creating the rail/road intermodal transport mode. A combination of more modes, for example, road/rail/inland waterways or sea shipping, has created multimodal freight transport mode(s). For example, the global express freight delivery companies such as DHL and UPS offer the integrated door-to-door delivery services by combining road, rail, and air transport mode.

2.4 CHARACTERISTICS OF TRANSPORT MODES AND THEIR SYSTEMS

2.4.1 INTRODUCTION

This section elaborates some characteristics of transport modes and their systems. The mass characteristics mainly include the topology of infrastructure and transport service networks, vehicles, and the numbers of passengers and freight shipments carried out during the specified period of time under given conditions. The other performances of transport modes and their systems are elaborated latter throughout the book.

2.4.2 URBAN AND SUB/URBAN ROAD AND RAIL-BASED TRANSIT SYSTEMS FOR PASSENGERS

2.4.2.1 Background

The road and rail mode include the road individual (car) and the road- and rail-based group (mass) transit systems for serving passengers in urban and suburban areas. The infrastructure networks of these systems operated by road and rail transport mode spreading over the urban and/or suburban areas are designed to enable frequent, fast, and relatively cheap transport services to their user—passengers. The layout of these networks is principally influenced by the factors such as spacing between lines and routes, their length, alignment, and interconnection (Vuchic, 2005). The spacing mainly depends on the trade-off between the walking distance to the lines and the transport service frequency there. The length of lines mainly depends on the size of urban and/or suburban areas and type of lines. The line alignment should enable rather substantive collection/distribution of passengers and their transport between origins and destinations. In the most cases, the lines of these networks include two lanes (road) and two tracks (rail) enabling vehicles' operations simultaneously in both directions, without mutual interference.

In general, transit networks can be classified into following basic types: radial, diametrical, tangential, circumferential, trunk with branches, trunk with feeder, and loops. These configurations can be recognized at almost all road and rail systems serving large urban and suburban agglomerations—buses including BRTs (bus rapid transit(s)) and trolleybuses, streetcars (tramways), LRTs, and subway (metro) systems. The latest two are also called rapid transit systems. Fig. 2.10 shows the simplified spatial layouts of some of these (generic) network configurations (Vuchic, 2005).

2.4.2.2 Bus system

The specific layout of urban bus networks is mainly influenced by the urban form and location of sufficient passenger demand. The bus lines follow the urban/city streets along the lanes used for mixed (road) traffic, lanes used exclusively for public transport (bus, streetcar, taxi), or completely dedicated lanes in the form of corridors such as those of the BRT systems. Fig. 2.11 shows a simplified scheme.

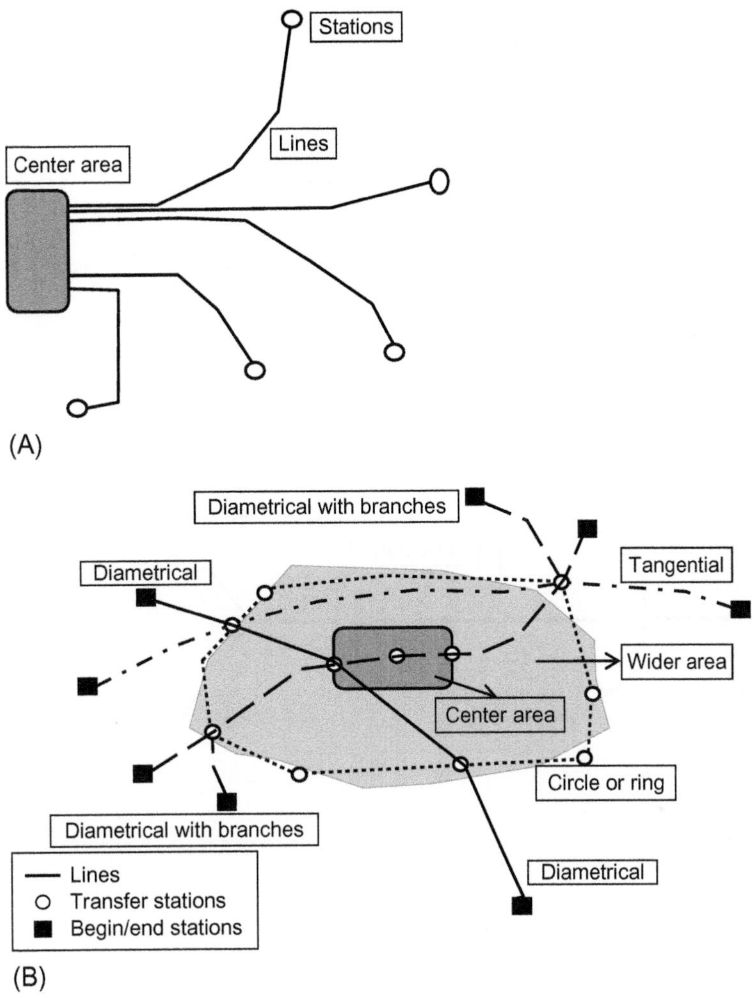

FIG. 2.10

Simplified layouts of networks of urban transit systems for passengers. (A) Radial, (B) diametrical, diametrical with branches, tangential, circle or ring,

Continued

Table 2.1 gives some characteristics of BRT systems round in the urban areas/cities round the world.

As can be seen, the BRT TransMilenio (Bogota, Columbia) has been the largest in terms of the daily number of passengers carried out. The BRT Mexico City Metrobus has the highest density of stations/stops, while the network of the BRT TransJakarta has been the longest.

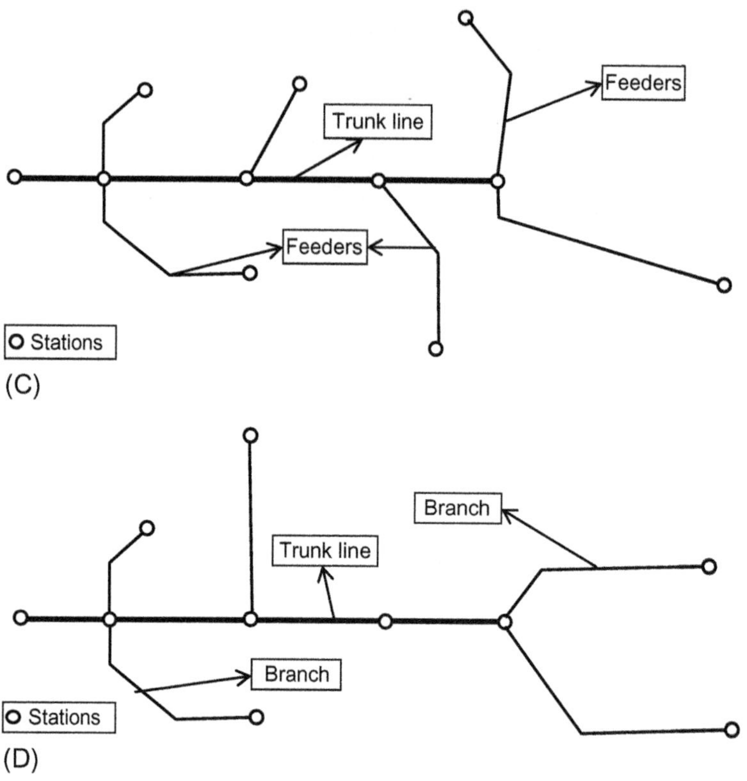

FIG. 2.10, CONT'D

(C) trunk line with feeder lines, and (D) trunk line with branch lines.

2.4.2.3 Streetcar (tramway) system

Similarly as at the urban bus network(s), the layout of infrastructure (track) networks of the streetcar (tramway) systems mainly is influenced by the size and composition of urban form.

The lines of these networks can share the same streets used by other traffic mixing with it. Such mixing of different categories of vehicles—streetcars (tramways) and cars—can often compromise punctuality and sometimes reliability of services provided by the former. This substantively improves if the system operates along the other traffic isolated-lanes similarly to corridors of the BRT system(s). In general the networks of streetcar (tramway) system(s) consist of several lines covering fully or partially given urban area. In some cases the network can consist only of a single line as shown in Fig. 2.12 for the city of Edinburgh (United Kingdom) (http://edinburghtrams.com/plan-a-journey/route-map).

As can be seen, this 14-km long line starts at the city center and ends at Edinburgh airport with 14 stations in between. In addition, there are 10 other stops along the line.

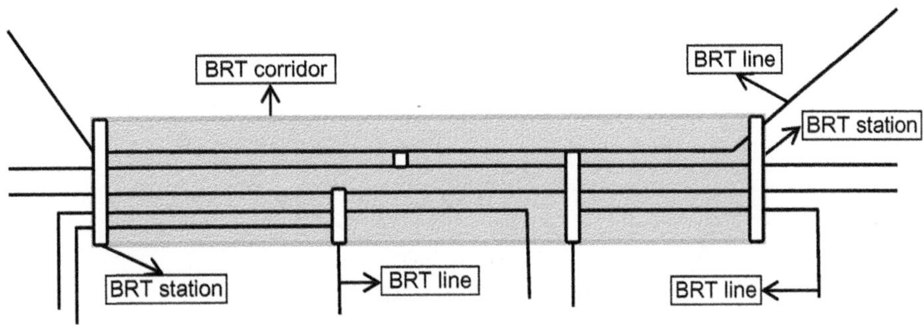

FIG. 2.11

Simplified scheme of the BRT corridor with few lines passing through it (Cervero, 2013; Weinstock et al., 2011).

Table 2.1 Characteristics of Some Largest BRT Systems (Cervero, 2013; https://en.wikipedia.org/wiki/List_of_bus_rapid_transit_systems)

Urban Area	System	Length (km)	No. of Stations	Station Density (km^{-1})	Passengers Per Day (10^6 day^{-1})
Ahmedabad	Janmarg	89	126	1.416	1.320
Bogota	TransMilenio	106	114	1.075	2.155
Guangzhou	Guangzhou Bus Rapid Transit	22	26	1.181	1.000
Curitiba, Brazil	Rede Integrada de Transporte	81	21	0.259	0.508
Mexico City, Mexico	Mexico City Metrobus	115	172	1.496	0.850
Istanbul	Metrobus (Istanbul)	52	45	0.865	0.800
Lahore	Metrobus (Lahore)	28	27	0.964	0.180
Tehran	Tehran Bus Rapid Transit	150	134	0.893	2.000
Jakarta	TransJakarta	208	223	1.072	0.350

2.4.2.4 LRT system

The LRT system has also been complementing or in some cases a predominant if not an exclusive mass urban and/or suburban transport system for passengers. In the former case, it has often operated as a complement to the urban metro systems and the regional rail systems as well. In the latter case, it has been as a backbone of the public transit system as shown in Fig. 2.13 (Kishimoto et al., 2007; Vuchic, 2007; https://www.google.nl/search?q=images+of+LRT+networks).

In general, the spatial layout of the LRT infrastructure networks have been mainly influenced on the area, size, and density of population, and the presence of the other urban transit modes serving the same urban and/or sub/urban areas. However, in many cases, it has not been easy to make a clear distinction

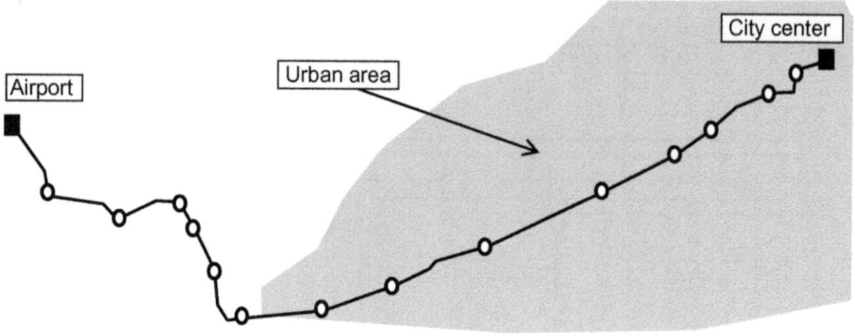

FIG. 2.12

Single line of the streetcar (tramway) system-case of Edinburgh (United Kingdom) (http://edinburghtrams.com/plan-a-journey/route-map).

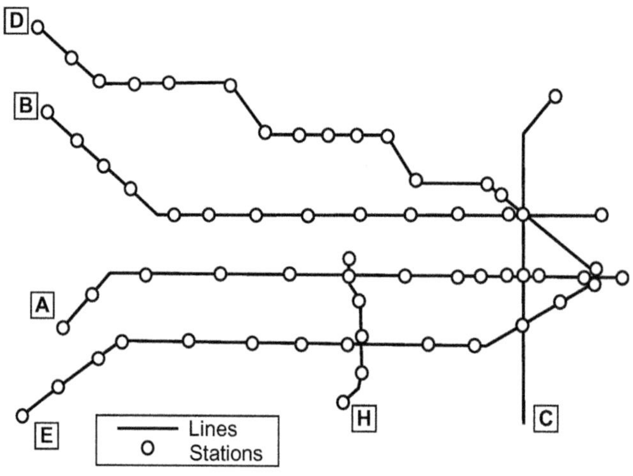

FIG. 2.13

Simplified layout of Toronto LRT system (Canada) (https://www.google.nl/search?q=images+of+LRT+networks).

between a streetcar (tramway) and LRT system. The main distinctive characteristic has related to the systems' infrastructure lines and networks. In the case of streetcars (tramways) share their rights-of-way with cars fully or partially, while LRT trains mainly operate along their right-of-ways. This enables providing higher quality of transit services in terms of travel speed, punctuality, and reliability. However, both systems have the similar vehicles with slightly distinctive capacities operating along the lines with different interstation distances in particular parts of a given urban/suburban area—the streetcar (tramway) with generally shorter, and the LRT system with generally longer—ones. Fig. 2.14 shows a simplified scheme of these characteristics.

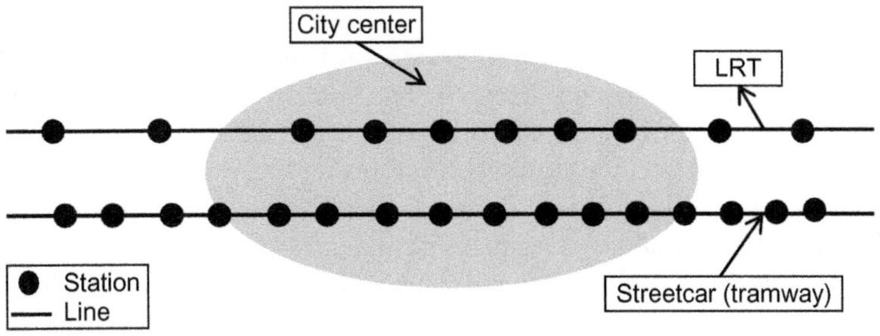

FIG. 2.14

Scheme of the lines and stations of streetcar (tramway) and LRT system.

As can be seen, in these cases, the streetcar (tramway) system tends to have more distant, ie, less frequent, stations along parts of its lines spreading out of the city center. The stations are closer to each other, ie, more frequent, along the same lines within the city center. The LRT usually has more distant, ie, less frequent, stations along the entire length of its lines.

The scale of operation of the LRT systems has often been considered together with that of the streetcar (tramway) systems, just because of the above-mentioned complexity of making a clear distinction between the two, at least in that statistical context. Therefore, the self-explanatory Table 2.2 gives characteristics—length of network, size of fleet, and the annual volume of served passengers by some largest LRT and streetcar (tramway) systems in Europe (UITP, 2015).

Table 2.2 Some Characteristics of the European largest LRT and Streetcar (Tramway) Systems (UITP, 2015; http://www.railway-technology.com/projects/category/light-rail-systems/)

Urban Area/City	Network Length (km)	Fleet (No. of Vehicles)	Served Demand (10^6 Pass/Year)
Prague	143	*920*	317
Moscow	181	919	252
St. Petersburg	*240*	833	312
Budapest	156	612	*396*
Warsaw	124	526	264
Vienna	178	520	363
Milan	172	481	n.a.[a]
Bucharest	145	483	322
Cologne/Bonn	195	382	210
Zurich	73	258	250

[a]*Not available.*

As can be seen, the longest network has been in St. Petersburg (Russia), the largest LRT/streetcar (tramway) fleet has been operated in Prague (Czech Republic), and the largest annual volume of satisfied passenger demand has been in Budapest (Hungary). It should always bear in mind that these figures relate to the integrated figures of both LRT and streetcar (tramway) systems (UITP, 2015).

2.4.2.5 Subway (metro) systems

The subway (metro) system represents the backbone of urban transportation system in many large urban areas around the world. The layout and length of its (mostly underground) infrastructure network can be different, mainly depending on the size of a given urban area, its population and intensity and type of particular activities, including the presence and scale of operations of other urban mass transport systems. In many cases, these other systems, as mentioned earlier, have complemented to the subway (metro) system. The spatial layout of the system's infrastructure networks can be one of those shown in Fig. 2.7B. The lines constituting the networks are the exclusive right-of-way enabling frequent, punctual, reliable, and fast transport services compared to other urban mass transport systems.

Some of the characteristics of these infrastructure networks are shown in Fig. 2.17. This is an example of the relationship between the length of network and the number of stations of 29 subway (metro) systems operating worldwide (UITP, 2014; http://www.railway-technology.com/features/featurethe-worlds-longest-metro-and-subway-systems-4144725/) (Fig. 2.15).

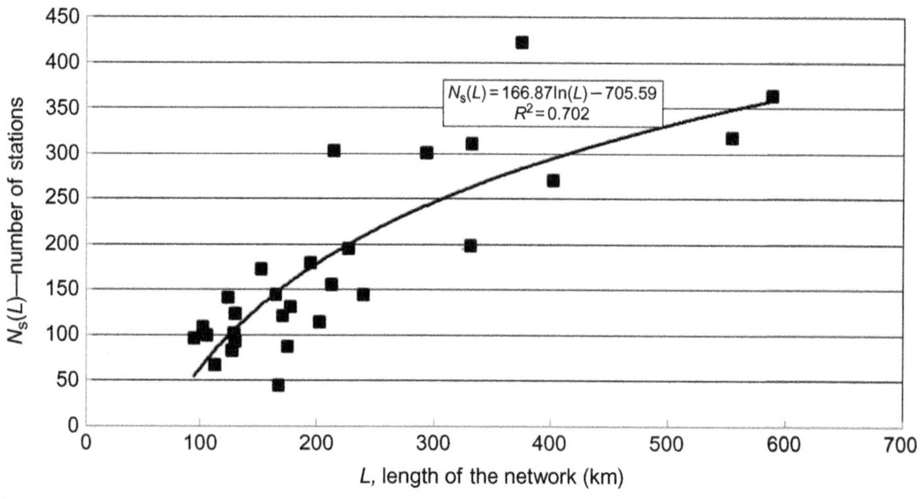

FIG. 2.15

Relationship between the number of stations and the length of network of selected subway (metro) systems (UITP, 2014; http://www.railway-technology.com/features/featurethe-worlds-longest-metro-and-subway-systems-4144725).

As can be seen, the number of stations increases with increasing of the network length at decreasing rate. One of the rather strong influencing factors is that the larger networks also cover the parts of larger urban areas with lower spatial density of population as the prospective demand, thus requiring less dense stations along the corresponding lines.

Some statistics indicate that the subway (metro) systems currently operate in 148 urban areas (cities) around the world with about 540 lines. The total daily passenger demand served by these systems has been about 150×10^6 passengers per day (UITP, 2014). Specifically in Europe, the total length of network of 2800 km in 45 cities and the fleet of 21,500 vehicles (cars) serves more than 30 million passengers per day.

2.4.3 URBAN AND SUB/URBAN TRANSPORT SYSTEMS FOR FREIGHT SHIPMENTS

The freight shipments as the semi- or final products for consumption and/or some other use are finally distributed to stores and final recipients located in urban and suburban areas. Usually, it is carried out by trucks of different size/payload capacity. In some other cases it is carried out by rail, and if convenient by inland waterways vessels/barges. The topology of these distribution networks mainly depends of type of freight, location of corresponding local depots, and the number and location of stores and final recipients. In particular, the level of inventories at stores, intensity of their consumption, and size/payload capacity of trucks deployed mainly influence the frequency and size of deliveries (Ehmke, 2012). In any case, the vehicles move along the urban street networks whose configuration determines the final topology of these distribution networks. A delivery from shippers to receivers can generally be direct and indirect. The latter is usually carried out through freight/goods distribution centers. Fig. 2.16 shows the simplified schemes of spatial network topology of delivering the freight/cargo/goods shipments to a given urban area with and without using the urban freight goods distribution center (EC, 2005).

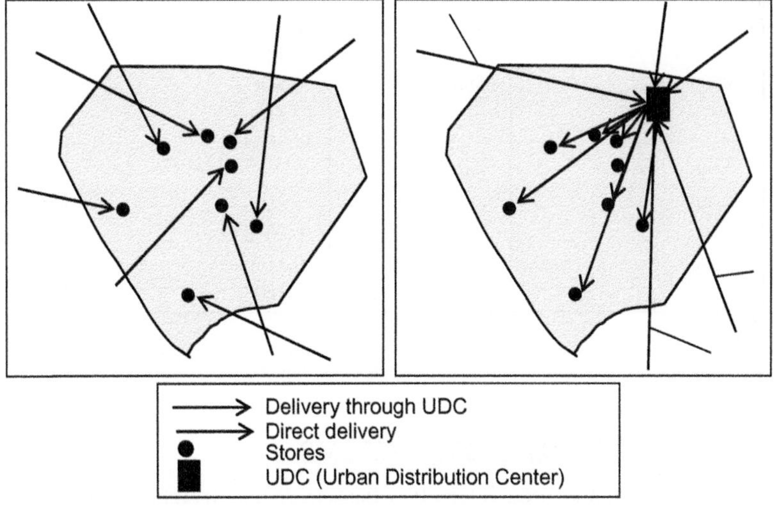

FIG. 2.16

A simplified spatial network topology of the freight/cargo/goods shipments entering a given urban area (EC, 2005).

As can be seen, the freight shipments can be delivered directly from the manufacturers as shippers to the stores as receivers, usually by the medium and/or heavy trucks. The latter are typically 5-axle trucks with the maximum gross weight of 36.32–39.95 ton in the United States and 40–44 ton in Europe. Alternatively, they can be first delivered from the manufacturers as shippers to the distribution center in urban area as an ultimate receiver by the above-mentioned medium or heavy trucks, and then further distributed locally to particular stores or the other final receivers, this time usually by medium and lighter single-unit 2- or 3-axles trucks and/or their combinations.

In order to get an idea about some characteristics of urban freight distribution activities, an information including the average population density, shop density (all commercial activities), size of loading/unloading area, and the intensity of deliveries for the selected large densely populated urban areas is given in Table 2.3 (Merchán et al., 2015; http://lastmile.mit.edu/km2; https://en.wikipedia.org/wiki/List_of_urban_areas_by_population).

Table 2.3 Some Characteristics of Urban Freight/Cargo/Goods Distribution in the Selected Urban Areas (Merchán et al., 2015; http://lastmile.mit.edu/km2; https://en.wikipedia.org/wiki/List_of_urban_areas_by_population)

Urban Area/City	Population Density (Inh./km^2)	Shop Density (km^{-2})	Loading/Unloading Area (10^3 m^2)	Intensity of Deliveries (h^{-1})
Beijing	5500	836	12.230	49
Bogota	18,300	733	5.800	24
Kuala Lumpur	3600	585	1.870p 47.700d	60
Madrid	4700	1420	10.740p 1.170d	51
Mexico City	9700	2580	0.584	35
Montevideo	6276	617	12.340p 84.350d	47
Quito	7200	1540	0	49
Rio de Janeiro	5800	2620	1.860p 0.260d	92
Santiago	6300	1800	0.415p 0.360d	22

p, public space; d, dedicated space.

2.4.4 INTERURBAN ROAD TRANSPORT SYSTEMS

The infrastructure network of the road interurban passenger and freight transport systems spreads between and connects urban and suburban areas in the given region, country, and more countries (USDT, 2005; Van de Velde, 2009). One of the examples of the global road networks is the TEN-T (Trans-European Transport Network), which includes, in addition to the infrastructure network of other transport modes, the so-called comprehensive road network 136,706 km long and the so-called core network 56,690 km long. The former network provides all European regions (including peripheral and outermost regions) with an accessibility that supports their further economic, social, and territorial development as well as the mobility of their citizens. The latter network is of the strategic importance for the major European passenger and freight/cargo/goods transport (http://ec.europa.eu/transport/themes/infrastructure/ten-t-guidelines/maps_en.htm). In the United States, the road network consists of different categories of roads such as paved and unpaved. The paved roads include major road system consisting of roads with less than four lands, roads with four or more lanes (highways), and local roads. The major part of the US road network in the interstate highway network about

75,000 km long. The similar major networks have been developed in the countries at other continents. Fig. 2.17 shows an example of development of the highway networks in the EU-27/28 Member States, United States, and China over time (EU-European Union) (EU, 2015; FHWA, 2013).

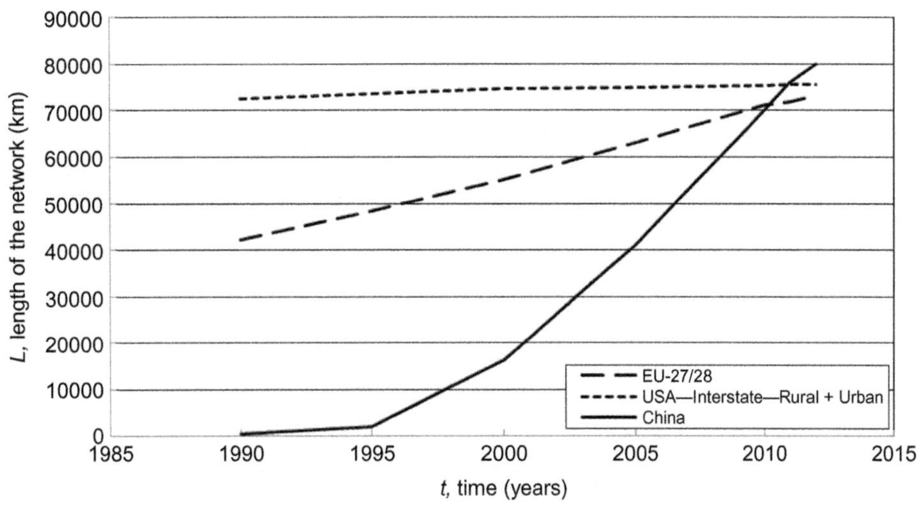

FIG. 2.17

Development of the highway/motorway network in the EU-27/28 Member States, United States, and China over time (1990–2012) (EU, 2015; FHWA, 2013).

As can be seen, in the EU-27/28 Member States the highway network has been continuously extended over the past two decades, just supporting growth of the road traffic and transport demand. The US interstate urban and rural highway network has been gradually extended (Fravel et al., 2011). However, in China, the highway network has been developing very fast over the past two decades and reached the length of EU-27/28 and United States. Fig. 2.18 shows the scheme of the US interstate road network connecting the capitals of particular provinces and states (http://www.mapsofworld.com/usa/usa-road-map.html).

2.4.4.1 Freight shipments

The freight (cargo, goods) road systems operate on the same road infrastructure network as the systems for passengers. The specificity is that their road trucks, in addition to highways for the long distance transport, also use local roads (and streets in urban and suburban areas) for approaching the shippers on one side, and receivers of freight shipments, on the other side of their delivery routes. This also includes local roads for operating around and at road freight consolidation terminals if necessary.

In general, freight/cargo can be delivered from shippers to receivers through different steps whose generic topology is shown in Fig. 2.19.

As can be seen, the initial step implies delivering raw materials from the suppliers to the manufacturing plant(s) where the semi- or the final products are made. In the former case, these are, after completion phase, transported, usually by heavy trucks, to the other manufacturing plants

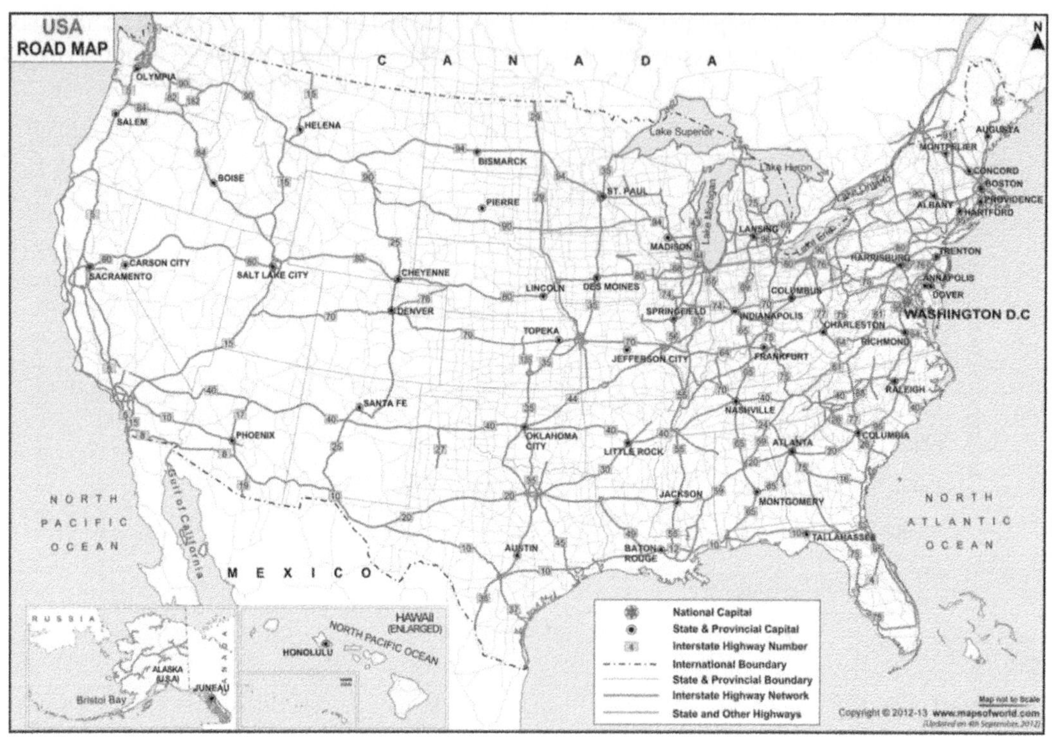

FIG. 2.18

Scheme of the US interstate highway network (http://www.mapsofworld.com/usa/usa-road-map.html).

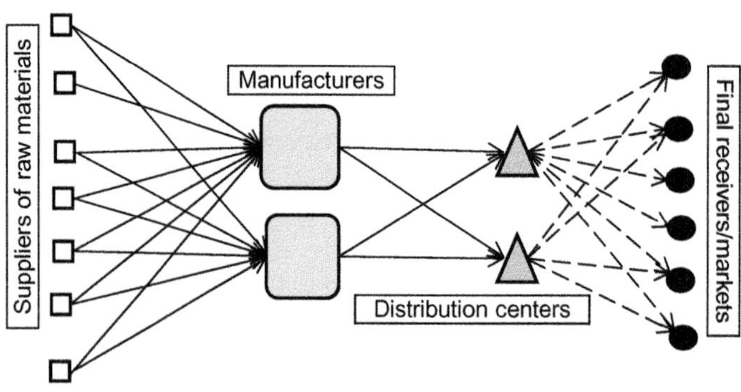

FIG. 2.19

Simplified scheme of topology of the road freight network.

for finalization. In the latter case, the final products are transported to the freight distribution centers usually located at the border, or just outside urban areas, usually by heavy or sometimes medium trucks. These distribution centers are also called road freight terminals, enabling consolidation, deconsolidation, and transshipment particular shipments between trucks of the same or different size/payload capacity. From the centers, the urban freight distribution takes place implying delivering the final products to the end receivers in a given urban area. Again, this is carried out by trucks of different size/payload capacity—light, medium, and sometimes heavy.

The road freight transport systems have dominated in terms of their market share in the total volumes of inland freight transport. The actual volumes of freight shipments in particular countries and wider regions transported by road have mainly depended on the overall economic development and supported by the adequate transport capacity. Fig. 2.20 shows the example of the relationship between the number of annual volumes of freight road haulage (t km) and the number of registered freight vehicles/trucks in the EU-27/28 Member States during the period 1995–2013 (t km—ton-kilometer) (EU, 2015).

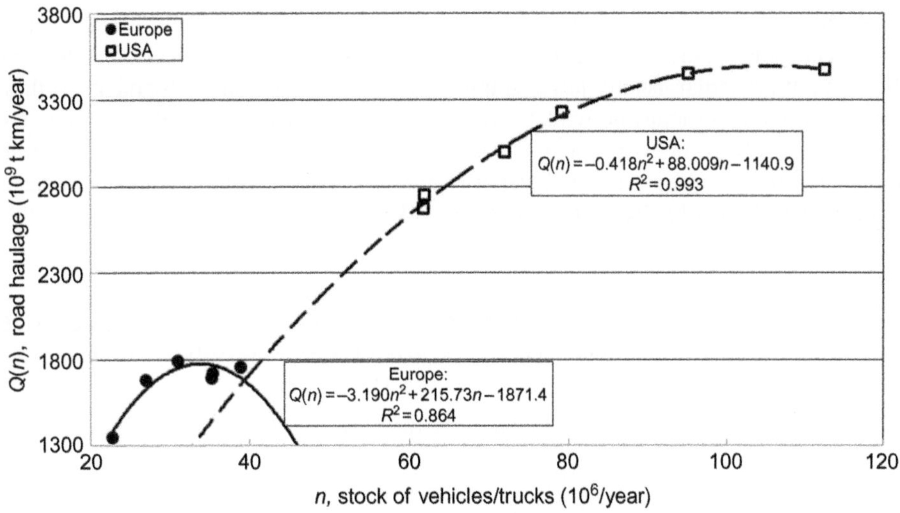

FIG. 2.20

Relationship between the annual volumes of freight/cargo/goods haulage by road and the number of vehicles/trucks in the EU-27/28 Member States and United States (1995–2013) (EU, 2015; USDT, 2013a,b,c).

As can be seen, the annual volumes of road haulage and the number of vehicles/trucks were for about 1.5–2.0 and 2–4 times, respectively, greater in the United States than in the EU-27/28 Member States. In both cases, the volumes of road haulage increased with increasing of the vehicle/truck stock at decreasing rate. In EU-27/28 Member States, the annual volumes also significantly stagnated despite further increasing of the vehicle/truck stock, which indicates its overall lower utilization. A similar but slighter situation happened in the United States during the observed period.

2.4.5 INTERURBAN RAIL TRANSPORT SYSTEMS

2.4.5.1 Introduction

Similarly as its road counterpart, the infrastructure network of the rail interurban passenger and freight transport systems spreads between and connects urban and suburban areas in the given regions and countries. An example of such rather global rail networks is the TEN-T (Trans-European Transport Network), which includes, in addition to the infrastructure network of other transport modes, the so-called comprehensive network of rail lines of 138,072 km and the so-called core network of these lines 68,915 km long. As in the case of the road infrastructure network, the former rail network provides all European regions (including peripheral and outermost regions) with an accessibility that supports their further economic, social, and territorial development as well as the mobility of their citizens. The latter rail network is of the strategic importance for the major European passenger and freight transport (http://ec.europa.eu/transport/themes/infrastructure/ten-t-guidelines/maps_en.htm).

Due to the historical reasons and latter specific local circumstances the rail infrastructure networks in different countries have usually been of different lengths. These infrastructure networks enable carrying out the passenger and freight/cargo/goods service networks. The nodes of the former are the corresponding stations usually located in the centers of urban areas where the transport services start and end. The nodes of the latter are the rail and rail/road intermodal terminals, rail shunting yards, and/or the doors of particular usually larger shippers and receivers of cargo shipments, if these being connected to the rail infrastructure network by the so-called industrial tracks. Table 2.4 gives the length of the 10 longest railway networks and their density in the corresponding countries (https://en.wikipedia.org/wiki/List_of_countries_and_dependencies_by_area; http://www.railway-technology.com/features/featurethe-worlds-longest-railway-networks-4180878/).

Table 2.4 Characteristics of the 10 Longest Rail Infrastructure Networks (https://en.wikipedia.org/wiki/List_of_countries_and_dependencies_by_area; http://www.railway-technology.com/features/featurethe-worlds-longest-railway-networks-4180878/)

Country	Area (10^6 km^2)	Length of Railway Network (10^3 km)	Network Density (km/km^2)
United States	9.162	250.0	0.0273
China	9.326	100.0	0.0107
Russia	16.378	85.5	0.0052
India	2.864	65.0	0.0227
Canada	9.094	48.0	0.0053
Germany	0.349	41.0	0.1174
Australia	7.634	40.0	0.0052
Argentina	2.764	36.0	0.0130
France	0.640	29.0	0.0453
Brazil	8.460	28.0	0.0033

As can be seen, the longest railway network is in the United States and the shortest in Brazil. However, the highest dense network is in Germany, followed by France and United States.

The lowest dense network is in Brazil. In these and other countries, the topology of railway networks has been influenced by location and distances between particular regions, urban areas/cities and towns to be connected. In general, these networks are used for both passenger and freight transport services with some exceptions, such as in the United States, where about 35,000 km of the railway lines are strictly dedicated to passenger transport and the rest to freight transportation.

Fig. 2.21 shows the simplified topology of the European railway "core network" for passengers with an indication of typical travel times between particular main urban areas/cities (http://www.eurail.com/plan-your-trip/railway-map).

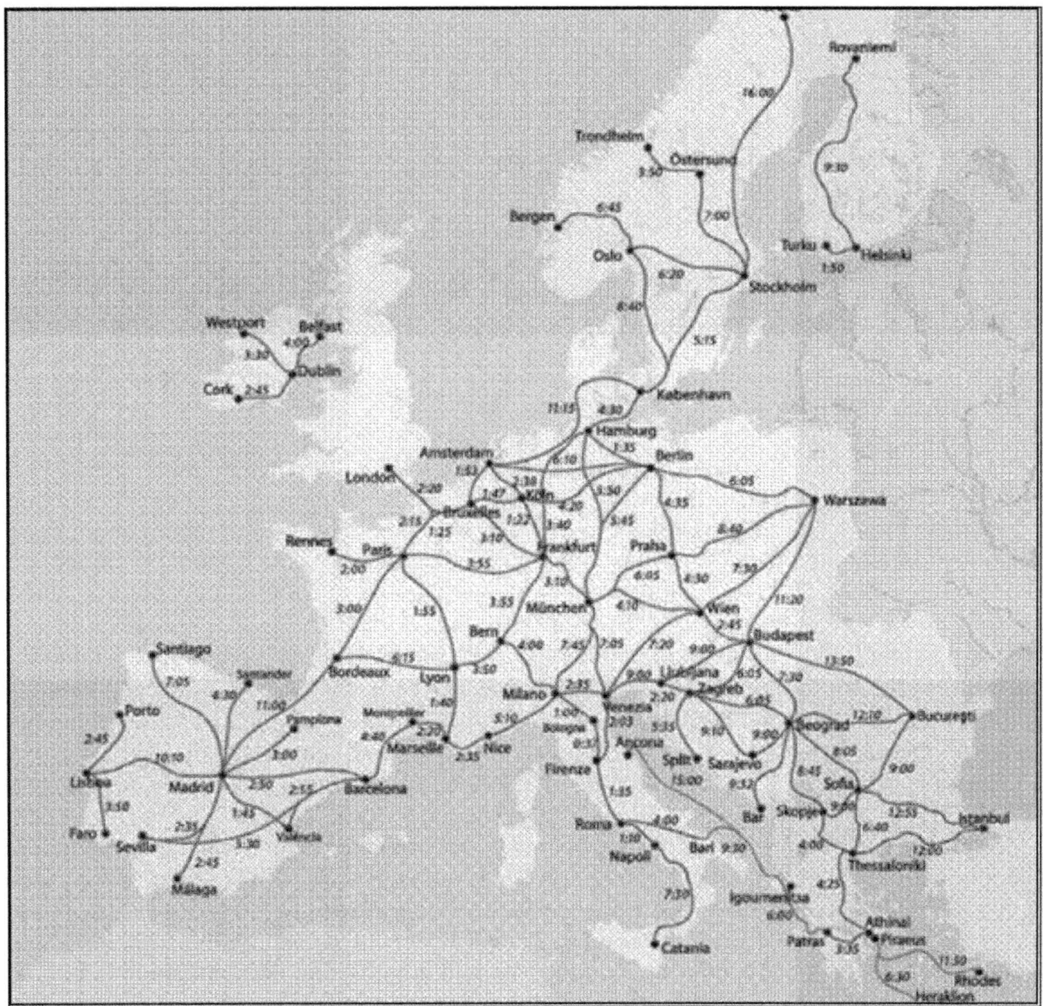

FIG. 2.21

European railway "core network" for passengers with typical travel times (http://www.eurail.com/plan-your-trip/railway-map).

The above-mentioned rail infrastructure has been used for the conventional passenger and freight trains.

2.4.5.2 Passengers

In addition to the above-mentioned conventional, the HSR transport networks developed in many countries worldwide has been fully dedicated for passenger services. These networks consist of lines with the rail tracks connecting the stations/stops along them and the end stations/terminuses. Topology of these networks has been mainly the country specific, but in general, three types can be distinguished as shown in Fig. 2.22 (Crozet 2013; http://www.johomaps.com/eu/europehighspeed.html).

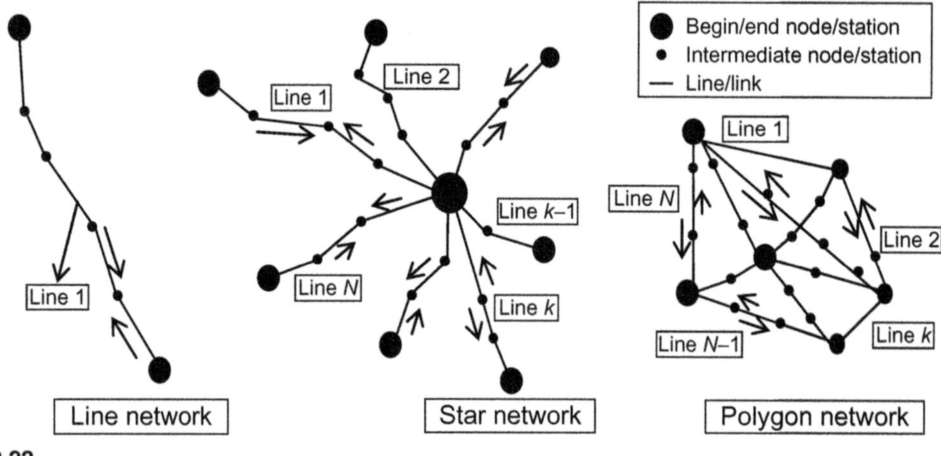

FIG. 2.22

Generic spatial topology of the HSR networks (Crozet, 2013; http://www.johomaps.com/eu/europehighspeed.html).

As can be seen, the spatial topology of the HSR networks has generally been as follows: line (for example, Italy), star (for example, France, Spain), and polygon (for example, Germany, China). In addition, most networks have the spatial topology combining these basic three. Fig. 2.23 shows the example of the star topology of Spain's HSR network (https://en.wikipedia.org/wiki/AVE).

Development of the HSR network infrastructure has been progressing in particular regions as given in Table 2.5.

As can be seen, the longest HSR network currently operating and being under construction is in Asia, mainly thanks to the fast developments in China (Ollivier et al., 2014), followed by that in Europe. The last are those in both Americas and Africa. Specifically, Table 2.6 gives some characteristic of the main grid (eight national backbone lines) of the HSR infrastructure network in China (Fu et al., 2015; Takagi, 2011; https://en.wikipedia.org/wiki/High-speed_rail_in_China/).

As can be seen, the main specificity of this (Chinese) compared to the other HSR rail networks, particularly those in Europe, is the length of lines, which varies between 1000 and 2400 km. In Europe, these lengths are much shorter and vary, for example, from 280 km between Berlin and Hamburg (Germany) to 770 km between Paris and Marseille (France) (UIC, 2014). However, the experience has shown that the

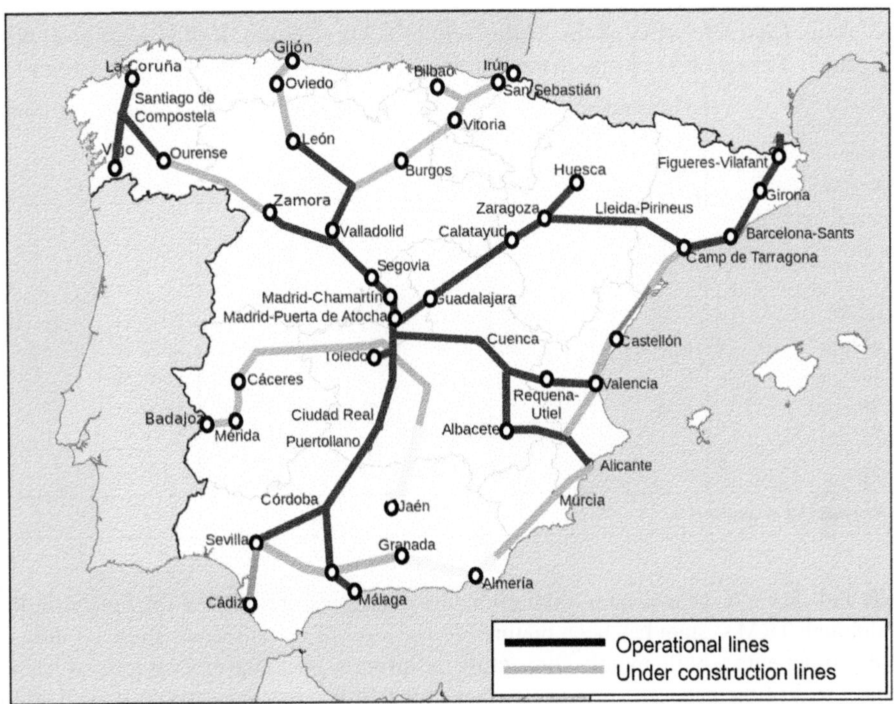

FIG. 2.23

Topology of the high-speed rail network in Spain (Europe) (https://en.wikipedia.org/wiki/AVE).

Table 2.5 Development of the HSR Networks at Particular Continents (CSP, 2014; Janić, 2016; UIC, 2014)

Status	Continent			World
	Europe	Asia	Others [a]	
In operation (km)	7351	15,241	362	22,954
Under construction (km)	2929	9625	200	12,754
Total (km)	*10,280*	*24,866*	*562*	*35,708*

[a] *Latin America, United States, Africa.*

average travel distances on some of these long Chinese lines have been about 560–620 km, which appears comparable to some of their (long) European counterparts (Fu et al., 2015).

The TRM (TransRapid MAGLEV—MAGnetic LEVitation) as an alternative HS (high speed) system is based on the Herman Kemper's idea of magnetic levitation dated from the 1930s. The magnetic levitations enables suspension, guidance, and propelling MAGLEV vehicles by magnets rather than by the mechanical wheels, axles, and bearings as at the HS (high speed) wheel/rail vehicles. Two

Table 2.6 Some Characteristics of the Main Grid of CRH (Chinese Rail High) Speed Network (Fu et al., 2015; Takagi, 2011; https://en.wikipedia.org/wiki/High-speed_rail_in_China/)

Relation	Orientation[a]	Length of Line (km)	Design Speed (km/h)
Beijing-Harbin	N-S	1800	350
Beijing-Shanghai	N-S	1318	350
Beijing-Hong Kong	N-S	2383	350
Hangzhou-Shenzhen	N-S	1499	250/350
Sublength:		*7000*	
Qingdao-Taiyuan	E-W	940	200/250
Xuzhou-Lanzhou	E-W	1434	250/350
Chengdu-Shanghai	E-W	2066	200/250
Kunming-Shanghai	E-W	2056	350
Sublength:		*6496*	
Total length:		*13,469*	

[a]N-S (north-south); E-W (east-west).

forces—lift and thrust or propulsion—both created by magnets are needed for operating the TRM vehicle. Although TRM system has been matured to the level of commercialization, its infrastructure has only been fragmentary built, mainly connecting the airport(s) with the city centers, which is still far from development of the network similarly as that of the HSR (Geerlings, 1998; Powell and Dunby, 2007). Table 2.7 gives the time milestones of developing the system (Janić, 2014).

Table 2.7 The Time Milestones of Developing TRM (TransRapid MAGLEV) System (Janić, 2014)

1970s	The Research on the Maglev Transportation Intensified (Japan, Germany)
1977	The first TRM (TransRapid Maglev) test line 7 km long built (the test speed achieved: 517 km/h) (Japan)
1993	The TRM test of 1674 km carried out (the achieved speed was 450 km/h) (Germany)
1990/1997	The Yamanashi TRM test line, which was 42.8 km long, had been constructed in the year 1990 and the first test carried out in the year 1997 (EDS (electro dynamic suspension)) (Japan)
2004	The first TRM line between Shanghai and its airport (China) was commercialized

In general, the TRM (TransRapid MAGLEV) system is characterized by: (i) infrastructure; (ii) rolling stock and operating speed; and (iii) commercial use.

Infrastructure

The main TRM infrastructure is a guideway consisting of the concrete (prefabricated) supporting piers and beams. The concrete piers depending on type of the guideway are located at different distances: 31 m—type I, 12 m—type II, and 6.19 m—type III. The beam laying on the solids can have different lengths: 62.92 m—type I, 24.78 m—type II, and 6.19 m—type III. The height of a beam can be 2 m—type I, 1 m—type II, and 0.4 m—type III. The total height of the guideway can be:

2.2–20.0 m—type I and II, and 1.35–3.5 m—type III. The beam carries the vehicle and provides the power to the entire system. Each type of beam constructed from steel, concrete, or hybrid (steel/concrete mixture) has the trapezoidal-like profile with the body and the track at the top. The above-mentioned three types of guideways can be single or double (or triple) track constructions. The scheme of double-track profile of the guideway is shown in Fig. 2.24 (Janić, 2014).

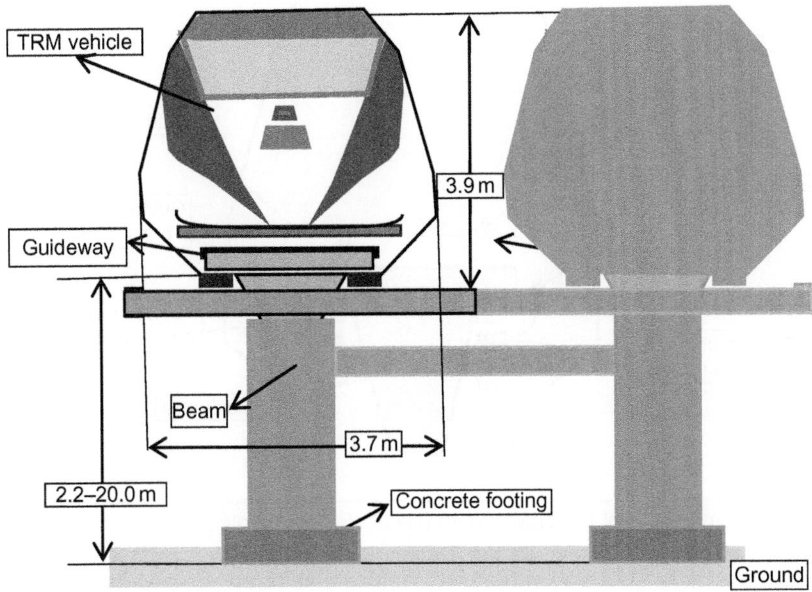

FIG. 2.24

A simplified scheme of the TRM system cross-section profile (Janić, 2014).

Rolling stock and operating speed

The TRM trains differ from the wheel/rail HS (high speed) trains in their operating. They levitate above the tracks supported by magnets and run on the principle of electromagnetism. As mentioned earlier, the magnets create two forces—lift and thrust or propulsion for operating the TRM vehicle. The lift force keeps the vehicle above the guideway at the distance of 10–15 mm during the trip. The propulsion force enables acceleration and deceleration as well as overcoming the air resistance during the cruising phase of the trip. Thanks to levitation, there is no friction between the tracks and the vehicle's wheels, thus enabling TRM trains to operate at the much higher speed. The electric energy is consumed for generating both lift and thrust force, but in much greater proportion for the latter than for the former force. In addition, a part of the energy spend during deceleration is returned to the network, thus also indicating possibilities for saving energy similarly as at the HSR.

Levitation and propulsion

Two technologies for TRM have been developed for full commercial implementation: EMS (electromagnetic Suspension) and EDS (electro dynamic suspension). They enable performing three basic functions of the TRM vehicles: (i) levitation above the track; (ii) propulsion enabling moving forward in terms of acceleration, cruise, and deceleration; and (iii) guidance implying maintaining stability along the guideway. Fig. 2.25 shows the main components enabling the above-mentioned movement (Janić, 2014).

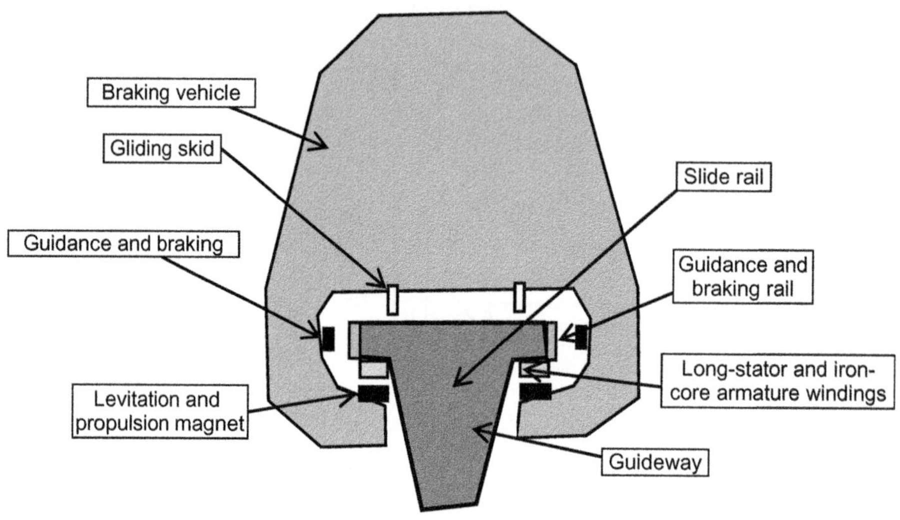

FIG. 2.25

Scheme of the main components of the TRM system (He et al, 1992; Janić, 2014).

Control systems

Contrary to the conventional rail and HSR, the TRM system independently on the technology does not have the outside signaling system. It is characterized by the fully automated and communication systems controlled by the computer system. This consists of the main computer in the system's command center and that on board the vehicle(s). These computers continuously communicate between each other thus providing the necessary monitoring and control/management of the driverless vehicles along the line(s).

In addition, the TRM trains change the tracks by using the bending switches, which consist of bending beams with the drive units installed on the every second solid of the bending switch. There are low-speed and high-speed switches. The former are used near and at the stations enabling passing between the tracks at the speed of about 100 km/h. The former are used along the main portion of the guideway enabling switching between the tracks at speed(s) of about 200 km/h.

Weight and energy consumption

The typical empty weight of the TRM train is about 50 ton, which with the weight of payload (passengers) of about 20 ton gives the total gross weight of 70 ton. A portion of the empty weight of a vehicle represents the weight of magnets on board. If the magnet force to maintain levitation of the vehicle is about 1–2 kW, then the total energy consumed for this part could be 70–140 kW. Table 2.8 gives the main technical/technological and operational characteristics of the TRM 07 train(s) (Janić, 2014).

Table 2.8 Technical/Technological and Operational Characteristics of TRM (TransRapid MAGLEV) 07 Train(s) (Janić, 2014)

Characteristic	Value
Carriages/sections per train	5/(2–10 possible)
Length (m)	128.3
Width (m)	3.70
Height (m)	4.16
Net weight of a train (ton)	247
Seating capacity (average) (seats)	446
Weight/seat ratio (average)	1.80
Axle load (tons/m)	1.93
Technical curve radius (m)	2825–3580
Maximum engine power (MW)	25
Lateral tilting angle (degree)	12–16
Operating speed (km/h)	400–500
Maximum acceleration/deceleration (m/s^2)	0.8–1.5

The TRM trains can operate at the speeds from about 10 to 15 km/h to about 400–500 km/h due to the lack of physical contact with the dedicated guide-way. This offers a couple of benefits compared to the other HS systems such as, for example, HSR. The first is the energy consumption as shown in Fig. 2.26 (Janić, 2014; Lukaszevicz and Anderson, 2009; TIG, 2012).

As can be seen, at both HSR and TRM trains, the specific energy consumption (SFC) increases with increasing of the operating speed more than proportionally, but remains lower at the latter (TRM) train at any speed. The other type of benefits is ability of the TRM to cover the travel distances between 500 and 800 km, also typical for the HSR and the short-haul air transport operations, in 1–2 h, respectively, which qualifies it as both competitive and complementary alternative to the other two HS systems. Consequently, the TRM will be able to connect centers and edges of the major cities with airports on the one hand, and provide an intermodal connection with the existing air passenger transport, HSR, and urban-mass transit systems, on the other. However, similarly as in the case of HSR, both national and/or international dedicated infrastructure networks for TRM may take decades to develop up to the level to have a significant impact on the existing and prospective transport market(s) (Powell and Dunby, 2007).

Commercial use

Up to date, the individual TRM lines have been constructed for different purposes. In particular, in addition to five testing tracks and four tracks (lines) under construction, only three lines have been commercialized for public use as follows (http://en.wikipedia.org/wiki/Maglev):

- The Shanghai Maglev Train as the German's TRM design, which has started operations in the year 2004 in Shanghai (China) by covering the distance of 30 km between the city of Shanghai and its airport in 7 min; this implies an average commercial speed of about 268 km/h.

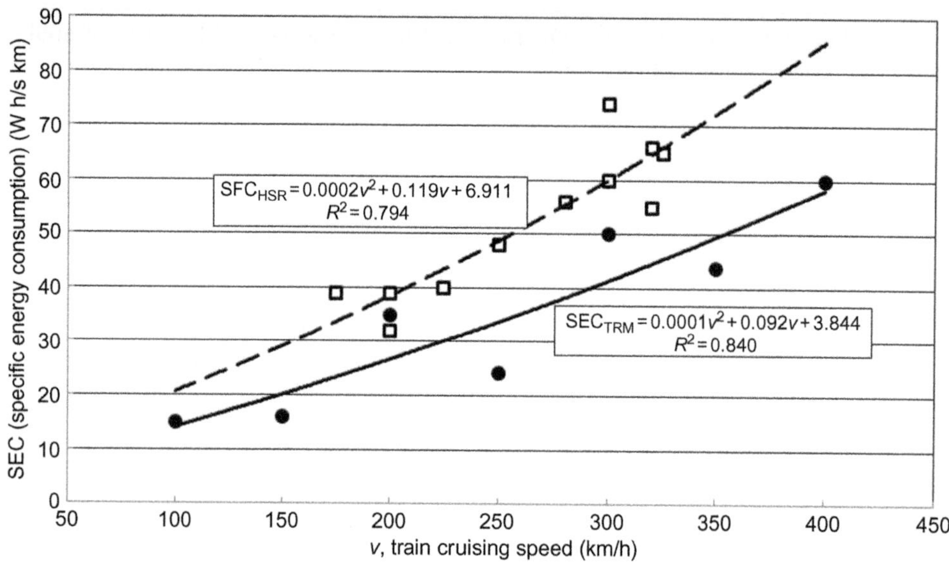

FIG. 2.26

Relationship between SFC (specific energy consumption) and cruising speed of TRM07 and HS train(s) (Janić, 2014; Lukaszevicz and Anderson, 2009; TIG, 2012).

- The "Urban MAGLEV," also known as the Tobu-Kyuryo Line, ie, the Linimo system, started operating in 2005 in Aichi (Japan); the length of line is 9 km with nine stations, a minimum curve radius of 75 m, and a maximum gradient of 6%; the trains almost free of noise and quite resistant to disruptions by bad weather operate at the maximum speed of 100 km/h, thus offering highly regular and reliable services (nearly 100%); consequently, almost 10 million passengers used this system during first 3 months of its operation.
- The urban MAGLEV system UTM-02 in Daejeon (South Korea) has started operations in the year 2008; the length of line is 1 km connecting Expo Park and National Science Museum.

2.4.5.3 Freight shipments

In many countries, the rail freight transportation services share the parts of infrastructure networks—rail lines—with their passenger counterparts, which connect the above-mentioned network nodes (the rail and rail/road intermodal terminals, rail shunting yards, and/or the doors of particular freight/cargo/goods shippers and receivers). Consequently, these service networks can have different topologies some of which area "point-to-point," "trunk line with collecting/distribution forks," "hub-and-spokes," and "line" ("ring") bundling networks. Some simplified schemes of these networks are shown in Fig. 2.27A–D.

Fig. 2.27A shows a topology of the "point-to-point" network, which serves regular and rather substantive volumes of cargo shipments between two terminals. These can be rail/rail consolidation/deconsolidation or rail/road intermodal terminals. In general, the network with rail/rail terminals operates as follows: the cargo shipments are delivered from their "local" origins—doors of their shippers to the origin terminal (*A*) by rail haulage. Then, they are loaded onto direct trains and transported to destination rail/rail terminal (*B*). From there, they are distributed (again by rail haulage) to their "final"

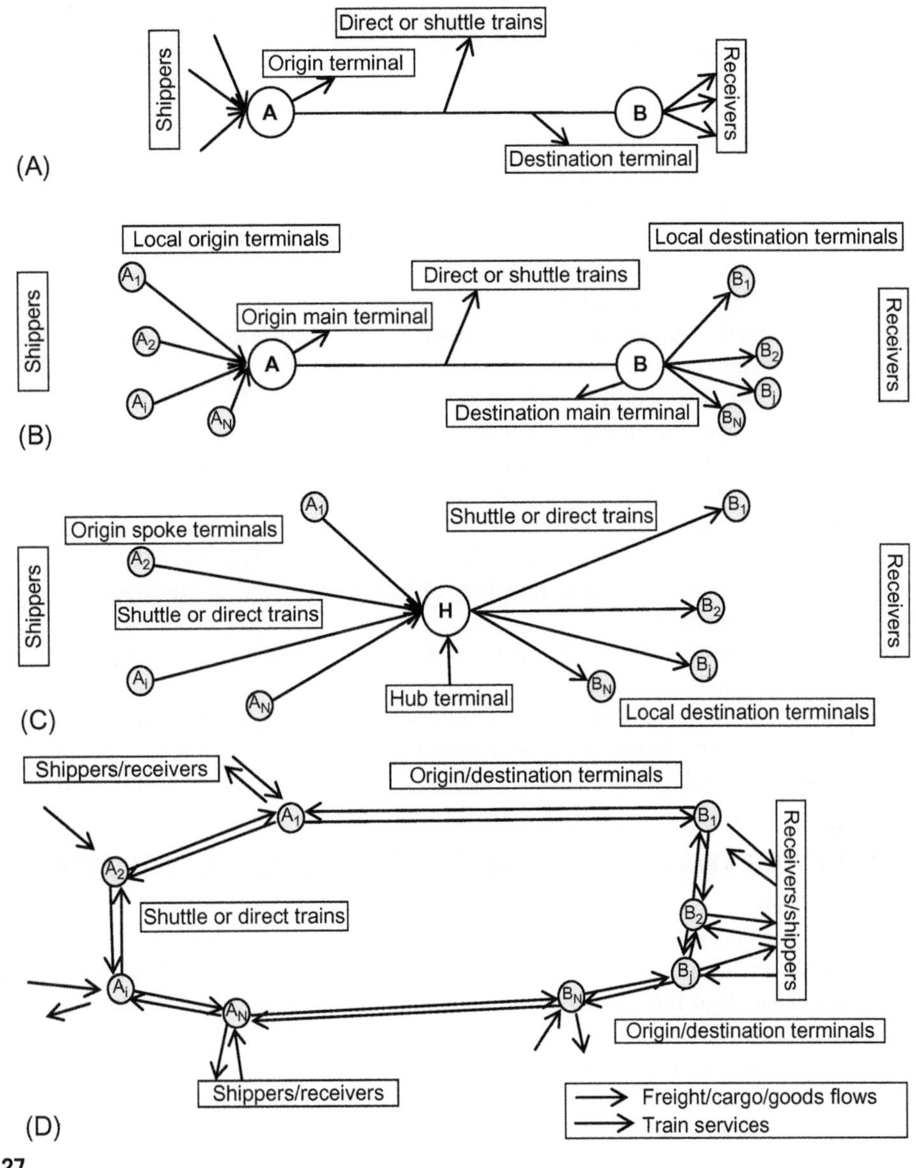

FIG. 2.27

Some topologies of the rail freight transport service networks (Janić et al., 1999). (A) Point-to-point, (B) trunk line with collecting/distribution forks, (C) hub and spokes, and (D) line or ring.

destinations—doors of their receivers. In the case of using the rail/road intermodal terminals, the local collection from the shippers and the local distribution to the receiver is carried out by road trucks.

Fig. 2.27B shows a topology of the "trunk line with collecting/distribution forks" network. It generally covers a wider area around the main origin and destination terminal(s) (*A*) and (*B*), respectively, of the "point-to-point" network. The network includes the "local" terminals (A_i, $i=1, 2, \ldots, N$; B_j, $j=1, 2, \ldots, M$) allocated to the main ones (A) and (B), respectively. On the one side, the "local" terminals are connected to the doors of shippers and receivers by rail, if being rail, and by road, if being the intermodal rail/road. In the former case, the main terminals (A) and (B) can also be rail shunting yards. On the other, thee "local" terminals are connected by rail to the main, this time rail terminals, which can also be large shunting yards. Handling of freight shipments in this network is carried out as follows: the shipments are transported between the "local" origin terminals and the main terminal (A) by direct or shuttle "feeder" trains. At the terminal (A) they are consolidated enabling scheduling the longer direct or shuttle trains along the (trunk) route (l_{AB}) to the destination main terminal (B). There these longer trains are deconsolidated and the shorter direct or shuttle trains dispatched toward the "local" destination terminals.

Fig. 2.27C shows that the "hub-and-spokes" network usually consists of a single hub central node—hub (H), and several peripheral nodes—spokes (S_i) ($i=1, 2, \ldots, N$). The hub is mainly the rail terminal. The spokes can be the "local" rail or the rail/road intermodal terminals. They are connected to the hub by direct or shuttle trains carrying the cargo shipments. At arriving from the shippers at the origin spoke terminals by the shorter trains or road trucks, the freight shipments are transshipped to the direct or shuttle trains, which are dispatched to the hub terminal (H). The shipments can pass through the hub (H) either by staying on the same carriage units (wagons) all the time (direct train) or by changing them. Then, the direct or shuttle trains are dispatched from the hub (H) to the destination spoke terminals, where freight shipments are again transshipped from these to the shorter trains or road trucks and then forwarded to their receivers.

Fig. 2.27D shows that the "line" or "ring" network has a line or ring configuration where the rail or rail/road intermodal terminals are located in line or ring in relation to the direction of flows of freight/cargo/goods shipments. The direct and/or shuttle train services usually connect these terminals. The exchange of freight shipments between two transport modes and their systems takes place at these terminals, which also can be locations of their entry and leave of the network (Janić et al., 1999).

2.4.6 INLAND WATERWAYS AND SEA SHIPPING SYSTEMS FOR CARGO SHIPMENTS

2.4.6.1 Introduction

The infrastructure networks of inland waterways and sea shipping systems for cargo shipments are, as expected, quite different. That of inland waterways systems consists of the natural navigable rivers and the artificial-built channels as links and the corresponding ports as nodes. As such they spread over a part or entire country and a continent. For example, in Europe, the inland waterways infrastructure network, which is 23,506 km long, represents a component of the above-mentioned TEN-T network (http://ec.europa.eu/transport/themes/infrastructure/ten-t-guidelines/maps_en.htm). These networks enable setting up the transport service systems by the vessel/barge operators and logistics companies involved.

The infrastructure network of sea shipping systems consists of two components. The first global includes the so-called sea lanes, sea roads, or shipping lanes, which are regularly used by the sea ships on oceans and seas as links, and the sea ports as nodes. This part of the network spreads globally between continents. The other part consists of the sea lanes closer to the coasts of the countries facing the sea and/or ocean and connecting the ports of a single or of the neighboring countries. In both cases, these sea lanes are commonly used for setting up the corresponding transport service systems. Fig. 2.28 shows the simplified scheme of the global sea lanes used for transporting containers.

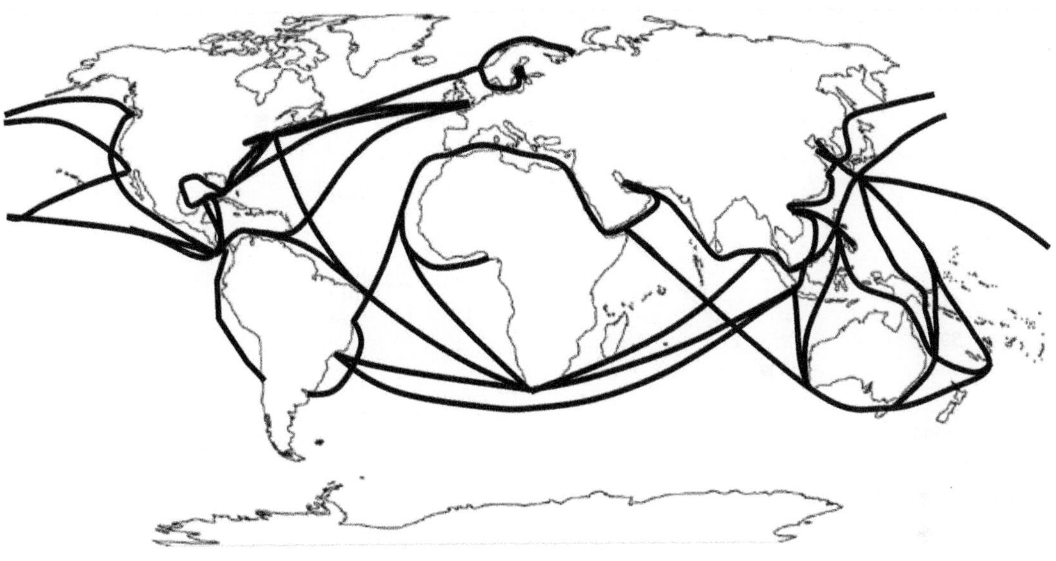

FIG. 2.28

Simplified scheme of the global sea lane network used for trade (WSC, 2013).

As can be seen, these lanes connect the main ports at the same and different continents, thus representing a global part of the infrastructure network of sea shipping systems.

2.4.6.2 Inland waterways

The topology of infrastructure network of the inland waterways is determined by the layout of navigable rivers and channels. The ports as nodes of these networks are usually located at in or close to the urban areas and towns where the pass through. In addition, these ports can be located within the area of sea large ports. The length of inland waterways networks varies across particular regions and countries. For example, the length of inland waterways in Europe consisting of navigable rivers, canals, and lakes regularly used for transport has been increasing over time and reached almost 42,000 km as shown in Fig. 2.29 (EU, 2015).

The extension of the above-mentioned inland waterways infrastructure network has been carried out by building new channels and regulating parts of the rivers such as Rhine and Danube. In general, the inland waterways in Europe are classified according the horizontal dimensions of the motor vessels and barges and pushed convoys, and especially by their width (ECMT, 1992). Table 2.9 gives an example of six categories of the motor vessels and barges as the basis for classification of these links.

As can be seen, the inland waterways are classified into seven classes respecting seven characteristics of the motor vessels and barges as such as: beam, length, draught, and tonnage. In some other countries the main inland waterways are rivers determining the infrastructure network topology. An illustrative example is the Mississippi river in the United States whose simplified layout with its tributaries is shown in Fig. 2.30 (Clark et al., 2005).

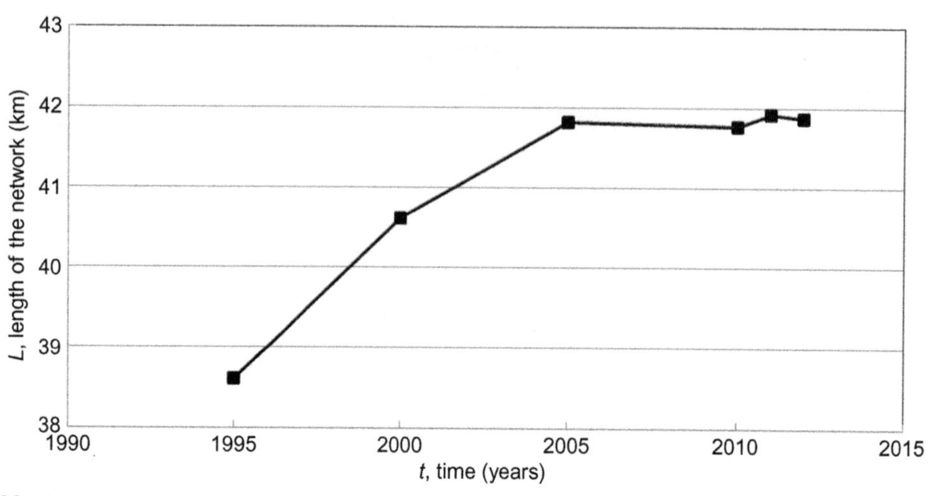

FIG. 2.29

Development of the European network of inland waterways (1995–2012) (EU, 2015).

Table 2.9 Categorization of the Inland Waterways Motor Vessels and Barges in Europe (ECMT, 1992)

Class of Navigable Waterways	Max. Length (m)	Max. Beam (m)	Draught (m)	Tonnage (ton)
I	38.50	5.05	1.80–2.20	250–400
II	50–55	6.60	2.5	400–650
II	67–80	8.20	2.50	650–1000
IV	80–85	9.50	2.50	1000–1500
Va	95–110	11.40	2.50–2.80	1500–3000
VIa	95–110[a]	11.40	2.50–4.50	3200–6000

[a]*Pushed convoys.*

As can be seen, topology of the given network is completely determined by the river's layout and the layout of parts of its navigable tributaries. The navigable length of the Mississippi river consists of three parts: the Upper Mississippi, which is about 1069 km long, including the series of man-made lakes located between Minneapolis and St. Louis (Missouri); the Middle Mississippi, which is 310 km long, spreading from St. Louis (Missouri) to its confluence with the Ohio River at Cairo (Illinois); and the Lower Mississippi, which is 1600 km long, between Cairo (Illinois) and its delta at the Gulf of Mexico. These make its total length as an inland waterway of 2979 km (the total length of the Mississippi river is 3734 km). In addition, the total length of the US inland waterways infrastructure network is about 41,000 km of which 18,000 is of depth <2.75 m (ECMT Class I), in China it is 110,000 km of which just 25,000 km with depth >2.5 m (ECMT Classes higher than I), and in Russia it is 145,000 km of which 90,000 km with depth <2.5 m (ECMT Class I) (The ECMT classification in

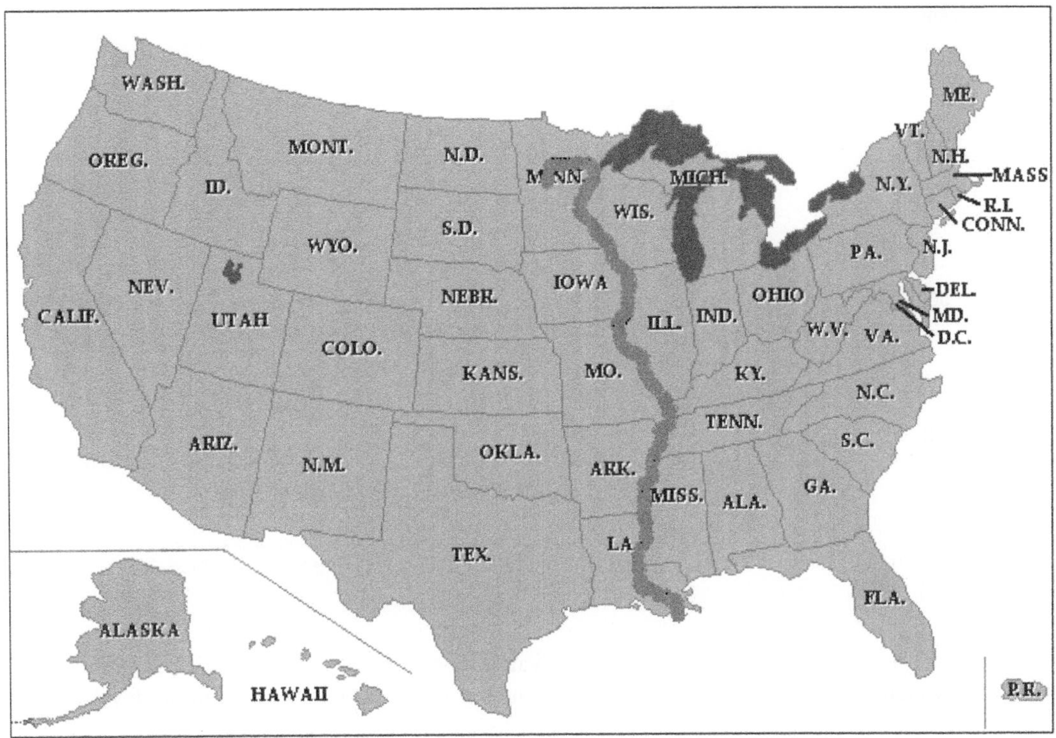

FIG. 2.30

Inland waterways network—the Mississippi river and its tributaries (United States) (Clark et al., 2005).

Table 2.9 is mentioned as a comparison (https://people.hofstra.edu/geotrans/eng/ch3en/conc3en/lengthwaterways.html)).

The topology of transport service networks operated by inland waterways systems is similar to the schemes shown for the freight railways. The differences are that the nodes are the inland waterway ports and the transport services between these ports are carried out by inland waterways vessels/barges of different size/capacity. The cargo shipments arrive from their inland origins to particular ports and depart to their inland destinations from these ports by road trucks or rail freight trains. Under such conditions, these ports operate as the multimodal transport nodes (Janić et al., 1999; An et al., 2015).

The inland waterways freight transport systems have been carried out rather substantive quantities of freight/goods shipments. Fig. 2.31 shows an example of the transported quantities of cargo shipments in the United States during the observed period (USACE, 2013).

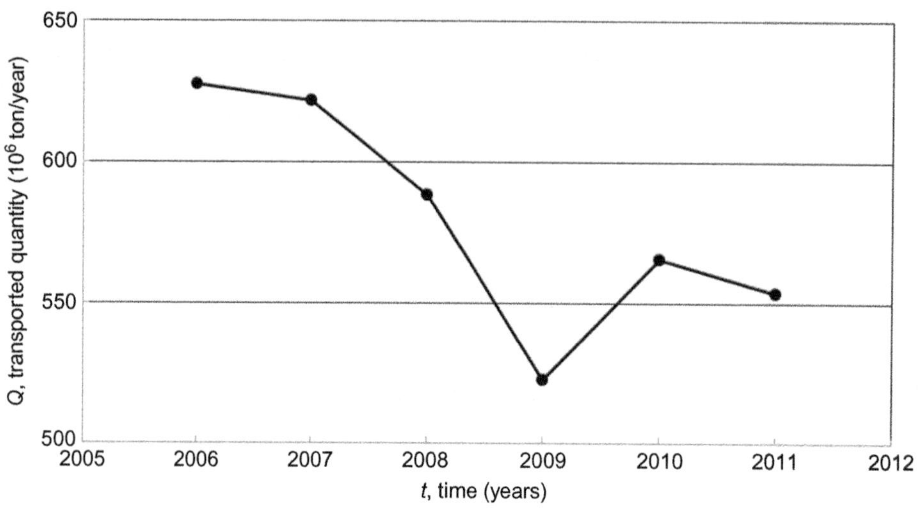

FIG. 2.31

Development of the US inland waterways freight transportation (excluding Great Lakes) (2006–2011) (USACE, 2013).

As can be seen, the transported quantities had varied over time with a substantive fall in the year 2009, after which the recovery started but over the next 4 years not reached the prefall levels.

2.4.6.3 Sea shipping

The topology of transport service networks of shipping lines depend on the shipping lines decisions. Each shipping line has been developing its own network. The choice of ports as the nodes of network of a given shipping line been mainly influenced by the volumes of expected demand between particular ports and the specific characteristics of these ports such as: geographical location, infrastructure and handling facilities and equipment, hinterland accessibility, efficiency, operational and economic performances of ground access systems, and overall efficiency of administration. An additional important factor has been imbalance between freight flows on the routes between particular ports. Under conditions of balancing between shippers' demands and supply of transport services, generally four different topologies of shipping line transport service networks have emerged as follows: (i) Line networks where transport services are carried out by the same ship(s) between any two ports as their and cargo shipments origins and destinations; (ii) Pick-up-and-delivery networks when transport services are carried out by the same ship(s) visiting the ports along the long sea lane/route(s) between the origin and destination port(s) and picking-up and delivering freight/cargo/goods shipments there; (iii) Hub-and-spoke networks where the hubs are the large sea ports at both ends of the long sea lane/route and the spokes are the sea ports close to these (hub) ports; the smaller ships provide feeder services between the spoke and hub ports, and vice versa, and the larger ships carry out the transport services between the hubs; (iv) Different combinations of the previous networks (Song and Panayides, 2015).

The sea ports as the nodes of sea shipping lines' networks generally operate as multimodal transport nodes facilitating the inland access systems such as road, rail, and inland waterways on the land and the shipping lines systems on the sea side. Actually, they enable transfer of freight shipments between their two sides and corresponding transport service systems. Over time, many sea ports, in addition to others, have also increasingly been developing container terminals in order to handle increased volumes of containerized cargo shipments. Fig. 2.32 shows an example of development of the number of handled containers at the six largest container ports during the specified period of time (2004–13) (http://www.iaphworldports.org/Statistics.aspx).

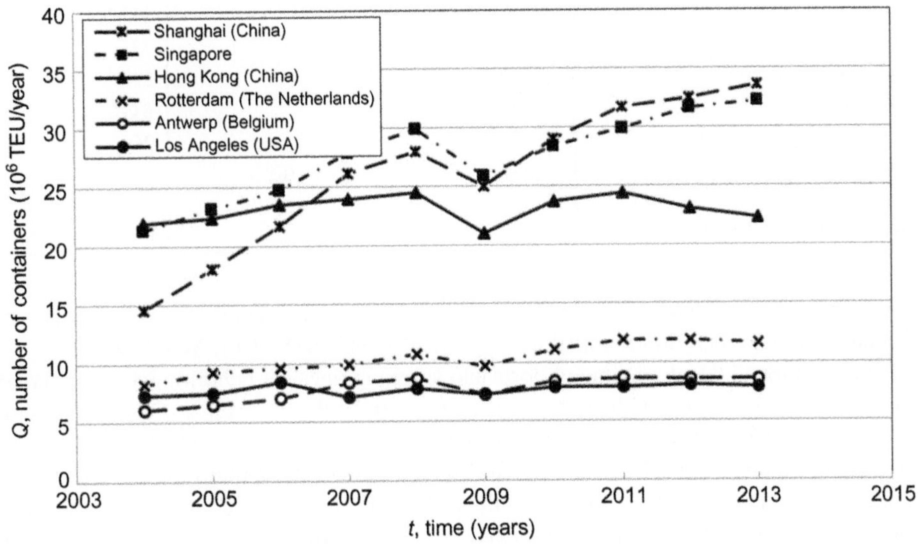

FIG. 2.32

Development of the container transport at selected ports (2004–2013) (http://www.iaphworldports.org/Statistics.aspx) (*TEU*, twenty foot equivalent unit).

As can be seen, the annual number of handled containers (TEUs) at three ports in the Far East have been about 2.5–3.0 time greater than that at two Western European and one US port during the observed period of time. At all ports, the number of handled containers was increasing, but with an observable fall in the year 2009, mainly due to the impact of global economic crisis. In addition, the first two Far East ports are the largest in the world, while the two European and one United States are the largest in the continent and the country, respectively.

The shipping lines have spread their service networks, which in most cases have global character. This implies that transport services are carried out between the ports located at different continents and their particular regions. In particular, as mentioned earlier, the container transport has been developing globally. Table 2.10 gives an example about such global traded container transport volumes between particular world's regions and transport service frequencies (WSC, 2013).

As can be seen, the most voluminous container trade was between Asia and North America and North Europe. However the transport service frequency was higher in the former than in the later cases, which could be an indication that the container ships of the smaller size/lower payload capacity were used. It should also be pointed out that about 500 shipping line service frequencies per week were

Table 2.10 The Volumes of Traded Containers Between Origin and Destination World's Regions (TEUs) (2013) (WSC, 2013)

Route	Traded Containers (Both Directions) (10^6 Year^{-1})	Transport Services (Week^{-1})
Asia-North America	23.125	73
Asia-North Europe	13.706	28
Asia-Mediterranean	6.739	31
Asia-Middle East	5.014	72
North Europe-North America	4.710	23
Australia-Far East	2.923	34
Asia-East Coast South America	2.131	26
North Europe/Mediterranean/East Coast South America	1.680	–
North America-East Coast South America	1.306	–

regularly scheduled on the routes between above-mentioned regions including also those to/from West Coast of South America, and South and West of Africa (UNCTAD, 2014; WSC, 2013).

The above-mentioned developments of ports and shipping lines for handling increasingly containerized volumes of cargo shipments have been possible also thanks to increasing the size of container ship fleets in terms of both number of ships and their payload capacity. In addition fewer and fewer ships have been equipped with their own "gear," which has required many ports to provide ships-to-shore cranes for loading and unloading them ("gear" handling crane on board the ship). Table 2.11 gives an example of the world's five largest shipping lines at the beginning of 2014 (UNCTAD, 2014).

Table 2.11 Some Characteristics of the Largest Container Shipping Lines (2014) (UNCTAD, 2014)

Shipping Line/Operator	Number of Ships	Total Capacity (10^6 TEU)	Average Ship Size (TEU/Ship)
Mediterranean Shipping Company S.A.	461	2.609	5660
Maersk Line	456	2.506	5495
CMA CGM S.A.	348	1.508	4333
Evergreen Line	229	1.102	4813
COSCO Container Lines Limited	163	0.880	5397

As can be seen the largest shipping line, Mediterranean Shipping Company S.A., operates the largest number of ships of the largest average capacity. In addition, Table 2.9 implicitly contains information that the most shipping lines usually operate fleets of ships of different size/payload capacity but on average: 1/3 of ships of 10,000 TEU or larger, 1/3 of ships of 5000–9999 TEU, and 1/3 of ships of under 4999 TEU. In general, transatlantic, transpacific, and Europe-Asia transport services are carried out by the ships of payload capacity between 5000 and 13,000 TEU and larger. The smaller ships with capacity lees than 5000 TEU are mainly used for the regional/continental transport services. Additionally, the average container ship size/payload capacity deployed per particular country has been increasing over time as shown in Fig. 2.33 (UNCTAD, 2014).

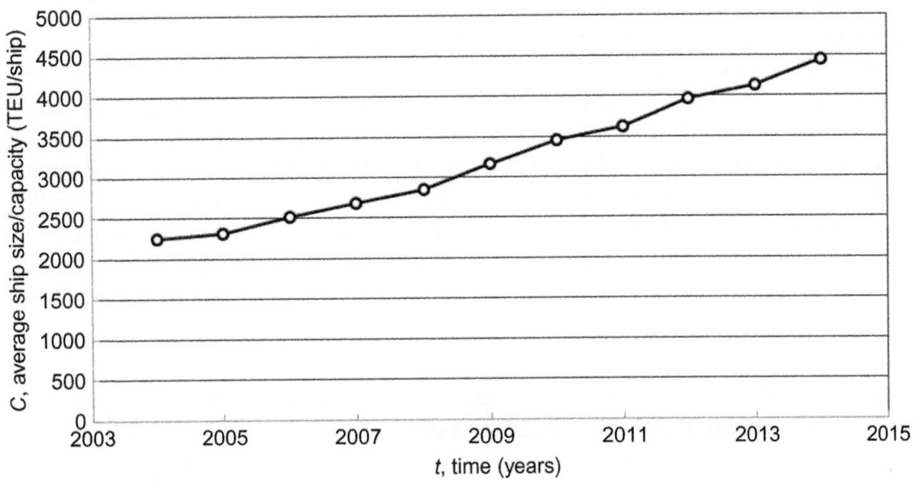

FIG. 2.33

Average ship size deployed per country (2004–2014) (UNCTAD, 2014).

As can be seen, the average ship size deployed by shipping lines in particular countries increased during the observed period. At the same time, the average fleet size was relatively constant: 130–135 ships/country. Also, it should be mentioned that the container ship fleet shared about 12.8% in the total dwt of the world's ship fleet in the year 2014, with increasing trend during the period 1980–2014. The share of dry bilk ship fleet increased from 27.2% in the year 1980 to 42.9% in the year 2014. The shares of the general cargo and oil tanker fleets decreased from 17.0% and 49.7%, respectively, in the year 1980 to 4.6% and 25.5%, respectively, in the year 2014 (dwt—deadweight ton, ie, a measure of the ship's payload capacity) (UNCTAD, 2014).

2.4.7 AIR TRANSPORT SYSTEM

2.4.7.1 Introduction

The air transport system generally includes airports, ATC (air traffic control) system, and airlines. The airports represent the ground part of the system's infrastructure handling the aircraft operated by different airlines transporting passengers and freight/cargo shipments. The organized and controlled airspace between airports represents the air part of the system's infrastructure. The ATC system provides guidance to aircraft while flying through the controlled airspace between airports and during their ground movements at the airports themselves. These aircraft are operated by airlines generally categorized into two classes: those, which primary transport passengers and to the limited extent cargo shipments; and those, which exclusively transport cargo shipments.

2.4.7.2 Airports

Airports located in a particular area, such as a country, group of neighboring countries, and/or continents with the controlled airspace and air routes through it, create the air transport system infrastructure network(s). The topology of this network in the horizontal plane depends on the micro location of airports within the above-mentioned area(s). Usually, they are located close to the urban areas and towns generating and attracting sufficient passenger and freight/cargo demand thus making themselves viable to be connected by airlines providing air transport services. Some larger urban areas can have, ie, are served by few airports in their vicinity. Consequently, it is not possible to extract and define generic topology of the airport network considered in the given context. As far as the particular airports are concerned, the layout of each of them is primarily dependent of the number and configuration of runways. The number of runways can be, for example, one at the airports up to six at the very large airports. In general, the size of the airports is expressed by the number of passengers and aircraft movements (atm) handled during the specified period of time, in this case during a year (1 atm is one aircraft landing or take-off). Table 2.12 gives the volumes of traffic in terms of the annual number of passengers and atms at the 15 world's largest airports (http://www.aci.aero/Data-Centre/Annual-Traffic-Data/Passengers/2013-final).

Table 2.12 The Traffic Volumes at the 15 Largest Airports in the World (2013) (http://www.aci.aero/Data-Centre/Annual-Traffic-Data/Passengers/2013-final)

Rank	Airport	Location	Atm (10^3)	Passengers (10^6)
1	Hartsfield-Jackson Atlanta International	Atlanta, Georgia, United States	911.074	94.43
2	O'Hare International	Chicago, Illinois, United States	883.28	66.88
3	Dallas/Fort Worth International	Coppell, Euless, Grapevine, and Irving, Texas, United States	678.06	60.44
4	Los Angeles International	Los Angeles, California, United States	614.92	66.70
5	Denver International	Denver, Colorado, United States	582.65	52.56
6	Beijing Capital International	Chaoyang-Shunyi, Beijing, China	567.76	83.71
7	Charlotte/Douglas International	Charlotte, North Carolina, United States	557.95	43.46
8	McCarran International	Las Vegas, Nevada, United States	520.99	41.86
9	George Bush Intercontinental	Houston, Texas, United States	496.91	39.87
10	Paris-Charles de Gaulle	Seine-et-Marne, Seine-Saint-Denis, Val-d'Oise, Île-de-France, France	478.31	62.05
11	Frankfurt	Frankfurt, Hesse, Germany	472.69	58.04
12	London Heathrow	London, United Kingdom	471.94	72.39
13	Phoenix Sky Harbor International	Phoenix, Arizona, United States	459.43	40.32
14	Amsterdam Schiphol	Haarlemmermeer, Netherlands	44.01	52.57
15	Philadelphia International	Philadelphia, Pennsylvania, United States	43.884	30.23

As can be seen, at the first five places were the US airports, followed by the Chinese one. The largest European airports—Paris Charles de Gaulle, Frankfurt, London Heathrow, and Amsterdam Schiphol—were in 10th, 11th, 12th, and 14th place, respectively. It is also evident that at some airports the number of atms and the number of passengers are not in the same order at given airports: some are ranked higher in terms of the number of atms and lower in terms of the number of passengers, and vice versa. In the former case, this implies that more atms (flights) are carried out by smaller aircraft. In the latter case, this implies opposite.

2.4.7.3 Transport service networks

The air transport service networks for both passengers and freight/cargo shipments are provided by airlines. At present, the passenger airlines operate according to two business models: conventional and LC (low cost). The conventional airlines usually use heterogeneous fleets consisting of different numbers of aircraft of different size, ie, capacity in terms of the number of seats and overall payload. The LC airlines also called LCC usually operate a single aircraft type of a given size/capacity. In addition, in order to guard market position and even strengthen it, many larger passenger airlines have created the airline alliances including themselves and several smaller airlines. Such development has influenced the topology of networks of these and other airlines, which can generally be "point-to-point," "hub-and-spokes," and a global "multi hub-and-spokes" as shown in Fig. 2.34.

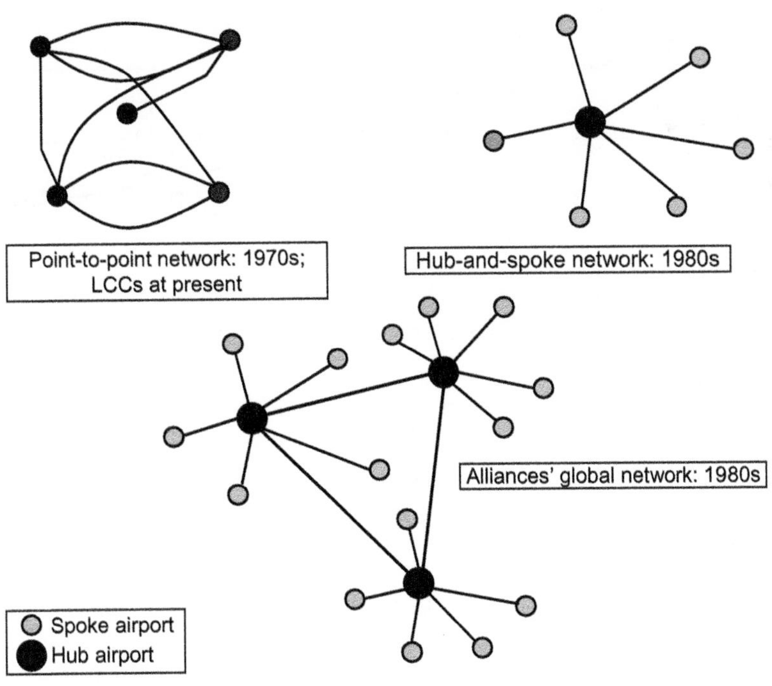

FIG. 2.34

Scheme of the topology of airline networks.

As can be seen, the particular networks can have different number of nodes-airports and links—routes connecting them. In general, for the given number of airports-nodes, the fully connected "hub-and-spoke" networks have the smaller number of links-routes than their "point-to-point" counterparts. This enables the airlines switching operations from the former to the later configuration of the network to schedule more flights on the smaller number of routes by the same number of aircraft in their fleets, ie, to concentrate its transport services. In the "point-to-point" networks, the airlines schedule direct flights between particular airports as origins and destinations of passengers. In the "hub-and-spoke" networks, the airlines schedule direct flights between the spoke airports and the hub. These direct incoming and outgoing flights are coordinated at the hub airport in order to enable transfer of those passengers whose origins and destinations are the spoke airports. In addition, some passengers also have the hub airport as origin and/or destination. The similar happens at the "multi hub-and-spokes" networks, which enable multi transfer of passengers at particular hubs while on routes between their origin and destination spoke airports.

The experience so far has shown that, generally, greater numbers of passengers have traveled on the shorter than on the longer routes. Fig. 2.35 shows an example of the relationship between the annual number of passengers and length of ten world's busiest air routes during the period 2011–2012 (https://www.iata.org/publications/pages/wats-passenger-carried.aspx; https://en.wikipedia.org/wiki/World%27s_busiest_passenger_air_routes).

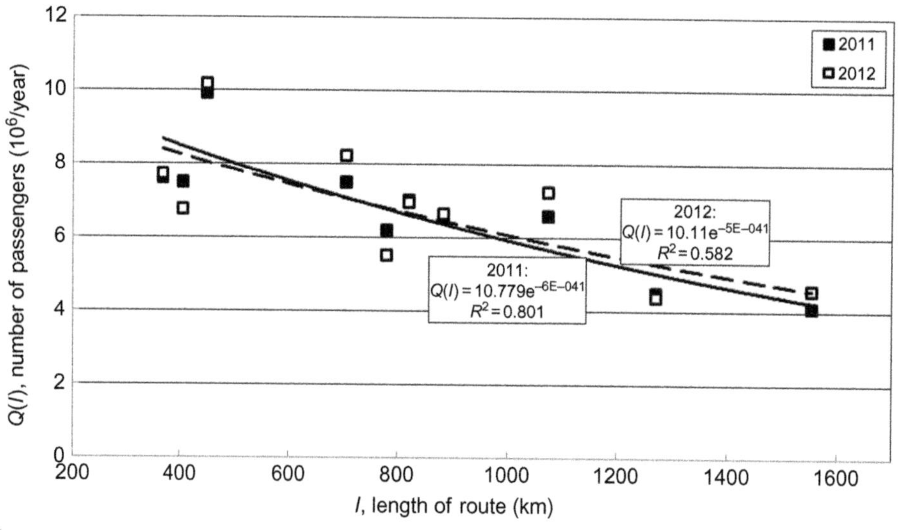

FIG. 2.35

Relationship between the annual number of passengers and the length of 10 busiest routes on the world (2011–2012) (https://en.wikipedia.org/wiki/World%27s_busiest_passenger_air_routes; https://www.iata.org/publications/pages/wats-passenger-carried.aspx).

As can be seen, the number of passengers was decreased more than proportionally when increasing the length of the route. In addition, Table 2.13 gives an example of the number of passengers and the volumes of p km (passenger kilometers) carried out by 10 world's largest passenger airlines in the year 2014 airlines (https://www.iata.org/publications/pages/wats-passenger-carried.aspx).

Table 2.13 The Number of Passengers and the Volumes of p km Carried Out by 10 World's Largest Airlines (International + Domestic) (2014) (https://www.iata.org/publications/pages/wats-passenger-carried.aspx)

Rank	Airline	Passengers (10^3)	p km[a] (10^9)
1	Delta Air Lines	129,433	290.862
2	Southwest Airlines	129,087	162.445
3	China Southern Airlines	100,683	112.247
4	United Airlines	90,439	287.547
5	American Airlines	87,830	208.046
6	Ryanair	86,370	96.324
7	China Eastern Airlines	66,174	120.461
8	easyJet	62,309	67.573
9	Lufthansa	59,850	143.403
10	Air China	54,577	112.247

[a]*Passenger-kilometer.*

As can be seen Delta Airlines was the largest in terms of both the annual number of passengers and the volume of p-km carried out followed by the LCC Southwest airline, but only in terms of the number of passengers. The European LCC Ryanair and easyJet were also among the first ten in terms of the number of passengers but not in terms of the volumes of p km. This indicates that they carried out the large number of passengers on the shorter (European) routes. In addition, Table 2.14 gives an example of the scheduled (offered) freight t km (ton-kilometers) offered by the world's 10 largest airlines in the year 2014 (https://www.iata.org/publications/Pages/wats-freight-km.aspx).

Table 2.14 Scheduled (Offered) Freight Ton-Kilometers by the World's 10 Largest Airlines (International and Domestic) (2014) (https://www.iata.org/publications/Pages/wats-freight-km.aspx)

Rank	Airline	t km[a] (10^9)
1	FedEx	16.020
2	Emirates	11.240
3	UPS Airlines	10.936
4	Cathay Pacific Airways	9.464
5	Korean Air Lines	8.079
6	Lufthansa	7.054
7	Singapore Airlines	6.019
8	Qatar Airways	5.997
9	Cargolux	5.753
10	China Airlines	5.266

[a]*Ton-kilometer.*

As can be seen, in addition to the specialized freight/cargo airlines such as FedEx, UPS Airlines, and Cargolux, the large passenger airlines were also offered the substantive volumes of freight/cargo capacity, which was quite comparable to that of their exclusively freight/cargo counterparts.

The above-mentioned airlines and many others operate fleets consisting of different number and size of aircraft. In general, the large conventional airlines have the total fleets of few tents and/or hundreds of aircraft usually of different size/capacity for the short, medium, and long-haul flights. For example, in the year 2014, the EU-27/28 Member States airlines operated the fleet of 3944 aircraft, of which 351 were with 50 or less seats, 1259 with 51 to 150 seats, 1816 with 151–250 seats, and 518 with 251 seats and more. The fleet of freight/cargo aircraft had 371 aircraft, of which 196 were with MTOW (maximum take-off weight) <45.3 ton, and 202 with MTOW >45.3 ton (EU, 2015). The LCCs have usually operated a single B737s/A320s aircraft type in their fleets containing up to several hundred units. As well, Fig. 2.36 shows an example of the total number of aircraft operated by the US civil aviation sector during the observed period (BTS, 2013).

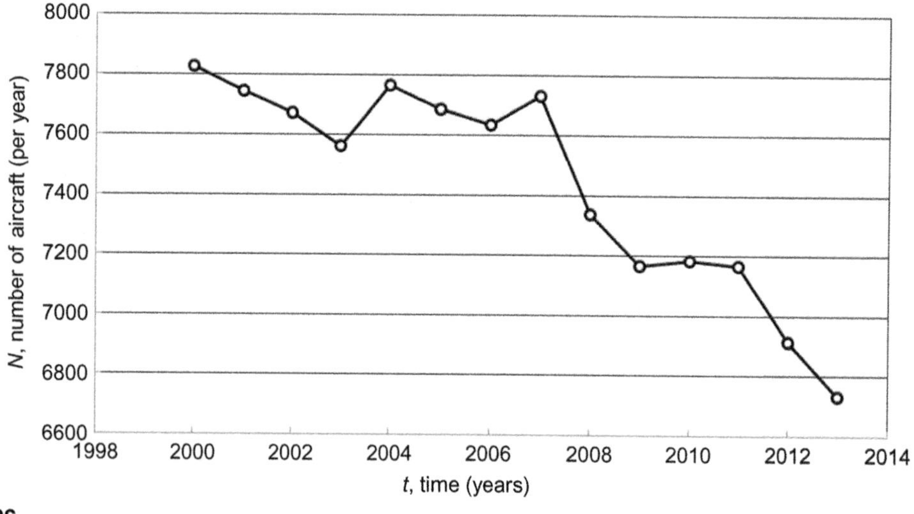

FIG. 2.36

The number of aircraft operated by the US civil aviation sector (2000–2013) (BTS, 2013).

As can be seen, the number of aircraft was continuously decreasing over time. At the end of the given period it was still for about two times greater than that above-mentioned in the EU-27/28 Member States.

2.5 TRANSPORTATION SYSTEMS TOPICS: PLANNING, CONTROL, CONGESTION, SAFETY, AND ENVIRONMENT PROTECTION

Every day, approximately, one third of the world trade by value travel by air. The total number of daily flights in the world is already >100,000, while the yearly number of air passengers is more than 3.5 billion.

The total length of the US National Highway System is more than 164,000 miles. The American I-80 freeway goes from the Atlantic to Pacific Ocean. According to US Department of Transportation, yearly travel on all roads and streets in the United States exceeds 2500 billion of vehicle miles of travel. Billions of tons of freight are transported every year within US National Highway System.

China has already about 20,000 km of High-speed rail (railway with train service at the speed of more than or equal to 200 km/h) track in service. The Beijing-Guangzhou High-Speed Railway line has length of ~2300 km. The Chinese trains are capable to reach operational speeds of up to 380 km/h.

Pedestrians, drivers, pilots, air traffic controllers, train drivers, captains of sailing and dispatchers are participants in traffic. Some of them travel, and others execute, or plan traffic operations. Safe and efficient transportation systems demand comprehensive transportation planning and sophisticated traffic control. Transportation planning and traffic control measures and actions should take care about level of congestion in the transportation system, safety, as well as environment protection.

The annual number of passengers between the city pairs, the number of passengers between the city zones, the annual number of requests for aircraft landing, etc., represents the basic input data necessary for transportation planning procedures. How to predict the total number of trips generated in specific urban zones? How to predict the traffic load at specific links in urban transportation network? Should we build another runway at the airport? Transportation planning methods and techniques help us to find the answers to these and similar questions. Comprehensive transportation planning must precede any significant decision regarding the transportation infrastructure development (construction of a new road, building the new runway, expanding the freight terminal at the port, etc.). Unfortunately, important decisions, concerning the development of transportation infrastructure, are still, in many countries of the world, taken without use of the sophisticated transportation planning techniques.

One of the main tasks of traffic engineers is to optimally use the existing transportation infrastructure, as well as to create efficient transportation systems. These objectives could be accomplished, above all, by developing and implementing various traffic control measures, methods and strategies. Word "efficiency" implies the vehicles' movement at the minimum, acceptable, or planned costs. Effectiveness assumes the vehicles' movement according to the planned/scheduled time. How to do green time allocation at the isolated intersection? How to control traffic along arterial streets? What are the best freeway traffic control measures? How to regulate movement of commercial airline aircraft between their origin and destination airports, and at these airports themselves? Traffic control techniques help us to obtain the answers to these and related questions.

The number of trips by private cars has considerably enlarged in recent decades in a lot of cities, and on many highways. Simultaneously, road network capacities have not kept up with this increase in travel demand. It has been widely documented that urban road networks in many countries are severely congested, resulting in increased travel times, increased number of stops, unexpected delays, greater travel costs, inconvenience to drivers and passengers, increased air pollution and noise level, and increased number of traffic accidents (Vuchic, 2008). The inhabitants of many big cities in the world already spend between 40 and 60 min of time when commuting to work. Traffic congestion has also been a problem at many airports all over the world. The air transportation operations at one specific airport are linked with the operations at many other airports. Consequently, delays at one airport have undesirable effects on aircraft delays at other airports. Great number of flights arrives at the busiest world's airport with a delay that is more than 15 min. At some airports, the average delay is equal to 50–60 min. The "Hours of delay per traveler" is a usual measure that is used to represent the congestion level in cities. Expanding traffic network capacities by building more roads, or more runways is very costly, as well as environmentally damaging. More efficient usage of the existing supply is crucial in order to maintain the rising travel demand. Planners, and transportation engineers have developed a variety of Travel Demand Management (TDM) techniques (Vickrey, 1969; Yang and Huang, 1999; Phang and Toh, 2004; Teodorovic and Edara, 2005), ie, different strategies that enlarge travel choices to travelers. TDM strategies consist of alternative mode encouragement strategies such as "Park-and-Ride facilities," "High Occupancy Vehicle (HOV) facilities," "Ride-sharing programs," "Telecommuting," "Alternative work hours," "Congestion Pricing," "Preferential parking to rideshare vehicles," among others.

Road traffic injuries are the primary cause of death for young people (World Health Organization, 2013). They are also the eighth most important cause of death worldwide. The 1.24 million people were killed on the world's roads in 2010. There are estimates that 20–50 million people sustain nonfatal injuries as a consequence of road traffic accidents. The average world road traffic fatality rate is 18 per 100,000 population. The main risk factors of traffic accidents are high speed, drink-driving, nonwearing of helmets, nonusage of seat-belts and low enforcement of child restraint laws.

Several countries have effectively decreased the number of deaths on their roads. The reduction in the total number of traffic accidents, and the total numbers of killed and injured people could be achieved through the development and implementation of the national road safety strategies, as well as building safer roads and safer vehicles. The specific actions that could reduce total number of traffic accidents include introducing and implementing drink-driving laws to meet the best practice (blood alcohol concentration of 0.05 g/dL or less), increasing seat-belt use, and increasing motorcycle helmet use. Since, half of all road traffic deaths are among pedestrians, cyclists and motorcyclists, the increasing motorcycle helmet use should be one of the ultimate goals. The significant contribution to the decrease of the total number of traffic accidents could be also achieved by reducing urban speeds, stronger enforcement of speed limits, implementation of the child restraint law, as well as law on mobile phones while driving.

Transportation systems are main consumers of energy, and the main polluters in modern world. The construction of new transportation infrastructure constantly leads to land take (green spaces, farming land), as well as to the extraction of transportation infrastructure construction materials (gravel, sand, iron, wood,...). Transportation systems are also the main consumers of crude oil (gasoline and other petroleum products).

Transportation has a large harmful impact on the natural environment. Transportation on a daily basis causes atmospheric and noise pollution. The most negative consequence of transportation activities is a climate change caused by emissions of CO_2 and other gases that cause global warming. There are numerous implications of climate change on biodiversity and human health. Transportation is also causing emissions of CO, lead, oxides of nitrogen (NOx), etc. The quality of life of the people that live in close proximity of highways, airports, railway stations and bus stations is also highly influenced by a noise level. Transportation is large generator of waste. Waste disposal (old vehicles, mechanical parts, vehicle tires, antifreeze, brake fluid, oil) in many countries in the world is not adequately controlled.

Transportation has various negative impacts on humans, flora and fauna. Technology development can partially decrease negative impacts of transportation on environment. There is no doubt, that the most important technological task is continuous reduction of the transportation dependency on fossil fuels.

When 30 solo-drivers of passenger cars park their cars and start to use the public bus, the pollution will be generated by only one car instead of 30. Various measures of transportation policies can lead to increased walking and cycling and increased public transport usage. When we walk, or ride a bike, we do not pollute the natural environment. We do not need fossil oils, we do not take green spaces and we do not destroy farming land. Many medical reports (Purcher et al., 2010) concluded that physical inactivity contributes to enlarged risk of a number of chronic diseases and health conditions (diabetes, hypertension, cardiovascular disease, gallstones, fatty liver, and some cancers). In addition, the medical research indicated that even 30 min per day of reasonable-intensity physical activity, if performed systematically, gives considerable health benefits.

2.6 PROBLEMS

1. Explain the milestones in developing transport sector from the middle century to the present day—infrastructure, services, and technologies.
2. Define and explain the structure of transport sector, transport modes, and transport systems.
3. What are the main components of particular transport modes and their systems?
4. What are the possible relationships between particular transport modes and their systems operating in the same area?
5. Itemize the main urban passenger transit systems, their components, and characteristics.
6. What are the main characteristics of urban freight transport systems?
7. Itemize the main interurban transport systems operated within particular transport modes, and explain their main characteristics.
8. Explain the main differences between the conventional rail, HSR, and MAGLEV system for passengers.
9. Define the water-based systems—inland waterways and sea shipping.
10. What are the types of freight shipments transported by the water-based transport systems?
11. Explain the main characteristics of container ships.
12. Itemize the main components of air transport systems.
13. What are the main topologies of airline networks?

REFERENCES

An, F., Hu, H., Xie, C., 2015. Service network design in inland waterway liner transportation with empty container repositioning. Eur. Transp. Res. Rev. 7(9). http://dx.doi.org/10.1007/s12544-015-0157-5.

Beckmann, M.J., McGuire, C.B., Winsten, C.B., 1956. Studies in the Economics of Transportation. Yale University Press, New Haven, CT.

Berechman, J., 2003. Transportation—economic aspects of Roman highway development: the case of Via Appia. Transp. Res. A 37, 453–478.

Blondé, B., 2010. At the cradle of the transport revolution? Paved roads, traffic flows and economic development in eighteenth-century Brabant. J. Transp. Hist. 31, 89–110.

Boyce, D.E., Mahmassani, H.S., Nagurney, A., 2005. A retrospective on Beckmann, McGuire, and Winsten's Studies in the Economics of Transportation. Pap. Reg. Sci. 84 (1), 85–103.

Braess, D., 1968. Uber ein Paradox der Verkerhsplannung. Unternehmenstorchung 12, 258–268.

BTS, 2013. Table 1–11: Number of U.S. Aircraft, Vehicles, Vessels, and Other Conveyances. Bureau of Transport Statistics, Office of the Assistant Secretary for Research and Technology, Washington, DC.

Cervero, R., 2013. Bus Rapid Transit (BRT): An Efficient and Competitive Mode of Public Transport, Report. 20thSAG.indd 3, University of California, Berkeley, CA.

Clark, J., 2015. Ships, clocks, and stars: the quest for Longitude: National Maritime Museum Greenwich, London. J. Transp. Hist. 36, 124–126.

Clark, C., Henrickson, E.K., Thoma, P., 2005. An Overview of the U.S. Inland Waterway System. IWR Report 1;05--NETS-R-12, U.S. Department of the Army Corps of Engineers, Institute for Water Resources, Alexandria, VA.

Crozet, I., 2013. High speed rail performance in France: from appraisal methodologies to ex-post evaluation. Discussion paper no. 2013-26, In: The Roundtable on Economics of Investments in High Speed Rails, International Transport Forum, 18–19 December 2013, New Delhi, India.

CSP, 2014. China Statistical Yearbook 2014. China Statistics Press, National Bureau of Statistics of China, Beijing.

EC, 2005. City Freight: Inter-and Intra-City Freight Distribution Networks, Final Report, Framework Programme, Energy, Environment and Sustainable Development, Key Action 4: City of Tomorrow and Cultural Heritage. European Commissions, Brussels, Belgium.

ECMT, 1992. Resolution no. 92/2 on new classification of inland waterway. In: European Conference of Ministers of Transport, Conference 11–12 June, Athens, Greece.

Ehmke, F.J., 2012. Integration of information and optimization models for routing in city logistics. International Series in Operations Research & Management Science, vol. 177. Springer, New York, NY.

EU, 2015. EU Transport in Figures: Statistical Pocketbook 2014. Publications Office of the European Union, Luxembourg.

FHWA, 2013. Highway Statistics 2012. Federal Highway Administration, Washington, DC.

Fitzpatrick, S.M., Callaghan, R., 2008. Magellan's crossing of the Pacific: using computer simulations to examine oceanographic effects on one of the world's greatest voyages. J. Pac. Hist. 43, 2.

Fravel, D.F., Barboza, R., Quan, J., Sartori, K.J., 2011. Toolkit for Estimating Demand for Rural Intercity Bus Services, TRCP Report 147, Transit Cooperative Research Program. Transportation Research Board, Washington, DC. www.TRB.org.

Fu, J., Nie, L., Meng, L., Sperry, R.B., He, Z., 2015. A hierarchical line planning approach for a large-scale high speed rail network: the China case. Transp. Res. A 75, 61–83.

Gazis, D., 2002. The origins of traffic theory. Oper. Res. 50 (1), 69–77. 50th Anniversary Issue (Jan.–Feb., 2002).

Geerlings, H., 1998. The rise and fall of new technologies: MAGLEV as technological substitution? Transp. Plan. Technol. 21, 263–286.

Harvey, E., 2010. Pavage grants and urban street paving in medieval England, 1249–1462. J. Transp. Hist. 31, 151–163.

He, J.L., Rote, D.M., Coffey, H.T., 1992. Survey of Foreign Maglev Systems. Center for Transportation Research, Energy Systems Division, Argonne National Laboratory, Argonne, IL.

Janić, M., 2014. Advanced Transport Systems: Analysis, Modelling and Evaluation of Performances. Springer, London.

Janić, M., 2016. A multidimensional examination of the performances of HSR (high speed rail) systems. J. Mod. Transp. 24 (1), 1–21.

Janić, M., Reggiani, A., Nijkamp, P., 1999. Sustainability of the European freight transport system: evaluation of the innovative bundling networks. Transp. Plan. Technol. 23 (2), 129–156.

Kishimoto, T., Kawasaki, S., Nagata, N., Tanaka, R., 2007. Optimal location of route and stops of public transportation. In: Proceedings of the 6th International Space Syntax Symposium, Istanbul, Turkey.

Knight, F.H., 1924. Some fallacies in the interpretation of social cost. Q. J. Econ. 38, 582–606.

Kohl, J.E., 1841. Der Verkehr and die Ansiedelungen der Menschen in Ihrer Abhängigkeit von der Gestaltung der Erdoberfläche. Arnold, Dresden, Leipzig.

Lei Luo, L., Wang, X., Liu, C., Guo, H., Du, X., 2014. Integrated RS, GIS and GPS approaches to archaeological prospecting in the Hexi Corridor, NW China: a case study of the royal road to ancient Dunhuang. J. Archaeol. Sci. 50, 178–190.

Lukaszevicz, P., Anderson, E., 2009. Green Train Energy Consumption: Estimation on High-Speed Rail Operations. KTH Engineering Science, KTH Railway Group, Stockholm.

Matthews Jr., K., 1960. The embattled driver in ancient Rome. Expedition 2, 22–27.

Merchán, E.D., Blanco, E.E., Bateman, H.E., 2015. Urban metrics for Urban logistics: building an Atlas for urban freight policy makers. In: Proceedings of the 14th International Conference on Computers in Urban Planning and Urban Management, July 7–10, 2015, Cambridge, MA, USA, p. 22.

Newman, P., Kenworthy, J., 1989. Cities and Automobile Dependence: An International Sourcebook. Gower, Aldershot, UK.

Nichols Busch, T., 2008. Connecting an empire eighteenth-century Russian roads, from Peter to Catherine. J. Transp. Hist. 29, 240–257.

Ollivier, G., Sondhi, J., Zhou, N., 2014. High-Speed Railways in China: A Look at Construction Costs. China Transport Topics No. 9, World Bank Office, Beijing, pp. 1–8.

Phang, S.Y., Toh, R.S., 2004. Road congestion pricing in Singapore: 1975 to 2003. Transportation 43, 16–25.

Pigou, A.C., 1920. The Economics of Welfare. Macmillan, London.

Powell, J., Dunby, G., 2007. Maglev: transport mode for 21st century. EIR Sci. Technol., 44–55.

Prigogine, I., Herman, R., 1971. Kinetic Theory of Vehicular Traffic. American Elsevier Publishing Co., New York, NY. pp. 17–54.

Purcher, J., Buehler, R., Bassett, D.R., Dannenberg, A.L., 2010. Walking and cycling to health: a comparative analysis of city, state, and international data. Am. J. Public Health 100 (10), 1986–1992.

Song, D.-W., Panayides, P., 2015. Developing liner service networks in container shipping. In: Decruet, C., Notteboom, T. (Eds.), Maritime Logistics: A Guide to Contemporary Shipping and Port Management. Kogan Page Publishers, London, pp. 125–134.

Takagi, K., 2011. Expansion of high speed rail services: development of high-speed railways in China. Jpn. Railw. Transp. Rev. 57, 36–41.

Teodorovic, D., Edara, P., 2005. Highway space inventory control system. In: Proceedings of ISTTT 16: Transportation and Traffic Theory: Flow. In: Dynamics and Human Interaction. Elsevier Publishers, MA, USA, pp. 43–62.

TIG, 2012. High Tech for Flying on the Ground. Transrapid International GmbH & Co, Berlin.

UIC, 2014. High Speed Lines in the World, Updated 1st September 2014. UIC High Speed Department, International Union of Railways, Paris.

UITP, 2014. World Metro Figures: Statistics Brief. UITP-Advancing Public Transport, Brussels.

UITP, 2015. Light Rail in Figures: Statistics Brief-Worldwide Outlook. UITP-Advancing Public Transport, Brussels.

UNCTAD, 2014. Review of maritime transport 2013. In: United Nations Conference on Trade and Development, New York, USA.

USACE, 2013. Waterborne Commerce of the United States. U.S. Army Corps of Engineers, New Orleans, LA.

USDT, 2005. Study of Intercity Bus Service, Report of the Department of Transportation to the United States Congress Pursuant to House Report 108-671. U.S. Department of Transportation, Washington, DC.

USDT, 2013a. Domestic Freight Activity by Mode: Table 5-2. U.S. Department of Transportation, Federal Highway Administration, Office of Freight Management and Operations, Washington, DC.

USDT, 2013b. Freight Analysis Framework. U.S. Department of Transportation, Federal Highway Administration, Office of Freight Management and Operations, Washington, DC.

USDT, 2013c. U.S. Water Transportation Statistical Snapshot. U.S. Department of Transportation, Maritime Administration, Washington, DC.

Van de Velde, D., 2009. Long-Distance Bus Services in Europe: Concessions or Free Market. Discussion Paper No. 2009-21 December, Joint Transport Research Centre, OECD/ITF (International Transport Forum), Paris.

Vickrey, W., 1969. Congestion theory and transport investment. Am. Econ. Rev. Pap. Proc. 59, 251–261.

Vuchic, V., 2005. Urban Transit: Operations, Planning, and Economics. John Wiley & Sons, Hoboken, NJ.

Vuchic, V., 2007. Urban Transit Systems and Technology. John Wiley & Sons, Inc., Hoboken, NJ

Vuchic, V., 2008. Transport systems and policies for sustainable cities. Therm. Sci. 12, 7–17.
Wardrop, J.G., 1952. Some theoretical aspects of road traffic research. Proc. Inst. Civ. Eng. II 1, 325–378.
Weinstock, A., Hook, W., Replogle, M., Cruz, R., 2011. Recapturing Global Leadership in Bus Rapid Transit A Survey of Select U.S. Cities. ITDP, Institute for Transport & Development Policy, New York, NY.
Williams, D.M., Armstrong, J., 2014. One of the noblest inventions of the age': British steamboat numbers, diffusion, services and public reception, 1812—c.1823. J. Transp. Hist. 35, 18–34.
World Health Organization, 2013. Global Status Report on Road Safety. WHO Press, Geneva.
WSC, 2013. Trade Routes. World Shipping Council, Washington, DC.
Yang, H., Huang, H.J., 1999. Carpooling and congestion pricing in a multilane highway with high-occupancy-vehicle lanes. Transp. Res. A 33A, 139–155.
Zuidweg, J.K., 1970. Automatic Guidance of Ships as a Control Problem (Ph.D. thesis). Technische Hogeschool Delft, Delft.

WEBSITES

http://www.britannica.com.
http://www.gojapango.com/travel/shinkansen_history.htm.
http://micanaldepanama.com/expansion/.
http://www.wikipedia.org.
http://en.wikipedia.org/wiki/Maglev.
https://people.hofstra.edu/geotrans/eng/ch3en/conc3en/lengthwaterways.html.
https://en.wikipedia.org/wiki/Stabilizer_%28aeronautics%29.
https://en.wikipedia.org/wiki/High-speed_rail_in_China/.
http://www.johomaps.com/eu/europehighspeed.html.
http://www.eurail.com/plan-your-trip/railway-map.
https://en.wikipedia.org/wiki/List_of_countries_and_dependencies_by_area.
http://www.railway-technology.com/features/featurethe-worlds-longest-railway-networks-4180878/.
https://www.iata.org/publications/Pages/wats-freight-km.aspx.
https://en.wikipedia.org/wiki/World%27s_busiest_passenger_air_routes.
https://www.iata.org/publications/pages/wats-passenger-carried.aspx.
https://en.wikipedia.org/wiki/World%27s_busiest_passenger_air_routes2015.
http://www.flightmanager.com/content/timedistanceform.aspx.
http://ec.europa.eu/transport/themes/infrastructure/ten-t-guidelines/maps_en.htm.
https://www.google.nl/search?q=images+of+LRT+networks/.
http://lastmile.mit.edu/km2/.
http://www.mapsofworld.com/usa/usa-road-map.html.
http://edinburghtrams.com/plan-a-journey/route-map.
http://www.railway-technology.com/projects/category/light-rail-systems/.
http://www.railway-technology.com/features/featurethe-worlds-longest-metro-and-subway-systems-4144725/.
http://www.rita.dot.gov/bts/sites/rita.dot.gov.bts/files/publications/national_transportation_statistics/index.html#chapter_1.
https://en.wikipedia.org/wiki/List_of_bus_rapid_transit_systems/.
https://en.wikipedia.org/wiki/Tram_and_light_rail_transit_systems/.
https://en.wikipedia.org/wiki/List_of_urban_areas_by_population/.
https://en.wikipedia.org/wiki/Transportation_in_the_United_States.
https://en.wikipedia.org/wiki/AVE.

Why do we use space-time diagrams? How could we find the shortest paths from one node to all other nodes in the traffic network? What are the main components of the mathematical description of the transportation problems? What is Linear Programming? Are number of cars on a link, and number of passengers on specific flight random variables whose values are unknown in advance? What are queueing systems? How do we imitate the behavior of the real transportation system? What are fuzzy sets? What is Fuzzy Logic?

TRAFFIC AND TRANSPORTATION ANALYSIS TECHNIQUES

3

3.1 OBJECT MOTION AND TIME-SPACE DIAGRAMS

Motion is the common characteristic for the pedestrian who walks, the bird that flies, the car that passes through an intersection, or the train that approaches a metro station. Movement through space and time is caused by people undertaking various economic, business, cultural, touristic, sport, and recreational activities. In every time point, when we drive a car, when we walk, or run, we occupy different position in the space. The trajectory of an aircraft that traveled from Paris to Washington, DC could be precisely described by time points and corresponding aircraft positions in the space. In the same way, our trip by car from home to university could be also described by time points and car positions. We denote by y the moving object's position. This position is a function of time, ie, $y=f(t)$. The notation $y(t)$ has exactly the same meaning. The functional relationship is called *equation of motion*. In some cases, it is very easy to determine this relationship. In some other cases, when the motion is very complex, determination of equation of motion could be also very complex, or even impossible. We use *space-time diagrams*, when we want to represent object (car, vessel, aircraft, crews, or passengers) movements through space and time. In these diagrams, space is represented in one dimension, and time is represented in the other dimension. Every node has two attributes—the first related to space, and the second related to time. Each node represents an event taking place in a specific intersection, city, airport, bus

Transportation Engineering. http://dx.doi.org/10.1016/B978-0-12-803818-5.00003-2

stop, or harbor at a specific time. Depending on the problem considered, time attribute could be related to the beginning of green light at the intersection, aircraft readiness for departure, train departure time, bus arrival time, etc. For example, if an object travels two distance units in every time unit, we write:

$$y = 2t \tag{3.1}$$

where y is the object's position and t is time.

The object's position y is measured taking into account a previously defined starting point. Fig. 3.1 contains a space-time diagram that shows the motion of an object as a function of time.

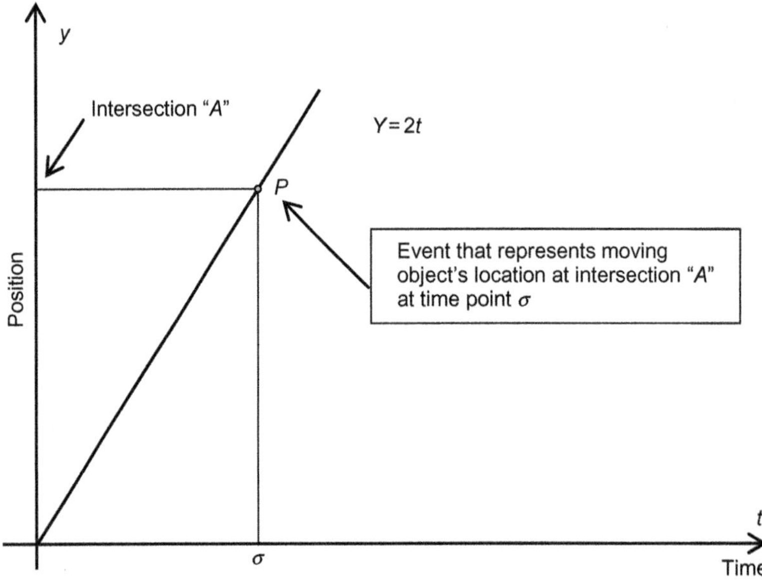

FIG. 3.1

Time-space diagram that shows the motion of an object as a function of time.

Time is shown on one of the space-time diagram axis (the x axis is frequently used). Every point on the curve represents an event. For example, point P shown in Fig. 3.1 denotes the following event: a car is passing the intersection "A" in time point σ. Time-space-time diagrams enable us to note acceleration and deceleration of a moving object, as well as time intervals during which object travels by a constant speed (Fig. 3.2). Studying Fig. 3.2, we conclude that the acceleration existed between point "O" and point "A." After that object traveled by a constant speed between point "A" and point "B." After point "B" object decelerated. The constant object's speed between point "A" and the point "B" equals:

$$v = \frac{\Delta y}{\Delta t} \tag{3.2}$$

where Δy is distance traveled and Δt is elapsed time.

Cars that travel from one place to another create traffic flows. When trying to study these flows we can use the macroscopic or the microscopic approach. Within the macroscopic approach, we do not recognize vehicles as entities. In other words, in this approach we do not identify vehicles in traffic flow individually. Traffic flow could be studied in this case as fluid-flow, or heat-transfer phenomena. The

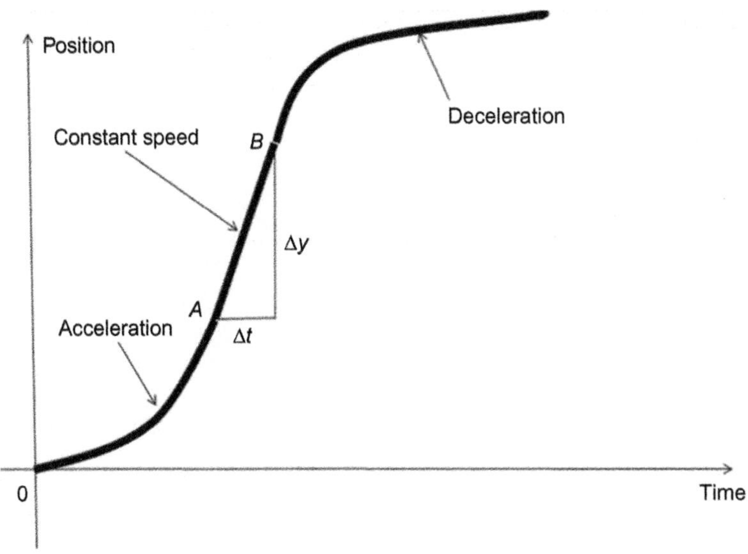

FIG. 3.2

Acceleration, deceleration, and constant speed.

microscopic approach could find its theoretical foundation in the fact that traffic flows are the result of the individual driver decisions to begin a trip at a certain time point and to choose a particular set of routes. Within the microscopic approach we consider every individual vehicle and study individual driving behaviors. We represent time on the horizontal axis and distance from the point on the vertical axis. We also plot vehicle trajectory on the time-space diagram. In other words, the movement of each driver through the traffic network could be described by the time-space diagram (Fig. 3.3). When equation of motion $y = f(t)$ is known, it is easy to analyze object's motion through space and time.

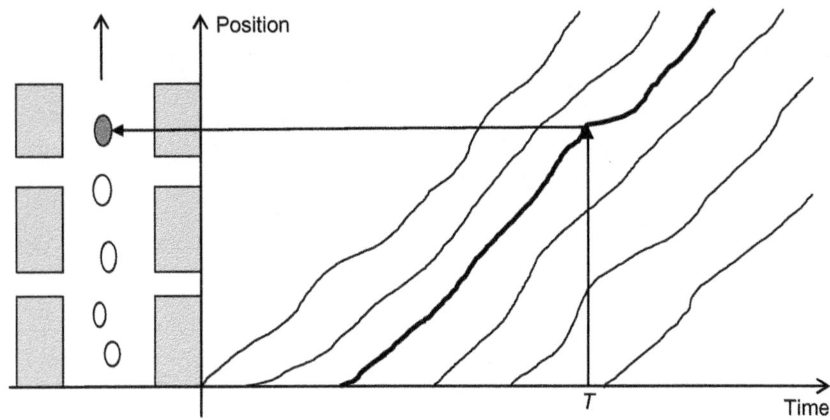

FIG. 3.3

Identification of the object's position at any time point.

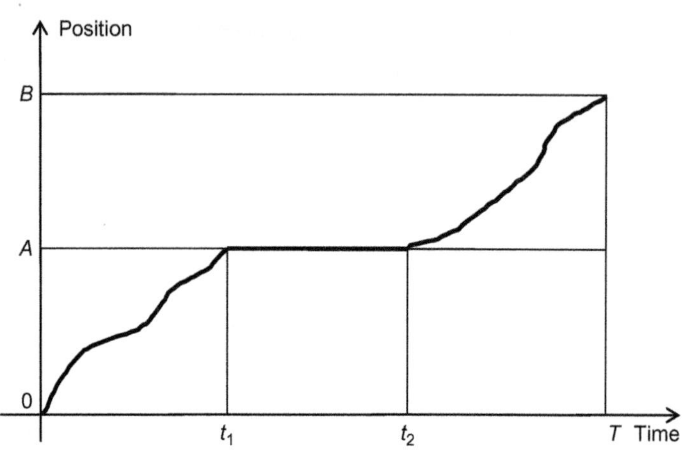

FIG. 3.4

Time-space time diagram that describes movement of individual car through space and time.

We observe the car's movement during the time interval $(0, T)$. We conclude from Fig. 3.4 that the car reached the intersection A at time point t_1. After waiting for the right-of-way at the intersection A, our car continued its trip at time point t_2. Finally, the car reached the intersection B at the time point T. We can also conclude from Fig. 3.4 that the car changed speed while traveling. Fig. 3.5 shows the movements of five cars.

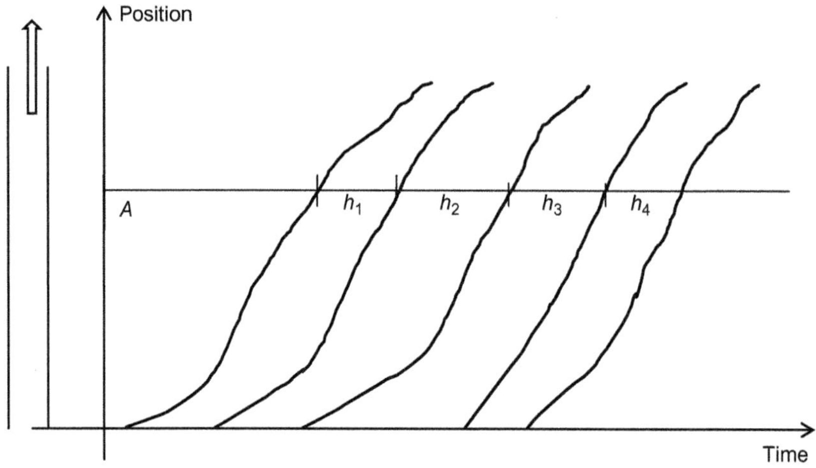

FIG. 3.5

Movements of five cars.

In the transportation analysis analysts use *headways* to measure the distance or time between vehicles. Fig. 3.5 shows headways h_1, h_2, h_3, and h_4. The first vehicle passed point A at the time point t_1. The second vehicle passed the point A, h_1 time units after the first vehicle. The third one passed the point A, h_2 time units after the second vehicle etc. The concept of headway is also used in the public transit and metro systems to describe frequency of service offered to the passengers. The shorter headways assume more bus departures along the bus line. The headway in many metro systems is in the range of 1–10 min. The smallest headways on a highway are approximately equal to 2 s.

Space-time diagrams are also used to illustrate traffic signal coordination. Fig. 3.6 shows traffic signals coordination along an arterial street.

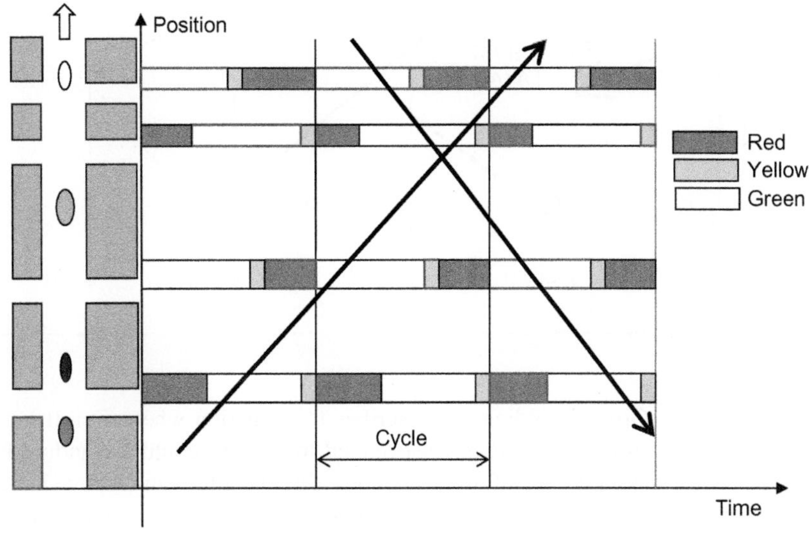

FIG. 3.6

Traffic signal coordination along an arterial street.

Successful signal coordination enables vehicles to pass through every intersection smoothly. The vehicle trajectories shown in Fig. 3.6 passage left to right along with time. At the same time, the vehicle distance passed through can be either northbound or southbound. The northbound is from the bottom to the top of Fig. 3.6, while the southbound is from the top to the bottom. Depending on the travel direction along an arterial street, vehicle trajectories can have a positive or negative slope.

The time-space diagrams are also used when designing airline schedule (Fig. 3.7). Fig. 3.7 shows daily assignments of different aircraft types (departure airports, arrival airports, times of departures, and flight times). The time-space diagram contains many flight "legs." Corresponding aircraft type and ("tail number") are shown for each flight leg. For example, the following information could be obtained by inspection of the space-time diagram (Fig. 3.7): Flight that departs at 11:00 am from city B to city D will be flown by Aircraft type 3 (for example, Boeing 747-400). The aircraft with the tail number "AKA17" will perform this flight. It is also possible to approximately determine needed number of aircraft and their corresponding routes using time-space diagrams.

Link that connects two nodes is usually called flight arc, since it connects aircraft departure node to aircraft arrival node. This is the usual way to represent trip, or flight in space-time network. Many vehicle, aircraft, and crew routing and scheduling models use so called ground holding arc, as well

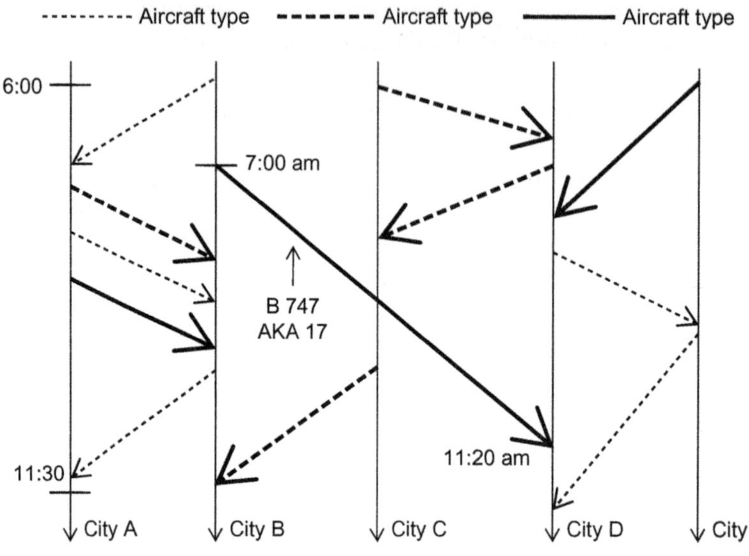

FIG. 3.7

Airline scheduling.

as overnight arc (Fig. 3.8). Ground holding arc describes the situation when aircraft is located at a particular airport for a certain period of time. Cost related to this arc could be landing fee, parking charges, etc. In the same, way, this type of arcs could describe the situation when vehicle is waiting at the intersection for a green time, or the situation when bus waits at the bus stop for the beginning of the trip. This type of arcs describes the situations when vehicle wait at the same position in space for a longer period of time. Overnight arc represents situations when bus, or aircraft is staying overnight at the specific city, or airport.

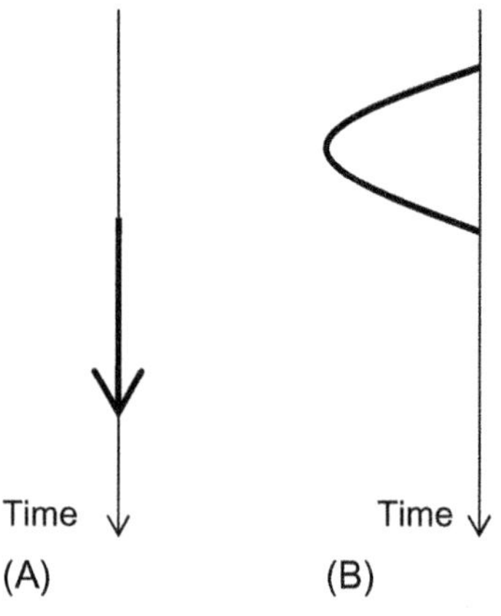

FIG. 3.8

(A) Ground holding arc and (B) overnight arc.

3.2 TRANSPORTATION NETWORKS BASICS

Transportation networks are represented by geometrical figures (Figs. 3.9–3.11). In these figures, some points (nodes) are mutually connected by lines (links). At the infrastructure networks, the nodes are generally intersections of lines. For example, at the road transport mode, the nodes are intersections of streets in urban, roads in suburban, and roads/highways in the interurban areas. These are also bus and road freight terminals. The links are streets and roads/highways converging to and diverging from the corresponding intersections and stations/terminals. At the rail transport mode, the nodes are junctions along the lines and the stations/terminals. The links are the rail lines/tracks connecting them. At the air transport mode, the nodes are intersections of airways and the links are the converging and diverging airways, and the airports as well. At the maritime transport mode, the nodes are the considered to be the sea ports, where the incoming and outgoing ships and the links are the maritime routes themselves. The service network consists of nodes and links too. For example, at the road and rail transport mode the nodes are passenger and freight/goods stations/terminals. The links are transport services scheduled between them to serve given volumes of passengers and freight demand, respectively. At the air and maritime transport mode, the nodes are airports and seaports, respectively, and the links are the airline flights and the ship transport services, scheduled between them to serve expected passenger and freight/cargo demand, respectively.

We denote transportation networks in a same way as *graphs* (Ford and Fulkerson, 1962; Busacker and Saaty, 1965; Newell, 1980). The notation $G=(N, A)$ of the transportation network refers to a set of *nodes* (or *vertices*) N, and a set of links (or *arcs* or *edges* or *branches*) A, that connect pairs of nodes. We denote by (i, j) link that connects node $i \in N$ to node $j \in N$. We usually assign one or more numerical characteristics to every link $(i, j) \in A$. Most frequently, these numerical characteristics represent "travel cost" c_{ij}, or link capacity u_{ij}. In the same way, numerical characteristics $b(i)$ are assigned to every node $i \in N$. The numerical characteristic could represent, for example, the total traffic generated at node i, or the yearly number of car accidents in node i. We use in this book terms network and graph interchangeably.

If all links in the transportation network are oriented, the network is called an *oriented network* (or *directed network*). In an oriented network, link (i, j) leads from node i to node j. In the opposite case, when none of the branches are oriented, the network is called *nonoriented*. If some of the branches in the network are oriented and some nonoriented, this is called a *mixed* network.

Figs. 3.9–3.11 provide examples of oriented, nonoriented, and mixed networks, respectively.

The *indegree of a node* in an oriented network represents the number of head endpoints adjacent to a node. Similarly, the *outdegree of a node* in an oriented network is defined as the number of tail endpoints adjacent to a node. The indegree of the node 5 in a network shown in Fig. 3.9 equals 1. The outdegree of the same node equals to 2. The *degree of a node* in a nonoriented network represents the number of links incident to the node. For example, the degree of the node 4 in a nonoriented network shown in Fig. 3.10 equals 3. *Path* leading from the node i to the node j is a sequence of all links and all nodes that should be passed when traveling from the node i to the node j. Path could be defined by a list of nodes, or by a list of links that should be passed. The sequence (3, 1, 2) denotes the path that we follow when traveling from node 3 through node 1 to node 2 (Fig. 3.10). This path could be also denoted as ((3, 1), (1, 2)), meaning that we traverse links (3, 1) and (1, 2) when traveling from node 3 to node 2. A path whose origin and destination nodes coincide is called a *cycle*. The path (3, 1, 5, 4, 3) is a cycle, since node 3 is the initial and the final node of the path (Fig. 3.10). The path is called *simple*, when all links appear only once in the path. The path is *elementary* when all nodes appear only once in the path.

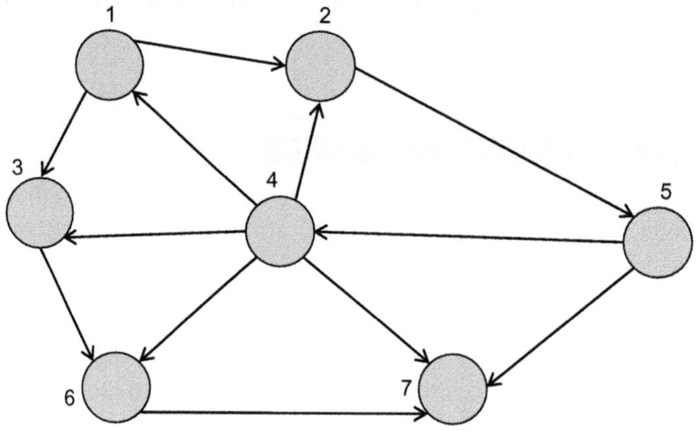

FIG. 3.9

Oriented network.

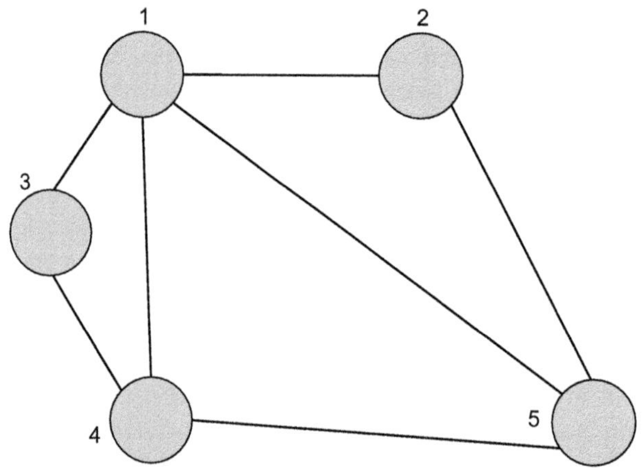

FIG. 3.10

Nonoriented network.

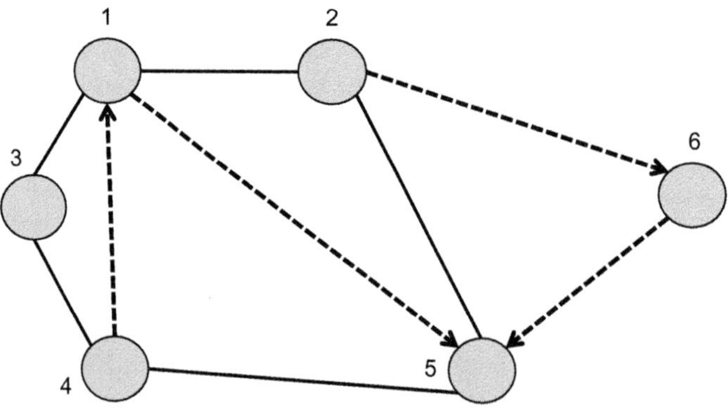

FIG. 3.11

Mixed network.

A node i is *connected* to node j if there is a path that leads from the node i to the node j. A nonoriented network is connected if there is a path between every pair of nodes $i, j \in N$. An oriented network is connected if a corresponding nonoriented network (the network that is created if orientation is removed from the oriented network) is connected. An oriented network that has paths between all pairs of nodes is called *strongly connected* oriented network. A nonoriented connected network is shown in Fig. 3.12.

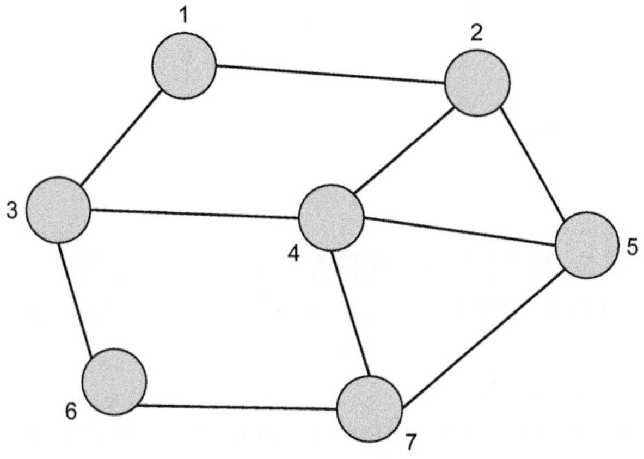

FIG. 3.12

Nonoriented connected network.

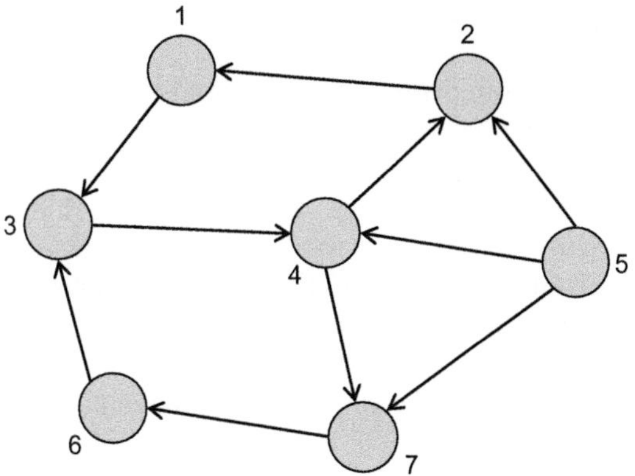

FIG. 3.13

Oriented connected network.

We oriented all links of the network shown in Fig. 3.12. In this way, we created the oriented network shown in Fig. 3.13. The network shown in Fig. 3.13 is an oriented connected network, since a corresponding nonoriented network (Fig. 3.12) is connected. There are no paths between some pairs of nodes in the network shown in Fig. 3.13 (there is no path between node 4 and node 5; there is no path between node 7 and node 5, etc.). Fig. 3.14 shows a strongly connected oriented network.

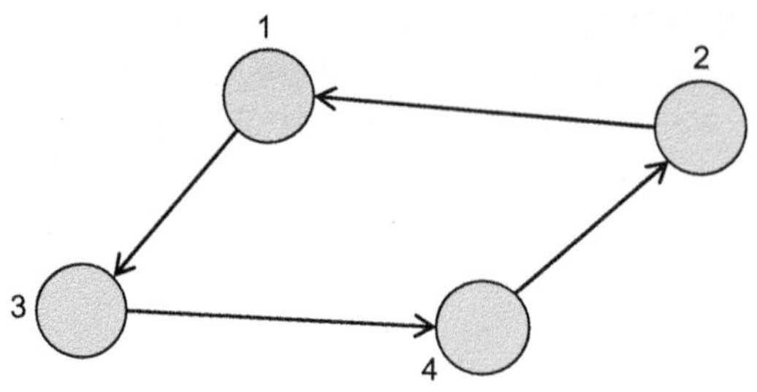

FIG. 3.14

Strongly connected oriented network.

3.3 OPTIMAL PATHS IN TRANSPORTATION NETWORKS

3.3.1 FINDING SHORTEST PATH IN A TRANSPORTATION NETWORK

When traveling through the network we are faced with the problem of finding the paths that are "optimal." In other words, paths that we are looking for must possess optimal properties. The problems of finding optimal paths in transportation networks are known as *shortest path problems* (Minty, 1957; Dantzig, 1960; Golden, 1976; Dial et al., 1979; Deo and Pang, 1984; Glover et al., 1984, 1985; Pallottino, 1984; Gallo and Pallottino, 1988). Depending of the context of the problem considered, the "shortest path" could be the shortest path, the fastest path, the most reliable path, the path with the greatest capacity, etc. network links are characterized by length. "Link length" could represent distance, travel time, travel cost, link reliability, etc. Link lengths are mostly treated as deterministic quantities.

Link lengths are treated in some problems as random variables. Most frequently, these link lengths represent travel times. There are random variations in travel times caused by weather conditions, randomness in traffic flows, traffic accidents, and other factors. In these cases, the *shortest paths in probabilistic network* should be determined. In some cases, when searching for the optimal path, we simultaneously try to optimize two or more objectives. For example, when searching for the best path, we could try simultaneously to take care of travel time as well as travel costs. In such cases, we are dealing with multicriteria *shortest path problems.*

We use the expression "*shortest path*" to denote optimal path. Depending on the context of the problem considered, the following variants of the shortest path problem could appear:

- shortest path between two specified nodes;
- shortest paths from a given node to all other nodes;
- *k* shortest paths from a given node to all other nodes;
- shortest paths between all pairs of nodes;
- *k* shortest paths between all pairs of nodes;
- shortest path between two specified nodes that must pass through some prespecified nodes; or
- shortest path between two specified nodes that must pass through some prespecified links.

3.3.2 DIJKSTRA'S ALGORITHM

Dijkstra (1959) developed one of the most efficient algorithms for finding shortest paths from one node to all other nodes in the network. This algorithm assumes that all the lengths $d(i, j)$ of all links in the network $G=(N, A)$ are nonnegative.

We denote by a the node for which we are to discover the shortest paths to all other nodes in the network. During the process of discovering these shortest paths, each node can be in one of the two possible states: in an open state if the node is denoted by a temporary label, or in a closed state if it is denoted by a permanent label. In the case of the permanent label, we are not sure whether the discovered path is the shortest path. Dijkstra's algorithm gradually changes temporary labels into permanent labels. The initial distances between any two nodes in the network are defined as follows. The distance between node a to node a is zero. The distance between two nodes is equal ∞ if there is no link between these two nodes. If there is a link that connects two nodes, the distance between these nodes is equal to the length of the link that connects them.

If there is a few links that connect two nodes, the distance between these nodes is equal to the length of the shortest link that connects them. Each node i in the network is denoted by the following two labels:

d_{ai}: the length of the shortest known path from node a to node i discovered so far,
q_i: predecessor node to node i on the shortest path from node a to node i discovered so far.

We denote by c the last node to be given a permanent label. We also denote by + node predecessor to node a. Dijkstra's algorithm is as follows:

Step 1: The process starts from node a. Set $d_{aa}=0$, $q_a=+$, $d_{ai}=\infty$ for $i \neq a$, and $q_i=-$ for $i \neq a$. The only node which is in a closed state is node a. Therefore, we write $c=a$.

Step 2: Explore all links (c, i) which exit from the last node that is in a closed state (node c). If node i is also in a closed state, we pass the examination on to the next node. If node i is in an open state, we obtain its first label d_{ai} based on the relation:

$$d_{ai} = \min\{d_{ai}, d_{ac} + d(c,i)\},$$

Step 3: Compare the values d_{ai} for all nodes that are in an open state and choose the node with the smallest d_{ai} value. Let this be some node j. Node j passes from an open to a closed state.

Step 4: Examine the lengths of all links (i, j) that lead from closed state nodes to node j until one is found such that the following relation is satisfied:

$$d_{aj} - d(i,j) = d_{ai}.$$

Let this relation be satisfied for some node t. Then set $q_j=t$.

Step 5: Node j is in a closed state. If all nodes are in a closed state, stop. If there are still some open state nodes, go to Step 2.

EXAMPLE 3.1

Using Dijkstra's algorithm, discover the shortest paths from node 0 to all other nodes in the transportation network shown in Fig. 3.15. The numbers next to the network links shown in Fig. 3.15 denote the link lengths.

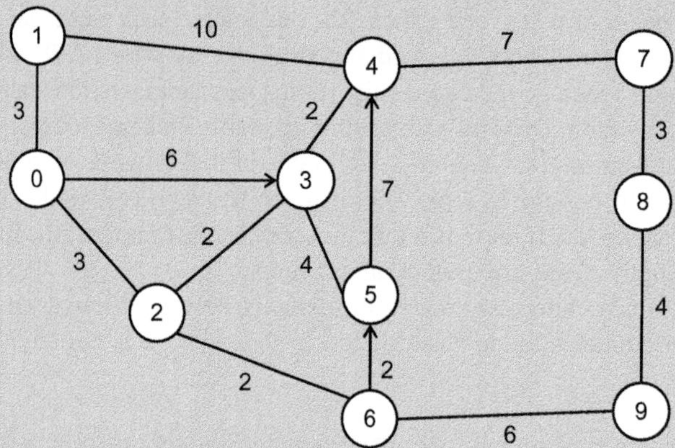

FIG. 3.15

Network in which shortest paths from node 0 to all other nodes should be discovered.

Solution

First Iteration

We start the process of discovering the shortest paths at node 0. The length of the shortest path from node 0 to node 0 is equal to zero ($d_{00}=0$). We denote by the symbol + the predecessor node to node 0 ($q_0=+$). The lengths of all other shortest paths from node 0 to all other nodes $i \neq 0$ are for the present unexplored, so for all other nodes $i \neq 0$ we set $d_{0i}=\infty$. Since predecessor nodes to nodes $i \neq 0$ are unknown, we set $q_i=-$ for all $i \neq 0$. The only node that is now in a closed state is node 0. Therefore, $c=0$. To the node 0 symbol we put the label (0, +), and add the symbol ' that indicates node 0 is in a closed state. This completes the first step of the algorithm.

The transportation network after the first algorithmic step is shown in Fig. 3.16.

We now move to the second step of the algorithm. By examining the lengths of all links leaving node 0 that are in a closed state we can write the following:

$$d_{01} = \min\{\propto, 0+3\} = 3$$

$$d_{02} = \min\{\propto, 0+3\} = 3$$

$$d_{03} = \min\{\propto, 0+6\} = 6$$

In step 3, we determine which node will be next in line to pass from an open to a closed state. Since

$$d_{01} = \min\{d_{01}, d_{02}, d_{03}\} = \min\{3, 3, 6\} = 3$$

EXAMPLE 3.1—cont'd

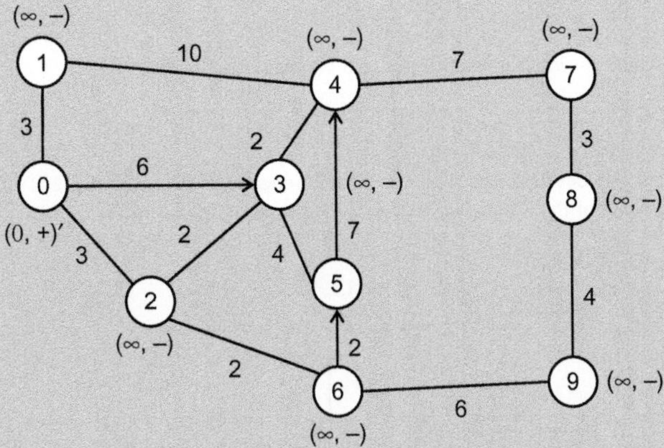

FIG. 3.16

Network after the first algorithmic step.

node 1 passes from an open to a closed state. In the same manner, since

$$d_{01} - d(0,1) = 3 - 3 = 0 = d_{00}$$

we can conclude in step 4 that node 0 is the immediate predecessor of node 1 on the shortest path, ie, $q_1 = 0$.

Now, in step 5 we note that there are still many nodes that are in an open state. Therefore, we have to return to step 2. The network, after going through all five steps of the algorithm for the first time is shown in Fig. 3.17.

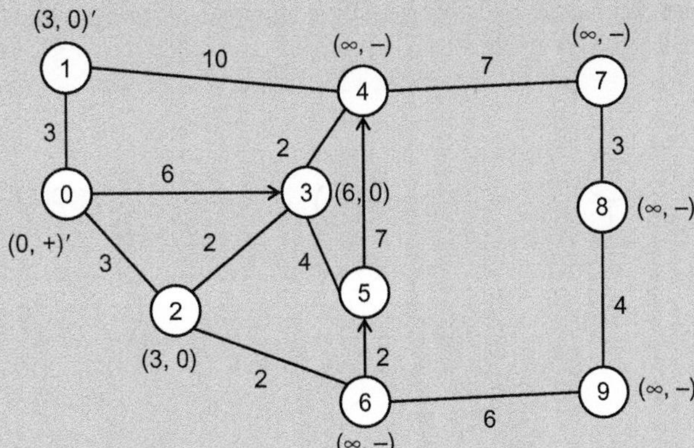

FIG. 3.17

Network after the first time through the algorithmic steps.

Second Iteration

Let us now return to the second step of the algorithm. The last node to go from an open to a closed state was node 1. This means that $c = 1$. Let us examine all links, leaving node 1 and going towards nodes that are in an open state. The following relation is satisfied:

(Continued)

EXAMPLE 3.1—cont'd

$$d_{04} = \min\{d_{04}, d_{01} + d(1,4)\} = \min\{\infty, 3 + 10\} = 13$$

The node 4 is the next node to switch from an open to a closed state. Since

$$d_{04} - d(1,4) = 13 - 10 = 3 = d_{01}$$

node 1 is the immediate predecessor of node 4, and $q_4 = 1$.

The network still contains many nodes in an open state, so we must once again return to step 2 of the algorithm. The network, after going through all five steps of the algorithm for the second time, is shown in Fig. 3.18.

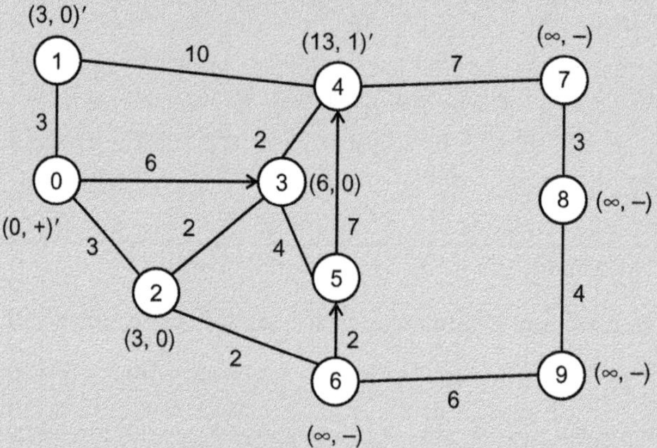

FIG. 3.18

Network after the second time through the algorithmic steps.

We obtained shortest paths from node 0 to all other nodes by going through few more iterations. The shortest paths from node a to all other nodes in the network are shown in Fig. 3.19.

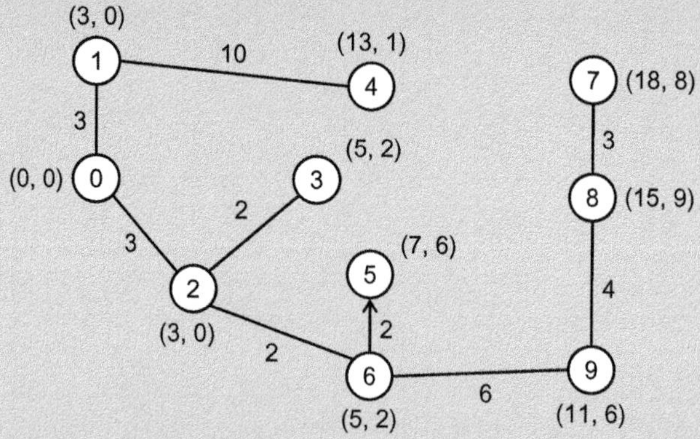

FIG. 3.19

Shortest paths from node 0 to all other nodes.

3.3.3 SHORTEST PATHS BETWEEN ALL PAIRS OF NODES

Let our problem be to find the shortest path between all nodes in transportation network $G=(N, A)$. *Floyd*'s algorithm (1962) is one of the best-known algorithms that can found shortest paths between all pairs of nodes. We denote all nodes of the network by positive whole numbers 1, 2, …., n.

We now introduce D_0 which is the beginning matrix of the shortest path lengths and Q_0 which is the predecessor matrix. We denote by d_{ij}^k the length of the shortest path from node i to node j which is found in the kth passage through the algorithm, and by q_{ij}^k the immediate predecessor node of node j on the shortest path from node i which is also discovered on the kth passage. Elements d_{ij}^0 of matrix D_0 are defined in the following manner:

If a branch exists between node i and node j, the length of the shortest path d_{ij}^0 between these nodes equals length $d(i, j)$ of branch (i, j) which connects them. Should there be several branches between nodes i and node j, the length of the shortest path d_{ij}^0 must equal the length of the shortest branch, ie,

$$d_{ij}^0 = \min\{d_1(i,j), d_2(i,j), \ldots, d_m(i,j)\} \tag{3.3}$$

where m is the number of branches between node i and node j.

It is clear that $d_{ij}^0 = 0$ when $i=j$. In the case when there is no branch between node i and node j, we have no information at the beginning concerning the length of the shortest path between these two nodes so we treat them as though they were infinitely far from each other, ie, that the following is true for such pairs of nodes:

$$d_{ij}^0 = \propto \tag{3.4}$$

Elements q_{ij}^0 of the predecessor matrix Q_0 are defined as follows:

First, we assume that $q_{ij}^0 = i$, for $i \neq j$, ie, that for every pair of nodes (i, j), for $i \neq j$, the immediate predecessor of node j on the shortest path leading from node i to node j is actually node i. Since we have defined the elements of matrixes D_0 and Q_0, we can now take a look at Floyd's algorithm, which contains the following steps:

Step 1: Let $k=1$.

Step 2: We calculate elements d_{ij}^k of the shortest path length matrix D_k found after the kth passage through algorithm using the following equation:

$$d_{ij}^k = \min\left\{d_{ij}^{k-1}, d_{ik}^{k-1} + d_{kj}^{k-1}\right\}$$

Step 3: Elements of predecessor matrix Q_k found after the kth passage through the algorithm are calculated as follows:

$$q_{ij}^k = \begin{cases} q_{kj}^{k-1} & \text{for } d_{ij}^k \neq d_{ij}^{k-1} \\ q_{ij}^{k-1} & \text{otherwise} \end{cases}$$

Step 4: If $k=n$ the algorithm is finished. If $k<n$, increase k by 1, ie, $k=k+1$ and return to Step 2.

Let us now look at the algorithm in a little more detail. In Step 2, each time we go through the algorithm we are checking as to whether a shorter path exists between nodes i and j other than the path we already know about which was established during one of the earlier passages through the algorithm. If we establish that $d_{ij}^k \neq d_{ij}^k$, ie, if we establish during the kth passage through the algorithm that the length of the shortest path d_{ij}^{k} between nodes i and j is less than the length of the shortest path d_{ij}^{k-1} known previous to the kth passage, we have to change the immediate predecessor node to node j.

Since the length of the new shortest path is:

$$d_{ij}^{k} = d_{ik}^{k-1} + d_{kj}^{k-1} \tag{3.5}$$

it is clear that in this case node k is the new immediate predecessor node to j, and therefore:

$$q_{ij}^{k} = q_{kj}^{k-1} \tag{3.6}$$

This is actually done in the third algorithmic step. It is also clear that the immediate predecessor node to node i does not change if, at the end of Step 2, we have established that no other new, shorter path exists. This means that:

$$q_{ij}^{k} = q_{ij}^{k-1} \text{ when } d_{ij}^{k} = d_{ij}^{k-1} \tag{3.7}$$

When we go through the algorithm n times (n is the number of nodes in the transportation network), elements d_{ij}^{k-1} of final matrix D_n will constitute the shortest path's lengths between pairs of nodes (i, j), and elements q_{ij}^{n} of matrix Q_n will enable us in to determine all of the nodes which are on the shortest path going from node i to node j.

EXAMPLE 3.2

Determine the shortest paths between all pairs of nodes on transportation network $T\ (N, A)$ shown in Fig. 3.20. Link lengths are shown in Fig. 3.20.

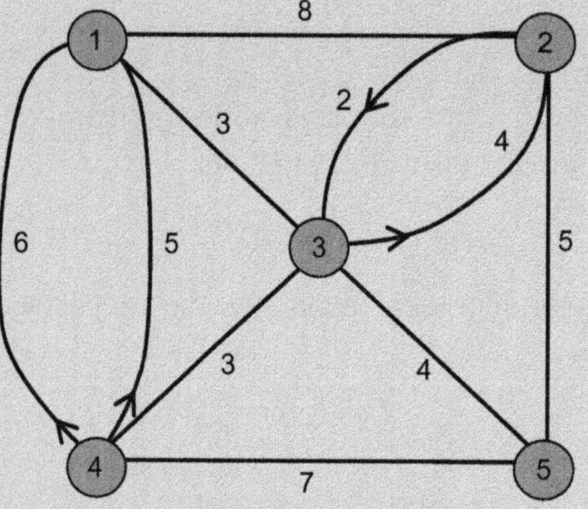

FIG. 3.20

Network in which shortest paths between all pairs of nodes should be discovered.

Solution

Starting matrix D_0 is as follows:

EXAMPLE 3.2—cont'd

$$D=\begin{array}{c}\\1\\2\\3\\4\\5\end{array}\begin{array}{c}\begin{array}{ccccc}1&2&3&4&5\end{array}\\\begin{bmatrix}0&8&3&5&\infty\\8&0&2&\infty&5\\\infty&1&0&3&4\\6&\infty&\infty&0&7\\\infty&5&\infty&\infty&0\end{bmatrix}\end{array}$$

All elements of the main diagonal of the matrix D_0 are equal to zero ($d_{ij}^{k-1}=0$ for $i=j$). Let us note the element $d_{1,2}^0$ of the matrix D_0. This element is equal to 8, since the length of the link that connects node 1 and the node 2 equals 8. Element $d_{3,1}^0$ is equal to ∞, since there is no link leading from node 3 to node 1 (there is only link oriented from node 1 to node 3). Element $d_{5,1}^0$ of the matrix D_0 is equal to ∞, since there is no link connected node 5 and node 1.

Initial matrix Q_0 reads:

$$Q_0=\begin{array}{c}\\1\\2\\3\\4\\5\end{array}\begin{array}{c}\begin{array}{ccccc}1&2&3&4&5\end{array}\\\begin{bmatrix}-&1&1&1&1\\2&-&2&2&2\\3&3&-&3&3\\4&4&4&-&4\\5&5&5&5&-\end{bmatrix}\end{array}$$

In the beginning, we consider node i as a predecessor to node j on the shortest path leading from node i to node j (for $i \neq j$). For example, the following relation is satisfied:

$$q_{2,1}^0=q_{2,3}^0=q_{2,4}^0=q_{2,5}^0=2$$

Let us start with a first algorithmic step ($k=1$). Let us calculate the elements of the matrix D_1. The following relations are satisfied:

$$d_{1,2}^1=\min\{d_{1,2}^0,d_{1,1}^0+d_{1,2}^0\}=\min\{8,0+8\}=8$$

$$d_{1,3}^1=\min\{d_{1,3}^0,d_{1,1}^0+d_{1,3}^0\}=\min\{3,0+8\}=3$$

$$d_{1,4}^1=\min\{d_{1,4}^0,d_{1,1}^0+d_{1,4}^0\}=\min\{5,0+5\}=5$$

$$d_{1,5}^1=\min\{d_{1,5}^0,d_{1,1}^0+d_{1,5}^0\}=\min\{\infty,0+\infty\}=\infty$$

$$d_{2,1}^1=\min\{d_{2,1}^0,d_{2,1}^0+d_{1,1}^0\}=\min\{8,8+0\}=8$$

$$d_{2,3}^1=\min\{d_{2,3}^0,d_{2,1}^0+d_{1,3}^0\}=\min\{2,8+3\}=2$$

$$d_{2,4}^1=\min\{d_{2,4}^0,d_{2,1}^0+d_{1,4}^0\}=\min\{\infty,8+5\}=13$$

$$d_{2,5}^1=\min\{d_{2,5}^0,d_{2,1}^0+d_{1,5}^0\}=\min\{5,8+\infty\}=5$$

$$d_{3,1}^1=\min\{d_{3,1}^0,d_{3,1}^0+d_{1,1}^0\}=\min\{\infty,\infty+\infty\}=\infty$$

(Continued)

EXAMPLE 3.2—cont'd

$$d^1_{3,2} = \min\{d^0_{3,2}, d^0_{3,1} + d^0_{1,2}\} = \min\{1, \infty + 8\} = 1$$

$$d^1_{3,4} = \min\{d^0_{3,4}, d^0_{3,1} + d^0_{1,4}\} = \min\{3, \infty + 5\} = 3$$

$$d^1_{3,5} = \min\{d^0_{3,5}, d^0_{3,1} + d^0_{1,5}\} = \min\{4, \infty + \infty\} = 4$$

$$d^1_{4,1} = \min\{d^0_{4,1}, d^0_{4,1} + d^0_{1,1}\} = \min\{6, 6 + 0\} = 6$$

$$d^1_{4,2} = \min\{d^0_{4,2}, d^0_{4,1} + d^0_{1,2}\} = \min\{\infty, 6 + 8\} = 14$$

$$d^1_{4,3} = \min\{d^0_{4,3}, d^0_{4,1} + d^0_{1,3}\} = \min\{\infty, 6 + 3\} = 9$$

$$d^1_{4,5} = \min\{d^0_{4,5}, d^0_{4,1} + d^0_{1,5}\} = \min\{7, 6 + \infty\} = 7$$

$$d^1_{5,1} = \min\{d^0_{5,1}, d^0_{5,1} + d^0_{1,1}\} = \min\{\infty, \infty + 0\} = \infty$$

$$d^1_{5,2} = \min\{d^0_{5,2}, d^0_{5,1} + d^0_{1,2}\} = \min\{5, \infty + 8\} = 5$$

$$d^1_{5,3} = \min\{d^0_{5,3}, d^0_{5,1} + d^0_{1,3}\} = \min\{\infty, \infty + 3\} = \infty$$

$$d^1_{5,4} = \min\{d^0_{5,4}, d^0_{5,1} + d^0_{1,4}\} = \min\{\infty, \infty + 5\} = \infty$$

Matrix D_1 reads:

$$D_1 = \begin{array}{c} \\ 1 \\ 2 \\ 3 \\ 4 \\ 5 \end{array} \begin{array}{c} \begin{array}{ccccc} 1 & 2 & 3 & 4 & 5 \end{array} \\ \begin{bmatrix} 0 & 8 & 3 & 5 & \infty \\ 8 & 0 & 2 & \mathbf{13} & 5 \\ \infty & 1 & 0 & 3 & 4 \\ 6 & \mathbf{14} & \mathbf{9} & 0 & 7 \\ \infty & 5 & \infty & \infty & 0 \end{bmatrix} \end{array}$$

The elements of the matrix D_0 whose values are changed are denoted by a bold letters.

The length of the shortest path from node 2 to node 4, after first algorithmic step, equals 13. In the initial matrix D_0 this length was equal to ∞. Since:

$$d^1_{2,4} = d^0_{2,1} + d^0_{1,4} = 13 < \infty = d^0_{2,4}$$

Node 1 is a new predecessor node to node 4 on the shortest path from node 2 to node 4. Matrix Q_1 after first passage through the algorithm reads:

$$Q_1 = \begin{array}{c} \\ 1 \\ 2 \\ 3 \\ 4 \\ 5 \end{array} \begin{array}{c} \begin{array}{ccccc} 1 & 2 & 3 & 4 & 5 \end{array} \\ \begin{bmatrix} - & 1 & 1 & 1 & 1 \\ 2 & - & 2 & \mathbf{1} & 2 \\ 3 & 3 & - & 3 & 3 \\ 4 & \mathbf{1} & \mathbf{1} & - & 4 \\ 5 & 5 & 5 & 5 & - \end{bmatrix} \end{array}$$

The matrices D_2 and Q_2 after second passage through the algorithm reads:

EXAMPLE 3.2—cont'd

$$D_2=\begin{array}{c|ccccc|} & 1 & 2 & 3 & 4 & 5 \\ \hline 1 & 0 & 8 & 3 & 5 & \mathbf{13} \\ 2 & 8 & 0 & 2 & 13 & 5 \\ 3 & \mathbf{9} & 1 & 0 & 3 & 4 \\ 4 & 6 & 14 & 9 & 0 & 7 \\ 5 & \mathbf{13} & 5 & \mathbf{7} & \mathbf{18} & 0 \end{array} \qquad Q_2=\begin{array}{c|ccccc|} & 1 & 2 & 3 & 4 & 5 \\ \hline 1 & - & 1 & 1 & 1 & \mathbf{2} \\ 2 & 2 & - & 2 & 1 & 2 \\ 3 & \mathbf{2} & 3 & - & 3 & 3 \\ 4 & 4 & 1 & 1 & - & 4 \\ 5 & \mathbf{2} & 5 & \mathbf{2} & \mathbf{2} & - \end{array}$$

Matrices D_3, Q_3, D_4, Q_4 i D_5, Q_5 are respectively equal:

$$D_3=\begin{array}{c|ccccc|} & 1 & 2 & 3 & 4 & 5 \\ \hline 1 & 0 & \mathbf{4} & 3 & 5 & \mathbf{7} \\ 2 & 8 & 0 & 2 & \mathbf{5} & 5 \\ 3 & 9 & 1 & 0 & 3 & 4 \\ 4 & 6 & \mathbf{10} & 9 & 0 & 7 \\ 5 & 13 & 5 & 7 & \mathbf{10} & 0 \end{array} \qquad Q_3=\begin{array}{c|ccccc|} & 1 & 2 & 3 & 4 & 5 \\ \hline 1 & - & \mathbf{3} & 1 & 1 & \mathbf{3} \\ 2 & 2 & - & 2 & \mathbf{3} & 2 \\ 3 & 2 & 3 & - & 3 & 3 \\ 4 & 4 & \mathbf{3} & 1 & - & 4 \\ 5 & 2 & 5 & 2 & \mathbf{3} & - \end{array}$$

$$D_4=\begin{array}{c|ccccc|} & 1 & 2 & 3 & 4 & 5 \\ \hline 1 & 0 & 4 & 3 & 5 & 7 \\ 2 & 8 & 0 & 2 & 5 & 5 \\ 3 & 9 & 1 & 0 & 3 & 4 \\ 4 & 6 & 10 & 9 & 0 & 7 \\ 5 & 13 & 5 & 7 & 10 & 0 \end{array} \qquad Q_4=\begin{array}{c|ccccc|} & 1 & 2 & 3 & 4 & 5 \\ \hline 1 & - & 3 & 1 & 1 & 3 \\ 2 & 2 & - & 2 & 3 & 2 \\ 3 & 2 & 3 & - & 3 & 3 \\ 4 & 4 & 3 & 1 & - & 4 \\ 5 & 2 & 5 & 2 & 3 & - \end{array}$$

$$D_5=\begin{array}{c|ccccc|} & 1 & 2 & 3 & 4 & 5 \\ \hline 1 & 0 & 4 & 3 & 5 & 7 \\ 2 & 8 & 0 & 2 & 5 & 5 \\ 3 & 9 & 1 & 0 & 3 & 4 \\ 4 & 6 & 10 & 9 & 0 & 7 \\ 5 & 13 & 5 & 7 & 10 & 0 \end{array} \qquad Q_5=\begin{array}{c|ccccc|} & 1 & 2 & 3 & 4 & 5 \\ \hline 1 & - & 3 & 1 & 1 & 3 \\ 2 & 2 & - & 2 & 3 & 2 \\ 3 & 2 & 3 & - & 3 & 3 \\ 4 & 4 & 3 & 1 & - & 4 \\ 5 & 2 & 5 & 2 & 3 & - \end{array}$$

We obtain from the matrices D_5 and Q_5 full information about shortest paths between all pairs of nodes. So, for example, the length of the shortest path from node 5 to node 4 is equal to 10. Node-predecessor to node 4 is node 3, since $q_{5,4}=3$. Node-predecessor to node 3 on the shortest path from node 5 to node 3, is node 2, since $q_{5,3}=2$. Since $q_{5,2}=5$, we conclude that the shortest path from node 5 to node 4 reads (5, 2, 3, 4).

3.4 MATHEMATICAL PROGRAMMING APPLICATIONS IN TRAFFIC AND TRANSPORTATION

Many real-life traffic and transportation problems can be relatively easily formulated in words. After such formulation of the problem, in the next step, engineers usually translate problem verbal description into a mathematical description. Main components of the mathematical description of the problem are *variables*, *constraints*, and the *objective*. Variables are also sometimes called *unknowns*. Some of the variables are under the control of the analyst. There are also variables that are not under the control of the analyst. Constraints could be physical, caused by some engineering rules, laws, or guidelines, or by various financial reasons. No one could accept more than 100 passengers for the planned flight, if the capacity of the aircraft equals 100 seats. This is typical example of a physical constraint. Financial constraints are usually related to various investment decisions. For example, no one could invest in road improvement more than $10,000,000 if the available budget equals $10,000,000. Variable values could be feasible, or infeasible. Variable values are feasible when they satisfy all the defined constraints. An objective represents the end result decision-maker wants to accomplish by selecting a specific program of action. Revenue maximization, cost minimization, or profit maximization are typical objectives in profit oriented organizations. Providing highest level-of service to the customers represents usual objective in a nonprofit organizations.

Mathematical description of a real-world problem is called a *mathematical model* of the real-world problem.

For example, if x_{ij}, $j = 1,2,\ldots,n$ are the n decision variables of the studied problem, and if the observed system is subject to m constraints, the general mathematical model can be written in the following way:

Optimize

$$y = f(x_1, x_2, \ldots, x_n) \tag{3.8}$$

subject to

$$h_i(x_1, x_2, \ldots, x_n) \leq b_i \quad i = 1,2,\ldots,m \tag{3.9}$$

$$x_1, x_2, \ldots, x_n \geq 0 \tag{3.10}$$

where $y = f(x_1, x_2, \ldots, x_n)$ is the objective function, $h_i(x_1, x_2, \ldots, x_n) \leq b_i \quad i = 1,2,\ldots,m$ is the constraint, and $x_1, x_2, \ldots, x_n \geq 0$ is the constraint (nonnegativity restrictions).

Optimization seeks the best value (*optimal value*) of the objective function. Optimization usually suggests the maximization or minimization of the objective function. *Optimal solution* to the model is the discovery of a set of variable values (feasible) which generate the optimal value of the objective function.

An *algorithm* represents some quantitative method used by an analyst to solve the defined mathematical model. Algorithms are composed of a set of instructions which are usually followed in a defined step-by-step procedure. Algorithm produces an optimal (the best) solution to a defined model. *Optimal solution* to the model is the discovery of a set of variable values (feasible) that generate the optimal value of the objective function. Depending on a defined objective function, optimal solution corresponds to maximum revenue, minimum cost, maximum profit, etc.

3.4.1 LINEAR PROGRAMMING IN TRAFFIC AND TRANSPORTATION

In many cases all variables are continuous variables. There is usually also only one objective function. Frequently, objective function and all constraints are *linear*, meaning that any term is either a constant or a constant multiplied by a variable. Any mathematical model that has one objective function, all continuous variables, linear objective function, and all constraints is called a *linear program* (LP). It has been shown through many years that many real-life problems can be formulated as linear programs. Linear programs are usually solved using widely spread Simplex algorithm (there is also an alternative algorithm called *Interior Point Method*).

EXAMPLE 3.3

We show the basics of the Linear Programming using the example related to the airline seat inventory control problem. The liberalization of airline tariffs has led to intensive competition among air carriers. In such conditions, an air carrier logically wants to sell the seats available in a way that maximizes profit. The liberalization of airline tariffs has also resulted in a large number of different tariffs existing on the same flight. Passengers paying lower tariffs (as a rule making private trips) often reserve seats before passengers paying higher tariffs (business passengers who decide to travel several days or hours before the flight), which is why a certain number of passengers who are prepared to pay a higher tariff cannot find a vacant seat on the flight they want. The simplest reservation system is often called distinct fare class inventories, indicating separate seat inventories for each fare class. Once a seat is assigned to a fare class inventory, it may be booked only in that fare class, or else remain unsold. In the case of a nested reservation system, the high fare request will not be rejected as long as any seats are available in lower fare classes. Let us consider the airline seat inventory control problem for a direct, nonstop flight. An aircraft capacity (the number of seats in the aircraft) equals 100. Let us assume that passengers are offered two tariff classes: \$200 and \$100. We assume that we are able to predict exactly the total number of requests in different passenger tariff classes. We expect 60 passenger requests in the first class, and 80 passenger requests in the second class. We decide to sell at least 10 seats to the passengers paying higher tariffs. We have to determine the total numbers of seats sold in different passenger tariff classes in order to reach the maximum airline revenue.

Solution

Since we wish to determine the total numbers of seats sold in different passenger tariff classes, the variables of the model can be defined as:

x_1 is the total number of seats planned to be sold in the first passenger tariff class;

x_2 is the total number of seats planned to be sold in the second passenger tariff class.

Since each seat from the first class sells for \$200, the total revenue from selling x_1 seats is $200 \cdot x_1$. In the same way, the total airline revenue from x_2 seats is equal to $100 \cdot x_2$. The total airline revenue equals the sum of the two revenues, ie, $200 \cdot x_1 + 100 \cdot x_2$.

From the problem formulation we conclude that there are specific restrictions on the seat selling and on demand. The seat selling restrictions may be expressed verbally in the following way:

- Total number of seats sold in both classes together must be less than or equal to the aircraft capacity.
- Total number of seats sold in any class must be less than or equal to the total number of passenger requests.
- Total number of seats sold in the first class must be at least 10.

Total number of seats sold in the second class cannot be less than zero (nonnegativity constraint).

The following is the mathematical model for airline revenue management problem:

Maximize

$$F = 200 \cdot x_1 + 100 \cdot x_2$$

subject to:

$$x_1 + x_2 \leq 100$$

$$x_1 \leq 60$$

$$x_2 \leq 80$$

$$x_1 \geq 10$$

$$x_2 \geq 0$$

(Continued)

EXAMPLE 3.3—cont'd

In our problem, we allow variables to take the fractional values (we can always round the fractional value to the closest integer value). In other words, all our variables are continuous variables. We also have only one objective function: We try to maximize the total airline revenue. Our objective function and all our constraints are linear. Since we have only two variables, we can also solve our problem graphically. Graphical method is impractical, or impossible for mathematical models with more than two variables. In order to solve problem graphically, we plot the feasible solutions (solution space) which satisfy all constraints simultaneously. Fig. 3.21 shows our solution space.

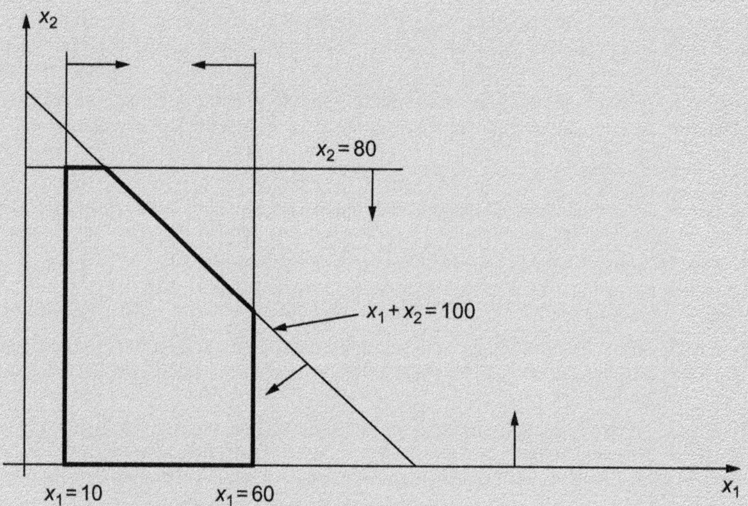

FIG. 3.21

Solution space of the airline seat inventory control problem.

All feasible values of the variables are located in the first quadrant. This is caused by the following constraints: $x_1 \geq 10$ and $x_2 \geq 0$. The straight-line equations $x_1=10$, $x_1=60$, $x_2=80$, $x_2=0$, and $x_1+x_2=100$ are obtained by substituting "$\leq$" by "$=$" for each constraint. Then, each straight-line is plotted. The region in which each constraint is satisfied when the inequality is put in power is indicated by the direction of the arrow on the corresponding straight line. The resulting solution space of the airline seat inventory control problem is shown in Fig. 3.21. Feasible points for the problem considered are all points within the boundary, or on the boundary of the solution space. The optimal solution is discovered by studying the direction in which the objective function rises. The optimal solution is shown in Fig. 3.22.

The parallel lines in Fig. 3.22 represent the objective function $F=200\cdot x_1+100\cdot x_2$. They are plotted by arbitrarily assigning increasing values to F. In this way, it is possible to make the conclusion about the slope and the direction in which the total airline revenue increases.

EXAMPLE 3.3—cont'd

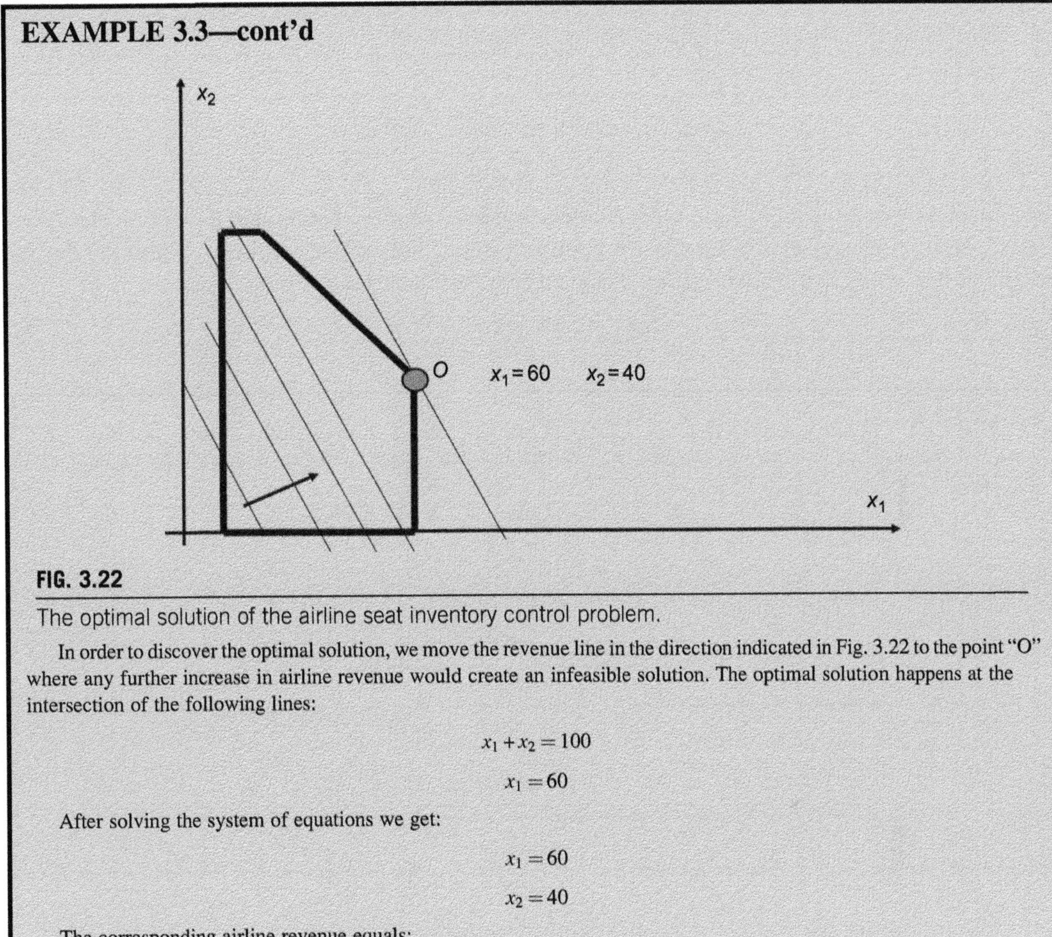

FIG. 3.22

The optimal solution of the airline seat inventory control problem.

In order to discover the optimal solution, we move the revenue line in the direction indicated in Fig. 3.22 to the point "O" where any further increase in airline revenue would create an infeasible solution. The optimal solution happens at the intersection of the following lines:

$$x_1 + x_2 = 100$$

$$x_1 = 60$$

After solving the system of equations we get:

$$x_1 = 60$$

$$x_2 = 40$$

The corresponding airline revenue equals:

$$F = 200 \cdot x_1 + 100 \cdot x_2 = 200 \cdot 60 + 100 \cdot 40 = 16{,}000$$

The problem considered is a typical *resource allocation* problem. *Linear Programming* (Taha, 1982; Hillier and Lieberman, 1990; Winston, 1994) help us to discover the best allocation of limited resources. The following is *Linear Programming* model:

$$F = c_1 \cdot x_1 + c_2 \cdot x_2 + \cdots + c_n \cdot x_n \tag{3.11}$$

$$a_{11} \cdot x_1 + a_{12} \cdot x_2 + \cdots + a_{1n} \cdot x_n \leq b_1 \tag{3.12}$$

$$a_{21} \cdot x_1 + a_{22} \cdot x_2 + \cdots + a_{2n} \cdot x_n \leq b_2 \tag{3.13}$$

$$\vdots$$

$$a_{m1} \cdot x_1 + a_{m2} \cdot x_2 + \cdots + a_{mn} \cdot x_n \leq b_m \tag{3.14}$$

$$x_1, x_2, \ldots, x_n \geq 0 \tag{3.15}$$

The variables describe level of various economic activities (number of seats sold to the first class passengers, duration of a green time for specific approach at the intersection, number of flights per day on specific airline route, number of vehicles assigned to a particular route, etc.).

EXAMPLE 3.4

There are three on-ramps on a freeway. The freeway is divided into three sections, each containing at most one on-ramp (Fig. 3.23).

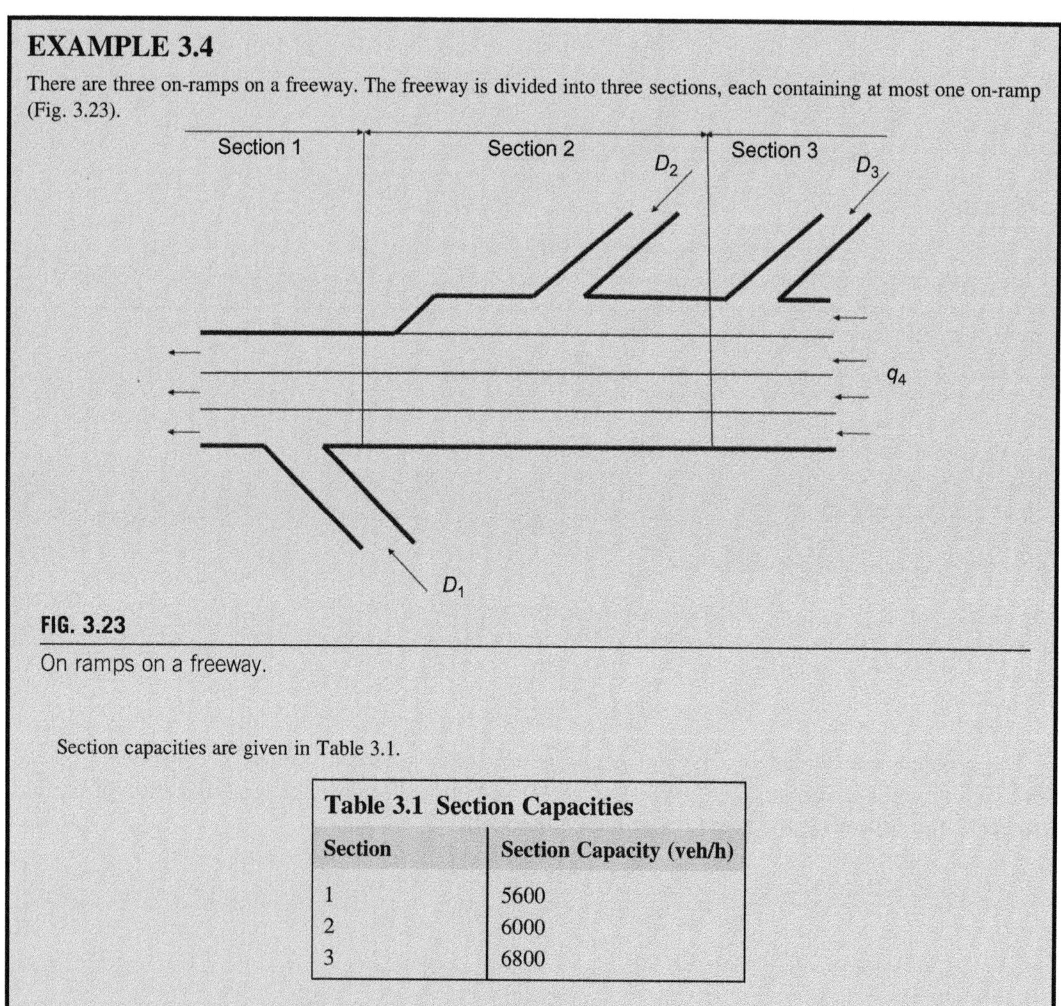

FIG. 3.23

On ramps on a freeway.

Section capacities are given in Table 3.1.

Table 3.1 Section Capacities

Section	Section Capacity (veh/h)
1	5600
2	6000
3	6800

EXAMPLE 3.4—cont'd

Ramp demands are given in Table 3.2.

Table 3.2 Ramp Demand

Ramp	Demand (veh/h)
1	700
2	800
3	900

Calculate the optimal number of vehicles allowed to enter the highway from every ramp.

Solution

The maximum number of vehicles allowed to enter any section must be less than or equal to the section capacity. At the same time, the number of vehicles allowed to enter the highway from any ramp must be less than or equal to demand on that ramp. We want to maximize the total number of vehicles that enter the highway. The mathematical formulation of our problem reads:

Maximize

$$q_1 + q_2 + q_3 \tag{3.16}$$

subject to

$$q_1 + q_2 + q_3 + 4000 \le 5600 \tag{3.17}$$

$$q_2 + q_3 + 4000 \le 6000 \tag{3.18}$$

$$q_3 + 4000 \le 6800 \tag{3.19}$$

$$0 \le q_1 \le 700 \tag{3.20}$$

$$0 \le q_2 \le 800 \tag{3.21}$$

$$0 \le q_3 \le 900 \tag{3.22}$$

We solve the problem (3.16)–(3.22) by using commercial package LINDO. The following are the obtained decision variables values:

$$q_1 = 0$$

$$q_2 = 700$$

$$q_3 = 900$$

The objective function value is equal to 1600.

3.4.2 INTEGER PROGRAMMING

Analysts frequently realize that some or all variables in the formulated linear program must be *integer*. This means that some variables or all take exclusively integer values. In order to make the formulated problem easier, analysts often allow these variables to take fractional values. For example, analyst knows that the number of first class passengers must be in the range between 30 and 40. Linear program could produce the "optimal solution" that tells us that the number of first class passengers equals 37.8. In this case, we can neglect the fractional part, and we can decide to protect 37 (or 38) seats for the first class passengers. In this way, we are making small numerical error, but we are capable to easily solve the problem. In some other situations, it is not possible for analyst to behave in this way. Imagine that you have to decide about new highway alignment. You must choose one out of numerous generated alternatives. This is kind of "yes/no" ("1/0") decisions: "Yes" if the alternative is chosen, "No," otherwise. In other words, we can introduce binary variables into the analysis. The variable has value 1 if the ith alternative is chosen, and value 0 otherwise, ie,

$$x_i = \begin{cases} 1 & \text{if the } i\text{th alternative is chosen} \\ 0 & \text{otherwise} \end{cases} \tag{3.23}$$

The value 0.7 of the variable means nothing to us. We are not able to decide about the best highway alignment if the variables take fractional values.

There are various logical constraints that should be taken into account when handling variables that take exclusively integer values. For example, in the case when decision-maker has to choose at most one alternative among n available alternatives, the constraint reads:

$$\sum_{i=1}^{n} x_i \leq 1 \tag{3.24}$$

In some situations, at least one alternative must be chosen among n alternatives. This constraint reads:

$$\sum_{i=1}^{n} x_i \geq 1 \tag{3.25}$$

When we solve problems similar to the highway alignment problem we work exclusively with integer variables. These kinds of problems are known as *integer programs*, and corresponding area is known as an *Integer Programming*. Integer programs usually describe the problems in which one, or more, alternatives must be selected from a finite set of generated alternatives. There are also problems in which some variables can take only integer values, while some other variables can take fractional values. These problems are known as *mixed-integer programs*. It is much harder to solve Integer Programming problems than Linear Programming problems.

The following is Integer Programming Model:

Maximize

$$F = \sum_{j=1}^{n} c_j x_j \tag{3.26}$$

subject to

$$\sum_{j=1}^{n} a_{ij}x_j \le b_i \quad \text{for} \quad i = 1,2,\ldots,m \tag{3.27}$$

$$0 \le x_j \le u_j \quad \text{integer for} \quad j = 1,2,\ldots,n \tag{3.28}$$

There are numerous software systems that solve linear, integer, and mixed-integer linear programs (CPLEX, Excel and Quattro Pro Solvers, FortMP, LAMPS, LINDO, LINGO, MILP88, MINTO, MIPIII, MPSIII, OML, OSL).

EXAMPLE 3.5

Many departing passengers significantly walk in airport terminal buildings between check-in desks and gates (the word gate is used in the literature to describe aircraft stands at the airport terminals, as well as off-pier stands at the apron). Simultaneously, the arriving passengers walking distances, between gates and the baggage claim area could be also considerable. Many transit passengers are exposed to significant walking between specific gates when changing the plane. The total passenger walking distance within airport terminal building may fluctuate depending on the specific assignment of aircraft to parking positions (Fig. 3.24).

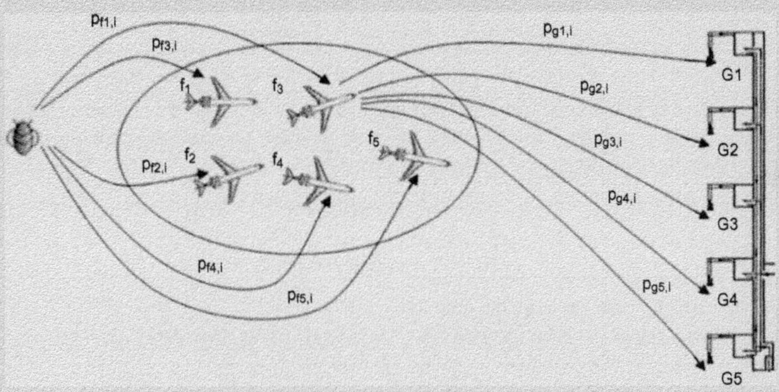

FIG. 3.24

Assignment aircraft to available gates.

The standard *Gate Assignment Problem* can be defined in the following way: For a given set of parking positions and a given set of aircraft which can use any of these parking positions, find a parking position assignment for aircraft that will minimize the total walking distance of all passengers arriving, transiting and departing by aircraft parked at set of parking positions. The decision-maker must assign aircraft to available gates and must determine the start and end time of serving aircraft at the gate it has been assigned.

(Continued)

EXAMPLE 3.5—cont'd

Airport gates are one of the greatest congestion points of the air transportation system. The total daily number of aircraft operations at big airports could be more than 1000, while the total number of gates is frequently more than 100.

Let us consider the toy example when we have to assign three aircraft to four available parking positions. Departure and arrival walking distances are given in Table 3.3.

Table 3.3 Departure and Arrival Walking Distances

Gate	Departure Walking Distance Between Check-in Desks and Gate (m)	Arrival Walking Distance Between Gate and the Baggage Claim Area (m)
1	70	85
2	75	95
3	120	110
4	100	130

Number of departing passengers and number of arriving passengers are given in Table 3.4.

Table 3.4 Numbers of Departing and Number of Arriving Passengers

Aircraft	Number of Departing Passengers	Number of Arriving Passengers
1	120	150
2	80	130
3	150	90

Aircraft 1 will land with 150 passengers, and will be assigned to one of the available positions. The same aircraft will depart from that parking position with the 120. Aircraft 2 will land with 130 passengers, etc. We denote by c_{ij} the total walking distance of all passengers if aircraft i is assigned to the gate j. The equals:

$$c_{ij} = n_i^a d_j^a + n_i^d d_j^d \tag{3.29}$$

where:

n_i^a is the number of arriving passengers in the aircraft i;
n_i^d is the number of departing passengers by the aircraft i;
d_j^a is the arrival walking distance between the gate j and the baggage claim area; and
d_j^d is the departure walking distance between check-in desks and the gate j.

The total walking distances of all passengers for every aircraft-gate pair are shown in Table 3.5.

Table 3.5 Total Walking Distance of All Passengers for Every Aircraft-Gate Pair

c_{ij}	Gate 1	Gate 2	Gate 3	Gate 4
Aircraft 1	21,150	23,250	30,900	31,500
Aircraft 2	16,650	18,350	23,900	24,900
Aircraft 3	18,150	19,800	27,900	26,700

EXAMPLE 3.5—cont'd

The Gate Assignment Problem could be formulated in the following way:
Minimize

$$Z=\sum_{i=1}^{3}\sum_{j=1}^{4}c_{ij}x_{ij}$$

subject to

$$\sum_{j=1}^{4}x_{ij}=1 \quad \forall\, i=1,2,3$$

$$\sum_{i=1}^{3}x_{ij}\leq 1 \quad \forall\, j=1,2,3,4$$

ie,
Minimize

$$\begin{aligned}Z=&21,150x_{11}+23,150x_{12}+30,900x_{13}+31,500x_{14}+16,650x_{21}+18,350x_{22}\\&+23,900x_{23}+24,900x_{24}+18,150x_{31}+19,800x_{32}+27,900x_{33}+26,700x_{34}\end{aligned}$$

subject to

$$x_{11}+x_{12}+x_{13}+x_{14}=1$$

$$x_{21}+x_{22}+x_{23}+x_{24}=1$$

$$x_{31}+x_{32}+x_{33}+x_{34}=1$$

$$x_{11}+x_{21}+x_{31}\leq 1$$

$$x_{12}+x_{22}+x_{32}\leq 1$$

$$x_{13}+x_{23}+x_{33}\leq 1$$

$$x_{14}+x_{24}+x_{34}\leq 1$$

We used commercial software to solve the problem. In the optimal solution decision variables x_{11}, x_{23}, and x_{32} take values 1. All other variables are equal to 0. We conclude that the aircraft 1 should be assigned to the gate 1, aircraft 2 to the gate 3, and aircraft 3 to the gate 2. The total walking distance in this case is equal to 64,850 m.

3.4.3 DIMENSIONALITY OF THE TRAFFIC AND TRANSPORTATION ENGINEERING PROBLEMS

A great number of practical real-world transportation problems was formulated and solved using Integer Programming, Dynamic Programming, and Graph Theory techniques during the last five decades. It is important to note, however, that the majority of real-world problems solved by some of the optimization techniques were of *small dimensionality*. Many engineering and management problems are combinatorial by their nature. Most of the combinatorial optimization problems are difficult to solve either because of the large dimensionality or because it is very difficult to decompose then into smaller subproblems. Typical representatives of this type of problems are the vehicle fleet planning and static and dynamic routing and scheduling of vehicles and crews for airlines, railroads, truck operations and public transportation services, designing transportation networks and optimizing alignments for highways and public transportation routes through complex geographic spaces, different locations problems, etc.

So, in many cases optimal solution cannot be discovered in a reasonable CPU time. Frequently, there is a combinatorial explosion of the promising combinations of the decision variables that could be optimal (for example, if we have in a problem considered 1000 binary variables that can take value 0, or 1, the total number of all possible solutions is equal to 2^{1000}). In some other cases, it could be very difficult to evaluate defined objective function. In other words, many discrete optimization problems are NP-complete. In order to overcome NP completeness various *heuristic algorithms* ('ευρισkω) were developed during last five decades. These algorithms are capable to produce good enough solution(s) in a reasonable amount of CPU time. On the other hand, heuristic algorithms that are based on experience and/or judgment cannot be guaranteed to generate the optimal solution. It could happen that in some problem instances heuristic algorithms discover optimal solution(s).

Metaheuristic algorithms (Simulated Annealing (SA), Genetic Algorithms (GA), Taboo Search (TS), Variable Neighborhood Search (VNS), Ant Colony Optimization (ACO), Particle Swarm Optimization (PSO), and Bee Colony Optimization (BCO)) are considered to be a general-purpose techniques capable to produce good solutions of a difficult discrete optimization problems in a reasonable computer time. Metaheuristic algorithms could be single-solution based (Simulated Annealing, Taboo Search), or population based (Genetic Algorithm). In the case of population based metaheuristic algorithms, as opposed to single-solution based metaheuristic algorithms, the search is run in parallel from a population of solutions. Numerous factors influence the usage of a specific heuristic or metaheuristic algorithm (the frequency of making decisions, the time available for generating problem solution, the number of decision variables, etc).

3.4.4 COMPLEXITY OF ALGORITHMS

In essence, all algorithms could be classified as exact, or heuristic. Heuristic algorithms that are capable to produce good enough solution(s) in a reasonable amount of CPU time could be described as a mixture of scientific methods, invention, experience, and intuition for problem solving. The complex engineering, management, and control problems are frequently solved by various heuristic algorithms. It is possible to develop various heuristic algorithms for a specific problem. The question logically arise which one of these heuristic algorithms is the best. In computer science, the goodness of an algorithm is mainly described by its complexity. The complexity of any algorithm is usually measured through the number of *elementary operations* (addition, subtraction, multiplication, division, comparison between two numbers, execution of a branching instruction) that have to be performed by the algorithm to reach the solution under the *worst-case conditions*.

Let us assume that the number of nodes in a transportation network equals n. We also assume that this number of nodes represents the dimensions of the problem considered. We further assume that the total number of elementary operations E, to be performed in order to execute the proposed algorithm is equal to:

$$E = 4n^4 + 5n^3 + 2n + 7 \tag{3.30}$$

The value of E is primarily determined by n^4 as n increases. It is usually to say that the complexity of the proposed algorithm is proportional to n^4. The other way is to say that the algorithm requires $O(n^4)$ time (under the assumption that each elementary operation requires one unit of time). The complexity of *polynomial algorithms* is proportional to, or bounded by a polynomial function of the dimension of the input. For example, the algorithm that requires $O(n^4)$ time is polynomial algorithm.

Nonpolynomial (exponential) algorithms break all polynomial limits (in a case of large sizes of the input). For example, the algorithm that requires $O(2^n)$ time is exponential. It is usual in computer science to consider polynomial algorithms as *good* algorithms. The exponential algorithms are considered as bad algorithms. The quality of the generated solution and the CPU time are two dominant criteria for evaluation of a specific algorithm. When evaluating heuristic algorithm it is also necessary to consider the simplicity of the algorithm and complexity of the algorithm implementation. The closer the objective function value produced by the algorithm to the optimal value, the better the proposed algorithm.

Worst case analysis assumes analysis of such numerical examples that will show the worst possible results that can be obtained by the proposed algorithm. As a rule, such numerical examples are rare within the problem considered. For example, we can more easily evaluate the proposed algorithm if we know that in the worst case algorithm produces solution that has objective function value 5% higher (in a case of minimization) than the optimal solution value.

Within the *average case analysis* the analysts usually generate great number of problem instances that can appear in a real-life, and perform statistical analysis about the algorithm performances. It is always very important to test the proposed heuristic algorithm on real-life examples.

3.5 PROBABILITY THEORY AND TRAFFIC PHENOMENA

Every trip-maker makes his/her own decision independently of all other trip-makers concerning the day, time, route, and transportation mode he/she wishes to travel. The numerous independent random factors affect various traffic phenomena (travel time, the total number of cars on a specific urban transportation network link, the total number of passengers on a specific flight, demand in nodes of a distribution system, demand (time and location) for emergency help from urban emergency services, etc). Travel times between specific nodes, waiting times at the intersections, number of cars on a link, number of passengers on a plane, etc., are *random variables* whose values are *unknown* in advance (Larson and Odoni, 1981). Let us consider the following example. There is left-turn bay at the intersection whose capacity equals 8 (Fig. 3.25).

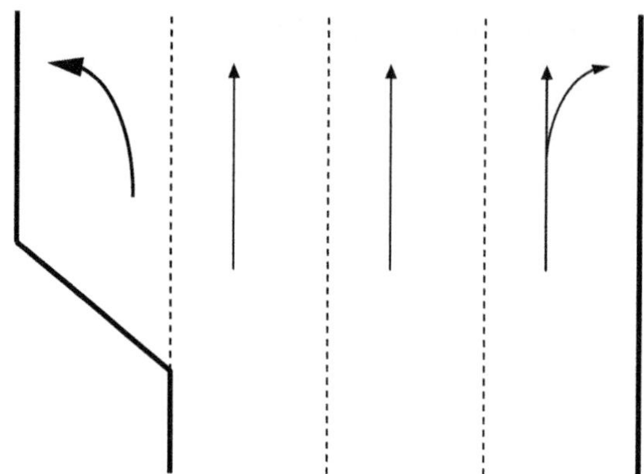

FIG. 3.25

Left-turn bay.

Let us perform the following experiment during 30 days. Every day, at 9:30 am we count the number of cars in the left turn bay. Daily *outcome* of our experiment is unknown. In other words, outcome of our experiment is subject to *chance*. It could happen that there is 0, 1, 2, 3, …, 7, or 8 cars in the left-turn bay. (When the number of left-turning cars is ≥ 8, cars fully occupy left-turn bay, as well as part of the through traffic lane). In or experiment, the number of possible outcomes is *finite*. The number of possible outcomes could be also *infinite*. For example, we can measure every day travel time between our home and the University. The outcomes in this case may take any nonnegative real value. Obviously, the number of potential outcomes in the case of travel time measurement is infinite. A *sample space* is composed of all possible experiment outcomes. For example, in our experiment of the number of cars counting in a left-turn bay, the sample space is $\{0,1,2,3,4,5,6,7,8\}$. In the case of travel time measurement, the sample space is composed from all values from the interval $\{0, \propto\}$. An *event* represents a collection of outcomes from the sample space. For example, the event could be "empty left-turn bay." The occurrence of this event is clearly related with the outcome "0." The event could be also "at least one car in the left-turn bay," "full left-turn bay," *etc*. Usually we use capital letters $A, B, C,\ldots$ to denote events. We denote by $P(A)$ the *probability* of event A. Probability Theory has its roots in the work of *Pierre de Fermat* and *Blaise Pascal* in the 17th century. *Andrey Kolmogorov*, "father" of the modern Probability Theory, presented the axiom system for the Probability Theory in 1933.

3.5.1 PROBABILITY THEORY BASICS

What is a probability? Probability is nonnegative real number not greater than one. Could probability be equal to zero? Yes. Could probability be equal to one? Yes. Could probability be greater than one? No. How could be calculate the probability of specific event? Usually, we repeat the experiment *many times*, and we count the number of trials that describe our event. Let us denote by n the total number of trials. By performing experiment we observe that m trials out of n trials describe our event. The probability of the event A equals:

$$P(A) = \lim_{n \to \propto} \frac{m}{n} \tag{3.31}$$

Probability of any event is always in between zero, and one:

$$0 \le P(A) \le 1 \tag{3.32}$$

When event A is *impossible*, then $P(A) = 0$. Impossible event never happens. For example, even in the cases of extremely high traffic demand, the number of cars in the left-turn bay will never be equal to 47, since the capacity of the left-turn bay equals 8. We denote by B the following event: "the number of cars in the left-turn bay equals 47." We can write that $P(B) = 0$. *Certain* events always happen. We denote by C the following event: "travel time by car between home and the university is greater than zero." We write that $P(C) = 1$, since event C is certain.

The *intersection* $A \cap B$ of event A and the event B means that both A and B are realized. The *union* $A \cup B$ of events A and B means that A or B, or both of them happens. The intersection is also denoted as "AB," while "$A+B$" is also used to denote union.

The *addition law* and the *conditional probability law* are the basic probability laws. The addition law (Fig. 3.26) is:

$$P(A+B) = P(A) + P(B) - P(AB) \tag{3.33}$$

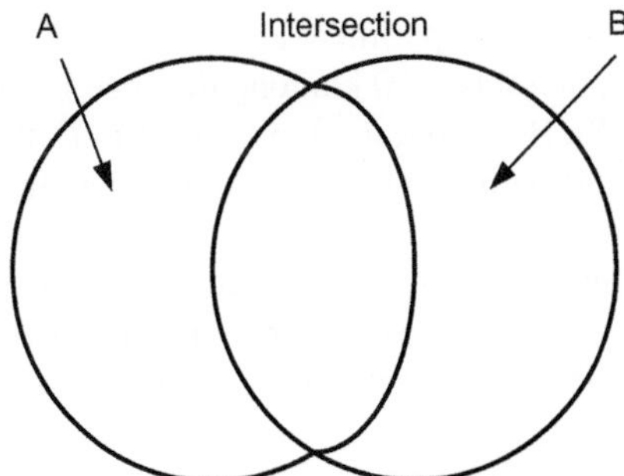

FIG. 3.26

Intersection of events.

Two events A, and B are *mutually exclusive*, if the occurrence of one event means nonoccurrence of the other. In other words, mutually exclusive events A and B cannot happen simultaneously (Fig. 3.27).

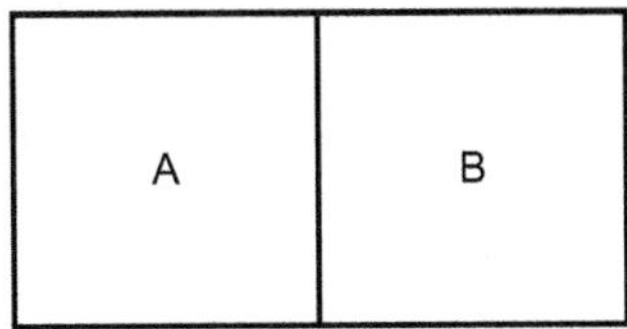

FIG. 3.27

Mutually exclusive events.

In the case of mutually exclusive events A and B, we have:

$$P(AB) = 0 \tag{3.34}$$

and

$$P(A+B) = P(A) + P(B) \tag{3.35}$$

When solving some problems we are facing so called *conditioning events* in a sample space. The probability that event A will happen knowing that event B already happened is usually denoted as $P(A/B)$. The conditional probability law helps us to compute the probability $P(A/B)$ of event A, given event B, ie,

$$P(A/B) = \frac{P(AB)}{P(B)} \tag{3.36}$$

Two events A and B are *independent* when:

$$P(A/B) = P(A) \tag{3.37}$$

In the case of independent events A and B we have:

$$P(AB) = P(A)P(B) \tag{3.38}$$

3.5.2 RANDOM VARIABLES AND PROBABILITY DISTRIBUTIONS

The numerical outcomes of the observed experiment are represented by a random variable. For example, let us assume that passengers, that travel 10 km long distance in a city, could choose for their trip private car (C), or public transit (P). By assigning 0 to C and 1 to P, the potential passengers' choices (outcomes of the experiment) could be presented as a random variable. A random variable could be *discrete*, or *continuous*. A discrete random variable takes on specific values at discrete points on the real line. In the case of left-turn bay at the intersection whose capacity equals 8, the number of vehicles *in* the bay could be 0, 1, 2, 3, 4, 5, 6, 7, or 8. The number of vehicles in the bay cannot be, for example, 4.32, or 6.17. In the case of continuous variables, the variable can take any value over continuous range of the real line.

There is a function $f(x)$ that assigns probability measure to the random variable values x. The function is called *probability density function* (pdf). Let us explain the concept of the probability density function by using the following example.

EXAMPLE 3.6

The number of cars waiting for a right of way through the intersection has been recorded at a specific time of a day during 365 days. Table 3.6 shows the distribution of number of cars waiting for a right of way through intersection.

Table 3.6 The Distribution of Number of Cars Waiting for a Right of Way Through Intersection

Number of cars	1–3	4–6	7–9	10–12	13–15	16–18	19–21	22–24
Number of days	38	52	70	55	45	40	40	25

The total number of days under observation equals 365. We transform all the data in Table 3.7 into probabilities by dividing by this total.

EXAMPLE 3.6—cont'd

Table 3.7 The Distribution of Number of Cars Waiting for a Right of Way Through Intersection

Number of cars	1–3	4–6	7–9	10–12	13–15	16–18	19–21	22–24
Number of days	0.11	0.14	0.19	0.15	0.12	0.11	0.11	0.07

From these data we can obtain the bar graph (Fig. 3.28). We call this bar graph *probability distribution histogram.*

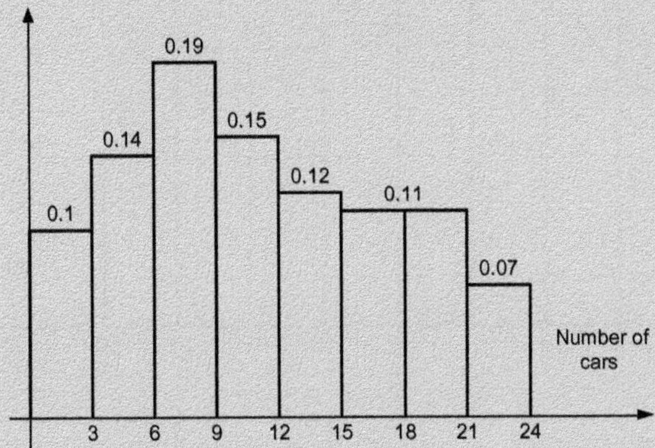

FIG. 3.28

Probability distribution histogram of the number of cars.

The probability distribution histogram enables us to calculate different probabilities. The probability that the number of cars waiting to pass through the intersection is in between 6 and 12 is shown in Fig. 3.29. In this way, we can calculate various probabilities by summing up corresponding areas.

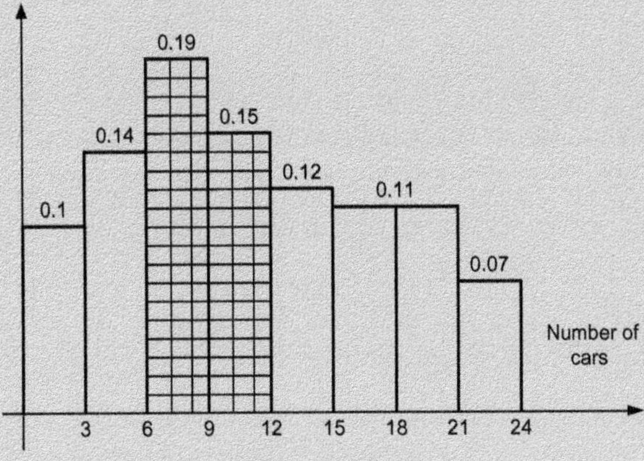

FIG. 3.29

Probability calculation by summing up corresponding areas.

The probability distribution histogram shown can be replaced by a continuous curve shown in Fig. 3.30. The curve shown in Fig. 3.30 is known as a *probability density function.*

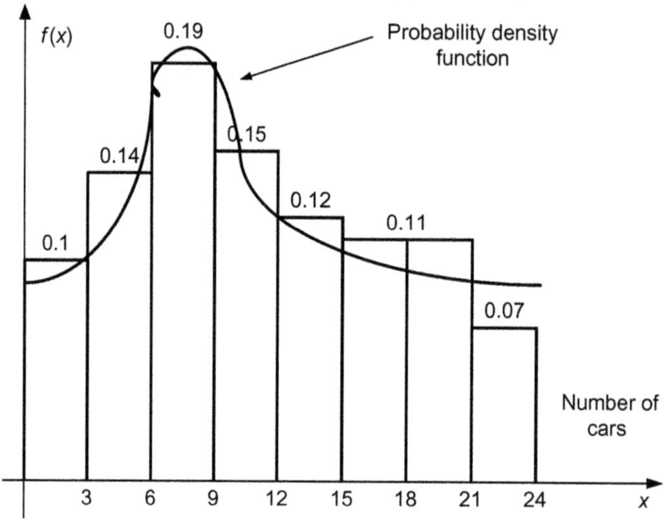

FIG. 3.30

Probability density function.

A probability density function (in the case of continuous random variable) has the following properties:

$$f(x) \geq 0 \ \forall x \tag{3.39}$$

$$\int_{-\infty}^{\infty} f(x)dx = 1 \tag{3.40}$$

In the case of discrete random variables, we denote the probability density function by $P(x)$. The pdf $P(x)$ defines the probability that x takes a given value. The $P(x)$ must satisfy the following:

$$P(x) \geq 0 \quad \text{for all } x \tag{3.41}$$

$$\sum_{\text{all } x} P(x) = 1 \tag{3.42}$$

The probability $P(a \leq x \leq b)$ that the continuous random variable X will take value from the interval $[a, b]$ equals (Fig. 3.31):

$$P(a \leq x \leq b) = \int_a^b f(x)dx \tag{3.43}$$

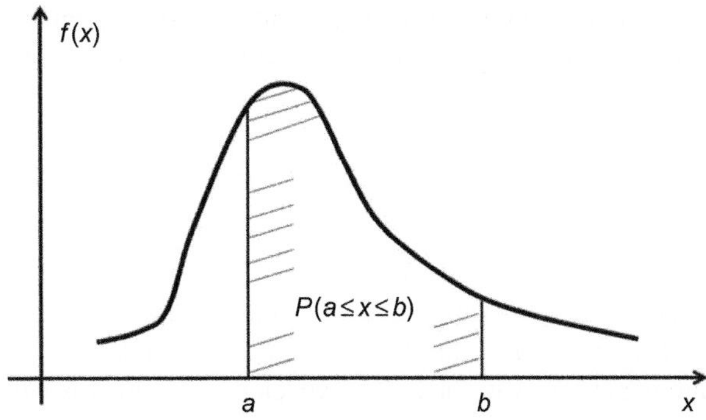

FIG. 3.31

The probability $P(a \leq x \leq b)$ that the random variable X will take value from the interval $[a, b]$.

The *cumulative density function* $F(x)$ is defined as the probability that the observed value of the random variable X will be less than or equal to x, ie,

$$F(x) = P(X \leq x) = \int_{-\infty}^{x} f(x)dx \tag{3.44}$$

Probability density function and the corresponding cumulative density function are shown in Fig. 3.32.

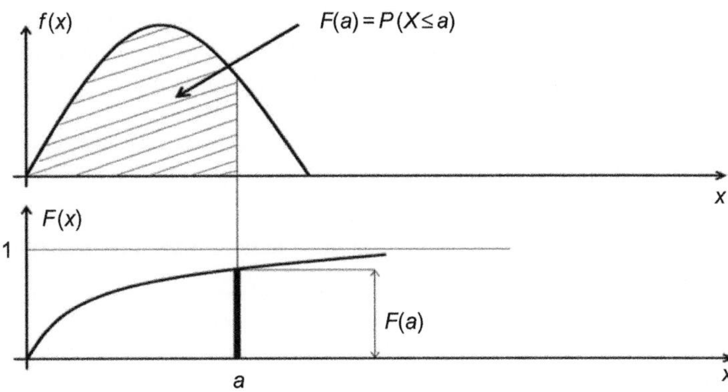

FIG. 3.32

Probability density function $f(x)$ and the corresponding cumulative density function $F(x)$.

The probability $P(a \leq x \leq b)$ that the random variable X will take value from the interval $[a, b]$ could be also calculated using cumulative density function:

$$P(a \leq x \leq b) = \int_{a}^{b} f(x)dx = \int_{-\infty}^{b} f(x)dx - \int_{-\infty}^{a} f(x)dx = F(b) - F(a) \tag{3.45}$$

We denote by $E(X)$ the *expected value* (mean value, mean) of the random variable X. The expected value measures the central tendency of the distribution. In the case of discrete random variables the expected value $E(X)$ equals:

$$E(x) = \sum_{i=1}^{n} x_i p_i \tag{3.46}$$

where:

x_i is the ith possible value of the random variable X;
p_i is the probability that the random variable X will take the value x_i; and
n is the total number of the possible values of the random variable X.

In the case of continuous random variables, the expected value $E(x)$ is calculated as follows:

$$E(x) = \int_{-\alpha}^{\alpha} x \cdot f(x) \cdot dx \tag{3.47}$$

where $f(x)$ is the probability density function of the random variable X.

The *variance* is a measure of dispersion of the distribution around its expected value. The variance is defined as:

$$\text{var}(x) = E\left\{(x - E(x))^2\right\} \tag{3.48}$$

It can be easily shown that variance equals:

$$\text{var}(x) = E(x^2) - (E(x))^2 \tag{3.49}$$

The *Poisson*, *Exponential*, and *Normal* distribution (Table 3.8) are the distributions that frequently appear in various traffic and transportation engineering problems. The Poisson distribution is discrete, while Exponential and Normal are continuous distributions (Table 3.8).

Table 3.8 Probability Density Functions that Frequently Appear in Traffic and Transportation

Name of the Distribution	Probability Density Function $f(x)$	Some Examples of the Random Variable Distributed According to the Probability Density Function $f(x)$
Poisson	$P(X=k) = \frac{\lambda^k}{k!} \cdot e^{-\lambda}$ $E(x) = \lambda$ $\text{var}(x) = \lambda$	The number of vehicle arrivals at the intersection during specific time interval The number of calls for emergency help from firefighters during specific time interval
Exponential	$f(x) = \lambda \cdot e^{-\lambda \cdot x}$ $E(x) = \frac{1}{\lambda}$ $\text{var}(x) = \frac{1}{\lambda^2}$	The vehicle interarrival times at the toll plaza Passenger interarrival time at the travel agent office
Normal	$f(x) = \frac{1}{\sigma\sqrt{2 \cdot \pi}} \exp\left[-\frac{(x-\mu)^2}{2 \cdot \sigma^2}\right]$ $E(x) = \mu$ $\text{var}(x) = \sigma^2$	The number of passengers in a bus The number of passengers in a plane

3.5.2.1 Poisson distribution

Measurements in many transportation systems show that the client arrivals pattern could be described by the *Poisson distribution*. It has been shown that Poisson distribution describes many real-life situations. Let us consider specific point at the highway (Fig. 3.33)

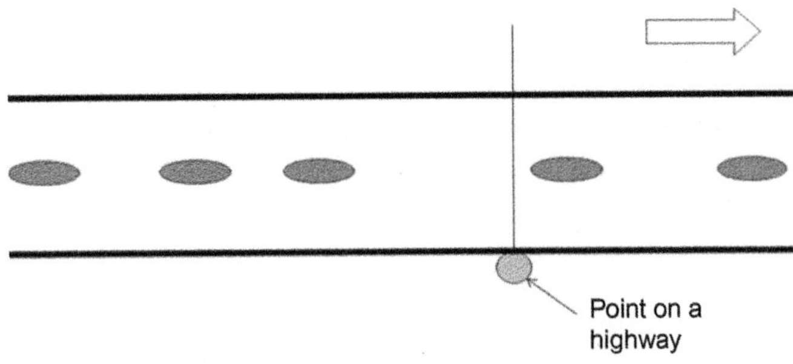

FIG. 3.33

Point on a highway.

Vehicles randomly show up and pass. We count every vehicle. The number of vehicle arrivals X during specific time interval represents random variable. In other words, it can happen that during specific time interval no vehicles arrive, one vehicle arrive, two vehicles arrive, etc. This random variable is distributed according to the *Poisson* distribution:

$$P(X=k)=\frac{(\lambda t)^k}{k!}\cdot e^{-\lambda t} \tag{3.50}$$

Relation (3.50) describes the probability $P(k)$ that the total number of vehicle arrivals X happening in a time interval of the length t is equal to k. The expected value $E(x)$ and the variance var(x) are equal in the case of Poisson distribution. By observing the collected statistical data, and by calculating mean and variance, one can easily get an impression about the observed traffic phenomena. In the case that the calculated mean and variance are approximately equal, there is a high chance that the Poisson distribution describes studied traffic phenomenon.

3.5.2.2 Exponential distributions

Let us again consider point at the highway (Fig. 3.33). The time interval between the appearances of successive vehicles (headway) could be, for example, 5 s, 10 s, 11 s, 14 s, etc. In other words, the time interval between vehicle arrivals is random variable that frequently has *exponential* distribution. This continuous random variable T has *exponential* distribution with the parameter λ:

$$f(t)=\lambda\cdot e^{-\lambda\cdot t} \tag{3.51}$$

Exponential distribution of the time between client arrivals is shown in Fig. 3.34.

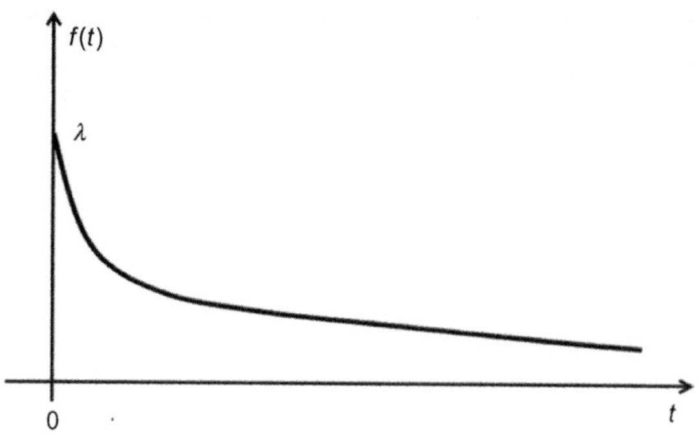

FIG. 3.34

Exponential distribution of the time between client arrivals.

As long as the vehicle interarrival time is *exponential*, the number of vehicle arrivals during time interval t is *Poisson*. The relationship between Poisson distribution and exponential distribution will be explained in more detail in Chapter 4, devoted to the traffic flow theory.

3.5.2.3 Normal distribution

The probability density function of the normal distribution equals:

$$f(x) = \frac{1}{\sigma\sqrt{2\cdot\pi}}\exp\left[-\frac{(x-\mu)^2}{2\cdot\sigma^2}\right] \quad -\infty < x < \infty \tag{3.52}$$

where parameter μ denotes mean, and σ^2 denotes variance of the distribution.

The cumulative density function of the normal distribution equals:

$$F(x) = \int_{-\infty}^{x}\frac{1}{\sigma\sqrt{2\cdot\pi}}\exp\left[-\frac{(y-\mu)^2}{2\cdot\sigma^2}\right]dy \tag{3.53}$$

Probability density function and the cumulative density function of the normal distribution are shown in Fig. 3.35.

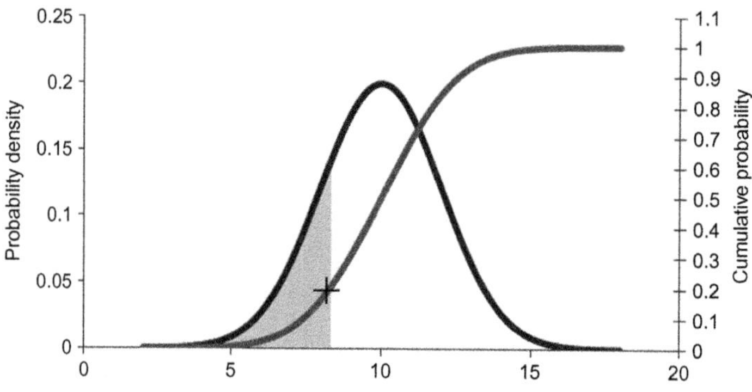

FIG. 3.35

Probability density function and the cumulative density function of the normal distribution.

EXAMPLE 3.7

The average interarrival time between two vehicles on a highway equals 20 (s). Assume that vehicle interarrival times are distributed according to the exponential distribution. Calculate the percentage of cases when interarrival time is <10 s.

Solution

The probability density function of vehicle interarrival times equals:

$$f(t) = \lambda \cdot e^{-\lambda \cdot t}$$

The expected interarrival time value equals:

$$E(t) = \int_{-\infty}^{\infty} t \cdot \lambda \cdot e^{-\lambda t} dt = \int_{0}^{\alpha} t \cdot \lambda \cdot e^{-\lambda t} dt$$

After solving the integral, we obtain the following:

$$E(t) = \frac{1}{\lambda}$$

Since the average interarrival time between two vehicles on a highway equals 20 (s), we have:

$$\frac{1}{\lambda} = 20$$

$$\frac{1}{\lambda} = 20$$

$$\lambda = 0.05\,(\text{veh/s})$$

The probability of event that the interarrival time is <10 s equals:

$$P(0 \le T \le 10) = \int_{0}^{10} f(t) \cdot dt = \int_{0}^{10} 0.05 \cdot e^{-0.05 \cdot t} \cdot dt$$

$$P(0 \le T \le 10) = 0.393$$

The probability of event that the interarrival time is <10 s represents the percentage of cases when interarrival time is <10 s. We conclude that in 39.3% of cases interarrival time will be <10 s.

EXAMPLE 3.8

The total of 720 vehicles that wanted to merge onto highway appeared during 1 h (Fig. 3.36). It has been shown that the vehicle arrival pattern could be described by the Poisson process. The assumption is the vehicle arrival pattern will be unchanged the following days.

Calculate:

(a) mean time between arrivals;

(b) Probabilities of having 3, 4, 5 vehicles during 30 s interval;

(c) Percentage of the 10 s intervals with no vehicles.

(Continued)

EXAMPLE 3.8—cont'd

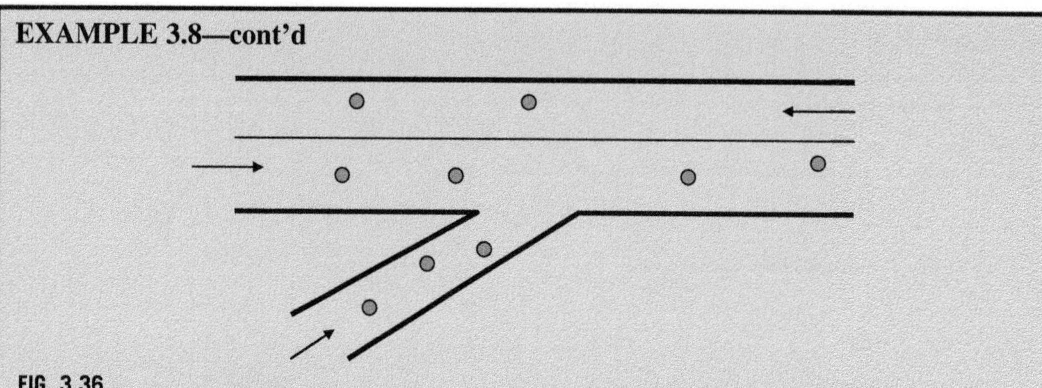

FIG. 3.36

Vehicles merging onto highway.

Solution

We express the average vehicle arrival rate in (veh/s), and the mean time between vehicle arrivals in (s). The average vehicle arrival rate equals:

$$q = \frac{720 \text{ vehicles}}{1 \text{ h}} = \frac{720 \text{ vehicles}}{3600\text{s}} = 0.2(\text{veh}/s)$$

The mean time between arrivals equal:

$$\frac{1}{q} = \frac{1}{0.2(\text{veh/s})} = 5\,\text{s}$$

Probability of having n vehicles during 30 s interval equals:

$$P(X=n) = \frac{(0.2(\text{veh/s})30\text{s})^n e^{-0.2(\text{veh/s})30\text{s}}}{n!}$$

$$P(X=n) = \frac{(6)^n e^{-6}}{n!}$$

$$P(X=3) = \frac{(6)^3 e^{-6}}{3!} = 0.089244$$

$$P(X=4) = \frac{(6)^4 e^{-6}}{4!} = 0.133866$$

$$P(X=5) = \frac{(6)^5 e^{-6}}{5!} = 0.1606392$$

Percentage of the 10 s intervals with no vehicles represents the probability of the event that no one vehicle will show up during 10 s. This probability equals:

$$P(X=0) = \frac{(0.2(\text{veh/s})30\text{s})^0 e^{-0.2(\text{veh/s})10\text{s}}}{0!}$$

$$P(X=0) = e^{-0.2(\text{veh/s})\,10\text{ s}} = e^{-2} = 0.13534$$

We conclude that in 13.534% cases during 10 s interval no one vehicle will show up.

EXAMPLE 3.9

East-westbound moving vehicles (vehicles moving along the major street) have right of way (Fig. 3.37). Vehicles coming from the minor street must wait for the acceptable gap in order to cross the major street. The intersection between the major and minor street is unsignalized intersection. The vehicle arrival pattern along the main street could be described by the Poisson distribution. Measurement shows that 900 vehicles pass through the intersection along the main street during 1 h.

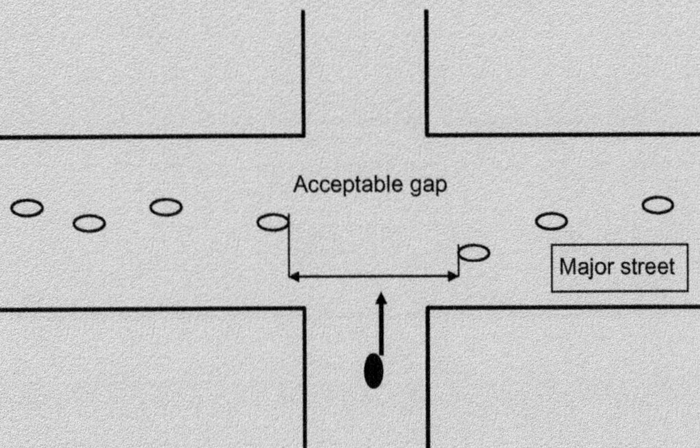

FIG. 3.37

Acceptable gap.

(a) Assume that the minimal acceptable gap equals 4 s and calculate the expected number of acceptable gaps during 1 h.
(b) The acceptable gap equals 8 s in the case of senior citizens. Calculate in this case the expected number of acceptable gaps during 1 h.
(c) The time between vehicle arrivals in a Poisson Process is random variable T that has exponential distribution with parameter q:

$$f(t) = q e^{-qt}$$

In our case, parameter q equals:

$$q = 900\,(\text{veh/h}) = 0.25\,(\text{veh/s})$$

The probability that the next vehicle arrival will happen after t equals e^{-qt}. The safe situation to cross the main street happens always when the gap in the major vehicles flow is >4 s. The probability that the random variable T takes the value >4 s equals:

$$P(T > 4) = e^{-0.25(4)} = e^{-1} = 0.367$$

(Continued)

EXAMPLE 3.9—cont'd

We know that 900 vehicles pass through the intersection along the main street during 1 h. This means that there are 899 gaps in the vehicle flow along the main street during 1 h. The expected number of acceptable gaps during 1 h equals:

$$899 \cdot P(T > 4) = 899 \cdot (0.367) = 330$$

We conclude there is an acceptable gap in the main flow in $\frac{330}{899}(100)\% = 36.7\%$ cases.

(d) The probability that the random variable T takes the value >8 s equals:

$$P(T > 8) = \mathrm{e}^{-0.25(8)} = \mathrm{e}^{-2} = 0.134$$

The expected number of acceptable gaps in this case equals:

$$899 \cdot P(T > 8) = 899 \cdot (0.134) = 120$$

The acceptable gap in the main flow in $\frac{120}{899}(100)\% = 13.3\%$ cases.

3.6 QUEUEING IN TRANSPORTATION SYSTEMS

Queueing is a part of our daily routine. In tall building we wait for the elevator. As pedestrians, we typically wait before crossing the street. Each day hundreds and thousands of cars are delayed at the intersections. In the cases when the airport's runways are busy, the aircraft are assigned to a holding pattern. Ships wait in ports, air passengers wait for the security checks at the airports, and many trucks wait to be loaded and unloaded. The queue appearance in any transportation system is the consequence of the fact that the transportation demand during specific time periods exceeds the capacity of the transportation system. Many clients (pedestrians, drivers, cars, aircraft, and ships) demand different services on a day-to-day basis (crossing the street, passing through the intersection, landing on runway, unloading at the dock, etc.).

Queueing theory represents the mathematical analysis of queues. The origin of queueing theory is related to the Danish engineer *Agner Krarup Erlang* (1878–1929). Erlang analyzed telephone traffic problems and published in 1909 the first paper on queueing theory. He showed in the paper that the Poisson distribution appears in a telephone traffic. Queueing theory has been used in modeling urban and road traffic, elevator traffic control, airport operations, air traffic control, crowd dynamics, emergency egress analysis, railway, telephone, and internet traffic (Larson and Odoni, 1981; Newell, 1982).

Queueing theory facilitates assessment of the level-of-service and operational performances of the transportation systems. Average waiting time a client spends in a queue, and the average number of clients in a queue, are traditional metrics for the level of transportation service. Utilization of the service facility has been regularly used as a metric for the system operational performance. Queueing theory has been used by the engineers and planners when designing future service facility (calculation of the number of lanes at intersection, the estimation of the length of left-turning bays, calculation of the size of the check-in area at airport, calculation of the required number of parking spaces, etc.). Queueing theory techniques help us to find the answers to the following questions: What is the level-of service-offered to the clients of the transportation system? What is the operational efficiency of the studied transportation system? Should transportation capacity be expanded in response to anticipated demand?

3.6.1 ELEMENTS OF QUEUEING SYSTEMS

Traffic intersections, airport runways, and elevators represent various queueing systems. All of them are characterized by queue existence and waiting times that clients spend in the system. The following are the basic characteristics of *every* queueing system:

(a) arrival process type;
(b) service process;
(c) number of servers;
(d) queue discipline; and
(e) queue capacity.

The arrivals in queueing system could be *deterministic*, or *random*. In the case of deterministic arrivals, the arrival rate is constant. Similarly, service time in a queueing system could be *deterministic service time*, or by service time, or it could be a *random variable*. In the case of stochastic queueing, clients' arrivals and service times are described by probability density functions. The expected number of clients in the system, the expected waiting time per client, and the percentage of time when server is busy are usual metrics, in the case of stochastic queueing.

The number of servers are one of the main characteristics of any queueing system. The number of servers is equal to one in the case of airport with one runway. The number of servers equals, for example, six when six toll booths are open on the highway. In the majority of queueing systems, queue discipline is *FIFO* (*First In-First Out*). The queue discipline *FIFO* is also frequently called *FCFS* (*First Come First Served*). There are also the other queue disciplines like queue discipline *Last In First Out* (*LIFO*), and *Service In Random Order* (*SIRO*).

There are no specific restrictions related to the allowed queue capacity in some queueing systems. In such cases, queue capacity is assumed to be equal to infinity. On the other hand, in some other queueing systems queue has specific capacity. For example, shock waves occur when left-turning vehicles are forced to slow down in the through lanes, in that way affecting through traffic to also slow down. Left-turn bays could considerably reduce the negative shock wave effect. The lengths of the left-turn bay (queue capacity) must be proper to meet left-turners loading requirements. The main queueing system characteristics are denoted by the following standardized format:

$$A/B/C$$

where A is arrivals distribution, B is service time distribution, and C is number of servers.

The uniform, deterministic distribution of arrivals or departures is denoted by D, while the exponential distribution is denoted by M. A "general distribution" is denoted by G. For example, the notation *M*/*M*/1 denotes the queueing system that is characterized by exponentially distributed interarrival times, exponentially distributed service times and the existence of one server, while notation *D*/*D*/1 describes queueing system with deterministic arrivals, deterministic departures and one server. Every queueing system could be graphically represented in the following way (Fig. 3.38):

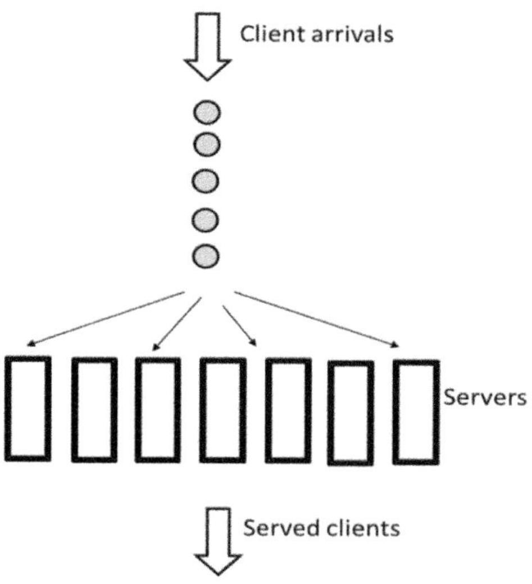

FIG. 3.38

Queueing system.

EXAMPLE 3.10

Tolls on toll roads are paid with a help of electronic toll collection equipment. The equipment communicates electronically with a car's transponder. In other words, toll collection points are unmanned on many modern highways. On the other hand, toll booths are still needed on many highways for the infrequent drivers who do not have a transponder. Such drivers must stop and pay the toll. Let us consider one toll booth (Fig. 3.39).

The vehicles are coming from the left. The shown toll booth could be treated as a queueing system. There is only one server in the system. The server is a toll both. The server is busy when driver is paying. Clients are vehicles that show up from the left. The service time is composed of a stop time, paying time, and a passage through the toll booth. The service time could be treated as a deterministic quantity, or as a random variable. In majority of cases, vehicle arrivals are random. In other words, vehicles show up from the left in random time points. Vehicle that requests service (pass through the toll booth) is immediately accepted if there is no queue of vehicles in a front of toll booth. If there is a queue, the newly arrived vehicle will join the queue. The queue discipline is the FIFO discipline.

EXAMPLE 3.10—cont'd

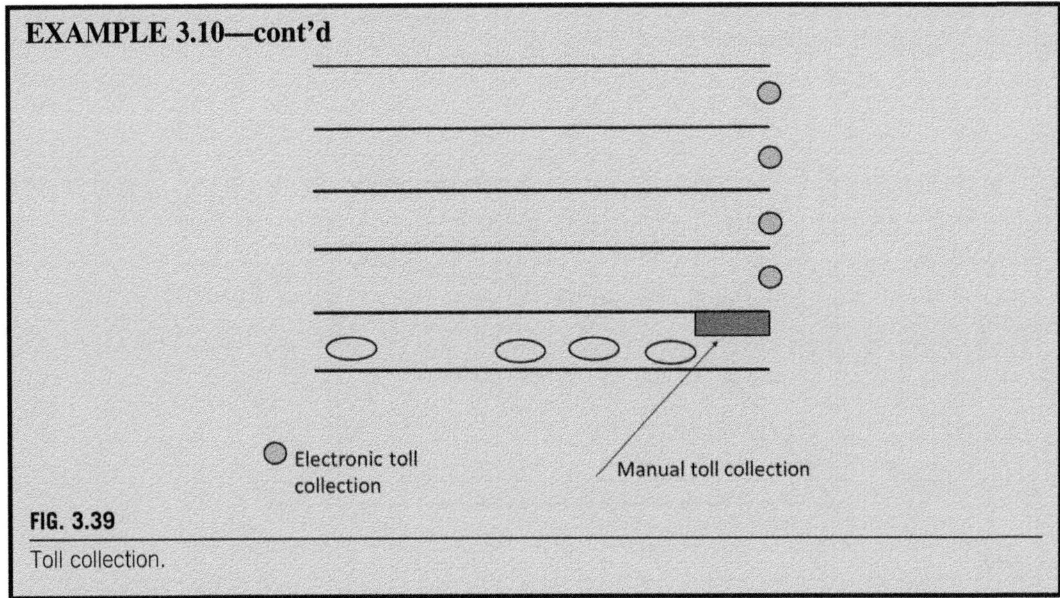

FIG. 3.39

Toll collection.

3.6.2 *D/D/*1 QUEUEING

There is no randomness in the case of deterministic arrivals. This means that the time points of arrival of the first, second, third,... client are accurately known (Fig. 3.40).

The total number of clients that will enter the queueing system during a specific time period is known in the case of constant arrival rate, as well as in the case of deterministic arrival rate that varies over time. In the case of $D/D/1$ queueing systems, service process is also characterized by deterministic service time. In the case shown in Fig. 3.40, the demand rate (clients/h) is known.

This rate rises from the beginning of the observation till time point t_1. The rate is constant between t_1 and t_3. The demand rate decreases between t_3 and t_4. In the end, behind t_4 the rate is constant. The service rate (clients/h) could be defined in a similar way. It is quite simple to create cumulative number of arrivals and cumulative number of departures for known arrival and departure times. These cumulative numbers deliver the information about the total number of arrived clients and the total number of departed clients till the certain time point. When we say, for example, that the cumulative number of arrived vehicles at the toll booth at 7:30 am equals 220, it means that by 7:30 am, a total of 220 vehicles arrived at the toll booth.

The main queueing concepts can be clearly understood after studying the simple $D/D/1$ queueing system. We use continuous lines to represent cumulative arrivals and cumulative departures in the $D/D/1$ queueing systems. It has been shown that these lines are very good approximation for the cumulative stepped lines that are real lines that represent arrivals and departures in the queueing system.

3.6.3 LITTLE'S LAW

Little's Law is the central result of queueing theory. This law is applicable for any queueing system that is in stable conditions. (Stable conditions do not assume, for example, the start of operations in the

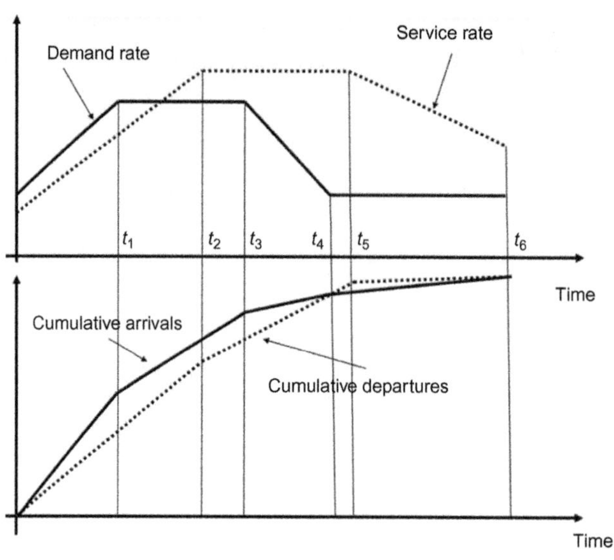

FIG. 3.40

Arrival rate, departure rate, cumulative number of arrivals, and cumulative number of departures.

system.) In the case of stochastic queueing, client arrivals and service times are described by probability distributions known as arrival and service time distributions. The relationship described by Little's Law requires no assumptions about probability distributions of the interarrival and service times. In some cases, queueing system could be composed of the queueing subsystems. Little's Law is valid to queueing subsystems, as well as for a whole queueing system.

The following is the explanation of Little's Law. Let us assume that N customers arrive in the queueing system during the time interval $(0, T)$. Cumulative number of arrivals and cumulative number of departures are shown in Fig. 3.41.

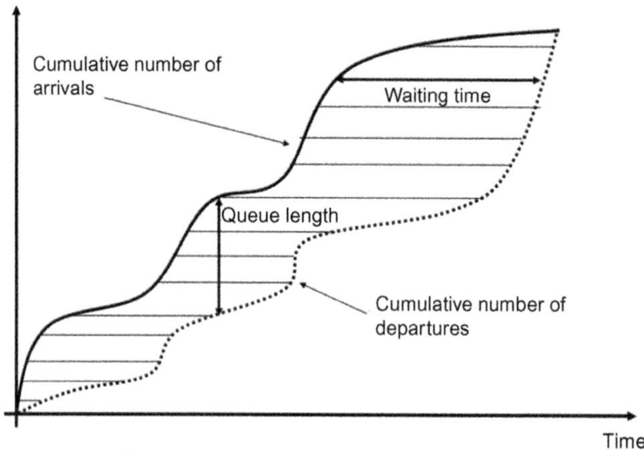

FIG. 3.41

Queue length and waiting time.

The queue length in any time point represents the maximum ordinate distance between the cumulative number of departures and cumulative number of arrivals curves. For example, let us assume that queueing system started with operations at 8:00 am. The total number of arrived customers by 10:00 am equals 900. The total number of served customers by 10:00 am equals 800. Clearly, the queue length at 10:00 am is equal to 100. It is also very easy to "read" from the figure, by visual inspection, the waiting time of every client. The area between cumulative arrivals curve and cumulative departures curve (shaded area in Fig. 3.41) represents the total waiting time of all clients. The average waiting time W represents the quotient of the total waiting time and the number of clients, ie,

$$W = \frac{\text{Area}}{N} \tag{3.54}$$

The shaded area also represents the total length of all queues in all time points. The average queue length L equals:

$$L = \frac{\text{Area}}{T} \tag{3.55}$$

We conclude the following:

$$W \cdot N = L \cdot T \tag{3.56}$$

$$L = \frac{N}{T} \cdot W \tag{3.57}$$

The ratio $\frac{N}{T}$ actually represents the arrival rate λ. In the end, we can write:

$$L = \lambda \cdot W \tag{3.58}$$

Relation (3.58) represents Little's Law. We interpret this relation in the following way: The average number of clients in a queueing system (during particular time interval) is equal to their average arrival rate multiplied by their average time in the system.

EXAMPLE 3.11

Traffic accident caused decreased road capacity. Vehicles travel through the area of traffic accident. After 80 min the road capacity is not decreased any more. We analyze the time period related to the decreased road capacity (80 min). After analyzing the statistical data, it has been concluded that both arrival and service rates were deterministic. Service rate is constant. Arrival rate varies over time. The arrival and departure (service) rates (veh/min) are respectively equal:

$$\lambda(t) = -\frac{1}{4} \cdot t + 20$$

$$\mu(t) = 10$$

where t is in minutes after beginning the observation of the queueing system. Cumulative number of arrivals $A(t)$ and cumulative number of departures $D(t)$ are respectively equal:

$$A(t) = \int_0^t \lambda(t)dt = \int_0^t \left(-\frac{1}{4}t + 20\right)dt = \left(-\frac{1}{4}\frac{t^2}{2} + 20t\right)\Big|_0^t = -\frac{1}{8}t^2 + 20t$$

(Continued)

EXAMPLE 3.11—cont'd

$$D(t) = \int_0^t \mu(t)dt = \int_0^t 10dt = 10t$$

Cumulative number of arrivals and cumulative number of departures are shown in Fig. 3.42.

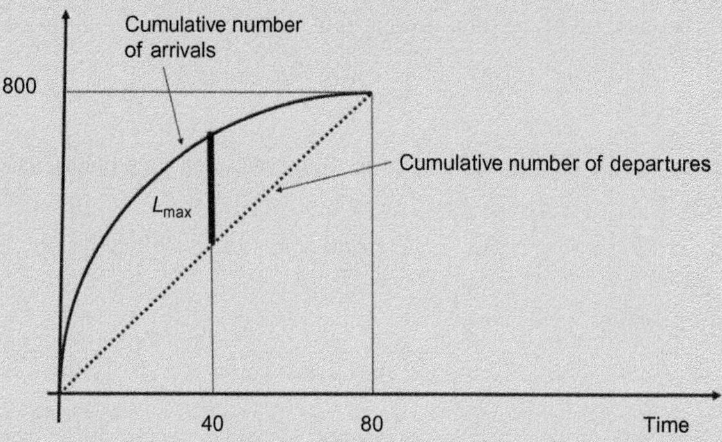

FIG. 3.42

Cumulative number of vehicle arrivals and cumulative number of vehicle departures in the area of traffic accident.

The queue will dissipate in the time point when the total number of cumulative arrivals equals the total number of cumulative departures, ie,

$$A(t) = D(t)$$

After substitution, we get the following equation:

$$-\frac{1}{8}t^2 + 20t = 10t$$

$$-\frac{1}{8}t^2 + 10t = 0$$

The solutions of the equation are $t_1 = 0$ and $t_2 = 80$. The cumulative number of arrivals $A(t)$ equals to the cumulative number of departures $D(t)$ for the first time at the beginning of our observation ($t = 0$). These cumulative numbers are equal for the second time when queue dissipates. We conclude that the queue will dissipate after 80 min. The total number of arrivals during 80 min equals:

$$A(100) = D(100) = 800$$

EXAMPLE 3.11—cont'd

The total delay D of all vehicles is represented by the area between cumulative number of arrivals and the cumulative number of departures. In other words:

$$D = \int_0^{80} A(t)dt - \int_0^{80} D(t)dt$$

$$D = \int_0^{80} \left[-\frac{1}{8}t^2 + 20t \right] dt - \int_0^{80} 10t dt$$

$$D = \left(-\frac{1}{8}\frac{t^3}{3} + 20\frac{t^2}{2} - 10\frac{t^2}{2} \right) \Bigg|_0^{80}$$

$$D = 10{,}667$$

The average delay d per one vehicle equals:

$$d = \frac{10{,}667}{800} = 13.33$$

Because of the traffic accident, average delay per vehicle equals 13.33 min. Queue length $L(t)$ in any moment t represents the difference between cumulative number of vehicle arrivals $A(t)$ by moment t, and cumulative number of vehicle departures $D(t)$ by moment t:

$$L(t) = A(t) - D(t)$$

$$L(t) = -\frac{1}{8}t^2 + 20t - 10t = -\frac{1}{8}t^2 + 10t$$

We determine the maximum queue length as follows:

$$\frac{d[L(t)]}{dt} = 0$$

$$\frac{d\left[-\frac{1}{8}t^2 + 10t \right]}{dt} = 0$$

$$-\frac{t}{4} + 10 = 0$$

$$t = 40$$

We conclude that the maximum queue length happens 40 min after the beginning of observation. The maximal queue length equals:

$$L_{\max} = -\frac{1}{8}40^2 + 10 \cdot 40$$

$$L_{\max} = 200 \text{ vehicles}$$

3.6.4 *M/M/*1 QUEUEING

A lot of queueing systems in transportation are *M/M/*1 queueing systems. The examples could be one open toll booth at the highway, one open check-in counter at the airport, vehicle inspection facility with one repairman working, etc. The *M/M/*1/queueing system has the following characteristics:

- Poisson arrivals (exponential interarrival times);
- Exponential service times;
- One server;
- FIFO queue discipline.

Let us introduce the following notation:

λ: mean arrival rate;
μ: mean service rate.

The following relations describe *M/M/*1 queueing:
Probability of having no customers in the queueing system equals:

$$p_0 = 1 - \frac{\lambda}{\mu} \tag{3.59}$$

Probability of having n customers in the queueing system equals:

$$p_n = \left(\frac{\lambda}{\mu}\right)^n \cdot p_0 \tag{3.60}$$

The average number of customers in the queue:

$$L_q = \frac{\lambda^2}{\mu \cdot (\mu - \lambda)} \tag{3.61}$$

The average number of customers in the queueing system:

$$L = L_q + \frac{\lambda}{\mu} = \frac{\lambda^2}{\mu \cdot (\mu - \lambda)} + \frac{\lambda}{\mu} = \frac{\lambda}{\mu - \lambda} \tag{3.62}$$

The average waiting time a customer spends in the queue:

$$W_q = \frac{L_q}{\lambda} = \frac{\frac{\lambda^2}{\mu \cdot (\mu - \lambda)}}{\lambda} = \frac{\lambda}{\mu \cdot (\mu - \lambda)} \tag{3.63}$$

The average waiting time a customer spends in the queueing system:

$$W = W_q + \frac{1}{\mu} = \frac{\lambda}{\mu \cdot (\mu - \lambda)} + \frac{1}{\mu} = \frac{1}{\mu - \lambda} \tag{3.64}$$

EXAMPLE 3.12

Let us analyze the operations of one toll booth on a highway. This toll booth enables drivers without transponder to stop and pay the toll (Fig. 3.38). Drivers pay in cash, or by credit card. Service time varies depending on the availability of the cash and coins, but on the average, toll booth attendant needs 18 s to collect the money. Average vehicles arrival rate equals 180 (veh/h). Treat toll booth as *M*/*M*/1 queueing system and calculate:

the average number of vehicles in the queue;
the average waiting time a client spends in the queue; and
the average waiting time a client spends in the queueing system.

Solution

The average arrival rate λ equals 180 (veh/h), ie, $\lambda = \frac{180\,\text{vehicles}}{3600\text{ s}} = 0.05\,(\text{veh/s})$.

The average service rate equals $\mu = \frac{1}{18} = 0.055\,(\text{veh/s})$.

The average number of clients in the queue equals:

$$L_q = \frac{\lambda^2}{\mu \cdot (\mu - \lambda)} = \frac{0.05^2}{0.055 \cdot (0.055 - 0.05)} \approx 9 \text{ vehicles}$$

The average waiting time a client spends in the queue equals:

$$W_q = \frac{\lambda}{\mu \cdot (\mu - \lambda)} = \frac{0.05}{0.055 \cdot (0.055 - 0.05)} \approx 182 \text{ s}$$

The average waiting time a client spends in the queueing system equals:

$$W = \frac{1}{\mu - \lambda} = \frac{1}{0.055 - 0.05} = 200s$$

3.6.5 *M*/*M*/s QUEUEING

In the case of the *M*/*M*/*s* queueing system, both arrivals and departures occur according to *Poisson* distribution. The total number of servers is equal to *s*. Consequently, the maximum *s* clients could be served simultaneously. Obviously, the system's service rate is much higher than in the case of one server. The examples of the *M*/*M*/*s* queueing system are: parking lot, where each parking place represents one server; airport operations in the case of multiple runways, or few toll booths on the highway. Utilization factor ρ of the facilities is defined in the following way:

$$\rho = \frac{\lambda}{s \cdot \mu} \tag{3.65}$$

The following relations describe *M*/*M*/*s* queueing:

Probability of having no customers in the queueing system equals:

$$p_0 = \frac{1}{\sum_{k=0}^{s-1} \frac{\left(\frac{\lambda}{\mu}\right)^k}{k!} + \frac{\left(\frac{\lambda}{\mu}\right)^s}{s!} \cdot \frac{1}{1 - \frac{\lambda}{s \cdot \mu}}} \tag{3.66}$$

Probability of having k customers in the queueing system equals:

$$p_k = \frac{\left(\frac{\lambda}{\mu}\right)^k}{k!} \cdot p_0 \quad \text{for } 0 \leq k \leq s \tag{3.67}$$

$$p_k = \frac{\left(\frac{\lambda}{\mu}\right)^k}{s! \cdot s^{k-s}} \cdot p_0 \quad \text{for } k > s \tag{3.68}$$

The average number of customers in the queue:

$$L_q = \frac{p_0 \cdot \left(\frac{\lambda}{\mu}\right)^s \cdot \rho}{s!(1-\rho)^2} \tag{3.69}$$

The average waiting time a client spends in the queue equals:

$$W_q = \frac{L_q}{\lambda} \tag{3.70}$$

The average number of clients in a queueing system equals:

$$L = \lambda \cdot W = \lambda\left(W_q + \frac{1}{\mu}\right) = L_q + \frac{\lambda}{\mu} \tag{3.71}$$

EXAMPLE 3.13

The airport terminal shown in the next figure has two security checkpoints for all passengers boarding aircraft. Each security check point has two X-ray machines. A survey reveals that on the average a passenger takes 45 s to go through the system (exponential distribution service time).

The arrival rate is known to be random (this equates to a Poisson distribution) with a mean arrival rate of one passenger every 25 s. In the design year the demand for services is expected to grow by 60% compared to the present one.

In order to properly plan further airport development, transportation engineers and planners are looking for answers to the following questions:

(a) What is the current utilization of the queueing system (ie, two X-ray machines)?
(b) What should be the number of X-ray machines for the design year of this terminal if the maximum tolerable waiting time in the queue is 2 min?
(c) What is the expected number of passengers at the checkpoint area on a typical day in the design year?
(d) What is the new utilization of the future facility?
(e) What is the probability that more than four passengers wait for service in the design year?

EXAMPLE 3.13—cont'd

Solution

(a) Utilization of the facility:

Note that this is a multiple server case with infinite source. The mean arrival rate equals one passenger on every 25 s, ie,

$$\lambda = \frac{1}{\frac{25}{3600}}(\text{pass/h}) = 144(\text{pass/h})$$

On the average it takes 45 s for a passenger to go through the system. The service rate equals:

$$\mu = \frac{1}{\frac{45}{3600}}(\text{pass/h}) = 80(\text{pass/h})$$

The utilization of the facility equals:

$$\rho = \frac{\lambda}{s \cdot \mu}$$

$$\rho = \frac{144}{2 \cdot 80} = 0.90$$

Other queueing parameters for a multiserver queueing system with infinite population are:
Idle probability:

$$p_0 = \frac{1}{\sum_{k=0}^{s-1} \frac{\left(\frac{\lambda}{\mu}\right)^k}{k!} + \frac{\left(\frac{\lambda}{\mu}\right)^s}{s!} \cdot \frac{1}{1 - \frac{\lambda}{s \cdot \mu}}}$$

$$p_0 = \frac{1}{\sum_{k=0}^{2-1} \frac{\left(\frac{144}{80}\right)^k}{k!} + \frac{\left(\frac{144}{80}\right)^s}{2!} \cdot \frac{1}{1 - \frac{144}{2 \cdot 80}}}$$

$$p_0 = 0.052632$$

Expected number of clients in the queue:

$$L_q = \frac{p_0 \cdot \left(\frac{\lambda}{\mu}\right)^s \cdot \rho}{s!(1-\rho)^2}$$

$$L_q = \frac{0.052632 \cdot \left(\frac{144}{80}\right)^2 \cdot 0.90}{2!(1-0.90)^2}$$

$$L_q = 7.6737$$

(Continued)

EXAMPLE 3.13—cont'd

Expected number of customers in the system:

$$L = L_q + \frac{\lambda}{\mu}$$

$$L = 7.6737 + \frac{144}{80}$$

$$L = 9.4737$$

Average waiting time in the queue:

$$W_q = \frac{L_q}{\lambda}$$

$$W_q = \frac{7.6737}{140} = 0.055\,(\text{h}) = 197\,(s)$$

Average waiting time in the system:

$$W = W_q + \frac{1}{\mu}$$

$$W = 197 + 45 = 242\,(s)$$

(b) The solution to this part is done by trial and error. As a first trial let us assume that the number of X-ray machines is $3(s=3)$.

In the design year the demand for services is expected to grow by 60% compared to that today. The average arrival rate in the design year will be:

$$\lambda = 144 \cdot 1.6 = 230\,(\text{pass/h})$$

Finding p_0 for the design year:
$p_0 = 0.0097$, or <1% of the time the facility is idle.
Finding the average waiting time in the queue:

$$W_q = 332\,(\text{s})$$

Since this waiting time violates the desired 2 min maximum it is suggested that we try a higher number of X-ray machines to expedite service (at the expense of cost). The following results show that four X-ray machines are needed to satisfy the 2-min operational design constraint:

$$P_o = 0.045$$

$$L_q = 1.16$$

$$W_q = 18\,(\text{s})$$

(c) The expected number of passengers in the system is (with $s=4$):
$L = 4.04$ passengers in the system on the average design year day.

(d) The utilization of the improved facility (ie, four X-ray machines) is:

$$\rho = \frac{\lambda}{s \cdot \mu}$$

$$\rho = \frac{230}{4 \cdot 80} = 0.72$$

EXAMPLE 3.13—cont'd

(e) The probability that more than four passengers wait for service is just the probability that more than eight passengers are in the queueing system, since four are being served and more than four wait.

$$P(n>8)=1-\sum_{k=0}^{8} p_k$$

where:

$$p_k=\frac{\left(\frac{\lambda}{\mu}\right)^k}{k!}\cdot p_0, \text{ for } 0\le k\le s$$

$$p_k=\frac{\left(\frac{\lambda}{\mu}\right)^k}{s!\cdot s^{k-s}}\cdot p_0, \text{ for } k>s$$

from where, $P(n>8)=0.0879$.

Note that this probability is low and therefore the facility seems properly designed to handle the majority of the expected traffic within the 2-min waiting time constraint.

3.6.6 QUEUEING THEORY AND INVESTMENTS IN TRANSPORTATION FACILITIES EXPANSION

Queueing theory techniques enable us to measure operational efficiency of the studied queueing system, as well as the level-of-service offered to the clients. By changing the number of servers in the queueing system we study the sensitivity of the service facility useless time, the sensitivity of the queue length and the average waiting time, as the number of servers increases. In many cases, we can considerably improve queueing system operations by adding more servers (more through lanes on a highway, new runway at the airport, expanding dock in a harbor). Expanding traffic network capacities is extremely costly, as well as environmentally damaging. In many cases, general public is not ready to accept new transportation projects that increase the transportation capacity. Transportation capacity increase projects are also highly correlated with complex land-use policy issues. To mitigate traffic congestion, traffic engineers, planners, and authorities should combine expansion of existing facilities, and construction of new transportation facilities with the use of various demand management strategies (congestion pricing, HOV lanes) and various advanced technologies (Intelligent Transportation Systems). Queueing theory techniques should always be used in the analysis of the potential transportation capacity increase.

Let us assume that we are in the stage of increasing the number of toll booths at the highway. We have to make the decision about the number of toll booths (Fig. 3.43).

It is unthinkingly clear that the queue length through the rush hour will decline with the rise in the number of toll booths. How many toll booths do we need? Using queueing theory techniques, we can calculate average queue length during rush hour, average waiting time per client, percentage of elevator

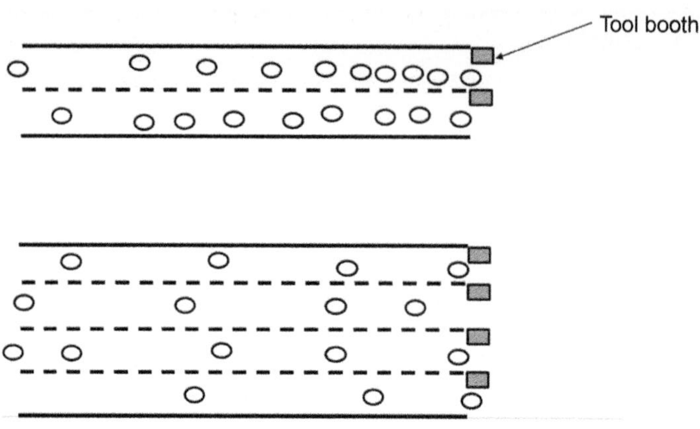

FIG. 3.43

Queue length as a function of the number of toll booths.

idle time, etc. The expected user waiting cost, and the total toll booths construction and maintenance cost are shown in Fig. 3.44.

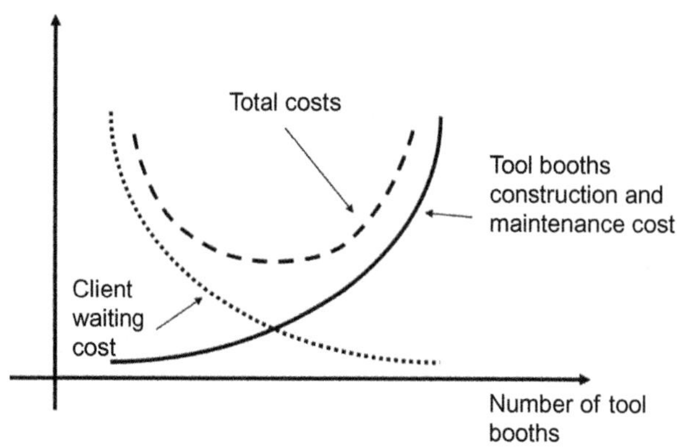

FIG. 3.44

User waiting cost, construction cost and total cost in the queueing system.

The higher the number of toll booths, the lower the drivers' waiting cost, and the higher the toll booths construction and maintenance cost. This is also valid for every other transportation facility (highway, airport, railway station, port). The curves shown in Fig. 3.44 characterize operations of all transportation facilities. High level-of-service (short waiting times, short queue lengths) is costly, but users' waiting costs are very low, and vice versa. It is clear that the "optimal" number of servers represents the compromise between queue lengths, user waiting times and construction and maintenance costs. Queueing theory assists us in performing comprehensive analysis of the queueing phenomenon, and to investigate the trade-off between the several service costs and the costs of waiting for the service.

3.7 SIMULATION

Elements of transportation systems interact among themselves all the time. Most often, we are not able to precisely predict changes in transportation system performances that arise as a result of changes in certain elements. How much will be the average travel time of network users if we make another bridge across the river? What will be the increase in number of transported passengers if air carrier significantly increases flight frequency on a specific route? Will the introduction of congestion pricing system and payment for entering the down-town significantly reduce traffic congestion in the city?

To properly answer these, and similar questions, analysts and traffic engineers usually develop *simulation models*. A simulation model has the main task to *imitate* the behavior of the real system. By studying the interaction among transportation system elements, simulation models allows us to estimate travel times, waiting times, the percentage of utilization of vehicles, utilization of crews, etc.

Simulation models enable performing of *statistical experiments*. These statistical experiments are executed with the help of computer. Therefore, instead of long-term observations of the real transportation system, traffic engineers often simulate transportation system behavior on a computer. Based on the large number of statistical experiments, appropriate statistical analysis is performed and conclusions are drawn about the transportation system performances.

3.7.1 THE MONTE CARLO SIMULATION METHOD

The Monte Carlo simulation method has been used in engineering applications from the late 1950s. In this method, random numbers are used to obtain samples from probability distributions. Sampling from whichever probability distribution is based on the utilization of the [0, 1] random numbers.

The [0, 1] random numbers are uniformly distributed in the interval [0, 1]. In other words, every one of the values in the interval [0, 1] has the same chance to happen. The [0, 1] random numbers are generated in an entirely random manner.

We illustrate the Monte Carlo method by the following example. There are two paths between node *A* and node *B* (Fig. 3.45).

We assume that there are equal chances of choosing left path (*L*), or right path (*R*) by the driver who travels from *A* to *B*. In other words, we assume that any driver chooses path between *A* and *B* with the following probabilities:

$$p(L) = 0.5 \quad p(R) = 0.5$$

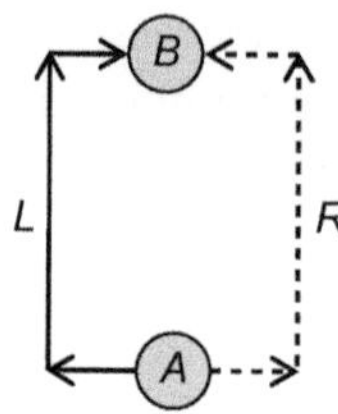

FIG. 3.45

Two paths between node *A* and node *B*.

We denote by r the generated random number from the interval [0, 1]. Given that the [0, 1] random numbers are uniformly distributed in the interval [0, 1], we formulate the following rules for determining driver's choice:

If $0 \leq r \leq 0.5$ the driver chooses L

If $0.5 \leq r \leq 1$ the driver chooses R

Let us study the route choice of first 10 drivers. The choices of first 10 drivers are equivalent to generating 10 random numbers from the interval [0, 1]. Let us assume that we generated the following random numbers:

0.051455 0.627205 0.084273 0.82207 0.298202 0.203535 0.535325 0.359749 0.701533 0.116597

The drivers' choices will be: $L, R, L, R, L, L, R, L, R, L$.

Let us consider the case when drivers can choose one among five routes. The routes are denoted respectively as 1, 2, 3, 4, and 5. We assume that there are equal chances of choosing any path by the driver who travels from A to B. In other words, we assume that any driver chooses path between A and B with the following probabilities:

$$p(1) = 1/5,\ p(2) = 1/5,\ p(3) = 1/5,\ p(4) = 1/5,\ p(5) = 1/5$$

We denote by x the outcome (driver's choice of the route). We also denote respectively by $p(x)$ and $F(x)$ probability density function and cumulative density function. The possible outcomes and the corresponding values of $p(x)$ and $F(x)$ are shown in Table 3.9, as well as in Fig. 3.46.

Table 3.9 The Possible Outcomes, and Corresponding Values of $p(x)$ and $F(x)$

x	1	2	3	4	5
$p(x)$	0.2	0.2	0.2	0.2	0.2
$F(x)$	0.2	0.4	0.6	0.8	1

The following rules determine driver's choice:

If $0 \leq F(x) \leq 0.2$ the driver chooses route 1

If $0.2 < F(x) \leq 0.4$ the driver chooses route 2

If $0.4 < F(x) \leq 0.6$ the driver chooses route 3

If $0.6 < F(x) \leq 0.8$ the driver chooses route 4

If $0.8 < F(x) \leq 1$ the driver chooses route 5

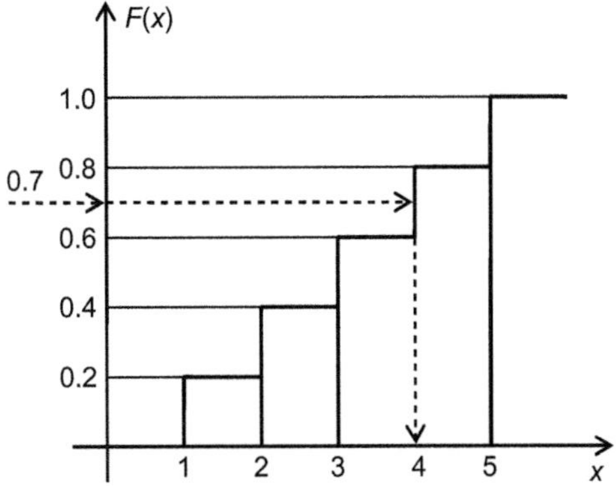

FIG. 3.46

Outcomes x and $F(x)$.

Fig. 3.46 also shows experiment in which we generated random number that is equal to 0.7. We assign this number to $F(x)$. We obtain the outcome (driver's route choice) by *inverting* $F(x)$. In our case, $r=0.7$ and $F(x)=0.7$, and consequently $x=4$. This method is called *method of inversion.* The method of inversion is used for all probability distributions. We illustrate the using of this method in the case when we have to perform sampling of the Exponential Distribution.

EXAMPLE 3.14

Vehicles arrive at the specific point at the highway according to the Poisson Process. The time between vehicle arrivals in a Poisson Process is random variable T that has exponential distribution with parameter λ:

$$f(t)=\lambda \mathrm{e}^{-\lambda t}$$

Let us assume that $\lambda=0.25\,(\mathrm{veh/s})$. The cumulative density function $F(x)$ equals:

$$F(x)=\int_0^t \lambda \cdot \mathrm{e}^{-\lambda t} dt = 1-\mathrm{e}^{-\lambda t}$$

We generate random number r from the interval [0, 1]. We get:

$$r=F(x)$$

$$r=1-\mathrm{e}^{-\lambda t}$$

$$t=-\frac{1}{\lambda}\cdot \ln(1-r)$$

Since r is a random number, $R=1-r$ is also random number, so we can write:

$$t=-\frac{1}{\lambda}\cdot \ln R$$

(Continued)

EXAMPLE 3.14—cont'd

We generated the following random numbers (R):

0.312230, 0.28341, 0.297506, 0.510998, 0.220226

The time intervals between vehicle arrivals are equal to:

$$t_1 = -\frac{1}{0.25} \cdot \ln(0.312230) = 4.66\,\text{s}$$

$$t_2 = -\frac{1}{0.25} \cdot \ln(0.283410) = 5.04\,\text{s}$$

$$t_3 = -\frac{1}{0.25} \cdot \ln(0.297506) = 4.85\,\text{s}$$

$$t_4 = -\frac{1}{0.25} \cdot \ln(0.510998) = 2.29\,\text{s}$$

$$t_5 = -\frac{1}{0.25} \cdot \ln(0.220226) = 6.05\,\text{s}$$

3.8 MULTIATTRIBUTE DECISION MAKING METHODS

Let us assume, for example, that we want to buy a new car. We need to rank the considered car models (set of alternatives), and to choose one car model (one alternative) from a set of possible alternatives. The criteria that we consider when choosing a new car model can be price, estimated future maintenance cost, fuel consumption, depreciation, safety, comfort, etc.

Government, industry, and/or traffic authorities frequently have to evaluate set of transportation projects (alternatives). The ranking of the alternatives is usually done according to a number of criteria that, as a rule, are mutually conflicting.

Multiattribute decision making (MADM) methods take into account different types of criteria with various dimensions (Hwang and Yoon, 1981; Roy and Vincke, 1981; Chen and Hwang, 1992). The MADM methods can be used to discover a single most favorite alternative, to rank the alternatives, or to make the distinction of acceptable from unacceptable alternatives.

We use term "alternative" to describe transportation project. The terms "option," "policy," "action," and "candidate" are also used in the literature. The alternatives are usually ranked according to few attributes. The number of the attributes depends on the nature of the problem considered. Considered attributes have different units of measurement. For example, the attributes and the units of measurements could be: price ($); fuel consumption (miles per gallon); waiting time (min); comfort (nonnumerical way (words)), etc.

By m we denote the total number of alternatives (transportation projects), and by n the total number of criteria according to which the considered alternatives are compared. In the decision matrix D, values x_{ij} are given that certain alternatives A_i ($i = 1, 2, \ldots, m$) take by particular attribute (criteria) X_j ($j = 1, 2, \ldots, n$):

$$D = \begin{matrix} & X_1 & X_2 & \cdots & X_n \\ A_1 & x_{11} & x_{12} & \cdots & x_{1n} \\ A_2 & x_{21} & x_{22} & \cdots & x_{2n} \\ \vdots & \vdots & \vdots & & \vdots \\ A_m & x_{m1} & x_{m2} & \cdots & x_{mn} \end{matrix} \quad (3.72)$$

More or less, all MADM need information about the relative importance (weight) of each attribute. Weights could be assigned by the analyst (decision-maker), or they could be calculated by various methods.

There are benefit attributes and cost attributes. In the case of benefit attributes, the greater the attribute value the more its preference (profit, revenue, fuel efficiency, ...). In the case of cost attributes, the greater the attribute value the less its preference (direct operation cost, passenger waiting time at hub, ...).

3.8.1 ATTRIBUTE WEIGHTS

In the case of cardinal weights of the attributes, numerical values (importance) are assigned to each attribute. All weights must be numerical values greater than or equal to zero, and smaller than or equal to one. The following relation must be satisfied:

$$\sum_{j=1}^{n} w_j = 1 \quad (3.73)$$

In other words, the total sum of all weights must be equal to one. Attributes could be also arranged in a simple rank order. In this case, we list the most important attribute first and the least important attribute last. The number of attributes (criteria) used for ranking of alternatives is equal to n. Analysts usually assign 1 to the most important criteria, and n to the least important attribute. The attribute weights are calculated as follows:

$$w_k = \frac{\frac{1}{r_k}}{\sum_{j=1}^{n} \frac{1}{r_j}} \quad (3.74)$$

where r_k is the rank of the kth attribute.

EXAMPLE 3.15

The analyst ranks alternative airport locations according to the following criteria:

X_1: total construction cost;
X_2: distance from the downtown; and
X_3: connectivity with highway and railway networks.

Let us assume that the rank order of the criteria is the following:

X_3: connectivity with highway and railway networks;
X_1: total construction cost; and
X_2: distance from the downtown.

The ranks are:

$$r_3 = 1$$

(Continued)

EXAMPLE 3.15—cont'd

$$r_1 = 2$$

$$r_2 = 3$$

The corresponding criteria weights are respectively equal:

$$w_1 = \frac{\frac{1}{r_1}}{\frac{1}{r_1}+\frac{1}{r_2}+\frac{1}{r_3}} = \frac{\frac{1}{2}}{\frac{1}{2}+\frac{1}{3}+\frac{1}{1}} = 0.272$$

$$w_2 = \frac{\frac{1}{r_2}}{\frac{1}{r_1}+\frac{1}{r_2}+\frac{1}{r_3}} = \frac{\frac{1}{3}}{\frac{1}{2}+\frac{1}{3}+\frac{1}{1}} = 0.181$$

$$w_3 = \frac{\frac{1}{r_3}}{\frac{1}{r_1}+\frac{1}{r_2}+\frac{1}{r_3}} = \frac{\frac{1}{1}}{\frac{1}{2}+\frac{1}{3}+\frac{1}{1}} = 0.547$$

When we determine weights from the ranks, the sum of all weights must be also equal to one, ie,

$$w_1 + w_2 + w_3 = 0.272 + 0.181 + 0.547 = 1$$

3.8.2 MINIMAX METHOD

The overall performance of an alternative is determined by its weakest or poorest attribute. Let us explain the concept of the *Minimax* method by considering the following example. *A*, *B*, *C*, *D*, and *E* are rural areas (Fig. 3.47). A joint fire-fighting brigade is to be designed for these five areas and the optimal location of the fire-fighting brigade must be determined. The optimal location must minimize the greatest distance between potential fire locations and the fire-fighting brigade station. The fire station can be in only one of the five areas.

Link lengths are shown in Fig. 3.47. We find the lengths of the shortest paths between all pairs of nodes. These lengths are shown in the matrix $[d_{ij}]$:

$$[d_{ij}] = \begin{array}{c} \\ A \\ B \\ C \\ D \\ E \end{array} \begin{array}{c} \begin{array}{ccccc} A & B & C & D & E \end{array} \\ \begin{bmatrix} 0 & 5 & 4 & 7 & 5 \\ 5 & 0 & 7 & 2 & 3 \\ 4 & 7 & 0 & 6 & 4 \\ 7 & 2 & 6 & 0 & 2 \\ 5 & 3 & 4 & 2 & 0 \end{bmatrix} \end{array}$$

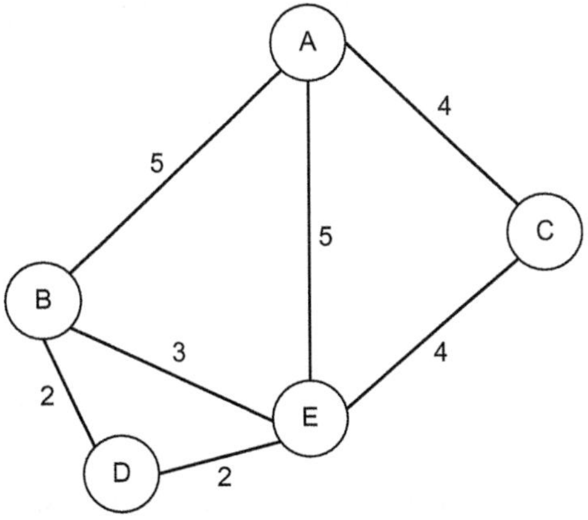

FIG. 3.47

Transportation network for which the location of the fire station has to be determined.

The node that has the minimum value of the maximum elements of its row is the optimal location for the fire station. In our case node E is the optimal location for the fire station.

3.8.3 MAXIMAX METHOD

The *Maximax* method selects an alternative by its best attribute rating. In the Maximax method only a single attribute represents an alternative. In the first step of the method the best attribute value for each alternative is identified. In the second step, the alternative with the maximum of the best values is selected.

EXAMPLE 3.16

The decision matrix D reads:

$$D = \begin{matrix} A_1 \\ A_2 \\ A_3 \end{matrix} \begin{bmatrix} 5 & 6 & \mathbf{7} \\ 2 & \mathbf{11} & 4 \\ 8 & 5 & \mathbf{9} \end{bmatrix}$$

All criteria are benefit criteria. The best attribute values for each alternative are respectively equal: 7, 11, and 9. The alternative with the maximum of the best values is the alternative A_2.

3.8.4 SIMPLE ADDITIVE WEIGHTING METHOD

The simple additive weighting (SAW) method is widely used MADM method in engineering and management. Various traffic, technical, economic, or environmental criteria are converted to a common scale before applying the SAW method. Within the SAW method, the score of each considered alternative is obtained by adding contributions from each attribute. The final score of the alternative is obtained by multiplying the rating for each attribute by the attribute weight and then summing these products over all the attributes.

The SAW method translates a multicriteria problem into a single-dimension. The weighted score V_i of the alternative A_i equals:

$$V_i = \sum_{j=1}^{n} w_j \cdot r_{ij} \tag{3.75}$$

where w_j is the weight of the criteria X_j and r_{ij} is the rating score for alternative A_i on criterion X_j.

The alternative with the highest weighted score is selected by the decision-maker.

3.8.5 TOPSIS

Technique for Order Preference by Similarity to Ideal Solution (TOPSIS) is based on the idea that the selected alternative should have the shortest distance from the *positive-ideal solution* and the longest distance from the *negative-ideal solution.*

In matrix D, values x_{ij} are given that certain alternatives A_i ($i=1, 2, \ldots, m$) take by particular criteria X_j ($j=1, 2, \ldots, n$):

$$D = \begin{matrix} & X_1 & X_2 & \cdots & X_n \\ A_1 & x_{11} & x_{12} & \cdots & x_{1n} \\ A_2 & x_{21} & x_{22} & \cdots & x_{2n} \\ \vdots & \vdots & \vdots & & \vdots \\ A_m & x_{m1} & x_{m2} & \cdots & x_{mn} \end{matrix}$$

By m we denote the total number of alternatives, and by n the total number of criteria according to which the considered alternatives are compared.

Normalized values r_{ij} are calculated as

$$r_{ij} = \frac{x_{ij}}{\sqrt{\sum_{i=1}^{m} x_{ij}^2}} \quad i=1, 2, \ldots, m, \; j=1, 2, \ldots, n \tag{3.76}$$

In the next step, each column's elements in matrix R are multiplied by weight w_j (significance of a criterion) corresponding to a particular column. In this manner, matrix V is obtained such that the values of its elements express the weights (significance) of individual criteria as well. Matrix V is found to be

$$V = \begin{bmatrix} v_{11} & \cdots & v_{1j} & \cdots & v_{1n} \\ \vdots & & & & \vdots \\ v_{i1} & \cdots & v_{ij} & \cdots & v_{in} \\ \vdots & & & & \vdots \\ v_{m1} & \cdots & v_{mj} & \cdots & v_{mn} \end{bmatrix} = \begin{bmatrix} w_1 r_{11} & \cdots & w_j r_{1j} & \cdots & w_n r_{1n} \\ \vdots & & & & \vdots \\ w_1 r_{i1} & \cdots & w_j r_{ij} & \cdots & w_n r_{in} \\ \vdots & & & & \vdots \\ w_1 r_{m1} & \cdots & w_j r_{mj} & \cdots & w_n r_{mn} \end{bmatrix} \tag{3.77}$$

On calculating the elements of matrix V, the positive ideal solution A^* and the negative ideal solution A^- are determined. These solutions are defined as:

$$A^* = \left\{ (\max_i v_{ij} \middle| j \in J), (\min_i v_{ij} \middle| j \in J') \middle| i = 1,2,\ldots,m \right\} \\ = \left\{ v_1^*, v_2^*, \ldots, v_j^*, \ldots, v_n^* \right\} \tag{3.78}$$

$$A^- = \left\{ (\min_i v_{ij} \middle| j \in J), (\max_i v_{ij} \middle| j \in J') \middle| i = 1,2,\ldots,m \right\} \\ = \left\{ v_1^-, v_2^-, \ldots, v_j^-, \ldots, v_n^- \right\} \tag{3.79}$$

where:

$$J = \{j = 1,2,\ldots,n | j \text{ belongs to the benefit criteria}\} \tag{3.80}$$

$$J' = \{j = 1,2,\ldots,n | j \text{ belongs to the cost criteria}\} \tag{3.81}$$

A positive ideal solution A^* represents the ideal alternative that takes the best values according to all criteria. The ideal solution usually does not exist in real life. The decision makers try to choose the alternative which is as close as possible to an ideal solution. The negative-ideal solution is composed of all worst attribute ratings (Fig. 3.48).

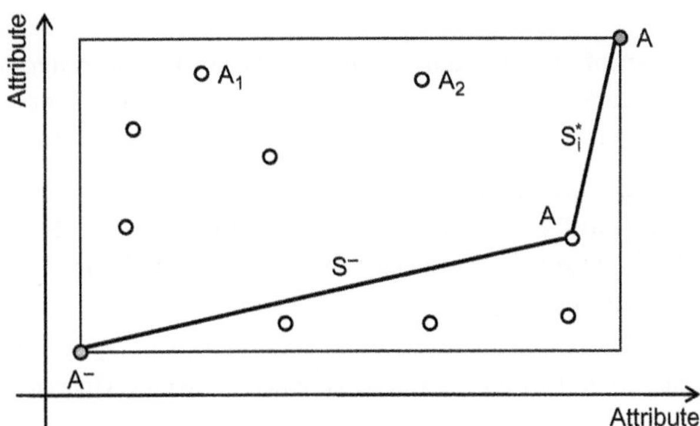

FIG. 3.48

Alternatives, positive ideal and negative-ideal solution in two-dimensional space.

Let us note that the benefit criteria are understood to be those by which an alternative is better if it takes greater values. As far as the cost criteria are concerned, an alternative is better if by these criteria it takes lower values. The distance S_i* of each alternative from the ideal alternative is:

$$S_i^* = \sqrt{\sum_{j=1}^{n} \left(v_{ij} - v_j^* \right)^2} \quad i = 1,2,\ldots,m \tag{3.82}$$

The distance S_i^- of each alternative from the negative ideal solution is:

$$S_i^- = \sqrt{\sum_{j=1}^{n} \left(v_{ij} - v_j^- \right)^2} \quad i = 1,2,\ldots,m \tag{3.83}$$

Relative closeness C_i* of the alternative A_i to the ideal solution A* is:

$$C_i^* = \frac{S_i^-}{S_i^* + S_i^-}, \quad 0 \le C_i^* \le 1 \quad i = 1,2,\ldots,m \tag{3.84}$$

Since $C_i^* = 1$ if $A_i = A^*$ and $C_i^* = 0$ if $A_i = A^-$, the alternative A_i is better if C_i* is closer to 1. It is clear that from the set of alternatives $A_1, A_2, \ldots, A_m$ the best alternative is A_i with the largest value of C_i*.

3.9 DATA ENVELOPMENT ANALYSIS (DEA)

Transportation engineers and analysts frequently face the problem of comparing the efficiency of airports, hubs, terminals, ports, airline routes and bus lines, as well as the problem of measuring their performances. Most frequently, when performing such an analysis, the engineers use *ratios*. Ratios are obtained by dividing some output measure (number of processed passengers, number of aircraft operations, cargo volumes, etc) by some input measure (number of runways, number of check-in desks, etc). Various ratios can produce different conclusions about the efficiency of the compared transportation facilities.

3.9.1 RATIOS

Let us analyze the efficiencies of the 10 airports. Number of runways, passenger terminal area in (thousands of m^2), cargo terminal area in (thousands of m^2), number of aircraft operations in (thousands), and the number of processed passengers in (millions) for 10 analyzed airports are shown in Table 3.10.

Table 3.10 Number of Runways, Passenger Terminal Area in (Thousands of m^2), Cargo Terminal Area in (Thousands of m^2), Number of Aircraft Operations in (Thousands), and the Number of Processed Passengers in (Millions)

Airport	Number of Runways	Passenger Terminal Area in (10^3 m^2)	Cargo Terminal Area in (10^3 m^2)	Number of Aircraft Operations in (10^3)	Number of Processed Passengers in (10^6)
A_1	2	180	65	270	30
A_2	3	120	55	160	13
A_3	3	255	96	300	17
A_4	5	250	84	430	33
A_5	3	72	135	220	14
A_6	1	28	5	40	2
A_7	2	780	810	183	28
A_8	1	40	73	44	3
A_9	2	142	45	186	19
A_{10}	3	675	43	300	34

We divide yearly number of aircraft operations by the number of runaways. We also divide yearly number of processed passengers by passenger terminal area. We obtain the ratios r_1 and r_2 shown in Table 3.11.

Table 3.11 Yearly Number of Aircraft Operations (in Thousands) Per Runway (r_1) and Yearly Number of Processed Passengers in Per m^2 of Passenger Terminal Area (r_2)

	Yearly Number of Aircraft Operations (in Thousands) Per Runway	Yearly Number of Processed Passengers in Per m^2 of Passenger Terminal Area
Airport	r_1	r_2
A_1	135	167
A_2	53	108
A_3	100	67
A_4	86	132
A_5	72	194
A_6	40	71
A_7	92	36
A_8	44	75
A_9	93	134
A_{10}	100	50

The calculated ratios r_1 and r_2 produce different conclusions about the efficiency of the compared airports. The airport A_1 has the highest yearly number of aircraft operations (in thousands) per runway, while the airport A_5 has the highest yearly number of processed passengers in per m^2 of passenger terminal area.

The simple graphical analysis could help us to clarify different ratios. We draw a horizontal line, from the y-axis to A_5. Then we connect A_5 with A_1, and finally, we draw a vertical line from A_1 to the x-axis (Fig. 3.49).

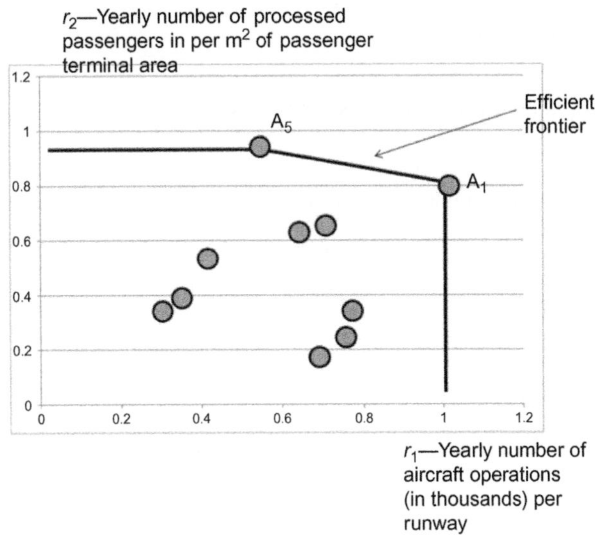

FIG. 3.49

Efficient frontier.

The line drawn is called the efficient frontier. The efficient frontier envelopes the data analyzed. The airports A_1 and A_5 are on the efficient frontier, and we say that the airports A_1 and A_5 have 100% efficiency. In the same way, all points that belong to the efficient frontier have efficiency equal to 100%. The other considered airports have efficiencies that are <100%.

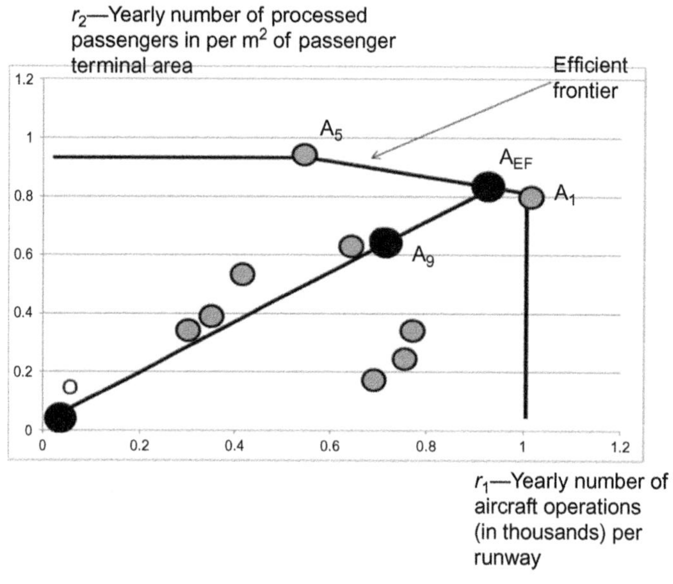

FIG. 3.50

Calculating the efficiency of the airport A_9.

For example, the efficiency of the airport A_9 (Fig. 3.50) equals:

$$\text{Efficiency of the } A_9 = \frac{\text{Length of the line from origin O to } A_9}{\text{Length of the line from origin O via } A_9 \text{ to efficient frontier}} \cdot 100\%$$

$$\text{Efficiency of the } A_9 = 77.6\%$$

The efficiencies of all analyzed airports are *relative efficiencies*, relative to the data analyzed.

3.9.2 DEA BASICS

The *Data Envelopment Analysis* (DEA) is a measurement technique that is used for evaluating the relative efficiency of decision-making units (DMU's). The DEA is also called *frontier analysis* (Charnes et al., 1978, 1979, 1981; Charnes and Cooper, 1985). By using the Data Envelopment Analysis (DEA) the analyst could evaluate the efficiency of any number of DMU's. The analyzed DMU's could have any number of inputs and outputs. The decision-making units in the area of traffic and transportation could be airports, airline routes, intersections, HOV lanes, ports, networks, park and ride facilities, etc.

The DEA was initially suggested by Charnes, Cooper, and Rhodes in 1978. Each DMU denotes the entity that changes inputs into outputs. The DEA defines the relative efficiency in the following way:

$$\text{Efficiency} = \frac{\text{Weighted sum of outputs}}{\text{Weighted sum of inputs}} \tag{3.85}$$

$$\text{Efficiency} = \frac{u_1 \cdot y_{1j} + u_2 \cdot y_{2j} + \cdots + u_n \cdot y_{nj}}{v_1 \cdot x_{1j} + v_2 \cdot x_{2j} + \cdots + v_m \cdot x_{mj}} \tag{3.86}$$

where:

u_i is the weight of the output I;
y_{ij} is the amount of output i from the unit j;
x_{ij} is the amount of input i to the unit j; and
v_i is the weight of the input i.

The DEA defines the efficiency for every DMU as a weighted sum of outputs divided by a weighted sum of inputs. The efficiency of any DMU is within the range [0, 1], or [0%, 100%]. Frequently, different DMU's have different goals. For example, some airports could try to maximize number of served passengers, while some other could try to maximize cargo volumes. When calculating the efficiency of a specific DMU by the DEA technique, the weights of a DMU are chosen to present the considered DMU in the best possible light.

The efficiency h_0 of the DMU j_0 could be obtained by solving the following fractional programming problem:

Maximize

$$h_0 = \frac{\sum_r u_r \cdot y_{rj_0}}{\sum_i v_i \cdot x_{ij_0}} \quad \forall_j \tag{3.87}$$

subject to:

$$\frac{\sum_r u_r \cdot y_{rj_0}}{\sum_i v_i \cdot x_{ij_0}} \leq 1 \tag{3.88}$$

$$u_r, v_i \geq \varepsilon \tag{3.89}$$

By solving the fractional programming problem, the analyst obtains the input weights v_i, as well as output weights u_r. In the case of n DMU to be evaluated, the analyst have to perform n optimizations (one for every DMU to be evaluated). The fractional program could be replaced by the following equivalent linear program:

Maximize

$$h_o = \sum_r u_r \cdot y_{rj_o} \tag{3.90}$$

subject to:

$$\sum_i v_i \cdot x_{ij_o} = 1 \tag{3.91}$$

$$\sum_r u_r \cdot y_{rj} \leq \sum_i v_i \cdot x_{ij} \quad \forall j \tag{3.92}$$

$$u_r, v_i \geq \varepsilon \tag{3.93}$$

The inputs and outputs in the DEA have various units. The DEA enable analyst to directly compare considered DMU with their peers.

EXAMPLE 3.17

We use DEA to evaluate efficiencies of ten airports. In other words, decision making units are airports. When evaluating airports we use the following inputs and outputs:

Inputs:

1. Input 1: Number of runways
2. Input 2: Passenger terminal area
3. Input 3: Number of gates

Outputs:

1. Output 1: Number of aircraft operations
2. Output 2: Number of processed passengers

The inputs and outputs are given in Table 3.12.

Table 3.12 Number of Runways, Passenger Terminal Area, Number of Gates, Number of Aircraft Operations, and Number of Processed Passengers

DMU	Input 1	Input 2	Input 3	Output 1	Output 2
A_1	2	180	61	270	30
A_2	3	120	50	160	13
A_3	3	255	95	300	17
A_4	5	250	108	430	33
A_5	3	72	60	220	14
A_6	1	28	12	40	2
A_7	2	780	67	183	28
A_8	1	40	16	44	3
A_9	2	142	93	186	19
A_{10}	3	675	149	300	34

The efficiencies obtained, by solving the linear program for each airport (DMU), are shown in Table 3.13.

EXAMPLE 3.17—cont'd

Table 3.13 Airport Efficiencies

Airport (DMU)	Efficiencies
A_1	1
A_2	0.76387
A_3	0.76527
A_4	0.95842
A_5	1
A_6	0.80137
A_7	0.86987
A_8	0.64985
A_9	0.79017
A_{10}	0.72001

3.10 COMPUTATIONAL INTELLIGENCE TECHNIQUES

Many traffic and transport parameters are characterized by uncertainty, subjectivity, imprecision, and ambiguity. Every day, dispatchers, drivers, air traffic controllers, operators, passengers, engineers and planners use subjective knowledge, approximately known parameter values, and linguistic information in a decision-making processes.

Some of the complex traffic and transportation problems can be successfully solved, by using various intelligent systems that are based on knowledge and techniques that belong to different scientific disciplines. These intelligent systems must be able to recognize different situations and to make appropriate decisions without explicitly known relationships between the individual variables. The new generation of intelligent systems that are used for transportation planning and traffic control traffic should be able to generalize, to adapt and to learn from experience, new knowledge and new information. Modern intelligent systems are based on computer techniques capable to count with words (Fuzzy Logic), to learn and to adapt (Artificial Neural Networks), and to perform in a systematic way stochastic search (Holland, 1975; Goldberg, 1989) and optimization (Genetic Algorithms). A set of these techniques, inspired by nature, is known as computational intelligence. Computational intelligence techniques deal with difficult real-world problems to which mathematical or traditional modeling can be inadequate. The traditional mathematical approaches to some of the complex transportation problems are sometimes inadequate for the following reasons: (a) the studied traffic phenomena might be too complex; (b) some of the important traffic parameters are characterized by uncertainty. Computational intelligence techniques have been used to solve a wide variety of traffic and transportation problems (urban traffic control, ramp metering, transportation facility location problems, traffic assignment problem, vehicle routing and scheduling, etc.). It is particularly important to use these techniques for solving real-time traffic problems, as well as for solving the problems characterized by uncertainty.

3.10.1 THE CONCEPT OF FUZZY SETS

In the classic theory of sets, very precise bounds separate the elements that belong to a certain set from the elements outside the set. In other words, it is quite easy to determine whether an element belongs to a set or not. For example, if we denote by A the set of signalized intersections in a city, we conclude that every intersection under observation belongs to set A if it has a signal. Element x's membership in set A is described in the classic theory of sets by the membership function $\mu_A(x)$, as follows:

$$\mu_A(x) = \begin{cases} 1, & \text{if and only if } x \text{ is member of A} \\ 0, & \text{if and only if } x \text{ is not member of A} \end{cases} \tag{3.94}$$

Fig. 3.51 presents set A and elements x, y, and z.

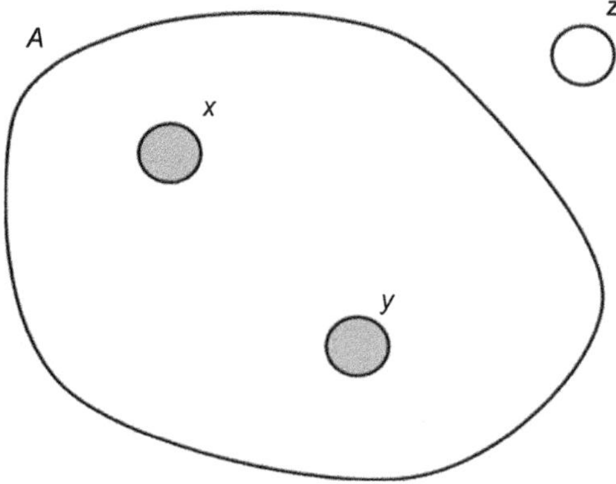

FIG. 3.51

Set A and elements x, y, and z.

It is clear from Fig. 3.51 that $\mu_A(x) = 1$, $\mu_A(y) = 1$, and $\mu_A(z) = 0$. Boolean logic utilizes razor-sharp divisions. It forces us to draw lines between members of a class and nonmembers. Many sets encountered in reality do not have precisely defined bounds that separate the elements in the set from those outside the set. Thus, it might be said that waiting time of a vehicle at a certain signal is "long." If we denote by **A** the set of "long waiting time at a signal," the question logically arises as to the bounds of such a defined set. In other words, we must establish which element belongs to this set. Does a waiting time of 30 s belong to this set? What about 15 or 80 s? The air traffic between two cities can be described as having "high flight frequency." Do flight frequencies of five flights a day, eight flights a day, three flights a day belong to "high flight frequency" category? Travel time between origin and destination is usually subjectively estimated as "short," "not too long," "long," "medium," "about 20 min," "around half an hour," and so on. Does a travel time of 45, 28, or 8 min belong to the set called "travel time of around half an hour"? We intuitively know that a travel time of 28 min belongs to the set called "travel time of around half an hour" "more" or "stronger" than a travel time of 8 min. In other words, there is more truth in the statement that travel time of 28 min is "travel time of around half an hour" than in the statement that travel time of 8 min is "travel time of around half an hour."

Fuzzy sets (Zadeh, 1965) and fuzzy logic try to reproduce the way how human beings think. In 1965, Lotfi Zadeh published his famous paper "Fuzzy sets." Zadeh (1965) initiated a new mathematical concept for using natural language terms. Fuzzy logic makes efforts to model human beings' meaning of words, and to replicate human being's decision making processes. The membership function for fuzzy sets can take any value from the closed interval [0,1]. Fuzzy set **A** is defined as the set of ordered pairs $\mathbf{A}=\{x, \mu_{\mathbf{A}}(x)\}$, where $\mu_{\mathbf{A}}(x)$ is the grade of membership of element x in set **A**. The greater $\mu_{\mathbf{A}}(x)$, the greater the truth of the statement that element x belongs to set **A**.

EXAMPLE 3.18

Let us note set $X=\{2, 5, 9, 18, 21, 25\}$, whose elements denote the number of vehicles waiting in line at a signal. Set **B** consists of the fuzzy set "small number of vehicles in line." Fuzzy set **B** can be shown as

$$\mathbf{B}=\frac{0.95}{2}+\frac{0.55}{5}+\frac{0.20}{9}+\frac{0.10}{18}+\frac{0.05}{21}+\frac{0.01}{25}$$

The grades of membership 0.95, 0.55, ..., 0.01 are subjectively determined and indicate the "strength" of membership of individual elements in fuzzy set **B**. For example, 2 with a grade of membership of 0.95 belongs to fuzzy set **B**, which comprises a "small number of vehicles in line" at the signal.

Fuzzy sets are often defined through membership functions to the effect that every element is allotted a corresponding grade of membership in the fuzzy set. Let us note fuzzy set **C**. The membership function that determines the grades of membership of individual elements x in fuzzy set **C** must satisfy the following inequality:

$$0 \leq \mu_{\mathbf{C}}(x) \leq 1 \quad \forall x \in X \tag{3.95}$$

EXAMPLE 3.19

Let us note fuzzy set **A**, which is defined as "travel time is approximately ~30 min." Membership function $\mu_{\mathbf{A}}(t)$, which is subjectively determined is shown in Fig. 3.52.

In this case, we have subjectively estimated that travel time between the two points can be within the limits of 25–35 min. A travel time of 30 min has a grade of membership of 1 and belongs to the set "travel time is approximately 30 min." All travel times within the interval of 25–35 min are also members of this set because their grades of membership are greater than zero. Travel times outside this interval have grades of membership equal to zero.

(Continued)

EXAMPLE 3.19—cont'd

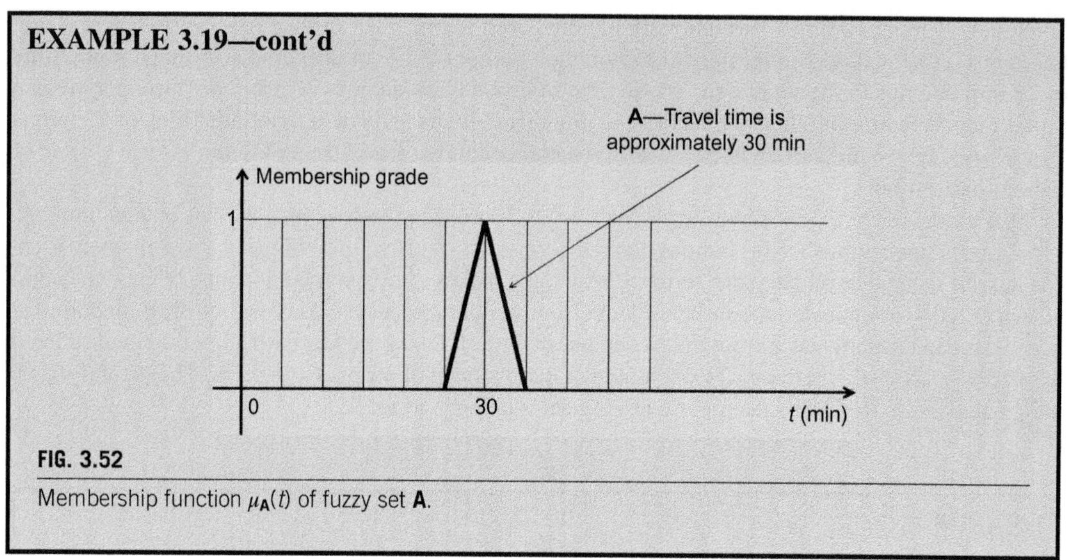

FIG. 3.52

Membership function $\mu_A(t)$ of fuzzy set **A**.

3.10.2 THE FUZZY SETS BASICS

Let us note fuzzy sets A and B defined over set X. Fuzzy sets A and B are equal (A=B) if and only if $\mu_A(x)=\mu_B(x)$ for all elements of set X.

Fuzzy set A is a subset of fuzzy set B if and only if $\mu_A(x)\leq\mu_B(x)$ for all elements x of set X. In other words, $A\subset B$ if, for every x, the grade of membership in fuzzy set A is less than or equal to the grade of membership in fuzzy set B (Fig. 3.53).

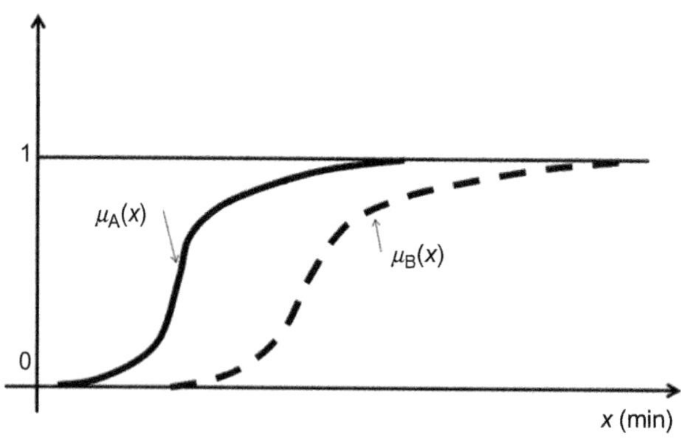

FIG. 3.53

Membership functions of fuzzy sets "long" and "very long" travel times.

We denote by **A** and **B**, respectively, the sets of "long" and "very long" travel times. The fuzzy set "very long" travel time is a subset of the fuzzy set "long" travel time since the following relation is satisfied for every x:

$$\mu_{\mathbf{B}}(x) \leq \mu_{\mathbf{A}}(x) \tag{3.96}$$

The intersection of fuzzy sets **A** and **B** is denoted by $\mathbf{A} \cap \mathbf{B}$ and is defined as the largest fuzzy set contained in both fuzzy sets **A** and **B**. The intersection corresponds to the operation "and." In classical set theory, an intersection between two sets includes the elements shared by these two sets. On the other hand, in fuzzy sets, an element may to a certain extent belong to both sets with different memberships. Membership function $\mu_{\mathbf{A} \cap \mathbf{B}}(x)$ of the intersection $\mathbf{A} \cap \mathbf{B}$ is defined as follows:

$$\mu_{\mathbf{A} \cap \mathbf{B}}(x) = \min \{\mu_{\mathbf{A}}(x), \mu_{\mathbf{B}}(x)\} \tag{3.97}$$

Fig. 3.54 presents the membership functions of sets **A**, **B**, and $\mathbf{A} \cap \mathbf{B}$.

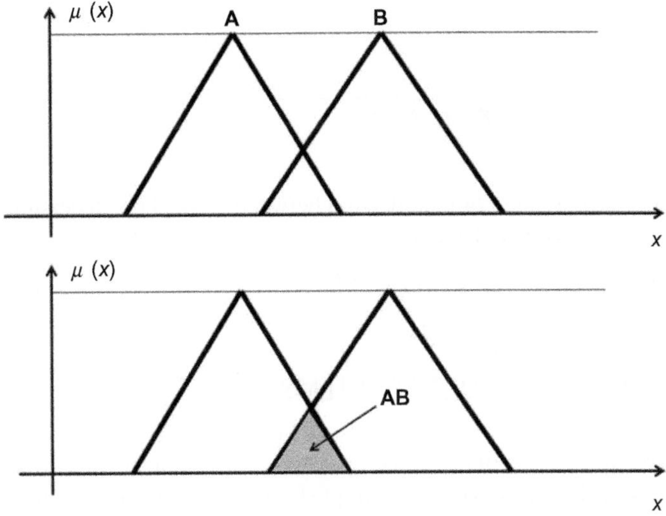

FIG. 3.54

Membership functions of fuzzy sets **A**, **B**, and $\mathbf{A} \cap \mathbf{B}$.

The union of fuzzy sets **A** and **B** is denoted by $\mathbf{A} \cup \mathbf{B}$ and is defined as the smallest fuzzy set that contains both fuzzy set **A** and fuzzy set **B**. The membership function $\mu_{\mathbf{A} \cup \mathbf{B}}(x)$ of the union $\mathbf{A} \cup \mathbf{B}$ of fuzzy sets **A** and **B** is defined as follows:

$$\mu_{\mathbf{A} \cup \mathbf{B}}(x) = \max \{\mu_{\mathbf{A}}(x), \mu_{\mathbf{B}}(x)\} \tag{3.98}$$

The union corresponds to the operation "or." Fig. 3.55 presents the membership functions of sets **A**, **B**, and $\mathbf{A} \cup \mathbf{B}$.

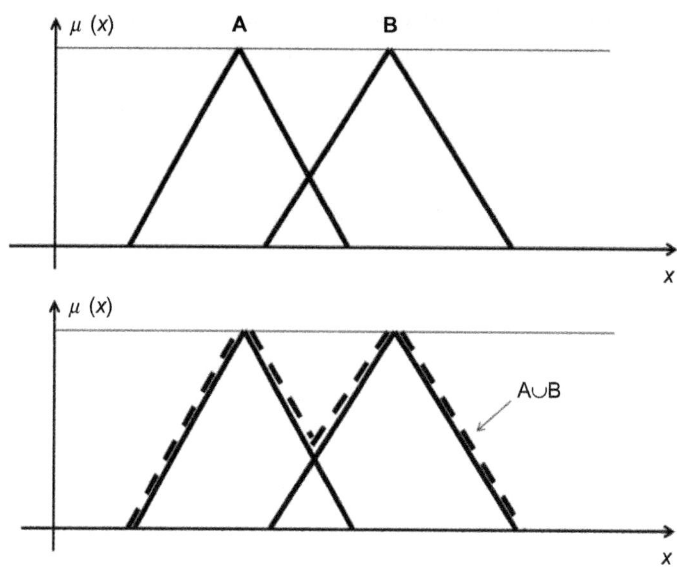

FIG. 3.55

Membership functions of fuzzy sets **A**, **B**, and **A**∪**B**.

A (whose elements have a grade of membership between 0 and 1) is understood to be fuzzy set $\bar{\mathbf{A}}$ whose membership function is calculated as

$$\mu_{\bar{\mathbf{A}}}(x) = 1 - \mu_{\mathbf{A}}(x) \tag{3.99}$$

3.10.3 BASIC ELEMENTS OF FUZZY SYSTEMS

Operators, dispatchers, air traffic controllers, pilots, and other rely on common sense when solving the problems. The control strategies of the drivers, operators and dispatchers can often be formulated in terms of numerous descriptive rules, which are simple for a manual processing and execution but complicated when it comes to the use of classical algorithms. These difficulties arise from the fact that, when describing different decisions made at various stages of a process, human beings prefer to use qualitative expressions instead of quantitative ones. The questions are how we can represent expert's knowledge (based on descriptive rules, and vague and ambiguous terms) to the computer. In 1973, Lotfi Zadeh proposed the new approach to the analysis of complex systems. Zadeh proposed the way to describe human knowledge by fuzzy rules.

Fuzzy logic systems, fuzzy expert systems, arise from the desire to model human experience, intuition, and behavior in decision making (Zimmermann, 1991). Fuzzy logic is not logic that is fuzzy. Fuzzy logic is the logic that we use to express fuzziness. The combination of imprecise logic rules in a single control strategy is called by Zadeh (1973) *approximate or fuzzy reasoning*.

Fuzzy rules include descriptive expressions such as small, medium, or large used to categorize the linguistic (fuzzy) input and output variables (Pappis and Mamdani, 1977; Self, 1990; Teodorović and

Vukadinović, 1998; Teodorović, 1999). A set of fuzzy rules, describing the control strategy of the operator, driver or a dispatcher forms a fuzzy control algorithm, that is, approximate reasoning algorithm, whereas the linguistic expressions are represented and quantified by fuzzy sets. The main advantage of this approach is the possibility of introducing and using rules from experience, intuition, heuristics, and the fact that a model of the process is not required.

The basic elements of each fuzzy logic system (Zadeh, 1972, 1973, 1975a,b, 1996; Wang and Mendel, 1992) are rules, fuzzifier, inference engine, and defuzzifier (Fig. 3.56).

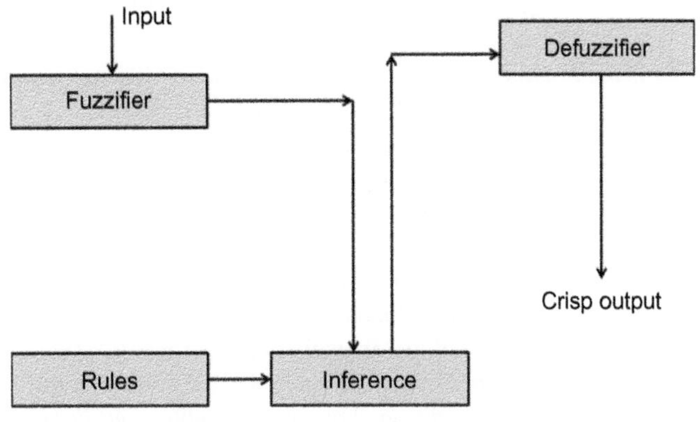

FIG. 3.56

Basic elements of a fuzzy logic system.

The input data are most commonly crisp values. The task of a fuzzifier is to map crisp numbers into fuzzy sets. Fuzzy rules can conveniently represent the knowledge of experienced experts (drivers, operators, passengers) used in control. These rules are arrived at either by verbalizing the operator's expertise or by conducting a carefully composed survey. The rules could be also formulated by using the observed decisions (input/output numerical data) of the operator.

Fuzzy rule (fuzzy implication) takes the following form:

$$\text{If } x \text{ is } \mathbf{A}, \text{ then } y \text{ is } \mathbf{B}$$

where **A** and **B** represent linguistic values quantified by fuzzy sets defined over universes of discourse X and Y.

The first part of the rule "x is **A**" is the premise or the condition preceding the second part of the rule "y is **B**" which constitutes the consequence or conclusion. The fact after "If" is called the *premise* or *hypothesis* or *antecedent*. From this fact, another fact called *conclusion* or *consequent* (the fact after "then") can be inferred.

Note the following simple rule:

$$\text{If variable } x \text{ is } \mathbf{Big}, \text{ then variable } y \text{ is } \mathbf{Small}$$

The membership functions of the fuzzy sets **Big** and **Small** are shown in Fig. 3.57.

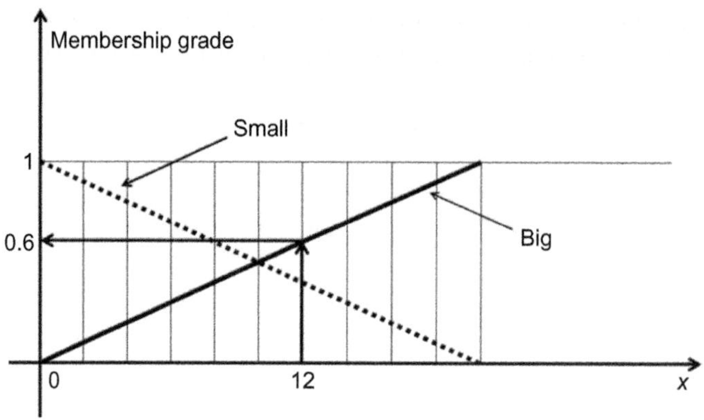

FIG. 3.57

Membership functions of fuzzy sets **Big** and **Small**.

As can be seen, the output of variable y is conditioned by the input of fuzzy variable x. Let us assume that we have obtained data or that we have estimated that the value of input variable $x = 12$. From Fig. 3.57, we can see that $x = 12$ corresponds to the grade of membership in fuzzy set **Big** of 0.6. This grade of membership is actually the "value of the truth" contained in the claim that the value of input variable $x = 12$ can be treated as **Big**. Since output variable y is conditioned by input variable x, we conclude that the claim that variable y is **Small** is only as true as the truth in the claim that input variable x is **Big**.

Fuzzy reasoning is an inference procedure, ie, the way of generating the conclusion from the premises when the linguistic expressions are quantified by fuzzy sets. The inference engine of the fuzzy logic system maps fuzzy sets into fuzzy sets. The inference engine "handles the way in which rules are combined" (Mendel, 1995). There are a number of various inferential procedures in literature.

The following set of rules which is called the fuzzy rule base is a typical example of an approximate reasoning algorithm:

If x is **Big**, then y is **Small**

or

If x is **Medium**, then y is **Medium**

or

If x is **Small**, then y is **Big**

Our known input variable value ($x = 8$) must go through all the above-defined rules: we must determine how much truth is contained in the claim that $x = 8$ is **Big**, how much truth in $x = 8$ is **Medium**, and, finally, how much truth there is in the claim that $x = 8$ is **Small**. After going through all the rules in the approximate reasoning algorithm, all the possible values of output variable y are associated with a grade of membership.

The last step in the approximate reasoning algorithm is defuzzification, ie, choosing one value for the output variable. Using the algorithm of fuzzy reasoning a fuzzy set is obtained as the output result with particular membership grades of possible numerical values of the output variable. By defuzzification the fuzzy information is compressed and given by a representative numerical information. The center of gravity of the resulting fuzzy set frequently represents the output numerical value.

An analyst or decision maker could also look at the grades of membership of individual output variable values, and chooses one of them according to the following criteria: "the smallest maximal value," "the largest maximal value," "center of gravity," "mean of the range of maximal values," and so on (Fig. 3.58).

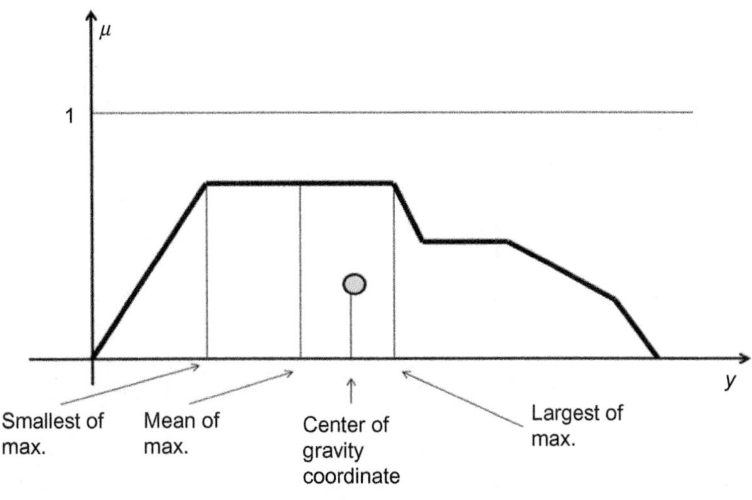

FIG. 3.58

Approximate reasoning using max-min composition for two rules.

In the case of fuzzy rule-based systems, the formation of the rule base can be made by human experts, according to numerical data or by combining numerical data and human experts. The development of models of fuzzy logic most often requires several iterations. The first step defines the set of rules and the corresponding membership functions of the input/output variables. After looking at the results, corrections are made to the rules and/or membership functions, if necessary. Then the model is tested once again with the modified rules and/or membership functions, and so on (Teodorović and Vukadinović, 1998).

A graphical interpretation of a fuzzy logic inference can considerably help us to reach a deeper understanding of the nature of fuzzy logic inference. Let us consider a set of fuzzy rules containing two input variables x_1, and x_2 and one output variable y.

$$\begin{gathered}\textit{Rule } 1: \text{ If } x_1 \text{ is } \mathbf{P}_{11} \text{ and } x_2 \text{ is } \mathbf{P}_{12} \\ \text{Then } y \text{ is } \mathbf{Q}_1\end{gathered}$$

or

$$\begin{gathered}\textit{Rule } 2: \text{ If } x_1 \text{ is } \mathbf{P}_{21} \text{ and } x_2 \text{ is } \mathbf{P}_{22} \\ \text{Then } y \text{ is } \mathbf{Q}_2\end{gathered}$$

or

$$\textit{Rule } k: \text{ If } x_1 \text{ is } \mathbf{P}_{k1} \text{ and } x_2 \text{ is } \mathbf{P}_{k2} \\ \text{Then } y \text{ is } \mathbf{Q}_k$$

The given rules are interrelated by the conjunction *or*. Such a set of rules is called a *disjunctive system of rules* and assumes the satisfaction of at least one rule. Let us note Fig. 3.59 in which our disjunctive

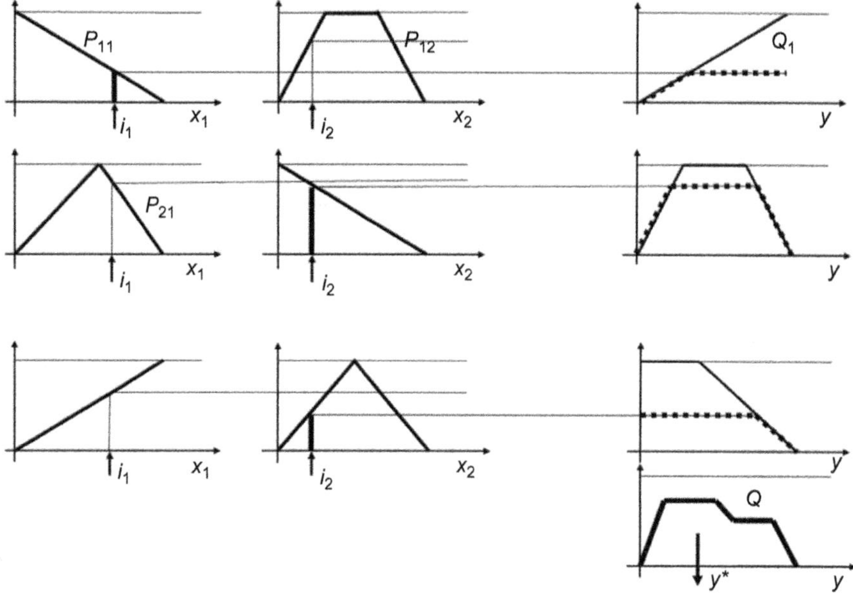

FIG. 3.59

Graphical interpretation of a disjunctive system of rules.

system of rules is presented.

Let the values i_1, and i_2 respectively, taken by input variables x_1, and x_2 be known. (Depending on the context of the problem considered, these values can be obtained in the basis of the collected data, through measurements, or by an expert evaluation.) In the considered case, the values i_1, and i_2 are crisp.

Fig. 3.59 also represents the membership function of output **Q**. This membership function takes the following form:

$$\mu_{\mathbf{Q}}(y) = \max_k \left\{ \min \left[\mu_{\mathbf{P}_{k1}}(i_1), \mu_{\mathbf{P}_{k2}}(i_2) \right] \right\} \quad k = 1, 2, \ldots, K \tag{3.100}$$

whereas fuzzy set **Q** representing the output is actually a fuzzy union of all the rule contributions $\mathbf{Y}_1$, $\mathbf{Y}_2$, ..., $\mathbf{Y}_k$, ie,

$$\mathbf{Q} = \mathbf{Y}_1 \mathrm{U} \mathbf{Y}_2 \mathrm{U} \ldots \mathrm{U} \mathbf{Y}_k \tag{3.101}$$

It is clear that

$$\mu_{\mathbf{Q}}(y) = \max\{\mu_{\mathbf{Y}_1}(y), \mu_{\mathbf{Y}_2}(y), \ldots, \mu_{\mathbf{Y}_k}(y)\} \tag{3.102}$$

Let us try to explain more thoroughly the manner in which, for the known values i_1, i_2, and i_3 of input variables x_1, x_2, and x_3, the corresponding value y^* of output variable y is calculated. Consider rule 1, which reads as follows:

If x_1 is $\mathbf{P}_{11}$ and x_2 is $\mathbf{P}_{12}$
Then y is $\mathbf{Q}_1$

The value $\mu_{\mathbf{P}11}(i_1)$ indicates how much truth is contained in the claim that i_1 equals $\mathbf{P}_{11}$. Similarly, value $\mu_{\mathbf{P}12}(i_2)$ indicates the truth value of the claim that i_2 equals $\mathbf{P}_{12}$. Value w_1, which is equal to $w_1 = \min\{\mu_{P_{11}}(i_1), \mu_{P_{12}}(i_2)\}$ indicates the truth value of the claims that, simultaneously, i_1 equals $\mathbf{P}_{11}$, and i_2 equals $\mathbf{P}_{12}$.

Since the conclusion contains as much truth as the premise, after calculating value w_1, the membership function of fuzzy set $\mathbf{Q}_1$ should be transformed. In this way, fuzzy set $\mathbf{Q}_1$ is transformed into fuzzy set $\mathbf{Y}_1$ (Fig. 3.59). Values w_2, w_3, …, w_k are calculated in the same manner leading to the transformation of fuzzy sets $\mathbf{Q}_2$, $\mathbf{Q}_3$, …., $\mathbf{Q}_k$ into fuzzy sets $\mathbf{Y}_2$, $\mathbf{Y}_3$, …., $\mathbf{Y}_k$.

As this is a disjunctive system of rules, assuming the satisfaction of at least one rule, the membership function $\mu_{\mathbf{Q}}(y)$ of the output represents the outer envelope of the membership functions of fuzzy sets $\mathbf{Y}_1$, $\mathbf{Y}_2$, …., $\mathbf{Y}_k$. The final value y^* of the output variable is arrived at upon defuzzification.

Fuzzy Logic Toolbox within MATHLAB—Math Works helps analysts to build fuzzy inference systems and view and analyze results.

EXAMPLE 3.20

Traffic congestion has been a problem in many cities in the world. The citizens of many big cities in the world by now spend between 40 and 60 min of time when commuting to work. The level of traffic congestion could be measured in a various way. (The procedure for measuring urban traffic congestion in this example is based on the work of Hamad and Kikuchi, 2002. This example is dedicated to the memory of Professor Shinya Kikuchi (1943–2012), who was one of the pioneers of applying fuzzy sets theory in transportation engineering.)

Let us first consider two possible measures of traffic congestion. In the next step, we combine these two measures into a single measure using fuzzy inference. The two measures are travel speed rate and proportion of time traveling at very low speed (below 5 mph) compared with total travel time.

When the traffic density is extremely low, our speed is influenced exclusively by performances of our vehicle, and by posted speed limits. This speed is known as a *free flow speed* (We shall study in more detail free flow speed in the chapter devoted to the traffic flow theory issues). Let us note Fig. 3.60.

The solid line in Fig. 3.60 denotes the path of the vehicle if it travels at the speed limit (the free-flow speed). The dashed line indicates actual vehicle's travel path. The time periods in which the vehicle's speed is ≤5 mph are denoted in Fig. 3.60 by S_i.

Travel speed rate could be defined as the rate of decrease in speed caused by congestion from the free-flow speed condition:

$$\text{Travel speed rate} = \frac{(\text{Free}-\text{flow speed} - \text{Average speed})}{\text{Free}-\text{flow speed}}$$

(Continued)

EXAMPLE 3.20—cont'd

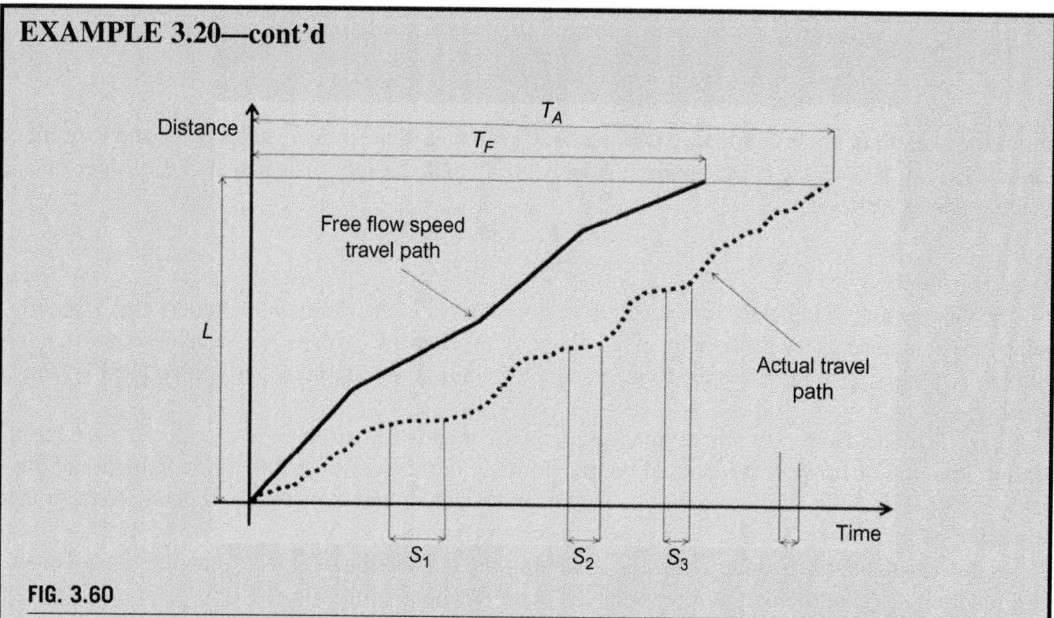

FIG. 3.60

Travel speed rate and very-low-speed rate.

$$\text{Travel speed rate} = \frac{\frac{L}{T_F} - \frac{L}{T_F}}{\frac{L}{T_F}}$$

Travel speed rate values are between 0 and 1. The value of 0 corresponds to the best condition, when the average speed is equal to the free-flow speed. The value 1 represents the worst condition, when average speed is near 0.

The very-low-speed rate is calculated on the basis of the percentage of time traveling at very low speed compared with the total travel time:

$$\text{Very low speed rate} = \frac{\sum_i S_i}{T_A}$$

The very-low-speed rate values are between 0 and 1. The value of 0 represents the best possible traffic condition, with no delay. The value of 1 represents the worst condition, with most of the travel time spent in delayed conditions. The delay is defined as the travel time at a speed of <5 mph.

The travel speed rate and the very-low-speed rate represent fuzzy (linguistic) input variables. They could be "Low," "Medium," "High," etc (Figs. 3.61 and 3.62). The congestion index values are between 0 and 1. The value of 0 represents the best possible traffic condition. The value of 1 represents the worst possible urban traffic congestion. The congestion index represent fuzzy (linguistic) output variable (Figs. 3.63).

EXAMPLE 3.20—cont'd

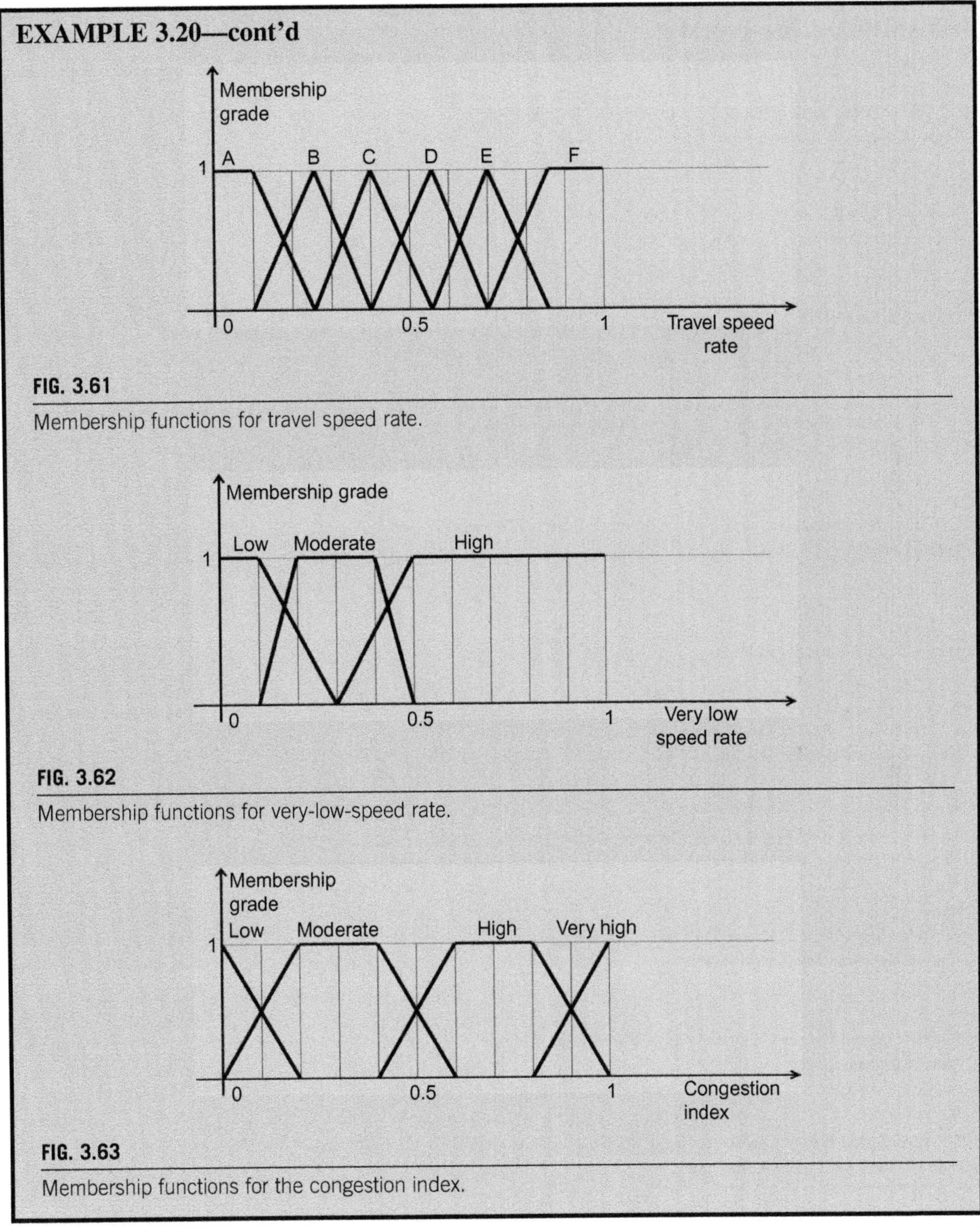

FIG. 3.61

Membership functions for travel speed rate.

FIG. 3.62

Membership functions for very-low-speed rate.

FIG. 3.63

Membership functions for the congestion index.

(Continued)

EXAMPLE 3.20—cont'd

Hamad and Kikuchi (2002) proposed the following approximate reasoning algorithm for calculating the congestion index:

If Travel Speed Rate is F and Very Low Speed Rate is High
Then Congestion Index is Very High

or
If Travel Speed Rate is F and Very Low Speed Rate is Moderate
Then Congestion Index is Very High

or
If Travel Speed Rate is F and Very Low Speed Rate is Low
Then Congestion Index is High

or
If Travel Speed Rate is E and Very Low Speed Rate is High
Then Congestion Index is Very High

or
If Travel Speed Rate is E and Very Low Speed Rate is Moderate
Then Congestion Index is High

or
If Travel Speed Rate is E and Very Low Speed Rate is Low
Then Congestion Index is High

or
If Travel Speed Rate is D and Very Low Speed Rate is High
Then Congestion Index is High

or
If Travel Speed Rate is D and Very Low Speed Rate is Moderate
Then Congestion Index is High

or
If Travel Speed Rate is D and Very Low Speed Rate is Low
Then Congestion Index is Moderate

or
If Travel Speed Rate is C and Very Low Speed Rate is High
Then Congestion Index is High

or
If Travel Speed Rate is C and Very Low Speed Rate is Moderate
Then Congestion Index is Moderate

EXAMPLE 3.20—cont'd

or
If Travel Speed Rate is C and Very Low Speed Rate is Low
Then Congestion Index is Moderate

or
If Travel Speed Rate is B and Very Low Speed Rate is High
Then Congestion Index is Moderate

or
If Travel Speed Rate is B and Very Low Speed Rate is Moderate
Then Congestion Index is Moderate

or
If Travel Speed Rate is B and Very Low Speed Rate is Low
Then Congestion Index is Low

or
If Travel Speed Rate is A and Very Low Speed Rate is High
Then Congestion Index is Moderate

or
If Travel Speed Rate is A and Very Low Speed Rate is Moderate
Then Congestion Index is Low

or
If Travel Speed Rate is A and Very Low Speed Rate is Low
Then Congestion Index is Low

The proposed approximate algorithm calculates congestion index value for the given Travel Speed Rate and Low Speed Rate values. The proposed approximate reasoning algorithm enables combining various traffic congestion measures.

3.10.4 ARTIFICIAL NEURAL NETWORKS

3.10.4.1 Introduction

People relatively easily perform a variety of complex tasks that are highly difficult to solve by computational techniques of traditional algorithms. The brain architecture largely differs from common serial computers. The researchers of artificial neural networks seek to enable computers to process data in a similar way like humans. Artificial neural networks are inspired by biology. They are composed of the elements that function similarly to a biological neuron. These elements are organized in a way that is reminiscent of the anatomy of a brain. In addition to this superficial similarity, artificial neural networks display a remarkable number of the brain's properties. For example, they are able to learn from experience, to apply to new cases generalizations derived from previous instances, and to abstract essential characteristics of input data that often contain irrelevant information (Hornik et al., 1989).

3.10.4.2 Biological neurons and artificial neurons

The human brain contains roughly 10^{11} neurons that apart from the characteristics that they share with other cells have unique capabilities to receive, process and transmit electrochemical signals to other neurons. Neurons communicate between themselves and are organized in the form of a neural network that is regarded as the brain's system of communication. Fig. 3.64 illustrates the structure of a biological neuron. Neurons consist of the cell body (*soma*) and several branches. The branches conducting information to the cell (*stimulus*) are called *dendrites*, and the branches conducting information out of the cell (*response*) are called *axons*. The emitted signals differ in frequencies, duration and amplitudes. The interaction between the neurons occurs at specific connection points called *synapses*.

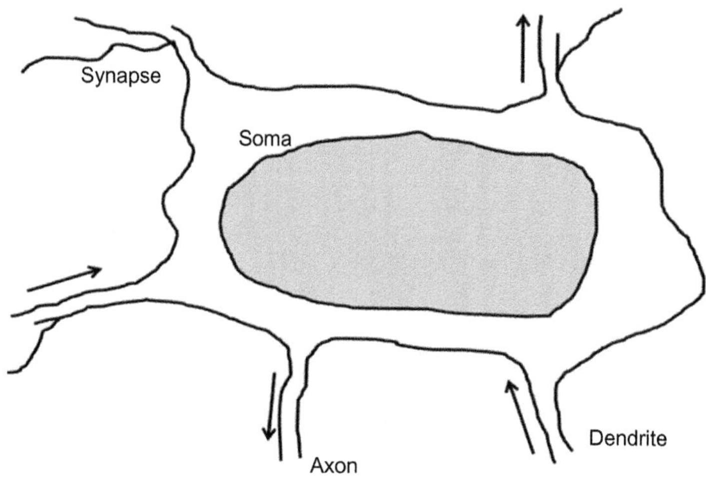

FIG. 3.64

Structure of a biological neuron.

Since the connections between neurons are of different strength, a stimulus is considered to be a weighted sum of the received inputs. If a stimulus exceeds a threshold, the neuron transmits the signals down the axon to other neurons. The brain contains over fifty different kinds of neurons relative to their shape and specialized functions. The brain is a complex communication system. The overall length of a network's branches approximates 10^{14} m. The presence of such a large number of connections establishes a high level of "massive parallelism." While a single neuron relatively slowly responds to outside stimuli, the brain as a whole solves complex problems in a remarkably short time—in a fraction of a second or in a couple of seconds. The brain can analyze complex problems and adequately cope with unforeseen situations due to the knowledge stored in synapses and the ability to adapt to new situations. The human brain can learn and generalize the acquired knowledge. The generalization of knowledge refers to similar stimuli inducing similar responses.

3.10.4.3 An artificial neuron

The first model of an artificial neuron that was proposed by McCulloch and Pitts (1943) was a binary device with a binary input, binary output, and fixed activation threshold.

The later model of an artificial neuron (Fig. 3.65) copied the properties of a biological neuron (Wasserman, 1989). The input signals $x_1, x_2, \ldots, x_n$, representing the output signals of other neurons, are multiplied by associated connection strengths, $w_1, w_2, \ldots, w_n$. Each connection strength (weight) corresponds to the strength of a biological synaptic connection. The output signal (*NET*) is equal to the weighted sum of input signals.

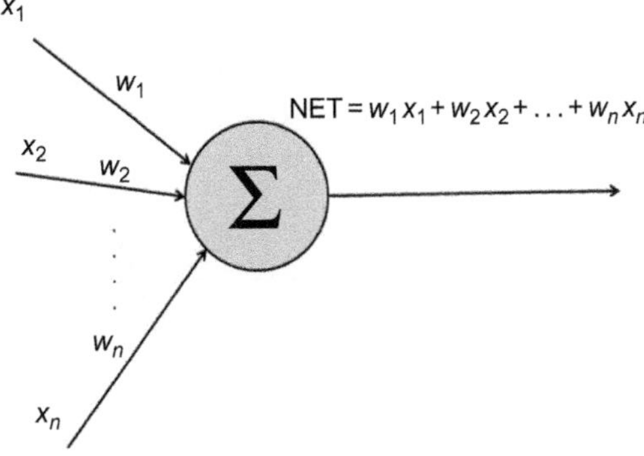

FIG. 3.65

Structure of an artificial neuron.

Fig. 3.66 represents an artificial neuron with an associated activation function that is, a typical processing element. The presented activation function is most commonly used: nonlinear, continuous, monotonously increasing, bounded, differentiable logistic function. The range of the weighted sum of

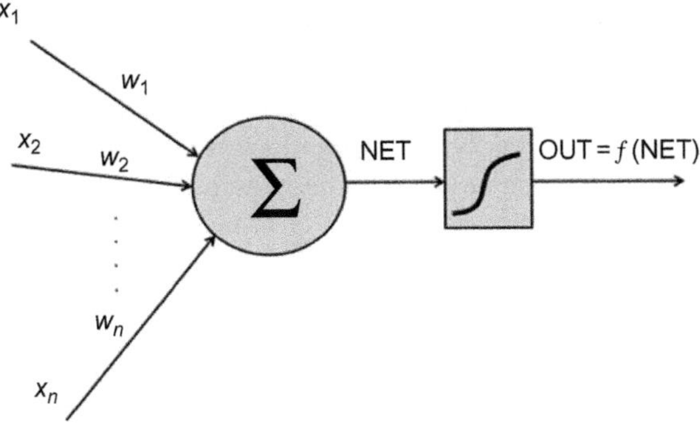

FIG. 3.66

Processing element: artificial neuron with activation function.

input signals, *NET*, is compressed by an "S" curve such that the value of the output signal, *OUT*, never exceeds a relatively low level regardless of the value of *NET*. The most commonly used activation functions are step function, sigmoid function, hypertangent function, and identity function (Fig. 3.67).

The transformation of input signals by a logistic curve enables the receiving and processing of very weak and very strong signals. When input signals are very strong (positive or negative), the curve slope is very small. When input signals are weak, the curve slope is large, which means that a usable output signal can be produced.

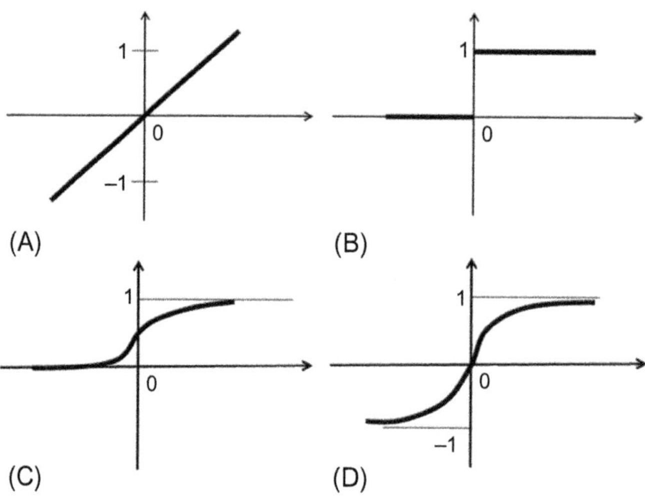

FIG. 3.67

The most commonly used activation functions: (A) identity function, (B) step function, (C) sigmoid function, and (D) hypertangent function.

3.10.4.4 Characteristics of neural networks

The present neural network architecture is based on a simplified model of the brain, the processing task being distributed over numerous neurons (nodes, units, or processing elements). Although a single neuron is able to perform simple data processing, the strength of a neural network is obtained as the result of the connectivity and collective behavior of these simple nodes.

Neural networks store the strength of synaptic connections (weights of the network's branches) by which the required data are reproduced. Neural networks can be trained by adjusting the connection strengths in order to abstract the relations between the presented input/output data.

A neural network is characterized by:

1. a set of processing elements;
2. connectivity of those elements;
3. the rule of signal propagation through the network;
4. activation or transfer functions; and
5. training algorithms (learning rules or learning algorithms).

A neural network configuration begins by defining a set of processing elements. Processing elements (nodes, neurons, units) are simple elements performing relatively simple signal processing in order to determine the output signal. In other words, each node receives input values from its nearest neighbors according to which it computes and transmits a single output value. A neural network is inherently parallel in the sense that a large number of processing elements make simultaneous computations. It can be said that a neural network performs parallel distributed data processing.

A neural network contains three types of nodes: input, output, and hidden. Input nodes receive input signals from outside sources that is, sources outside the network. Output nodes transmit signals that is, output values outside the network. Each node transmits signals of different strengths to its neighbors (the nodes to which it is connected).

Many of the currently used neural models have a fixed network structure. A network structure that is, the number of nodes and their connection types can also be very flexible. An artificial neural network is completely determined (capabilities of modeling, representation, and generalization) once we have established the network's structure, activation function of each node, and learning rule.

How a network's nodes are connected is of vital importance. The synaptic connectivity of a network can be total or partial (for example, only connections between different layers are allowed). Also, if, for example, a neural network is intended to support control of a traffic dispatcher, its structure can mimic the dispatcher's decision process (special connectivity). Each branch of the network is associated with a weight (positive or negative value) modifying the strength of a signal. The absolute value of a branch's weight represents the connection strength. In order to calculate the output signal of a node, the weighted sum of input signals is modified by an activation function.

3.10.4.5 A multilayered feedforward neural network

A multilayered feedforward neural network is a network in which the input signal extends forward through several layers, while it is being processed to estimate the network's output signal. Each layer contains a certain number of nodes. Each node is a processing element associated with the corresponding activation function by which the weighted sum of input values is transformed to determine the output value. To each node's input only the outputs of nodes from a previous layer are supplied and the output signal is transmitted to the nodes of the next layer (Fig. 3.68). Each node is associated with a weight vector by which the input vector is linearly transformed. The node connectivity is generally total except in some transformations when it is partial.

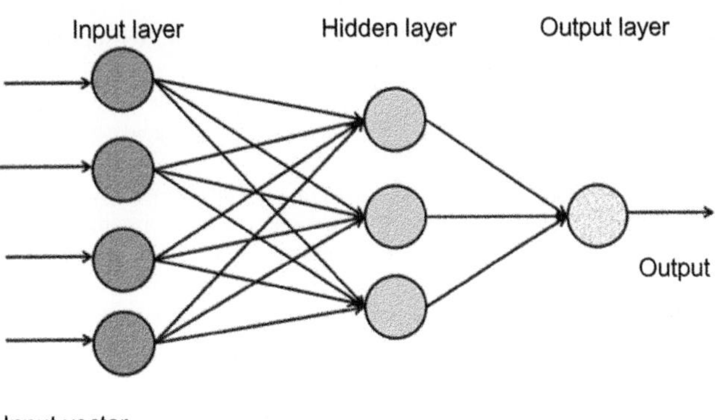

FIG. 3.68

Feedforward neural network.

3.10.4.6 Training of a neural network

Artificial neural networks are capable of modifying their behavior in response to the environment. When presented with a set of input data (most frequently with the desired output data) they self-adapt to incite appropriate responses. A wide range of training algorithms has been developed, each having its own advantages and disadvantages.

The chosen structure of a neural network model can influence the convergence rate of a training algorithm and even determine the type of learning to be used. Training algorithms are very simple mechanisms for adapting the weights of a network's branches, requiring for each branch only locally available data. Their implementation generally does not involve complex computations and consequently no powerful computation configurations are needed.

3.10.4.7 Validation of a neural model

The most important part of any model design procedure is validation of the given model. The tests are performed to estimate the extent to which the network has learnt the training data and how able it is to generalize (interpolate and extrapolate) to unforeseen cases. Most training algorithms are able to successfully learn a set of training data. The performance of the trained network must be fully understood; a network cannot simply be regarded as a black box.

There are a number of ways in which the performance of a trained network can be assessed, the simplest one is the assessment of the network's performance in reproducing the training data. A better approach is to divide the available data into a training set and a test set and to use the testing data to evaluate the final model. In other words, any set of data presented to a learning algorithm must be split into a part that is used to train the network and another part used for estimating the network's performance by testing its ability to generalize correctly. One heuristic that works well in practice is to use two-thirds for training and one-third for testing. This approach is reasonable since a learning algorithm cannot perform well if there are insufficient training data, but the output results cannot be assessed accurately if the test set is too small.

EXAMPLE 3.21

The transportation demand between two regions depends on socioeconomic characteristics of the given regions, as well as the characteristics of the connecting transportation system. (Chapter 8 will cover in more detail travel demand analysis). The output data of transportation demand models usually consist of an estimated number of the daily vehicle miles of travel (in individual road traffic), number of miles of travel (in air, railway and international coach travel), used telephone impulses (in communication traffic), and so on.

The input data for planning transportation networks and transportation facilities and for building new and reconstructing the existing traffic and transportation facilities include the data characterizing transportation demand. The transportation demand forecasting is of outstanding importance for the functioning of different traffic and transportation systems, as well as their further development.

Chin et al. (1992) used a multilayered, feedforward neural network for the travel demand forecasting. Their model is based on socioeconomic and highway operation data for 339 urban areas in the United States from 1982 to 1988. One of the neural networks proposed by Chin et al. (1992) is presented in Fig. 3.64.

EXAMPLE 3.21—cont'd

As seen from Fig. 3.69, the authors used a three-layered neural network. Between input and hidden layers there is a full connectivity. The authors used the proposed network to estimate the daily vehicle miles of travel (DVMT). The data supplied to the input layer are traffic-generation-related, traffic-attraction-related, and network-supply-related. As Chin et al. (1992) noted, "to account for the three above-mentioned explicit relationships as well as two unknown potential relationships, a total of five nodes are used in the hidden layer." As already pointed out, the author experimented with several different neural networks. These neural networks differed in the number of hidden layers, as well as the number of nodes in the hidden layers. The authors demonstrated that the average absolute relative errors between the neural network results and real data are within 4.77% and 7.09%, which can certainly be considered satisfactory.

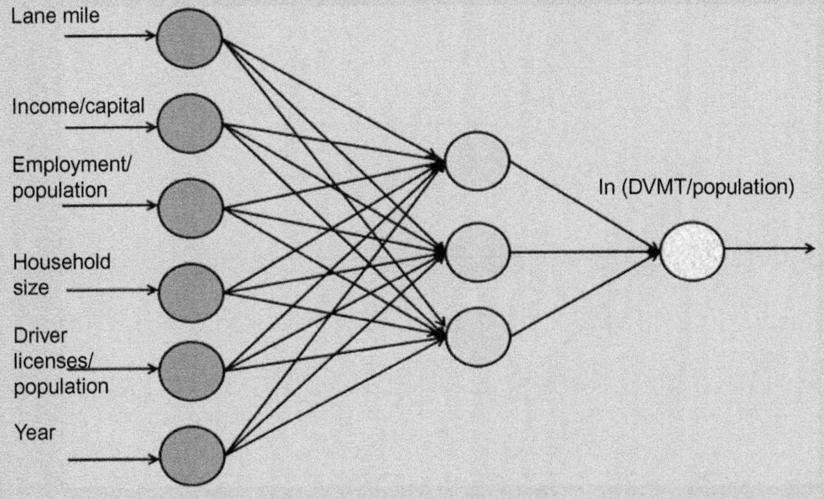

FIG. 3.69

Neural network daily vehicle miles of travel (DVMT) forecasting model.

3.11 PROBLEMS

1. Using Dijkstra's algorithm, find the shortest paths from node a to all other nodes in the transportation network shown in Fig. 3.70.
2. Using Dijkstra's algorithm, find the shortest paths from the node 1 to all other nodes in the transportation network shown in Fig. 3.71.
3. Consider the traffic network shown in the previous example. The travel demands between node a and all other nodes are given in Table 3.14.

   Assume that the travelers always travel along the shortest path. Calculate the link flows.

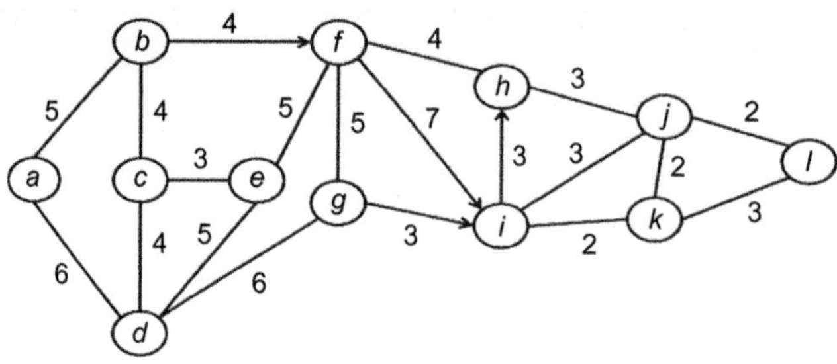

FIG. 3.70

Network in which shortest paths from node *a* to all other nodes should be discovered by Dijkstra's algorithm.

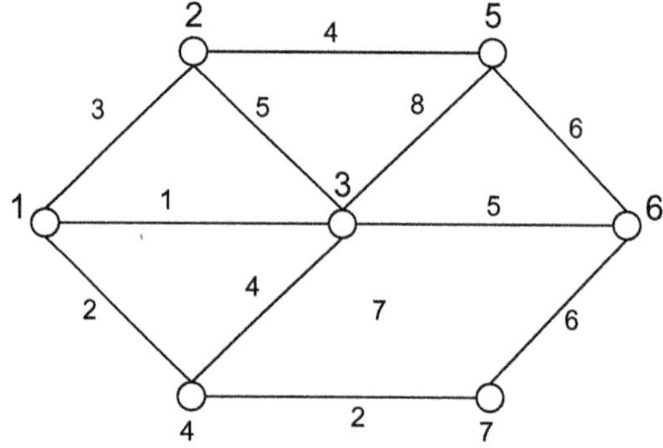

FIG. 3.71

Network in which shortest paths from node 1 to all other nodes should be discovered by Dijkstra's algorithm.

Table 3.14 The Travel Demands Between Node *a* and All Other Nodes

Pair of Nodes	Travel Demand
(*a*,*b*)	100
(*a*,*c*)	50
(*a*,*d*)	30
(*a*,*e*)	120
(*a*,*f*)	20
(*a*,*g*)	20
(*a*,*h*)	40
(*a*,*i*)	130
(*a*,*j*)	50
(*a*,*k*)	60
(*a*,*l*)	180

4. By using Floyd's algorithm find the shortest paths between all pairs of nodes in the transportation network shown in Fig. 3.72.

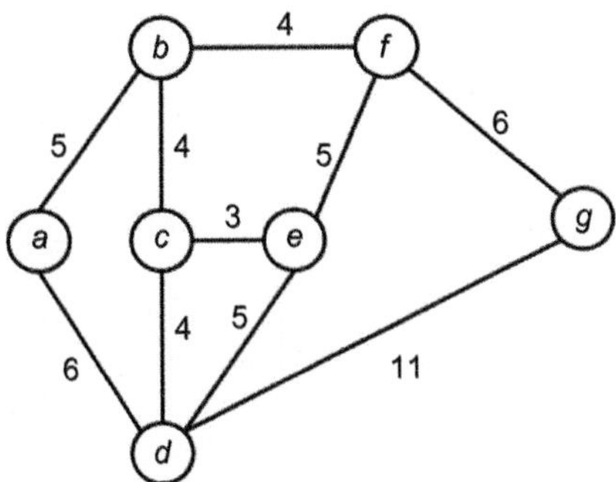

FIG. 3.72

Network in which the shortest paths between all pairs of nodes should be discovered by Floyd's algorithm.

5. Compute the shortest paths and path lengths between all pairs of nodes in the network shown in Fig. 3.73 using Floyd's algorithm.

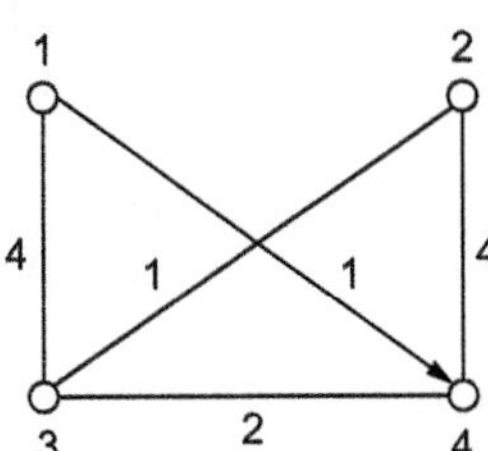

FIG. 3.73

Network in which the shortest paths between all pairs of nodes should be discovered by Floyd's algorithm.

6. In the case of the network shown in Fig. 3.74 determine the shortest path between node 1, and node 4 and between node 2, and node 4 using Floyd's algorithm.

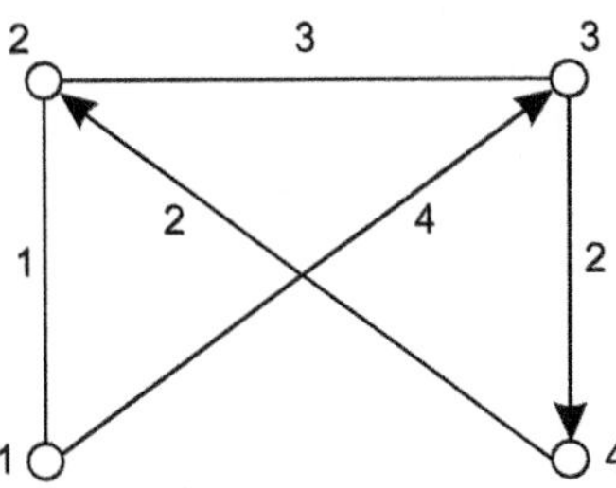

FIG. 3.74

Network in which the shortest paths between node 1, and node 4 and between node 2, and node 4 should be discovered by Floyd's algorithm.

7. The coal mines CM_1 and CM_2 deliver coal to the coal trade depots A, B, and C. The coal mine CM_1 is capable to deliver 500 tons of coal to these three coal trade depots. The coal mine CM_2 can deliver 800 tons. The demands at the coal trade depots A, B, and C are respectively equal to 500, 400, and 400 tons. The transportation costs (in monetary units per one ton of coal) are given in Table 3.15:

Table 3.15 Transportation Costs (in Monetary Units Per One Ton of Coal)

	A	*B*	*C*
CM_1	8	5	5
CM_2	4	6	8

Generate the optimal plan of coal transportation (calculate quantities of coal to be transported between coal mines and coal trade depots that will generate minimal total transportation costs).

8. Vehicle types in traffic flow are divided into the following categories: cars (C), SUVs (S), vans (V), trucks (T), buses (B), and emergency vehicles (E). The observer recorded vehicles that appeared in the 5-min interval. Explain (in words) the following events: SVT, EBT, CCC, V, SVC, C.
9. The outcome of the traffic experiment is represented by the number of left-turning cars in the left-turn bay. The capacity of the left-turn bay equals 10 vehicles. Define the associated sample space in terms of all outcomes.
10. Traffic engineers studied the violators in the HOV lane. It was discovered 250 violators among the 10,000 observed cars. Out of these 250 violators 190 were younger than 25 years. Calculate the following probabilities:
    - **a.** The probability the driver in traffic flow will be traffic violator.
    - **b.** The probability that the traffic violator is younger than 25 years.
    - **c.** The probability that the traffic violator is older than 25 years.
11. There are three intersections on the driver's chosen path. The intersections are not coordinated. The probabilities of the red lights on these intersections are respectively equal 0.3, 0.4, and 0.5. Calculate the following probabilities:

    The driver will pass through the intersections with no delay.

    The driver will be stopped by at least one traffic light.

    The driver will be stopped by two traffic lights.

    The driver will be stopped by at all three intersections.
12. The river divides city into two parts. There are two bridges on the river (Fig. 3.75).

    The probability that during any winter month the severe traffic accident happen that completely block the Bridge 1 equals 0.001. In the case when Bridge 1 fails, the traffic load on the bridge 2 is unusually high. In such a situation, there is the probability equal to 0.004 that the bridge 2 will also fail. Calculate the probability that there will be no traffic between two parts of the city.

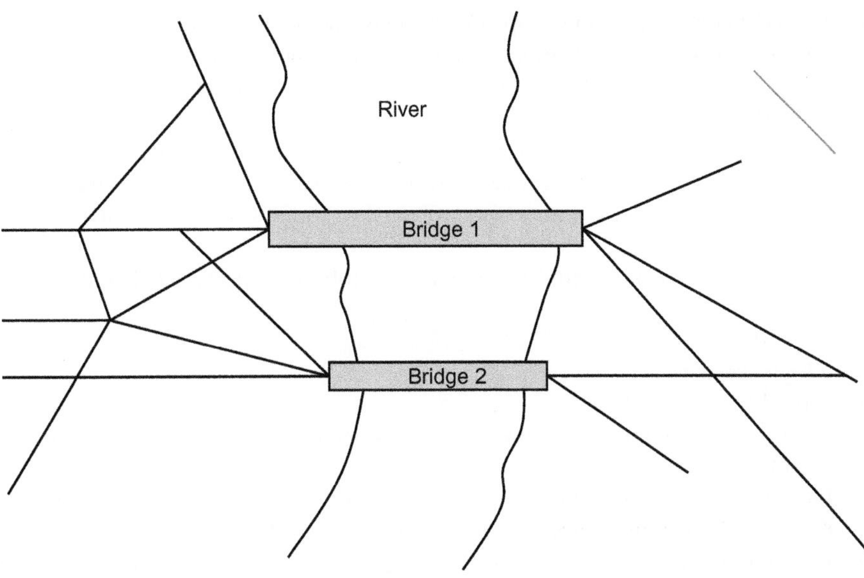

FIG. 3.75

Two bridges on the river.

13. Check if the following function satisfies the conditions to be the probability density function.

$$f(x) = \begin{cases} a, & \text{for } 0 \leq x \leq 5 \\ 0, & \text{otherwise} \end{cases}$$

Calculate the value of the parameter a. Show the function $f(x)$ graphically.

14. Consider the following probability density function:

$$f(x) = \begin{cases} \frac{1}{5}, & \text{for } 0 \leq x \leq 5 \\ 0, & \text{otherwise} \end{cases}$$

Define the cumulative density function (CDF) and show the CDF graphically. What is the probability that the random variable X will be in the range $2 \leq x \leq 4$?

15. The cumulative density function reads:

$$F(x) = \begin{cases} 1 - e^{-3x}, & x \geq 0 \\ 0, & \text{otherwise} \end{cases}$$

Define the probability density function. What is the probability that the random variable X will be in the range $3 \leq x \leq 6$?

16. Drivers pay the highway toll in cash, or by credit card. On the average, toll booth attendant needs 20 s to collect the money from the driver. Average vehicles arrival rate equals 240(veh/h). Treat toll booth as $M/M/1$ queueing system and calculate:

the average number of vehicles in the queue;

the average waiting time a client spends in the queue; and
the average waiting time a client spends in the queueing system.
What will be the benefits of introducing the electronic pay system that will need 3 s per driver to pay the toll?

17. Vehicles arrive at the specific point at the highway according to the Poisson Process. Let us assume that $=0.2(\text{veh/s})$. Determine by simulation the time intervals between six successive vehicle arrivals. Determine the probability that 20 vehicles will arrive within 2-min time interval.

18. Develop a reasonable membership functions for the following fuzzy sets based on the speed measured in km/h:
a. "high speed";
b. "very high speed";
c. "low speed."

19. Describe the linguistic term "approximately 10 bus departures per day" by reasonable membership function.

20. Develop a reasonable membership functions for the fuzzy sets " number of vehicles in a queue is ≥ 10" and "number of vehicles in a queue is about 10."

21. If the speed limit in an urban area is equal to 50 km/h, determine the membership functions of the fuzzy sets "permitted speed" and "nonpermitted speed."

22. Consider the following fuzzy sets:

$$A=\frac{1}{2}+\frac{0.6}{3}+\frac{0.3}{4}+\frac{0.7}{5}+\frac{0.2}{6} \text{ and } B=\frac{0.4}{2}+\frac{0.2}{3}+\frac{0.6}{4}+\frac{0.8}{5}+\frac{0.3}{6}$$

For these two fuzzy sets find the following:
(a) Union; (b) Intersection.

23. Propose the fuzzy logic system that will control a ramp at the entrance to the highway. The possible input variables are: density of traffic flow and speed of vehicles on the highway. The possible output variable is the number of vehicles that will be the acceptable to enter the highway in a time unit.

REFERENCES

Busacker, R.G., Saaty, T.L., 1965. Finite Graphs and Networks: An Introduction with Applications. McGraw Hill, New York, NY.

Charnes, A., Cooper, W.W., 1985. Preface to topics in data envelopment analysis. Ann. Oper. Res. 2, 59–94.

Charnes, A., Cooper, W.W., Rhodes, E., 1978. Measuring the efficiency of decision making units. Eur. J. Oper. Res. 2, 429–444.

Charnes, A., Cooper, W.W., Rhodes, E., 1979. Short communication; measuring the efficiency of decision making units. Eur. J. Oper. Res. 3, 339.

Charnes, A., Cooper, W.W., Rhodes, E., 1981. Evaluating program and managerial efficiency: an application of data envelopment analysis to program follow through. Manag. Sci. 27, 668–697.

Chen, S.-J., Hwang, C.-L., 1992. Fuzzy Multiple Attribute Decision Making. Springer-Verlag, Berlin, Heidelberg.

Chin, S.M., Hwang, H.L., Miaou, S.P., 1992. Transportation demand forecasting with a computer simulated neural network model. In: Ritchie, S.G., Hendrickson, C. (Eds.), Proceedings of the International Conference on

Artificial Intelligence Applications in Transportation Engineering. Institute of Transportation Studies, University of California, Irvine, San Buenaventura, CA.

Dantzig, G.B., 1960. On the shortest route problem through a network. Manag. Sci. 6, 187–190.

Deo, N., Pang, C., 1984. Shortest path algorithms: taxonomy and annotation. Networks 14, 275–323.

Dial, R., Glover, F., Karney, D., Klingman, D., 1979. A computational analysis of alternative algorithms and labeling techniques for finding shortest path trees. Networks 9, 215–248.

Dijkstra, E.W., 1959. A note on two problems in connection with graphs. Numer. Math. 1, 269–271.

Floyd, R.W., 1962. Algorithm 97—shortest path. Commun. ACM 5, 345.

Ford Jr., L.R., Fulkerson, D.R., 1962. Flows in Networks. Princeton University Press, Princeton, NJ.

Gallo, G.S., Pallottino, S., 1988. Shortest path algorithms. Ann. Oper. Res. 7, 3–79.

Glover, F., Glover, R., Klingman, D., 1984. Computational study of an improved shortest path algorithm. Networks 14, 25–36.

Glover, F., Klingman, D., Phillips, N., Schneider, R.F., 1985. New polynomial shortest path algorithms and their computational attributes. Manag. Sci. 31, 1106–1128.

Goldberg, D., 1989. Genetic Algorithms in Search, Optimization, and Machine Learning. Addison Wesley Publishing Company Inc., New York, NY

Golden, B., 1976. Shortest path algorithms: a comparison. Oper. Res. 24, 1164–1168.

Hamad, K., Kikuchi, S., 2002. Developing a measure of traffic congestion: fuzzy inference approach. Transp. Res. Rec. J. Transp. Res. Board 1802, 77–85.

Hillier, F., Lieberman, G., 1990. Introduction to Operations Research. McGraw Hill Publishing Company, New York, NY.

Holland, J., 1975. Adaptation in Natural and Artificial Systems. University of Michigan Press, Ann Arbor, MI.

Hornik, K., Stinchcombe, M., White, H., 1989. Multilayer feedforward networks are universal approximators. Neural Netw. 2, 359–366.

Hwang, C.L., Yoon, K., 1981. Multiple Attribute Decision Making. Springer-Verlag, Berlin.

Larson, R., Odoni, A., 1981. Urban Operations Research. Prentice Hall, Englewood Cliffs, NJ.

McCulloch, W.S., Pitts, W., 1943. A logical calculus of the ideas immanent in nervous activity. Bull. Math. Biophys. 5, 115–133.

Mendel, J.M., 1995. Fuzzy logic systems for engineering: a tutorial. Proc. IEEE 83, 345–377.

Minty, G.J., 1957. A comment on the shortest route problem. Oper. Res. 5, 724.

Newell, G.F., 1980. Traffic Flow on Transportation Networks. MIT Press, Cambridge, MA.

Newell, G.F., 1982. Applications of Queueing Theory, second ed. Chapman and Hall, London.

Pallottino, S., 1984. Shortest-path methods: complexity, inter-relations and new propositions. Networks 14, 257–267.

Pappis, C., Mamdani, E., 1977. A fuzzy controller for a traffic junction. IEEE Trans. Syst. Man Cybern. ASMC 7, 707–717.

Roy, B., Vincke, P., 1981. Multicriteria analysis: survey and new directions. Eur. J. Oper. Res. 8, 207–218.

Self, K., 1990. Designing with fuzzy logic. IEEE Spectr. 105, 42–44.

Taha, H., 1982. Operations Research. MacMillan Publishing Co., Inc., New York, NY.

Teodorović, D., 1999. Fuzzy logic systems for transportation engineering: the state of the art. Transp. Res. 33A, 337–364.

Teodorović, D., Vukadinović, K., 1998. Traffic Control and Transport Planning: A Fuzzy Sets and Neural Networks Approach. Kluwer Academic Publishers, Boston, MA/Dordrecht/London.

Wang, L.X., Mendel, J., 1992. Generating fuzzy rules by learning from examples. IEEE Trans. Syst. Man Cybern. 22, 1414–1427.

Wasserrnan, P.D., 1989. Neural Computing: Theory and Practice. Van Nostrand Reinhold, New York, NY.

Winston, W., 1994. Operations Research. Duxbury Press, Belmont, CA.

Zadeh, L., 1965. Fuzzy sets. Inf. Control. 8, 338–353.
Zadeh, L., 1972. A fuzzy-set-theoretic interpretation of linguistic hedges. J. Cybern. 2, 4–34.
Zadeh, L., 1973. Outline of a new approach to the analysis of complex systems and decision processes. IEEE Trans. Syst. Man Cybern SMC 3, 28–44.
Zadeh, L., 1975a. The concept of a linguistic variable and its application to approximate reasoning: I. Inf. Sci. 8, 199–249.
Zadeh, L., 1975b. The concept of a linguistic variable and its application to approximate reasoning: II. Inf. Sci. 8, 301–357.
Zadeh, L., 1996. Fuzzy logic = computing with words. IEEE Trans. Fuzzy Syst. 4, 103–111.
Zimmermann, H.J., 1991. Fuzzy Set Theory and Its Applications. Kluwer, Boston, MA.

What are traffic flow, speed, and density? Who was Bruce Greenshields? How do we measure the basic flow variables? What are headways? Is the distribution of headways Exponential, Normal, or Pearson Type III distribution? What is Annual Average Daily Traffic (AADT)? What is a fundamental diagram of traffic flow? Why do shock waves appear on a highway? What is a Network Flow Diagram (NFD)?

TRAFFIC FLOW THEORY 4

4.1 TRAFFIC FLOW PHENOMENON

Thousands of drivers and passengers that move from one place to another form various traffic flows. Flows of cars on streets and highways, flows of bicycles on streets, flows of peoples in metro stations and shopping malls, flows of pedestrians on pedestrian crossings, and flows of aircraft on airport's taxiways vary over the time of day, day of the week, and month of the year. Traffic flows are essentially different from flows in other engineering areas. Traffic flows appear as the consequences of human decisions, and they are determined by rules of human behavior. Rate flows (which we shall call simply "flows") are expressed in the number of units per unit of time. Units could be cars, aircraft, pedestrians, vessels, or containers. The unit of time could be a minute, hour, day, month, or year. Traffic engineers frequently use units like vehicles per hour, vehicles per day, passengers per hour, aircraft per 15 min, etc. Traffic phenomena are very complex. Researchers from the Los Alamos National Laboratory, United States made the following statement after 4 years of work developing the traffic and transportation planning software package called TRANSINMS: "Modeling traffic phenomena has proven to be more difficult than predicting and modeling subatomic level reactions inside the atom-for nuclear warhead simulations."

Flow measurements and modeling of traffic phenomena are required in order to estimate the capacities of transportation facilities, as well as to make appropriate decisions related to the further expansion and development of the transportation facilities.

Flow measurements and analysis facilitate prediction of future queueing at the highway, and understanding of traffic flow propagation in space and time. Flow measurements also help us to estimate the queue lengths, level-of-service, and to perform actions that will mitigate traffic congestion.

Traffic flow theory (Greenshields, 1935; Drew, 1968; Gerlough and Huber, 1975; Herman and Prigogine, 1979; Herman and Ardekani, 1984; Williams et al., 1987; May, 1990; Ross, 1991; Kerner,

Transportation Engineering. http://dx.doi.org/10.1016/B978-0-12-803818-5.00004-4

2009; Elefteriadou, 2014) describes the essential characteristics of traffic flows (flow, speed, and density). Traffic flows can be described using various parameters. One possibility for studying traffic flows is to study the behavior of *individual* vehicles (speed changes, headway changes, etc.) over time. More frequently, we are interested in *macroscopic* approach that assumes description of traffic flow conditions by various aggregate parameters like mean traffic speed, mean headway, traffic density, etc.

The majority of traffic and transportation activities are characterized by uncertainty related to time of occurrence. It is practically impossible to predict exactly the time of occurrence of any particular traffic activity. Traffic phenomena are stochastic processes (described by random variables).

Traffic flow theory has its roots in 1933, when Bruce Greenshields mathematically described traffic flow characteristics based on his pioneering studies and observations of flows on American highways. Greenshields measured traffic flow, traffic density, and speed using photographic measurement methods for the first time.

4.2 MEASUREMENTS OF THE BASIC FLOW VARIABLES

Traffic flows fluctuate over space and over time. Traffic flows changes over time can be determined by collecting appropriate statistical data. The flow (vehicles per unit time), speed (distance per unit time), and density (vehicles per unit distance) are the basic flow variables that are measured by the traffic engineers (Greenshields et al., 1933; Greenshields, 1935; Kühne, 2011). Traffic engineers also measure the occupancy that represents the percent of time a specific point on the highway is occupied by vehicles. Frequently, within traffic studies and projects, we analyze the distribution of the time headway between vehicles.

Measurements of the traffic flow variables could be performed at a specific highway point, over length of a highway, or simultaneously from a number of vehicles in a wide area.

Inductive loop detector technology is most frequently used when performing measurement at a specific point on a highway. This technology started to be used for detecting traffic in the 1960s. The electronics are buried into the highway. When a moving vehicle reaches the loop, the vehicle metal disrupts the magnetic field over the loop. As a consequence of this disturbance, the loop inductance is changed (Fig. 4.1).

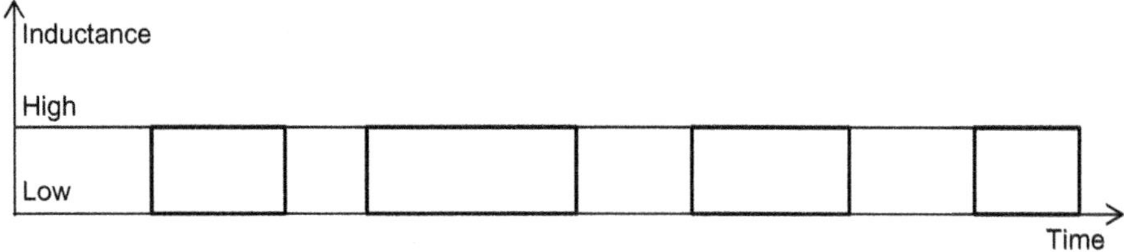

FIG. 4.1

The loop inductance changes over time.

This change is recorded. Single loops measure the occupancy and the traffic volume (Fig. 4.2).

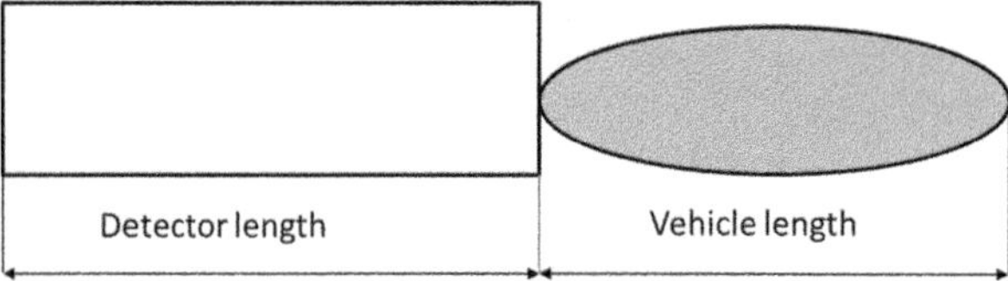

FIG. 4.2

Speed estimation by the loop detector.

When a vehicle comes into the detection zone, the loop inductance is changed. The inductance stays changed until the vehicle leave the detection zone. The time interval when the detector is activated is equal to the time necessary for the vehicle to travel a distance equal to the detector length plus the distance equal to its length. The loop detector can estimate speed in the following way (Fig. 4.2).

$$\text{Speed} = \frac{\text{Detector length} + \text{Vehicle length}}{\text{Occupancy time}} \tag{4.1}$$

4.3 VEHICLE HEADWAYS AND FLOW

Let us consider a specific point at the highway (Fig. 4.3).

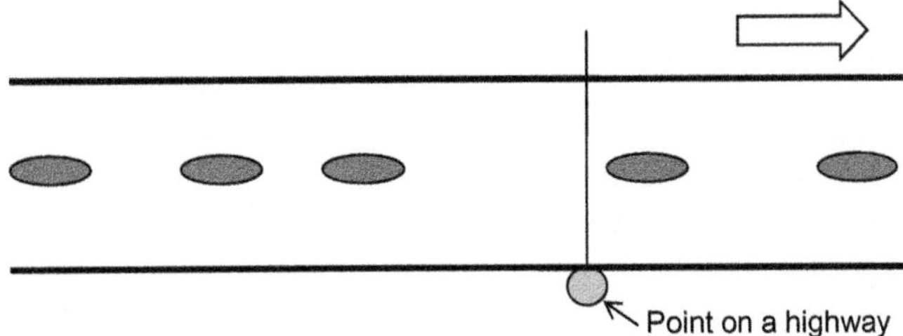

FIG. 4.3

Flow measurement.

In order to estimate the flow, we must measure (count) vehicles over time. We denote by N the total number of vehicles counted during the time period T. Flow q (veh/h) is defined as

$$q = \frac{N}{T} \tag{4.2}$$

As we can see, the flow q represents the number of vehicles that pass specific point during the time period T. Traffic flow values vary over time. The flow is expressed in vehicles per unit time. Transportation engineers frequently use annual average daily traffic (AADT) in the transportation analysis. The AADT practically represents the total number of vehicles that passed the specific point in a year divided by 365. The period of observation and measurements is often less than 1 year. In this case, the average daily traffic (ADT), or average weekly traffic are the basic traffic flow indicators.

Peak hour volume gives us the information related to the highest hourly traffic volume through a day.

The distance (space headway) between following vehicles (FVs) in a traffic flow is measured from front bumper to front bumper (Fig. 4.4).

FIG. 4.4

The distance (space headway) between following vehicles.

Headway represents the time (in seconds) between successive vehicles, as their front bumpers pass a specific point. We denote respectively by $h_1, h_2, h_3, \ldots$ first, second, third,...headway. The total counting time T actually represents the sum of the recorded headways, ie,

$$T = \sum_{i=1}^{N} h_i \tag{4.3}$$

We can substitute an expression for T in Eq. (4.3). We get

$$q = \frac{N}{T} = \frac{N}{\sum_{i=1}^{N} h_i} = \frac{1}{\frac{1}{N}\sum_{i=1}^{N} h_i} \tag{4.4}$$

$$q = \frac{1}{\bar{h}} \tag{4.5}$$

where $\bar{h}$ represents average headway.

We conclude from Eq. (4.5) that the flow has reciprocal relationship with the average headway. The shorter the average headway, the higher the flow, and vice versa. Some characteristic flow and average headway values are shown in Table 4.1.

Table 4.1 Some Characteristic Flow and Average Headway Values

Flow (veh/h)	Average Headway (s)
1800	2
900	4
360	10
180	20

4.4 POISSON DISTRIBUTION OF THE NUMBER OF ARRIVALS AND THE EXPONENTIAL DISTRIBUTION OF HEADWAYS

Depending on the traffic flow conditions, various distributions could be used to describe headways. Practically, there is no interaction between the arrivals of any two vehicles in the case of low traffic volumes. In this case, vehicles randomly show up and pass. The time interval between the appearances of successive vehicles (headway) could be, for example, 5, 10, 11, 14 s, etc. In other words, the time interval between vehicle arrivals is a random variable that frequently has *exponential* distribution (Fig. 4.5).

The probability density function in the case of exponential distribution equals

$$f(x) = \lambda e^{-\lambda x} \tag{4.6}$$

where $\lambda > 0$ is the average arrival rate.

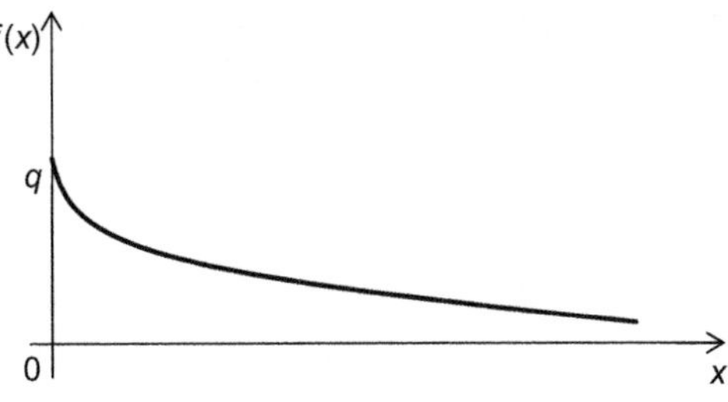

FIG. 4.5

Exponential distribution.

The average arrival rate λ is expressed in vehicles per second. Since flow q is expressed in vehicles per hour, we can write the following:

$$\lambda = \frac{q}{3600} \tag{4.7}$$

The exponential distribution can be derived from the distribution of the number of vehicles that appear during specified time interval. If the distribution of the number of vehicles that appear during specified time interval is *Poisson* distribution, the *exponential* random variable will represent the time between two successive vehicles. Poisson process is characterized by the following four postulates:

(1) The probability that at least one vehicle arrives during a small time interval Δt is approximately equal to $\lambda\ \Delta t$. (For example, when flow of vehicles q equals 360 veh/h, the probability that at least one vehicle arrives during a small time interval $\Delta t = 3$ s equals (360 veh/3600 s) $\times$ 3 s $= 0.3$.)
(2) The number of vehicle arrivals in any prespecified time interval does not depend on the starting time point of the interval. The number of vehicle arrivals also does not depend on the total number of vehicle arrivals observed prior to the interval.
(3) The numbers of vehicle arrivals in disjoint time intervals are mutually independent random variables. (The number of vehicle arrivals between 9:00 am and 9:05 am does not depend, for example, on the number of vehicle arrivals between 10:05 am and 10:10 am.)
(4) Two or more vehicle arrivals cannot happen simultaneously.

The probability $P(k)$ that the total number of vehicle arrivals happening in a time interval of the length t is equal to k is calculated in the case of *Poisson* process as:

$$P(k) = \frac{(\lambda t)^k e^{-\lambda t}}{k!} \tag{4.8}$$

Let us consider the case when no vehicles arrive in a time interval t. The probability of this event equals

$$P(0) = \frac{(\lambda t)^0 e^{-\lambda t}}{0!} = e^{-\lambda t} \tag{4.9}$$

We conclude that if no vehicles arrive in a time interval t, the headway h is equal to or greater than t, ie,

$$P(0) = P(h \geq t) = e^{-\lambda t} \tag{4.10}$$

$$1 - P(h < t) = e^{-\lambda t} \tag{4.11}$$

$$P(h < t) = 1 - e^{-\lambda t} \tag{4.12}$$

$$F(t) = 1 - e^{-\lambda t} \tag{4.13}$$

$$f(t) = \frac{d}{dt}F(t) = \frac{d}{dt}(1 - e^{-\lambda t}) = \lambda \cdot e^{-\lambda t} \tag{4.14}$$

In this way, we proved that the *exponential* random variable represents the time between two successive vehicles in the case when the distribution of the number of vehicles that appear during specified time interval is *Poisson* distribution.

4.5 NORMAL DISTRIBUTION AND PEARSON TYPE III DISTRIBUTION OF HEADWAY

The higher the traffic flow value, the more interactions there are between vehicles in traffic flow. In the case of medium traffic volumes, some vehicles go freely, while some other vehicles have interactions with other vehicles in the traffic flow. It has been shown that the Pearson Type III distribution could successfully describe headway distribution in such cases.

There is a very high interaction among the vehicles in the traffic flow in the cases of high traffic volumes. All headway values in such a situation are similar. In other words, the mean and the standard deviation of the headways are very low. It has been shown that the normal distribution (Fig. 4.6) could in an appropriate way describe the headway distribution in the case of relatively high traffic volumes. The probability density function in the case of normal distribution equals

$$f(x) = \frac{1}{\sigma \cdot \sqrt{2\pi}} e^{-\frac{(x-\mu)^2}{2\sigma^2}} \tag{4.15}$$

where μ and σ are given parameters.

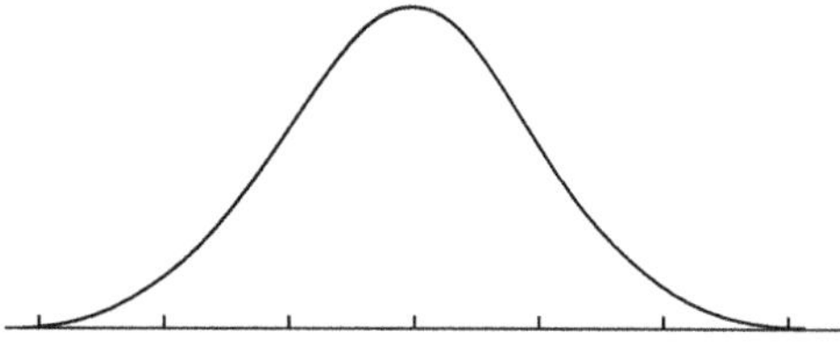

FIG. 4.6

Normal distribution.

In the case when traffic flow value comes close to capacity, all vehicles are interacting, and all headways are (more or less) constant.

4.5.1 SPEEDS

Imagine that we also recorded speeds of all vehicles that we observed and counted at our point on the highway (Fig. 4.7).

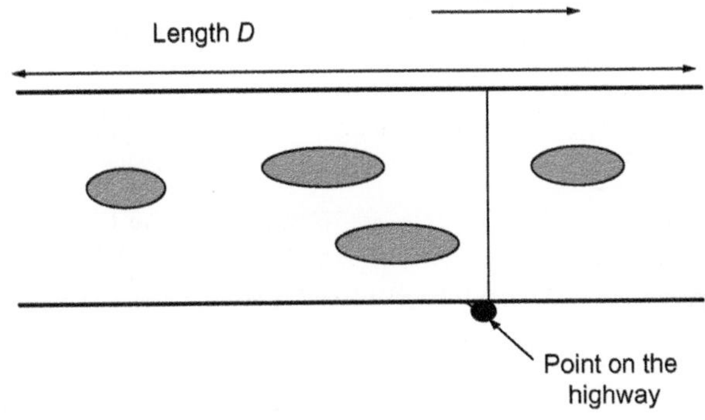

FIG. 4.7

Calculation of the time-mean speed and space-mean speed.

We can record speeds with certain level of accuracy using various equipment (radar, microwave, inductive loops). There are two ways to describe speeds on a highway: (a) the time-mean speed and (b) the space-mean speed.

The time-mean speed $\bar{u}_t$ is defined in the following way:

$$\bar{u}_t = \frac{1}{N}\sum_{i=1}^{N} u_i \tag{4.16}$$

where u_i represents recorded speed of the ith vehicle.

We see that the time-mean speed can be calculated by calculating the arithmetic time mean speed. The space-mean speed is the average speed that has been used in the majority of traffic models. Let us note the section of the highway whose length equals D. We denote by t_i the time needed by the ith vehicle to travel along this highway section. The space-mean speed $\bar{u}_s$ is defined in the following way:

$$\bar{u}_s = \frac{D}{\frac{1}{N}\sum_{i=1}^{N} t_i} = \frac{D}{\bar{t}} \tag{4.17}$$

The expression $\frac{1}{N}\sum_{i=1}^{N} t_i$ represents average travel time $\bar{t}$ of the vehicles traveling along the observed highway section.

EXAMPLE 4.1

Measurement points are located at the beginning and at the end of the highway section whose length equals 1 km (Fig. 4.8). The recorded speeds and travel times are shown in Table 4.2.

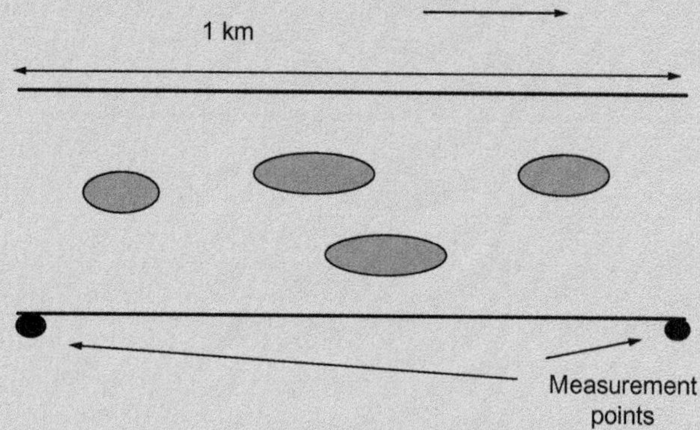

FIG. 4.8

Measurement points.

Table 4.2 Recorded Speeds and Travel Times

Vehicle Number	Speed at Point *A* (km/h)	Travel Time Between Point *A* and Point *B* (s)
1	80	45
2	75	50
3	62	56
4	90	39
5	70	53

Speeds of the five vehicles are recorded at the beginning of the section (point *A*). The vehicle appearance at points *A* and point *B* were also recorded. Calculate the time-mean speed and the space-mean speed.

Solution

The time-mean speed $\bar{u}_t$ at point *A* is

$$\bar{u}_t = \frac{1}{N}\sum_{i=1}^{N} u_i$$

$$\bar{u}_t = \frac{1}{5}(80+75+62+90+70) = \frac{1}{5}(377) = 75.4\,(\text{km/h})$$

The mean-space speed represents measure of the average traffic speed along the observed highway section. The mean-space speed is

$$\bar{u}_s = \frac{D}{\frac{1}{N}\sum_{i=1}^{N} t_i}$$

EXAMPLE 4.1—cont'd

The total travel time of all five vehicles equals

$$45+50+56+39+53=342\,(s)=\frac{243}{3600}\,(\text{h})=0.0675\,(\text{h})$$

The mean-space speed equals

$$\bar{u}_s=\frac{1}{\frac{1}{5}(0.0675)}\,(\text{km/h})=74.07\,(\text{km/h})$$

4.6 SPEED-DENSITY RELATIONSHIP

Imagine that we are in the helicopter. We can observe the traffic along specific highway section and make the photographs from the helicopter. Later, we can count the number of cars occupying section of the highway whose length is D. Traffic density k represents the number of vehicles occupying observed highway section at a specific time point, ie,

$$k=\frac{N}{D} \tag{4.18}$$

The traffic density is expressed in (veh/km). So-called *jam density* describes traffic conditions when cars on a freeway are bumper to bumper. At this density, traffic stops.

When the density is extremely low (eg, at 3:30 am) we do not worry about other drivers, since we are practically alone on a freeway. In other words, the other drivers (if any) have no influence on our speed. Our speed is influenced exclusively by performances of our vehicle, and by posted speed limits. This speed is known as a *free flow speed*. Over time, in early morning, traffic density increases, and in some cases creates traffic jam (Fig. 4.9).

At the beginning of our early morning observation, traffic density is close to zero. In this case, the vehicles' speed is free flow speed. Slowly, traffic density increases. The higher traffic density becomes, the lower the average speed will be. Finally, very high traffic density will cause a traffic jam. Vehicles will practically not move and the speed will be equal to zero. A typical speed-density relationship is shown in Fig. 4.10. Free-flow speed and the jam density are indicated in Fig. 4.10. Speed-density relationship shown in Fig. 4.10 is linear, ie,

$$u=u_{\text{f}}\left(1-\frac{k}{k_j}\right) \tag{4.19}$$

Linear speed-density relationship is approximation of the reality. This relationship is known as the *Greenshields* model. Many field measurements showed that speed-density relationship is nonlinear especially in the areas of low and high-speed traffic density. On the other hand, linear relationship is simply, and helps us to better understand complex issues in traffic flow.

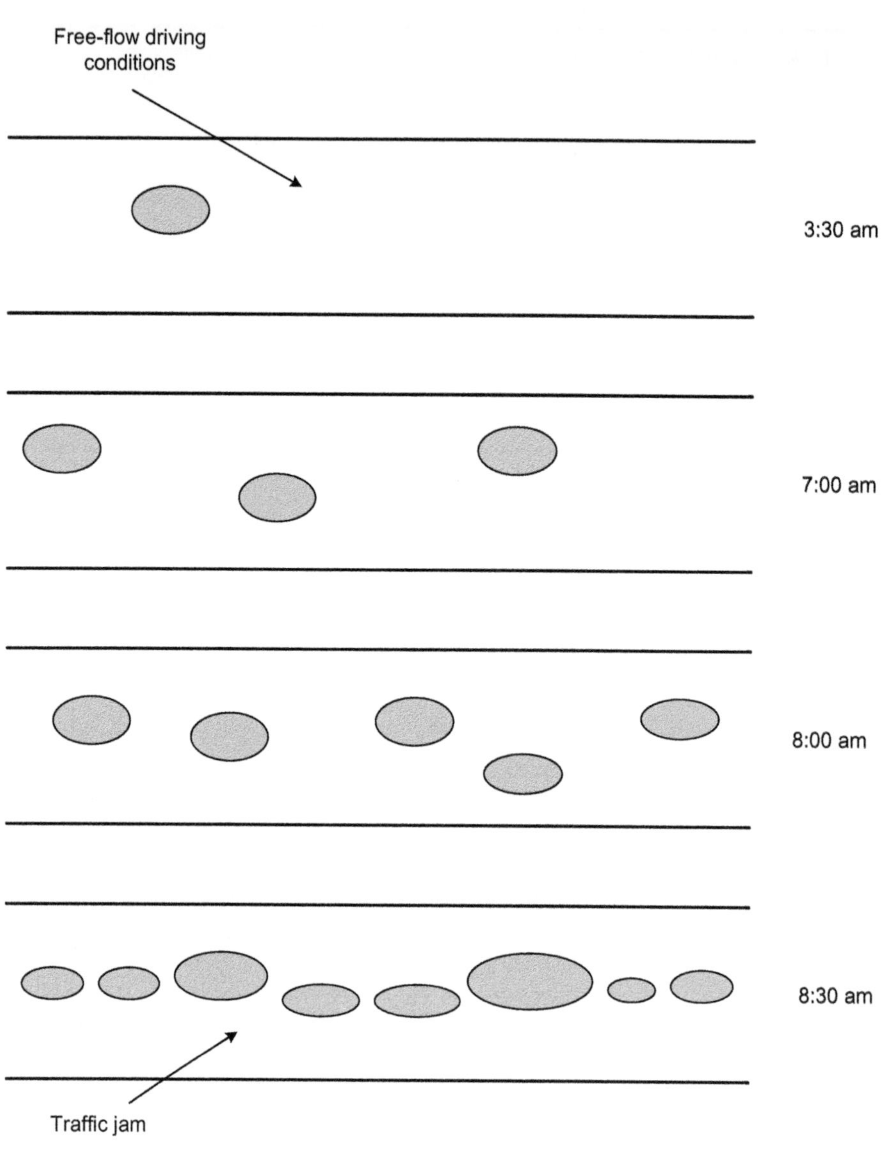

FIG. 4.9

Increase in traffic density over early morning.

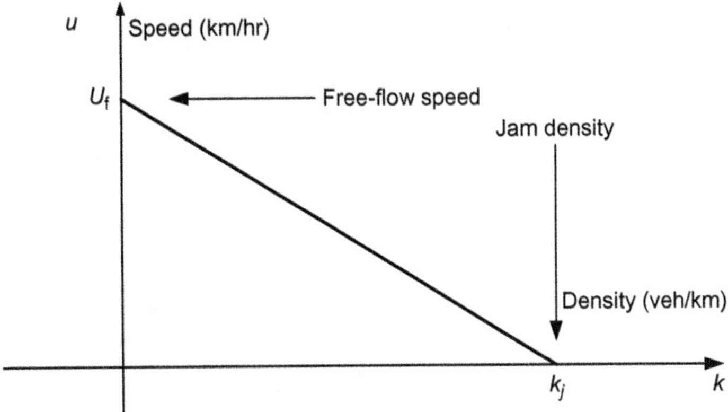

FIG. 4.10

Speed-traffic density relationship.

4.7 FLOW-DENSITY RELATIONSHIP

Let us multiply density k by speed u. If we analyze the units, we conclude that we are multiplying (veh/km) by (km/h). In this way, we get

$$\left(\frac{\text{veh}}{\text{km}}\right)\left(\frac{\text{km}}{\text{h}}\right)=\left(\frac{\text{veh}}{\text{h}}\right)$$

Obviously the units $\left(\frac{\text{veh}}{\text{h}}\right)$ are flow units. We see that we can obtain flow units by multiplying units of density by units of speed. This means that

$$q=uk \tag{4.20}$$

The flow is equal to the density multiplied by the speed. Using the linear relationship between speed and density, we get

$$q=uk=u_f\left(1-\frac{k}{k_j}\right)k \tag{4.21}$$

ie,

$$q=u_f\left(k-\frac{k^2}{k_j}\right) \tag{4.22}$$

From relation (4.22), we conclude that the flow q is quadratic function of the density k.

The graph of this function (parabola) is shown in Fig. 4.11. As density increases, flow increases from zero to the maximum value q_m. The value q_m represents *highway capacity*. This value is sometimes also called "traffic flow at capacity." Obviously, k_m is the density that corresponds to the highway capacity. We denote also by u_m the corresponding speed. With further increase in density, flow starts to decrease. Finally, at jam density, flow is equal to zero (all vehicles are stopped and there is no flow). It is very important to properly study density at capacity k_m. This density determines the border between stable traffic conditions (densities between zero and k_m) and the unstable traffic conditions (densities higher than k_m). Obviously, additional vehicles that enter the highway and create density higher than k_m decreases highway flow.

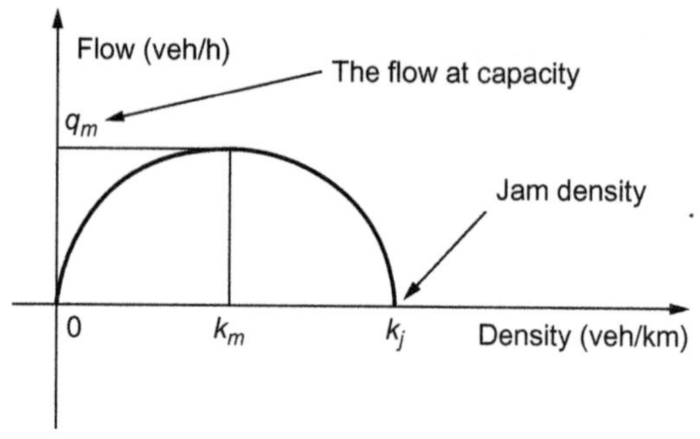

FIG. 4.11

Flow q (the quadratic function of the density k).

All parabolas are symmetric with respect to the axis of symmetry. A parabola intersects its axis of symmetry at a point called the vertex of the parabola. In our case, vertex of the parabola has coordinates $(k_{\mathrm{m}}, q_{\mathrm{m}})$. These coordinates are calculated in the following way. At maximum flow, first derivative $\frac{dq}{dk}$ must be equal to zero, ie,

$$\frac{dq}{dk}=0 \tag{4.23}$$

$$\frac{d\left[u_{\mathrm{f}}\left(k-\frac{k^2}{k_j}\right)\right]}{dk}=0 \tag{4.24}$$

$$\frac{d\left[u_{\mathrm{f}}k-u_{\mathrm{f}}\frac{k^2}{k_j}\right]}{dk}=0 \tag{4.25}$$

$$u_{\mathrm{f}}-u_{\mathrm{f}}\frac{2k}{k_j}=0 \tag{4.26}$$

$$u_{\mathrm{f}}\left(1-\frac{2k}{k_j}\right)=0 \tag{4.27}$$

The product $u_{\mathrm{f}}\left(1-\frac{2k}{k_j}\right)$ is equal to zero. The first term of this product is different than zero since u_{f} is the free-flow speed. This means that

$$1-\frac{2k}{k_j}=0 \tag{4.28}$$

By solving this equation, we get

$$k_{\mathrm{m}}=\frac{k_j}{2} \tag{4.29}$$

The density that corresponds to the maximal flow (highway capacity) equals one half of the jam density. The speed u_{m} that corresponds to the maximal flow equals

$$u_{\mathrm{m}} = u_{\mathrm{f}}\left(1 - \frac{k_{\mathrm{m}}}{k_j}\right) \tag{4.30}$$

$$u_{\mathrm{m}} = u_{\mathrm{f}}\left(1 - \frac{(k_j/2)}{k_j}\right) = u_{\mathrm{f}}\left(1 - \frac{k_j}{2k_j}\right) = \frac{u_{\mathrm{f}}}{2} \tag{4.31}$$

The flow at capacity q_{m} equals

$$q_{\mathrm{m}} = u_{\mathrm{m}} k_{\mathrm{m}} \tag{4.32}$$

$$q_{\mathrm{m}} = \frac{u_{\mathrm{f}}}{2}\frac{k_j}{2} \tag{4.33}$$

$$q_{\mathrm{m}} = \frac{u_{\mathrm{f}} k_j}{4} \tag{4.34}$$

The flow at capacity is equal to the one quarter of the free flow speed multiplied by the jam density.

Let us note points A and B (Fig. 4.12). Corresponding slopes shown in the figure represent speeds related to the points A and B. For example, the slope related to the point B represents the space-mean speed of all vehicles taken into account to calculate density k_{b}.

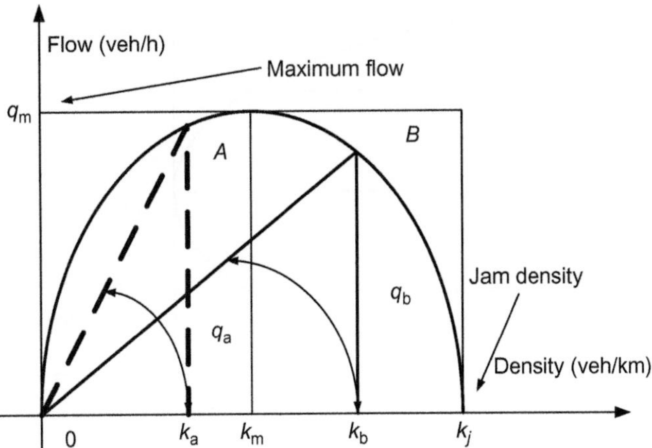

FIG. 4.12

Slopes representing speeds.

4.8 SPEED-FLOW RELATIONSHIP

We consider linear speed-density relationship, ie,

$$u = u_{\mathrm{f}}\left(1 - \frac{k}{k_j}\right) \tag{4.35}$$

Using this relation, we can express density k as a function of speed u, ie,

$$k = k_j\left(1 - \frac{u}{u_f}\right) \tag{4.36}$$

Since:

$$q = uk = uk_j\left(1 - \frac{u}{u_f}\right) \tag{4.37}$$

we get

$$q = k_j\left(u - \frac{u^2}{u_f}\right) \tag{4.38}$$

The relationship between speed and flow is shown in Fig. 4.13.

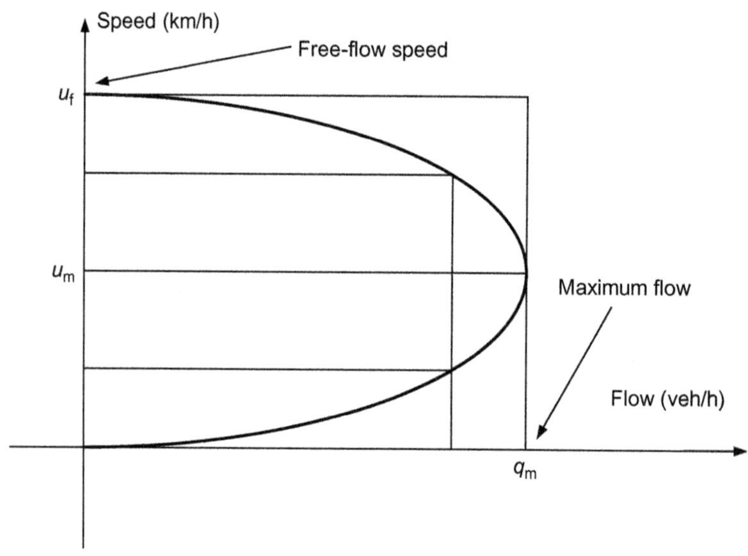

FIG. 4.13

Speed-flow relationship.

In this case, the vertex of the parabola has coordinates (q_m, u_m). We see that specific flow value always corresponds to the two speed values (except in the case of flow at capacity q_m). Speeds in the range from free-flow speed to the speed at capacity characterize stable traffic conditions (Fig. 4.14). Unstable traffic conditions characterize highly congested highways. Average space speed in unstable traffic conditions is below the speed at capacity (below u_m). Unstable traffic conditions are frequently described as a "stop-and-go" traffic.

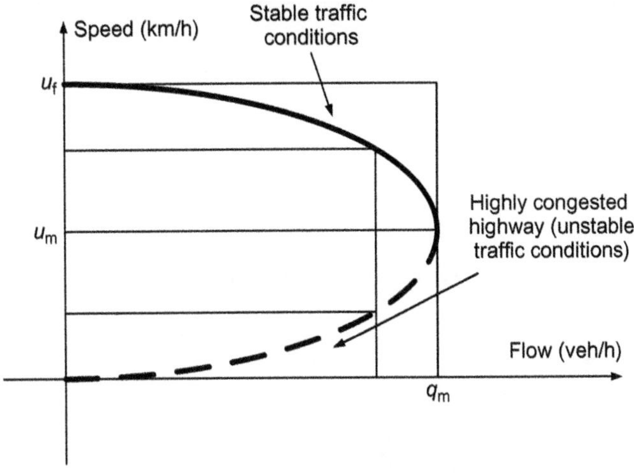

FIG. 4.14

Stable and unstable traffic conditions.

EXAMPLE 4.2

The following equation describes a flow-speed relationship:

$$q = 48k - 0.32k^2$$

Calculate the flow at capacity, and jam density.

Solution

At maximum flow, first derivative $\frac{dq}{dk}$ must be equal to zero, ie,

$$\frac{dq}{dk} = 0$$

$$\frac{d[48k - 0.32k^2]}{dk} = 0$$

$$48 - 0.64k = 0$$

$$k_m = 0.75$$

Since $k_j = 2k_m$, we get

$$k_j = 2 \cdot 75 = 150(\text{veh/km})$$

Maximum flow equals

$$q_m = 48k - 0.32k^2 = 48 \cdot 75 - 0.32 \cdot 75^2 = 1800(\text{veh/h})$$

EXAMPLE 4.3

The jam density, and the capacity are respectively equal $k_j = 160(\text{veh/km})$ and $q_m = 1600(\text{veh/h})$. Assume linear speed-density relationship (Greenshields model). Mathematically describe relationships in traffic flow.

Solution

The density at maximum capacity k_m equals

$$k_m = \frac{k_j}{2} = \frac{160}{2} = 80(\text{veh/km})$$

We denote by A and B, respectively, two characteristic points on the parabola. The coordinates of these points are denoted in Fig. 4.15.

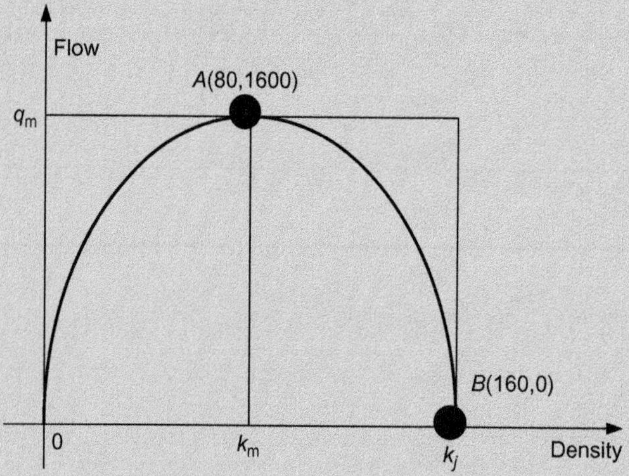

FIG. 4.15

Parabola describing traffic conditions when $k_j = 160(\text{veh/km})$ and $q_m = 1600(\text{veh/h})$.

The flow-density relationship is described by the following equation:

$$q = a \cdot k + b \cdot k^2$$

Points A and B are points located at the parabola. This means that the following must be satisfied:

$$1600 = a \cdot 80 + b \cdot 80^2$$

$$0 = a \cdot 160 + b \cdot 160^2$$

After solving this system of equations, we get

$$a = 40; \; b = -0.25$$

The flow density relationship reads

$$q = 40k - 0.25k^2$$

The maximum flow equals

$$q_m = \frac{u_f \cdot k_j}{4}$$

EXAMPLE 4.3—cont'd

From this equation we get

$$u_f = \frac{4 \cdot q_m}{k_j} = \frac{4 \cdot 1600}{160} = 40\,(\text{km/h})$$

The speed-density relationship reads

$$u = u_f\left(1 - \frac{k}{k_j}\right) = 40\left(1 - \frac{k}{160}\right)$$

$$u = 40 - \frac{1}{4} \cdot k$$

The speed-flow relationship reads

$$q = k_j\left(u - \frac{u^2}{u_f}\right) = 160\left(u - \frac{u^2}{40}\right)$$

$$q = 160 \cdot u - 4 \cdot u^2$$

EXAMPLE 4.4

The following is one of the best known nonlinear speed-density relationships:

$$u = c \cdot \ln\left(\frac{k_j}{k}\right)$$

The relationship is known as Greenberg model. (a) Determine the constant c; (b) jam density equals 170 (veh/km). Speed at capacity equals 80 (km/h). Determine the speed and the flow in the case when density equals 120 (km/h).

Solution

(a) In this model, the flow equals

$$q = k \cdot u = k \cdot c \cdot \ln\left(\frac{k_j}{k}\right)$$

At maximum flow, first derivative $\frac{dq}{dk}$ must be equal to zero, ie,

$$\frac{dq}{dk} = 0$$

$$\frac{dq}{dk} = \frac{d\left[k \cdot c \cdot \ln\left(\frac{k_j}{k}\right)\right]}{dk} = 0$$

$$c \cdot \ln\left(\frac{k_j}{k}\right) + k \cdot c \cdot \frac{1}{\left(\frac{k_j}{k}\right)} \cdot \left(-\frac{k_j}{k^2}\right) = 0$$

(Continued)

EXAMPLE 4.4—cont'd

$$c \cdot \ln\left(\frac{k_j}{k}\right) - c = 0$$

$$\ln\left(\frac{k_j}{k}\right) = 1$$

$$\frac{k_j}{k} = e$$

We finally get that the density that corresponds to the maximum flow equals

$$k_m = \frac{k_j}{e}$$

The speed at maximum flow u_m equals

$$u_m = c \cdot \ln\left(\frac{k_j}{k_m}\right) = c \cdot 1 = c$$

We now can rewrite Greenberg's model as:

$$u = u_m \cdot \ln\left(\frac{k_j}{k}\right)$$

(b) The speed equals

$$u = u_m \cdot \ln\left(\frac{k_j}{k}\right)$$

The jam density equals 170 (veh/km), and the speed at capacity equals 80 (km/h). In the case when density equals 120 (km/h), the speed equals

$$u = 60 \cdot \ln\left(\frac{170}{120}\right)$$

$$u = 27.9\,(\text{km/h})$$

The corresponding flow value is

$$q = u \cdot k = 27.9 \cdot 120 = 3348\,(\text{veh/h})$$

EXAMPLE 4.5

Free-flow speed and the capacity of the highway section are respectively equal

$$u_f = 85\,(\text{km/h}) \text{ and } q_m = 3200\,(\text{veh/h})$$

(a) Calculate the jam density.
(b) Calculate the density that corresponds to the flow value $q = 800\,(\text{veh/h})$.
(c) Calculate the space-mean speed in the case when flow equals $q = 1600\,(\text{veh/h})$. Assume that relationships between traffic stream variables could be described by the Greenshields model.

EXAMPLE 4.5—cont'd

Solution

(a) The flow at capacity equals

$$q_m = \frac{u_f \cdot k_j}{4}$$

The jam density is

$$k_j = \frac{4 \cdot k_m}{u_f} = \frac{4 \cdot 3200}{85} = 151\,(\text{veh/km})$$

(b) The flow-density relationship reads

$$q = u_f \cdot \left(k - \frac{k^2}{k_j}\right)$$

By substituting corresponding values we get

$$800 = 85 \cdot \left(k - \frac{k^2}{151}\right)$$

ie,

$$\frac{85}{151} \cdot k^2 - 85 \cdot k + 800 = 0$$

The rounded solution of the quadratic equation are

$$k_1 = 10\,(\text{veh/km});\;\; k_2 = 141\,(\text{veh/km})$$

(c) Speed-flow relationship reads

$$q = k_j \cdot \left(u - \frac{u^2}{u_f}\right)$$

By substituting corresponding values we get

$$1600 = 151 \cdot \left(u - \frac{u^2}{85}\right)$$

ie,

$$\frac{151}{85} \cdot u^2 - 151 \cdot u + 1600 = 0$$

The rounded solutions of the quadratic equation are

$$u_1 = 12.4\,(\text{km/h});\;\; u_2 = 72.6\,(\text{km/h})$$

4.9 FUNDAMENTAL DIAGRAM OF TRAFFIC FLOW

In Fig. 4.16, we can see *all three* major variables *simultaneously*. This is the reason that researchers call this diagram *fundamental diagram* of traffic flow. The other diagrams that we use to describe traffic stream show only two major traffic variables.

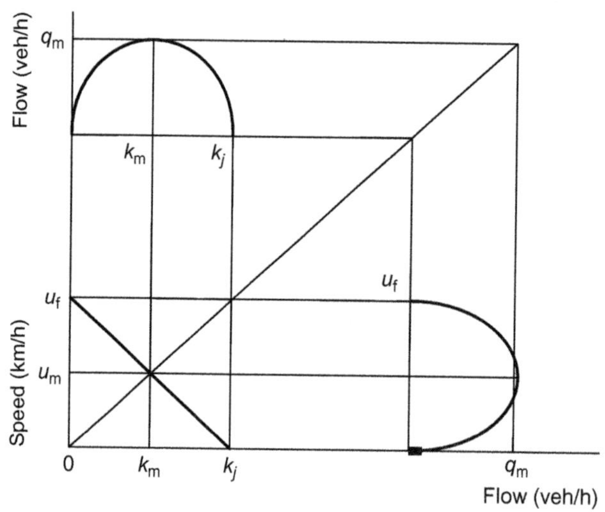

FIG. 4.16

Fundamental diagram of traffic flow.

4.10 SHOCK WAVES

Shock wave in fluids could be generated by the sudden, intense disturbance of a fluid. This disturbance could be, for example, powerful explosion, or flight of the supersonic aircraft. Supersonic aircraft produce shock waves that propagate through the atmosphere at a speed higher than the speed of sound. Shock waves also frequently occur on many highways on a daily basis. Imagine that the highway has reduction in the capacity caused by the work zone. For example, because of the work zone, the number of highway lanes is reduced from three to one (Fig. 4.17).

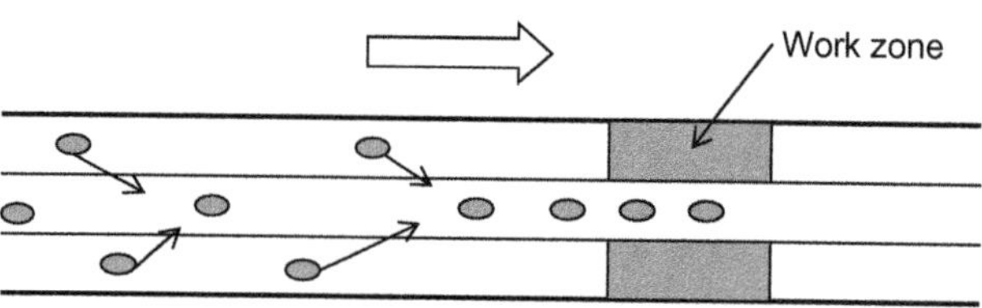

FIG. 4.17

Reduction in the number of highway lanes.

A reduction in highway capacity causes (in the case of large flow and density) reduction in vehicle speeds. Imagine that more and more vehicles continue to approach to the bottleneck. The point where reduction in speed happens will start to move upstream. In other words, this point travels in the opposite direction of traffic. The phenomenon in traffic stream that we described is known as a *shock wave* (Fig. 4.18).

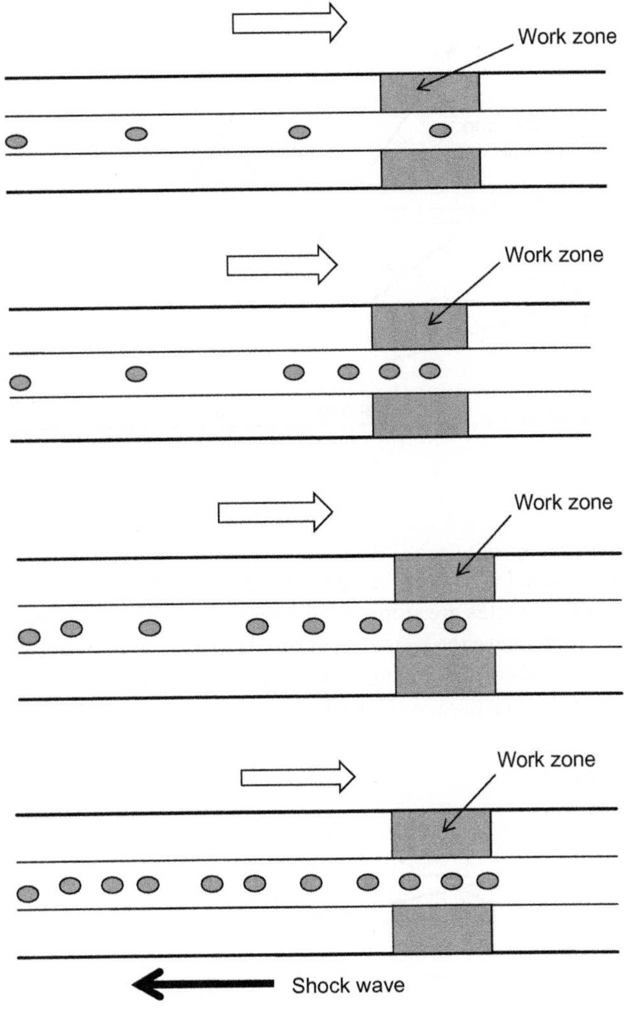

FIG. 4.18

Platoon composition in the case of work zone.

The shock wave can also travel in the direction of traffic stream (it happens when vehicles can again increase their speeds). Platoon composition in the case of a work zone is shown in Fig. 4.18. Queues formed in this way could be, in some cases, very long. Frequently, slow-moving trucks that travel along rural roads create shock waves and significant delays of other vehicles from the traffic stream. Vehicle trajectories are shown in Fig. 4.19. Fig. 4.19 shows the case, when no vehicle overtaking can happen.

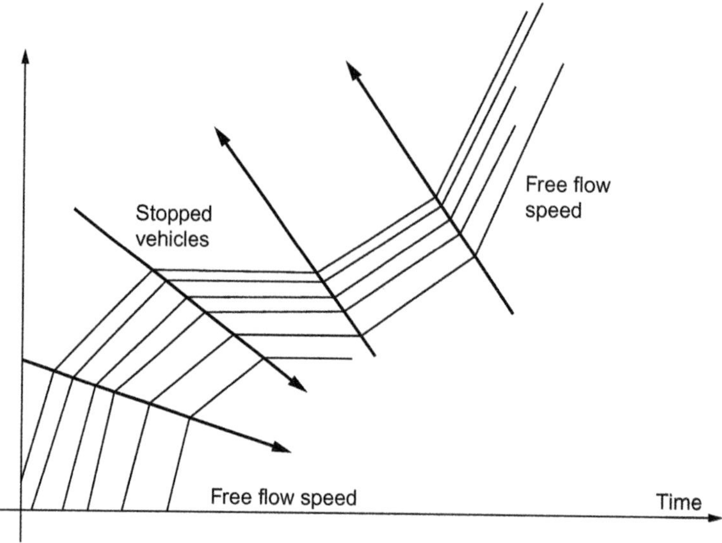

FIG. 4.19

Vehicle trajectories.

At the very beginning, vehicles travel in a free-flow traffic conditions. After that, vehicles start to decrease speeds, and finally vehicles are stopped. Vehicles spend some time in the area of completely stopped traffic, then start to increase the speeds, and finally again travel in free flow traffic conditions. Four different shock waves and corresponding shock wave speeds are shown in Fig. 4.20. The sign of the shock wave speed could be positive, or negative depending on the shock wave direction of travel.

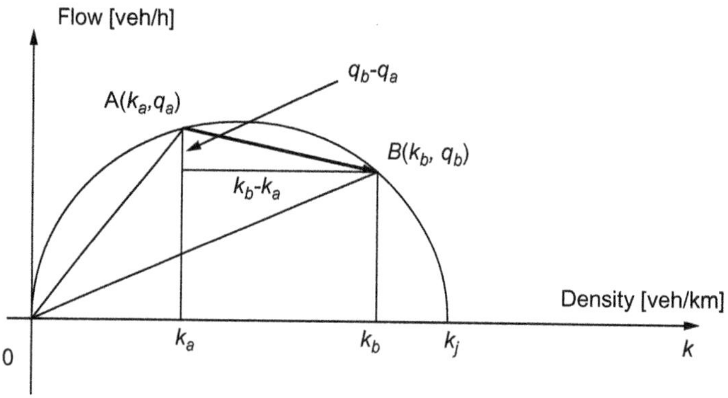

FIG. 4.20

The slope of the lane *AB* that represents the shock wave speed.

Fig. 4.20 shows point *A* (relatively normal traffic conditions) and the point *B* (bottleneck road section). The bottleneck road section is characterized by the significantly lower flow q_b, and the significantly higher traffic density k_b. Vehicles that approach the road bottleneck area are forced to decelerate and reduce their speed from the speed $\frac{q_a}{k_a}$ to the speed $\frac{q_b}{k_b}$. The shock wave propagates upstream by the speed represented by the vector $\overrightarrow{AB}$ (Fig. 4.20). The shock speed value can be calculated by calculating the slope of the lane that connects the point *A* and the point *B*, ie,

$$u_w = \frac{q_b - q_a}{k_b - k_a} \tag{4.39}$$

The sign of the shock wave speed could be positive, or negative, depending on the values of q_a, q_b, k_a, and k_b. The negative sign is related to the situation when shock wave travels in the upstream direction. The positive sign denotes the situation when shock wave travels in the direction of traffic stream.

EXAMPLE 4.6

A road work zone is causing significant vehicle delays. We denote by q_1, q_2, q_3 and u_1, u_2, u_3, respectively, lows and speeds in the approaching area to the work zone, in the work zone, and downstream of the work zone. The measured flow and speed values are shown in Table 4.3.

Table 4.3 The Measured Flow and Speed Values in the Approaching Area to the Work Zone

Road Section	Flow (veh/h)	Speed (km/h)
1—Approaching area to the work zone	1200	65
2—Work zone	1000	10
3—Downstream of the work zone	1100	80

Calculate the speeds of the shock waves generated by the work zone.

Solution

The work zone generates the shock wave at the tail of the platoon of vehicles entering the work zone, as well as the shock wave at the end of work zone. The corresponding densities are respectively equal

$$k_1 = \frac{q_1}{u_1} = \frac{1200}{65} = 18.46(\text{veh/km})$$

$$k_2 = \frac{q_2}{u_2} = \frac{1000}{10} = 100(\text{veh/km})$$

$$k_3 = \frac{q_3}{u_3} = \frac{1100}{80} = 13.75(\text{veh/km})$$

(Continued)

EXAMPLE 4.6—cont'd

The shock wave speed u_{w1} at the tail of the platoon of vehicles entering the work zone equals

$$u_{w1} = \frac{q_2 - q_1}{k_2 - k_1} = \frac{1000 - 1200}{100 - 18.46} = -2.45\,(\text{veh/h})$$

The shock wave speed u_{w2} at the end of work zone equals

$$u_{w2} = \frac{q_3 - q_2}{k_3 - k_2} = \frac{1100 - 1000}{13.75 - 100} = 1.16\,(\text{veh/h})$$

4.11 MICRO-SIMULATION TRAFFIC MODELS

Today, traffic systems are frequently studied and analyzed by using traffic simulation techniques. A greater number of traffic simulation models have been developed during the last two decades. The development of these models was significantly influenced by the development of personal computers, as well as by computer science achievements. Macro-simulation models attempt to describe the traffic on networks by using speed, flow, and density as the main entities in the model. On the other hand, micro-simulation models try to simulate network traffic starting from the behavior of individual drivers and individual vehicles.

Traffic simulation models significantly help traffic engineers and analysts in predictive travel time calculations, dynamic route guidance, urban congestion management, dynamic emergency vehicle routing, emissions management, evacuation management, etc.

The micro-simulation models take into account gap-acceptance behavior, speed adjustment, lane-changing, overtakes, and car-following behavior (Fig. 4.21).

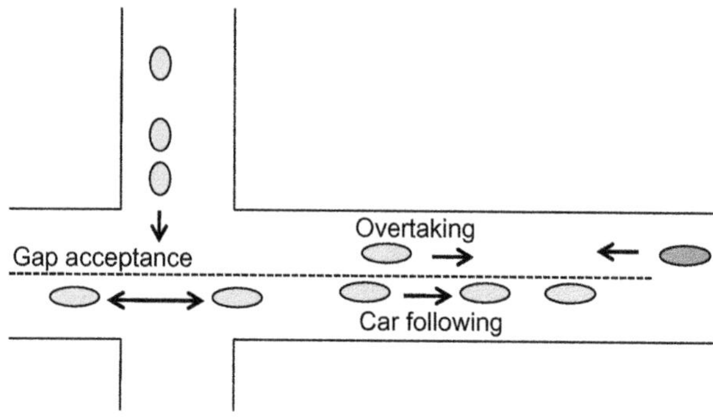

FIG. 4.21

Micro-simulation model.

There are submodels, within micro-simulation models, devoted to all these phenomena. The gap acceptance models describe driver behavior at the unsignalized intersections and other merging situations. Lane-changing models explain drivers' behavior when changing lane, while overtake models describe drivers' overtaking behavior. The car-following models describe drivers' behavior in an uninterrupted traffic flow.

The road network topology, the origin-destination matrix (matrix that gives the information about the number of trips going from each origin to each destination) and turning percentages (percentages of left turning, straight moving, and right turning vehicles) at every intersection are the main input to micro-simulation models. The input could be also intersection control plans.

The main output of the traffic simulation models could be average travel time between two points, average speed, number of stopped vehicles at the intersections, traffic condition predictions (future network flow patterns), congestion levels, emission levels, etc.

4.12 CAR FOLLOWING MODELS

Longitudinal spacing of vehicles is one of the very important issues that have to be analyzed in traffic systems studies. It has high influence on traffic safety, capacity and a level-of-service. The traffic simulation models (AIMSUN, MITSIM, VISSIM, etc.) are composed of a few submodels. One of the submodels is car-following model. The car-following models (Edie, 1961; Gazis et al., 1961) attempt to study longitudinal spacing of vehicles, and to describe drivers' behavior in an uninterrupted traffic flow. These models are related to single-lane traffic, with no passing. The car-following models are created, with an assumption that the density of the observed traffic flow is high. Interactions between individual vehicles disappear in the cases of low traffic density, so in these cases the car—the following models should not be used. The essential equation of the car-following models is based on the assumption that every driver responds to a given stimulus, ie,

$$\text{Response} = \text{Sensitivity} \times \text{Stimulus}$$

According to this theoretical concept, every driver responses to a stimulus from the car ahead of and/or behind her/him. Various parameters are used in various models to describe response, sensitivity, and stimulus. The stimulus could depend of one or more variables (distance headway, vehicle speed, relative speeds, etc.). The acceleration of the vehicle represents the logical response. By applying the gas and brake pedals, by accelerating or decelerating, every driver responses to a stimulus. Car-following models describe the way in which every vehicle follows another vehicle in an uninterrupted traffic flow (Fig. 4.22).

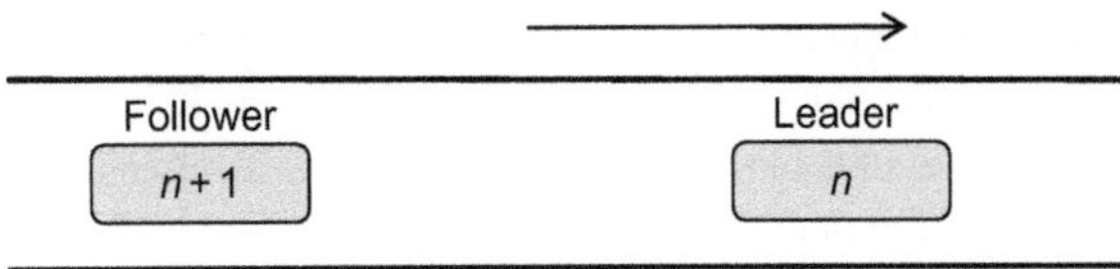

FIG. 4.22

Leader vehicle and follower vehicle.

Let us introduce the following notation:

n is the leader vehicle
$n+1$ is the following vehicle
x_n^t is the location of the leader vehicle at time t
v_n^t is the speed of the leader vehicle at time t
x_{n+1}^t is the location of the follower vehicle at time t
v_{n+1}^t is the speed of the follower vehicle at time t

There are numerous car-following models in the literature. The widely used model is the general motors (GM) model, also known as a Gazis-Herman-Rothery Model. GM has developed five generations of the car-following models. The most general has the following form:

$$\ddot{x}_{n+1}(t+\Delta t)=\left\{\frac{\alpha(l,m)\cdot(\dot{x}_{n+1}(t+\Delta t))^m}{(x_n(t)-x_{n+1}(t))^l}\right\}\cdot[\dot{x}_n(t)-\dot{x}_{n+1}(t)] \tag{4.40}$$

where

$x_n(t)$ is the distance from some arbitrary point
$\dot{x}_n(t)$ is the speed of the nth vehicle at time t
$\ddot{x}_n(t)$ is the acceleration/deceleration of the nth vehicle at time t
Δt is the perception reaction time
l is the distance headway exponent (in the range from +4 to −1)
m is the speed exponent (in the range from −2 to +2)
$\alpha(l,m)$ is the sensitivity coefficient

The model parameters l, m, and $\alpha(l,m)$ require calibration. The relative speed between follower vehicle and leader vehicle (the term within the brackets) represents a stimulus for a follower vehicle in the GM model. The term within the braces is called the sensitivity term.

4.12.1 THE CAR-FOLLOWING MODEL BASED ON FUZZY INFERENCE RULES

Car-following represents a control process. During this control process, the driver of the FV makes an effort to preserve a safe distance between his/her car and the leading vehicle (LV). The driver of the FV accelerates or decelerates in response to the actions of the vehicle in front. It is important to know that "drivers do not *completely* follow any deterministic behavior" (Ceder, 1976). There is vagueness present in the perception of stimuli. Vagueness is also present in the rules that driver applies, as well as in an action performed by a driver. Chakroborty and Kikuchi (1999) started from the assumption that human "perceives the environment, uses his/her knowledge and experience to infer possible actions, and responds in an *approximate* manner." The car-following model proposed by Chakroborty and Kikuchi (1999) consists of a set of fuzzy inference rules (the basic elements of the fuzzy logic systems are described in Chapter 3). The graphical representation of the fuzzy-inference based model is given in Fig. 4.23.

The fuzzy inference rules connect a specific driving environment (the existing relative speed, distance headway, and actions of the LV) at time t to an appropriate action by FV at time $t+\Delta t$. The rules are of the form:

If (at time t): Distance headway (DS) is A_i; relative speed (RS) is B_j; and acceleration of LV is C. Then (at time $t+\Delta t$): Acceleration/deceleration of FV should be D_l.

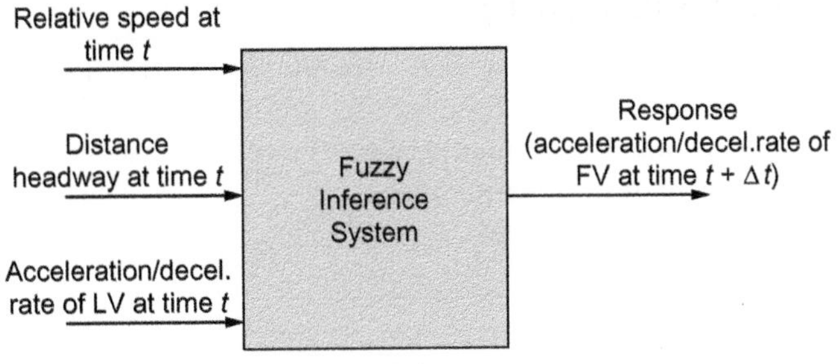

FIG. 4.23

A fuzzy-inference based model of car-following.

The car-following model based on fuzzy inference rules tries to explain an approximate nature of the car-following behavior. It has been shown that drivers usually react differently when decelerating and when accelerating. Various experiments showed that "the reaction of a driver to positive relative speed of a given magnitude is not the same in magnitude as to a negative relative speed of the same magnitude. In other words, the drivers response is asymmetric with respect to relative speed" (Chakroborty and Kikuchi, 1999). The car-following model based on fuzzy inference rules can describe asymmetric response of the drivers.

The car-following model based on fuzzy inference rules was evaluated using the same evaluation scheme that is used for the GM models. (The data were obtained by the instrumented test vehicle.) The performed comparison demonstrated that the fuzzy inference model can overcome many limitations of the GM based car-following models.

4.13 NETWORK FLOW DIAGRAM

The relationships between flow and density, and speed, which were described in the previous discussion, are related to highways. The logical question is: does a similar *network-wide* relationship exist? In other words, the question is whether similar relationships could describe flows in traffic networks in urban areas. Researchers have attempted to describe the relations between flow, density, and speed at the network level during the last few decades (Smeed, 1966, 1968; Thomson, 1967; Wardrop, 1968; Mahmassani et al., 1984, 1987, 2013; Ardekani and Herman, 1987; Mahmassani and Peeta, 1993; Cassidy and Bertini, 1999; Daganzo, 2007; Daganzo and Geroliminis, 2008; Daganzo et al., 2011; Boyaci and Geroliminis, 2011; Saberi and Mahmassani, 2012; Saberi et al., 2014). It has been shown (Geroliminis and Daganzo, 2008; Saberi et al., 2014), after a field experiment in Yokohama (2001), that a Macroscopic Fundamental Diagram that connects space-mean flow, density, and speed exists in a large urban area. The experiment in Yokohama, performed in 2001, used a combination of link-based measurements and trajectory-based measurements. The link-based measurements were performed by using fixed 500 ultrasonic and loop detectors. The trajectory-based measurements were carried out by using mobile sensors (140 taxis equipped with Global Positioning System, GPS).

The expression "Network Fundamental Diagram" is also used in the literature, and we shall continue to use this expression. The seminal work of Edie (1965), in which he proposed generalized definitions of traffic flow, density, and speed, together with the field experiments from Yokohama, enabled deriving conclusions about traffic flow characteristics in urban areas.

4.13.1 LINK-BASED MEASUREMENTS

Link measurements are frequently used to compute network-wide traffic flow variables. Link-based measurements assume that we collect traffic data from the sensors. Sensors are installed on individual links of the transportation network. The average network flow Q, the average network density K, and the average network speed V are, respectively, equal

$$Q = \frac{\sum_{i=1}^{M} l_i \cdot q_i}{\sum_{i=1}^{M} l_i} \tag{4.41}$$

$$K = \frac{\sum_{i=1}^{M} l_i \cdot k_i}{\sum_{i=1}^{M} l_i} \tag{4.42}$$

$$V = \frac{\sum_{i=1}^{M} l_i \cdot v_i}{\sum_{i=1}^{M} l_i} \tag{4.43}$$

where

l_i is the length of lane-link i, $i = 1,2,\ldots,M$
M is the total number of lane-links
q_i is the flow on lane-link i, $i = 1,2,\ldots,M$
k_i is the density on lane-link i, $i = 1,2,\ldots,M$
v_i is the speed on lane-link i, $i = 1,2,\ldots,M$

4.13.2 GENERALIZED TRAFFIC FLOW VARIABLES

Edie (1965) proposed generalized definitions of traffic flow, density, and speed. Let us explain this generalized definitions in the case of the region A (Fig. 4.24). Fig. 4.24 shows the trajectory of six vehicles passing through the region A. In the considered case, the region A is in the form of a rectangle with the sides are L and T. The thin horizontal rectangle and thin vertical rectangle are also shown in Fig. 4.24 (these rectangles are shaded). The thin, horizontal rectangle in Fig. 4.24 is related to a fixed surveillance point at the highway.

The flow at the surveillance point is equal to $\frac{m}{T}$, where m is the number of vehicles. In the case shown in Fig. 4.24, $m = 4$. By multiplying both nominator and denominator by dx, the flow could be expressed as

$$\frac{m \cdot dx}{T \cdot dx}$$

The denominator $T \cdot dx$ represents the area of the thin horizontal rectangle. The denominator is expressed in units of distance × time.

The nominator $m \cdot dx$ represents the total distance traveled by all vehicles in this thin horizontal rectangle. The flow represents the ratio of the distance traveled by vehicles in a region

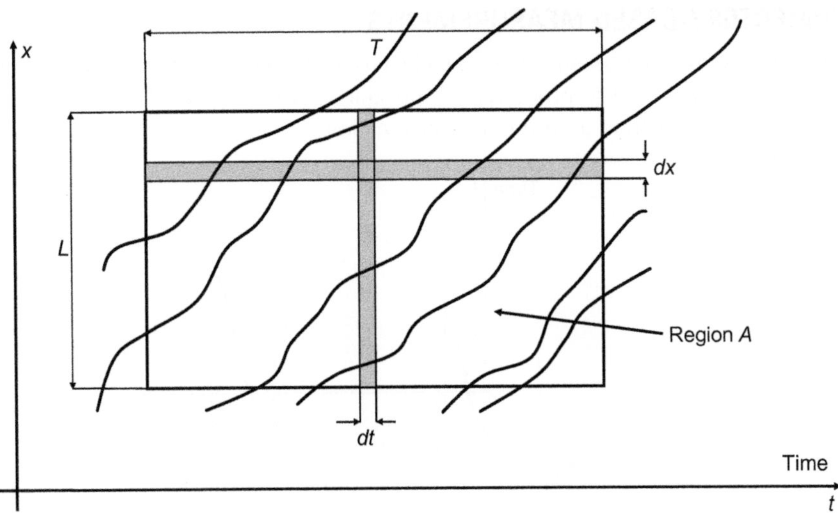

FIG. 4.24

Trajectories of six vehicles.

to the region's area. Any time-space region is composed of thin, elementary rectangles. We conclude that flow $q(A)$ in *any region A* represents the distance $d(A)$ traveled by vehicles in a region A to the region's area $|A|$ A, ie,

$$q(A) = \frac{d(A)}{|A|} \tag{4.44}$$

The thin, vertical rectangle shown in Fig. 4.24 is related to an instant in time. The density is equal to $\frac{n}{L}$, where n is the number of vehicles. In the case shown in Fig. 4.24, $n=3$. By multiplying both nominator and denominator by dt, the density could be expressed as:

$$\frac{n \cdot dt}{L \cdot dt}$$

The nominator $n \cdot dt$ represents the total time spent by all vehicles in this thin vertical rectangle, while the denominator $L \cdot dt$ represents the area of the thin vertical rectangle.

We conclude that density $k(A)$ in *any region A* represents the time $t(A)$ spent by vehicles in a region A to the region's area $|A|$ A, ie,

$$k(A) = \frac{t(A)}{|A|} \tag{4.45}$$

The speed (velocity) $v(A)$ in *any region A* is equal

$$v(A) = \frac{q(A)}{k(A)} = \frac{d(A)/|A|}{\frac{t(A)}{|A|}} = \frac{d(A)}{t(A)} \tag{4.46}$$

Edie's generalized traffic flow variables along a highway are recently extended to the network level.

4.13.3 TRAJECTORY-BASED MEASUREMENTS

Trajectory-based measurements assume that we collect traffic data from individual vehicle trajectories. In order to analyze network traffic properly, we introduce 3D time-space diagram. Fig. 4.25 shows movements of four vehicles from an origin to a destination.

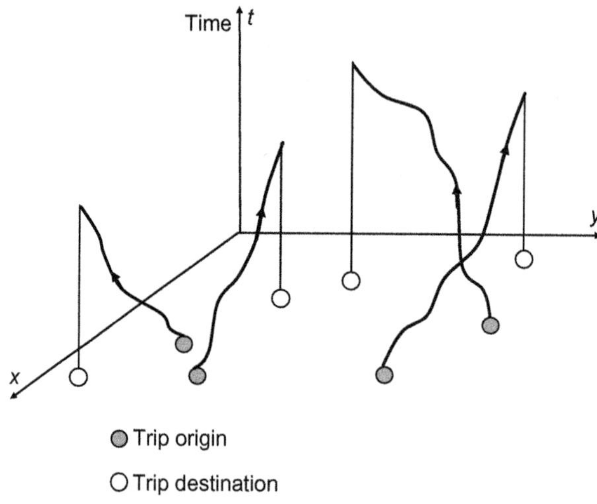

FIG. 4.25

3D space-time diagram: movements of four vehicles.

In order to estimate network-wide flow, density, and speed, analogously to the region A in the 2D time-space diagram, Saberi et al. (2014) proposed a closed 3D shape. Fig. 4.26 shows a closed 3D shape (cube). The shape could be any form. The transportation network structure is laid down on the x–y plane (Fig. 4.26).

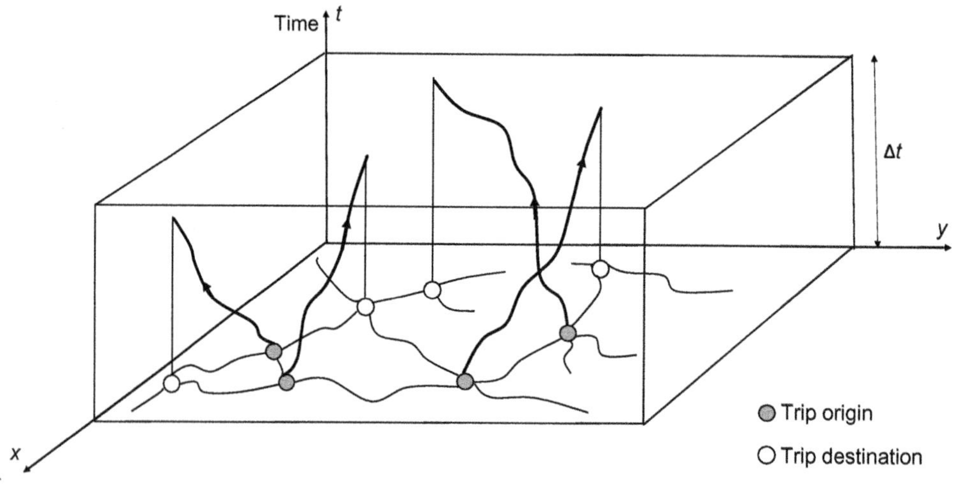

FIG. 4.26

3D shape.

Saberi et al. (2014) expressed the generalized network-wide traffic flow variables (based on the extended definitions of Edie) in the following way:

$$Q(B) = \frac{d(B)}{L_{xy}(B) \cdot \Delta t} \tag{4.47}$$

$$K(B) = \frac{t(B)}{L_{xy}(B) \cdot \Delta t} \tag{4.48}$$

$$v(B) = \frac{d(B)}{t(B)} \tag{4.49}$$

where

$Q(B)$ is the network-wide average flow for the specified shape B
$K(B)$ is the network-wide average density for the specified shape B
$v(B)$ is the network-wide average speed for the specified shape B
$L_{xy}(B)$ is the total length (in lane-miles) of the network on the x–y plane associated with the shape B
Δt is the time height of the shape B

4.14 PROBLEMS

1. The distribution of the number of vehicles that appear at a specific highway location during specified time interval is *Poisson* distribution. The flow at the observed highway location equals 720 veh/h. Calculate the average arrival rate λ. Calculate the probabilities of having 0, 1, 2, and 3 vehicles over a 30-s interval.
2. The distribution of the number of vehicles that appear at a specific highway location during specified time interval is *Poisson* distribution. The flow at the observed highway location equals 540 veh/h. Calculate the average arrival rate λ. Calculate the probabilities of having 6 or more vehicles over a 25-s interval.
3. The distribution of the number of vehicles that appear at a specific highway location during specified time interval is *Poisson* distribution. The flow at the observed highway location equals 360 veh/h. Calculate the probability of having headway being greater than or equal to the 30-s interval.
4. The distribution of the number of vehicles that appear at a specific highway location during specified time interval is *Poisson* distribution. The flow at the observed highway location equals 360 veh/h. Calculate the probability of having headway being greater than or equal to the 30-s interval.
5. Traffic engineers counted the number of vehicles that appear at a specific highway location during a time interval of 30 s. In total, vehicle counts are performed in 100 time intervals. Statistical analysis showed that the distribution of the number of vehicles that appear at the observed location during the specified time interval is *Poisson* distribution. No cars arrived in 10 of the observed time intervals. Calculate the average arrival rate λ. Calculate the probabilities of having 1, 2, and 3

vehicles over a 30-s interval. Calculate the number of time intervals in which 1, 2, and 3 vehicles arrived.

6. Vehicles appear at a specific highway location at an average of 360 vehicles per hour. Calculate the probability that none passes in a given minute. What is the expected number of vehicles in 3 min?
7. On average, there are of 250 incoming vehicles to toll booths during 8:00 am and 9:00 am on Monday. The experience shows that the existing staff can handle up to 500 vehicles in an hour. What is the probability that 600 vehicles will arrive next Monday between 8:00 and 9:00 am? Do traffic authority have enough staff?
8. The distribution of the number of traffic accidents that happen at a specific highway section during specified time interval is *Poisson* distribution. It has been observed that the average number of traffic accidents requiring medical help on the observed highway section between 9:00 and 10:00 am on Friday is 1. Calculate the probability that there will be a need for exactly 2 ambulances on the highway section during 9:00 and 10:00 am next Friday.
9. Vehicles arrive at a toll booth according to Poisson process at a mean rate of 60 vehicles per hour. What is the probability that the toll booth operator has to wait more than 2 min for the next student?
10. The number of miles that a specific car can run before its battery wears out is exponentially distributed with an average of 10,000 miles. The owner of the car has to take a 5000-mile trip. What is the probability that she will be capable to finish the trip without having to substitute the car battery?
11. The following is the speed-density relationship:

$$u = u_f\left[1 - \left(\frac{k}{k_j}\right)^2\right]$$

The jam density k_j is equal to 145 veh/km. Calculate the following quantities:
(a) free-flow speed
(b) space-mean speed of the traffic at capacity
12. The following equation describes flow-density relationship:

$$q = 46k - 0.30k^2$$

Calculate flow at capacity, and jam density.
13. The jam density, and the capacity are respectively equal $k_j = 150(\text{veh/km})$ and $q_m = 1500(\text{veh/h})$. Assuming the linear speed-density relationship, mathematically describe relationships in traffic flow.
14. The speed-density relationship equals

$$u = c \cdot \ln\left(\frac{k_j}{k}\right)$$

Jam density equals 160 (veh/km). Speed at capacity equals 80 (km/h). Determine the speed and the flow in the case when density equals 100 (km/h).

15. The following is the flow-speed relationship:

$$q = 250u - 65u \ln u$$

Calculate flow at capacity, and free flow speed.

16. The speed-density relationship equals

$$u = 25 \cdot \ln\left(\frac{150}{k}\right)$$

Calculate flow at capacity, and jam density.

REFERENCES

Ardekani, S.A., Herman, R., 1987. Urban network-wide traffic variables and their relations. Transp. Sci. 24, 1–16.

Boyaci, B., Geroliminis, N., 2011. Estimation of the network capacity for multimodal urban systems. Procedia Soc. Behav. Sci. 16, 803–813.

Cassidy, M.J., Bertini, R.L., 1999. Some traffic features at freeway bottlenecks. Transp. Res. B 33, 25–42.

Ceder, A., 1976. A deterministic traffic flow model for the two regime approach. Transp. Res. Rec. 567, 16–30.

Chakroborty, P., Kikuchi, S., 1999. Evaluation of the general motors based car-following models and a proposed fuzzy inference model. Transp. Res. C 7, 209–235.

Daganzo, C.F., 2007. Urban gridlock: macroscopic modeling and mitigation approaches. Transp. Res. B 41 (1), 49–62.

Daganzo, C.F., Geroliminis, N., 2008. Analytical approximation for the macroscopic fundamental diagram of urban traffic. Transp. Res. B 42, 771–781.

Daganzo, C.F., Gayah, V., Gonzales, E., 2011. Macroscopic relations of urban traffic variables: bifurcations, multivaluedness and instability. Transp. Res. B 41, 278–288.

Drew, D., 1968. Traffic Flow Theory and Control. McGraw-Hill, New York.

Edie, L.C., 1961. Car following and steady-state theory for non-congested traffic. Oper. Res. 9 (1), 66–76.

Edie, L.C., 1965. Discussion of traffic stream measurements and definitions. In: Almond, J. (Ed.), Proceedings of the Second International Symposium on the Theory of Traffic Flow. OECD, Paris, pp. 139–154.

Elefteriadou, L., 2014. An Introduction to Traffic Flow Theory. Springer, New York.

Gazis, D.C., Herman, R., Montroll, E.W., Rothery, R.W., 1961. Nonlinear follow-the-leader models of traffic flow. Oper. Res. 9, 545–560.

Gerlough, D.L., Huber, M.J., 1975. Traffic Flow Theory: A Monograph TRB Special Report 165. Transportation Research Board, Washington, DC.

Geroliminis, N., Daganzo, C.F., 2008. Existence of urban-scale macroscopic fundamental diagrams: some experimental findings. Transp. Res. B 9, 759–770.

Greenshields, B.D., 1935. A study of traffic capacity. Highway Res. Board Proc. 14, 448–477.

Greenshields, B.D., Thompson, J.T., Dickinson, H.C., Swinton, R.S., 1933. The photographic method of studying traffic behavior. Highway Res. Board Proc. 13, 382–399.

Herman, R., Ardekani, S.A., 1984. Characterizing traffic conditions in urban areas. Transp. Sci. 18, 101–140.

Herman, R., Prigogine, I., 1979. A two-fluid approach to town traffic. Science 204, 148–151.

Kerner, B.S., 2009. Introduction to Modern Traffic Flow Theory and Control: The Long Road to Three-Phase Traffic Theory. Springer, New York.

Kühne, R.D., 2011. Greenshields' legacy highway traffic. In: 3-10 in Transportation Research Circular Number E-C 149 (75 Years of the Fundamental Diagram for Traffic Flow Theory-Greenshields Symposium). Transportation Research Board of the National Academies, Washington, DC.

Mahmassani, H.S., Peeta, S., 1993. Network performance under system optimal and user equilibrium dynamic assignments: implications for ATIS. Transp. Res. Rec. J. Transp. Res. Board 1408, 83–93.

Mahmassani, H.S., Williams, J.C., Herman, R., 1984. Investigation of network-level traffic flow relationships: some simulation results. Transp. Res. Rec. J. Transp. Res. Board 971, 121–130.

Mahmassani, H.S., Williams, J.C., Herman, R., 1987. Performance of urban traffic networks. In: Proceedings of the 10th International Symposium on Transportation and Traffic Theory. Elsevier Science Publishing, New York, pp. 1–20.

Mahmassani, H.S., Saberi, M., Zockaie, A., 2013. Urban network gridlock: theory, characteristics, and dynamics. Soc. Behav. Sci. 80, 79–98.

May, A., 1990. Traffic Flow Fundamentals. Prentice-Hall, Upper Saddle River, NJ.

Ross, P., 1991. Some properties of macroscopic traffic models. Transp. Res. Rec. 1194, 129–134.

Saberi, M., Mahmassani, H.S., 2012. Exploring the properties of network-wide flow-density relations in freeway networks. Transp. Res. Rec. J. Transp. Res. Board 2315, 153–163.

Saberi, M., Mahmassani, H., Hou, T., Zockaie, A., 2014. Estimating network fundamental diagram using three-dimensional vehicle trajectories: extending Edie's definitions of traffic flow variables to networks. Transp. Res. Rec. J. Transp. Res. Board 2422, 12–20.

Smeed, R.J., 1966. Road capacity of city centers. Traffic Eng. Control 8 (7), 455–458.

Smeed, R.J., 1968. Traffic studies and urban congestion. J. Transp. Econ. Policy 2 (1), 33–70.

Thomson, J.M., 1967. Speeds and flows of traffic in central London: 2. Speed-flow relations. Traffic Eng. Control 8 (12), 721–725.

Wardrop, J.G., 1968. Journey speed and flow in central urban areas. Traffic Eng. Control 9 (11), 528–532.

Williams, J.C., Mahmassani, H.S., Herman, R., 1987. Urban traffic network flow models. Transp. Res. Rec. J. Transp. Res. Board 1112, 78–88.

What is the influence of the number of lanes, the width of the lanes and shoulders, and percentage of heavy vehicles in the traffic flow, on the freeway capacity? How do we calculate the capacity of a single rail line? What is the capacity of a rail station? How could we balance the capacity of the seaside with the capacity of the landside area in a maritime port? How do we calculate airport capacity? What is the capacity of the airspace?

CAPACITY AND LEVEL OF SERVICE 5

5.1 INTRODUCTION

Capacity is considered as the maximum capability of a given transportation mode or its particular component to serve a certain volume of demand (passenger and/or freight), during a specified period of time, under given conditions. We describe, in this chapter, the capacity and level of service of different transportation modes. The capacity of a given modal component depends on its size, technical/technological performances, and rules and procedures of its operation. These three attributes are interrelated. Given conditions are usually characterized by the constant demand for service during the specified period of time—"ultimate" capacity, or by the service level provided to each unit of demand while getting service—"practical" capacity.

Consequently, the "ultimate" capacity of a given modal component can be defined as the maximum number of served entities during given period of time under conditions of constant demand for service. The "practical" capacity of the same modal component can be defined as the maximum number of entities served during a given period of time when each is provided the specified service level. In the given context, the service level is usually expressed by the waiting time/delay while getting service. In addition, it can be expressed by additional attributes such as the available space (ie, internal and external comfort), accessibility, and reliability of service characterized for each particular transportation mode. In such a way, particular attributes of service level are interrelated. In this chapter, the service level, expressed by the average waiting time/delay of a user, while receiving the service, will be particularly under focus.

Transportation Engineering. http://dx.doi.org/10.1016/B978-0-12-803818-5.00005-6

Dealing with the capacity and level of service assumes that each transportation mode consists of the infrastructure network. The network of transportation services is operated by transport operators/companies. For the infrastructure network, the capacity is considered for serving transport vehicles, thus implying the traffic capacity. For the service network, the capacity is considered for serving passengers and freight/goods carried by transport vehicles, thus implying the transport capacity.

Each transportation network generally consists of the nodes and links. Both are characterized by the capacity and corresponding level of service. The networks spread in urban, sub-urban-regional (rail and road), and interurban areas (rail, road, air, and sea). The transport services of all transportation modes can be carried out by the vehicles of different size (carrying capacity—payload) and other technical/technological and operational characteristics. They operate in the infrastructure networks according to the specified rules and procedures.

The traffic and transport "ultimate" and "practical" capacity and corresponding level of service of particular components of the infrastructure and service networks are described. These descriptions are given for road, rail, air, and maritime transport mode and their systems, serving both passenger and freight transport demand in urban, sub-urban, and inter-urban areas.

5.2 HIGHWAY CAPACITY AND LEVEL OF SERVICE

One of the main issues faced by traffic engineers refers to the estimation of the traffic volume that can be serviced by the transportation facility. How many aircraft could land and take-off from the airport for an hour, or how many cars could pass along road segments are questions that need to be answered by transportation experts when analyzing and planning future development of transportation facilities. A logical question that also arises is under which operating conditions the expected traffic volume could be served. Traffic engineers calculate *capacity* of highway, street and other transportation facilities in order to implement appropriate traffic control, as well as to derive conclusions about potential expansion of transportation facilities. Airport capacity and rail systems capacity largely depends on applied control strategies. In other words, operators and dispatchers can, by their actions, significantly increase, or reduce airport capacity, or rail system capacity.

On the other hand, the greatest influence on highway capacity is driver behavior. Relations speed-flow and flow-density existing in traffic flows are studied, both theoretically and empirically, over the last few decades. The *Highway Capacity Manual* (HCM), published by the Transportation Research Board of the National Academies (TRB), combines the theoretical and empirical knowledge and makes recommendations regarding the highway capacity computation, as well as the assessment of the level of service offered to the users. HCM also offers analysis and recommendations for the traffic management at unsignalized and signalized intersections, and traffic operations along freeways, multilane highways, and two-lane highways. The first *Highway Capacity Manual* was published in 1950. The presentation in this book, related to highway capacity and level of service, is based on the HCM.

Highway capacity depends on a number of various factors. Geometric conditions (number of lanes, the width of the lanes and shoulders, horizontal and vertical alignment, grades, etc.), traffic conditions (directional distribution, lane distribution, percentage of heavy vehicles (trucks, buses, and recreational vehicles) in the traffic flow, percentage of turning movements, etc.), and traffic control strategies (established speed limits, traffic signals, location of STOP and YIELD signs, etc.) highly influence the freeway capacity. Capacity of transportation facility *is not* a permanent number. If we are willing

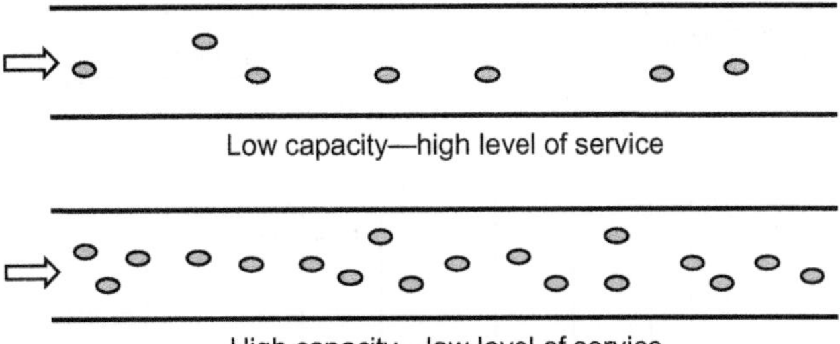

FIG. 5.1

Capacity and level of service.

to accept higher level of congestion, we can allow more vehicles to enter the highway, and vice versa (Fig. 5.1).

The density in the upper case, shown in Fig. 5.1, is much lower than the density in the lower case, and the transportation facility provides services to a small number of users, that enjoy very high level of service.

5.2.1 HIGHWAY CAPACITY AND TRAFFIC DEMAND VARIATIONS

There are monthly, daily, and hourly variations in traffic demand along a freeway. When studying freeway capacity and a level of service, analysts should take into account all these variations.

Seasonal and monthly variations in traffic demand are higher on highways serving resorts, beaches, national parks, historical places, botanical gardens, etc. Highways with significant intercity traffic have lower monthly variations.

The traffic demand differences by day of the week are also associated with the highway type. For example, traffic weekday volumes are significantly higher than the traffic weekend volumes on many urban freeways in the world. The situation is completely reverse for freeways that serve recreational traffic.

It is also well documented that traffic flow rates usually fluctuate over the course of a day and over the course of an hour.

It is usual to use *volume* and *flow rate* to measure the number of vehicles passing a point over a given point or section of a lane during a given time interval. Volume is the number of vehicles that pass a point for the duration of one hour. On the other hand, flow rate denotes the number of vehicles that pass a point through a time interval of less than 1 h (usually 15 min), but expressed as an *equivalent* hourly rate. For example, 125 vehicles recorded in a 15-min time period corresponds to an hourly rate of 500 vehicles/h:

$$\left(\frac{125\,\text{vehicles}}{15\,\text{min}} = \frac{125\,\text{vehicles}}{0.25\,h} = 500\,\text{vehicles/h}\right)$$

Transportation planners and engineers must decide on the value of the traffic flow for which they want to design transportation facility. Let us consider Fig. 5.2.

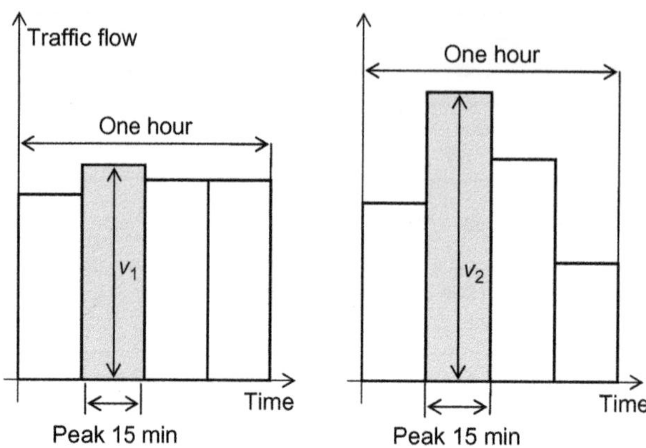

FIG. 5.2

Peak flow rates υ_1 and υ_2.

Fig. 5.2 shows 15 min peak flow rates and a hour volumes. Let us consider the situation when the hourly volume is equal in both cases that are shown in Fig. 5.2. The 15-min peak flow rate υ_1 is significantly higher than the 15 min peak flow rate υ_2. Transportation facilities are not designed to accommodate the hourly traffic volume, since it could produce an oversaturated traffic conditions for a considerable part of the hour. In other words, we would make a significant design mistake if we used in our calculations hourly volume instead of 15 min peak flow rate.

The HCM recommendation to the traffic engineers is to analyze the peak 15 min of flow during the analysis hour. The HCM exploits the peak hour factor (*PHF*) to change hourly volume into a peak 15-min flow rate. The *PHF* represents the ratio of total hourly volume to the peak flow rate inside the peak hour, ie:

$$PHF = \frac{\text{Hourly volume}}{\text{peak flower rate(within the peak hour)}} \tag{5.1}$$

The *PHF* value indicates traffic flow variations. In the case of lower *PHF* values, traffic flow has high variability, while the high *PHF* values indicate less flow variation within the hour. When 15 min periods are used, the *PHF* may be calculated in the following way:

$$PHF = \frac{V}{4 \cdot V_{15}} \tag{5.2}$$

where:

PHF is peak hour factor;
V is hourly volume (veh/h); and
V_{15} is volume during the peak 15 min of the analysis hour (veh/15 min).

The hourly volume could be converted into a peak flow rate, by using the known value of the *PHF*, ie:

$$\upsilon = \frac{V}{PHF} \tag{5.3}$$

where υ represents flow rate for a peak 15-min period, expressed in vehicles per hour.

5.2.2 FREEWAYS

HCM defines freeways as "separated highways with full control of access and two or more lanes in each direction dedicated to the exclusive use of traffic." The freeway capacity analysis helps us in estimating current level of service. We also perform the freeway capacity analysis in order to calculate number of lanes needed for target LOS, as well as to predict when freeway capacity will be exceeded.

Freeways are composed of merge, diverge, weaving, and basic freeway segments. In the case of merge segments, two or more traffic streams mix and create a single traffic stream. Along a diverge segment, a single traffic stream divides and create two or more individual traffic streams. According to HCM, the span of a merge segment is from the point where the boundaries of the travel lanes of the merging roadways meet to a point 1500 ft downstream of that point. The span of a diverge segment is from the point where the boundaries of the travel lanes of the merging roadways meet to a point 1500 ft upstream of that point (Fig. 5.3).

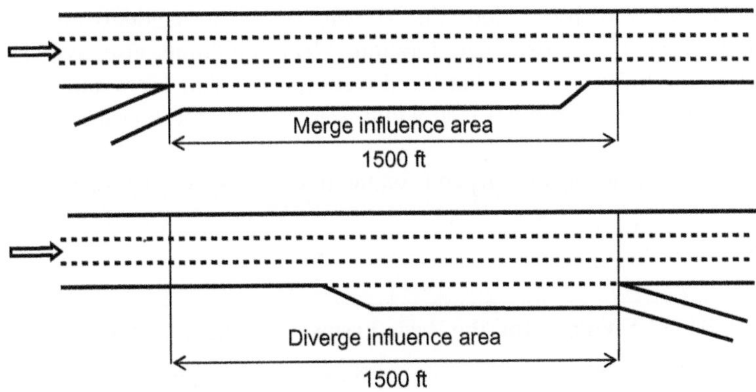

FIG. 5.3

Merge and diverge influence areas.

Weaving segments are formed when a diverge segment is close to a merge segment. In such a situation, two or more traffic streams cross paths along a sizeable length of freeway without help of traffic control devices. Weaving segments are characterized by entry and exit points. These points are defined as the points where the appropriate edges of the merging and diverging lanes meet. The length of a weaving segment is composed of the base length of the weaving segment, 500 ft upstream of the entry point to the weaving segment, and 500 ft downstream of the exit point from the weaving segment (Fig. 5.4).

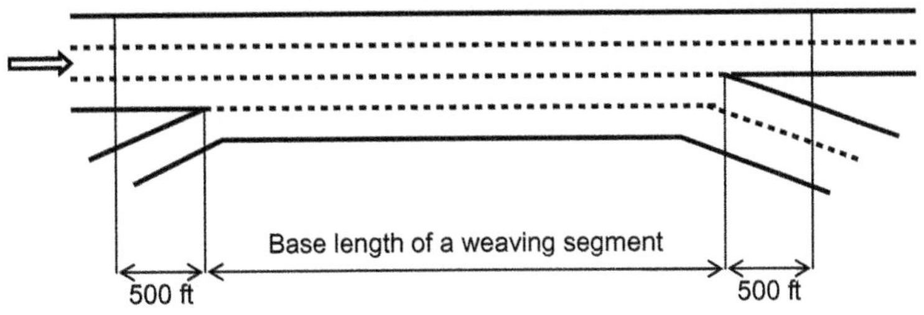

FIG. 5.4

Weaving influence area.

Basic freeway segments are segments along the freeway that are not within merge, diverge and weaving influence areas.

The *Highway Capacity Manual* (HCM, 2010) defines capacity in the following way:

> Capacity is the maximum sustainable hourly flow rate at which persons or vehicles reasonably can be expected to traverse a point or a uniform section of a lane or roadway during a given time period under prevailing roadway, environmental, traffic, and control conditions.

The capacity could be expressed in vehicles per hour, passenger-car equivalents (PCEs) per hour, or persons per hour, depending on the type of transportation facility element and/or the type of transportation study being performed.

The HCM calculates the capacity of the basic freeway segments under *base conditions* which include good weather, good visibility along the freeway, no traffic incidents or accidents, no work zone doings, and no severe pavement deterioration. Ideal conditions also assume that there are no heavy vehicles in the traffic stream, that the drivers are frequent users and commuters that are familiar with the freeway, that minimum lane width is 12-ft lane, and that the right-side clearances are 6 ft.

The basic freeway segment capacity depends on the free-flow speed (FFS). Table 5.1 shows basic freeway segment capacity values and corresponding FFS values.

Table 5.1 *FFS* Values and the Basic Freeway Segment Capacity Values

Free Flow Speed (mi/h)	Base Capacity (pc/h/ln)
75	2400
70	2400
65	2350
60	2300
55	2250

pc/h/ln, *passenger cars per hour per lane.*

The *Highway Capacity Manual* (HCM, 2010) defines *level of service* for freeways in the following way: "A quality measure describing operational conditions within a traffic stream, generally in terms of such service measures as speed and travel time, freedom to maneuver, traffic interruptions, and comfort and convenience."

The HCM recommends the following rating scale A–F for the level-of-service (Table 5.2), which indicate best to poorest traffic operations:

Denoting the level of service by letters from A to F has the same meaning as denoting with the help of linguistic expression like "excellent," "good," "average," etc (Fig. 5.5).

Table 5.2 The Rating Scale A–F for the Level of Service

Level of Service	Traffic Flow Characteristics
Level of service *A*	Flow is characterized by the free flow traffic conditions. Vehicles are not prevented to maneuver within the traffic flow. Any potential traffic incident is absorbed without difficulty.
Level of service *B*	Traffic flow is reasonably free, and vehicles' ability to maneuver is to some extent limited. Drivers enjoy high overall level of physical and psychological comfort. Potential traffic incidents are absorbed without difficulty.
Level of service *C*	Traffic flow is stable, but vehicles' freedom to maneuver (lane changes) is limited. Minor traffic incidents are absorbed, but more serious incidents could create queues of vehicles.
Level of service *D*	Traffic conditions are approaching to unstable flow. Vehicle speeds are decreased, and a vehicles' freedom to maneuver is significantly limited. Traffic incidents cannot be absorbed, and even small traffic incidents generate vehicle queues.
Level of service *E*	Traffic flow is unstable. Vehicle in a flow are narrowly spaced. Vehicles' ability to maneuver is enormously limited, and drivers experience extensive physical and psychological discomfort. Even a lane change, or a vehicle that enter from a ramp can cause a disruption wave and delays.
Level of service *F*	Traffic flow has breakdown. Vehicle queues establish instantly behind points where arrival flow rate momentarily exceeds the departure rate (point of small incident, on-ramps, off-ramps). Vehicles stop and go in cycles, and vehicle queues appear and dissipate in cycles.

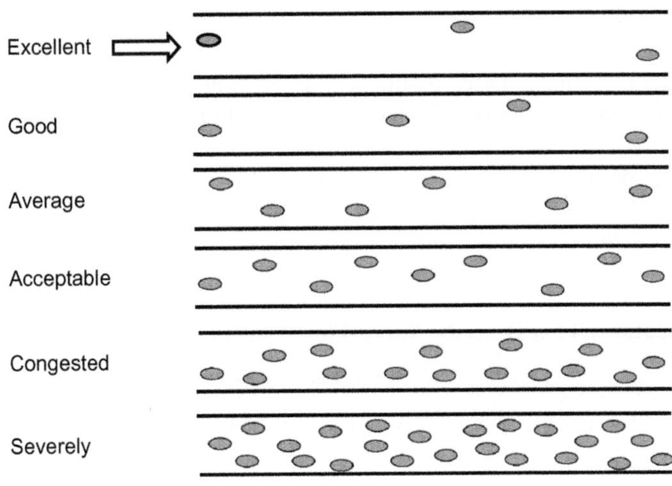

FIG. 5.5

Denoting the level-of service by the linguistic expressions.

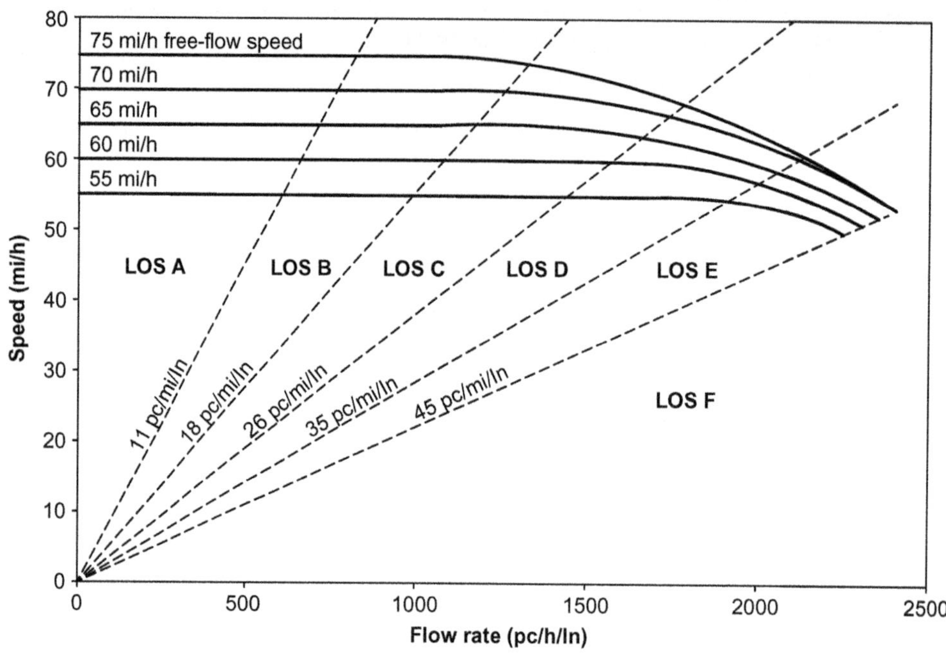

FIG. 5.6

Level of service for basic freeway segments.

From Transportation Research Board, 2010. Highway Capacity Manual. *National Research Council, Washington, DC. Special Report.*

Fig. 5.6 shows the LOS that is based on the base speed-flow curves. The lines of constant slopes beginning at the origin represent density. Each of the defined LOS is characterized by a specific range on the speed-flow curves.

The equations that define each of the curves shown in Fig. 5.6 are given in Table 5.3.

Table 5.3 The Equations That Describe Speed-Flow Curves

		Flow Rate	
FFS **(mi/h)**	**Breakpoint (pc/h/ln)**	**[0, Breakpoint)**	**(Breakpoint, capacity]**
75	1000	75	$75-0.00001107\cdot(v_p-1000)^2$
70	1200	70	$70-0.00001160\cdot(v_p-1200)^2$
65	1400	65	$65-0.00001418\cdot(v_p-1400)^2$
60	1600	60	$60-0.00001816\cdot(v_p-1600)^2$
55	1800	55	$55-0.00002469\cdot(v_p-1800)^2$

The HCM recommends not doing interpolating between the basic curves. The *FFS* should be always rounded to the nearest 5 mi/h. The levels of service, in the case of basic freeway segments are based on density (Table 5.4).

Table 5.4 Level of Service for Basic Freeway Segments

LOS	Density (pc/mi/ln)
A	≤ 11
B	$>11-18$
C	$>18-26$
D	$>26-35$
E	$>35-45$
F	Demand exceeds capacity >45

5.2.3 METHODOLOGY FOR THE CAPACITY ANALYSIS, LOS, AND THE LANE REQUIREMENTS

The HCM recommends the following methodology (Fig. 5.7) to estimate the level of service, as well as to determine the number of lanes required to provide a target LOS for a known demand volume:

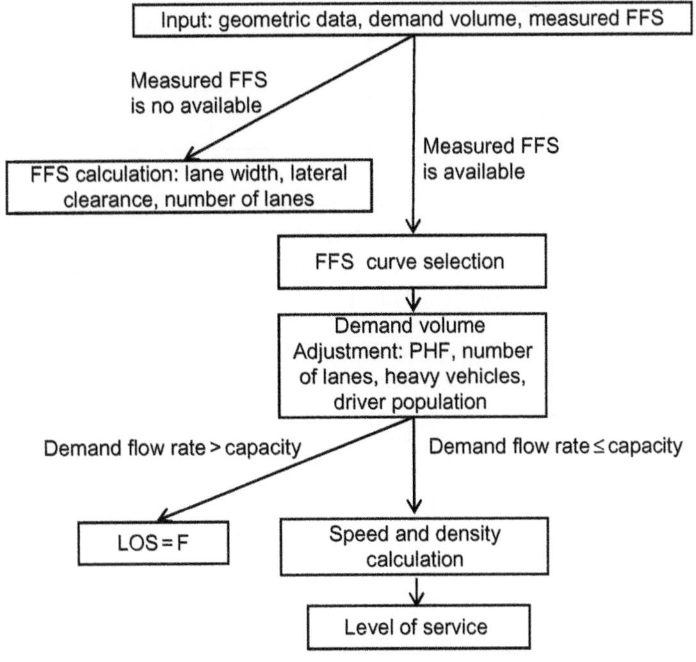

FIG. 5.7

The methodology to estimate level of service, as well as to determine the number of lanes required to provide a target LOS for a known demand volume.

Let us explain in more detail the HCM methodology shown in Fig. 5.7. The HCM suggests using the following relation for the estimation of the *FFS* (when *FFS* is not available):

$$FFS = 75.4 - f_{LW} - f_{LC} - 3.22 \cdot TRD^{0.84} \quad (5.4)$$

where:

FFS is free flow speed of the basic freeway segment (mi/h);
f_{LW} is adjustment for lane width (mi/h);
f_{LC} is adjustment for right-side lateral clearance (mi/h); and
TRD is total ramp density (ramps/mi).

The empirical investigations showed that there is a significant impact of the merging and diverging vehicles on *FFS*. This impact is measured by the total ramp density *TRD*, which is defined as the total number of on ramps and off ramps located between 3 miles upstream and 3 miles downstream of the midpoint of the basic freeway segment, divided by 6 miles.

The HCM defines the base condition for lane width is 12 ft or greater. There are negative effects (reduction in *FFS*) in the cases when the average lane width across all lanes is less than 12 ft. Table 5.5 shows the adjustments that are related to the effects of narrower average lane width.

Table 5.5 The Adjustments (Reduction in *FFS*) That Are Related to the Effects of Narrower Average Lane Width

Average Lane Width (ft)	Reduction in *FFS*, f_{LW} (mi/h)
≥ 12	0
$\geq 11-12$	1.9
$\geq 10-11$	6.6

The base condition for right-side lateral clearance is 6 ft or greater. The lateral clearance is measured from the right edge of the travel lane to the closest lateral obstruction. The base condition assumes that the right-side lateral clearance must be 6 ft or greater. Table 5.6 shows the adjustments to the base *FFS* that should be made because of the existence of obstructions closer than 6 ft to the right travel lane edge.

Table 5.6 Adjustment to *FFS* for Right-Side Lateral Clearance, f_{LC} (mi/h)

	Lanes in One Direction			
Right-Side Lateral Clearance (ft)	2	3	4	≥ 5
≥ 6	0	0	0	0
5	0.6	0.4	0.2	0.1
4	1.2	0.8	0.4	0.2
3	1.8	1.2	0.6	0.3
2	2.4	1.6	0.8	0.4
1	3.0	2.0	1.0	0.5
0	3.6	2.4	1.2	0.6

When estimating level of service, the demand volume under prevailing conditions V (veh/h) should be converted into the demand flow rate under equivalent base conditions v_p (pc/h/ln) in the following way:

$$v_p = \frac{V}{PHF \cdot N \cdot f_{HV} \cdot f_p} \tag{5.5}$$

where:

v_p is demand flow rate under equivalent base conditions (pc/h/ln);
V is demand volume under prevailing conditions (veh/h);
PHF is peak-hour factor;
N is number of lanes in analysis direction;
f_{HV} is adjustment factor for presence of heavy vehicles in traffic stream; and
f_p is adjustment factor for unfamiliar driver populations.

The HCM defines heavy vehicle "as any vehicle with more than four wheels on the ground during normal operation." Trucks, intercity buses, public transit buses, school buses, self-contained motor homes, small trucks with trailers for boats, and terrain vehicles belong to the class of heavy vehicles. The heavy-vehicle adjustment factor f_{HV} is computed as follows:

$$f_{HV} = \frac{1}{1 + P_T \cdot (E_T - 1) + P_R \cdot (E_R - 1)} \tag{5.6}$$

where:

f_{HV} is heavy vehicle adjustment factor;
P_T is proportion of trucks and buses in the traffic stream;
P_R is proportion of recreational vehicles in the traffic stream;
E_T is PCE for trucks and buses in traffic stream; and
E_R is PCE for recreational vehicles in traffic stream.

The HCM considers the following three categories of general terrain: *Level terrain*, *Rolling terrain* and *Mountainous terrain*. The Level terrain usually includes short grades of no more than 2%. This terrain enables heavy vehicles to keep the same speed as passenger cars. Rolling terrain affects heavy vehicles to decrease their speed to a large extent under those of passenger cars. On the other hand, Rolling terrain does not cause heavy vehicles to operate at *crawl* speeds (the maximum speed that trucks can maintain on an extended upgrade). Mountainous terrain causes heavy vehicles to operate at crawl speed for considerable distances.

Table 5.7 shows the PCEs for trucks and buses and RVs in general terrain segments.

Table 5.7 PCEs for Trucks and Buses and RVs in General Terrain Segments

	PCE by Type of Terrain		
Vehicle	**Level**	**Rolling**	**Mountainous**
Trucks and buses, E_T	1.5	2.5	4.5
RVs, E_R	1.2	2.0	4.0

The adjustment factor f_p values for unfamiliar driver populations range from 0.85 to 1.00. The HCM recommends that the analyst should use a value of 1.00 which is related to the assumption that the familiarized drivers are the majority of freeway users.

The HCM methodology suggests that the demand flow rate in passenger cars per hour per lane under equivalent base conditions should be compared with the base capacity of the basic freeway segment. In the case when this demand rate goes beyond the capacity, the LOS is F, and a breakdown has been recognized. In the opposite case, the analysis of the level of service continues.

In the next step, we estimate estimated speed and density of the traffic stream. For known *FFS* and v_p values, we determine the expected mean speed *S* and the density *D* of the traffic stream. The density of the traffic stream *D* equals:

$$D = \frac{v_p}{5} (\text{pc/mi/ ln}) \tag{5.7}$$

where:

v_p demand flow rate (pc/h/ln)
S mean speed of traffic stream under base conditions (mi/h).
For known density we are capable to determine level of service.

EXAMPLE 5.1

Determine the expected LOS for the freeway during the worst 15 min of the peak hour. The freeway is four-lane freeway that has two lanes in each direction. The following are freeway characteristics:

Lane width = 11 ft
Right-side lateral clearance = 3 ft
Freeway is used primarily by commuters
Peak-direction demand volume during peak hour = 1800 veh/h
Traffic composition: 4% trucks, 0% recreational vehicles
$PHF = 0.93$
Rolling terrain
TRD = 4 ramps/mi

Solution

We calculate *FFS* by using the following relation:

$$FFS = 75.4 - f_{LW} - f_{LC} - 3.22 \cdot TRD^{0.84}$$

We select the adjustment for the lane width f_{LW}. Since the lane width = 11 ft, the adjustment f_{LW} is equal to 1.9 mi/h.

We select the adjustment for right-side lateral clearance. Since the right-side lateral clearance = 3 ft, the adjustment for right-side lateral clearance f_{LC} is equal to 1.8 mi/h.

The *FFS* equals:

$$FFS = 75.4 - 1.9 - 1.8 - 3.22 \cdot 4^{0.84}$$

EXAMPLE 5.1—cont'd

$$FFS = 61.4\text{mi/h}$$

Since the *FFS* calculated value is in the interval (57.5 mi/h, 62.5 mi/h), we shall use the 60-mi/h speed-flow curve for further analysis.

In the next step, we convert the demand volume under prevailing conditions V (veh/h) into the demand flow rate under equivalent base conditions υ_p (pc/h/ln). The υ_p equals:

$$\upsilon_p = \frac{V}{PHF \cdot N \cdot f_{HV} \cdot f_p}$$

The demand volume V is equal to 1800 veh/h, while the *PHF* is 0.93. The number of lanes N in each direction is equal to 2. Since freeway is used primarily by commuters, the driver population factor f_p is equal to 1.00. The υ_p equals:

$$\upsilon_p = \frac{1.800}{0.93 \cdot 2 \cdot f_{HV} \cdot 1}$$

The heavy vehicle adjustment factor f_{HV} equals:

$$f_{HV} = \frac{1}{1 + P_T \cdot (E_T - 1) + P_R \cdot (E_R - 1)}$$

There are 4% trucks in the traffic stream ($P_T = 0.04$). There are no recreational vehicles in the traffic stream ($p_R = 0$). We have to determine a heavy-vehicle adjustment factor in the case of rolling terrain, We conclude that the PCE for trucks is equal to $E_T = 2.5$ for rolling terrain.

The heavy vehicle adjustment factor f_{HV} equals:

$$f_{HV} = \frac{1}{1 + 0.04 \cdot (2.5 - 1) + 0} = 0.943$$

The υ_p equals:

$$\upsilon_p = \frac{1.800}{0.93 \cdot 2 \cdot f_{HV} \cdot 1} = \frac{1.800}{0.93 \cdot 2 \cdot 0.943 \cdot 1} = 1026\text{pc/h/1}$$

The base capacity for a freeway with *FFS* = 60 mi/h is equal to 2300 pc/h/ln. Since $1026 \leq 2300$, we conclude that the level of service *F* does not exist. We continue our analysis.

Let us use Fig. 5.6. In our case, *FFS* = 60 mi/h, and the demand flow rate is 1026 pc/h/l, which is less than 1600 pc/h/l. We see from Fig. 5.6 that the speed is equal to 60 mi/h. The density of the traffic stream D equals:

$$D = \frac{\upsilon_p}{5}$$

$$D = \frac{1026}{60} = 17.1\text{pc/mi/ ln}$$

The density of 17.1 pc/mi/ln matches the level of service *B*.

5.2.4 THE NUMBER OF LANES REQUIRED TO DELIVER THE TARGET LOS

The speed-flow curves enable us to calculate maximum service flow rate for the targeted level of service. For example, if we want to offer to the drivers *FFS* that is equal to 70 (mi/h) and the level of service *B*, the maximum service flow rate is equal to 1250 passenger cars per hour per lane. The maximum service flow rates for the targeted level of service are shown in Table 5.8.

Table 5.8 Maximum Service Flow Rates in Passenger Cars Per Hour Per Lane for Basic Freeway Segments Under Base Conditions

	Target Level of Service				
FFS (mi/h)	A	B	C	D	E
75	820	1310	1750	2110	2400
70	770	1250	1690	2080	2400
65	710	1170	1630	2030	2350
60	660	1080	1560	2010	2300
55	600	990	1430	1900	2250

Maximum flow rate MSF_i is related to the base conditions. A service flow rate SF_i is related to the prevailing conditions and represents the maximum rate of flow that can be while LOS i is sustained through the 15-min analysis period under prevailing conditions. The SF_i is equal to:

$$SF_i = MSF_i \cdot N \cdot f_{\text{HV}} \cdot f_{\text{p}} \tag{5.8}$$

where:

N is the number of lanes required to deliver the target LOS.

The important task faced by traffic engineers is determining number of lanes required to deliver target LOS. In other words, we have to specify target LOS in the future, and to determine number of lanes required to achieve defined LOS. In the first step we have to predict future demand volume, as well as to specify lane width and lateral clearance.

The number of lanes N required to deliver the target LOS is calculated as:

$$N = \frac{v}{MSF_i} \tag{5.9}$$

$$N = \frac{V}{MSF_i \cdot PHF \cdot f_{\text{HV}} \cdot f_{\text{p}}} \tag{5.10}$$

where:

MSF_i is the maximum service flow rate for LOS I (the MSF_i values are shown in Table 5.8).

In the case when the calculated value of N is fractional, it must be rounded to the next-higher value. This rounding could, in some cases, creates level of service different than the targeted level of service.

EXAMPLE 5.2

Determine the number of lanes needed to provide LOS C during the worst 15 min of the peak hour, taking into account the following facts:

The estimated demand volume in one direction is equal to 3600 vet/h.
The general terrain is Level terrain.
There are 10% trucks, and 2% recreational vehicles in traffic stream.
The lane width of 12 ft. will be provided.
The lateral clearance of 6 ft will be provided.
The freeway will be used primarily by regular users.

$$PHF = 0.90$$

Ramp density is equal to 3 ramps/mi.

Solution

We first calculate *FFS*. The *FFS* equals:

$$FFS = 75.4 - f_{LW} - f_{LC} - 3.22 \cdot TRD^{0.84}$$

$$FFS = 75.4 - 0 - 0 - 3.22 \cdot 3^{0.84} = 67.3 \text{mi/h}$$

The quantities f_{LW} and f_{LC} are equal to zero, since the lane width and lateral clearance that will be provided on the new freeway are respectively equal to 12 ft and 6 ft. The calculated *FFS* is in the interval (62.5 mi/h, 67.5 mi/h). We use the 65-mi/h speed-flow curve for further analysis.

We use Table 5.8 to select the maximum service flow rate for *FFS* of 65 mi/h and LOS C. This value is 1630 pc/h/ln. The *PHF* is equal to 0.9. Since commuters will use future freeway, the driver population factor is equal to 1.00.

We use Table 5.7 to determine PCEs for 10% of heavy vehicles and 2% of recreational vehicles in traffic stream. For level terrain, we get $E_T = 1.5$ and $E_R = 1.2$.

$$f_{HV} = \frac{1}{1 + p_T \cdot (E_T - 1) + P_R \cdot (E_R - 1)}$$

$$f_{HV} = \frac{1}{1 + 0.1 \cdot (1.5 - 1) + 0.02(1.2 - 1)} = 0.949$$

The number of lanes N required to deliver the target LOS is equal:

$$N = \frac{V}{MSF_i \cdot PHF \cdot f_{HV} \cdot f_p}$$

$$N = \frac{3600}{1.630 \cdot 0 \cdot 9 \cdot 0.949 \cdot 1} = 2.586 \text{lanes}$$

In order to provide LOS C during the worst 15 min of the peak hour we should build 2.586 lanes. Since it is not possible to build 2.51 lanes, we should build three lanes in each freeway direction. In this way, we shall build a six-lane freeway. Let us explore what will be the speed and the density in the case of six-lane freeway.

The actual demand flow rate per lane v_p under equivalent base conditions is equal:

$$v_p = \frac{V}{PHF \cdot N \cdot f_{HV} \cdot f_p}$$

(Continued)

EXAMPLE 5.2—cont'd

$$\upsilon_p = \frac{3600}{0.9 \cdot 3 \cdot 0 \cdot 949 \cdot 1} \approx 1405 \text{pc/h/ ln}$$

The expected speed S of the traffic stream is equal:

$$S = 65 - 0.00001418 \cdot (\upsilon_p - 1400)^2$$

$$S = 65 - 0.00001418 \cdot (1405 - 1400)^2$$

$$S = 64.999 \text{mi/h}$$

The density D is equal:

$$D = \frac{\upsilon_p}{S}$$

$$D = \frac{1405}{64.999} = 21.615 \text{pc/mi/ ln}$$

The density value corresponds to the LOS C.

5.3 "ULTIMATE" AND "PRACTICAL" CAPACITY OF BUS STATIONS

The bus stop stations enable stopping of buses performing transport of passengers between their ultimate origins and destinations. In general, the size and nature of a bus stop stations can be different ranging from a roadside bus stop with a little in not at all facilities for passengers to the purposely bus stop station providing a rather wide range of facilities and services to buses, their crews, and users of transport services—passengers, and their accompanies. In cases of the very low intensity of the arriving and departing buses (vehicles/unit of time) and their passengers, the roadside stop station will be sufficient to accommodate them efficiently and effectively. The first implies provision of sufficient space for buses to stop and passengers to disembark and embark them in the short time. The other implies provision of the sufficient space for passengers to wait for the bus arrival(s) and for crossing the stop after disembarking the buses. If this demand is high, the off-road bus stop station will be needed in order to accommodate both buses and passengers efficiently by not causing the road traffic congestion. The other difference is that most bus routes/lines only transit through the stop station(s). Most of them, however, terminate and begin at the off-road stop stations, ie, terminal (s) or bus station(s). This station(s) is usually located in the central urban areas and connected by the local urban public transport systems enabling accessibility for passengers. In addition, they can be located very close to the railway stations, thus enabling passengers to change transport mode—from bus to rail, and vice versa, on their medium to long-haul trips between origins and destinations.

The buses arrive at the bus station, disembark the arriving passengers, board the departing passengers, perform the necessary formalities, and then depart. In some cases, this time may include the "buffer" time between arrival and departure, which can take few hours. In cases of such an even longer-term parking needs buses are usually parked at the depots with facilities and equipment for their servicing and cleaning.

The bus station consists of the bus-side area and the user/passenger-side area. The bus-side area includes access roads and lines to platforms for parking buses. These platforms are faced with the bus passenger terminal/building accommodating the arriving and departing passengers. A simplified layout of a bus station is shown in Fig. 5.8.

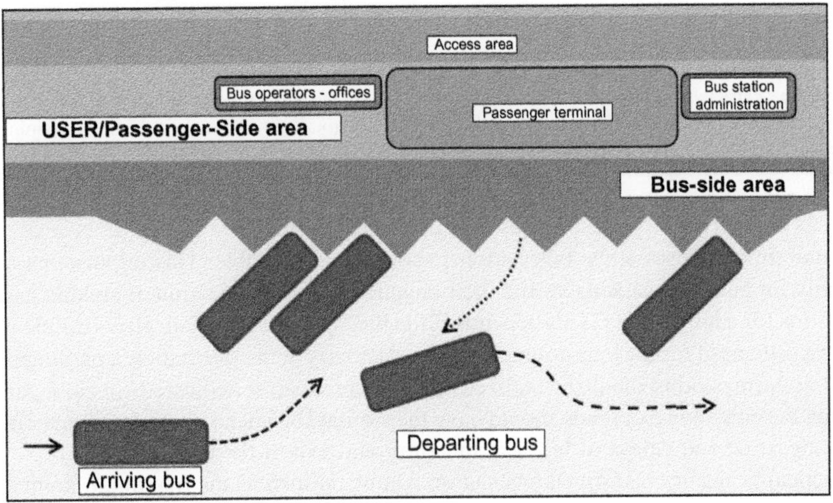

FIG. 5.8

A simplified scheme of a bus station—angle parking places/platforms.

The "ultimate" capacity of the bus-side area is that of the parking places/platforms. It can be expressed by the maximum number of buses that can be served there during a given period of time under given conditions specified by the constant demand for service. This capacity μ can be estimated as follows:

$$\mu = N/\tau \tag{5.11}$$

where:

N is the number of available parking places-platforms at the bus station; and
τ is the average bus turnaround time at the bus station's parking place/platform (h).

As well, the number of required parking places-platforms N can be estimated means by the steady-state queuing system theory, ie, Little's formula,[1] as follows:

$$N = \lambda \cdot \tau \tag{5.12}$$

where

λ is the constant intensity of flow of the arriving buses at the bus station during the time (τ).

[1]This is one of the most general and versatile laws in queueing theory. If used in an appropriate and clever ways, it can lead to remarkably simple derivations (Newell, 1982).

The bus turnaround time at the bus station τ in relations (5.11) and (5.12) includes the time for the bus's entering the platform, disembarking the arriving passengers and their baggage, short cleaning, the necessary administration if any, embarking the departing passengers and their baggage, and departing the platform.

EXAMPLE 5.3

Let the given bus station has: $N=20$ parking places/platforms. The average bus turnaround time is: $\tau=30$ min, ie, 0.5 h.

The "ultimate" capacity of this station will be: $\mu=20/0.5=40$ buses/h. In addition, let the intensity of arriving buses at the given bus station is: $\lambda=20$ buses/h, and the average turnaround time of each of them: $\tau=30$ min. The required number of parking places/platforms will be: $N=20\cdot(30/60)=10$.

The turnaround time of buses at the bus station τ should be reasonable. This implies that it should be spent primarily for handling passengers and their baggage with minimization of parking at the parking places/platforms for a longer time. This depends if the bus operators use their fleets efficiently. In such case, there is a little need for the long-term parking particularly at the bus station's parking places. Otherwise, the long-term parking should be realized at the rather separate dedicated but close parking areas.

In addition the entry and exit roads should have the similar (balanced) "ultimate" capacities in order to prevent congestion and delays of buses at the entry and exit of the station.

The "practical" capacity of a given bus station can be defined as the maximum number of buses, which can be accommodated during a given period of time (usually 1 h) under conditions of the specified maximum average delay of their acceptance. This delay depends on the ratio between the intensity of the arriving buses and their service rate, ie, the above-mentioned "ultimate" capacity of the bus-side area. If this ratio is lower than 1.0, this delay is usually practically negligible. It becomes more substantive when this ratio is greater than 1.0, ie, in this case the intensity of arriving buses is greater than the "ultimate" capacity of the bus-side area of a given bus station. According to the above-mentioned notation, assuming that: $\lambda(t)>\mu(t)$ (ie, $\rho(t)=\lambda(t)/\mu(t)>1.0$), the number of buses $n(t)$ waiting for accommodation in the bus-side area at some time t, can be estimated as follows:

$$n(t)=\max\{0;n(t-\Delta t)+[\lambda(t)-\mu(t)]\Delta t\} \tag{5.13}$$

where:

Δt is the time interval between successive observation of the bus queue (min); and
$n(t-\Delta t)$ is the number of buses in the queue waiting to enter the bus station at *time* $(t-\Delta t)$.

The other symbols are analogous to those in the previous relations. The waiting time/delay $w(t)$ of the last bus in the queue at time t at the entry of bus station, is equal to:

$$w(t)=n(t)/\mu(t) \tag{5.14}$$

In addition to an average waiting time, this waiting time $w(t)$ can be used as an indicator of service quality provided to the buses in the bus-side area of a given bus station. If limiting it, the maximum acceptable arrival flow of buses under given conditions can be defined as the "practical" capacity of the bus-side area under given conditions. For example, if the maximum waiting time is limited to: $w(t) = w^*$, then the maximum acceptable bus arrival rate at the bus-side area of a given bus stations, ie, its "practical" capacity will be equal as follows:

$$\lambda^*(t) = \mu(t) \cdot (w^* + 1) \tag{5.15}$$

For example, if in Eq. (5.13): $n(t - \Delta t) = 0$, if: $w(t) = w = 10$ min, and if the "ultimate" capacity of the bus-side area of a given bus station is: $\mu(t) = 40$ buses/h, its maximum acceptable arrival rate, ie, the "practical" capacity at time t will be: $\lambda(t) = 40 \cdot [(10/60) + 1] \approx 47$ buses/h.

5.4 RAIL INTER-URBAN TRANSPORT SYSTEMS

5.4.1 GENERAL

The capacity of inter-urban rail traffic and transport system for serving the passenger demand relates to that of infrastructure consisting of the rail lines and stations, and vehicles—rolling stock/trains. Both "ultimate" and "practical" traffic and transport capacity is considered for the single and multi (double)-track rail lines and the line/transit and central stations/terminals. This assumes that the capacity and service level are calculated for the given system's components.

5.4.1.1 *Capacity and service level of infrastructure*

"Ultimate" capacity of a single-track line(s)

This section deals with the "ultimate" traffic and transport capacity of a single-track line. The single-track lines are usually divided into several segments by stations (sidings). In the case of two-way traffic on these single-track lines, because of meets and overtakes, trains have to wait or not for passing each other to at intermediate stations/sidings along the line and thus adhere to priority rules. A simplified scheme of these stations/sidings and track configuration there, which enables trains meetings and overtakes without stopping, is shown in Fig. 5.9.

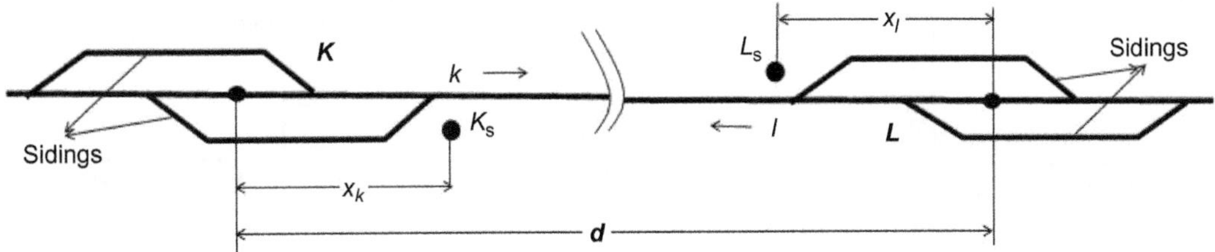

FIG. 5.9

Scheme of the bottleneck segment of the given single rail line.

Usually, the trains with a lower priority have to wait for those with a higher priority to pass first.

Depending on the type/category of trains and service they perform, these priority rules are built in the given line's timetable (during the day, week, month, and year).

In the case of mixed traffic on the line, the freight trains usually have to wait for the passenger trains. The local passenger trains have to wait for the inter-city trains. The direct freight trains are usually prioritized over the local freight collection/distribution trains. In addition, on the lines with the mixed traffic, operations of passenger and freight trains can be separated in time, the former during the day, and the latter during the night. As well, the trains are spaced along the particular segments of the line implying that only one train can be found on a particular section of the line at time. When different categories of trains run on a line (for example, passenger and freight), it is possible to calculate, for each category, the average occupancy time of the line's "bottleneck" segment including the anticipated delay(s). In this case, the "bottleneck" segment is defined by the longest distance or the longest running time between two of the stations of the line. Then, means by these values of time, the "ultimate" traffic capacity of the given single-track line can be defined and estimated as the maximum number of trains, which can pass the "bottleneck" segment and the entire line during the specified period of time (usually 24 h) under given conditions, ie, constant demand for service (Hunt and Wyman, 2010; Janić, 1984).

The main characteristics of the particular categories of trains, which are important for estimating the capacity of the single rail line(s) are their average speeds along particular segments of the given line and length. The average speeds are determined by characteristics of the track and trains themselves. The maximum lengths of trains, which permit safe and undisturbed movement along the line, are limited by the minimum sidings length (arrival-departure track) of one of the stations of the line.

Each segment of the given line is divided into the smaller sections called fixed block sections, by wayside signals, which are located at block boundaries (ie, automatic block section). Regulation of wayside signals, which indicate whether a train is allowed to head out onto a certain track section, is carried out automatically when trains pass signals. Fig. 5.10 shows a simplified scheme of operating the signaling system along the single line for the trains moving in the same direction.

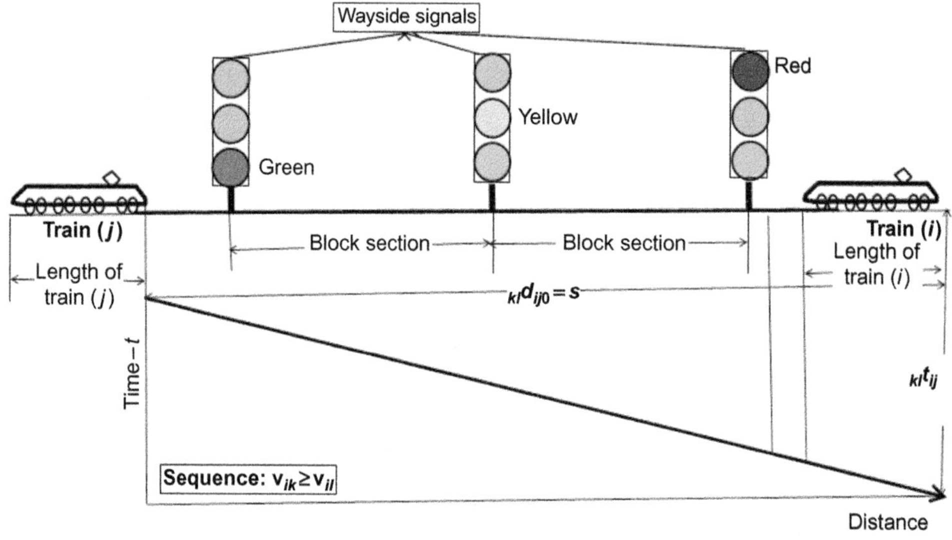

FIG. 5.10

Scheme of the signaling system of a given single rail line.

The standard characteristics of the above-mentioned railway signaling system is the three-aspect (green, yellow, red) fixed block signaling. In the case of a pair of trains i and j, a train j may enter a block section only after the train i ahead has completely cleared the block section and is protected by a stop (red light) signal. A red signal aspect means that the subsequent block section is either occupied by another train or out of service, a yellow signal aspect means that the subsequent block section is empty but the following block section is still occupied by another train, and a green signal aspect indicates that the next two block sections are empty as shown in Fig. 5.10.

The train j is allowed to enter the next block section if the signal aspect is either green or yellow, but the latter requires deceleration and stop before the next signal if this remains red. In addition, the given rail line is equipped with facilities for centralized traffic control which enable the train dispatcher to monitor and control the movement of trains. On the basis of knowledge of the position of trains along the line and the occupancy of particular sections, the dispatcher may instruct the train drivers independently of the scheduled time and requiring speed regulation.

Under the above-mentioned operating conditions of a given single-track line, the model of its "ultimate" capacity implies that the trains passing the bottleneck segment during the specified period of time (usually 24 h) are counted. One modification of the UIC (The International Union of Railway) single-track line capacity model is the one in which the capacity is not related to time-tables. In this model, the maximum number of trains (denoted by μ), which can be served on the bottleneck segment of the line, can be computed using following formula (Janić, 1984):

$$\mu = \frac{T}{t_{\text{fm}} + \bar{t} + t_{\text{zn}}} \tag{5.16}$$

where:

T is the time interval for which capacity is to be computed;
$\bar{t}$ is the average minimum inter-arrival time between entering of successive trains into the bottleneck segment of the line; and
t_{fm}, t_{zn} are the time buffers, which are determined empirically.

In relation (5.17), the minimum average inter-arrival time $\bar{t}$ can be estimated as follows:

$$\bar{t} = \sum_{ijkl} p_k \cdot q_{ik} \cdot {}_{kl}t_{ij} \cdot p_l \cdot q_{jl} \tag{5.17}$$

where:

i, j is the index of the train categories operating on the given single-track line ($i, j = 1, 2, \ldots, N$);
k, l is the index of directions from which trains enter the bottleneck segment of a given single-track line ($k, l = 1,2$);
${}_{kl}t_{ij}$ is the minimum time between entering the bottleneck segment of the trains of categories i and j from the directions k and l, respectively;
p_k, p_l is the relative traffic intensity from the directions k and l, respectively; and
q_{ik}, q_{jl} is the probability that the trains of categories i and j enter the bottleneck segment from the directions k and l, respectively.

When the trains i and j enter the bottleneck segment d long from the same direction ($k = l$), three combinations of trains sequences (ij) can occur respecting their operating speed: (i) the sequence of trains with equal speed ($v_{ik} = v_{lj}$); (ii) the sequence where the train with a higher speed (the leading one) i moves in front of the train with lower speed (the following one) j ($v_{ik} > v_{jl}$); and (iii) the sequence where the train with lower speed (the leading one) i moves in front of the train a higher speed j($v_{ik} < v_{jl}$). These cases are presented by simplified time-distance diagrams in Fig. 5.11A–C, respectively. The sequence of trains (ij), which move in opposite directions ($k \neq l$), independently on their speeds, is shown in Fig. 5.12.

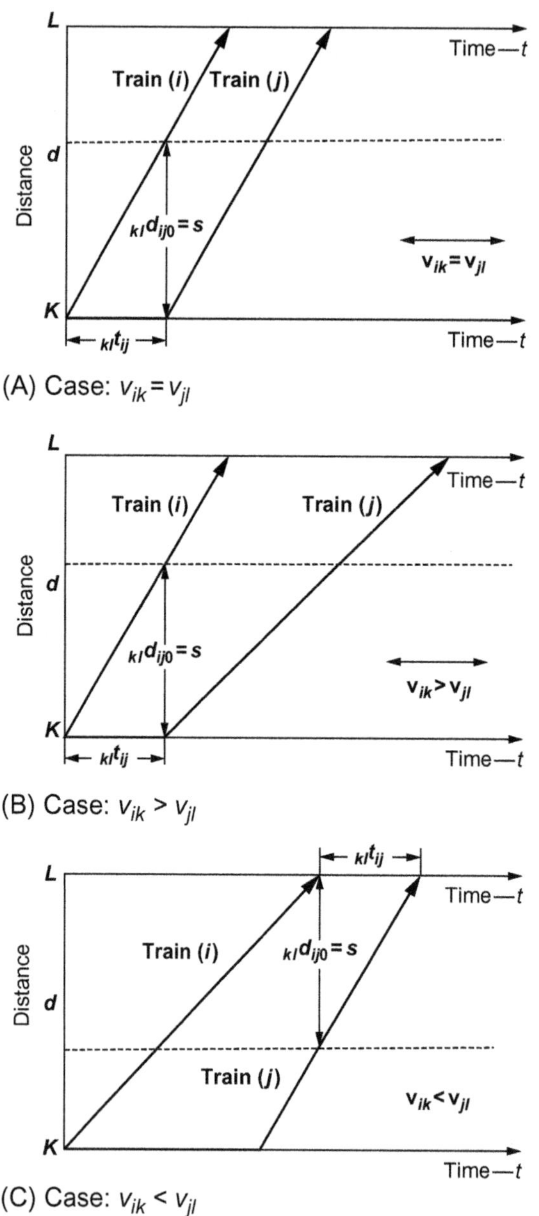

FIG. 5.11

Simplified time-distance diagram for different combinations of trains entering the bottleneck segment of a single-track line from the same direction ($k = l$). (A) Case: $v_{ik} = v_{jl}$. (B) Case: $v_{ik} > v_{jl}$. (C) Case: $v_{ik} < v_{jl}$.

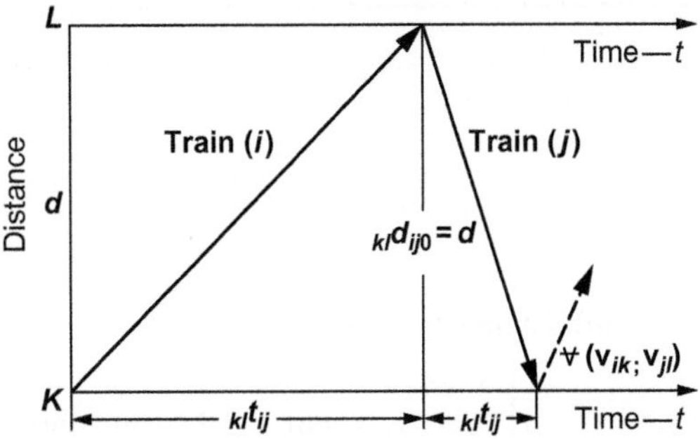

FIG. 5.12

Simplified time-distance diagram for different combinations of trains entering the bottleneck segment of a single-track line from different directions ($k \neq l$).

Based on Figs. 5.11 and 5.12, the minimum time ($_{kl}t_{ij}$) in Eq. (5.18), can be estimated as follows (Janić, 1984):

$$
_{kl}t_{ij} = \begin{cases} \begin{bmatrix} \dfrac{l_{ik}/2 + x_k + _{kl}d_{ij0}}{v_{ik}} + t_{\text{fm}} + t_{\text{zn}};\ v_{ik} = v_{jl} \\ \dfrac{l_{ik}/2 + x_k + _{kl}d_{ij0}}{v_{ik}} + d\left(\dfrac{1}{v_{jl}} - \dfrac{1}{v_{kl}}\right) + t_{\text{fm}} + t_{\text{zn}};\ v_{ik} > v_{jl} \end{bmatrix} k = l \\ \dfrac{_{kl}d_{ij0}}{v_{jl;v}} + t_{\text{fm}} + t_{\text{zn}};\ \begin{bmatrix} v_{ik} < v_{jl}, k = l \\ \forall(v_{ik}; v_{jl}, k \neq l] \end{bmatrix} \end{cases} \tag{5.18}
$$

where:

l_{ik} is the average length of train category i coming from direction k;
x_k is the distance between axis and exit signal of the station/siding k;
$_{kl}d_{ij0}$ is the minimum spacing between the trains i and j entering the bottleneck segment d from the moving from directions k and l, respectively;
v_{ik}, v_{jl} is the average speed of train i coming from direction k and of the train j coming from direction l, respectively.

The minimum spacing $_{kl}d_{ij0}$ between the trains i and j is equal to:

$$
_{kl}d_{ij0} = \begin{bmatrix} s; k = l, v_{ik} \geq v_{jl} \\ d; k \neq l, other - i, j, k, l \end{bmatrix} \tag{5.19a}
$$

where:

s is the minimum permitted distance between trains i and j, measured along the line (Fig. 5.10); and
d is the length of bottleneck segment of the line, between the axis of the stations/sidings k and l.

EXAMPLE 5.4

For example, the length of bottleneck segment of the given single-track line is: $d=12.0$ km; the distance between axes and exit signals of end bottleneck stations/sidings are approximately equal to: $x_1=x_2=0.8$ km. The bottleneck segment d is divided by the wayside signals into the shorter block sections, each of approximately equal lengths of 2 km (see Fig. 5.10). The train categories and their characteristics are given in Table 5.9.

Table 5.9 Train Categories and Their Characteristics

Train Category i	Description	Average Speed v_i (km/h)	Average Length l_i (m)	Proportion in the Mix q_i (%)
1	All freight	53	500	50
2	Local passenger	61	150	35
3	International passenger	66	275	15

From Eq. (5.19), the matrix of minimum train spacing intervals is determined as follows:

$$_{kl}d_{ij0}=\begin{bmatrix} s=4\text{km}; & k=l, v_{ik}\geq v_{jl} \\ d=12\text{km}; & k\neq l, other-i,j,k,l \end{bmatrix} \tag{5.19b}$$

The time buffers are set up to be: $t_{fm}=t_{zn}=2.5$ min. By using the above-mentioned values of inputs in Eqs. (5.16)–(5.19), the "ultimate" capacity of the given single-track line is calculated as follows: $\mu=78$ trains/24 h if: $p_k=p_l=100\%$, ie, if the trains are moving only in the single direction. It is: $\mu=63$ trains/24 h if: $p_k=p_l=50\%$, ie, if the trains are moving in the opposite directions in the equal proportions. If the length of the bottleneck segment increases, the capacity will decrease, and vice versa. In addition, if the proportion of the slower trains in the traffic mix increases, the "ultimate" capacity of the single-track line will decrease, and vice versa (Janić, 1984). As well, the "ultimate" capacity of double-track rail line can be calculated by multiplying that of the single-track line by 2, if all conditions along both tracks are identical. In our example, this will imply: $p_k=p_l=100\%$ (k and l are along the separate tracks), which gives: $\mu(2)=2\cdot78$ trains/24 h$=156$ trains/24 h.

By selecting the proportions of the particular train categories, the "ultimate" capacity can be calculated for the single- and double-track line(s) accommodating exclusively passenger and/or freight trains. Analogously, this capacity for the high speed rail (HSR) lines can be calculated.

"Practical" capacity of a single rail line(s) and service level

In addition to the above-mentioned "ultimate" capacity, the additional important factor to be considered in planning and carrying out operations of trains along a single-track line is its "practical" capacity. In this context, this capacity is expressed by the maximum number of trains accommodated on the "server," ie, bottleneck segment of the line during a given period of time, when each of these trains is imposed the maximum average delay. This delay is generally caused by two reasons: (i) the randomness of real times in which particular trains arrive and request entering the bottleneck segment can compromise the minimum distance separations between them and thus require delaying some of them (in this case the intensity of demand for entering the bottleneck segment as the server is lower than its "ultimate" capacity, thus indicating trains' delays as rather stochastic); and (ii) persistence of train dispatchers to maintain the planned priorities of individual trains operating along the given line, ie, its bottleneck segment, also by imposing delays on some of them. Under such conditions, the average delay per train requesting service at the bottleneck segment of the given single-track line operating under steady-state conditions during a given period of time can be estimated as follows:

$$w = \frac{\rho}{2\mu(1-\rho)} \tag{5.20}$$

This is a well-known simple formula from *M/D/*1 queuing system model where $\rho < 1.0$ (M=Poisson arrival flow; D=Deterministic service time; 1=single server). The parameter $\rho = \lambda/\mu$ represents the rate of utilization of the server, ie, the bottleneck segment. In this context, it is called level of service (LOS) ratio. The parameter λ represents the intensity of train demand for service assumed to be according to Poisson law. The average service time of each train at the bottleneck segment is assumed to be deterministic, ie, equal to: $1/\mu$. Fig. 5.13 shows an example of typical relationship between the average delay and the utilization rate of the "ultimate" capacity of the given single-track line.

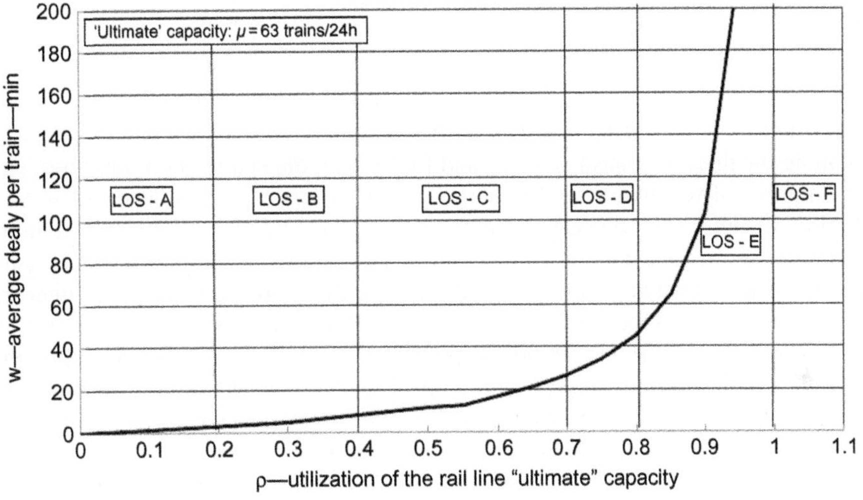

FIG. 5.13

An example of the relationship between the average delay per train and the rate of utilization of the "ultimate" capacity of a given single-track line.

As can be seen, the average delay increases more than proportionally, ie, exponentially with increasing of the utilization of the "ultimate" capacity of the given single-track line ρ. In addition, six grades of LOS provided to trains based on the values of ρ can be distinguished as described in Table 5.10.

Then if, for example, the average delay per train is adopted to be: W=20 min, the LOS will be at the grade "C" and the utilization rate: ρ=0.6. Given the "ultimate" capacity of: μ=63 trains/24 h, the "practical" capacity enabling the above-mentioned average delay per train will be: $\lambda = \rho \cdot \mu = 0.6 \cdot 63 = 38$ trains/24 h. The "practical" capacity of the double-track line can be similarly calculated by setting from the above-mentioned case: $\mu(2) = 2 \cdot 78$ trains/24 h = 156 trains/24 h. If: ρ=0.6 for both tracks, this will be: $\lambda(2) = \rho \cdot \mu(2) = 0.6 \cdot 156 = 4$ trains/24 h. As well, as at the "ultimate" capacity, the "practical" capacity of the single and double rail line(s) accommodating exclusively passengers and/or freight trains can be calculated similarly.

Table 5.10 An Example of Gradation of the Level of Service at U.S. Railways (CS, 2007; Hunt and Wyman, 2010)

Grade of LOS	Description		$\rho = \lambda/\mu$
A B C	Below capacity	Low to moderate train intensity with capacity to recover from incidents and enable maintenance	0.0–0.2 0.2–0.4 0.4–0.7
D	Near capacity	Heavy train intensity with moderate capacity to recover from incidents and enable maintenance	0.7–0.8
E	At capacity	Very heavy train intensity with the very limited capacity to recover from incidents and enable maintenance	0.8–1.0
F	Above capacity	Unstable train intensity; conditions for breaking down of services	>1.0

"Ultimate" and "practical" capacity of rail station(s), and service level

Passenger stations. The "ultimate" capacity of the passenger rail stations(s) can be calculated for the stations along the lines, ie, transit stations, and for those at the end/begin of rail lines, ie, terminal stations. The number of required tracks in the given rail passenger station depends on the intensity of traffic (trains per unit of time) and the average trains' stopping time. The transit stations located on a single-track line can have only two tracks. Those located on the double-track lines usually have two through tracks and also two or more stopping tracks. The terminal stations of either track lines where the trains start and end their transport services usually have several tracks with the associated platforms for passengers waiting before boarding and passing through after dis-boarding their trains when starting and ending their journeys, respectively. The common width of the platforms at the contemporary rail stations is 10–12 m and the length in proportion to the length of the longest trains handled there (Teixeira, 2011). Fig. 5.14A and B shows the simplified scheme of the layout of tracks at two above-mentioned types of rail passenger stations—transit and terminal.

In the case of the transit stations, the incoming trains stop directly on the main tracks. Their stopping/dwell time should generally be shorter than the inter-arrival time of the incoming trains. For example, at German railways, this dwell time typically amounts to 2–3 min (Clausen, 2011). In the case of the terminal stations usually located in the city centers the incoming trains from different lines/directions stop at the greater number of the main tracks with the associated platforms. After disembarking and embarking passengers, these trains can continue their journey in the same or different directions compared to those they previously arrived. In addition, there may be more additional tracks for the longer parking of trains while being out of operation. In any case, the required number of tracks at either transit or terminal station can be estimated as follows:

$$N_t(T) = 2 + \lambda(T) \cdot t_s \tag{5.21}$$

$\lambda(T)$ is the intensity of arriving trains at the given station (trains/h);
t_s is the average dwell (stopping) time of a train at the given station (h); and
T is the time unit (usually 1 h).

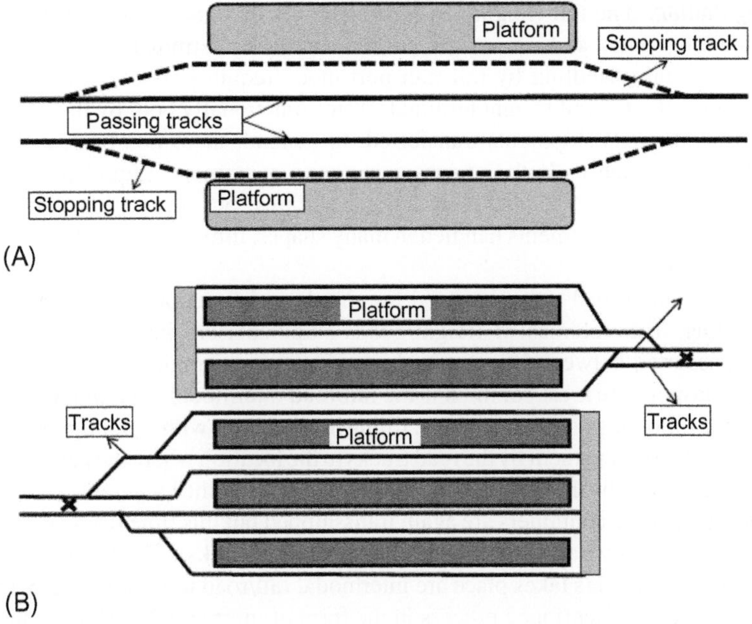

FIG. 5.14

Simplified scheme of the possible layout of tracks in the rail passenger stations. (A) Passing by stations along the double-track line (Anderson and Lindvert, 2013). (B) Begin/end station of Tokyo Shinkansen HSR (Nishiyama, 2010).

The first term in relation (5.21) indicates the minimum required number of tracks anyway. The second term indicates the required number of tracks for handling the given volume of train traffic. Consequently, the "ultimate" capacity of the given rail station can be determined as:

$$\mu_s(T) = \frac{N_t(T) - 2}{t_s} \tag{5.22}$$

For example, the central station in Tokyo (Japan), which accommodates the Shinkansen rail passenger trains (Fig. 5.14B) has: $N_t = 5$ tracks for handling traffic (not counting 2 additional tracks). The average train turnaround time includes the time for dis-boarding arriving passengers (about 4 min), time for cleaning and inspecting the train (12 min) and the time of boarding departing passengers (about 4 min), which gives: $t_s = 20$ min. This gives the station's "ultimate" capacity during the period of $T = 1$ h of: $\mu_s(T) = 5/(20/60) = 15$ trains/h. For the average train's turnaround time of: $t_s = 12$ min, the "ultimate" capacity of the station will be: $\mu_s(T) = 5/(12/60) = 25$ trains/h.

In any case, the "ultimate" capacities of the rail line and the transit and terminal stations on it need to in balance.

The "practical" capacity of the rail stations can be estimated similarly as that of the rail lines (relation 5.20). In the case of the Tokyo Shinkansen station, if the LOS is at the grade: "C–D," ie, $\rho = 0.8$, and the "ultimate" capacity is: $\mu_s(T) = 15$ trains/h, the average delay per arriving train at the station will be: $W = [(1/2)0.8/(1 - 0.8)/15] \cdot 60 = 8.0$ min. Consequently, the "practical" capacity will be: $\lambda(T) = \rho \cdot \mu_s(T) = 0.8 \cdot 15 = 12$ trains/h (or 282 trains/24 h).

Freight handling station. The rail freight handling stations can generally be categorized into freight terminals, rail shunting yards, and intermodal rail/road container terminals.

Freight terminals. Freight handling by rail transport mode requires specific loading and unloading equipment. Consequently, the rail freight terminals differ functionally both by the transport modes involved and the freight/goods shipments transferred. In general, there is a basic difference between freight/goods shipments such as bulk, general cargo, and containers. Liquid or dry bulk refers to freight/goods handled unpacked in relatively large quantities, but in uniform dimensions. General cargo refers to freight/goods shipments handled in many shapes, dimensions, and weights, thus making their handling labor-intensive.

Containers are standard load units designed for simple and functional, ie, efficient and effective mechanized handling. The volumes of freight/goods shipments are increasingly consolidated into containers and transported between shippers and receivers—by a single or combination of transport modes. In inland transport, the most common is the rail/road intermodal combination. This usually implies carrying out the following phases/activities: filling in containers with freight/goods shipments at the shippers' doors and transporting them by the road trucks to the location where they are transshipped to the rail container train; transport by the rail container train usually along the longer distance line, which ends with the location where these containers are again transshipped but this time from rail container train to the road trucks, and then delivered to the doors of receivers (by trucks). The locations where road/rail/road transshipment of containers takes place are intermodal rail/road terminals. Fig. 5.15 shows a simplified scheme of the above-mentioned process in the form of intermodal rail/road transport network.

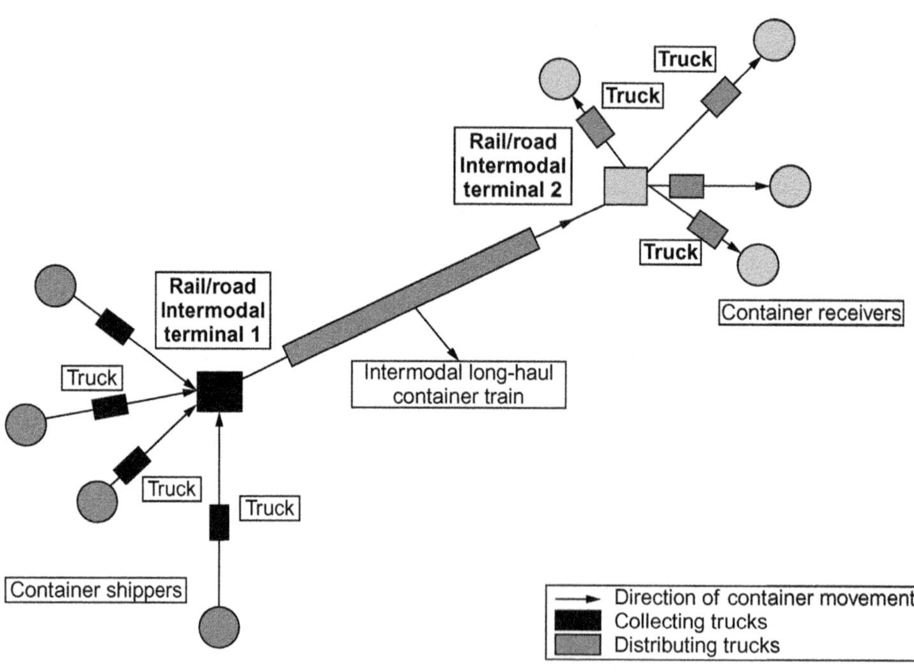

FIG. 5.15

Simplified scheme of the rail/road intermodal transport network.

Each type of the above-mentioned rail terminals including the container ones is equipped with the rail tracks enabling handling of the arriving trains where their unloading and loading takes place. This implies that a given train is first unloaded and then loaded while staying on the same track. During these processes the composition of the trains stays fixed without shunting decomposing and composing operations. The number of tracks accommodating the arriving trains depends on their intensity of arrivals and the corresponding average service, ie, unloading time, as follows (Rallis, 1977):

$$N_{\mathrm{ft}}(T) = \lambda_{\mathrm{f}}(T) \cdot t_{\mathrm{sf}} \tag{5.23}$$

$\lambda_f(T)$ is the intensity of arriving freight trains at the given station (trains/h);
t_{fs} is the average service, ie, unloading, time of a freight train at a given rail freight station (h); and
T is the time unit (1 h).

In the case of the rail container terminal handling the intermodal container trains, the unloading time (t_{sf}) of a train in relation (5.23) depends on the number and unloading/loading rate of container unloading devices. At the same terminal, two main types of these devices can be used exclusively or together: gantry cranes (rail mounted or rubber tire) and side-loaders (forklifts and/or reachstackers) (Kozan, 2006; Mocuta and Ghita, 2007).

$$t_{\mathrm{sf}} = (Q_{\mathrm{ac}} + Q_{\mathrm{dc}})/(n_{\mathrm{c}} \cdot m_{\mathrm{c}}) + t_{\mathrm{m}} \tag{5.24}$$

where:

Q_{ac}, Q_{dc} is the number of unloaded and loaded containers per a train (containers/train)
n_c is the number of unloading devices engaged for unloading a train (number/train); and
m_c is the unloading/loading rate per unloading/loading device (containers/h); and
t_m is the train maneuvering time in the terminal (h).

EXAMPLE 5.5

The arrival rate of trains at the container terminal is: $\lambda_f(T) = 5$ trains/h ($T = 1$ h), the number of containers per train: $Q_{ac} = Q_{dc} = 35$ containers/train, the number of loading/unloading devices: $n_c = 2$/train, the loading/unloading rate per device: $m_c = 20$ containers/h, and the train maneuvering time: $t_m = 1.5$ h. The processing (loading/unloading) unloading time per train is: $t_{sf} = (35+35)/(2 \cdot 20) + 1.5 = 3.25$ h.

The required number of tracks for handling these trains is: $N_{ft} = 5$(trains/h)$\cdot 3.25$ h ≈ 17 tracks. In addition, if the number of tracks at the given container terminal is, for example, $N_{ft} = 30$, and the average service time of a train is: $t_{sf} = 3.25$ h, the "practical" capacity will be equal: $\mu_{tc}(T) = N_{ft}/t_{sf} = 30/3.25 \approx 9$ trains/h. If the terminal operates during the entire day, this gives the daily capacity of: $\mu_{ct}(T) \approx 222$ trains/24 h ($T = 24$ h). If each train carries an average of 35 containers/train, the "ultimate" capacity of the terminal in terms of the number of processed containers will be: $9 \cdot 35 = 315$ containers/h and $222 \cdot 35 = 7700$ containers/24 h (day).

If, for example, the LOS grade is specified to be: $\rho = 0.6$, then the "practical" capacity of the given rail container terminal can be estimated as follows: $\mu^*_{ct}(T) = \rho \cdot \mu_{ct}(T) = 0.6 \cdot 9 \approx 5$ trains/h ($T = 1$ h) and 120 trains/24 h ($T = 24$ h). Under such conditions, the average waiting time of a train for starting its processing can be estimated by relation (5.20) as: $W = 0.6/[(2 \cdot 9) \cdot (1 - 0.6)] \cdot 60 = 5\,\mathrm{min}$.

(Continued)

EXAMPLE 5.5—cont'd

Similar consideration of the service quality can be made for trucks while being served in the intermodal rail/road terminal during unloading incoming and loading outgoing containers. In addition to the waiting time, reliability, flexibility, safety and security, and terminal accessibility during the day can also be considered as the attributes of service quality from the aspects of terminal users—trains and trucks (Ballis, 2003).

Rail shunting yard. The rail shunting yards play an important role in railway freight operations. The main function of the shunting yard is to uncouple trains and then reassemble them according to their common destinations. The experience so far has shown that the largest and most effective type of rail shunting yards is the hump/gravity yards. There, the arrived trains are generally decomposed and their wagons rolled down in order to get reassembled into the departure trains. This process requires that these yards have a specific layout and consist of four different components: (i) the receiving yard with sidings/tracks whose number depends on the intensity of arriving trains and their average track/ siding occupancy time; (ii) hump with the height of 2.5–5.5 m compared to the level of the classification yard enabling rolling down wagons or wagon groups from the decomposed train(s) towards the specified track(s)/siding(s) of the classification yard; the pushing speed of wagons/trains towards the hump is 5–10 km/h; the wagons' rolling down speed is regulated by automatic brakes; (iii) the classification yard with the number of tracks/sidings clustered into groups of typically 4–8, whose total number depends on the number of directions for which the new trains are assembled (typically not less than 16 but also 24, 36, 48, or even more); and (iv) the departing yard with sidings/tracks whose number depends on the intensity of assembled departing trains and their waiting time for departure (Yagar et al., 1983). In addition, the receiving and classification yard are connected by tracks enabling the shunting locomotive(s) to maneuver. Fig. 5.16 shows a typical layout of a hump/gravity yard consisting of the four different components.

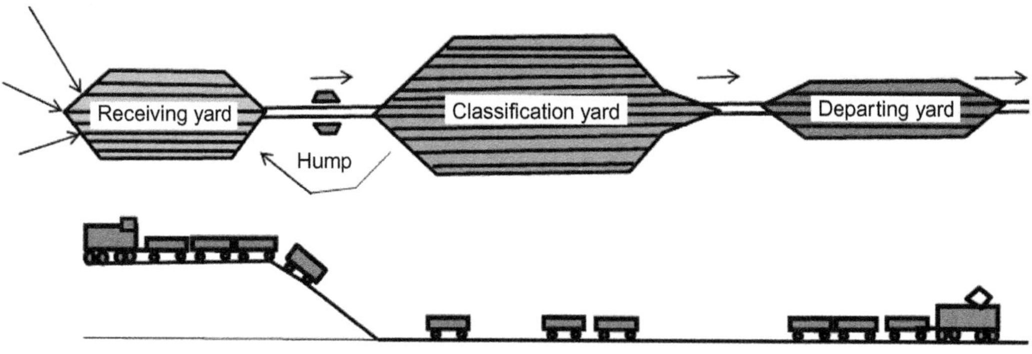

FIG. 5.16

Simplified layout of the rail hump/gravity shunting yard.

In order to estimate the "ultimate" and "practical" capacity of the above-mentioned components of a given shunting yard including the corresponding level of service to the arriving and departing trains it is needed to analyze the shunting process. This is carried out as follows: an arrived train handled at the receiving track(s) siding(s) is first inspected; then, the locomotive is uncoupled; after these, the wagons wait on the receiving track until they come on top of the queue; then, they are gradually pushed up the hump by an (usually diesel) engine while being uncoupled into the wagons and wagon groups; after passing the edge of the hump, they roll down the hump thanks to gravity and get into the specified track (s)/siding(s) of the classification yard—all wagons and wagon groups with the common destination(s). They wait there until assembling the train by the specific shunting engine(s). The last step is pulling the assembled train to the departure yard and its preparation for departure. Finally, the train departs.

The "ultimate" and "practical" capacity and service level of the above-mentioned components of a given shunting yard can be determined as follows:

(a) Receiving yard

The "ultimate" capacity of the receiving yard can be estimated as follows:

$$\mu_{ar} = \frac{N_{ar}}{\tau_{ar}} \tag{5.25}$$

where:

N_{ar} is the number of tracks/sidings in the receiving yard; and

τ_{ar} is the average minimum time, which arrived trains spend at the receiving yard, before starting their disassembling (h/train).

The time (τ_{ar}) in relation 5.25 consists of the time of train's entering the station, inspection, and finalization of documentation before disassembling. For example, if the number of receiving tracks/sidings is: $N_{ar} = 10$, and the average time of train's staying there: $\tau_{ar} = 0.5$ h, the "ultimate" capacity of the receiving yard will be: $\mu_{ar} = 10/0.5 = 20$ trains/h.

If the maximum intensity of the arriving trains during the given period of time (T) is: $\lambda_{ar}(T)$, and the average time of staying of each of them at the receiving yard is: (τ_r), the required number of tracks/sidings at the receiving yard can be estimated as follows: $N_{ar} = \lambda_{ar}(T)^*\tau_{ar}$. For example, if the intensity of arriving trains is: $\lambda_{ar}(T) = 5$ trains/h ($T = 1$ h), and the train occupancy time of the: $\tau_{ar} = 0.5$ h/train, the number of required track/siding in the receiving yard will be: $N_{ar} = 5 \cdot 1 = 5$.

The "practical" capacity of receiving yard can be estimated as follows:

$$\rho_{ar}(\text{LOS}_{ar}, T) = \frac{\mu^*_{ar}(T) \cdot \tau_{ar}}{N_{ar}} \quad \text{and} \quad \mu^*_{ar}(T) = \frac{(\rho_{ar}, \text{LOS}_{ar}) \cdot N_{ar}}{\tau_{ar}} \tag{5.26}$$

where:

$\mu_{ar}(T)$ is the "practical" capacity of the receiving yard at the given LOS (trains/h); and

LOS_{ar} is the specified level of service at the receiving yard (≤ 1.0).

The other symbols area analogous to those in relation (5.25). For example, if the above-mentioned grade LOS of "C" is specified as: $\rho_{ar}(0.7, T) = 0.7$ ($T = 1$ h), the number of receiving tracks/sidings as: $N_{ar} = 10$, and the track/siding occupancy time as: $\tau_{ar} = 0.5$ h, then the "practical" capacity of receiving yard based on relation (5.26) will be: $\mu_{ar} = 0.7 \cdot 10/0.5 = 14$ trains/h.

(b) Hump

The "practical" capacity of hump is expressed by the maximum number of trains and/or trains, which can be decomposed during a given period of time T under conditions of constant demand for service. It can be estimated as follows:

$$\mu_h(T) = \frac{T}{\tau_h} \quad \text{and} \quad \mu_{hw}(T) = \frac{T}{\tau_h} m \tag{5.27}$$

where:

τ_h is the technological time interval of the hump (h/train).

m is the average number of wagons per train to be disassembled in the receiving yard.

The first term of relation (5.27) expresses the hump's "ultimate" capacity by the number of processed trains/h and the second one by the number of processed wagons/h. The technological time interval of hump (τ_h) is defined as the time between the starting of successive disassembling of two trains at the hump after being taken from the receiving yard. This time includes the times of empty ride of the shunting (usually diesel) locomotive pushing a train to the hump, disassembling of a train into wagons and wagon groups over the hump and then compressing them again into a train in the classification yard, and eventually the final formatting of a train.

For example, if: $\tau_h = 0.5$ h, the average number of wagons per train: $m = 30$, and the operational time of the shunting yard: $T = 12$ h, the "ultimate" capacity of hump will be: $\mu_h(12) = 12/0.5 = 24$ trains/12 h or 2 trains/h, and $\mu_{hw}(12) = (12/0.5) \cdot 30 = 720$ wagons/12 h or 60 wagons/h.

The corresponding "practical" capacity for the specified LOS (level of service) is determined as:

$$\mu_h^*(T) = \rho_h(\text{LOS}_h) \cdot \mu_h(T) \quad \text{and} \quad \mu_{hw}^*(T) = \rho_{hw}(\text{LOS}_{hw}) \cdot \mu_{hw}(T) \tag{5.28}$$

where all symbols are analogous as in the previous equations.

If the grade "C" of LOS is specified, ie, $\rho_h(\text{LOS}_h) = \rho_{hw}(\text{LOS}_{hw}) = 0.7$, the corresponding "practical" capacities will be: $\mu_h(12) = 0.7 \cdot 24 \approx 17$ trains/12 h or ≈ 1 trains/h, and $\mu_{hw}(12) = 0.7 \cdot (12/0.5) \cdot 30 = 504$ wagons/12 h or 42 wagons/h.

If $n_{ar}(T)$ is the number of trains simultaneously waiting in the receiving yard, the average waiting time of a train for starting its disassembling can be estimated as follows:

$$w_{ar}(T) = (1/2) \cdot n_{ar}(T)/\mu_h(T) \tag{5.29}$$

For example, if: $n_{ar}(T) = 10$ and $\mu_h = 2$ train/h, the average waiting time of a train will be: $w_{ar}(T) = (1/2) \cdot (10/2) = 2.5$ h/train.

By summing the times $w_{ar}(T)$ and τ_{ar} as the total train's spending time in receiving yard, the corresponding "ultimate" and "practical" capacity and the required number of tracks/sidings in the received yard should be recalculated by Eqs. (5.26) and (5.27).

(c) Classification yard

The number of tracks/sidings in the classification yard is equal to the number of directions the trains are formatted. The maximum intensity of wagons and wagon groups arriving at the classification yard and its particular tracks/siding during the specified period of time T are proportional/equal to the hump "ultimate" or "practical" capacity. In the former case, it can be estimated as follows:

$$\lambda_{cw}(T) \equiv \mu_{hw}(T) = \frac{T}{\tau_h} \cdot m \tag{5.30}$$

$$\lambda_{cw}(T) \equiv \mu_{hw}(T) \cdot p_k(T) = \frac{T}{\tau_h} \cdot m \cdot p_k(T) \tag{5.31}$$

$$\sum_{k=1}^{K} p_k(T) = 1.0 \tag{5.32}$$

where:

$p_k(T)$ is the proportion of wagons or wagon groups in the arriving trains during time T, which will be in the trains(s) assembled for the direction k;

K is the number of directions for which trains are assembled in the classification yard.

Eq. (5.31) also indicates the intensity of collecting wagons and wagon groups for the given direction k. If the number of wagons per train formatted during time T in the classification yard for the direction k is $M_k(T)$, the time for collecting them is equal from Eq. (5.32) to:

$$\tau_{ck}(T) = M_k(T)/\lambda_{cw}(T) \tag{5.33}$$

Then, the average waiting time of a train and/or a wagon in it is:

$$w_{ck}(T) = (1/2)M_k(T)/\lambda_{cw}(T) \tag{5.34}$$

For example, if: $\lambda_{cw}(T)=2$ trains/h, the number of wagons per train: $m=30$, the proportion of wagons in the arriving train for the direction k, $p_k=0.1$, and the number of wagons in the departing train for the direction k, $M_k(T)=30$, the total waiting time will be: $\tau_{ck}(T)=30/(2\cdot 25\cdot 0.1)=6.0$ h. The average waiting of a train is: $w_{ck}(T)=(1/2)\cdot 6.0=3.0$ h. If this is considered as the service time, the "ultimate" capacity of the classification yard is:

$$\mu_c(T) = N_c(T)/w_{ck}(T) \tag{5.35}$$

Similarly, the "practical" capacity is equal to:

$$\mu_c^*(T) = \rho_c(LOS_c)\cdot\mu_c(T) \tag{5.36}$$

where:

N_c is the number of tracks/sidings in the classification yard.

The other symbols are as in the previous equations. Fig. 5.17 shows the relationship between the "ultimate" and "practical" capacity and the number of tracks/sidings in the classification yard.

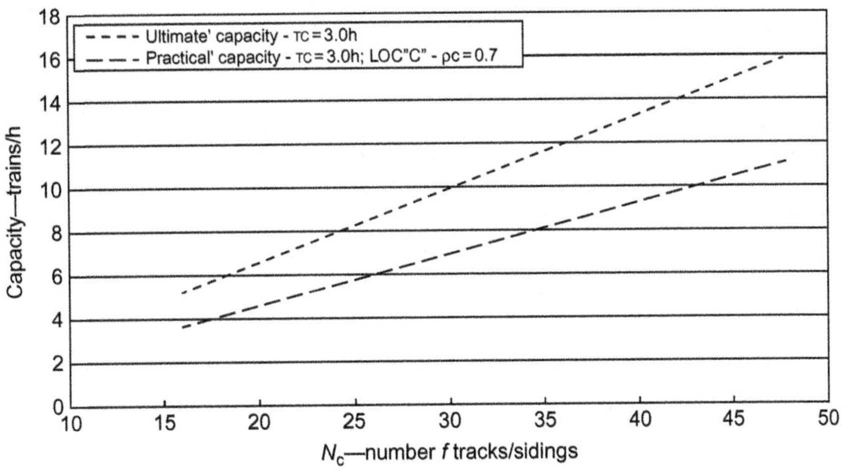

FIG. 5.17

Dependence of the capacity of classification yard on its number of tracks/sidings.

As expected, both "ultimate" and "practical" capacity of the classification yard increases linearly with increasing of the number of tracks/sidings there given their average occupancy time. It is clear that these capacities will increase with shortening of the track/siding occupancy time and decreasing of grade of LOS, and vice versa.

(d) Departing yard

The "ultimate" capacity of departing yard depends on the number of tracks/sidings there and the minimum time for the final preparation of train(s) for departure. This can be estimated as follows:

$$\mu_{\mathrm{d}}(T) = \frac{N_{\mathrm{d}}}{\tau_{\mathrm{d}}} \tag{5.37}$$

where:

N_{d} is the number of tracks/siding in the departing yard;

τ_{d} is the average minimum time for preparing trains for departure (h/train).

The "practical" capacity of receiving yard is estimated as follows:

$$\mu_{\mathrm{d}}^{*}(T) = \rho_{\mathrm{d}}(LOS_{\mathrm{d}}) \cdot \mu_{\mathrm{d}}(T) \tag{5.38}$$

where all symbols are analogous to those in the previous equations. The similar examples can be made as in the previous cases

(e) Entire shunting yard

The "ultimate" and "practical" capacity of a given shunting yard consisting of the receiving yard, hump, classification and departing yard operating in the serial order can be determined as the minimum among their capacities as follows:

- Ultimate capacity:

$$\mu(T) = \min\left[\mu_{\mathrm{ar}}(T); \mu_{\mathrm{h}}(T); \mu_{\mathrm{c}}(T); \mu_{\mathrm{d}}(T)\right] \tag{5.39}$$

- "Practical" capacity

$$\mu^{*}(T) = \min\left[\mu_{\mathrm{ar}}^{*}(T); \mu_{\mathrm{h}}^{*}(T); \mu_{\mathrm{c}}^{*}(T); \mu_{\mathrm{d}}^{*}(T)\right] \tag{5.40}$$

Therefore, in order to enable efficient and effective operations of a given shunting yard, the capacities of its particular components should be balanced: those of receiving, classification, and departing yard by providing the sufficient number of tracks/sidings and effectively and efficiently performing corresponding activities by their automation; those of the hump mainly by automating and consequently speeding up the particular sub-operations of the trains' disassembling process carried out there.

5.4.1.2 Capacity and service level of the vehicle fleet

The rail vehicle fleet or rolling stock generally consists of passenger and freight trains both composed of the corresponding wagons/cars and electric and/or diesel-powered locomotives pulling them as shown in Fig. 5.18A and B.

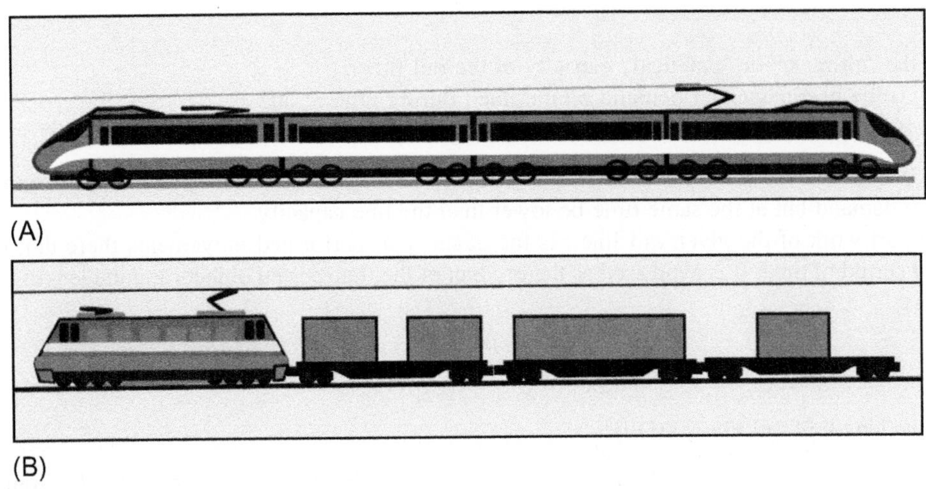

(A)

(B)

FIG. 5.18

The railways rolling stock. (A) Passenger train. (B) Freight train.

Passenger trains

The passenger trains are scheduled along the rail lines to serve expected user/passenger demand. The scheduling implies designing the timetable specifying itineraries of particular trains of the given seat capacity, which include their departure frequency, time of departure, dwell time at origin and destination terminal and transit stations along the line, and the travel time between them.

The offered transport capacity of line (i) ($TC_i(T)$) of the rail network during the specified time interval (T) (peak or off-peak period of passenger demand) can be estimated as follows:

$$TC_i(T) = f_i(T) \cdot s_i(T) \tag{5.41}$$

where:

$f_i(T)$ is the departure frequency of trains on the line (i) during time (T)(dep/TU) (TU = time unit); and
$s_i(T)$ is the average seat capacity of a train per departure on the line (i) during time (T) (seats/train).

In case when the railway network consists of (N) lines, its total offered transport capacity can be is equal to:

$$TC(T) = \sum_{i=1}^{N} TC_i(T) = \sum_{i=1}^{N} f_i(T) \cdot s_i(T) \tag{5.42}$$

The train departure frequency $f_i(T)$ ($i = 1,2,\ldots, N$) should be set up as follows:

$$f_i(T) = \min\left[\mu_i; \frac{Q_i(T)}{\lambda_i(T) \cdot s_i(T)}\right] \tag{5.43}$$

where:

μ_i is the "ultimate" or "practical" capacity of the rail line i;
$Q_i(T)$ is the user/passenger demand on the line i during time T; and
$\lambda_i(T)$ is the average load factor of trains scheduled on the line i during time T.

Eq. (5.43) indicates that the departure frequency on a given line should be set up to satisfy the expected passenger demand but at the same time be lower than the line capacity.

Transport work of the given rail line i is the quantity of performed movements there during the specified period of time. It is expressed as the product of the transported objects and the length of line as follows:

$$TW_i(T) = f_i(T) \cdot s_i(T) \cdot l_i \tag{5.44}$$

where:

l_i is the length of rail line (i) (km).

Productivity of the rail line (i) and of the network with (N) lines represents the transport work performed there per unit of time during the revenue service. It is estimated, respectively, as follows:

$$TP_i(T) = f_i(T) \cdot s_i(T) \cdot v_i(T) \tag{5.45}$$

and

$$TP(T) = \sum_{i=1}^{N} TP_i(T) = \sum_{i=1}^{N} f_i(T) \cdot s_i(T) \cdot v_i(T) \tag{5.46}$$

where

v_i is the average train operating speed along the line (i) (km/h).

The size of rolling stock ie, train fleet engaged on particular lines of the given rail network depends on the scheduled departure frequencies and the train turnaround times along them. This turnaround time for the line i can be estimated as follows:

$$\tau_i = t_{io} + 2 \cdot \left[\sum_{k=1}^{K_i} t_{ik} + \sum_{k=1}^{K_i - 1} \frac{d_{ik}}{v_{ik}} \right] + t_{id} \tag{5.47}$$

t_{io}, t_{id} is the train's dwell time at the origin and destination terminal of the line i (min);
t_{ik} is the dwell time at the transit station k of the line i (min);
d_{ik} is the length of segment k of the line i (km);
v_{ik} is the train's operating speed along the segment k of the line i (km/h);
K_i is the number of transit station along the line i.

The required train fleet to be scheduled along the line (i) and in the given network during time (T) is determined as follows:

$$n_i(T) = f_i(T) \cdot \tau_i \tag{5.48}$$

$$n(T) = \sum_{i=1}^{N} n_i(T) = \sum_{i=1}^{N} f_i(T) \cdot \tau_i \tag{5.49}$$

Under the above-mentioned conditions, the level of service provided to users/passengers on the line i during time period T can be expressed by the scheduled delay as follows:

$$sd_i(T) = (1/2) \cdot \frac{T}{f_i(T)} \tag{5.50}$$

EXAMPLE 5.6

An illustrative example of the above-mentioned transport capacity and service level of rail passenger lines and rolling stock is Japanese high speed Tokaido Shinkansen line connecting Tokyo and Shi-Osaka terminals (CJRC, 2014). There are 15 transit stations along the line with three types of train services: Nozomi (stops at 6 transit stations), Hikari (stops at 6 transit stations), and Kodama (stops at 15 transit stations). The length of the line is: 514.4 km.

The train scheduled travel time along the line in the single direction including stops at transit stations is about 2 h and 25 min, which gives the average travel speed of: $v_i = 213$ km/h. If the dwell time at the origin and destination terminal is: $t_{io} = t_{id} = 20$ min, the total train's turnaround time will be: $\tau_i = 20 + 2 \cdot (145) + 20 = 330$ min, ie, 5.5 h.

The scheduled departure frequency is: $f_i(T) = 7$ dep/h per direction ($T = 1$ h). If each train set has the seat capacity of: $s_i(T) = 1323$ seats/train (type N700A), the total offered capacity along the line in both directions will be equal to: $TC_i(T) = 14 \cdot 1323 = 18{,}522$ seats/h.

The average passenger demand has been: $Q_i(T) = 16{,}083$ passenger/h, which divided by the offered seat capacity gives the average load factor of: $\lambda_i(T) = 0.87$.

In addition, the required train fleet will be: $n_i(T) = 7 \cdot 5.5 \approx 39$ sets N700A.

The transport work and productivity of the line are estimated as:

(i) Transport work: $TW_i(T) = 18{,}522 \cdot 514.4$ km $= 9527716.8$ seat-km;
(ii) Productivity: $TP_i(T) = 18{,}522 \cdot 213 = 3{,}945{,}186$ seat-km/h.

As well, the average schedule delay based on will be: $sd_i(T) = (1/2) \cdot (60/7) = 4.3$ min. The departure and arrival delays due to all other reasons (weather, failure of the system components, etc.) have been between 0.6 and 0.9 min/service. It should be pointed out that this line operates with the above-mentioned performances during the entire day (24 h) (CJRC, 2014).

Freight trains

The rail freight rolling stock also consists of locomotives and freight wagons. These can be of different type depending on the type of goods/freight intended to transport such as, for example, open, covered, flat, with opening roof, special, tank, etc. (https://en.wikipedia.org/wiki/UIC_classification_of_goods_wagons).

In general, the rail freight companies operating in a given area use diverse fleets of these wagons. The wagons of each type (except being at inspection and maintenance) are distributed over the network operated by a given rail operator. Theoretically and often practically, each wagon passes through the following operations taking time, which constitute its turnaround time as follows: loading at the origin freight terminal, moving in train between origin and destination terminal, unloading at destination freight terminal, and moving empty in train to the next loading/origin terminal. The time in train can consists of the running time in train and dwell times at transit freight terminals including shunting yard(s). Under such conditions, the required number of wagon-days[2] $n_{fw}(T)$ of given type to be loaded in different terminals of the network during time T can be estimated as follows:

$$n_{fw}(T) = \left[\frac{Q_f(T)}{\lambda_{fw} \cdot PL}\right] \cdot t_{tr} \tag{5.51}$$

where:

$Q_f(T)$ is the quantity of goods/freight to be loaded on the wagons of given type during time (T);
λ_{fw} is the average load factor of a given wagon (≤ 1.0);
PL is the average payload capacity of a given wagon (tons, TEU); and
t_{tr} is the average turnaround time of a given wagon (days).

In Eq. (5.51), the term in brackets actually represents the number of wagons, which need to be loaded during time (T). For example, if: $T=1$ day, $Q_f(T)=1000$ tons, $PL=20.0$ tons, $\lambda_{fw}=0.8$, and $t_{tr}=4$ days, the required number of wagon-days will be: $n_{fw}(1)=[(1000)/(0.8 \cdot 20.0)] \cdot 4=250$, ie, 250 wagons need to be supplied to be loaded.

In addition, let the block freight trains[3] operate along a given line i at given service frequency.[4]

Capacity. Similarly as in the case of passenger trains and lines, the line transport capacity is determined as follows:

$$TC_{i/f}(T) = f_{i/f}(T) \cdot n_i(T) \cdot PL_{i/pl}(T) \tag{5.52}$$

where:

$f_{i/f}(T)$ is the frequency of block freight trains on the line i during time T(dep/TU) (TU = time unit);
$n_i(T)$ is the number of wagons in the block freight train operating on the line i during time T; and
$PL_{i/pl}(T)$ is the average payload of a wagon included in block train scheduled on the line (i) during time T (ton/wagon).

In case when the block freight trains operate on the network consisting of N lines, its total transport capacity is estimated as follows:

$$TC_f(T) = \sum_{i=1}^{N} TC_{i/f}(T) \tag{5.53}$$

Transport work. The transport work on the line i, is equal to:

$$TW_{i/f}(T) = f_{i/f}(T) \cdot n_i(T) \cdot PL_{i/pl}(T) \cdot L_i \tag{5.54}$$

where:

L_i is the length of line i (km).

---

2 This implies availability of each wagon during the day.
3 Block train has the fixed composition, ie, the number of wagons between origin and destination rail terminals.
4 Similarly as at the passenger trains, this is set up to satisfy given volume(s) of goods/freight demand respecting the train's payload capacity and average load factor.

The transport work in the network of N lines is equal to:

$$TW_{\mathrm{f}}(T) = \sum_{i=1}^{N} TW_{i/\mathrm{f}}(T) \tag{5.55}$$

Productivity. The productivity of the line (i) and that of the rail network with (N) lines handling the above-mentioned block freight trains is determined, respectively, as follows:

$$TP_{i/\mathrm{f}}(T) = f_{i/\mathrm{f}}(T) \cdot n_i \cdot PL_{i/\mathrm{pl}}(T) \cdot v_{i/\mathrm{f}}(T) \tag{5.56}$$

$$TP_{\mathrm{f}}(T) = \sum_{i=1}^{N} TP_{i/\mathrm{f}}(T) \tag{5.57}$$

where

$v_{i/\mathrm{f}}$ is the average train operating speed along the line i (km/h).

The other symbols are as in the previous equations.

Size of rolling stock. The size of rolling stock ie, fleet of block trains engaged on the above-mentioned rail line (i) depends on the scheduled departure frequencies and the train turnaround time there. Let the fully loaded block trains operate in both directions of the line (i) constrained by the origin and destination terminal. These terminals are at the same time loading for trains in one and unloading for trains in opposite direction. Under such conditions, the number of block trains required on the line is estimated as follows:

$$N_{i/\mathrm{f}}(T) = f_{i/\mathrm{f}}(T) \cdot \left[t_{i/\mathrm{ot}} + \frac{2L_i}{v_{i/\mathrm{f}}(L)} + t_{i/\mathrm{dt}} \right] \tag{5.58}$$

where:

$t_{i/\mathrm{ot}}$, $t_{i/\mathrm{dt}}$ is the average time of train's spending at the origin and destination terminal of the line (i) (unloading/loading and unloading/loading, respectively) (h, days);
L_i is the length of line (km); and
$v_{i/\mathrm{f}}(L)$ is the average train speed along the line (i) equal in both directions (km/h).

The number of wagons in the block trains is equal to:

$$n_{i/\mathrm{f}}(T) = m_{i/\mathrm{f}} \cdot N_{i/\mathrm{f}}(T) \tag{5.59}$$

where

$m_{i/\mathrm{f}}$ is the number of wagons per block train operating along the line i.

Under such conditions, the service level provided to users—freight/goods shippers and receivers—can be expressed again by the average schedule delay the goods/freight shipments experience if uniformly arrive between successive trains' departures. The delays along the line should also be considered as an indicator of service quality particularly in the case of the time-sensitive freight/goods shipments.

EXAMPLE 5.7

An illustrative example is the Trans-Siberian Railway line between Moscow and Vladivostok via northern Mongolia, which is about 9300 km long, as shown in Fig. 5.19.

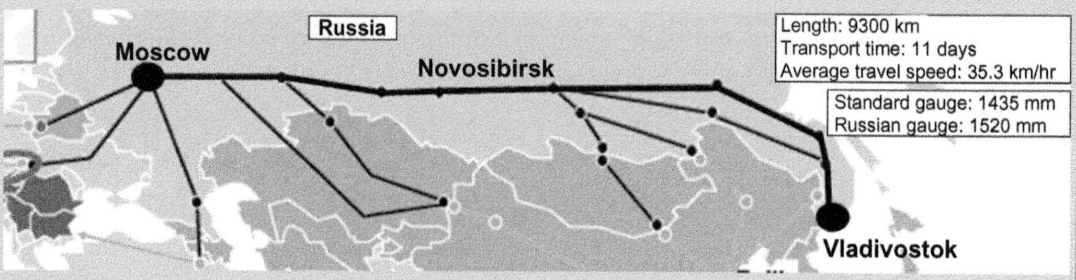

FIG. 5.19

A simplified scheme of the Trans-Siberian Railway via northern Mongolia.

The transport capacity of the line depends on the characteristics of freight trains and their scheduled frequency. Let the block container trains are scheduled at the frequency of: $f_c(T)=3$ trains/wk ($T=1$ wk; wk = week). Each train is composed of 39 wagons (flat cars), each of the payload capacity of 3 containers of the TEU[5] size and tare weight of: $w_w=29$ ton. The weight of the full TEU is adopted to be about 18 ton (ie, 2.5 ton tare plus 15.5 of payload). These give the total payload capacity of a train of: $PL(\text{TEU})=39\cdot 3=117$ TEU/train, or: $PL(\text{weight})=39\cdot 3\cdot 15.5=1813.5$ ton/train. In addition, the transport capacity of the line under given conditions is: $TC_{pl}(\text{TEU})=3\cdot 39\cdot 3=351$ TEU/wk, and $TC_{pl}(\text{weight})=3\cdot 39\cdot 3\cdot 15.5=5440.5$ ton/wk. Consequently, the transport work of the line will be: $TW(\text{TEU/wk})=351\cdot 9300=2{,}929{,}500$ TEU-km, or $TW(\text{weight/wk})=5440.5\cdot 9300=50{,}596{,}650$ ton-km. The corresponding gross weight of a train consisting of the given number of loaded wagons and locomotive is: $W_t=39\cdot 3\cdot 18+39\cdot 29+400=3637$ ton. When the length of each flat wagon is 19.8 m and that of the locomotive 32.0 m, the total length of the train will be: $L_t=39\cdot 19.8+32.0=804.2$ m (In most estimations, this length is considered to be 850 m.) (Song and Na, 2012; Viohl, 2015).

If the average train travel time is about 11 days, the average travel speed in one direction is: $v_t=9300/(11\cdot 24\text{ h})\approx 35.3$ km/h. Then, the line productivity will be: $TP(\text{TEU/wk})=351\cdot 35.3=12390.3$ TEU-km/h and $=5440.5\cdot 35.3=192049.65$ ton-km/h.

If the above-mentioned trains spend at both end of the given line: $t_{i/ot}=t_{i/dt}=3$ days, the required size of the train fleet is determined as follows: (a) number of trains: $N_{i/f}(T)=(3/7)\cdot(3+22+3)=12$ trains; and (b) number of flat cars: $n_{i/f}(T)=12\cdot 39=468$ wagons/ flatcars. The required number of locomotives is equal to the number of trains—12. The continuous rating/power of the locomotives pulling these and much heavier trains (weighing about 9000 tons and consisting of more than 100 cars) is 12–13 MWh (mega-watt hours), which enables operating speed of 51 km/h (type 2 × 2ES5K or 4ES5K) (Kinzhigaziev and Zadorozhny, 2014). In addition, the schedule delay of freight shipment is: $sd_F=(1/2)\ (7/3)=1.16$ days.

5.5 INLAND WATERWAY FREIGHT/CARGO TRANSPORTATION SYSTEM

5.5.1 GENERAL

Similarly to transport systems operated by the other transport modes, the inland waterway freight/cargo transport system (further—IWT (Inland Waterway Transportation)) consists of the infrastructure and transport service network. The IWT infrastructure network generally consists of links—inland

[5]TEU—twenty foot equivalent unit—is based on the volume of a 20-foot-long (6.1 m) intermodal container (a standard-sized metal box) easy transferred between different transport modes road trucks, freight trains, and container ships.

waterways such as the man-built channel and the rivers enabling operation of the freight/cargo vessels and barge fleet of given size—and the nodes—inland ports located on the banks of channels and rivers, where IWT services start and end. Only rivers of sufficient size—width and depth—can be used. Under such conditions, the additional factor for regular operations of IWT services is the state, ie, the level of water generally depending on the weather. For example, under conditions of persistent draught, the water depth can decrease to compromise use of the vessels or barge fleets of a given (planned) size. The inland ports enable transfer of freight/goods shipments between the IWT and rail and road transport mode, thus enabling their delivery from the shippers' to the receivers' doors, just by combining services these different transport modes. In such case the IWT operates as the main mode of an intermodal transport networks where the rail and road are used for collecting and distributing freight/goods shipments from the shippers and receivers located in the gravitational zones of the origin and destination port as the intermodal transfer locations, respectively. The IWT service network consists of routes along the rivers and/or channels where transport services (as links) are carried out between particular ports (as nodes), under given conditions. These transport services can be carried out by fleets consisting of different number, size and type of vessels and/or barges. For example, the number of vessels or barges in the fleet can vary from one to few hundreds. The size expressed by the vessel or barge payload capacity (ton, TEU) can also vary substantially—from hundreds to thousands of tons. As well, the type of these vessels and barges depends on the type of cargo they can carry such as general, bulk, liquid, and container cargo. The ITW services can be carried out only along the channels and rivers, which make them less flexible than that of rail and road.

The IWT infrastructure network is characterized by the capacity and service level of its components—inland ports as nodes and rivers and man-built channels as links. The IWT service network is characterized by the physical size (the number of links and nodes) and spatial coverage conditioned by the available infrastructure network, and transport service frequency, fleet size and type, transport work and productivity, all dependent on the volumes of demand (freight/cargo) to be transported under given conditions.

5.5.2 CLASSIFICATION

The IWT infrastructure networks and their components—ports as nodes, and rivers and channels as links—are classified based on the vessel classification. For example, in Europe, the Classification of European Inland Waterways is carried out in order to set up the standards for interoperability of large navigable waterways forming part of the Trans-European Inland Waterway network within Continental Europe and Russia. This classification implies that the size of each waterway is limited by the dimensions of the structures including the locks and boat lifts on the route (ECMT, 1992). Table 5.11 gives basic characteristics of this classification.

In Table 5.11, the classes of inland waterways I–III are of the regional, and the classes IV–VII of the international importance. Classes I–III refer to the individual motor vessels and barges. The classes IV–VII refer to the pushed convoys consisting of the towboat or tug and non-propelled barges.

Fig. 5.20 gives the relationship between the average area ($L \times B$) and the capacity (PL) of the above-mentioned classes of vessel, barges, and pushed convoys. The same relationship is provided for the IWT fleet operating along the Mississippi river in the United States (CECW-CP, 2004).

Table 5.11 Classification of the IWT Infrastructure in Europe (ECMT, 1992)—Motor Vessels, Barges, and Pushed Convoys—Maximum Characteristics (ECMT, 1992)

Classes of Navigable Waterways	Type of Vessel, Barges, and Pushed Convoys General Characteristics				
	Capacity PL (ton)	Length L (m)	Beam B (m)	Draught D (m)	Number of Barges in Convoy
I	250–400	38.5	5.05	1.80–2.20	0
II	400–650	50.0–55.0	6.60	2.50	0
III	650–1000	67.0–80.0	8.20	2.50	0
IV	1000–1500	80.0–85.0	9.50	2.50	1
V_a	1500–3000	95.0–110.0	11.40	2.50–4.50	1
V_b	3200–6000	172.0–185.0	11.40	2.50–4.50	2
VI_a	3200–6000	95.0–110.0	22.80	2.50–4.50	2
VI_b	6400–12,000	185.0–195.0	22.80	2.50–4.50	4
VI_c	9600–18,000 9600–18,000	270.0–280.0 195.0–200.0	22.80 33.00–34.20	2.50–4.50 2.50–4.50	6 6
VII	14,500–27,000	285.0	33.00–34.20	2.50–4.50	9

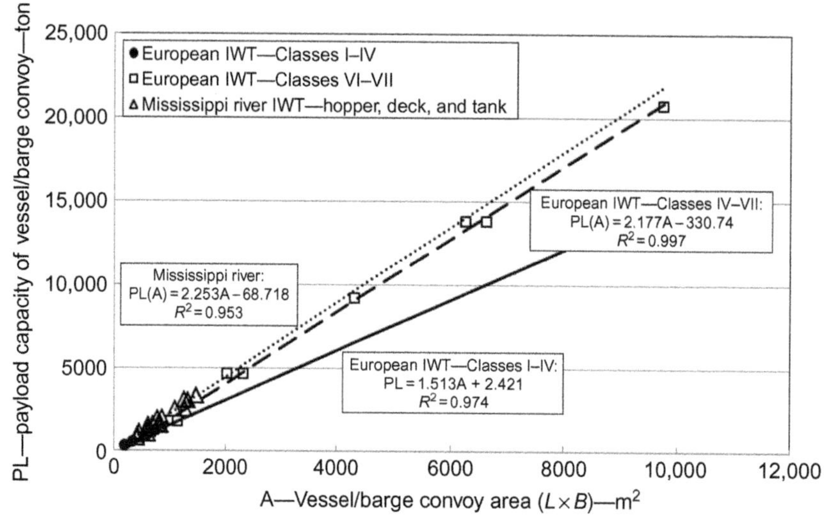

FIG. 5.20

Examples of the relationship between the average physical dimensions (layout) of the IWT barge/vessels and their payload capacity (CECW-CP, 2004; ECMT, 1992).

As can be seen, the Mississippi river and the European IV–VII classes of vessels, barges, and pushed convoys have the similar area per unit of payload capacity despite the absolute size (both area and payload capacity) of that in Europe is much larger. The absolute size of European I–IV class vessels/barges is similar to that at the Mississippi river. In addition, the European I–IV class vessels and barges have much lower area per unit of payload capacity, ie, 1.513 versus 2.177–2.253 m^2/ton, which indicates their higher spatial efficiency.

In some other countries outside Europe classification of the components of IWT infrastructure is carried out according to their physical characteristics—dimensions, which in turn dictate the size of vessel/barge/convoy they serve. This relates to the rivers and man-built channels, and ports classified according to the vessel/barge dimensions they can safely accommodate. One such classification is carried for IWT along the Yangtze River in China. The river is 6300 km long with 25 inland ports classified at four levels, as shown in Table 5.12.

Table 5.12 Classification of Inland Ports Along the Yangtze River (China) (Deming)

Segment	Location/Type of Port	Length of Segment(km)	Water Depth (m)	Max Vessel/Barge Size (DWT[a])
1	Lower Yangtze—deep draft Nanjing-the estuary—port of Shanghai	250	10.5	25,000–30,000
2	Lower Yangtze-shallow draft Wuhan—Nanjing	700	4.0–4.5	5000–7000
3	Middle Yangtze—Shallow draft Yichang—Wuhan	630	2.9	3000
4	Upper Yangtze—Shallow draft Chongqing—Yichang	660	2.9	1500

[a]*Dead weight tonnage (DWT) or deadweight as the payload capacity expressed in tons not including the weight of a vessel or a barge.*

As can be seen, the size of vessels and barge fleet decreases with decreasing of the water depth, which in the given case happens upstream of the river. Thus, the water depth along the first and second segment of the river enables operation and docking of the oceangoing vessels and barge fleets of the specified size. The third and fourth segment enable operations and docking of the river vessels and barge fleet of the specified size.

In addition, the inland ports can be further classified according to the volumes of handled freight/cargo during the specified period of time (usually one year).

5.5.3 INFRASTRUCTURE NETWORK

5.5.3.1 Ports

Components, operations, and capacity

The IWT ports are generally located on the bank of rivers, lakes, and man-made channels. These ports consist of the water side and land side area. At the water side area, the main component is the quay area with berths for anchoring vessels and barges. The important factor for handling particular vessels and/or barge of particular size is the water depth along the quay. The main components of the land side area are terminals dedicated to type of freight/cargo, as mentioned above. Each terminal consists of the bank platform/space with fixed or mobile (along the tracks on the quay) lift cranes or shiploaders (depending on the type of cargo container, general, bulk) for unloading and loading vessels and barges, and terminal yard or warehouse where the freight/cargo is stored after being unloaded from the vessels and/or barges and before being loaded onto the trucks or trains, and vice versa. The movement of freight/cargo shipments between the bank platform and terminal yard can be carried out by wagons loaded/unloaded by the tire crane or forklifts. In cases when the freight/cargo shipments go directly from/to their shippers/receivers, respectively, they do not pass through the terminal yard, but are directly transhipped between truck/rail and vessel/barge and vice versa, respectively. Anyway, the additional component on the port's land side area are its ground access systems—rail and road lines with space for their loading and unloading. Specifically, in case of liquid cargo (crude oil, fuel oil, petrochemical liquids), the corresponding terminals are characterized by the pipeline transportation, where liquids flow through the pipelines driven by the pumps. The liquid tanks are located at the yard connected to the vessels/barges at the berths by the mobile arms/pipes. Fig. 5.21 shows the simplified general layout of the IWT port.

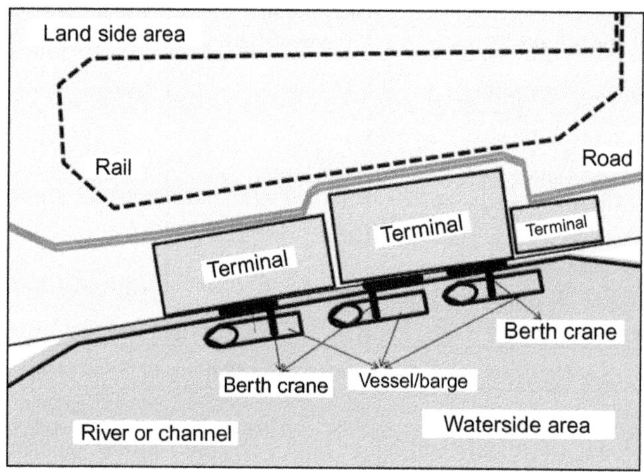

FIG. 5.21

Simplified layout of the IWT port.

The quays with berths at a given IWT port dedicated to the terminals handling particular type of freight/cargo transported by different inland vessel/barge operators are characterized by their "ultimate" and "practical" capacity.

Capacity and service level of the water side area

The capacity of the waterside area of a given IWT port can be "ultimate" and "practical." They both are expressed by the maximum number of motor vessels or barge convoys (towboat with few attached non-propelled barges), which can be accommodated/served at the given number of berths during a given period of time (day, week, month, year) under given conditions. For the former capacity, given conditions are constant demand for service during given period of time. For the latter capacity given conditions are the average delay(s) imposed on each vessel or barge convoy while waiting for getting service, ie, docking.

"Ultimate" capacity. The "ultimate" capacity of a given quay area with $N_i(\tau)$ available berths during the period of time τ can be estimated as follows:

$$\mu_{b/i}(\tau) = N_i(\tau)/\tau_{b/i} \tag{5.60}$$

where:

i is the type i berth in front of the corresponding terminal handling the corresponding freight/cargo type i;
$\mu_b(\tau)$ is the "ultimate" capacity of a given quay area with $N_i(\tau)$ available berths during the time period τ (vessels or barge convoys/h or day); and
$\tau_{b/i}$ is the average service time of the vessel or barge convoy at berth type i (hours, days).

If in Eq. (5.60), $\mu_{b/i}(\tau) = \lambda_{b/i}(\tau)$, where $\lambda_{b/i}(\tau)$ is the intensity of arriving vessels or barge convoys of type (i) during the given period of time τ, the required number of berths can be estimated as: $N_i(\tau) = \lambda_{b/i}(\tau) \cdot \tau_{b/i}$.

In general, the vessel's or barge convoy's service time $\tau_{b/i}$ in Eq. (5.60) depends on the payload to be unloaded after arriving and loaded before departing of a given motor vessel or a barge convoy and the loading/unloading service rate of deployed devices as follows:

$$\tau_{b/i} = \frac{(\theta_{1/i} + \theta_{2/i}) \cdot PL_i}{m_i \cdot r_i \cdot u_i} \tag{5.61}$$

where:

$\theta_{1/i}, \theta_{2/i}$ is the average load factor of the arriving and departing vessel or barge convoy, respectively, freight/cargo type (i) (≤ 1.0);
PL_i is the average payload capacity of a vessel or barge convoy carrying freight/cargo type (i) (ton, TEU);
m_i is the number of cranes deployed for unloading/loading a vessel or barge convoy carrying the freight/cargo type (i);
r_i is the service rate of a crane unloading and loading the freight/cargo type (i) (tons/h; TEU/h); and
u_i is the utilization rate of a crane unloading and loading the freight/cargo of type (i) (≤ 1.0).

By summing up the capacities of particular categories/types of berths, the total "ultimate" capacity of a given port can be estimated.

EXAMPLE 5.8

The quay area in front of the container terminal in a given IWT port has a length of $l = 600$ m. The water depth is 5 m, thus enabling, according to Table 5.11, handling all categories of vessels and barge convoys (Europe). The number of container vessels or barge convoys, which can be simultaneously docked in this area is approximately equal to: $N = l/1.2\,L$ where L is the length of a vessel or a barge convoy (factor 0.2 is used for determining the separation between vessels or barge convoys along the quay) (see Fig. 5.22). From Table 5.11 follows, for example, that: $N = 600/1.2 \cdot 55 = 9$ berths for docking the vessels of length: $L = 55$ m (Class IV) or $N = 600/1.2 \cdot 270 = 1.85 \approx 2$ berths for docking the barge convoys of length: $L = 270$ m (Class VI$_c$) can be simultaneously docked.

(Continued)

EXAMPLE 5.8—cont'd

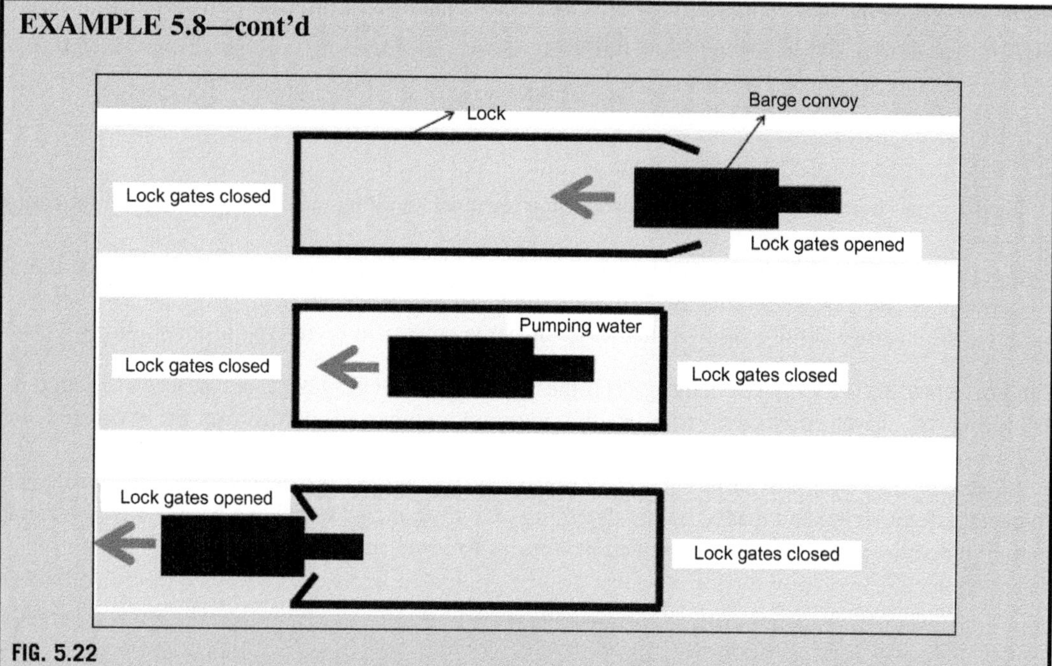

FIG. 5.22

Simplified scheme of the lock operations.

The payload capacity of the smaller (IV class) vessels is: $PL=1500$ ton. If one TEU weights 14.3 ton (12 ton payload and 2.3 ton of tare), the payload capacity of the vessel in terms of the number of TEUs is: $PL=1500/14.3\approx 105$ TEU. The vessels' average load factor at arrivals and departures is assumed to be: $\theta_1=\theta_2=0.85$. In addition, to each vessel: $m=1$ crane with the service rate of: $r=15$ TEU/h and the utilization rate of service rate: $u_i=1.0$ is deployed. Then, the average service time per vessel is equal to: $\tau=[(0.85+0.85)\cdot 105]/1\cdot 15=11.9$ h. As a result, the capacity of the quay area will be equal to: $\mu=9/11.9=0.756$ vessels/h or 18 vessels/day (1 day $=24$ h).

The payload capacity of the larger (VI$_c$ class) barge convoy is: $PL=18{,}000$ ton. In terms of TEU, this capacity amounts: $PL=18{,}000/14.3=1259$ TEU. The average load factor of each convoy's is adopted to be: $\theta_1=\theta_2=0.85$. In addition, each convoy is served by: $m=3$ cranes, each with the loading/unloading rate of: $r=15$ TEU/h, and the utilization rate of service rate of: $u_i=1.0$. This gives the average service time per a barge convoy of: $\tau=[(0.85+0.85)\cdot 1259]/3\cdot 15=47.7$ h. Then, the resulting capacity of the quay area with 2 berths will be: $\mu=2/47.7=0.042$ barge convoys/h or ≈ 1 barge convoy/day.

These examples show that the "ultimate" capacity of the water side of a given IWT port(s) depends not only on its physical characteristics (length of quay and the number of berths) but also of the productivity of loading and unloading and size of vessels or barge convoys. This can be an example how the characteristics of demand, in this case vessels and barge convoys influence the capacity of fixed (available) infrastructure.

"Practical" capacity and service level. The average delay imposed on a vessel or barge convoy carrying freight/cargo of type (i) before its docking at one of the berths in front of the corresponding terminals of a IWT port can be estimated by the formula from the theory of steady-state queuing systems (model M/G/1/–) as follows:

$$W_i = \frac{\lambda_i(\sigma_i^2 + 1/\mu_i^2)}{2 \cdot (1 - \lambda_i/\mu_i)} \tag{5.62}$$

where:

- λ_i is the average demand/arrival rate of ships, ie, port calls (vessels or barge convoys/day;)
- μ_i is the "ultimate" capacity of the quay/berth (vessels or barge convoys/day); and
- σ_i is the standard deviation of the vessels or barge convoy's service time at the quay/berth area (h).

The "ultimate" capacity μ_i can be estimated from Eq. (5.62). Consequently, by specifying the maximum average delay W_i· as the service level, the "practical" capacity of the given quay/berth area λ_i· can be estimated as follows:

$$\lambda_i^* = \frac{2^*W_i}{\left[2^*W_i/\mu_i + (\sigma_i^2 + 1/\mu_i^2)\right]} \tag{5.63}$$

where all symbols are as in the previous equations. For example, if the maximum delay per vessels or barge convoy is: $W_i = 2$ h, the "ultimate" capacity: $\mu_i = 0.756$ vessels or barge convoys/h, and standard deviation of their service time: $\sigma_i = 1$ h, the corresponding "practical" capacity will be: $\lambda_i = 2 \cdot 2/[2 \cdot 2/0.756 + (1 + 1/0.756^2)] = 0.568$ vessels or barge convoys/h or $0.568 \cdot 24\ \text{h} = 13.63 \approx 14$ per day (1 day = 24 h). With increasing of the maximum delay, the "practical" capacity will approach closer to its "ultimate" counterpart under given conditions.

Capacity of and service level of the land side area

The "ultimate" and "practical" capacity, and the corresponding service level of the land side area of given IWT port mainly relate to its terminals and ground access systems.

"Ultimate" capacity. The "ultimate" capacity of a terminal of a given IWT port can be defined as the maximum number of freight/cargo shipments, which can be processed during a given period of time under given conditions, ie, constant demand for service. Since the dedicated space is allocated to the incoming (from the water side) and outgoing (from the land side) freight/cargo shipments, the corresponding capacities can be estimated as follows:

(a) For the incoming freight/cargo shipments:

$$\mu_1 = \frac{1}{\Delta_1} \cdot \frac{A_1}{s_1} \tag{5.64}$$

where:

- μ_1 is the "ultimate" capacity of terminal yard for the incoming freight/cargo shipments (ton/h, day, year; TEU/h, day, year);
- Δ_1 is the average time of passing of the unit of incoming freight/cargo shipment through the terminal (h, days);
- A_1 is the available area for storing the incoming freight/cargo shipments (m^2); and
- s_1 is the average size of the footprint of a unit of freight/cargo shipment (m^2).

In order to prevent that the incoming freight/cargo shipments wait for storing/passing through the terminal, the following condition needs to be fulfilled:

$$\mu_1 \geq m_1 \cdot r_1 \cdot u_1 \tag{5.65}$$

where:

m_1 is the number of quay/berth cranes deployed to unload the arrived motor vessels or barge convoys (−);

r_1 is the unloading/loading rate per deployed crane (ton/h; TEU/h); and

u_1 is the average utilization rate of a crane (≤1.0).

Since the area of the terminal yard A_1, average footprint of freight/cargo unit s_1, and the crane unloading/loading rate are constant, the balancing between the two capacities can be achieved by adjusting the average time of passing through the terminal Δ_1 and the number of deployed cranes m_1.

(b) For the outgoing freight/cargo shipments:

$$\mu_2 = \frac{1}{\Delta_2} \cdot \frac{A_2}{s_2} \tag{5.66}$$

where all symbols area analogous to those in Eq. (5.64), but relate to the characteristics of the outgoing freight/cargo shipments and corresponding part of the terminal yard.

In order to prevent waiting of the outgoing freight/cargo shipment for passing through the terminal, the following condition needs to be fulfilled:

$$\mu_2 \geq \frac{n_{2/1}}{\tau_{2/1}} \cdot PL_{2/1} \cdot \theta_{2/1} + \frac{n_{2/2}}{\tau_{2/2}} \cdot PL_{2/2} \cdot \theta_{2/2} \tag{5.67}$$

where:

$n_{2/1}$ is the number of parking places for trucks bringing the outgoing freight/cargo shipments from the shippers to the IWT port's terminal;

$\tau_{2/1}$ is the average time of unloading a truck with the outgoing freight/cargo shipments (h);

$PL_{2/1}$ is the average payload capacity of a truck with the outgoing freight/cargo shipments (ton; TEU);

$\theta_{2/1}$ is the average load factor of a truck with the outgoing freight/cargo shipments (≤1.0);

$n_{2/2}$ is the number of rail tracks for trains bringing the outgoing freight/cargo shipments from the shippers to the IWT port's terminal;

$\tau_{2/2}$ is the average time of unloading a train with the outgoing freight/cargo shipments (h);

$PL_{2/2}$ is the average payload capacity of a train with the outgoing freight/cargo shipments (ton; TEU); and

$\theta_{2/2}$ is the average load factor of a train with the outgoing freight/cargo shipments (≤1.0).

In this case, the balancing between the capacities of the IWT port's terminal ground access modes and the corresponding terminal yard can be carried out by adjusting the intensity of trucks and trains with the outgoing freight/cargo shipments and the time of their passing through the corresponding part of terminal yard.

EXAMPLE 5.9

Let the freight/cargo shipments be TEUs. If the area of the terminal for the incoming TEUs is: $A_1 = 240{,}000\ m^2$, the footprint of 1 TEU, $s_1 = (6.1\ m \times 2.44\ m) = 14.844\ m^2$, and average time of TEU's passing through the terminal: $\Delta_1 = 2$ days.

The corresponding "ultimate" terminal capacity will approximately be: $\mu_1 = (1/3) \cdot 240{,}000/14.844 \approx 8062$ TEU/day or $8062/24 = 336$ TEU/h. If the average unloading rate of each deployed crane is: $r = 3 \cdot 15$ TEU/h, and the utilization rate: $u_1 = 1.0$, the maximum number of deployed cranes will need to be: $m \leq 7$, just to prevent congestion of the unloaded TEUs at the quay/berth platform in front of the terminal. The similar calculation can be performed for the outgoing TEUs, but respecting the specificities of the capacities of the ground access systems.

"Practical" capacity and service level. The "practical" capacity of the IWT port's a terminal(s) and ground access systems can be estimated similarly as for the water side area of the port. In such case, the necessary modifications are needed respecting the relevant inputs. However, as mentioned above, since the "ultimate" capacities of the water side and landside area have been balanced, the maximum delays of the incoming and outgoing freight/cargo shipments specifying the corresponding "practical" capacities are usually "absorbed" within their time of passing through the terminal(s).

5.5.3.2 Rivers and man-built channels

The rivers and man-built channels similarly as the IWT ports are characterized by their "ultimate" and "practical" capacity. They both are expressed by the maximum number of motor vessels or barge convoys (towboat with few attached non-propelled barges), which can pass through the selected reference location for their counting during a given period of time (day, week, month, year) under given conditions. For the "ultimate" capacity, the given conditions are constant demand for service during a given period of time. For the "practical" capacity, the given conditions are the average delay(s) imposed on each motor vessel or barge convoy while waiting to pass through the reference location. The "reference location" can be the critical cut (line or segment) or the most constraining lock,[6] both along the given river or channel where all vessels or barge convoys moving in the same direction pass through.

"Ultimate" capacity. The "ultimate" capacity of a given river or channel for traffic moving in the same direction can be estimated as follows:

$$\mu_r = 1/\overline{t_r} \tag{5.68}$$

where:

μ_r is the river or channel capacity estimated for the selected reference location (vessels or barge convoys/min, h, day, year);

$\overline{t_r}$ is the average service time of a motor vessels or a barge convoy at the reference location.

---

[6] A lock is actually a chamber/box holding the water and accommodating vessels or barge convoys either lower on raise them to a lower or higher level, respectively. Therefore, a lock is needed to follow the level of the ground. A vessel or barge convoy enters the lock. Then the crew shuts the gate(s) behind it. If it is going down, the water is let out at the other end until the same level is reached, then the gate(s) are opened and the vessel can proceed on the lower level. If it is going up, the water from the higher level is let in the lock until the water in the lock is leveled with that above, then the vessel can proceed on that higher level (CS, 1998).

In particular, if the convoy consists of k barges and the reference location is the lock, its "ultimate" capacity, ie, the number of barges or motor vessels served during a given period of time will be:

$$\mu_r(k) = \frac{k}{\bar{t}_r(k)} \cdot u_r \tag{5.69}$$

k is the number of motor vessels or barges in the convoy, which can simultaneously be accommodated at the lock;

$\bar{t}_r(k)$ is the average lock's cycle time (min, h); and

u_r is the rate of utilization of a lock during the given period of time ($\leq$1.0).

The other symbols are as analogous to that in the previous equations.

The number of barges or motor vessels k in Eq. (5.69) can be approximated as follows:

$$k = \left| \frac{l \cdot w}{L \cdot B} \right| \tag{5.70}$$

l, w is the length and width of the given lock, respectively (m, m); and

L, B is the average length and beam of a motor vessel or barge convoy including the virtual safety belt around the footprint in the given context (m, m).

The average lock's cycle time ($\bar{t}_r(k)$) in Eq. (5.69) consists of three components coinciding with three general steps of the lock's operations shown in Fig. 5.22 (Kooman and de Bruijn, 1975).

The motor vessel or barge convoy approaches to the entry gate of the lock, which is opened to accept it. The exit gate on the other side of the lock is closed. When the motor vehicle or barge convoy is in the lock, the gates on its both ends are closed, and the water is pumped to adjust the level of water in the lock to that downstream, ie, behind the lock's exit gate. When it is carried out, the exit gate is opened and the motor vessel or barge convoy leaves the lock. Consequently, the components of the lock's cycle time are: (i) the entry time of the motor vessel(s) or barge convoy(s), (ii) the lock's operating time, and (iii) the exit time of motor vessel(s).

For the critical cut along a river or a channel, the time ($\bar{t}_r$) in Eq. (5.69) is equivalent to the average time between passing of the successive vessels or barge convoys through it. In order to estimate this time, let (ij) be the pair of vessels or barge convoys passing through the given cut. In this pair, the vessel or barge convoy i is the leading and j is the following in the sequence as shown in Fig. 5.23.

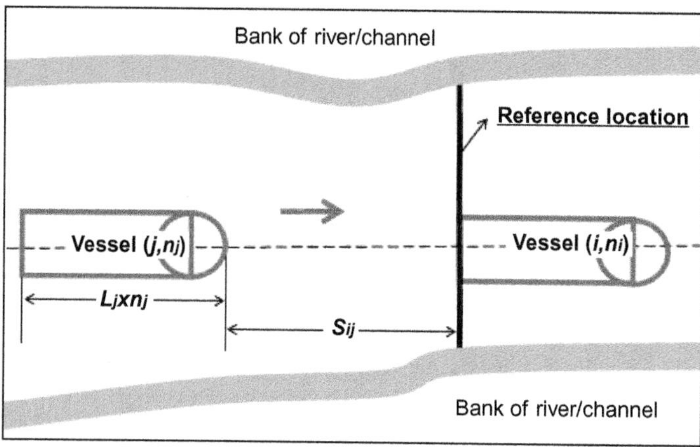

FIG. 5.23

Scheme for estimating the time between passing two successive vessels or barge convoys through the reference location on a river or channel.

The time between passing of the pair of vessels or barge convoys i and j through the reference location can be estimated as follows:

$$t_{ij} = \frac{s_{ij}}{v_j} \tag{5.71}$$

s_{ij} is the mutual distance between the stern of leading vessel or barge convoy i and the bow of the trailing vessel or barge convoy j (km, m); and
v_j is the average sailing speed of the vessel or barge convoy j (km/h).

In Eq. (5.71), the motor vessels or barge convoys differentiate according to the operating speed at the cut. The average time of passing of the different combinations of pairs of the motor vessels or barge convoys through the reference location, ie, selected cut along the river or channel can be estimated as follows:

$$\overline{t_r} = \sum_{i=1}^{M}\sum_{j=1}^{M} p_i \cdot t_{ij} \cdot p_j \tag{5.72}$$

where:

p_i, p_j is the proportion of motor vessels and/or barge convoys of the category i and j, respectively in the traffic mix; and
v_j is the average speed of the motor vessel or barge convoy of the category j (km/h).

Then, the capacity of the given cut can be calculated by Eq. (5.68). In addition, the average payload of the motor vessels or barge convoys passing through the reference location can be estimated as follows:

$$\overline{PL_r} = \sum_{i=1}^{M} p_i \cdot n_i \cdot PL_i \cdot \theta_i \tag{5.73}$$

where:

i is the category of motor vessel or barge convoy regarding type of freight/cargo it carries;
M is the number of types of motor vessels or barge convoys respecting the freight/cargo type, which pass through the lock;
p_i is the proportion of vessels or barge convoys of the category i in the traffic mix;
n_i is the average number of barges in the convoy of category i (for vessels of the category i, $n_i = 1$);
PL_i is the payload capacity of a barge in the convoy of category i or of the motor vessel of category i (ton, TEU); and
θ_i is the average load factor of a barge in the convoy of category i or of the motor vessel of category i (≤ 1.0).

From Eqs. (5.71), (5.72), and (5.73), the capacity of a lock or of some other critical cut of the given river or channel in terms of quantity of freight/cargo processed during the specified period of time can be estimated as follows:

$$Q_r = \mu_r \cdot \overline{PL_r} \tag{5.74}$$

where all symbols are analogous to those in the previous equations.

EXAMPLE 5.10

One illustrative example is the lock's capacity of locks on the Mississippi River (United States), which is about 4100 km long. Its elevation is 450 m above sea level, which drops to 0 m above sea level at the Gulf of Mexico. In particular, the upper Mississippi River is a 1067 km segment of the Mississippi River, which extends from Minneapolis (Minnesota), to its juncture with the Missouri River near St. Louis (Missouri). At this segment, there are 28 locks located at 16–75 km distance intervals. The lift high of these locks varies from 1.5 to 15 m. Three locks have chambers that are: $l=366$ m long and $w=33.5$ m wide. The others are: $l=183$ m long and $w=33.5$ m wide. The typically used barges are $L=59.5$ m long and $B=10.7$ m wide. This implies that the storage capacity of the larger locks in terms of simultaneously accommodating barge convoys is equal to: $k=(366\times 33.5)/(59.5\times 10.7)\approx 19$ units (towboat + 18 barges), and that of the smaller locks: $k=(183\times 33.5)/(59.5\times 10.7)\approx 10$ units (towboat+9 barges) (the virtual safety belt around the footprint is not considered in both cases). This implies that the convoys with a towboat pushing 18 barges of typical size need to be split up in order to pass through the smaller locks, which makes their service time there typically $\bar{t}_r=1$–1.5 h. The average service time of the same convoys at the larger locks is $\bar{t}_r=0.5$ h. Based on Eq. (5.69), this gives the capacity of the smaller locks of: $\mu_r=(24)\cdot(1/1$–$1.5)=16$–24 barge convoys/day. The capacity of the larger locks is: $\mu_r=(24)\cdot(1/0.5)=48$ barge convoys/day. In addition, if the utilization of the lock time is, for example, $u_r=0.8$, the total annual number of barges served at each smaller and larger lock, respectively is equal to: $\mu_r=(24)\cdot(9/1$–$1.5)\cdot 0.8\cdot 365=42{,}048$–$63{,}072$ barges/year, and $\mu_r=(24)\cdot(18/0.5)\cdot 0.8\cdot 365=252{,}288$ barges/year. If each barge has the average capacity and load factor: $\bar{PL}_r=1500$ ton and $\theta=0.75$, respectively, then, the capacity of smaller and larger lock in terms of quantity of freight/cargo processed during a year will be equal to: $Q=(42{,}048$–$63{,}072)\cdot 1500\cdot 0.75=474{,}730{,}400$–$70{,}956{,}000$ ton/year, and $Q=252{,}288\cdot 1500\cdot 0.75=283{,}824{,}000$ ton/year, respectively [In Europe, the size of the barge convoy is regulated to the customary 4 barges pushed by a towboat and exceptionally to 6 barges pushed by a towboat along 270 km long segment of the river Rhine Koblenz and Emmerik (Vergeij et al., 2008)].

The other example is the capacity of the line cut of a European river or channel. For example, if the motor vessels of the same category (ie, $p_i=p_j=1.0$) operate along the given river or channel in the same direction at an average sailing speed of: $v_j=12$ km/h, and if the distance between each pair is: $s_{ij}=1.45\cdot(L_j\cdot n_j)$ where $L_j=110$ m, and $n_j=1$, then the inter arrival time at the selected reference location—cut from Eq. (5.71) will be: $t_{ij}=(1.45\cdot 0.110/12)\cdot 60\approx 0.8$ min (Groenveld et al., 2006; Solar, 2012). Based on Eq. (5.69), this gives the "ultimate" capacity of the cut of: $\mu_r=60/0.8\approx 112$ vessels/h. If each of them has the payload capacity of: $PL=208$ TEU and load factor: $\theta=0.8$, the "ultimate" capacity of the given cut in terms of the quantity of freight/cargo processed will be: $Q_r=112\cdot 208\cdot 0.8\approx 18{,}720$ TEU/h. If the average sailing speed is: $v_j=20$ km/h, the average inter arrival time at the cut will be: $t_{ij}=(1.45\cdot 0.110/20)\cdot 60\approx 0.5$ min, and the corresponding "ultimate" capacity of the cut: $\mu_r=60/0.5=120$ vessels/h. The capacity in terms of the quantity of freight/cargo will be: $Q_r=120\cdot 208\cdot 0.8=19{,}968$ TEU/h.

"Practical" capacity and service level. The "practical" capacity of the rivers and channels as inland waterways is also determined respecting the reference location, which is the selected lock(s) or cut(s). Usually, it is assumed that the reference location operate as the service channel with general distribution of the service time of motor vessels or barge convoys as customers arriving according to Poisson process. Under such condition, the average waiting time of the vessel of barge convoy can be estimated as in Eq. (5.72). In addition, the total time of passing through the selected reference location (lock or cut) can be estimated as:

$$\bar{t}_h = W_h + \Delta_h = \frac{\lambda_h\left[\sigma_h^2 + (\Delta_h/u_u)^2\right]}{2[1-\lambda_h\cdot(\Delta_h/u_h)]} + \Delta_h \tag{5.75}$$

where:

W_h is the average waiting time of a motor vessel or barge convoy for service at the selected reference location (lock or cut) (h);
λ_h is the average intensity of arriving motor vessels or barge convoys at the selected reference location (lock or cut) (arr/h);
σ_h is the standard deviation of the service time of motor vessels or barge convoy at the selected reference location (h);
Δ_h is the average service time of a motor vessel or barge convoy at the selected reference location (lock or cut) (h); and
u_h is the probability that the reference location is not operational as the service channel during given time period of time independently on the causes (≤ 1.0).

Fig. 5.24 shows an example of the relationship between the average delay of delayed barge convoy and the overall utilization of the capacity of locks L18–L25 on the upper the Mississippi River (Campbell et al., 2007.

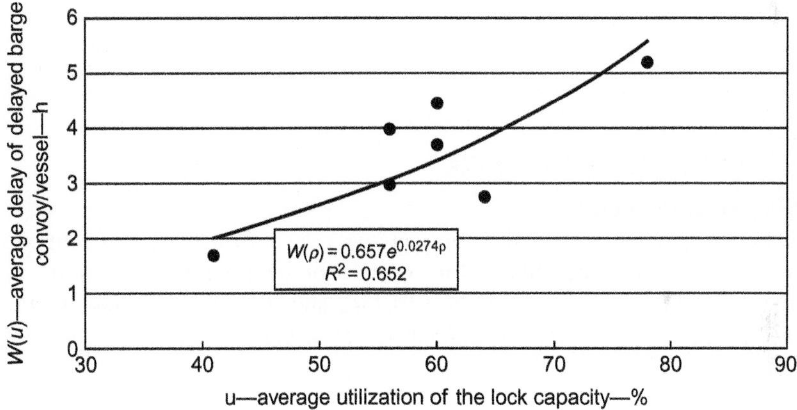

FIG. 5.24

Relationship between the average delay per barge convoy/vessel and utilization of capacity of Locks L18-L25 at the upper Mississippi river (period: 1980–99).

If both these delays are set up to the certain level, for example, at: $W = 2$ h, the utilization of the "ultimate" capacity will be: $u = 0.40$ (ie, 40%). This gives the "practical" capacity of the smaller lock of: $\mu = 0.40 \cdot (16\text{–}24) \approx 6\text{–}10$ barge convoys/day, and of the larger lock of: $\mu = 0.40 \cdot 48 \approx 19$ barge convoys/day. If the average delay is: $W = 5$ h, the corresponding utilization of the "ultimate" capacity will be: $u = 0.8$, and the practical capacity: $\mu = 0.80 \cdot (16\text{–}24) \approx 13\text{–}19$ barge convoys/day for the smaller and $\mu = 0.80 \cdot 48 \approx 38$ barge convoys/day for the larger lock. In addition, Fig. 5.25 shows the example of the average accumulated delays of barge convoys while passing through the sequences of locks in the given example. As can be seen, these delays increased with increasing of the number of successive locks passed through. In addition, the differences between their average, minimum, and maximum values were noticeable.

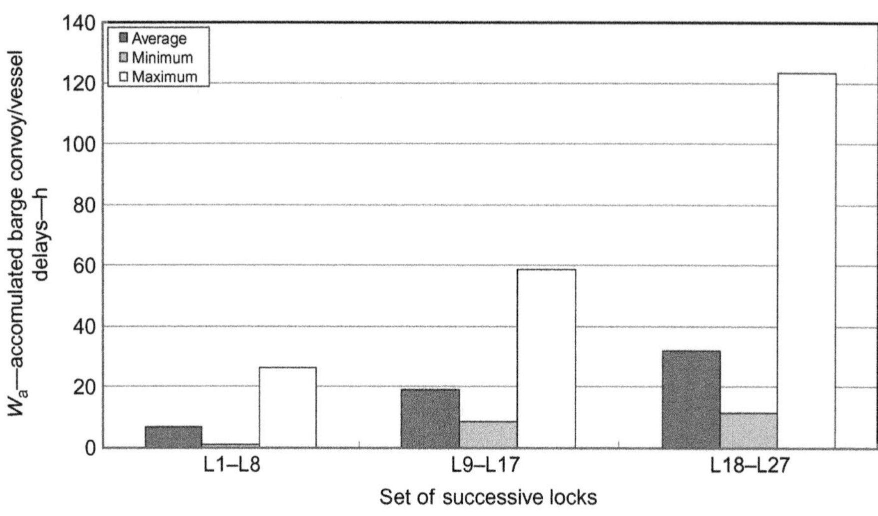

FIG. 5.25

Accumulated barge convoy/vessel delays at particular sets of locks along the upper Mississippi river (Period: 1980–99).

5.5.4 TRANSPORT SERVICE NETWORK

5.5.4.1 Rolling stock/vehicles

The characteristics of motor vessels and barge convoys relevant for the given context are given in Table 5.11 for Europe. In the United States, for example, on the river Mississippi, the typically used barges have the length: $L=59.5$ m, width: $B=10.7$ m, and the payload capacity: $PL=1500$ ton. Fig. 5.26 shows a scheme of the motor container vessel of the category I–IV operating in Europe, as mentioned in Table 5.11.

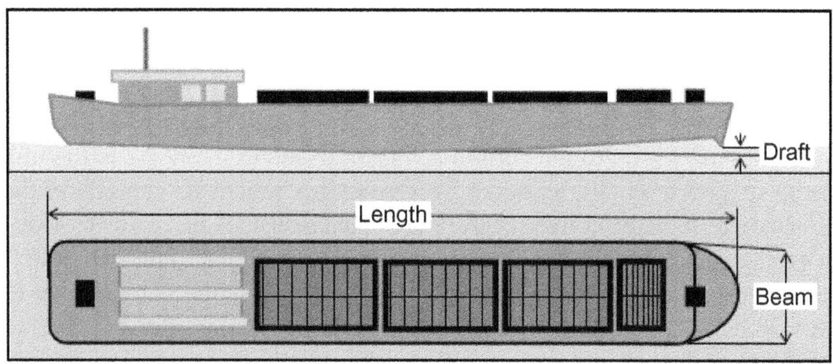

FIG. 5.26

Scheme of the IWT motor container vessel operating in Europe.

5.5.4.2 Route and network

Transport service frequency

The service frequency along the route (*ij*) during time (τ) carried out by the given type of motor vessel or barge convoy can be estimated as follows:

$$f_{ij}(\tau) = \frac{Q_{ij}(\tau)}{\theta_{ij}(\tau) \cdot PL_{ij}} \tag{5.76}$$

$Q_{ij}(\tau)$ is the volume/quantity of freight/cargo during the time (τ) to be transported on the route (*ij*) (ton, TEU);
PL_{ij} is the payload capacity of a barge convoy or the motor vessel operating on the route (*ij*)(ton, TEU); and
θ_{ij} is the load factor of a barge convoy or the motor vessel operating on the route (*ij*) (≤ 1.0).

The required barge convoy or motor vessel fleet to be deployed on the route (*ij*) during time τ can be estimated as follows:

$$N_{ij}(\tau) = f_{ij}(\tau) \cdot \overline{t_{ij}(\tau)} \tag{5.77}$$

where:

$\overline{t_{ij}(\tau)}$ is the average turnaround time of a barge convoys or the motor vessels deployed on the route (*ij*) during time (τ) (h, days).

The turnaround time $\overline{t_{ij}(\tau)}$ in Eq. (5.77) can be estimated as follows:

$$\begin{aligned} \overline{t_{ij}(\tau)} = {} & \frac{\theta_{ij}(\tau) \cdot PL_{ij}}{r_{l/i}} + \sum_{h=1}^{H-1} \frac{d_{ij/h}}{v_{ij/h}(\tau)} + \sum_{h=1}^{H} [W_{ij/h}(\tau) + \Delta_{ij/h}(\tau)] + \frac{\theta_{ij}(\tau) \cdot PL_{ij}}{r_{u/j}} + \\ & + \frac{\theta_{ji}(\tau) \cdot PL_{ji}}{r_{l/j}} + \sum_{h=1}^{H-1} \frac{d_{ji/m}}{v_{ji/h}(\tau)} + \sum_{h=1}^{H} [W_{ij/h}(\tau) + \Delta_{ji/h}(\tau)] + \frac{\theta_{ji}(\tau) \cdot PL_{ji}}{r_{u/i}} \end{aligned} \tag{5.78}$$

$r_{l/i}$, $r_{u/j}$ is the loading and unloading rate of motor vessels or barge convoys at the origin port *i* and destination port *j*, respectively (ton/h, TEU/h);
$r_{l/j}$, $r_{u/i}$ is the loading and unloading rate of motor vessels or barge convoys at the origin port *j* and destination port *i*, respectively (ton/h, TEU/h);
H is the number of locks along the route (*ij*);
$d_{ij/h}(\tau)$, $d_{ji/h}(\tau)$ is the distance of the segment *h* of the given route in the direction (*ij*) and (*ji*), respectively (km);
$v_{ij/h}(\tau)$, $v_{ji/h}(\tau)$ is the average sailing speed of a barge convoy or motor vessel along the segment *h* of the given route in the direction (*ij*) and (*ji*), respectively (km/h);

$w_{ij/h}(\tau)$, $w_{ji/h}(\tau)$ is the average delay of motor vessel or barge convoy while being served at the lock h of the given route in the direction (ij) and (ji), respectively (km/h); and
$\Delta_{ij/h}(\tau)$, $\Delta_{ji/h}(\tau)$ is the average service time of motor vessel or barge convoy at the lock h of the given route in the direction (ij) and (ji), respectively (km/h).

In Eq. (5.78), the first term represents the loading time at the origin port i, the second term the sailing time along particular segments, the third term the delay and service time at the locks, and the fourth term the unloading time at the destination port (j), all in the direction (ij) of the route for the given motor vessel or barge convoy. The last four terms are analogous to the first four for operating the given motor vessel or barge convoy in the direction (ji) of the given route.

Transport work

The transport work on the route (ij) based on Eq. (5.76) can be estimated as follows:

$$TW_{ij}(\tau) = Q_{ij}(\tau) \cdot d_{ij} = f_{ij}(\tau) \cdot \theta_{ij}(\tau) \cdot PL_{ij} \cdot d_{ij} \tag{5.79}$$

where all symbols are as in the previous equations.

Productivity

The productivity of the given route (ij) based on Eq. (5.76) is estimated as follows:

$$PR_{ij}(\tau) = Q_{ij}(\tau) \cdot v_{ij} = f_{ij}(\tau) \cdot \theta_{ij}(\tau) \cdot PL_{ij} \cdot v_{ij} \tag{5.80}$$

Network

By summing up the number of required fleet, transport work, and productivity for all routes (ij), the corresponding totals for a given barge operator's network can be obtained under conditions that the fixed fleet is exclusively allocated to particular routes.

EXAMPLE 5.11

The example of the busiest IWT route is the segment of Lower Rhine River sphere connecting Rotterdam (the Netherlands) and Duisburg (Germany) (Europe). The length of route is: $d_{ij} = 192$ km, the average width: $w = 150$ m, and debt: $b = 2.80$ m (van Donselaar and Carmigchelt, 2001; Solar, 2012). According to Table 5.11, all seven categories and motor vessels can operate along the route but those of the category V–VII need always adjustments of the total weight. The barge operators provide an extensive range of services, of which the relevant in the given context are the "shuttle" "point-to-point" (without the intermediate stops) services of containers between the two inland ports. Let the total quantity of containerized freight/cargo to be transported in one direction, for example, between Rotterdam and Duisburg, is: $Q(\tau) = 5 \cdot 10^6$ ton/year ($\tau = 1$ year).

If the average weight of 1 TEU is 14.3 ton (12 ton of cargo and 2.3 ton of tare), this gives: $Q(\tau) = 5 \cdot 10^6/14.3 \approx 350 \cdot 10^3$ TEU/year. The container motor vessels of the payload capacity of: $PL = 208$ TEU/vessel are used. The average load factor of the vessel is: $\theta = 0.85$. Then, the required service frequency will be: $f(\tau) = (350 \cdot 10^3)/(208 \cdot 0.85) = 1978$ dep/year

EXAMPLE 5.11—cont'd

or 1978/52 = 38 dep/week or 38/7 = 5 dep/day. If the average sailing speed of the container vessels upstream is: $v_{ij} = 12$ km/h and downstream: $v_{ji} = 20$ km/h, the corresponding sailing times will be: 192/11 = 17.5 h and 192/20 = 9.6 h, respectively.

If the 2-crane loading/unloading rate at both end ports is: $r = 2 \cdot 15$ TEU/h, the vessels' average loading/unloading time will be: $2 \cdot (208 \cdot 0.85/2 \cdot 15) \approx 6.0$ h. Based on Eq. (5.78), the vessel's turnaround time on the given route is equal to: $t_{ij} = 6.0 + 17.5 + 6.0 + 9.6 = 39.1$ h. Then the required fleet of container vessels is equal to: $N = (5/24) \cdot 39.1 \approx 8$ vessels. The total transport work carried out in both directions is equal to: $TW_{ij} = 2 \cdot 5 \cdot 208 \cdot 0.85 \cdot 192 = 385{,}456$ TEU-km/day. The productivity of the given route is equal to: $PR_{ij} = (5/24) \cdot 208 \cdot 0.85 \cdot 11 = 405.17$ TEU-km/h in the direction (*ij*), and $PR_{ji} = (5/24) \cdot 208 \cdot 0.85 \cdot 20 = 736.7$ TEU-km/h in the direction (*ji*).

5.6 MARITIME FREIGHT/CARGO TRANSPORT SYSTEM

5.6.1 GENERAL

The maritime or sea transport mode consists of the infrastructure and service networks. The infrastructure network includes seaports as nodes and maritime shipping lines/routes as links connecting them. Since the ports are usually at the coasts of different countries at the same and/or at different continents, these links can be very long. The transport service networks are established by transport operators—shipping liners—operating vehicles—ships—to serve the expected passenger and freight/cargo demand between particular seaports under given conditions. These ports and transport services between them are the nodes and links, respectively, of these service networks. In this context, the capacity and service level of ports and those of shipping liners transporting freight/cargo are only considered.

5.6.2 PORTS

5.6.2.1 Configuration—layout

A maritime port consists of seaside and landside area. The seaside area generally includes the ship anchorage area and quay/berth area. In some cases, the waterways connecting the port with its inland area are considered as parts of its seaside area. The landside area contains the terminal yard with dedicated terminals handling particular type of freight/cargo such as containers, liquid, dry, and break bulk, and the port's access transport modes such as road and rail freight/cargo systems. The port's seaside and landside operations meet at the quay/berth area facing terminal yard, where freight/cargo is unloaded from the ships and transferred to the terminal, and vice versa, ie, transferred from the terminal yard and loaded on the ships. In such case the port container terminals are actually multimodal ones since facilitate different transport modes: on the one side it is maritime/sea and on the other inland freight/cargo transport modes such rail, road, and in some cases inland waterways/barge mode. Fig. 5.27 shows the simplified scheme of the port layout where with container terminal served by rail and truck transport mode from its inland side.

In this context, the static and dynamic capacity of particular components of the port seaside and landside area can be estimated. The static capacity is defined as the maximum number of objects (ships, containers, other freight/cargo shipments), which can be accommodated/stored on the currently available area of a given port component (at sea or on land) under given conditions. These areas are defined

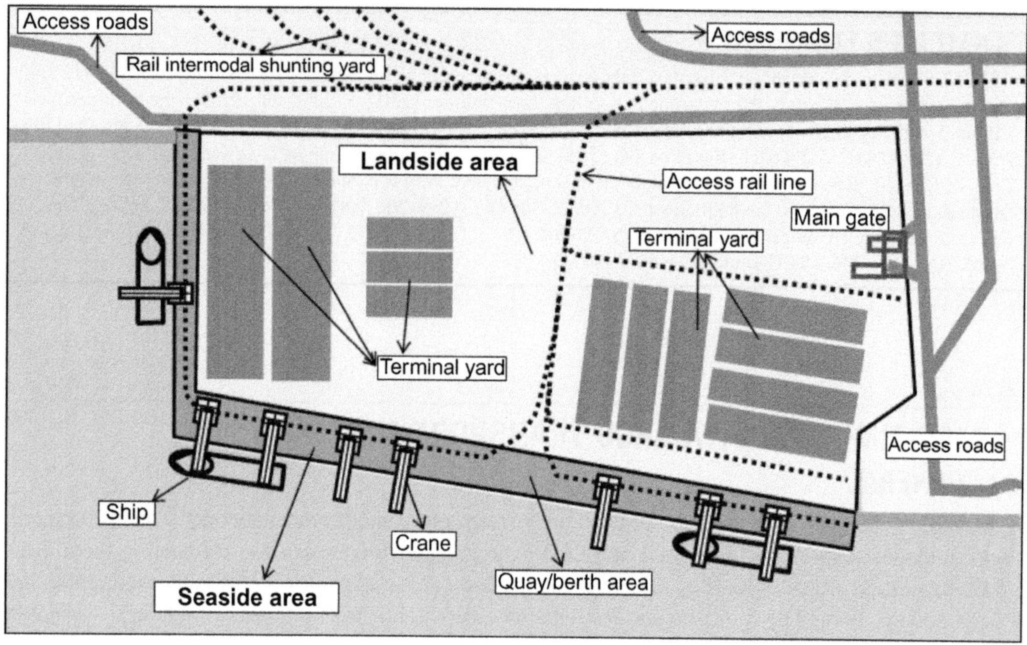

FIG. 5.27

Scheme of the port with the seaside (quay/berth) and the landside (container terminal and ground access systems) area.

by the available space at the certain point in time according to accommodating/storing operational rules and procedures (Lagoudis and Rice, 2011). The dynamic capacity is defined as the maximum number of objects (ships, containers, other freight/cargo shipments), which can be served by a given port component during the specified period of time under given conditions. These can be constant demand for service ("ultimate" capacity) and the average delay while getting service ("practical" capacity) (Salminen, 2013). Both types of capacities for the quay/berth area and the terminal yard area with container terminals are further described.

5.6.2.2 Capacity—seaside area

The seaside area generally contains the anchorage and quay/berth area. The static "ultimate" capacity of this component (they are almost identical for container, dray and break bulk ships), can be expressed by the maximum number of ships (N_q), which can be simultaneously anchored there as follows:

$$N_q = \frac{L_q}{(l_s + s_s)^s} \tag{5.81}$$

where:

L_q is the length of quay/berth area (m);
l_s is the length of a ship (m); and
s_s is the safe longitudinal separation between ships anchored in line at berths along the quay (m) (in general, $s_s = 0.15\text{–}0.20 \cdot (\max l_s)$).

If the average beam/width of each ship is (w_s), the minimum area of anchorage (quay/berth) component shown in Fig. 5.28 can be determined as follows:

$$S_q = L_q \cdot w_s \tag{5.82}$$

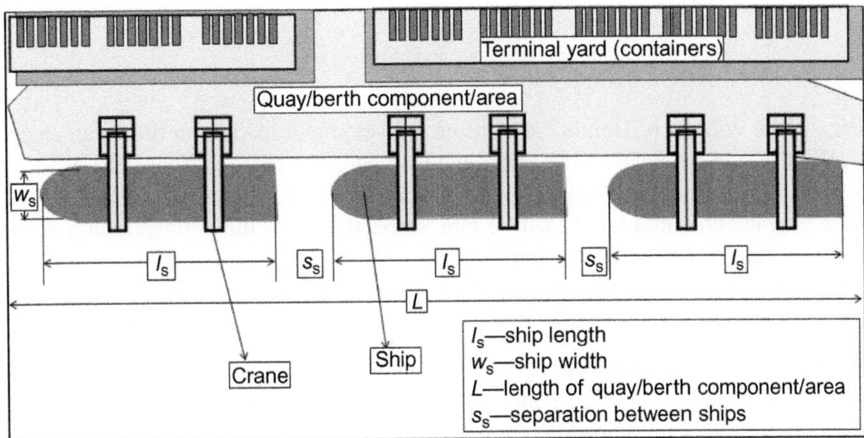

FIG. 5.28

Scheme of the ships anchored at berths along the quay.

The dynamic "ultimate" capacity expressed by the maximum number of ships, which can be handled at the given quay/berth component/area during a given period of time (ships/day) can be estimated as follows:

$$\mu_q = \frac{N_q}{\tau_q} \tag{5.83}$$

where

τ_q is the average ship's turnaround time at the quay/berth area (h; days).

If the carrying (payload) capacity of each ship is PL_s, the berth's "ultimate" capacity in terms of the handled payload (tons/day or TEU/day) will be equal:

$$M_q = \mu_q \cdot PL_s \tag{5.84}$$

The ship's turnaround time τ_q in Eqs. (5.83) and (5.84) consists of its anchoring, freight/cargo unloading, refueling, food and water supplying, freight/cargo loading, and berth leaving time. As such, it can be estimated as follows:

$$\tau_q = t_a + t_u + t_{rf/fws} + t_l + t_f \tag{5.85}$$

where:

t_a is the ship's anchoring time (h);
t_u is the ship's unloading time (days);
$t_{rf/fws}$ is the ship's refueling, and food and water supplying time (days);
t_l is the ship's loading time (days); and
t_f is the ship's berth leaving time (h).

The ship unloading and loading times t_u and t_l in Eq. (5.85) depend on the volumes of the corresponding freight/cargo and related unloading and loading rates. These rates depend on the number of engaged devices (cranes) and their specific handling rates. Consequently, these times can be calculated as follows:

$$t_u = \frac{PL_{s/u}}{n_u \cdot r_u \cdot \theta_u} \quad \text{and} \quad t_l = \frac{PL_{s/l}}{n_l \cdot r_l \cdot \theta_l} \tag{5.86}$$

$PL_{s/u}$, $PL_{s/l}$ is the volume of freight/cargo to be unloaded and loaded on the given ship, respectively; this can be equal to the ship's payload capacity $PL_{s/max}$ (tons, TEU);
n_u, n_l is the number of engaged unloading and loading devices (quay/berth cranes), respectively;
r_u, r_l is the average unloading and loading rate, respectively, of the corresponding device (quay/berth cranes) (tons/h; TEU/h); and
θ_u, θ_l is the utilization rate of the unloading and loading devices (quay/berth cranes), respectively (≤ 1.0).

For example, let the length of quay/berth component in a given port is: $L_q = 3000$ m, the average ship length: $l_s = 294$ m (Panamax 4000 TEU container ship), and the longitudinal separation between ships: $s_s = 100$ m. The static "ultimate" capacity of the quay will be, based on Eq. (5.83), equal to: $N_q = 3000/(294 + 100) \approx 8$ ships. Since the width of each of these ships is: $w_s = 32$ m, the minimum size of anchorage area will be: $S_q = 3000 \cdot 32 = 96{,}000$ m^2 (ie, 9.6 ha (ha = hectare; 1 ha = $10 \cdot 10^3$ m^2)). It should be mentioned that the minimum draft in anchorage area for these ships should be at least 13.5 m.

Fig. 5.29 shows an example of the relationship between total length of a berth/quay area and the number of berths/quays at 20 world's largest container ports (PHK, 2006).

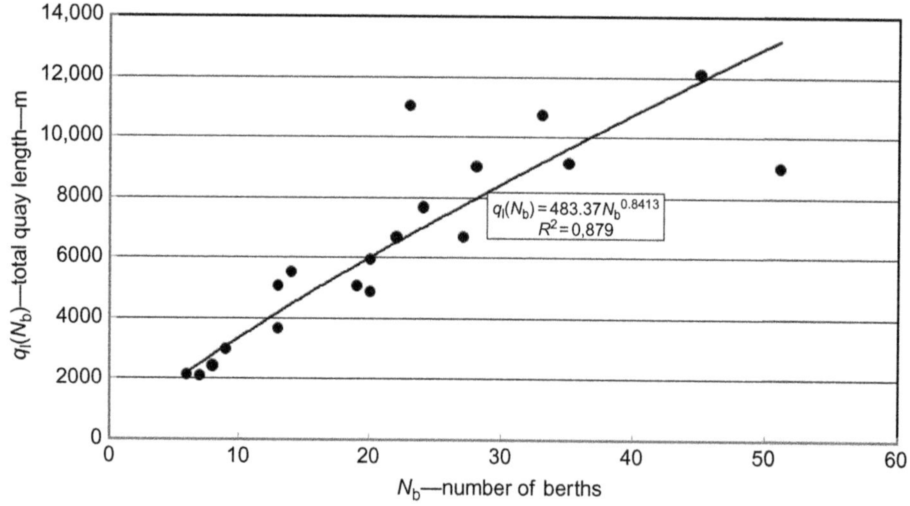

FIG. 5.29

Example of the relationship between the total berth/quay length and the number of berths in the 20 world's largest container ports (PHK, 2006).

As can be seen, the total length of berths/quays increases less that proportionally with increasing of the number of quays/berths. The average length of a berth/quay is about 483 m, which allows anchoring one container ship of type Panamax (4000 TEU, length 294 m, draft 13.3 m) or Triple E Maersk (18,000 TEU, length 399 m, draft 14.5 m). In all these ports, the maximum alongside water depth is equal or greater than 15 m, which enables handling the above-mentioned two ships.

Further, let the unloading and loading volume of containerized freight/cargo of this ship is: $PL_{s/u} = PL_{s/l} = 4000 \cdot 0.80 = 3200$ TEU (ie, load factor (θ) is: 0.80, ie, 80%), the number of unloading and unloading devices (gantry cranes): $n_u = n_l = 4$, each with the corresponding unloading and loading rates of: $r_u = r_l = 25$ TEU/h, their utilization: $\theta_u, = \theta_l = 1.0$ (100%), and the sum of ship's anchoring, refueling, food/water supplying, and berth leaving time 24 h (1 day). Then, the ship's turnaround time will be: $\tau_q = 24 + (2 \cdot 3200)/(4 \cdot 25 \cdot 1) = 64 + 24 = 88$ h (ie, 3.7 days). From Eq. (5.83), the quay/berth dynamic capacity will be equal: $\mu_q = 8/3.7 \approx 2$ ships/day. If the port operates during 250 days/year and 12 h/day, the annual dynamic "ultimate" capacity will be: $\mu_q = 2 \cdot 250 \cdot 16/24 \approx 333$ ships/year.

In addition, the container terminals in large ports worldwide have been faced with handling the smaller number of larger ships requesting the specified turnaround time as an indicator of service quality. For example, in order handle the ship coming with the payload of: $PL_{u/s} = PL_{u/l} = 18{,}000$ TEU (Triple E Maersk) in the required time of $t_u = t_l = 24$ h (total $2 \cdot 24$ h $= 2$ days), the required unloading and loading rate will have to be: 750 TEU/h. If the loading and unloading rate of quay/berth cranes is: $r_u = r_l = 25$ TEU/h and their utilization during the period: $\theta_u, = \theta_l = 1.0$ (100%), the number of engaged cranes have to be: $n_l = n_u = 30$ (continuously operating over two days).

5.6.2.3 Service level—seaside area

If ships arrive at a given port's quay with N_q berths, each with the dynamic capacity μ_q (Eq. 5.83), according to the Poisson process, then the average delay/waiting time W_q (h) in anchoring can be estimated by the well-known formula from the theory of steady-state queuing system (model M/G/1/–), as follows:

$$W_q = \frac{\lambda_q\left(\sigma_q^2 + 1/\mu_q^2\right)}{2(1 - \lambda_q/\mu_q)} \tag{5.87}$$

where:

λ_q is the average demand/arrival rate of ships, ie, port calls (ships/day);
μ_q is the dynamic "ultimate" capacity of the quay/berth component/area (ships/day); and
σ_q is the standard deviation of the ship's service (turnaround) time at the quay/berth component/area (day).

Consequently, by specifying the maximum average delay, the dynamic "practical" capacity of the quay/berth area can be estimated similarly as at the port of WT (inland waterway transport), but with the necessary specification of the input parameters.

5.6.2.4 Capacity—landside area

The static and dynamic "ultimate" and "practical" capacity of the terminal yard and that of the port access modes—rail and road—as components of a given port's landside area are elaborated. In particular, the capacities of terminal yard are separately estimated for the terminal yard and for the facilities and equipment handling freight/cargo (containers) there.

Terminal yard/area

The static "ultimate" capacity of the port's terminal yard/area can be expressed by the size of available space for storing freight/cargo shipments under given conditions. These conditions are specified by the available space, size of footprint of a unit of freight/cargo, and stacking policy, the latest in the case of the container terminals. Fig. 5.30 shows an example of the relationship between the total area of container terminals and their available storage capacity at 20 largest world container ports (PHK, 2006).

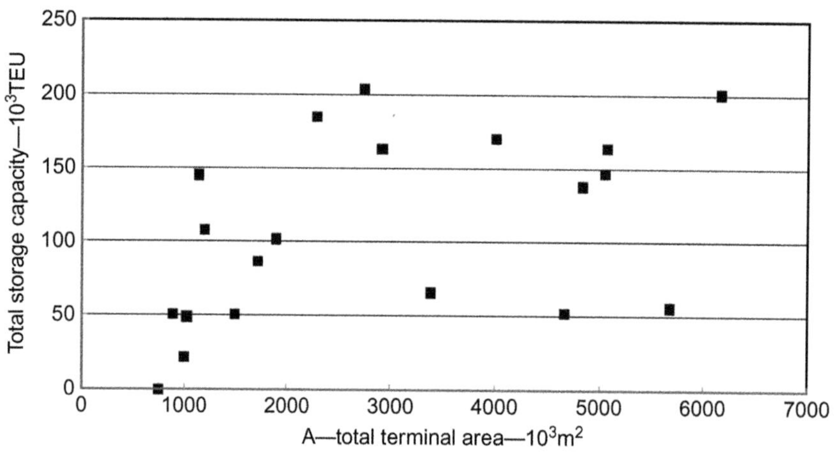

FIG. 5.30

An example of the relationship between the total storage capacity and total terminal area at 20 largest container ports (PHK, 2006).

As can be seen, in this case there is not particular causal relationship between the size of available container terminal space and its storage capacity mainly due to the above-mentioned reasons of handling and storing containers in particular port terminals. Under such conditions, the "static" capacity μ_s (containers = TEU) for a given container terminal yard/area can be estimated as follows:

$$\mu_s = \frac{A_c}{a_c} \cdot m_c \cdot u_c \tag{5.88}$$

where:

A_c is the available area of the yard (m^2);
a_c is the area of container footprint (m^2);
m_c is the number of container stacking levels; and
u_c is the yard's utilization rate regarding congestion.

Based on Eq. (5.88), the dynamic "ultimate" capacity of a given container terminal yard/area (containers-TEU/h, or per day) can be calculated as follows:

$$\mu_d = \frac{\mu_s}{\tau_c} \tag{5.89}$$

where

τ_c is the average dwell time of a container in a given terminal yard/ area (h/TEU; days/TEU).

For example, let the yard's available area is: $A_c = 10{,}000\ m^2$ (ie, 1 ha), the container footprint: $a_c = 14.884\ m^2$ (this is the standard 20 foot container: length 6.1 m, width 2.44 m, and height 2.59 m), the number of stacking levels: $m_c = 5$, and the yard's utilization rate: $u_c = 0.70$. The distance between containers in the block ranges from 0.35 to 0.5 m. Then the static "ultimate" capacity will be equal to: $\mu_s = [10{,}000/6.1 \cdot (2.44 + 0.5)] \cdot 5 \cdot 0.70 \approx 1952$ TEUs. If the container average dwell time in the yard is: $\tau_c = 5$ days, ie, 36 h, then the yard's dynamic "ultimate" capacity will be: $\mu_d = 1952/5 \approx 390$ TEU/day or $1390/(5 \cdot 24) \approx 16$ TEU/h. In addition, the terminal yard/area "practical" capacity can be calculated similarly as that for the berth/quay area. However, this time, the incoming containers wait on board the arrived ships for their unloading. The outgoing containers wait for their unloading in front of the terminal yard/area on the trains and/or the road trucks. In both cases, their short temporary storage can be provided at both ends of the yard.

Freight/cargo handling equipment

The freight/cargo handling equipment in the port container terminals include different types of devices enabling handling and manipulating with containers from the quay/berth area to the yard, within the yard, and from the yard to terminal ground access systems, and vice versa. These are: quay/berth cranes for moving containers from ships to berths, the vehicles for transporting containers between berths and terminal yard, terminal yard cranes for stacking containers, straddle carriers as an equipment for transporting and stacking containers in terminal yard, internal transport vehicles generally used for transporting containers within the terminal and moving containers between the terminal truck entry/exit gate or rail intermodal yard to the terminals stacking cranes, and the rail/road (horizontally and vertically) transshipment equipment [for example, the forklift trucks, automated guided vehicles (AGVs), etc.].

In general, the dynamic "ultimate" capacity of any kind of the above-mentioned equipment can be calculated as follows:

$$\mu_e = n_e \cdot r_e \cdot u_e \tag{5.90}$$

n_e is the number of engaged equipment units (for example, cranes); and
r_e is the average handling rate per engaged equipment unit (crane) (tons/h; TEU/h); and
u_e is the utilization rate of equipment unit (crane) (<1.0).

For example, if the number of stacking cranes in a given terminal yard is: $n_e = 3$, the container handling and utilization rate of each of them: $r_e = 15$ TEU/h and $u_e = 0.70$, respectively, the dynamic "ultimate" capacity will be: $\mu_e = 3 \cdot 15 \cdot 0.70 \approx 32$ TEU/h. In the hypothetical case of completely empting the above mentioned yard with: $N_c = 1952$ TEUs, the empting time will be: $1952/32 \approx 61$ h or 2.54 days. The average container dwell time will be: $\tau_c = (1/2) \cdot 61 = 30.5$ h or ≈ 1.3 days. By specifying the maximum dwell time of a container in the yard, the "practical" capacity of the handling equipment can be determined. In addition, let the full train transporting containers to and from the yard consists of 25 flat wagons each with the payload capacity of 2 TEU (ie, the full train payload capacity is 50 TEU). The loading/unloading device (for example, the forklift track) has the rate of about:

$\mu_e = 1$ TEU/min. If: $m_e = 3$ three devices are engaged, the train's loading/unloading rate will be: $\mu_e = 3 \cdot 1 \cdot 0.70 \approx 2.1$ TEU/min, and its unloading or loading time will be: $t_{l/u} = 50/2.1 \approx 24$ min. The total time a train spends at the terminal excluding the time for administration will be equal to: $\tau_t = 2 \cdot 24 = 48$ min. The average waiting time of an incoming container for unloading or an outgoing container for the train departure is equal to: $w_c = (1/2) \cdot t_{l/u} = (1/2) \cdot 24 = 12$ min.

By specifying this waiting time in advance regarding the train characteristics (length, payload), the "practical" capacity of the loading/unloading equipment of the given characteristics to guarantee just such performances can be determined.

Faced with increased demand, numerous container terminals around the world ports have adopted the automated above-mentioned container handling equipment. In this case the automation enables remote operations, ie, remote control of handling containers through the terminal—from the ship-to-quay/berth unloading/loading phase, stacking in the terminal yard, and preparing for handling by the ground access systems, as well as remote monitoring of automatic gates, where human intervention is the exception. The efficiency of operations can be further increased through the horizontal integration of these systems and equipment. In order to be particularly attractive for the large container ships an automated terminal should possess an efficient quay/berth cranes, intelligent automatic stacking cranes, integrated terminal equipment from ship to exit/entry gate, and remote operations from the control room (PEMA, 2014). For example, one of the most advanced is the Euromax Terminal Rotterdam (Rotterdam, the Netherlands) especially designed for the efficient, effective, and safe handling of the largest container ships. These are anchored alongside the quay/berth area in about one hour. The cranes in the quay/berth area are semi-automated. AGVs are used to transport containers between the quay/berth area stacks in the terminal yard. In the stacking lanes, the automated rail mounted gantry cranes are used. The ground access systems include rail intermodal, road truck, and inland waterways/barge transport mode. Trains are handled at the on-site rail terminal with two cranes and six tracks. This rail terminal is directly connected to the Port Railway Line and the Betuweroute. The road trucks takeover and deliver containers at the terminal gate. The barges take over and deliver containers at both ends of the quay/berth area means by separate cranes (http://www.ect.nl/en/content/euromax-terminal-rotterdam).

5.6.2.5 Capacity—access modes

A port can linked with its hinterland by road, rail, and in some cases inland waterway/barge transport mode. They are characterized by the static and dynamic "ultimate" and "practical" capacity.

Rail

The static "ultimate" capacity of the port's rail access mode can be defined as the required number of tracks for simultaneous accommodation of the trains serving given terminal yard. This number (N_t) can be estimated as follows:

$$N_t = \mu_t \cdot \tau_t \tag{5.91}$$

where:

μ_t is the dynamic "ultimate" capacity of the rail lines connecting the terminal yard and the port's hinterland (trains/h or trains/day); and
τ_t is the average track's occupancy time by an incoming or outgoing train (h/train).

The dynamic "ultimate" capacity of the rail tracks, ie, rail network can be expressed by the maximum number of freight/cargo units (containers, TEUs), which can be served by trains operating on the lines connecting the terminal yard and the port's hinterland. This capacity (M_t) in can be calculated as follows:

$$M_t = n_t \cdot \mu_t \cdot PL_t \tag{5.92}$$

where:

n_t is the number of incoming and outgoing rail lines connecting given terminal yard with the port's hinterland; and
PL_t is the maximum average payload capacity of a train operating between given terminal yard and the port's hinterland (TEU/train).

The other symbols are analogous to those in the previous equations.

For example, if train spends about: $\tau_t = 4.8$ h on the tracks serving the given terminal yard, and if the capacity of rail lines connecting this terminal yard and the port's hinterland is: $\mu_t = 4$ trains/h, the required number of tracks will be:

$N_t = 4 \cdot 4.2 \approx 17$. In addition, if the number of incoming and outgoing rail lines is: $n_t = 2$, the dynamic "ultimate" capacity of each of the: $\mu_t = 2$ trains/h, and the capacity of each train: $PL_t = 25$ wagons $\cdot$ 2 TEU/wagon $= 50$ TEU, the dynamic "ultimate" capacity of the rail tracks, ie, of the rail network serving given terminal yard will be equal: $M_t = 4 \cdot 2 \cdot 50 = 400$ TEU/h. If the rail mode operates 16 h/day, its daily capacity will be: $M_t = 16 \cdot 400 = 2400$ TEU/day. Over the 250 working days/year, this capacity will be: $M_t = 400 \cdot 250 = 100{,}000$ TEU/yr.

Road

The static "ultimate" capacity of the port's road access mode can be defined as the required number of parking places for simultaneous accommodation of road trucks while serving given terminal yard, ie, while being unloaded and/or loaded by containers. This number (N_r) can be estimated as follows:

$$N_r = \lambda_r \cdot \tau_r \tag{5.93}$$

where:

λ_r is the intensity of truck's demand for accessing given terminal yard (trucks/h or trucks/day); and
τ_r is the average occupancy time of a parking place by an incoming or outgoing truck (h/truck).

The dynamic "ultimate" capacity of the road network serving given terminal yard can be expressed by the maximum number of freight/cargo units (containers, TEUs), which can be served by trucks operating on the lines connecting the terminal yard and the port's hinterland. This capacity (M_r) can be calculated as follows:

$$M_r = n_r \cdot \mu_r \cdot PL_r \tag{5.94}$$

where:

N_r is the number of incoming and outgoing road lanes connecting given terminal yard with the port's hinterland;
μ_r is the dynamic "ultimate" capacity of road lane connecting given terminal yard with the port's hinterland (trucks/h); and
PL_r is the maximum average payload capacity of a truck operating between given terminal yard and the port's hinterland (TEU/truck).

For example, if the intensity of truck's demand for accessing given terminal yard is: $\lambda_r = 100$ trucks/h, the truck unloading/loading time: $\tau_r = 2/15$ (TEU/TEU/h) $= 0.4$ h, the required number of parking places will be: $N_r = 40$. In addition, if the number of road lanes connecting given yard with the port's hinterland is: $n_r = 2$, each with the dynamic "ultimate" capacity of: $\mu_r = 60$ trucks/h, and if the payload capacity of each truck is: $PL_r = 2$ TEU/truck, the dynamic "ultimate" capacity of the road network in the given case will be: $M_r = 2 \cdot 60 \cdot 2 = 240$ TEU/h. If this road network operates 16 h/day, its daily capacity will be: $M_r = 240 \cdot 16 = 3840$ TEU/day. For 250 working days per year, this capacity will be: $M_r = 3840 \cdot 250 = 960{,}000$ TEU/year.

Total capacity

The total dynamic "ultimate" capacity of the ground access modes serving given terminal handling containers can be calculated as by summing up the dynamic "ultimate" capacity of both rail and rail transport mode as follows:

$$M_{gas} = M_t + M_r \tag{5.95}$$

In Eq. (5.95), it is assumed that both modes operate simultaneously under given above-mentioned conditions.

Balancing the seaside and landside capacity and the overall service quality

In order to enable efficient and effective operation of the above-mentioned port's seaside and landside area, the dynamic "ultimate" capacities of their particular components positioned in the serial order for handling incoming and outgoing container flows need to be in balance, ie, equal under given operating conditions. Since it rarely happens, the "critical" or "bottleneck" capacity is the smallest among those of particular components. For example, for an incoming container, it will be:

$$\mu_{i/c} = \min\left[M_q; n_{q/t}/t_{q/t}; \mu_c; n_{t/t}/t_{t/t}; M_{gas}\right] \tag{5.96}$$

where:

$n_{q/t}$, $t_{q/t}$ is the number of engaged vehicles and the average time, respectively, for moving container from the quay/berth to the terminal yard (min/TEU); and
$n_{t/t}$, $t_{t/t}$ is the number of vehicles and the average time, respectively, for moving container from the terminal yard to the loading position on the train or/ot truck (min/TEU).

The other symbols are as in the previous equations. In addition, the total average time of passing an incoming container through the port (ie, the total port's dwell time) consisting of its service (handling) time and the waiting time/delay for service can be calculated under the above-mentioned conditions as follows:

$$\tau_{i/c} = (1/2) \cdot \frac{PL_{s/max} \cdot \theta_s}{n_{q/u} \cdot r_{q/u} \cdot u_{q/u}} + \frac{1}{r_{q/u} \cdot u_{q/u}} + t_{q/t} + \tau_c + t_{t/t} + \frac{1}{r_e \cdot u_e} + (1/2) \cdot \frac{PL_t \cdot \theta_t}{n_e \cdot r_e \cdot u_e} \tag{5.97}$$

where:

$u_{q/u}$ is the utilization rate of unloading devices (cranes) at quay/berth; and
θ_t is the average load factor of the outgoing train.

The other symbols area analogous to those in the previous equations. As an illustration, for example, if: $PL_{s/max} = 4000$ TEU/ship, $\theta_s = 0.80$, $n_{q/u} = 4$, $r_{q/u} = 25$ TEU/h, $u_{q/u} = 0.70$, $t_{q/t} = 15$ min, $\tau_c = 5$ days,

$t_{t/t}=15$ min, $PL_t=50$ TEU/train, $\theta_t=1.00$, $n_e=3$, $r_e=15$ TEU/h, and $u_e=0.70$ the container's total time through port will be equal as: $\tau_{i/c}=(1/2)\cdot\frac{4000\cdot 0.80}{4\cdot 25\cdot 0.70}+\frac{1}{25\cdot 0.70}+15/60+5\cdot 24$ $+15/60+\frac{1}{15\cdot 0.70}+(1/2)\cdot\frac{50\cdot 1.00}{3\cdot 15\cdot 0.70}=144.30$ h, or 6.01 days. This time for an outgoing container can be calculated similarly, but respecting its relevant handling operations and values of particular parameters.

Generalization

The above-mentioned capacity and level of service have been considered for the port's container terminal(s). The similar approach, but with the necessary modifications of particular variables/parameters, can be applied to considering the port's liquid, dry, and/or break bulk terminal(s) (Salminen, 2013).

5.6.3 SHIPPING LINES

The liner shipping is the maritime transport service provided by the shipping line companies operating the high-capacity, ocean-going ships. These ships sail along the regular ocean routes usually according to the fixed scheduled service frequencies serving the expected freight/cargo demand.

5.6.3.1 Route

Capacity

For a given line/route connecting two ports at different continents, the service frequency satisfying given volumes of expected freight/cargo demand during the specified period of time can be determined as follows (Janić, 2014c):

$$f(\tau)=\frac{Q(\tau)}{\theta(\tau)\cdot PL_{s/max}} \tag{5.98}$$

where:

- τ is the period of time for which the service frequency is scheduled (week); duration of the chain's production/consumption cycle (TU);
- $Q(\tau)$ is the quantity of freight/cargo demand transported on the given route connecting two ports during the time period (τ) (tons; TEUs/TU);
- $PL_{s/max}$ is the payload capacity of ships operating on the route between two port (%; tons, m^3, or TEU/ship);and
- $\theta(\tau)$ the average load factor of ships operating on the route between two ports during time (τ) (<1.0).

In Eq. (5.98), it is assumed that all ships are approximately of the same size/payload capacity. In addition, if the service frequency is already given, the total quantity of freight/cargo shipments (tons; TEUs), which can be transported between two ports during time (τ) will be as follows:

$$Q(\tau)=f(\tau)\cdot\theta(\tau)\cdot PL_{s/max} \tag{5.99}$$

The size of deployed ship fleet (the number of vehicles/ships) on the given route under given conditions is equal to:

$$N(\tau)=f(\tau)\cdot\tau(l,v) \tag{5.100}$$

where

$\tau(l, v)$ is the ship's turnaround time while operating on the route of length l at an average sailing cruising speed v (days) d (nm); v (kt); nm = nautical mile; 1 nm = 1.852 km; kt = knot; 1 kt = 1 nm/h.

If each ship sails relatively full in both directions of the given route, its average turnaround time $\tau(l, v)$ (days) in Eq. (5.100) can be estimated as follows:

$$\tau(d,v) = \Delta_{11} + \frac{\theta_1 \cdot PL_{s/max}}{u1_{q/l} \cdot r1_{q/l}} + \frac{l_1}{v_1(l_1)} + D_1 + \Delta_{12} + \frac{\theta_1 \cdot PL_{s/max}}{u1_{q/u} \cdot r1_{q/u}} + \Delta_{21} + \frac{\theta_2 P^2 L_{s/max}}{u2_{q/l} \cdot r2_{q/l}} + \frac{l_2}{v_2(l_2)} + D_2 + \Delta_{22} + \frac{\theta_2 P^2 L_{s/max}}{u2_{q/u} \cdot r2_{q/u}} \quad (5.101)$$

where:

1, 2 is the route direction: 1 = one direction; 2 = opposite/return direction;
Δ_{11}, Δ_{12} is the time between starting ship's loading at the origin and unloading at the destination port while operating in the direction (1) (h);
Δ_{21}, Δ_{22} is the time between starting ship's loading at the origin and unloading at the destination port while operating in the direction (2) (h);
θ_1, θ_2 is the average ship's load factor in the directions (1) and (2) of the given route, respectively, 20 and the payload capacity, respectively, of a vehicles serving the chain (*ij*) (ton, m^3, or TEU per vehicle);
$r1_{q/l}$, $u1_{q/l}$ is the ship's loading rate and its utilization, respectively, at the origin port of the direction (1) (TEU/h; $\leq$1.0);
$r1_{q/u}$, $u1_{q/u}$ is the ship's unloading rate and its utilization, respectively, at the destination port of the direction (1) (TEU/h; $\leq$1.0);
$r2_{q/l}$, $u2_{q/l}$ is the ship's loading rate and its utilization, respectively, at the origin port of the direction (2) (TEU/h; $\leq$1.0);
$r2_{q/u}$, $u2_{q/u}$, is the ship's unloading rate and its utilization, respectively, at the destination port of the direction (2) (TEU/h; $\leq$1.0);
l_1, l_2 is the length of route in the direction (1) and (2), respectively (nm);
$v_1(l_1)$, $v_2(l_2)$ is the average sailing cruising ship's speed in the direction of the route (1) and (2), respectively (kt) (kt = knots); and
D_1, D_2 is the average delay per transport service due to the traffic conditions on the route in the direction (1) and (2), respectively (h, days).

For the route direction (1), the particular terms in Eq. (5.101) denote the following: the ship's preparation and loading time at the origin port, sailing time and eventual delay along the route, and preparation and unloading time at the destination port. The particular terms for the direction (2) are analogous. In addition, Eq. (5.101) indicates that operating conditions at origin and destination ports and along the route in opposite directions can be different.

The transport work (ton-nm; TEU-nm) on the given route in the single direction, based on Eq. (5.99), is calculated as follows:

$$TW(\tau) = Q(\tau) \cdot l \quad (5.102)$$

The productivity of a given route (ton-nm/h; TEU-nm/h) in the single direction, based on Eqs. (5.99) and (5.101) is equal to:

$$TP(\tau) = Q(\tau) \cdot \overline{v(l)} = Q(\tau) \cdot \frac{2l}{\tau(l, v)} \tag{5.103}$$

where all symbols are analogous to those in previous equations.

Service level

The service level provided along the given line/route can be expressed by the schedule delay the freight/cargo shipments weight at the origin port to be transported and the delay along the route if caused by the internal-ship related causes. The schedule delay (h, days) during time (τ) can be determined as follows:

$$sd(\tau) = (1/2) \cdot \frac{\tau}{f(\tau)} \tag{5.104}$$

The delay along the route D (h, days) can be determined as the difference between the scheduled and actual travel time.

EXAMPLE 5.12

The above-mentioned models are illustrated for the route connecting North Europe and Far East Asia served by the liner container shipping. The route connects the port of Rotterdam—APM Terminals Rotterdam (the Netherlands) and the port of Shanghai—Yangshan Deepwater Port Phases 1/2 or 3/4 (People Republic of China). Currently, this is one of the world's busiest sea trading routes,[7] schematically shown in Fig. 5.31.

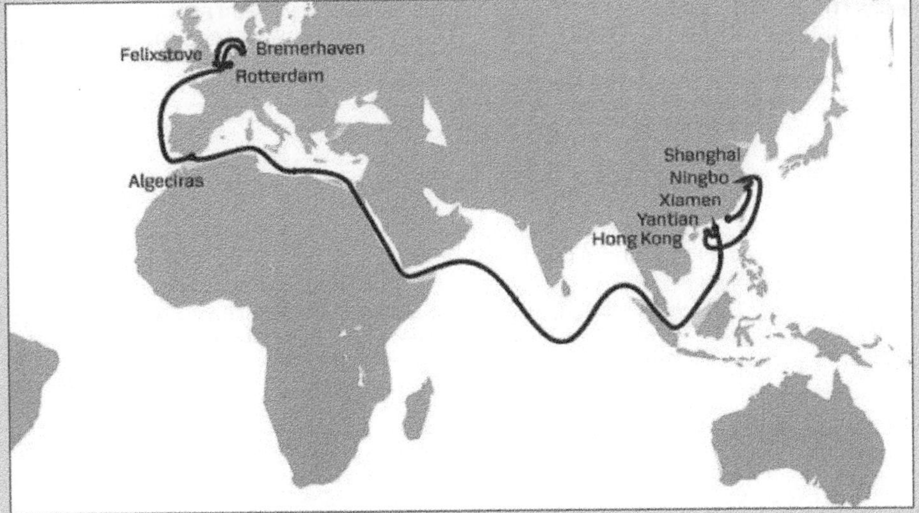

FIG. 5.31

Scheme of geography of the given supply chain—the liner shipping route Rotterdam-Shanghai (Janić, 2014b). Arrows indicate directions.

(Continued)

[7]This sea trading route, included in the World Container Index together with other 10 most voluminous world's container sea trading routes, shares about 35% of their total volumes (TEUs) (http://www.worldcontainerindex.com/).

EXAMPLE 5.12—cont'd

Its length is: $l = 10{,}525$ nm (http://www.sea-distances.org/). The container terminals at both ports enable access and operation of the large container ships including the currently largest Triple E Maersk. The collection and distribution of goods/ freight shipments (TEUs) at both ports are carried out by rail/intermodal, road, inland waterway (barge), and feeder (including short-sea) vessel transport modes (Janić, 2014b; Zhang et al., 2009).

Two scenarios of operating the given route in the single direction are considered: an exclusive use of the container ships of capacity of 4000 TEU (the current Panamax); and exclusive use of the container ships of capacity of 18,000 TEU (Neo Panamax, ie, Triple E Maersk ship). The average load factor of both ships is: $\theta = 0.8$. (AECOM/URS, 2012; http://www.worldslargestship.com/;http://en.wikipedia.org/wiki/List_of_largest_container_ships).

Schemes of both ships are shown in Fig. 5.32. In both scenarios, the ships performing transport services are assumed to operate at the typical slow steaming speed of 20 kts (knots) and the supper slow steaming speed[8] of 15 kts (SCG, 2013). The waiting time for beginning of loading and unloading at both ports is assumed to be: $\Delta_{11} = \Delta_{12} = 1$ day, and the average delay along the route: $D = 0.0$ days.

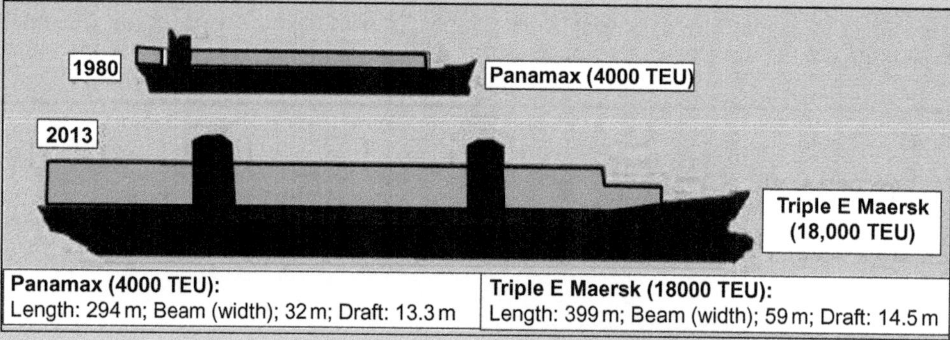

FIG. 5.32

Scheme of container ships operating on the given route (Janić, 2014b).

The loading and unloading rate in both ports are adopted to be: $r_{q/l} = r_{q/u} = 92$ TEU/h (3–4 canes) for the ship of 4000 TEU and $r_{q/l} = r_{q/u} = 215$ TEU/h (7–8 cranes) for the ship of 18,000 TEU. The utilization rates of loading and unloading are assumed equal for both ships in both ports, ie, $u_{q/l} = u_{q/u} = 1.0$. If the total annual demand on the route in the single direction is: $Q(\tau) = 750{,}000$ TEUs, the frequency for the ship transport services is determined from Eq. (5.98), as follows: $f(\tau) = [750{,}000\ /(4000 \cdot 0.8)]/52 \approx 5$ dep/wk for the smaller and $f(\tau) = [750{,}000/(18{,}000 \cdot 0.8)]/52 \approx 1.0$ dep/wk for the larger ship. If the sailing conditions are the same in both directions, the ship's turnaround time on the route is calculated from Eq. (5.101) as follows:

For the smaller ship: $\tau(d, v) = 2 \cdot \left(24 + \frac{0.8 \cdot 4000}{92} + \frac{10525}{15} + 0 + 24 + \frac{0.8 \cdot 4000}{92}\right) = 1638.5$ h or 68.3 days for the super slow steaming and 1463 h or 61 day for the slow steaming sailing policy. The corresponding fleet size will be from Eq. (5.100) equal to: $N(\tau) = 5 \cdot 68.3/7 \approx 48$ ships for the super slow steaming and $N(\tau) = 5 \cdot 61/7 \approx 44$ ships for the slow steaming sailing policy. The transport work from Eq. (5.102) will be: $TW(\tau) = (750{,}000/52) \cdot 10{,}525 =$ 151,802,885 TEU-nm/week. The route productivity from Eq. (5.103) will be: $TP(\tau) = (750{,}000/52) \cdot (10{,}525/$

[8]This operating policy has been practiced by the shipping line companies in order to reduce fuel consumption and related emissions of green house gases.

EXAMPLE 5.12—cont'd

(1638.5/2)) ≈ 185,294 TEU-nm/wk for the super slow steaming and $TP(\tau) = (750{,}000/52) \cdot (10{,}525/(1463/2)) \approx 207{,}523$ TEU-nm/wk for slow steaming sailing policy.

For the larger ship: $\tau(d,v) = 2 \cdot \left(24 + \frac{0.8 \cdot 18000}{215} + \frac{10525}{15} + 0 + 24 + \frac{0.8 \cdot 18000}{215}\right) = 1767.24\text{h}$ or 73.6 days for the super slow steaming speed and 1592 h or ≈ 66 days for the slow steaming sailing policy. The corresponding fleet size will be from Eq. (5.100) equal to: $N(\tau) = 1 \cdot 73.6/7 \approx 11$ ships for the super slow steaming and $N(\tau) = 1 \cdot 66/7 \approx 9$ ships for the slow steaming policy. The transport work from Eq. (5.102) will be: $TW(\tau) = (750{,}000/52) \cdot 10{,}525 = 151{,}802{,}885$ TEU-nm/week. The route productivity from Eq. (5.103) will be: $TP(\tau) = (750{,}000/52) \cdot (10{,}525/(1767.24/2)) \approx 171{,}797$ TEU-nm/wk for the super slow steaming and $TP(\tau) = (750{,}000/52) \cdot (10{,}525/(1592/2)) \approx 190{,}707$ TEU-nm/wk for the slow steaming policy. The corresponding average schedule delays of freight/cargo shipments from Eq. (5.104) will be: $sd = 7/5 = 1.4$ days for the smaller and: $sd = 7/1 = 7$ days for the larger ship (Janić, 2014c).

5.6.3.2 Network

Capacity

The above-mentioned route is a part of the transport service network of a shipping line (s). The routes between other ports also constitute this network. If a given shipping line company transport freight/cargo shipments between (*N*) ports, then its network consists of these ports as the nodes and $N(N-1)$ routes, ie, network links. In this case, the required fleet in terms of the number of ships to operate this network during the specified time (τ) can be determined as follows:

$$N(\tau) = \sum_{i=1}^{N} \sum_{\substack{j=1 \\ j \neq i}}^{N} f_{ij}(\tau) \cdot \tau_{ij}\left[l_{ij}; v_{ij}\left(l_{ij}\right)\right] \tag{5.105}$$

The offered transport capacity (tons, TEU) will be:

$$PL(\tau) = \sum_{i=1}^{N} \sum_{\substack{j=1 \\ j \neq i}}^{N} f_{ij}(\tau) \cdot \tau_{ij}\left[l_{ij}; v_{ij}\left(l_{ij}\right)\right] \cdot PL_{s/\max//ij} \tag{5.106}$$

where:

$f_{ij}(\tau)$ is the transport service frequency on the route (*ij*) and (*ji*), ie, between the ports *i* and *j*, and return (dep/wk);
$\tau_{ij}[l_{ij}; v_{ij}(l_{ij})]$ is the ship's turnaround time on the route l_{ij} and l_{ji}, while operating at the speed $v_{ij}(l_{ij})$ (days);
$v_{ij}(l_{ij})$ is the average ship's operating speed along the route l_{ij}, and return (kts);
l_{ij} is the length of route between the ports *i* and *j* (nm); and
$PL_{s/\max/ij}$ is the payload capacity of ships operating along the route (*ij*) (tons, TEU).

It is assumed, in Eqs. (5.105) and (5.106), that the transport service frequency, length of routes, ship's sailing speed, and payload capacity are the same in both directions of the particular routes. By replacing the service frequency from Eq. (5.98) into Eq. (5.105), the ship fleet satisfying the expected freight/cargo demand under given conditions can be estimated.

Based on Eqs. (5.105) and (5.106), the offered transport work (ton-nm; TEU-nm) in the network is equal to:

$$TW(\tau) = \sum_{i=1}^{N} \sum_{\substack{j=1 \\ j \neq i}}^{N} f_{ij}(\tau) \cdot 2 \cdot l_{ij} \cdot PL_{s/\max//ij} \tag{5.107}$$

The offered productivity of the network (ton-nm/h or TEU-nm/h) will be equal:

$$TP(\tau) = \sum_{i=1}^{N} \sum_{\substack{j=1 \\ j \neq i}}^{N} f_{ij}(\tau) \cdot PL_{s/\max//ij} \cdot v_{ij}\left(l_{ij}\right) \tag{5.108}$$

where all symbols are as in the previous equations.

Service level

In addition to the number and diversity of origin and destination ports, the shipping lines consider the service quality offered to their users—freight cargo shippers and receivers—through attributes such as the offered service frequency between particular ports during given period of time, punctuality (no or acceptable delays), reliability (realized as agreed), and safety and security (no damages of freight/goods shipments). In particular, the service frequency on the particular routes influences the schedule delay, which could be very important for the time-sensitive freight/goods shipments, if there. For the given network, this can be determined on the route-by Eq. (5.104).

EXAMPLE 5.13

Shipping line can be very large. Table 5.13 gives an indication of the size and related capacity of ten largest shipping lines in the world.

Table 5.13 Characteristics of the Available Capacity of Ten Largest Shipping Line Companies (http://www.alphaliner.com/top100/)

Rank	Name of the Company	Capacity (TEU)	Number of Ships	Average Ship Capacity PL_{max} (TEU/ship)
1	APM-Maersk	3.056.603	607	5036
2	Mediterranean Shg Co	2.685.524	508	5286
3	CMA CGM Group	1.789.032	475	3766
4	Hapag-Lloyd	958.358	180	5324
5	Evergreen Line	948.815	201	4720
6	COSCO Container L.	866.260	166	5218
7	CSCL	704.837	139	5071
8	Hanjin Shipping	622.190	102	6100
9	Hamburg Süd Group	615.902	131	4702
10	OOCL	589.491	110	5359

EXAMPLE 5.13—cont'd

As can be seen, some companies operate the larger total fleet capacity but with the smaller number of bigger ships, and vice versa. In this case, the typical average ship size varies between 3000 and 6000 TEU. By multiplying the available company's capacity by the average route length and the average sailing speed (which is typically between 15 and 20 kts), the corresponding transport work and productivity at the company's level can be estimated by relations (5.102) and (5.103).

5.7 AIR TRANSPORT SYSTEM

5.7.1 GENERAL

Similarly as others, the air transport system consists of demand and supply component. The demand component includes the users-air passengers and air freight/cargo shipments send between shippers and receivers. The supply component embraces airports, airlines, and air traffic control (ATC). This time, the capacity and service level of airports, ATC, and airlines are analyzed under conditions of given demand. Specifically, those of the ATC are constrained only to the airspace while the rest related to the capacity of ATC controller(s) are elaborated in Chapter 6. In this case, the air transport infrastructure consists of airport as nodes and the controlled airspace between them organized into airways (air routes) as links of the air transport infrastructure network. The airlines operating their flights in this infrastructure network form the transport service network handling the expected passenger and freight/cargo demand under given conditions.

5.7.2 AIRPORTS

5.7.2.1 Background

An airport consists of the airside and landside area. The airside area includes the runway system with approach airspace in its vicinity, taxiways, and the apron/gate complex. The runway system and close airspace enable the aircraft final approach and landing, and taking-off. The apron/gate complex is used for aircraft parking after landing and for their preparation before take-off (departure). Taxiways enable aircraft to taxi between runways and the apron/gate complex after landing and before taking-off. The apron/gate complex enables parking the aircraft after landing and their preparation for take-off. This includes disembarking of arriving passengers and unloading of freight/goods shipments, refueling, cleaning, supplying with food and water, and embarking the departure passengers and freight/goods shipments, all taking the aircraft turn-around time. The airport landside area consists of airport passenger and cargo terminals and ground access systems. The airport terminals enable users of air transport services—air passengers and air cargo shippers—access to the airlines and their aircraft as providers of these services. These are elaborated in more detail in Chapter 8. The ground access systems enable users of air transport service access to the airport from its gravitational area usually spreading around the nearest big city or urban agglomeration. These can generally be individual cars and public transport systems such as bus, metro, light rail, and regional and long-distance conventional and HSR. In the latter two cases, the gravitational area of an airport can be much wider than the area around the nearest big city (urban agglomeration), ie, a part or entire country. The capacity and service quality of these public airport access systems are very similar to those of the above-elaborated urban public transport systems and therefore will not be particularly considered. Nevertheless, it should bear in mind that they always need to be put in the given context, in this case as the airport ground access systems. The simplified scheme of an airport main components is shown in Fig. 5.33.

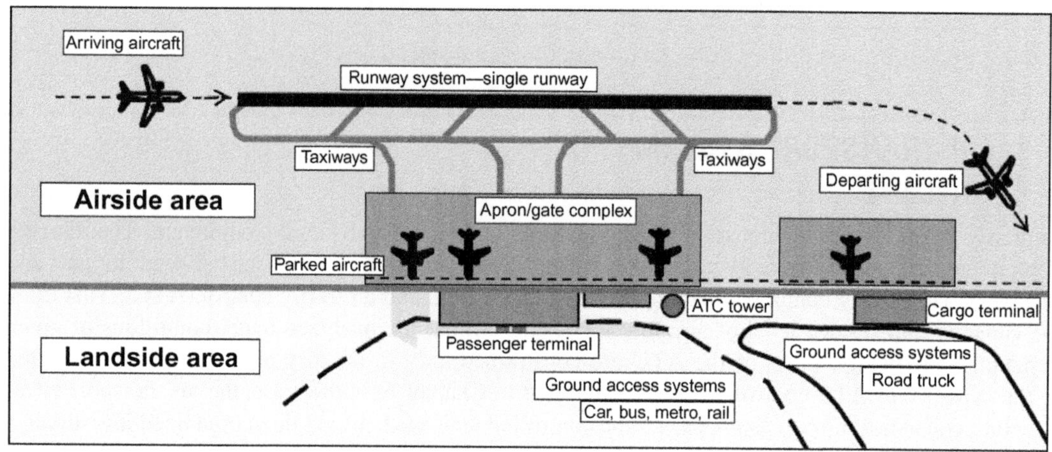

FIG. 5.33

Simplified scheme of an airport components.

5.7.2.2 "Ultimate" capacity

The "ultimate" capacity of an airport can be considered for three components of its airside area—runway system, taxiways, and apron/gate complex. This capacity for particular components of the airport landside area such as passenger and freight/cargo terminal(s) and that of ground access systems is elaborated in Chapter 8. The "ultimate" capacity a given airport airside component can be expressed by the maximum number of aircraft operations (movements), which can be carried out there during a given period (usually 1 h) under conditions of constant demand for service. This capacity is dependent on many fixed and changing/variable factors. The most important fixed factor is the airport design characterized by the number and directions of runways, taxi-ways, runway instrumentation, the number and type of apron/gate parking stands, etc.

The most important changing, relatively variable factors are: (i) meteorological conditions prevailing at given airport, which in combination with the wind direction dictate application of the ATC separation rules between landing aircraft and the runway in use, respectively; the meteorological conditions can be instrument meteorological conditions (IMC) and visual meteorological conditions; the former are used round the world and both in the United States; Table 5.14 gives the IMC separation rules for landing aircraft in the sequence leasing-trailing (ICAO, 1978; Janić, 2000); (ii) characteristics of the aircraft fleet operating at the airport distinguished subject to the aircraft maximum take-off weight (MTOW), approach/departure speed, the runway occupancy time during landing and/or taking-off, etc. In particular, the MTOW is used to classify the aircraft into five wake vortex categories as given in Table 5.15; and (iii) The ATC tactics applied in handling the aircraft landings and taking-offs under given conditions usually depending on the prevailing demand.

Table 5.14 The ATC Separation Standards for Landings (ICAO, 1978)

	Trailing A/C		
Leading A/C	**L**	**M**	**H**
L	3	3	3
M	4	3	3
H	6	5	4

U.S. FAA (Federal Aviation Administration) recommends 2.5 nm instead of 3 nm between the corresponding aircraft sequences at busy airports.
IFR, *instrument flight rules;* nm, *nautical mile;* L, *light;* M, *medium;* H, *heavy;* A/C, *aircraft.*

Table 5.15 Classification of Aircraft According Their MTOW and Landing Speed (ICAO, 1978; Janić, 2000)

Aircraft Class	Type	MTOW (Maximum Take-Off Weight) (kg)	Landing Speed (kt)[a]
A, B	Dornier 228, Cessna Citation	≤7000	70–90 and 91–120
C	A319, A320, B737	7000–136,000	121–140
D, E	B747, B767, B777, B787 A330, A340, A350, A380	>136,000	141–165 and >165

[a]*Knots.*

Runway system

The "ultimate" capacity of the runway system of a given airport is expressed by the maximum number of handled aircraft during the specified period of time (usually 1 h) under conditions of constant demand for service. Additional conditions are given aircraft fleet mix requesting service, the length of the final approach and/or departure path where aircraft spend time just before landing and after taking-off, respectively, and the ATC separation rules. These rules should prevent the aircraft coming closer to each other than prescribed, either airborne or on the ground. Essentially, they ensure that only a single aircraft, either landing and/or taking-off, is occupying the runway at time. This capacity can be computed when the runway system is exclusively used for landing, taking-off and mixed operations.

In the well-known analytical models, the "ultimate" capacity of the airport runway system is computed as the reciprocal of the average aircraft inter-arrival time through the "reference location," defined as the point in space where all operations should pass and as such are counted. This "location" is the runway threshold (Blumstein, 1960; Janić, 2000). Fig. 5.34 shows that, according to the above-mentioned definition, the arriving aircraft are assumed to follow a common three-dimensional approach path connecting the final approach gate F (the system's entry point) and runway threshold T. This path is defined by instrument landing system.[9]

[9]This is a ground-based instrument approach system that provides precision lateral and vertical guidance to an aircraft approaching and landing on a runway (https://en.wikipedia.org/wiki/Instrument_landing_system).

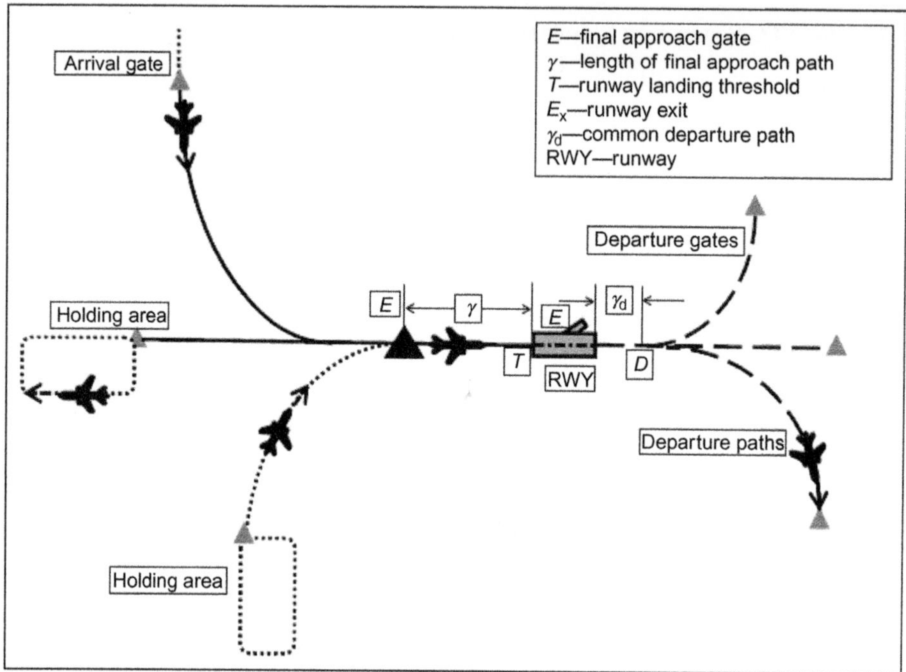

FIG. 5.34

Simplified scheme of the arriving and departing flight paths to/from a single runway(Newell, 1979; Janić, 2000).

The departing aircraft start take-off at the runway threshold *T*, fly over the opposite end of the runway and climb along the departure path to the altitude assigned by the ATC. The aircraft are assumed to maintain constant speeds while on the final approach and departure path(s), which are dependent on the aircraft category (see Table 5.15). However, the differences in the aircraft approach speeds produce four distinct pairs of landing sequences: two consecutive Slow aircraft—*SS*; leading aircraft is Slow and trailing aircraft is Fast—*SF*; leading aircraft is Fast and trailing aircraft is Slow *FS*; and both aircraft are Fast—*FF*. Classification of the particular types of aircraft into two groups (Fast and Slow) is both relative since the one with the higher speed final approach speed is always considered "Fast" and the other "Slow." Similar combinations of sequences occur at the departure aircraft. The ATC minimum separation rules in Table 5.14 should established at the final approach gate (point *E* in Fig. 5.34) for the sequences *SS*, *FS* and *FF* and at the runway threshold (point *T* in Fig. 5.35) for the sequence *SF*. Applying the ATC separation rules in this way causes increasing of the horizontal distance between aircraft in the sequence *FS*, its decreasing in the sequence *SF*, and being constant in the sequences *SS* and *FF*.

Fig. 5.35 shows time-space diagram for the aircraft operations at a single runway. In addition, it shows that one or more successive take-offs can be carried out between particular pairs of the consecutive landings.

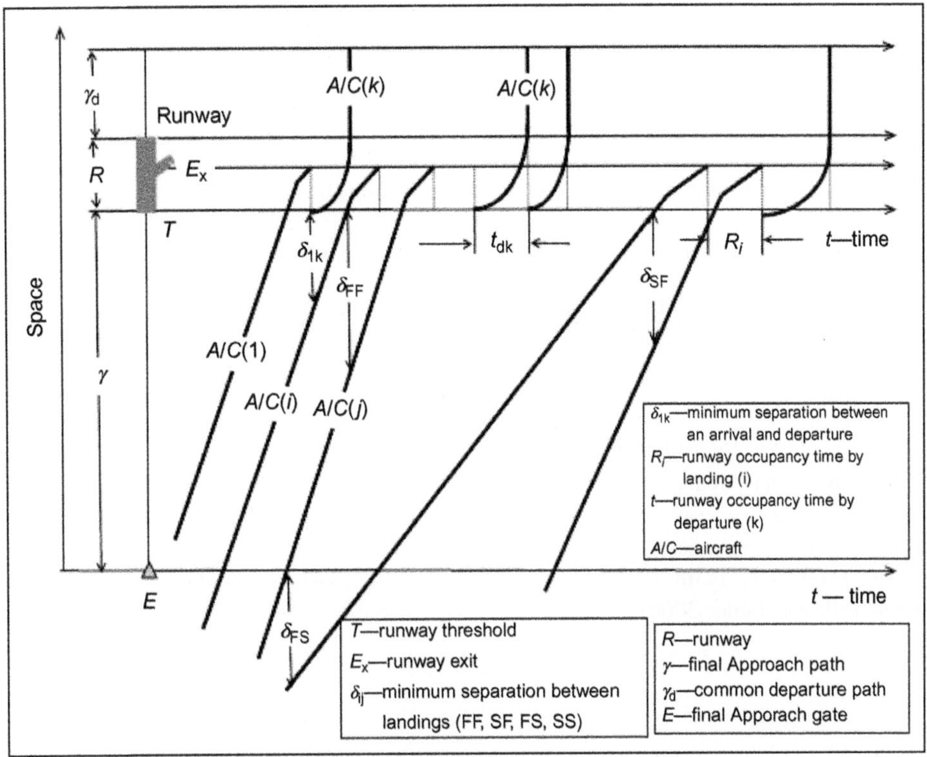

FIG. 5.35

Time-space diagram of using a single runway (Newell, 1979; Janić, 2000).

Landing capacity. The model of runway landing capacity is based on the calculation of the average service time of the landing aircraft during given period of time τ (usually 1 h). This time is equivalent to the arithmetic mean of the inter-event times adjusted to take their minimum allowed values under given conditions (Newell, 1979). As mentioned above, four types of inter-event times relating to landings are possible at the reference location (eg, landing threshold). These are *SS*, *FF*, *SF* and *FS*. If the inter-event time between the leading aircraft *i* and trailing aircraft *j* in the sequence (*ij*) is ($_at_{ij}$), and if the proportions of aircraft of type *i* and *j* in traffic mix are p_i and p_j, respectively, then if *i* and *j* are assumed to be the independent events, the average minimal inter-event time at the runway threshold for all combinations of landing sequences (*ij*) will be determined as follows (Blumstein, 1960; Janić, 2000):

$$t_A = \sum_{ij} p_i \cdot {}_at_{ij} \cdot p_j \tag{5.109}$$

and

$${}_at_{ij} = \begin{Bmatrix} \max\left[R_{ai}; \delta_{ij}/v_j, \quad \text{for } v_i \leq v_j\right] \\ \delta_{ij}/v_j + \gamma \cdot \left(1/v_j - 1/v_i\right), \quad \text{for } v_i > v_j \end{Bmatrix} \tag{5.110}$$

where:

R_{ai} is the runway occupancy time by landing aircraft i;
δ_{ij} is the minimum longitudinal separation between the aircraft i and j measured along the path of aircraft j; it is applied either on the runway threshold (sequence $v_i \leq v_j$) or at the final approach gate (sequence $v_i > v_j$);
γ is length of the final approach path common for all aircraft; and
v_i, v_j is the final approach speed of the aircraft i and j, respectively, assumed to be constant along distance γ in closed case where ($v_i \leq v_j$), and distances δ_{ij} and γ in an open case where ($v_i > v_j$).

From Eqs. (5.109) and (5.110), the runway landing capacity can be computed as:

$$\mu_A = \tau / t_A \tag{5.111}$$

where

τ is the time interval for which the capacity is calculated.

Take-off capacity. Analytical models for calculating the runway "ultimate" take-off capacity have the similar structure as the above-mentioned for calculating the runway landing capacity. The only difference is in the expressions for determining the inter-event times between successive aircraft movements (take-offs). Consequently, in this case, the time between successive take-off of the aircraft i and j can be computed as follows (Janić, 2000):

$$_d t_{ij} = \left[{_d t_{ij0}};{_d t_{ijmin}} - \left(R_{dj} - R_{di}\right) - \gamma_d \cdot \left(1/v_{jd} - 1/v_{id}\right)\right] \tag{5.112}$$

where:

${_d t_{ij0}}$ is the ATC minimum time-based separation rule applied to the aircraft i and j at the take-off threshold;
${_d t_{ijmin}}$ is the minimum required time separation between the aircraft i and j at the point where they both leave the airport zone (point D in Fig. 5.35);
R_{di}, R_{dj} is the runway occupancy time of the taking-off aircraft i and j, respectively;
γ_d is the length of common departure path for the aircraft i and j; and
v_{id}, v_{jd} is the average speed of the aircraft i and j, respectively, on the distance γ_d.

The probability of occurrence of sequence (ij), assumed to be the realization of a pair of independent events is equal to ${_d p_{ij}} = {_d p_i} \cdot {_d p_j}$. The average inter-departure time for all combinations of departure sequences (ij) is equal to:

$$t_D = \sum_{ij} {_d p_{ijij}} \cdot {_d t_{ijij}} \tag{5.113}$$

The taking-off capacity is equal to:

$$\mu_D = \tau / t_D \tag{5.114}$$

Capacity for mixed operations. According to the time-space scheme shown in Fig. 5.35, a take-off between two successive landings can be realized if the following condition is fulfilled at the runway (Janić, 2000):

$$_{ad} t_{ij} \geq R_{ai} + \frac{\delta_{jk}}{v_j} \tag{5.115}$$

where:

$_{ad}t_{ij}$ is the minimum time gap allowing a take-off between landings i and j;
R_{ai} is the runway occupancy time by leading aircraft i in the landing sequence (ij);
δ_{jk} is the minimum distance of landing aircraft j from the runway threshold allowing a safe take-off of aircraft k; and
v_j is average speed of the arriving aircraft j along the distance δ_{jk}.

If the probability of occurrence of the time gap $_{ad}t_{ij}$ in time τ is p_d, the total runway capacity can be computed as:

$$\mu = (1 + p_d)\mu_A \tag{5.116}$$

An illustration of calculating the "ultimate" capacity of an airport runway system using the above-mentioned analytical approach is the case of Frankfurt Main airport (Frankfurt, Germany). A simplified layout is shown in Fig. 5.36.

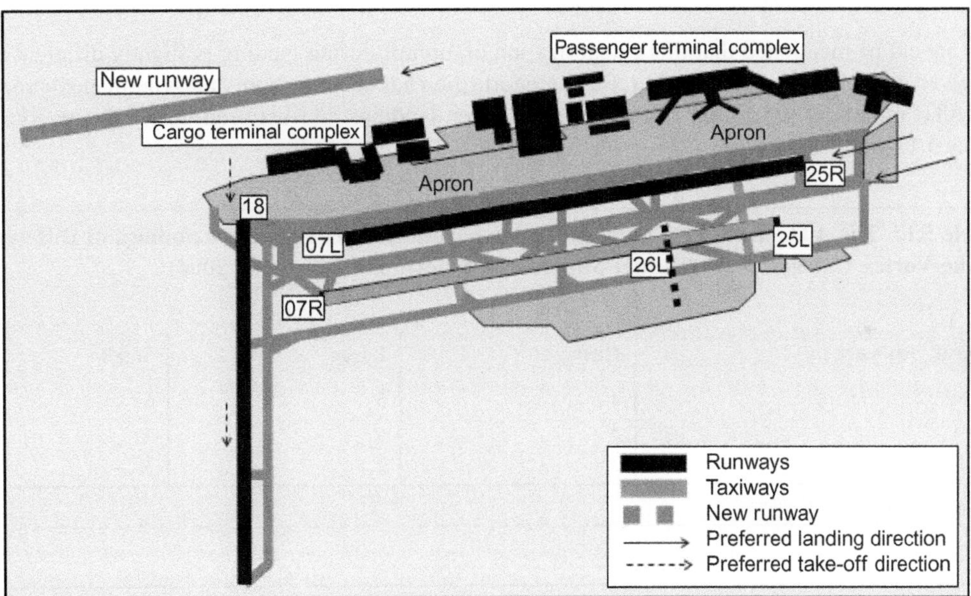

FIG. 5.36

Simplified scheme of Frankfurt Main airport (Frankfurt, Germany) (Fraport, 2004).

The airport operates the runway RWY07 L/R and RWY25 L/R 4000 m long for landings and take-offs, the new runway 2500 m long for landings, and the runway RWY 26 L 2500 m long for landings. Physically, this is a part of RWY 25 L with displaced landing threshold for 1500 m. The runway RWY18 is used only for take-offs and the new runway, each 4000 m long. The runways RWY 07 and 25 L/R and RWY 26 L are laterally separated for $d = 1700$ ft (518 m), which prevents them to be operated independently. The length of the common final approach path is: $\gamma = 6$ nm. Characteristics of the aircraft fleet using the airport are given in Table 5.16.

Table 5.16 Characteristics of the Aircraft Fleet at Frankfurt Airport (Fraport, 2004)

A/C Category	Type	Proportion (%)	Approach Speed (kt)[a]	Runway Landing Occupancy Time (s)[b]
Super Heavy	A380	10	150	60
Heavy	A300–600; A330; A340; B767 B777; B747	10	140	60
Large	B737; A320, 321 s	60	130	55
Small	ATR42,72; Avrojet; Dash8	20	110	45

[a]*Knots.*
[b]*Seconds.*

It should be mentioned that this categorization of aircraft during landing is slightly different than that given in Table 5.15 since the aircraft re-categorization has become more or less permanent process. The ATC minimum separation rules applied between landings and between take offs are given in Tables 5.17 and 5.18.

Table 5.17 The ATC Minimum Longitudinal Separation Rules Between Landings of Different Wake-Vortex Categories (Frankfurt Main airport-Germany) (Fraport, 2004)

	Category		
Aircraft Sequence (*i/j*)	**Heavy**	**Large**	**Small**
Heavy	4	5	6
Large	3	4	4
Small	3	3	3

All values in nm (nautical miles).

Table 5.18 The ATC Minimum Separation Rules Between Successive Departures of Different Wake-Vortex Categories (Frankfurt Main Airport-Germany) (Fraport, 2004)

	Category		
Aircraft Sequence (*i/j*)	**Heavy**	**Large**	**Small**
Heavy	90	120	120
Large	90	90	120
Small	45	45	45

All values are in seconds.

In addition, the distance separation between landing and taking-off aircraft is: $\delta_{jk} = 2$ nm.

Both landing and taking-off aircraft are handled at runways according to the FIFO (first-in-first-out) service discipline and the following scenario: RWY 25R/L—26 L or RWY 18 are used as a single runway for both landings and taking-offs; the ATC applies only longitudinal separation rules between landings and taking-offs. Fig. 5.37 shows the resulting so-called "capacity envelope" indicating the relationship between the landing and taking-off "ultimate" capacity.

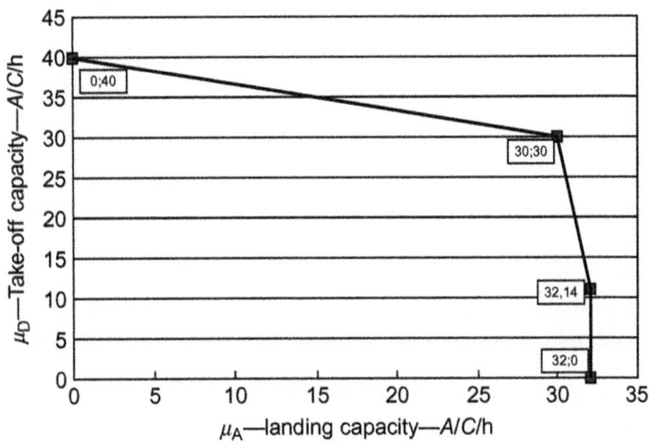

FIG. 5.37

Capacity coverage curves for given scenario of using the runway system at Frankfurt Main airport—Germany (Janić, 2006).

As can be seen, the landing capacity—horizontal axis—is the greatest when there are not take-offs between landings. The same is happening with the take-off capacity—vertical axis. With increasing of the landing capacity, the take-off capacity decreases, and vice versa. This is because landings have ultimate priorities over take-offs in the given case.

Taxiways

The "ultimate" capacity of taxiways depends on their layout/configuration, the aircraft fleet mix characterized by individual aircraft taxing speeds, longitudinal separation between aircraft taxing in the same direction either after landings or before taking-offs, separation between aircraft at the taxiway intersections, etc. In this case, the "ultimate" capacity can be estimated by calculating the average time between passing successive taxing aircraft through a prior selected "reference location" for counting capacity and then by taking its reciprocal of for the specified period of time (usually 1 h). The average time between passing successive taxing aircraft can be estimated by dividing the average longitudinal separation between them by their average taxing speed. The experience so far has indicated that the "ultimate" capacity of taxiway system has not been the capacity "bottlenecks'" of the entire airport airside area.

Apron/gate complex

The "ultimate" capacity of apron/gate complex is defined as the maximum number of aircraft handled on the given number of parking stands/gates during some period of time under a condition of constant demand for service. The parking stands/gates are designed to accommodate particular aircraft types regarding their dimension (length, wing span, ie, footprint). Consequently, two cases can happen: (i) all aircraft types operated by the same or different airlines can use all available parking stands; and (ii) only a certain types of aircraft and airlines can use the specific stands/gates.

If all aircraft types may use all available parking stands/gates, the apron/gate complex "ultimate" capacity μ_g can be calculated as follows (Janić, 2000):

$$\mu_g = \frac{U_g \cdot N_g}{\tau_g} \text{ and } N_g = \lambda_g \cdot \tau_g / U_g \tag{5.117}$$

where:

U_g is the utilization factor of stand/gate as the ratio between the time it is occupied and the total available time (<1.0);
λ_g is the intensity of aircraft arrivals at the apron/gate complex (aircraft/h);
N_g is the number of available parking stands/gates at the apron/gate complex; and
τ_g is the average stand/gate occupancy time averaged over various aircraft types (classes) (h).

The average stand/gate occupancy time (τ_g) in Eq. (5.117) is the time between the aircraft's wheel stop at the stand/gate and the time of the aircraft's moving out from the gate. This time can also include the so-called "separation time" consisting of the aircraft push-out time, the time needed for departing aircraft to "clear" the apron area, and the time needed for a new (arriving) aircraft to come from the apron's entrance to the assigned stand/gate. The intensity of arrivals at the apron/gate complex λ can be equal to the greater between the "ultimate" capacity of the runway and taxiway system(s) or to the actual arrival rate.

For example, if: $N_g = 30$, $\tau_g = 1$ h, and $U_g = 0.8$, the "ultimate" capacity will be: $\mu_g = 30 \cdot 0.8/1 = 24$ aircraft/h. Or, if: $\lambda_g = 30$ aircraft/h, $\tau_g = 0.75$ h (ie, 45 min), and $U_g = 0.8$, the required number of parking stands will be: $N_g \approx 28$ stand/gates.

If all aircraft types operated by the same or different airlines cannot use all available gates, the "ultimate" capacity of each type of parking stands/gates by Eq. (5.117) and then the overall capacity of the apron/gate complex should be calculated as follows (Janić, 2000):

$$\mu_g = \sum_{i=1}^{M} \frac{U_{gi} * N_{gi}}{\tau_{gi}} \text{ and } N_g = \sum_{i=1}^{M} (\lambda_{gi} * \tau_{gi/})/U_{gi} \tag{5.118}$$

where:

U_{gi} is the utilization factor of the parking stand/gate by the aircraft type (i);
λ_{gi} is the intensity of arrivals of aircraft of type (i) at the apron/gate complex (A/C/h);
N_{gi} is the number of stands/gates, which can accommodate the aircraft type (i); and
τ_{gi} is the average gate occupancy time by the aircraft type (i) (h/A/C).

For example, let three aircraft types use the given apron/gate complex with: $N_{g1} = 15$, $N_{g2} = 25$, and $N_{g3} = 10$ parking stands/gate. The occupancy time of these stands/gates is: $\tau_{g1} = 0.3$ h (ie, 20 min),

$\tau_{g2}=0.75$ h (ie, 45 min), and $\tau_{g3}=1$ h (ie, 60 min). The corresponding utilization factors are: $U_{g1}=0.8$, $U_{g2}=0.8$, and $U_{g3}=0.7$. The capacity of the apron gate complex will be: $\mu_g=(0.8\cdot 15)/0.3+(0.8\cdot 25)/0.75+(0.7\cdot 10)/1=63$ aircraft/h, and vice versa.

5.7.2.3 "Practical" capacity and service level

The "practical" capacity of particular components of the airport airside area—runway system, taxiways, and apron/gate complex can be expressed by the maximum number of aircraft served during a given period of time (usually 1 h) when each of them is imposed an average delay specified in advance. If this delay is used as an indicator of service quality, then the "practical" capacity is related to it. This delay can be imposed aircraft before landing and taking-off. The delay has also be imposed on aircraft while taxiing (maneuvering) on the apron/gate complex and along taxiways. Models to estimate the aircraft delays and consequently an airport runway system "practical" capacity are based on the queuing system theory (). In these cases the airport runway system with adjacent airspace is always assumed to operate as a single server system. When two or more runways are used independently, the system is considered to operate as a multi- (two-) server queuing system (de Neufville and Odoni, 2003; Hurdle, 1991; Janić, 2000). The landing and taking-off aircraft represent two classes of customers simultaneously requesting service. The server's service rate is the above-mentioned runway system's "ultimate" capacity. The number of waiting places/positions where the landing and/or taking-off aircraft queue is assumed to be unconstrained. The network of taxi-ways has always been treated as a component connecting the runway system and apron/gate complex and usually been modeled in a scope of modeling of the whole airport complex operating as the service system. The apron/gate complex has always been modeled as the multi-server queuing system where the parking stands/gates are considered as servers with unlimited waiting space. The processing rate has again been the "ultimate" capacity of each of them. The landing queues and related delays take place in the holding stacks located in the vicinity of landing runway(s), which are at least theoretically, unlimited. The taking-off queues and related delays are realized on the ground, either at the stands/gates or along taxiways just before take-offs, again using theoretically unlimited space.

Queues, ie, congestion and delays arise whenever the rate of demand for service exceeds the service rate, ie, the "ultimate" capacity of given airside component. Consequently, the component becomes over-saturated for a shorter or longer time. Dependent on the rate and intensity of the over-saturation, two typical situations may occur.

- The demand rate is overall lower that the service rate, ie, ultimate capacity during the longer period of time (let's say 1 h):

  In this case, despite there is a plenty of capacity overall, congestion and delays happen because the instant (short-time) demand rate exceeds the component's service rate, ie, "ultimate" capacity. When this demand rate and capacity come closer to each other, the frequency of demand's exceeding the capacity significantly increases causing congestion and delays to grow disproportionately faster than the demand rate (de Neufville and Odoni, 2003). In this case, based the queuing system theory, for example, the average delay of an aircraft requesting landing at a single runway can be computed by the following equation (Horonjeff and McKelvey, 1983):

$$W_a = \frac{\lambda_a(\sigma_s^2 + 1/\mu_A^2)}{2(1-\lambda_a/\mu_A)} \tag{5.119}$$

where:

W_a is the average delay to arriving aircraft (min);

λ_a is the average demand/arrival rate (A/C/h);

μ_A is the "ultimate" landing capacity of given runway (A/C/h); ($\mu_A = T/t_a$), where t_A is the average service time for landings (min); $T = 1$ h (60 min) (Eq. 5.111); and

σ_s is the standard deviation of the service time for landings (min).

Eq. (5.119) is well known formulae from the theory of "equilibrium queues" developed for the conditions when the arrival flow of customers requesting service is Poisson process and the service time has general probability distribution (ie, this is M/G/1/- queuing system) (Newell, 1982). The average delay of taking-off aircraft can be similarly calculated. Fig. 5.38 shows an example of development of the average landing delays at London Heathrow airport (UK).

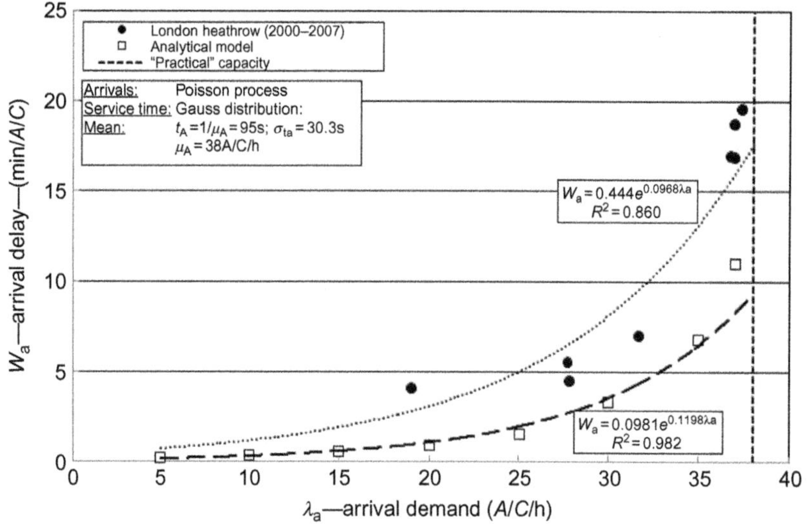

FIG. 5.38

Relationship between demand and average delay per landing aircraft at London Heathrow airport (ACL, 2007).

As can be seen, the average landing delay of an aircraft increases more than proportionally, ie, exponentially with increasing of the intensity of demand for landings. Both practice and the above-mentioned model confirm this. The airport has declared the "practical" landing capacity of $\mu_a = 38$ A/C/h, which regarding the intensity of demand generates the average landing delay of about: $W_a = 20$–25 min/A/C. This pattern of operations sustains at the airport during the entire day. If the average delay is set up to be: $W_a = 10$ min, the "practical" landing capacity to be declared will be: $\mu_a = 32$ A/C/h. Etc.

- The demand rate is overall greater that the service rate, ie, "ultimate" capacity, during the longer period of time (let's say greater than 1 h):

  In this case, congestion and delays can develop in three phases. The first one is when the demand rate is lower than or equal to the capacity actually preventing registration of congestion and related delays. The component's throughput is equal to the rate of demand. The second is when the demand rate exceeds the component's capacity causing queues to build up and grow over time. The system's throughput is equal to the component's capacity. Finally, when the demand rate falls again below the component's capacity the throughput remains at the level of this capacity until all demand is cleared from the queue. After that, the throughput is equal to the demand rate again. These developments are usually modeled by the deterministic queuing systems (de Neufville and Odoni, 2003; Newell, 1982). These models do not provide information about stochastic variations in queues and delays of customers. It actually implies that queues and delays exist only when the server is over saturated. The process of congestion of the airport airside area or some of its components, for example, the runway system, can be presented by the typical graphical form, This is shown in Fig. 5.39 for the NY La Guardia airport (New York, United States) on 30 June 2001. The horizontal axis represents the time of day and the vertical axis the cumulative.

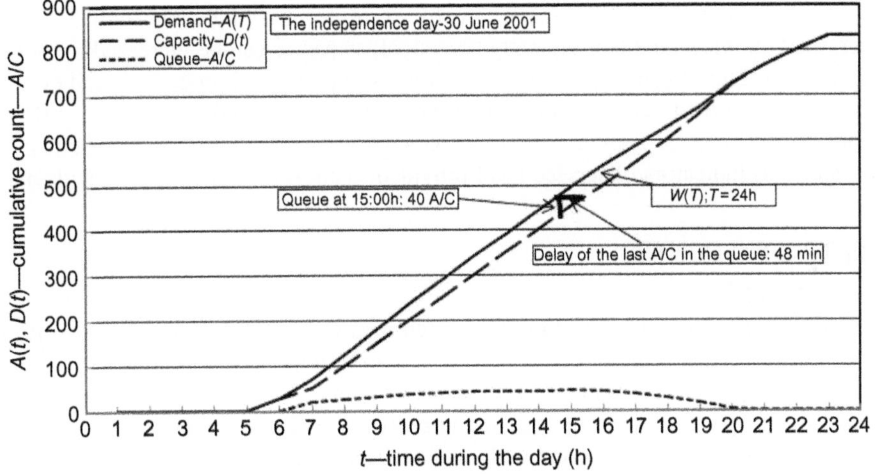

FIG. 5.39

Cumulative counts of demand, capacity, congestion, and delays at NY La Guardia (PANYNJ (2003).

Fig. 5.39 counts of total demand (curve $A(t)$) and capacity (curve $D(t)$), ie, the aircraft arrivals to and departures from the queue, respectively. Actually, both cumulative curves are non-decreasing step functions of time, which increase by one whenever a new aircraft arrives and/or departs the queue. The area between the curves $A(t)$ and $D(t)$ denoted by $W(T)$ represents the total waiting time of all aircraft requesting service in the airport airside area during time (in this case $T = 24$ h). These delays are measured by the total aircraft hours. The length of the queue $Q(t)$ at some time t is determined as the vertical distance between the curves $A(t)$ and $D(t)$ as follows:

$$Q(t) = \max[0; A(t) - D(t)] \tag{5.120}$$

As can be seen in Fig. 5.39, this queue was the greatest at 15:00 h—44 aircraft were there. Delay of an aircraft waiting for service in the queue is determined as the horizontal distance between two curves, ie, it is the difference between the time when an aircraft joins the queue and the time when it starts to be served (FCFS discipline is assumed to be applied). This delay is equal as follows:

$$w(t) = Q(t)/\mu(t) \tag{5.121}$$

where

$\mu(t)$ is the service rate, ie, "ultimate" capacity being constant during a given (short) period of time (A/C/h).

In the case shown in Fig. 5.39, this delay was the longest at 15:00 h—48 min. In addition, by dividing the total area $W(T)$ by the number of aircraft requesting service during the time period T, the average delay per an aircraft can be obtained indicating an average service level provided under given conditions. However, this service level based on the demand/capacity ratio is: $\rho(t) = A(T)/day(t) > 1$ most of the time and as such is outside of the above-mentioned gradation of LOS when: $\rho(t) < 1.0$. This raises the question of matching the capacity to demand in order to prevent occurrence of such extreme congestion and delays at particular airports. On the one side, the intensity of demand can be maintained at or just below the level of the "practical" capacity by, for example, charging congestion (Janić, 2005a,b). On the other, the "ultimate" and consequently "practical" capacity can be increased by, for example, introducing the new technologies supporting the innovative operational procedures and even by building additional airside infrastructure (for example, new runway(s)) (Janić, 2014a,b,c, 2015). Finally, both previous options can be applied aiming at balancing the "ultimate" and "practical" capacity and related delays as indicators of service level provided to aircraft in the airport airside area

5.7.3 AIR TRAFFIC CONTROL

5.7.3.1 Background

The ATC system is established over some designated area (airspace) to provide safe, efficient, and effective air traffic. That means that the aircraft, as the system's users, should be served without any mutual conflicts and the surrounding terrain as well (safety). In addition, they have to be provided flying through the airspace between the origin and destination airports along their fuel/time optimal flight paths (trajectories) (efficiency), and without significant delays caused by the ATC system itself (regularity).

In order to fulfil these operational attributes, the large controlled airspace is divided into smaller parts—airport zones, terminal areas, and low and high altitude en-route areas. These are then divided into the smaller parts called the "ATC sectors," each sector assigned to one or more ATC controllers with responsibility to monitor and control the air traffic there.

The airport zones are established around the airports to provide efficient the aircraft landings, ground movements, and taking-offs. The terminal airspace is established around large airports with substantive traffic. It spreads around them for about 40–50 nm, with the vertical boundary on FL (Flight Level) 100 (10,000 ft) (each FL is defined by 1000 ft; 1 ft = 0.305 m) (ICAO, 1978). The aircraft fly through this area along the prescribed arrival and departure paths (trajectories) defined either by the radio-navigational facilities or by the ATC radar vectoring of aircraft. They are separated by the ATC horizontal distance-based separation rules.

The low altitude area is established around and the above-mentioned terminal airspace and horizontally spread over a larger space. Vertically, they cover range of altitudes between FL100 and FL245 middle sea level (Horonjeff and McKelvey, 1983). The aircraft climbing and descending trajectories to and from the cruising FLs are located in this airspace. They continue to the departure and approach trajectories in the terminal airspace. Since the aircraft permanently change the altitude while in this area, they are separated by the ATC horizontal distance- or time-based separation rules.

The high-altitude area lies above the FL245. The aircraft use this area for cruising along the airways on the specific FLs where they are vertically separated by1000 ft, up to FL 290 (29,000 ft) and 2000 above FL 290. This means that the closest traffic flows moving in the same direction along the same airway/route are vertically separated by 2000 ft, if they are on the levels up to FL 290 and by 4000 ft, if they are on the flight levels above FL 290. In addition, the aircraft on the same flight levels of a given airway or its part are horizontally separated by the ATC longitudinal distance- or time-based separation rules (ICAO, 1978; Janić, 2000).

The main operational characteristics of the above-mentioned particular parts of the airspace are their "ultimate" and "practical" capacity. These include the capacities of airspace and that of the ATC controller(s) having jurisdiction over this airspace, the latter elaborated in Chapter 6.

5.7.3.2 "Ultimate" capacity

The "ultimate" capacity of a given type of airspace can be is expressed by the maximum number of aircraft, which can be served during given period of time (usually 1 h) under given conditions. These are characterized by constant demand for service and the structure of the aircraft fleet mix (Janić, 2000). Estimation of this capacity is illustrated for the case of the long high-altitude airway where the aircraft fly at FLs at approximately constant the same or different cruising speeds as shown in Fig. 5.40.

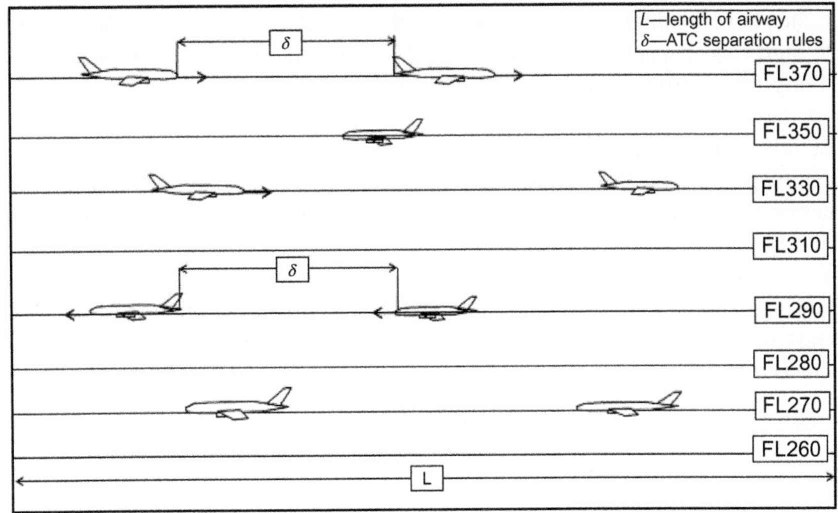

FIG. 5.40

Simplified scheme of an airway with cruising aircraft at particular flight levels.

In cases of different cruising speeds of the aircraft flying along the given airway at the same FL in the same direction, the potential overtaking conflicts can happen, which are usually resolved by the ATC controller(s)—either by changing the FL, course, or both of one of the aircraft. In order to calculate the capacity, the "reference location" for counting' operations is usually defined as an arbitrarily chosen point along the given airway where all aircraft pass through. In an ideal (hypothetical) situation, if the ATC establishes the prescribed separation rules between aircraft entering the given airway at several flight levels, the inter-arrival time between entering successive aircraft at one of them, (i), can be calculated as follows:

$$t_{FL/i} = \delta_i / v_i \tag{5.122}$$

where:

δ_i is the ATC minimum longitudinal distance-based separation rules between successive aircraft entering FL (i) of the given airway (nm); and
v_i is the average aircraft cruising speed at the FL (i) (kts).

Then, the "ultimate" capacity of FL (i) during some time (τ) and that of the airway with (N) FLs, respectively, is equal to:

$$\mu_{FL/i} = 1/t_{FL/i} = v_i/\delta_i \tag{5.123}$$

and

$$\mu_{AR} = \sum_{i=1}^{N} \mu_{FL/i} \tag{5.124}$$

For example, if: $\delta_i = 30$ nm and $v_i = 380$ kts for $\forall i \in N$, and if: $N = 5$, the airway capacity will be: $\mu_{AR} = 5 \cdot [1/(30/380)] \approx 63$ A/C/h (see also Chapter 6).

However, the above-mentioned ideal scenario rarely takes place particularly along the straight-line airways/route segments. This is mainly due to the ATC acting to prevent the potential overtaking conflicts by unifying the air traffic sub-flows at particular flight levels respecting their cruising speeds. Unification of these sub-flows can be realized by "sifting" the original flow, ie, by requesting some aircraft to use other alternative airways/routes, other free FLs on the original airway/route segment, or carry out some holding pattern before entering the airway/route (this latest can be carried out in advance either on the ground or just before in the en-route airspace located nearby the required airway/route segment). Consequently, each of the above-mentioned "flow-sifting" actions reduces the intensity of the"sifted" sub-flows remaining on the particular FLs of the given airway/route segment. Under such traffic scenario, it is really to assume the "sifted" flow is a Poisson process, thus implying that the aircraft inter-arrival times at the entrance of the given airway can be represented by the continuous random variable with the shifted exponential distribution (Dunlay, 1975) as follows:

$$f(t) = \left\{ \begin{array}{l} \lambda e^{-\lambda(t-\theta)}, \text{ for } t > 0 \\ 0, \text{ otherwise} \end{array} \right\} \tag{5.125}$$

where:

λ is the intensity of the original (non-"sifted") flow of air traffic (A/C/h);
θ is the minimum time separation rules applied by ATC between the aircraft in the "sifted" flow (min); and
t is the time assumed to be the continuous variable (h).

The average inter-arrival time of the aircraft in the "sifted" flow at the "reference location," ie, the entry of the given airway is determined as follows (Alfredo and Tang, 1975):

$$\bar{t}_s = \int_{\theta}^{\infty} t\lambda e^{-\lambda(t-\theta)}dt = 1/\lambda + \theta \tag{5.126}$$

Based on Eq. (5.126), the intensity of "sifted" flow, eg, the actually achievable capacity of the given airway is equal to:

$$\lambda_s = 1/\bar{t}_s = \frac{1}{(1/\lambda)+\theta} = \frac{\lambda}{1+\lambda^*\theta} \tag{5.127}$$

The ATC minimum time separation rules (θ) may be taken as the constant value or as a realization of the random variable dependent on the skills and abilities of the ATC controller(s) to establish the minimum separation between each pair of the aircraft in the "sifted" traffic flow. When the distance-based separation rules are applied, the average value of (θ) can be determined as in Eq. (5.122).

Fig. 5.41 shows an (hypothetical) example of the dependence of the intensity of "sifted" aircraft flow on the intensity of the original (non-"sifted") flow when two TC longitudinal distance-based separation rules are applied: $\delta = 30$ nm, which is still a part the ATC practice (ICAO, 1978), and $\delta = 10$ nm, which is increasingly used thanks to developing the new ATC technologies. The average speed of "sifted" flow is: $v = 360$ kts (knots) implying that the flow consists of medium and heavy commercial aircraft flying in the low altitude area.

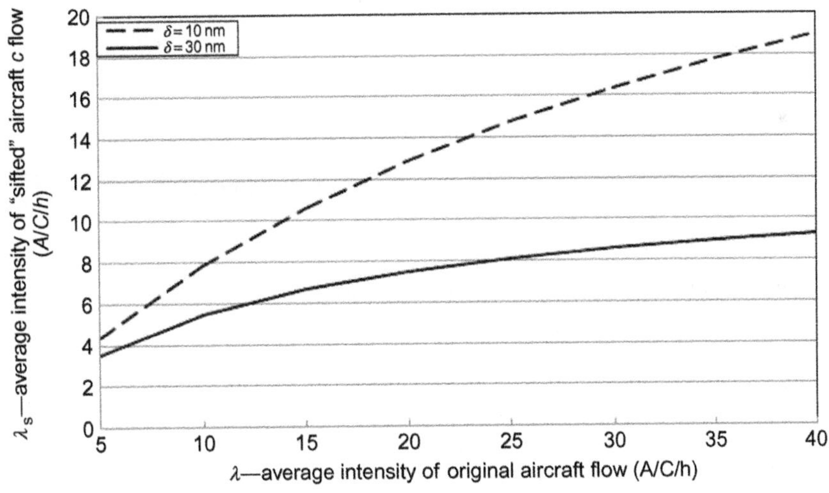

FIG. 5.41

Dependence of the "sifted" and original aircraft flow in the given example.

As can be seen, the intensity of "sifted" aircraft flow will increase with increase in the intensity of original flow and with reduction of the ATC minimum separation rules, as intuitively expected. The models for calculating the "ultimate" capacity of other parts of the ATC airspace such as the intersection of airways, the high altitude sector, and terminal airspace can be found in other reference sources (Janić, 2000).

5.7.3.3 "Practical" capacity and service level

If the "ultimate" capacity of a given part of airspace is temporary saturated, the newly arriving aircraft will have to wait for entering it. For example, the maximum number of aircraft, which can occupy FL (i) of the given airway/route segment of the length (L) in the given airspace can be determined as:

$$n_{\max/i} = L_i/(\delta_i + 1) \tag{5.128}$$

In addition, under conditions of constant demand for service, this number is also equal to:

$$n_i = \lambda_i{}^* \tau_i = \lambda^* (L_i/v_i) \tag{5.129}$$

where all symbols are analogous to those in the previous equations.

From Eqs. (5.128) and (5.129), the maximum intensity of aircraft flow on the given FL(i) of the given airway/rote segment, ie, its "ultimate" capacity, is equal to: $\lambda_{\max/i} \equiv \mu_i = v_i/(\delta_i + 1)$, which is analogous to Eq. (5.123).

By summing up the capacities for all flight levels of all airways in the given airspace, its total "ultimate" capacity (μ) as the service rate can be obtained. The number of waiting aircraft to enter this airspace at some time (t) can be calculated by the theory of deterministic queues as follows (Janić, 2000):

$$n(t) = \max\,[0; n(t - \Delta t) + (\lambda - \mu) \cdot \Delta t] \tag{5.130}$$

where:

- $n\ (t - \Delta t)$ is the number of aircraft waiting to enter the given airspace at time $(t - \Delta t)$ (A/C);
- λ is the intensity of the aircraft demand to enter the given airspace during the time Δt; and
- μ is the "ultimate" capacity of the given airspace during time Δt.

The waiting time of the last aircraft requesting entering the given airspace at time t can be calculated as:

$$w(t) = n(t)/\mu \tag{5.131}$$

Consequently, the waiting time in Eq. (5.131) can be used in some way as the measure of the service quality provided to users/aircraft under given conditions. By setting up this maximum acceptable waiting time/delay as the level of service quality, the "practical" capacity of a given airspace expressed by the maximum intensity of demand, which can be accepted under given conditions can be calculated from Eq. (5.130) as follows:

$$\lambda_{\max} = \mu^* \left[\frac{w^*_{\max}}{\Delta t} + 1\right] \text{ when } n(t - \Delta t) = 0 \tag{5.132}$$

Fig. 5.42 shows an example.

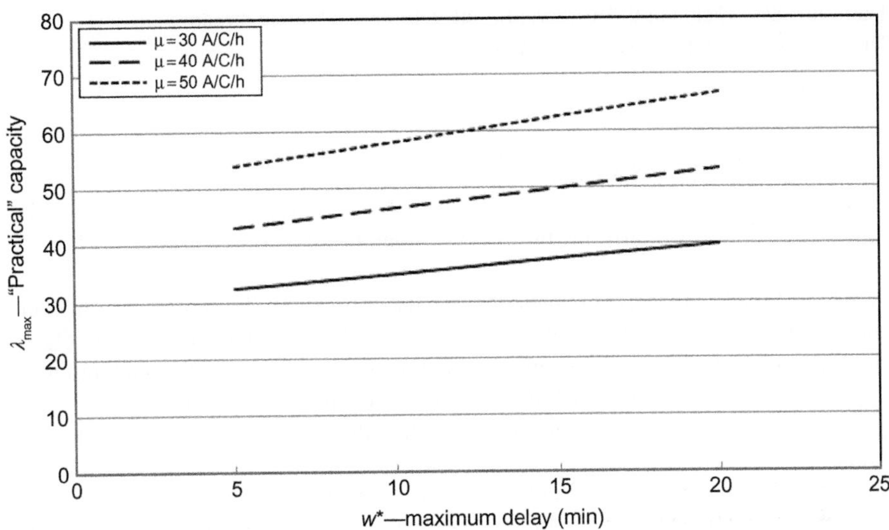

FIG. 5.42

Relationship between the maximum allowable delay, "ultimate," and "practical" capacity of the given airspace.

As can be seen, with increasing of the "ultimate" capacity and the maximum allowable delay as an indicator of service quality before entering the given airspace, the "practical" capacity can also be linearly increased. In order to achieve this, the ATC needs to apply the strategic and pre-tactical ATFM procedures on the one hand and to respect the constrains on the capacity of ATC controller(s) having jurisdiction over the given airspace on the other (see Chapter 6) (Janić, 2000).

5.8 PROBLEMS

1. Determine the expected LOS for the freeway during the worst 15 min of the peak hour. The freeway is four-lane freeway that has two lanes in each direction. The following are freeway characteristics:
   a. Lane width = 11 ft
   b. Right-side lateral clearance = 3 ft
   c. Freeway is used primarily by commuters
   d. Peak-direction demand volume during peak hour = 1700 veh/h
   e. Traffic composition: 5% trucks, 0% recreational vehicles
   f. *PHF* = 0.95
   g. Rolling terrain
   h. *TRD* = 4 ramps/mi
2. Determine the number of lanes needed to provide LOS C during the worst 15 min of the peak hour, taking into account the following facts:
   a. The estimated demand volume in one direction is equal to 3400 vet/h.
   b. The general terrain is Level terrain.

c. There are 8% trucks, and 3% recreational vehicles in traffic stream.
d. The lane width of 12 ft. will be provided.
e. The lateral clearance of 6 ft will be provided.
f. The freeway will be used primarily by regular users.
g. $PHF = 0.95$
h. Ramp density is equal to 3 ramps/mi.

3. The bus station has: $N = 25$ parking places/platforms. The average bus turnaround time is: $\tau = 30$ min. The intensity of arriving buses at the given bus station is: $\lambda = 25$ buses/h. Calculate the "ultimate" capacity of the bus station.
4. The number of tracks at the given container terminal is $N_{ft} = 25$. The arrival rate of trains at the container terminal is: $\lambda_f(T) = 4$ trains/h ($T = 1$ h), the number of containers per train: $Q_{ac} = Q_{dc} = 30$ containers/train, the number of loading/unloading devices: $n_c = 2$/train, the loading/unloading rate per device: $m_c = 20$ containers/h, and the train maneuvering time: $t_m = 1.5$ h.
   a. Calculate the processing time per train. Calculate the required number of tracks for handling trains.
5. The quay area in front of the container terminal in a given IWT port has a length of $l = 600$ m. The water depth is 5 m thus enabling handling all categories of vessels and barge convoys. The number of container vessels or barge convoys, which can be simultaneously docked in this area is approximately equal to: $N = l/1.2\,L$ where L is the length of a vessel or a barge convoy (factor 0.2 is used for determining the separation between vessels or barge convoys along the quay).
   a. Propose feasible combination of the Class IV, Class V, and Class VI vessels that could be simultaneously docked.
6. Define the concept of "ultimate" and "practical" capacity of nodes and links of transport infrastructure networks.
7. The bus station has 20 parking places-platforms. Each bus spends about 20 min at each place-platform. What would be the "ultimate" capacity of the given bus station.
8. The buses are assumed to arrive at a bus station every 10 min during the peak-hour. Each bus is expected to spend about 15 min at the parking place-platform. Determine the number of required parking places-platforms at the station.
9. Determine the "practical" capacity of a given bus station if the "ultimate" capacity is 25 buses/h and the maximum waiting time to occupy the parking place-platform is 5 min.
10. Define the "bottleneck" segment of the single-track rail line.
11. Define the main elements for estimating the "ultimate" capacity of the single rail line.
12. Explain time-distance diagram and different combinations of trains entering the "bottleneck" segment of a single-track line from the same and different directions.
13. How is the minimum spacing between trains determined? Same direction; Different directions?
14. How can the train service level be graded?
15. Describe the main concept of rail passenger stations.
16. What are the main factors influencing the "ultimate" capacity of the rail passenger stations.
17. The trucks arrive at the freight/cargo terminal every 10 min during the period of two hours. Each truck is unloaded/loaded in 25 min. Determine the required number of ramps/places for handling trucks under the above-mentioned conditions.
18. The freight/cargo terminal has 40 ramps/places for handling trucks. If each truck spends 30 min at the dock, what is the "ultimate" capacity of the ramp/places area of given freight /cargo terminal?

19. What are the main components and operations of the rail shunting yard?
20. Explain the difference between the traffic and transport capacity of a given rail line.
21. The rail operator has scheduled passenger trains each with the capacity of 300 seats every 5 min during an hour. What is the transport capacity of the given rail line? If the rail network has 10 such identical lines, what is its transport capacity?
22. The demand on a given rail line amounts 5000 passengers per day. The rail operator offers the train sets with the capacity of 200 seats/train expecting the average load factor of 60%. At which service frequency will these trains operate along the given line during the day?
23. If the train's turnaround time on a given rail line is 6 h and if the transport service frequency is 5 trains/h, estimate the required number of trains to operate on the line at the given service frequency.
24. What are the main criteria for classification of the IWT (Inland waterways) infrastructure in Europe?
25. What are the main components of inland waterways ports?
26. Define the waterside capacity inland waterways ports and itemize the main influencing factors.
27. Define the landside capacity of inland waterways ports and itemize the main influencing factors and their relationship.
28. What is the river/channel lock? Define its capacity.
29. What are the transport work and productivity of barges, each with the capacity of 3000 ton and load factor of 90%, scheduled at the frequency of 5 departures/week sailing at the speed of 20 km/h along the route of length of 500 km?
30. Itemize the main factors influencing the capacity of seaside area of the sea ports?
31. Itemize the main components of the sea port landside area and factors influencing their capacity. What is the relationship between particular factors?
32. Explain what a shipping line is.
33. Explain whether it is more convenient to use the smaller number of larger than the larger number of smaller ships on the given route to transport the same quantity of freight/goods shipments during the specified period of time.
34. What are the main components of an airport?
35. Aircraft continuously arriving at an airport runway are separated by ATC by the distance of 5 nm (nm = nautical mile). The average speed of each aircraft is 130 kts (kts = knots). Estimate the "ultimate" capacity of the given runway.
36. The continuously departing aircraft from an airport runway are mutually separated for 1.5 min by the ATC. What is the corresponding runway capacity of the given runway for the period of an hour?
37. Explain the concept of the airport runway capacity coverage curve.
38. The aircraft arrive at the apron/gate complex of an airport every 3 min. They spend at the gate (turnaround time) on average 45 min. What is the required number of parking stands/gates at the given apron/gate complex?
39. The airport apron/gate complex has 100 parking stand/gates where the aircraft spend on average 35 min. What is the "ultimate" capacity of a given apron/gate complex?
40. The airport has declared the "ultimate" capacity of 38 ac/h. If the intensity of demand is 28 ac/h, what will be the average delay of an aircraft?
41. Explain the concept of organization of airspace for the purpose of managing and controlling air traffic.

42. An controlled airspace has a single airway route with 10 FL (flight levels). The aircraft enter each level every 15 min. What is the "ultimate" capacity of the given route during an hour?

43. If the aircraft enter the given airway every 15 min at 10 FLs and spend for about 6 h flying there, what is the number of aircraft simultaneously being along the airway?

44. Explain the main differences between the "ultimate" and practical' capacity of a given airspace.

REFERENCES

ACL, 2007. Peak Periods at Heathrow Airport, Report Submitted to CAA, Attachment 1. Airport Co-ordination Limited, London.

AECOM/URS, 2012. NC Maritime Strategy: Vessel Size vs. Cost, Prepared for North Carolina Department of Transportation. AECOM Technology Corporation/URS Corporation, Los Angeles.

Alfredo, H.S.A., Tang, H.W., 1975. Probability Concepts in Engineering Planning and Design: Basic Principles, Probability Concepts in Engineering. Wiley, New York.

Anderson, T., Lindvert, D., 2013. Station Design on High Speed Railway in Scandinavia: A Study of How Track and Platform Technical Design Aspects Are Affected by High Speed Railway Concepts Planned for the Oslo Göteborg Line. MSc Thesis, Chalmers University of Technology, Göteborg.

Ballis, A., 2003. Introducing Level of Service Standards for Intermodal Freight Terminals. In: TRB 2003 Annual Meeting CD-ROM. Transportation Research Board, Washington, DC.

Blumstein, A., 1960. An Analytical Investigation of Airport Capacity. Cornell Aeronautical Laboratory Inc., New York. Report No TA-1358-8-1.

Campbell, F.C., Smith, D.L., Sweeney II, C.D., Mundy, R., Nauss, M.R., 2007. Decision tools for reducing congestion at locks on the Upper Mississippi River. In: Proceedings of the 40th Hawaii International Conference on System Sciences, Honolulu, Hawaii, USA.

CECW-CP, 2004. Shallow Draft Vessels Operating Costs—Fiscal Year 2004, Economic Guidance Memorandum, 05-06. U.S. Army Corps of Engineers Civil Works, Vicksburg, MS.

CJRC, 2014. Visitor Guide 2014. Central Japan Railway Company, Tokyo.

Clausen, U., 2011. Logistik-und Verkehrsmanagement, ÖPNV I. Dortmund. ITL TU Dortmund, Dortmund.

CS, 1998. Multimodal Corridor and Capacity Analysis Manual. Cambridge Systematics, Transportation Research Board, National Research Council, Washington, DC. NCHRP Report 399.

CS, 2007. National Rail Freight Infrastructure Capacity and Investment Study. Prepared for Association of American Railroads. Cambridge Systematics, Cambridge, MA.

de Neufville, R., Odoni, A., 2003. Airport Systems: Planning, Design, and Management. McGrow Hill, New York.

Dunlay Jr., J.W., 1975. Analytical models of perceived air traffic control conflicts. Transp. Sci. 9 (2), 149–164.

ECMT, 1992. Resolution No. 92/2 on New Classification of Inland Waterways, [CEMT/CM(92)6/FINAL]. In: European Conference of Ministers of Transport, Athens, Greece.

Fraport, 2004. EDDF-SOP—Standard Operating Manual. Virtual Frankfurt Airport SOP V9.6, Frankfurt Main Airport. http://www.vacc-sag.org (accessed 15.04.04).

Groenveld, R., Vergeij, H.J., Stolker, C., 2006. Capacity of Inland Waterways, Lecture Notes CT 5306. Department of Hydraulic Engineering, Faculty of Civil Engineering and Geosciences, Delft University of Technology, Delft.

Highway Capacity Manual, 2010. Highway Capacity Manual. Transportation Research Board of the National Academies, Washington, DC.

Horonjeff, R., McKelvey, X.F., 1983. Planning and Design of Airports, third ed. McGraw Hill, New York.

Hunt, D., Wyman, O., 2010. Modelling Rail Capacity. AASHTO SCORT (American Association of State Highway and Transportation Officials, Standing Committee on Rail Transportation), Washington, DC.

Hurdle, F.V., 1991. Queuing Theory Application. In: Papageorgiu, M. (Ed.), Concise Encyclopaedia of Traffic and Transportation Systems. Pergamon Press, Oxford, pp. 337–341.

ICAO, 1978. Rules of the Air and Air Traffic Services, Doc 4444/RAC 501, eleventh ed. International Civil Aviation Organization, Montreal.

Janić, M., 1984. Single track line capacity model. Transp. Plan. Technol. 9, 135–151.

Janić, M., 2000. Air Transport System Analysis and Modeling: Capacity, Quality of Services and Economics, Transportation Studies, vol. 16. Gordon & Breech, Amsterdam.

Janić, M., 2005a. Modelling performances of intermodal freight transport networks. Logist. Sustain. Transp. 1 (1), 19–26.

Janić, M., 2005b. Modelling airport congestion charges. Transp. Plan. Technol. 28 (1), 1–26.

Janić, M., 2006. Model of the ultimate capacity of dual-dependent parallel runways. Transp. Res. Rec. 1951, 76–85.

Janić, M., 2014a. Advanced Transport Systems: Analysis, Modelling, and Evaluation of Performances. Springer, London.

Janić, M., 2014b. Modelling the effects of different air traffic control (ATC) operational procedures, separation rules, and service priority disciplines on runway landing capacity. J. Adv. Transp. 48, 556–574.

Janić, M., 2014c. Modelling performances of the supply chain(s) served by the mega freight transport vehicles. In: Proceedings of ICTTE. International Conference on Traffic and Transport Engineering, Belgrade, November 27–28, 2014, Belgrade, Serbia.

Janić, M., 2015. A multi-criteria evaluation of solutions and alternatives for matching capacity to demand in an airport system: the case of London. Transp. Plan. Technol. http://dx.doi.org/10.1080/03081060.2015.1059120.

Kinzhigaziev, V., Zadorozhny, V., 2014. World's most powerful S5K electric locomotive. Mag. Transmashholding Partners 4, 10–15.

Kooman, C.I., de Bruijn, P.A., 1975. Lock Capacity and Traffic Resistance of Locks, Rijkswaterstaat, Traflic Engineering Division, Shipping Branch. Government Publishing Office, The Hague.

Kozan, E., 2006. Optimum capacity for intermodal container terminals. Transp. Plan. Technol. 29 (6), 471–482.

Lagoudis, I., Rice Jr., J., 2011. Revisiting Port Capacity: A Practical Method for Investment and Policy Decisions. Unpublished manuscript.

Mocuta, E.G., Ghita, E., 2007. The analysis of the transhipment capacity of an intermodal terminal. In: International Conference on Economic Engineering and Manufacturing Systems, October 25–26, 2007, Brasov, Romania.

Newell, F.G., 1979. Airport capacity and delay. Transp. Sci. 13 (3), 201–241.

Newell, G., 1982. Applications o Queuing Theory, Monographs on Statistics and Applied Probability. Chapman and Hall, New York.

Nishiyama, T., 2010. High-Speed Rail Operations in Japan, International Practicum on Implementing High Speed Rail in the United States. APTA, New York.

PANYNJ, 2003. La Guardia Airport: Traffic Statistics. The Port Authority of NY&PJ, New York. Report.

PEMA, 2014. Container Terminal Yard Automation. A Pema Information Paper. Port Equipment Manufacturer Association, Brussels.www.pema.org.

PHK, 2006. Port Benchmarking for assessing Hong Kong's Maritime Services and Associated Costs With Other Major International Ports. Maritime Department, Planning Development and Port Security Branch, Port of Hong Kong, Hong Kong.

Rallis, T., 1977. Intercity Transport: Engineering and Planning. The McMillan Press, London.

Salminen, J.B., 2013. Measuring the Capacity of a Port System: A Case Study on a Southeast Asian Port. MSc Thesis, MIT, Cambridge, MA.

SCG, 2013. Global supply chain news: maersk triple E cost advantages are too great to ignore. In: Supply Chain Digest. http://www.scdigest.com/index.php.

Solar, R.M., 2012. Inland Waterway Transport in the Rhine River: Searching for Adaptation Turning Points. MSc Thesis, Earth System Science Group, Wageningen University, Wageningen.

Song, J.-Y., Na, H.S., 2012. A Study on the intercontinental transportation competitiveness enhancement plan between northeast Asia and Europe using the trans-Siberian railway. IACSIT—Int. J. Eng. Technol. 4 (2), 208–212.

Teixeira, P., 2011. Engenharia Ferroviária—Slides da cadeira. Instituto Superior Técnico, Lisbon.

van Donselaar, P., Carmigchelt, H., 2001. Container Transport on the Rhine Marginal Cost: Case Study Infrastructure, Environmental and Accident Costs for Rhine Container Shipping, Competitive and Sustainable Growth Program (European Commission). Workpackage 5/8/9, Version 2. NEI B.V, Netherlands.

Vergeij, H.J., Stolker, C., Groenveld, R., 2008. Inland Waterways: Ports, Waterways, and Inland Navigation, Lecture Notes CT 4330. Department of Hydraulic Engineering, Faculty of Civil Engineering and Geosciences, Delft University of Technology, The Netherlands, Delft.

Viohl, B., 2015. Euro Asian transport links: transport flows and non-physical barriers. In: Informal document WP.5/GE.2 (2015) No.1, Economic Commission for Europe, Inland Transport Committee, Working Party on Transport Trends and Economics Group of Experts on Euro-Asian Transport Links, Thirteenth Session, June 9–10, 2015, Dushanbe, Tajikistan, p. 49.

Yagar, S., Saccomanno, F., Shi, Q., 1983. An efficient sequencing model for humping in rail yard. Transp. Res. A 17 (4), 251–262.

Zhang, M., Wiegmans, B., Tavasszy, L.A., 2009. A comparative study on port hinterland intermodal container transport: Shanghai and Rotterdam. In: The Fifth Advanced Forum on Transportation of China (AFTC), October 17, Beijing, China.

WEBSITES

http://www.alphaliner.com/top100.
http://www.britannica.com/place/Greater-London.
https://www.dhl-discoverlogistics.com/cms/en/glossary/buchstabe_s.jsp.
http://www.eurolines.com/en.
http://dot.gov/airconsumer.
http://www.delta.com/content/www/en_US/traveling-with-us/where-we-fly/routes/downloadable-route-maps.html.
http://www.ect.nl/en/content/euromax-terminal-rotterdam.
https://www.google.nl/search?q=southwest+airlines+route+map.
http://www.interfreight.co.za/container_information.html.
www.londonlorrycontrol.com/routes.
https://www.rta.ae.
http://www.ryanair.com/en.
http://www.sea-distances.org.
http://www.ship.gr/news6/hanjin28.htm.
http://en.wikipedia.org/wiki/Container_ship.
https://en.wikipedia.org/wiki/Instrument_landing_system.
http://en.wikipedia.org/wiki/List_of_largest_container_ships.
https://en.wikipedia.org/wiki/Low-cost_carrier.
https://en.wikipedia.org/wiki/UIC_classification_of_goods_wagons.
http://www.worldcontainerindex.com.
http://www.worldslargestship.com.

What is fixed-time control at the isolated intersection? How to calculate vehicle delays at signalized intersections? How to perform green time allocation? What are Intelligent Transportation Systems? What will be the most appropriate traffic control in the case of traffic flows dominated by autonomous vehicles? What are the best freeway traffic control measures? What are Demand Management Strategies? How to control separation distances between successive trains? What are the main components of the air traffic control system?

TRAFFIC CONTROL 6

6.1 INTRODUCTION

The number of trips by cars has considerably grown in latest decades in many cities, and on many highways in the world. At the same time, road network capacities have not kept up with this increase in travel demand. Urban road networks in many countries are harshly congested, resulting in increased travel times, increased number of stops, unexpected delays, greater travel costs, inconvenience to drivers and passengers, increased air pollution and noise level, and increased number of traffic accidents (Triantis et al., 2011). Increasing traffic network capacities by building more roads is very costly, as well as, environmentally destructive.

Traffic congestion has also been a problem at numerous airports all over the world. Delays at one airport have unlikable effects on aircraft delays at other airports. Big number of flights land at the busiest world's airport with a delay that is more than 15 min. At several airports, the average delay is equal to 50–60 min.

Traffic engineers and planners are trying to optimally use the existing transportation infrastructure, as well as to create efficient transportation systems. These goals could be achieved primarily by developing and implementing a variety of traffic control measures, methods, and strategies.

This chapter describes the main components and principles of operations of the road, rail, and air traffic control system. Traffic control techniques should enable transportation system to operate efficiently, effectively, and safely. Efficiency implies the vehicles' movement at the minimum, acceptable, or planned costs. Effectiveness is related to the vehicles' movement according to the planned/scheduled time, while the safety assumes vehicles' movement without conflicts causing traffic incidents and accidents.

Transportation Engineering. http://dx.doi.org/10.1016/B978-0-12-803818-5.00006-8

– In particular, the road traffic control regulates and manages traffic flows in urban streets, and suburban, and interurban roads and highways.

The air traffic control (ATC) system regulates movement of commercial airline aircraft between their origin and destination airports and at these airports themselves. In some cases, the movement of the military aircraft is also regulated. However, this is not considered in this chapter. The ATC system consists of components such as infrastructure, supportive facilities and equipment (on the ground and on-board the aircraft), and staff including the ATC controllers and ultimately the aircraft pilots.

The rail traffic control regulates movement of passenger and freight trains between their origin and destination stations. The control system includes components such as rail lines and stations (with tracks), supportive facilities and equipment (signaling system), and staff—the train dispatchers and ultimately train drivers.

6.2 TRAFFIC CONTROL AT SIGNALIZED INTERSECTIONS

Majority of road intersections in the world are not signalized. The reasons for this are low traffic volumes and/or acceptable sight distances. Higher traffic volumes and/or frequency and harshness of traffic accidents at the intersections create the needs for traffic control. Traffic control represents a supervision of the vehicles and pedestrians movements, with an intention of securing maximum efficiency and safety of conflicting traffic movements. Traffic lights or traffic signals are the basic devices used in traffic control. Traffic signals warn, control and/or direct traffic movements at an intersection. They are located at road intersections and/or pedestrian crossings. The first traffic light was installed even before the automobile traffic. It happened in London on December 10, 1868. The current traffic lights were invented in United States (Salt Lake City, 1912; Cleveland, 1914; New York and Detroit, 1920). Adequately designed and located traffic signals provide orderly traffic movement in all directions, increase intersection capacity, decrease delay, and decrease number of crashes at the intersection (Webster, 1958; Little, 1966; Allsop, 1971, 1976; Hunt et al., 1982; Papageorgiou, 1983; Mirchandani and Head, 1998; Gazis, 2002; Papageorgiou et al., 2003; Kotsialos and Papageorgiou, 2000, 2004). At the same time traffic signals help pedestrians to cross the street, and help side-street vehicles to move into the main traffic stream. In the case of arterial streets, traffic signals enable progressive flow of traffic. On the other hand, improper traffic signals could increase vehicle delays, and number of stops, change traffic assignment in the network, increase fuel consumption and increase harmful emissions. The engineering practice shows that many of the traffic signals in the world could be advanced by updating the timing plans. The signal timing plan should be updated from time to time, since travel demand patterns modify after a while. The traffic signal retiming is one of the best ways to mitigate traffic congestion.

Green, yellow, and red colors are used as colors for traffic lights. Green time represents the time interval during which the traffic signal has the green indication. The green *indication* gives the *right-of-way* to specific group of vehicles and/or pedestrians at the intersection for the certain length of time (Fig. 6.1).

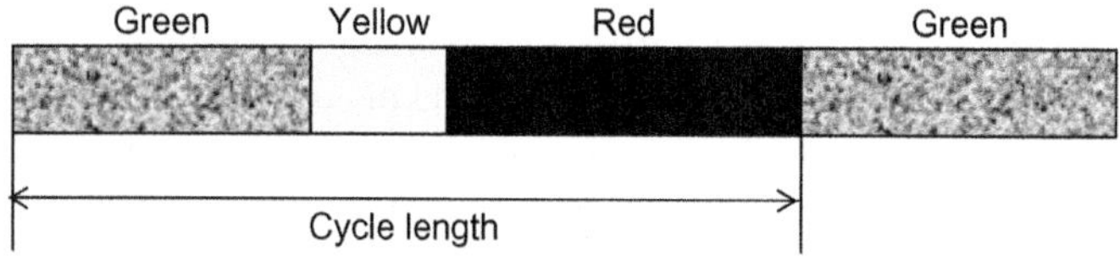

FIG. 6.1

Signal timing.

The right-of-way is finished by the yellow time interval. During this interval yellow signal is displaced. The purpose of the yellow light is to make possible protected transfer of right-of-way from one movement to a new one. The yellow signal is followed by a red signal. The meaning of a red light is the same as the meaning of the STOP sign. *Cycle length* represents the total time needed for the signal to finish one sequence of signal indications. The term *cycle time* is also used to describe the cycle length.

Drivers move toward the intersection from different approaches. Every intersection is composed of a number of approaches and the crossing area (Fig. 6.2).

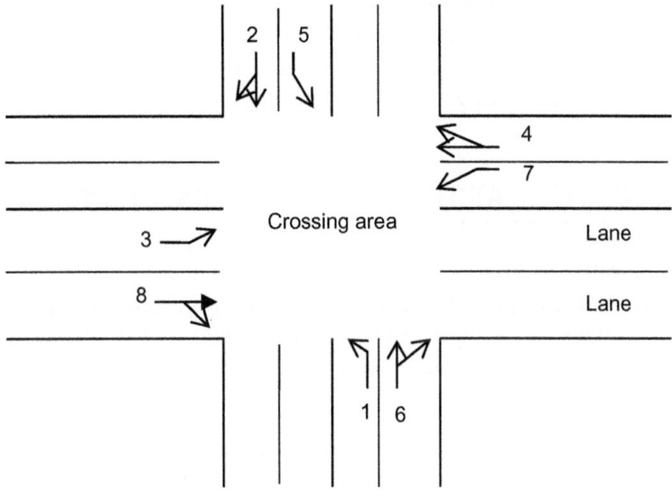

FIG. 6.2

Intersection: approaches, crossing area, and NEMA numbering scheme.

Approach can have one or more lanes. The American National Electrical Manufacturers Association (NEMA) standardized numbering scheme for the intersections. According to this standardization, left turns are every time given odd numbers (Fig. 6.2). Through movements are always given even numbers. In this scheme, right movements are not numbered alone. There are also some other numbering schemes used in different countries.

During the green time, vehicles from the observed approach can leave the stop line and cross the intersection. If the upstream demand is, in this case, extensive, and if the downstream link is not congested, vehicles continuously cross the stop line. The corresponding average flow rate of vehicles that cross the stop line is known as a *saturation flow*. During red time no vehicle departures happen. In many cases, queues of vehicles are established exclusively during the red phases, and are terminated during the green phases. Such traffic conditions are known as an *undersaturated traffic conditions*. An intersection is considered as an unsaturated intersection when all of her approaches are undersaturated. Traffic conditions, in which queue of vehicles can arrive at the upstream intersection, are known as an *oversaturated traffic conditions*.

Traffic engineers apply various traffic control strategies (Fig. 6.2) in order to minimize the total delay at the intersection. Isolated traffic control strategies are used to control isolated intersection.

Many isolated intersections operate under the *fixed-time control strategies*. These pretimed strategies are independent of the existing traffic. The phases, cycle length, and all green times are preset, and all cycles of the signals are the same.

At the same time, some other intersections operate under *demand actuated signals*. Demand actuated signals use detectors and corresponding control logic to respond to the traffic demand at the intersection. These signals could be further divided into *semiactuated signals* and *fully actuated signals*. A semiactuated signal is usually installed at the intersection of the high traffic volume major road, and the low traffic volume minor road. A semiactuated signal involves an installation (magnetic-loop detector in the pavement) on the minor road that detects when vehicle is present. This detection changes the green phase to the minor road to allow vehicle to cross the intersection. A fully actuated signal involves installation on all intersection approaches. This installation detects the volume of existing traffic. Fully actuated signal timing (cycle length, green split among competitive approaches) is directly influenced by the volumes of existing traffic, and every cycle is different from another.

6.2.1 FIXED-TIME CONTROL AT THE ISOLATED INTERSECTION

Isolated intersection is not coordinated with adjacent intersections. Coordinated control strategies simultaneously control traffic on many intersections. Many isolated intersections in the world operate under the fixed-time control strategies. These strategies assume existence of the *signal cycle* that represents one execution of the basic sequence of signal combinations at an intersection. A *phase* represents part of the signal cycle, during which one set of movements has right of way simultaneously (Fig. 6.3).

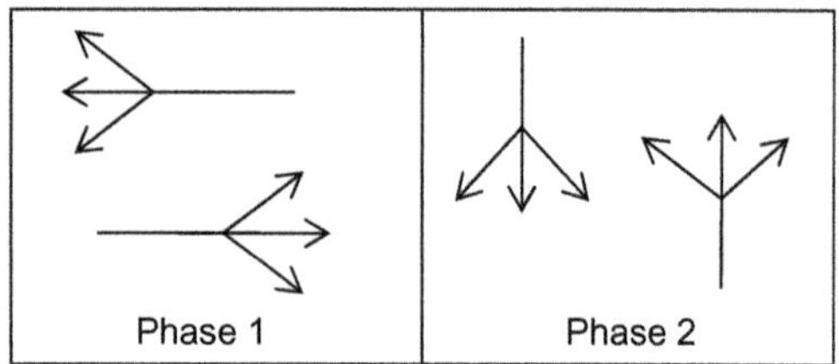

FIG. 6.3

Two-phase traffic operations.

Fig. 6.3 shows two-phase traffic operations for the intersection shown in Fig. 6.1. Cycle shown in Fig. 6.3 contains only two phases. Phase 1 is related to the movement of the east-westbound vehicles through the intersection. Phase 2 represents the movement of the north-southbound vehicles. The cycle length represents the duration of the cycle measured in seconds. The sum of the phase lengths represents the cycle length. For example, in the case shown in Fig. 6.4, the cycle length could be 90 s, length of the Phase 1 could be 50 s, while the length of the Phase 2 could be equal to 40 s (Fig. 6.4).

Fig. 6.5 shows intersection with three approaches and six possible movements.

Fig. 6.6 shows a potential phase diagram for the intersection shown in Fig. 6.5.

A higher number of phases is usually caused by traffic engineer's wish to protect some movements (usually left-turning vehicles). "Protection" assumes avoiding potential conflicts with the opposing

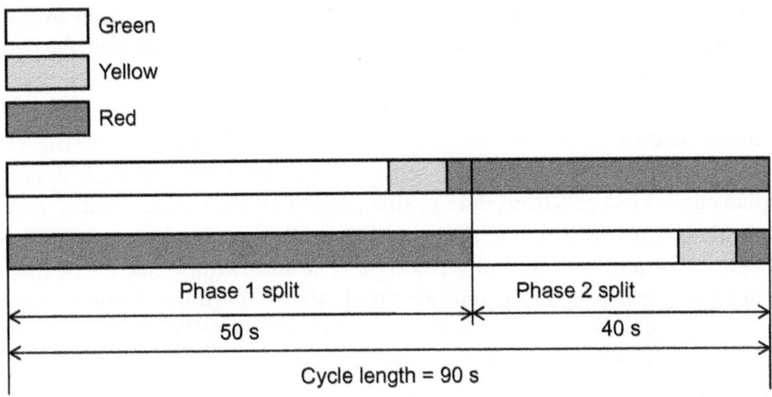

FIG. 6.4

Two phases: the cycle length is equal to 90 s.

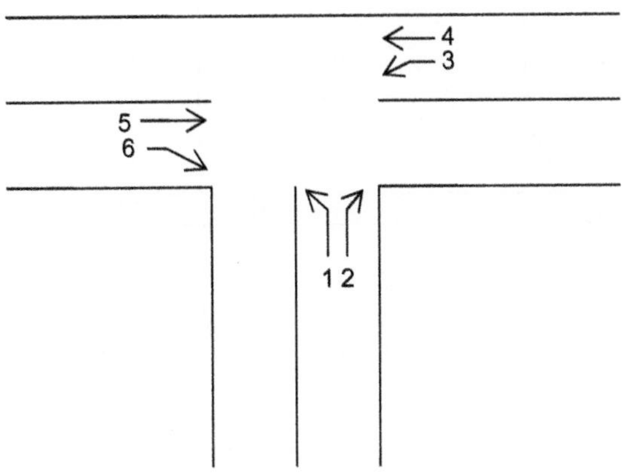

FIG. 6.5

Intersection with three approaches and six possible movements.

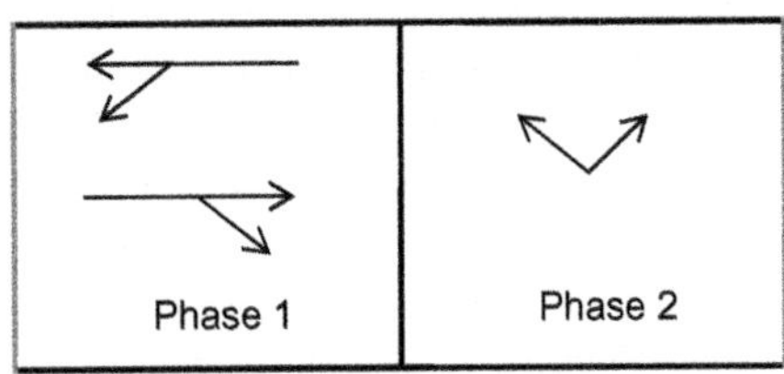

FIG. 6.6

A potential phase diagram for the intersection shown in Fig. 6.5.

traffic movement, and/or pedestrians. On the other hand, there is always certain amount of lost time (few seconds) during phase change. Obviously, the higher the number of phases, the better the protection, and the higher the value of the lost time associated with a phase change. Fixed time traffic control plans are based on the past traffic counts. Different time periods during a day (morning, afternoon, etc.) have different traffic control plans. As we already mentioned, the updated traffic counts must be obtained from time to time (usually once in 2 years), since traffic flow changes over time. The new traffic control plans are developed from time to time, based on the updated traffic counts.

Traffic signals are control devices. They alternatively order vehicles to stop and continue at various intersections by means of red, yellow, and green light. The following sequence of lights at the intersection approach is frequent in many countries: "Red, Red All, Green, Yellow, Red, Red All,..." (Fig. 6.7).

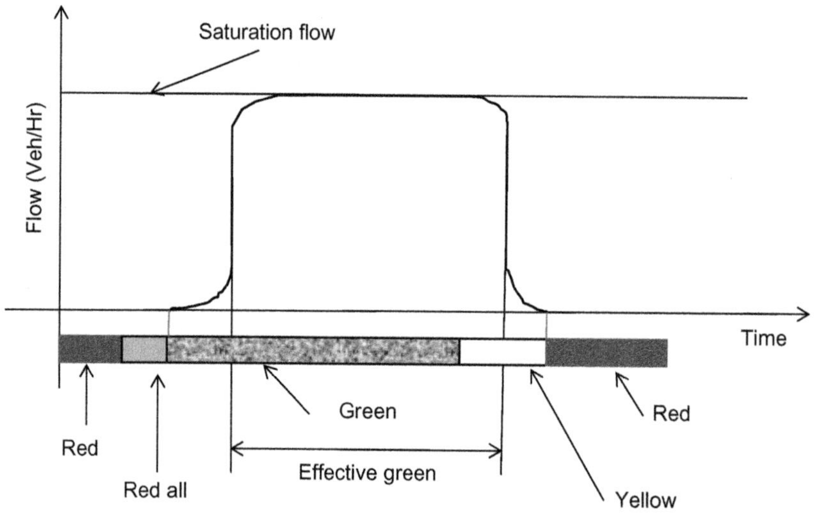

FIG. 6.7

Green and effective green.

"Green time," "effective green," "red time," and "effective red" are linguistic expressions frequently used by the traffic engineers. Theoretically, all drivers should cross the intersection during the green light. In reality, no one driver starts his/her car exactly in a moment of the green light appearance. Similarly, at the end of a green light, some drivers speed up, and cross the intersection during the amber light. With the assumptions that the upstream demand is extensive, and that the downstream link is not congested, vehicles continuously cross the stop line during green light. Few vehicles also cross the intersection during the amber light. Vehicle flow as a function of time is shown in Fig. 6.4. The saturation flow rate represents the maximum rate at which vehicles can go through a given point in an hour in existing conditions. Let us assume that vehicles depart from a queue with an average headway of 2 s. This means that saturation flow rate s equals

$$s = \frac{3600}{2} = 1800 \text{ vehicles per hour per lane}$$

"Green time" represents the time interval within the cycle when observed approach has green *indication*. On the other hand, "effective green" represents the time interval during which observed vehicles are *crossing* the intersection. "Green time," and "effective green" have similar numerical values, but they are not equal. There is also similar difference between "red time," and "effective red."

The capacity for an individual movement at the intersection depends on the saturation flow rate and the percentage of time during which vehicles may go into the intersection. Capacity of an individual movement is shown in Fig. 6.8.

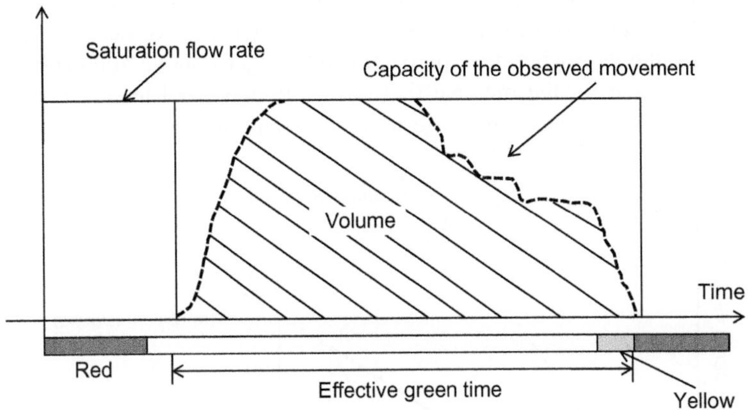

FIG. 6.8

Capacity of the observed movement.

The capacity of the observed movement represents the area of a rectangle whose sides are saturation flow rate and the effective green time. The area shaded in the figure represents volume. The capacity equals

$$c = s \cdot \frac{g}{c} \tag{6.1}$$

where

c is the capacity
s is the saturation flow rate
g is the effective green time for the movement (s)
C is the cycle length (s)

The degree of saturation for each movement is used in analyzing isolated intersections. This degree α is represented by the volume-to-capacity ratio:

$$\alpha = \frac{v}{c} = \frac{v}{s \cdot \frac{g}{C}} = \frac{v \cdot C}{s \cdot g} \tag{6.2}$$

Traffic movements that have volume-to-capacity ratios less than 0.85 have satisfactory capacity. In such a situation, signalized intersection is operating under the capacity, and vehicles are not suffering significant delays. In other words, observed signalized intersection is characterized by stable traffic operations. Movements with a volume-to-capacity ratio between 0.85 and 1.00 are characterized by the less stable traffic flow conditions, as well as queue appearance. When volume-to-capacity ratio exceeds 1.00 vehicle queue becomes longer and could, in some cases, disturbs normal operations at neighboring intersections.

6.2.2 VEHICLE DELAYS AT SIGNALIZED INTERSECTIONS

Success of any traffic signal control strategy highly depends, among other factors, on traffic engineers' capabilities to exactly calculate vehicle delays at signalized intersections. Let us first calculate the average vehicle delay in the case when the approach capacity goes above approach arrivals. Because of simplicity, let us assume for the moment that observed signalized intersection could be treated as a *D*/*D*/1 queueing system. We assume uniform arrivals, and uniform departure rate (Fig. 6.9).

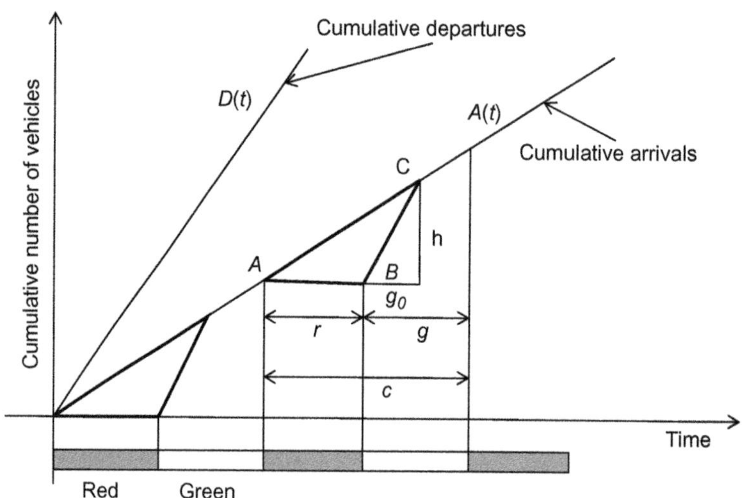

FIG. 6.9

Cumulative vehicle arrivals at and cumulative vehicle departures from the signalized intersection approach.

Let us denote by λ vehicles arrival rate, and by μ vehicles departure rate during the green time period. In the deterministic case, the cumulative number of arrivals $A(t)$ and the cumulative number of departures $D(t)$ are respectively equal (Fig. 6.9):

$$A(t) = \lambda t \tag{6.3}$$

$$D(t) = \mu t \tag{6.4}$$

We also introduce the following notation:

C is the duration of the signal cycle
r is the effective red
g is the effective green

The duration of the signal cycle equals

$$C = r + g \tag{6.5}$$

Formed queue is the longest at the beginning of effective green. Queue start to decrease at the beginning of effective green. We denote by g_0 the time necessary for queue to dissipate (Fig. 6.9). Queue must dissipate before the end of effective green. In the opposite case, queue would escalate indefinitely. In other words, queue dissipation will happen in every cycle if the following relation is satisfied:

$$g_0 \leq g \tag{6.6}$$

Relation (6.6) will be satisfied if the total number of vehicle arrivals during cycle length C is less than or equal to the total number of vehicle departures during effective green, ie,

$$\int_0^c \lambda dt \leq \int_0^g \mu dt \tag{6.7}$$

$$\lambda c \leq \mu g \tag{6.8}$$

Finally, we get

$$\frac{\lambda}{\mu} \leq \frac{g}{C} \tag{6.9}$$

Let us note the triangle ABC (Fig. 6.9). Vehicles arrive during the time period $(r + g_0)$. Vehicles depart during the time period g_0. The total number of arrived vehicles equals the total number of departed vehicles, ie,

$$\lambda \cdot (r + g_0) = \mu \cdot g_0 \tag{6.10}$$

$$(\mu - \lambda) \cdot g_0 = \lambda \cdot r \tag{6.11}$$

The time period g_0 required for queue to dissipate equals

$$g_0 = \frac{\lambda \cdot r}{\mu - \lambda} \tag{6.12}$$

We divide both numerator and denominator by μ. We get

$$g_0 = \frac{(\lambda/\mu) \cdot r}{1 - \frac{\lambda}{\mu}} \tag{6.13}$$

Since $\rho = \frac{\lambda}{\mu}$, we can write

$$g_0 = \frac{\rho \cdot r}{1 - \rho} \tag{6.14}$$

The area $A_{\Delta ABC}$ of the triangle ABC represents the total delay d of all vehicles arrived during the cycle. This area equals

$$A_{\Delta ABC} = \frac{1}{2} \cdot r \cdot h \tag{6.15}$$

where h is the triangle height.

The ratio $\frac{h}{r+g_0}$ represents the slope λ, ie,

$$\frac{h}{r+g_0}=\lambda \tag{6.16}$$

The height h equals

$$h=\lambda(r+g_0) \tag{6.17}$$

The area of the triangle ABC equals

$$A_{\Delta ABC}=\frac{1}{2}\cdot r\cdot h=\frac{1}{2}\cdot r\cdot\lambda(r+g_0)=\frac{\lambda\cdot r}{2}(r+g_0) \tag{6.18}$$

The total delay d of all vehicles arrived during the cycle equals

$$d=\frac{\lambda\cdot r}{2}(r+g_0)=\frac{\lambda\cdot r}{2}\left(r+\frac{\rho\cdot r}{1-\rho}\right)=\frac{\lambda\cdot r^2}{2}\left(1+\frac{\rho}{1-\rho}\right) \tag{6.19}$$

ie,

$$d=\frac{\lambda\cdot r^2}{2\cdot(1-\rho)} \tag{6.20}$$

The average delay per vehicle $\bar{d}$ represents the ratio between the total delay d and the total number of vehicles per cycle. The total number of vehicles per cycle equals $\lambda\cdot C$. This means that the average delay per vehicle $\bar{d}$ equals

$$\bar{d}=\frac{d}{\lambda\cdot C} \tag{6.21}$$

ie,

$$\bar{d}=\frac{(\lambda\cdot r^2)/(2\cdot(1-\rho))}{\lambda\cdot C} \tag{6.22}$$

Finally, the average delay per vehicle equals

$$\bar{d}=\frac{r^2}{2\cdot C\cdot(1-\rho)} \tag{6.23}$$

EXAMPLE 6.1

The cycle length at the signalized intersection equals 90 s. The considered approach has the saturation flow of 2200 (veh/h), the green time duration of 27 s, and flow rate of 600 (veh/h). Analyze traffic conditions in the vicinity of the intersection. Calculate average delay per vehicle. Assume that the $D/D/1$ queueing system adequately describes considered intersection approach.

Solution

The corresponding values of the cycle length and the green time are

$$C=90\,(\text{s});\quad g=27\,(\text{s})$$

EXAMPLE 6.1—cont'd

The flow rate and the service rate are respectively equal

$$\lambda = 600\,(\text{veh/h}) = \frac{600}{3600}\,(\text{veh/s}) = 0.167\,(\text{veh/s})$$

$$\mu = 2200\,(\text{veh/h}) = \frac{2200}{3600}\,(\text{veh/s}) = 0.611\,(\text{veh/s})$$

Traffic intensity ρ equals

$$\rho = \frac{\lambda}{\mu} = \frac{0.167\,(\text{veh/s})}{0.611\,(\text{veh/s})} = 0.273$$

The duration of the red light for the considered approach equals

$$r = C - g = 90 - 27 = 63\,(\text{s})$$

The number of arrived vehicles per cycle equals

$$\lambda \cdot C = 0.167\,(\text{veh/s}) \cdot 90\,(\text{s}) = 15.03\,(\text{veh})$$

The number of departed vehicles during green light equals

$$\mu \cdot g = 0.611\,(\text{veh/s}) \cdot 27\,(\text{s}) = 16.497\,(\text{veh})$$

We conclude that the following relation is satisfied:

$$\lambda \cdot C \le \mu \cdot g \tag{6.24}$$

This means that the traffic conditions in the vicinity of the intersection are undersaturated traffic conditions. The average delay per vehicle equals

$$\bar{d} = \frac{r^2}{2 \cdot C \cdot (1 - \rho)} \tag{6.25}$$

$$\bar{d} = \frac{63^2}{2 \cdot 90 \cdot (1 - 0.273)} = 30.33\,(\text{s})$$

EXAMPLE 6.2

The cycle length at the signalized intersection equals 60 s. The considered approach has the saturation flow of 2200 (veh/h), the green time duration of 15 s, and flow rate of 400(veh/h). Analyze traffic conditions in the vicinity of the intersection. Assume that the $D/D/1$ queueing system adequately describes considered intersection approach. Calculate: (a) the average delay per vehicle; (b) the longest queue length; and (c) percentage of stopped vehicles.

Solution

(a) The corresponding values of the cycle length and the green time are

$$C = 60\,(\text{s}); \quad g = 20\,(\text{s})$$

The red time equals

$$r = C - g = 60 - 20 = 40\,(\text{s})$$

The flow rate and the service rate are respectively equal

$$\lambda = 400\,(\text{veh/h}) = \frac{400}{3600}\,(\text{veh/s}) = 0.111\,(\text{veh/s})$$

(Continued)

EXAMPLE 6.2—cont'd

$$\mu = 2200\,(\text{veh/h}) = \frac{2200}{3600}\,(\text{veh/s}) = 0.611\,(\text{veh/s})$$

The utilization ratio ρ for the queue equals

$$\rho = \frac{\lambda}{\mu} = \frac{0.111\,(\text{veh/s})}{0.611\,(\text{veh/s})} = 0.182$$

The average delay per vehicle equals

$$\bar{d} = \frac{r^2}{2 \cdot C \cdot (1-\rho)} \tag{6.26}$$

$$\bar{d} = \frac{40^2}{2 \cdot 60 \cdot (1-0.182)} = 16.3\,(\text{s})$$

(b) The longest queue length $L_{\max}$ happens at the end of a red light (Fig. 6.9). The quantity $L_{\max}$ equals

$$L_{\max} = \lambda \cdot r = 0.111\,(\text{veh/s}) \cdot 40\,(\text{s}) = 4.44\,(\text{veh})$$

(c) Vehicles arrive all the time during the cycle. The total number vehicles A arrived during the cycle equals

$$A = \lambda \cdot C = 0.111\,(\text{veh/s}) \cdot 60\,(\text{s}) = 6.66\,(\text{veh})$$

All vehicles that arrive during time interval $(r + g_0)$ are stopped. The total number of stopped vehicles S equal

$$S = \lambda \cdot (r + g_0) \tag{6.27}$$

The time period g_0 required for queue to dissipate equals

$$g_0 = \frac{\lambda \cdot r}{\mu - \lambda} \tag{6.28}$$

We get

$$S = \lambda \cdot (r + g_0) = \lambda \cdot \left(r + \frac{\lambda \cdot r}{\mu - \lambda}\right) = 0.111 \cdot \left(40 + \frac{0.111 \cdot 40}{0.611 - 0.111}\right)$$

$$S = 5.43\,(\text{veh})$$

The percentage of stopped vehicles equal

$$P = \frac{S}{A} \cdot 100 = 5.43 \cdot 100 = 81.53\,(\%)$$

Traffic flows are characterized by random fluctuations. The delay that a specific vehicle experiences depends on the probability density function of the interarrival times, as well as on signal timings and the time of a day when the vehicle shows up. Obviously, individual vehicles experience at a signalized approach various delay values. In other words, there is significant variability of vehicle delays.

Let us consider simple intersection approach that consists of a single through lane. The cumulative number of arrivals and the cumulative number of departures are shown in Fig. 6.10.

Let us calculate the delay D for the vehicle arriving at time τ (Fig. 6.10). The overall delay D is composed of the *uniform delay d* and the *overflow delay* d_R, ie,

$$D = d + d_R \tag{6.29}$$

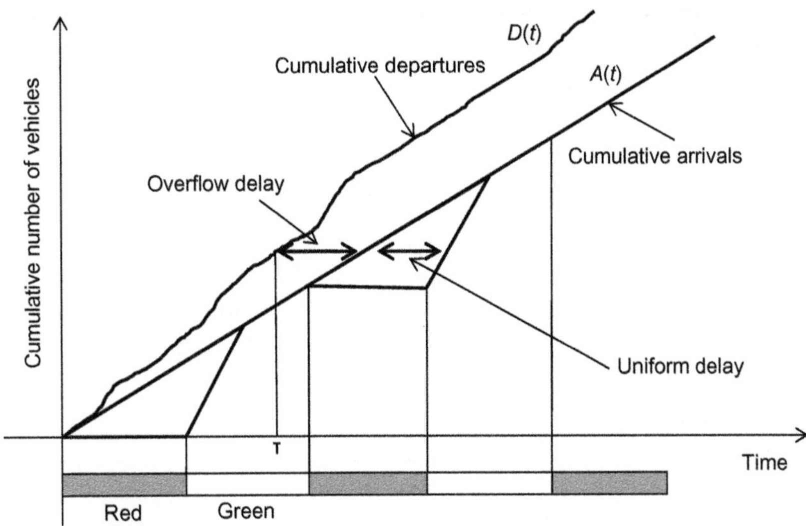

FIG. 6.10

Overflow delay and uniform delay.

The uniform delay *d* represents delay that would be experienced by a vehicle when all vehicle arrives uniformly and when traffic conditions are unsaturated. Because of the random nature of vehicle arrivals, arrival rate during some time periods can go over the capacity, causing overflow queues. The overflow delay represents the delay that is caused by short-term overflow queues.

The practical and theoretical knowledge showed that it has been difficult to offer the exact formula for the delay that is applicable at various intersections, in various traffic conditions. Traffic engineers widely use approximate delay formulas. The following two formulas (derived by combining analytical and simulation approaches) for the average vehicle delay calculation were proposed and widely used:

Webster's formula

$$D=\frac{r^2}{2\cdot C\cdot(1-\rho)}+\frac{\alpha^2}{2\cdot\lambda\cdot(1-\alpha)}-0.65\left(\frac{C}{\lambda^2}\right)^{1/3}\cdot\alpha^{\left(2+\frac{5\cdot g}{c}\right)} \quad (6.30)$$

Allsop's formula

$$D=\frac{9}{10}\cdot\left[\frac{r^2}{2\cdot C\cdot(1-\rho)}+\frac{\alpha^2}{2\cdot\lambda\cdot(1-\alpha)}\right] \quad (6.31)$$

EXAMPLE 6.3

The cycle length at the signalized intersection equals 80 s. The considered approach has the saturation flow of 2200(veh/h), the green time duration of 30 s, and flow rate of 600(veh/h). Calculate the average vehicle delay by using Webster's formula.

Solution

(a) The corresponding values of the cycle length and the green time are

$$C=80(\text{s}) \quad g=30(\text{s})$$

(Continued)

EXAMPLE 6.3—cont'd

The red time equals

$$r = C - g = 80 - 30 = 50\,(\text{s})$$

The flow rate and the service rate are respectively equal

$$\lambda = 600\,(\text{veh/h}) = \frac{600}{3600}\,(\text{veh/s}) = 0.167\,(\text{veh/s})$$

$$\mu = 2200\,(\text{veh/h}) = \frac{2200}{3600}\,(\text{veh/s}) = 0.611\,(\text{veh/s})$$

The utilization ratio ρ for the queue equals

$$\rho = \frac{\lambda}{\mu} = \frac{0.167\,(\text{veh/s})}{0.611\,(\text{veh/s})} = 0.273$$

The volume to capacity ratio equals

$$\alpha = \frac{\lambda \cdot C}{\mu \cdot g} = \frac{0 \cdot 167 \cdot 80}{0 \cdot 611 \cdot 30} = 0.73$$

The average delay per vehicle equals

$$D = \frac{r^2}{2 \cdot C \cdot (1-\rho)} + \frac{\alpha^2}{2 \cdot \lambda \cdot (1-\alpha)} - 0.65\left(\frac{C}{\lambda^2}\right)^{1/3} \cdot \alpha^{\left(2+\frac{5 \cdot g}{C}\right)}$$

$$D = \frac{50^2}{2 \cdot 80 \cdot (1-0.273)} + \frac{0.73^2}{2 \cdot 0.167 \cdot (1-0.73)} - 0.65\left(\frac{80}{0.167^2}\right)^{1/3} \cdot 0.73^{\left(2+\frac{5 \cdot 30}{80}\right)}$$

$$D = 21.49 + 5.91 - 2.73 \approx 25\,(\text{s})$$

6.2.3 THE DETERMINATION OF TIMING FOR FIXED-TIME SIGNALS

Fixed time traffic control plans are based on the past traffic counts. Updated traffic counts must be obtained from time to time (usually once in 2 years), since traffic flow changes over time. Traffic engineers are facing the following questions when developing the new fixed traffic signals timing: What is the optimal cycle length? What is the optimal number of phases in the cycle? What is the best way to allocate available green time among conflicting traffic movements? Could pedestrians cross the street safely? The following are usual steps when developing fixed traffic signals timing:

Signal phasing selection
Volume adjustment (calculation of equivalent straight-through passenger cars)
Critical lane volumes selection
Change interval calculation
Cycle length calculation
Green time allocation
Pedestrian crossing time check
Signal timing plan

6.2.4 SIGNAL PHASING SELECTION

Different traffic streams at the intersection have different turn movements. This mean, that some of them could be frequently in conflict. We use traffic lights to separate conflicting vehicles in time. Typical crossing, merging and diverging conflicts are shown in Fig. 6.11.

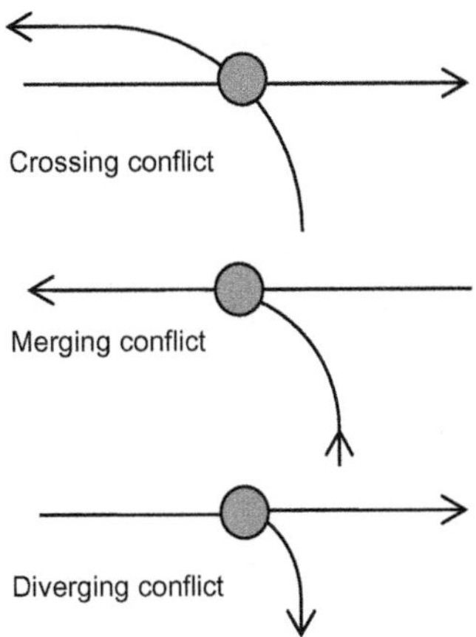

FIG. 6.11

Various conflict types.

The number of phases and their composition depends on many factors (chosen treatment of left turns, chosen treatment of right turns, location of crosswalks, and pedestrian requirements, etc.). For example, crash history could indicate significant hazard at the intersection that could be reduced by introducing protected or exclusive pedestrian phases. Some vehicle movements cannot go together at the same time. This means that these movements cannot be together in the same phase. Higher number of phases is usually caused by traffic engineer's wish to protect some movements (usually left-turning vehicles). "Protection" assumes avoiding potential conflicts with the opposing traffic movement, and/or pedestrians. Left-turning vehicles are in conflict with opposing through traffic, and pedestrians, while right-turning vehicles are frequently in conflict with pedestrians (Fig. 6.12).

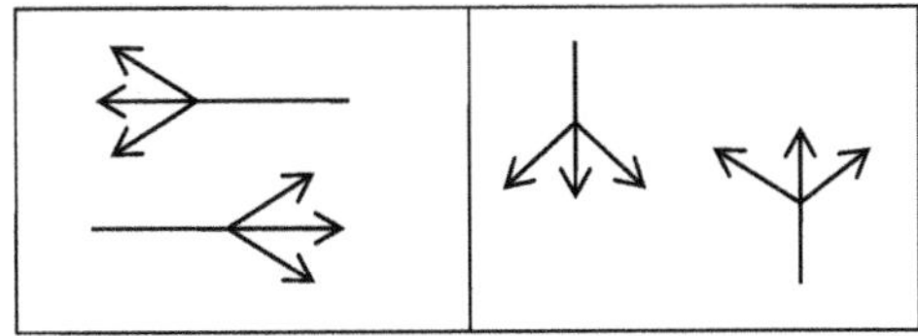

FIG. 6.12

Two-phase traffic operations.

Three-phase traffic operations are usually introduced in the case of high pedestrian volumes, as well as in the case of high left-turning traffic volume in one, or both of the streets (Fig. 6.13). Four-phase traffic operations are introduced in the cases of high left-turning traffic volumes in both intersecting streets (Fig. 6.14).

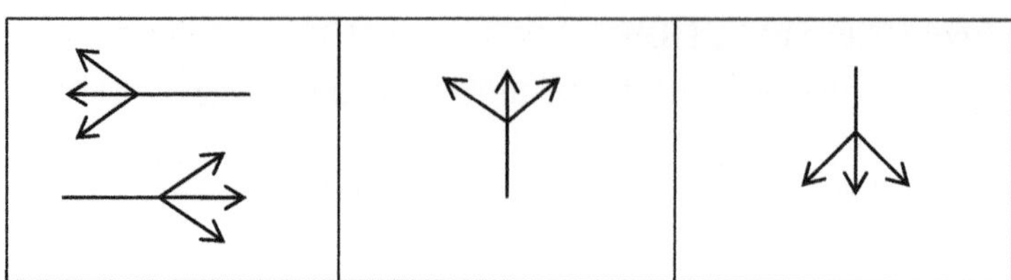

FIG. 6.13

Three-phase traffic operations.

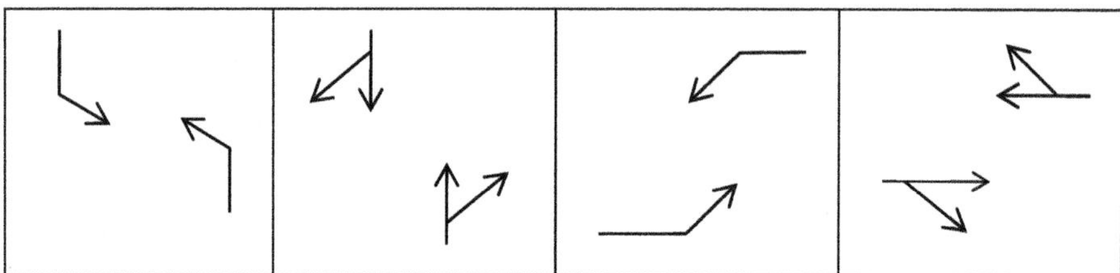

FIG. 6.14

Four-phase traffic operations.

There is always certain amount of lost time (few seconds) during phase change. Obviously, the higher the number of phases, the better the protection, and the higher the value of the lost time associated with a phase change. The number of phases and their composition highly influence intersection capacity and efficiency. Final decision about number of phases is based on the established rules for introducing individual left-turning phase, as well as on common sense, and engineering judgment. According to the United States Department of Transportation (Federal Highway Administration), the left-turn phasing should be advised if any one of the following criteria is satisfied:

(a) There are minimum of 2 left-turning vehicles per cycle and the product of opposing and left turn hourly volumes exceeds 50,000 in the case of one opposing lane, and 100,000 in the case of two opposing lanes.
(b) The left-turning movement crosses 3 or more lanes of opposing through traffic.
(c) The posted speed of opposing traffic exceeds 70 km/h.
(d) Analysis of crash history for a 12-month period shows 5 or more left-turn collisions.

6.2.5 VOLUME ADJUSTMENT (CALCULATION OF EQUIVALENT STRAIGHT-THROUGH PASSENGER CARS)

Vehicles performing different turning movements behave in a different way. For example, a left-turning passenger car usually does not accelerate as fast as a straight-moving passenger car. It is also important to notice that different vehicle types (passenger cars, trucks, buses) behave in a different way, and need different amount of time to perform specific maneuver (buses and trucks move more slowly than passengers cars). In order to properly calculate fixed traffic signal parameters, it is necessary to perform all calculations with only one vehicle type (straight-through passenger car). In the first step, all nonthrough traffic movements and all vehicle types are transformed into straight-through passenger cars. In the second step, fixed traffic signal parameters are calculated using corresponding traffic data expressed in (straight-through passenger cars) units. The adjustment factors that are used by the traffic engineers to convert actual traffic flows into straight-through passenger cars are given in Table 6.1 (*Highway Capacity Manual*).

Table 6.1 Conversion of Actual Flows into Straight-Through Passenger Cars

Vehicle Type/Movement	Adjustment Factor
Buses and trucks (heavy vehicles)	1.5 (straight-through passenger cars)
Right-turn vehicles	1.4 (straight-through passenger cars)
Left-turn vehicles	1.6 (straight-through passenger cars)

For example, every heavy vehicle in the flow should be counted as 1.5 straight-through passenger cars. Every right-turn vehicle should be counted as 1.4 straight-through passenger cars. Bus that makes left-turn should be counted as straight-through passenger cars, etc.

EXAMPLE 6.4

Numbers of left-turning, straight-moving, and right-turning passenger cars and heavy trucks are indicated in Fig. 6.15. Transform all nonthrough traffic movements and corresponding vehicle types into straight-through passenger cars.

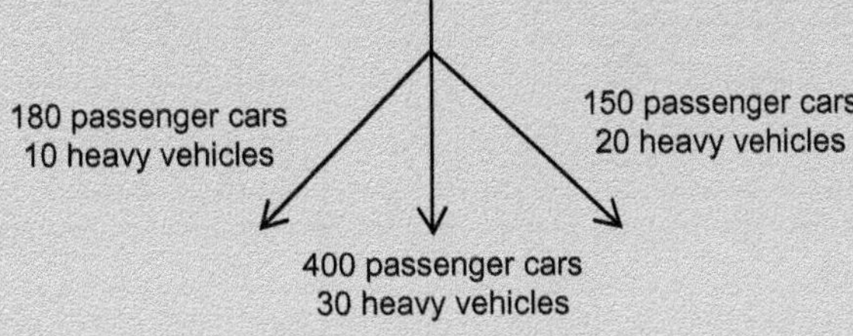

FIG. 6.15

Traffic volumes adjustment.

Solution

Calculated numbers of equivalent straight-through passenger cars are shown in Fig. 6.16.

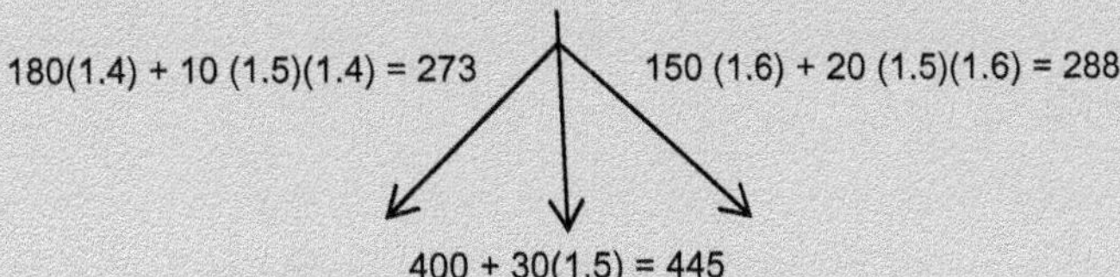

FIG. 6.16

Calculated numbers of equivalent straight-through passenger cars.

6.2.6 CRITICAL LANE VOLUMES SELECTION

Traffic movements within the same phase have different traffic volumes. The maximum traffic volume per lane among all traffic volumes per lane within the same phase is known as the critical lane volume. Obviously, the quantity of time needed for each signal phase is determined by this lane volume, since all other lane movements in the same phase need less time than the critical lane movement.

The critical lane volumes are determined in this way for every approach. In order to calculate critical lane volume, traffic engineer must distribute the total approach volume among approach lanes. This should be done according to the obtained field measurements data (some lanes are more used than some others). In the absence of such data, approximate lane volumes could be obtained by dividing the total approach volume by the number of lanes. There are also some other recommendations in the literature how to perform this distribution.

EXAMPLE 6.5

The traffic signal cycle is composed of two traffic phases. The east-westbound phase is shown in Fig. 6.17.

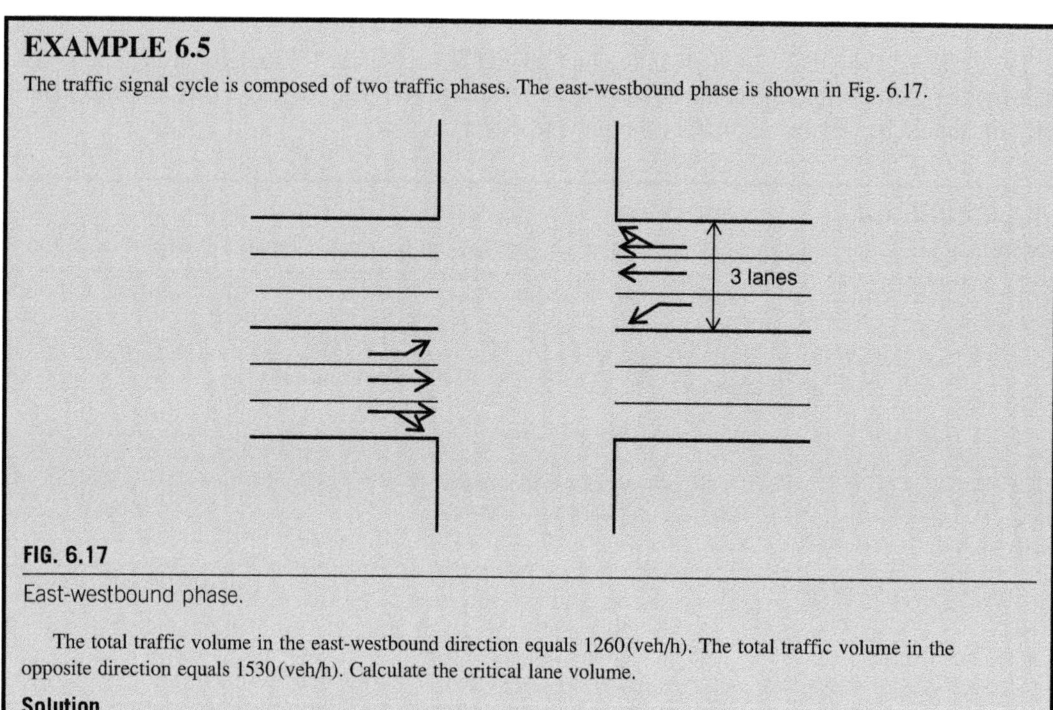

FIG. 6.17

East-westbound phase.

The total traffic volume in the east-westbound direction equals 1260 (veh/h). The total traffic volume in the opposite direction equals 1530 (veh/h). Calculate the critical lane volume.

Solution

We distribute approach volume equally among the lanes. The volumes per lanes are respectively equal

East-westbound direction: $\frac{1260}{3} = 420$

West-eastbound direction: $\frac{1530}{3} = 510$

The critical lane volume v_{cl} for the considered traffic phase equals

$$v_{cl} = \max\{420, 510\} = 510$$

6.2.7 CHANGE INTERVAL CALCULATION

A yellow time always ends a green time. The length of the yellow change interval must be adequate to avoid generating a dilemma zone. Dilemma zone is defined as a situation under which the driver can neither stop nor move through the intersection safely (Fig. 6.18).

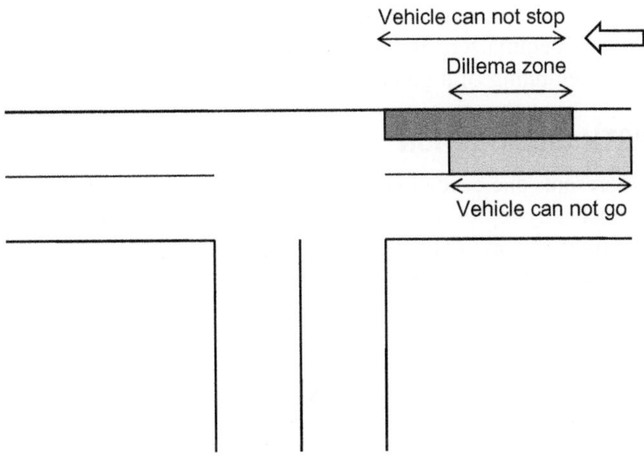

FIG. 6.18

Dilemma zone.

The minimum length of yellow time to avoid the dilemma zone equals

$$YT = t + \frac{v}{2 \cdot \alpha} \tag{6.32}$$

where

t is the driver's perception-reaction time in (s)
v is the approach speed in (m/s)
α is the deceleration rate in (m/s^2)

Usually, is rounded to the nearest 0.5 s. A yellow time less than the value described by relation (6.32) will generate a possible situation in which the driver may not be capable to stop before getting in the intersection, or enter the intersection before the end of the yellow time. The yellow time is usually in the range between 3 and 5 s. The yellow time must be succeeded by an all red time of adequate length. In this way, a vehicle that has entered the intersection at the end of the yellow time interval is able to clear that intersection before the conflicting movements are given right of way. The length of the all red time interval AR is calculated as:

$$AR = \frac{w + l}{v} \tag{6.33}$$

where

w is the width of intersection to be crossed
l is the length of the vehicle
v is the approach speed

The total change interval represents the sum of the yellow time and the all red time. The length of the total change interval must be greater than or equal than the sum of YT and AR (relations 6.32 and 6.33). In this way, creation of the dilemma zone will be prevented, and the vehicles will not be trapped within the intersection after the beginning of a green display for conflicting movements.

6.2.8 CYCLE LENGTH CALCULATION

The signal cycle that represents one execution of the basic sequence of signal combinations at an intersection. Cycle length practically represents the length of the time interval from the time point when a movement receives the right of way, until the time point when the same movement receives right of way again. Cycle length is usually in the range of 45–190 s. The cycle length c is calculated using the following Webster's formula:

$$c = \frac{1.5 \cdot LT + 5}{1 - \sum_{i=1}^{n} y_i} \tag{6.34}$$

where

LT is the total lost time during the cycle (approximately equal to the sum of total yellow, and all red times)
y_i is the ratio of the critical lane volume to the saturation flow per lane for the ith signal phase

The seconds are the units of the cycle length. Calculated cycle length is always rounded to the closest 5-s increment.

EXAMPLE 6.6

Traffic volumes in the case of four-way intersection are indicated in Fig. 6.19. The corresponding saturation flows are given in Table 6.2.

Traffic at the intersection is regulated by the two-phase traffic operations (Fig. 6.20).

The lost time per phase change equals 4 s. Calculate the cycle length at the intersection.

Solution

The ratios of the critical lane volume to the per lane saturation flow are respectively equal

Phase 1:

$$y_1 = \max\left\{\frac{650}{1700}, \frac{700}{1700}\right\} = 0.41$$

Phase 2:

$$y_2 = \max\left\{\frac{450}{1650}, \frac{550}{1700}, \frac{600}{1700}, \frac{500}{1650}\right\} = \max\{0.27, 0.32, 0.35, 0.30\} = 0.35$$

The total lost time LT per cycle equals

$$LT = 2 \cdot 4 = 8\,(\mathrm{s})$$

EXAMPLE 6.6—cont'd

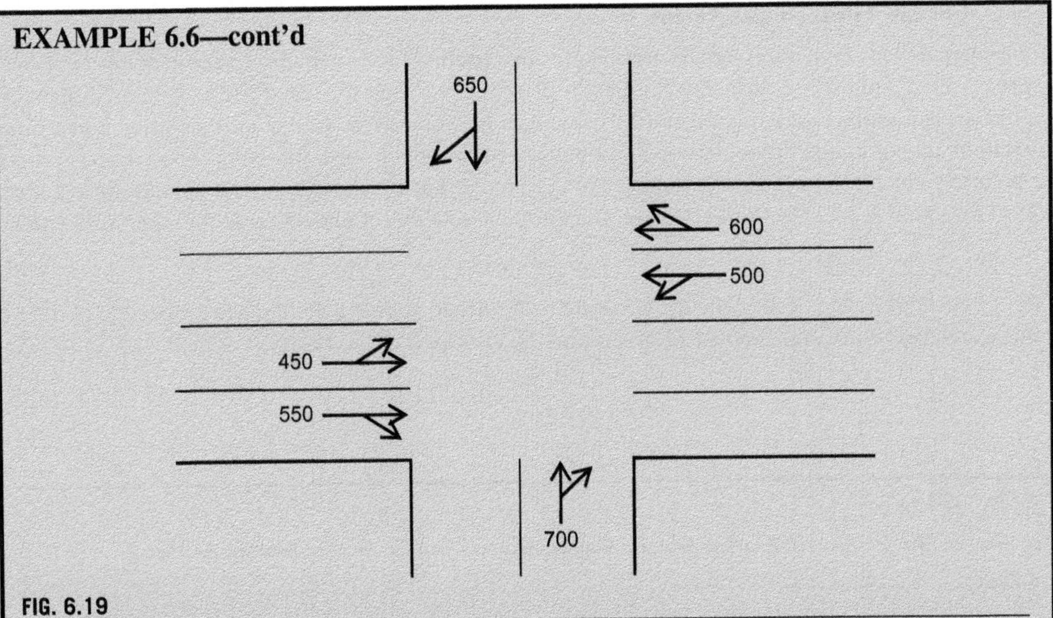

FIG. 6.19

Traffic volumes in the case of four-way intersection.

Table 6.2 Saturation Flows

Movements in the Lane	Saturation Flow (Vehicles Per Hour of Green Light)
Through	1800
Through and left-turns	1650
Through and right-turns	1700

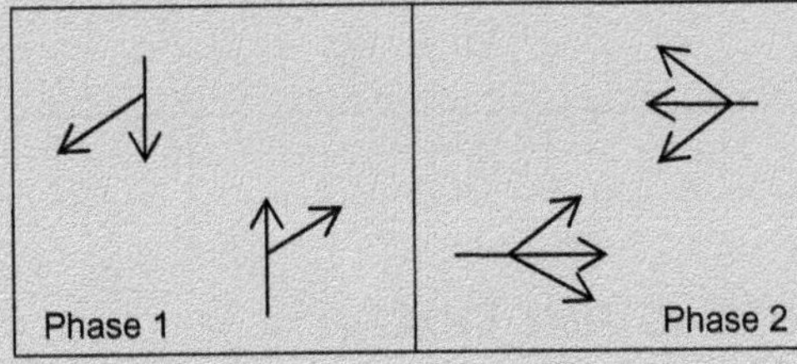

FIG. 6.20

Two-phase traffic operations.

The cycle length c equals

$$c = \frac{1.5 \cdot LT + 5}{1 - \sum_{i=1}^{n} y_i} = \frac{1.5 \cdot 8 + 5}{1 - (0.41 + 0.35)} = \frac{17}{0.24} = 70.83$$

After rounding to the nearest 5-s increment, we get

$$c = 70(\text{s})$$

6.2.9 GREEN TIME ALLOCATION

If one approach is given a red signal for too long, an extremely long vehicles queue can appear in that approach. It is intuitively clear that the length of the green time allocated to the specific approach should depend more or less on the traffic volume in that approach. Green split assumes green time allocation among competitive phases. The duration of the green time for every defined phase must be precisely calculated within the defined cycle. The cycle length equals c. The amount of lost time within the cycle is LT. The total effective green time available for allocation within the cycle equals

$$G = c - LT \tag{6.35}$$

The effective green time within the cycle should be allocated among the phases in proportion to their y values. In other words, the amount of green time G_i allocated to the ith phase equals

$$G_i = \frac{y_i}{y_1 + y_1 + \cdots + y_n} \cdot G \tag{6.36}$$

EXAMPLE 6.7

Allocate the effective green time among phases in the case of intersection shown in the Example 6.6.

Solution

The cycle length and the lost time are respectively equal

$$c = 70\,(\mathrm{s})$$

$$LT = 8\,(\mathrm{s})$$

The effective green time to be allocated equals

$$G = c - LT = 70 - 8 = 62\,(\mathrm{s})$$

The amount of green time G_1 that should be allocated to the first phase equals

$$G_1 = \frac{y_1}{y_1 + y_2} \cdot G = \frac{0.41}{0.41 + 0.35} \cdot 62 = 33.45$$

The amount of green time G_2 that should be allocated to the second phase equals

$$G_2 = G - G_1 = 62 - 33.45 = 28.55\,(\mathrm{s})$$

6.2.10 PEDESTRIAN CROSSING TIME CHECK

The allocated green time for the specific phase must be compared with the minimum green time that should be given to pedestrians. This is done within the pedestrian crossing time check. Pedestrian crossing time check must be performed in order to prevent pedestrians to be trapped within the intersection after the beginning of a green display for conflicting traffic movements. The usual pedestrian walking speed equals 1.2 (m/s). Pedestrian reaction time is equal to 0.7 s.

A pedestrian clearance time starts momentarily after the appearance of the "Walk" signal indication. Pedestrians cross the street during a pedestrian change interval, and some of them during the yellow time and even during the all red time. This means that the sum of pedestrian green time PGT, yellow time YT, and the all red time AR must be equal to the sum of pedestrian reaction time and the time necessary for a pedestrian to cross the street, ie,

$$0.7+\frac{w}{PWS}=PGT+YT+AR \tag{6.37}$$

The pedestrian green time equals

$$PGT=0.7+\frac{w}{PWS}-YT-AR \tag{6.38}$$

There are some intersections where a lot of pedestrians who walk slower than usual or pedestrians who use wheelchairs, regularly cross the street. In such cases, a lower walking speed 0.9 (m/s) should be planned.

If the allocated green time for the specific phase is less than the green time that should be given to pedestrians, the allocated green time must be increased.

6.2.11 ACTUATED SIGNAL CONTROL

Fixed (pretimed) traffic control is not sensitive to traffic flow variations. In other words, pretimed traffic control does not take into account the actual traffic demand. Frequently, some approaches receive more green time than they really need. At the same time, green allocated to some other approaches is not sufficient. This result in inadequate usage of intersection capacity, increased delays and number of stopped vehicles.

An actuated controller receives information by the actuation of vehicle detectors (Fig. 6.21). This type of controller overcomes the problem of inadequate usage of intersection capacity by operating signals based on actual traffic demands. Depending on the actual vehicle flows, the approach green times varies between minimum and maximum lengths. Vehicle actuation of loop detectors also causes adjustment of the cycle length and phases. Even various actuated control strategies try to adjust signal phase lengths to the actual traffic flows, they do not try to find the optimal phase lengths based on obtained information. The basic timing of actuated green times is shown in Fig. 6.21.

An actuated signal control works very well during nonpeak periods. It has been shown that this type of traffic control is successful at the intersections that are characterized by modest traffic flows, and that are relatively distant from other intersections.

Let us explain the basic principles of the actuated control using simple intersection shown in Fig. 6.21. The detector is placed in the minor street. The intersection shown in Fig. 6.21 is characterized by two-phase traffic operations. Traffic engineer must prescribe the minimum and maximum possible duration of the green times for both approaches.

For example, traffic engineer could decide that the green time in the minor street, under any circumstances, will never be less than 10 s, nor greater than 30 s. In there is no actuation in the minor street, the major street will continue to have the right-of-way. The major street green time can be broken up in the case when major-street minimum green time already expired and there are minor-street actuations. In such situation, the minor street will get the right of way. The duration of the minor-street green time depends on the minor street minimum green time and the passage time extensions. Whenever vehicle in the minor street is detected, minor street green time has prolongation that enable detected vehicle to cross the intersection (Fig. 6.21). The more vehicles appear in the minor street, the green time will be more extended. The total amount of the green time allocated to the minor street

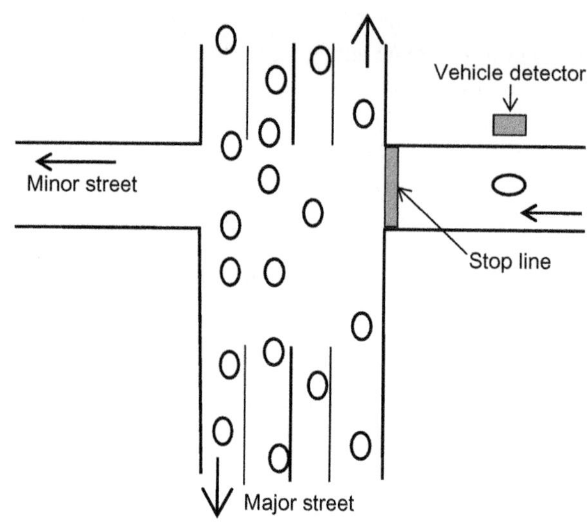

FIG. 6.21

Actuation of vehicle detector.

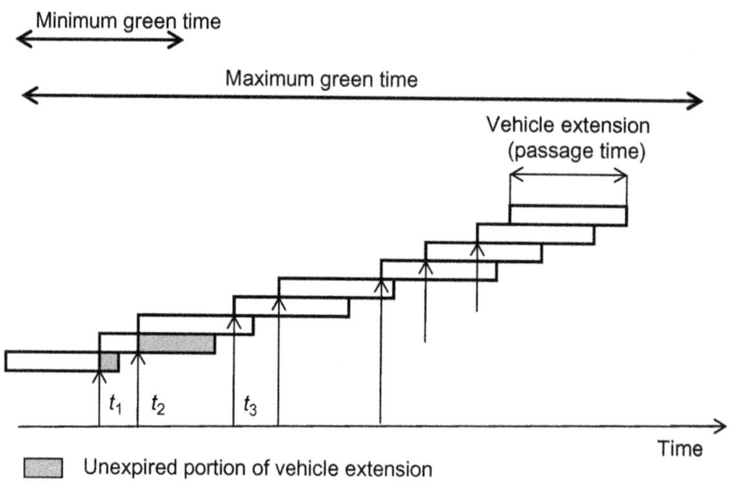

FIG. 6.22

The basic timing of actuated green times.

cannot be greater than the minor-street maximum green time (Fig. 6.22). Traffic engineer should also decide about vehicle passage time. An actuated signal control can be either *fully-* or *semiactuated.* In the case of a fully actuated signal control, detectors are placed on all intersection approaches. In a semi-actuated traffic signal control (Fig. 6.21), one (or more) approaches to an intersection do not have a detector. The approach without the detector is usually the major street.

6.3 TRAFFIC CONTROL FOR ARTERIAL STREETS

Drivers like to drive on an arterial street through sequence of intersections where the traffic light turns green just ahead of them. This can be achieved through traffic signal coordination (Little et al., 1981; Gartner et al., 1991; Papageorgiou, 1994). Traffic coordination actually means that traffic signals along arterial street "work together." Traffic signal synchronization should provide smooth traffic flow through the arterials minimizing at the same time vehicle delays in the side streets. In other words, the series of pretimed traffic signals is usually coordinated, in order to keep vehicles away from being stopped at each intersection along the arterial street. Intersections are usually coordinated if the distances between them are less than or equal to 400 m. In this case, the arrival flow at an intersection is still in the shape of "platoon," and coordination of traffic signals makes sense. In the opposite case, when the distances between adjacent intersections are longer than 400 m, vehicles arrive at an intersection less or more individually. When links between the intersections are very short, all movements for a considered direction should receive green at the same time. Fig. 6.23 shows coordination of a one-way arterial street.

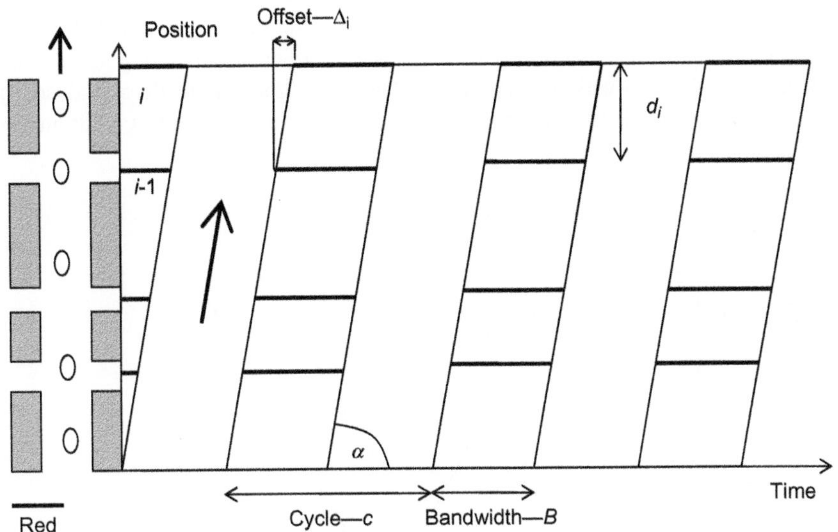

FIG. 6.23

Coordination of the one-way arterial street.

The distance between $(i-1)$st and ith intersection is denoted by d_i. The *offset* represents the number of elapsed seconds until a beginning of green at a ith traffic signal happens after the beginning of green of an adjacent $(i-1)$st traffic signal. Offset helps traffic signals to adjust the beginning of the green time to the beginning of the green time on adjacent signal. Grouped vehicles that travel in the same direction at the same approximate speed form platoon. We denote platoon speed by v. This sped equals (Fig. 6.23):

$$v = \tan\alpha = \frac{d_i}{\Delta_i} \tag{6.39}$$

where

d_i is the distance between $(i-1)$st and ith intersection
Δ_i is the offset

Platoon speed along an arterial street depends on the signal spacing, as well as on the cycle length at traffic signals. Uniformly spaced traffic signals provide very good conditions for forming vehicle platoons, as well as for smooth traffic flow through the arterial street. It is less likely that vehicle platoon will be formed in the cases of uneven distances between intersection, or relatively closely spaced traffic signals. Platoon speeds are lesser in the cases of closely spaced traffic signals and shorter cycle length, and vice versa.

The offset Δ_i equals

$$\Delta_i = \frac{d_i}{v} \tag{6.40}$$

The green bandwidth B is also shown in Fig. 6.23. The two parallel lines that determine bandwidth borders represent the constant speed trajectories (*speed of progression*) of the first, and the last vehicle in platoon of vehicles. This platoon of vehicles can clear all the intersections without stopping.

The properly calculated offsets and bandwidth create a "green wave" along an arterial. Fig. 6.24 shows coordination of the two-way arterial street. The basic goal of traffic signal coordination in the case of two arterial streets is to maximize the number of vehicles traveling without stops in both directions.

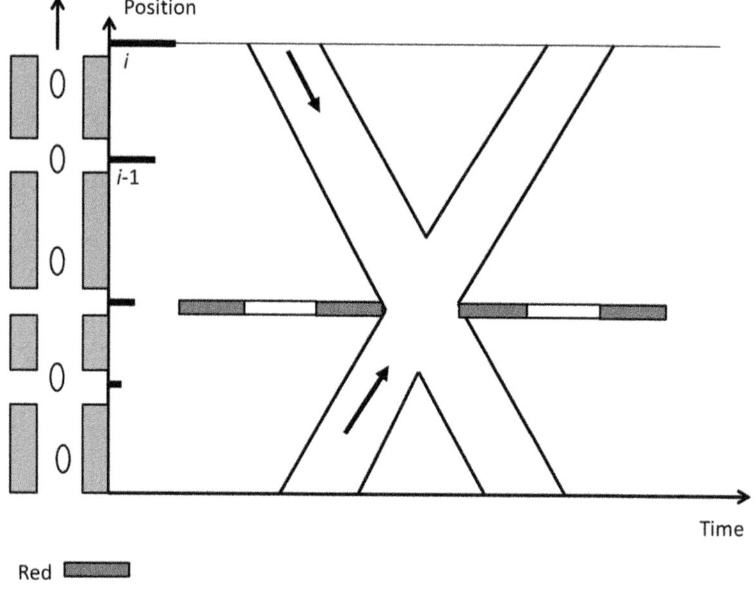

FIG. 6.24

Coordination of a two-way arterial street.

Cycle length is the same for all intersections that are together in the system. In the first step, cycle length is calculated independently for each intersection. The largest cycle length is usually chosen to be the cycle of the whole system. It has been shown in theory, as well as in practical implementations that proper traffic signal coordination could provide optimal travel speeds, and could reduce average travel times, the number of vehicle stops, the number of vehicle accidents, fuel consumption, and level of vehicle emissions.

6.3.1 ADAPTIVE CONTROL STRATEGIES

Some arterial streets are characterized by significant fluctuations in traffic demand. In order to properly control traffic along such arterial streets, traffic engineers use various adaptive control strategies (Lowrie, 1982; Teodorovic et al., 2006). Minimization of vehicle delays and the number of stops are the main goals of the adaptive traffic control strategies. Adaptive control strategies make decisions about green time allocation, offset, phase length, and phase sequence based on current traffic conditions (real-time information).

This means that the new traffic control parameters are calculated for every cycle. Pavement loop detectors provide basic traffic information, while local traffic controllers and central traffic controller communicate all the time.

The following are some of the most important adaptive traffic control systems:

SCATS (Sydney coordinated adaptive traffic system)
SCOOT (split, cycle, offset optimization technique)
ATSC (automated traffic surveillance and control)
OPAC (optimized policies for adaptive control)
RHODES (real-time hierarchical optimized distributed effective system)
RTACL (real-time traffic adaptive control logic)

SCATS, SCOOT, and ATSC have been already implemented in different cities (Table 6.3).

Table 6.3 Some Examples of Implemented Adaptive Control Systems

System	City	Number of Intersections
ATSC	Los Angeles, CA	1170
SCATS	Oakland County, MI	350+
SCATS	Hennepin County, MN	71
SCOOT	Arlington, VA	65
SCOOT	Minneapolis, MN	60
SCOOT	Anaheim, CA	20

Source: http://www.itsdocs.fhwa.dot.gov

Adaptive control strategies are powerful tools for arterial traffic management. It is better to implement these strategies than the traditional fixed-time control strategies, primarily in the situations, when it is difficult to predict future traffic accurately.

6.4 AREA-WIDE TRAFFIC CONTROL SYSTEMS

Area-wide control systems consider simultaneously all, or a major fraction, of traffic signals in a considered area (Robertson, 1969; Papageorgiou, 1990). There are offline area-wide control systems, as well as the area-wide traffic responsive control systems (online systems). The area-wide traffic responsive control systems act in response to traffic changes on a system-wide basis. These systems respond very quickly, typically at the next phase of the traffic cycle.

The widely used TRANSYT model is offline system. The TRANSYT calculates the average delays that could be expected from given average flows. The first version of the TRANSYT was developed in 1968 by Robertson of the UK Transport and Road Research Laboratory (TRRL). The latest versions of the TRANSYT are based on the genetic algorithm optimization of cycle length, phasing sequence, splits, and offsets. TRANSYT has two major modules. Within the first module, the traffic model is utilized to calculate the performance index for a specified set of signal timings. The second element represents the optimizing process. This process creates changes to the settings and decides whether they advance the defined performance index (most frequently, the "performance index" represents a weighted combination of the total amount of delay and the total number of stops experienced by vehicles), or not. The TRANSYT model calculates the "performance index" value for the given traffic network, for any fixed-time signal plan and set of average traffic flows. The gained computational experience shows that the genetic algorithm needs longer computer running times when weigh against the other optimization techniques. On the other hand, there is the possibility in genetic algorithm to specify the "number of generations." By specifying a small number of generations, the analyst can obtain a reasonable good timing plan. By specifying a large number of generations, the analyst could obtain optimal or near-optimal solution.

EXAMPLE 6.8

Fig. 6.25 shows the area that should be controlled. The network shown in Fig. 6.25 and the data given in Table 6.4 are modified from the example proposed by Gartner (1983).

The solution obtained by TRANSYT 7F is given in Table 6.5.

EXAMPLE 6.8—cont'd

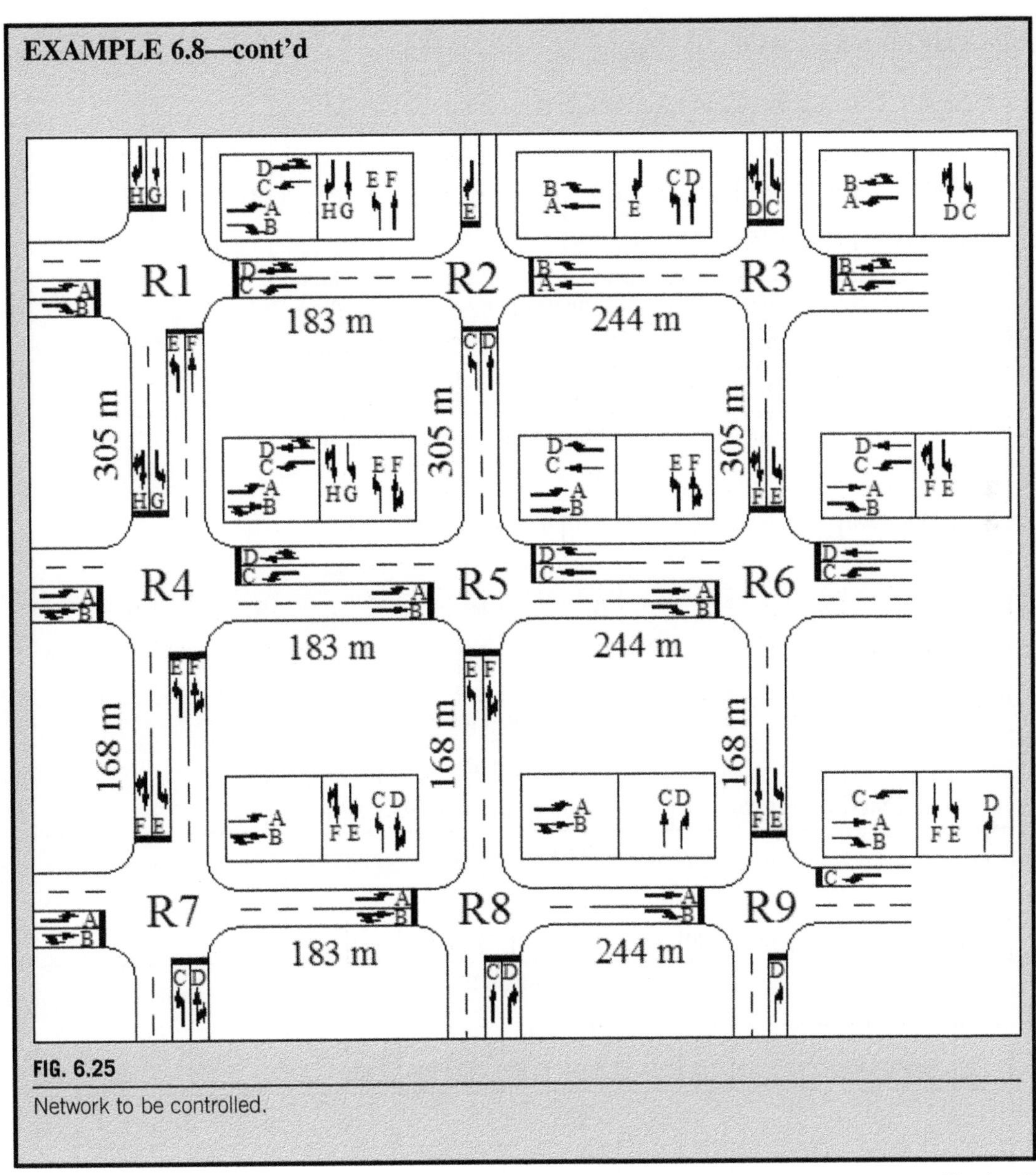

FIG. 6.25

Network to be controlled.

(Continued)

EXAMPLE 6.8—cont'd

Table 6.4 Lane Flows q_k and Lane Saturations Flows s_k (veh/h)

Intersection R_1

k	A		B		C		D		E		F		G		H	
q_k/s_k	170	900	103	1500	177	1125	370	1450	146	900	250	1600	350	1600	280	1500

Intersection R_2

k	A		B		C		D		E	
q_k/s_k	335	1500	212	2120	130	1500	302	2120	82	1500

Intersection R_3

k	A		B		C		D	
q_k/s_k	150	1500	537	1490	145	1500	570	1490

Intersection R_4

k	A		B		C		D		E		F		G		H	
q_k/s_k	108	1005	520	1350	180	900	367	1550	84	1005	454	1330	218	1125	412	1370

Intersection R_5

k	A		B		C		D		E		F	
q_k/s_k	120	900	672	2120	357	1600	190	1500	190	1500	242	1330

Intersection R_6

k	A		B		C		D		E		F	
q_k/s_k	600	1600	192	1500	140	765	415	2120	200	1500	430	1350

Intersection R_7

k	A		B		C		D		E		F	
q_k/s_k	168	1500	425	1140	112	1005	600	1330	250	900	380	1430

Intersection R_8

k	A		B		C		D	
q_k/s_k	172	1500	680	1550	260	2120	200	1500

Intersection R_9

k	A		B		C		D		E		F	
q_k/s_k	600	1600	200	1500	117	765	190	1500	130	1500	500	2120

EXAMPLE 6.8—cont'd

Table 6.5 Solution Obtained by the TRANSYT 7F[a]

Cycle Length = 107 (s)

Green Time (s)

R_1		R_2		R_3		R_4		R_5		R_6		R_7		R_8		R_9	
g_1	g_2	g_1	g_2	g_1	g_2	g_1	g_2	g_1	g_1	g_2	g_1	g_2	g_1	g_2	g_1	g_2	g_1
50	47	54	43	48	49	50	47	65	32	53	44	32	65	73	24	50	37

Offset (s)

Link 1–4	Link 2–1	Link 3–2	Link 3–6	Link 4–1	Link 4–5	Link 4–7	Link 5–2	Link 5–4	Link 5–6	Link 6–5	Link 6–9	Link 7–4	Link 7–8	Link 8–5	Link 8–9
33	34	103	104	74	21	70	19	86	20	87	13	37	81	84	10

[a]*The following GA parameters are used: crossover probability—30%, mutation probability—4%, convergence threshold—0.01%, population size—20, and the maximum number of generations—30.*

The widely used SCOOT model is online. The TRRL in Great Britain developed SCOOT in 1973, and the first SCOOT implementation happened in 1979 in Glasgow.

6.5 TRAFFIC CONTROL SIGNAL NEEDS STUDIES

Traffic engineers try to keep vehicles away from conflicts by disjointing them in time and space. We separate vehicles in space by building interchanges. We use traffic lights to separate conflicting vehicles in time. The following question frequently appears in engineering studies: Whether we should install traffic control signal at a specific location? To properly answer this question, we should perform detailed traffic engineering study of the local traffic conditions, pedestrian characteristics, as well as physical characteristics of the considered location. The Manual on Uniform Traffic Control Devices for Streets and Highways (2009 Edition, US Department of Transportation, Federal Highway Administration) offers the following traffic signal warrants and other factors related to existing operation and safety at the study location:

Warrant 1: Eight-Hour Vehicular Volume.
Warrant 2: Four-Hour Vehicular Volume.
Warrant 3: Peak Hour.
Warrant 4: Pedestrian Volume.
Warrant 5: School Crossing.
Warrant 6: Coordinated Signal System.
Warrant 7: Crash Experience.
Warrant 8: Roadway Network.
Warrant 9: Intersection Near a Grade Crossing.

Warrant 1

The need for a traffic control signal shall be considered if initial study discoveries that minimally prescribed traffic volumes exist for each of any 8 h of an average day on the major street and the minor-street approaches.

Warrant 2

This warrant applies when for each of any 4 h of an average day, the sum of the vehicles per hour on the major street in both directions and the corresponding vehicles per hour on the higher volume minor-street approach all fall above specified values.

Warrant 3

This signal warrant should be used in the cases that include office complexes, manufacturing plants, industrial complexes, high-occupancy vehicle facilities that attract or generate big numbers of vehicles over a short time, etc.

Warrant 4

This warrant applies when the pedestrian volume crossing the major street at an intersection in an average day is higher than the prescribed number. This warrant should not be applied at locations where the distance to the adjacent traffic control signal along the major street is quite small.

Warrant 5

The need for a traffic control signal shall be considered when study shows that the number of satisfactory gaps in the traffic stream during the period when the children (elementary through high school students) are using the crossing is less than the recommended number.

Warrant 6

This warrant applies in order to provide a progressive movement along the arterial streets. The proposed and neighboring traffic control signals will jointly provide better degree of platooning of vehicles.

Warrant 7

The need for a traffic control signal shall be considered if an engineering study finds that specific number of reported crashes have occurred within a 12-month period, each crash characterized by personal injury or substantial property damage.

Warrant 8

This warrant applies when the intersection has a total entering traffic volume higher than the recommended value, and when a major intersection route is an element of the street or highway system that is the main roadway network for through traffic flow, or contains rural or suburban highways outside, entering, or traversing a city.

Warrant 9

This warrant applies when the nearness to the intersection of a grade crossing on an intersection approach controlled by a STOP or YIELD sign is the main reason to consider installing a traffic control signal.

6.6 INTELLIGENT TRANSPORTATION SYSTEMS

Intelligent transportation systems (ITS) use range of technologies to monitor, evaluate and manage transportation systems to improve safety and efficiency. They link transportation infrastructure and vehicles by using information and communications technology (computers, electronics, and sensing technologies).

Intelligent Transportation Society of America (ITS America) was founded in 1991. Today, ITS America includes over 450 public agencies, private sector companies, and academic and research institutions. Thus far, more than thirty user services have been developed in cooperation by US Department of Transportation and ITS America. ERTICO—ITS Europe was also founded in 1991. The ERTICO consists of over a hundred partners that work in various sectors. The ITS Australia was established in 1992, while the Vehicle, Road and Traffic Intelligence Society (VERTIS (2)) was inaugurated in Japan in January, 1994.

The ITS concepts have been applied to different transportation modes. There are numerous successful applications of the ITS in the areas of traffic safety, traveler information, traffic operations, commercial vehicle operations, public transit operations, tolling, congestion pricing, etc. (Whelan, 1995; Chowdhury and Sadek, 2003; Gordon, 2009). The global positioning system (GPS) is a satellite-based navigation system. The GPS is composed of a number of satellites placed into orbit. It is worthy to note that GPS performs in every weather conditions. The GPS could be used everywhere in the world, 24 h a day. Weather information systems monitor the weather information, while Bus Information System provides real-time bus stop information. Electronic toll collection (ETC) significantly decreased queues at toll plazas. These systems enabled vehicles to go all the way through toll gates at traffic speed. ETC systems are also used within congestion pricing schemes. The ITS applications also include pretrip travel information, en-route driver information, route guidance, incident management, etc.

The ITS technologies offered novel services to users. Simultaneously, the ITS technologies enabled transportation services users to be better informed. The gained experience have shown that the ITS technologies have helped in traffic congestion mitigation, number of crashes decrease, and pollution reduction. They also enabled users to make safer and smarter utilization of transportation facilities and services.

The phrase "intelligent transportation system," and especially word "system" within this phrase, means that there is a group of agents that have mutual objective and that they execute tasks according to specific regulations. The word "transportation" is related to the fact that these agents collaborate when moving passengers, or transporting cargo. Finally word "intelligent" assumes that agents have the ability to use available information, to learn, and adapt to new situations.

6.6.1 ITS ARCHITECTURE

The ITS architecture should serve as a framework for the design of ITS. The US national ITS architecture is shown in Fig. 6.26.

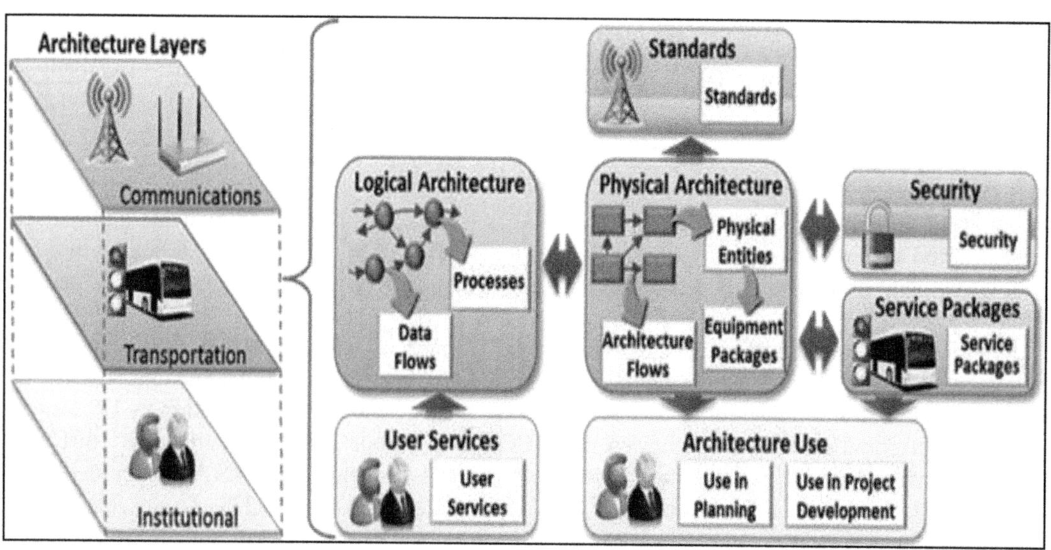

FIG. 6.26

National ITS architecture (http://www.iteris.com/itsarch/html/menu/hypertext.htm).

The shown architecture includes three layers (Fig. 6.26). The institutional layer (ITERIS, 2013) consists of the institutions, policies, funding mechanisms, and processes that are needed for successful implementation, operation, and maintenance of an ITS. The transportation layer defines the transportation services in terms of the subsystems and interfaces. User services explain what the system will perform from the user's perspective. The communications layer describes the communication services needed to maintain these transportation services.

The National ITS architecture is a reference framework for the development of Standards (US DOT, http://www.iteris.com/itsarch/html/menu/hypertext.htm).

6.6.2 ITS USER SERVICES

The following are the user services enabled by the ITS technologies:
Travel and traffic management:

Pretrip travel information
En-route driver information
Route guidance
Ride matching and reservation
Traveler services information
Traffic control
Incident management
Travel demand management

Public transportation management:

Public transportation management
En-route transit information
Personalized public transit
Public travel security

Electronic payment:

Electronic payment services

Commercial vehicle operations:

Commercial vehicle electronic clearance
Automated roadside safety inspection
On-board safety and security monitoring
Commercial vehicle administrative processes
Hazardous materials security and incident response
Freight mobility

Emergency management:

Emergency notification and personal security
Emergency vehicle management
Disaster response and evacuation

Advanced vehicle safety systems:

Longitudinal collision avoidance
Lateral collision avoidance
Intersection collision avoidance
Vision enhancement for crash avoidance
Safety readiness
Precrash restraint deployment
Automated vehicle operation

Information management:

Archived data

Maintenance and construction management:

Maintenance and construction operations

Due to the limited space, we do not describe in detail all developed ITS services. In order to illustrate the ITS technology achievements we just offer the description of a few service packages (US DOT, http://www.iteris.com/itsarch/html/mp/mpindex.htm):

Transit vehicle tracking

> This service package monitors current transit vehicle location using an Automated Vehicle Location System. The location data may be used to determine real time schedule adherence and update the transit system's schedule in real-time. Vehicle position may be determined either by the vehicle (eg, through GPS) and relayed to the infrastructure or may be determined directly by the communications infrastructure. A two-way wireless communication link with the Transit Management Subsystem is used for relaying vehicle position and control measures.

Demand response transit operations

> This service package performs automated dispatch and system monitoring for demand responsive transit services. This package monitors the current status of the transit fleet and supports allocation of these fleet resources to service incoming requests for transit service while also considering traffic conditions. The Transit Management Subsystem provides the necessary data processing and information display to assist the transit operator in making optimal use of the transit fleet.

Transit signal priority

> This service package determines the need for transit priority on routes and at certain intersections and requests transit vehicle priority at these locations. The signal priority may result from limited local coordination between the transit vehicle and the individual intersection for signal priority or may result from coordination between transit management and traffic management centers. Coordination between traffic and transit management is intended to improve on-time performance of the transit system to the extent that this can be accommodated without degrading overall performance of the traffic network.

Dynamic route guidance

> This service package offers advanced route planning and guidance that is responsive to current conditions. The package combines the autonomous route guidance user equipment with a digital receiver capable of receiving real-time traffic, transit, and road condition information, which is considered by the user equipment in provision of route guidance.

Dynamic ridesharing

This service package provides dynamic ridesharing/ride matching services to travelers. This service could allow near real time ridesharing reservations to be made through the same basic user equipment used for Interactive Traveler Information. This ridesharing/ride matching capability also includes arranging connections to transit or other multimodal services.

HOV lane management

This service package manages HOV lanes by coordinating freeway ramp meters and connector signals with HOV lane usage signals. Preferential treatment is given to HOV lanes using special bypasses, reserved lanes, and exclusive rights-of-way that may vary by time of day. Vehicle occupancy detectors may be installed to verify HOV compliance and to notify enforcement agencies of violations.

6.6.3 AUTONOMOUS VEHICLES

Technological revolution in transportation happened at the beginning of the 20th century. Carriages drawn by horses were gradually replaced by the first cars. At the same time, people began to fly planes. It seems that we are nowadays on the beginning of a new technological revolution in transportation. Petrol cars began to receive significant competition in electric cars, hydrogen vehicles, and hybrid cars. Autonomous vehicles are the most important achievement of the new technological revolution. The DARPA Grand Challenge (a competition held by the United States Defense Advanced Research Projects Agency) initiated the research and development of the fully autonomous ground vehicles. The first Grand Challenge was held in 2004. Vehicles that were in competition navigated around 150 miles in the Mojave Desert area in California. The third Grand Challenge happened in an urban environment with real-life conditions such as other vehicles on the road. All competitors were obliged to follow all driving laws.

Today, all leading car manufacturers are developing prototypes of autonomous vehicles. Autonomous cars (self-driving cars, driverless cars, robotic cars) are already around us. Autonomous cars are capable of performing the major transportation abilities of a conventional car. Self-driving cars are equipped with GPS combined with tachometers, altimeters, and gyroscopes. Cars also have radar sensors and ultrasonic sensors that can measure position of objects very close to the vehicle (parked vehicles and curbs). Lidar (light detection and ranging) helps in identifying lane markings and road edges. Self-driving cars have video cameras that detect traffic lights and read road signs. These cameras also look out for other vehicles, pedestrians, and road obstacles. Car central computer analyses information from all sensors. The implemented software takes care about steering, acceleration and brakes.

Self-driving cars have the capability to perceive their neighboring environment. Autonomous cars can navigate themselves and perform path planning without human assistance. They are equipped with various sensors, actuators, and computers. The software powering Google's cars is called *Google Chauffeur*. Washington, District of Columbia, Nevada, Florida California, Michigan and Idaho recently allowed testing of driverless cars in public roads. In 2014, Google presented a new prototype of their self-driving car without a steering wheel and pedals.

It is expected that autonomous vehicles can significantly contribute to reduction of the number and consequences of traffic accidents. At the same time, there is also an expectation that self-driving vehicles could reduce fuel consumption and the negative effects on the environment. Reduction in average travel time, as well as reducing the space required for parking also represent potential benefits of using self-driving vehicles. The use of autonomous vehicles in taxi services, dial-a-ride systems, and distribution systems are also explored.

Increasing the reaction time of self-driving cars is one of the major problems that should be solved in the future. There is also a great lack of public confidence in autonomous vehicles.

The development of autonomous vehicle opens up a whole range of technical, legal, and ethical issues. How fast the autonomous vehicles will became part of everyday life? What will be the most appropriate traffic control in the case of traffic flows dominated by autonomous vehicles? Whether the usual types of intersections will slowly disappear? Will the number of traffic accidents be significantly reduced? Who will be responsible in the case of a traffic accident? Will the drivers who spend a great deal of time in traffic congestion get a more free time? Ii is not easy to find the answers today on these and similar questions.

The development of the fully autonomous vehicle requests the new approaches to the intersection control.

6.6.4 AUTONOMOUS INTERSECTION MANAGEMENT

Traffic signals and stop signs are used to control traffic at many intersections in the world. Regardless of the continuous improvement of methods for the traffic management at intersections, the large number of intersections in the world is characterized by significant vehicle delays, and a large number of stopped vehicles. At the same time, a very large number of traffic accidents happen in the intersection areas. Researchers from the University of Texas at Austin, leaded by Peter Stone are today the pioneers in the development of autonomous intersection management (AIM), an intersection control protocol designed for autonomous vehicles. Preliminary research on simulation models, as well as field experiments with autonomous vehicles have shown, that in the case of AIM, a large number of vehicles could pass through the intersection without stopping. Therefore, preliminary studies indicate that aeronautical information services (AIS) can become traffic control for the future that will enable the reduction level of traffic congestion, decrease fuel consumption and vehicle emissions.

Let us explain briefly how AIS works. AIM is based on the computer programs called *driver agents* that control the vehicles. Simultaneously an agent called an *intersection manager* is located at each intersection.

The vehicle approaches to an intersection. The vehicle wants to turn left, and requests from the server permission to enter the intersection (Fig. 6.27).

In its request, the vehicle (driver agent) specifies its size, estimated arrival time, velocity, acceleration, and arrival and departure lanes. The server (intersection agent) simulates the trajectory of the vehicle through the intersection, and assesses whether there are potential vehicles collisions with other vehicles. If there are no collisions, the server sends a confirmation to the vehicle, so the vehicle can enter the intersection. The vehicle is responsible to arrive at, and move through the intersection as specified.

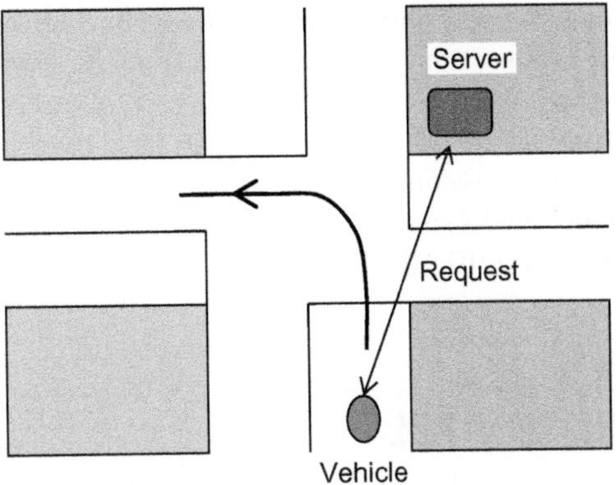

FIG. 6.27

Vehicle requests from the server permission to enter the intersection.

Let us now consider the situation when two vehicles are approaching intersection on a collision course. Let us also assume that a straight-moving vehicle first sent the request to enter the intersection (Fig. 6.28).

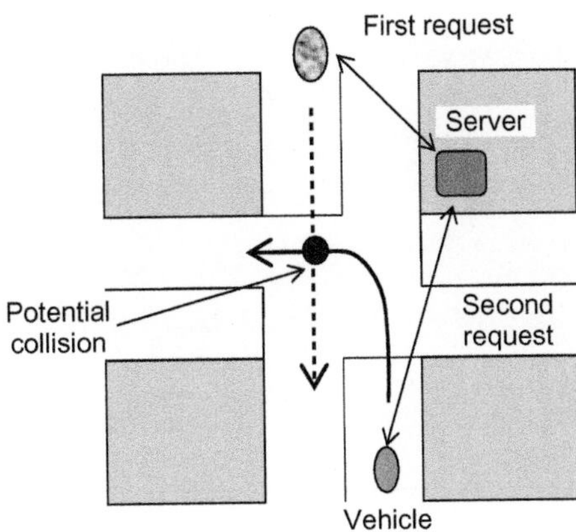

FIG. 6.28

Potential collision at the intersection.

The request of the vehicle that wants to turn left will be rejected by the server due to the potential collision at the intersection. The intersection manager suggests to the left-turning vehicle to make later reservation. Left-turning vehicle must slow down and send a new request to enter the intersection. If there is no potential collision this vehicle also safely passes through the intersection at a later time.

No one car can enter the intersection without approved reservation. Every car after departing from the intersection reports to the intersection manager that its move through the intersection is finished. The performed numerical experiments have shown that the application of AIM can significantly reduce vehicle delay at the intersections. It is also shown that if any message between vehicles and intersection manager is dropped that safety is not compromised, but it can only be a small increase in vehicle delay. In other words, it has been shown that AIM is robust to these types of disturbances.

Researchers from the University of Texas at Austin also recently proposed the SemiAIM, which allows human-driven cars and semiautonomous cars, to also make reservations and enter an intersection. The performed experiments shows significant decrease in vehicle delays with an increase of percentage of autonomous vehicles in traffic flow.

6.7 FREEWAY TRAFFIC CONTROL

Freeways are characterized by the complete control of access. All freeways have at least two lanes for the traffic in each direction. Opposite directions of travel are separated by traffic barriers. There are no intersections, stop signs, or traffic signals that regulate traffic flows on freeways. The flows on freeways are regulated only by vehicle to vehicle interactions. In other words, traffic flows on freeways are uninterrupted flows. Overpasses and underpasses across the freeway help to solve potential conflicts with other roads or railways. The congestion appears on freeways due to various reasons such as meteorological conditions, high density values, traffic accidents, etc. (Fig. 6.29).

FIG. 6.29

Freeway.

The number of trips by private cars has considerably increased in latest decades on many freeways. At the same time, road network capacities have not kept up with this raise in travel demand. Freeways and traffic networks in many countries are severely congested, resulting in increased travel times, increased number of stops, unexpected delays, greater travel costs, inconvenience to drivers and passengers, increased air pollution and noise level, and increased number of traffic accidents. Recurrent congestion occurs daily during rush hours; nonrecurrent congestion is caused by traffic incidents.

Freeway congestion appears in a situation when demand exceeds the capacity. The congestion could be mitigated by increasing the freeway capacity or by decreasing the travel demand. Expanding traffic network capacities by building more roads is extremely costly as well as environmentally damaging. Various freeway traffic control techniques can reduce traffic congestion problems caused by traffic incidents, various bottlenecks, and very high traffic demand.

6.7.1 FREEWAY TRAFFIC CONTROL MEASURES

The following are usual freeway traffic control measures:

- Ramp metering
- Link control
- Driver information and guidance systems
- Congestion pricing

Ramp metering is based on installed traffic signals at freeway on-ramps. Link control is performed through variable speed limits, keep-lane instructions, warnings, etc. Driver information and guidance systems assumes usage of radio services, television, internet, and road side variable message signs (VMS) with an idea to provide real-time information to drivers.

6.7.2 RAMP METERING

One of the most important freeway traffic control techniques is ramp metering (Papageorgiou et al., 1990; Haj-Salem and Papageorgiou, 1995; Kotsialos and Papageorgiou, 2001; Papageorgiou and Kotsialos, 2002). Ramp metering is based on traffic signals at freeway on-ramps that control the flow rate of vehicles entering the freeway (Fig. 6.30).

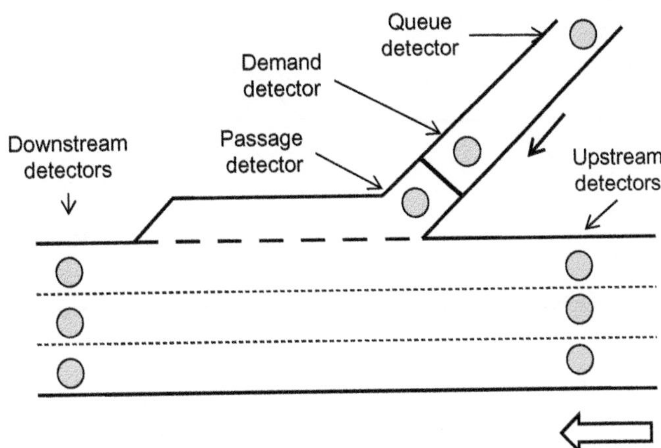

FIG. 6.30

Ramp metering.

It has been shown that various ramp metering strategies could significantly improve freeway traffic operations. The properly implemented ramp metering strategies could increase mainline throughput, reduce traffic accidents caused by merging conflicts, and enable more efficient incident response. The ramp metering rate could be fixed, or it could be based on current traffic conditions in the ramp area.

Fixed metering rates are calculated using historical traffic data. Ramp metering strategy could be local, or coordinated. Coordinated ramp metering strategy assumes coordination between adjacent freeway ramps. In other words, there are three categories of the ramp metering systems:

- Fixed time ramp metering operations
- Local traffic responsive ramp metering operations
- System-wide traffic responsive ramp metering operations

In the case of fixed time ramp metering operations, vehicles enter the freeway one at a time. The metering rate is fixed (obtained after statistical analysis of historical data). This fixed rate is calculated offline without taking into account any real-time measurements. In other words, fixed ramp metering strategy does not respond to traffic variations. This could cause in some cases significant ramp queuing and delays. Local traffic responsive ramp metering strategies are based on current traffic conditions in the ramp area. There are various algorithms that chose an appropriate metering rate by analyzing the traffic flow data from the ramp and freeway detectors (upstream detectors, downstream detectors, passage detector, presence detector, queue detector).

Local traffic responsive ramp metering systems are more costly than the systems based on fixed metering rates. On the other hand, these systems produce shorter delays and queues. The best known local traffic responsive ramp metering strategy is the demand-capacity strategy (Fig. 6.31) and the ALINEA (Papageorgiou et al., 1991, 1997; Fig. 6.32).

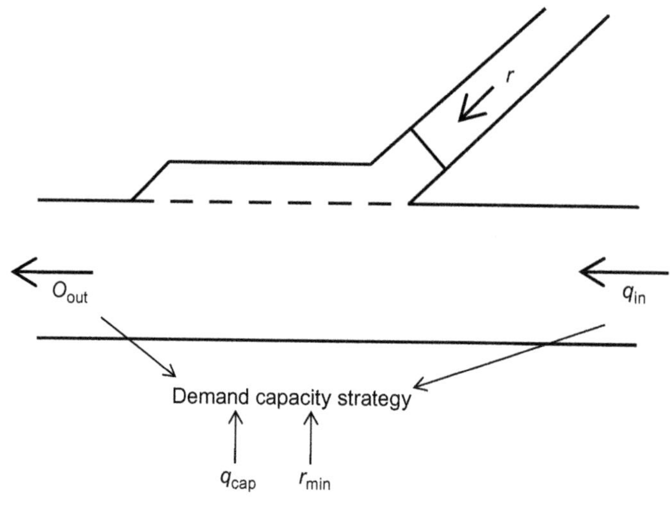

FIG. 6.31

Demand capacity strategy.

A discrete-time representation of traffic variables with discrete time index $k = 1, 2, \ldots$ and time interval T is usually used in ramp metering models. We denote by $q(k)$ the traffic flow (veh/h) obtained as the ratio between the number of vehicles crossing specified location during the time interval

$[k \cdot T, (k+1) \cdot T]$. In a similar way, we denote by $r(k)$ the ramp flow value during the time interval $[k \cdot T, (k+1) \cdot T]]$.

The demand-capacity strategy could be defined in the following way:

$$r(k) = \begin{cases} q_{\text{cap}} - q_{\text{in}}(k-1) & \text{if } o_{\text{out}}(k) \leq o_{\text{cr}} \\ r_{\min} & \text{otherwise} \end{cases} \quad (6.41)$$

where

q_{cap} is the freeway capacity downstream of the ramp
q_{in} is the freeway flow measurement upstream of the ramp
o_{out} is the freeway occupancy measurement downstream of the ramp
o_{cr} is the critical occupancy (at this occupancy the freeway flow is maximal)
$r_{\min}$ is the prespecified minimum ramp flow value

Demand-capacity strategy tries to add to the measured upstream flow q_{in} as much ramp flow as possible to achieve the downstream freeway capacity q_{cap}. When measurements of the downstream occupancy show that occupancy becomes critical (at this occupancy the freeway flow is maximal, and traffic congestion can start), ramp flow $r(k)$ is cut down to the minimum flow to escape potential congestion (Fig. 6.32).

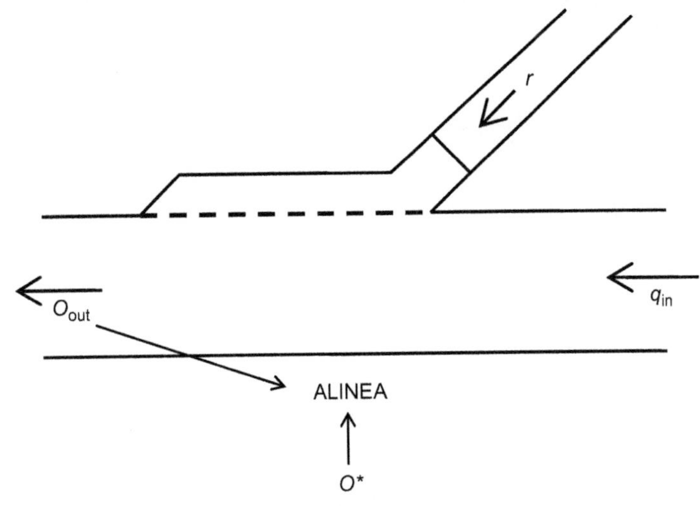

FIG. 6.32

ALINEA.

The ALINEA ramp metering strategy is defined in the following way:

$$r(k) = r(k-1) + K_{\text{R}} \cdot (o^* - o_{\text{out}}(k)) \quad (6.42)$$

where $K_{\text{R}} > 0$ is a regulator parameter and o^* is a desired value for the downstream occupancy.

It has been shown in field experiments that ALINEA has not been very sensitive to the value of the regulator parameter K_{R}. From equations (6.41) and (6.42), one can conclude the following.

The demand-capacity strategy acts to extreme occupancies only after a threshold value O^* is exceeded. The reaction is very crude, meaning that the ramp flow value is suddenly decreased to the prespecified minimum ramp flow value r_{min}. On the other hand, ALINEA reacts smoothly even to small differences between desired value for the downstream occupancy o^* and the freeway occupancy measurement downstream of the ramp O_{out}. In this way, ALINEA prevents potential congestion securing at the same time high values of traffic flow.

The ramp volume r is converted into a green-phase duration in the following way:

$$g = \frac{r}{r_{sat}} \cdot c \quad (6.43)$$

where

g is the green time duration
c is the cycle length
r is the ramp volume
r_{sat} is the ramp saturation flow

Many ramps operate on the principle "one-car-per-green realization." This means that the ramp volume in such cases is controlled by changing the red-phase duration. In the case of system wide traffic responsive ramp metering operations, a traffic control center equipped with centralized computer overlooks few ramps (Fig. 6.33).

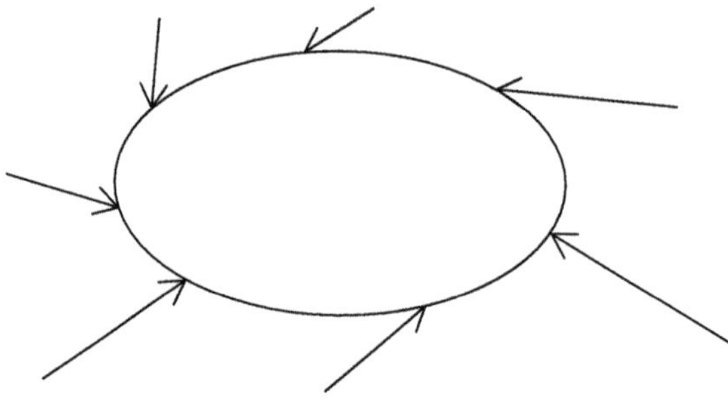

FIG. 6.33

System wide traffic responsive ramp metering.

The metering rate at any ramp depends on local traffic conditions, as well as on traffic conditions at other locations within the controlled highway section. Comprehensive ramp metering could provide total control of the network traffic flows and could contribute to the better utilization of the overall highway infrastructure. System wide traffic responsive ramp metering systems are especially useful in the cases of freeway incidents.

6.7.3 DRIVER INFORMATION AND GUIDANCE SYSTEMS

The advanced traveler information systems have the task to inform the drivers about the expected travel times, traffic accidents, as well as the expected congestion on the freeway and possible alternative routes (Smulders, 1990; Hall, 1993; Mammar et al., 1996). In other words, by providing appropriate information to the driver we attempt to influence drivers' route choice. Modern traveler information systems try to direct drivers to less congested alternative routes, and in this way to relieve congestion in the part of traffic network (Fig. 6.34).

FIG. 6.34

Traveler information system.

Help in trip planning is usually based on various telephone and web-based traffic information. VMS provide traffic information during the trip. Real-time traffic measurements, traffic accident information, as well as information on general traffic conditions are today available via smart phones.

6.8 TRANSPORTATION DEMAND MANAGEMENT

Transportation congestion in cities, on highways, and at the airports could be attacked by supply-side or demand-side strategies. Building more roads is the main supply-side strategy. The Braess paradox, which can happen in some networks, is the best example that expanding transportation capacities is not always the best solution for traffic congestion problems.

The transportation strategies that try to decrease the demand for existing transportation systems represent an alternative to strategies that promote building and expanding transportation infrastructure. Planners, engineers, and economists have introduced various demand management methods in an attempt to reduce the fast growing traffic congestion (park-and-ride, high-occupancy vehicle (HOV) lanes, high-occupancy toll (HOT) lanes, ridesharing and transit use, road pricing, parking pricing). *Transportation demand management* (TDM), is a common term for various activities that advocate a decrease in the demand for existing transportation systems. Demand management actions also force transportation network users to travel and use transportation facilities more during off-peak hours. Some of the demand management strategies could advance the transportation choices accessible to users. Some other demand management strategies generate alterations in departure time, route choice, destination or mode choice. Successfully planned and implemented demand management strategies can result in significant toll revenues, decrease in total number of vehicle trips, decrease in total number of vehicle trips during peak periods, increase in number of vehicle trips during off-peak periods, increase in ridesharing, rise in public transit ridership, and in some cases increased cycling, walking, and telework. Some of the demand management strategies have been already successfully implemented. The following are the best known demand management strategies (majority of them have not been widely implemented).

6.8.1 RIDESHARING (CARPOOLING)

Ride sharing is one of the most extensively used TDM techniques that assumes the involvement of two or more travelers that together share a vehicle when traveling from few origins to few destinations (Teodorović and Dell'Orco, 2008; Šelmić et al., 2011). The benefit of ride sharing is evident: ride sharing considerably decreases the total number of trips. By sharing the ride with just one other commuter, one can reduce commuting everyday expenditure by 50%. Simultaneously, it is possible, while ride sharing, to use HOV lanes, to develop social life, and even create new friendships. Participants in ride sharing decide by themselves about various ride sharing operational issues (vehicle schedule, pick-up and drop-off points, maximum waiting time, music playing, smoking policy) (Fig. 6.35).

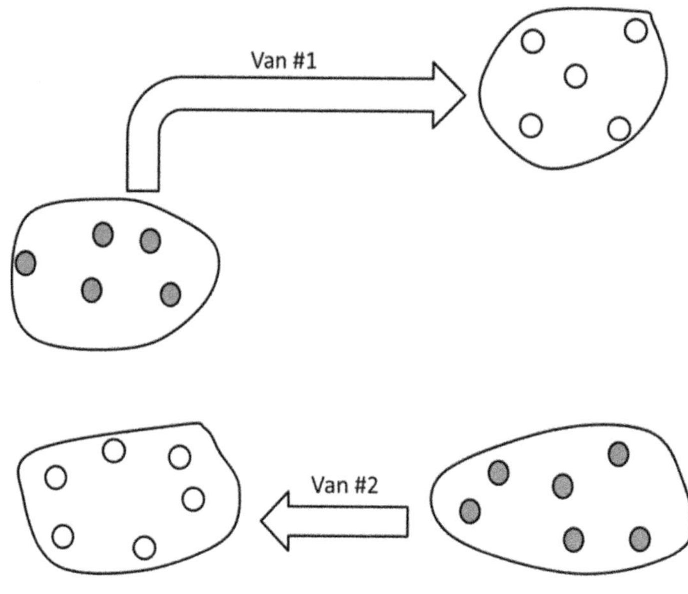

FIG. 6.35

Ride sharing.

Ride share programs use a wide range of traveler databases to match commuters who live and/or work in close proximity to each other for carpools and vanpools. Depending on the number of commuters in the group, the carpool or the vanpool will be proposed and formed. Carpooling is a widespread type of ride sharing. The participants in carpooling are neighbors who work at different companies located only a short distance away from each other, who also have similar work hours. The participants are frequently also staff of a single company who lives next to each other. In some cases, the same traveler drives all the time, while the other commuters participate in sharing the cost. In some other cases, travelers alternate in driving. Vanpooling is also a well-known type of ride sharing.

A vanpool is usually composed of 5–15 commuters. Vans are leased or purchased by individuals that participate in vanpooling, by third party, or by employer, or a group of employers. The vanpool participants define the vanpool schedule and route. Most frequently, ride share programs put a new commuter into one of the vacant vanpools. The fares are based on the van type, and the mileage traveled.

Ridesharing is very effective demand management strategy, especially in regions that are not well served by public transportation.

6.8.2 REMOTE PARKING AND PARK AND RIDE

Remote Parking encourages drivers to use parking facilities located a few blocks from a down town. Drivers continue their trips from the parking by public transportation.

6.8.3 IMPROVED WALKABILITY

Improving sidewalks, crosswalks and paths, and creating new pedestrian shortcuts could lower driving.

6.8.4 TELEWORK

Telework describes working activities of employees who could work from home or another location in order to decrease commute travel. Distance learning, tele-shopping, tele-banking, electronic government, internet business and other telecommunications based (email, websites, telephone, fax, video connections, etc.) activities significantly decrease demand for existing transportation systems.

6.8.5 CONGESTION PRICING

The high traffic congestion level in many cities in the world could be explained by the phenomenon called the "*tragedy of commons.*" Tragedy of commons refers to situations where a free public good is available to the general public, where each user tries to maximize her/his own utilities, by continually incrementally increasing his/her use of that good. Finally, the public good become overloaded and of reduced, or no value to any user. The term "*tragedy of the commons*" was introduced by ecologist Garrett Hardin (1968). The resources could be atmosphere, oceans, rivers, fish stocks, etc. Hardin's (1968) research has its roots in the two lectures on population (1833) of English mathematician and economist *William Forster Lloyd*. Lloyd (1833) offered an example of herders sharing a common parcel of land on which they are each entitled to let their cows graze. He pointed out that with each additional sheep herder get benefits, while the commons is damaged.

This is exactly the situation with the use of roads and streets in many cities in the world. Length of roads and streets in each city is limited. The use of roads and streets is open to any user. When considering travel costs, drivers usually perceive only their own costs (fuel, time cost, etc.). Drivers are not able to look at wider social cost (costs of traffic congestion), resulting from the individual decisions. Drivers are usually not paying for these social costs. In other words, the drivers are not seeing enough the fact that the decision, to travel to down-town, as a solo driver in private car, contribute to the increase of traffic congestion level. When there are long lines of cars moving slowly, and with many stops and starts (bumper-to bumper traffic), the free public good (roads and streets) has a reduced value to all drivers.

Traffic planners, engineers, economists, and city authorities have extensively accepted the concept of "*congestion pricing*" in an effort to relieve the congestion during peak periods. The economist Pigou (1920) argued for a tax on congestion in his book *The Economics of Welfare*. Congestion pricing practically represents basic economic concepts of supply and demand. *William Vickrey*, winner of the Nobel Prize for Economics, is considered among researchers as the "father" of modern congestion pricing concepts. Congestion pricing already exists in many different sectors. Airlines have been offered off-peak discounts for many years. Usually, in many countries, hotel rooms cost more for the duration of peak tourist seasons. Telephone companies have been applying congestion pricing concepts for many years.

In some countries road-use charges vary with the level of congestion (Singapore, United States, Europe). Thus, it is possible to reallocate some trips to off-peak times, less-congested routes, or other transportation. The basic idea of congestion pricing (value pricing) is to charge road users with diverse fees through different traffic conditions. Different fees or tolls that change with a location in the network, time of day and/or level of traffic congestion have been proposed (Vickrey, 1955, 1963, 1969, 1994; Verhoef et al., 1995, 2002a,b; Sullivan and El Harake, 1998; Yang and Huang, 1998, 1999; De Palma and Lindsey, 2002; Teodorović and Lučić, 2006; Teodorović and Edara, 2007; Larson and Sasanuma, 2010). In other words, drivers should pay for exploiting particular road, corridor, bridge, tunnel, or for entering specific area in some time periods (area-based charging). The simple idea behind the concept of congestion pricing is to push drivers to travel and use transportation facilities more during off-peak hours and less during peak hours, as well as to intensify the use of underutilized routes (Fig. 6.36).

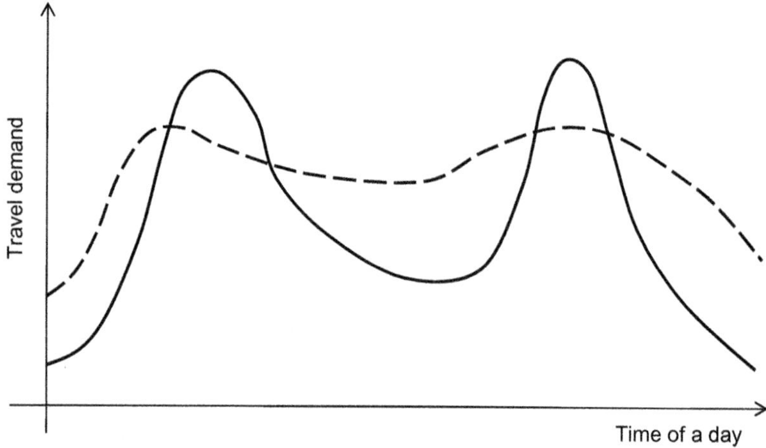

FIG. 6.36

More evenly distributed passenger demand.

Congestion pricing projects that are well planned and successfully implemented could result in significant toll revenues, decreased total number of vehicle trips, decreased total number of vehicle trips during peak periods, increased number of vehicle trips during off-peak periods, increase in ridesharing, greater number of passengers in public transit, and in some cases increased cycling and walking (Fig. 6.37).

The fist congestion pricing scheme in the world was applied in the city of Singapore in 1975. The traffic authorities introduced, at that time, a paper license scheme, that was replaced by an ETC system in 1998. The Singapore's congestion pricing scheme decreased the total peak-period traffic each working day by 45%.

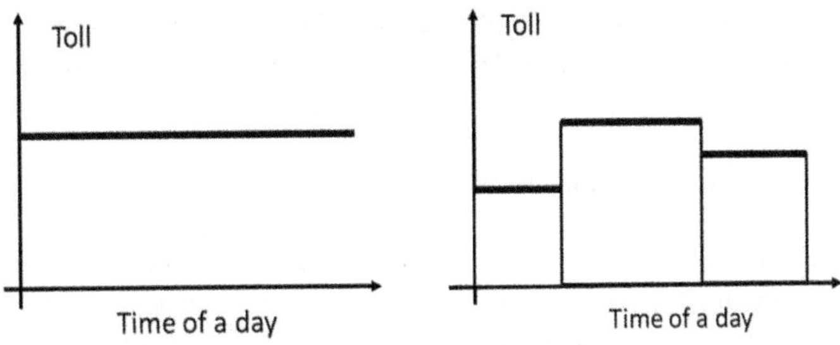

FIG. 6.37

Toll as a function of a time of a day.

Congestion pricing (cordon-tolls) started in London on February 17, 2003 (Litman, 2006). The London's authorities charged 5 UK pounds every vehicle per day. All drivers were obliged to pay the charge if they entered a defined congestion-charging zone between 7 am and 6:30 pm weekdays. By using more than 600 cameras, the traffic authorities were capable to capture the license plate number of every entering vehicle. The total number of vehicles entering central London through charging hours failed approximately 20–30%. There are also significant decreases in vehicle delays, as well as in carbon dioxide emissions. The London's authorities spent congestion pricing net revenue in improving bus services. As a consequence, bus passengers entering the charging zone in morning rush hour during the first year raised about 35%.

Stockholm introduced congestion pricing scheme on January 3, 2006. After the initial 6 months work of the system, the referendum was organized, and citizens voted on whether to continue with a congestion pricing scheme. The majority of citizens voted "Yes" at the referendum. After introducing the congestion pricing scheme, the total traffic volumes were reduced 25%. Simultaneously, the public transit ridership is increased by 40,000 per day.

In addition to road pricing, the congestion pricing can be applied also as a parking pricing. Japan has significantly changed the Road Traffic Law in 2006, by practically removing free parking on the streets. Authorities started to apply high penalty fees to parking violators, resulting in significant decrease of traffic jams duration, as well as decrease in average travel times.

6.8.6 CONGESTION CHARGES

Drivers and passengers recognize only their own travel time costs. Correspondingly, bus operators and airlines perceive, above all, their own operating costs. They do not take into account the additional delay their trips impose on other passengers. Noticing these facts, Nobel Prize laureate William Vickrey (1955, 1963, 1969) claimed that "Charges should reflect as closely as possible the marginal social cost of each trip in terms of the impacts on others. There is no excuse for charges below marginal social cost." In other words, it appears very rational that users of a transportation infrastructure should pay a price equal to the delay cost they impose on others (Andreatta and Odoni, 2003). In this way, travel patterns could probably be changed. The probable consequence could be efficiency improvement of transportation capacity utilization, and probably the decrease of upcoming transportation capacity requirements. There are many opponents to the congestion pricing idea. It appears that there are no simple solutions of traffic congestion problems.

Let us denote by c the delay cost per unit time per client. The total cost of delay per unit time equals

$$C = c \cdot L_q \tag{6.44}$$

where L_q is the average number of clients in the queue.

We conclude, based on the *Little's Law* that the total cost of delay per unit time equals

$$C = c \cdot \lambda \cdot W_q \tag{6.45}$$

where λ is the average arrival rate and W_q is the average waiting time a client spends in the queue.

More vehicles on a highway, higher is the congestion level, and higher are vehicle delay costs. The more aircraft is in the air, the higher is aircraft waiting time for landing, and the higher are aircraft operating costs. Marginal cost indicates the additional increase in queueing cost with every additional customer. Marginal delay cost *MC* imposed by an additional client equals (Andreatta and Odoni, 2003):

$$MC = \frac{dC}{d\lambda} \tag{6.46}$$

ie,

$$MC = c \cdot W_q + c \cdot \lambda \cdot \frac{dW_q}{d\lambda} \tag{6.47}$$

The last equation indicates that the marginal delay cost *MC* has two terms. The term $c \cdot W_q$ represents the internal cost experienced by the additional customer. The second term $c \cdot \lambda \cdot \frac{dW_q}{d\lambda}$ represents the "external" cost imposed. The majority of customers in queueing systems in transportation do not realize the existence of these "external" costs. In other words, when we enter the congested highway, we are exposed to the significant delay and increase in travel cost, but, at the same time, we must realize that, by our entering the congested highway, we further increase delay costs of all other drivers and passengers. Aircrafts trying to land during the high airport congestion are sent to the holding pattern. In this way, their operating costs are increased. At the same time, these aircraft also increase operating costs to all other aircrafts that try to land.

The marginal delay cost in the case of *M/M/*1 queueing system equals

$$MC = c \cdot W_q + c \cdot \lambda \cdot \frac{dW_q}{d\lambda} \tag{6.48}$$

In the case of *M/M/*1 queueing system, the average waiting time W_q equals

$$W_q = \frac{\lambda}{\mu \cdot (\mu - \lambda)} \tag{6.49}$$

Marginal delay cost *MC* equals

$$MC = c \cdot \frac{\lambda}{\mu \cdot (\mu - \lambda)} + c \cdot \lambda \cdot \frac{d\left[\frac{\lambda}{\mu \cdot (\mu - \lambda)}\right]}{d\lambda} \tag{6.50}$$

$$MC = \frac{c \cdot \lambda \cdot (2 \cdot \mu - \lambda)}{\mu \cdot (\mu - \lambda)^2} \tag{6.51}$$

EXAMPLE 6.9

There are two alternative routes between point A and point B. The work zone is located on the shorter route (Fig. 6.38). Because of the work zone, this route is characterized by the high congestion level.

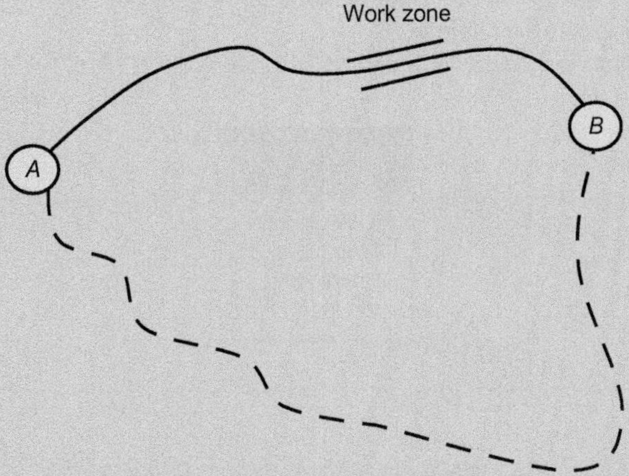

FIG. 6.38

Congestion in the work zone neighborhood.

Traffic engineers concluded that the route with the work zone should be treated as $M/M/1$ queuing system. The average arrival rate equals: 150 (veh/h), ie, $\lambda = \frac{150 \text{ vehicles}}{3600 \text{s}} = 0.0417 \,(\text{veh/s})$. The average service rate equals: $\mu = \frac{1}{20} = 0.05 \,(\text{veh/s})$. In order to decrease congestion level, the traffic authorities can introduce congestion pricing. Calculate congestion price (marginal delay cost) that should be charged to drivers that use congested route.

Solution

Marginal delay cost MC imposed by an additional vehicle equals

$$MC = \frac{c \cdot \lambda \cdot (2 \cdot \mu - \lambda)}{\mu \cdot (\mu - \lambda)^2} = \frac{0.0417 \cdot (2 \cdot 0.05 - 0.0417)}{0.05 \cdot (0.05 - 0.0417)^2} \cdot c = 706 \cdot c$$

$$MC = 706 \cdot c$$

Table 6.6 shows congestion price MC (\$) (marginal delay cost) as a function of the delay cost per unit time per driver c.

Table 6.6 Congestion Price *MC* (\$) (Marginal Delay Cost) as a Function of the Delay Cost Per Unit Time Per Driver *c*

c (\$/h)	*c* (\$/s)	*MC* (\$)
18	0.005	3.53
15	0.00417	2.942
12	0.00333	2.353
9	0.0025	1.765

6.9 HOV FACILITIES

By congestion pricing concepts we practically change freeway operational capacity. The freeway capacity could be also changed by dynamically changing speed limits over freeway sections. Introducing limitations to the vehicle types permitted to use specific freeway lanes is one of the widely accepted and implemented congestion pricing techniques.

HOV facilities exist in many metropolitan areas in the world (Fig. 6.39).

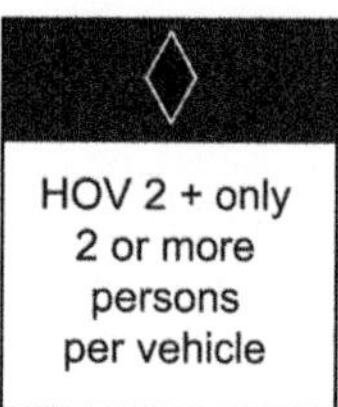

FIG. 6.39

HOV traffic sign.

The basic idea behind the HOV facilities is to increase the *number of people* traveled along the freeway. The existence of the HOV lanes promote shared rides over solo driving concept. HOV lanes could be used exclusively by the HOVs. The minimum number of passengers in the vehicle could be two, three, or four. These lanes could be also used by buses.

The HOV lanes are transformed on some freeways into HOT lanes. There are also completely new HOT lanes on many freeways. The HOT lanes could be used by all vehicles that meet prescribed occupancy requests for an HOV lane, as well as by all other vehicles if they pay a toll. The gained experience showed that HOT lanes increased the capacity utilization on some formerly underutilized HOV lanes. The HOT lanes significantly help in decreasing congestion on the regular freeway lanes.

6.10 HIGHWAY SPACE INVENTORY CONTROL SYSTEM

Over the last decade, researchers have begun to consider the idea of *highway booking* concept. The essential idea of the highway booking concept is that all road users, at some roads, have to book *in advance*. Airline industry, hotels, car rental, rail, and many other industries are already using reservation systems and revenue management concepts when selling their products. It is usual in many industries (airlines, hotels) to use the term *inventory* to represent aircraft seats, hotel rooms, etc. These inventories are specific, in a sense that they are *perishable*. The highway inventories are also perishable. Highway spaces not utilized through some time interval are lost. Highways also have limited resources. The existing highway capacity is limited and cannot be increased in a small time period. There is also market segmentation associated with the highway usage. The following are some of the possible categories of potential highway users: (a) single-occupant cars; (b) transit vehicles; (c) carpools; (d) trucks; and (e) low emission vehicles (LEV).

The highway operator could offer to prospective customers through the same time interval a great number of different tariffs (or assign different priorities) based on a set of defined criteria

(Edara and Teodorović, 2008; Edara et al., 2011). This is due to the highway operator's potential wish to reach the highest possible number of passenger miles along the highway, to spread peak over time and space, to accept the greatest possible number of LEV and/or vehicles equipped with ETC units, to reach the highest possible total revenue, etc. Highway operators should offer various options to the highway clients, for example, users willing to pay only low tariffs should be permitted to reserve the highway space well in advance. Highway users paying higher tariff should receive certain benefits like open date and/or time for come back, various cancelation opportunities, or possibility for planned trip change. On the other hand, users paying lower tariff should accept, for example, to make advanced booking, to pay penalty in the case of cancelation, etc. In this way, the highway operator would be able to minimize the unused highway space.

Reservation requests can be made via telephones or electronically. Driver should specify entry time (within time interval whose width could be 5–10 min) on ramp, trip, and off ramp and inform the highway operator (by telephone or electronically). Requests would be usually made several days, hours, or minutes (last-minute request) before the planned trip. In this way, highway operator would know planned driver itineraries, as well as planned departure times.

6.11 AUCTIONS

Auctions also represent one of the possible demand-side strategies for attacking traffic congestion. Auctions have been applied in air transportation, but it is possible to assume that, in years to come, auctions will be used in other transportation areas (Teodorović et al., 2008).

Airline schedules are market-driven and take into account travel time preferences of air passengers. Airspace systems in many countries are overloaded. The long-term supply side solution is to build new airports and to add new runways, taxiways, and other airport facilities to meet increasing demand and to reduce the delays.

Slot auction assumes auction of landing rights at specific airport. Airlines at some airports bid on an available landing for a particular time of a day.

The *auction* represents a *market-based* procedure in which an item or a collection of items is sold on the basis of *bids* (prices offered by the auction participants for the item being auctioned). The auction participants are also called *bidders*, or *agents*. Post stamps, old coins, paintings, old automobiles are typical items that are frequently auctioned. Recently, many services started also to be auctioned. Depending on the items being auctioned, auctions could be performed online or offline.

An airline's request for a take-off slot at a flight's origin is not independent of its demand for a landing slot at the flight's destination. In other words, airlines need matching slots at airports in the network. Airport slot auctions belong to the class of combinatorial auctions where bidders compete to buy many different but related items. Every auction participant makes one or more bids for any of the possible item combinations.

6.12 RAIL TRAFFIC CONTROL

6.12.1 BACKGROUND

The rail traffic control regulates the train operations along the railway lines, at stations, and consequently through the given railway network. The main objective is to simultaneously provide safe, efficient, and effective running of trains. In this context, "safe" implies running trains without incidents and accidents. "Efficient" means running trains at the acceptable prices for users—passengers and/or freight shippers and at, as low as possible, if not minimal, costs for the rail/train operators. "Effective" implies delivering the above-mentioned users of transport services between their origins and destinations as planned, usually on time specified by the timetable.

Timetable is the way of allocating the available transport resources to serve the expected user demand over given period of time and lines of the railway network. For a given line, the timetable specifies the following: (a) the departure and arrival time of each train at the beginning and ending station of the line; (b) the running time between intermediate stations, and the time of stop there; and (c) the train capacity (the number of seats per train for passengers and the payload capacity for freight). The trains running elements are balanced, in order to provide their conflict-free paths respecting the availability of the infrastructure along the given line and at the stations. The prioritizing of particular categories of trains on the lines with the mixed—passenger/freight—traffic is also specified.

Safe, efficient, and effective running of trains should be achieved at least at the planning level. There, on a daily basis, various irregularities and deviations from the plan due to many reasons. The timetable needs also to be operationalized, ie, realized accordingly. This requires the real-time rail traffic control. In order to understand the main characteristics and operations of the rail traffic control, it is necessary first to analyze the characteristics of the railway infrastructure, supporting facilities and equipment, railway operations, and the workload/capacity of the train dispatcher(s). The main principles of the rail traffic control do not differentiate too much between the rail-based systems, operated in urban areas (metro, light rail), regional (light rail), and the interurban conventional rail and HSR (high speed rail). The former two carry mainly passengers. The latter two carry both passengers and freight. What makes the differences between, for example, conventional and HSR are the supportive facilities and equipment, partially influencing the way of operating the staff. In this last case, the fully automated railway systems (mainly metro systems operating in urban areas) are not included.

6.12.2 INFRASTRUCTURE

The railway infrastructure generally consists of the stations and rail lines with tracks connecting the stations. The railway lines can be categorized according to different criteria such as the width of tracks, importance and volume of traffic, type of traffic, and the number of tracks. Regarding the widths of tracks, for example, in Europe, three categories of rail lines exist: (i) normal line with the width of tracks of 1435 mm; (ii) narrow lines with the width of tracks smaller than 1435 mm; and (iii) wider lines with the tracks greater than 1435 mm. Regarding the importance and volume of traffic, the railway line can be generally of the first, second, and third category. For example, the rail lines of the first category in Europe (CEC, 2001, 2011) are the RNE (rail net Europe) corridors and the RFCs (rail freight corridor(s). The same can be said for the HSR lines/networks in, for example, Japan, Europe, and China (EC, 1996a,b; UIC, 2010).

The railway lines can also be differentiated regarding the predominant type of traffic—for example, those carrying out passengers, and those carrying out freight. The examples are the above-mentioned HSR and RFC, respectively.

As far as the number of tracks is concerned, the railway lines can be single, double, and multitrack. The latest can have 3, 4, 5, or even 6 tracks. At the single-track lines, trains are running in opposite directions and "meet" at the "passing place(s)." The "passing places" stations, or sidings along the line, defined by the timetable, where trains wait for each other. Neither train is permitted to move before the other has arrived at the given "passing" or "meeting" place. This implies that the segments of the line between the "meeting places" can be occupied exclusively by a single train at time. At double-track lines, trains are running in the opposite directions on the different tracks. In this case, the traffic can be organized on the left and right side. The right side traffic is when the trains run on the right track in the direction of running train. The left side traffic is when the trains run on the left track is the direction of running train. This organization is usually the country's specific. In addition, the trains running on the same track in the same direction are prevented to occupy the same section of the line at the same time. The similar is provided at the multitrack lines.

6.12.3 SUPPORTIVE FACILITIES AND EQUIPMENT

The train running along the lines is regulated by timetable and supporting facilities and equipment. While performing operations according to the planned timetable, trains are prevented to come into conflicts, or collide with each other. In particular, timetable is a plan of allocation of the available rolling stock, ie, transport capacity to serve expected passenger and/or freight demand during the specified period of time (week, day, hour). Since, the railway lines are divided into sections called blocks, the timetable at the same time provides that only one train is permitted in each block at a time. In such a way, the safety of railway operations is provided. At the level of railway network, or its part (for example, a line with intensive traffic), the timetable realization is monitored, and controlled by the centralized traffic control system. Occupation of the single block exclusively by a single train, ie, maintaining the trains either moving in the opposite (single-track) or in the same (double-track) direction in the real time is provided by the signaling, interlocking, and ATP (automatic train protecting) system. These systems enable safe, efficient, and effective rail traffic by imposing a minimum safety separation between trains, setting up the conflict-free routes, and enforcing the speed restrictions on running trains. In particular, signals are located before every junction as well as along the lines and inside the stations. The ATP system ensures safe rail operations by, for example, automatic braking if the train ignores the valid speed restrictions (in the case of technical or human failures).

6.12.3.1 Signaling systems

The blocks are the segments of the railway lines that could be occupied exclusively by a single train at time. In general, the block systems can be divided into the fixed and moving block systems.

Fixed block systems

The fixed block systems can generally be the station-based (between two stations) and ATP system-based (between two signals of ATP) blocks as shown in Fig. 6.40A and B).

Fig. 6.40A shows that the station-based block is characteristic for the single-track railway lines. If train (1) has already started running from the station *A* in the direction *AB*, then it faces green signaling aspect along the entire segment *AB*. In such case, the train (2) must wait at the station *B* for the train (1) to arrive at the station *B* before starting run in the direction *BA*. Under such condition the train (2) is faced with the red signaling aspect forbidding it to leave the station *B*. The time of occupation of the block *AB* (or *BA*) is equal to the running time of trains (1) and (2) there, respectively (depending on the block length and train operating speed), and the time for setting up their conflict-free paths.

Fig. 6.40B shows two trains running in the same direction on the double-track line with the ATP system-based blocks. This case is called 3-aspect signaling system with red, yellow, and green aspects.

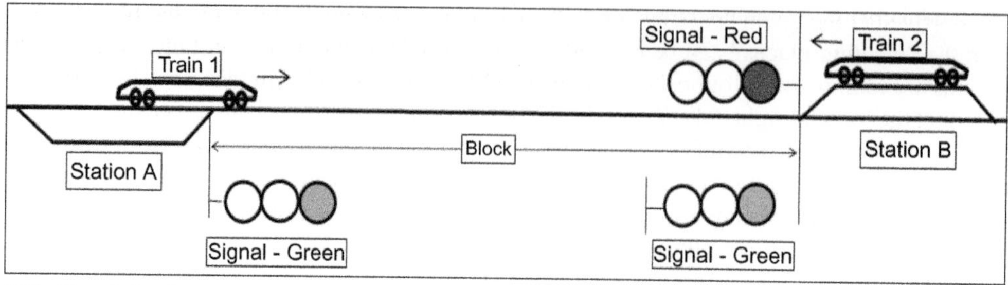

(A) The station-based blocks—single track line

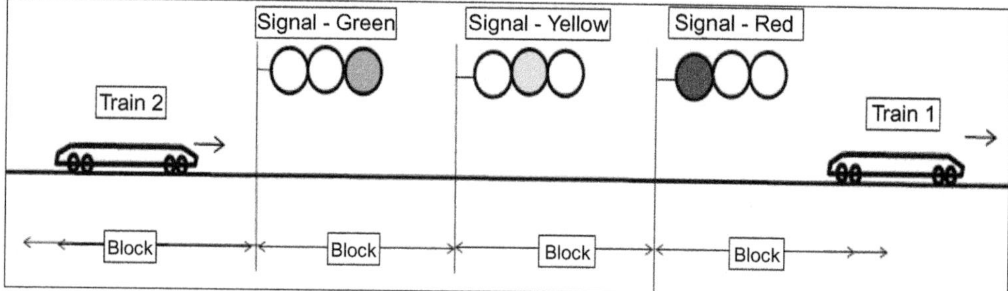

(B) The ATP system-based blocks—double or multitrack lines

FIG. 6.40

Simplified schemes of the different block systems: (A) the station-based blocks—single-track line and (B) the ATP system-based blocks—double or multitrack lines.

As can be seen, train 2 faces green aspect in front of the clear block it is just to enter. At the entry of the successive block ahead is yellow aspect warning that at the entry of the next (third) successive block the red signal can be expected. This is because the Train 1 set up the red aspect at the entrance of this block after entering it. Evidently, train 2 faced with the green aspect is separated by at least two clear blocks from train 1, which enables it to maintain the maximum allowed speed until facing the yellow signal. In such way, train 2 have at least two blocks of distance for the safe braking.

In addition, a 4-aspect signaling system is used is some countries (for example, the UK). It operates similarly to the 3-aspect signaling system except that two warnings—a double yellow and a single yellow—are used before a red aspect. This system provides early warnings of a red signal aspect for higher speed trains, allows better track occupancy by shortening the length of blocks, provides an advanced warning of red aspects for HS (high speed) trains, and closer running of the lower speed trains.

The above-mentioned signaling systems are currently operated by the track circuits, which enable monitoring of trains. The low voltage currents applied to the rails cause the signal, via a series of relays (originally) or electronics (more recently) to show an aspect at the signal. If the current flow is to be interrupted by the wheels of a train or any other cause of the current interruption on the forbidden block the aspect of the signal will protect that block by the red aspect command. The green or yellow aspect, ie, proceed signal will only be displayed if the current does smoothly flow. Thus, the track circuits enable the above-mentioned signals with three aspects to operate automatically or semiautomatically.

The remaining question is how to determine the length of block on the particular railway lines. For the single-track line(s), it is clearly distance between particular stations or the train "passing/meeting" locations. At the double and multitrack lines with the ATP system-based blocks, the minimum length of a block mainly depends on the length of the longest trains to be accommodated there and the length of the "overlapping" distance. Under such conditions, it can be estimated as follows (Kovačević, 1988):

$$S_{min} = n_{max} \cdot l_c + N \cdot l_e + \Delta \qquad (6.52)$$

where

S_{min} is the minimum length of a block
n_{max} is the maximum number of railcars in the longest train
l_c is the average length of a car (m)
N is the number of engines(locomotives) in the train
l_e is the length of locomotive (m)
Δ is the length of the "overlapping" distance (m)

For example, if the number of railcars per train is: $n_{max} = 30$, the average length of each wagon: $l_c = 20$ m, the number of engines/locomotives: $N = 1$, the length of engine/locomotive: $l_e = 15$ m, and the length of the "overlapping" distance: $\Delta = 100$ m, the length of the train and consequently the minimum length of the block will be: $S_{min} = 30 \cdot 20 + 1 \cdot 15 + 100 = 715\text{m}$

Why the "overlapping distance"? The actual length of the block in the given case and consequently the spacing/distance between the signals are influenced by the factors such as: (a) the line speed as the maximum permitted speed over the given line-section; (b) train speed as the maximum speed of different categories of traffic—conventional, HS, passenger, freight; (c) line/segment gradient needed to compensate for longer or shorter braking distances; (d) the train braking characteristics, different for different categories of trains such as freight, conventional passenger, HS passenger; (e) sighting indicating how far ahead a driver can see a signal; and (f) the driver's reaction time. In the given context, the train braking characteristics in terms of the required minimum braking distance respecting the category (type) of train is of particular importance. Since this distance, as we shall see below, depending on the train initial speed and deceleration rate could be substantive, the blocks based exclusively on the maximum length of train(s) are extended for the "overlapping" distance(s). It is located at the certain distance from the signal, at the entrance of the section it protects. In most cases, its length is 100–185 m. As such, it helps the train to stop even it has passed the red/stop aspect of the signal. Consequently, at most conventional railway lines in Europe and elsewhere in the world, the typical length of a block amounts about 1000 m (for the HSR lines this is about 1500 m). Consequently, for 3-aspect signaling system, this gives the spacing between the signals of about 3000 m. This enables trains to run at the maximum operating cruising speed $S(v_0, a^-)$, which can be estimated as follows:

$$S(v_0, a^-) = \frac{v_0^2}{2a^-} \qquad (6.53)$$

Consequently:

$$v_0 = \sqrt{S(v_0, a^-) \cdot 2a^-} \qquad (6.54)$$

where

v_0 is the train's operating cruising speed at which deceleration and braking starts (km/h)
a^- is a constant deceleration rate during the braking (m/s^2)
$S(v_0, a^-)$ is braking distance from the speed (v_0) and deceleration (a^-) equal to the spacing between two signals (m)

For example, if: $S(v_0, a^-) = 1000$ m and $a^- = -0.5$ m/s^2, then the train's maximum operating cruising speed enabling the safe braking will be

$$v_0 = \sqrt{1000 \cdot 2 \cdot 0.5} = 31.62\text{m/s} = 31.62 \cdot 3.6\text{km/h} \approx 114\text{km/h}$$

It is obvious that with increasing of the deceleration rate, the maximum operating cruising speed will increase, but at decreasing rate given the fixed spacing between signals in the given case.

ERTMS (European rail traffic management system)

The ERTMS has been recommended by the EC (European Commission) and gradually implemented in particular countries in addition to the already existing national signaling systems (EC, 1996a,b, 2001a, b; Jarašūnienė, 2005). The main objectives of the ERTMS are expected to be improvement of technical interoperability, safety, train operating performances, and availability/reliability.

Improvement of technical interoperability by unified signaling system/equipment is expected to stimulate opening the rail transport markets for more rail operators. Improvement of safety is to be provided by designing the system according to the given/specified standards. The train operating performances are expected to be improved. The improvement will be possible by enabling safe operations of both passenger and freight trains at the higher speeds in the same direction, separated at much shorter time intervals than those specified by the national (above-mentioned 3-aspect) signaling systems. Availability/reliability is expected to be achieved by reducing the quantity of equipment along the rail lines, and probability of the component failures.

The ERTMS consists of two primary components: (i) the ETCS (European train control and command system), which is an automatic control system that control the speed limits of a train by communicating with the driver, and (ii) the GSM-R (global system for mobile communications—railway, ie, the radio communications system to enable exchange of information between the train (driver) and the traffic management center.

The ERTMS is designed at three levels:

The ERTMS Level 1 uses Eurobalises installed under the tracks, the existing trackside signals, and track circuits. The Eurobalises are the electronic beacons or transponders installed usually below the ties of tracks at the distance of 3 m and represent a part of the current ATP system. The track signals are the same as those at the conventional railways. The track circuits as devices enable collecting information on the train integrity and its position. As such, the system can continuously supervise current and generate prospective safe train's speed. The LEU (lineside equipment unit) located by side of the tracks generates the movement authorities (ie, the safe train's path(s)) and the track description data based on the information received from the trackside signals and track circuits (the latter on the train's integrity and position). The movement authorities are transmitted to the train through balises. Then, the on-board computer system calculates the dynamic speed profile ahead (the actual speed and the maximum allowed speed) by taking into account the train's braking characteristics. In addition, it also monitors and controls the indicators in front of the driver. In this case, use of the trackside signals is necessary.

The ERTMS Level 2 is based on the radio based ATC (automatic train control) system, which provides continuous information and supervision of the train speed towards fixed points of the line (ends of block sections, restrictions of speed, etc.). In addition, the RBC (radio block centre) generates messages on the movement authorities, state of tracks, current speed, eventual restrictions, and emergency based on the information transmitted from the HS train(s), external interlocking system, and the track circuits. The RBC usually covers and manages about 100 km of double-track line. The messages constituting the movement authorities are transmitted between RBC and the train(s) means by GSM-R system. Its basis is BTSs (base transceiver station(s)) positioned on the approximate distance from each other of about 3–4 km. The GSM-R operates as follows: at the moment when HS train passes over a Eurobalise, it transmits its new position and its speed to RBC. Then, it receives back agreement (or disagreement) to enter the next section of the track and its new maximum speed. Then, the on board computer system calculates the train's dynamic speed profile by taking into account braking characteristics and other commands to be eventually used. In this case, the side track signals can also be used, but optionally.

Under such conditions, the trains running in the same direction (usually on the double or multitrack lines) can come close to each other just at the minimum braking distance of the following train, which is still safe. Fig. 6.41 shows the simplified scheme of the principle of controlling the train speed by the ATC (automatic train control) system.

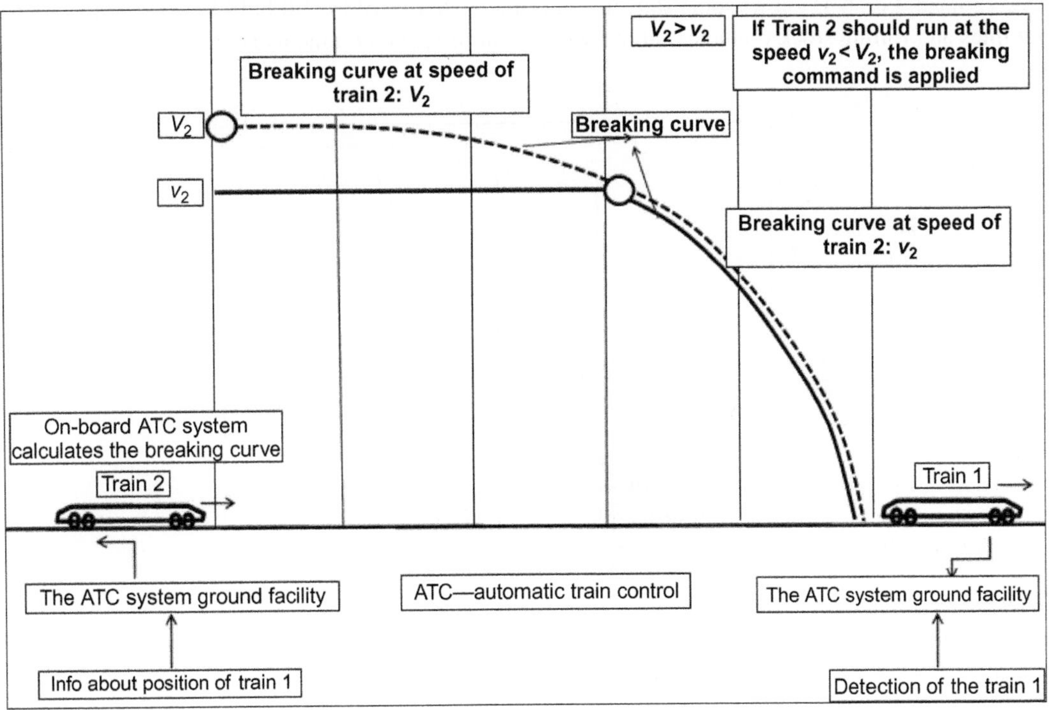

FIG. 6.41

Simplified scheme of controlling separation distances between successive trains by controlling the braking distance their speeds.

As can be seen, in this example, the system permanently detects the position of train (1) and transmits it to the train (2) whose on-board system/computer calculates its braking curve and distance as its safe zone/segment under given conditions. (The system similarly creates the safe zone/segment for the train (1).) If the speed of train (2) is higher than the minimum braking distance to maintain its safe separation from the train (1), the command for braking and reducing speed to the required level is automatically applied.

The ERTMS Level 3 has the very similar characteristics as the ERTMS Level 2 with some technical and functional differences. The former are: the necessity for having on-board equipment for checking the train's integrity, which is then used by RBC for generating the movement authority. The track circuitries for the train detection are not needed. The latter is that the preceding train is considered as the moving block and target while specifying the minimum time/distance intervals between the trains' separation (UNIFE, 2014a,b,c,d, http://demo.oxalis.be/unife/ertms/?page_id=42).

The European ERTMS deployment plan predicts that the ERTMS equipment will be implemented in about 25,000 km by 2020. A number of key European freight lines have also been identified for ERTMS deployment but six corridors initially prioritized have been as follows: Corridor A: Rotterdam–Genoa, Corridor B: Stockholm–Naples, Corridor C: Antwerp–Basel, Corridor D: Budapest–Valencia, Corridor E: Dresden–Constanta, and Corridor F: Aachen–Terespol. In addition, the ERTMS has been implemented in other countries round the world such as China, India, South Korea, Taiwan, Saudi Arabia, UAE, Morocco, Algeria, Turkey, Kazakhstan, Indonesia, Brazil, Mexico, Australia, and New Zealand. Fig. 6.41 shows some statistics of implementation of the ERTMS Level 2 related to the rail lines and rolling stock (vehicles) at the railways at different continents (Fig. 6.42).

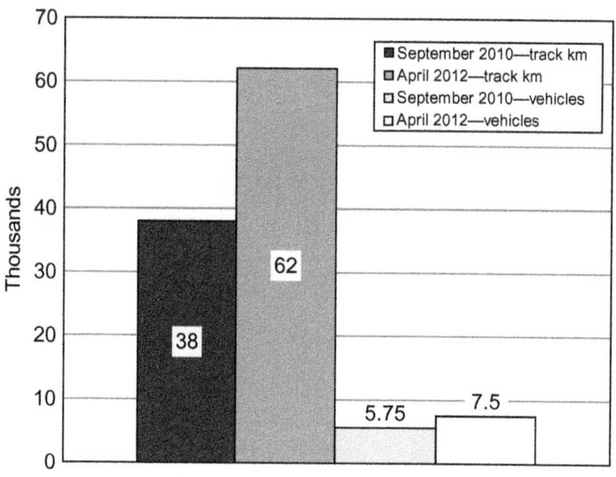

FIG. 6.42

Some statistics of implementing the ERTMS round the world (UNIFE, 2014d, http://demo.oxalis.be/unife/ertms/?page_id=42).

As can be seen, the number of both track km and on board units has been increasing during the observed period, for about 63% and 32%, respectively.

Moving block system

General: As mentioned above, the fixed block signaling system shows constraints when the maximum train operating cruising speed needs to be increased. With increasing of this speed, the braking distance will be increased too, thus requiring increasing the length of blocks and spacing between successive signals along the given line (all other factors are assumed to be constant). Such repositioning the signals would be very expensive to justify the gains by increased train speeds. Consequently, the solution has been found in developing the moving block system, where computers calculate a "buffer" or "safe" zone, ie, a segment or a block of the rail line to be used by a single train. This implies that no other train is permitted to enter the zone-segment. The determination of the length of a zone/segment is based on knowledge of the precise location and speed and direction of movement of a train, which are determined by using several sensors: active and passive markers along the tracks and tachometers and speedometers onboard the train (it should be mentioned that GPS cannot be used because it does not work in tunnels). Under such circumstances, the created virtual zone/segments are moving along the line together with the trains, thus making the above-mentioned line side signals unnecessary. The instructions on the speed control and potential collision risk with other trains including avoiding actions are passed directly to the trains. While the moving, block system is already used at some regional rail lines with homogenous train fleet/traffic (Vancouver's *Skytrain*, London's *Docklands Light Railway*, New York City's *BMT Canarsie Line*, and London's Jubilee Line) and at the HSR lines again with the homogenous trains/traffic, it is not sufficiently developed to be applied to the rail lines with heterogeneous trains/traffic, such as passenger and freight. Nevertheless, further development represents a part of the above-mentioned ERTMS Level 3, which will enable trains to follow each other exactly at the current braking distance(s). This implies reduction of spacing of trains moving in the same direction, which will contribute to increasing of the capacity of the given rail line(s).

TVM (transmission vole machine): The TVM signaling system is applied exclusively to the HSR lines in France. The system is based on ATP system, which distributes information on the train speeds from about 270 to 360 km/h, depending on the version. The ground-based components of the system are TCCs (trackside control centre(s)) located approximately every 15 km along each track of the line. They are linked to the line's centralized traffic control center while directly controlling about ten blocks of track, each equipped with its own track circuit. In addition, the TVM system exclusively relies on cab-signaling, which implies that it operates without trackside signaling. The main characteristic of the cab-signaling system is that the signaling information is transmitted through the tracks as electrical signals, which are picked up by antennas under the train, ie, it is continuously transmitted through the track circuits as the track-to-train transmission. Four these antennas, two on each end are mounted underneath the train, but the two are used in the direction of travel. The track circuits in both tracks are used to transmit the signaling information to the train's on-board computers, as well as fixed inductive loop beacons. In addition, TVM is a fixed block system. This means that the track is subdivided into fixed segments each of which has a particular state. Only one train may occupy any block at one time under regular operating conditions. The length of a block is about 1500 m. The blocks are shorter than the HS train's braking distance, so a braking safe separation interval spreads over the several blocks whose number depends on the maximum operating speed and the maximum train's deceleration rate. For example, it is usually 4 block for the speeds between 270 and 300 km/h, 5 blocks for the speed of 300 km/h, and 6 blocks for the speeds of

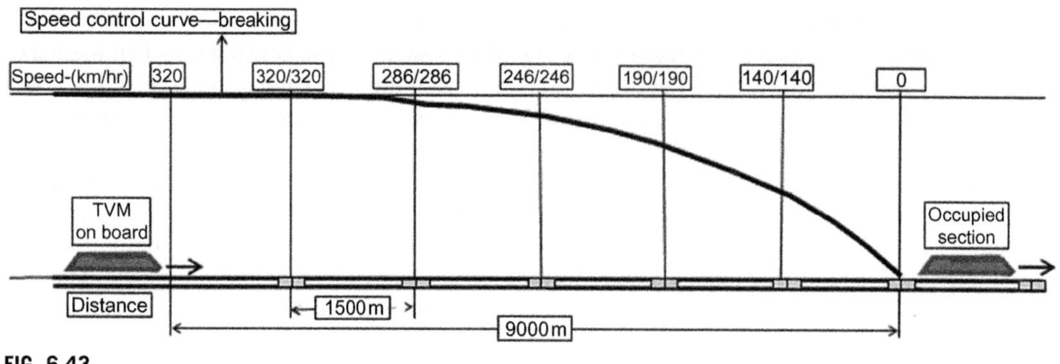

FIG. 6.43

A scheme of braking pattern of HS (high speed) train—speed control curve by the TVM signaling system.

320–360 km/h. Each block possesses the constant relevant properties for the train occupying it such as length, gradient, and the maximum operational safe speed. In addition, the variable property is the train's target speed as the speed at which the train should exit the current and enter the next block. Fig. 6.43 shows the simplified generic scheme of the braking pattern of the HS train based on the TVM signaling system.

As can be seen, in the given case, the HS train will take the distance of about 9000 m (9 km) to stop if starting braking at the operating cruising speed of about 320 km/h at a constant deceleration rate of 0.46 m/s^2. All above-mentioned information is transmitted by TVM system to the train's computers and the cab displays where the driver can monitor them. Specifically, he/she monitors the target speeds for the current and subsequent blocks (displayed in km/h), full line speed, and the speedometer continuously indicating varying target and current speed (with precision of about 2%). In addition, in order to mitigate the driver's workload the speeds are displayed over several blocks ahead of the train. As well, since the system itself cannot adapt to the irregular operating conditions, the human operator-driver is kept in the control loop. For example, if the train exceeds the specified maximum speed of 300 km/h the computer will undertake an action to reduce it and establish again the regular operations. This implies that TGV (Train à Grande Vitesse) trains are carried out manually, but the safety has been provided by the automated signaling system. Last but not least, the digital recording system based on the desktop computer system monitors and records every action of the driver (electro motors' operating regime, activation/deactivation of brakes, pantographs, etc.) as well as the above-mentioned signaling information (ABB, 2014).

6.12.3.2 Rail traffic control/management system

In general, the contemporary rail traffic control/management system on the conventional rail lines/networks usually includes the following components: CTC (centralized traffic control), communication facility, and remote control of power supply facilities. The CTC main function is to monitor the status of train operations on the conventional lines and carry out their control while being under jurisdiction. The communication facility enables the real-time transmission of traffic control data and provides connections between the center of the system and particular locations. The remote control of power supply facilities enables control of the power supply facilities in order to guarantee reliable functioning of the entire center. At the HSR lines/networks the rail traffic/control management systems is fully computer-supported and can include, such as the Japanese Shinkansen railway network, the following main components[1]: TOC—train operation controller, PC—power controller, STC—signal and telecommunication controller, CCC—crew and car utilization controller, PSC—passenger service controller, and TSMC—track and structure maintenance controller. Fig. 6.44 shows the simplified scheme of the house/spatial organization of the above-mentioned components (JR, 2011).

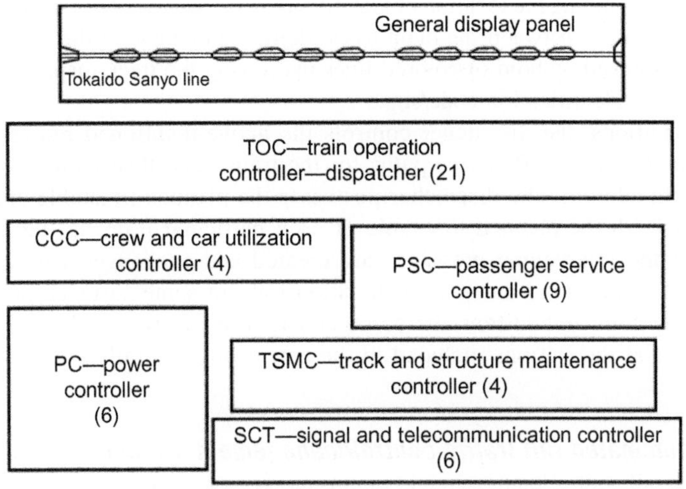

FIG. 6.44

Simplified layout of organization of the Tokaido-Sanyo Shinkansen General Control Center (the number of consoles/workstations in given in parentheses) (JR, 2011).

The core of the system is COMTRAC (computer-aided traffic control), which enables permanent online monitoring of the status and control of all currently operating Shinkansen trains (JR, 2011).

The main activity of each rail traffic control/management system (in the further text "CTC" (computer train control)) is to control and monitor realization of the timetable and provide online instructions for mitigating deviations from it, which can happen due to any reason. The CTC unit is accommodated in the dedicated office where the train dispatcher carries out these. The office is equipped with the control panel(s) (screen(s)) providing the graphical representation of the part of the rail line/network under his/her jurisdiction. The size of the part of the line/network depends on many factors such as: type of the rail line (single, double, multitrack), the number and configuration of stations on the line/network, traffic intensity in terms of the number of trains during the specified period of time, the number and time for executing necessary tasks by the dispatcher, and the required productivity of the dispatcher. Consequently, the given rail network can have several CTC units usually

[1]These are the components of the Tokaido-Sanyo Shinkansen General control center considered as the most advanced operating control/management system ensuring safety, efficiency, and effectiveness of tarins' operations (JR, 2011).

located at the large stations or rail yards. In addition, the office is equipped with telecommunication devices and links enabling communication of the dispatcher with the train drivers and dispatcher at stations of its part of the network and other CTC units. They also enable the dispatcher to communicate with other services including maintenance of tracks and rolling stock, as well as with passenger and freight service units.

In the modern computer-supported CTC unit, the panels/screens generally display rather simplified schemes of the tracks including the location of signals and powered switches usually located at the end of sidings and at crossovers between the main track along the line and at the stations. The occupied tracks are usually displayed by bold or colored lines overlaying the track display. The trains are displayed as tags with the identification code/sign.

Under such conditions, the dispatcher controls the above-mentioned red-and-green-aspect of the signals, creates the safe conflict-free paths for the trains, and then monitors their progressing along these paths. In addition, the dispatcher possesses the planned timetable enabling him/her to compare the actual with the planned timetable of each particular train. In the case of deviations due to any reason, the intervening messages are created and exchanged between the dispatcher and the train drivers either via voice communication link, in some cases automatically (as mentioned above), or by both. In addition, the communication is carried out between the central and the local dispatchers at the stations included in the part of the centrally controlled/managed line/network.

6.12.3.3 Fully automated rail traffic control/management system

Development of the rail control/management systems has been moving towards more computer-based operations, which in most cases have maintained a man in the control loop either as the train driver, ground-based dispatcher/controller, or both. This has been the case at the current conventional rails but also at the HSR such as the above-mentioned French TGV (the driver remained in the control loop) and Japanese Shinkansen HSR systems (the controllers monitor operations on trains on the given line). The additional cases are many urban rail-based systems—metros and light rail. Up to date, the fully automated rail-based systems have shown to be metros operating in 25 urban areas/cities worldwide. The networks in these cities are of different length—the longest one is in Dubai (UAE—United Arab Emirates) (80 km) and the shortest one in Hong Kong (5 km). The longest automated metro line is the Red line, 52.1 km long, again in Dubai (UITP, 2012).

Concept of metro automation

At metro systems, automation generally implies transferring processes and responsibility for train operations and management from the man-driver to the train control/management system.

In general, there can be different levels of automation in the given case specified by the type and number of basic functions of train operations allocated to the staff and to the train control system itself. These basic functions for distinguishing the particular levels of automation are: (1) setting the train in motion; (2) door closure; (3) stopping train; and (4) handling disruptions. Consequently, at first Level 0, the above-mentioned functions are carried out by the train driver and train attendee(s). At the last Level 4 of automation all these functions are carried out by the train control system. At the two intermediate levels, particular functions are divided between train driver and attendant on the one side, and the train control system on the other. Fig. 6.45 shows the simplified scheme (UITP, 2012).

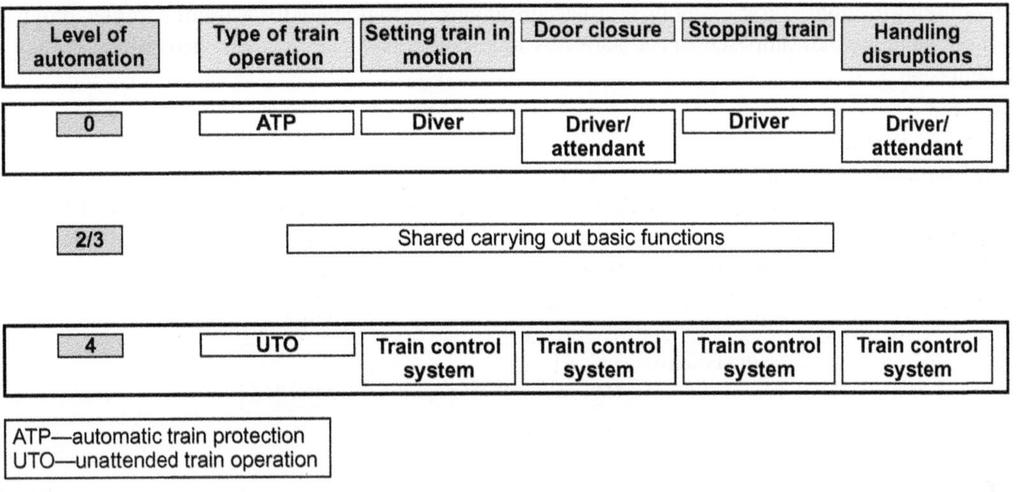

FIG. 6.45

Scheme of the different levels of automation of metro system (UITP, 2012).

Components of automated metro system

The main components of the fully automated (ATC—automatic train control) metro system are: ATP, ATO (automatic train operation), and ATS (automatic train supervision) (UITP, 2012; Mohan, 2012).

- The ATP provides basic level of safety by preventing collisions, red signal overrunning, and exceeding the speed limits. In any of these cases, the train brakes are automatically activated. In particular, the ATP ensures and maintains a safe separation between successive trains by established the buffer safety zone around them, sets up the maximum allowable speed, and required deceleration rate, both depending on the train's location along the line and other (traffic) conditions, detects eventual obstacles along the train path(s), and secure routes and switch Interlocking.
- The ATO performs all activities/functions, which would otherwise be performed by the driver. In particular, the ATO controls train's acceleration, deceleration, and stop at the station(s) including the speed control, in order to optimize the train's running time and comfort between stations and along the entire route from the aspects of users—passengers. In addition, it controls the time of train's stops at stations, and adjust simultaneous closing of the platform screen doors and the train doors, followed by an audio and visual signal for passenger information. As well, it performs other functions such as route selection.
- The ATS (automatic train supervision) automatically sets up the routes for trains, and regulate their operations according to the timetable, but within the tolerable deviations. In particular the ATC executes instructions received from the CCO (central control operator(s), ie, train controller(s). Thus, it enables trains to skip particular stations, the station office to hold a train in a station including dispatching and adjustment of the station stop time, station interlocking and computation of train schedules,

monitoring the train's progress along the route, displaying the train status to the train controller, provision of output to the indicators/screens at platforms and/or other passenger/management information media, etc.

Signaling and train control

The signaling and train control of the ATC metro system is organized at three hierarchical levels as follows (top-down): (i) management; (ii) operational; and (iii) activation level. Fig. 6.46 shows the simplified scheme (Mohan, 2012).

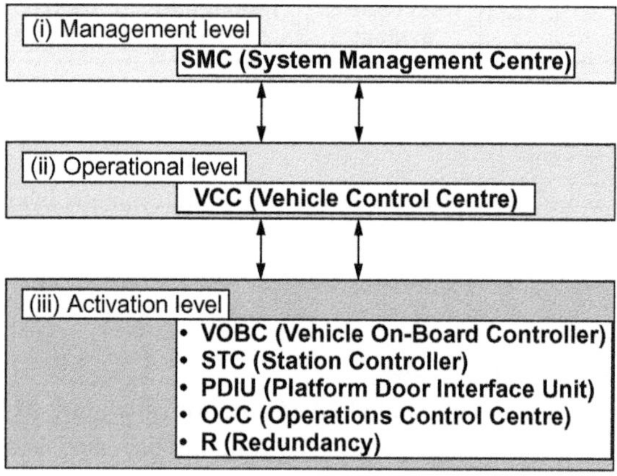

FIG. 6.46

Simplified scheme of the configuration of signaling and train control of Dubai automated metro system—red line (Mohan, 2012).

- **(i)** *Management level*: The SMC (system management center) performs the functions at the strategic/management level. These generally include managing the train operations in the network. Actually, the SMC monitors and controls the entire network with running trains and provides the required ATS (automatic train supervision) function. In addition, it provides information to the CCOs) on the position and status of all trains and status of the equipment within the ATC system. As well, the commands on the train and guideway control are entered at CCO workstations.
- **(ii)** *Operational level*: The VCC (vehicle control center) performs the system's function at the operational level. In general, the system is responsible for safe operations by providing the minimum separation between trains including performing all interlocking functions. Thanks to the permanent contact with all trains and wayside devices in the network, and use of the real-time information, the VCC can generate and pass the movement authority to each train. In this way, VCC initiate commands for positioning the wayside signals accordingly in order to realize the conflict-free movements under given conditions. For example, at Dubai metro system with two lines, there are 6VCC on red and 4VCC on green line. (The system has two lines.)

(iii) *Activation level*: The components of the system at this level are: VOBC (vehicle on-board controller), STC (station controller), PDIU (platform door interface unit), ILC (inductive loop communications), OCC (operations control centre), and R (redundancy).

The VOBC provides the on-board ATO and ATP functions to the vehicles by constantly communicating with VCC and is primarily responsible for the control of propulsion, maximum speed, brakes, and train doors under ATP constraints. These all is needed to ensure that train operates according to the given moving authority.

The STC (*station controller*) carries out control and supervision of automatic switch of all switches including reporting of their current status. In addition, it provides control and monitoring of the platform screen doors.

The PDIU supervises, ie, monitors and reports for the intrusions status of the network infrastructure components. The ILC equipment located in the equipment rooms and at trackside enabling exchanging information between VCC and VOBC for each train.

The OCC as the "nerve" center for train operations houses the relevant controller workstations and consoles, central visual control panel/screen, and the required communication and security facilities. In addition to the central panel/screen, the controllers of OCC can have an overview of location of trains in the entire network on their workstations.

The R (*redundancy*) component contains duplicated all important subsystems. They should prevent disruptions of the regular train services in the case of failures of particular subcomponents (computers, microprocessors, and power supply subsystems).

6.12.4 THE WORKLOAD AND CAPACITY OF TRAIN DISPATCHER(S)

As mentioned above, except in the fully automated systems, the train dispatchers directly control train operations in the areas of their jurisdiction. They seat in the railway control center(s), which can simultaneously handle single (for example, HS or a mixture of different (for example, long distance passenger and freight) trains. In general, the control center is situated in a large room equipped by several working stations for the same number of dispatchers operating in parallel, each with assigned responsibility for different adjoining territories, and possibility to directly orally communicate between each other. Each workstation can contain few video display terminals: one usually displays radio and phone communication information and the others mainly the train-related information. One of these latter shows schematics of portions of the track, ie, portions of the territory controlled by a single dispatcher. The second displays information related to the trains' status. A PC (Personal Computer) is also there for carrying out the administrative tasks. In addition, a large wall-projected overview display shows a schematic of the entire configuration of rail tracks controlled from a given dispatcher center.

Together with the train drivers, the train dispatchers represent some of the key players in carrying out the train operations. Their main objectives are to enable efficient (the lowest cost), effective (on time without substantive delays), and safe (conflict and accident free) train traffic along the particular tracks/lines and stations of the given rail network. These objectives can be fulfilled by carrying out generally monitoring and communicating activities as five different tasks such as: monitoring trains, controlling trains by preparation of routes, communicating by short messages, communicating by (longer lasting) consultation, and adjusting the plan(s). Performing these tasks under regular, and particularly under irregular disruptive operating conditions, generate the mental workload[2] at the train dispatcher(s), which needs to be quantified in order to obtain its capacity (Andersson et al., 1997; Lenior, 1993). Quantification of the train dispatcher's workload can be carried out by measuring his/her subjective mental and objective workload, both under specified/given conditions, ie, within the given working environment (Volpe, 1999; Zeilstra et al., 2012). In particular, some projects have dealt with measuring of the objective workload by weighting, ie, allocating a certain number of points to each of the above-mentioned train dispatcher's task with respect to its inherent complexity, frequency, and execution time, all for a single train and with respect to its current and prospective interactions with other trains within the area, ie, track/line segment, under dispatcher's jurisdiction. Consequently, each the tasks with assigned points have been summed-up in order to calculate the dispatcher's mental workload. The train dispatcher's workload WL in monitoring train traffic could be calculated in the following way (Zeilstra et al., 2012):

$$WL = \left[\left(\frac{1}{2} + I_c\right) \cdot \lambda + 4 \cdot \lambda_s\right] + (1 \cdot N_s + 0.1 \cdot \lambda) + 0.4 \cdot I_c \lambda_s \tag{6.55}$$

where

WL is the train dispatcher's workload (points)
I_c is the measure of the rail track/line infrastructure complexity
λ is the intensity of regular train traffic entering or leaving the dispatcher's area (trains/h)
λ_s is the intensity of special train traffic entering or leaving the dispatcher's area (trains/h)

In Eq. (6.55), three terms reflect the workload of detection of a train, interpretation of the train status, and anticipation of possible actions. However, the calculated number of workload points (WL) always needs to be compared with the standards for acceptable workload (W^*). This standard can be defined by the train dispatchers and/or some railway safety authority (ISO, 1991). As an example, made by the subjective judgment of the train dispatchers under conditions of regular operations during the peak time: $WL < 150$ implies low, $250 < WL < 350$ good, and $350 < WL \leq 400$ acceptable workload (but the latest lasting up to 3 h). Consequently, for example, by setting up in Eq. (6.55), that the above-mentioned acceptable workload (WL) is equal to the maximum allowed workload (WL^*), ie, $WL = WL^*$, the monitoring capacity of the train dispatcher expressed by the intensity of regular and special trains, handled during the given period of time can be calculated as follows:

$$\lambda^* = \frac{WL^* - \lambda_s(5 + 0.4 I_s)}{(0.6 + I_c)} + \lambda_s \tag{6.56}$$

where all symbols are as in the previous equations. Fig. 6.47 shows an example of the calculated relationship between the train's dispatcher's monitoring capacity, his/her allowed workload during the peak hour, and the complexity of tracks/rail lines in the area of his/her jurisdiction.

As can be seen, the capacity of train dispatcher in terms of the intensity of handled trains during a given period of time linearly increases with increasing of the allowable workload and decreases with increasing of the rail track/line infrastructure complexity in the area of his/her jurisdiction. Fig. 6.48 shows the relationship between the train dispatcher's capacity and the rail track/line complexity in the area of his/her jurisdiction.

[2]Several terms have been frequently used for expressing the term "workload": working pressure, workload, task-load, mental strain, stress, etc. However, a clear definition is still lacking.

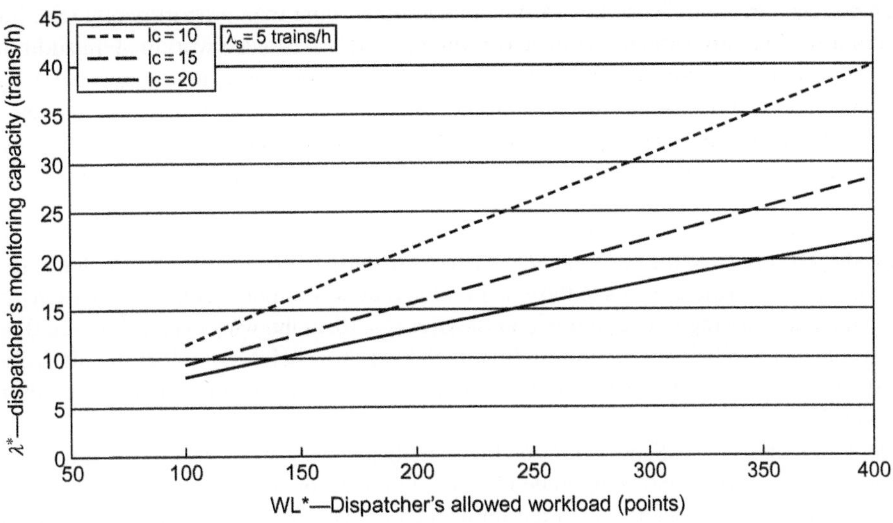

FIG. 6.47

Relationship between the train dispatcher's monitoring capacity and allowed workload, and the rail track/line infrastructure complexity in the area of jurisdiction.

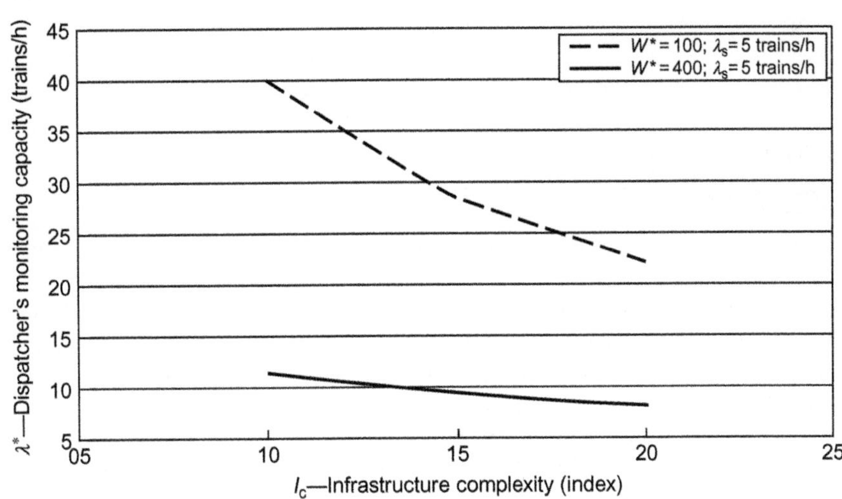

FIG. 6.48

Relationship between the train dispatcher's monitoring capacity, the rail track/line infrastructure complexity in the area of jurisdiction, and his/her allowed workload.

As can be seen, the train dispatcher's capacity decreases more than proportionally with increasing with the rail track/line infrastructure complexity for any level of allowed workload. In addition, based on the train dispatcher's capacity under given conditions (level of workload) and the average time the trains spend in the area of his/her jurisdiction, the maximum number of trains simultaneously being under monitoring can be estimated by known Little' s formula from queuing system theory as follows (Newell, 1982):

$$N^* = \lambda^* \cdot t_r \tag{6.57}$$

where t_r is the average time a train spends in the area of jurisdiction of a train dispatcher (h).

The other symbols are analogous to those in previous equations. For example, if the train dispatcher operates in the area with the infrastructure complexity: $I_c = 10$, at the workload of $WL = 350$ points, its monitoring capacity will be: $\lambda^* = 35$ trains/h (of which 5 are the special ones). If each train spends in the area: $t_r = 0.5$ h (30 min), the number of trains simultaneously being under dispatcher's monitoring will be, $N^* = 35 \cdot 0.5 \approx 18$.

In order to obtain more realistic figures on the train dispatcher's workload and related capacity, the above-mentioned models should be upgraded by taking into account the other tasks he/she carries out. Frequently, these are those dealing with the unplanned demand for multiple use of particular tracks (maintenance, inspection, in addition to trains), and the need for dynamic re-planning of the timetable aiming at mitigating the train delays and track outages. The intensive radio communication with the train drivers, neighboring centers/dispatchers, and other services needs also to be taken into account (Volpe, 1999).

It should be mentioned that the two capacities always need to be considered while planning (timetable) and executing the train operations in the given area (part of the railway network): the capacity of infrastructure (lines, stations), and the capacity of train dispatcher(s), who monitors and controls the trains operating in his/her area of jurisdiction. In such case, the lower capacity value is always taken as relevant.

6.13 AIR TRAFFIC CONTROL

6.13.1 BACKGROUND

The air transport system consists of three main components: airports, ATC (air traffic control) system, and airlines. The airports accommodate the airline aircraft at the beginning and end of their flights between them, serving the passengers and freight cargo demand. The ATC monitors and controls flights at the airports and in the air between them in order to provide their safety, efficiency, and effectiveness. The ATC should prevent conflicts that cause air traffic incidents and accidents (safety), significant aircraft delays (effectiveness), and additional airline and passenger costs (efficiency).

The ATC system is characterized by the components such as infrastructure, supportive facilities and equipment, and staff (ATC controllers). All of them function within a given organization, and according to the specified operational rules and procedures. The organization and operational rules and procedures are influenced by the characteristics of infrastructure, supportive facilities and equipment and aircraft/flight demand which needs to be served, under given conditions.

In the most general sense, the ATC system infrastructure includes airports and airspace with air routes connecting them. The supportive facilities and equipment are the radio-navigational aids on the ground (beacons) and in the air (satellites), the ground and air-based surveillance equipment—radars, avionics on board the aircraft, computing facilities processing the flight (and supportive) data for the system's different planning and operational purposes, and operating staff (ATC controllers) (Graves, 1998; Horonjeff and McKelvey, 1994). The operational rules and procedures are applied in two ways: (i) as those related to execution of particular monitoring and controlling activities by the ATC controllers; and (ii) as those related to separation of aircraft/flights while being under jurisdiction of given ATC unit in order to enable fulfillment of the main objectives in terms of safety, efficiency, and effectiveness.

6.13.2 INFRASTRUCTURE

6.13.2.1 Airspace

General: In general, there is the international agreement how to regulate and divide airspace, which is used for commercial air transportation (flights). (In this context the airspace organization and division for the military purposes is not considered.) These agreements are facilitated through ICAO (International Civil Aviation Organization) whose members are all countries of the world. The headquarters of the organization are in Montreal (Canada).

The ICAO issues the guiding documents called International Standards and Recommended Practices, known as ICAO annexes. There are 18 annexes and three PANS (procedures for air navigation services). The two annexes related to division and categorization of airspace Annex 2 (Rules of the Air) and Annex 11 (ATC Services) (ICAO, 2001).

According to these documents, the airspace is generally divided into controlled and uncontrolled. In the controlled airspace, the services to commercial and military flights are provided by the ATC. Depending on the treatment and provision of separation of IFR (instrument flight rules) and VFR (visual flight rules) flights, the airspace is categorized into classes A, B, C, D, and E. In the uncontrolled airspace the aircraft/flights of both above-mentioned categories receive the ATC advices on request.

Specific: The above-mentioned actually large parts of the airspace are organized into smaller parts, which usually cover the territory of the country. In general, this airspace is further divided into the parts under jurisdiction of the centralized ARTCC (air traffic control centers), such as, for example, the US airspace. The European airspace is divided into so-called FABs (functional airspace blocks). Both are shown in Fig. 6.49A and B, respectively.

In particular, the main objective of establishing the FABs in Europe is to overcome the constraints imposed by the countries' borders and consequently improve the current capacity, efficiency, and effectiveness on ATC services, while at the same enhancing safety. This also should substantively contribute to implementing the concept of SES (Single European Sky) (EC, 2012).

In addition, the airspace of the country or the wider functional blocks is also divided into the vertical plan into the airport zone as shown in Fig. 6.50 for the UK airspace (Graves, 1998).

As can be seen, the airspace is vertically but also at the same time horizontally divided into the airport zone(s), terminal airspace(s), low altitude airspace, and high altitude (upper)

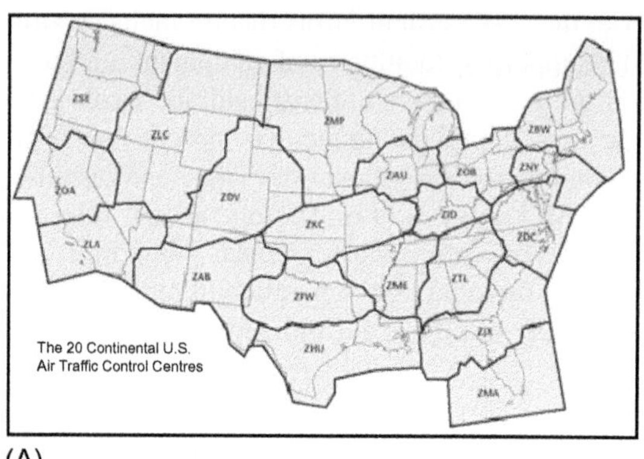

(A)

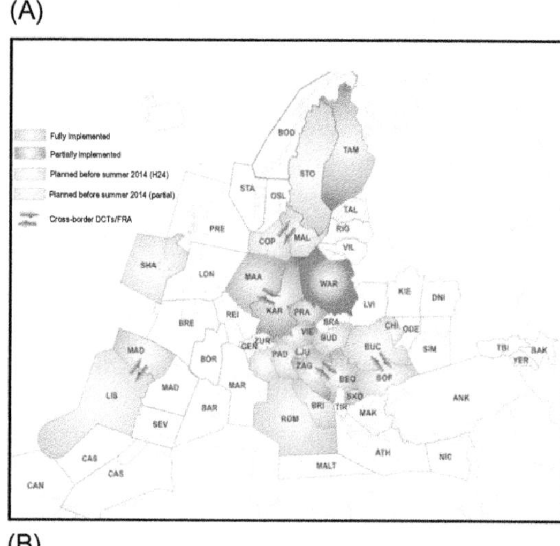

(B)

FIG. 6.49

Scheme of division of airspace (the horizontal plane): (A) US airspace (TRACON location map: www.artofanderson.com) and (B) European airspace—FABs (functional airspace block(s)) (EC, 2012).

- *Airport zone* is established around an airport with a usual radius of 2.0—2.5 nm (nm—nautical mile; 1 nm = 1852 m) and attitude of 2000 ft (600 m) MSL (middle sea level) to enable control and management of landing and taking-off air aircraft (1 ft = 0.305 m). In the horizontal plane, it can have different shapes—circle, semicircle, ellipse, square, rectangle, trapezoid, etc. depending on the local conditions. One example is shown in Fig. 6.51.

  (The arriving aircraft use information from the radio navigational facilities such as ILS (instrument landing system) or MLS (microwave landing system) while carrying out the final approach and landing).
- *Terminal airspace* spreads around and above the airport zone. It is established around a single busy or several airports of the same airport system with a radius of 40–50 nm (nautical miles) and altitude

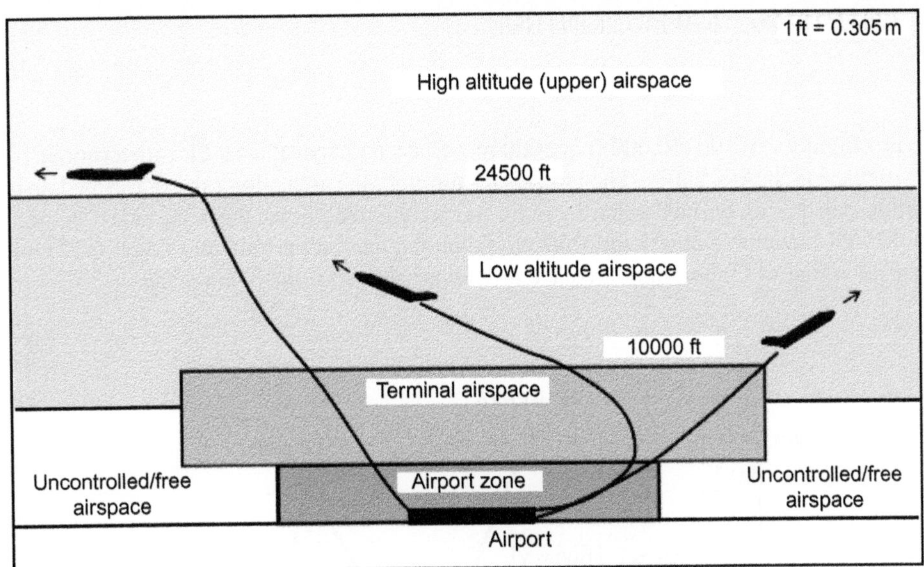

FIG. 6.50

Scheme of division of airspace in the vertical plane (Graves, 1998).

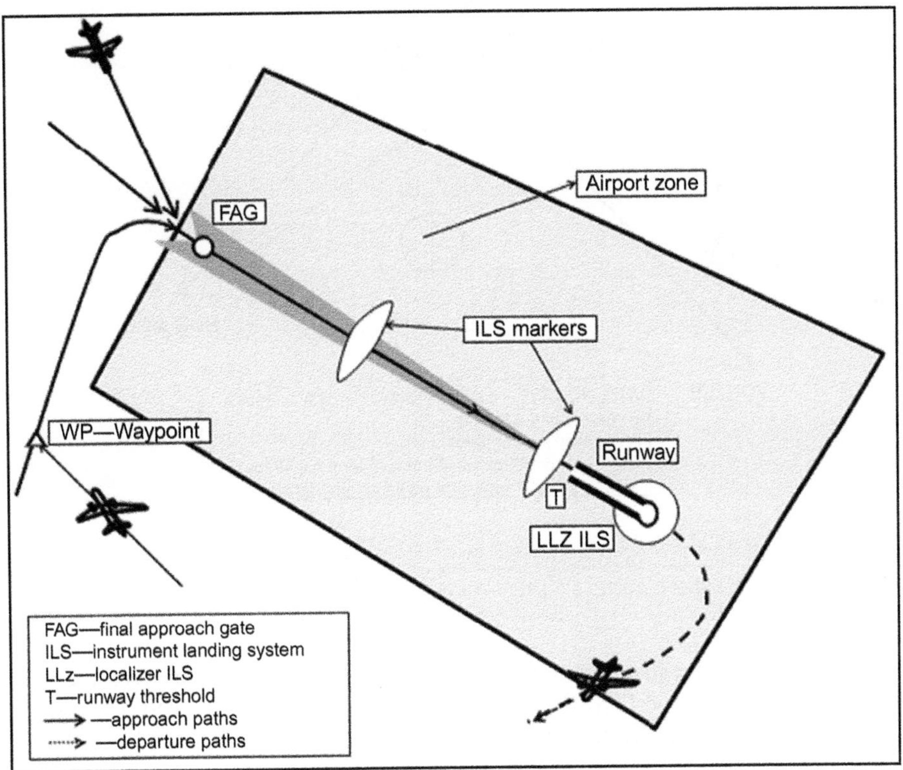

FIG. 6.51

Scheme of the horizontal layout of an airport zone (SMATSA. 2011).

up to FL (flight level) 100 (10,000 ft or 3000 m—Class B airspace; each FL is determined by 10^3 ft (1 ft ~ 0305 m)) (ICAO, 2007). The aircraft fly through this area, along the prescribed arrival and departure trajectories, defined either by radio-navigational facilities, the ATC radar vectoring, and/or by RNAV (area navigation) and RNP (required navigation performance). Fig. 6.52 shows the horizontal layout of the terminal airspace around an airport of the average size.

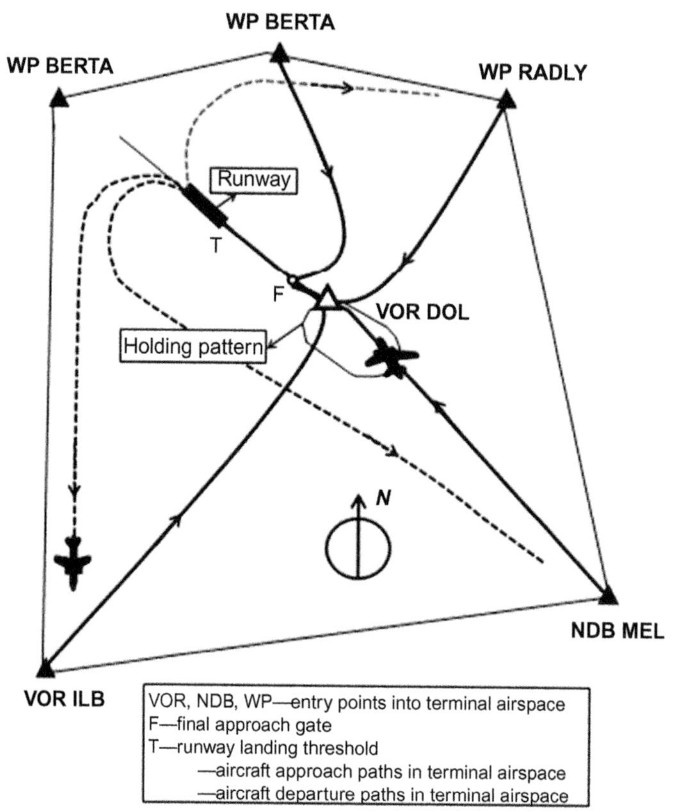

FIG. 6.52

Scheme of the horizontal layout of terminal airspace—airport Ljubljana (Republic of Slovenia) (Class E airspace).

- *Low altitude airspace* spreads around and above the terminal airspace between the altitudes between FL100 (10,000) and FL180 (18,000 ft) (or FL245—24,500 ft) MSL (Horonjeff and McKelvey, 1994; ICAO, 2007). The FLs can be odd and even. The odds are FL110, 130, 150, etc. The evens are FL120, 140, 160, etc. The departing aircraft from the given airport use this airspace for climbing to the cruising altitude, which is usually in the high altitude (upper) airspace. Their trajectories continue to those established in the terminal airspace. The arriving aircraft, after

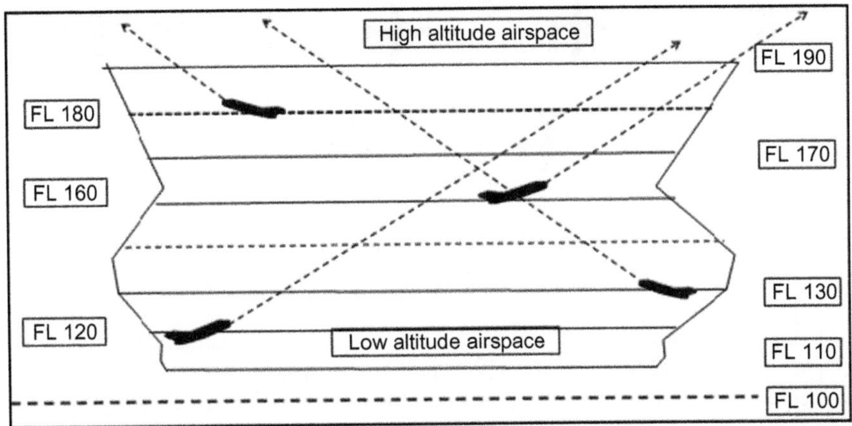

FIG. 6.53

Scheme of the vertical profile and common traffic pattern in the low altitude airspace (Class E airspace).

leaving their cruising altitudes in the high altitude (upper airspace) descend through this airspace along the trajectories, which precede to those arrival trajectories in the terminal area. In this airspace, the aircraft trajectories are defined similarly as those in the terminal airspace—either by radio-navigational facilities, the ATC radar vectoring, and/or RNAV and RNP. Fig. 6.53 shows a scheme of possible traffic pattern in this airspace within the range of altitudes between FL100 and FL180.

- *High altitude (upper) airspace* spreads covers the altitudes between FL180 (or FL245) and FL600 MSL (Class A airspace) (FL600—60,000 ft). In this airspace, the aircraft perform cruising phase of their flights (ICAO, 2007). Fig. 6.54A and B shows the horizontal and vertical profile of this airspace and the possible traffic pattern over the North Atlantic airspace where the aircraft trajectories as tracks in the horizontal plane are defined by the sets of so-called WPs (way point(s)), each defined with its geographical coordinates—longitudes and latitudes. In the vertical plane, these trajectories have 12 FLs in both directions, between FL285 and FL410. The aircraft being in the constant HF (high frequency) radio-communication with ATC follow this trajectories thanks to GPS, ADS-C (automatic dependent surveillance—contract), CPDLC (controller-pilot data link communications), RNAV and RNP systems. As can be seen, the aircraft on the same tracks at the same FL are separated by the time separation minima of 10 min. Recently, on the very busy tracks, based on these 10 min minima have been reduced to 5 min. The minimum vertical separation between aircraft amounts 1000 ft (NATS, 2011).

6.13.2.2 Airports

Airports represent the ground-based infrastructure of the ATC system. They handle the aircraft before their departure and during take-off phase of their flights, and during landing and after arrival. The ATC monitors and controls the aircraft movements in the airport airside area consisting of runways, taxiways, and apron-gate complex. Fig. 6.55 shows a simplified scheme of an airport airside area under jurisdiction of the ATC airport unit.

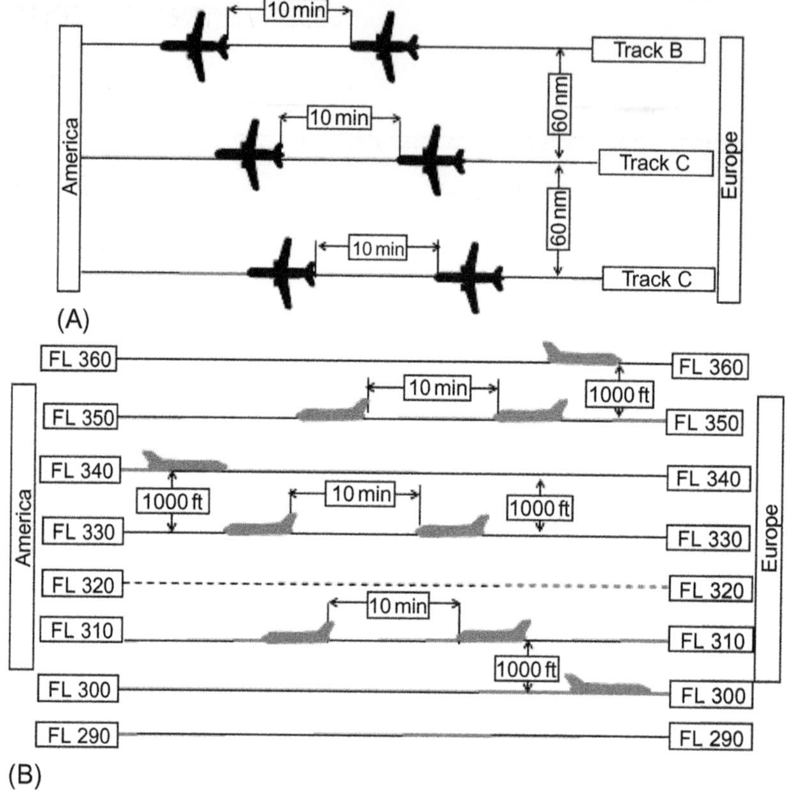

FIG. 6.54

Scheme of the high altitude (upper) airspace over North Atlantic between Europe and America: (A) horizontal layout and (B) vertical profile.

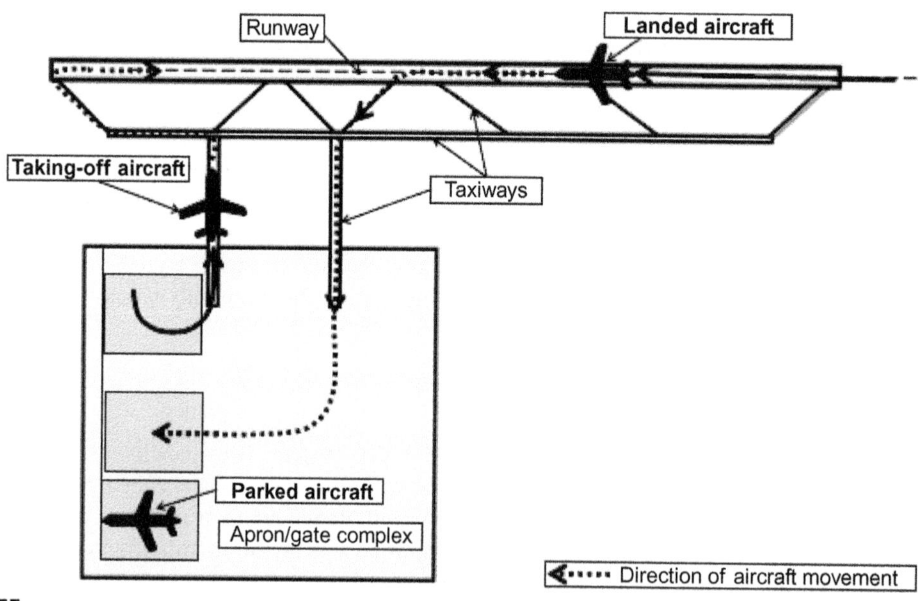

FIG. 6.55

Simplified layout of an airport airside area under jurisdiction of ATC airport unit.

6.13.3 SUPPORTIVE FACILITIES AND EQUIPMENT

The supportive facilities and equipment of the ATC system include: (i) communication equipment; (ii) navigation facilities; and (iii) surveillance equipment—radars, both with the components on the ground, at aircraft, and in the (air) space; and (iv) ANS (air navigation services) system provider.

- **(i)** Communication equipment includes: communication channels for transmission of information between (1) pilots and ATC controllers (VHF—very high frequency/UHF—ultra high frequency air/ground voice and nonvoice communication links); (2) particular ATC's control units; and (3) the ATC and the outside environment. The forthcoming communication systems under development within the EU (European Union) SESAR (Single European Sky ATM Research) and US NextGen (Next Generation) initiatives imply mostly the automated nonvoice transmission of information throughout the ATC system through all three above-mentioned "channel" (http://www.sesarju.eu/; http://www.faa.gov/nextgen/).
- **(ii)** Navigational facilities include the ground aids and space satellites both used for the aircraft primary navigation. The ground aids are: NDB (nondirectional (radio) beacon), VOR (VHF omni directional radio range), DME (distance measuring equipment), VOR/DME, and RNAV system (the information about distance and azimuth by VORTAC stations as input); the overwater en-route OMEGA, LORAN-C; DNS (doppler navigation system); INS (inertial navigation system); the terminal airspace/airport ILS (instrumental landing system) and MLS (microwave landing system), and the airport navigational equipment including the approach lighting systems, slope indicators, the surface detection equipment, etc. The forthcoming navigational procedures are to be primarily based on RNAV and RNP system, which will enable flying along the most preferred routes thus contributing to increased flexibility of using airspace and consequently higher efficiency and effectiveness of flights (Graves, 1998).
- **(iii)** Surveillance equipment consists of the primary and the beacon (secondary) radars enabling the ATC controllers to monitor and control separation between aircraft and make appropriate decisions, which are sent to the pilots by the A/G (air-ground) communications link (Horonjeff and McKelvey, 1994; Graves, 1998). The forthcoming ADS-B (automatic dependent surveillance) in combination with CDTI (cockpit display traffic information) system will enable the aircraft/pilots self-monitoring of their mutual separation, thus partially relieving the ATC controllers' workload (http://www.sesarju.eu/; http://www.faa.gov/nextgen/).
- **(iv)** ANS system provides services to the aircraft during all phases of their flights. These services are categorized as follows (ICAO, 2011): (1) ATM—air traffic management; (2) CNS—communication, navigation, and surveillance; (3) MET—meteorological; (4) AIS; and (5) SAR—search and rescue. Fig. 6.56 shows the scheme.

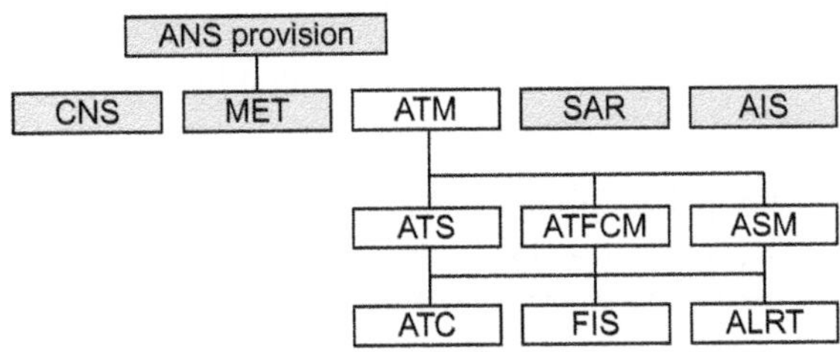

FIG. 6.56

Categorization of ANS (air navigation services) (ICAO, 2011).

In Fig. 6.55, the component CNS provide services at airports, MET services about the weather, SAR the corresponding services and activities, AIS a range of databases on the air transport components and operations, and ATM the real-time functioning of ATC providing: (a) a safe separation between particular aircraft according to the prescribed rules and (b) a safe, efficient, and effective movement of traffic flows under given conditions. For such purpose, as can be seen from Fig. 6.54, the ATM system subfunctions at the first level are: ATS (air traffic services), ATFM (air traffic flow control/management), and ASM (airspace management) (ICAO, 2007, 2012, 2013; Janić, 2000). They possess the characteristics as follows:

- *ATS* as a core subfunction of the ATM function continuously provides conditions for safe, efficient, and effective movement of air traffic flows;
- *ATFM* balances the air traffic flows as the demand for service and the available ATC system capacity aiming at preventing the long ground/airport and air/airspace aircraft/flight delays, and possible safety implications. The ATFM can be applied to the strategic and tactical level. At the strategic level, it provides planning of handling the aircraft/flights respecting their preferred four-dimensional trajectories several hours and/or even several months in advance. At the tactical level, the ATFM sets-up the acceptance rates for the specific en-route points and/or origin/destination airports accompanied by the allocation of departure slots The tactical ATFM operates online implying updating of the aircraft/flights position regularly in the very short time spots (1–3 min) (EEC, 2007, 2008; Odoni, 1987).
- *ASM* focuses on the availability of airspace for the aircraft/flight optimal trajectories. Let us consider the above-mentioned example of the currently very busy North Atlantic airspace: The aircraft traffic flows are kept on tracks (airways), the ATC applies the minimum longitudinal separation between aircraft on the same FL(s) of 10 min, the vertical separation of 2000 ft, and the lateral separation between tracks (airways) of 60 nm. By reducing the vertical separation to 1000 ft (also thanks to the ADS-B) for the aircraft flying at the altitudes FL290-FL410, six additional cruising levels can be provided. In addition, reducing the lateral separation from the current 60 to 30 nm, which will be possible thanks to RNAV, proportionally more additional tracks (airways) can be opened. Consequently, the airspace capacity could be doubled as compared to its current counterpart (http://www.sesarju.eu/; http://www.faa.gov/nextgen/).

The other ATM functions are: ATC, FIS (flight information services), and ALRT (air traffic alert), all providing services to flights during their planning and realization (ICAO, 2011).

6.13.4 THE ATC SEPARATION RULES AND PROCEDURES

The main task of the ATC system (and ATC controller(s)) is to prevent collisions between aircraft while in the air and on the ground, and their collisions with the ground. For such purpose, the ATC controllers establish, monitor, and control the safe separation between aircraft/flights by using the above-mentioned facilities and equipment.

- In the airport zone and terminal airspace, the ATC horizontal distance-based separation rules are applied between the landing aircraft on the same runway depending on the sequences in order to protect trailing aircraft from the wake vortices of leading aircraft. These are: 2.5, 3, 4, 5, and 6 nm.

The ATC time-based separation rules of 1 and 2 min are applied between the successive taking-off aircraft from the same runway depending on the aircraft taking-off sequence. In addition, only one aircraft can occupy the runway at time (ICAO, 2007).

- In the low and high (upper) altitude airspace, both the ATC vertical and horizontal time- or distance-based separation rules are applied. The vertical separation rules of 1000 ft are applied to the aircraft/flights cruising at FLs up to FL290 (29,000 ft) and 2000 ft above. In some controlled airspaces, the vertical separation of 1000 ft is applied up to FL410 (41,000 ft) and 2000 ft above. In addition, the vertical separation rules can be applied between two climbing/descending aircraft when one reach the required FL before the other arrives on its previous FL. Otherwise, the ATC horizontal separation rules are applied between these and the aircraft cruising on the same FL and along the same route. The distance-based longitudinal separation rule is 15 nm (28 km). In the case of the radar/ADS-B/CDTI surveillance, the minimum horizontal separation rules between the aircraft/flights on the same FL is 5 nm. The minimum time-based separation rules are 3, 5, 10, and 15 min, depending on the configuration of air route(s), the navigational aids used, and the traffic situation (for example, crossing routes at different angles/courses, climbing and descending along the same route, etc.). If RNP system is applied, the minimum lateral separation rules between the aircraft flying on the same route and on the same FL need to be for at least 4 times greater than the desired routes accuracy limits (ICAO, 2007; Janić, 2014).

6.13.5 THE ATC STAFF—CONTROLLER AND PILOTS

The ATC staff includes the ATC controllers and ultimately the pilots. In particular, the ATC controllers are responsible for monitoring and controlling the air traffic in the airspace of their jurisdiction—this is usually a sector within the terminal, low, and/or high (upper) altitude airspace. The ATC controllers are usually organized as teams of two or three persons with the precise division of tasks and responsibilities. The latter actors—pilots participate in the ATC processes by following, in addition to their own, instructions and guidance from the ATC controllers. These instructions mainly relate to maintaining the allocated four-dimensional trajectories and maintaining separation from other aircraft/traffic and the obstacles on the ground (in the vicinity of and at the airports). In the contemporary ATC systems, the ATC controllers use the workstations equipped by the synthetic radar screens and consoles with the information on the aircraft/flights, which are currently under their jurisdiction, and those which are expected to be so soon at the entry of their sector(s) (in the next half or an hour). At the beginning, the information about the aircraft/flight based on the flight plan is passed to the ATC unit through AFTN (Aeronautical Fixed Telecommunication Network). The ATC registers the received information about the flight, which generally contains: the radar identification code, aircraft type, planned origin and destination airport, required route and FLs along it, ground speed, type of the avionics onboard, etc. In total, in strip-based system, each flight is described by 21 descriptors, each with its field on the strip. In stripless-automated system, the "strip" has digital form with the fields for 16 flight descriptors. Each aircraft/flight is characterized by five states while being handled in a given ATC sectors as follows: (1) notified, (2) coordinated, (3) assumed, (4) transfer initiated, and (5) redundant state. In terms of time, the airline flights are planned 3–6 months in advance. The ATC capacity needed to handle them is planned per day and hours of the day. The planning of handling flights in given sector is planned 5–20 min in advance, ie, before they arrive, and then planning how to control of flights in the sector is planned about 5 min in advance. Finally, the flights are navigated and guided through the sector in time windows less than 5 min. It should be mentioned that the contemporary ATC system are equipped with the radar coverage (primary and secondary radar) of the airspace in their jurisdiction. This enables the ATC controller(s) to have two-dimensional picture of the traffic situation in the sector. The following are information on the radar screen, related to the flight/aircraft: identification sign/code of the aircraft/flight, its position in the given airspace/sector, flying altitude, and the time at the reporting points. Fig. 6.57 shows the very simplified scheme.

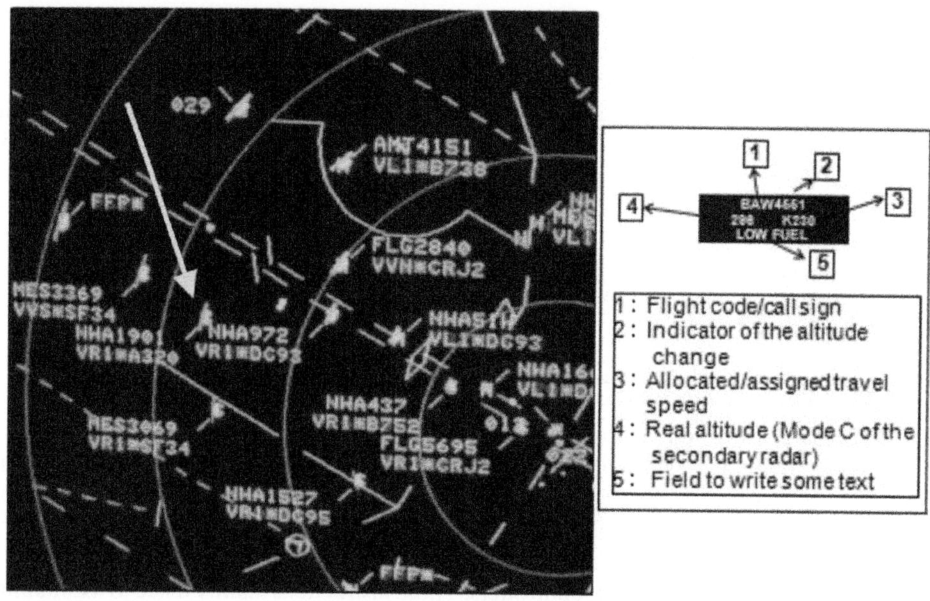

FIG. 6.57

Simplified scheme of the ATC radar screen (Janić, 2000).

Consequently, the question may be how a given flight is realized? An illustrative example can be the flight JP438 between Ljubljana (Slovenia) and Amsterdam (the Netherlands) in Europe. At the departure gate of the origin airport (Ljubljana), the pilot of A319 aircraft communicates with the ATC tower controller at the Ljubljana airport by confirming the planned characteristics of its flight such as altitude, speed, route, and estimated flight time. After the flight gets the departure clearance, the pilot receives taxiing instructions, leaves the apron/gate complex, and proceeds to the departing runway threshold. After arriving at the departure threshold of the (single) runway and being ready for take-off, the pilot once again contacts the ATC tower controller, who using his own view from the tower clears the aircraft for taking-off. One nm (nautical mile) away from take-off location, the controller at Ljubljana tower transfers the responsibility (by voice communication) for the flight to a terminal airspace controller also at Ljubljana airport, who directs the pilot to the proper course/trajectory for the first leg of the flight (see Fig. 6.56). About 25 nm farther along the aircraft/flight climbing trajectory, the terminal controller transfers the responsibility by instructing the pilot to contact the ATC controller of the en-route (low altitude) Austrian ATC Center-sector (again by controller-controller vice communication link). The controller of the Austrian Center-sector tracks the plane as it continues to climb to approximately 24,000 ft. Then he/she hands over the flight to another controller of the en-route (high-upper) altitude German ATC center—sector (again by voice communication), who handles the flight reaching the cruising altitude of 33,000 ft. At this moment, the aircraft is about 100 nm from the departure airport (Ljubljana). After some time, the German ATC Center passes responsibility to the Netherlands en-route ATC center, which instructs pilot to start descending procedure. The plane continues its descent until the ATC terminal controller takes over the responsibility for the flight and lines up the plane for its final approach to Amsterdam Schiphol airport. About 5 nm from the assigned landing runway, the responsibility is again passed to the ATC approach controller of Schiphol airport, where he/she monitors the aircraft/flight instrument final approach and landing on the radar screen and also visually. The last handoff of the flight takes place from the approach to the tower/ground control, which directs the aircraft/flight to its assigned arrival gate at the apron/gate complex. The duration of this flight is usually about 2 h. As well, the flight is continuously monitored at the radar screen(s) covering the particular ATC areas of jurisdiction/sectors while passing through.

6.13.6 AUTOMATION

In general, automation can be defined as an application of devices and/or systems partially or completely for realizing the functions and/or activities, which have previously been carried out by the human operators (Wickens, 1988). The main objectives of implementation of the automation in the ATC system are raising of the level of safety, and increasing efficiency and effectiveness of the air traffic under jurisdiction of particular ATC controllers(s), while at the same time reducing his/her corresponding workload.

Up to date, 10 levels of automation of the given system have been identified. The first level corresponds to the complete absence of automation—the man carries out all functions and activities. In this case, the machine does not provide any support to the man. The last, 10th level corresponds to the systems whose all activities and functions are carried out by machine(s). In this case, the machine(s) decides on everything and operates completely autonomously. At the intermediate levels, the performance and functions are divided between the man and machine(s), with increasing role of the latter on the account of the former at the higher automation levels. In the ATC system, automation relates to three interrelated and linked activities of the ATC controller: (1) generation of the right decisions and selection of the related actions; (2) collection of information necessary for carrying out monitoring and controlling activities and processes; and (3) implementation of the selected actions (Wickens, 1988). In general, at the working place of ATC controller(s), the automated tools are the self-explaining: CDT (conflict detection tools), MA (monitoring aids), and SYSCO (system supported co-ordination). In addition, more intensive automation of the ATC system's operations and processes has recently started through the European SESAR and US NextGen research and development programs (Erzberger, 2004; http://www.sesarju.eu/; http://www.faa.gov/nextgen/). Their main objectives are to develop a new generation ATC system, which will be able to cope with growing demand by increasing the system capacity, while at the same time improve safety, efficiency, and effectiveness of flights, the latter two by improving the fuel/energy/cost efficiency and reducing the aircraft/flight delays, and consequently the overall environmental and social friendliness of air transportation (FAA, 2011).

The main ATC components to be developed with slight variations in both above-mentioned programs depend on the phase of flight as given in Table 6.7 for the US NextGen (Next Generation) program.

Table 6.7 Components of the US ATC NextGen Development Initiative (FAA, 2013, http://www.faa.gov/nextgen)

Flight Planning
Ground ATC facilities/equipment
CSS-Wx Data comm. ERAM Modernized AIM system NWP SWIM TFDM TFMS
Push Back, Taxi, and Departure
Ground ATC facilities/equipment
ADS-B ground stations ASDE X GSS-Wx Data comm. Integrated departure and arrival coordination system Modernized AIM system NWP SBAS STARS SWIM. TFDM. TEMS
Avionics
ADS-B in and out with associate displays like CDTI RNAV and RNP Data comm.
Climb and Cruise
Ground ATc facilities/equipment
ADS-B ground stations Advanced Technologies and Oceanic Procedures GSS-Wx, data comm., ERAM, NVP, TBFM TFMS

Table 6.7 Components of the US ATC NextGen Development Initiative (FAA, 2013, http://www.faa.gov/nextgen)—cont'd
Avionics
ADS-B in and out with associate displays CDTI Data comm. including integration with FMS Future air navigation system in oceanic airspace RNAV and RPN
Descent and Approach
Ground ATC facilities/equipment
ADS-B ground stations ASDE-X Data comm. GSS-Wx NWP, SBAS STARS enhancements, TBFM, TEDM, TFMS
Avionics
ADS-B in and out with associate displays like CDTI EFVS Data comm. RNAV and RNP, VN
Landing, Taxi, and Arrival at the Gate
Ground ATC facilities/equipment
ADS-B ground stations ASDE-X CSS-W Data comm. GBAS Integrated departure and arrival coordination system Modernized AIM system SBAS, STARS enhancements, WIM, TBFM, TFDM, TFMS
Avionics
ADS-B in and out with associate displays like CDTI EFVS, data comm., GBAS
AIM, aeronautical information management; *ASDE X*, airport surface detection equipment—model X; *ADS-B*, automatic dependent surveillance broadcast; *CDTI*, cockpit display of traffic information; *CSS-Wx*, common support services-weather; *Data comm.*, data communication; *EFVS*, enhanced flight vision system; *ERAM*, en route automation modernization; *FMS*, flight management system; *GBAS*, ground based augmentation system; *NWP*, NextGen weather processor; *RNAV*, area navigation; *RNP*, required navigation performance; *SBAS*, satellite based augmentation system; *STARS*, standard terminal automation replacement system; *SWIM*, system wide information management, *TBFM*, time based flow management; *TFDM*, terminal flight data manager; *TFMS*, traffic flow management system; *VN*, vertical navigation.

The particular components in Table 6.7 should enable:

- Improved access to different information about the state of airspace through common network during the flight planning.
- More precise planning of the actual departure times including improvements in the departure performances by using RNAV and RNP supported multiple departure paths particularly over the metropolitan areas.
- Improved surveillance enabling reduction of the aircraft separation minima during climbing and cruising phase of flight (the responsibilities will be shared between the ATC controllers and pilots, thanks to ADS-B and CDTI).
- Planning well in advance arrivals and departures along the optimal single and/or multiple 4D trajectories (supported by RNAV and RNP) between the end/beginning of cruising and the runway landing/departure phase of flight, respectively, in terms required safety, fuel consumption (and related emissions of GHG (green house gases)), and noise. These trajectories will also end/begin at the runways depending on the current and predicted congestion and delays. As well, during the arrivals, ground maneuvering, and departures, independently on the weather conditions (good or bad), both ATC controllers and pilots will be able to monitor the aircraft position relative to the surrounding traffic on the associated screens (CDTI).
- Reducing the aircraft/flight separation minima thanks to the surveillance improved precision.
- Transmitting most of the information/data automatically thus reducing quantity and intensity of the A/G voice communications independently on the content (routine, exceptional, emergency, weather, etc.), and consequently increasing the ATC controller's capacity at the same workload.

The European SESAR has the very similar components as those in Table 6.7 and the quantitative targets as follows (EEC, 2013; Griffiths, 2013; http://www.sesarju.eu/):

- Decreasing the ATM-related charges for the airspace users for about 50%.
- Reducing the impacts of aircraft/flights on the environment for about 10%.
- Increasing the ATC system capacity for about 200%, which will enable reducing both ground and airborne delays.
- Improving the safety level by 10%.

The development of eth above-mentioned system and components is in progress. Very likely, they will be gradually implemented, implying parallel functioning of existing and new components until complete replacement of the former with the later.

6.13.7 THE WORKLOAD AND CAPACITY OF ATC CONTROLLER(S)

In order to enable moving of air traffic efficiently (as low as possible aircraft flight cost), effectively (without significant aircraft/flight delays), and safely (without incidents/accidents due to known reasons), the controlled airspace of a given area or a country is usually divided into the smaller three-dimensional parts called "ATC sector(s)." Each sector and its traffic are assigned to one or a team (usually two) air traffic (ATC) controllers. They generally perform monitoring and executing tasks for each aircraft before and while in the sector under their jurisdiction. These two categories of tasks consist of six specific tasks: monitoring identified aircraft, coordination (external and between sectors), standard R/T (radio/telephony) communications, flight data management, conflict search and detection, and radar conflict monitoring and resolution. In the automated ATC system, some of the above-mentioned tasks such as standard R/T coordination, and support in conflict detection and resolution are executed automatically, which leaves the ATC controller(s) to carry out mainly task of monitoring of the identified aircraft and eventually take part in resolving the potential conflict(s).

The main operational parameter of an ATC sector is its capacity, which is usually expressed by the maximum number of aircraft/flights processed through the sector under given conditions during the specified period of time. These given conditions are usually constant demand for service ("ultimate" capacity) and the average delay per aircraft/flight ("practical" capacity). Alternatively, these capacities can be expressed by the maximum number of aircraft of aircraft/flights simultaneously being in the sector under monitoring and controlling by the ATC controller(s). Over the past 40 years, different analytical and simulation models have been developed for estimating the ATC sector capacity. The most common have been those based on the capacity of R/T communication channel and the ATC controller's workload. For example, at the latter models, the ATC controller's (mental) workload (*WL*) generated by the arriving traffic at the airspace sector of its jurisdiction during the specified period of time (τ) (usually ¼, ½, or 1 h), has been estimated as follows (Janić, 1989, 2000):

$$WL \approx \lambda_c \tau_r + \lambda_c^2 \tau_c \tag{6.58}$$

where

- λ_c is the capacity of the given terminal airspace based on the ATC controller's workload (ac/h) (ac—aircraft)
- τ_r is the average time of executing a routine control task for an aircraft while in the terminal airspace (s)
- τ_c is the average time for detecting and resolving potential conflict (including the task's decision-making and execution time) (s)

The workload (*WL*) in Eq. (6.58) is measured by time, which the ATC controller spends for carrying out the above-mentioned control tasks. In addition, Eq. (6.58) indicates that the ATC controller workload (*WL*) increases with the square of number of aircraft handled in the sector mainly due to at least two aircraft could participate in a single conflict. If: $WL = \tau$, then after the necessary transformations of Eq. (6.58), the "ultimate" capacity (λ^*_c) of the given sector can be calculated as the nonnegative root of the quadratic equation:

$$A\lambda_c^2 + B\lambda_c + C = 0 \tag{6.59}$$

where

$$A = 1;\ B = \tau_r/\tau_c;\ C = -(\tau \cdot U)/\tau_c \tag{6.60}$$

and

$$\lambda_c^* = \frac{-\tau_r/\tau_c \pm \sqrt{\left(\frac{\tau_r}{\tau_c}\right)^2 + 4\left(\frac{\tau \cdot U}{\tau_c}\right)}}{2} \tag{6.61}$$

The parameters (τ_r) and (τ_c) in Eqs. (6.58)–(6.61) can be estimated by measurements of time of executing particular monitoring and control tasks, by interviewing the ATC controllers in order to get their subjective judgments about the weights/complexity and the time of carrying out particular control and monitoring tasks, and/or by both simultaneously. Fig. 6.58 shows an example of the relationships between the ATC controller workload and the arriving traffic in the given sector estimated by Eq. (6.61) for the conditions of conflict-free passing through the sector.

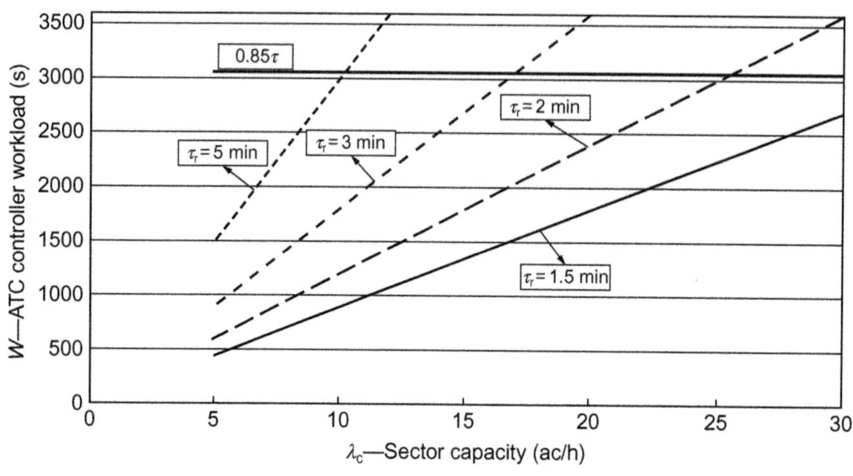

FIG. 6.58

Dependence of ATC controller workload on the intensity of arriving traffic and the average time of execution of ATC control tasks (Janić, 1989, 2000).

As can be seen, the ATC controller's workload increase linearly with increase of the arriving traffic and the time for executing control task per each aircraft while in the sector. If this workload is limited, for example, to $U=85\%$ of time of an hour of the ATC controller's operation, the intensity of arriving traffic can be constrained, thus representing the sector capacity based just on his/her allowed workload. Under such conditions, if, for example, $\tau_r=1.5$ min/ac, the capacity will be more than 30 ac/h; if: $\tau_r=3$ min/ac, the capacity will be 17 ac/h, etc.

In addition, based on Eq. (6.61), the sector capacity expressed by the number of aircraft simultaneously being in the given sector can be estimated using the Little's formula as follows (Janić, 1989, 2000; Newell, 1982):

$$N=\lambda_c^* \cdot t_f \tag{6.62}$$

where N is the number of aircraft simultaneously being in the given sector and t_f is the average aircraft flying time through the sector.

The other symbols are as in the previous equations. Fig. 6.59 shows an example of the relationships between the sector capacity (N) and the aircraft flying time through the sector (t_f).

As can be seen, for the given intensity of arriving traffic, which could be determined by the above-mentioned ATC controller's workload, the sector capacity in terms of the number of aircraft simultaneously handled there will increase with increasing of their average flying time through the sector. In addition, this capacity will increase with increase of the intensity of arriving traffic given the average flying time through the sector (Janić, 1989, 2000; Newell, 1982).

It should be mentioned that the two capacities always need to be considered while planning (timetable) and carrying the air traffic operations: the capacity of airspace and the capacity of ATC controller having jurisdiction over traffic in this airspace. In such case, the lower capacity is usually considered as relevant.

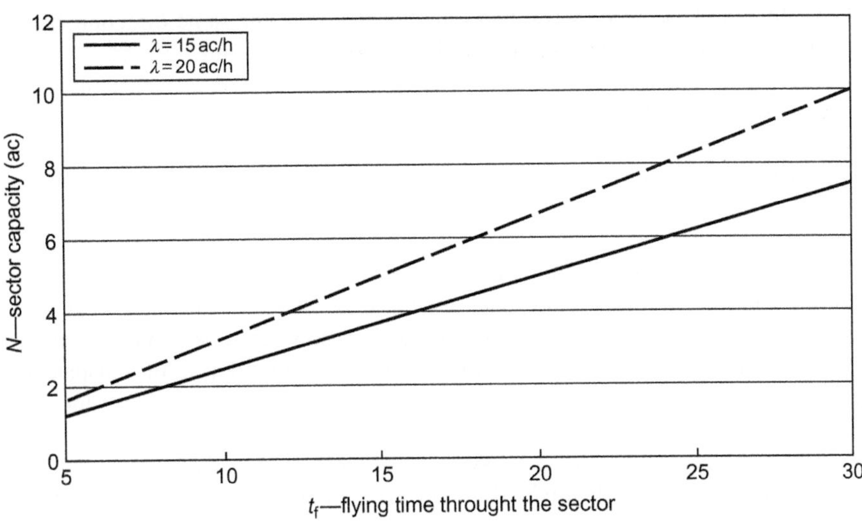

FIG. 6.59

Dependence of ATC sector capacity on the aircraft flying time in the sector and the intensity of arriving traffic (Janić, 1989, 2000).

6.14 PROBLEMS

1. The cycle length at the signalized intersection equals 80 s. The considered approach has the saturation flow of 2200 (veh/h), the green time duration of 25 s, and flow rate of 500 (veh/h). Analyze traffic conditions in the vicinity of the intersection. Calculate average delay per vehicle. Assume that the $D/D/1$ queueing system adequately describes considered intersection approach.
2. The cycle length at the signalized intersection equals 70 s. The considered approach has the saturation flow of 2200 (veh/h), the green time duration of 20 s, and flow rate of 450 (veh/h). Analyze traffic conditions in the vicinity of the intersection. Assume that the $D/D/1$ queueing system adequately describes considered intersection approach. Calculate: (a) the average delay per vehicle; (b) the longest queue length; and (c) percentage of stopped vehicles.
3. The cycle length at the signalized intersection equals 90 s. The considered approach has the saturation flow of 2200 (veh/h), the green time duration of 25 s, and flow rate of 500 (veh/h). Calculate the average vehicle delay by using Webster's formula.
4. The cycle length at the signalized intersection equals 60 s. The green time duration of the observed approach is equal to 20 s. The considered approach has the saturation flow of 1500 (veh/h), and flow rate of 500 (veh/h). Calculate the average vehicle delay using Webster's formula and Allsop's formula.
5. Explain the characteristics of the train signaling system and its operations.
6. How to determine the length of blocks on the particular railway line(s)?

7. Calculate minimum spacing between two trains operating in the same direction at the speed of 300 km/h, and the ability of decelerating at the rate of 0.5, 1.0, and 1.5 m/s^2.
8. Calculate the maximum operating speed at which a train should start braking at the rate of 1.5 m/s^2 along the distance of 3000 m.
9. Explain the *ERTMS*.
10. Explain functioning of TVM system at the HSR.
11. Itemize the main components of the rail traffic control/management system and explain their main functions.
12. Describe the concept of metro automation system, its components, and operations.
13. Explain the workload of a train dispatcher and the main factors influencing it.
14. What is the monitoring capacity of the train dispatcher, how can it be expressed, and calculated?
15. Calculate the number of trains simultaneously being under monitoring of a dispatcher if his/her monitoring capacity is 20 trains/h and if each train spends in the area of his/her jurisdiction about 45 min.
16. What are the main components of ATC system and their characteristics?
17. Explain how the airspace can be divided in both horizontal and vertical plane for the purpose of controlling air traffic.
18. What is the principal difference between an airport zone, terminal area, and low and high altitude area? What is the pattern of flights carried out there?
19. Describe the structure and role of the ATC's supporting facilities and equipment.
20. Itemize and explain the ATC operations and the aircraft/flight separation rules and procedures in particular parts of the controlled airspace.
21. What are the main objectives of increasing the level of automation in the ATC system?
22. What are the main research and developing programs dealing with increasing of automation in ATC system? What are the main ATC system's components to be automated?
23. Itemize the main expected effects from increased automation in the TAC system.
24. Describe the main tasks of an ATC controller and their influence on his/her workload.
25. Calculate the sector capacity based on the ATC controller workload if the utilization time, ie, the time of active operation of the controller during an hour is 90%, the average time for carrying out a routine task 1 min, and the time for resolving potential conflict between the aircraft 1.5 min.
26. Calculate the capacity of an ATC sector based on the number of aircraft simultaneously being under jurisdiction of the ATC controller if his capacity based on the workload is 20 ac/h and the average time the aircraft spend in the sector 40 min.

REFERENCES

ABB, 2014. Powering the World's High Speed Rail Networks. ABB ISI Rail, Geneva.

Allsop, B.R., 1971. SIGSET: a computer program for calculating traffic capacity of signal-controlled road junctions. Traffic Eng. Control 12, 58–60.

Allsop, B.R., 1976. SIGCAP: a computer program for assessing the traffic capacity of signal-controlled road junctions. Traffic Eng. Control 17, 338–341.

Andersson, A.W., Sandblad, B., Hellström, P., Frej, I., Gideon, A., 1997. A systems analysis approach to modelling train traffic control. In: Proceedings of WCRR '97, November 16–19, Florence, Italy.

Andreatta, G., Odoni, A.R., 2003. Analysis of market-based demand management strategies for airports and en route airspace. In: Ciriani, T. (Ed.), Operations Research in Space and Air. Kluwer Academic Publishers, Boston, MA.

CEC, 2001. European transport policy for 2010: time to decline. White Paper, Commission of the European Communities COM (2001) 370 Final, European Communities, Brussels.

CEC, 2011. Roadmap to a single european transport area—towards a competitive and resource efficient transport system. White Paper, Commission of the European Communities COM (2011) 144 Final, European Communities, Brussels.

Chowdhury, M., Sadek, A., 2003. Fundamentals of Intelligent Transportation Systems Planning. Artech House, Boston.

De Palma, A., Lindsey, R., 2002. Private roads, competition, and incentives to adopt time-based congestion tolling. J. Urban Econ. 52, 217–241.

EC, 1996a. Directive 96/48/EC of 23 June 1996 on Interoperability of Trans-European System of High-Speed Railways. European Commission, Brussels.

EC, 1996b. Interoperability of the Trans-European High Speed Rail System, Directive 96/48/EC. European Commission, Brussels. p. 155.

EC, 2001a. Directive 2001/16/EC of 19 March 2001 on Interoperability of Trans-European System of Conventional Railways. European Commission, Brussels.

EC, 2001b. Decision 2001/260/EC of 21 March 2001 on Basic Parameters of Control System Containing Specifications ERTMS/ETCS and ERTMS/GSM-R. European Commission, Brussels.

EU, 2012. The Roadmap for Sustainable Air Traffic Management: European ATM Master Plan. European Union, Brussels, Belgium. www.atmmasterplan.eu.

Edara, P., Teodorović, D., 2008. Model of an advance-booking system for highway trips. Transp. Res. C 16, 36–53.

Edara, P., Teodorović, D., Triantis, K., Natarajan, S., 2011. A simulation-based methodology to compare the performance of highway space inventory control and ramp metering control. Transp. Plan. Technol. 34, 705–715.

EEC, 2007. Capacity Assessment & Planning Guidance: An Overview of the European Network Capacity Planning Process, September 2007 ed. EUROCONTROL, Brussels.

EEC, 2008. European Medium-Term ATM Network Capacity Plan Assessment 2009–2012, DMEAN—Dynamic Management of the European Airspace Network, September 2008 ed. EUROCONTROL, Brussels.

EEC, 2013. Challenges of Growth 2013: Task 4: European Air Traffic in 2035. European Organization for the Safety of Air Navigation (EUROCONTROL), Brussels.

Erzberger, H., 2004. Transforming the NAS: the next generation air traffic control system. In: 24th ICAS (International Congress of the Aeronautical Sciences), September, Yokohama, Japan.

FAA, 2011. FAA Aerospace Forecast: Fiscal Years 2012–2035. US Department of Transportation, Aviation Policy and Planning, Federal Aviation Administration, Washington, DC.

FAA, 2013. NextGen Implementation Plan. Office of the NextGen, Federal Aviation Administration, US Department of Transportation, Washington, DC.

Gartner, H.N., 1983. OPAC: a demand-responsive strategy for traffic signal control. Transp. Res. Rec. 906, 75–81.

Gartner, H.N., Assmann, F.S., Lasaga, F., Hom, L.D., 1991. A multiband approach to arterial traffic signal optimization. Transp. Res. B 25, 55–74.

Gazis, C.D., 2002. Traffic Theory. Kluwer, Boston, MA.

Gordon, R., 2009. Intelligent Freeway Transportation Systems Functional Design. Springer, New York, NY.

Graves, D., 1998. UK Air Traffic Control: A Layman's Guide, third ed. Airlife Publishing Ltd., Shrewsbury.

Griffiths, P., 2013. SES Performance Targets and Achievements. EUROCONTROL, Brussels.

Haj-Salem, H., Papageorgiou, M., 1995. Ramp metering impact on urban corridor traffic: field results. Transp. Res. A 29, 303–319.

Hall, W.R., 1993. Non-recurrent congestion: how big is the problem? Are traveller information systems the solution? Transp. Res. C 1, 89–103.

Hardin, G., 1968. The tragedy of the commons. Science 162 (3859), 1243–1248.

Horonjeff, R., McKelvey, X.F., 1994. Planning & Design of Airports, third ed. McGraw-Hill, New York, NY.

Hunt, B.P., Robertson, L.D., Bretherton, D.R., 1982. The SCOOT on-line traffic signal optimization technique. Traffic Eng. Control 23, 190–192.

ICAO, 2001. Air Traffic Services: Annex 11 to the Convention on International Civil Aviation, 13th ed. International Civil Aviation Organization, Montreal. July.

ICAO, 2007. Air Traffic Management, Doc. 4444, 15th ed. International Civil Aviation Organisation, Montreal.

ICAO, 2011. Rules of the Air: Annex 2 to the Convention on International Civil Aviation, 10th ed. July, International Civil Aviation Organization, Montreal, Canada.

ICAO, 2012. ICAO's Policies on Charges for Airports and Air Navigation Services, Doc 9082, ninth ed. International Civil Aviation Organization, Montreal.

ICAO, 2013. Manual on Air Navigation Services Economics, Doc 9161, fifth ed. International Civil Aviation Organization, Montreal.

ISO 10075, 1991, Ergonomic principles related to mental work-load, general terms and definitions. https://www.iso.org/obp/ui/.

ITERIS, 2013. Key Concepts of the National ITS Architecture 7.0. http://www.iteris.com/itsarch/documents/keyconcepts/keyconcepts.pdf (accessed April 19, 2014).

Janić, M., 1989. Terminal area capacity model—a problem concerning air traffic controller workload. Transp. Plan. Technol. 13 (3), 205–216.

Janić, M., 2000. Air Transport System Analysis and Modelling: Capacity, Quality of Services, and Economics. Gordon and Breach Science Publisher, Amsterdam.

Jarašūnienė, A., 2005. General description of European railway traffic management system (ERTMS) and strategy of ERTMS implementation in various railway managements. Transp. Telecommun. 6 (5), 21–27.

JR, 2011. Data Book 2011. Central-Japan Railway Company, Tokyo.

Kotsialos, A., Papageorgiou, M., 2000. The importance of traffic flow modelling for motorway traffic control. Netw. Spat. Econ. 1, 179–203.

Kotsialos, A., Papageorgiou, M., 2001. Efficiency versus fairness in network-wide ramp metering. In: 2001 IEEE Intelligent Transportation Systems Conference Proceedings. Oakland, CA, USA, August 25–29, pp. 1190–1195.

Kotsialos, A., Papageorgiou, M., 2004. Motorway network traffic control systems. Eur. J. Oper. Res. 152, 321–333.

Kovačević, S., 1988. Eksploatacija zeleznica I and II. Zavod za novinsko-izdavacku i propagandnu delatnost, Belgrade.

Larson, R., Sasanuma, K., 2010. Urban vehicle congestion pricing: a review. J. Ind. Syst. Eng. 3, 227–242.

Lenior, T.M.J., 1993. Analyses of cognitive processes in train traffic control. Ergonomics 36 (11), 1361–1368.

Litman, T., 2006. London Congestion Pricing: Implications for Other Cities. Victoria Transport Policy Institute, Victoria, BC, Canada. http://www.vtpi.org/london.pdf.

Little, D.C.J., 1966. The synchronization of traffic signals by mixed integer-linear-programming. Oper. Res. 14, 568–594.

Little, D.C.J., Kelson, D.M., Gartner, H.N., 1981. MAXBAND: a program for setting signals on arteries and triangular networks. Transp. Res. Rec. 795, 40–46.

Lloyd, W.F., 1833. Two Lectures on the Checks to Population. Oxford Univ. Press, Oxford, England. Reprinted (in part) in: Hardin, G. (Ed.), Population, Evolution, and Birth Control. Freeman, San Francisco, 1964, p. 37.

Lowrie, P.R., 1982. SCATS: the Sydney co-ordinated adaptive traffic system—principles, methodology, algorithms. In: Proc. IEE Int. Conf. Road Traffic Signalling, London, pp. 67–70.

Mammar, S., Messmer, A., Jensen, P., Papageorgiou, M., Haj-Salem, H., Jensen, L., 1996. Automatic control of variable message signs in Aalborg. Transp. Res. C 4, 131–150.

Mirchandani, P., Head, L., 1998. RHODES—a real–time traffic signal control system: architecture, algorithms, and analysis. In: TRISTAN III (Triennial Symp. Transport. Analysis), vol. 2, San Juan, Puerto Rico, June 17–23.

Mohan, S., 2012. Dubai Metro Signalling & Train Control System, Aspect 2012—100 Years. Institute of Railway Signal Engineers, London. p. 10.

NATS, 2011. AIRE lot 3.3—reduced longitudinal separation minimum in the North Atlantic. D2—Phase 2 Deliverable Package, A2RLG/D2/01, NATS En Route Plc, National Air Traffic Control Services, Whiteley.

Newell, G., 1982. Application of Queuing Theory, second ed. Chapman and Hall, Routledge, London.

Odoni, A.R. (Ed.), 1987. The flow management problem in air traffic control. In: Flow Control of Congested Networks. NATO ACI Series, vol. F38. Springer-Verlag, Berlin, Heildelberg.

Papageorgiou, M., 1983. Application of Automatic Control Concepts to Traffic Flow Modeling and Control. Springer-Verlag, New York, NY.

Papageorgiou, M., 1990. Dynamic modeling, assignment, and route guidance in traffic networks. Transp. Res. B 24, 471–495.

Papageorgiou, M., 1994. An integrated control approach for traffic corridors. Transp. Res. C 3, 19–30.

Papageorgiou, M., Kotsialos, A., 2002. Freeway ramp metering: an overview. IEEE Trans. Intell. Transp. Syst. 3, 271–281.

Papageorgiou, M., Blosseville, J.-M., Hadj-Salem, H., 1990. Modeling and real-time control of traffic flow on the southern part of Boulevard Périphérique in Paris—part II: coordinated on-ramp metering. Transp. Res. A 24, 361–370.

Papageorgiou, M., Hadj-Salem, H., Blosseville, J.-M., 1991. ALINEA: a local feedback control law for on-ramp metering. Transp. Res. Rec. 1320, 58–64.

Papageorgiou, M., Haj-Salem, H., Middelham, F., 1997. ALINEA local ramp metering: summary of field results. Transp. Res. Rec. 1603, 90–98.

Papageorgiou, M., Diakaki, C., Dinopoulou, V., Kotsialos, A., Wang, Y., 2003. Review of road traffic control strategies. Proc. IEEE 91, 2043–2067.

Pigou, A.C., 1920. The Economics of Welfare. Macmillan and Co., London.

Robertson, I.D., 1969. TRANSYT method for area traffic control. Traffic Eng. Control 10, 276–281.

Šelmić, M., Macura, D., Teodorović, D., 2011. Ride matching using K-means method: case study of Gazela bridge in Belgrade. J. Transp. Eng. 138 (1), 132–140.

SMATSA, 2011. Zbornik Vazduhoplovnih podataka Republike Srbije. Aeronuatical Information Publication—AIP, Belgrade.

Smulders, S., 1990. Control of freeway traffic flow by variable speed signs. Transp. Res. B 24, 111–132.

Sullivan, E.C., El Harake, J., 1998. California route 91 toll lanes—impacts and other observations. Transp. Res. Rec. 1649, 55–62.

Teodorović, D., Dell'Orco, M., 2008. Mitigating traffic congestion: solving the ride-matching problem by bee colony optimization. Transp. Plan. Technol. 31, 135–152.

Teodorović, D., Edara, P., 2007. A real-time road pricing system: the case of two-link parallel network. Comput. Oper. Res. 34, 2–27.

Teodorović, D., Lučić, P., 2006. Intelligent parking systems. Eur. J. Oper. Res. 175, 1666–1681.

Teodorović, D., Varadarajan, V., Popović, J., Chinnaswamy, M.R., Ramaraj, S., 2006. Dynamic programming—neural network real—time traffic adaptive signal control algorithm. Ann. Oper. Res. 143, 121–129.

Teodorovic, D., Triantis, K., Edara, P., Zhao, Y., Mladenovic, S., 2008. Auction-based congestion pricing. Transp. Plan. Technol. 31, 399–416.

Triantis, K., Sarangi, S., Teodorović, D., Razzolini, L., 2011. Traffic congestion mitigation: combining engineering and economic perspectives. Transp. Plan. Technol. 34, 637–645.

UIC, 2010. Necessities for future high speed rolling stock. Report January 2010, International Union of Railways, Paris.

UITP, 2012. Metro Automation Facts, Figures and Trends: A Global Bid for Automation: UITP Observatory of Automated Metros Confirms Sustained Growth Rates for the Coming Years. The International Association of Public Transport (UITP), Brussels.www.uitp.org.

UNIFE, 2014a. The ERTMS Memorandum of Understanding a Cross-Sector Agreement to Ensure ERTMS' Success. The European Railway Industry, Brussels.

UNIFE, 2014b. The ERTMS Levels: Different Levels to Match Customer's Needs. The European Railway Industry, Brussels.

UNIFE, 2014c. The ERTMS from the Drivers' Point of View: How ERTMS Facilitates Train Operations for Drivers. The European Railway Industry, Brussels.

UNIFE, 2014d. The ERTMS Deployment Statistics—Overview. The European Railway Industry, Brussels.

Verhoef, E.T., 2002a. Second-best congestion pricing in general networks: algorithms for finding second-best optimal toll levels and toll points. Transp. Res. B 36, 707–729.

Verhoef, E.T., 2002b. Second-best congestion pricing in general static transportation networks with elastic demands. Reg. Sci. Urban Econ. 32, 281–310.

Verhoef, E., Nijkamp, P., Rietveld, P., 1995. Second-best regulation of road transport externalities. J. Transp. Econ. Policy 29, 147–167.

Vickrey, W., 1955. A proposal for revising New York's subway fare structure. J. Oper. Res. Soc. Am. 3, 38–68.

Vickrey, W., 1963. Pricing in urban and suburban transport. Am. Econ. Rev. Pap. Proc. 53, 452–465.

Vickrey, W., 1969. Congestion theory and transport investment. Am. Econ. Rev. Pap. Proc. 59, 251–261.

Vickrey, W., 1994. Statement to the joint committee on Washington, DC, metropolitan problems. J. Urban Econ. 36, 42–65.

Volpe, J., 1999. Understanding how train dispatchers manage and control trains. Results of a preliminary cognitive task analysis, human factors in railroad operations. Final Report, DOT/FRA/ORD-99/XX, DOT-VNTSC-FRA-99-XX. US Department of Transportation, Federal Railroad Administration, Massachusetts Institute of Technology (MIT), the Volpe National Transportation Systems Center, Cambridge, MA.

Webster, V.F., 1958. Traffic signal settings. Road Res. Tech. Paper No. 39, Road Research Laboratory, London.

Whelan, R., 1995. Smart Highways, Smart Cars. Artech House Publishers, Boston, MA.

Wickens, C., 1988. Automation issues in air traffic management. In: Future of Air Traffic Control: Human Operators and Automation. National Academies Press, Washington, DC, pp. 11–45.

Yang, H., Huang, H.J., 1998. Principle of marginal-cost pricing: how does it work in a general road network? Transp. Res. A 32, 45–54.

Yang, H., Huang, H.J., 1999. Carpooling and congestion pricing in a multilane highway with high-occupancy-vehicle lanes. Transp. Res. A 33, 139–155.

Zeilstra, M., De Bruijn, D., Van der Weide, R., 2012. Development and implementation of a predictive tool for optimizing workload of train dispatchers. In: Wilson, R.J., Mills, A., Clarke, T., Rajan, J., Dadashi, N. (Eds.), Rail Human Factors around the World: Impacts on and of People for Successful Rail Operations. CRC Press, Taylor & Francis Group, London, UK.

WEBSITES

http://demo.oxalis.be/unife/ertms/?page_id=42.

http://en.wikipedia.org/wiki/European_Rail_Traffic_Management_System.

http://ertico.com/vision-and-mission.
http://www.faa.gov/nextgen.
http://www.iteris.com/itsarch/html/menu/hypertext.htm.
http://www.its-jp.org/english/what_its_e.
http://www.railway-technology.com/projects/european-rail-traffic-management-system-ertms.
http://www.sesarju.eu.

When did organized urban public transportation and rail intercity transportation appear? What are demand-responsive transportation systems? What is the relationship between the average waiting time per passenger at the station and the vehicle headway? How to generate timetable? How to design public transit network? What is airline supply and what is airline capacity? What is Revenue Management concept?

PUBLIC TRANSPORTATION SYSTEMS

7

7.1 INTRODUCTION

Public transportation is a *shared* transportation service. There are *urban public transit* and *intercity public transportation*. The main transportation modes in urban public transit are buses, trams, trolley-buses, trains, and metro. Ferries also appear in some cities in the world as an urban transportation mode. Airlines, buses, intercity rail, and, in some countries, high-speed rail are the transportation modes that appear in intercity public transportation. In the majority of cases, all forms of urban public transit and intercity public transportation are offered to the general public.

Organized urban public transportation and rail intercity transportation appeared in the 19th century. In 1885, Europe had already more than 50 tramway operators. Today, rapid transit rail systems in New York, Moscow, Tokyo, Beijing, Seoul, and other big cities in the world deliver millions of rides on weekdays. The New York rapid transit system has close to 500 transit stations in operation. Forty million passengers use the rail system daily within Greater Tokyo. Only these numbers are sufficient to indicate the extent to which the life in big cities depends on public transit. Today, public transit operators are one of the key factors in improving quality of life by promoting sustainable transportation in urban areas.

Scheduled commercial aviation started in 1914, when a single paying passenger was transported on a flight across Tampa Bay, Florida. This happened on New Year's Day 1914. A hundred years later, air transportation has created an opportunity for the people to be practically any place in the world in 24 h. The world's major carriers made about 30 million commercial flights in 2013, carrying more than 3 billion passengers.

Transportation Engineering. http://dx.doi.org/10.1016/B978-0-12-803818-5.00007-X

Public transportation problems are very complex. When designing public transit network, determining the bus frequencies and defining bus schedules it is necessary to take into account the city zones coverage, passenger interests, the operator interests, the available number of vehicles, the drivers' workload and working regulations, the available budget, etc.

There are transit operators in the world that operate a few thousand buses that serve a few hundred bus routes. They employ thousands of drivers.

The largest world's airline serves approximately 100 airports in 41 countries by a fleet composed of nearly 700 aircraft. Planning and maintaining operations in such big systems is not an easy task.

7.2 NUMBER OF TRANSPORTED PASSENGERS VERSUS NUMBER OF SERVED VEHICLES

Traffic engineers try by various techniques and measures to enable as many vehicles as possible to pass, during a specific period of time, through a traffic intersection or through a road section. In other words, traffic experts try to maximize the *number of vehicles* that are served during a certain period of time. On the other hand, when it comes to public transit, we are trying to maximize the *number of transported passengers*. The number of passengers that can be served during the observed time interval represents *transit line capacity*. Similarly, the number of vehicles that can be served during the observed time interval represents the *vehicle line capacity*.

Let us note the freeway lane shown in Fig. 7.1.

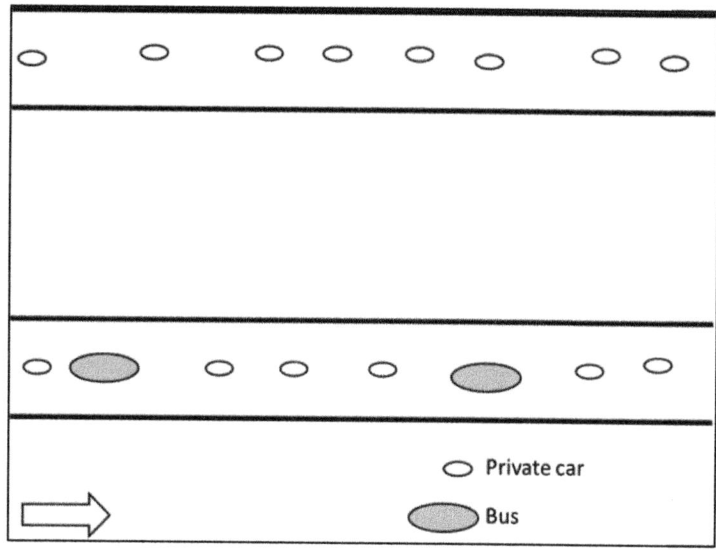

FIG. 7.1

Freeway lane.

Let us assume that the freeway lane can serve 2200 vehicles per hour (the freeway lane capacity equals 2200 vehicles per hour). The upper part of Fig. 7.1 shows the situation when only private cars used freeway lane. The lower part of Fig. 7.1 refers to the situation when some buses also participate in freeway lane traffic. We assume that average number of passengers in bus and private car (average occupancies) are, respectively, equal to 50 and 1.4. We also assume that instead on any two private cars we can allow one bus to participate in freeway lane traffic.

We denote by C, B, and P, respectively, the total number of cars, total number of buses, and total number of transported passengers. If we allow, for example, 50 buses to enter the freeway lane, we need to reduce the total number of private cars for $50 \cdot 2 = 100$ private cars. In this case, the total number of transported passengers P by private cars and buses equals:

$$P = C \cdot 1.4 + B \cdot 50 = 2100 \cdot 1.4 + 50 \cdot 50 = 5440 \text{ passengers}$$

Table 7.1 and Fig. 7.2 show increase in the total number of transported passengers with the increase of the number of buses engaged.

Table 7.1 Number of Cars per Hour, Number of Buses per Hour, and the Total Number of Transported Passengers

Number of Cars per Hour (C)	Number of Buses per Hour (B)	Number of Passengers per Hour (P)
2200	0	3080
2100	50	5440
2000	100	7800
1900	150	10,160
1800	200	12,520

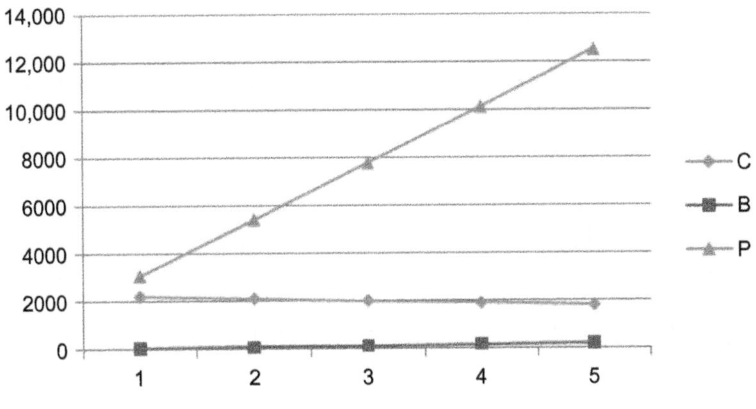

FIG. 7.2

Number of cars per hour, number of buses per hour, and the total number of transported passengers.

Fig. 7.2 illustrates the effects that public transit can achieve. With a very small percentage share of public transport vehicles in the total number of vehicles, it is possible to increase dramatically the total number of passengers carried. To achieve such effects in reality, it is necessary constantly to increase the public transit attractiveness.

7.3 URBAN PUBLIC TRANSIT

Urban public transit enables people movement in cities and in many cases also in suburban areas. They include *demand-responsive* and *mass transit systems*. The former are taxi systems and dial-a-ride systems. Taxi system uses operating individual passenger cars and sometimes small vans. They are the real-time demand-responsive systems, which provide transit services starting from the location/origin of the person's request/call and ending at his/her desired destination. They use the urban streets and regional roads as the infrastructure network (Fig. 7.3).

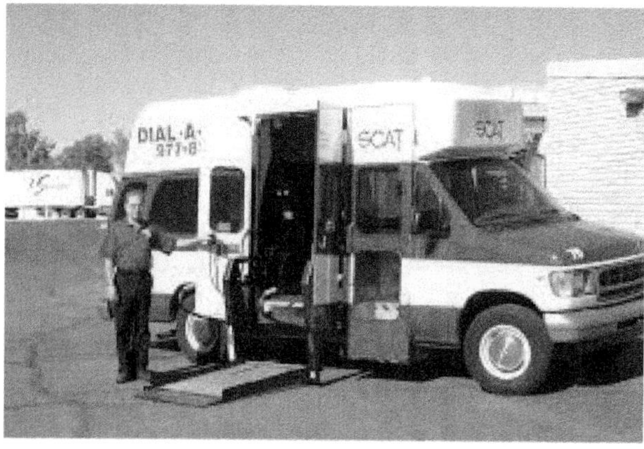

FIG. 7.3

Dial-a-ride system.

Dial-a-ride systems exist in some areas with lower population densities. The planned trip has previously to be announced, that it is necessary to make reservation. Such systems are also often used to transport elderly and citizens with disabilities.

There are also public transportation forms that combine fixed route system with demand-responsive transportation (DRT) system.

The mass transit systems operate according to the schedule by transporting larger number of persons/users/passengers simultaneously being onboard of the larger vehicles—generally buses and trains (Vuchic, 2005). It is a shared transportation service, since many passengers are carried in the same vehicle. The phrases "mass transit," "public transportation," and "public transport" are also used to describe this type of transportation service (Vuchic, 1981; Desaulniers and Hickman, 2007). The fixed-route system is the most common public transportation system in many cities in the world. It is characterized by high values of the number of requests for travel between the nodes in the transportation network. The vehicle routes are fixed, and the vehicle schedule is the same every working day.

There are two groups of public transportation users. The first group consists of people that have also other opportunities to travel. These are vehicle owners who choose public transportation, due to savings in transportation costs, and/or potential problems with parking. Most frequently, drivers become passengers in public transportation with an idea to avoid traffic jams. In big cities like New York, the majority of morning commuters arrive down-town by public transportation.

The second group of public transit users represents users who cannot drive private cars. Even in many developed countries, up to a third of the population cannot drive a private car. The very young, the elderly, citizens without a valid driving license, citizens who do not own a car, and citizens with disabilities need to use public transportation services.

Public transportation uses group travel technologies (trolleybuses, buses, trams, trains, ferries, etc.). Public transport can play an extremely important role in decreasing traffic congestion and air pollution reduction in many cities globally.

It is well-known that, in most cases, it is not easy for the public transit operators to make profit from this activity. Asian public transit operators are mainly privately owned companies, while the transit operators in USA and Canada are usually run by municipal transit authorities. The European public transit operators are run by municipal transit authorities, as well as by state-owned companies.

The authorities in charge of public transportation try primarily to increase the mobility of the urban population. The city authorities attempt to provide to the passengers the best possible level of services within the available budget. Public transportation operators are interested, first of all, in serving all the planned trips with the minimum possible number of buses and/or the minimum transportation costs. Together with passengers, the agencies and operators are interested in small waiting times at bus stops and at transfer points, satisfactory comfort in vehicles, etc. Passengers are also very interested in a high service reliability that represents the degree to which buses provide punctual service.

Public transportation vehicles serve groups of unconnected passengers. It is not an easy task to collect trips in space and time efficiently. Consequently, one of the most important problems in public transportation that has to be solved is the problem of matching transport capacities and passenger demand. Public transit network topology, vehicle frequencies, as well as distribution of departure times on specific routes represent the manner in which passenger demand and transport capacities are matched.

The horse-drawn omnibus are the first known public transit system. The transportation service with the horse-drawn omnibus started in France in 1828. Rails in the street, cable car, steam and electric trains, and elevated rail transit lines appeared in the 19th century (Fig. 7.4).

FIG. 7.4

Horse-drawn omnibus (http://www.istockphoto.com/).

An electrified rapid transit train system appeared in the second part of the 19th century in London. Today, there are more than 160 metro systems in the world. Car pools, fixed bus routes, light rail, heavy rail, aerial tramway, cable car, trolleybus systems, and automated guideway transit (AGT) are some modern public transportation systems (Fig. 7.5).

FIG. 7.5

Bangkok Mass Transit System (http://www.everystockphoto.com/).

Some of these forms share right-of-way (ROW) with private cars, while some other have dedicated ROW. The public transportation forms also differ due to the technology used for guidance, size of the cabin used, the type of a fuel used, and the type of routes and vehicle schedules.

Taxis and dial-a-ride systems have low capacity. The medium capacity is related to buses, trolley buses, and trams, while light rail transit and rapid rail transit are considered to be high-capacity transportation modes.

The mass transit systems can be the road- and rail-based transit systems consisting of the infrastructure and the service networks generally characterized by the technical/technological characteristics. The technological characteristics are as follows: ROW category (any or exclusively intended path for operating public transport vehicles), support (type of vehicle's wheels—rubber tires or steel), guidance (steered or guided), propulsion (internal combustion engine, electric), control (manual, semi- and fully automatic), and transit unit (small, medium, and large vehicles, and short and long trains). Public transit is accessible to general public in many cities in the world. The elements of the urban public transit are vehicles (buses, trolleybuses, trams), ways (streets, tracks, guide ways), stops (stops, stations, terminals), garages, power supply systems, and control systems.

Road transport modes include: Regular Buses (RB), Trolleybuses (TB), and Semi-Rapid Buses (SRB), and their derivatives/modifications. The rail transport mode includes: Streetcar (STC) or tramways, Light Rail Transit (LRT), Rail Rapid Transit (RRT) or Metro, Regional Rail (RGR), and their upgrades/modifications. It should be pointed out that the LRT and PRT systems and their particular liens/routes have been increasingly semi- and/or fully automated, ie, driverless, in many urban areas round the world (Vuchic, 1981, 2005).

7.3.1 ROAD-BASED URBAN TRANSIT SYSTEMS

7.3.1.1 Regular buses

This road-based medium-capacity urban transport system consists of buses operating along fixed routes and according to the fixed, announced in advance, schedule with relatively frequent/dense stops along the lines/routes. The public transportation systems based on bus operations have the lowest investment cost per line length. Simultaneously these systems have the lowest performances. Buses serve passengers in many cities in the world. The capacity of buses in terms of spaces (seats and stands) for passengers can vary from that of the minibus (25–30) and standard bus (80) to that of the long (articulated) bus (120–125) (EU, 2011). This transit system is very flexible due to the ability of buses to operate along almost all streets and roads in urban and suburban areas. In addition, they can change the layout of line/route by changing particular streets and roads they follow due to any reason (Fig. 7.6) (Vuchic, 1981).

FIG. 7.6

Typical bus used in public transit (http://www.everystockphoto.com/).

7.3.1.2 Trolleybuses

This is another road-based medium-capacity urban transport system very similar to the above-mentioned bus system including the capacity of vehicles-trolleybuses. The exception is that trolleybuses are propelled by electric power obtained from two overhead wires spreading along their lines/routes. Such ultimate energy-power dependency reduces the spatial flexibility of the system's lines/routes to adapt to the spatial changes due to any reason. The system also operates the transport service network along the streets and roads in urban and, in some cases, also in the sub urban areas, respectively (Fig. 7.7) (Vuchic, 1981).

FIG. 7.7

Trolleybus used in public transit (http://www.everystockphoto.com/).

7.3.1.3 Semi-rapid buses

This road-based high-performance system operates the high-performance buses of the similar capacity as its regular counterpart, usually along the dedicated (reserved) paths—lanes—along the streets and roads in urban and suburban areas. The transport services are carried out according to the fixed schedule. The number of stops along the lines/routes is smaller than that at the above-mentioned RB system. An improvement of the Regular and particularly SRB system has been the development of the BRT (Bus Rapid Transit) system. As a flexible rubber-tired road rapid transit system, it combines stations, vehicles, services, running ways, and ITS (Intelligent Transport System) into an integrated system with a strong positive image and identity. In particular, the system uses the concept of HOV (High-Occupancy Vehicles) and the new/innovative vehicles-buses compared to those of its conventional counterparts. They operate along the strictly reserved existing and new-built bus lanes on streets, bus/pedestrian malls, and other roads while having priority treatment at intersections. In many respects, BRT can be considered as a rubber-tired LRT-like system but with greater operating flexibility and potentially lower capital and operating costs (Levinson et al., 2003). The system has shown flexibility in terms of feasibility of implementation in the urban areas with a population of between 0.2 and 10 million. As such, in many transit corridors/routes, it has represented a test-bed before implementing a rail-based urban transit system such as LRT (Fig. 7.8) (Janić, 2014; Vuchic, 2005).

FIG. 7.8

Bus Rapid Transit.

7.3.2 RAIL-BASED URBAN TRANSIT SYSTEMS

7.3.2.1 Streetcars or tramways

This is a rail-based medium-capacity urban transit system operating the electrically powered vehicles usually in composition of 1–3 units, with the capacity of 100–300 spaces for passengers. The layout of line/route is determined by alignment of the rail tracks located mainly along the dedicated lanes of particular streets and roads in urban areas. The electricity is provided by the single wire above the line/route, ie, tracks. However, in many cases, these lines/routes have shared the same streets and road lanes with other transit modes and individual cars, which often caused congestion and considerable friction with each other, particularly with individual car traffic. The transit services are provided according to the fixed schedule with the number of stops along the lines/routes similar to that of the RB system. As such, this system has been competing and later on complementing to the RB systems in many urban areas (Fig. 7.9).

7.3.2.2 Light rail transit

This is the rail-based high-performance urban transit system operating trains along predominantly reserved grade-separated ROWs—tracks. The trains are electrically powered, consisting of 1–4 vehicles/cars providing the capacity of a train of 110–600 spaces for passengers. The services are provided according to the fixed schedule at stops/stations, which are rarer that those at bus and tramway system. This system possesses some advantages and disadvantages regarding the spatial flexibility of its lines/routes: on the one hand it can run on the grade-crossing tracks, but also on the streets, which increases its spatial flexibility; on the other, the layout of its lines/routes remains ultimately inflexible following the alignment of tracks. In addition, the system has originated as an substantive upgrade of Streetcar or Tramway system, but it also possesses the ability to be upgraded into rapid transit system, such as Light Rail Rapid Transit (LRRT) or RRT. After being fully automated, ie, driverless, the LRRT system has also become known as Automated Light Rail Transit (ALRT) (Fig. 7.10) (Vuchic, 2005).

FIG. 7.9

Tramway used in public transit (http://www.morguefile.com/).

FIG. 7.10

Light Rail Transit (http://www.everystockphoto.com/).

7.3.2.3 Rail rapid transit or subway or metro

The London Underground, the first metro system in the world, was opened in the second part of the 19th century. Nowadays, there are approximately 160 metro systems in 55 countries in the world. Metro has the highest investment cost per line length. This is the rail-based high-performance urban transit system operating trains along the dedicated lines/routes with rail tracks usually spreading underground, ie, with the tunnel alignment in the large densely populated urban areas. The electrically powered trains are composed of 1–10 vehicles/cars with the capacity of 140–2000 spaces for passengers. The transit services are provided according to a fixed schedule with relatively close stops at underground stations in dense urban areas, and fewer stations in suburban areas. Compared with its above mentioned LRT counterparts, the RRT system provides much higher transit capacity, travel speed, internal comfort, reliability, punctuality, and safety of services. Similarly as in the case of LRRT systems, the particular lines/routes or the entire network/system has been also increasingly semi- or fully automated (Fig. 7.11) (Vuchic, 2005).

FIG. 7.11

Metro (http://www.morguefile.com/).

7.3.2.4 *Regional rail*

This is the rail-based high-performance suburban transit system, which operates trains along the rail lines/routes spreading between urban and suburban areas. These usually electrically powered trains are composed of 1–10 vehicles/cars with the capacity of 140–1800 spaces for passengers. The transport services are provided according to the fixed schedule, at lower service frequency and the rarer stops at stations on the longer lines/routes. Thanks to the longer lines/routes and rarer stops, the travel speed of RGR trains is higher compared to that the above-mentioned RRT and LRT counterparts (Fig. 7.12) (Vuchic, 1981; 2005).

7.3.3 COMPLEMENTARITY OF THE SYSTEMS

The above-mentioned systems can operate in particular urban and/or suburban areas individually/exclusively or together, manly depending on their size—area, number, and density of population. The additional influencing factors can be availability of individual passenger cars, ie, motorization rate,

FIG. 7.12

Regional rail.

habits of inhabitants of using public transport systems, and the offered performances of these systems, usually compared to those of the individual passenger car use under given conditions. In smaller less densely populated urban areas, RB system usually operates. In larger and higher density populated areas, Regional and Semi-rapid bus, Trolleybus, and Streetcar (Tramway) system can simultaneously operate. In many cases, transit services complement each other by enabling passengers' passing from one to other system at dedicated stops/stations. In the large very densely populated urban and suburban areas, in addition to the above-mentioned medium-capacity transit systems, the high-performance LRT and RRT system scan operate exclusively or simultaneously. In such case, the medium-capacity systems can provide collection and distribution of passenger flows for these high-performance systems. The stops/stations of all these systems are close to each other enabling fast passing between systems. The RGR system is also considered as some kind of spreading of LRT and RRT system transit services towards the periphery of large urban areas, again by locating the stations close to each other, even on the common location(s).

7.4 INFRASTRUCTURE OF URBAN TRANSIT SYSTEMS

The infrastructure networks of urban transit systems consists of stops/stations as the network nodes and depending on the ROW category of the street/road lanes and the rail lines/tracks as the network's links connecting them. Fig. 7.13 shows three public transit lines.

Every line is characterized by the line length in one direction L. In other words, the line length represents distance from the beginning line terminal to the end terminal of the line. This length is expressed in kilometers or miles. Lines #1 and #2 are partially overlapping. Because of the overlap of individual lines, the total length of all lines in the network is greater than the total length of alignment along which lines go. If there is no overlapping in the network, these two quantities are equal. The transit line terminals, as well as the individual stops are also denoted in Fig. 7.13. At transit stops,

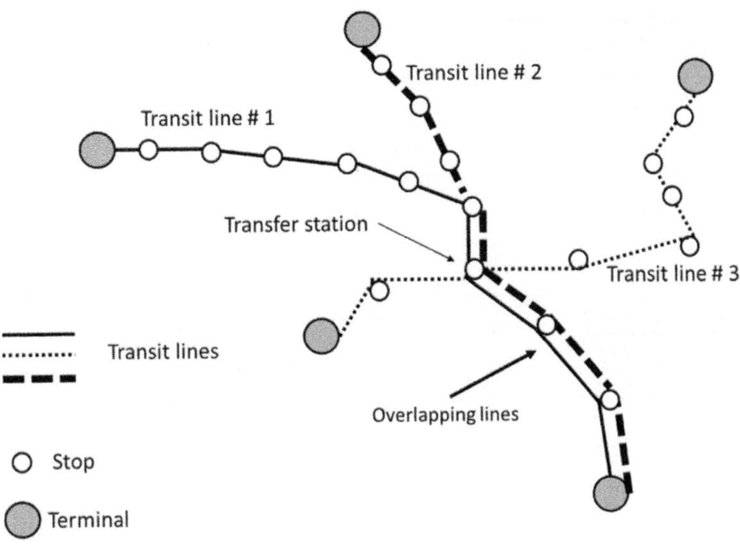

FIG. 7.13

Fixed-route system.

vehicles pick up and/or drop off passengers. At some of stops, passengers can make transfer among transit lines (Fig. 7.13). The term station is also used to denote line stop.

7.4.1 STOPS/STATIONS IN URBAN TRANSIT SYSTEMS

The stops stations of urban transit systems should be placed according to several criteria: reducing walking distance in order to achieve minimum passenger travel time (eg, the common access time of an ideally located stop/station should be about 5 min), providing maximum area coverage in order to attract reasonably high passenger demand, achieving the minimum investment costs, all under conditions of having the convenient space for waiting, embarking, disembarking the vehicles, and passing between different lines safely, including minimal disturbance of other traffic (Kikuchi and Vuchic, 1982; Vuchic, 2005).

At the road-based RB, TB, and SRB systems and the rail-based STC systems, the stops/stations are located respecting the above-mentioned criteria usually on streets behind intersection(s) in the direction of vehicle movement at the minimum distance of 25 m. The stations of the lines in different directions are located at a minimum distance of 50 m (Banković, 1982). Fig. 7.14 shows the simplified scheme.

As can be seen, the dimension of RB, TB, and SRB stops/stations depends on the vehicle typical length and width and the number of vehicles expected to occupy the stop/station simultaneously. These are usually the vehicles from different lines sharing the same stop/station.

The streetcar or tramway system stops/stations can generally be micro-located in two ways. If the lines/tracks are positioned on the side lanes of streets, the stops/stations are located similarly as those of buses. If they are located in the middle lanes of streets/roads, they look like and operate as some kind of isolated islands. In such cases, passengers need to cross the lanes intended to other traffic in order to access these stops/stations. Fig. 7.15 shows a simplified scheme.

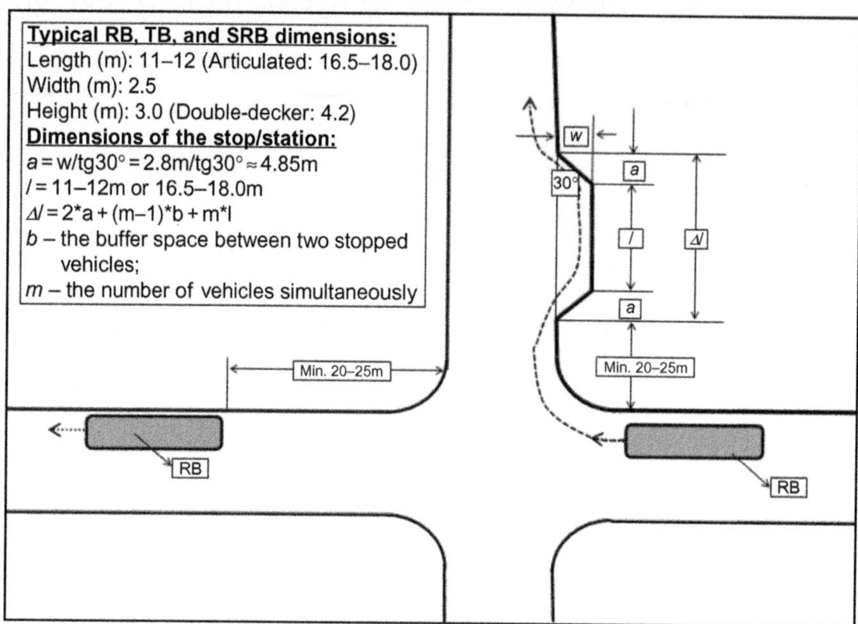

FIG. 7.14

Examples of micro location of a bus stop/station.

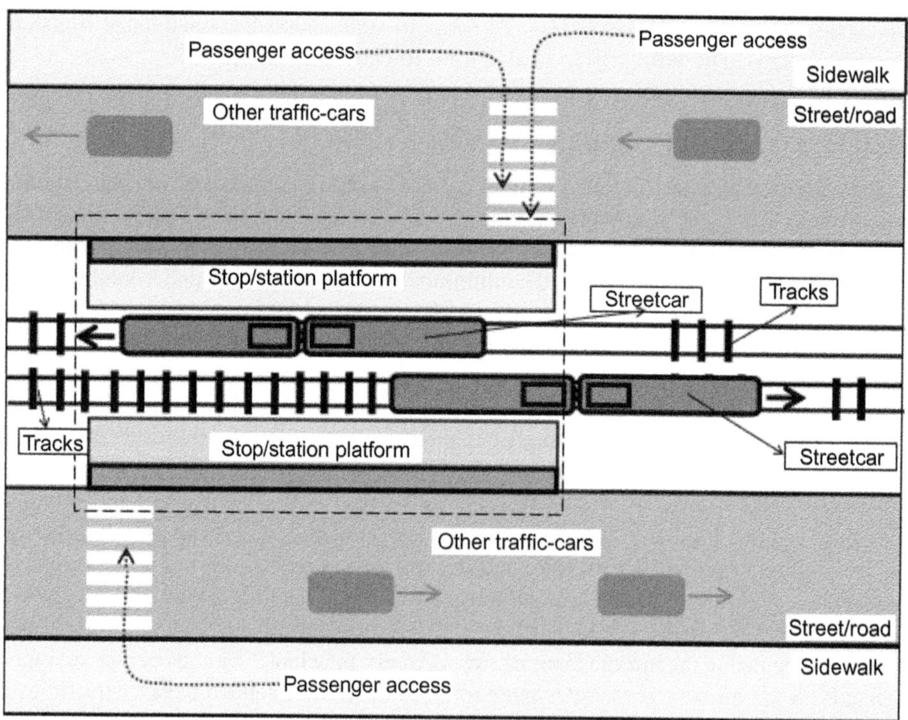

FIG. 7.15

Simplified scheme of the micro location of a streetcar (tramway) stop/station-lateral platforms.

As can be seen, the streetcar moving in one direction has stopped and the other moving in the opposite direction has just left the stop/station. The width of double track lane is: 6.087.0 m (Vuchic, 1981).

At the rail-based LRT, RRT, and RGR systems, the double-tracks spread along the isolated both surface and underground (tunnel) lanes of typical with of 7.0–8.0 m. The stops/stations are micro located on the track sides with usually lateral platforms of the sufficient dimensions (length and width) for accommodating trains and providing space for relatively fast and absolutely safe passenger waiting, embarking, and disembarking the trains. The typical height of these platforms relative to the level of tracks is: 0.5–1.0 m. Their length varies from 120 to 270 m, thus enabling accommodation of a single train composed of 10 vehicles/cars of the typical length of 14.6–26 m (Vuchic, 1981).

In addition, the spacing, ie, the distance between neighboring stops/stations along particular lines/routes has been particularly important. This spacing divided by the line/route length represents the density of stops/stations, which in turn reflects the coverage of the given line/route. In general, the stop/station density should be based on the trade-off between two components of passenger travel time: access and on line travel time. This travel time mostly depends on the vehicle operating speed, which in turn, particularly at the RB and SRB systems is influenced by spacing between stops/stations. For example, at European SRB systems in 35 cities and related urban areas, there has been strong correlation between the vehicle operating speed and spacing between stops/stations, as follows: $V=0.0192\Delta S+10$ (km/h) ($R^2=0.923$; $n=35$) (ΔS is the average spacing between stops/stations—(m)) (EU, 2011). As can be seen, on the one hand, increasing of the stop/station distances contributes to increasing of the operating speed, which raises attractiveness of the system. However, on the other hand, increased spacing also increases the walking distance to/from the stops/stations, which deteriorates the attractiveness of the system. Therefore some balance should always be established. The detailed models of determining the stop/station density on the lines of urban transit systems can be found in the relevant references (Banković, 1982; Vuchic, 2005).

Fig. 7.16 shows examples of the relationship between the average stop/station spacing and the length of the network for RRT-metro and RGR systems in the selected world's cities/urban and suburban areas.

As can be seen, at both systems there is not strong correlation between the average stop/station spacing and the length of the network. At the RRT systems the average stop/station spacing varies between 600 and 1200 m, but most frequently it is between 800 and 1000 m. At the RGR systems the average stop/station spacing varies between 1400 and 1800 m.

7.4.2 URBAN TRANSIT SYSTEMS LINKS AND INDICATORS OF NETWORK SIZE

Links of the urban transit networks are represented by infrastructure depending on the system's ROW category. For the road-based system these are the city and urban area streets with lanes of width 6.0–6.5 m enabling operation of the system's vehicles—buses—in both directions simultaneously (EU, 2011). For the rail-based systems, these are (usually standard 1435 mm wide) double tracks of the above-mentioned width of 7.0–8.0 m enabling operation of the system's vehicles—trains—simultaneously in both directions. These street and road lanes and the rail tracks with the corresponding stops/stations form the infrastructure networks of urban transit systems. These networks are characterized by quantitative elements of spatial performances such as: configuration/geometry, size, topology, and relationship with the urban areas they operate. These performances can be measured by different

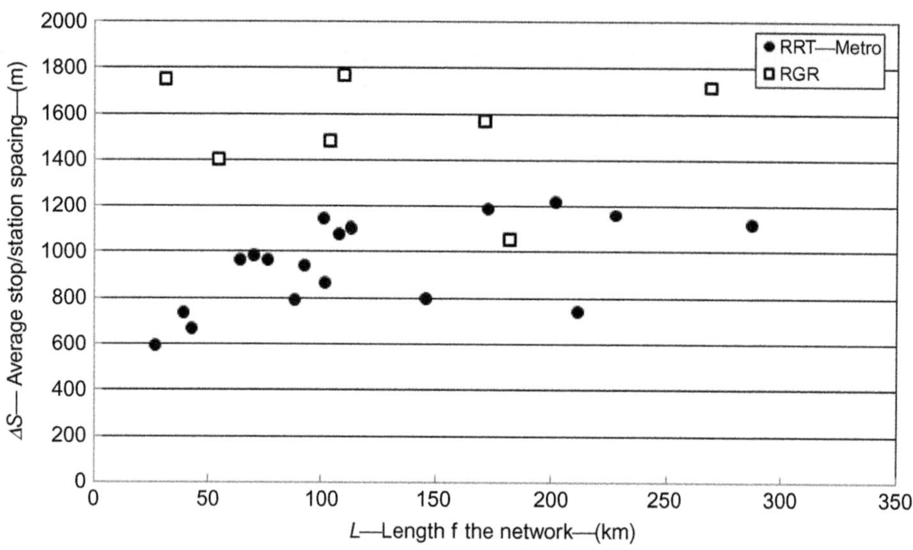

FIG. 7.16

Relationships between the average spacing of stops/stations and the length of network for the selected RRT and RGR systems (Vuchic, 2005).

indicators, which have shown to be particularly useful for analyzing, comparing, and/or planning the RRT or metro systems.

The size of a RRT system infrastructure network can be measured by different (simplified) indicators. Some of them are given in Table 7.2 (Vuchic, 2005).

Table 7.2 Some Indicators of the Size of a RRT System Infrastructure Network

Definition	Symbol	Expression
Number of lines in the network (counts)	N_l	N_l
Number of stations per line (i)[a] (counts)	n_i	n_i
Number of stations in the network (counts)	N_s	$N_s = \sum_{i=1}^{N} n_i$
Number of interstation spacings on line i (counts)	a_i	$a_i = n_i - 1$
Number of interstation spacings in the network (counts)		$A = \sum_{i=1}^{N} a_i$
Length of line i (km)	l_i	l_i
Length of the network (km)	L	$L = \sum_{i=1}^{N} l_i$
The number of possible O/D[b] paths (counts)	OD	$OD = (1/2) \cdot N_s \cdot (N_s - 1)$

[a]*Not counting multiple stations common for two or more lines.*
[b]*Possible station-to-station O/D (Origin-Destination) paths.*

The number of lines in the network N_l (Table 7.2) is an indicator representing in some sense the network's size, ie, more lines—larger the network. In such case, the smallest network has only one line. Number of stations per line n_i indicates in some sense its accessibility, ie, more stations—higher accessibility of given line, and vice versa. The similar can be applied to the number of stations of the network N_s. The numbers of interstation spacings on the line and in the network, a_i and A, reflect their accessibility on one hand, and also inherently traveling at lower speeds on the other. This means that if the number of interstation spacings is greater, the number of stations is also greater and the given RRT system network is more accessible. On the other hand, a greater number of interstation spacings implicitly indicates that they are shorter, and as such affects the operating and travel speed of trains.

The length of line and of the network l_i and L, respectively, explicitly indicate their spatial scale and implicitly the size of urban area they serve. The number of possible OD paths indicates the maximum possible number of passenger demand flows and related transit services in the network under an assumption that each station is at the same time their origin and destination. As can be seen, this number increases with increasing of the number of stations in the network, thus reflecting in some sense the network complexity from on its demand size.

7.4.2.1 Topology and relationship to the urban area/city

Similarly as in the case of the RRT system network size, its topology and relationship to the urban area/city can be quantified by some, again simple, indicators given in Table 7.3 (Vuchic, 2005).

Table 7.3 Some Indicators of the Topology and Relationship to Urban Area of a RRT System Infrastructure Network

Definition	Symbol	Expression
Topology		
Average interstation spacing (km)	S	$S=L/A$
Network complexity (–)	α	$\alpha=A/N_s$
Relationship to urban area		
Network density (km/km^2)	D	$D=L/S_u$
Network extensiveness (km/10^6 population)	E	$E=L/P$
Network area coverage (%)	AC	$AC=(N_s * S_O)/S_u$

where:
S_u is the size of urban area served by the given RRT system (km^2);
P is the population in given urban area (million—10^6); and
S_O is the size of area around the station of a given RRT system usually of the radius of 400–500 m (walking distance) (km^2).

In Table 7.3, the average interstation spacing (S) represents how good the urban area is covered by the system's stations. If it is low, the stations are closer thus reducing operating and travel speed, and vice versa. The network complexity ratio α expresses the number of spacings to the number of stations in the given RRT system network. Its minimum value is 0.5, indicating just a single line with two end stations. With increasing the number of stations along the lines, this indicator increases thus showing potentially lower operating and travel speed along the particular lines and in the network. The network density D indicates how the network of a given RRT system is extensive compared to the size of urban area it serves.

The network extensiveness E indicates the relative importance of the given RRT system network for population in the urban area it serves. In this case, the greater is the better, and as such more important. The network area coverage AC indicates the quality of accessibility of a given RRT system throughout the urban area it serves. As such, it represents one of the most important indicators for planning the RRT system lines and networks in urban areas.

It should be mentioned that the above-mentioned set of indicators has not been exhausted. A more complete and detailed list can be found in the relevant literature (Vuchic, 2005).

EXAMPLE 7.1

We illustrate the above-mentioned indicators is the RRT—using the example of the metro network is Dubai Automated metro System (Dubai, UAE—United Arab Emirates). The attribute "automated" indicates that the system is fully automated, ie, driverless. Fig. 7.17 shows the network's simplified layout.

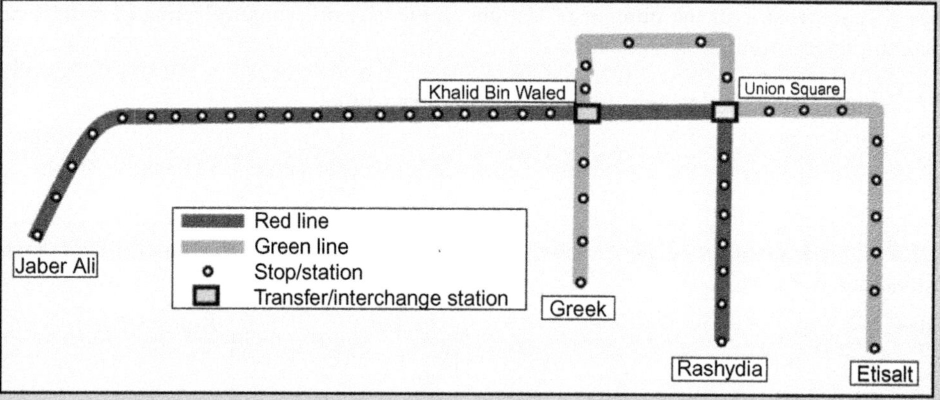

FIG. 7.17

Simplified layout of RRT—metro network in Dubai (UAE) (https://www.rta.ae).

Dubai is situated in Persian Gulf and is one of seven United Arab Emirates with the area $S_u = 4114\ km^2$ and population: $P = 2{,}106{,}177$ inhabitants. The Metro network has started operation in September 2009. The network consists of: $N_l = 2$ lines called Red Line and Green Line. The stations of the network are accessible on foot, by car and public bus system. The parking places for cars and the stops/stations for buses are provided at the end and transfer system's stations. The main indicators of the network's size, topology, and relation to the urban area, that are calculated by using relations from Tables 7.2 and 7.3, are given in Table 7.4.

Table 7.4 Some Indicators of the Spatial Performances of Dubai RRT—Metro Infrastructure Network (https://www.rta.ae)

Indicator	Red Line ($i = 1$)	Green Line ($i = 2$)	Network
	Size		
N_l (−)	1	1	2
N_s (−)	29	20	49
A (−)	28	19	47
L (km)	52.1	22.5	74.6

EXAMPLE 7.1—cont'd

Table 7.4 Some Indicators of the Spatial Performances of Dubai RRT—Metro Infrastructure Network (https://www.rta.ae)—cont'd

Indicator	Red Line ($i=1$)	Green Line ($i=2$)	Network
OD (−)	406	190	1176
	Topology		
S (km)	1.86	1.18	1.58
α (−)	–	–	0.959
	Relation to urban area		
D (km/km^2)	–	–	0.0183
E (km/10^6 pop)			35.42
AC (%)[a]	–	–	0. 935

[a]$S_0=0.785$ km^2 *(The radius around each station is assumed to be 500 m).*

Similarly, performances of infrastructure networks of other urban transit systems can be estimated.

7.5 PUBLIC TRANSPORTATION AVAILABILITY

In order to increase the number of passengers in public transportation, it is necessary to give due consideration to public transportation availability.

The first prerequisite for public transportation to become a possible alternative for the user is the existence of the public transit in a user's neighborhood. This spatial availability assumes existence of the public transportation system near the trip origin and the trip destination, within acceptable walking distance. Few studies and gained experience from real life show that majority of transit passengers accept to walk about 400 m (approximately one-quarter mile) to public transportation stops. This walking distance usually assumes 5 min of walking. In the case of rail transit, the maximum accepted walking time is about 10 min. The public transit stop spacing should be done in accordance with acceptable walking distances (Fig. 7.18).

In some countries, public transportation becomes possible alternative for bicycle users, in the case when there is a connection between bicycle and public transportation. In the case of bicycle riders, the catchment area of public transportation is not limited to 400 m. Good connections to transit stops, and available bicycle parking at transit stops, attract bicycle riders to public transportation and increase the number of passengers in public transit. In the case of bicycle riders, the catchment area could have the radius up to 2 km.

Drivers choose public transportation, due to savings in transportation costs, and/or potential problems with parking in down-town. Manu users of commuter rail service in lower density areas in the USA and Canada are vehicle owners. Park-and ride lots attract these group of passengers to public transportation (Fig. 7.19).

FIG. 7.18

Bicycle parking at a transit stop.

FIG. 7.19

Park and ride lot.

At park-and-ride facilities, drivers make the transfer from cars to public transit. These intermodal transfer facilities are usually located at suburbs. Parking costs down-town are usually significantly higher than parking costs at park-and-ride lots, and this highly motivates drivers to choose public transportation.

Public transportation is attractive to potential passengers if the service is provided throughout the day. In some cities, the service is also offered during the night hours (with reduced frequency). In other words, company should also provide temporal availability of the public transportation service to the passengers.

Capacity availability is another aspect of availability that should be taken into account. Public transit operators should provide a sufficient number of seats (spaces) to potential customers. This primarily means that a traveler can enter the vehicle that came to the station and find an available seat or enough space in the vehicle.

7.6 PASSENGER FLOWS IN PUBLIC TRANSPORTATION

Passenger flows vary considerably in public transportation. During the working day, in most cities in the world, the daily numbers of passengers on most transit lines are relatively uniform. This is due to the fact that many passengers, each working day, use the same transportation mode and the same route when going to work. During a weekend the number of passengers in public transit is significantly smaller. Fig. 7.20 shows daily variations in passenger volume.

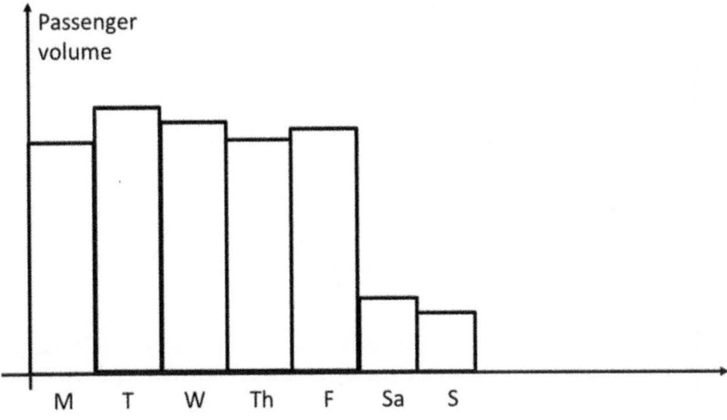

FIG. 7.20

Daily variations in passenger volume.

Fig. 7.21 shows hourly variations in passenger volume. The variations shown in Fig. 7.21 are typical for many cities in the world. There are morning and evening peaks when people go, and when people come back from work. The differences in hourly passenger volumes could be very high by hours of a day.

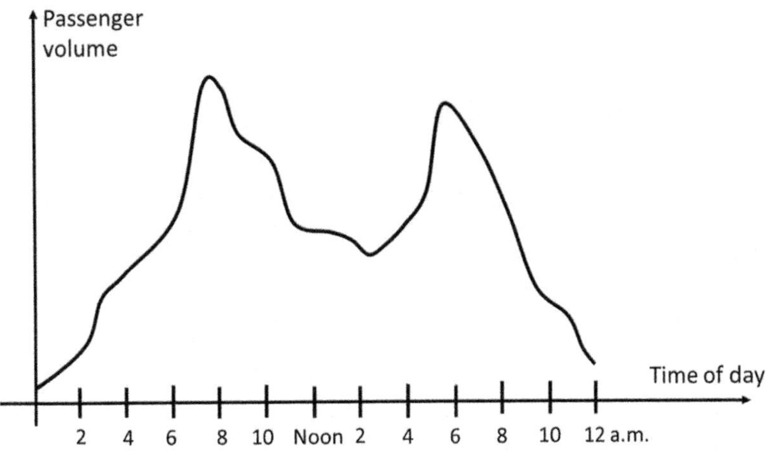

FIG. 7.21

Hourly variations in passenger volume.

Hourly variations in passenger volume have, as a consequence, different number of vehicle departures from the terminals during certain time intervals (Fig. 7.22). In this way, hourly variations in passenger volume require the engagement of different number of vehicles during certain time periods. The number of engaged vehicles is much higher during the rush-hours (Fig. 7.22). Outside the peak periods the transit operator has a surplus of vehicles and drivers. The transit operator, therefore, meets with a range of organizational problems that have to be solved ("empty" vehicle trips to garage and from garage, drivers working hours divided in two shifts, vehicle maintenance planning, etc.).

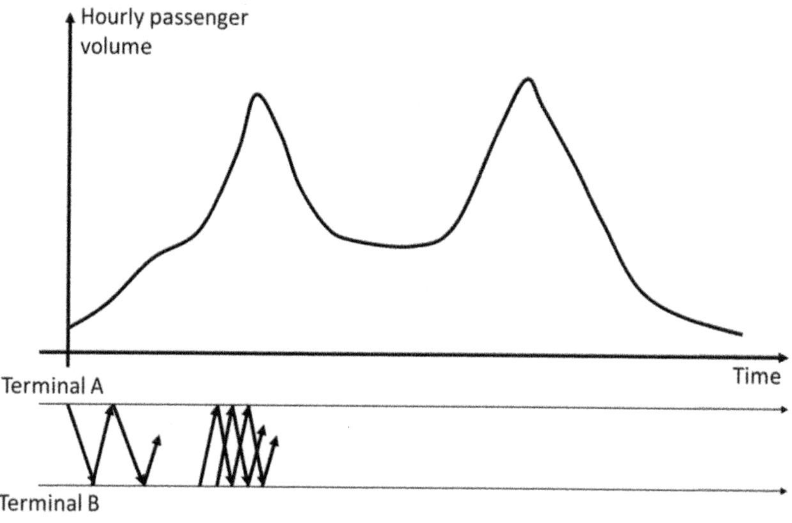

FIG. 7.22

Different number of vehicle departures from the terminals during certain time intervals.

7.7 PASSENGER FLOWS ALONG A TRANSIT LINE

Headway in public transportation operations represents the time interval between vehicles past a specific point (Fig. 7.23). Headways are expressed in minutes. It is essential to study passenger flow along the transit line, in order to determine the appropriate transit line headway.

Let us assume that we have data on the number of boarding passengers and number of alighting passengers on individual line stops and *terminals* (end stations on a transit line) (Table 7.5). The transit line has 5 stations. The terminals are denoted respectively by *A* and *B*.

The last column of Table 7.5 shows the numbers of passengers in the vehicle, after departing from bus stops. Thus, for example, after leaving the station 2, there are 21 passengers in the vehicle that travels along line section between stop 3 and stop 4. The number of boarding passengers, number of alighting passengers and the number of passengers in the vehicle for any line section are shown in Fig. 7.23.

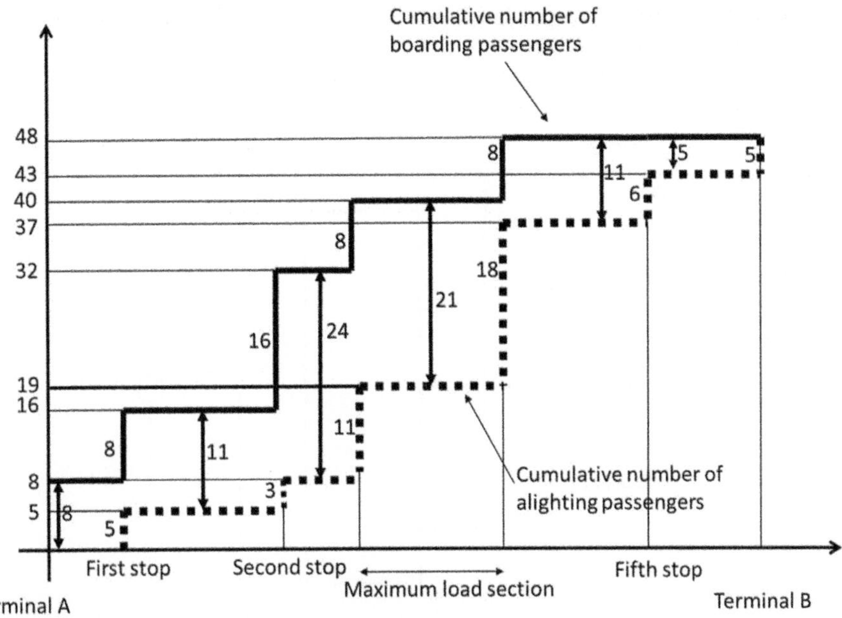

FIG. 7.23

The number of boarding passengers, number of alighting passengers and the number of passengers in the vehicle for any line section.

Table 7.5 Number of Boarding Passengers, Number of Alighting Passengers and the Number of Passengers in the Vehicle

Bus Stop	Number of boarding Passengers	Number of Alighting Passengers	Number of Passengers in the Vehicle
Terminal A	8	0	8
1	8	5	11
2	16	3	24
3	8	11	21
4	8	18	11
5	0	6	5
Terminal B	0	5	0

The numbers of passengers on certain line sections respectively equal 8, 11, 24, 21, 11, and 5. The *maximum passenger volume* equals $\max\{8, 11, 24, 21, 11, 5\} = 24$ and corresponds to the section between stop 2 and stop 3. We call section between stop 2 and stop 3 the *maximum load section* (MLS).

The data shown in Table 7.5 and Fig. 7.23 are related to one vehicle trip. In real-life applications, 1 h is the basic time unit that is used when describing cumulative number of alighting and boarding passengers, as well as the maximum passenger volume on MLS. In other words, the maximum passenger volume is expressed in passengers per hour. The passenger volume profile is calculated for both directions of the transit line (Fig. 7.24).

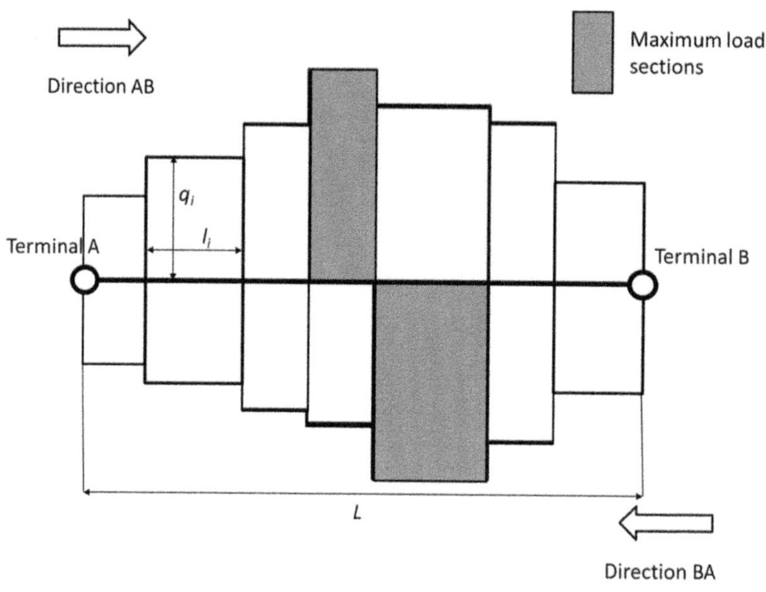

FIG. 7.24

Passenger volume profile.

We denote by L in Fig. 7.24 the *line length*. This length represents the one-way distance between the line terminals along the line alignment. The line length is measured in miles or kilometers. We denote by q_i passenger volume on the ith section. The length of the ith transit section is denoted by l_i. The MLS in one direction is usually different from the MLS in another direction. The maximum passenger volumes in both directions should be taken into account when determining the number of vehicles to be engaged on the transit line.

7.8 SERVICE FREQUENCY AND HEADWAYS

One of the most important problems encountered by public transit operators is how to match transportation supply and passenger demand on individual transit line. The matching problem is far more complex over the entire route network than on individual routes. Service frequencies (Bowman and Turnquist, 1981; Furth and Wilson, 1982; Ceder, 1984) and vehicle departure times on transit lines in the network reflect the manner in which transportation supply and passenger demand are matched. Service frequencies and vehicle departure times depend on passenger volume profile and on the number and type of vehicles in the fleet. The number of passengers that decide in the end to use public transit on a particular transit line depends, to the highest degree, on service frequency and vehicle departure times. For example, if service frequency is low or if vehicle departure times during the day are not convenient, a number of potential passengers will instead choose other modes of transportation.

Let us note the bus line shown in Fig. 7.25. Vehicles move from Terminal A to Terminal B. On the way to the Terminal B, vehicles stop at pre-defined bus-stops, where passengers enter and exit the vehicle. On arrival at the Terminal B, the driver rests for a while, and then the vehicle travels to Terminal A. On the way to Terminal A, car stops at bus-stops where passengers enter and exit the vehicle.

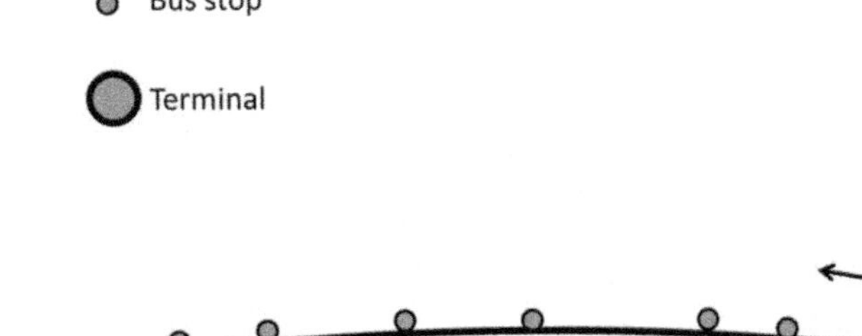

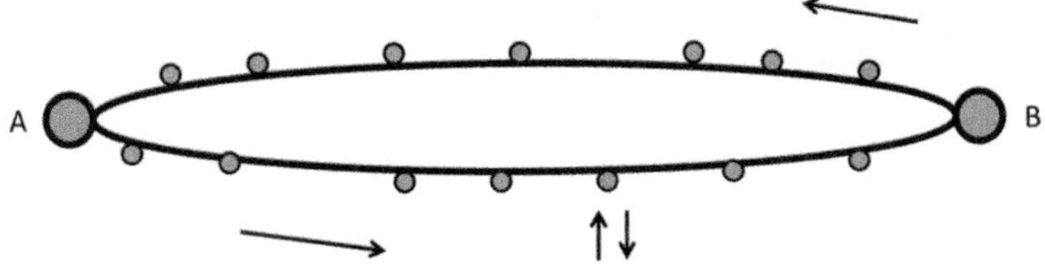

FIG. 7.25

Public transit line.

We denote by T turnaround time. This time is the time that elapses from the moment when vehicle leaves Terminal A to the moment when vehicle returns to Terminal A. Let us assume that we have on our disposal N vehicles that we can assign to the bus line. The service frequency represents the number of vehicles per time unit past a specific point in the same direction. The frequency equals:

$$f = \frac{N}{T}\left[\frac{vah}{h}\right] \tag{7.1}$$

The frequency is expressed in the number of vehicle per hour. *Headway* h in public transportation operations represents the time interval between vehicles past a specific point (Fig. 7.26). Headways are expressed in minutes. Since the frequency represents the number of vehicles per time unit past a specific point in the same direction, we conclude that the frequency is the inverse of the headway, ie:

$$f = \frac{1}{h} \tag{7.2}$$

When calculating and rounding headways, it is desirable to obtain a so called *clock headway*. Clock headways have a feature that enables the generation of timetable that is repeated every hour, starting on the hour. Thus, for example, in the case when the headway is equal to 15 min, it is possible to have the vehicle departures from the terminal in 8:00, 8:15, 8:30, 8:45, 9:00, 9:15, 9:30, 9:45, 10:00, etc.

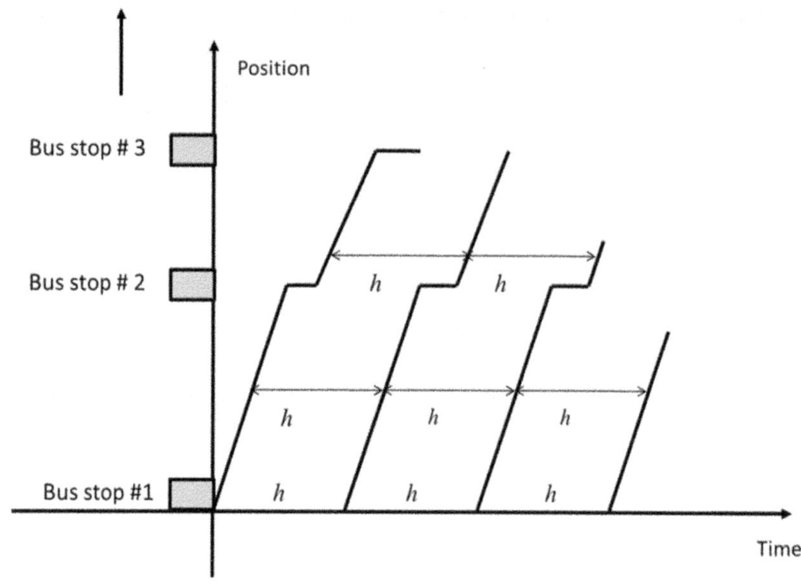

FIG. 7.26

Bus headways.

7.8.1 THE MAXIMUM SERVICE FREQUENCY

The maximum service frequency is defined by the maximum number of transit vehicles passing through a given point of a line/route i in one direction during a given period of time (usually 1 h) under prevailing operating conditions. This can be estimated as follows:

$$f_{\max/i}(\tau) = \tau / h_{\min/i} \tag{7.3}$$

where:

τ is the given period of time (h); and
$h_{\min/i}$ is the minimum headway, ie, the time interval between the successive transit vehicles passing through a given point of a line/route i in the same direction (min).

The minimum headway $h_{\min/i}$ in Eq. (7.3) can be determined according to different criteria, but in many cases the prevailing factors are characteristics of the system such as technology and way of operations along the line and at the stations. These factors influence the minimum headway for a given line line/route and for the stops/stations along it. Most frequently the stop/station headway is as much greater than that of the line(s)/route(s). In addition, the headway $h_{s/i}$ at the stop/station on a given line/route i, should be greater than the vehicle stop time $t_{s/i}$ at that stop/station. Consequently, the "ultimate" capacity of this stop/station will be:

$$C_{ss/i}(\tau) = \tau / \max\left[h_{s/i}, t_{s/i}\right] \tag{7.4}$$

In addition, the following must be satisfied for all stops/stations on the line/route i:

$$f_{max/i}(\tau) \leq C_{ss/i}(\tau) \tag{7.5}$$

In other words, the maximum number of transit vehicles passing through a given stops/stations cannot exceed the "ultimate" capacity of any of these stops/stations.

7.8.2 PASSENGER WAITING TIME

Passengers' walk to a stop and passenger waiting time are the basic attributes of the public transit level of service. Walk to stop in the range of 400–800 m is considered acceptable for public transit users.

In order to estimate the average passenger waiting time, let us first consider the situation when the bus arrives at the bus stop regularly, according to the published timetable. We also assume that all passengers at the bus stop can enter the vehicle, and that the passengers appear at the bus stop in random moments of time. It has been shown that, in this case, the average waiting time per passenger at the station w is equal to the one half of the vehicle headway h, ie:

$$w = \frac{h}{2} \tag{7.6}$$

The average waiting time per passenger could be longer in the case of irregular bus arrivals. In the case of irregular bus arrivals, vehicle headway is not any more deterministic quantity. In this case, the vehicle headway is a random variable. If the planned headway equals, 10 min, in the case of irregular arrivals headway values could be, for example, 8, 9, 12, 15,.... minutes. It has been shown, that in the case of irregular vehicle arrivals at the stop, the average passenger waiting time equals:

$$E(W) = \frac{E(H)}{2} + \frac{\text{var}(H)}{2 \times E(H)} \tag{7.7}$$

where:

$E(H)$ is the expected value of the random variable H; and
var(H) is the variance of the random variable H (variance represents the square of standard deviation).

7.8.3 HEADWAY DETERMINATION BY "SQUARE ROOT FORMULA"

Transit operator cost and passenger cost depend on chosen headway. Passenger cost, in the case when vehicles arrive regularly, has linear increase with headway. The greater the headway, the greater the passenger waiting time and passenger cost. On the other hand, greater headway means for transit operator smaller number of departures and lower costs (Fig. 7.27).

Z is the total cost per hour;
c is the transit operator cost per bus hour;
v is the value of passenger waiting time per hour;
r is the total number of passengers on line per hour (ridership per hour);
N is the number of vehicles assigned to the bus line; and
h is the headway.

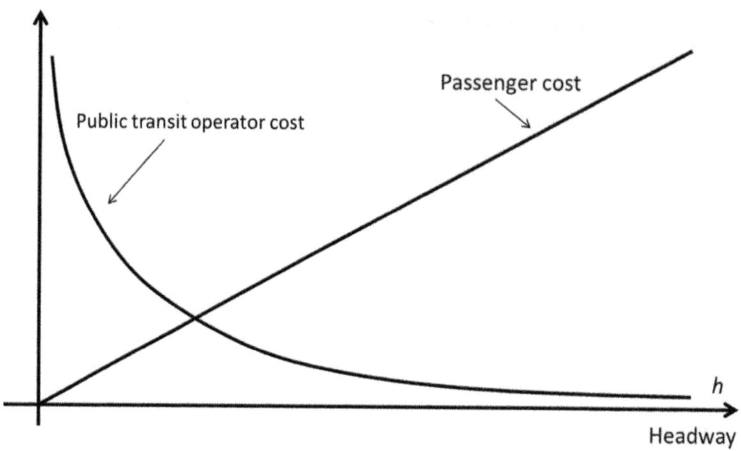

FIG. 7.27

Dependence of the transit operator cost and passenger cost of headway.

The transit operator cost per hour is equal to $N \cdot c$. We assume regular vehicle arrivals. In this case, the average waiting time per passenger at the station w is equal to the one half of the vehicle headway h, ie:

$$w = \frac{h}{2} \tag{7.8}$$

The waiting cost of all passengers is equal to the $v \cdot r \cdot \frac{h}{2}$. The total cost is equal:

$$Z = N \cdot c + v \cdot r \cdot \frac{h}{2} \tag{7.9}$$

Since $N = \frac{T}{h}$, we can write:

$$Z = c \cdot \frac{T}{h} + v \cdot r \cdot \frac{h}{2} \tag{7.10}$$

The optimal headway is found by setting the derivative of Z with respect to h equal to zero:

$$\frac{\mathrm{d}Z}{\mathrm{d}h} = -c \cdot \frac{T}{h^2} + \frac{v \cdot r}{2} = 0 \tag{7.11}$$

The optimal headway equals:

$$h = \sqrt{\frac{2 \cdot c \cdot T}{v \cdot r}} \tag{7.12}$$

Eq. (7.12) represents the "square root formula" for optimizing headway and service frequency. Minimal headway values in real-life are usually between 2 and 3 min. Maximal headways values are between 15 and 30 min. Outside of peak periods, on some transit lines, maximum headway values reach 60 min.

The service frequency f is the inverse of the headway, ie:

$$f = \frac{1}{h} = \frac{1}{\sqrt{\frac{2 \cdot c \cdot T}{v \cdot r}}} = \sqrt{\frac{v}{2 \cdot c \cdot T}} \cdot \sqrt{r} \tag{7.13}$$

Service frequency dependence of the total number of passengers on line per hour is shown in Fig. 7.28.

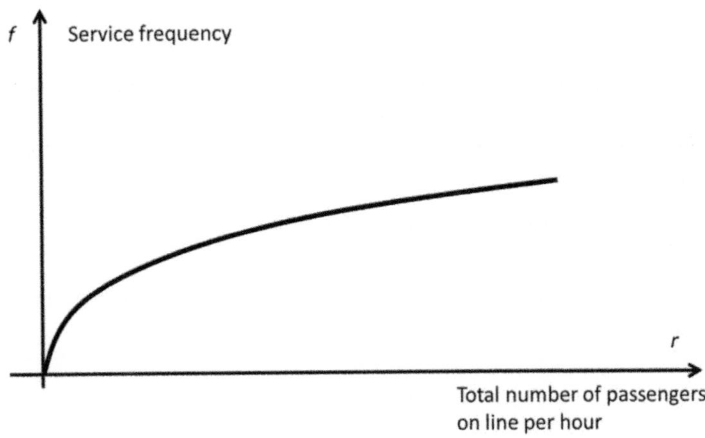

FIG. 7.28

Service frequency dependence of the total number of passengers on line per hour.

EXAMPLE 7.2

Let us assume that transit line parameters are respectively equal:

$$c = 120\$ \text{ per hour}$$

$$v = 10\$ \text{ per passenger hour}$$

$$r = 1200 \text{ passengers per hour}$$

$$T = 90\,\text{min} = 1.5\text{h}$$

The optimal headway equals:

$$h = \sqrt{\frac{2 \cdot c \cdot T}{v \cdot r}} = \sqrt{\frac{2 \cdot 120 \cdot 1.5}{10 \cdot 1.200}} = 0.173h \tag{7.14}$$

$$h \approx 10\,\text{min}$$

7.8.4 HEADWAY DETERMINATION BY MAXIMUM LOAD METHOD

When determining headways, transit operators try to provide enough space (especially during peak hours) to meet passenger demand. Majority of transit operators also define maximum headways on transit lines. For example, operator could define that maximum headway on a specific route, is equal to 30 min. Maximum headways guarantee a minimum service frequency offered to the passengers. The prescribed maximum headway is usually called *policy headway* and denoted by h_p.

The Maximum load method (that can have few variations) is based on counting passengers on the transit stop that is at the beginning of the MLS. Depending on the time of a day, the location of the MLS could change. For example, the transit stop, that is at the beginning of the MLS, could be Stop #5 between 10:00 am and 11:00 am, while Stop #7 could be at the beginning of the MLS between 5:00 pm and 6:00 pm Frequently, passenger counting is performed at the bus stop that has the highest *daily* passenger volume. The location of this transit stop is usually well known to the transit operator. The counting interval (whole day, between 7:00 am and 10:00 am, between 4:00 pm and 7:00 pm, etc.) is different for different transit operators and different cities.

We denote by P_{max} the average value of the maximum daily passenger volume. For example, transit operator monitored during seven days period the daily number of passengers that departed from the station # 6. The following 7 values were recorded: 1262, 1348, 1439, 1285, 1290, 1391, and 1287. The P_{max}, in this case is equal to:

$$P_{max} = \frac{1262 + 1348 + 1439 + 1285 + 1290 + 1391 + 1287}{7} = 1329$$

The service frequency f that should be offered in order to satisfy maximum passenger volume and desired vehicle occupancy is equal to:

$$f = \frac{P_{max}}{\alpha \cdot C_{car}} \tag{7.15}$$

where:

C_{car} is maximum number of passengers per car; and
$0 \leq \alpha \leq 1$ is load factor.

The load factor α is related to the concept of desired vehicle occupancy. The product $\alpha \cdot C_{car}$ defines the desired vehicle occupancy during the observed time period.

The corresponding headway is equal to:

$$h = \frac{1}{f} = \frac{1}{\frac{P_{max}}{\alpha \cdot C_{car}}} = \frac{\alpha \cdot C_{car}}{P_{max}} \tag{7.16}$$

7.9 TIMETABLE

The transit line timetable is generated at one point (usually terminal). In the next step, by using information about average travel times between transit stops, timetable is generated for all transit stops. The transit operator's timetable contains information about vehicle departure times at all transit stops.

The basic input data for the timetable construction represent service frequency values in certain intervals of time during the day. Timetable can easily be generated in the following way. Let the abscissa represents time. Based on the known values of service frequency in certain intervals of time, we draw a cumulative frequency (Fig. 7.29).

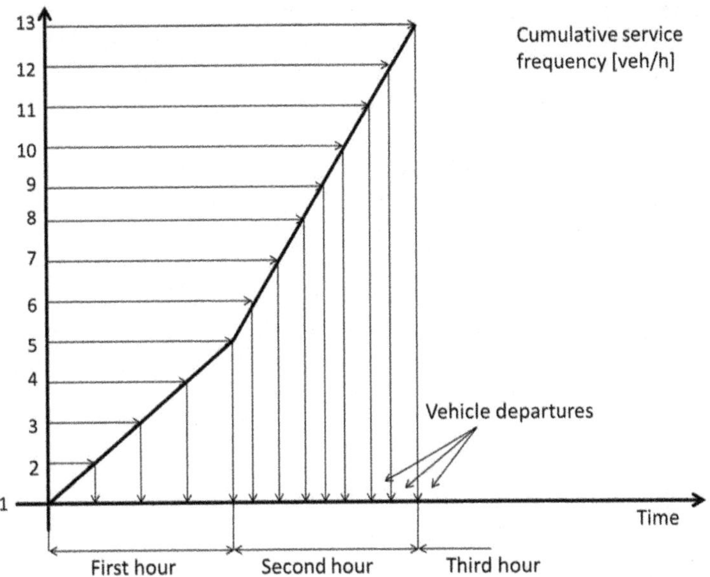

FIG. 7.29

Constant vehicle headways within time intervals.

The cumulative frequency shown in Fig. 7.29 is related to the case when service frequency within first hour is equal to 4, and service frequency within second hour is equal to 8. Let us first vehicle departure happen at the beginning of the first hour. We can go horizontally for every vehicle departure, until intersecting the cumulative curve (Fig. 7.29). From the point of intersection we go vertically. The intersection of this vertical line and abscissa represents vehicle departure time. In this way, we generate timetable at specific point. In this way, we generate constant headways during specific time intervals.

Because of demand variability, constant headways frequently do not generate equal number of passengers at every vehicle departure. In order to achieve even-load at every vehicle departure, we draw cumulative loads on the transit stop that is at the beginning of the MLS (Fig. 7.30).

Unequal number of passengers at specific vehicle departures can cause overcrowding in some vehicles, long boarding time at some transit stops, "bunching" of vehicles, and a decrease in the level of service offered to the passengers. On the other hand, equal load is related to unequal headways which could be inconvenient for passengers.

Trips between nodes in public transit networks may be made with or with no making transfers. Transfers generally cause inconvenience to passengers. Given that inadequately coordinated transfers can increase waiting times considerably, it is particularly important (when constructing timetables) to synchronize schedules cautiously in cases of larger headways. Unsuccessfully coordinated transfers can also reduce the number of passengers using public transit as a result of switching to competitor modes. When designing synchronized schedules, it is essential to try to minimize the total waiting times of all passengers at transfer nodes in a transit network.

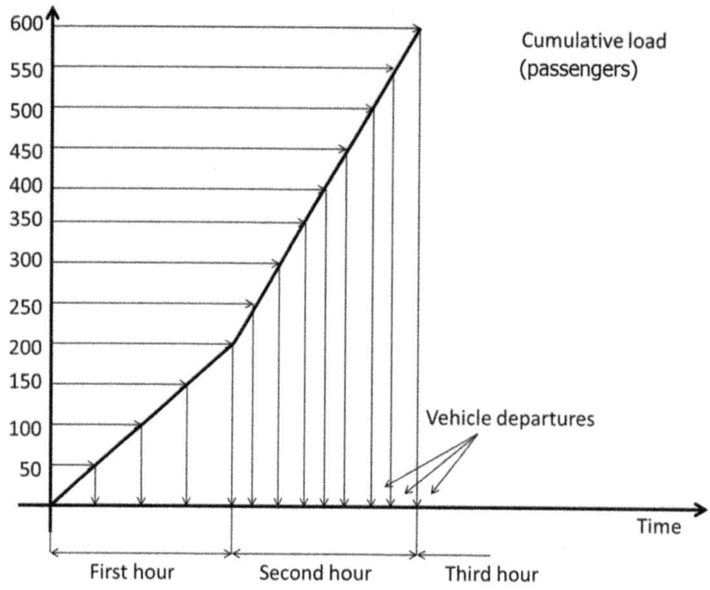

FIG. 7.30

Equal number of passengers at every vehicle departure.

7.10 TRANSIT LINE CAPACITY

The transit line is the basic element of public transit system. The transit route lengths in one direction are usually between 40 and 90 min, while stop spacing in urban areas are in the range from 120 to 400 m. It is desirable that transit route intersects few other transit routes. In this way, transfer points are generated that enable passengers to create various itineraries when making a trip. The waiting time at transfer points, which is less than 8 min, is considered as an acceptable waiting time. The transit line operating hours by weekdays are usually between 5:00 am and midnight. The spacing between transit lines, in the majority of cities, is in the range of 700–1000 m.

We denote by C_{car} the maximum number of passengers per car. Bus capacity represents the sum of number of seated passengers and legal standees. Public transit agencies and operators usually assume six passengers per square meter, as legal standees. In some countries, this figure could be higher.

The line/route (offered) "ultimate" capacity $C_{max}(\tau)$ is expressed by the maximum number of spaces, which can be transported in one direction during a given period of time τ (usually 1 h) under prevailing operating conditions. In other words, the "ultimate" capacity of a public transportation lines represents the product of the maximum frequency of the service $f_{max/i}(\tau)$ and the maximum number of passengers in the transit vehicle.

$$C_{max}(\tau) = f_{max}(\tau) \cdot N_{car} \cdot C_{car}$$

The practical capacity C_{line} of a public transit line represents the product of the offered service frequency f and the maximum number of passengers in the transit vehicle.

$$C_{\text{line}} = f \cdot N_{\text{car}} \cdot C_{\text{car}} \tag{7.17}$$

where:

N_{car} is the number of cars in the train; and
C_{car} is the maximum number of passengers per car.

The number of vehicles N_{car} per service frequency f can be different. For buses, trolleybuses, and trams it is usually one to two and for LRT, RGR, and RRT systems it is usually 5–10. In the case of bus operations, $N_{car} = 1$, and the capacity of a public transportation line C_{line} equals:

$$C_{\text{line}} = f \cdot C_{\text{car}} \left[\frac{\text{spaces}}{h}\right] \tag{7.18}$$

The turnaround time equals:

$$T = \frac{2 \cdot L}{u} \tag{7.19}$$

where:

L is the distance between Terminals A and B (due to simplicity we assume that distance from A to B is equal to the distance from B to A); and
u is the average vehicle speed.

$$C_{\text{line}} = f \cdot C_{\text{car}} \tag{7.20}$$

$$C_{\text{line}} = \frac{N}{T} \cdot C_{\text{car}} \tag{7.21}$$

$$C_{\text{line}} = \frac{N}{\frac{2 \cdot L}{u}} \cdot C_{\text{car}} \tag{7.22}$$

$$C_{\text{line}} = \frac{N \cdot u \cdot C_{\text{car}}}{2 \cdot L} \left[\frac{\text{spaces}}{h}\right] \tag{7.23}$$

As we can see, the line capacity C_{line} depends on the number of engaged vehicles N, the average travel speed u, vehicle capacity C_{car} and the length of the line L. By changing some of these quantities, it is possible to change line capacity.

EXAMPLE 7.3

The public transit line length equals 10 km in one direction. The average bus speed on a city heavy traffic equals 20 km/h. The total of 12 buses is assigned to the line. The capacity of every vehicle equals 50. Calculate the line capacity, frequency, and headway.

Solution

The line capacity equals:

$$C_{\text{line}} = \frac{N \cdot u \cdot C_{\text{car}}}{2 \cdot L}$$

(Continued)

EXAMPLE 7.3—cont'd

$$C_{\text{line}} = \frac{12.20\left[\frac{\text{km}}{\text{h}}\right]\cdot 50\,[\text{spaces}]}{2.10\,[\text{km}]}$$

$$C_{\text{line}} = 600\left[\frac{\text{spaces}}{h}\right]$$

In our case, frequency f and headway h are respectively equal:

$$f = \frac{N}{T}\left[\frac{veh}{h}\right]$$

$$f = \frac{12}{1} = 12\left[\frac{veh}{h}\right]$$

$$h = \frac{1}{f} = \frac{1}{12\left[\frac{veh}{h}\right]} = \frac{1}{\frac{12\,veh}{60\,\text{min}}}$$

$$h = 5\,\text{min}$$

The approximate capacity values of some transit modes are given in Table 7.6.

Table 7.6 The Approximate Capacity Values of Transit Modes

Transit Mode	Capacity (passengers/h)
Bus in mixed traffic	500–1500
Bus on HOV lane	4000–8000
Bus Rapid Transit	7000–30,000
Light Rail Transit with exclusive right-of-way	8000–25,000
Heavy Rail	14,000–60,000

7.10.1 TRANSIT LINE CAPACITY UTILIZATION

The transit line capacity represents the number of spaces offered to passengers that pass a specific point in one direction during 1 h. Transit operator offers specific number of spaces to the passengers during specific period of time. It could happen that the offered capacity is insufficient in certain situations. On the other hand, is it possible that the offered capacity is underutilized. Therefore, there is a need to measure the utilization of the offered capacity. Fig. 7.31 shows transit line capacity and passenger volume profile on the transit line between terminal A and terminal B.

The *transportation work* w_i, made by the transit operator when carrying q_i passengers along the section that has length equal to l_i, equals:

$$w_i = q_i \cdot l_i \tag{7.24}$$

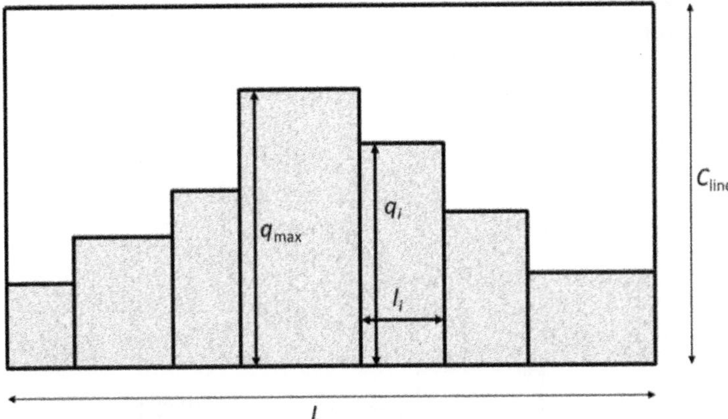

FIG. 7.31

Transit line capacity and passenger volume profile.

Transit operator offers to the passengers the capacity that is equal to C_{line}. The transportation work that is possible to make is equal to the area of a rectangle with sides L and C_{line}. The realized transportation work is equal to the sum of the areas of shaded rectangles (Fig. 7.31). The average transit line capacity utilization α is equal to:

$$\alpha = \frac{\sum_{i=1}^{n} q_i \cdot l_i}{C_{line} \cdot L} \tag{7.25}$$

where n is the number of line sections.

EXAMPLE 7.4

The public transit line length equals 5 km in one direction (Fig. 7.32). The average bus speed on a city heavy traffic equals 20 km/h. The total of 10 buses is assigned to the line. The capacity of every vehicle equals 50. The passenger volume profile of the line is given in Fig. 7.32.

Calculate the turnaround time, service frequency, headway, line capacity, and the average transit line capacity utilization α.

Solution

The turnaround time equals:

$$T = \frac{2 \cdot L}{u}$$

$$T = \frac{2.5\text{km}}{20\left[\frac{\text{km}}{\text{h}}\right]} = 0.5[\text{h}] = 30[\text{min}]$$

(Continued)

EXAMPLE 7.4—cont'd

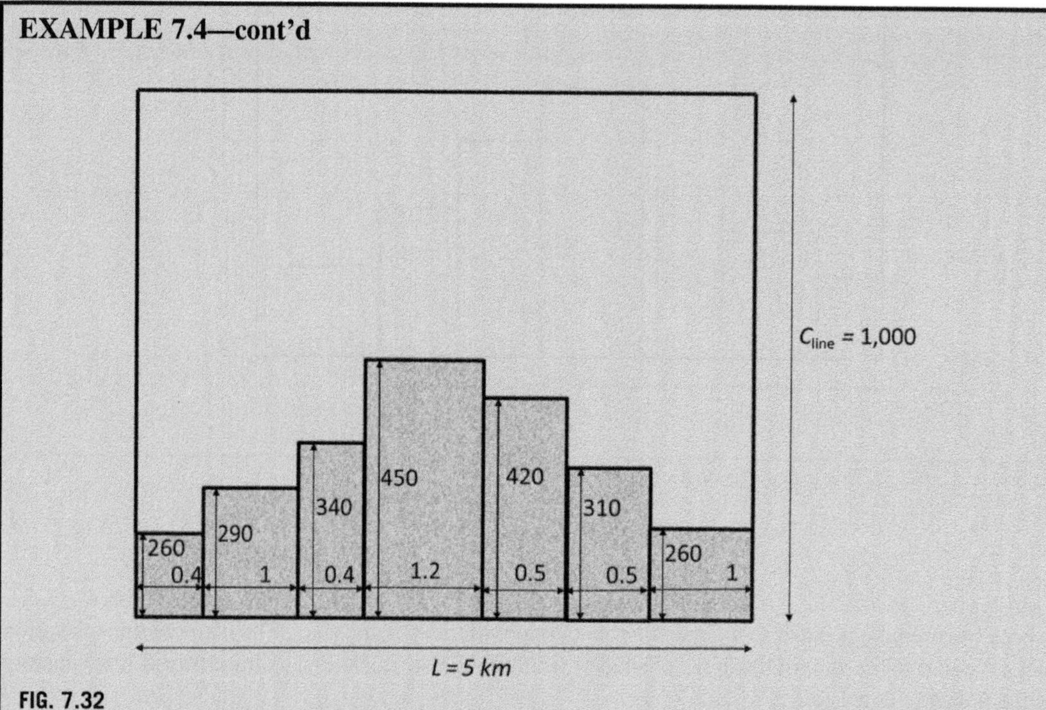

FIG. 7.32

Passenger volume profile in the case of line whose length equals $L=5$km.

The service frequency and the headway are respectively equal:

$$f=\frac{N}{T}\left[\frac{veh}{h}\right]$$

$$f=\frac{10}{0.5}\left[\frac{veh}{h}\right]=20\left[\frac{veh}{h}\right]$$

$$h=\frac{1}{f}=\frac{1}{20\left[\frac{veh}{h}\right]}=\frac{\frac{1}{20}}{60\,\text{min}}$$

$$h=3\,\text{min}$$

The line capacity equals:

$$C_{\text{line}}=\frac{N\cdot u\cdot C_{\text{car}}}{2\cdot L}$$

$$C_{\text{line}}=\frac{10.20\left[\frac{\text{km}}{\text{h}}\right]\cdot 50[\text{spaces}]}{2.5[\text{km}]}$$

EXAMPLE 7.4—cont'd

$$C_{\text{line}} = 1000\left[\frac{\text{spaces}}{h}\right]$$

The average transit line capacity utilization equals:

$$\alpha = \frac{\sum_{i=1}^{n} q_i \cdot l_i}{C_{\text{line}} \cdot L}$$

$$\alpha = \frac{260\left[\frac{\text{pass}}{\text{h}}\right]\cdot 0.4[\text{km}] + 290\left[\frac{\text{pass}}{\text{h}}\right]\cdot 1[\text{km}] + 340\left[\frac{\text{pass}}{\text{h}}\right]\cdot 0.4[\text{km}] + 450\left[\frac{\text{pass}}{\text{h}}\right]\cdot 1.2[\text{km}] + 420\left[\frac{\text{pass}}{\text{h}}\right]\cdot 0.5[\text{km}] + 310\left[\frac{\text{pass}}{\text{h}}\right]\cdot 0.5[\text{km}] + 260}{1000\left[\frac{\text{spaces}}{\text{h}}\right]\cdot 5[\text{km}]}$$

$$\alpha = 0.339$$

The average transit line capacity utilization is relatively low. In order to increase, the average line capacity utilization, operator could decrease the number of buses engaged. If, for example, transit operator operates with $N = 5$ buses, the basic transit line parameters would be:

$$f = 10\left[\frac{veh}{h}\right]$$

$$h = 6\text{min}$$

$$C_{\text{line}} = 500\left[\frac{\text{spaces}}{\text{h}}\right]$$

$$\alpha = 0.678$$

7.11 THE PERFORMANCES OF THE URBAN TRANSIT NETWORK

The "ultimate" capacity of urban transit network during a given period of time τ, $C_{\text{network}}(\tau)$ can be determined as the sum of the capacities of its particular lines/routes as follows:

$$C_{\text{network}}(\tau) = \sum_{l=1}^{N_l} C_{\text{line}\,i}(\tau) \tag{7.26}$$

where N_l is the number of lines/routes in the network.

The other symbols are analogous to those in the previous relations.

The transport work is the quantity of transport services offered or utilized on a given line/route or in the entire transit network during a given period of time under prevailing operating conditions. For the given public transit line i, the transport work carried out during a given period of time τ (eg, 1 h), (spaces-km and pax-km) can be estimated as follows:

$$TW_i(\tau) = C_{\text{line}\,i} \cdot l_i \tag{7.27}$$

The transport work $TW_{\text{network}}(\tau)$ in the public transit network, carried out during a given period of time τ (eg, 1 h), (spaces-km and pax-km) equals:

$$TW_{\text{network}}(\tau) = \sum_{i=1}^{N_l} TW_{\text{line}\,i}(\tau) \tag{7.28}$$

where l_i is the length of a given line/route i (km).

The productivity $TP_i(\tau)$ of urban transit links and the network $TP_{\text{network}}(\tau)$ reflects the speed of carrying out the transport work. It is calculated as the product of the number of offered spaces and the speeds of transit vehicles (spaces-km/h) on a given line/route and in the entire network:

$$TP_i(\tau) = C_{\text{line } i} \cdot u_i \tag{7.29}$$

$$TP_{\text{network}}(\tau) = \sum\nolimits_{i=1}^{N_i} TP_{\text{line}\, i}(\tau) \tag{7.30}$$

where u_i is the average speed of transit vehicles along a given line/route i (km/h).

The fleet size of an urban transport operator/company is defined by the number of vehicles needed to operate on the particular lines/routes of the entire network. This fleet usually consists of the vehicles needed for the regular operations, reserve, and maintenance (Vuchic, 2005). For a given line/route i, the fleet of vehicles required to operate during a given period of time τ can be calculated as follows:

$$m_i(\tau) = f_i(\tau) \cdot \tau_{ci} = f_i(\tau) \cdot [2 \cdot (\tau_{oi}) + t_{ti}] \tag{7.31}$$

where:

$f_i(\tau)$ is the scheduled service frequency on the line/route i during the time period τ (dep/h);
τ_{ci} is the vehicle turnaround or cycle time on the line/route i (min or h);
τ_{oi} is the one-way vehicle's operating time on the line/route i (min or h); and
t_{ti} is the time, which the vehicles spend at the end stations (terminals) of the line/route (i) (min).

The one-way vehicle's operating time on the line/route τ_{oi} in Eq. (7.31) can be determined as follows:

$$\tau_{oi} = (N_i - 1) \cdot \frac{\Delta S_i}{v_i} + (N_i - 2) \cdot t_{si} \tag{7.32}$$

where:

N_i is the number of stops/stations on the line/route i;
ΔS_i is the average spacing between stops/stations on the line/route i (m);
v_i is the average vehicle's operating speed between successive stops/stations (km/h); and
t_{si} is the average vehicle's stop time at stops/stations on the line/route (i) excluding the end stops/stations (sec, min).

Fig. 7.33 shows the time-space diagram of typical turnaround time of a transit vehicle on the route.

As can be seen, this time consists of the vehicle's stop time at stops/stations and riding time between them.

The required number of vehicles to operate in the network consisting of N_l lines/routes during a given period of time is equal:

$$M(\tau) = \sum_{i=1}^{N_l} m_i(\tau) \tag{7.33}$$

where all symbols are analogous to those in the previous relations.

The offered service level of service can be expressed by different indicators relevant for users such as: access time, schedule delay, comfort on board transit vehicle(s), travel speed on the line/route,

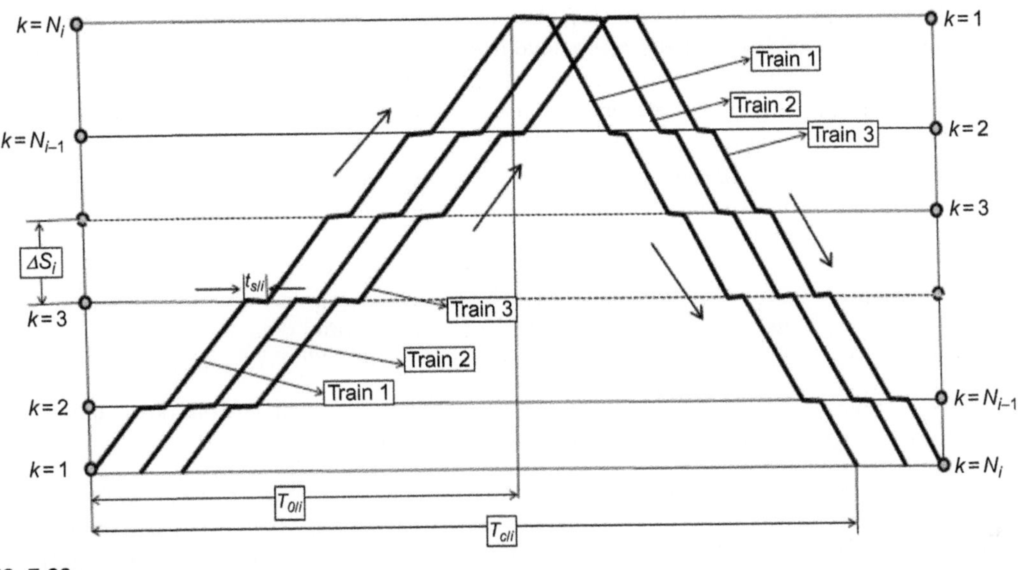

FIG. 7.33

Time-space diagram of typical turnaround time of vehicle(s) on a route of an urban transit system.

travel time on the line/route, transfer time, punctuality and reliability of transit services, and door-to-door or origin-destination travel time (Vuchic, 1981; 2005).

The access time t_a is the time a user-passenger takes for accessing the stop/station of chosen transit system from its origin, or vice versa, from the stop/station his/her destination.

The schedule delay represents the time interval between passenger' arrival at given stop/station and departure of the first arriving transit vehicle. If the scheduled service frequency $f_i(\tau)$ on the given line i is relatively high (several per hour), the passengers are supposed to uniformly arrive during the headway, ie, between two successive vehicle departures. Under such conditions, the schedule delay $sd_i(\tau)$ (min) can be estimated as follows:

$$sd_i(\tau) = \frac{1}{2} \cdot \frac{\tau}{f_i(\tau)} \tag{7.34}$$

In cases of rare scheduled service frequencies, passengers are supposed to know the schedule and arrive closer to the desired service departure time. Consequently, the schedule delay in such case is shorter.

Comfort on board transit vehicles are measured by the offered space on board transit vehicle(s) to seating and standing passengers. In general there are five levels of comfort as follows: (1) >1.0 m^2/pax, implying comfortable standing and free circulation; (2) 0.33–0.5 m^2/pax, some contacts and slightly compromised circulation; (3) 0.25 m^2/pax, extensive contacts, high compromised circulation; (4) 0.20 m^2/pax, difficult—pressed standing and extremely compromised circulation; and (5) 0.15 m^2/pax, crowd, forced circulation. In addition, the design standard for RB and SRB buses in Europe is: 4 pax/m^2. This gives 0.25 m^2/pax, what is equivalent to level "3" of the comfort onboard (EU, 2011). It should be mentioned that these levels of comfort are dependent, in addition to the area of floor of transit vehicles indicating their size, also on the current relationship between the utilized and offered capacity, ie, load factor, under given conditions.

Travel speed on the line/route is the speed a passenger experiences on a given line/route i while traveling between its origin and destination. It is equivalent to the vehicle operating speed V_i (km/h) offered to passengers by transit operator/company along a given line/route i.

Travel time t_{ti} on the line/route is duration of passenger travel on board the vehicle. It can be expressed by the ratio of spacing between passenger origin and destination stop/station and travel speed. For the line/route i, it is as follows:

$$t_{ti} = \frac{S_i}{V_i} \tag{7.35}$$

where:

S_i is the average interstation spacing on the line/route (i) of the given network (km); and
V_i is the travel speed on the line/route i of the given network (km/h).

Transfer time $t_{\mathrm{tr}\ ij}$ is the time needed for passengers to pass from one line/route to another operated by the same or different transit system. This time starts from disembarking vehicle of one line/route until time of boarding vehicle of another line/route. It mainly depends on the distance between platforms of the corresponding lines/routes and walking speed and can be calculated for passing from the line/route i to the line/route j of the same or different transit systems as follows:

$$t_{\mathrm{tr}\ ij} = \frac{d_{ij}}{v_{ij}} \tag{7.36}$$

where:

d_{ij} is the average (passing) distance between the lines/routes i and j (m); and
v_{ij} is the average walking speed on the distance d_{ij} (m/s).

In general, in cases of coordinated schedule, the headways on departing lines/routes should be greater than the transfer times from arriving lines/routes.

Punctuality of service is measured by delays of particular transit services as the difference between the actual and planned arrival/departure time(s) of vehicles of given transit system at the passenger origin and destination stop(s)/station(s). For the road-based systems operating under prevailing (rather regular) conditions, these delays can be more stochastic and longer due to influence of traffic sharing the same streets and road lanes as the given lines/routes. For the rail-based systems, except for streetcars (tramways), they are less stochastic and relatively shorter due to the above-described nature of operations mainly on the isolated routes/lines. It should be mentioned that the delays W_i do not happen due to discrepancies between the demand for service represented by the scheduled service frequency and the capacity of infrastructure components—segments of road lanes, rail lines, and corresponding stops/stations.

Reliability of service is measured as the ratio between actually realized and scheduled/planned transit services on particular lines/routes during a given period of time (day, month, year). This implies that some scheduled transit services can be canceled due to many reasons. The reliability of services on the given line/route and of the entire network during a given period of time can be estimated, respectively, as follows:

$$r_i(\tau) = \frac{F_{ai}(\tau)}{f_i(\tau)} \cdot 100\% \tag{7.37}$$

$$r(\tau) = \frac{\sum_{i=1}^{N_i} F_{ai}(\tau)}{\sum_{i=1}^{N_i} f_i(\tau)} \cdot 100\% \tag{7.38}$$

where $F_{ai}(\tau)$ is the number of actually realized transit services on the line/route i during the time period τ (dep/h).

The other symbols area analogous to those in the previous relations. On the longer time scale (month, year), this indicator is more relevant for transit operator(s)/company(s). On the short time scale (few hours, or a day), this indicator is highly relevant for passengers since cancelation of particular services can substantively prolong their door-to-door or origin-destination travel time mainly due to prolonging schedule delay.

Door-to-door or Origin-Destination (OD) travel time t_{OD} is the time of which passengers spend from the door at their origin to the door at their destination. This time for a given pair of doors can be calculated as follows:

$$t_{OD} = t_{a1} + \frac{sd}{p} + t_t + \gamma \cdot t_{tr} + \chi \cdot W + t_{a2} \tag{7.39}$$

where:

t_{a1}, t_{a2} is the average access time of a passenger at his/her origin and destination stop(s)/station(s), respectively (min);
p is the proportion of actually scheduled transit services on the line/route between given passenger origin and destination ($0 < p < 1.0$);
γ, χ is the coefficient, which takes the value "1" if the transfer and the delay, respectively, take place, and the value "0," otherwise; and
W is the average delay between a given pair of origin and destination.

The other symbols are analogous to those in the previous relations.

In the case when $p = 1.0$, all scheduled transit services will be realized with a corresponding schedule delay(s). If $p = 0.0$, none of the scheduled transit services will be realized thus infinitely prolong the corresponding schedule delay(s) and consequently OD travel time(s). This indicates that trips are not realized.

EXAMPLE 7.5

The above-mentioned indicators of performances of service networks—capacity and level of service of urban transit systems are illustrated by the service network of RRT—metro system in Dubai (UAE) (Fig. 7.17). The indicators of performances such as capacity and service level based on the scheduled services are calculated and given in Table 7.7 together with the characteristics of rolling stock/trains.

(Continued)

EXAMPLE 7.5—cont'd

Table 7.7 Some Indicators of Performances of the RRT—Metro Service Network in Dubai—supply side (UAE) (https://www.rta.ae)

Indicator	Red Line ($i=1$)	Green Line ($i=2$)	Network
	Rolling stock		
s (spaces/car)	128	128	128
m (cars/train)	5	5	5
s_f (spaces/train)	640	640	640
Length (m/train)	85.5	85.5	85.5
	Capacity		
τ (h)	1	1	1
h_s (min)[a]	3	3	3
$f_s(\tau)$ (dep/h)[a]	20	20	20
$\mu(\tau)$ (spaces/h) (2 directions)	25,600	25,600	25,600
$TW(\tau)$ (spaces-km) (2 directions)	1,310,720	576,000	1,886,720
v (km/h)	43	35	39
$TP(\tau)$(spaces-km/h) (2 directions)	1,111,040	933,680	2,012,160
τ_c (h)	2.4	1.4	3.8
$m(\tau)$ (trains)	48	28	76
	Service level		
t_a (min)	5	5	5
sd (min)	1.5	1.5	1.5
$\theta(-)$[b]	0.81	0.81	0.81
Comfort (m^2/pax)[c]	0.35	0.35	0.35
V (km/h)	43	35	39
t_t (h)[d]	1.2	0.7	–
t_{tr} (min)	–	–	–
W (%)	–	–	–
r (%)	–	–	–
t_{OD} (h)[e]	1.4	0.9	–

[a]*The scheduled headway between departures and the related scheduled frequency; the minimum headway can be:* $h_{min}=1.5$ *min.*
[b]*Based on 500 000 transported passengers per day (24 h).*
[c]*Category "2" of the comfort onboard the train.*
[d]*In single direction between the end stations of the line.*
[e]*In single direction between the end stations of the line with 100% reliable transit services* ($p=1.0$) *and no delays* ($\chi=0.0$).

Similarly, performances of transit service networks of other urban transit systems can be estimated.

7.12 PUBLIC TRANSIT NETWORK TYPES

In general, the urban transit networks based on the ROW category of streets/roads differ significantly from the others using the ROW category of rail tracks. The former are characterized for the road and the later for the rail-based modal systems. The difference is mainly due to: (1) the road-based systems' transit lines/routes must strictly follow alignment of urban streets, while those of LRT or RRT system should; (2) the network of road-based systems is denser with close stops attracting mostly shorter trips, while that of the rail-based systems is characterized and performs quite opposite; and (3) the performances of road-based systems such as capacity, travel time, reliability and punctuality of services, and safety are lower than those of its rail-based counterparts mainly because of the dedicated/isolated ROWs and technology dictating the nature of vehicle operations along the routes/lines. Consequently, contrary to the urban road-based transit systems, the infrastructure networks of the rail-based systems have some typical/recognizable configuration/geometry in the typical form such as: radial, radial/circumferential, rectangular or grid, ubiquitous network.

We denote the public transit network by $G=(N, A)$, where N is the set of nodes, and A set of links (street segments). Nodes represent potential bus stops (intersections, zone centroids). Any path used by transit passengers is defined by a sequence of nodes, and links. Public transit networks have emerged in modern cities since the mid-19th century. These networks have evolved over time with the development of cities and the development of transport technologies. Public transit networks usually represent a combination and/or modification of the basic types shown in Fig. 7.34.

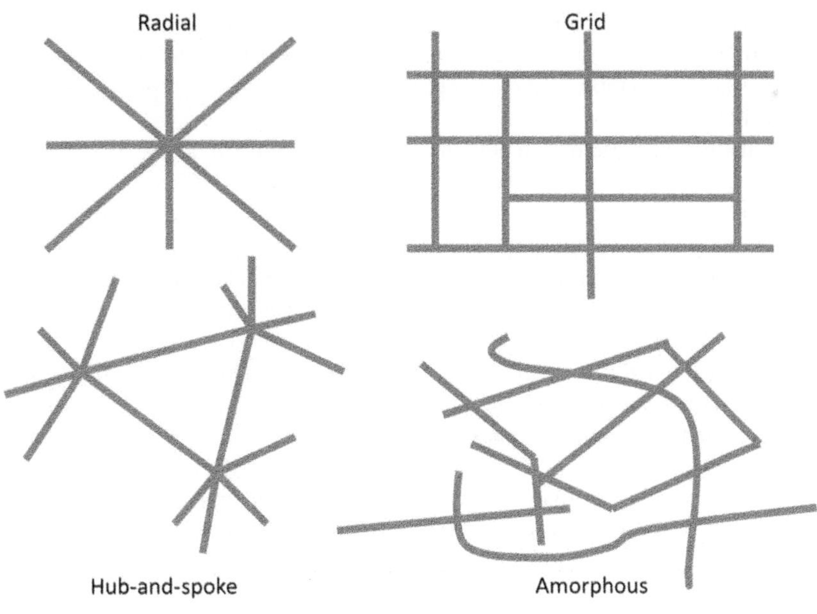

FIG. 7.34

Radial, grid, hub-and-spoke, and amorphous transit network.

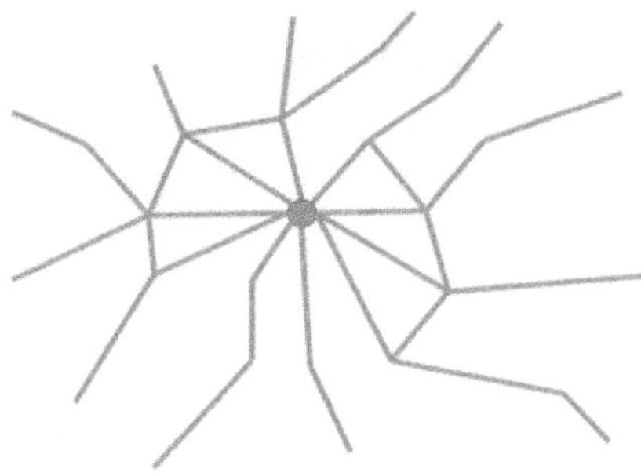

FIG. 7.35

Radial/circumferential transit network.

Radial transit networks have been developed, primarily in the cities, where the majority of activities (work, shopping, business, school, etc.) were concentrated in a single node (usually central business district). Some transit lines in radial networks start at the suburb and end in the city center. Some other lines start on the suburb, pass through the city center and end at some other suburb. These networks usually do not have transfer stations between particular lines. Any passenger trip (between any two nodes), in a radial network, is possible to do without transfer, or with at most one transfer. The radial networks consist primarily of the radial rail (LRT, RRT, or RGR) lines converging towards the relatively small area (center) of a given urban area. Typical examples are Chicago Regional Rail System-Metra (Chicago) network, and the regional rail networks in Paris, Munich, and Oslo.

The radial/circumferential LRT or RRT networks (Fig. 7.35) contain the radial lines, but also a circle, circumferential, and/or tangential lines. These intersect with the radial lines at the transfer stations thus enabling passenger transfer between themselves and the radial lines. In some cases, they provide collection and distribution of passengers for the radial lines. A typical example is the RRT Moscow Metropolitan network.

In most cases, the phrase "grid transit network" (Fig. 7.34) refers to a rectilinear grid network that contains the east-west and north-south lines. This type, as well as modifications of this basic type of public transit network is mainly seen in some American cities. The rectangular or grid LRT, RRT networks have the rectangular spatial pattern with lines intersecting at rather right angles with each other at stations enabling passenger transfer between them. These networks provide uniform coverage of a given urban area, thus being suitable for the large urban areas with relatively uniform intensity of population activities. A typical example is the Boston CDB network.

The ubiquitous usually RRT networks have become an increasing trend among urban transport planners. They consist of radial lines without focusing on the small central area but on almost the entire urban area. Their stations are dispersed along the entire urban area thus providing a wider spatial accessibility and coverage than the radial networks. The particularities are that most lines split into two of more branches towards suburban area(s) they connect to. In addition to serving high passenger demand by the radial lines, these networks enable good accessibility to the center and adequate coverage of both center and periphery of given urban areas, good connectivity and transfer between particular lines, and reasonable connections to other urban and suburban transit systems. A typical example is the Paris Metro network (Fig. 7.36) (Vuchic, 2005).

In some cities, public transit operators organize a hub-and-spoke transit network (Fig. 7.34). Hubs are usually down-towns, rail stations, big shopping centers, etc. At hubs, passengers make transfers.

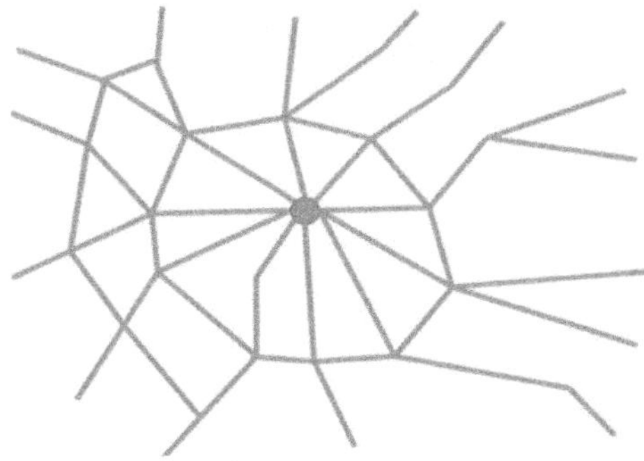

FIG. 7.36

Ubiquitous transit network.

Passengers from suburbs are transported to hubs. After making the transfer at hub, some of them continue trip by using another transit line that leads to another hub. Practically, one can claim that hub-and-spoke network contains few smaller radial networks. Hub-and-spoke networks are characterized by timed transfers operations. The arrivals of vehicles to hubs are carefully planned, and vehicles usually wait a few minutes at hubs for passengers to transfer.

Amorphous transit networks (Fig. 7.34) could appear in the cities when there are few transit operators that do not have appropriate coordination. The network is just a collection of the few smaller transit networks developed by individual transit operators independently. In an amorphous transit network, passengers usually make more transfers and have longer travel and waiting times.

7.13 THE PUBLIC TRANSIT NETWORK DESIGN

The public transit network design problem is the main strategic problem in public transportation. It is one of the most significant problems faced by the transit operators and city authorities in the world. Let us consider the road network shown in Fig. 7.37.

We denote this network by $G=(N,A)$, where N is the set of nodes, and A set of links (street segments). Nodes represent potential bus stops (intersections, zone centroids). Any path used by transit passengers is defined by a sequence of nodes, and links. We study the transit network design problem in the case of connected undirected street networks. Connected street network assumes that any two nodes in the network are connected by at least one path. Within transit network design problem we search for the best possible set of routes R. In other words, we make the decision about the links from the set A to be included in the set of routes R, as well as the decision how to bring together chosen links into the fixed transit routes. In addition to this, authors in some cases also determine frequency of transit service on each of the defined routes. One of the possible public transportation networks for the road network shown is shown in Fig. 7.38.

Public transit network design problem belongs to the class of difficult combinatorial optimization problem, whose optimal solution is difficult to discover (Lampkin and Saalmans, 1967; Mandl, 1979; Newell, 1979; Banković, 1982; Pattnaik et al., 1998; Bielli et al., 2002; Zhao and Zeng, 2007; Mauttone and Urquhart, 2009; Ceder, 2015). The bus network shape, as well as bus frequencies, highly depend on both passenger demand, and on the number and type of available buses (fleet size), and/or available budget (Baaj and Mahmassani, 1992, 1995; Ceder and Wilson, 1986; Israeli and Ceder, 1989, 1995;

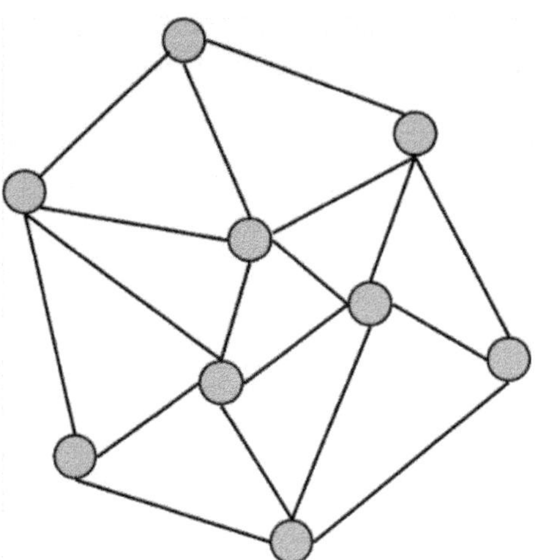

FIG. 7.37

Road network.

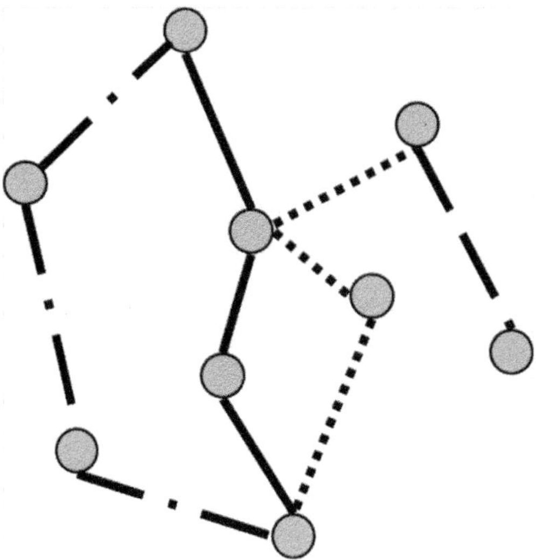

FIG. 7.38

One of the possible public transportation networks for the road network.

Chakroborty and Dwivedi, 2002; Chakroborty, 2003; Guan et al., 2003; Fan and Machemehl, 2006, 2008; Mauttone and Urquhart, 2009). Poorly designed bus network can cause very long passengers' waiting times, and/or inexactness in bus arriving times. In addition, inadequately designed network can show high inappropriateness among the designed bus routes and paths of the majority of users. Many of the factors that should be taken into account when designing bus network are mutually in conflict. For example, the shorter passengers waiting times, the higher the number of buses needed, etc. When designing the bus network, the interests of both the operator and the passenger must be taken into account. Due to the conflicting nature of these interests, we treat the bus network design problem as a multi-criteria decision-making problem.

The main indicator that planners use to describe the level of transit service is the total travel time spent by the users of transit service. The total travel time (frequently is also used the expression "the generalized cost of travel") has the following components: access time, waiting time, in-vehicle time, transfer time, and egress time, ie:

$$t = w_1 \cdot t_a + w_2 \cdot t_w + w_3 \cdot t_{iv} + w_4 \cdot t_t + w_5 \cdot t_e \tag{7.40}$$

where:

$w_1, w_2, \ldots, w_5$ is the travel time component weights ($0 \le w_i \le 1 \quad i = 1,2,\ldots,5$);
t is the total travel time;
t_a is the access time;
t_w is the waiting time;
t_{iv} is the in-vehicle time;
t_t is the transfer time; and
t_e is the egress time.

When calculating the total travel time, the analysts assign different weights to the travel time components.

When measuring the quality of the solution generated, we take into account the total number of transfers, as well as the total number of unsatisfied passengers, since transfers and unsatisfied demand keep back passengers to use transit. In some public transit network design models, the number of passenger transfers and the total number of unsatisfied passengers are also converted into time. Obviously, the total number of transfers and the total number of unsatisfied passengers may be decreased by optimizing the configuration of the transit network.

We denote by Q the origin-destination matrix:

$$Q = \begin{bmatrix} q_{11} & \cdots & q_{1n} \\ \vdots & \ddots & \vdots \\ q_{n1} & \cdots & q_{nn} \end{bmatrix} \tag{7.41}$$

where q_{ij} is the number of travelers that want to travel from the zone i to the zone j.

The more details related to the estimation of the number of travelers, and other travel demand analysis issues will be given in Chapter 8 devoted to the travel demand analysis.

The following are typical objective functions used in public transit network design models:

Minimization of the total travel time of all passengers
Maximization of the transit operator's profit (in the case of private transit operators)
Maximization of the area covered by public transportation

The transit network design problem could be defined in the following way:

> For a given set of n nodes, known origin-destination matrix Q that describes demand among these nodes, and known travel time matrix TR, generate set of transit routes on a network in such a way to optimize defined objective function.

7.13.1 SIMPLE GREEDY ALGORITHM FOR PUBLIC TRANSIT NETWORK DESIGN

Let us consider bus line *l* whose terminals are located in the nodes *i* and *j* respectively (Fig. 7.39). Bus line *l* contains all nodes that belong to the shortest path between *i* and *j*. Let us denote with N_l the set of nodes connected by the line *l*.

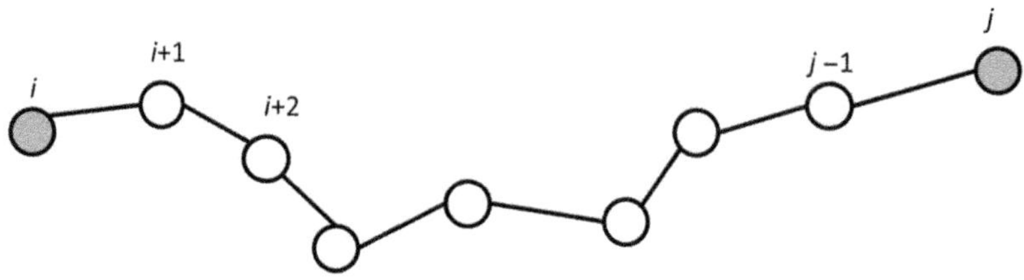

FIG. 7.39

Bus line whose terminals are located in the nodes *i* and *j*.

This bus line could be used by the passengers that enjoy direct service, as well as by passengers that have to make the most two transfers during their trip. The total number of passengers ds_{ij} that enjoy the direct service along this bus line *l* equals:

$$ds_{ij} = \sum_{m \in N_l} \sum_{n \in N_l} d_{mn} \tag{7.42}$$

We denote by *DS* the corresponding matrix that contains information about the number of passengers that enjoy the direct service:

$$DS = \{ds_{ij} | i, j \in [1, 2, \ldots, |N|]\} \tag{7.43}$$

Nikolić and Teodorovic (2013) proposed a simple greedy algorithm to generate the public transit network routes. In this algorithm, the authors tried to connect, by the direct service, pairs of nodes that have high ds_{ij} values. In this way, the number of passengers that enjoy the direct service is increased. The algorithm is composed of the following steps:

Step 1: Prescribe the total number of bus lines *NBL* in the network. Denote the set of bus lines by *Y*. Set $Y = \varnothing$. Let $m = 1$.
Step 2: Find the pair of nodes that has the highest ds_{ij} value. Let this pair is the pair of nodes (a, b). The nodes *a* and *b* are the terminals of the new bus line. Find the shortest path between these two nodes. The nodes that belong to the shortest path represent stations in the bus lines.
Add line *l* in set *Y*.
Step 3: Update the matrix *DS*, without taking into account passenger travel demands that is already satisfied.
Step 4: If $m = NBL$, stop; otherwise, set $m = m + 1$ and return to Step 2.

EXAMPLE 7.6

Fig. 7.40 shows a road network. This hypothetical transportation network, which was proposed by Mandl (1979) often serves as a benchmark example for testing various methods for transit networks design.

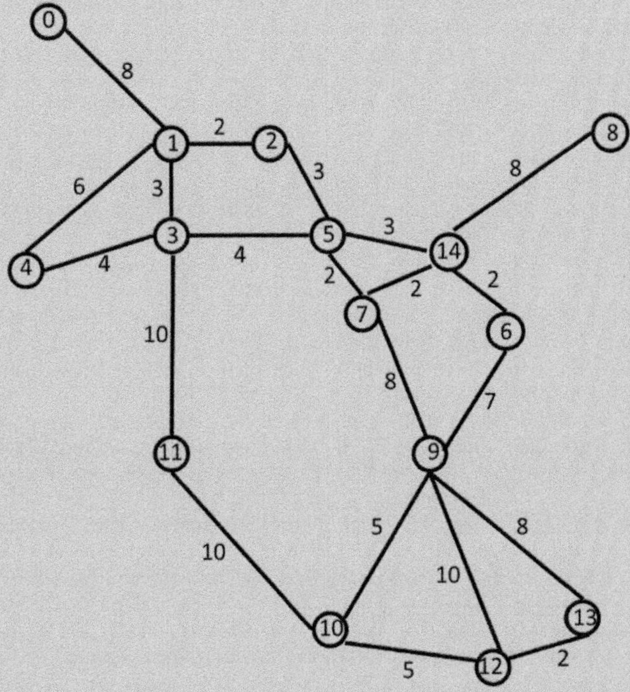

FIG. 7.40

Mandl's road network.

The Origin-Destination matrix reads:

	1	2	3	4	5	6	7	8	9	10	11	12	13	14	15
1	0	400	200	60	80	150	75	75	30	160	30	25	35	0	0
2	400	0	50	120	20	180	90	90	15	130	20	10	10	5	0
3	200	50	0	40	60	180	90	90	15	45	20	10	10	5	0
4	60	120	40	0	50	100	50	50	15	240	40	25	10	5	0
5	80	20	60	50	0	50	25	25	10	120	20	15	5	0	0
6	150	180	180	100	50	0	100	100	30	880	60	15	15	10	0
7	75	90	90	50	25	100	0	50	15	440	35	10	10	5	0
8	75	90	90	50	25	100	50	0	15	440	35	10	10	5	0
9	30	15	15	15	10	30	15	15	0	140	20	5	0	0	0
10	160	130	45	240	120	880	440	440	140	0	600	250	500	200	0
11	30	20	20	40	20	60	35	35	20	600	0	75	95	15	0
12	25	10	10	25	15	15	10	10	5	250	75	0	70	0	0
13	35	10	10	10	5	15	10	10	0	500	95	70	0	45	0
14	0	5	5	5	0	10	5	5	0	200	15	0	45	0	0
15	0	0	0	0	0	0	0	0	0	0	0	0	0	0	0

(Continued)

EXAMPLE 7.6—cont'd

Design the public transit network composed of $NBL=6$ bus lines by using the described simple greedy algorithm.

Solution

The total number of bus lines NBL equals 6. We denote the set of bus lines by Y. At the beginning, $=\varnothing$. We also let $m=1$.

We calculate ds_{ij} values for all pair of nodes. The matrix DS reads:

	0	1	2	3	4	5	6	7	8	9	10	11	12	13	14
0	0	800	1300	1160	1000	2320	3030	3030	2500	6340	7870	1280	7500	6790	2320
1	800	0	100	240	40	820	1380	1380	940	4370	5840	310	5460	4820	820
2	1300	100	0	420	260	360	740	740	450	3470	4900	510	4540	3910	360
3	1160	240	420	0	100	200	500	500	290	3620	5090	50	4690	4060	200
4	1000	40	260	100	0	400	750	750	510	4110	5620	180	5190	4550	400
5	2320	820	360	200	400	0	200	200	60	2840	4230	280	3890	3270	0
6	3030	1380	740	500	750	200	0	100	30	880	2150	600	1900	1290	0
7	3030	1380	740	500	750	200	100	0	30	880	2150	600	1900	1290	0
8	2500	940	450	290	510	60	30	30	0	1190	2500	380	2210	1600	0
9	6340	4370	3470	3620	4110	2840	880	880	1190	0	1200	1850	1000	400	880
10	7870	5840	4900	5090	5620	4230	2150	2150	2500	1200	0	150	190	310	2150
11	1280	310	510	50	180	280	600	600	380	1850	150	0	480	600	280
12	7500	5460	4540	4690	5190	3890	1900	1900	2210	1000	190	480	0	90	1900
13	6790	4820	3910	4060	4550	3270	1290	1290	1600	400	310	600	90	0	1290
14	2320	820	360	200	400	0	0	0	0	880	2150	280	1900	1290	0

We go to step 2 of the algorithm. The pair of nodes (0, 10) has the highest ds_{ij} value. This value equals 7870. The nodes 0 and 10 are the terminals of the first bus line. The shortest path between terminal 0 and terminal 10 reads: 0—1—2—5—7—9—10. The nodes 1, 2, 5, 7, and 9 are the stations of the first transit line. The first transit line reads: 0—1—2—5—7—9—1. The set Y reads: $Y=\{1\}$.

We go to step 3 of the algorithm and update the matrix DS (we do not take into account any more the passenger travel demands that is already satisfied by the transit line 1). The updated matrix DS reads:

	0	1	2	3	4	5	6	7	8	9	10	11	12	13	14
0	0	0	0	360	200	0	710	0	180	0	0	480	1160	450	0
1	0	0	0	240	40	0	560	0	120	0	0	310	1090	450	0
2	0	0	0	320	160	0	380	0	90	0	0	410	1070	440	0
3	360	240	320	0	100	200	500	300	290	780	860	50	1850	1220	200
4	200	40	160	100	0	400	750	550	510	1270	1390	180	2350	1710	400
5	0	0	0	200	400	0	200	0	60	0	0	280	1050	430	0
6	710	560	380	500	750	200	0	100	30	880	950	600	1900	1290	0
7	0	0	0	300	550	0	100	0	30	0	0	400	1020	410	0
8	180	120	90	290	510	60	30	30	0	1190	1300	380	2210	1600	0
9	0	0	0	780	1270	0	880	0	1190	0	0	650	1000	400	880
10	0	0	0	860	1390	0	950	0	1300	0	0	150	190	310	950
11	480	310	410	50	180	280	600	400	380	650	150	0	480	600	280
12	1160	1090	1070	1850	2350	1050	1900	1020	2210	1000	190	480	0	90	1900
13	450	450	440	1220	1710	430	1290	410	1600	400	310	600	90	0	1290
14	0	0	0	200	400	0	0	0	0	880	950	280	1900	1290	0

We go to step 4 of the algorithm. We check if $m=NBL$. Since $m=1<6=NBL$, we set $m=2$ and return to step 2. The pair of nodes (4, 12) has the highest ds_{ij} value. This value equals 2350. The nodes 4 and 12 are the terminals of

EXAMPLE 7.6—cont'd

the second bus line. The shortest path between terminal 4 and terminal 12 reads: 4—3—5—7—9—12. The nodes 3, 5, 7, and 9 are the stations of the second transit line. The second transit line reads: 4—3—5—7—9—12. The set Y reads: $Y = \{1, 2\}$.

The set of public transit lines generated by the described algorithm is shown in Table 7.8.

Table 7.8 The Generated Set of the Public Transit Lines by the Greedy Algorithm

Route Number	Route Description
1	0, 1, 2, 5, 7, 9, 10
2	4, 3, 5, 7, 9, 12
3	8, 14, 6, 9, 13
4	0, 1, 2, 5, 14, 6
5	9, 10, 11
6	0, 1, 3, 11

The generated public transit lines are also shown in Fig. 7.41.

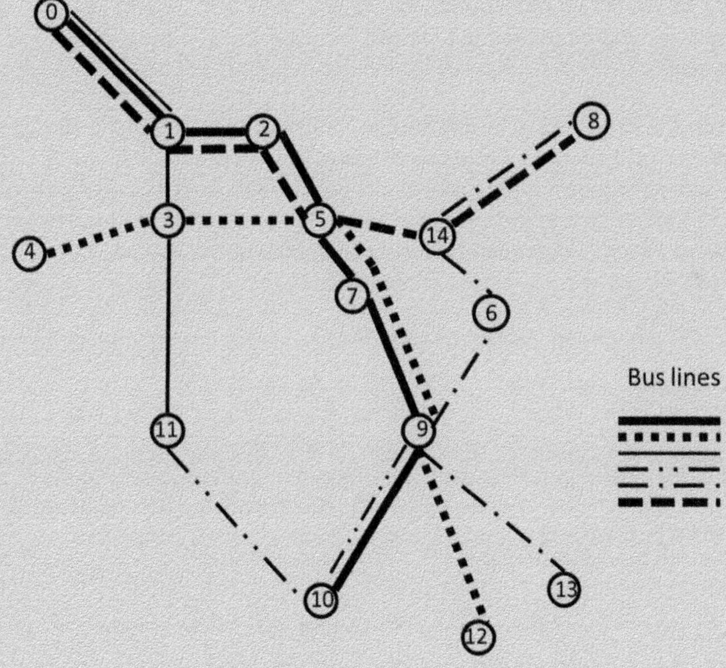

FIG. 7.41

The solution (generated set of the public transit lines) obtained by the greedy algorithm.

The more complex models devoted to the public transit network design problem (that are beyond the scope of this book) are based on the assumption that public transit demand depends on the transit network configuration, as well as on the service frequencies of the routes. This realistic assumption leads to the creation of complex mathematical models that capture relationships between passenger route choice mechanism, passenger flows along offered routes, service frequency values and optimal transit network configuration.

7.14 SERVICE FREQUENCIES DETERMINATION IN TRANSIT NETWORK

The service frequencies are determined after the set of transit routes is generated. The usual objective function, to be minimized, represents the total passenger travel time. Mathematically, service frequencies determination problem could be formulated in the following way:

Minimize

$$\sum_{i\in N}\sum_{j\in N} q_{ij} t_{ij}(f) \tag{7.44}$$

subject to:

$$\sum_{r\in R} T_r \cdot f_r \leq FS \tag{7.45}$$

where:

q_{ij} is the number of travelers that want to travel from the node i to the node j;
$t_{ij}(f)$ is the travel time from node i to node j that is function of service frequencies;
f_r is the service frequency on transit route r;
T_r is the turnaround time on transit route r; and
FS is the fleet size.

Travel time between any two nodes in the network is composed of access time, waiting time, in-vehicle time, transfer time, and egress time. Passenger waiting time between node i and node j depends on service frequency along the transit route that connect these nodes (in the case of possible direct service between i and j), as well as on frequencies of all routes that connect origin i and transfer nodes on the shortest path between i and j. Consequently, travel time from node i to node j, $t_{ij}(f)$ is the function of service frequencies f.

EXAMPLE 7.7

There are four routes in Mandl's network (see Fig. 7.42 and Table 7.9). An analyst has to make the decision on service frequency values along these routes. In a previous analysis an analyst defined for each route three potential frequency values. In a final decision, an analyst has to choose one frequency value for each route. There are 86 buses available. Determine the route frequency values along routes in such a way to minimize total travel time of all passengers in the network. The total travel time includes in vehicle time and waiting time.

Solution

In order to solve the problem, we use following mathematical formulation given by Martinez et al. (2014):

$$\min \sum_{k\in K}\left(\sum_{a\in A} c_a v_{ak} + \sum_{n\in N^P} w_{nk}\right)$$

s.t.

$$\sum_{l\in L}\sum_{f\in 1,\ldots,m} \theta_f y_{lf} \sum_{a\in l} c_a \leq B$$

EXAMPLE 7.7—cont'd

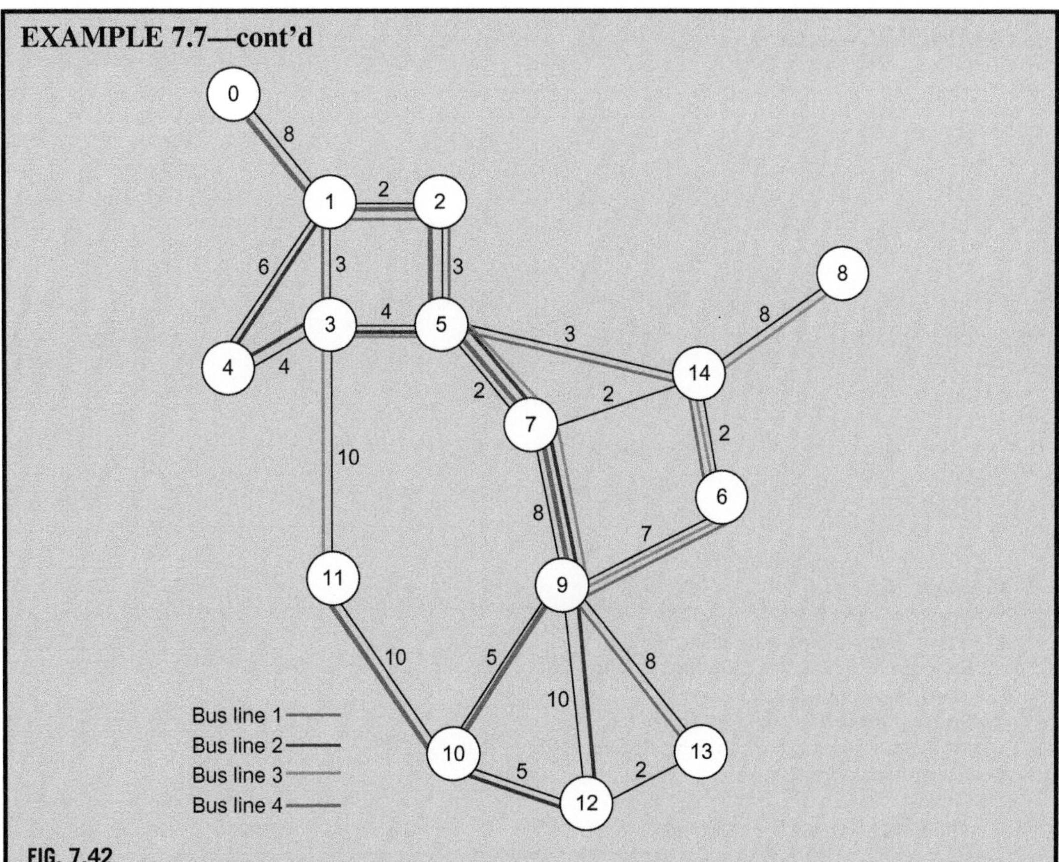

FIG. 7.42

Set of routes in the Mandl's network.

Table 7.9 Characteristics of The Transit Routes in the Mandl's Network

The Route Number	Route	Turnaround Time (min)	Number of Buses	Frequency (min^{-1})
1	0–1–2–5–7–9–10–11		35	0.461
		76	36	0.474
			37	0.487
2	1–4–3–5–7–9–12–10		22	0.282
		78	23	0.295
			24	0.308
3	8–14–6–9–7–5–3–11		15	0.183
		82	16	0.195
			17	0.207
4	3–1–2–5–14–6–9–13		8	0.143
		56	9	0.161
			10	0.179

(Continued)

EXAMPLE 7.7—cont'd

$$\sum_{f\in 1,\dots,m} y_{lf}=1 \quad \forall l\in L$$

$$\sum_{a\in A_n^+} v_{ak}-\sum_{a\in A_n^-} v_{ak}=b_{nk} \quad \forall n\in N^P, a\in A_n^+, k\in K$$

$$v_{ak}\le \theta_{f(a)}w_{nk} \quad \forall n\in N^P, a\in A_n^+, k\in K$$

$$v_{ak}\le \delta_k y_{l(a)f(a)} \quad \forall a\in A^B, k\in K$$

$$v_{ak}\ge 0 \quad \forall a\in A, k\in K$$

$$w_{nk}\ge 0 \quad \forall n\in N^P, k\in K$$

$$y_{lf}\in\{0,1\} \quad \forall l\in L, f=1,\dots,m$$

where:
N^P (N^S) is the set of stop nodes;
A is the set of arcs;
A^T is the set of travel arcs;
A^B (A^L) is the set of boarding (alighting) arcs;
A_n^+ (A_n^-) is the set of outgoing (incoming) arcs from (to) node n;
L is the set of line with generic element l;
Θ is the set of frequencies with generic element θ_f;
y_{lf} is the variable which indicates whether frequency θ_f is set to line l;
B is the upper limit of fleet size;
c_a is the cost of arc a;
v_a is the amount of demand flowing through arc a;
f_a is the frequency value of the line corresponding to boarding arc a;
$f(a)$ is the index in Θ of the frequency which represents arc a;
$l(a)$ is the index in L of the line corresponding to arc a;
K is the set of OD pairs with generic element k;
O_k (D_k) is the origin (destination) node of OD pair k;
δ_k is the amount of trips of OD pair k;
w_n is the waiting time multiplied by the demand at stop node n; and
b_n is the a value equal to δ_k if $n=O_k$,—δ_k if $n=D_k$ and 0 otherwise.

It was obtained, by solving mix-integer program, based on the proposed mathematical formulation, that the transit routes should have the following frequencies (buses):

- route 1: 0.487 min^{-1} (37 buses)
- route 2: 0.282 min^{-1} (22 buses)
- route 3: 0.207 min^{-1} (17 buses)
- route 4: 0.179 min^{-1} (10 buses)

The numbers of passengers that travel through the network links are given in Table 7.10. Passengers waiting times in the network nodes are given in Table 7.11. The total passengers travel time (the total in vehicle time + total waiting time) is 220,313 min.

EXAMPLE 7.7—cont'd

Table 7.10 Numbers of Passengers That Travel Through Specific Links

Link	Number of Passengers	Link	Number of Passengers	Link	Number of Passengers
(0,1)	1436	(5,2)	1270	(9,13)	297
(1,0)	1460	(5,3)	829	(10,9)	1591
(1,2)	1589	(5,7)	2276	(10,11)	405
(1,3)	227	(5,14)	266	(10,12)	757
(1,4)	160	(6,9)	660	(11,3)	115
(2,1)	1644	(6,14)	1006	(11,10)	417
(2,5)	1198	(7,5)	2321	(12,9)	295
(3,1)	240	(7,9)	2066	(12,10)	690
(3,4)	389	(8,14)	310	(13,9)	298
(3,5)	792	(9,6)	658	(14,5)	279
(3,11)	127	(9,7)	2123	(14,6)	985
(4,1)	160	(9,10)	1639	(14,8)	318
(4,3)	381	(9,12)	215		

Table 7.11 Passengers Waiting Times in the Nodes

Node	Waiting Time (min^{-1})	Node	Waiting Time (min^{-1})	Node	Waiting Time (min^{-1})
0	2950.18	5	7230.16	10	6484.48
1	7224.39	6	4317.86	11	1411.05
2	4100.50	7	4458.55	12	2448.05
3	4688.78	8	1495.29	13	1666.00
4	1916.83	9	8794.95	14	4758.67

7.15 VEHICLE SCHEDULING IN PUBLIC TRANSIT

The scheduling phase is performed after a timetable is determined (Salzborn, 1972; Ceder, 1986; Ceder et al., 2001). Within the phase of vehicles and crews scheduling, the operator defines the *use* of vehicles and crews. As already mentioned, passenger flows in public transit varies considerably throughout the day. The passenger flows during peak periods are often several times higher than passenger flows during off peak periods (Fig. 7.42). Most often, the time period of public transit operations is divided into several smaller *scheduling periods*. In the next step, the vehicle scheduling is performed within each of the scheduling periods (Fig. 7.43).

Fig. 7.44 shows bus operations in the case of the existence of three depots (depots 1, 2, and 3). Solid lines indicate planned trips planned to be performed. Dotted lines in the figure indicate deadheads (bus

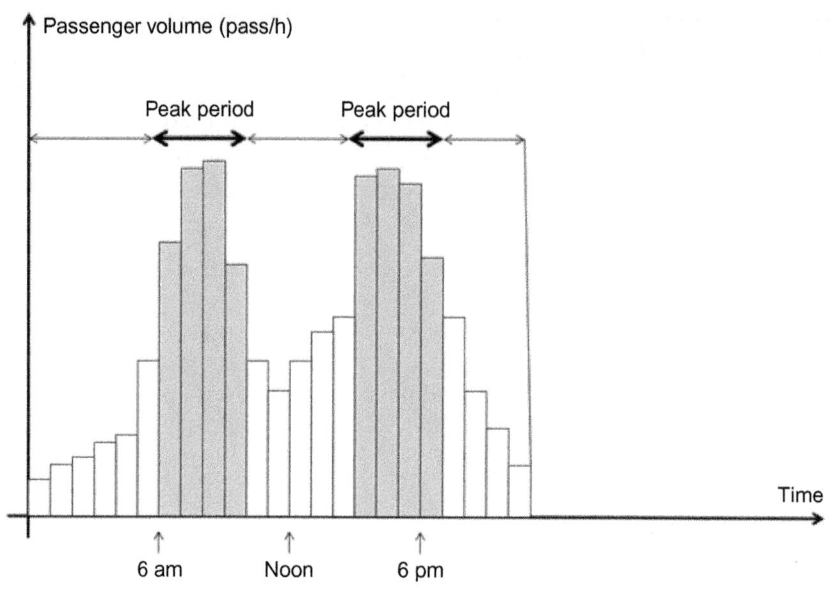

FIG. 7.43

Scheduling periods.

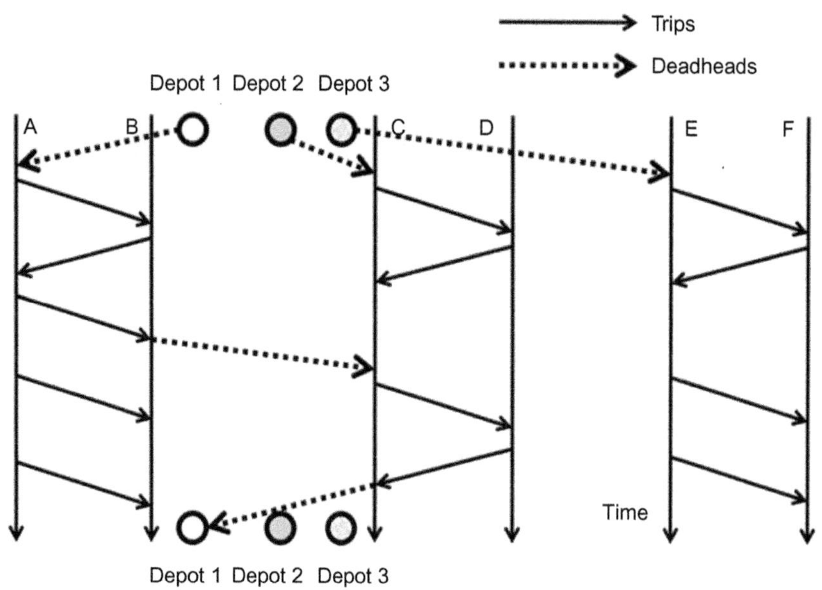

FIG. 7.44

Graphical representation of the vehicle scheduling problem.

trips without passengers from the depots to the terminals, from the terminals to the depots, or between individual terminals). The figure shows one specific case of bus operations. One of the buses operates between Terminals *A* and *B*. In one moment in time, this bus is sent (without passengers) from Terminal B to Terminal C. This is indicated by the dashed arrow in the figure. After reaching Terminal C, the bus continues to operate between Terminal C and Terminal D. It was also possible to keep this bus all the time to operate between Terminals *A*, and *B*, to send bus, in some moment in time, from Terminal B to Terminal E, etc. The vehicle scheduling problem is combinatorial by its nature. The vehicle scheduling problem could be defined in the following way: Assign available vehicles to the set of planned timetabled trips in such a way to minimize the total costs, or the total number of vehicles needed.

The sequence of trip that vehicle makes is usually called *vehicle rotation* (term *block* is also used). The example of timetabled trips and deadhead trips is shown in Table 7.12. The solution of the vehicle

Table 7.12 Vehicle Timetabled and Deadhead Trips

Vehicle Activity	Start Time	Starting Location	Ending Location
Pull-out	5:00 am	Garage	Terminal A
Trip	5:15 am	Terminal A	Terminal B
Trip	6:00 am	Terminal B	Terminal A
Trip	6:40 am	Terminal A	Terminal B
Trip	7:25 am	Terminal B	Terminal A
...	...	...	...
...	...	...	...
Trip	11:05 pm	Terminal B	Terminal A
Pull-in	11:45 pm	Terminal A	Garage

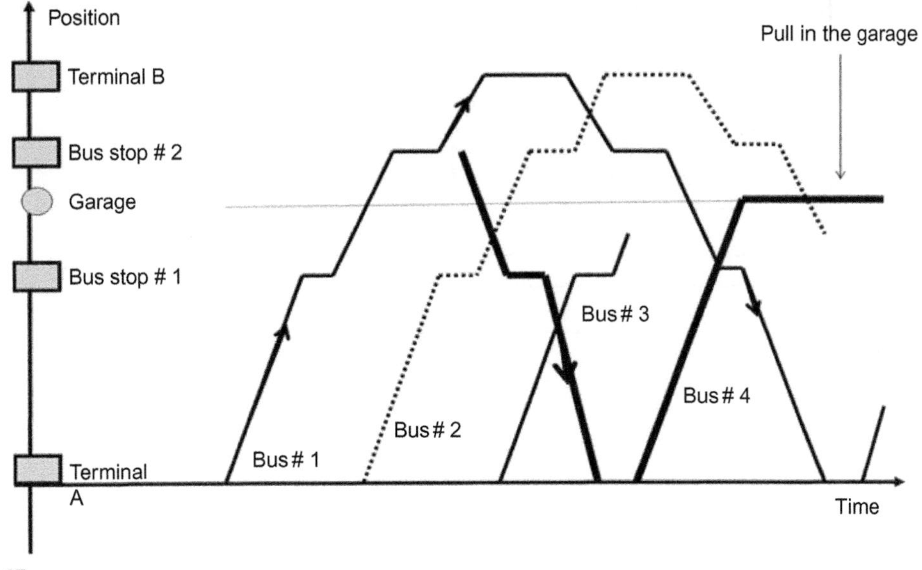

FIG. 7.45

Graphical schedule of a transit line.

scheduling problem represents the set of vehicle rotations (Daduna and Paixão, 1995). Every rotation represents the sequence of vehicle timetabled and deadhead trips (Table 7.12 and Fig. 7.45).

The vehicle scheduling problem is solved by Mathematical programming techniques, or by various heuristic algorithms. The successfully solved real-life problems show that substantial saving in the total number of buses needed could be achieved (up to 20% of vehicles). Once the vehicle schedule is generated, transit operators calculate different schedule performance data such as vehicle-kilometers, work-hours, pay-hours, etc. These data could contribute significantly to the estimate of schedule efficiency.

7.16 CREW SCHEDULING IN PUBLIC TRANSIT

Crew costs are very high in public transit. In some cases, these costs represent nearly 70% of the total costs. Crews are, in many companies in the world, paid for 8 h of work. Higher wages/salaries could be given for 12 h spread time interval of work, and especially for overtime, working shifts during a night, work on Saturdays, and Sundays, etc.

In public transit, as in other fields of transportation, the vehicle schedule is usually designed first, and then the crew is scheduled based on it (Smith and Wren, 1988; Wren and Rousseau, 1995). There are also models that simultaneously perform vehicle and crew scheduling (Haase et al., 2001).

When assigning crews to individual trips, various constraints regarding the crews' work (legal and union rules) should be taken into account (authorized number of work hours, required breaks between trips, regulated payment for overtime, etc.).

The *crew scheduling* problem is solved separately for every depot of the public transit operator. The crew scheduling (the term *duty scheduling* is also used) consists of determining drivers duties while covering all vehicle rotations that are assigned to a specific depot.

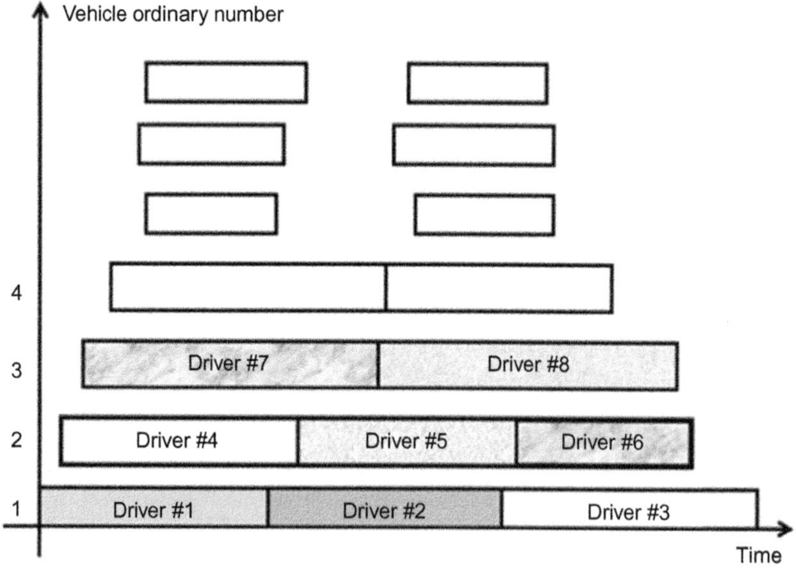

FIG. 7.46

Run-cutting.

Vehicle trips (runs) are shown in Fig. 7.46. Some vehicle runs are spread along the whole day. Some other vehicle runs are composed of two pieces (vehicles that operate only during rush hours). Every vehicle run is cut into the pieces. Every piece is assigned to one driver (Fig. 7.46). The crew scheduling problem consists in assigning crews to a given vehicle schedule in such a way to minimize total costs, while taking care about all operational and legal requirements and constraints. Crew scheduling problems (the term *run-cutting* is also used) in public transit are often of large dimensions. Typically, the number of drivers in the transit company is 2–3 times higher than the number of available buses. The number of daily trips to be performed is 10–20 times higher than the number of buses. The number of buses could be up to few thousands. The transit operator usually has few depots and different bus types.

The solution of the crew scheduling problems is represented by *duties* and *rosters*.

The crew scheduling problems are combinatorial optimization problems by their nature, and fit in to the class of NP-hard problems. A variety of optimization and heuristic techniques that can generate solutions of reasonably good quality in an acceptable amount of computer time have been used when solving crew scheduling problems in public transit.

7.17 DISRUPTION MANAGEMENT IN PUBLIC TRANSIT

In many public transport systems buses leave their terminals at fixed intervals. Because of disturbances, headways become irregular as the vehicle moves along the bus route. There are many causes of these variations. The most important are the level of congestions, the number of passenger at specific transit stops, the number of signalized intersections at which the vehicle must stop, traffic incidents, and the driver's characteristics.

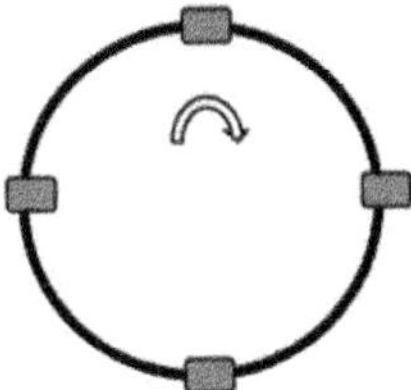
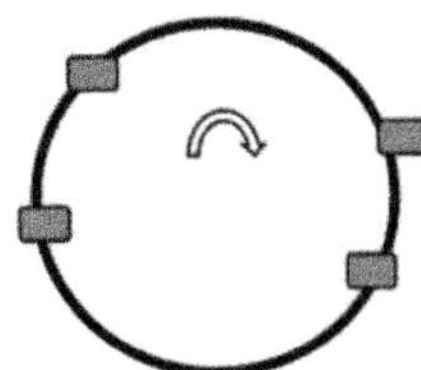
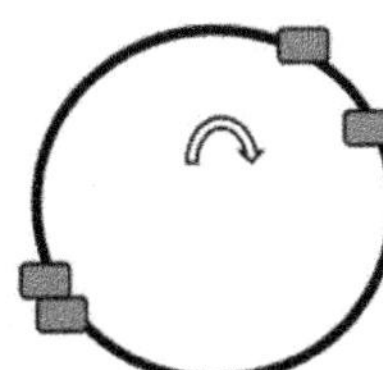

FIG. 7.47

Bus "bunching".

The bus *bunching* happens more frequently in the case of higher values of service frequency (Hickman, 2001; Eberlein et al., 2001; Dessouky et al., 2003). When previous vehicle is behind the schedule, and the following vehicle runs ahead of planned schedule, it could happen that two vehicles form a pair of buses that operate as one bus (Fig. 7.47). In order to mitigate negative effects of bus "bunching," and to again create more equal headways, the dispatchers in charge of vehicle operations usually apply one of the following strategies: (a) advising some drivers to speed up or slow down; (b) holding some vehicles at transit stops; (c) allowing some vehicles to omit some transit stops. The applied dispatcher's strategy could be also some combination of these basis strategies.

In many cases, lower service reliability can be influenced by the fact that some planned buses are out of operation due to different technical reasons. Companies in less developed countries who operate with older buses are often faced with this type of a problem. In extreme cases it is possible that 30–40% of the planned buses are out of operation due to technical reasons. A smaller number of buses in operation, than it is planned, can also be caused by the shortage of drivers. This is the case in some developed countries. Operators in many countries are every day faced with the problem that they must operate with a smaller number of buses than it was planned. What can they do in such a situation? Are they able to transport all passengers? Will they cut some lines? How will they assign available buses to bus lines to minimize the total passenger waiting time?

In the cases of shortage of planned buses, dispatchers in charge of operations decrease service frequencies on many transit lines (Fig. 7.48).

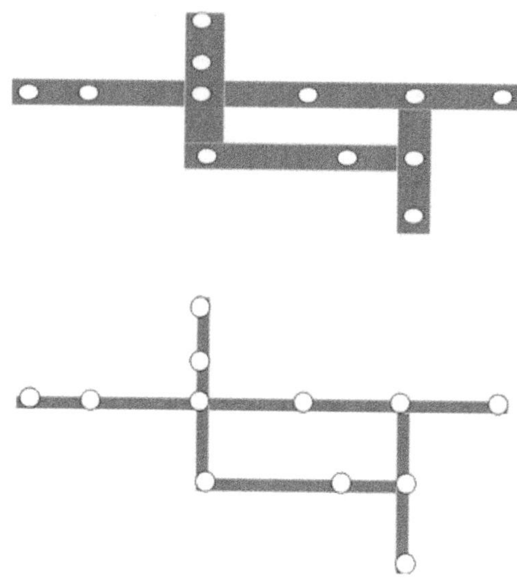

FIG. 7.48

Service frequencies changes in public transit network.

The thickness of the public transit network links shown in Fig. 7.48 reflects the service frequency values.

In many companies some bus lines or some parts of some lines are sometimes shortened and/or cut in the case of shortage of buses. The remaining bus lines in such cases can be served as planned.

Fig. 7.49 shows one possible modification of the transit network shown in Fig. 7.41, caused by the shortage of buses.

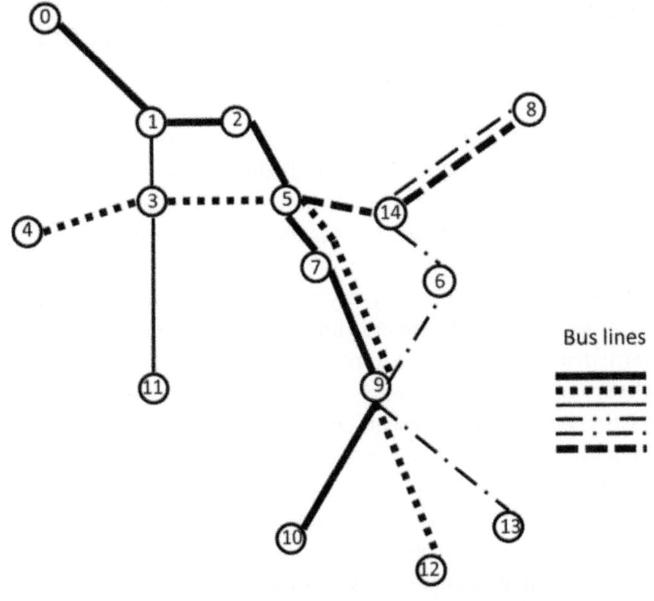

FIG. 7.49

Modification of the transit network shown in Fig. 7.41, caused by the shortage of buses.

7.18 PUBLIC TRANSIT PLANNING PROCESS

The public transit planning process is graphically shown in Fig. 7.50.

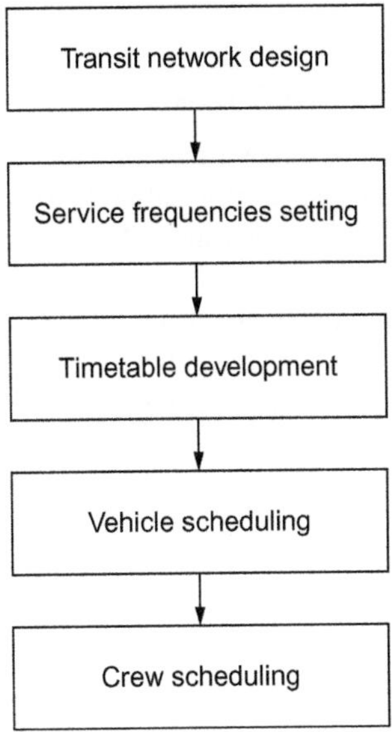

FIG. 7.50

Public transit planning process.

7.19 DEMAND-RESPONSIVE TRANSPORTATION SYSTEMS

DRT is the transit mode that uses passenger cars, vans and/or small buses. After a call from passengers, the operator's dispatcher dispatches vehicles to pick up the passengers and transport them to their requested destinations. This, shared-ride service is characterized by flexible routing and scheduling. The other passengers may be picked up or dropped off during the ride of particular passenger. As a rule, within DRT, vehicles do not run over a fixed-route or on a fixed-schedule. Demand-Responsive Transit, Dial-a-Ride and Flexible Transport Services are the expressions also used in the literature to describe DRT.

Various versions of DRT are found in every day practice; transportation of people in rural and low density areas, transportation of elderly and people unable to use public transportation (people that have a permanent or long-term disability), services during off-peak times (late evenings and weekends), and parcel pick-up and delivery service in urban areas are some of the examples. There are more than 1500 DRT rural systems in USA, and about 400 urban DRT systems. There are also hundreds of urban DRT systems. There are numerous DRT systems in Germany, Austria, the United Kingdom, Australia, Canada, Italy, Japan, and other countries.

The DRT service complements the traditional public transit systems by enabling transportation to people unable to use existing public transportation. When giving a service to the people unable to use public transportation, drivers help passengers to get to and from the vehicle. Passengers usually have a right to bring one person with them, as well as any number of children if they have. DRT service typically provides a collective journey on a vehicle, with other passengers. Since Dial-a-Ride is a shared-ride service, passenger can expect longer travel time than by private car or a taxi service. In the majority of cases, service users travel within a service area without transferring to another vehicle. Dial-a-Ride service could be free, or service fares could vary by service area and distance traveled. Good cooperation between public transit operator and the health, education, and other local organizations can significantly contribute to the low cost, and high level of service DRT.

7.19.1 TYPE OF ROUTING AND SCHEDULING IN DRT

Fixed schedule in public transportation assumes timetable with specified vehicle departure times. All vehicle stops within fixed schedule ate previously defined stops. The *Route deviation* system is the DRT form in between fixed schedule and flexible routes. This system has fixed route between two terminals, but deviations from the fixed route are allowed when there is request for transportation (Fig. 7.51).

Fig. 7.51 shows the case when vehicle deviates from the fixed route in order to pick up on demand passengers (marked by the triangle in Fig. 7.51). Contrary to this, the majority of DRT forms are characterized by the flexible routes within the area of operations. There are various DRT forms, depending on the type of routing and scheduling. The major DRT forms are "Many to many," "Many to one," "One to many," etc.

In the case of "Many to many," there are many different pick-ups going to many different destinations (Fig. 7.53).

The "Many to one" case describes the situations when group of passengers is transported from various origins to just one destination (Fig. 7.52). Fig. 7.52 shows the case when vehicle pick-up senior citizens and drive them to the regional hospital.

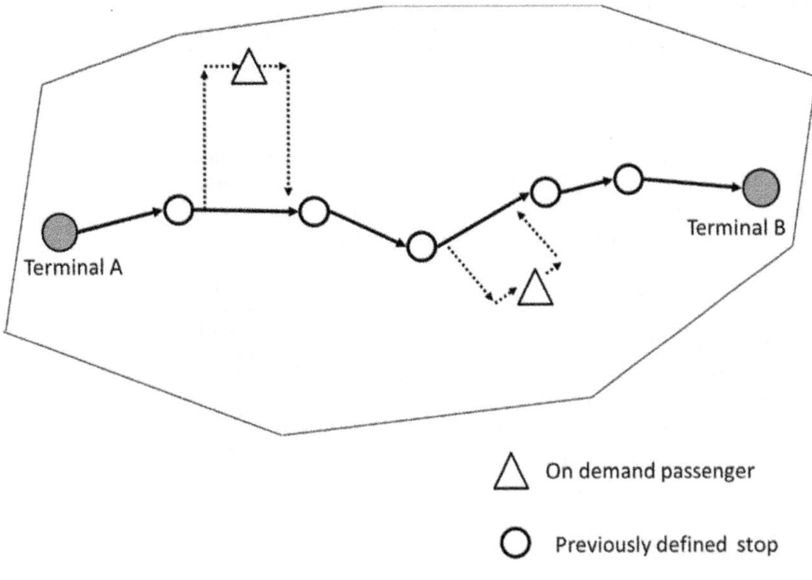

FIG. 7.51

Route deviation.

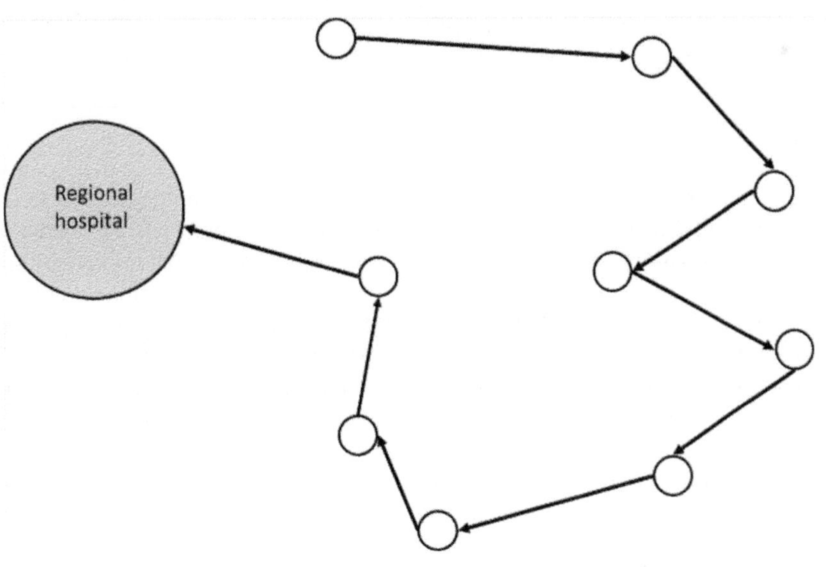

FIG. 7.52

Many to one.

7.19.2 DIAL-A-RIDE

The "Many to many" case describes the situations when group of passengers is transported from various origins to various destinations (Fig. 7.53). The phrase "Dial-a-ride" is frequently used instead of the phrase "Many to many." The operator, which is running the transportation service, receives calls for the transportation. The origin, destination, and preferred beginning and/or end of the service characterize every request for transportation (Fig. 7.53).

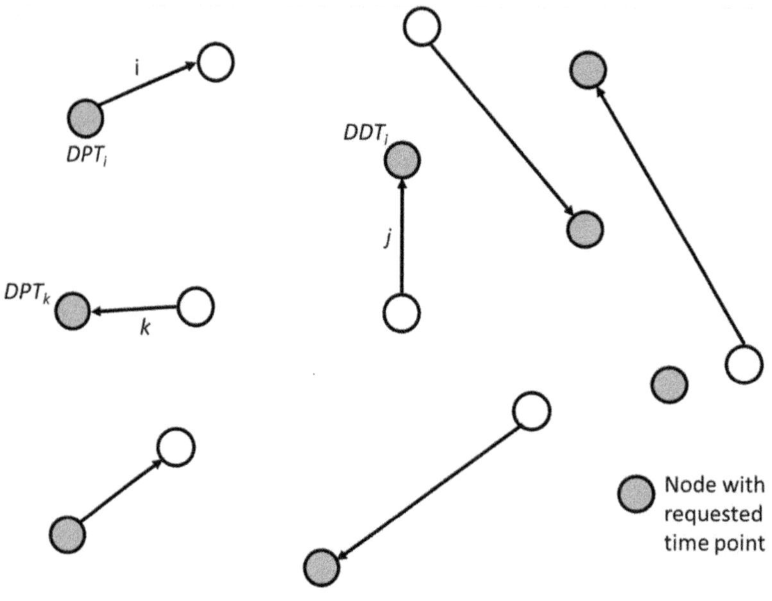

FIG. 7.53

Static version of a Dial-a-Ride problem.

Every client who wants to be served must specify to the central dispatching center desired moment when she/he wants to be picked up at the trip origin or desired moment when she/he wants to be dropped off at the destination. Usually, the vehicles are dispatched to pick up several passengers at different locations, previous to transporting them to their requested destinations. In some cases, the planned trip could be interrupted en route to planned destinations to pick up more passengers.

A carrier has to design vehicle routes so as to achieve the greatest number of requested trips, minimize the total traveling distance, reduce the number of vehicles needed, minimize the detour time, etc. In the *static* version of the Dial-A-Ride problem (Fig. 7.53), it is usual to collect the requests for transportation the day prior to the beginning of the service.

In the *dynamic* version of the problem, the vehicle routing and scheduling must be performed in real time so that the customers' demand for the requested service is satisfied (Fig. 7.54).

At the time point, when a new passenger's request appears, some of the earlier passenger requests have already been satisfied, or were rejected from the service, so they are not related any more to any upcoming dispatching decisions. The group of the remaining former passengers has also been assigned

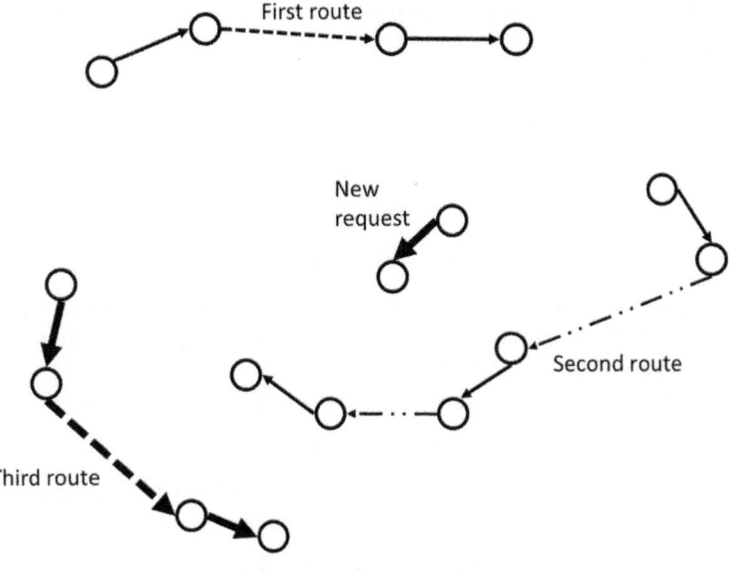

FIG. 7.54

Dynamic version of the dial-a-ride problem.

to particular vehicle routes. Some of them are waiting to be picked-up, and some of them are already in the vehicle on their way to their destinations.

In some countries, DRT system clients have the right to request service with very short call ahead time. This generates a necessity for the transit operator to have very fast heuristic algorithm for vehicle schedule design.

We use the following notation:

N is the number of clients that request service;
n is the number of available vehicles;
i^+ is the origin of the i-th passenger (the node where the i-the passenger should be picked up);
i^- is the destination of the i-th passenger (the node where the i-the passenger should be dropped off);
DPT_i is the desired pick-up time of customer i;
DDT_i is the desired delivery time of customer i;
EPT_i is the earliest pick-up time for customer i;
EDT_i is the earliest delivery time for customer i;
LPT_i is the latest pick-up time for customer i;
LDT_i is the latest pick-up delivery time for customer i;
APT_i is the actual pick-up time for customer i;
ADT_i is the actual delivery time for customer i;
$D(x,y)$ is the vehicle travel time from point x to point y using the shortest route between x and y;
DRT_i is the direct ride time of the ith passenger (the time needed to transport passenger directly from the node i^+ to the node i^-);

MRT_i is the maximum ride time of the ith passenger (maximum allowed time that the ith passenger could spend in the vehicle); and
DV_i is the actual time deviation for customer i from his desired pick-up (delivery) time.

In the case of clients that specify desired pick-up time, the DV_i equals:

$$DV_i = APT_i - DPT_i \tag{7.46}$$

In the case of clients that specify desired delivery time, the DV_i equals:

$$DV_i = DDT_i - ADT_i \tag{7.47}$$

WS_i is the maximum acceptable deviation of customer i from his desired pick-up or delivery time ($DV_i \leq WS_i$) All defined operational constraints must be checked for all already scheduled passenger requests that are influenced by the newly inserted passenger request.

The following relation must be satisfied for all clients that request service:

$$ADT_i - APT_i \leq MRT_i \tag{7.48}$$

In other words, no one passenger can spend in the vehicle more than MRT_i time units. The following relation must be satisfied in the case of passengers that specify desired pick-up time:

$$DPT_i \leq APT_i \leq DPT_i + WS_i \tag{7.49}$$

No one passenger who specifies a desired pick-up time can have a deviation from her/his desired pick-up time greater than WS_i. In the case of passengers who specify the desired delivery time, the following must be satisfied:

$$DDT_i - WS_i \leq ADT_i \leq DDT_i \tag{7.50}$$

No one passenger who specifies a delivery time can have a deviation from her/his desired delivery time greater than WS_i. The maximum allowed time MRT_i that ith passenger could spend in the vehicle could be defined as:

$$MRT_i = a + b \cdot DRT_i \tag{7.51}$$

where a and b are pre-specified constants.

Jaw et al. (1986) considered the static version of the dial-a-ride problem and proposed the general procedure of inserting a new request in the current schedule. The proposed algorithm consists of the following three steps:

Step 1: For each vehicle route, generate all feasible positions to insert a new request. Calculate the change in the objective function value caused by insertion.
Step 2: Insert the new request into the position that has the lowest incremental cost.
Step 3: If no feasible insertion exists, reject the request.

The dial-a-ride problem is characterized by existence of the following constraints: (a) maximum ride time; (b) time window; (c) waiting time; and (d) vehicle capacity. All defined operational constraints must be checked for all already scheduled passenger requests that are influenced by the newly inserted passenger request.

DRT systems are usually evaluated according to the following criteria:

- total number of vehicle-hours;
- total number of vehicle-miles;

- total number of passenger trips;
- total operating expense;
- total number of Accidents/safety incidents; and
- total number of on-time trips.

7.20 INTERURBAN ROAD TRANSPORT SYSTEMS

7.20.1 INTRODUCTION

In this section, the intercity medium and long-haul bus system serving passenger demand is described. The bus transport system consists of the infrastructure network including the bus stations/terminals as the network's nodes and the roads connecting them as the network's links. The transport service network consists of the bus transport services scheduled between particular terminals during given period of time under given conditions. The intercity buses use local and regional roads within and highways between rather distant regions and cities, and streets in the urban areas around the bus stations. Their "ultimate" and "practical" capacity and the corresponding level of service in this context are analogous to those analyzed previously. Therefore, in the following text they will not be particularly considered.

7.20.2 SERVICE NETWORKS IN AN INTERURBAN ROAD TRANSPORTATION

The "ultimate"' capacity of a road passenger transport vehicle-bus is specified by its seat capacity. Some of its characteristics are shown in Fig. 7.55.

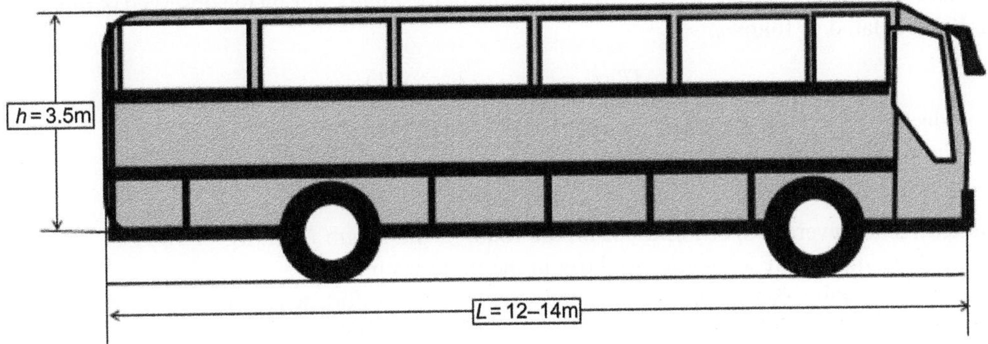

FIG. 7.55

Scheme of an intercity bus (SCANIA, 2009).

In general, GVM (Gross Vehicle Mass)[1] of these buses is about 18–22 tons (GVM). Their seat-capacity is from 50 to 55 seats, length 12–14 m, width 2.6 m, and height 3.5 m (SCANIA, 2009).

The load factor $\lambda_i(\tau)$ on a given route i during a given period of time τ equals:

[1]This is the maximum operating weight/mass of a bus as specified by the manufacturer including the vehicle's empty weight, fuel, passengers and their baggage (SCANIA, 2009).

$$\lambda_i(\tau) = \frac{Q_i(\tau)}{F_i(\tau) \cdot s_i} \tag{7.52}$$

where:

$Q_i(\tau)$ is the expected passenger demand on the route i during time period τ (pax/day, week);
s_i is the number of seats per departure frequency on the route i (seats/dep); and
$F_i(\tau)$ is the service frequency on the route i during time τ (departures/day).

The transport work on route i, carried out during a given period of time τ (in both directions of the route) can be calculated as follows:

$$TW_i(\tau) = F_i(\tau) \cdot s_i \cdot \lambda_i \cdot 2 \cdot l_i \tag{7.53}$$

The transport work $TW(\tau)$ in the whole transport service network equals:

$$TW(\tau) = \sum\nolimits_{i=1}^{N} TW_i(\tau) = \sum\nolimits_{i=1}^{N} F_i(\tau) \cdot s_i \cdot \lambda_i \cdot 2 \cdot l_i \tag{7.54}$$

where:

$F_i(\tau)$ is the service frequency on the route i during time τ (departures/day or week);
l_i is the length of route i (km); and
N is the number of routes in the transport service network of the given bus operator/company.

It is assumed that the service frequencies, bus seat-capacity per frequency, the average load factor, and the length of route are equal in both directions for all routes of the network.

The productivity of a route and the productivity of a service network indicate the time efficiency of use of the above-mentioned capacity. The productivity of the route i, during the observed period of time τ can be calculated as follows:

$$TP_i(\tau) = F_i(\tau) \cdot s_i \cdot \lambda_i(\tau) \cdot v_i(l_i) \tag{7.55}$$

The productivity $TP(\tau)$ in the whole transport service network equals:

$$TP(\tau) = \sum\nolimits_{i=1}^{N} TP_i(\tau) = \sum\nolimits_{i=1}^{N} F_i(\tau) \cdot s_i \cdot \lambda_i(\tau) \cdot v_i(l_i) \tag{7.56}$$

where $v_i(l_i)$ is the average speed of a bus on the route of the length i (km/h).

The other symbols are analogous to those in the previous relations.

The service quality can be expressed by the internal bus comfort, schedule delay, punctuality, and reliability of services related to the particular routes. The internal bus comfort is expressed by its air-conditioned cabin, the size, ie, the number of usually reclining seats, comfortable legroom, on board toilets and Wi-Fi. The schedule delay depends on the service frequency on a given route and can be determined for the route i of the given bus network as follows:

$$SD_i(\tau) = (1/2) \cdot \frac{\tau}{F_i(\tau)} \tag{7.57}$$

where all symbols are as in the previous relations.

The punctuality of services implies delays in arriving at destination terminal. They can be expressed as the positive difference between the actual and scheduled arrival time. Reliability of services represents the actually realized transport services compared to the planned (scheduled) ones.

The size of fleet operated by a bus company generally depends on the number of routes in the service network, the service frequency scheduled on the particular routes during a given period of time, and the length of particular routes, ie, the bus turnaround time along them. Consequently, for the network consisting of N nodes/bus terminals, the required fleet carrying out punctual transport services during a given period of time can be estimated as follows:

$$M(\tau) = \sum_{i=1}^{N} F_i(\tau) \cdot 2 \cdot \left[t_i + \frac{l}{v_i(l_i)} \right] \tag{7.58}$$

where t_i is the average time which a bus spends at both end terminals of a given route i (h).

The other symbols are analogous to those in the previous relations.

EXAMPLE 7.8

The seat-capacity of the intercity uses is usually 50 seats/vehicle. One of the illustrative cases is *Eurolines* bus system operating the international transport services throughout Europe starting from the year 1985. Under this system, 29 large bus companies from different European countries have been operating together and continuously developing of the tarns-continental bus service network consisting of more than 700 nodes/terminals usually located in the city centers or nearby them in 32 European countries. The longest route is that of 4500 km between Lisbon (Portugal) and Moscow (Russia). The system has served more than 10 million passengers per year (ECMT, 2001; Van de Velde, 2009; http://www.eurolines.com/en). Fig. 7.56 shows an example of the relationship between the weekly service frequencies and the length of route/line for the selected routes/lines of the Eurolines bus system. As can be seen, the service frequencies generally decrease more than proportionally with increasing of the route/line length. In addition, Fig. 7.57 shows that the average travel speed on these routes is about 60 km/h.

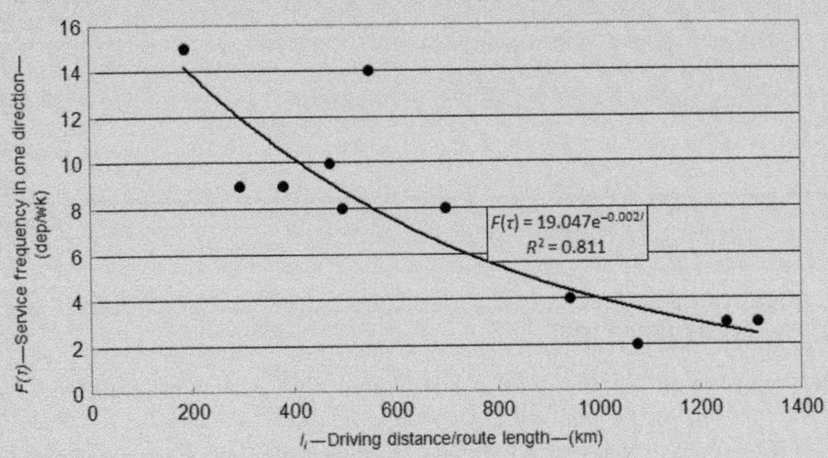

FIG. 7.56

Relationship between the service frequency and the length of route/line for the selected routes of the Eurolines bus system (one direction) (http://www.eurolines.com/en).

(Continued)

EXAMPLE 7.8—cont'd

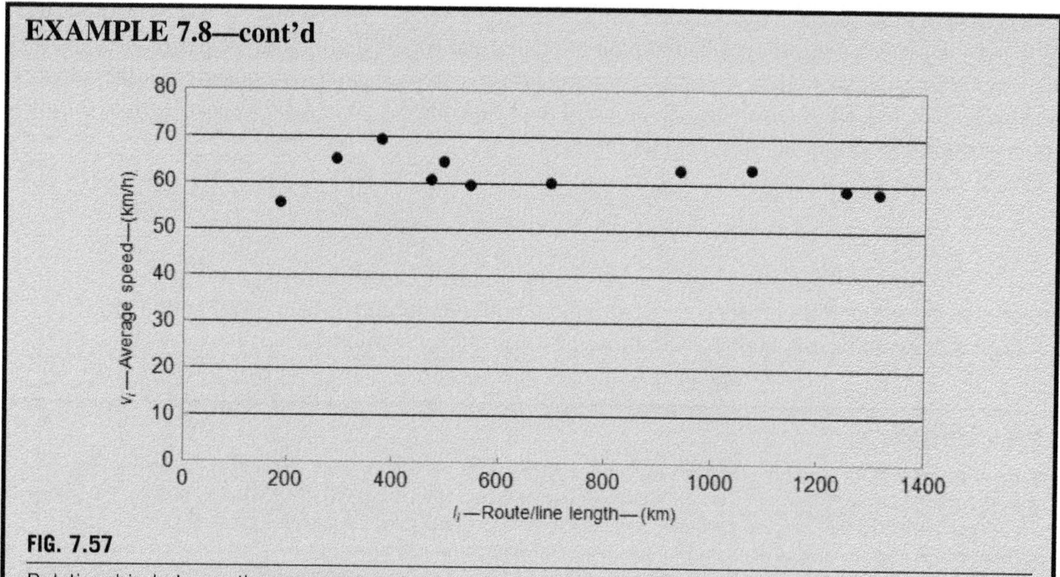

FIG. 7.57

Relationship between the average travel speed and the length of route/line for the selected routes of the Eurolines bus system (http://www.eurolines.com/en).

An additional example relates to the route/line between London (UK) and Amsterdam (the Netherlands) (excluding passing though Channel Tunnel). Its length is: $l=547$ km, where the weekly service frequencies by the buses of seat-capacity of: $s=50$ seats/bus is: $F=15$ dep/week in each direction are carried out. The corresponding supply/capacity transport work in both direction is: $TWS=2\cdot15\cdot50\cdot547=820{,}500$ seat-km. If the load factor is: $\lambda=0.8$, the demand/passenger carried transport work in both directions will be equal to: $TWD=2\cdot15\cdot50\cdot0.80\cdot547=656{,}400$ pax-km. The average bus travel speed on this route/line is about: $v=60$ km/h as shown in Fig. 7.60. Consequently, the productivity of the route/line in a single direction on the supply/capacity side and the demand/passenger carried side will be equal to: $TPS=1550\cdot60/(6\cdot24)=312.5$ seat/km-h and $TPS=15\cdot50\cdot0.8\cdot60/(6\cdot24)=250$ pax/km-h (1 week = 6 days; 1 day = 24 h). As well, if the average turnaround time on this route in both directions including the time spend at both terminals is: $2\cdot(t_i+l/v)=2\cdot(3+574/60)\approx22$ h and if the daily service frequency is: $F(\tau)=3$ dep/day, ($\tau=1$ day), the required fleet (the number of buses) will be equal to: $M(\tau)\approx(3/24)\cdot22\approx3$ buses.

7.21 AIR TRANSPORTATION

During the hundred years of its existence, scheduled commercial aviation has experienced a great expansion. According to the International Air Transport Association (IATA) in 2014, airlines transported 3.3 billion passengers. Air carriers also transported 50 million metric tons of cargo. The world air transportation network was characterized in 2014 by nearly 50,000 air routes. There are approximately 1400 airlines in the world that operate a total fleet of over 25,000 aircraft.

Air transportation has allowed rapid exchange of goods, people, ideas, cultures, and values. Commercial aviation has significantly contributed to the shaping of the modern world into a "global village" (Fig. 7.58).

FIG. 7.58

Airplane (http://www.imagebase.net/).

The airlines are companies scheduling flights between particular airports and through airspace between them by different aircraft in order to serve expected passenger and freight/cargo demand under given conditions. In such a way, the airports connected by flights represent the nodes and the flights themselves the links of airline transport service network. Since the flights can be temporal, the airline service networks have inherently temporal character. In such a context, the questions are: what are the capacity and service level of airlines and how can they be expressed?

The airline industry is highly competitive. Traditional air carriers, low-cost carriers (LCC), and various joint ventures in international flights try to maximize profit, as any other industry. Every airline tries to maintain or increase market share. In the first step, every air carrier is facing the problem of *market selection*. In other words, the air carrier is facing the simple question—where to fly? Should we fly between New York and Paris? Should we offer flights between London and Belgrade? Should we establish service to New Zealand? Once the market is selected, the air carrier should determine *flight frequency*. The air carrier must determine how often will fly in the chosen market. Will air carrier fly three times per week, or every day? In the next step airline must decide about *departure* and/or *arrival times*. Should we depart at 9:15 am, or 11:20 am? Air carrier is also facing the *fleet selection* problem. In other words, the airline must decide which aircraft type to use between specific airports at a specific time. High competition among airlines forces airlines to develop various *pricing strategies*. Airlines must decide about the number of different products and fares offered to every market they serve. Finally, in order to maximize revenue, airlines must develop *a* revenue management policy related to the number of accepted bookings by fare type for every market.

7.21.1 AIR TRANSPORTATION DEMAND

Air passenger flow represents the number of air passengers in a unit of time traveling from one city to another. The potential passenger flow comprises the number of air passengers in a unit of time that wish to travel from one city to another. For one reason or another, a difference is very often found between the actual and potential passenger flow. For example, due to low frequency in flights between two cities, an air carrier operating without competition is not able to satisfy the market demand which means that the actual passenger flow is below its potential. Air passenger flow is a value that changes over time. Changes are noticed by month, by week, by day in a week, and finally by hour in a day.

Transportation flows changes over time can be determined by collecting appropriate statistical data, as well as by conducting passengers' surveys. It is extremely important to monitor air passenger flows by day in the week and particularly by hour in the day in order to adequately determine flight frequency and departure times.

The *Revenue Passenger Mile* (RPM) is also used to describe passenger demand. The RPM is achieved when one paying passenger is transported 1 mile. *Yield* represents the average fare paid by passengers, per mile flown.

Airline statistics over years show that business demand for flights are higher on weekdays, during spring, and during falls. On the other hand, the leisure demand is higher on weekends and during summer. As a consequence, airlines change flight frequencies and/or departure times based on season and day in the week.

There are, every day, on many flights leisure passengers who go on vacation, and who bought ticket 2 months before the take-off, and business passengers whose ticket was purchased just before the take-off. The company paying for the ticket of business travelers primarily takes care about the possibility that a company representative catch a sudden business meeting. On the other hand, a large number of tourists plan vacation on time, buy tickets well in advance and takes much care about the ticket price. In other words, air transportation is characterized by market segments with different willingness to pay for a trip (Fig. 7.59).

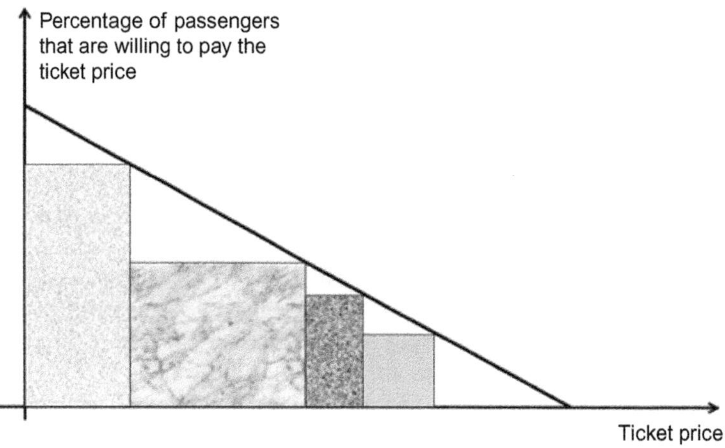

FIG. 7.59

Percentage of passengers who are willing to pay the ticket price as a function of a ticket price.

Business passengers are, generally, less sensitive to ticket price changes than leisure passengers. At the same time, business passengers usually have less flexibility to reschedule or abandon their trip than leisure travelers. All airlines apply various strategies of *differential pricing*, meaning that all airlines charge different prices to different passenger types. Airlines sell seats, at the same time, to various market segments.

Passenger paying cheaper tickets, as a rule, have various restrictions (3 weeks advance purchase, Saturday night stay, etc.). In other words, air carries apply various rules that prevent sale to business passengers.

It has been shown that, by using differential pricing, airlines are capable to increase the revenue and achieve higher passenger load factors, than in the case of a single fare that is applied to all groups of passengers.

7.21.2 AIRLINE SUPPLY AND AIRLINE CAPACITY

In general, the capacity of an airline can be static and dynamic, both reflecting the airline capability to produce the given volumes of transport services during a given period of time (day, week, month, year) under given conditions. In general, an airline static capacity can be considered to be its available aircraft fleet and the air route network consisting of airports and routes connecting them. The airline dynamic capacity can be considered to be the level of using the elements of the static capacity to serve the expected passenger and freight/cargo demand during given period of time under given operating conditions.

The available resources used by an airline as inputs for the service production process reflecting its static capacity can be divided into physical and non-physical. In this context, the main physical input is the aircraft fleet and the supportive facilities and equipment-buildings, repair/overhaul and maintenance equipment, computer and communication facilities, and airport, aircraft, passenger and baggage service facilities. The main non-physical input is the airline network with routes and scheduled flights between particular airports based on the flying rights at corresponding airports and airspace between them.

Airlines use aircraft of different types to provide air transport services in their networks. In general, as mentioned above these aircraft can be of different size influencing their MTOW (Maximum Take-Off Weight) including OEW (Operating Empty Weight), ie, the weight of payload consisting of passengers, their bags, and freight/cargo load, and trip fuel consisting of trip fuel and fuel reserve (the latter for flying to the alternate airport in the case of not being able to land at original destination airport). While considering the aircraft static capacity, it is usually seat and freight-cargo capacity. While analyzing this capacity, the common approach is to consider the so-called payload diagram. An example of the B737–800 aircraft including its simplified layout is shown in Fig. 7.60A and B (Boeing, 2011).

The payload-range diagram indicates payload-range capabilities and determines how much payload (passengers and their bags, and air cargo) the aircraft can transport over different distances under given operational constraints. As can be seen, the aircraft MTOW (Maximum Take-off Weight) (the weight of empty aircraft itself + payload—passengers and their bags and freight cargo + fuel) is about 172 thousand lb (78.1 ton). The OWE (Operating Empty Weight) is 90 thousand lb (41 ton) and the full payload is about 48 thousand lb (20.8 ton). The rest of 36 thousand lb (16.3 tons) is trip fuel (not including fuel reserve for flying to the alternative airport and delaying before landing) (1 lb = 0.453 kg). With the full payload, the aircraft range can be about 2000 nm (3704 km) (point A in Fig. 7.60). The payload of 35.6 thousand lb (16.1 ton) enables the aircraft range of about 3065 nm (5676 km) (point B in Fig. 7.60). Finally, without any payload but fully fueled, the aircraft can reach the distance of about 3800 nm (7038 km) (point C in Fig. 7.60). This means that the aircraft range can be extended by increasing the amount of trip fuel, but only on the account of payload.

The airline fleet consists of aircraft of different types and seat-capacity. The number of different aircraft types and the number of aircraft of each type depend on the characteristics of the airline network and the flight frequencies on its particular routes (Swan, 1979; Teodorović, 1988). These are set up to satisfy passenger and freight/cargo demand under given conditions. Given the number of aircraft

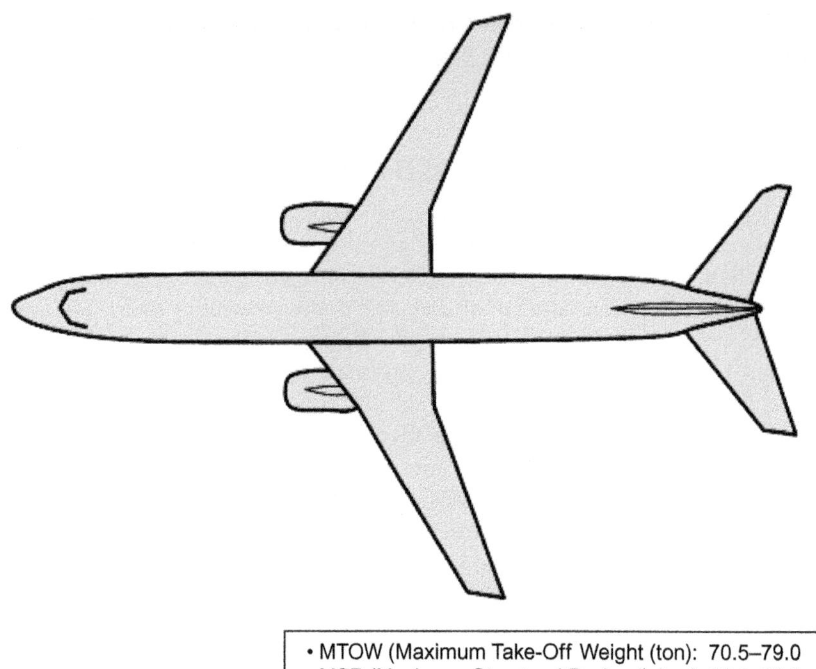

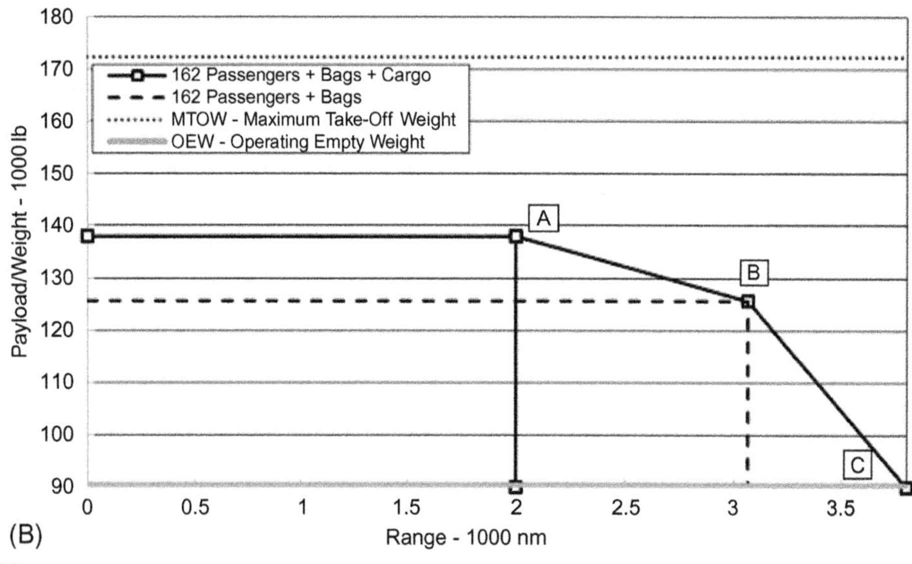

FIG. 7.60

Basic characteristics of the Boeing B737–800 aircraft (Boeing, 2011). (A) Layout and (B) payload-range diagram.

of each type and their seat-capacity, the airline static in terms of the total number of available seats S capacity can be calculated as follows:

$$S = \sum_{i=1}^{N} n_i \cdot s_i \tag{7.59}$$

where:

n_i is the number of aircraft of type i in the airline fleet;
s_i is the average seat-capacity of aircraft of type i; and
N is the number of different aircraft type in an airline fleet.

Table 7.13 shows these characteristics of one of the largest air carriers in the USA—Delta Airlines.

Table 7.13 Some Characteristics of the Aircraft Fleet of Delta Airlines

Aircraft Type	Number	Average Seat-Capacity	Sub-Fleet Seat-Capacity
B717	71	134	9514
B737s	124	150	18,600
B747–400	13	416	5408
B757s	138	210	28,980
B767s	95	244	23,180
B777s	18	350	6300
A319s	57	140	7980
A320s	69	165	11,385
A330s	34	315	10,710
MD88/90	181	145	26,245
Total:	800		148,302

As scan be seen, the airline operates the fleet of 800 aircraft consisting of 10 different types with different modifications. Based on relation (7.59), the static capacity of this airline is computed to be: 148,302 seats. Similar calculations can be made by taking into account also the freight/cargo capacity of this fleet. In addition, the world's largest low-cost carrier[2] Southwest Airlines operates the fleet of 692 B737s aircraft. If each has about 150 seats, the airline static capacity is: $S = 692 \cdot 150 = 103{,}800$ seats.

In general, the number of different aircraft types depends on the total size of the airline fleet as shown in Fig. 7.61 for the U.S. conventional airlines.

As can be seen, the number of different aircraft types generally increases with increase in the fleet size at decreasing rate. In this case, the average number of different aircraft types is five, but some airlines also operate the fleets of 10 and 11 different aircraft types. But, as mentioned above, southwest operates only one aircraft type.

[2]These are the airlines with a lower operating cost than their competitor conventional airlines. These costs enable low cost carriers to offer lower fares to user/air passengers on the account of reduced comfort during handling at airports and while onboard the aircraft. The additional elements of comfort are additionally (extra) charged. The airlines usually operate the fleets of single or most two aircraft types (usually Boeing B737s or Airbus A320s) on the networks consisting the regional or secondary airports connected by direct (non-stop) short-haul routes and flights (https://en.wikipedia.org/wiki/Low-cost_carrier).

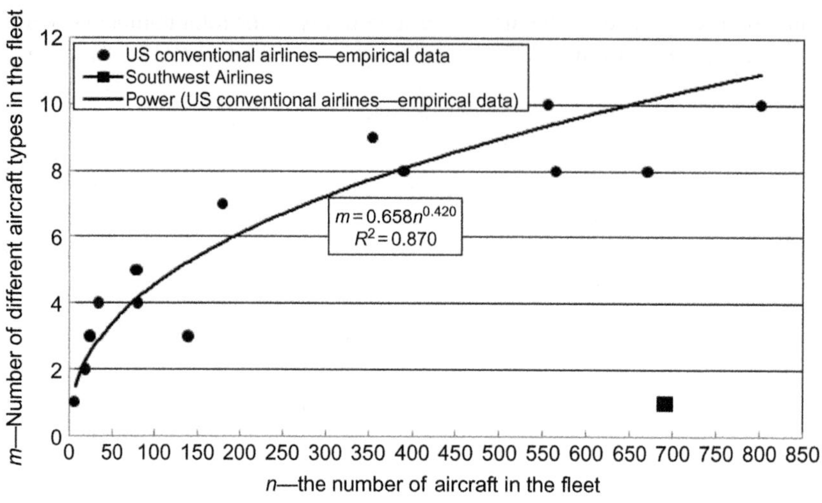

FIG. 7.61

An example of the relationship between the number of different aircraft types and size of the airline fleet of some US airlines (Janić, 2000).

Flight leg represents non-stop operation between two airports (Fig. 7.62). The flight leg is characterized by the origin airport, destination airport, departure time, arrival time and the aircraft type flying. The flight leg is also called *flight segment*.

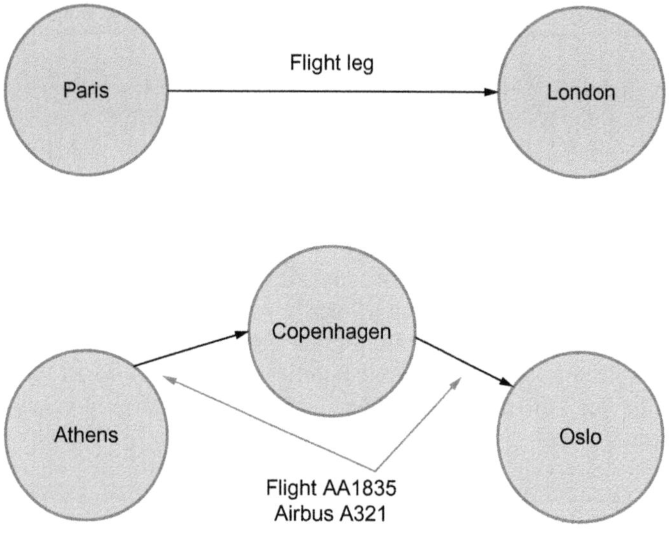

FIG. 7.62

Flight leg Paris—London and Flight Athens—Copenhagen—Oslo.

Flight is composed of one or more flight legs (Fig. 7.62). A flight is most frequently operated by a single aircraft. Every flight is labeled by a *flight number* (*flight code*). A particular flight could be easily identified by an airline name, flight number, and date. The flight AA 1835 shown in the figure is a two-leg flight that connects Athens (Greece) with Oslo (Norway) via Copenhagen (Denmark). The flight is operated by the Airbus A321.

When traveling between two cities in the network, passengers make various combinations of flight legs. In other words, *passenger paths* are composed of one or more flight legs.

As we showed in the case of Delta Airlines, each airline has in the fleet several different aircraft types that have various capacities. Within the fleet planning, airline makes decisions on number and type of aircraft required. These decisions are strategic in character and have a direct impact on both the operating costs of the carrier, and the airline's ability to serve specific routes and markets. The majority of aircraft are capable of operating economically for more than 30 years.

Airline supply is expressed in *Available Seat Miles* (ASM). The seat-mile is achieved when one aircraft seat is flown 1 mile. The aircraft seat could be empty, or occupied by passenger. The airline's Average Load Factor (ALF) equals:

$$ALF = \frac{RPM}{ASM} \tag{7.60}$$

The average load factor is the basic indicator of the airline capacity utilization. According to the International Air Transport Association (IATA), which represents 240 world airlines, the average passenger load factor of these airlines, in recent years, is between 75% and 85%.

EXAMPLE 7.9

The Boeing 747–400 aircraft that has 416 seats flies 7257 miles between Los Angeles and Hong Kong. There are 330 passengers on the flight.

The available seat miles (ASM) equals:

$$ASM = 416 \text{ seats} \cdot 7257 \text{ miles} = 3{,}018{,}912 \text{ seat miles}$$

The achieved RPM equals:

$$RPM = 330 \text{ passengers} \cdot 7257 \text{ miles} = 2{,}394{,}810 \text{ passenger miles}$$

The average load factor (ALF) equals:

$$ALF = \frac{RPM}{ASM} = \frac{2{,}394{,}810}{3{,}018{,}912} = 0.79$$

7.22 AIR TRANSPORTATION NETWORKS

An air transportation network can be defined as a set of nodes and a set of links on which air transportation activities are carried out. Depending on problem considered, nodes can signify regions, cities, airports, airport terminals, gates, runway thresholds, navigational points, *etc*. Nodes are connected by links that can signify airline flights, airways, or taxiways. A Geographic Information System (GIS) is a convenient way of storing information about air transportation network. In all air transportation networks request arise in certain nodes to transport goods or passengers. Infrastructure costs, direct operating costs, travel times, and the level of service all highly depend on the air transportation networks' design.

Air transportation networks are denoted in the same manner as graphs. The notation represents a network containing set of nodes N, and a set of links between these nodes denoted by A. The notation denotes link that connects node with node. Air transportation network could be represented as *spatial network*, or as a *space-time network*. In a spatial network, nodes represent airports, departure fixes, arrival fixes, en route fixes, intersection airways points, runway thresholds, or gates. Links can represent runways, taxiways, parts of airways, etc. (Fig. 7.63).

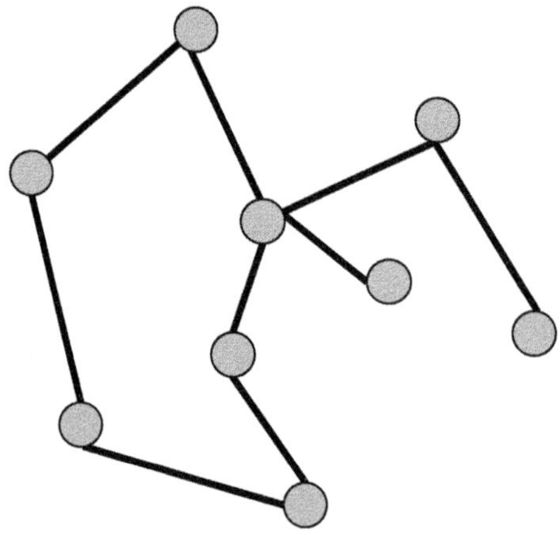

FIG. 7.63

Spatial Network.

We use Space-Time networks when we want simultaneously to represent aircraft, crews, or passengers movements through space and time. In these networks, space is represented in one dimension, and while time is represented in the other dimension. Every node has two attributes—first related to space, and second related to time. In other words, each node represents an event taking place in a specific airport, gate, runway threshold, or fix at a specific time. Depending on the problem considered, time attribute could be related to the beginning of boarding, aircraft readiness for departure, aircraft departure time, aircraft arrival time, etc. For example, node with attributes (Los Angeles, 10:00 am) could represent situation when specific aircraft makes take-off from Los Angeles at 10:00 am (Fig. 7.64).

The flight leg (link) connects node (Los Angeles, 10:00 am) with the node (San Francisco, 11:15 am). Leg is also called *flight arc*, since it connects aircraft departure node to aircraft arrival node. This is the usual way to represent flight in Space-Time network. Many aircraft and crew routing and scheduling models use so called *ground holding arc*, as well as *overnight arc* (Fig. 7.65). Ground holding arc describes the situation when aircraft is located at a particular airport for a certain period of time. Cost related to this arc could be landing fee, parking charges, etc. Overnight arc represents situations when aircraft or aircrew is staying overnight at the specific airport. The "cost" associated with overnight arcs usually represent the number of aircraft staying overnight at the considered airport.

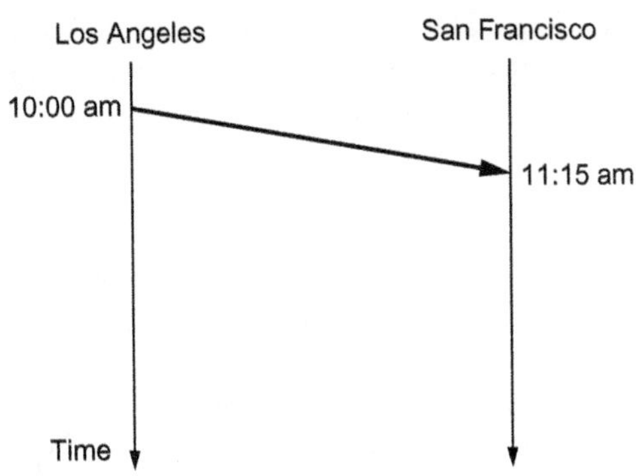

FIG. 7.64

Space-Time Network (Flight arc).

FIG. 7.65

Ground holding arc and overnight arc.

The airports within airline networks are connected by routes, where the airlines schedule their flights at certain frequencies during the specified period of time. The routes and flight frequencies are based on the obtained air traffic rights. The number of airports, within network, can explicitly express and measure the size of an airline network and implicitly indicate the size of covered geographical area. The spatial layout of airline networks can generally be different such as point-to-point, hub-and-spoke, and multi-hub-and-spoke as shown in Fig. 7.66.

Until 1978, many airline services in the U.S.A. were arranged on a *point-to-point* basis. The U.S.A. deregulated domestic air transportation in 1978 (Airline Deregulation Act). American domestic air transportation market has become highly competitive (Carter and Morlok, 1975; Pollack, 1982; Teodorović and Krčmar-Nožić, 1989; Chou, 1990; Teodorović et al., 1994; Teodorović and Kalić, 1995). Hub and spoke system was introduced by American Airlines and immediately adopted also by other major American carriers. Most airlines started to use strategically located airports (the hubs) as air passenger exchange points for flights to and from remote cities (the spokes). Many flights that served one single market in a point-to-point system, serve many different markets in a hub-and-spoke system. Number of flights requiring passengers to change aircraft and/or airlines at an intermediate stop has been significantly increased.

Today, in many world regions, air travel is characterized by the existence of hundreds and thousands of small markets. As a rule, the demand within these markets does not support existing of the *point-to-point* non-stop flights (Fig. 7.66A). The *hubbing* phenomenon happens when air carriers use one or more airports as collection-distribution centers for their passengers (Fig. 7.66B). Hub-and-

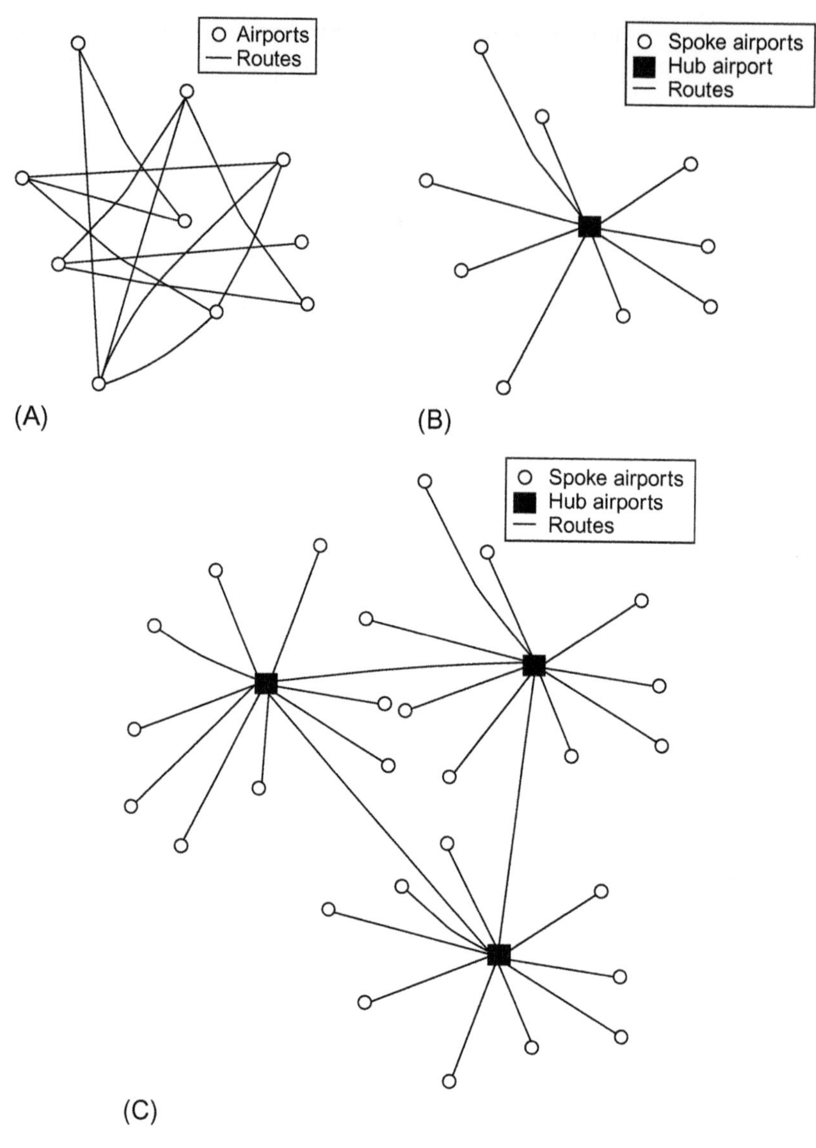

FIG. 7.66

Schemes of different spatial configurations of airline networks. (A) Point-to-point network, (B) single hub-and-spoke network, and (C) multi-hub-and-spoke network.

spoke airline networks are inspired by the benefits of increased flight frequencies and by the economies of operating bigger aircraft (Gordon and de Neufville, 1973; Kanafani, 1981, 1983; Kanafani and Ghobrial, 1982, 1985; Ghobrial, 1983; Ghafouri and Lam, 1986). In the hub-and-spoke networks, passengers travel between origin and destination through the hub airports. Airlines consolidate passengers through airport hubs, and operate between hubs by larger aircraft with increased flight frequencies. It is well documented that hub-and spoke networks can provide higher frequency of larger aircraft at lower cost per passenger. On the other hand, hub-and spoke networks cause additional circuity of travel, and additional passenger waiting time at hubs. Trying to meet passengers' transportation requests and to enable successful transfers, most of the airlines densely packed their schedules (Figs. 7.67 and 7.68).

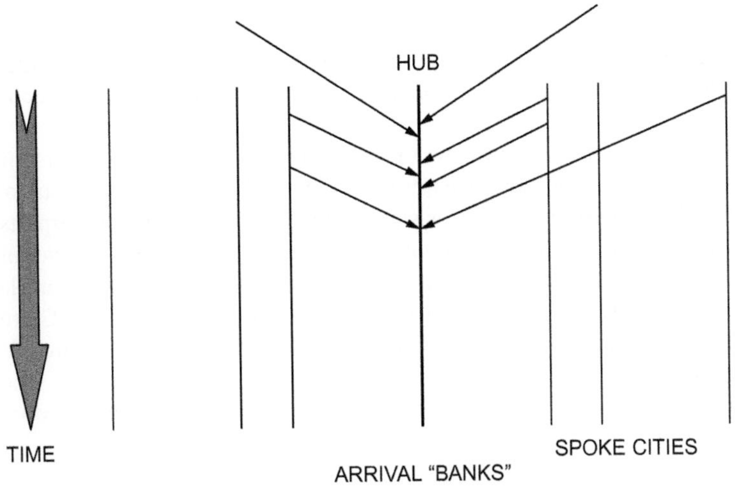

FIG. 7.67

Arrival "banks".

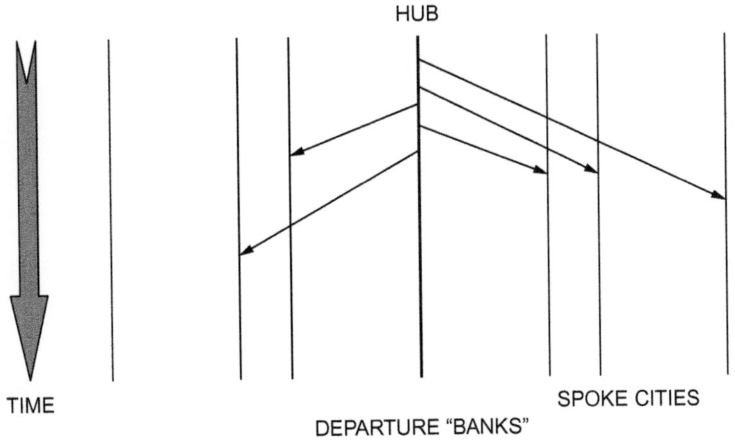

FIG. 7.68

Departure "banks".

The hub-and-spoke and multi-hub-and-spoke networks are operated by the larger conventional airlines. Passengers from the flights of arriving banks can connect to the flights of departing banks on their way between origin and destination spoke airports. The hub airport(s) can also be an origin and destination of passengers. Such operational pattern reduces the number of routes needed to fully connect the given number of airports of the airline network compared to the corresponding number of routes in the equivalent point-to-point network. In the case of the single hub-and-spoke network with N airports, the number of routes fully connecting them is: $(N-1)$. Arrival "banks" consist of incoming flights to hub from many spoke cities landing at approximately same time. Departure "banks" consist of outgoing flights from a hub to many spoke cities departing at approximately the same time (Fig. 7.67).

Bad meteorological conditions at hub airports could, during a winter season, could cause hundreds of canceled and delayed flights, greater airline and airport operating costs, and inconvenience to air passengers.

The *point-to-point networks* are nowadays operated mainly by low-cost carriers. In this case, some airports serve as the main base airports where the majority of carrier's fleet is stationed. The others are also base airports if at least one aircraft is stationed there. These and other airports of the network are usually connected by direct (non-stop) flights. For example, the number of routes in such fully connected network with N airports is: $N \cdot (N-1)$. Some of these networks can be very large such as for example that of the largest world's low-cost carrier Southwest Airlines (US). Currently the airline serves 95 airports in 41 countries by the above-mentioned fleet of 692 B737s aircraft. The 10 largest U.S. airports are used as the base airports where between 150 and 250 flights are scheduled to depart daily. Fig. 7.69 shows the scheme of spatial layout of the airline domestic network operated in the U.S. (SA, 2015).

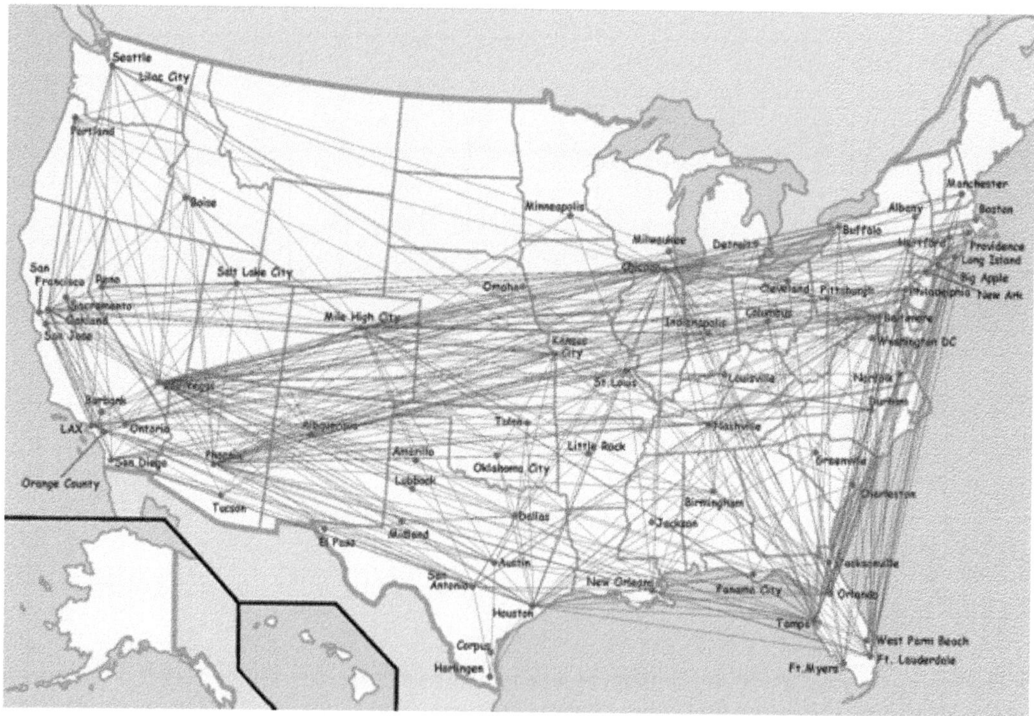

FIG. 7.69

Scheme of the air route network of Southwest airline (SA, 2015; https://www.google.nl/search?q=southwest+airlines+route+map).

In Europe, Ryanair is the largest low-cost carrier, serving 179 airports by 318 B737s aircraft stationed in the two main largest airports (http://www.ryanair.com/en).

As we already mentioned, the hub-and-spoke and multi-hub-and-spoke networks are operated by the larger conventional airlines. Fig. 7.70 shows the scheme of air route network of Delta Airlines. In the U.S., with the fleet of 800 aircraft the Delta Airlines operates 9 major hubs: Atlanta, Boston, Detroit, Los Angeles, Minneapolis/St. Paul, New York-JFK, New York-LaGuardia, Salt Lake City, and Seattle (http://news.delta.com/stats-facts).

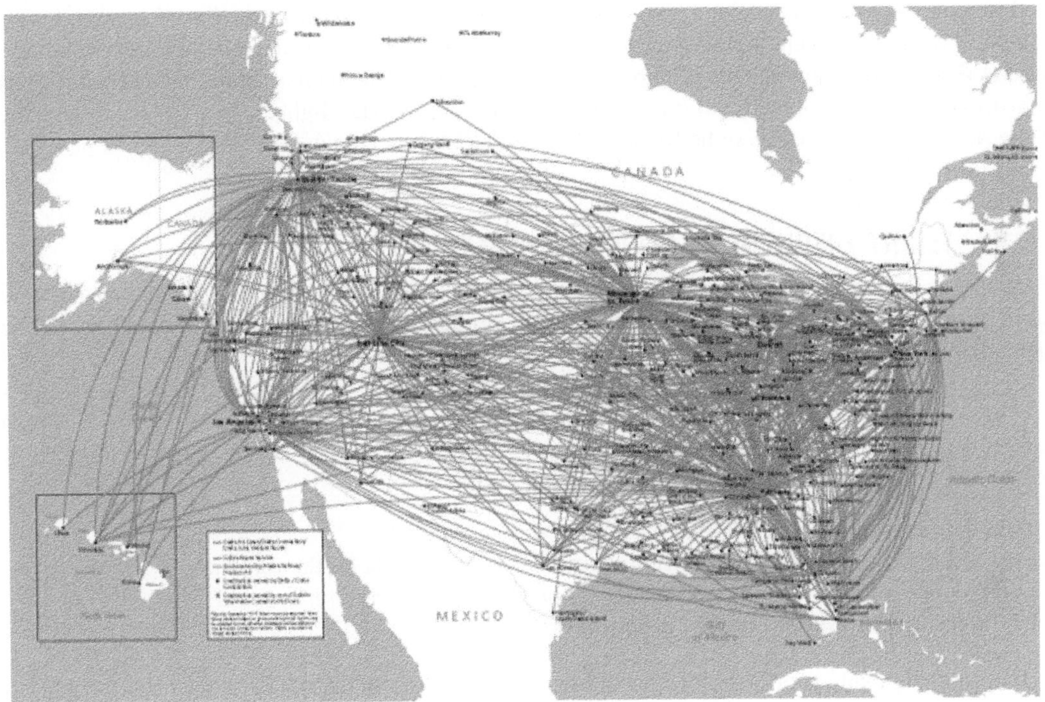

FIG. 7.70

Scheme of the air route network of Delta Airlines (http://www.delta.com/content/www/en_US/traveling-with-us/where-we-fly/routes/downloadable-route-maps.html).

7.23 FLIGHT FREQUENCIES

The airline dynamic capacity expresses the intensity and volume of utilization of the available airline fleet carrying out flights at given frequencies during a given period of time (hour, day, season, year) on particular routes of its air route network under given operating conditions.

Each flight is specified by aircraft type, distance and scheduled time between given pair of airports, and the departure and arrival times from/to the particular airports. At a scheduled airline, the routes and flights are announced in the timetable. In that sense, the timetable can be considered as the airline's "production" plan, which may change several times during the year dependent on the expected variations of the volumes of passenger and freight/cargo demand. Several flights can be carried out on the same routes during the specified period of time. As such, they represent the flight frequencies there.

In general, the airlines schedule flight frequencies on particular routes according to different criteria (Pollack, 1982; Teodorović, 1983). At least theoretically, these can be: satisfying the passenger (and/or freight/cargo) demand, sustaining the existing airline market share, gaining the airline market share particularly in cases of new entries, minimizing the total route costs, maximizing the route profits as difference between the obtained revenues and costs, and maximizing the quality of service. In this context, the flight frequency satisfying passenger (and/or freight/cargo) demand, gaining the airline market share, and minimizing the total route costs is analyzed. The main assumption is that all flights are carried out by the same aircraft type, ie, of the seat-capacity. The flight frequencies determined by using the remaining criteria can be found in relevant references (Janić, 2000).

7.23.1 FLIGHT FREQUENCY SATISFYING DEMAND

Flight frequency satisfying demand is represented by the number of flights scheduled on a given route during given period of time τ to satisfy expected passenger (and/or freight/cargo) demand. It can be estimated as follows:

$$f_i(\tau) = \frac{Q_i(\tau)}{\lambda_i(\tau) \cdot n_{ik}} \quad \text{for } i \in M \tag{7.61}$$

where:

$\lambda_i(\tau)$ is the average load factor on the route i during time τ;
n_{ik} is the seat-capacity of the aircraft of type k carrying flights on the route i;
$Q_i(\tau)$ is the expected number of passengers on the route i during time τ; and
M is the number of routes in the airline network.

In this case, the load factor is defined as a proportion between the realized/occupied and the offered seats on a given route under given conditions.

7.23.2 FLIGHT FREQUENCY GAINING MARKET SHARE

Flight frequency gaining market share enables an airline to gain some market share by entering a given route/market for the first time. This market share is defined as a proportion of the total passenger (and/or cargo) demand served by particular airlines operating the route including the new entrant during given period of time. For example, on the route i, where K airlines compete among themselves, the j-th airline market share M_{ji} can be expressed as follows (Teodorović and Krčmar-Nožic, 1989):

$$M_{ji} = \frac{(f_j \cdot n_j)^{\alpha}}{\sum_{k=1}^{K} (f_k \cdot n_k)^{\alpha}} \tag{7.62}$$

where α is an empirical constant ranging between 1 and 2.

In Eq. (7.62), it is assumed that both the aircraft seat-capacity and the flight frequencies are used as the airline competitive tool in order to gain market share on the given route. The other symbols are analogous to those in (7.61).

Fig. 7.71 shows the various relationships between the market share and the capacity share for the airline k ($k \in K$). It can be seen that when the value of α is greater than one, the curve representing the market share of the corresponding airline will take some S-shape. This can be explained by considering expected and actual behavior of air passengers. First, the passengers most frequently think about the dominant airline. Then they make their seat-reservation at that airline expecting the preferred-best departure time (Janić, 2000).

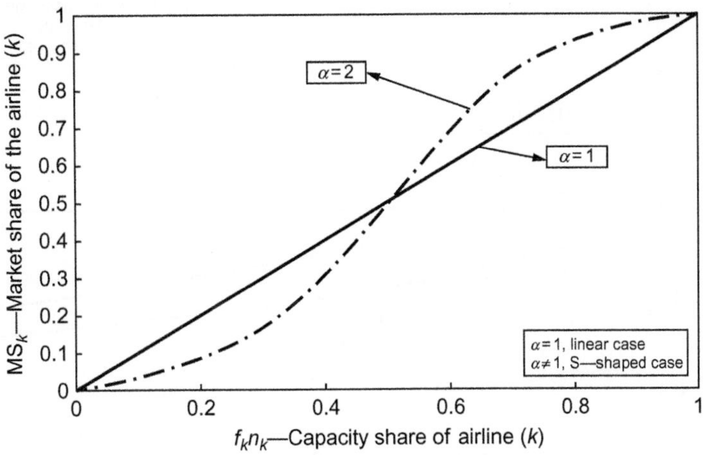

FIG. 7.71

Relationship between the market share and the capacity share of the airline k on a route given route (Janić, 2000).

An inherent tendency of the airlines operating in competitive markets is to over-schedule. This may also influence the shape of the market share-capacity curve. Especially the new entrants have the short-term tactics to supply the extra capacity in an effort to obtain an adequate market share in a relatively short time. The other airlines already operating in the same market(s) always respond to such challenge and expand their capacity supply. In order to analyze such capacity-supply policy let's consider two airlines competing on a single route. They are assumed to supply the flights with given number of seats in order to maintain their current market shares while satisfying the expected passenger demand. They schedule their flights according to the following scenario: if the airline *No.1* increases the number of flights by 1 flight, the airline *No.2* will immediately respond by supplying additional flight, and vice versa. The new market shares of both airlines and the route load factor can be determined as follows:

$$Q(\tau) = [f_1(\tau) \cdot n_1 + f_2(\tau) \cdot n_2] \cdot \lambda(\tau) \tag{7.63}$$

$$\lambda(\tau) = Q(\tau) / [f_1(\tau) \cdot n_1 + f_2(\tau) \cdot n_2] \tag{7.64}$$

where all symbols area analogous to those in previous relations.

Fig. 7.72 shows an example of the dependence of the average route load factor and the airlines' market shares on the number of extra flights supplied by both competing airlines. If both airlines simultaneously supply the same number of extra flights, their market shares will not change. The average route load factor will decrease indicating that both airlines fly increasing number of empty seats (the other factors influencing demand are assumed constant). In addition, any decrease in demand additional decrease in the load factor. Under such conditions, the airlines usually stop scheduling extra flights but also retreat some of the previously existing in order to maintain the acceptable load factor.

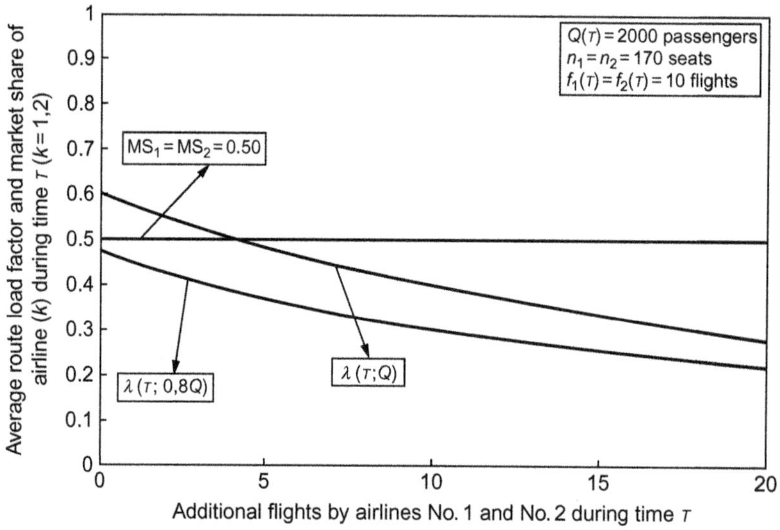

FIG. 7.72

Dependence of the airline load factor and market share on the number of additional flights scheduled on a route (Janić, 2000).

7.23.3 FLIGHT FREQUENCY MINIMIZING THE TOTAL ROUTE COST

Flight frequency minimizing the total route cost implies setting it up to minimize the total cost consisting of the passenger cost and the airline operating cost on the route. The passenger cost includes the cost of their lost while traveling but not the cost of airfare. The passenger time consists of the time waiting for the first convenient departure (ie, scheduled delay or defer time) and in-aircraft time. The airline operating costs include the cost of operating flights on a given route during time τ. The total route cost $C(\tau)$ can be expressed as follows (Janić, 2000):

$$C(\tau) = Q(\tau)\left[\frac{\alpha(\tau)}{2\cdot f(\tau)} + \beta(\tau)\cdot\left(\frac{L}{v(n,L)} + D(\tau)\right) + a_0\right] + c_f(n,L)\cdot f(\tau) \qquad (7.65)$$

where:

$Q(\tau)$ is the number of passengers traveling on the given route in time τ (pax);
$\alpha(\tau)$, $\beta(\tau)$ are the unit cost of passenger time while waiting for departure (eg, schedule delay cost) and traveling (eg, in-aircraft time) during time τ, respectively (min/pax);
L is the length of a route (nm);
$v(n, d)$ is the average block speed of the aircraft of seat-capacity n on the route L (kts);
$f(\tau)$ is the number of flights scheduled on the route during time τ(flights);
$D(\tau)$ is the average anticipated delay per flight carried out in time τ(min);
a_o is the airline fixed cost per passenger ($US/pax); and
$c_f(n, L)$ is the cost per flight carried out by the aircraft of seating capacity n on the route L ($US/flight).

By deriving the cost function $C(\tau)$ by $f(\tau)$ and equalizing the obtained results to zero, the minimum of function $C(\tau)$ can be found as follows:

$$\frac{\partial C(\tau)}{\partial f(\tau)} = -\frac{1}{2}\frac{Q(\tau)\cdot\tau\cdot\alpha(\tau)}{[f(\tau)]^2} + c_f(n,L) = 0 \text{ and } \frac{\partial^2 C(\tau)}{\partial f^2(\tau)} = \frac{Q(\tau)\cdot\tau\cdot\alpha(\tau)}{[f(\tau)]^3} > 0 \tag{7.66}$$

From Eq. (7.66), the flight frequency which minimizes the total route cost is equal as follows:

$$f^*(\tau) = \left[\frac{Q(\tau)\cdot\tau\cdot\alpha(\tau)}{2\cdot c_f(n,L)}\right]^{\frac{1}{2}} \tag{7.67}$$

From (7.67) follows that the optimal flight frequency on a given route will increase at decreasing rate (0.5) by increasing of the demand consisting of more "expensive" passengers in terms of the value of their time (ie, business passengers) and decrease with increasing of the flight cost. By combining (7.61) and (7.67), the optimal aircraft size in terms of its seat—capacity to be engaged on the route during time τ can be determined as follows:

$$n^*(\tau) = \frac{1}{\lambda(\tau)}\left[\frac{2\cdot Q(\tau)\cdot c_f(n,L)}{\tau\cdot\alpha(\tau)}\right]^{\frac{1}{2}} \tag{7.68}$$

where all symbols are analogous to the corresponding ones in previous expressions. If $n^*(\tau) > n_{\max}$ ($n_{\max}$ is the maximum seat—capacity of the available aircraft in the airline's fleet), two or more aircraft should be engaged per single departure.

Fig. 7.73 shows an example of determining the optimal flight frequency and aircraft size on the route.

As can be seen, the optimal frequency decreases and the aircraft optimal seat-capacity increases with increasing of the route length. Both the optimal frequency and the aircraft seat-capacity increase with growing demand on the route during time τ (day).

The validity of the above-mentioned theoretical were tested on a real-life example of Lufthansa Airlines (Germany) scheduling flights from its main hub airport—Frankfurt Main (Germany) as shown in Fig. 7.74.

As can be seen, Figs. 7.73 and 7.74 are analogous. Overall, the similarity between particular relationships in theory and practice also indicates that the theoretical model(s) could be useful in explaining these phenomena in the contemporary airline industry.

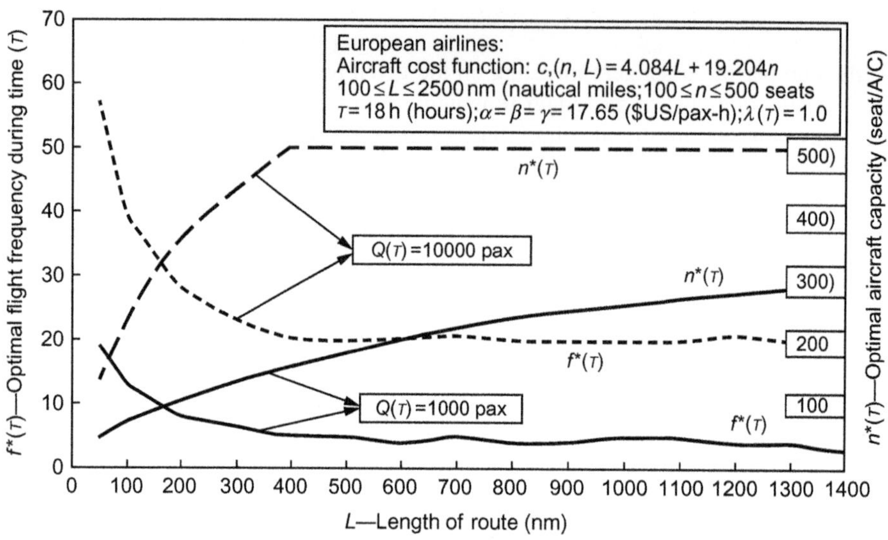

FIG. 7.73

An example of the relationship between the optimal flight frequency, the aircraft seat-capacity, and the volume of demand on a given route (Janić, 2000).

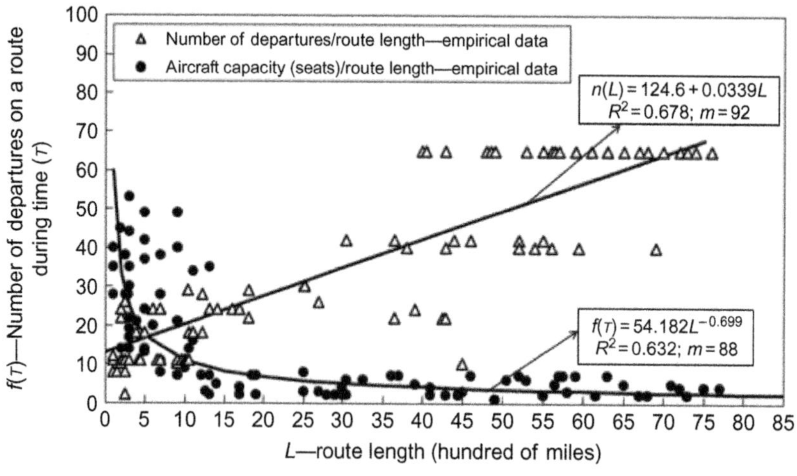

FIG. 7.74

Relationships between the flight frequency, aircraft seat-capacity and route length of the Lufthansa flights departing from Frankfurt Main Airport (Janić, 2000).

7.24 AIRLINE TRANSPORT WORK AND PRODUCTIVITY

The transport work and productivity of particular routes and of the entire (given) airline network generally indicate the volume of offered capacity and its utilization.

Transport work on airline route includes that based on the supplied capacity and that realized by transporting/serving passenger (and/or freight/cargo) demand. Similarly as in the previously mentioned cases of other transport modes, the supply-based (seat-nm or seat-km) and the demand-based transport work (pax-mi or pax-km) on the route i during the specified period of time τ can be calculated as follows:

$$TWS_i(\tau) = f_i(\tau) \cdot n_i(\tau) \cdot L_i \tag{7.69}$$

$$TWD_i(\tau, \lambda) = f_i(\tau) \cdot n_i(\tau) \cdot \lambda_i(\tau) \cdot L_i \tag{7.70}$$

where all symbols are analogous to those in the previous relations (mi—mile; 1 mile = 1.609 km).

For the route i of an airline network, the productivity in terms of seat-km/h or seat-mi/h, and pax-km/h or pax-mi/h, respectively, can be calculated as follows:

$$TPS_i(\tau) = f_i(\tau) \cdot n_i(\tau) \cdot v_i(n_i, L_i) \tag{7.71}$$

$$TPD_i(\tau, \lambda) = f_i(\tau) \cdot n_i(\tau) \cdot \lambda_i(\tau) \cdot v_i(n_i, L_i) \tag{7.72}$$

where $\lambda_i(\tau)$ is the load factor on the route i.

The other symbols are analogous to those in the previous relations. In calculating the productivity of particular flights a concept of "break-even" load factor can be applied to determine the acceptable level of their commercial feasibility (Janić, 2000).

Transport work of airline network can be calculated by summing up the transport work on particular routes of the network, based on relations (7.69) and (7.70), as follows:

$$TWS(\tau) = \sum_{i=1}^{M} TWS_i(\tau) = \sum_{i=1}^{M} f_i(\tau) \cdot n_i(\tau) \cdot L_i \tag{7.73}$$

$$TWD(\tau) = \sum_{i=1}^{M} TWD_i(\tau, \lambda) = \sum_{i=1}^{M} f_i(\tau) \cdot n_i(\tau) \cdot \lambda_i(\tau) \cdot L_i \tag{7.74}$$

where M is the number of considered routes in the airline network.

The productivity of airline network can be calculated by summing up the productivity on particular routes constituting the network. Based on relations (7.71) and (7.72), it is as follows:

$$TPS(\tau) = \sum_{i=1}^{M} TPS_i(\tau) = \sum_{i=1}^{M} f_i(\tau) \cdot n_i(\tau) \cdot v_i(n_i, L_i) \tag{7.75}$$

$$TPD(\tau) = \sum_{i=1}^{M} TPD_i(\tau, \lambda) = \sum_{i=1}^{M} f_i(\tau) \cdot n_i(\tau) \cdot \lambda_i(\tau) \cdot v_i(n_i, L_i) \tag{7.76}$$

In relations (7.75) and (7.76), the route block speed can be estimated as:

$$v_i(n_i, L_i) = \frac{L_i}{\tau(L_i)} \tag{7.77}$$

where $\tau(L_i)$ is the average aircraft/flight block time on the route i of the length L_i.

The average aircraft/flight block time is usually estimated in advance by taking into consideration the local traffic and weather conditions on the route, as well as the performance of the engaged aircraft. This time is the time difference between leaving the gate at the origin airport until arriving at the gate of destination airport. Delays imposed on particular flights by the ATC (Air Traffic Control) prolong the aircraft/flight block time and decrease the expected block speed. For example, if the average delay on route i, with the length L_i, is $w(L_i)$, the actual block speed will be:

$$v_i(n_i, L_i) = \frac{L_i}{[\tau(L_i) + w(L_i)]} \tag{7.78}$$

Generally, reduction of the block speed due to delays and any other reasons diminishes the productive capacity of particular routes and consequently the airline network. Introducing a greater number of the longer routes, engagement of the larger aircraft and scheduling more flights are all the factors contributing either partly or together to increasing of the productivity of the airline network.

A specific indicator of airline productivity is the average overall load factor. This is defined as the ratio between the total demand-based transport work—the volume of realized revenue passenger and/or freight/cargo-kilometers (miles) and the total supply/capacity-based transport work—the volume of offered seat and/or freight/cargo-kilometers (miles) during a given period of time. The overall load factor can be calculated as follows:

$$\theta(\tau) = \frac{TWD(\tau)}{TWS(\tau)} \tag{7.79}$$

This implies that as a greater percentage of the offered seats are filled, the overall airline productivity will be higher, and vice versa.

Fig. 7.75A–C shows some of the above-mentioned metrics—transportation work at the supply and demand side, corresponding airline load factor, and the average route length (the average route length for the European large conventional airline—Lufthansa and low-cost carrier—Ryanair.

As can be seen, these indicators have been quite different at both airlines as intuitively expected. The scale of transport work at Lufthansa has been for about twice higher than that of Ryanair. It has generally increased at decreasing rate while that of Ryanair has decreased. This latter has indicated exposure of the carrier to stronger competition and the necessity to squeeze the scale of its operations in order to maintain the required (break-even) load factor. As well, Lufthansa has also permanently adjusted its supply capacity in order maintain the required (break-even) load factor. Overall, the average load factor for both airlines has been about 80% over the observed period. By dividing the transport work by the number of transported passengers, the average route length has been calculated. It has been again for about twice longer at Lufthansa (about 2000–2500 km) mainly thanks to its long-haul intercontinental flights. At Ryanair it has gradually increased from about 1000 to 1250 km, mainly due to operating exclusively the short- and medium-haul European flights (Lufthansa, 2014; Ryanair, 2014).

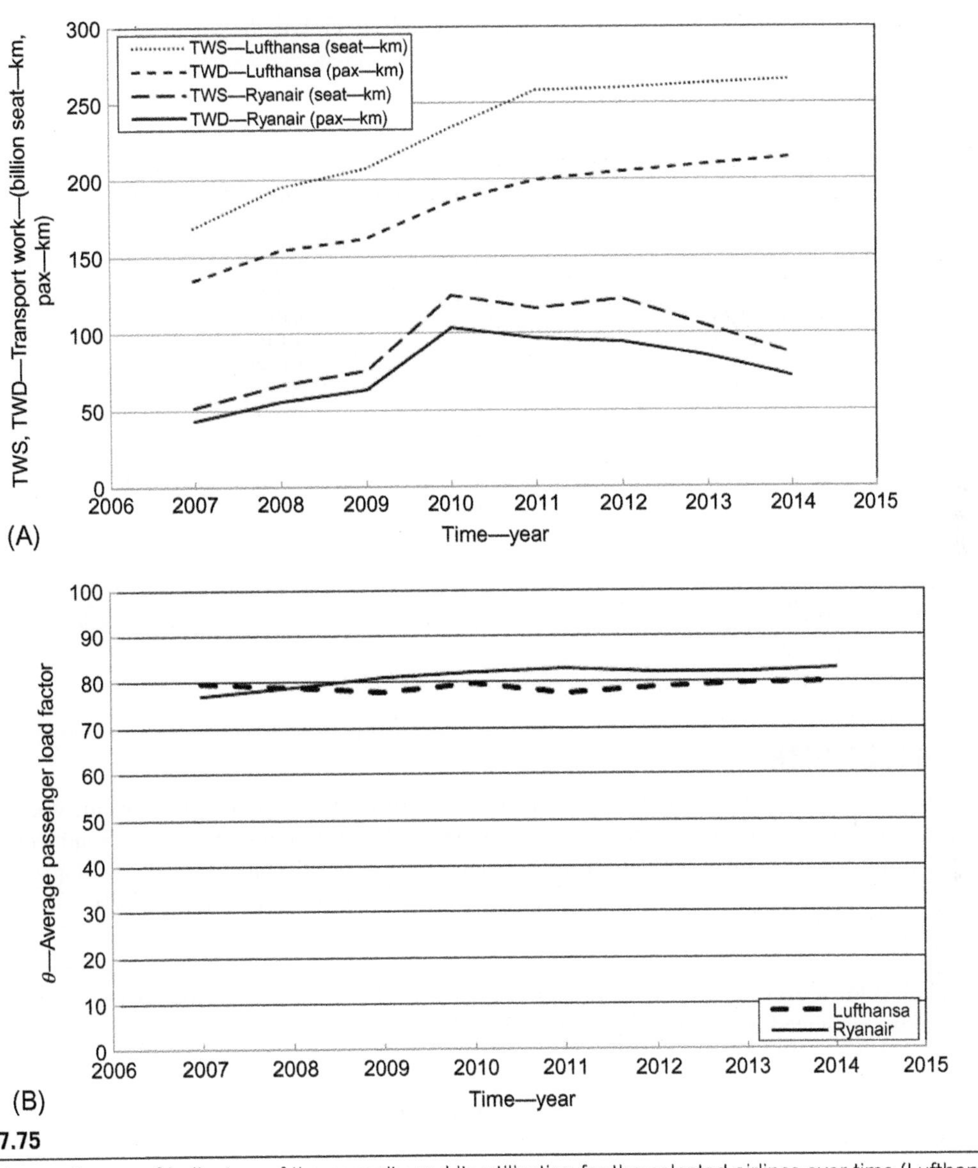

FIG. 7.75

Development some of indicators of the capacity and its utilization for the selected airlines over time (Lufthansa, 2014; Ryanair, 2014). (A) Transport work, (B) average load factor, and

(Continued)

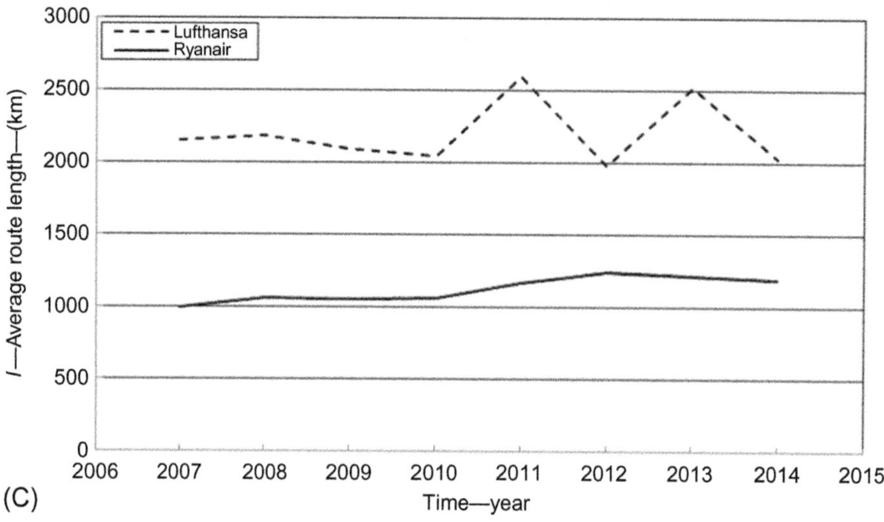

FIG 7.75, CONT'D

(C) average route length.

7.25 FLEET SIZE

The airline fleet size is expressed by the number of aircraft operated by the given airline during the specified period of time (as mentioned before, it is 800 by Delta Airlines and 692 by Southwest Airlines). The main factors influencing the size of airline fleet are: the size and complexity of airline network expressed by the number of airports and routes connecting them, planned flight frequencies on particular routes, and the estimated the above-mentioned block times on particular routes. In order to calculate the number of aircraft in a fleet, it is assumed that an airline operates only one aircraft type (for example, it is low-cost carrier). This number to be operated during the time τ (day, week, season, and year) can be calculated as follows (Janić, 2000):

$$A(\tau) \geq \frac{1}{U(\tau)} \cdot \sum_{i=1}^{M} f_i(\tau) \cdot t_i(L_i) \tag{7.80}$$

where:

$U(\tau)$ is the average utilization of an aircraft of the airline fleet during time period τ; and
$t_i(L_i)$ is the aircraft turnaround time on the route i.

The other symbols are analogous to those in the previous expressions. The aircraft turnaround time on the route $t_i(L_i)$ includes the above-mentioned block time $\tau_i(L_i)$ and a part of the aircraft turnaround time at the airport gate spent for its preparation for the flight on the route i. Relation (7.80) indicates that the stock of aircraft increases with increasing individually or together of the number of routes in the network, the number of flights scheduled there, and the aircraft turnaround time on these routes. The other methods for estimating the size of an airline fleet can be found in other references (Janić, 2000).

7.26 LEVEL OF SERVICE

The level of service at commercial airlines can be considered by different actors/stakeholders involved. These are mainly airlines themselves using it as a competitive tool and—air passengers expecting these airlines to satisfy their travel needs under given conditions. As far as users-air passengers are concerned, service quality can be defined "level of their satisfaction" or "continuity in satisfaction of customer requirements." This service quality may have the external and internal dimension influencing choice of the air transport mode and then particular airline. The external dimension is generally represented by attributes such as the size of airline network, available fleet and its schedule of particular routes, accessibility and punctuality of services. The internal dimension becomes relevant when users-air passengers enter the airline system. Its main attributes are quality of service provided while on the ground and onboard the aircraft, ie, during the flight (Janić, 2000). Consequently, estimating the service quality provided by airlines has appeared to be the very complex task. One of the most systematic and continuous efforts has been development and use of the AQR (Airline Quality Rating) system (Bowen and Headley, 2015). The system has been developed for evaluating and comparing service quality of the 12 largest U.S. airlines means by the standard set of 4 attributes/criteria, each with the assigned weight ranging from "0"—"no importance" to "10"—"great importance." In addition, each attribute is assigned a specific sign reflecting direction of its impact on the overall AQR. The positive sign "+" denotes that the corresponding attribute is more favorable and the negative sign "−" does just opposite. The attributes, weights, and their signs are as follows: OT—On-Time, Weight—8.63, Sign "+"; DB—Denied Boarding, Weight—8.03, Sign "−"; MB—Mishandled Baggage, Weight—7.92, Sign "−"; CC—Customer Complaints, Weight—7.17, Sign "−." This last "Customer Complaints" contains attributes/criteria such as: Flight Problems, Over Sales, Reservations, Ticketing, and Boarding, Fares, Refunds, Baggage, Customer Service, Disability, Advertising, Discrimination, Animals, and Other. Data for these criteria are obtained from the U.S. Department of Transportation's monthly "Air Travel Consumer Report" (http://dot.gov/airconsumer). Its basic structure is as follows:

$$AQR = \frac{\sum_{i=1}^{4} w_i F_i}{\sum_{i=1}^{4} w_i} = \frac{(+8.63 \cdot OT) + (-8.03 \cdot DB) + (-7.92 \cdot MB) + (-7.17 \cdot CC)}{(8.63 + 8.03 + 7.92 + 7.17)} \quad (7.81)$$

where:

w_i is the weight of attribute i of service quality; and
F_i is the attribute i of service quality.

Table 7.14 gives the results from application of the AQR system to the U.S. airlines in the year 2014.

Table 7.14 Ranking of the Service Quality of U.S. Airlines in 2014 (AQR System) (Bowen and Headley, 2015)

Airline	AQR Score	Rank
Alaska	−0.65	5
American	−1.35	7
Delta	−0.60	3
Envoy/Am Eagle	−2.83	12
ExpressJet	−2.12	11
Frontier	−1.48	8
Hawaiian	−0.53	2
JetBlue	−0.61	4
SkyWest	−1.84	10
Southwest	−1.22	6
United	−1.62	9
Virgin America	−0.30	1
Industry—total:	−1.24	

As can be seen, the AQR score for all considered airlines has been negative, implying that the criteria DB (Denied Boarding), MB (Mishandled Baggage), and CC (Consumer Complaints) have dominated over the criteria OT (On-Time). The worst ranked has been Envoy/Am Eagle (12th), followed by ExpressJet (11th). The best ranked has been Virgin America (1st) and Hawaiian (2nd). In addition, Fig. 7.78A and B shows AQR of the two U.S. airlines—Delta and Southwest Airlines—over time. As can be seen from Fig. 7.78A, during the period 2010–14, Delta has improved its AQR score—from about (−1.2) to (−0.6). During the period 2007–12, Southwest has also improved its AQR score—from about (−1.6) to (−0.8), but after that time its score has deteriorated—from (−0.8) to (−1.2) in the year 2014. At the same time, the overall AQR score of the airline industry has been improving—from about (−2.25) in the year 2007 to about (−1.1) in the year 2014. Fig. 7.76B shows that the rank of Delta has improved during the period 2010–14 from 7th to 3rd. The rank of Southwest has deteriorated during the period 2007–13 from 3rd to 8th, but then recovered to 6th in the year 2014. It should be mentioned that these figures should be considered in the relative terms and also be compared to the AQR scores and ranks of other 10 airlines during the observed period (Bowen and Headley, 2015).

In addition, service quality of the above-mentioned U.S. airlines can also be considered by taking into account sub-attributes/criteria in the scope of criterion "Customer Complaints" if the relevant data are available. In addition, the AQR system is sufficiently generous to be applied to other non-U.S. airlines, but with the necessary modifications and based on the availability of relevant data. In such case, the set of standard attributes/criteria, sign of their impacts and weights need to be determined.

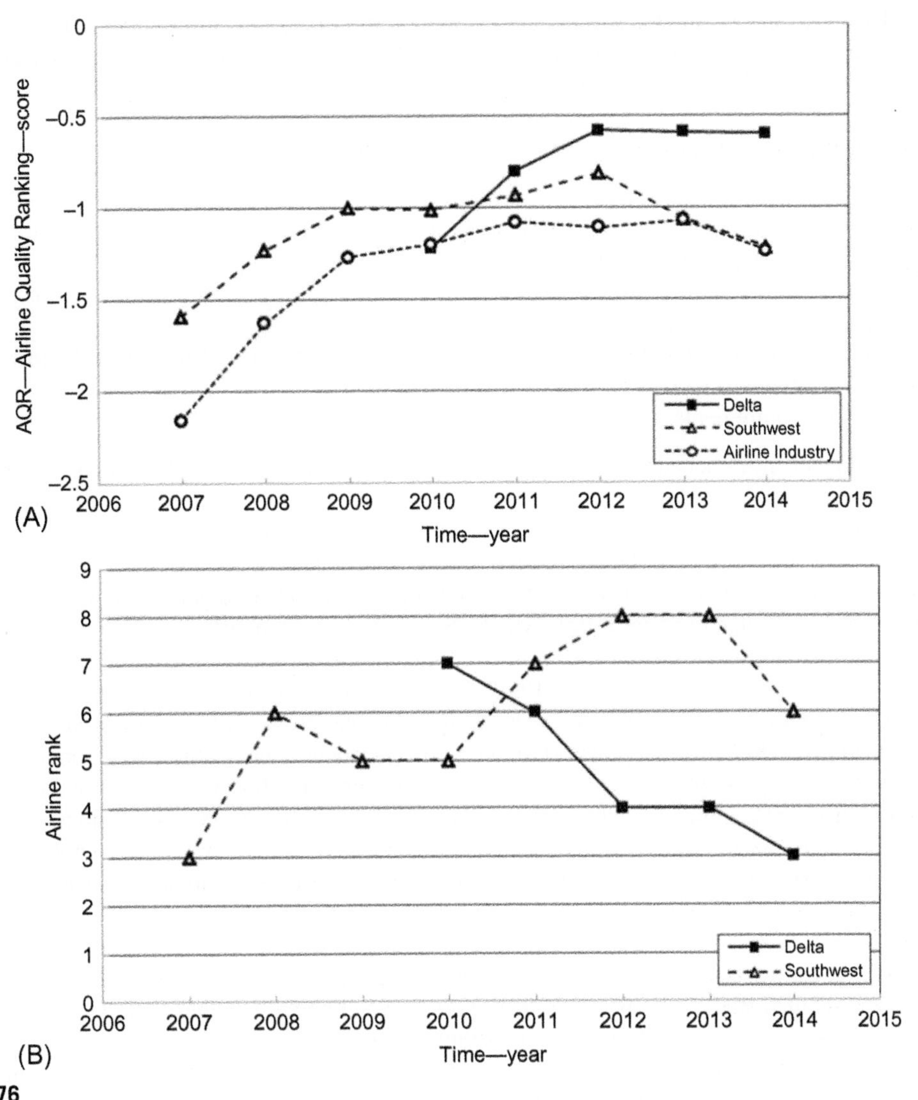

FIG. 7.76

Results of ranking of the selected U.S. airlines over time by AQR system. (A) Ranking score and (B) rank.

7.27 AIRLINE SCHEDULING

Airlines' schedules could be considered as the main airlines' product that airlines offer to the customers. Elaborating a network airline schedule is combinatorial by nature: of the large number of possible alternatives, those must be chosen that satisfy to the greatest extent the interests of the air carrier, the passengers and operational constraints.

Passengers are interested in the greatest possible flight frequency, departure times that are adapted to their desires, a high probability of finding a vacant seat on a particular flight, short waits to continue traveling at transfer points, few canceled and late flights, etc. On the other hand, the carrier's interests are in a high load factor, high annual aircraft utilization, low operating costs and high profit. When creating a schedule a lot of various operational requirements should be carefully studied and taken into account. The following are the most important factors:

- airline maintenance department requirements (aircraft must appear at airline maintenance facilities during certain time windows for inspection and maintenance);
- aircraft and crew availability (numbers of various aircraft types, gates availability, pilots, flight attendants, ground service personnel, customer service personnel);
- crew working hours regulations (crew work time must be within limits prescribed by low and/or by airline internal rules); and
- crew training requirements.

When designing an airline schedule, one must enable to each aircraft from the fleet to stay overnight at a maintenance facility according to the aircraft maintenance schedule (Within aircraft maintenance schedule, frequency and duration of aircraft service are prescribed). For reasons of safety, aircraft must have certain parts inspected or replaced at specific times at specific airports that are most often the air carrier's technical base. In addition, when greater technical work must be done, aircraft are removed from operations for a longer period of time which also has an important effect on the choice of airline schedule.

When scheduling aircrews (pilots, and flight attendants), care must be taken about legal, airline, and union rules that precisely regulate crew working hours and crew working duties. Similarly, it is necessary to plan carefully the manpower (ground service personnel that handle aircraft on the ground, and customer service personnel that serve passengers in the terminal) for every station in the network.

Another constraint to remember when designing the airline schedule is the working hours of airports in the network. Some airports are open 24 h a day. Most airports have prescribed working hours and only accept aircraft in emergency situations outside of these hours. Most international air carriers are interested in completing take-off and landing operations within a fixed time interval during the day. A result of this desire is peak overload periods at airports that considerably hinder the work of airport services.

In long-haul traffic, different time zones present special operational constraints. Due to different time zones, a detailed study must be made of airport working hours, whether there is transportation between the airport and the city during different times of day, etc.

For each planned flight, the type of meal service must be defined, depending on flight status (domestic, or international) and the duration of the flight.

Meteorological conditions prevailing at the airports in the network also act as a limiting factor when choosing a flight schedule (Teodorović, 1985). During the winter season, some regions often have airports that are closed to take-off or landing operations which can result in many canceled and delayed flights (Teodorović and Guberinić, 1984; Teodorović and Stojković, 1995).

A result of coordinating the carrier's interests, and passenger requests and the large number of operational constraints is that many departures can only be made during fixed time window during the day. Some schedule variations suit the passengers better while others are more satisfactory to the carrier's interests. The choice of schedule alternative is most certainly the result of the air carrier's overall transportation policy on a specific transportation market.

The oldest and most frequent airline schedule type is one that is prepared beforehand and is valid for a set period of time. Aircraft departure times for every day of the week are known for all routes on the network and are widely publicized (publications intended for the transportation market, web pages, advertisements in newspapers, etc.) (Table 7.15).

Table 7.15 Basic Information Contained in Airline Schedule

Flight Number	Departure Airport	Arrival Airport	Flight Departure Time	Flight Arrival Time	Days in a Week
AB 22	BEL	PAR	6:40 am	9:15 am	1234567
AB 33	PAR	LON	8:50 am	9:05 am	1234567
.....	.....	.....	.....	.....	

For example, flight AB 22 departs from Belgrade at 6:40 am and arrives in Paris at 6:40 am local time. The airline provides this service every day in the week (last column). Flight AB 33 departs every day at 8:50 am local time from Paris, at arrives at London at 9:05 am local time. An airline schedule can also contain information regarding equipment type to be used on particular flights.

Passengers adjust their desired departure times to those given in advance for this type of schedule and most passengers purchase their tickets in advance and reserve a seat before departure time. The air carrier must keep to the announced schedule and flights must be made at the specified time regardless of the number of passengers in the plane. This type of schedule gives passengers complete information on the possibilities of making a trip.

7.28 AIRLINE SCHEDULE PLANNING PROCESS

Airline scheduling process begins few months, or even 1 year before the airline schedule execution. This process is graphically shown in Fig. 7.77.

Airline marketing department provides basic information related to various airline markets to the airline scheduling department. Within the phase of flight frequencies determination, flight frequencies are determined between particular city pairs. During flight scheduling process, the origin airport, the destination airport, as well as the departure time, and the arrival time of each leg are determined. In other words, in this stage, possible departure times are determined taking into account both passenger requests and convenient times to arrive at important transfer points in order to allow passengers to continue their trip without much waiting (Soumis et al., 1981). When departure times are established, the proposed airline schedule could be tested in terms of operational constraints. Airlines usually operate with more than one aircraft type. In other words, airlines usually have few fleets. Particular fleet is composed of the aircraft of the same type. In the fleet assignment phase, each leg defined in the flight scheduling phase is assigned to a specific fleet (Lohatepanont and Barnhart, 2004; Barnhart et al., 2002). Aircraft of different capacities and performances can be assigned to the network in different ways. The basic question that airline tries to answer in this planning phase is the following one: Which aircraft type should fly each leg? When doing this assignment, airline must take care about the available number of aircraft of each type, as well as about the cost of assigning certain aircraft type to a particular

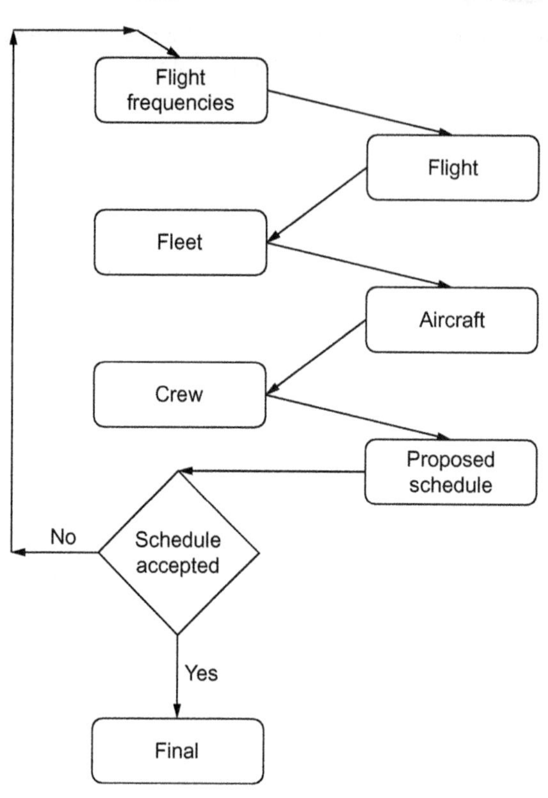

FIG. 7.77

Airline scheduling process.

leg. Obviously, airline tries to "cover" all planned flight legs in such a way to achieve total maximum profit or minimum "coverage" costs. The expected number of passengers that could appear on each leg is assumed to be known.

Capacity of particular aircraft types play significant role in fleet assignment problem, since assigning aircraft with a small capacity could cause lost revenue for the airline. On the other hand, assigning too big aircraft could be very expensive and could also cause extremely low or even negative airline profit. Some of the major world carriers perform more than 2000 flights per day. They serve between 100 and 200 cities, have between 7 and 10 various aircraft type. The total number of aircraft, owned by the biggest airlines, ranges between 300 and 500 aircraft. Obviously, the fleet assignment problem is characterized by big dimensions. Various optimization techniques and heuristic algorithms have been developed in order to solve this complex problem.

An aircraft rotation could be defined a sequence of legs flown by the aircraft of the same type (aircraft that belong to the same fleet). Rotation duration could be one, two, three, or a few days (Fig. 7.79).

Fleet assignment models help us to assign aircraft types to each of the planned flight legs. These models indicate for each flight leg only the aircraft type, and do not assign specific aircraft from the fleet (tail number) to flight legs. Aircraft routing models help us to assign specific aircraft (tail number)

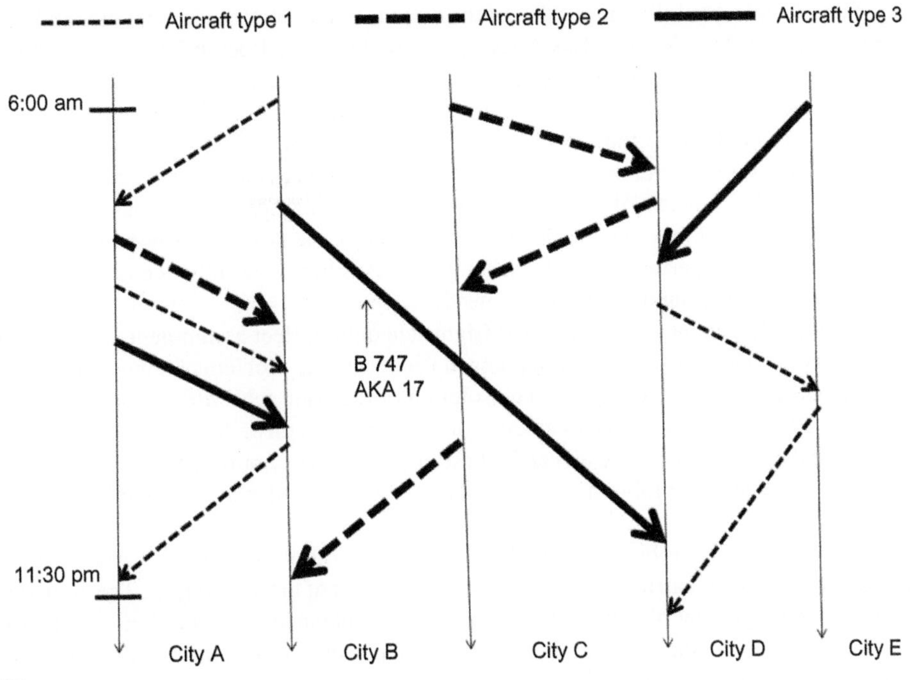

FIG. 7.78

Fleet assignment.

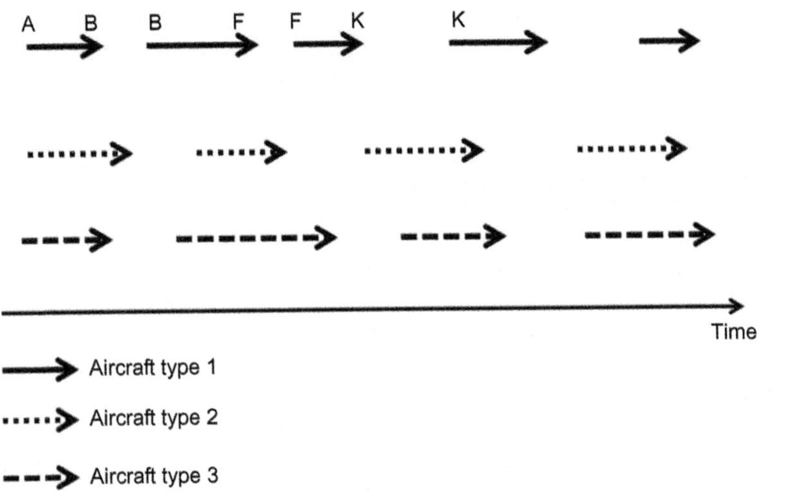

FIG. 7.79

Aircraft rotations.

to flight leg. For example, the following is one of the outputs of the fleet assignment model: Flight that departs at 3:00 pm from New York to Los Angeles will be flown by Boeing 747–400. Using aircraft routing model, we assign, for example, from the Boeing 747–400 fleet, aircraft with the tail number "AKA17" to perform flight this flight. As we can see, aircraft routing models give us more specific information. They also determine the needed number of aircraft and their routes during a fixed time period (usually 1 week). When creating set of aircraft rotations, one must take care about airline maintenance department requirements. We will use the expression *aircraft route* to describe part of the aircraft rotation composed of legs that are flown within 1 day. In the next step, since aircraft have been routed, an analysis could be performed of the schedule's flexibility, ie, whether certain changes in the schedule might decrease the number of aircraft needed.

Once the airline schedule has been designed (fight scheduling, fleet assignment, and aircraft rotation) scheduling crews and assigning them to planned flights are the problems facing air carriers in the next planning phase (Barnhart et al., 1999). The solution of the *crew scheduling problem* represents a set of crew rotations. A crew rotation is a sequence of legs on consecutive days that are made by a crew that begin and end their trip in the same air carrier base. The majority of pilots fly one aircraft type. This means that the crew scheduling problem could be subdivided into smaller crew scheduling problems by aircraft type.

Once the crew scheduling problem is solved, that is, a set of rotations has been defined to be carried out by the crew members, air carriers are faced with the problem of crew rostering. The crew rostering problem entails the assignment of different crew members to planned crew rotations. In other words, the crew rostering problem includes the construction of personalized monthly schedules (rosters). When the crew rostering problem has been solved, each crew member will be assigned rotations to be made during the following month.

During peak periods, aircraft departure and arrival rate are substantially high. This means that during peak periods many aircraft are on the ground at once requesting parking positions. Current airport terminal building designs that have finger piers and air bridges occasionally result in considerable walking distances for passengers. For a given set of aircraft parking positions and a given set of aircraft (with known numbers of arriving and departing passengers), it is necessary to assign aircraft to parking positions (gate assignment problem) in such a way to minimize passengers walking distances.

7.29 AIRLINE REVENUE MANAGEMENT

The liberalization of airline tariffs has led to extremely strong competition among world air carriers. The difference in ticket prices for the same economy-class round-trip could be few hundreds, or even more than one thousand dollars. Airlines seats are only physically identical. They are commercially different, since airlines sell different seats to different passenger's types (market segments) under different circumstances. In the 1970s, the Australian carrier Qantas was one of the pioneers in promoting and introducing "Tourist Class," "Business Class," holiday packages, special fares for young travelers, etc. Air carriers want to sell the available seats in a way that maximizes profits. The majority of airlines use revenue management concepts that could be described as a group of different scientific techniques of managing the airline's revenue when trying to deliver the right product to the right passenger at the right price at the right time. Today, airline industry, hotels, car rental, rail, cruise, healthcare, broadcast industry, energy industry, golf, equipment rental, restaurant, and other industries are utilizing revenue management concepts when selling their products (Cross, 1997). The expression "revenue management" has been simultaneously used with the expressions "yield management," "revenue optimization," and "seat inventory control" (Rothstein, 1971, 1975, 1985; Vickrey, 1972; Littlewood, 1972; Nagarajan, 1979; Allstrup et al., 1986; Belobaba, 1987, 1989; Teodorović, 1988). All airlines use automated reservation systems. The first automated reservation system was introduced by American Airlines in the early 1960s. Today, global distribution systems serve hundreds of thousands of travel agencies on all continents.

Passengers make their requests at random moments of time. A certain number of passengers cancel their reservations before flight departure. These cancelations are also made at random moments of time. Also, a certain number of passengers do not appear for flights for which they have a confirmed reservation and purchased ticket. Although these "no-show" passengers can considerably decrease the air carrier's profits, they usually do not suffer any serious financial consequences (their reasons for not appearing are often of an objective nature). Some passengers appear right before departure looking for an empty seat on the flight, even though they do not have a confirmed reservation (*goshow* passengers). Frequently, the number of passengers appearing for a flight is larger than the available number of seats in the plane (for which situation there can be a variety of reasons for this overbooking). Majority of airlines deliberately sell more seats than available in order to compensate the negative consequences of passenger cancelations and no-shows. Without overbooking policy, many airlines would depart with significant number of empty seats, even in the cases when aircraft is "fully booked." Airlines usually offer various benefits (additional mileage, free pass, etc.) to compensate overbooked passengers. Airline's overbooking policy has significant influence on airline's image and reputation, as well as passengers' loyalty.

Large airlines make more than 1000 flights per day and they usually serve few thousands origin-destination markets. As a result, the number of possible fare class/origin-destination combinations could be extremely large for every seat on each flight leg. Every of these combinations have different contribution to the airline's revenue. The management of an air carrier's revenues represents an extremely complex problem.

It is common in airline business to use word "inventory" to denote aircraft seats. These inventories are specific, in the sense that they are perishable. Airline inventories cannot be stored like inventories in some other industries. Every empty seat at the departure is definitely lost for the airline. Any revenue management technique that can increase an airline's load factor, and total revenue achieved makes a lot of sense. Airlines have limited resources. They cannot enlarge aircraft capacity. They also cannot lease the new aircraft in a short period of time.

Airline decision-makers are practically facing the following dilemma all the time: do we accept early reservations from passengers paying low ticket prices, or do we wait for passengers who are ready to pay higher ticket prices? The simplest reservation system is called *distinct fare class inventories* (Fig. 7.80A), indicating separate seat inventories for each fare class; once the seat is assigned to a fare class inventory, it may be booked only in that fare class or else remain unsold. In the case of a *nested reservation system*, the high fare request will not be rejected as long as any seats are available in lower fare classes. For example, if we have four fare classes, then there is no booking limit for class 1, but there are booking limits (BLi, $i=2,3,4$) for each of the remaining three classes (Fig. 7.80B).

As we can see from Fig. 7.80, all seats are always available to class 1. There are always a certain number of seats protected for class 1, certain number of seats protected for classes 1 and 2 and certain number of seats protected for classes 1, 2, and 3. If we make a request-by-request revision of booking limits, there is no longer a difference between distinct and nested reservation system.

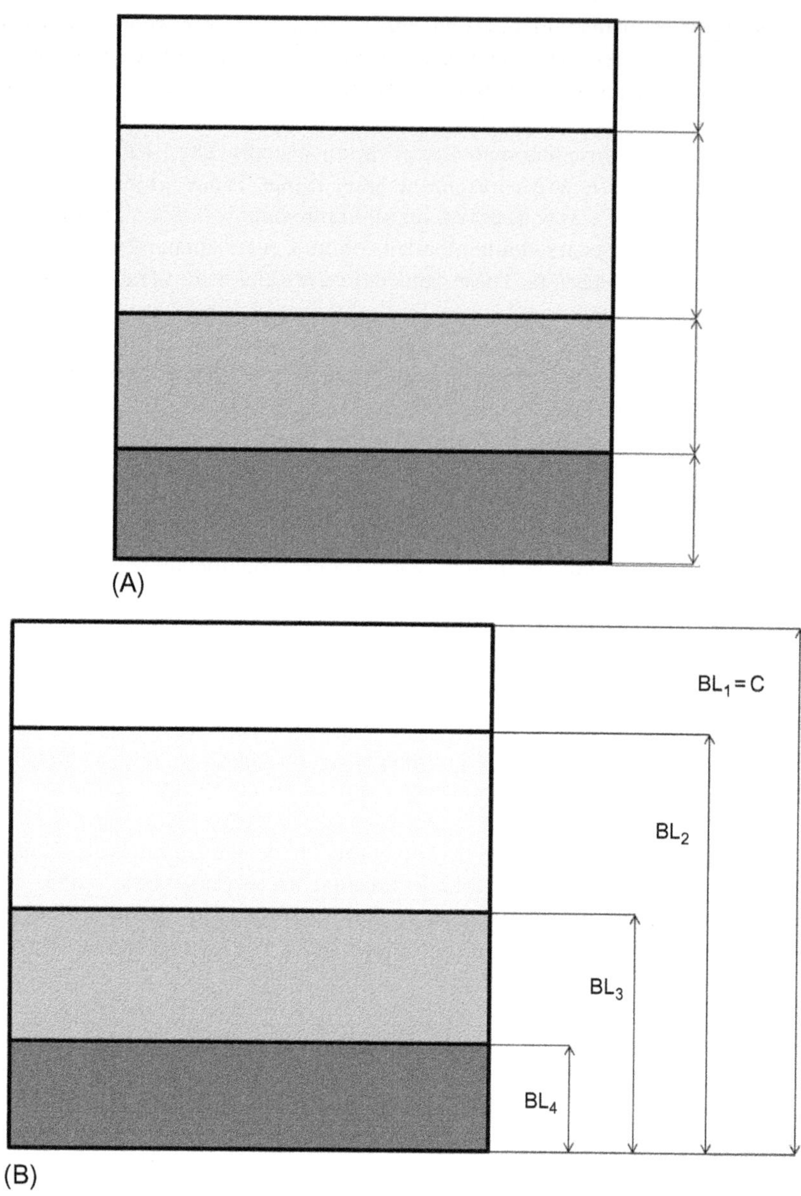

FIG. 7.80

(A) Distinct fare class inventories; (B) Nested fare class inventories.

The internet provides the information on ticket prices to millions of users. Together with e-commerce Internet has opened the new stage in revenue management concepts and practice. Internet airline ticket sales have risen quickly. New forms of airline ticket sales like On-line travel agents, air carriers' web sites, and various forms of auctions and "last minute sales" appeared on the market in the late 1990s. Using web sites of online travel agents, passengers explore various travel options. The majority of web sites are user-friendly and allow potential passengers to take into account different factors (ticket prices, departure times, number of connections, etc.) when making travel decision. Airlines already sell a significant percentage of their tickets through their own web sites. The era of dynamic airline ticket pricing has already started creating a lot of complex problems for practitioners and air transportation researchers.

7.30 PROBLEMS

1. Itemize urban transit systems, explain their main characteristics, similarities, and differences.
2. Itemize and define the main components of urban public transit systems.
3. What are the main indicators of size and topology of an urban transit network?
4. Itemize and define the main characteristics of passenger demand in urban public transit systems-volume, intensity, time pattern.
5. Define the urban transit service frequency and its estimation depending on the most influencing factors.
6. Define the passenger waiting time and costs due to service frequency. What does balancing of passenger and transport operators costs mean?
7. Define a timetable, its main characteristics, and principles of design.
8. What is transit line capacity and its utilization, and how do you estimate them both?
9. Define the main types and performances of urban transit network. How these performances can be expressed depending on type of the network?
10. What are the main characteristics of disrupted transit service networks in a given urban area?
11. Itemize and explain the main steps of planning process of urban transit system(s).
12. Itemize and explain the main characteristics of the demand-responsive transit systems.
13. What are the main characteristics of road inter-urban transport systems and their service networks?
14. The number of boarding passengers, number of alighting passengers and the number of passengers in the vehicle are given in the following table:

Bus Stop	Number of Boarding Passengers	Number of Alighting Passengers	Number of Passengers in the Vehicle
Terminal A	9	0	9
1	10	6	13
2	14	4	23
3	9	11	21
4	7	17	11
5	0	7	4
Terminal B	0	4	0

Show graphically the number of boarding passengers, number of alighting passengers and the number of passengers in the vehicle for all line sections.

15. Taking into account the data given in the previous example, show graphically passenger load profile and identify the MLS.
16. Let us assume that transit line parameters are respectively equal:
    - $c = 120$ \$ per hour
    - $v = 12$ \$ per passenger hour
    - $r = 1400$ passengers per hour
    - $T = 120$ min $= 2$ h

    Calculate the optimal headway by using the "square root formula."
17. The transit operator monitored during 10 days period the daily number of passengers that departed from the station #5. The following 10 values were recorded: 1160, 1245, 1440, 1280, 1180, 1380, 1220, 1358, 1178 and 1382. The maximum number of passengers per car is equal to 100. The transit operator would like to achieve the load factor that is equal to 0.75. Calculate the average value of the maximum daily passenger volumes. Determine the service frequency f that should be offered in order to satisfy maximum passenger volume and the desired vehicle occupancy.
18. The public transit line length equals 12 km in one direction. The average bus speed on a city heavy traffic equals 25 km/h. The total of 14 buses is assigned to the line. The capacity of every vehicle equals 60. Calculate the line capacity, frequency, and headway.
19. A public transit operator wants to provide service frequency between 6:00 and 7:00 that is equal to 4. Between 7:00 am and 8:00 am the operator wants to offer service frequency equals to 6. The operator provides service frequency that is equal to 8 in the time interval between 8:00 am and 9:00 am. Generate vehicles timetable that will provide constant vehicle headways within time interval from 6:00 am to 9:00 am. Show the solution graphically.
20. Itemize and explain the main characteristics of air transport system.
21. What are the relationships between the capacity and demand at airlines?
22. What are the main factors influencing the fleet size of an airline?
23. Specify and explain the main characteristics of the quality of services provided by an airline.
24. Explain the processes of airline scheduling, schedule planning, and yield management.

REFERENCES

Alstrup, J., Boas, S., Madsen, O.B.G., Vidal, R.V.V., 1986. Booking policy for flights with two types of passengers. Eur. J. Oper. Res. 27, 274–288.

Baaj, M.H., Mahmassani, H.S., 1992. Artificial intelligence-based system representation and search procedures for transit route network design. Transp. Res. Rec. 1358, 67–70.

Baaj, M.H., Mahmassani, H.S., 1995. Hybrid route generation heuristic algorithm for the design of transit networks. Transport. Res. C 3 (1), 31–50.

Banković, R., 1982. Javni gradski putnički prevoz. Naučna knjiga, Beograd (in Serbian).

Barnhart, C., Johnson, E.L., Nemhauser, G.L., Vance, P.H., 1999. Crew scheduling. In: Hall, R.W. (Ed.), Handbook of Transportation Science. Kluwer Academic Publisher, Norwell, MA, pp. 493–521.

Barnhart, C., Kniker, T., Lohatepanont, M., 2002. Itinerary-based airline fleet assignment. Transp. Sci. 36 (2), 199–217.

Belobaba, P.P., 1987. Airline yield management: an overview of seat inventory control. Transp. Sci. 21, 63–73.

Belobaba, P.P., 1989. Application of a probabilistic decision model to airline seat inventory control. Oper. Res. 37, 183–197.

Bielli, M., Caramia, M., Carotenuto, P., 2002. Genetic algorithms in bus network optimization. Transport. Res. C 10 (1), 19–34.

Boeing, 2011. Boeing commercial airplanes: a better way to fly, the boeing company, Seattle, Washington, USA.

Bowen, B.D., Headley, D.E., 2015. Airline quality rating 2015: the 25th year reporting airline performance. Retrieved from http://commons.erau.edu/aqrr/25.

Bowman, L.A., Turnquist, M.A., 1981. Service frequency, schedule reliability and passenger wait times at transit stops. Transport. Res. A 15 (6), 465–471.

Carter, E., Morlok, E., 1975. Planning air transport network in Appalachia. Transport. Eng. J. ASCE 101, 569–588.

Ceder, A., 1984. Bus frequency determination using passenger count data. Transp. Res. A 18 (5–6), 439–453.

Ceder, A., 1986. Methods for creating bus timetables. Transport. Res. A 21 (1), 59–83.

Ceder, A., 2015. Public Transit Planning and Operation: Modeling, Practice and Behavior, second ed. CRC Press, Taylor & Francis Group, London, UK.

Ceder, A., Wilson, N.H.M., 1986. Bus network design. Transport. Res. B 20 (4), 331–344.

Ceder, A., Golany, B., Tal, O., 2001. Creating bus timetables with maximal synchronization. Transport. Res. A 35, 913–928.

Chakroborty, P., 2003. Genetic algorithms for optimal urban transit network design. Comput. Aided Civil Infrastruct. Eng. 18, 184–200.

Charkroborty, P., Dwivedi, T., 2002. Optimal route network design for transit systems using genetic algorithms. Eng. Optim. 34 (1), 83–100.

Chou, Y.H., 1990. The hierarchical-hub model for airline networks. Transp. Plan. Technol. 14, 243–258.

Cross, R., 1977. Revenue Management. Broadway Books, New York.

Daduna, J.R., Paixão, J.M.P., 1995. Vehicle scheduling for public mass transit—an overview. In: Daduna, J.R., Branco, I., Paixão, J.M.P. (Eds.), Computer-Aided Transit Scheduling. In: Lectures Notes in Economics and Mathematical Systems, vol. 430. Springer-Verlag, Heidelberg, pp. 76–90.

Desaulniers, G., Hickman, M.D., 2007. Public transit. In: Barnhart, C., Laporte, G. (Eds.), Handbook in OR & MS, vol. 14. Elsevier, New York, pp. 69–127.

Dessouky, M., Hall, R., Zhang, L., Singh, A., 2003. Real-time control of buses for schedule coordination at a terminal. Transport. Res. A 37, 145–164.

Eberlein, X.-J., Wilson, N.H.M., Bernstein, D., 2001. The holding problem with real-time information available. Transp. Sci. 35 (1), 1–18.

ECMT, 2001. Annual report 2000, European conference of ministers of transport. Paris Cedex, France.

EU, 2011. Buses with high level of service: fundamental characteristics and recommendations for decision-making and research—results from 35 European Cities, Final report—COST (COoperation in Science and Technology). Action TU0603, RTD Framework Programme, Brussels, Belgium

Fan, W., Machemehl, R.B., 2006. Optimal transit route network design problem with variable transit demand: genetic algorithm approach. J. Transport. Eng. ASCE 132, 40–51.

Fan, W., Machemehl, R.B., 2008. Tabu search strategies for the public transportation network optimizations with variable transit demand. Comput. Aided Civil Infrastruct. Eng. 23, 502–520.

Furth, P.G., Wilson, N.H.M., 1982. Setting frequencies on bus routes: theory and practice. Transp. Res. Rec. 818, 1–7.

Ghafouri, M., Lam, T., 1986. Accessibility in the deregulated domestic airline network. Transp. Res. Rec. 1094, 10–17.

Ghobrial, A., 1983. Analysis of the air network structure: the hubbing phenomenon. Ph.D. thesis, University of California, Berkeley.

Gordon, S., Neufville, R., 1973. Design of air transportation networks. Transp. Res. 7, 207–222.

Guan, J.F., Yang, H., Wirasinghe, S.C., 2003. Simultaneous optimization of transit line configuration and passenger line assignment. Transp. Res. B 40 (10), 885–902.

Haase, K., Desaulniers, G., Desrosiers, J., 2001. Simultaneous vehicle and crew scheduling in urban mass transit systems. Transp. Sci. 35 (3), 286–303.

Hickman, M., 2001. An analytic stochastic model for the transit vehicle holding problem. Transp. Sci. 35 (3), 215–237.

Israeli, Y., Ceder, A., 1989. Designing transit routes at the network level. In: Proceedings of the First Vehicle Navigation and Information Systems Conference. IEEE Vehicular Technology Society, pp. 310–316.

Israeli, Y., Ceder, A., 1995. Transit route design using scheduling and multiobjective programming techniques. In: Daduna, J.R., Branco, I., Piaxão, J. (Eds.), Computer-Aided Transit Scheduling. Lecture Notes in Economics and Mathematical Systems, vol. 430. Springer-Verlag, Heidelberg, pp. 56–75.

Janić, M., 2000. Air transport system analysis and modeling: capacity, quality of services and economics. In: Transportation Studies, vol. 16. Gordon & Breech, Amsterdam, The Netherlands.

Janić, M., 2014. Advanced Transport Systems: Analysis, Modelling, and Evaluation of Performances. Springer-Verlag, London, UK.

Jaw, J., Odoni, A., Psaraftis, H., Wilson, N., 1986. A heuristic algorithm for the multi-vehicle advance-request dial-a-ride problem with time windows. Transport. Res. 20B, 243–257.

Kanafani, A., 1981. Aircraft technology and network structure in short-haul air transportation. Transp. Res. 15A, 305–314.

Kanafani, A., 1983. Transportation Demand Analysis. McGraw-Hill, New York.

Kanafani, A., Ghobrial, A., 1982. Aircraft evaluation in air network planning. Transport. Eng. J. ASCE 108, 282–300.

Kanafani, A., Ghobrial, A., 1985. Airline hubbing-some implications for airport economics. Transp. Res. 19A, 15–27.

Kikuchi, S., Vuchic, V., 1982. Transit vehicle stopping regimes and spacings. Transp. Sci. 16, 311–331.

Lampkin, W., Saalmans, P.D., 1967. The design of routes, service frequencies, and schedules for a municipal bus undertaking: a case study. Oper. Res. Q. 18 (4), 375–397.

Levinson, S.H., Zimmerman, S., Clinger, J., Gast, J., Rutherford, S., Smith, L.R., Cracknell, J., Soberman, R., 2003. Bus rapid transit, volume 1: case studies in bus rapid transit, Report, TCRP (Transit Cooperative Research Program), TRB (Transportation Research Board), Washington, D.C., USA.

Littlewood, K., 1972. Forecasting and control of passenger bookings. In: Proceedings of the XII AGIFORS Symposium, Nathanya, Israel.

Lohatepanont, M., Barnhart, C., 2004. Airline schedule planning: integrated models and algorithms for schedule design and fleet assignment. Transp. Sci. 38 (1), 19–32.

Lufthansa, 2014. First choice—annual report. Deutsche, Lufthansa, Cologne, Germany.

Mandl, C.E., 1979. Evaluation and optimization of urban public transportation network. Eur. J. Oper. Res. 5, 396–404.

Martínez, H., Mauttone, A., Urquhart, M.E., 2014. Frequency optimization in public transportation systems: formulation and metaheuristic approach. Eur. J. Oper. Res 236, 27–36.

Mauttone, A., Urquhart, M.E., 2009. A route set construction algorithm for the transit network design problem. Comput. Oper. Res. 36, 2440–2449.

Nagarajan, K.V., 1979. On an auction solution to the problem of airline overbooking. Transp. Res. 13A, 111–114.

Newell, G., 1979. Some issues relating to the optimal design of bus routes. Transp. Sci. 13, 20–35.

Nikolić, M., Teodorovic, D., 2013. Transit network design by Bee Colony Optimization. Expert Syst. Appl. 40, 5945–5955.

Pattnaik, S.B., Mohan, S., Tom, V.M., 1998. Urban bus transit network design using genetic algorithm. J. Transp. Eng. 124 (4), 368–375.

Pollack, M., 1982. Airline route-frequency planning: some design trade-offs. Transp. Res. 16A, 149–159.

Rothstein, M., 1971. Airline overbooking: the state of the art. J. Trans. Econ. Policy 5, 96–99.
Rothstein, M., 1975. Airline overbooking: fresh approaches are needed. Transp. Sci. 9, 169–173.
Rothstein, M., 1985. OR and the airline overbooking problem. Oper. Res. 33, 237–248.
Ryanair, 2014. Annual report 2014, http://www.ryanair.com/.
SA, 2015. Southwest Airlines fleet details and history, Planespotters.net Just Aviation, planespotters.net, Retrieved Aug 19, 2015.
Salzborn, F.J.M., 1972. Optimum bus scheduling. Transp. Sci. 6 (2), 137–148.
SCANIA, 2009. Annual report 2008: pride quality driver appeal productivity sustainable development. Södertälje, Sweeden. www.scania.com.
Smith, B.M., Wren, A., 1988. A bus crew scheduling system using a set covering formulation. Transport. Res. A 22, 97–108.
Soumis, F., Ferland, J.A., Rousseau, J.M., 1981. MAPUM: a model for assigning passengers to a flight schedule. Transp. Res. 15A, 155–162.
Swan, W.M., 1979. A systems analysis of scheduled air transportation networks. Report FTL-R79-5, MIT, Cambridge, MA.
Teodorović, D., 1983. Flight frequency determination. J. Transp. Eng. 109, 747–757.
Teodorović, D., 1985. Model for designing meteorologically most reliable airline schedule. Eur. J. Oper. Res. 21, 156–165.
Teodorović, D., 1988. Airline Operations Research. Gordon and Breach Science Publishers, New York.
Teodorović, D., Guberinić, S., 1984. Optimal dispatching strategy on an airline network after a schedule perturbation. Eur. J. Oper. Res. 15, 178–183.
Teodorović, D., Kalić, M., 1995. A fuzzy route choice model for air transportation networks. Transp. Plan. Technol. 19, 109–119.
Teodorović, D., Krčmar-Nožic, E., 1989. Multicriteria model to determine flight frequencies on an airline network under competitive conditions. Transp. Sci. 24, 14–27.
Teodorović, G., Stojković, G., 1995. A model to reduce airline schedule disturbances. J. Transp. Eng. 121, 324–331.
Teodorović, D., Kalić, M., Pavković, G., 1994. The potential for using fuzzy set theory in airline network design. Transp. Res. 28B, 103–121.
Van de Velde, D., 2009. Long-distance bus services in europe: concessions or free market?, International Transport Forum—OCDE/ITF, Discussion Paper No. 2009-21, Dec 2009.
Vickrey, W., 1972. Airline overbooking: some further solutions. J. Trans. Econ. Policy 6, 257–270.
Vuchic, V., 1981. Urban Public Transportation: Systems and Technology. Prentice Hall Inc, Englewood Cliffs, New Jersey. IUSA NC.
Vuchic, V., 2005. Urban Transit: Operations, Planning, and Economics. John Willey & Sons, Inc., Hoboken, New Jersey, USA.
Wren, A., Rousseau, J.-M., 1995. Bus driver scheduling—an overview. In: Daduna, J.R., Branco, I., Paixão, J.M.P. (Eds.), Computer-Aided Transit Scheduling. In: Lecture Notes in Economics and Mathematical Systems, vol. 430. Springer-Verlag, Heidelberg, pp. 173–187.
Zhao, F., Zeng, X.G., 2007. Optimization of transit route network, vehicle headways and timetables for large-scale transit networks. Eur. J. Oper. Res. 186, 841–855.

How to predict the total number of trips generated in specific urban zones? How to predict the traffic load at specific links in urban transportation network? What is Four step planning procedure? What is Gravity model? How do travelers choose transportation mode? How do drivers choose route? When traffic network is in user equilibrium conditions? What are Wardrop's principles? What is Logit model? What are Activity-based travel demand models?

TRANSPORTATION DEMAND ANALYSIS

8

8.1 INTRODUCTION

People travel for various reasons. There is always distance between trip origin and trip destination, and distance between goods supply and goods demand. This distance should be beat at a transportation cost. Passenger and freight flows are the consequences of spatial interaction between various regions. Various socioeconomic activities in a society, as well as land uses, highly influence transportation demand. Highways, streets, airports, ports, etc. should be capable of meeting transportation demand and providing an adequate level of service to users. Unfortunately, in many countries in the world, transportation network capacities have not kept up with the raise in travel demand. Consequently, many urban road networks and many airports are severely congested resulting in increased travel times, increased transportation costs, and increased air pollution. There are also quite opposite examples when transportation system experiences overdesign (transportation capacity goes beyond transportation demand). Such transportation systems are economically inefficient.

Vital decisions concerning the design and improvement of transportation infrastructure (highways, airports, sea ports, railway stations, freight terminals, public transit networks, network of airways, etc.), as well as matching transportation demand and transportation supply should be made with the help of various transportation planning techniques. In the first step of the transport planning process, it is necessary to define precisely the considered transportation problem (building a new highway, expansion of

Transportation Engineering. http://dx.doi.org/10.1016/B978-0-12-803818-5.00008-1

the existing airport, construction of a new port, etc.). In the next steps, engineers and planners define the goals they want to achieve, generate possible alternatives and perform their evaluation (Steenbrink, 1974; Kanafani, 1983; Sheffi, 1985; de Ortuzar and Willumsen, 1990).

8.2 TRANSPORTATION DEMAND AND TRANSPORTATION SUPPLY

The yearly number of passengers between city pairs, the daily number of passengers between city zones, and the daily number of requests for landing represent some of the input data for the transportation facilities planning (expansion and reconstruction). By monitoring the number of vehicles driven on different network links, or passengers flown on different routes during a day, week and month, certain patterns are noted that characterize the demand for transportation services. In other words, transportation flows change over time. Changes are noticed by month, by week, by day in a week, and finally by hour in a day. Fig. 8.1 shows typical flow changes over time of day on many links in urban transportation networks. Transportation flows changes over time can be determined by collecting appropriate statistical data, as well as by conducting passenger surveys. Demand could be expressed in various ways. For example, demand could be expressed as the number of drivers wanted to drive between certain city zones in a unit of time. Demand could also be expressed, for example, as a number of tons of freight that should be transported in a unit of time (day, week, month) between two regions, as a number of truck loads per week, or as the number of aircraft requesting landing during certain time period.

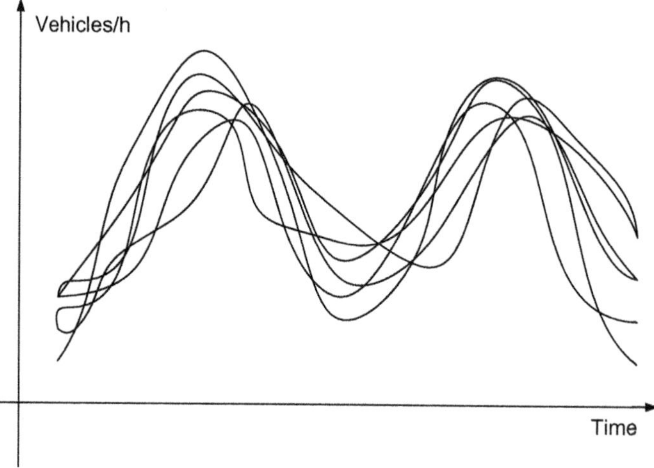

FIG. 8.1

Traffic flow changes over hours during a day.

From the point of view of a highway operator, transportation supply could be treated as a set of facilities available to highway users (entrances, ramps, exits, lanes, message signs). From the point of view of a bus operator, transportation supply represents the number of seats in the buses provided to the passengers at certain time of a day. In a similar way, transportation supply could be treated from the point of view of Air Traffic Control, as a set of facilities available to airspace users (airports, runways, gates, fixes). Various transportation service attributes (non-physical attributes) like price for using the road, metro ticket price, bus schedule, or signal timing should be also considered as the elements of the transportation supply.

The level of service offered to users highly depends on the relationship between the demand and the supply. For example, if a bus operator's supply is inadequate (small number of departures, small number of offered seats) certain number of passengers will not get a vacant seat on the bus during the proffered time period (Fig. 8.2).

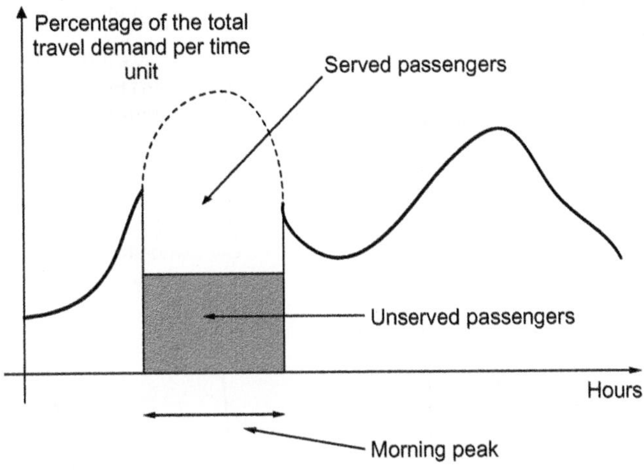

FIG. 8.2

Inadequate supply of the bus operator.

The supply of transportation services has also an essential effect on transportation demand. For example, by increasing bus frequency, better adjusting the bus schedule to passenger requests, reducing tariffs, or some other promotional measure, public transportation demand can be increased to a certain extent and thereby the number of passengers transported on certain routes. In the same vein, a decline in demand is often caused by an inadequate supply.

8.3 TRANSPORTATION DEMAND MODELING

The prediction of travel demand constitutes an essential ingredient in transportation planning. All transportation planners try to predict travel demand as precisely as possible. The main goal of the transportation demand analysts is to develop user friendly, accurate, and reliable models that can provide various information to planners, traffic engineers, and other decision makers (Smeed, 1968; Robertson, 1969; Ben-Akiva, 1974; Florian and Nguyen, 1978; Gartner, 1980; Jansen and Bovy, 1982; Fisk, 1988; Safwat and Magnanti, 1988; Goncalves and Uysseaneto, 1993; Teodorović and Vukadinović, 1998; Teodorović and Trani, 2002; Hensher et al., 2005; Jara-Díaz, 2007). For example, the following questions should be answered in the urban planning context:

- What is the total number of trips generated in specific urban zones?
- What is the number of trips between all urban pair zones?
- What is the traffic load at specific links in urban transportation network?
- What is the number of passenger kilometers that can be achieved?

When traveling, passengers and drivers individually decide on transportation mode, departure time, route choice, etc. This means that we should try to describe, analyze, and predict choice behavior of the individual passengers and drivers. When doing this, we should take into account individual traveler's experience, age, behavior, etc. Most frequently, due to the lack of data, high survey costs, and computational complexity, it is not possible to perform transportation analysis on the individual or household level. Transportation planners usually divide observed region into geographical entities of, more or less, uniform socio-economic characteristics. These entities are usually called transportation analysis zones. Fig. 8.3 shows map with a three zones—*A*, *B*, and *C*. Each zone is symbolized by a *centroid*. The main activities of the zone are usually spatially concentrated in centroid. Centroids generate and attract traffic. The centroids that generate traffic flows are called *origin*, and the centroids that attract traffic flows are called *destination*.

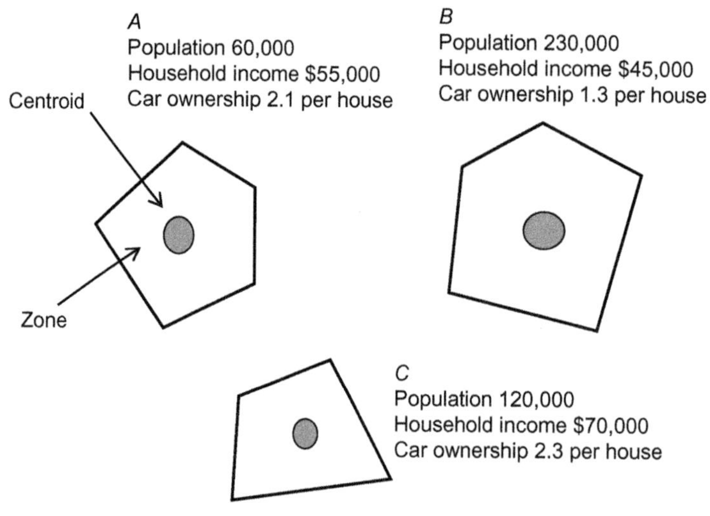

FIG. 8.3

Zones *A*, *B*, and *C*.

The following is trip interchange matrix (Table 8.1) in the case of zones *A*, *B*, and *C*:

Table 8.1 Trip interchange matrix in the case of zones *A*, *B*, and *C*

Origin zone/Destination Zone	A	B	C
A	–	*A-B*	*A-C*
B	*B-A*	–	*B-C*
C	*C-A*	*C-B*	–

The Origin-Destination matrix (*O-D* matrix) contains information about traffic flow values between all pairs of centroids (Cascetta, 1984; Cascetta and Nguyen, 1988). Origin-Destination tables that contain information about traffic flows during short period of time are usually unbalanced (the total traffic from the centroid is not equal to the total traffic to the centroid). For example, if we study traffic flows between 7:00 am and 9:00 am, the total number of trips from the centroid that represent residential area, will be significantly higher than the total number of trips coming to the residential area. The situation is quite opposite in the time period between 5:00 pm and 7:00 pm When we study traffic flows during whole day, the total number of trips from the centroid is approximately equal to the total traffic to the centroid that represents residential area. In this case, O-D matrix would be balanced.

In the case of example with three zones, we are interested in estimating the total number of trips generated in the zones A, B, and C. We are also interested in estimating current, as well as in predicting future number of trips between A and B, A and C, and B and C. Our final goal could be, for example, to accurately estimate current and predict future number of trips along various roads that connect zones A, B, and C. In order to make prediction of the number of trips, transportation planners and traffic engineers develop various transportation planning models. The following is one of the first questions that analysts face: what drives the number of trips? Obviously, the number of trips is highly influenced by the number of persons per household, number of cars per household, income levels, road infrastructure density (lane-km or road per square kilometer), as well as many other factors. Many of the transportation demand models establish the relationships between the number of trips, and all other relevant factors that influence the number of trips.

Macroscopic transportation demand models are used to estimate the development of transportation in a certain country or region. These models are used to estimate the number of trips in urban transportation, the number of passenger kilometers, or the number of passengers in air transportation, and the number of aircraft operations. The number of passenger kilometers is arrived at by multiplying the number of trips by the length of each trip. The number of passenger kilometers in an indicator of effectuated transportation.

Microscopic models estimate the number of passengers along a specific route when there are several different routes, the number of passengers in each transportation mode when there are various transportation modes, or the number of passengers in each class when there are various tariffs on a route.

8.4 TRANSPORTATION DEMAND FORECASTING TECHNIQUES

In general, transportation demand could have trend component, cyclical component, seasonal peaks, and random variations (Fig. 8.4). A trend represents permanent, general increasing, or general decreasing pattern. A pattern could be caused by various factors. The seasonal component represents usual pattern of up and down oscillations. For example, oscillations happen inside a single year on many touristic routes in air transportation. Cyclical component represents repeating up and down oscillations. Cyclical component could be also caused by various reasons. Random variations happen due to unpredicted events. These variations are usually short and they do not repeat.

Forecasting techniques for prediction of future events are based on information from the past. Consequently, no forecast is any better than the information included in the past data. The analyst usually has at his/her disposal a set of evenly spaced numerical data. These data are collected by observing the demand at regular time periods (day, month, year).

Experience gained in prediction of various phenomena has shown that all forecasts are always incorrect. The fundamental question is how big the forecast error is. Any forecast is more precise for shorter than longer time periods.

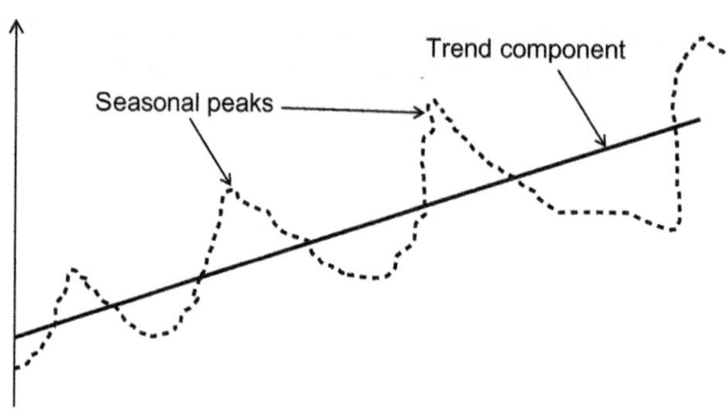

FIG. 8.4

Components of transportation demand.

There are qualitative and quantitative forecasting techniques (Fig. 8.5).

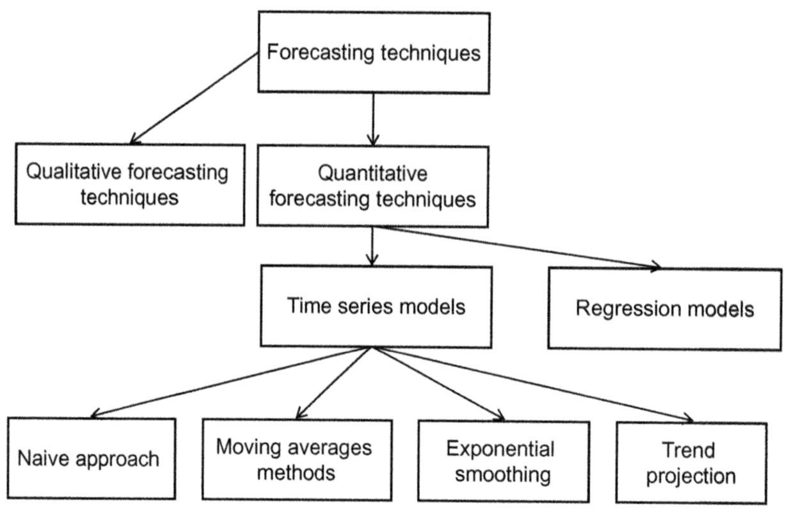

FIG. 8.5

Forecasting techniques.

Qualitative methods are based on the intuition and experience of the analysts. These methods are most frequently used when not much data exist. The Delphi method is one of the widely used qualitative methods. Within the Delphi method, a group of experts is queried iteratively until the agreement is reached.

There are two main classes of quantitative forecasting techniques:

- the extrapolation method (time series models) and the explanatory method (regression models).

Within the extrapolation method, future transportation demand is predicted having as a starting point the past characteristics of the demand over a period of time. When using the explanatory models, the future transportation demand is predicted on the basis of identified and quantifiable factors that influence the transportation demand (socio-economic characteristics, and characteristics of the transportation system). The extrapolation method is usually more difficult to apply and validate than the explanatory method. The extrapolation method can be primarily used to predict the future transportation demand under unchanged economic and social environment. Causal models (Regression models) are very convenient tools that enable us to explore various "what-if" scenarios.

Time series models include the naive approach, moving averages, exponential smoothing, and trend projection.

The naive approach is based on the assumption that demand in the next time period is identical to demand in most recent time period.

8.4.1 TIME SERIES MODELS

Time series forecasting is based exclusively on data from the past. The basic assumption in time series forecasting is that the dominant factors from the past will maintain influence in the future.

The moving average model utilizes previous t periods in order to predict demand in the time period $t+1$. Analysts use simple moving average and weighted moving average method. The forecast value in the simple moving average method equals:

$$F_{t+1} = \frac{A_t + A_{t-1} + \cdots + A_{t-n}}{n} \quad (8.1)$$

where:

t is the present time period;
F_{t+1} is the forecast for the subsequent period;
N is the forecasting horizon; and
A_i is the actual demand in the ith time period.

As we can see, the forecast, within the moving average method, represents linear combination of the four demand from the past time periods.

EXAMPLE 8.1

The data on passenger traffic at Belgrade Airport, Serbia, from 1970 to 1977 are given in Table 8.2.

Predict the number of passengers in 1975 by using a 3-year moving average.

Solution

$$F_{1975} = \frac{A_{1974} + A_{1973} + A_{1972}}{3}$$

$$F_{1975} = \frac{1,688,247 + 1,434,454 + 1,155,166}{3}$$

$$F_{1975} = 1,425,955$$

(Continued)

EXAMPLE 8.1—cont'd

Table 8.2 Passenger Traffic at Belgrade Airport, Serbia, 1970–77

Year	Passenger Traffic at Belgrade Airport
1970	838,156
1971	1,036,311
1972	1,155,166
1973	1,434,454
1974	1,688,247
1975	2,020,291
1976	2,047,016
1977	2,280,972

The weighted moving average method allows analyst to give more importance to some of the historical data. The forecast value in the weighted moving average method equals:

$$F_{t+1} = w_t \times A_t + w_{t-1} \times A_{t-1} + \cdots + w_{t-n} \times A_{t-n} \quad (8.2)$$

$$w_t + w_{t-1} + \cdots + w_{t-n} = 1 \quad (8.3)$$

where:

t is the present time period;
F_{t+1} is the forecast for the subsequent period;
n is the forecasting horizon;
A_i is the actual demand in the ith time period; and
w_i is the importance (weight) the analyst gives to the ith time period.

EXAMPLE 8.2

Predict the number of passengers in 1975 at Belgrade Airport, by using the weighted moving average method. Use a 3-year moving average, and the following set of weights:

$$w_{1974} = 0.6,\ w_{1973} = 0.3,\ w_{1972} = 0.1$$

Solution

$$F_{1975} = w_{1974} \times A_{1974} + w_{1973} \times A_{1973} + w_{1972} \times A_{1972} + w_{1971} \times A_{1971}$$

$$F_{1975} = 0.6 \times 1,688,247 + 0.3 \times 1,434,454 + 0.1 \times 1,155,166$$

$$F_{1975} = 1,558,800$$

The actual number of passengers in 1975 was 2,020,291. The simple moving average method predicted 1,425,955 passengers, while the weighted moving average method forecasted 1,558,800 passengers. We decreased the forecast error by giving more importance to the events that happened recently.

The weighted moving average method is more useful for the analysis of the transportation demand because of the capability to modify the time periods' weights. An analyst can determine the set of weights based on the significance that she/he thinks that the data from the past have.

Exponential smoothing is based on the idea of assigning higher weights to more recent observations than to the observations from the far-away past. In other words, in the case of exponential smoothing, the weights decrease exponentially with the past. In this way, the lowest weights are given to the most distant observations. The forecast value in the exponential smoothing equals:

$$F_{t+1} = \alpha \times A_t + (1-\alpha) \times F_t \tag{8.4}$$

ie:

$$F_{t+1} = F_t + \alpha \times (A_t - F_t) \tag{8.5}$$

$$F_{t+1} = F_t + \alpha \times e_t$$

where:

- t is the present time period;
- F_t is the forecast for the period t;
- A_i is the actual demand in the ith time period;
- e_t is the forecast error for the ith time period; and
- α is the smoothing constant.

Exponential smoothing is based on the idea that forecast should depend on the latest observation, as well as on the forecast error of the latest forecast. As we can see from Eq. (8.5), the new forecast is equal to the old forecast, plus the correction for the error that happened in the last forecast. The smoothing constant α articulates analyst's reaction to forecast error. The lower the α value, the lower reaction to forecast error. The higher the α value, the higher analyst's reaction to the difference between actual and forecasted value.

EXAMPLE 8.3

Predict the number of passengers at Belgrade Airport, by using the exponential smoothing method. The actual numbers of passengers are given in Table 8.3. Use the smoothing constant $\alpha = 0.8$. Assume that $F_1 = A_1$.

Table 8.3 The Number of Passengers at Belgrade Airport (1970–77)

Year, t	Actual Number of Passengers, A_t
1 (1970)	838,156
2 (1971)	1,036,311
3 (1972)	1,155,166
4 (1973)	1,434,454
5 (1974)	1,688,247
6 (1975)	2,020,291
7 (1976)	2,047,016
8 (1977)	2,280,972

(Continued)

EXAMPLE 8.3—cont'd

Solution

Since $F_1 = A_1$, we get:

$$F_1 = A_1 = 838,156$$

The first forecast error e_1 equals:

$$e_1 = A_1 - F_1 = 838,156 - 838,156 = 0$$

The second forecasted value F_2 equals:

$$F_2 = F_1 + \alpha \times e_1 = 838,156 + 0.8 \cdot 0 = 838,156$$

The second forecast error e_2 equals:

$$e_2 = A_2 - F_2 = 1,036,311 - 838,156 = 198,155$$

The third forecasted value F_3 equals:

$$F_3 = F_2 + \alpha \times e_2 = 838,156 + 0.8 \cdot 198,155 = 996,680$$

The third forecast error e_3 equals:

$$e_3 = A_3 - F_3 = 1,155,166 - 996,680 = 158,486$$

The fourth forecasted value F_4 equals:

$$F_4 = F_3 + \alpha \times e_3 = 996,680 + 0.8 \times 158,486 = 1,123,468$$

All forecasted values and all forecast errors are given in Table 8.4.

Table 8.4 Forecasted Values of the Number of Passengers at Belgrade Airport

Year, t	Actual Number of Passengers, A_t	Forecasted Number of Passengers, F_t	Forecast Error, $e_t = A_t - F_t$
1 (1970)	838,156	838,156	0
2 (1971)	1,036,311	838,156	198,155
3 (1972)	1,155,166	996,680	158,486
4 (1973)	1,434,454	1,123,468	310,985
5 (1974)	1,688,247	1,372,256	315,991
6 (1975)	2,020,291	1,625,048	395,242
7 (1976)	2,047,016	1,941,241	105.774
8 (1977)	2,280,972	2,025,860	255,111

8.4.2 TREND PROJECTION

A trend should be understood as a tendency of data to increase or decrease gradually over time. Long-term trend is most frequently modeled as a linear, quadratic or exponential function. An extrapolation is simply made of trends noted by the number of passengers transported, or the number of passenger kilometers effectuated, and a forecast on this basis is made of the number of passengers, or passenger kilometers in the next period. This is known as the independent estimation method, since factors that affect the number of passengers are not taken into consideration when estimating the number of users.

An increase in the number of users is an essential property of many transportation systems. We denote time by t, and the number of trips that changes over time by $D(t)$. We also denote by $F(t)$ the forecast for the transportation demand $D(t)$ at some future time t.

The most frequently used trend models of the transportation demand in a period t are given in Table 8.5.

Table 8.5 The Most Frequently Used Trend Models of the Transportation Demand

Model Name	Mathematical Description
Linear trend	$F(t) = a + b \times t$
Exponential curve	$F(t) = a \times b^t$
Modified exponential curve	$F(t) = k + a \times b^t$
Gompertz curve	$F(t) = k \times a^{b^t}$
Logistic function	$F(t) = \frac{k}{1 + b \times e^{-at}}$

In the case of linear trend, it is assumed that the growth in demand for transportation is constant within a unit of time (Fig. 8.6). Model calibration, ie, estimation of the parameters a, and b is made using the least squares method.

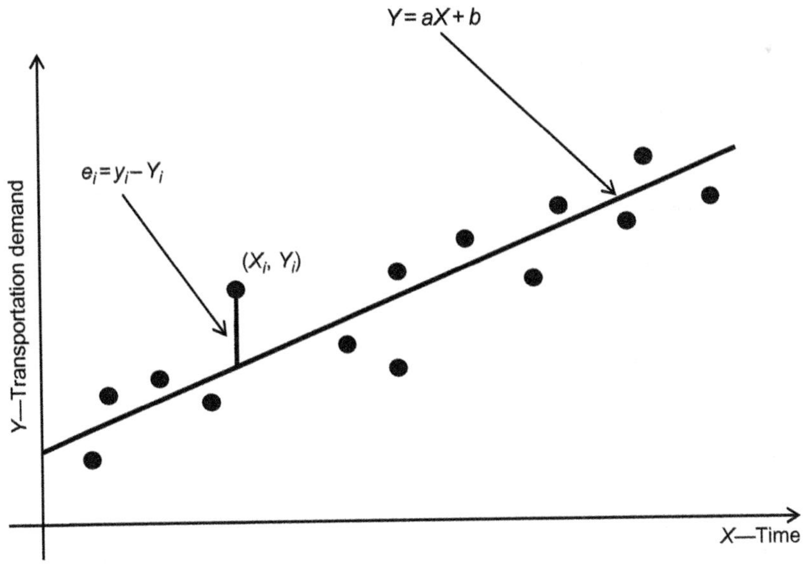

FIG. 8.6

Linear trend model.

The parameters a and b are determined to best fit a data set. In other words, by the least squares method we try to minimize the distance between the line and the points. Let us assume that the data set consists of n data pairs $(x_t, y_t),\ t=1,2,3,\ldots,n$ where x_t is an independent variable (time) and y_t is dependent variable (transportation demand). In the case of linear trend, the forecasted line is:

$$Y = a \times X + b \tag{8.6}$$

The difference e_i between the actual value of the transportation demand y_i and the value of the transportation demand predicted by the linear model Y_i (the ith error equals):

$$e_i = y_i - Y_i \tag{8.7}$$

where:

e_i is the ith error;
y_i is ith the actual value; and
Y_i is ith the predicted value.

The sum of squared errors S equals:

$$S = \sum_{i=1}^{n} (e_i)^2 = \sum_{i=1}^{n} (y_i - Y_i)^2 \tag{8.8}$$

$$S = \sum_{i=1}^{n} (e_i)^2 = \sum_{i=1}^{n} (y_i - a \times X_i - b)^2 \tag{8.9}$$

The optimal values of the parameters a and b are obtained when the sum S is minimal. The minimum value of the sum S is obtained by setting the gradient to zero, ie:

$$\frac{\partial s}{\partial a} = \sum_{i=1}^{n} 2 \times (y_i - a \times X_i - b) \times (-X_i) = 0 \tag{8.10}$$

$$\frac{\partial s}{\partial b} = \sum_{i=1}^{n} 2 \times (y_i - a \times X_i - b) \times 1 = 0 \tag{8.11}$$

After solving Eqs. ()–() the following is obtained:

$$a = \frac{\sum_{i=1}^{n} (x_i - \bar{x}) \times (y_i - \bar{y})}{\sum_{i=1}^{n} (x_i - \bar{x})^2} \tag{8.12}$$

$$b = \frac{\sum_{i=1}^{n} y_i}{n} - a \times \frac{\sum_{i=1}^{n} x_i}{n} \tag{8.13}$$

where:

$$\bar{x} = \frac{\sum_{i=1}^{n} x_i}{n}$$

$$\bar{y} = \frac{\sum_{i=1}^{n} y_i}{n}$$

A nonlinear trend model is shown in Fig. 8.7. Growth phenomena in transportation can be successfully described by the Gompertz curve, as well as by the logistic curve (Fig. 8.6). Many growth phenomena in transportation show an "S" shaped pattern. Initially the growth is slow. In the second stage growth speeds up, and finally in the third stage growth slows down and approaches a limit. As t increases, the populations approaches k. The logistic curve is typically S-shaped, or sigmoidal. A typical logistic curve is shown in Fig. 8.8.

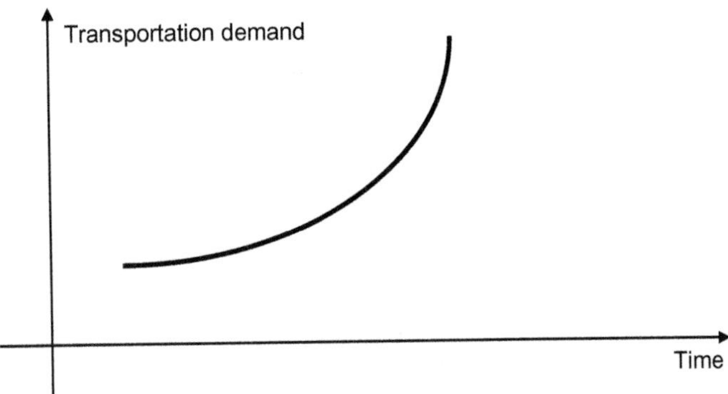

FIG. 8.7

Nonlinear trend model.

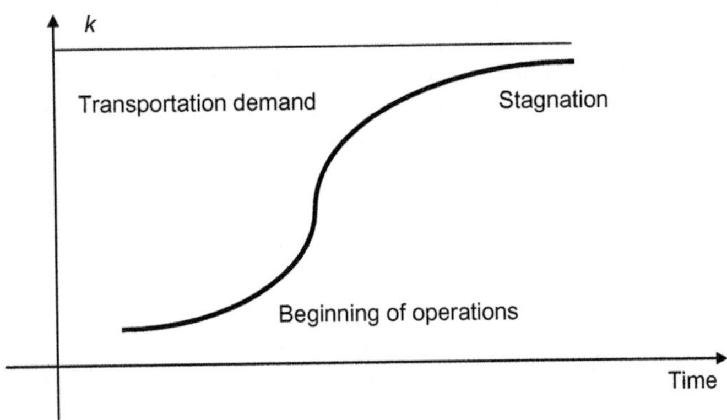

FIG. 8.8

Logistic curve.

The total passenger traffic at Belgrade Airport, Serbia, has been described as a function of time using a logistic curve (Grilihes, 1977). Data on passenger traffic from 1962 to 1978 are given in Table 8.6.

The following logistic equation was obtained:

$$F(t) = \frac{9,023,394}{1 + 39.88e^{-0.176t}} \tag{8.14}$$

Table 8.6 Data on Passenger Traffic at Belgrade Airport, Serbia, 1962–78

Year	Number of Passengers
1962	220,726
1963	282,873
1964	329,619
1965	405,191
1966	335,999
1967	399,066
1968	462,919
1969	602,257
1970	838,156
1971	1,036,311
1972	1,155,166
1973	1,434,454
1974	1,688,247
1975	2,020,291
1976	2,047,016
1977	2,280,972

Logistic curve could be very convenient tool to describe number of tourists, as well as number of air passenger demand on touristic routes. Lundtorp (www2.dst.dk/internet/4thforum/docs/c2-5.doc) developed the tourism destination life cycle model based on logistic curve. Using statistical data related to the Danish package tours to Portugal in the period 1976–94, Lundtorp showed that the following logistic curve explain these touristic trips:

$$AP = 13{,}000 + \frac{52{,}000}{1 + e^{-0.54(t-1986)}} \tag{8.15}$$

where AP represents number of Danish air passengers to Portugal.

Prediction of airport growth could be successfully done using logistic curve. It it reasonable to assume that the population of passengers will not grow without end. As in any other market or environment, there is always eventually certain airport "saturation" level. As many natural populations compete for scarce food resources or living space, and as many new technologies compete in a limited market, in the same way various destinations compete for travelers. Even if airline tickets become very cheap, nobody will travel to Miami for vacation seven times per month. In other words, there is always a limit (saturation level) to how many times a passenger wants to travel to specific destination.

The Mean Forecast Error and the Mean Absolute Deviation are usual measure for measuring the accuracy of the forecast. The Mean Forecast Error (MFE) represents the average error in the observations. The MFE equals:

$$MFE = \frac{\sum_{i=1}^{n}(A_i - F_i)}{n} \tag{8.16}$$

The Mean Absolute Deviation (MAD) represents the average absolute error in the observations. The MAD equals:

$$MAD = \frac{\sum_{i=1}^{n} |A_i - F_i|}{n} \tag{8.17}$$

The smaller the *MFE* and *MAD* values, the better the forecast.

8.5 FOUR-STEP PLANNING PROCEDURE

Many transportation agencies, consultants, analysts, and researchers used approach known as the "four-step process" when performing transportation demand analysis. The four-step process consists of the following four procedures:

- trip generation;
- trip distribution;
- modal split; and
- route choice.

The main task within the first step is to determine the number of passengers that will happen in the city zone or region that is observed (Fig. 8.9). The first planning step is called trip generation, and it estimates the total number of passengers from each of the considered regions. Different trip purposes could be modeled (business trips, or leisure trips).

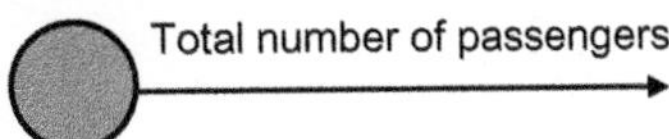

FIG. 8.9

Trip generation.

Within the second planning step (trip distribution), we try to estimate the number of trips between particular regions. For example, the number of leisure trips produced by the New York region is matched with leisure trip attractions through the U.S. The outputs are estimated numbers of leisure trips between New York region and Miami region, between New York region and San Francisco region, etc. (Fig. 8.10).

Within the third planning step, we try to answer the following question: which transportation mode will passengers use when traveling to their destinations? The third step of the four-step planning process is known as modal split (Fig. 8.11).

In intercity transportation, passengers are assumed to choose from the following transportation modes:

- air transportation (airlines);
- general aviation (private aircraft);
- intercity bus;
- rail;

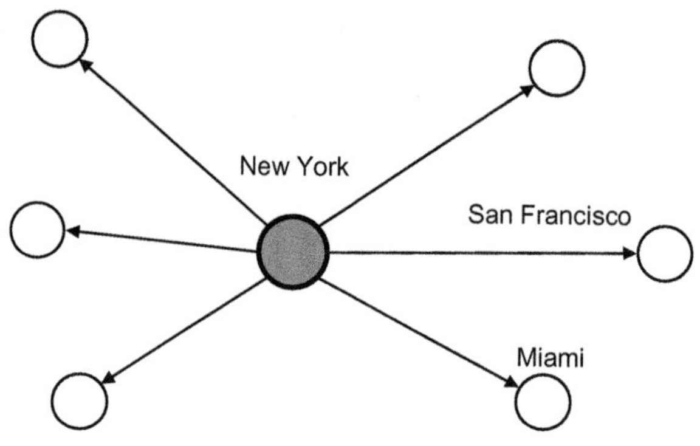

FIG. 8.10

Trip distribution.

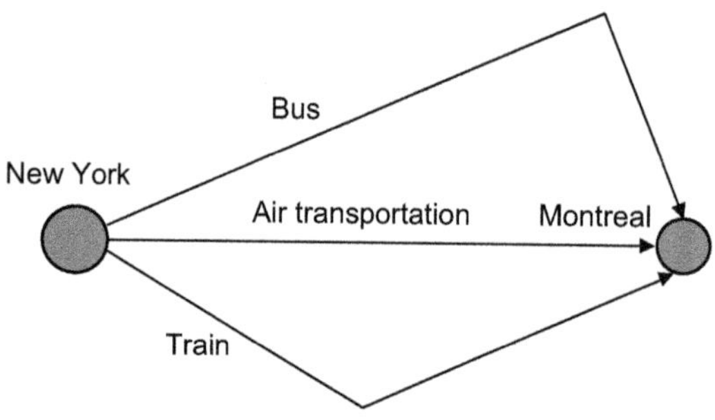

FIG. 8.11

Modal split.

- high-speed rail;
- feeder bus/rail; or
- car.

Mode choice predicts the percentage of person-trips selecting each mode of transportation while traveling between two zones in the region of interest. Within the fourth step, we determine the routes users choose to arrive at their final destinations, as well as corresponding link loads (Fig. 8.12).

This step is also known as traffic assignment. The outputs of this planning step represent the number of vehicles, passengers, or aircraft that will appear on each transportation network link. The route choice mechanism is the most important part of the traffic assignment models. The traveler-decision maker is the starting point in many route-choice models. Modeling traveler's behavior when choosing route represents the main challenge when developing route choice model.

Besides the sequential modeling approach there is also simultaneous modeling. Simultaneous modeling refers to the simultaneous prediction of trip generation, trip distribution, and modal split.

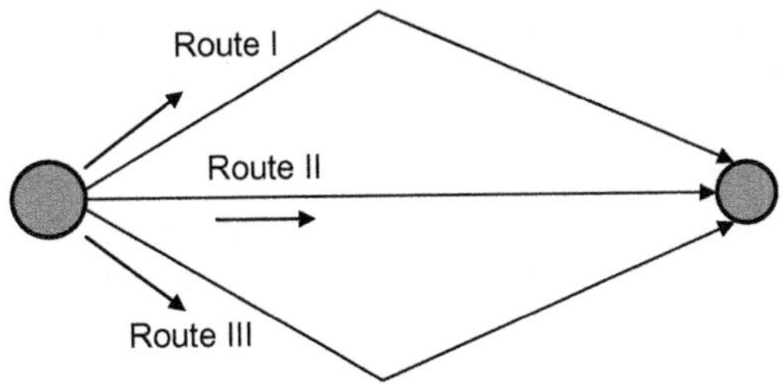

FIG. 8.12

Route choice.

8.5.1 TRIP GENERATION

The usual phrases used within the trip generation procedure are "trip production" and "trip attraction." Region- or zone-produced trips are region residents' trips to other regions. A region's attracted trips are trips originated in other regions and finished in the observed region. The outputs of the trip generation procedure are estimated numbers of the trips produced and attracted to each of the studied regions.

The cross classification tables are frequently used within trip generation phase (Table 8.7). These tables provide a picture of potential trips per household. The usual units used in the trip rate tables are [trips per household per day].

Table 8.7 Trip Rate Table for Urban Areas

Persons per Household	Vehicles per Household		
	0	1	2 or More
1	1.02	1.9	2.1
2	2.12	3.25	3.7
3	2.15	3.75	3.9
4 or more	3.96	5.02	6.54

Trip rate tables are primarily obtained through the surveys. Trip generation output could be shown in the form of a trip matrix. The elements of the trip matrix are all predicted trip attractions and trip productions (Table 8.8). The usual units used in the trip matrix are [trip-persons per day].

Table 8.8 Trip Matrix of Trip Attractions and Trip Productions in the Case of Zones *A*, *B*, and *C*

Zone	Production	Attraction
A	230,000	200,000
B	400,000	590,000
C	360,000	200,000

For example, zone B can attract 590,000 trip-persons per day. Trip attractions depend on variables like employment, retail floor space, etc. (Depending on the context of the problem studied, trip attraction depend on various factors. For example, in the case of touristic trips in air transportation, trip attractions could depend on variables like number of rooms in hotels.)

The trip generation procedure, when finished, generates information about the global level of transportation activities. Transportation demand is a function of socio-economic characteristics. Dependent variable is the number of trips, number of passengers, the number of operations, or the number of passenger kilometers. Independent variables are chosen from the set of socio—economic characteristics and characteristics of the transportation system. Multiple regression analysis has been widely used to predict transportation demand. In many cases nonlinear regression was also used for prediction. Nonlinear regression does not assume a linear relationship between variables. Artificial Neural Networks (ANN) have been successfully used to predict the future transportation demand values.

The following are two basic inputs to many trip generation models:

- *SE* is the set of socio-economic variables (population (current and forecasted), income, employment, volume of trade, average level of education...).
- *LOS* is the set of level-of-service variables (service frequencies, total travel times, departure and arrival schedule, routing, waiting times, fares, travel costs, schedule reliability, perceived level of comfort, perceived level of safety, carrier reputation...).

Models where demand is a function of socio-economic characteristics and the characteristics of the transportation system can be written in the following general form:

$$D(t) = a\prod_{i=1}^{m} S_{it}^{b_i} \prod_{j=1}^{n} T_{jt}^{c_j} \tag{8.18}$$

where:

m is the total number of socio-economic characteristics;
n is the total number of transportation system characteristics;
$D(t)$ is the number of passenger in year t;
S_{it} is the value of the ith socio-economic characteristic in year t;
T_{jt} is the value of the jth transportation system characteristic in year t; and
a, b_i, c_j are the parameters to be estimated.

EXAMPLE 8.4

Behbehani and Kanafani (1980), in analyzing the traffic between North America and the South Pacific from 1970 to 1976, established the following functional dependence between the number of passengers, income the trade volume and transportation tariffs:

$$Y_t = e^{-12.6} S_{1t}^{2.99} S_{2t}^{0.2} e^{-0.05 T_{1t}} \tag{8.19}$$

where:

Y_t is the number of trips in year t;
S_{1t} is the income in year t;
S_{2t} is the trade volume in year t; and
T_{1t} is the transportation tariff in year t.

8.5.2 TRIP DISTRIBUTION

When the total number of trips that a region, or city zone can generate has been established, the trips are then distributed. This distribution establishes the number of trips between individual regions. Let us denote by m the total number of origins, and by n the total number of destinations. The estimation of origin-destination patterns involves the estimations of the number of trips f_{ij} ($i=1, 2,\ldots, m$, $j=1, 2,\ldots, n$) between particular regions when the values of trip generated G_i ($i=1, 2,\ldots, m$) and trip attracted A_j ($j=1, 2,\ldots, n$) are known (Fig. 8.13).

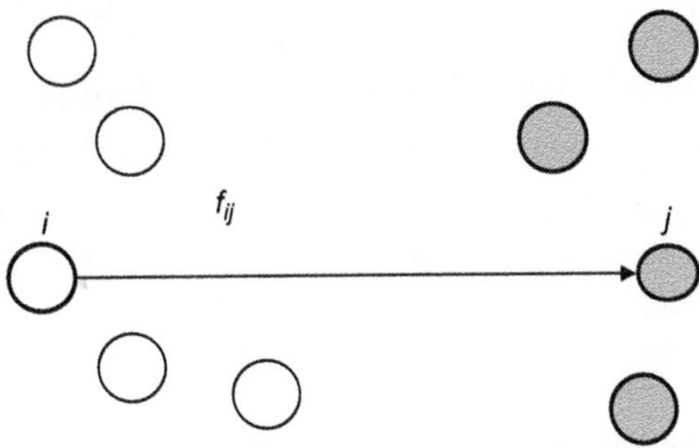

FIG. 8.13

m origins and *n* destinations.

The number of trips G_i generated by node i, and the number of trips A_j attracted to node j are, respectively, equal to:

$$G_i = \sum_{j=1}^{n} f_{ij} \quad i = 1, 2, \ldots, m \tag{8.20}$$

$$A_j = \sum_{i=1}^{m} f_{ij} \quad j = 1, 2, \ldots, n \tag{8.21}$$

The total number of trips T made from all origins to all destinations is:

$$T = \sum_{i=1}^{m} G_i = \sum_{j=1}^{n} A_j \tag{8.22}$$

The general form of the trip distribution model is:

$$f_{ij} = f\left(SE_i, SE_j, C_{ij}\right) \tag{8.23}$$

where:

SE_i is the function of the socio-economic characteristics of trip origin I;
SE_j is the function of the socio-economic characteristics of trip destination j; and
C_{ij} is the function representing the "resistance" when travelling from i to j (eg, the "resistance" is greater the greater the transportation tariff).

The gravity model is the most frequently used trip distribution model. The gravity model is trip distribution model that expresses interaction between two regions (countries, cities) as a function of the "magnitude" of the two regions and the distance between them (Fig. 8.14). The name "Gravity model" came from model's analogy to gravitational force between two bodies in space.

8.5.3 GRAVITY MODEL

Various gravity models assume that the total number of trips between two regions increases with the "size" (population, employment) of each region, and decreases with distance. For example, the total attraction of the region B is 590,000 [trip-persons per day] (Table 8.5). The attraction of the region B is much higher than the attraction of A (200,000 [trip-persons per day]). One could assume that region B will attract $\frac{590{,}000}{200{,}000}=2.95$ times more trips from the C than region A. On the other hand, the distance between A and C is 10 km, while the distance between B and region C is 20 km. Intuitively, is clear that the region B will not attract 2.95 times more trips from C than region A, because region B is on a further distance from region C than region A.

The gravity model assumes that the number of trips between two regions (zones) is directly proportional to the number of productions and attractions in the regions, and inversely proportional to the spatial separation between the regions. The basic input data for all gravity models are productions and attractions for each region and a matrix of interregional travel impedances.

The following is the basic form of the gravity model:

$$f_{ij}=k\frac{P_iA_j}{d_{ij}^2} \tag{8.24}$$

where:

f_{ij} is the total number of trips from region i to region j;
P_i is the magnitude (production, population, number of employees, etc.) of region i;
A_j is the magnitude (attraction, population, retail store areas, number of hotel rooms, etc.) of region j;
d_{ij} is the distance between region i and region j (distance may be measured by the Euclidean distance (distance along the straight line), or real travel distance); and
k is the parameter to be calibrated.

The following is a more general form of the gravity model:

$$f_{ij}=k\frac{P_iA_j}{d_{ij}^{\alpha}} \tag{8.25}$$

where a is the parameter to be calibrated.

The form of the gravity model described by relation () can be easily modified in the following way. The total production of the ith region equals:

$$P_i=\sum_{j=1}^{n}f_{ij} \tag{8.26}$$

After substituting Eq. (8.25) into Eq. (8.26) we get:

$$P_i = \sum_{j=1}^{n} f_{ij} = \sum_{j=1}^{n} k\frac{P_i A_j}{d_{ij}^{\alpha}} = kP_i \sum_{j=1}^{n} \frac{A_j}{d_{ij}^{\alpha}} \tag{8.27}$$

After solving for k, we get:

$$k = \frac{P_i}{P_i \sum_{j=1}^{n} \frac{A_j}{d_{ij}^{\alpha}}} = \frac{1}{\sum_{j=1}^{n} \frac{A_j}{d_{ij}^{\alpha}}} \tag{8.28}$$

Finally, after substituting (8.13) into (8.10), the classical form of the gravity model is obtained:

$$f_{ij} = k\frac{P_i A_j}{d_{ij}^{\alpha}} = \frac{1}{\sum_{j=1}^{n} \frac{A_j}{d_{ij}^{\alpha}}} \frac{P_i A_j}{d_{ij}^{\alpha}} \tag{8.29}$$

$$f_{ij} = P_i \frac{\frac{A_j}{d_{ij}^{\alpha}}}{\sum_{j=1}^{n} \frac{A_j}{d_{ij}^{\alpha}}} \tag{8.30}$$

The gravity model belongs to the class of spatial interaction models, and can be used for estimation (Hyman, 1969) of the spatial interactions in many different areas as migration, telephone traffic, internet traffic, or "retail gravitation" (It is logical to assume, for example, that the total number of shoppers from a surrounding areas increases with the size of the city and/or shopping center and decreases with distance from the city and/or shopping center.) The constant k in gravity models "takes care" of various dimensions of the variables. At the same time, various forms of spatial interaction (migration, telephone traffic) are in different countries and in different time periods "captured" and described by different constants.

The calibrated gravity model can help us in the following situations:

(a) to estimate the current amount of interaction (number of trips) between zones, regions, or cities.
(b) to estimate future amount of interaction once we have forecasted zone or city "magnitudes" in the future.

The outputs of the trip distribution are usually shown in a trip interchange matrix (Table 8.9). From the interchange table, we obtain full information about the number of trips that go between specific pair of zones.

Table 8.9 Trip Interchange Matrix

Origin Zone/Destination Zone	*A*	*B*	*C*
A	–	*A-B*	*A-C*
B	*B-A*	–	*B-C*
C	*C-A*	*C-B*	–

EXAMPLE 8.5

Cities 1 and 2 generate touristic trips towards cities 3, 4, and 5 (Fig. 8.12). The weekly number of tourists from city i towards city j is described by the following relation:

$$f_{ij} = k\frac{P_i R_j}{d_{ij}^2} \tag{8.31}$$

where:
P_i is the population of the city i;
R_j is the number of rooms in hotels in city j; and
d_{ij} is the distance in [km] between city i and city j.

The constant k is equal to 2. The distances between cities, city populations, and the number of hotel rooms in the cities are given in Table 8.10, 8.11, and 8.12, respectively.

Table 8.10 Distances Between Cities

City pair	Distance [km]
(1,3)	1500
(1,4)	2000
(1,5)	1500
(2,3)	2000
(2,4)	2000
(2,5)	1300

Table 8.11 City Populations

Region	Population
1	500,000
2	1,000,000

Table 8.12 Number of Hotel Rooms

Region	Number of Hotel Rooms
3	1000
4	200
5	1200

Calculate the weekly number of tourists that can travel between cities 1, and 2 towards cities 3, 4, and 5.

Solution

The numbers of tourists are, respectively, equal to:

$$f_{12} = 2\frac{(500,000)(1000)}{1500^2} = 444$$

EXAMPLE 8.5—cont'd

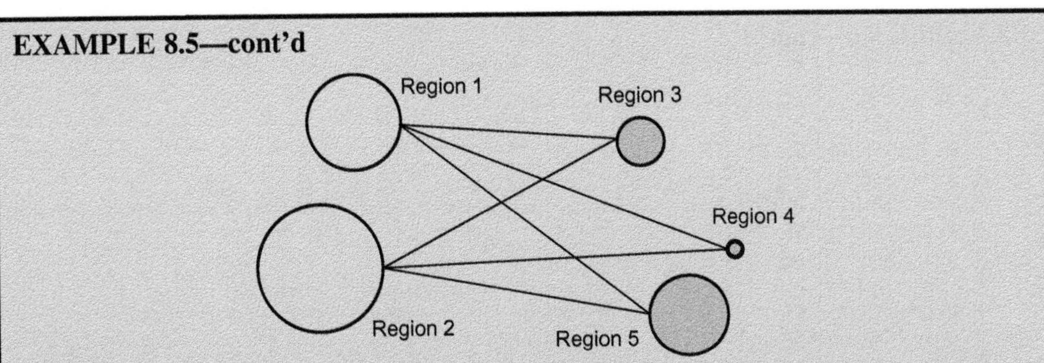

FIG. 8.14

Cities 1 and 2 generate touristic trips towards cities 3, 4, and 5.

$$f_{14} = 2\frac{(500,000)(200)}{2000^2} = 50$$

$$f_{15} = 2\frac{(500,000)(1200)}{1500^2} = 534$$

$$f_{23} = 2\frac{(1,000,000)(1000)}{2000^2} = 500$$

$$f_{24} = 2\frac{(1,000,000)(200)}{2000^2} = 100$$

$$f_{25} = 2\frac{(1,000,000)(1200)}{1300^2} = 1420$$

EXAMPLE 8.6

In 1978, air transportation in the U.S. was deregulated. Before 1978, many airline services were arranged on a point-to-point basis. After the deregulation, the hub-and-spoke system was adopted by major American airlines. Most airlines started to use strategically located airports (the hubs) as air passenger exchange points for flights to and from remote cities (the spokes).

Quality of service offered to the passengers has a significant influence on air transportation demand. Ghobrial and Kanafani (1995) were one of the first who developed the model for estimating air passenger demand that uses post-regulation data. They analyzed passengers flying directly between origin airport and destination airport. ("Direct" flight is defined either as a non-stop flight, or as a multi-stop flight without changing the plane.) Ghobrial and Kanafani (1995) proposed the following model that includes some of the level-of-service attributes (flight frequency during peak and off-peak periods, aircraft size (number of seats offered), and travel time):

$$T_{ij} = \alpha P_{ij}^{\beta} I_{ij}^{\gamma} FR_{ij}^{\varphi} FP_{ij}^{\mu} FO_{ij}^{\eta} SP_{ij}^{\lambda} SO_{ij}^{\phi} TM_{ij}^{\theta} h_{ij} \tag{8.32}$$

where:

$$h_{ij} = e^{(\omega TR_{ij} + \psi HUB_{ij})\varepsilon}$$

$$\forall FP_{ij} > 0 \;\; \forall FO_{ij} > 0$$

(Continued)

EXAMPLE 8.6—cont'd

where:
T_{ij} is the daily passenger demand who fly directly in market i, j;
P_{ij} is the product of populations of cities i and j;
I_{ij} is the product of income per capita of cities i and j;
FR_{ij} is the weighted average airfare by class type in market i, j;
FP_{ij} is the number of daily direct flights between city i and city j during peak periods (sum of morning (6:00–9:00 am) and evening (4:00–7:00 pm) peak flights);
FO_{ij} is the number of daily direct off-peak flights between city i and city j;
SP_{ij} is the weighted average aircraft size (number of seats) during peak periods between city i and city j;
SO_{ij} is the weighted average aircraft size (number of seats) during off-peak periods between city i and city j;
TM_{ij} is the average travel time in hours between cities i and j;
TR_{ij} is the a dummy variable for tourist markets (equal to one if city i or j is located in Florida, Hawaii, or Las Vegas; equal to zero otherwise);
HUB_{ij} is the a dummy variable for capacity constrained airport (equal to one if city i or j is capacity constrained airport (O'Hare, La Guardia, Logan, etc); equal to zero otherwise)
$\alpha,\beta,\gamma,\varphi,\mu,\eta,\lambda,\phi,\theta$ are the coefficients to be estimated; and
ε is the error term of estimation.

Since dummy variables *TR* and *HUB*, by their definition, could be, in some cases, equal to zero, they were expressed in an exponential form. In order to calibrate the model, Ghobrial and Kanafani (1995) used data related to the 100 top airport pairs in 1986. The estimated coefficients are shown in Table 8.13.

Table 8.13 Estimated Coefficients

Variable	Coefficient	Estimated Coefficient
Constant	α	11.180
Population	β	0.116
Per income capita	γ	0.139
Airfare	φ	−1.314
Peak flights	μ	0.436
Off-peak flights	η	0.296
Peak aircraft size	λ	0.786
Off-peak aircraft size	ϕ	0.700
Travel time	σ	0.359
Dummy for tourist markets	ω	0.058
Dummy for congested hubs	ψ	−0.231

After analyzing Ghobrial and Kanafani's model and results obtained, one can conclude that the air transportation demand is elastic with respect to fare, and highly dependent on flight schedule and travel time.

8.5.4 MODAL SPLIT

Within modal split analysis (modal share, mode split) we try to answer the following questions: how many trips between two specific nodes will be by car? How many trips between these two nodes will be by transit, etc.? Modal split models enable us to predict the percentage of travelers using a specific transportation mode. Why do we need answers to these questions? The transportation systems are dynamical systems. Various changes often occur in these systems. Frequently, traffic authorities in cities desire to raise the percentage of trips made by sustainable modes (public transit, cycling, walking). Public transit operators often want to change the topology of the public transit network, or some service elements (frequencies of service, ticket price, transfer system, etc.). Airlines regularly examine the possibility of opening a new route. City and traffic authorities could consider opening High Occupancy Lines (HOV). Various transportation policy changes (congestion pricing, new parking policy, etc.) lead to a different distribution of passenger flows by transportation modes (modal shift). In all these cases, it is necessary to estimate what will be the number of passengers that will use the new transit line, a new airline route, or a new HOV lane. To perform modal split analysis properly, it is first necessary to answer the following question: what has an effect on people's mode choice? There are three major groups of factors that influence a traveler's mode choice: (a) traveler characteristics; (b) trip characteristics; and (c) service characteristics. Income, car availability, gender, age, and family size are the most important traveler characteristics. Trip purpose, trip length, and time of day are the most important trip characteristics. The most main service characteristics are frequency of service, schedules, fares, travel time, schedule reliability, parking availability, etc.

Teleconferencing, video conferencing, and other emergent technologies will influence modal split in intercity and urban transportation in years to come. It is difficult to predict now the magnitude of this influence.

Distance to be traveled represents one of the most important issues in modal choice. There is a so-called one-day driving threshold which is about 700–900 km of driving. Usually, further than this point, air transportation obviously dominates in intercity transportation. Practically, further than this point, air transportation is the only promising transportation mode. Modal split models estimate the probability of selecting each transportation mode, given the characteristics of both the transportation mode and the passenger. These kinds of models are also used to estimate market share within a specific transportation mode.

One of the classical models to estimate the number of passengers choosing various transportation modes is the abstract mode model, developed by Quandt and Baumol (1966). The model has been applied in many transportation studies. Numerous modifications of this model have also been developed. We shall herewith present the basic assumptions of the abstract mode model using a suitable numerical example.

Let there be three different transportation modes operating between two cities. Costs and travel time of these transportation modes are given in Table 8.14.

Table 8.14 Travel Time and Costs of Three Transportation Modes

Transportation Mode, M_i	Travel Cost, cM_i	Travel Time, tM_i
M_1	1	10
M_2	2	4
M_3	3	2

Transportation mode M_1 has the lowest travel cost. However, the mode M_3 has the shortest travel time. Relative travel costs, or time of these transportation modes are defined as the ratio between each mode's travel costs or time and the least possible travel costs or time on the route under observation. In other words, the relative characteristics of transportation modes M_1, M_2, and M_3 are:

$$cM_i^r = \frac{cM_i}{\min\{cM_j\}} \quad j=1,2,3 \tag{8.33}$$

$$tM_i^r = \frac{tM_i}{\min\{tM_j\}} \quad j=1,2,3 \tag{8.34}$$

A transportation mode's relative attributes are calculated by dividing each of its attributes by the best possible value of that attribute. In the case of travel costs and time, the best value is the least one, while for frequency the best value would be the greatest one.

The relative travel costs of the transportation mode M_1 equal:

$$cM_1^r = \frac{1}{\min\{1, 2, 5\}} = 1$$

The relative travel costs and time for the three transportation modes are given in Table 8.15. The relative frequencies of the transportation modes are obtained as the ratio between the frequency of the mode in question and the greatest frequency on the route under observation, ie:

Table 8.15 Relative Travel Costs and Time for the Three Transportation Modes

Transportation Mode, M_i	Relative travel Cost, cM_i	Relative Travel Time, tM_i
M_1	1	5
M_2	2	2
M_3	3	1

The transportation mode frequencies and their relative values are given in Table 8.16. Quandt and Baumol (1966) stated with the assumption that the passenger's choice of transportation

Table 8.16 Relative Values of the Transportation Mode Frequencies

Transportation Mode, M_i	Frequency (Daily Number of Departures)	Relative Frequency
M_1	6	0.5
M_2	12	1
M_3	4	0.33

mode, or the number of passengers choosing a certain mode, depends on both the best attribute values of all transportation modes (least costs, least travel time, greatest frequency) and on the relative attribute values of the individual mode. The model also has a built in assumption that the total number of trips by all modes of transportation between two cities depends on the socio-economic characteristics of those cities. The original Quandt-Baumol model reads:

$$T_{kij} = \alpha_0 P_i^{\alpha_1} P_j^{\alpha_2} Y_i^{\alpha_3} Y_j^{\alpha_4} M_i^{\alpha_5} M_j^{\alpha_6} N_{ij}^{\alpha_7} TB_{ij}^{\beta_0} TR_{kij}^{\beta_1} CB_{ij}^{\gamma_0} CR_{kij}^{\gamma_1} FB_{ij}^{\delta_0} FR_{kij}^{\delta_1} \tag{8.35}$$

where:

T_{kij} is the number of trips between city i and city j by transportation mode k;
P_i, P_j is the populations in cities i and j;
Y_i, Y_j is the gross national income per capita in cities i and j;
M_i, M_j is the percentage of the population employed in industry in cities i and j;
N_{ij} is the number of different transportation modes that operate between cities i and j;
TB_{ij} is the least possible travel time between cities i and j;
TR_{kij} is the relative travel time between cities i and j by transportation mode k;
C_{ij}^b is the least possible travel cost between cities i and j;
CR_{kij} is the relative travel cost between cities i and j by transportation mode k;
FB_{ij} is the greatest frequency of service between cities i and j;
FR_{kij} is the relative frequency between cities i and j by transportation mode k; and
$\alpha_0, \alpha_1, \alpha_2, \alpha_3, \alpha_4, \alpha_5, \alpha_6, \alpha_7 \beta_0, \beta_1, \gamma_0, \gamma_1, \delta_0, \delta_1$, are the parameters estimated statistically.

In the context of freight transportation, modal split models predict the percentages of goods to be transported from the port to the hinterland by specific transportation modes (rail, inland waterways, road transport). Modal shift in freight transportation in the future assumes decrease in transported goods by road transportation and increase in transported goods by rail and inland waterways.

There are aggregate and disaggregate modal split models. The aggregate modal split models study the group of travelers, while disaggregate modal split models represents the behavior of individual travelers.

8.5.5 ROUTE CHOICE AND TRAFFIC ASSIGNMENT

Let us consider the simplest case when the driver is to choose one of three possible paths between the origin and destination of movement (Fig. 8.15). The origin is denoted by A, the destination by B. We assume that we know the traffic volume q_{AB} between A and B. Possible paths are marked by I, II, and III. The number of possible paths between two nodes in a network is most often very large, but in practice network users usually consider only two or three alternative paths.

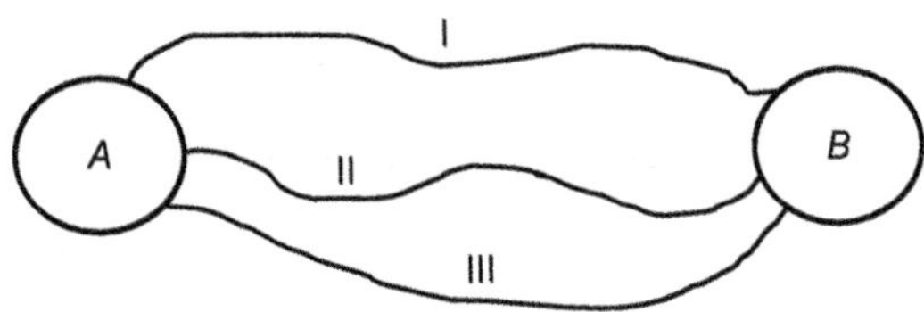

FIG. 8.15

Choosing one of three possible paths between the origin and the destination.

The traveler-decision maker is the starting point in many route-choice models. Modeling traveler's behavior when choosing route represents the main challenge when developing route choice model. Even the route choice problems have been extensively studied during last 5 decades (Wardrop, 1952; Dial, 1971; Ben-Akiva, 1974; Domencich and McFadden, 1975; Daganzo, 1979; Ben-Akiva and Lerman, 1985; Sheffi, 1985; de Ortuzar and Willumsen, 1990; Teodorović and Kikuchi, 1990; Bierlaire, 1998; Ben-Akiva and Bierlaire, 1999; Hensher et al., 2005; Bovy and Fiorenzo-Catalano, 2007), there have still been a lot of open questions related to these problems. The most important are the following:

- How do drivers and passengers choose?
- What are the most important factors that influence their decision-making?
- Do drivers and passengers always behave rationally when deciding about their trips?
- Do they always act and choose in the same manner?
- How do the characteristics of competitive routes influence route choice?
- How do travelers' characteristics influence route choice?
- How much confidence does the passenger have in information received from different sources (travel agents, Internet, advertisements)?
- How do travelers perceive the information they receive?
- What is the extent of previously gained traveler experience, and how does it affect route choice?

Travel time represents one of the most important parameters influencing route choice decision. In many transportation systems route travel time is a function of route flow. The most obvious example is urban traffic. Travel time along links in urban traffic is increasing function of link flow. The functional relationship between link travel time and link flow is described by the performance function. Typical performance function is shown in Fig. 8.16.

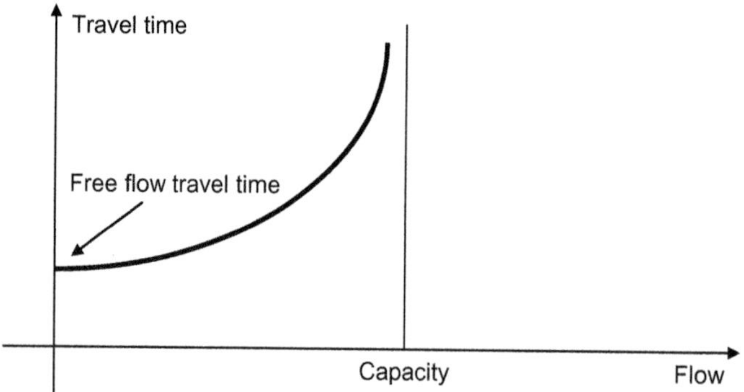

FIG. 8.16

Performance function.

Free-flow travel time corresponds to the free-flow traffic conditions. The more vehicles travel along the link, the higher the level of congestion, and the higher the link travel time. Transportation network is described by its set of nodes, set of links, link orientation, node connections, and link performance functions (Fig. 8.14). All these elements comprise transportation supply. Transportation demand is described by the Origin-Destination matrix. The traffic assignment problem could be defined in the following way:

> For defined transportation network's supply and for known Origin-Destination matrix, calculate link flows and link travel times.

In other words, through the traffic assignment procedure we try to answer the following question: how are networks users distributed through the transportation network? Obviously, networks users could be distributed in many different ways through the network routes and links. Traffic assignment procedures that have been developed so far use certain principles when distributing users. The best known principles are user-optimal principle, and system-optimal principle. We shall elaborate these principles in more details in subsequent sections. Traffic assignment scheme is shown in Fig. 8.17.

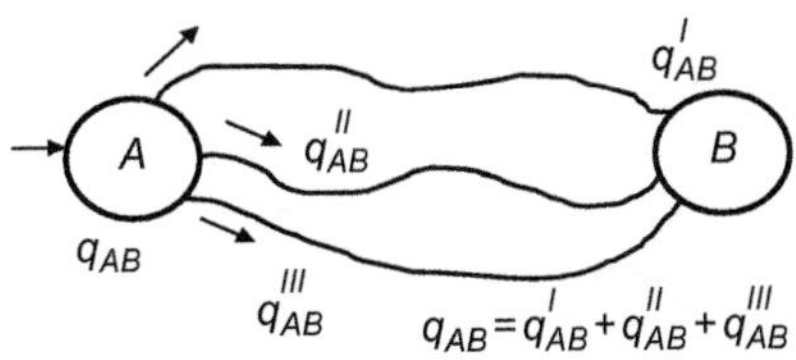

FIG. 8.17

Traffic Assignment Scheme.

The problems of route choice and traffic assignment are certainly among the most important transportation planning problems. Different variations of this problem appear in city, intercity, air, and rail transportation (Hall et al., 1980; Harker, 1987; Spiess and Florian, 1989).

The most important deciding factors for drivers when choosing among routes in an urban network are perceived travel time, perceived number of stops, perceived level of congestion, etc. Deciding factors when choosing among routes in an airline network include travel time, the cost of transport on different routes, flight frequency, and the number of stopovers. Many non-business passengers will usually chose the least expensive path, that frequently may contain and or more transfer points. Many business travelers will usually choose the "shortest" path.

Route choice and traffic assignment problem also appear in the area of air traffic control. A lot of pilots prefer (choose) the identical air routes, overloading in this way many air traffic control sectors, and airways. The main task of Air Traffic Flow Management (in the phase of air traffic planning that happened few hours before planned flights) is to change original aircraft routes (solve traffic assignment problem), and create such traffic assignment that will significantly reduce air traffic control sector congestion, potential aircraft conflicts, and controllers workload.

8.6 USER EQUILIBRIUM AND SYSTEM OPTIMUM

Let us note Fig. 8.18, which represents a network with five nodes that are the origins and destinations of movement.

The volumes between individual origins and destinations are given in matrix F:

$$F = \begin{array}{c} \\ 1 \\ 2 \\ 3 \\ 4 \\ 5 \end{array} \begin{array}{c} \begin{array}{ccccc} 1 & 2 & 3 & 4 & 5 \end{array} \\ \begin{pmatrix} - & 10 & 30 & 10 & 10 \\ 10 & - & 20 & 10 & 70 \\ 10 & 20 & - & 10 & 50 \\ 10 & 20 & 50 & - & 50 \\ 30 & 10 & 50 & 20 & - \end{pmatrix} \end{array}$$

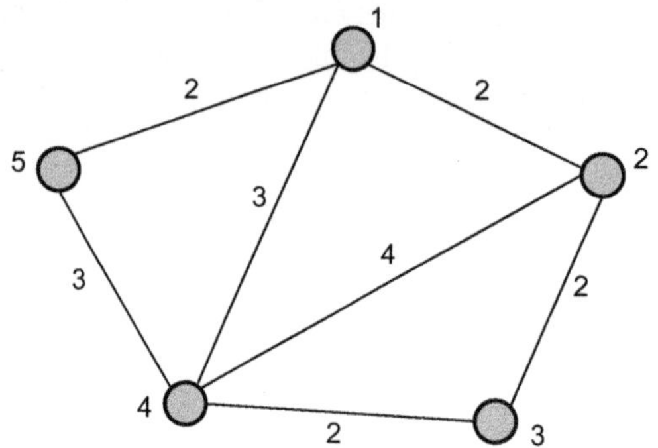

FIG. 8.18

Transportation network whose expected volumes on its links must be determined.

Travel "costs" along the links are noted in Fig. 8.16. Travel "costs" could represent real travel cost, passenger's travel time, aircraft flight time, etc. It is assumed that travel costs are constant and independent of the volume assigned to a link. Assuming that the network user will choose to travel along a path exclusively according to the criterion of travel cost, all network users will choose the shortest (cheapest) paths between the origins and destinations of travel. The shortest path between node 1 and node 3 is shown in Fig. 8.19.

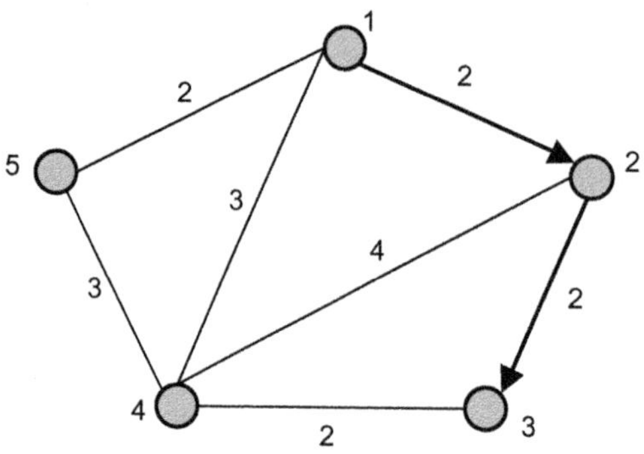

FIG. 8.19

Shortest path between node 1 and node 3.

Table 8.17 shows the paths that were selected by network users.

Using data shown in Table 8.14, we calculate the total traffic volume for each link in the network. For example, link (1, 2) is contained in the path (1, 2), as well as in the paths (1, 2, 3) and (5, 1, 2). The volume along the link (1, 2) equals $10+30+10=50$. Table 8.18 shows the total traffic volume for each link in the network.

The traffic assignment technique is based on the assumption that all network users are oriented exclusively to the shortest path (Dijkstra, 1959) between pairs of nodes is known as "all or nothing." The all or nothing principle does not take into account traffic congestion in the network. The more

Table 8.17 Paths Selected by the Network Users

Origin-Destination	Shortest Path Chosen by the Network Users	The Length of the Shortest Path	The Traffic Volume Between the Origin and the Destination
(1, 2)	(1, 2)	2	10
(1, 3)	(1, 2, 3)	4	30
(1, 4)	(1, 4)	3	10
(1, 5)	(1, 5)	2	10
(2, 1)	(2, 1)	2	10
(2, 3)	(2, 3)	2	20
(2, 4)	(2, 4)	4	10
(2, 5)	(2, 1, 5)	4	70
(3, 1)	(3, 2, 1)	4	10
(3, 2)	(3, 2)	2	20
(3, 4)	(3, 4)	2	10
(3, 5)	(3, 4, 5)	5	50
(4, 1)	(4, 1)	3	10
(4, 2)	(4, 2)	4	20
(4, 3)	(4, 3)	2	50
(4, 5)	(4, 5)	3	50
(5, 1)	(5, 1)	2	30
(5, 2)	(5, 1, 2)	4	10
(5, 3)	(5, 4, 3)	5	50
(5, 4)	(5, 4)	3	20

Table 8.18 Link Volumes

Link	Volume	Link	Volume
(1, 2)	50	(3, 4)	60
(1, 4)	10	(4, 1)	10
(1, 5)	80	(4, 2)	20
(2, 1)	20	(4, 3)	50
(2, 3)	50	(4, 5)	100
(2, 4)	10	(5, 1)	40
(3, 2)	30	(5, 4)	20

sophisticated assignment approaches try to incorporate the influence of traffic congestion on route choice and traffic assignment. This means that such approaches assume that link travel time is function of the volume assigned to the link.

When performing urban traffic assignment, the basic assumption is that travel time is the dominant factor influencing users in their choice of path. The similar assumption could be made for business travelers in air transportation. When travel time along a link (or some other factor that influences the user) is a function of the volume assigned to the link, assuming that every network user tries to minimize his/her travel time, the network will be in stable conditions when no user is able to decrease his/her travel time (business travelers) or his/her transportation cost (non-business travelers) through the network by changing his/her path. These conditions are known in the literature as user equilibrium conditions (UE), and has been known as Wardrop's first principle Wardrop (1952): "The travel times on all used paths between an origin and a destination point are equal, and less than those which would be experienced by a single vehicle on any unused path," or "No traveler can improve his travel time by unilaterally changing routes," or, "Every traveler follows the minimum travel time path" (According to M. Patriksson (Algorithms for Urban Traffic Network Equilibria, Linkoping, Sweden, 1991), mathematician J.E. Kohl (1841) and an economist A.C. Pigou (1920) studied these traffic phenomenon before Wardrop). Wardrop's user-equilibrium corresponds to a Nash equilibrium (after the Nobel Prize-winning mathematician and economist John Nash) in game theory. Generally, the Nash equilibrium is the state where no individual can advance his/her situation by making a different choice.

Every user wishes to minimize individual travel time (travel cost). There is a competition between network users who all are looking for the best route for themselves independently of each other. Wardrop's first principle is deterministic by its nature. It assumes that all network users, that are autonomous in their decisions, have perfect information about all route travel times and travel costs. This principle implicitly assumes that all network users always make rational decisions, always choosing the shortest or cheapest route. The main components of user equilibrium models are traffic assignment based on the principle "all or nothing" (users always choose the "shortest" path between the origin and destination) and a specific performance function that describes functional dependence between travel time along a link and the traffic volume assigned to the link.

In many user equilibrium models, a very strict distinction is made between real and perceived travel time (or real and perceived travel costs, real and perceived flight frequency, and so on). Improved user equilibrium models are based on the assumption that the choice of path depends on perceived, and not on observed, travel time, flight frequency, transportation cost, schedule reliability, etc. Perceived travel time is treated in such models as a random variable, whereby every network user corresponds to a specific value of perceived travel time. Stable conditions in this case arise when no network user believes that any change in his path could decrease his travel time. Such conditions are known in the literature as stochastic user equilibrium (Sheffi, 1985; Patriksson, 1994), and corresponding models are called stochastic user equilibrium models. A stochastic model to "load" the network and a functional dependence between travel time on a link and the volume assigned to the link are the main components of stochastic user equilibrium models (Daganzo and Sheffi, 1977). The system optimization approach to assigning traffic in a network endeavors to assign user flows so that the total travel time of all users through the network is minimized.

User equilibrium conditions do not provide minimum value of the total travel time (travel cost) in the air transportation network. Wardrop's second principle (the travel times on all used paths between an origin and a destination point are of equal marginal travel time, and less than those which would be experienced by a single vehicle on any unused path) deals with the minimization of the total travel time in the network (system optimization (SO) formulation). System optimization formulation is particularly important when dealing with traffic congestion problems. SO formulation provides smaller total number of hours traveled than the UE formulation. The optimum value of the total number of the number of hours traveled in the network can be achieved only through cooperation between users (In this case some network users would have longer travel times than in the UE case).

8.6.1 FORMULATION OF THE USER EQUILIBRIUM PROBLEM

Let us elaborate in more details user equilibrium problem using the example shown in Fig. 8.20. There are two alternative routes 1 and 2. The drivers are free to choose route 1 or route 2 when they leave node r. The total number of trips that should be performed between node r and node s is known, and equal to q_{rs}.

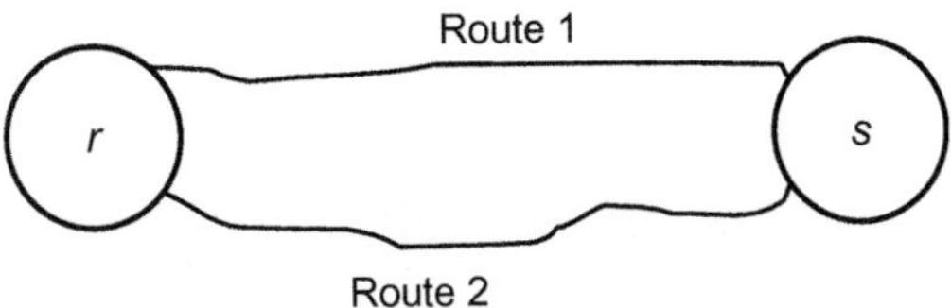

FIG. 8.20

User equilibrium problem in the case of two alternative routes.

Let us denote by x_1 and x_2, respectively, number of trips along routes 1 and 2. Therefore,

$$q_{rs} = x_1 + x_2$$

The minimum values of the total travel times along routes 1 and 2 are, respectively, equal to:

$$T_1(0) = 25 \text{ min}$$

$$T_2(0) = 20 \text{min}$$

It is not possible to make the trip for a smaller amount of time. We call these travel times *free-flow travel times*. The travel times along these routes are increasing functions of the number of trips performed along the routes. We assume that the route performance functions equal:

$$T_1(x_1) = 25 + 0.02 \times x_1$$

$$T_2(x_2) = 20 + 0.03 \times x_2$$

where:

T_1 and T_2 are total travel times along considered routes; and
x_1 and x_2 are number of trips along considered routes.

We also assume that:

$$q_{rs} = 1500$$

Free-flow travel times along routes 1 and 2 are, respectively, equal to:

$$x_1 = 0 \Rightarrow T_1(0) = 25$$

$$x_2 = 0 \Rightarrow T_2(0) = 20$$

We first assume that all trips use exclusively route 1. In this case, total travel time along route 1 equals:

$$x_1 = 1500 \Rightarrow T_1(1500) = 55$$

We now assume that all trips use exclusively route 2. In this case, total travel time along route 2 equals:

$$x_2 = 1500 \Rightarrow T_2(1500) = 65$$

Because:

$$T_1 = (1500) \Rightarrow 55 > T_2(0) = 20$$

and

$$T_2 = (1500) \Rightarrow 65 > T_1(0) = 25$$

we conclude that both routes will be used.

Wardrop's user equilibrium gives:

$$T_1(x_1) = T_2(x_2)$$

or:

$$25 + 0.02 \times x_1 = 20 + 0.03 \times x_2$$

Flow conservation reads:

$$x_1 + x_2 = 1500$$

We obtain:

$$x_1 = 800$$

$$x_2 = 700$$

Travel times are equal to:

$$T_1(800) = 25 + 0.02 \times 800 = 41\,\text{min}$$

$$T_2(700) = 20 + 0.03 \times 700 = 41\,\text{min}$$

The summary of the problem considered is given in Table 8.19.

Table 8.19 The Summary of the Problem Considered

Route	Performance Function	Free-Flow Travel Time [min]	Number of Trips in User Equilibrium Conditions	Travel Time in User Equilibrium Conditions [min]
1	$25 + 0.02x_1$	25	800	41
2	$20 + 0.03x_2$	20	700	41

Obviously, user equilibrium problem becomes more complex in the case of many alternative routes. Let us introduce the following notation:

R is the set of origins;
S is the set of destinations;
A is the set of network links;
P_{rs} is the set of all paths between node r and node s;
$r \in R$ is the origin;

$s \in S$ is the destination;
$p \in P_{rs}$ is the path leading from node r to node s;
q_{rs} is the total number of travelers from node r to node s;
x_a is the flow along link a;
f_p^{rs} is the flow along path p that leads from node r to node s;
t_a is the travel time along link a;
t_p^{rs} is the travel time along path p that leads from node r to node s; and
$t_a(w)$ is the travel time (travel cost) along link a, that depends on link flow w.

Beckman et al. (1956) and Sheffi (1985) proposed to obtain user equilibrium flows by solving the following problem:

Minimize

$$F = \sum_a \int_0^{xa} t_a(w)dw \tag{8.36}$$

subject to:

$$\sum_p f_p^{rs} = q_{rs} \text{ for } \forall r, s \tag{8.37}$$

$$f_p^{rs} \geq 0 \text{ for} \forall p, r, s \tag{8.38}$$

$$x_a = \sum_r \sum_s \sum_p f_p^{rs} \delta_{ap}^{rs} \text{ for } \forall\ a \tag{8.39}$$

where $\delta_{ap}^{rs} = 1$ when link a belong to the path p. In the opposite case, $\delta_{ap}^{rs} = 0$.

The proposed objective represents sum of integrals of the link performance functions. Objective function does not have any traffic, behavioral, economic, or any other interpretation. Proposed objective function is only mathematically convenient. Beckman et al. (1956) and Sheffi (1985) created this objective function in order to solve the user equilibrium problem.

The first derivative $\frac{\partial x_a}{\partial f_l^{mn}}$ equals:

$$\frac{\partial_{xa}}{\partial f_l^{mn}} = \frac{\partial \left[\sum_r \sum_s \sum_p f_p^{rs} \delta_{ap}^{rs} \right]}{\partial f_l^{mn}} = \delta_{ap}^{rs} \tag{8.40}$$

Since $\frac{\partial_{xa}}{\partial f_l^{mn}} = 0$ for $(r, s) \neq (m, n)$ or $\neq l$, we conclude that the first derivative $\frac{\partial x_a}{\partial f_l^{mn}}$ is equal to 1 in the case when link a belongs to the path l. In the opposite case, this derivative equals 0. We need the derivative $\frac{\partial x_a}{\partial f_l^{mn}}$ in order to perform the following analysis. Since we need to find minima of a function subject to constraints, we introduce a new variable u_{rs} called a *Lagrange multiplier*. The new function L to be minimized reads:

$$L = \sum_a \int_0^{xa} t_a(w)dw + \sum_{rs} u_{rs} \left(q_{rs} - \sum_p f_p^{rs} \right) \tag{8.41}$$

The new problem to be solved reads:

Minimize:

$$L=\sum_a \int_0^{xa} t_a(w)dw+\sum_{rs} u_{rs}\left(q_{rs}-\sum_p f_p^{rs}\right) \tag{8.42}$$

subject to:

$$f_p^{rs} \geq 0 \text{ za } \forall r,s,p. \tag{8.43}$$

when the following conditions are satisfied:

$$f_p^{rs}\frac{\partial L}{\partial f_p^{rs}}=0 \ \forall r,s,p \tag{8.44}$$

$$\frac{\partial L}{\partial f_p^{rs}} \geq 0 \ \forall r,s,p \tag{8.45}$$

$$\frac{\partial L}{\partial u_{rs}}=0 \ \forall r,s, \tag{8.46}$$

ie:

$$f_p^{rs}\left(t_p^{rs}-u_{rs}\right)=0 \text{ for } \forall,r,s,p \tag{8.47}$$

$$\left(t_p^{rs}-u_{rs}\right) \geq 0 \text{ for } \forall,r,s,p \tag{8.48}$$

$$\sum_p f_p^{rs}=q_{rs} \text{ for } \forall r,s \tag{8.49}$$

$$f_p^{rs} \geq 0 \text{ for } \forall r,s,p \tag{8.50}$$

What is the interpretation of the obtained conditions? Relation (8.49) represents conservation of flows. It shows that the sum of all flows along paths that lead from node r to node s is equal to the total flow from node r to node s. Relation shows that all flow values should be non-negative. The Lagrange multiplier u_{rs} represents the smallest travel time among all travel times along paths leading from node r to node s. Let us clarify this statement. If $f_p^{rs}=0$, than $f_p^{rs}\left(t_p^{rs}-u_{rs}\right)=0$ and $t_p^{rs}-u_{rs} \geq 0$. We conclude that in the case when $f_p^{rs}=0$, travel time t_p^{rs} along path p (that leads from node r to node s) is greater than or equal to the smallest possible travel time u_{rs}. It is also $f_p^{rs}\left(t_p^{rs}-u_{rs}\right)=0$ when $f_p^{rs} \neq 0$ and $\left(t_p^{rs}-u_{rs}\right)=0$. In this case, travel time t_p^{rs} along path p (that leads from node r to node s) is equal to the smallest possible travel time u_{rs}. This produces positive value of the flow f_p^{rs}. We can conclude that positive flow values appear along the paths that have travel time equal to the smallest possible travel time u_{rs}. There are no flows along the paths that have travel time greater than the smallest possible travel time u_{rs} (Fig. 8.21). There are obviously two subsets of paths. Network users travel exclusively along paths that belong to the first subset of paths. They do not use at all paths from the second subset. This represents the basic characteristic of the user equilibrium. The users do not use paths from the second subset, since the achieved travel time could not be decreased when using these paths.

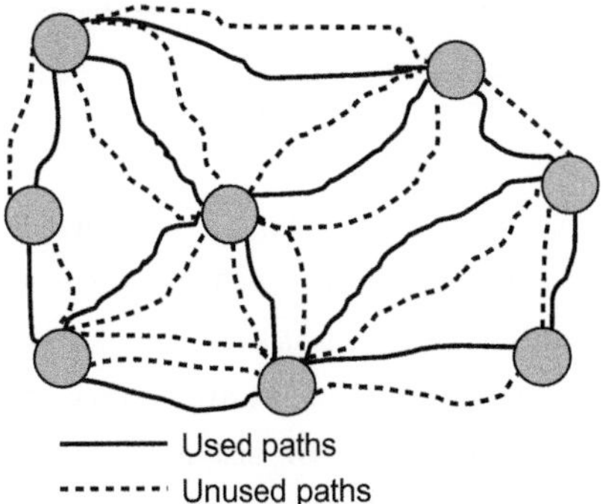

FIG. 8.21

User equilibrium.

EXAMPLE 8.7

There are three alternative routes connecting node *A* and node *B* (Fig. 8.22). The drivers are free to choose route 1, 2, or 3 when they leave node *A*. The total number of trips that should be performed between node *A* and node *B* is known, and equal to 5000.

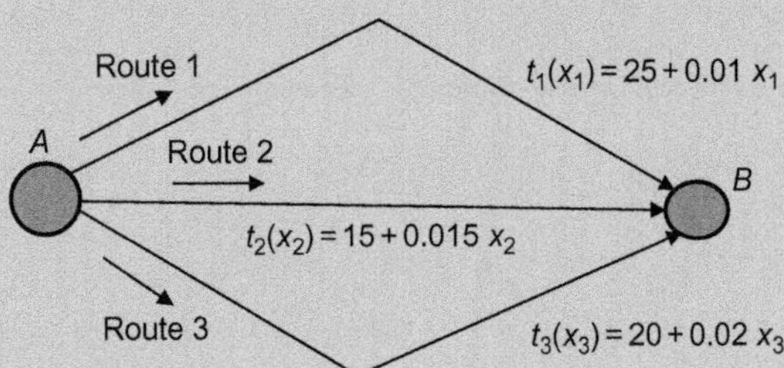

FIG. 8.22

User equilibrium and system optimum in the case of three alternative routes.

The route performance functions (Fig. 8.20) are, respectively, equal to:

$$t_1(x_1)=25+0.01\times x_1$$

$$t_2(x_2)=15+0.015\times x_2$$

$$t_3(x_3)=20+0.02\times x_3$$

where:

x_1, x_2, x_3 are the flows along routes 1, 2, and 3; and

$t_1(x_1)$, $t_2(x_2)$, $t_3(x_3)$ are the travel times along routes 1, 2, and 3.

(a) Determine the user equilibrium flows;

(b) Analyze route usage by drivers depending on the total flow value.

Solution

(a) Free-flow travel times along routes 1, 2, and 3 are, respectively, equal to:

$$x_1=0\Rightarrow t_1(0)=25$$

(Continued)

EXAMPLE 8.7—cont'd

$$x_2 = 0 \Rightarrow t_2(0) = 15$$

$$x_3 = 0 \Rightarrow t_3(0) = 20$$

We first assume that all trips use exclusively route 1. In this case, total travel time along route 1 equals:

$$x_1 = 5000 \Rightarrow t_1(5000) = 25 + 0.01 \times 5000 = 75$$

We now assume that all trips use exclusively route 2. In this case, total travel time along route 2 equals:

$$x_2 = 5000 \Rightarrow t_2(5000) = 15 + 0.015 \times 5000 = 75$$

In the case that all trips exclusively use route 3, total travel time along route 3 equals:

$$x_3 = 5000 \Rightarrow t_3(5000) = 20 + 0.02 \times 5000 = 120$$

The following inequalities are satisfied:

$$t_1(5000) = 75 > t_2(0) = 15$$

$$t_1(5000) = 75 > t_3(0) = 20$$

$$t_2(5000) = 75 > t_1(0) = 25$$

$$t_2(5000) = 75 > t_3(0) = 20$$

$$t_3(5000) = 120 > t_1(0) = 25$$

$$t_3(5000) = 120 > t_2(0) = 15$$

We conclude that all three routes will be used. Wardrop's user equilibrium gives:

$$t_1(x_1) = t_2(x_2) = t_3(x_3)$$

ie:

$$25 + 0.01 \times x_1 = 15 + 0.015 \times x_2$$

$$25 + 0.01 \times x_1 = 20 + 0.02 \times x_3$$

Flow conservation reads:

$$x_1 + x_2 + x_3 = 5000$$

After solving the system of Eqs. ()–(), we obtain the following integer flow values:

$$x_1 = 1884 \left[\frac{veh}{h}\right]$$

$$x_2 = 1923 \left[\frac{veh}{h}\right]$$

$$x_3 = 1193 \left[\frac{veh}{h}\right]$$

Corresponding travel times are:

$$t_1(1884) = 25 + 0.01 \times 1884 = 43.8\,\text{min}$$

$$t_2(1923) = 15 + 0.015 \times 1923 = 43.8\,\text{min}$$

$$t_3(1193) = 20 + 0.02 \times 1193 = 43.8\,\text{min}$$

The total travel time of all users equal:

$$T = 5{,}000 \cdot 43.8 = 219{,}000\ \textit{minutes}$$

EXAMPLE 8.7—cont'd

(b) The route performance functions are, respectively, equal to:

$$t_1(x_1) = 25 + 0.01 \times x_1$$

$$t_2(x_2) = 15 + 0.015 \times x_2$$

$$t_3(x_3) = 20 + 0.02 \times x_3$$

When the flow rate q_{AB} between node A and node B is very small, all drivers will choose route 2, since route 2 has the lowest free-flow travel time. The corresponding flows are:

$$x_1 = 0$$

$$x_2 = q_{AB}$$

$$x_3 = 0$$

A small increase in the total flow q_{AB} will increase travel time along route 2. However, travel time along route 2 will be still lower than the free-flow travel time along routes 1 and 3. With further increase in the total flow, in one moment, travel time along route 2 will become equal to the free-flow travel time along route 3. The flow q_{AB} that produces this equality must satisfy the following equation:

$$15 + 0.015 \times q_{AB} = 20$$

$$q_{AB} = 334 \left[\frac{veh}{h}\right]$$

Routes 2 and 3 will be used by drivers when the total flow is higher than $334 \left[\frac{veh}{h}\right]$. With further increase in the total flow, in one moment, travel times along routes 2 and 3 will become equal to the free-flow travel time along route 1. The flows on routes 2 and 3 that create this situation must satisfy the following equations:

$$15 + 0.015 \times x_2 = 25$$

$$20 + 0.02 \times x_3 = 25$$

After solving the system of equations, we get the following flow rate values:

$$x_2 = 667$$

$$x_3 = 250$$

The total flow equals:

$$q_{AB} = x_2 + x_3 = 667 + 250 = 917$$

All three routes will be used when the total flow is higher than $917 \left[\frac{veh}{h}\right]$. The results of the analysis are summarized in Table 8.20.

Table 8.20 Usage of Alternatives Routes as a Function of the Total Flow

Total Flow [veh/h]	Travel Time in User Equilibrium [min]	Routes Used
$q_{AB} \leq 334$	$15 \leq t \leq 20$	2
$334 \leq q_{AB} \leq 917$	$20 \leq t \leq 25$	2 and 3
$q_{AB} \geq 917$	$t \geq 25$	1, 2, and 3

8.7 HEURISTIC ALGORITHMS FOR FINDING USER-EQUILIBRIUM FLOW PATTERN

The problem of finding the user-equilibrium flow pattern is defined as a mathematical programming problem. The problem is usually solved by the mathematical programming techniques. On the other hand, planners and traffic engineers have developed, during the last few decades, few relatively simple heuristic algorithms for finding the user-equilibrium flow pattern. These algorithms have been applied in many transportation studies. The most important are the capacity restraint algorithm, the FHWA algorithm, and the incremental assignment algorithm. Let us introduce the following notation:

$t_a^0 = t(0)$ is the free-flow travel time on link a;
t_a^i is the travel time on link a in the ith iteration of the algorithm;
x_a^i is the flow on link a in the ith iteration of the algorithm; and
x_{rs} is the flow between origin r and destination s.

8.7.1 CAPACITY RESTRAINT ALGORITHM

The capacity restraint algorithm contains the following steps:

Step 1: Treat travel time on all links as free-flow travel time ($t_a^0 = t_a(0), \forall a$). Make all-or-nothing assignment. Get a set of link flows $\{x_a^0\}$. Set iteration counter $i = 1$.
Step 2: Set $t_a^i = t_a\left(x_a^{i-1}\right), \forall a$.
Step 3: Make all-or-nothing assignment based on travel times $\{t_a^i\}$. Get a set of link flows $\{x_a^i\}$.
Step 4: If: $\max_a\left\{\left|x_a^i - x_a^{i-1}\right| \leq c\right\}$ stop. The set of link flows $\{x_a^i\}$ represents the user-equilibrium link flows. Otherwise, set $= i+1$, and return to step 1.

EXAMPLE 8.8

There are three alternative routes connecting node A and node B (Fig. 8.23). The drivers are free to choose route 1, 2, or 3 when they leave node A. The total number of trips that should be performed between node A and node B is known, and equal to 5000.

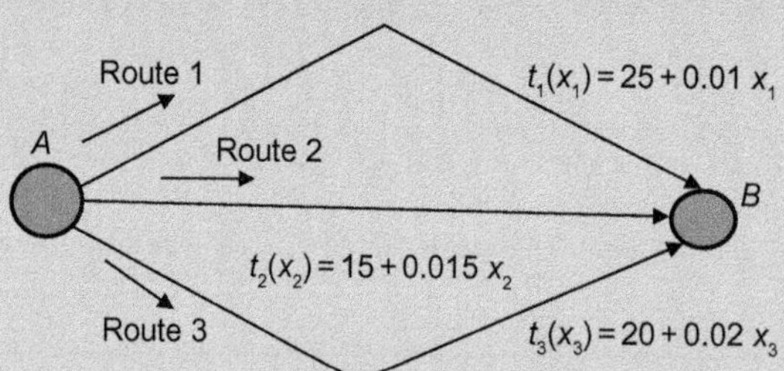

FIG. 8.23

Solving traffic assignment problem by the capacity restraint algorithm.

The route performance functions (Fig. 8.21) are, respectively, equal to:

$$t_1(x_1) = 25 + 0.01x_1$$

$$t_2(x_2) = 15 + 0.015x_2$$

EXAMPLE 8.8—cont'd

$$t_3(x_3) = 20 + 0.02x_3$$

where:

x_1, x_2, x_3 are the flows along routes 1, 2, and 3; and

$t_1(x_1)$, $t_2(x_2)$, $t_3(x_3)$ are the travel times along routes 1, 2, and 3.

Find the user-equilibrium flow pattern by the capacity restraint algorithm. Stop the algorithm if the maximum change in link flow between successive iterations is less than 100.

Solution

The steps of the capacity restraint algorithm are shown in Table 8.21.

Table 8.21 The Steps of the Capacity Restraint Algorithm

Iteration Number	Step of the Algorithm	Link 1	Link 2	Link 3
0	0	$t_1^0(0) = t_1(0) = 25$ $x_1^0 = 0$	$t_2^0(0) = t_2(0) = 15$ $x_2^0 = 5000$	$t_3^0(0) = t_3(0) = 20$ $x_3^0 = 0$
1	1	$t_1^1 = t_1(x_1^0) = t_1(0)$ $= 25 + 0.01 \cdot 0 = 25$	$t_2^1 = t_2(x_2^0) =$ $= t_2(5000) =$ $= 15 + 0.015 \times 5000 = 90$	$t_3^1 = t_3(x_3^0) = t_3(0)$ $= 20 + 0.02 \times 0 = 20$
	2	$x_1^1 = 0$	$x_2^1 = 0$	$x_3^1 = 5000$
	3	The maximum change in link flow between successive iterations is equal to 5000		
2	1	$t_1^2 = t_1(x_1^1) = t_1(0) =$ $= 25 + 0.01 \times 0 = 25$	$t_2^2 = t_2(x_2^1) = t_2(0)$ $= 15 + 0.015 \times 0 == 15$	$t_3^2 = t_3(x_3^1) == t_3(5000)$ $= 20 + 0.02 \times 5000 = 120$
	2	$x_1^2 = 0$	$x_2^2 = 5000$	$x_3^2 = 0$
	3	The maximum change in link flow between successive iterations is equal to 5000		
3	1	$t_1^3 = t_1(x_1^2) = t_1(0) =$ $= 25 + 0.01 \times 0 = 25$	$t_2^3 = t_2(x_2^2) = t_2(5000)$ $= 15 + 0.015 \times 5000 = 90$	$t_3^3 = t_3(x_3^2) = t_3(0)$ $= 20 + 0.02 \times 0 = 20$
	2	$x_1^3 = 0$	$x_2^3 = 0$	$x_3^3 = 5000$
	3	The maximum change in link flow between successive iterations is equal to 5000		

As can be seen, after the first three iterations, the capacity restraint algorithm does not converge. Links 2 and 3 receive alternately entire flow, while the link 1 remains constantly without traffic flow.

In order to overcome the shortcomings of the capacity restraint algorithm, the FHWA modified the capacity restraint algorithm and proposed the FHWA algorithm.

8.7.2 FHWA ALGORITHM

The FHWA algorithm consists of the following algorithmic steps:

Step 0: Treat travel time on all links as free-flow travel time ($t_a^0 = t_a(0),\ \forall a$). Make all-or-nothing assignment. Get a set of link flows $\{x_a^0\}$. Set iteration counter $i=1$.
Step 1: Set $\tau_a^i = t_a\left(x_a^{i-1}\right),\ \forall a$.
Step 2: Set $t_a^i = 0.75 \times t_a^{i-1} + 0.25 \times \tau_a^i,\ \forall i$.
Step 3: Make all-or-nothing assignment based on travel times $\{t_a^i\}$. Get a set of link flows $\{x_a^i\}$.
Step 4: If: $i=1$ go to Step 5. Otherwise, set $i=i+1$ and go to Step 1.
Step 5: Compute the average values $x_a^8,\ \forall a$ of the link flow values obtained through the iterations. Finish the algorithm. The set of link flows $\{x_a^8\}$ represents the user-equilibrium link flows.

The FHWA algorithm contains a smoothing step and an averaging step. Step 2 represents a smoothing step, while step 5 represents an averaging step. The algorithm also contains stopping rule instead of convergence test (I is the number of iterations prescribed by the analyst).

EXAMPLE 8.9

Apply the FHWA algorithm in the case of the traffic network shown in Fig. 8.21. Finish the algorithm after $I=5$ iterations.

Solution

The steps of the FHWA algorithm are shown in Table 8.22.

Table 8.22 The Steps of the FHWA Algorithm

Iteration Number	Step of the Algorithm	Link 1	Link 2	Link 3
1	0	$t_1^0(0)=25$ $x_1^0=0$	$t_2^0(0)=15$ $x_2^0=5000$	$t_3^0(0)=20$ $x_3^0=0$
1	1	$\tau_1^1=t_1\left(x_1^0\right)=t_1(0)$ $=25+0.01\times 0=25$	$\tau_2^1=t_2\left(x_2^0\right)=$ $=t_2(5000)=$ $=15+0.015\times 5000=90$	$\tau_3^1=t_3\left(x_3^0\right)=t_3(0)$ $=20+0.02\times 0=20$
	2	$t_1^1=0.75\times 25+0.25\times 25$ $t_1^1=25$	$t_2^1=0.75\times 15+0.25\times 90$ $t_2^1=33.75$	$t_3^1=0.75\times 20+0.25\times 20$ $t_3^1=20$
	3	$x_1^1=0$	$x_2^1=0$	$x_3^1=5000$
2	1	$\tau_1^2=t_1\left(x_1^1\right)=25$	$\tau_2^2=t_2\left(x_2^1\right)=15$	$\tau_3^2=t_3\left(x_3^1\right)==120$
	2	$t_1^2=0.75\times 25+0.25\times 25$ $t_1^1=25$	$t_2^2=0.75\times 33.75$ $+0.25\times 15$	$t_3^2=0.75\times 20+0.25\times 120$ $t_3^1=45$
	3	$x_1^2=5000$	$t_2^1=29.0625$ $x_2^2=0$	$x_3^2=0$

EXAMPLE 8.9—cont'd

Table 8.22 The Steps of the FHWA Algorithm—cont'd

Iteration Number	Step of the Algorithm	Link 1	Link 2	Link 3
3	1	$\tau_1^3 = t_1(x_1^2) = 75$	$\tau_2^3 = t_2(x_2^2) = 15$	$\tau_3^3 = t_3(x_3^2) = 20$
	2	$t_1^3 = 0.75 \cdot 25 + 0.25 \cdot 75$ $t_1^1 = 37.5$	$t_2^3 = 0.75 \times 29.0625$ $+0.25 \times 15$ $t_2^1 = 25.546875$	$t_3^3 = 0.75 \times 45 + 0.25 \times 20$ $t_3^1 = 38.75$
	3	$x_1^3 = 0$	$x_2^3 = 5000$	$x_3^3 = 0$
4	1	$\tau_1^4 = t_1(x_1^3) = 25$	$\tau_2^4 = t_2(x_2^3) = 90$	$\tau_3^4 = t_3(x_3^3) = 20$
	2	$t_1^4 = 0.75 \times 37.5 + 0.25 \times 25$ $t_1^4 = 34.375$	$t_2^4 = 0.75 \times 25.546875$ $+0.25 \times 90$ $t_2^4 = 31.66015625$	$t_3^4 = 0.75 \times 38.75 + 0.25 \times 20$ $t_3^4 = 34.0625$
	3	$x_1^4 = 0$	$x_2^4 = 5000$	$x_3^4 = 0$
5	1	$\tau_1^5 = t_1(x_1^4) = 25$	$\tau_2^5 = t_2(x_2^4) = 90$	$\tau_3^5 = t_3(x_3^4) = 20$
	2	$t_1^5 = 0.75 \times 34.375 + 0.25 \times 25$ $t_1^5 = 32.03125$	$t_2^5 = 0.75 \times 31.66015625$ $+0.25 \times 90$ $t_2^5 = 36.2451171$	$t_3^5 = 0.75 \times 34.0625$ $+0.25 \times 20$ $t_3^5 = 30.546875$
	3	$x_1^5 = 0$	$x_2^5 = 0$	$x_3^5 = 5000$
6	Average link flow values	$x_1^8 = 1000$	$x_2^8 = 2000$	$x_3^8 = 2000$

8.7.3 INCREMENTAL ASSIGNMENT ALGORITHM

The incremental assignment algorithm contains the following steps:

Step 0: Split each origin-destination flow into I equal shares, ie:

$$q_{rs}^i = \frac{qrs}{I}$$

Set iteration counter $i = 1$ and $x_a^0 = 0, \ \forall a.$

Step 1: Set $t_a^i = t_a(x_a^{i-1}), \ \forall a.$

Step 2: Make all-or-nothing assignment of shares q_{rs}^i, based on travel times $\{t_a^i\}$. Get a set of link flows $\{y_a^i\}$.

Step 3: Set $x_a^i = x_a^{i-1} + y_a^i, \ \forall a.$

Step 4: If $i = 1$, finish the algorithm. Otherwise, set $i = i + 1$, and return to step 1.

EXAMPLE 8.10

Apply the incremental assignment algorithm in the case of the traffic network shown in Fig. 8.21. (a) Finish the algorithm after $I=5$ iterations. (b) Finish the algorithm after $I=20$ iterations.

Solution

(a) We shall first split the total flow q_{AB} into 5 shares, ie, $q^i_{AB}=\frac{5000}{5}=1000 \quad i=1,2,\ldots,5.$

The steps of the incremental assignment algorithm are shown in Table 8.23.

Table 8.23 The Steps of the Incremental Assignment Algorithm in the Case of Five Shares

Iteration Number	Step of the Algorithm	Link 1	Link 2	Link 3
1	0	$x_1^0=0$	$x_2^0=0$	$x_3^0=0$
	1	$t_1^1=25$	$t_2^1=15$	$t_3^1=20$
	2	$y_1^1=0$	$y_2^1=1000$	$y_3^1=0$
	3	$x_1^1=0$	$x_2^1=1000$	$x_3^1=0$
2	1	$t_1^2=25$	$t_2^2=30$	$t_3^2=20$
	2	$y_1^2=0$	$y_2^2=0$	$y_3^2=1000$
	3	$x_1^2=0$	$x_2^2=1000$	$x_3^2=1000$
3	1	$t_1^3=25$	$t_2^3=40$	$t_3^3=40$
	2	$y_1^3=1000$	$y_2^3=0$	$y_3^3=0$
	3	$x_1^3=1000$	$x_2^3=1000$	$x_3^3=1000$
4	1	$t_1^4=35$	$t_2^4=40$	$t_3^4=40$
	2	$y_1^4=1000$	$y_2^4=0$	$y_3^4=0$
	3	$x_1^4=2000$	$x_2^4=1000$	$x_3^4=1000$
5	1	$t_1^5=45$	$t_2^5=30$	$t_3^5=40$
	2	$y_1^5=0$	$y_2^5=1000$	$y_3^5=0$
	3	$x_1^5=2000$	$x_2^5=2000$	$x_3^5=1000$

(b) We split the total flow q_{AB} into 20 shares, ie, $q^i_{AB}=\frac{5000}{20}=250 \quad i=1,2,\ldots,5.$ The steps of the incremental assignment algorithm are shown in Table 8.24.

Table 8.24 The Steps of the Incremental Assignment Algorithm in the Case of 20 Shares

Iteration Number	Step of the Algorithm	Link 1	Link 2	Link 3
1	0	$x_1^0=0$	$x_2^0=0$	$x_3^0=0$
	1	$t_1^1=25$	$t_2^1=15$	$t_3^1=20$
	2	$y_1^1=0$	$y_2^1=250$	$y_3^1=0$
	3	$x_1^1=0$	$x_2^1=250$	$x_3^1=0$
2	1	$t_1^2=25$	$t_2^2=27.5$	$t_3^2=20$
	2	$y_1^2=0$	$y_2^2=0$	$y_3^2=250$
	3	$x_1^2=0$	$x_2^2=250$	$x_3^2=250$
3	1*	$t_1^3=25$	$t_2^3=18.75$	$t_3^3=25$
	2	$y_1^3=0$	$y_2^3=250$	$y_3^3=0$
	3	$x_1^3=0$	$x_2^3=500$	$x_3^3=250$

EXAMPLE 8.10—cont'd

Table 8.24 The Steps of the Incremental Assignment Algorithm in the Case of 20 Shares—cont'd

Iteration Number	Step of the Algorithm	Link 1	Link 2	Link 3
4	1	$t_1^4=25$	$t_2^4=22.5$	$t_3^4=25$
	2	$y_1^4=0$	$y_2^4=250$	$y_3^4=0$
	3	$x_1^4=0$	$x_2^4=750$	$x_3^4=250$
5	1	$t_1^5=25$	$t_2^5=26.25$	$t_3^5=25$
	2*	$y_1^5=125$	$y_2^5=0$	$y_3^5=125$
	3	$x_1^5=125$	$x_2^5=750$	$x_3^5=375$
6	1	$t_1^6=26.25$	$t_2^6=26.25$	$t_3^6=27.5$
	2*	$y_1^6=125$	$y_2^6=125$	$y_3^6=0$
	3	$x_1^6=250$	$x_2^6=875$	$x_3^6=375$
7	1*	$t_1^7=27.5$	$t_2^7=28.125$	$t_3^7=27.5$
	2	$y_1^7=125$	$y_2^7=0$	$y_3^7=125$
	3	$x_1^7=375$	$x_2^7=875$	$x_3^7=500$
8	1	$t_1^8=28.75$	$t_2^8=28.125$	$t_3^8=30$
	2	$y_1^8=0$	$y_2^8=250$	$y_3^8=0$
	3	$x_1^8=375$	$x_2^8=1125$	$x_3^8=500$
9	1	$t_1^9=28.75$	$t_2^9=31.875$	$t_3^9=30$
	2	$y_1^9=250$	$y_2^9=0$	$y_3^9=0$
	3	$x_1^9=625$	$x_2^9=1125$	$x_3^9=500$
10	1	$t_1^{10}=31.25$	$t_2^{10}=31.875$	$t_3^{10}=30$
	2	$y_1^{10}=0$	$y_2^{10}=0$	$y_3^{10}=250$
	3	$x_1^{10}=625$	$x_2^{10}=1125$	$x_3^{10}=750$
11	1	$t_1^{11}=31.25$	$t_2^{11}=31.875$	$t_3^{11}=35$
	2	$y_1^{11}=250$	$y_2^{11}=0$	$y_3^{11}=0$
	3	$x_1^{11}=875$	$x_2^{11}=1125$	$x_3^{11}=750$
12	1	$t_1^{12}=33.75$	$t_2^{12}=31.875$	$t_3^{12}=35$
	2	$y_1^{12}=0$	$y_2^{12}=250$	$y_3^{12}=0$
	3	$x_1^{12}=875$	$x_2^{12}=1375$	$x_3^{12}=750$
13	1	$t_1^{13}=33.75$	$t_2^{13}=35.625$	$t_3^{13}=35$
	2	$y_1^{13}=250$	$y_2^{13}=0$	$y_3^{13}=0$
	3	$x_1^{13}=1125$	$x_2^{13}=1375$	$x_3^{13}=750$
14	1	$t_1^{14}=36.25$	$t_2^{14}=35.625$	$t_3^{14}=35$
	2	$y_1^{14}=0$	$y_2^{14}=0$	$y_3^{14}=250$
	3	$x_1^{14}=1125$	$x_2^{14}=1375$	$x_3^{14}=1000$
15	1	$t_1^{15}=36.25$	$t_2^{15}=36.625$	$t_3^{15}=40$
	2	$y_1^{15}=0$	$y_2^{15}=250$	$y_3^{15}=0$
	3	$x_1^{15}=1125$	$x_2^{15}=1625$	$x_3^{15}=1000$

(Continued)

EXAMPLE 8.10—cont'd

Table 8.24 The Steps of the Incremental Assignment Algorithm in the Case of 20 Shares—cont'd

Iteration Number	Step of the Algorithm	Link 1	Link 2	Link 3
16	1	$t_1^{16}=36.25$	$t_2^{16}=39.375$	$t_3^{16}=40$
	2	$y_1^{16}=250$	$y_2^{16}=0$	$y_3^{16}=0$
	3	$x_1^{16}=1375$	$x_2^{16}=1625$	$x_3^{16}=1000$
17	1	$t_1^{17}=38.75$	$t_2^{17}=39.375$	$t_3^{17}=40$
	2	$y_1^{17}=250$	$y_2^{17}=0$	$y_3^{17}=0$
	3	$x_1^{17}=1625$	$x_2^{17}=1625$	$x_3^{17}=1000$
18	1	$t_1^{18}=41.25$	$t_2^{18}=39.375$	$t_3^{18}=40$
	2	$y_1^{18}=0$	$y_2^{18}=250$	$y_3^{18}=0$
	3	$x_1^{17}=1625$	$x_2^{17}=1875$	$x_3^{17}=1000$
19	1	$t_1^{19}=41.25$	$t_2^{19}=43.125$	$t_3^{19}=40$
	2	$y_1^{19}=0$	$y_2^{19}=0$	$y_3^{19}=250$
	3	$x_1^{19}=1625$	$x_2^{19}=1875$	$x_3^{19}=1250$
20	1	$t_1^{20}=41.25$	$t_2^{20}=43.125$	$t_3^{20}=45$
	2	$y_1^{20}=250$	$y_2^{20}=0$	$y_3^{20}=0$
	3	$x_1^{20}=1875$	$x_2^{20}=1875$	$x_3^{20}=1250$

8.8 SYSTEM OPTIMAL ROUTE CHOICE

System optimal route choice results in the minimum total travel costs of all users. In the case of the system optimal route choice, some network users are able to decrease their travel time (cost) by unilaterally changing routes. System optimal flows are not stable (some users will always try to change routes in order to improve their own travel costs). The system optimal route choice problem is described by the following mathematical program:

Minimize

$$F=\sum_a x_a t_a \tag{8.51}$$

subject to:

$$\sum_p f_p^{rs}=q_{rs} \quad \forall r,s \tag{8.52}$$

$$f_p^{rs}\geq 0 \quad \forall p,r,s \tag{8.53}$$

where t_a represents link "cost" when link flow equals x_a.

The traffic assignment that is result of this program usually cannot be achieved in practice. The objective function represents the total travel time (total travel costs) of all users spent in the transportation network.

EXAMPLE 8.11

There are two alternative routes: 1 and 2 (Fig. 8.24). Route 1 starts from node r, goes through intersection 1, and finishes at node s. Route 2 also starts from node r, goes through intersection 2, and finishes at node s. The total number of trips that should be performed between node r and node s is known, and equal to 1500. The route performance functions equal:

$$T_1(x_1) = 25 + 0.02x_1$$

$$T_2(x_2) = 20 + 0.03x_2$$

where $x_{1,}, x_2$ are the number of trips along routes A and B.

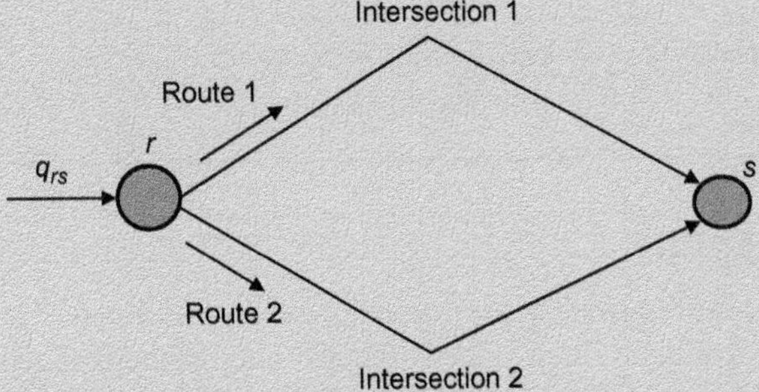

FIG. 8.24

System optimal route choice in the case of two alternative routes.

Assign flows to routes 1, and 2, in such a way to produce minimal total travel time of all network users.

Solution

The system optimal route choice problem is described by the following mathematical program:

Minimize:

$$F = (25 + 0.02x_1)x_1 + (20 + 0.03x_2)x_2$$

subject to:

$$x_1 + x_2 = 1500$$

$$x_1 \geq 0 \quad x_2 \geq 0$$

After substitution, we obtain:

$$F = [25 + 0.02(1500 - x_2)](1500 - x_2) + (20 + 0.03x_2)x_2$$

$$F = 0.05x_2^2 - 65x_2 + 82,500$$

After the first derivative is set to zero

$$\frac{\partial F}{\partial x_2} = 0$$

we obtain:

$$0.1x_2 - 65 = 0$$

which solving for x_2 gives:

$$x_2 = 650 \quad x_1 = 850$$

(Continued)

EXAMPLE 8.11—cont'd

Corresponding travel times are equal to:

$$T_1(850)=25+0.02(850)=42$$

$$T_2(650)=20+0.03(650)=39.5$$

System optimum (total minimum travel time of all air network users) will be achieved when 850 trips are assigned to route 1, and 650 trips are assigned to route 2. The total travel time of all network users equals:

$$T=850\times 42+650\times 39.5=61,375\,\text{min}$$

In the case of user equilibrium, the total travel time of all users equals:

$$T=1500\times 41=61,500\,\text{min}$$

8.9 PRICE OF ANARCHY

There is a significant difference in total travel time of all drivers between the "every driver for himself" policy and the societal optimum. There is no difference only in the case when travel times do not depend on link flows, and when we assume that all users always follow the corresponding shortest paths (all-or-nothing assignment). In this case the travel times are:

$$t_a(x_a)=t_a \tag{8.54}$$

where:

t_a is the travel time along link a that is constant and independent of the link flow; and
x_a is the flow along link a.

We examine every O-D pair (r, s). Let us assign the flow between node r and node s to every link that belongs to the shortest path between node r and node s.

The flows along other paths that connect node r, and node s are equal to zero. We denote by F_1 the objective function in the case of user equilibrium. When travel times do not depend on link flows, the following relation is satisfied:

$$F_1=\sum_a\int_0^{x_a} t_a(w)dw=\sum_a\int_0^{x_a} t_a dw=\sum_a x_a t_a \tag{8.55}$$

where the sum $\sum_a x_a t_a$ represents the objective function F_2 in the case of system optimum route choice problem, ie:

$$F_2=\sum_a x_a t_a \tag{8.56}$$

We conclude that when travel times do not depend on link flows, the objective function value in the case of user equilibrium is identical to the objective function value in the case of system optimum route choice problem, ie:

$$F_1=F_2=\sum_a x_a t_a \tag{8.57}$$

In all other cases, the values of these objective functions differ. Usually, the "every driver for himself" behavior produces significantly longer travel times than if the drivers had been optimally routes through the network. *The price of anarchy* (Roughgarden, 2005; Youn et al., 2008) is defined as the ratio of the total cost (total travel time) of the Wardropian user equilibrium to the total cost (total travel time) of the social optimum. The price of anarchy is the measure of the inefficiency of decentralization. For example, if the price of anarchy is equal to 1.35, drivers waste 35% of minimum possible travel time for not being coordinated. One of the main tasks of traffic engineers and planners is to propose and implement a proper transportation policy to reduce the price of anarchy.

8.10 BRAESS PARADOX AND TRANSPORTATION CAPACITY EXPANSIONS

The number of trips by private cars has significantly increased in recent decades in many cities, and on many highways. At the same time, road network capacities have not kept up with this increase in travel demand. Urban road networks in many countries are severely congested, resulting in increased travel times, increased number of stops, unexpected delays, greater travel costs, inconvenience to drivers and passengers, increased air pollution and noise level, and increased number of traffic accidents. Expanding traffic network capacities by building more roads is extremely costly as well as environmentally damaging. At the same time, it has been shown that building more roads does not necessarily cause decrease in total travel time in the transportation network. This phenomenon is known as *Braess' paradox* (Braess, 1968). Braess' paradox happens in some transportation networks because the equilibrium of such network is not necessarily optimal. Building new roads can sometimes slow traffic! Let us illustrate Braess' paradox using the original Braess example (Fig. 8.25).

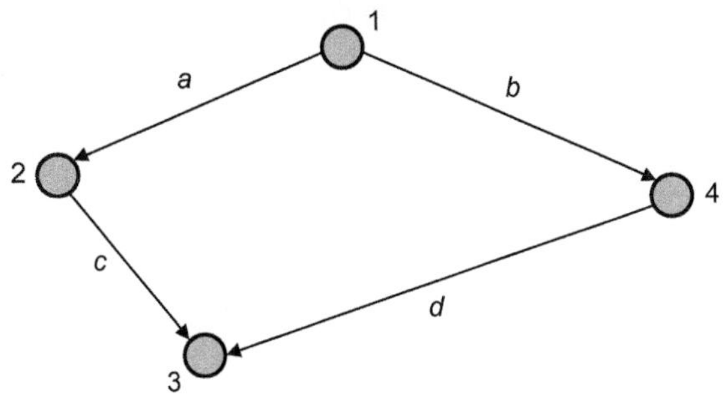

FIG. 8.25

Transportation network before improvement.

Fig. 8.23 shows simple transportation network that contains four nodes (1, 2, 3, 4) and four links (a, b, c, d). The performance functions of the links are, respectively, equal to:

$$t_a(x_a) = 10 \times x_a$$

$$t_b(x_b) = x_b + 50$$

$$t_c(x_c) = x_c + 50$$

$$t_d(x_d) = 10 \times x_d$$

where:

x_a, x_b, x_c, x_d are the flows on links *a*, *b*, *c*, *d*; and
$t_a(x_a), t_b(x_b), t_c(x_c), t_d(x_d)$ is the travel time on links *a*, *b*, *c*, *d*.

The total number of network users wanted to travel from node 1 to node 4 equals 6 (thousand). Network users traveling from 1 to 4 can choose one of the following two paths:

Path 1: (*a*, *c*)
Path 2: (*b*, *d*)

We denote by x_1 and x_2, respectively, flows on paths 1, and 2. By simple visual inspection, we conclude that the equilibrium link flows are:

$$x_a = x_b = x_c = x_d = 3$$

The equilibrium path flows are:

$$x_1 = x_2 = 3$$

Travel times in equilibrium are, respectively, equal to:

$$t_a(x_a) = 10 \times x_a = 10 \times 3 = 30$$

$$t_b(x_b) = x_b + 50 = 3 + 50 = 53$$

$$t_c(x_c) = x_c + 50 = 3 + 50 = 53$$

$$t_d(x_d) = 10 \times x_d = 10 \times 3 = 30$$

The total travel times t_1 and t_2 on paths 1 and 2 in equilibrium equal:

$$t_1 = t_a(x_a) + t_c(x_c) = 30 + 53 = 83$$

$$t_2 = t_b(x_b) + t_d(x_d) = 53 + 30 = 83$$

In this situation, no one network user has incentive to change his/her path. Path change would increase a user's travel time. Let us assume that in order to improve traffic conditions, we build the new link *e* that goes from node 2 to node 3 (Fig. 8.26).

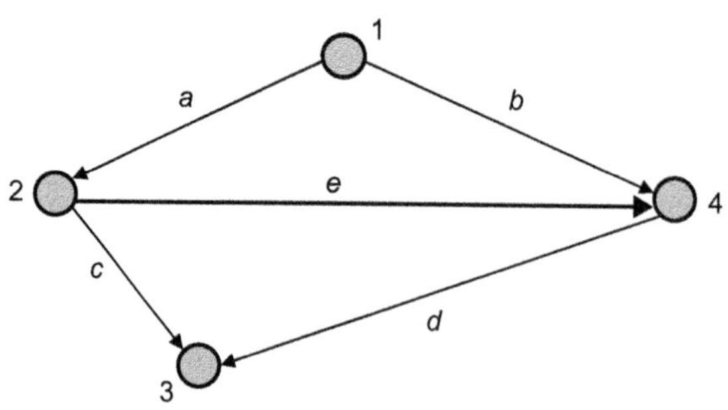

FIG. 8.26

Adding link *e*.

The link e performance function reads:

$$t_e(x_e) = x_e + 10$$

In modified network users traveling from 1 to 4 can choose one of the following three paths:

Path 1: (a, c)
Path 2: (b, d)
Path 3: (a, e, d)

Let us assume, for the moment, that nobody will use the new road (a, e, d). The path travel times in this situation would be:

$$t_2 = t_b(3) + t_d(3) = 53 + 30 = 83$$

$$t_2 = t_a(3) + t_e(0) + t_d(3) = 30 + 10 + 30 = 70$$

Network users would immediately realize that the new road travel time t_3 is lower than the travel times t_1 and t_2 on existing roads. In other words, travel time on unused path would be lower than travel times on used paths. Transportation network would not be any more in equilibrium. Users would also start to use path 3, and the new user equilibrium would be created. The new equilibrium link flows are:

$$x_a = 4, \quad x_b = 2, \quad x_c = 2, \quad x_d = 4, \quad x_e = 2$$

The new equilibrium path flows are:

$$x_1 = x_2 = x_3 = 2$$

Travel times in new equilibrium are, respectively, equal to:

$$t_a(x_a) = 10 \times x_a = 10 \times 4 = 40$$

$$t_b(x_b) = x_b + 50 = 2 + 50 = 52$$

$$t_c(x_c) = x_c + 50 = 2 + 50 = 52$$

$$t_d(x_d) = 10 \times x_d = 10 \times 4 = 40$$

$$t_e(x_e) = x_e + 10 = 2 + 10 = 12$$

The total travel times t_1, t_2, and t_3 on paths 1, 2, and 3 in new equilibrium equal:

$$t_1 = t_a(x_a) + t_c(x_c) = 40 + 52 = 92$$

$$t_2 = t_b(x_b) + t_d(x_d) = 52 + 40 = 92$$

$$t_3 = t_a(x_a) + t_e(x_e) + t_d(x_d) = 40 + 12 + 40 = 92$$

The new equilibrium travel time equals 92. This is greater than the old equilibrium travel time that is equal to 83. The addition of the link increased travel time in the network.

At first glance, the obtained result seems counterintuitive. It is difficult to accept the statement that adding more roads could sometimes *slow traffic*! The Braess paradox explanation is related to the essence of the user equilibrium and system optimal traffic assignment concepts. We built the new link e that goes from node 2 to node 3 in order to improve traffic conditions. This means that we wanted to decrease the total travel time of all users (system optimum objective function). On the other hand, network users distribute themselves through the network according to the user equilibrium principles. Every network user only wishes to minimize his/her individual travel time, not taking into account the interests of other users. Individual drivers are not interested in decreasing the total travel time. They are always only interested in decreasing their own travel time. Individual drivers are only concerned about the number of vehicles in the queue in a front of them. They do not care about vehicles in the queue behind them! The flows that are distributed according to the user equilibrium principles cannot decrease system optimal objective function. It is important to note that the addition of the new link in the wrong location would not increase the total travel time, if flows were distributed through the network according to the system optimum principles. The best examples are some countries that recently significantly increased expressway networks, and simultaneously increased average commute time. The opposite, in some cases, may also be true—removing roads may even advance traffic conditions.

8.11 DYNAMIC TRAFFIC ASSIGNMENT

Transportation network characteristics that change over time significantly affect the traveler's route and start time choice. Traffic congestion, as well as the various information provided to drivers often influence the change in the planned start time, as well as the choice of alternative routes. Most drivers are characterized by a dynamic route choice behavior. Dynamic traffic assignment (DTA) techniques describe the existing relationship between the transportation network characteristics and users' dynamic behavior (Mahmassani, 2001; Peeta and Ziliaskopoulos, 2001; Ziliaskopoulos et al., 2004). DTA models use the standard static assignment assumptions, but treat the time-varying flows. DTA techniques are very useful when analysts consider various demand management strategies (congestion pricing, HOV and HOT lanes), make decisions related to the construction of additional transportation capacities, analyze work zones impacts, deploy variable message signs (VMS), organize special events, manage traffic incidents, and perform emergency evacuation modeling.

DTA models use various types of route choice models. These models map the Origin-Destination matrix into path flows. In the next step, based on path flows, analysts determine the link flows.

The Origin-Destination matrix (O-D matrix) represents the basic input data in the case of static traffic assignment problem (Robillard, 1975). The O-D matrix contains information about traffic flow values between all pair of nodes in the network. Within the DTA context, the analysts use a time-dependent origin-destination demand. In the case of the dynamic traffic assignment problem, the (O-D) is replaced by the (O-D-T), where letter T is related to time. In the dynamic assignment modeling framework, the time is divided into small increments (several seconds, 1, 5, 10 or 15 min in length), and it is necessary to know the O-D matrix for each of numerous small intervals. Estimating and predicting time-dependent OD matrix is among the most complex tasks.

Travel times along various routes in the network *vary* in time increments. Consequently, there is no unique shortest path per O-D pair. In the case of dynamic traffic assignment, shortest path have to be discovered for each origin-destination-departure time increment combination.

DTA assumes that "the experienced travel time for all used routes is the same for travelers departing at the same time" (Sloboden et al., 2012). The equilibrium based DTA procedure discover shortest paths, assign trips to paths, load trips on the assigned paths, and evaluate traffic conditions in the transportation network.

The DynaMIT (Ben-Akiva et al., 1998; Wen et al., 2006b; Balakrishna et al., 2008), DYNASMART (Mahmassani and Hawas, 1997; Mahmassani et al., 2004), VISTA (Ziliaskopoulos et al., 2004), DynusT (Chiu et al., 2008), AIMSUN (Barcelo and Casas, 2002a,b, 2006), TRANSCAD (Caliper Corporation, 2009), INTEGRATION (Aerde et al., 1996) and METROPOLIS (de Palma and Marchal, 2002) are some of the most valuable dynamic traffic assignment (DTA) software for online traffic prediction and/or for offline traffic operations planning. The gained experience showed that the developed DTA models are capable of reproducing real traffic situations.

8.12 TRANSPORTATION DEMAND ANALYSIS BASED ON DISCRETE CHOICE MODELS

Numerous trips are made on a daily basis in intercity transportation, and each trip is linked to a certain number of decisions that must be made. Trip makers decide whether or not to make a certain trip. They choose the destination and the mode of transportation, choose the departure time, choose the carrier and route. Drivers choose the route in the city, while air passengers make the choice of tariff, and the class of transportation (business, first, and tourist class in airplanes). All these decisions are short-term travel decisions. During the past five decades, a large number of models have been developed that describe trip generation, trip distribution, modal split, and route choice. Some models are based on the socio-economic characteristics of a city, region, or zone and/or the characteristics of the transport system.

Discrete choice models start with the passenger as the trip's decision maker. A decision—maker is faced with the problem of choice of one alternative from a finite set of mutually exclusive alternatives (Daganzo, 1979; Sheffi et al., 1982, Ben-Akiva and Lerman, 1985). These models are disaggregate by their nature. This means that these models try to describe, analyze, and predict choice behavior of the individual passengers or other decision makers (organization, household, shipper, etc.). Discrete choice models are capable of giving an explanation as to why individual decision-makers make specific choices in a particular situation (Luce, 1959; McFadden, 1975; Manski, 1977; Ortúzar, 1982). These models are also capable of predicting changes in choice behavior due to changes in a decision-maker's characteristics and changes in alternatives' attributes. For example, in many situations it is worthy to analyze to what limit the probability of choosing a transportation mode will decrease, if the charge for that mode raise by a specific amount.

In addition to the frequency, travel time, and cost of different modes of transportation, numerous other factors are present whose effect on decision making process cannot be quantifies without pooling the passenger population. Thus, without surveys it is impossible to determine the effect of comfort, the passenger's feeling of safety during the trip, or schedule reliability on transportation mode.

Pools concerning the importance passengers give to non-demographic factors typically use a scale that enables the passenger to indicate his/her feelings. For example, the feeling of safety could be determines by the following:

Place the sign on the scale below to indicate your feeling of safety when flying from city *A* and city *B* (Fig. 8.27).

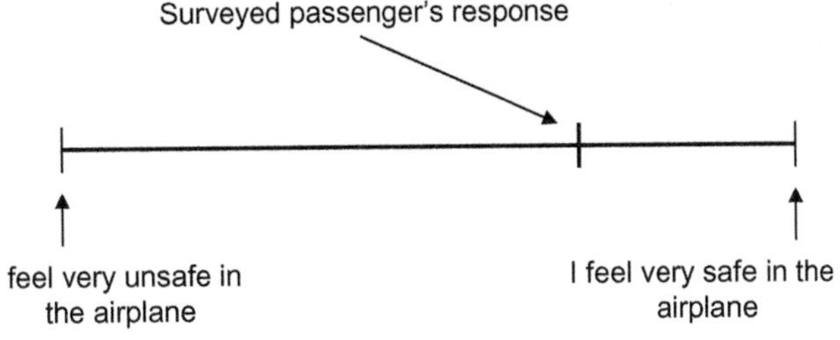

FIG. 8.27

Safety evaluation scale.

If the scale is 10 cm long, for example, a simple measurement of the length from left to right, from the beginning of the scale to the sign marked by surveyed passenger, will give a numerical value to the feeling of safety when flying. When taking the survey, precisely defined travel scenarios should be offered to participants. This means that they should indicate whether this is a business trip or a private trip, whether the trip is being made in summer or winter, etc. To this effect due to the frequent possibility of flights being canceled or late during winter, passengers will certainly give a poorer rating to the air transportation as a reliable means in terms of executing the schedule on time. In the same vein, the choice of transportation mode vitally depends on the purpose of the trip, making it essential to mention this purpose within the offered travel scenario. The survey questionnaire should therefore contain sentences of this type: "Imagine that you are taking a business trip, alone, during the summer...." Fig. 8.28 shows another possible type of scale to evaluate the non-demographic factors.

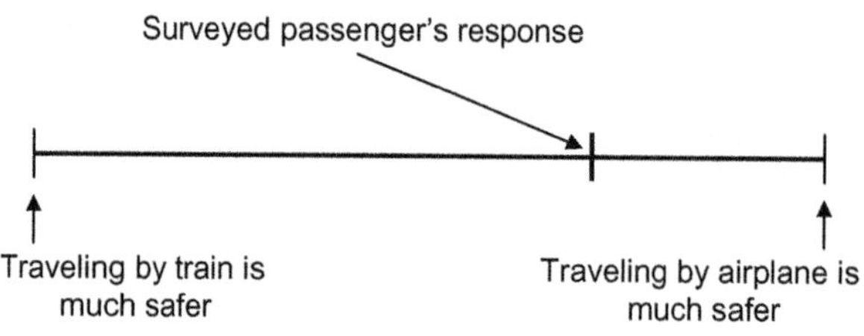

FIG. 8.28

A possible scale to evaluate the feeling of safety when traveling.

A good number of models used in travel behavior analysis are based on utility theory. Utility represents the measure of the alternative attractiveness. Utility depends on the individual that makes the decision. Utility maximization discrete choice models assume that the decision-maker selects the alternative having the highest utility. These, deterministic choice models have been extensively used in transportation engineering. Deterministic choice models assume that the process of choosing a mode of transportation, for example, is essentially deterministic and that the individual always chooses the mode of transportation in the same manner. In other words, it is assumed that if the individual choose the train once when traveling from city A to city B, he/she will also choose the train in subsequent situations. Many real life examples showed that the assumption about choosing the alternative having the highest utility has not been always correct. This caused development of the choice models that include also probabilistic aspect. Stochastic choice models assume that the choice process is subject to many random effects that cannot be precisely perceived. Stochastic models more realistically describe the process of passenger choice than deterministic models.

Let us note the passenger q that travels between two cities. We denote by A_i one of the possible alternatives to make this trip. We also denote by choice set C_q that represent set of considered alternatives. To every alternative we join function V_{iq} called the choice function. The choice function is a function of the alternative attributes and the characteristics of the passenger. The choice function is most of the form:

$$V_{iq} = \sum_j a_i x_{iqj} \tag{8.58}$$

where:

x_{iqj} are variables affecting the choice of mode that relate to alternative A_i (For example, these variables could be total travel time, total travel cost, number of transfers, walking distance, schedule reliability, etc.); and

a_i are parameters estimated when calibrating the choice model that indicate the effect of variables x_{iqj} on the choice of alternative A_i.

The choice process is subject to many random effects. The reason for this is that passengers give different evaluations to the utility they have by opting for a specific mode of transportation. For example, passengers often do not have complete information on the frequency of competitive modes of transportation, or on the departure schedules. In some cases, passengers do not know precisely the cost of some of the modes. In the same vein, some passengers do not opt for a certain alternative even though it would be very logical to do so because of its advantages compared to other alternatives.

We denote by U_{iq} the utility that passenger q connects with alternative A_i. The utility U_{iq} is given by:

$$U_{iq} = V_{iq} + \varepsilon_{iq} \tag{8.59}$$

where V_{iq} is the deterministic part of the utility, and ε_{iq} is the random term. The random term ε_{iq} is a *random* variable that reflects the stochastic nature of the choice of the alternatives.

There is usual assumption that decision maker always choose the alternative with the highest utility. The probability $P(i)$ that alternative A_i is chosen by passenger q from choice set C_q equals:

$$P(i) = P\left[U_{iq} \geq U_{jq} \quad \forall j \in C_q\right] \tag{8.60}$$

For a different probability density functions of the random term, different stochastic choice models are obtained.

8.13 LOGIT MODEL

One of the best known stochastic choice models is the *logit* model (McFadden, 1973; Domencich and McFadden, 1975; Koppelman, 1976; Ortúzar, 1983; Daly, 1987; Daganzo and Kusnic, 1993; Bhat, 1997; Bierlaire, 1998; Koppelman and Wen, 1998, 2000; Hunt, 2000; Wen and Koppelman, 2001; Hensher and Greene, 2003; Bhat and Guo, 2004; Koppelman and Bhat, 2006). This model was first introduced in the case of binary choice. The *multinomial logit* represents the generalization of the original model to more than two alternatives. The family of logit models was extensively used in various transportation studies. The logit model is based on the assumption that random variables representing random terms are independent and distributed by *Gumbel*'s probability density function.

The probability that the alternative A_i will be chosen equals:

$$P(i) = \frac{e^{V_i}}{\sum_k e^{V_k}} \tag{8.61}$$

Eq. (8.48) is known as a *logit* model. The standard utility function of the logit model is linear. In some cases, the choice function is not linear so the least squares method or the multiple regression technique can only be used to estimate the parameters of the choice function in special cases. The maximum likelihood method is used to estimate parameters of choice functions. Choice models can be based on disaggregated or aggregated data. Choice models based on disaggregated data presume that each passenger evaluates the advantages and defects of each alternative differently, so that the passenger population must be polled before estimating the parameters of the choice function. In this case, each pooled passenger corresponds to a different choice function value. For choice models based on aggregate data, all passengers correspond to the same choice function value, since it is assumed that the variables in the choice function are equal for all passengers. Therefore, the probability of choosing a certain alternative for aggregated data is equal for all passengers.

Let us consider the following example. Passengers traveling between node A and node B can use private cars, or public transit (Fig. 8.29).

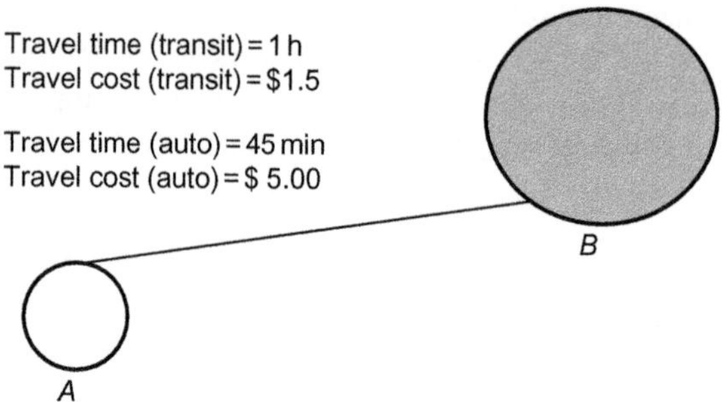

FIG. 8.29

Logit model: passengers traveling between node A and node B can use private cars, or public transit.

A modal split has been calibrated using the maximum likelihood technique (an advanced statistical method). The following equations describing utilities have been obtained:

$$U_{\text{Auto}} = 2.2 - 0.25C - 0.02T \tag{8.62}$$

$$U_{\text{Transit}} = 0.3 - 0.25C - 0.02T \tag{8.63}$$

where:

C is the out-of-pocket cost [$]; and
T is the travel time [minutes].

Calibrated logit models helps us to answer the following questions: How many trips will be between A and B by auto? How many trips will be between A and B by transit? The following are characteristics of the competitive modes:

- Travel time (transit) = 1 h;
- Travel cost (transit) = $1.5;
- Travel time (auto) = 45 min; and
- Travel cost = $5.00 (including parking).

The utilities are:

$$U_{\text{Auto}} = 2.2 - 0.25(5) - 0.02(45) = 0.05$$

$$U_{\text{Transit}} = 0.3 - 0.25(1.5) - 0.02(60) = -1.275$$

Estimated probabilities of travel by competitive modes are:

$$P_{\text{Auto}} = \frac{e^{\text{Auto}}}{e^{\text{Auto}} + e^{\text{Transit}}} = \frac{e^{0.05}}{e^{0.05} + e^{-1.275}} = 0.79 \tag{8.64}$$

$$P_{\text{Transit}} = \frac{e^{\text{Transit}}}{e^{\text{Auto}} + e^{\text{Transit}}} = \frac{e^{-1.275}}{e^{0.05} + e^{-1.275}} = 0.21 \tag{8.65}$$

How should we interpret the obtained results? We can express obtained probabilities in percentages and say: (a) The probability that a traveler from *A* to *B* uses auto is 79%; (b) The probability that a traveler from *A* to *B* uses transit is 21%. Similar interpretation is that 79% of travelers between *A* and *B* use auto, and 21% of travelers use transit. Why is this important? The Logit model and similar models are valuable tools in the transportation planning process. Logit model enable us to perform sensitivity analysis. Let us imagine that the auto cost is extremely low and equal to $1.00. In this case, the model predicts a ridership of 9% for the bus (compared to 21% previously). The bus still captures a small fraction of the riders. If the auto cost is $20.00, the model predicts a ridership of 9% for the auto mode. The cost of auto is quite high and forces many decision makers to "walk away" from auto mode and take the bus.

There are a lot of variations, modifications and improvements of the original logit model (multinomial logit model, nested logit model, cross-nested logit model, *C*-logit).

EXAMPLE 8.12

Assume that passengers taking a private trip on a certain route choose between two possible modes of transportation solely on the basis of the travel cost. We assume that the following choice function has been calibrated:

$$V_i = -0.1694c_i \tag{8.66}$$

where V_i is the utility of the *i*th alternative and c_i is the travel cost of the *i*th alternative. Data on travel costs are given in Table 8.25.

Table 8.25 Data on Travel Costs

Transportation Mode	Travel Cost
A_1	5
A_2	10

The probabilities that the competitive transportation modes will be chosen are, respectively, equal to:

$$p_1 = \frac{e^{5(-0.1694)}}{e^{5(-0.1694)} + e^{10(-0.1694)}} = 0.7$$

$$p_2 = \frac{e^{10(-0.1694)}}{e^{5(-0.1694)} + e^{10(-0.1694)}} = 0.3$$

We assume that a new mode of transportation joins the route in question. The cost c_3 of a new mode equals 7. We then calculate the market share between three modes of transportation. The probabilities of different modes of transportation being used are:

$$p_1 = \frac{e^{5(-0.1694)}}{e^{5(-0.1694)} + e^{10(-0.1694)} + e^{7(-0.1694)}} = 0.47$$

$$p_2 = \frac{e^{10(-0.1694)}}{e^{5(-0.1694)} + e^{10(-0.1694)} + e^{7(-0.1694)}} = 0.20$$

$$p_3 = \frac{e^{7(-0.1694)}}{e^{5(-0.1694)} + e^{10(-0.1694)} + e^{7(-0.1694)}} = 0.33$$

As we can see, by introducing a third mode of transportation on the route in question, the share of the first mode decreases from 70% of all passengers transported to 47%, and the second mode goes from 30% to 20%.

EXAMPLE 8.13

Two air carriers fly between New York and Mexico City (Fig. 8.30).

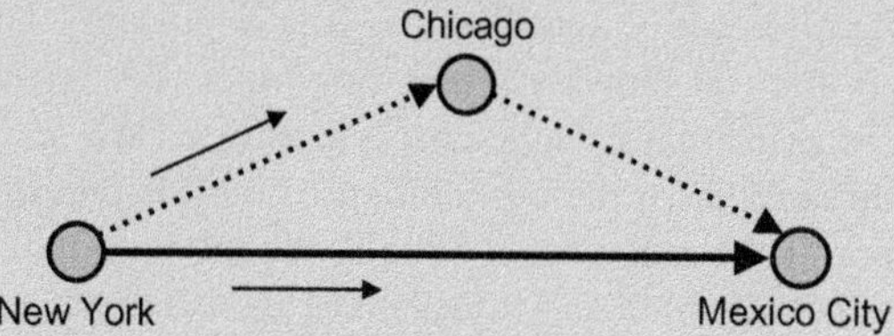

FIG. 8.30

Route choice in air transportation.

Carrier 1 flies non-stop between New York and Mexico City. The average ticket price equals \$434 (round trip). The flight time equals 5 h 5 min (305 min). When flying from New York to Mexico City, carrier 2 makes a stopover in Chicago. The average ticket price equals \$374. Flight time (including change of the plane in Chicago) equals 8 h 10 min (490 min). Air passengers can choose a non-stop flight, or less expensive flight with one stopover. The following choice function has been calibrated:

$$V_i = -0.05 \times c_i - 0.02 \times t_i \tag{8.67}$$

(a) Calculate the market share of every air carrier. (b) Carrier 1 wants to increase the average ticket price to \$470. Estimate the new market shares.

Solution

(a) The utilities are:

$$V_1 = -0.05 \times 434 - 0.02 \times 305 = -27.8$$

$$V_2 = -0.05 \times 374 - 0.02 \times 490 = -28.5$$

The probabilities that the competitive air carriers will be chosen are, respectively, equal to:

$$p_1 = \frac{e^{-27.8}}{e^{-27.8} + e^{-28.5}} = 0.668188$$

$$p_2 = \frac{e^{-28.5}}{e^{-27.8} + e^{-28.5}} = 0.33812$$

These probabilities represent carriers' market shares. In other words, carrier 1 can attract approximately 67% of the total market, while carrier 2 captures 33%.

(b) The new market shares are:

$$V_1 = -0.05 \times 470 - 0.02 \times 305 = -29.6$$

$$V_2 = -0.05 \times 374 - 0.02 \times 490 = -28.5$$

The new values of probabilities are, respectively, equal to:

$$p_1 = \frac{e^{-29.6}}{e^{-29.6} + e^{-28.5}} = 0.24974$$

$$p_2 = \frac{e^{-28.5}}{e^{-29.6} + e^{-28.5}} = 0.75026$$

We conclude that after introducing the new ticket prices, air carrier 1 would significantly decrease market share (from approximately 67% of the market to 25% of the total market).

8.13.1 INDEPENDENCE OF IRRELEVANT ALTERNATIVES PROPERTY

One of the most important characteristics of the multinomial logit model (MNL) is its independence from irrelevant alternatives (IIA) property. The premise is that additional alternatives are irrelevant to the decision of deciding among the two alternatives in the pair. Let us clarify this property by using the following example. Passengers on a certain route choose between train, bus, and private cars. The probabilities of different modes of transportation being used are:

$$P(\text{Train}) = \frac{e^{V(\text{Train})}}{e^{V(\text{Train})} + e^{V(\text{Bus})} + e^{V(\text{Car})}}$$

$$P(\text{Bus}) = \frac{e^{V(\text{Bus})}}{e^{V(\text{Train})} + e^{V(\text{Bus})} + e^{V(\text{Car})}}$$

$$P(\text{Car}) = \frac{e^{V(\text{Car})}}{e^{V(\text{Train})} + e^{V(\text{Bus})} + e^{V(\text{Car})}}$$

The ratios of each pair of these probabilities are, respectively, equal to:

$$\frac{P(\text{Train})}{P(\text{Bus})} = \frac{\dfrac{e^{V(\text{Train})}}{e^{V(\text{Train})} + e^{V(\text{Bus})} + e^{V(\text{Car})}}}{\dfrac{e^{V(\text{Bus})}}{e^{V(\text{Train})} + e^{V(\text{Bus})} + e^{V(\text{Car})}}} = \frac{e^{V(\text{Train})}}{e^{V(\text{Bus})}} = e^{V(\text{Train}) - V(\text{Bus})}$$

$$\frac{P(\text{Train})}{P(\text{car})} = \frac{\dfrac{e^{V(\text{Train})}}{e^{V(\text{Train})} + e^{V(\text{Bus})} + e^{V(\text{Car})}}}{\dfrac{e^{V(\text{Car})}}{e^{V(\text{Train})} + e^{V(\text{Bus})} + e^{V(\text{Car})}}} = \frac{e^{V(\text{Train})}}{e^{V(\text{Car})}} = e^{V(\text{Train}) - V(\text{Car})}$$

$$\frac{P(\text{Bus})}{P(\text{Car})} = \frac{\dfrac{e^{V(\text{Bus})}}{e^{V(\text{Train})} + e^{V(\text{Bus})} + e^{V(\text{Car})}}}{\dfrac{e^{V(\text{Car})}}{e^{V(\text{Train})} + e^{V(\text{Bus})} + e^{V(\text{Car})}}} = \frac{e^{V(\text{Bus})}}{e^{V(\text{Car})}} = e^{V(\text{Bus}) - V(\text{Car})}$$

As can be seen from relations (), the ratio of probabilities for any pair of alternatives depends purely on the attributes of those alternatives. This ratio does not depend on the attributes of the third alternative, fourth alternative, etc. This ratio is unchanged in any case (whether the third, or the fourth transportation mode is available or not).

There is a criticism of the MNL, for its independence of irrelevant alternatives (IIA) property. The unwanted characteristic of the IIA property means that the establishment of a new transportation mode on a considered route will decrease the probability of existing transportation modes proportionally to their probabilities prior to the change.

Some transportation modes could be similar. They could share attributes that are not contained in the utility function. For example, buses and trams in public transit have similar fare structure, similar level of privacy, etc. The Nested Logit (NL) model (Daly and Zachary, 1979; Williams, 1977) put together transportation modes (alternatives) that are more similar to each. In other words the NL model forms groups (nests) of similar alternatives. In this way, the unwanted characteristic of the IIA property could be prevailed.

8.13.2 LOGIT MODEL ESTIMATION

Let us study travelers that have to choose transportation mode. Let us denote by I the set of individuals, and by J the set of alternatives (transportation modes) to be chosen. We collect the data related to the travelers' choices of transportation modes, as well as the data related to the values of the choice function's variables. We note the i-th traveler, and the j-th alternative (transportation mode). The number of probabilities p_{ij} that are to be calculated in Logit model based on disaggregated data (in the case of disaggregated data, the values of the variables in the choice function are different for all travelers) is very large. This number equals the product of the number of observed travelers and the number of competitive transportation modes. In the case of a logit model based on aggregated data, we assume that the values of the variables in the choice function are equal for all travelers. Consequently, the probability of choosing specific transportation mode is constant, and does not vary from traveler to traveler. In other words, the following is satisfied in the case of aggregate logit model:

$$p_{ij} = p_j \quad i = 1, 2, \ldots, |I| \tag{8.68}$$

Parameters of the logit model are estimated by the maximum likelihood method. This method discovers parameters that maximize the likelihood that the sample was produced from the model with the chosen parameter values. In other words, the maximum likelihood method finds the values of parameters that are most likely to produce the choices detected in the sample. The likelihood function $L(\alpha)$ for a sample composed of $|I|$ individuals, and $|J|$ alternatives is defined in the following way:

$$L(\alpha) = \prod_{\forall i \in I} \prod_{\forall j \in J} (P_{ij}(\alpha))^{\delta_{ij}} \tag{8.69}$$

where:

$$\delta_{ij} = \begin{cases} 1, & \text{if } i\text{th individual chooses } j\text{th alternative} \\ 0, & \text{otherwise} \end{cases} \tag{8.70}$$

The values of the parameters that maximize the likelihood function are obtained by equating the first derivative of the likelihood function to zero. The log of a likelihood function produces the same maximum as the likelihood function. We introduce into the analysis log of a likelihood function, as a replacement for the likelihood function itself, since it is easier to differentiate log of a likelihood function. Log-likelihood function $LL(\alpha)$ equals:

$$LL(\alpha) = \mathrm{Log}\,(L(\alpha)) \tag{8.71}$$

$$LL(\alpha) = \prod_{\forall i \in I} \prod_{\forall j \in J} \delta_{ij} \times \ln\left(P_{ij}(\alpha)\right) \tag{8.72}$$

By equating the first derivative of the likelihood function $LL(\alpha)$ to zero we obtain the values of the parameters that maximize the likelihood function.

EXAMPLE 8.14

Drivers taking a trip between two nodes choose between two possible routes on the basis of perceived travel time. Data on the perceived travel time are given in Table 8.26.

Table 8.26 Data on the Perceived Travel Time

Route	Perceived Travel Time
1	5
2	10

A traffic engineer observed the decisions of 1000 drivers. The total of 700 drivers chose the first route. The remaining 300 drivers chose the second route. Calibrate the binomial aggregate logit model based on these data.

Solution

Since we have to calibrate the binomial aggregate logit model, we assume that the probability of choosing specific transportation mode is constant, and does not vary from traveler to traveler. The choice function equals:

$$V(i) = a \times t_i \quad i = 1,2 \tag{8.73}$$

where:
a is the parameter to be estimated; and
t_i is the perceived travel time of the ith route.

The probabilities of route 1 and route 2 being chosen, by any driver, are, respectively, equal to:

$$P_1 = \frac{e^{5a}}{e^{5a} + e^{10a}} \tag{8.74}$$

$$P_2 = \frac{e^{10a}}{e^{5a} + e^{10a}} \tag{8.75}$$

Log-likelihood function $LL(a)$ equals:

$$LL(a) = \prod\nolimits_{\forall i \in I} \prod\nolimits_{\forall j \in J} \delta_{ij} \times \ln(P_{ij}(a)) \tag{8.76}$$

$$LL(a) = 700 \times \ln(P_1) + 300 \times \ln(P_2) \tag{8.77}$$

$$LL(a) = 700 \times \ln\left(\frac{e^{5a}}{e^{5a} + e^{10a}}\right) + 300 \times \ln\left(\frac{e^{10a}}{e^{5a} + e^{10a}}\right) \tag{8.78}$$

$$LL(a) = 700 \times 5a + 300 \times 10a - 1000 \times \ln(e^{5a} + e^{10a}) \tag{8.79}$$

By equating the first derivative $\frac{dLL(a)}{da}$ of the likelihood function $LL(a)$ to zero we get the following equation:

$$1500 \times e^{5a} - 3500 \times e^{10a} = 0 \tag{8.80}$$

The solution of this equation equals:

$$a = -0.1694$$

The probabilities of route 1, and route 2 being chosen are, respectively, equal to:

$$P_1 = \frac{e^{5(-0.1694)}}{e^{5(-0.1694)} + e^{10(-0.1694)}} = 0.7$$

$$P_2 = \frac{e^{10(-0.1694)}}{e^{5(-0.1694)} + e^{10(-0.1694)}} = 0.3$$

EXAMPLE 8.15

Business travelers traveling on a specific route during a winter choose between the airplane and train uniquely on the basis of the schedule reliability. Let there be a possible evaluation from 0 to 10, with 0 denoting a totally unreliable transportation mode, and 10 an exceptionally reliable mode. The transportation planner has to make the travelers' choice prediction between air transportation and train. The following are assumed utility functions of the competitive transportation modes:

$$V_{\text{Air}} = a \times r_{\text{Air}} \tag{8.81}$$

$$V_{\text{Train}} = a \times r_{\text{Train}} \tag{8.82}$$

where:

r_{Air} is the perceived air transportation schedule reliability; and

r_{Train} is the perceived train schedule reliability.

Transportation planner assumes that the values of the variables in the choice function are different for all travelers. She observed the mode choice of seven travelers. The estimated schedule reliability and performed choices of seven passengers are shown in Table 8.27.

Table 8.27 Estimated Schedule Reliability and Performed Choices of Seven Travelers

Traveler	Estimated Airplane Schedule Reliability	Estimated Train Schedule Reliability	Chosen Transportation Mode
1	9	7	Air transportation
2	8	6	Air transportation
3	9	8	Air transportation
4	7	10	Train
5	8	9	Train
6	5	7	Train
7	9	10	Train

Calculate the value of the parameter a that maximizes the likelihood function.

Solution

We treat air transportation as the first transportation mode ($j=1$), and train as the second transportation mode ($j=2$). For the considered sample of travelers, the log-likelihood reads:

$$LL(a) = \prod\nolimits_{\forall i \in I} \prod\nolimits_{\forall j \in J} \delta_{ij} \times \ln\left(P_{ij}(a)\right) \tag{8.83}$$

$$LL(a) = 1$$

$$\ln P_{11} + 0 \times \ln P_{12} + 1 \times \ln P_{21} + 0 \times \ln P_{22} + 1 \times \ln P_{31} + 0 \times \ln P_{32} + 0 \times \ln P_{41} + 1 \times \ln P_{42} + 0 \times \ln P_{51} + 1 \times \ln P_{52} + 0 \times \ln P_{61} + 1 \times \ln P_{62} + 0 \times \ln P_{71} + 1 \times \ln P_{72} \tag{8.84}$$

$$LL(a) = \ln P_{11} + \ln P_{21} + \ln P_{31} + \ln P_{42} + \ln P_{52} + \ln P_{62} + \ln P_{72} \tag{8.85}$$

The probabilities $P_{11}, P_{22}, P_{31},$ and P_{42} are, respectively, equal to:

$$P_{11} = \frac{e^{9a}}{e^{9a} + e^{7a}} \tag{8.86}$$

$$P_{21} = \frac{e^{8a}}{e^{8a} + e^{6a}} \tag{8.87}$$

EXAMPLE 8.15—cont'd

$$P_{31} = \frac{e^{9a}}{e^{9a} + e^{8a}} \tag{8.88}$$

$$P_{42} = \frac{e^{10a}}{e^{7a} + e^{10a}} \tag{8.89}$$

$$P_{52} = \frac{e^{9a}}{e^{8a} + e^{9a}} \tag{8.90}$$

$$P_{62} = \frac{e^{7a}}{e^{5a} + e^{7a}} \tag{8.91}$$

$$P_{72} = \frac{e^{10a}}{e^{9a} + e^{10a}} \tag{8.92}$$

By equating the first derivative $\frac{dLL(a)}{da}$ of the likelihood function $LL(a)$ to zero, we get the following solution:

$$a \approx 10.4$$

8.14 APPLICATION OF THE COMPUTATIONAL INTELLIGENCE TECHNIQUES FOR THE PREDICTION OF TRAVEL DEMAND

The majority of the transportation planning models developed to date are based on analytical relations. Traffic flows are calculated based on these relations. For the period of the last decade, it has been shown that the computational intelligence techniques (fuzzy logic, artificial neural networks, and genetic algorithms) can effectively be utilized in predicting travel demand. Computational intelligence represents a set of nature-inspired computational methodologies capable of calculating with words (fuzzy logic), to learn and adapt (artificial neural networks), and perform stochastic search and optimization (genetic algorithms). The possibility of and justification for using fuzzy logic has gained in importance in the light of a mathematical proof that fuzzy systems are universal approximators.

Contrasting to statistical methods, artificial neural networks and fuzzy systems estimate the functions without specifying the mathematical model that describes the way in which output results depend on input data (artificial neural networks and fuzzy systems are often described as "models without a model"). The models based on computational intelligence techniques are able to make the correct prediction without knowing the functional relationships in effect between individual variables. As in other intelligent systems, these models are also able to generalize, adapt and learn based on new knowledge and new information.

EXAMPLE 8.16

There are 4 industrial towns, denoted by A, B, C, and D, respectively. Passengers depart from industrial towns to tourist cities 1, 2, 3, 4, 5, 6, and 7 (Fig. 8.31). The model based on the combination of fuzzy logic and genetic algorithm (Kalić and Teodorović, 2003) was used to estimate yearly number of air passengers between the observed industrial towns and tourist resorts.

(Continued)

EXAMPLE 8.16—cont'd

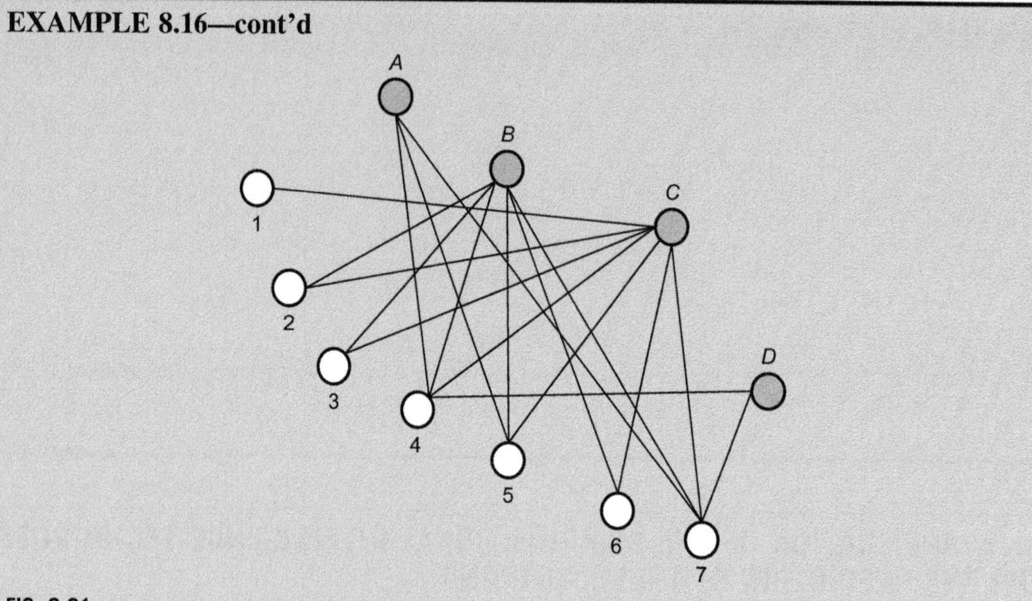

FIG. 8.31

Industrial cities and tourist resorts connected by air transportation.

The air transportation network consists of 4 industrial towns, 7 tourist resorts, and 18 links. The flows were estimated based on the number of passengers generated by industrial towns and the number of passengers attracted by tourist resorts. The example considered is based on real data collected in 1989 and 1990. The developed approximate reasoning algorithm used to estimate yearly number of passengers was composed of the rules of the following type:

If the number of passengers departing from an industrial town to tourist resort is LARGE and the number of passengers from industrial towns arriving in a tourist resort is LARGE

Then the number of passengers between the observed pair industrial town-tourist resort is LARGE

The model developed has been tested on the real data, and the obtained results are given in Table 8.28 and Fig. 8.32.

Table 8.28 Comparison of Real and Estimated Passenger Flows

Pair of Cities	Real Value of Passenger Flow	Estimated Value of Passenger Flow Obtained by Fuzzy Logic and Genetic Algorithm
(C,5)	106,332	98,839
(B,5)	67,607	67,975
(A,5)	4619	4110
(C,7)	16,296	19,639
(B,7)	3180	2902
(A,7)	468	1264
(D,7)	585	1264
(C,1)	24,006	21,996
(B,1)	3524	3448

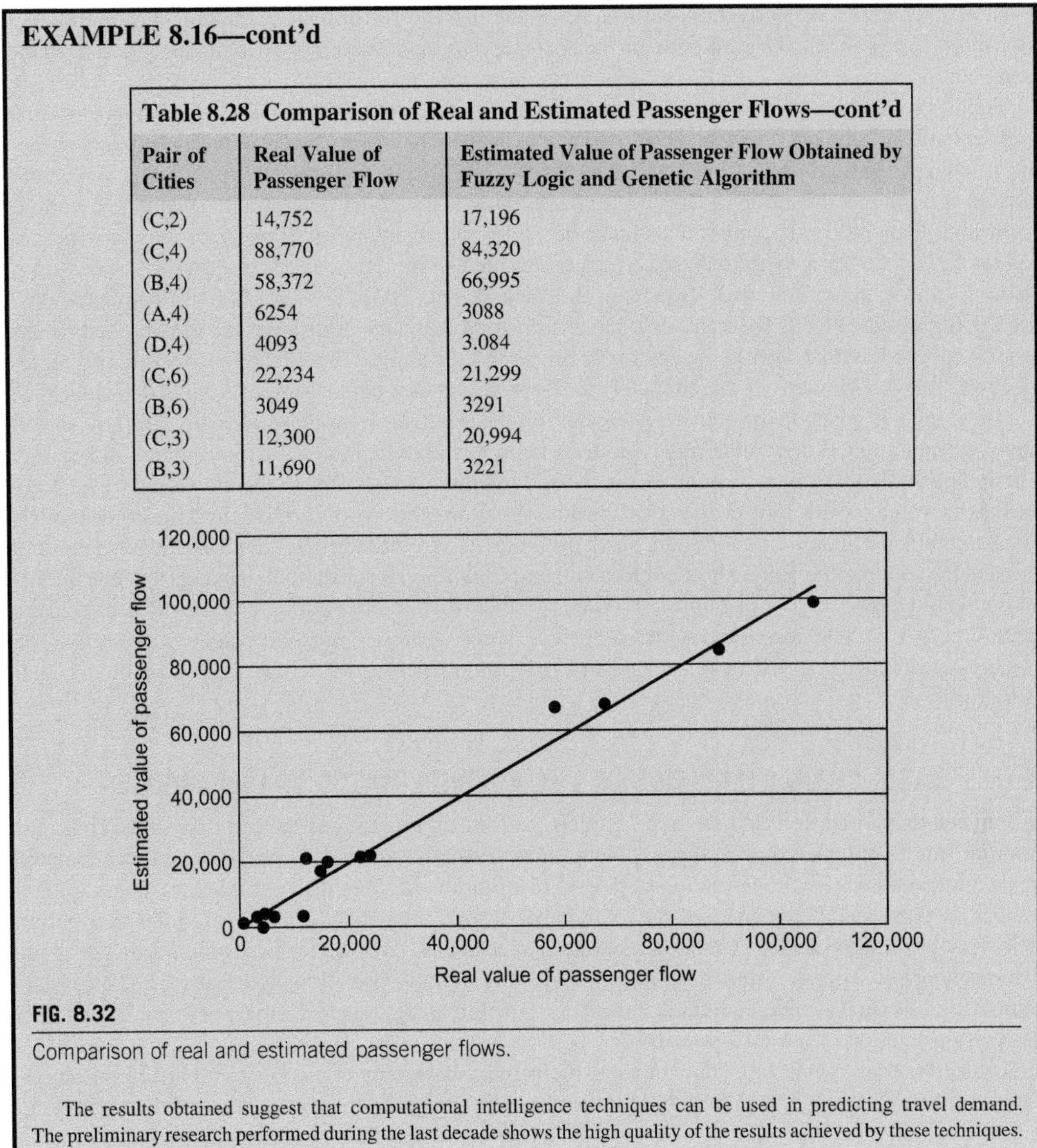

EXAMPLE 8.16—cont'd

Table 8.28 Comparison of Real and Estimated Passenger Flows—cont'd

Pair of Cities	Real Value of Passenger Flow	Estimated Value of Passenger Flow Obtained by Fuzzy Logic and Genetic Algorithm
(C,2)	14,752	17,196
(C,4)	88,770	84,320
(B,4)	58,372	66,995
(A,4)	6254	3088
(D,4)	4093	3.084
(C,6)	22,234	21,299
(B,6)	3049	3291
(C,3)	12,300	20,994
(B,3)	11,690	3221

FIG. 8.32

Comparison of real and estimated passenger flows.

The results obtained suggest that computational intelligence techniques can be used in predicting travel demand. The preliminary research performed during the last decade shows the high quality of the results achieved by these techniques.

8.15 ACTIVITY-BASED TRAVEL DEMAND MODELS

In previous sections, we explained the "four-step" models. All these models (trip generation models, trip distribution models, modal split models, and route choice models) are *trip-based* models. These models use the individual person *trip* as the unit of analysis. The "four-step" models work at the zone level. In other words, we study the specific zones in the city and perform the prediction of the number of trips that will be generated by this zone. In the next step, the models determine trip distribution from the zone, etc.

People on a daily basis go to work and go to school, go with friends for a dinner, go shopping during the weekend, visit movies, theaters, and go to sport events. In other words, people have needs and wishes to participate in various *activities*. The most frequently, the activities in which we participate are located outside of our home, resulting in our need to travel outside the home. Taking into account the various constraints that could exist, individuals make decisions where and when to participate in activities, as well as the decision how to get to these activities. During the last three decades, the *activity-based travel demand models* have emerged (Bowman and Ben-Akiva, 2001; Jovičić, 2001; Bowman, 2009; Castiglione et al., 2015). These models have been used in many transportation studies. Unlike the "four-step" models, that work at the aggregate zone level, the activity-based models work at a disaggregate person-level. The activity-based travel models represent how persons travel through the whole day.

These models generate the *activities* and identify the destinations of these activities. Activity-based travel demand models determine travel modes and perform prediction of the routes that will be used. These models also take into account the spatial and temporal constraints of the individual traveler activities, as well as of the multiple persons in a household. In other words, unlike the Trip-based models that assume that all trips are made independently, the activity-based travel demand models take into account the cooperation among household members. Children put substantial demands and constraints on household members. For example, a mother or father often, before going to work, drive the kids to school, on a trip excursion usually goes the whole family together, etc. We could conclude that the activity-based travel demand models do not analyze trips independently of other trips made by the other individuals.

8.15.1 BASIC CHARACTERISTICS OF THE ACTIVITY-BASED TRAVEL MODELS

In the first step, traditional "four-step" models perform aggregate estimate of demand. Thus, for example, for a particular zone in the city we predict the total number of trips that could be generated by the observed zone. In the following steps (trip distribution, modal split, route choice) the traditional models disaggregate the estimate of the total demand (total number of trips). Within the distribution step, the total number of trips generated is disaggregated to O-D pairs. In the next step, within each O-D pair the number of trips is further disaggregated to particular transportation modes, etc. The basic characteristic of the traditional models is that they first perform an aggregate estimate of demand, and then do a disaggregation of the total demand.

Unlike the traditional "four step" models, the activity-based travel models first perform disaggregate estimates of demand, and then these estimates are aggregated by space, day, and time. Fig. 8.33 shows a hypothetical daily activity itinerary. The traveler's tour shown contains five trips.

The five trips to be performed are also shown in Fig. 8.34.

The traditional trip-based model would model all five trips independent of the other trips. Contrary to this, the activity-based travel demand models consider the activities "work," "lunch at the restaurant," and "shopping at supermarket" and associated five trips as a part of the same decision-making process. The work at the workplace, work at home, attending school, attending university, shopping, travel as escort passenger, lunch/dinner outside of home, medical check-ups, visiting friends, and recreation represent typical activities.

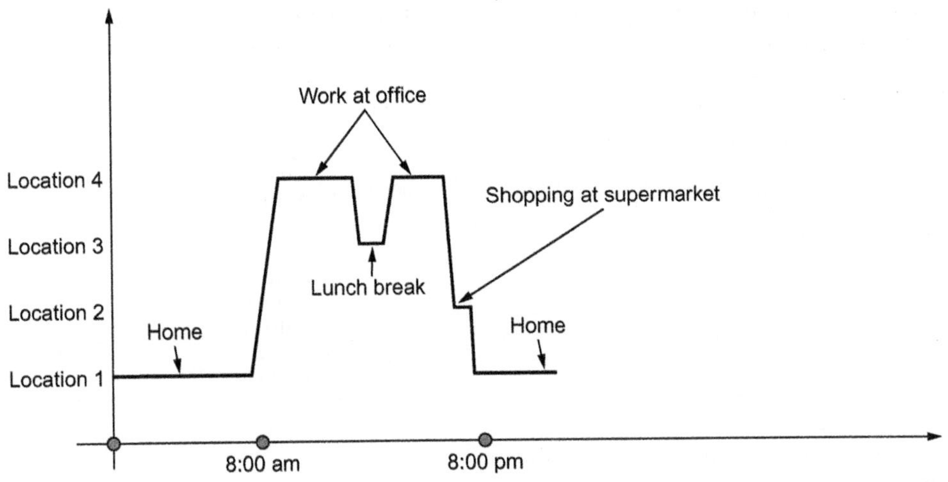

FIG. 8.33

Hypothetical daily activity itinerary (tour that contains five trips).

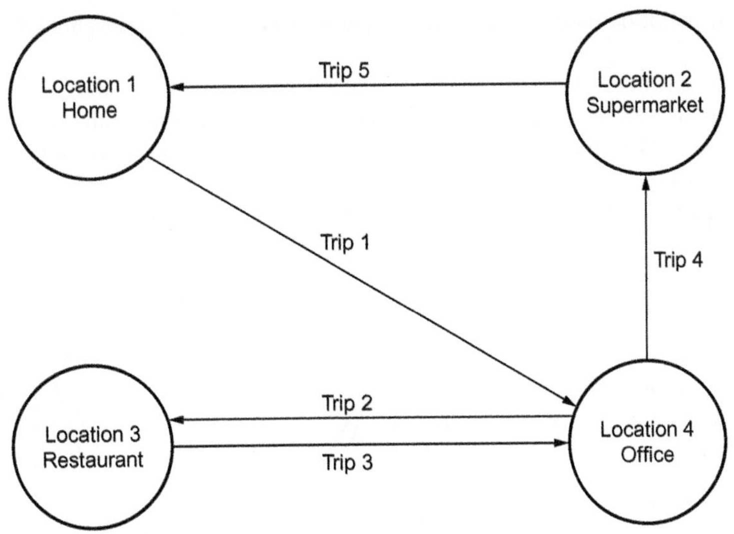

FIG. 8.34

Five trips performed during a hypothetical day.

The activity-based travel demand models take as inputs individual attributes, household attributes, attributes, and actions of other household members, as well as transportation networks characteristics. The following are typical household attributes that are used in the activity-based travel demand models:

- residential location;
- number of persons in the household;

- age of every family member;
- total household income;
- total number of vehicles owned;
- number of employees in the household; and
- number of students in the household.

The following are a person's attributes that are used in the models:

- relationship to householder;
- gender;
- age;
- worker status;
- number of hours worked per week;
- student status;
- grade in school;
- sponsored parking at work; and
- transit pass ownership.

Using the designated inputs, the activity-based models select activities to be carried out, specifying the beginning and the duration of these activities, the destination to be reached in order to perform the activities, transportation mode to be used and participants from the household that participate in the activities.

The activity-based travel demand models assume that the travel demand is a derived demand from participation in various activities. These models focus on the chain of various activities that are spread over a 24-h time period. In essence, the activity-based models simulate the activity-travel decisions. Usually, these models simulate decisions within 30–60-min decision time intervals. The decision makers could be individual members of the household, or the whole household (Fig. 8.35).

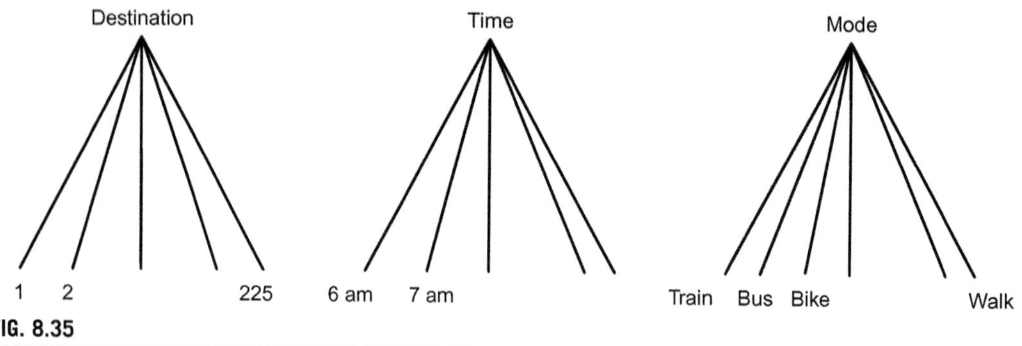

FIG. 8.35

Choices of space, time and mode.

Activity-based models use discrete choice models (*logit* model) to describe the choice of destination, starting time, and mode of travel (Fig. 8.31). Some choices are completely independent, while some others are interdependent. For example, frequently travelers choose among combinations of transportation modes and destinations, while number of cars (and cars usage) in a household highly depend on the locations where household members work, etc.

The activity-based models use simulation techniques to predict future transportation activities.

8.16 PROBLEMS

1. Explain the macroscopic transportation demand models and microscopic transportation demand models.
2. Explain qualitative and quantitative forecasting techniques.
3. Define and explain transportation demand components: trend component, cyclical component, seasonal peaks, and random variations.
4. Explain time series models.
5. The actual numbers of passengers at Airport X are given in Table 8.29.

Table 8.29 The Number of Passengers at Airport X

Year, t	Actual Number of Passengers, A_t
1	838,156
2	1,036,311
3	1,155,166
4	1,434,454
5	1,688,247
6	2,020,291
7	2,047,016
8	2,280,972

Predict the number of passengers in 9th year at Airport X, by using the weighted moving average method. Use a 3-year moving average, and the following set of weights:

$$w_8 = 0.7, \quad w_7 = 0.2, \quad w_6 = 0.1$$

6. Predict the number of passengers at Airport Y, by using the exponential smoothing method. The actual numbers of passengers are given in Table 8.30. Use the smoothing constant $\alpha = 0.7$. Assume that $F_1 = A_1$.
7. Explain the logistic curve.
8. Explain the usual measures for measuring the accuracy of the forecast.
9. Consider the following x and y values (Table 8.31):

   Calculate the least square regression line.

10. Explain the four-step process.
11. Explain the route choice and traffic assignment problems.
12. Explain user equilibrium, system optimum and price of anarchy.
13. Explain Braess' paradox.

Table 8.30 The Number of Passengers at Airport *Y*

Year, t	Actual Number of Passengers, A_t
1	825,159
2	1,084,422
3	1,195,246
4	1,487,653
5	1,692,358
6	2,043,291
7	2,089,214
8	2,295,876

Table 8.31 *X* and *Y* Values

X	Y
70	4.2
71	4.6
72	4.8
74	5.1
75	5.4
77	5.8
79	6.5
82	7.4

14. There are two alternative routes 1 and 2 (Fig. 8.36). Route 1 begins from node r, goes through intersection 1, and terminates at node s. Route 2 also begins from node r, proceeds through intersection 2, and ends at node s. The total number of trips that should be performed between node r and node s is known, and equal to 2000. The route performance functions equal:

$$T_1(x_1) = 30 + 0.03 \times x_1$$

$$T_2(x_2) = 25 + 0.04 \times x_2$$

where x_1, x_2 is the number of trips along routes 1 and 2.
Assign flows to routes 1 and 2 in such a way to make minimal total travel time of all traffic network users.
15. Explain dynamic traffic assignment.
16. Explain independence of irrelevant alternatives property.
17. Drivers that take a trip between two nodes decide between two achievable routes on the basis of perceived travel time. Data on the perceived travel time are given in Table 8.32.

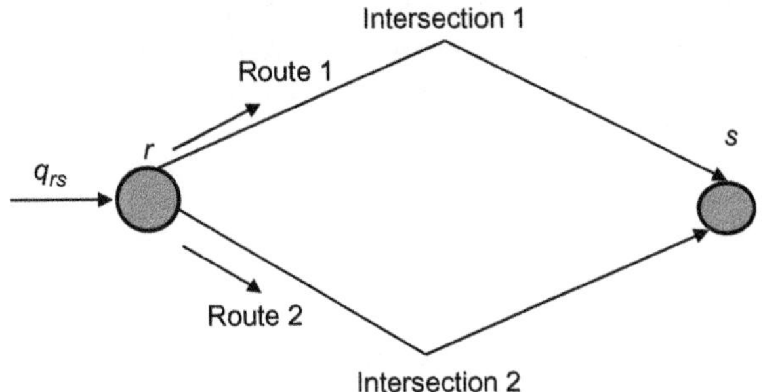

FIG. 8.36

Two alternative routes.

Table 8.32 Perceived Travel Times

Route	Perceived Travel Time
1	10
2	20

An analyst examined the decisions of 2000 drivers. A total of 1500 drivers chose the first route. The remaining 500 drivers chose the second route. Calibrate the binomial aggregate logit model based on these data.

18. Explain the potential applications of the computational intelligence techniques for the prediction of travel demand.
19. Explain the activity-based travel demand models.
20. Which household attributes are used in the activity-based travel demand models?

REFERENCES

Aerde, M.V., Hellinga, B., Baker, M., Rakha, H., 1996. INTEGRATION: an overview of current simulation features. In: Transportation Research Board 75th Annual Meeting Compendium of Papers, Washington, DC, January 7–11.

Balakrishna, R., Wen, Y., Ben-Akiva, M., Antoniou, C., 2008. Simulation-based framework for transportation network management for emergencies. Transp. Res. Rec.: J. Transp. Res. Board 2041, 80–88.

Barcelo, J., Casas, J., 2002a. Dynamic network simulation with AIMSUN. In: International Symposium Proceedings on Transport Simulation, Yokohama. Kluwer Academic Publishers.

Barcelo, J., Casas, J., 2002b. Heuristic dynamic assignment based on microscopic simulation. In: Proceedings of the 9th Meeting of the Euro Working Group on Transportation, Bari, Italy, June 10–13.

Barcelo, J., Casas, J., 2006. Stochastic heuristic dynamic assignment based on AIMSUN microscopic traffic simulator. Transp. Res. Rec.: J. Transp. Res. Board 1964, 70–80.

Beckman, M.J., McGuire, C.B., Winsten, C.B., 1956. Studies in the Economics of Transportation. Yale University Press, New Haven.

Behbehani, R., Kanafani, A., 1980. Demand and supply models of air traffic in international markets. Research report, ITS-WP-80-5, Institute of Transportation Studies, University of California at Berkeley, Berkeley.

Ben-Akiva, M.E., 1974. Structure of passenger travel demand models. Transp. Res. Rec. 526, 26–42.

Ben-Akiva, M., Bierlaire, M., 1999. Discrete choice methods and their applications to short-term travel decisions. In: Hall, R. (Ed.), Hanbook of Transportation Science. Kluwer Academic Publishers, Boston/Dordrecht/London.

Ben-Akiva, M., Bierlaire, M., Koutsopoulos, H., Mishalani, R., 1998. DynaMIT: a simulation-based system for traffic prediction. Paper Presented at the DACCORD Short Term Forecasting Workshop, February, Delft, The Netherlands.

Ben-Akiva, M.E., Lerman, S.R., 1985. Discrete Choice Analysis: Theory and Application to Travel Demand. MIT Press, Cambridge, MA.

Bhat, C.R., 1997. A nested logit model with covariance heterogeneity. Transp. Res. B 31, 11–21.

Bhat, C.R., Guo, J., 2004. A mixed spatially correlated logit model: formulation and application to residential choice modelling. Transp. Res. 38B, 147–168.

Bierlaire, M., 1998. Discrete choice models. In: Labb, M., Laporte, G., Tanczos, K., Toint, P. (Eds.), Operations Research in Traffic and Transportation Management. NATOASI Series, Series F: Computer and Systems Sciences, vol. 166. Springer Verlag, Berlin.

Bovy, P.H.L., Fiorenzo-Catalano, S., 2007. Stochastic route choice set generation: behavioural and probabilistic foundations. Transportmetrica 3, 173–189.

Bowman, J.L., 2009. Historical development of activity-based models: theory and practice. Traffic Eng. Control 50, 314–318.

Bowman, J.L., Ben-Akiva, M.E., 2001. Activity-based disaggregate travel demand model system with activity schedules. Transp. Res. 35A, 1–28.

Braess, D., 1968. Uber ein Paradox der Verkerhsplannung. Unternehmenstorchung 12, 258–268.

Caliper Corporation, 2009. A dynamic traffic simulation model on planning networks. TRB Planning Application Conference, Houston, TX, May 20.

Cascetta, E., 1984. Estimation of trip matrices from traffic counts and survey data: a generalised least squares approach estimator. Transp. Res. 18B, 289–299.

Cascetta, E., Nguyen, S., 1988. A unified framework for estimating or updating origin/destination matrices from traffic counts. Transp. Res. 22B, 437–455.

Castiglione, J., Bradley, M., Gliebe, J., 2015. Activity-Based Travel Demand Models: A Primer. Transportation Research Board, Washington DC.

Chiu, Y.C., Zheng, H., Villalobos, J.A., Peacock, W., Henk, R., 2008. Evaluating regional contra-flow and phased evacuation strategies for Texas using a large-scale dynamic traffic simulation and assignment approach. J. Homeland Secur. Emerg. Manag. 5 (1), Article 34.

Daganzo, C.F., 1979. Multinomial Probit: The Theory and Its Applications to Demand Forecasting. Academic Press, New York.

Daganzo, C.F., Kusnic, M., 1993. Two properties of the nested logit model. Transp. Sci. 27, 395–400.

Daganzo, C.F., Sheffi, Y., 1977. On stochastic models of traffic assignment. Transp. Sci. 11 (3), 253–274.

Daly, A., 1987. Estimating "tree" logit models. Transport. Res. B 21 (4), 251–268.

Daly, A.J., Zachary, C., 1979. Improved multiple choice models. In: Hensher, D., Dalvi, Q. (Eds.), Identifying and Measuring the Determinants of Mode Choice. Teakfield, London, pp. 335–357.

de Ortuzar, J.D., Willumsen, L.G., 1990. Modelling Transport. John Willey & Sons, New York.

de Palma, A., Marchal, F., 2002. Real cases applications of the fully dynamic METROPOLIS tool-box: an advocacy for large-scale mesoscopic transportation systems. Netw. Spatial Econ. 2, 347–369.

Dial, R.B., 1971. A probabilistic multipath traffic assignment model which obviates path enumeration. Transp. Res. 5, 83–111.

Dijkstra, E.W., 1959. Note on two problems in connection with graphs (spanning tree, shortest path). Numer. Math. 1, 269–271.

Domencich, T., McFadden, D., 1975. Urban Travel Demand: A Behavioural Analysis. North-Holland, Amsterdam.

Fisk, C.S., 1988. On combining maximum entropy trip matrix estimation with user optimal assignment. Transp. Res. 22B, 69–73.

Florian, M., Nguyen, S., 1978. A combined trip distribution modal split and trip assignment model. Transp. Res. 12, 241–246.

Gartner, N.H., 1980. Optimal traffic assignment with elastic demands: a review. Transp. Sci. 14, 192–208.

Ghobrial, A., Kanafani, A., 1995. Quality of service model of intercity air travel demand. J. Transp. Eng. 121, 135–140.

Goncalves, M.B., Uysseaneto, I., 1993. The development of a new gravity opportunity model for trip distribution. Environ. Plan. A 25, 817–826.

Grilihes, D., 1977. Bor Airport. Graduation thesis, University of Belgrade, Faculty of Transport and Traffic Engineering, Belgrade (in Serbian).

Hall, M.D., Van Vliet, D., Willumsen, L.G., 1980. SATURN—a simulation assignment model for the evaluation of traffic management schemes. Traffic Eng. Control 21, 168–176.

Harker, P.T., 1987. Predicting Intercity Freight Flows. VNU Science Press, Utrecht.

Hensher, D.A., Greene, W.H., 2003. The mixed logit model: the state of practice. Transportation 30, 133–176.

Hensher, D.A., Rose, J.M., Greene, W.H., 2005. Applied Choice Analysis: A Primer. Cambridge University Press, Cambridge.

Hunt, G.L., 2000. Alternative nested logit model structures and the special case of partial degeneracy. J. Reg. Sci. 40, 89–113.

Hyman, G.M., 1969. The calibration of trip distribution models. Environ. Plan. 1, 105–112.

Jansen, G.R.M., Bovy, P.H.L., 1982. The effect of zone size and network detail on all-or-nothing and equilibrium assignment outcomes. Traffic Eng. Control 23, 311–317.

Jara-Díaz, S.R., 2007. Transport Economic Theory. Elsevier Science, Amsterdam.

Jovičić, G., 2001. Activity Based Travel Demand Modelling—A Literature Study, Note 8. Danmarks Transport Forksning, Copengagen.

Kalić, M., Teodorović, D., 2003. Trip distribution using fuzzy logic and genetic algorithm. Transp. Plann. Technol. 26, 213–238.

Kanafani, A., 1983. Transportation Demand Analysis. McGraw Hill Book Company, New York.

Koppelman, F.S., 1976. Guidelines for aggregate travel prediction using disaggregate choice models. Transp. Res. Rec. 610, 19–24.

Koppelman, F., Bhat, C., 2006. Self Instructing Course in Mode Choice Modeling: Multinomial and Nested Logit Models. U.S. Department of Transportation Federal Transit Administration.

Koppelman, F.S., Wen, C.-H., 1998. Alternative nested logit models: structure, properties and estimation. Transport. Res. B 32, 289–298.

Koppelman, F.S., Wen, C.H., 2000. The paired combination logit model: properties, estimation and application. Transp. Res. 34B, 75–89.

Luce, R., 1959. Individual Choice Behavior: A Theoretical Analysis. John Wiley and Sons, New York.

Mahmassani, H.S., 2001. Dynamic network traffic assignment and simulation methodology for advanced system management applications. Netw. Spat. Econ. 1 (3/4), 267–292.

Mahmassani, H.S., Hawas, Y., 1997. Data requirements for development, calibration of dynamic traffic models, and algorithms for ATMS/ATIS. In: Proceedings of the 76th Annual Meeting of the Transportation Research Board, Washington, DC.

Mahmassani, H., Qin, X., Zhou, X., 2004. DYNASMART-X evaluation for real time tmc application: Irvine test bed: TrEPS phase 1.5b final report. Technical report. Maryland Transportation Initiative, University of Maryland, College Park, Maryland.

Manski, C., 1977. The structure of random utility models. Theor. Decis. 8, 229–254.

McFadden, D., 1973. Conditional logit analysis of quantitative choice behavior. In: Zaremmbka, P. (Ed.), Frontier of Econometrics. Academic Press, New York.

Ortúzar, J. de D., 1982. Fundamentals of discrete multimodal choice modelling. Transp. Rev. 2, 47–78.

Ortúzar, J. de D., 1983. Nested logit models for mixed-mode travel in urban corridors. Transp. Res. 17A, 283–299.

Patriksson, M., 1994. The Traffic Assignment Problem: Models and Methods. VSP, Utrecht.

Peeta, S., Ziliaskopoulos, A.K., 2001. Foundations of dynamic traffic assignment: the past, the present and the future. Netw. Spat. Econ. 1, 233–265.

Quandt, R.E., Baumol, W.J., 1966. The demand for abstract modes: theory and measurements. J. Reg. Sci. 6, 13–26.

Robertson, D.I., 1969. TRANSYT: a traffic network study tool. TRRL Report LR 253, Transport and Road Research Laboratory, Crowthorne.

Robillard, P., 1975. Estimating the O–D matrix from observed link volumes. Transp. Res. 9, 123–128.

Roughgarden, T., 2005. Selfish Routing and the Price of Anarchy. MIT Press, Cambridge, MA.

Safwat, K.N.A., Magnanti, T., 1988. A combined trip generation, trip distribution, modal split and trip assignment model. Transp. Sci. 22, 14–30.

Sheffi, Y., 1985. Urban Transportation Networks. Prentice Hall, Englewood Cliffs, NJ.

Sheffi, Y., Hall, R., Daganzo, C.F., 1982. On the estimation of the multinomial probit model. Transp. Res. 16A, 447–456.

Sloboden, J., Lewis, J., Alexiadis, V., Chiu, Y.-C., Nava, E., 2012. Traffic Analysis Toolbox Volume XIV: Guidebook on the Utilization of Dynamic Traffic Assignment in Modeling. U.S. Department of Transportation Federal Highway Administration, Washington, DC.

Smeed, R.J., 1968. Traffic studies and urban congestion. J. Transport Econ. Policy 2, 2–38.

Spiess, H., Florian, M., 1989. Optimal strategies: a new assignment model for transit networks. Transp. Res. 23B, 82–102.

Steenbrink, P.A., 1974. Optimisation of Transport Networks. John Wiley & Sons Inc., New York.

Teodorović, D., Kikuchi, S., 1990. Transportation route choice model using fuzzy inference technique. In: Proceedings of ISUMA'90, The First Symposium on Uncertainty Modelling and Analyses, University of Maryland, College Park, Maryland, pp. 140–145.

Teodorović, D., Trani, A., 2002. Introduction to Transportation Engineering. Virginia Tech, Blacksburg/Falls Church. Unpublished lecture notes.

Teodorović, D., Vukadinović, K., 1998. Traffic Control and Transport Planning: A Fuzzy Sets and Neural Networks Approach. Kluwer Academic Publishers, Boston/Dordrecht/London.

Wardrop, J.G., 1952. Some theoretical aspects of road traffic research. Proc. Inst. Civil Eng. II 1, 325–362.

Wen, C.-H., Koppelman, F.S., 2001. The generalized nested logit models. Transport. Res. B 35 (7), 627–641.

Wen, Y., Balakrishna, R., Ben-Akiva, M., Smith, S., 2006a. Online deployment of dynamic traffic assignment: architecture and run-time management. IEE Proc. Intell. Transp. Syst. (now IET Intell. Transp. Syst.) 153 (1), 76–84.

Wen, Y., Balakrishna, R., Gupta, A., Ben-Akiva, M., Smith, S., 2006b. Deployment of DynaMIT in the city of Los Angeles. Technicalreport, Massachusetts Institute of Technology and Volpe National Transportation Systems Center.

Williams, H.C.W.L., 1977. On the formation of travel demand models and economic evaluation measures of user benefit. Environ. Plann. A 9, 285–344.

Youn, Y., Gastner, M.T., Jeong, H., 2008. Price of anarchy in transportation networks: efficiency and optimality control. Phys. Rev. Lett. 101, 128701.

Ziliaskopoulos, A., Waller, S., Li, Y., Byram, M., 2004. Large-scale dynamic traffic assignment: implementation issues and computational analysis. J. Transp. Eng. 130 (5), 585–593.

How companies should synchronize production, dispatching and transportation processes at different locations? What is freight transportation and logistics? How many warehouses do we need in a distribution network? Where facilities in a network should be located? How should demand for facilities service be allocated to facilities? How to organize distribution of goods from warehouses to shops and supermarkets, newspapers distribution, emergency services, waste collection, street cleaning and sweeping in one city?

FREIGHT TRANSPORTATION AND LOGISTICS

9

9.1 LOGISTICS SYSTEMS BASICS

The cars we drive are composed of a large number of components. Many different companies participate in their production. Some companies provide the necessary raw materials, other produce specific parts, etc. It is similar with the production of computers, TVs, washing machines, and many other products. Toyota, one of the leading car producers in the world, has manufacturing facilities, and joint venture, licensed and contract factories on all continents. The GSK (GlaxoSmithKline), the biggest pharmaceutical company in the world, has 84 manufacturing sites in 36 countries that make medicines, vaccines and consumer healthcare products. In 2014, the GSK distributed more than 800 million doses of vaccines around the world. Chiquita Brands International Inc. that sells bananas, ready-made salads, and health foods operated in 2014 in 70 countries.

In order to produce high-quality products, accepted by the market, and to make profits, companies must synchronize production, dispatching and transportation processes at different locations. For example, supplying the world's retail chains with fresh fruits, vegetables, and cut flowers is certainly a complex task that involves detailed planning and coordination of producers, transportation companies, and trade enterprises.

As companies involve in global competition, transportation costs turn out to be even more important. Airfreight, motor carriers, ocean transportation, railroad, multi-modal transport operators, and couriers are the main transportation operators that appear, together with different suppliers, manufacturers, distributors, and retailers, in the stages of production, storage, transportation, distribution and sale of goods (Fig. 9.1).

Transportation Engineering. http://dx.doi.org/10.1016/B978-0-12-803818-5.00009-3

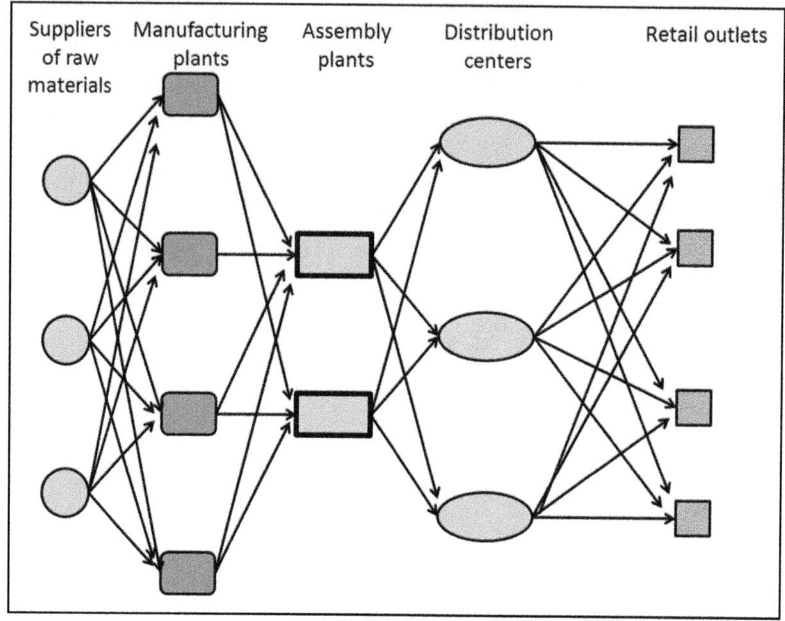

FIG. 9.1

Stages of production, storage, transportation, distribution, and sale of goods.

All of them are interconnected by various information and transportation tasks to be performed. In every production and sales process, there are material flows from raw material sources through factories, distribution centers, wholesalers, retailers to final customers. There are also the reverse flows of information and materials from the clients to the suppliers.

Logistics has a task to synchronize material flows and information flows, to meet customer expectations, and to provide on-time delivery (Daganzo, 2005; Don Taylor, 2007; Simchi-Levi et al., 2014; Taniguchi, 2014, 2016). Procurement, inventory management, transportation management, warehouse management, materials handling, and distribution are the main logistical subsystems. Logistics has the task to merge these subsystems and, in that way, provide clients with the right products, at the right place, at the right time.

The logistical system is composed of a set of various facilities (factories, distribution centers, transportation terminals, wholesalers, retailers, etc.) that are connected by transportation activities. Goods to be transported are usually consolidated into pallets and/or containers. In this way, goods are better protected. Simultaneously, it is much easier to handle goods at freight terminals. The typical pallet sizes are: 80×100 cm, 100×120 cm, 90×110 cm, and 120×120 cm. There are also various container types (closed, with upper opening, ventilated, refrigerated, etc.).

In order to provide low transportation costs and high level-of-service to the clients, logistic systems need to be appropriately configured and managed. Configuration of the distribution system assumes the problem of determining the locations of distribution centers and warehouses in a space. Accessibility, capability, total travel time, transit time, door-to-door transportation costs, frequency of service, service reliability, damage and freight loss, and security are some of the major attributes of the logistics/transportation system. The main task of the distribution logistics is to deliver final products to the customers.

Partially due to the high level of freight vehicle traffic in urban areas, urban road networks in many cities are severely congested, resulting in increased travel times, increased number of stops, sudden delays, bigger travel costs, increased air pollution and noise level, and an increased number of traffic accidents. Special branch of logistics is called City Logistics, which is defined as "the process for totally optimizing the logistics and transport activities by private companies with support of advanced information systems in urban areas considering the traffic environment, the traffic congestion, the traffic safety and the energy savings within the framework of a market economy" (Taniguchi et al., 2001).

The distribution network configuration is one of the most important logistical problems. How many distribution centers do we need in one region, or in one city, and where should we locate these centers? Fig. 9.2 shows the dependence of the inventory costs and transportation costs on the number of warehouses in a specific region.

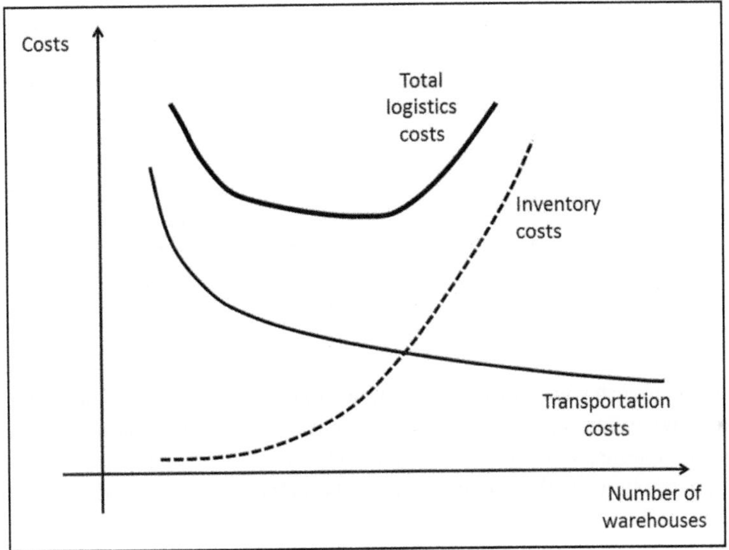

FIG. 9.2

Total logistics costs as a function of the number of warehouses.

The higher the number of warehouses, the higher the inventory costs, and the lower the transportation costs. When trying to solve these and similar problems one must take into account existing demand, available financial resources, equipment required by the distribution center, material handling procedures, as well as various environmental and transportation issues. There are various distribution strategies in practice. Direct shipment assumes that the final products are distributed from production plants directly to the retailers (Fig. 9.3).

The strategy of direct shipment is usually not used when there are numerous clients that require small amount of goods (small shipment size). This strategy is used when it is possible to have high truck utilization factor, as well as in the cases of perishable goods. When logistic system contains one or more warehouses, the goods are, in the first step, dispatched from the production plants to the warehouses. Items are stored and when warehouse receives customers' orders, items from the warehouse are retrieved, packed and shipped to the clients.

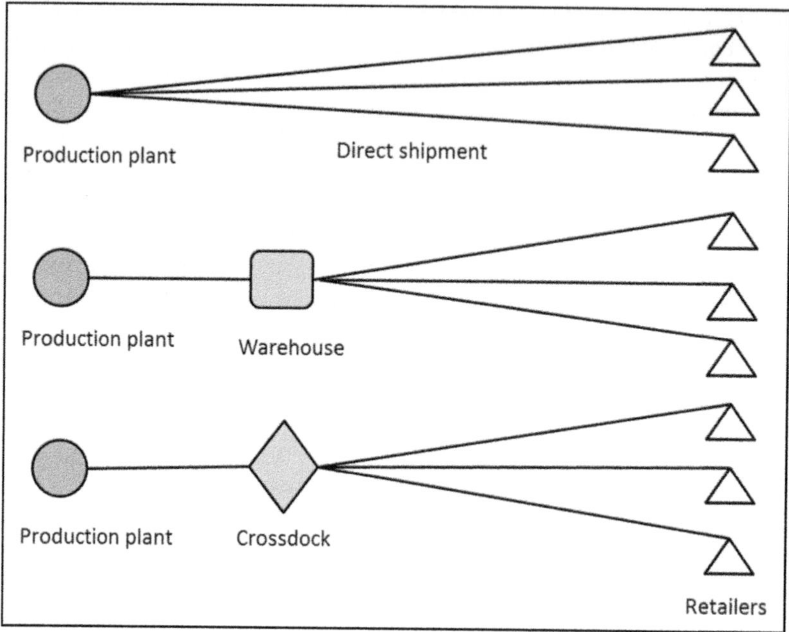

FIG. 9.3

Various distribution strategies.

Cross docking is a modern warehouse management concept. The items are delivered to a warehouse by inbound trucks at receiving docks. All delivered items are, without a delay, sorted out by destinations on the sortation systems, and routed and loaded into outbound trucks on the shipping docks. The outbound trucks delivery items to the clients. Practically, in the case of cross-docking, the items are not held in inventory at the warehouse (Fig. 9.4).

The cross-docking concept usually uses companies that distribute great volumes of goods and supply a large number of stores (Bartholdi and Gue, 2004). This concept enables reduction of the turnaround times for customer orders. The practice also shows reduced inventory costs, as well as smaller warehouse space needed. The cross docking concept requires very careful planning of all operations, especially trucks scheduling, as well as allocation of the inbound trucks items to the outbound trucks.

Distribution of goods from warehouses to shops and supermarkets cannot be well organized without appropriate vehicle routing and scheduling. Location analysis and vehicle routing and scheduling are two important areas where logistics and transportation intersect. In this chapter we outline the basics of the location analysis, as well as vehicle routing and scheduling problems.

9.1.1 REVERSE LOGISTICS

Reverse logistics is relate to the return flows of manufactured goods, materials or equipment back from the purchasers to the logistics network. For example, manufactured goods could go from the client to the distributor or to the manufacturer. These goods, materials, and/or equipment could be remanufactured, reused, recovered, or recycled. Millions of personal computers in the world became outdated.

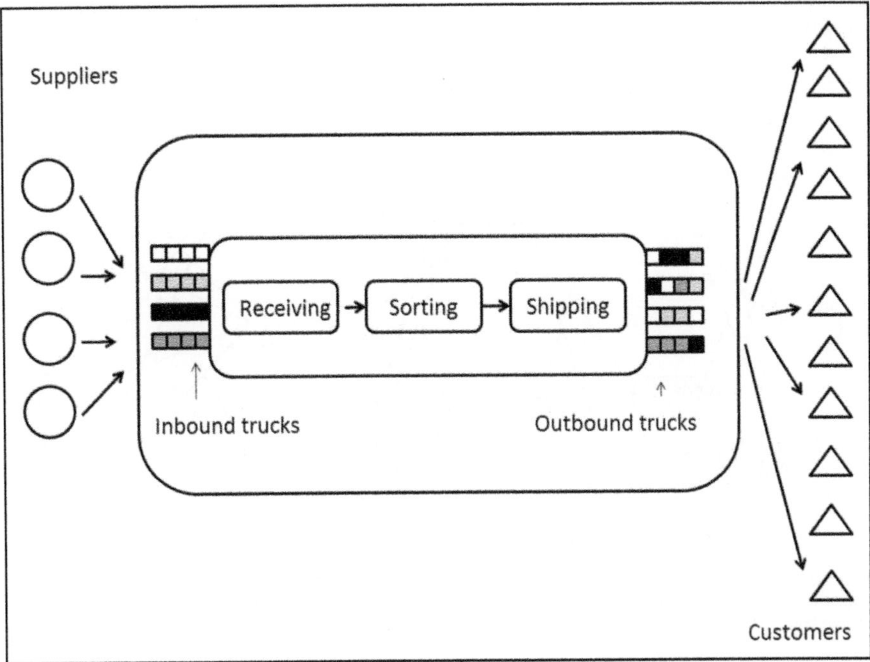

FIG. 9.4

Cross docking operations.

Obviously, there are great opportunities to reuse these computers and create the new values. E-waste contains cell phones, computers, TV sets, audio equipment, batteries, etc. E-waste contains aluminum, lead, copper, plastics, glass, etc. Within automotive industry waste could include engines, alternators, starters, transmissions, etc. Reverse logistics is already one of the most important issues related to magazine publishers, computer manufacturers, printers, auto industry parts, consumer electronics, household chemicals, etc. Fig. 9.5 shows logistic and reverse logistic flows. Reverse logistic flows are denoted in Fig. 9.5 by dashed lines.

There are various reasons why return flows exist. They could be customer service oriented (return of defective products, return of unsold goods, etc.), environmental (various green initiatives and government regulations), or economic. Large world companies already capable of handling reverse logistics issues include DHL, UPS, Fedex, Ryder, etc. Reverse logistics is becoming increasingly important in the modern world, due to the increasing attention to the protection of the environment.

9.2 ROAD FREIGHT TRANSPORT INFRASTRUCTURE

The road freight transport infrastructure network consists of the road, rail/road, and road/port freight terminals as the primary and doors of freight/goods shippers and receivers as secondary network nodes, and roads/highways connecting them. The road trucks use the local and regional roads within and highways between regions/cities, and streets in the urban areas/cities to access doors of particular

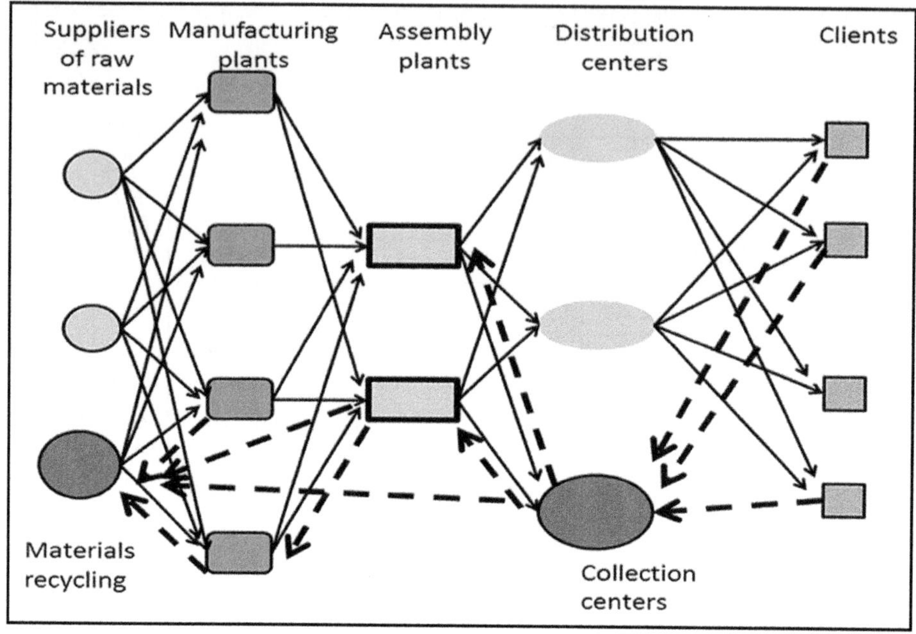

FIG. 9.5

Logistics flows and reverse logistic flows.

freight/goods shippers and receivers, road, rail/road, and port/road freight terminals, in the latter case if the freight/goods shipments are containerized.

9.2.1 "ULTIMATE" AND "PRACTICAL" CAPACITY AND SERVICE LEVEL OF ROAD TRUCK ROADS

The road trucks share the same roads and highways as buses and individual/private vehicles/cars. Their "ultimate" and "practical" capacity and corresponding service levels in this context are analogous to those analyzed in the previous sections. Therefore, similarly as in the case of the intercity bus transport, they will not be particularly considered in the following text.

EXAMPLE 9.1

The Brenner Motorway Tunnel through Alps as one of the main European trunk routes, which connects Innsbruck in Austria and Modena in northern Italy (Europe). Its length is: $l=36$ km. If the maximum allowed speed of heavy trucks is: $v_f=80$ km/h and the average separation distance between successive trucks moving in the same direction: $s_f=150$ m, the "ultimate" capacity of the lane full of trucks will be: $\mu_f=v_f/s_f=80{,}000/150\approx 533$ trucks/h. In addition, the number of trucks simultaneously being in the tunnel and moving in the same direction will be equal to: $n_f=\mu_f\times(l/v_f)=533\times(36/80)\approx 244$ trucks (Lauber, 2001).

9.2.2 "ULTIMATE" AND "PRACTICAL" CAPACITY OF ROAD FREIGHT TERMINALS AND THEIR LEVEL-OF-SERVICE

The road freight terminals usually operate as the centrally located hubs in the networks of road freight/goods operators. They are connected between themselves and the doors of particular shippers and receivers with the numerous inbound and outbound connections/road freight/goods transport services carried out by trucks. In this case, these terminals can represent the freight/goods interchanging locations within the same modal (road) system ensuring continuity of the freight/goods flows. These flows can be consolidated differently as bulk, general, and containerized shipments. In the case of a given terminal, after arriving by an incoming truck, the goods/freight shipments are unloaded, stored temporarily in the terminal, then loaded on the outgoing trucks and departed from the terminal. The loading and unloading of tracks is carried out by the dedicated corresponding equipment depending on the type of consolidation of freight/goods. In such cases, the "ultimate capacity" of the terminal consists of three components: that of the incoming truck unloading and the outgoing truck loading bay area, and the freight/goods storage space within the terminal. Fig. 9.6 shows the simplified layout of such a terminal.

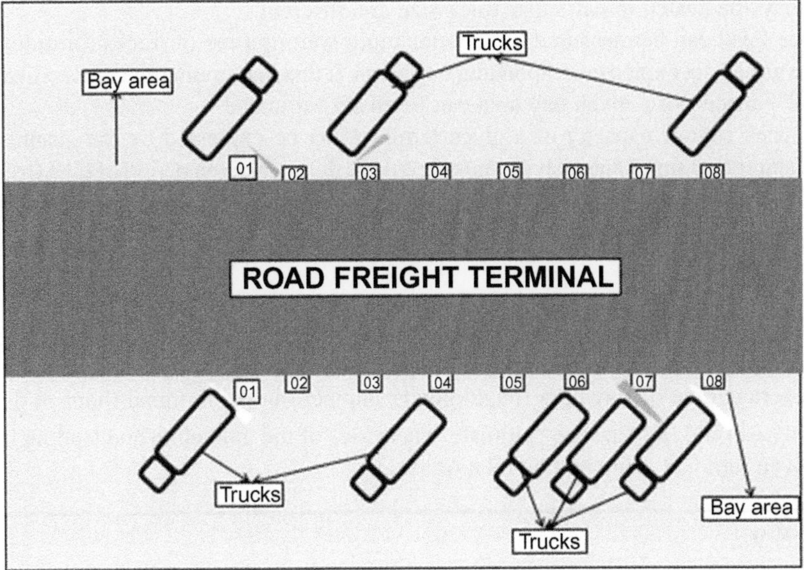

FIG. 9.6

Scheme of a road freight/goods terminal.

For example, the "ultimate" capacity of the incoming unloading/loading bay area can be defined as the quantity/number of freight/goods shipments, which can be handled there during a given period of time (usually 1 h) under conditions of constant demand for service. For the specified type of consolidated freight/goods shipments transported to a given truck size, this capacity (μ_f) can be estimated by the Little's formulae from the queuing theory, as follows:

$$\mu_f = (N_f/\tau_f) \times PL_f \tag{9.1}$$

and

$$N_f = \lambda_f \times \tau_f \tag{9.2}$$

where:

N_f is the number of loading/unloading bays of a given type at a given terminal for accommodating trucks of the payload capacity (PL_f);
τ_f is the average turnaround time of the truck(s) of payload capacity (PL_f) at an unloading/loading bay (this time consists of the unloading, intermediate/"buffer," and loading time);
PL_f is the payload capacity of a truck requesting handling (unloading and loading) at the corresponding bay area (ton/truck); and
λ_f is the average intensity of demand (trucks) of payload capacity (PL_f) requesting unloading or loading at corresponding bay areas (trucks/h).

The trucks operating at given terminal can be of different size. For example, they are smaller if collecting and distributing freight/goods shipments within the gravitational areas of their origin and destination hub terminals, respectively, and larger if transporting freight/goods shipments between hub terminals. In such cases, the hub terminals consolidate freight/goods flows. In addition, the dimensions of particular bays dedicated to particular truck size are different.

The service level can be measured by the maximum waiting time of trucks for unloading and/or loading due to already occupied corresponding bay areas. It this maximum time is specified in advance, the "practical" capacity of a given bay area can be easily estimated.

The "ultimate" storage capacity of a given terminal can be expressed by the quantity/number of freight/goods shipments simultaneously being/stored there during a given period of time (from few hours to few days). If a given terminal is assumed to be empty, this capacity can be estimated as follows:

$$q_{f/max} = \max\left[0; \left(\mu_{f/e} - \mu_{f/l}\right) \times \Delta_f\right] \tag{9.3}$$

where:

$\mu_{f/e}$, $\mu_{f/l}$ is the maximum intensity of entering and leaving of the freight/goods shipments from a given terminal, respectively (freight/goods shipments/day); and
Δ_f is the average time of staying a freight/goods shipment in the terminal (hour or days).

The capacities ($\mu_{f/u}$) and ($\mu_{f/l}$) are the "ultimate" capacities of the unloading and loading bays, respectively, of a given terminal estimated by relation (9.1).

EXAMPLE 9.2

Let the number of bays of a given terminal intended to handle trucks of payload capacity of: $PL_f = 26$ tons/truck, is: $N_f = 20$, and the average truck's turnaround time: $\tau_f = 1$ h. Then, the "ultimate" capacity of the terminal's bay area will be: $\mu_f = (20/1) \times 26 = 520$ tons/h. If terminal operates 12 h/day at the full hourly "ultimate" capacity, its daily capacity will be: $520 \times 12 = 6240$ tons/day (entering and leaving the terminal). Or, if the intensity of arriving trucks of given payload capacity is: $\lambda_f = 30$ trucks/h, and if the truck's handling time is: $\tau_f = 0.75$ h, the required number of bays will be: $N_f = 30 \times 0.75 \approx 23$. As well, if the maximum intensity of the freight/goods shipments entering the above-mentioned terminal is: $\mu_{f/e} = 6240/2 = 3120$ tons/day, the maximum intensity of their leaving: $\mu_{f/l} = 2000$ tons/day, and if the average time of their staying in the terminal: $\Delta_f = 3$ days, the terminal's "ultimate" storage capacity will be: $q_{f/max} = \max[0; (3120 - 2000) \times 3] = 3360$ tons. If this freight/goods shipment are containerized and if each container is full with the gross weight of 24[1]tons, the stacking space in the terminal or around $3360/24 = 140$ containers needs to be provided.

[1]This is the gross weight (tare + payload) of TEU (20 Foot Equivalent) load unit-container (STC, 2015).

In addition to the maximum waiting time of trucks for being handled, the service quality of road terminals can be measured by the quality of storing of freight/goods in terms of the internal ambient, security, and safety in handling.

9.3 SERVICE NETWORKS OF THE ROAD FREIGHT TRANSPORT OPERATORS

The road freight transport operators/companies operate their transport service networks on the selected infrastructure network covering certain geographical area/territory. In general, these service networks can be of the so-called point-to-point and hub-and-spoke shape (terminology similarly used for the intermodal freight transport and air transport networks). The "point-to-point" network consists of direct transport connections (the network links) of doors of the freight/goods shippers and receivers (the network nodes). The "hub-and-spoke" network consists of direct connections (the secondary network links) of doors of the freight/goods shippers and receivers (the secondary network nodes) with consolidating hub terminal(s) (the primary network nodes) and direct connections (the primary network links) of the hub terminals themselves. Fig. 9.7A and B shows the simplified layout of both network schemes.

9.3.1 CAPACITIES AND SERVICE LEVEL OF THE ROAD FREIGHT TRANSPORT SERVICE NETWORKS

The capacities and service level of the road freight transport service networks are generally expressed by the size of the network, ie, the area it covers and the number of nodes and links, ie, shippers and receivers, the size and payload capacity of road trucks engaged and their service frequencies on the particular links of the networks. The latest two depend on the quantity/volume of freight/goods transport demand to be served during a given period of time under given conditions.

Transport companies can use trucks of different size and payload capacity to operate in their service networks. For the medium—to long-haul interurban transportation the most common configurations consist of a towing unit, either a truck or tractor, and one or more trailers or semitrailers. For example, in Europe, since 1992, the size of the road freight trucks has been limited: their maximum gross weight (the weight of empty truck, trailer, and cargo), number of axles, and length of these road trucks have been harmonized and standardized to 40–44 tons, 5–6 axles, and 18.75 m, respectively. This gives a maximum axle load of these vehicles of 40–44/5, ie, between 8 and 8.8 tons/axis (ASECAP, 2010). Fig. 9.8 shows a typical configuration of this truck.

The capacity of this truck is expressed by its payload capacity, which is 26 tons. This gives the ratio payload/gross weight of: 26/40 = 0.65. In addition, the maximum allowed gross weight of 3-axle trucks is 26 tons and of 2-axles truck 18 tons (EU, 2015). In the United States, the maximum size of the 5-axle road trucks in different configurations varies across the particular States. However, the maximum length of a typical truck-trailer configuration is: 14.63–19.81 m, width 2.6 m, and height 4.11–4.3 m. Its typical maximum allowable gross weight is: 80,000 lb (36 tons), of which tare is 33,600 lb (16 tons), and payload 46,400 lb (20 tons) (1 ton = 453 kg). The payload/gross weight ratio is: 20:36 = 0.56. In addition, the maximum gross weight of the 9-axle road truck-trailer combination is

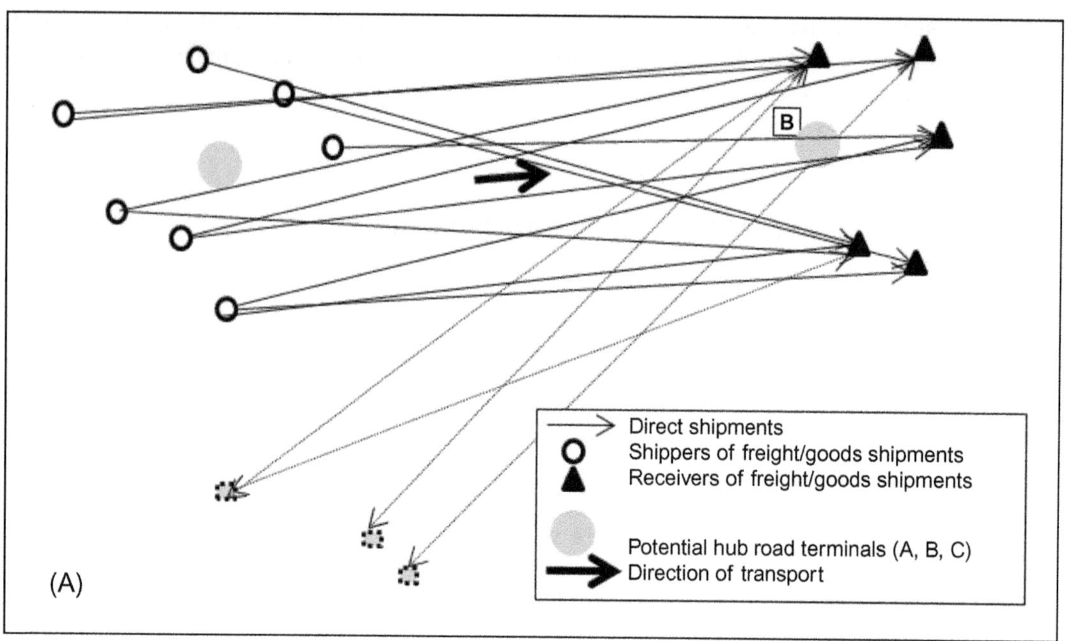

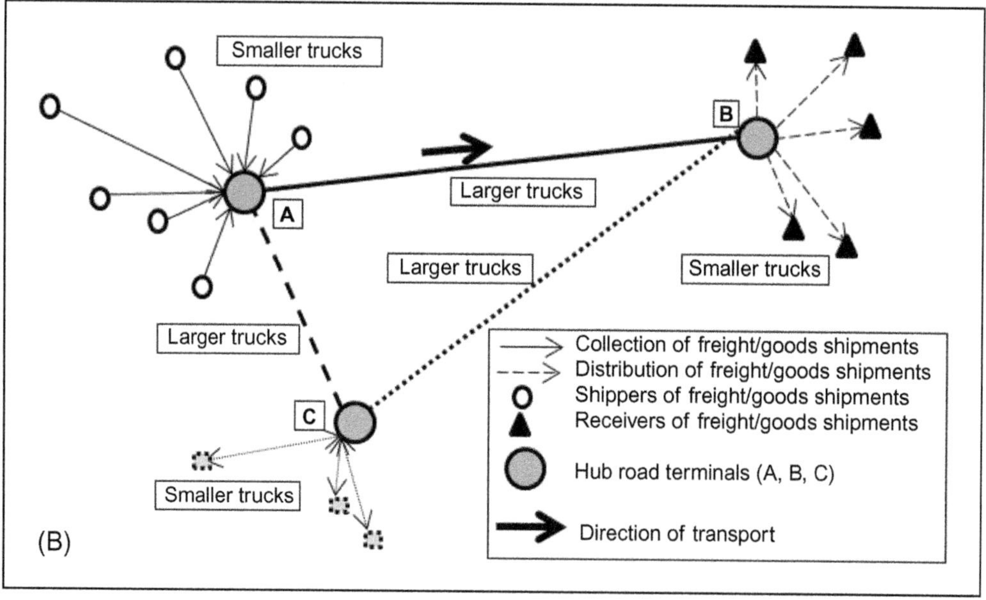

FIG. 9.7

Simplified layouts of the road freight transport service networks. (A) "Point-to-point" network and (B) "Hub-and-spoke" network.

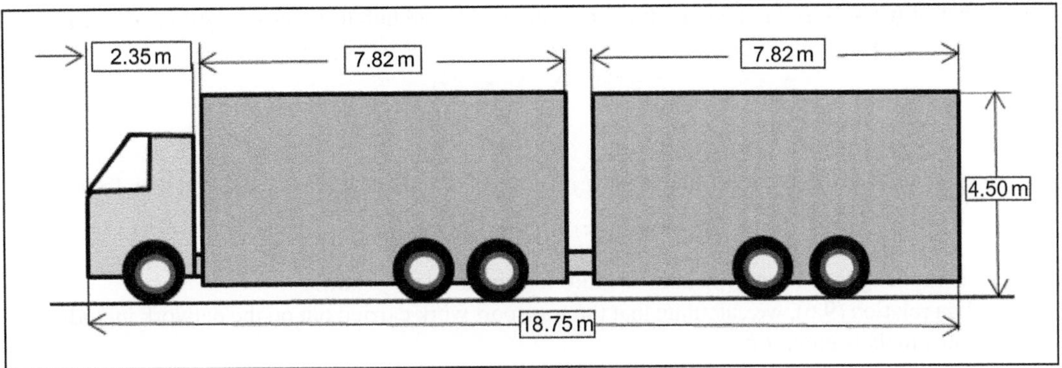

FIG. 9.8

Typical configuration of a standard European truck (Fraunhofer, 2009).

110,000 lb (≈50 tons), with tare 42,800 lb (19.5 tons), and payload 67,200 lb (≈30.5 tons). The ratio payload/gross weight is 30.5/50 = 0.61 (USDT, 2004).

In general, the European and U.S. road networks are designed to meet the existing standards on the weight and length of these heaviest standard trucks. In particular, there is evidence that the European road network is not completely suitable to efficiently, effectively, and safely accommodate trucks heavier and longer than the above-mentioned standard ones. This implies that the more widespread use of heavier and longer (ie, mega) trucks, with a maximum gross weight of 60 tons and length of 25.25 m, will likely require modification of the existing road infrastructure as follows (ASECAP, 2010; UIC, 2008).

It is assumed that the road trucks serving given network consisting of (*M*) road hub terminals (as shown in Fig. 9.6). We assume that trucks have the same payload capacity and the average route load factor. The required number of trucks $n_{f/ij}(\tau)$ operating on the route between a given pair of road hub terminals (*A* and *B* shown in Fig. 9.7) can be estimated as follows:

$$n_{f/ij}(\tau) = \frac{Q_{ij}(\tau)}{\theta_{f/ij} \times PL_{f/ij}} \tag{9.4}$$

where:

$Q_{ij}(\tau)$ is the quantity of freight/goods shipments to be transported during the time period τ on the route (i,j) connecting the hub road terminals i and j (tons);
$\theta_{f/ij}$ is the average load factor of trucks operating on the route (i,j) between hub terminals i and j (≤ 1.0); and
$PL_{f/ij}$ is the payload capacity of trucks operating on the route (i,j) between hub terminals i and j (tons/truck).

The number of trucks serving the network consisting of M road hub terminals is estimated as follows:

$$n_{\mathrm{f}}(\tau) = \sum_{i=1}^{M} \sum_{\substack{j=1 \\ j \neq i}}^{M} n_{\mathrm{f}/ij}(\tau) \tag{9.5}$$

The transport work carried out on the route (i,j) during the time period τ is calculated as follows:

$$TW_{ij}(\tau) = Q_{ij}(\tau) \times l_{ij} = n_{\mathrm{f}/ij} \times PL_{\mathrm{f}/ij} \times l_{ij} \tag{9.6}$$

where l_{ij} is the length of route (i,j) between hub terminals i and j (km).

Based on relation (9.6), we calculate that the transport work carried out on the network including M road hub terminals is equal to:

$$TW(\tau) = \sum_{i=1}^{M} \sum_{\substack{j=1 \\ j \neq i}}^{M} TW_{ij}(\tau) \tag{9.7}$$

In addition, productivity of a given route connecting a given pair of road terminals can be calculated as follows:

$$TP_{ij}(\tau) = Q_{ij}(\tau) \times v_{ij}(l_{\mathrm{AB}}) = n_{\mathrm{f}/ij} \times PL_{\mathrm{f}/ij} \times v_{ij}(l_{ij}) \tag{9.8}$$

where $v_{ij}(l_{ij})$ is the average speed of trucks on the route (i,j) between hub terminals i and j (km).

In addition, productivity of the network including M road hub terminals is equal to:

$$TP(\tau) = \sum_{i=1}^{M} \sum_{\substack{j=1 \\ j \neq i}}^{M} TP_{ij}(\tau) \tag{9.9}$$

where all symbols are analogous to those in the previous equations.

The service quality can be expressed by the punctuality and reliability of services, ie, on-time and realized (as guaranteed) transport between the doors of shippers and receivers of the freight/goods shipments.

For the route between the terminals i and j, the required number of trucks of the given characteristics is equal to:

$$N_{\mathrm{f}/ij}(\tau) = n_{\mathrm{f}/ij} \times \left(\tau_{\mathrm{f}/i} + 2 \times \frac{l_{ij}}{v_{\mathrm{f}/ij}(l_{ij})} + \tau_{\mathrm{f}/j} \right) \tag{9.10}$$

where $\tau_{\mathrm{f}/i}$, $\tau_{\mathrm{f}/j}$ is the truck's loading and unloading time at the origin and destination terminal i and j, respectively (h).

The other symbols are analogous to the symbols in previous relations. The required number of trucks (fleet size) to serve the network of M road hub terminals under given conditions is determined as follows:

$$N(\tau) = \sum_{i=1}^{M} \sum_{\substack{j=1 \\ j \neq i}}^{M} N_{\mathrm{f}/ij}(\tau) \tag{9.11}$$

where all symbols are analogous to those in the previous expressions.

EXAMPLE 9.3

Let's assume that the road route/line is the above mentioned Brenner tunnel through Alps. Fig. 9.9 shows the relationship between the annual quantity of freight/goods and the number of heavy trucks during the specified period of time (EU, 2015).

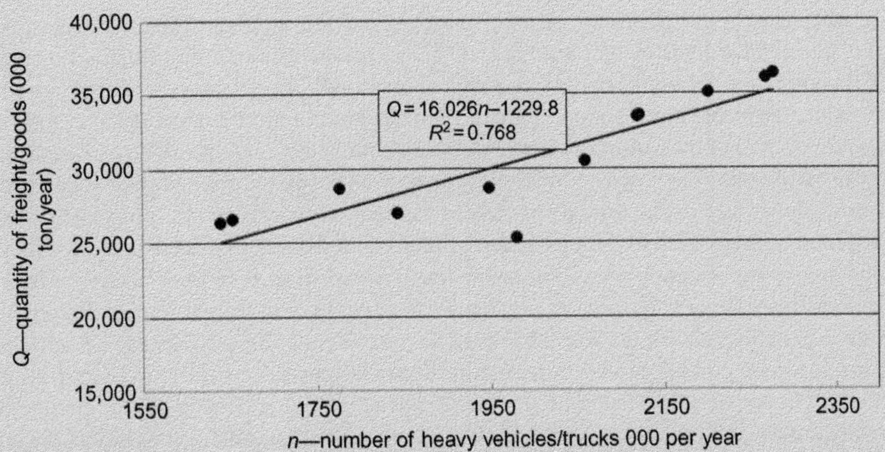

FIG. 9.9

Relationship between the quantity of freight/goods and the number of heavy trucks through Alpine Brenner Tunnel (Europe) (Period 1999–2012) (EU, 2015).

From the simple regression equation, it can be seen that the average payload per truck has been 16 tons. If the payload capacity of truck is: $PL_f = 26$ tons, the average load factor has been: is $\theta_f = 16/26 \approx 0.62$. Given the length of a tunnel of: $l = 36$ km, the transport work carried out per a truck has been: $TW = 16 \times 36 = 576$ tons-km/truck. In addition, since the truck's speed has been limited to: $v = 80$ km/h, its productivity through the tunnel has been: $TP = 16 \times 80 = 1280$ tons-km/h. If it is assumed that the quantity of gods to be transported in both directions through the tunnel is: $Q = 300$ tons/day, the transport frequency will be: $n_f = 300/16 \approx 19$ services/day. If the truck loading and unloading time at both ends of the tunnel is: $\tau_{fl1} = \tau_{fl2} = 1.0$ hr, the time of passing through the tunnel in both directions: $2 \times (36/80) = 0.9$ h. Then the required truck fleet shuttling through the tunnel will be: $N_f = 19 \times (1.0 + 0.9 + 1.0) \approx 54$ trucks.

Similarly, for the given road freight/goods service network, the transport work, productivity, and the required fleet size, based on the quantity of freight/goods to be transported during a given period of time can be calculated.

9.4 CITY LOGISTICS

In 2014, the total population of cities made up over 50% of the total world population. It is expected, in years to come, that there will be an increase in city populations at a rate of 1–2% yearly. In the near future, even in developing countries, the largest number of inhabitants will live in cities. A large number of cities in the world, inhabited by millions of people, are facing very large urban traffic congestion, as well as the significant air quality problems. Urban freight transportation significantly contributes to the high level of traffic congestion in the cities. At the same time, urban freight transportation is a large producer of greenhouse gas emissions. Urban freight transportation represents the movement of freight vehicles whose main function is to transport goods into, out of and within cities. Construction in cities, retail, courier and post services, hotel, restaurant and catering activities, as well as waste collection generate significant transportation flows in every city. Freight vehicles characteristically participate with 10–15% in total traffic flows in cities.

The supply of city residents and the quality of life in world cities highly depends on efficient and sustainable urban freight transportation. In many cities in the world, urban freight transportation is inefficient. Freight vehicle load factors are relatively low, and there are many freight vehicles empty running. There are many deliveries made to individual locations. For decades, goods for supermarkets, grocery stores, department stores, shopping malls, bookstores, hotels and restaurants, in most cities in the world, were delivered by various companies. These companies, independently of each other, organized goods deliveries. They determined the type of vehicle that carried supplies, determined the delivery frequency, as well as the time points where the delivery is carried out. Independent deliveries by many different operators led to the inefficient utilization of delivery vehicles in many cities. Urban freight vehicles also significantly contribute to the traffic congestion, emissions and noise level. In the past, city authorities were not significantly involved in solving urban freight transportation problems. City authorities in many cities in the world have not developed, over the past decades, neither planning strategies nor adequate control strategies related to urban freight transportation. There have been, in some cities, regulatory measures such as limited access hours for freight vehicles, restricted access to specific city zones, and parking restrictions. The increase in urban traffic congestion, air pollution, noise level, number of traffic accidents, and travel times, led to new approaches in solving urban freight transportation problems.

City logistics represents a relatively new concept that tries to reduce economic, social and environmental costs of goods movement into, out and within cities. City logistics is defined as "the process for totally optimizing the logistics and transport activities by private companies with support of advanced information systems in urban areas considering the traffic environment, the traffic congestion, the traffic safety and the energy savings within the framework of a market economy" (Taniguchi et al., 2001). City inhabitants, city authorities, shippers, and freight carriers are the main stakeholders that are involved in city logistics issue.

9.4.1 URBAN FREIGHT TRANSPORT BASICS

Urban freight transport includes moving freight/goods shipments in, within, and out of urban areas. City logistics is based on the principle of consolidation of several shipments of various shippers and carriers by the same vehicle. City logistics concepts assume coordination of shippers and carriers.

Global supply chains deliver freight/goods shipments very often from the far way manufacturers to the retailers in a given urban area. In general, this supply chain(s) consists of the following components: (i) transport network infrastructure and the freight transport service networks of different transport modes—rail, road, inland waterways/barges, air, and sea/maritime and their vehicles of different size (weight and payload capacity); (ii) the centers/buildings for collecting, warehouses, and distributing goods shipments. Moving of goods shipments through such supply chain generally consists of the following steps:

(1) Transporting freight/goods products/commodities from the manufacture(s) warehouse of the final products to Regional Logistics/Distribution Center (RLDC) by different transport modes—usually road trucks and freight trains; before, they can also be transported by cargo aircraft and/or freight/cargo ships to the airport and port cargo terminal(s), respectively, where they are taken over by trucks or trains; in many cases, such combination of different modes is used implying use of intermodal transport;

(2) Warehousing of these products/commodities at RLDC where waiting for the further step(s);

(3) Transporting these products/commodities from RLDC to Urban Local Distribution Center (ULDC) mostly by trucks and in some cases by inland waterway barges where they are again warehoused while waiting for the succeeding step. City could have one or more ULDC, depending on a city size. A single RLDC can supply few ULDCs in the same urban area. Freight is consolidated in a ULDC (the phrase urban consolidation center is also used). ULDC offers storage, sorting, consolidation, and deconsolidation facilities. The goods are transported from a ULDC to the clients in smaller, environment friendly vehicles.

(4) Distributing products/commodities from the ULDCs to particular retailers. In some cases, freight/goods shipments are transported directly from RLDC to the retailers. Fig. 9.10 shows a simplified scheme.

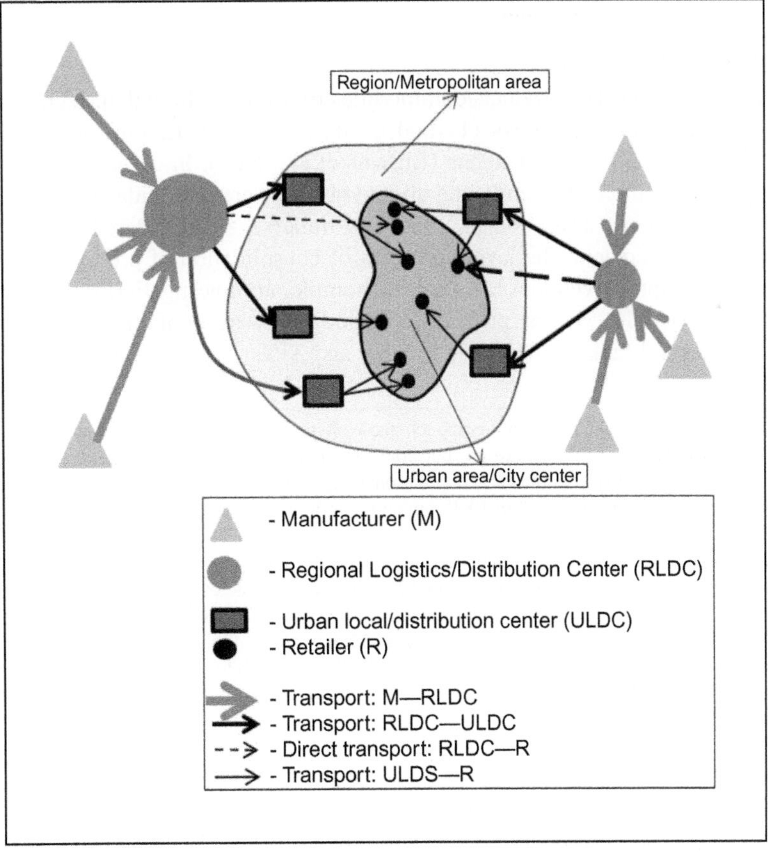

FIG. 9.10

Simplified scheme of supply chain including urban freight/goods distribution.

It is also possible the existence of urban local distribution centers organized into two levels. Large urban distribution centers are located on the outskirts of the city, while satellite distribution centers are located in the urban area. Final distribution could be performed by small, environment-friendly vehicles located in the satellite distribution centers.

The introduction of a limited number of permits for urban distribution, together with the ban for large vehicles to travel through the city center can significantly force the shippers to use the urban distribution center services.

City logistics concept could significantly reduce number of freight vehicles operating within city, and the total number of empty vehicle kilometers, decrease the level of traffic congestion, emissions, and noise.

9.4.2 URBAN FREIGHT DISTRIBUTION SYSTEMS

In general, the urban freight/goods distribution systems can be divided into conventional and advanced. The former are based on the existing infrastructure, technology, and related operations. The latter (advanced) uses the advanced infrastructure, technology, and related operations.

9.4.2.1 Conventional systems

Freight/goods consolidation: The products/commodities transported through the supply chains ending in urban areas can be classified as follows: (1) food, drink, and tobacco; (2) bulk products; (3) chemical, petrol, and fertilizers; and (4) miscellaneous (Browne et al., 2004). In order to move some of them through the supply chains more efficiently and effectively, they are consolidated usually at two levels: pallets,[2] and then containers,[3] swap bodies,[4] and semi-trailers,[5] all considered as the standard intermodal load units. In this context, the level and form of consolidation of products/commodities also implies use of the appropriate vehicles/trucks. For example, an investigation showed that in Greater London area[6] (London, UK), 3% of products/commodities were containerized, 0% swap-bodied, 24% palletized, 36% solid bulked, 4% liquid bulked, and 33% as others (Browne et al., 2004).

---

[2]This is a structure supporting freight/goods in a compact stable shape, thus enabling their manipulation by the loading/unloading devices. They are mostly made of wood, but could be made of some other materials. Dimensions of pallets are different in different world regions, but the ISO (International Organization for Standardization) considers 6 dimensions as standard. In general their width is from 800 to 1000 mm and length of 1000 to 1219 mm (https://en.wikipedia.org/wiki/Pallet#ISOpallets).

[3]A container is load unit designed and built for intermodal freight/goods transport. They can be exchanged between different freight transport modes—truck/train/ship without the need for unloading and reloading their content—freight/goods shipments. They can be of different size but the standard ones are 20, 40, and 45 foot containers. The basic one called TEU (20 Foot Equivalent Unit) is 20-foot container of length of 6.058 m, width 2.438 m, and height 2.591 m (http://www.iso.org/iso/iso_catalogue/catalogue_ics/catalogue_ics_browse.htm?ICS1=55&ICS2=180&ICS3=10).

[4]A swap body (or swop body) is the non-stackable standard load unit, which can be interchanged between the road trucks and the freight trains (DHL, 2008; DFDS, 2014).

[5]Semi-trailer as a container like load unit—box does not have a front axle so a tractor unit supports most of its weight. Instead of axle on the front side it has "legs" enabling its stable standing when it is not coupled to the tractor. They could be of different size—dimensions. The most common width is 2.44–2.6 m and length, for example in the U.S., from 8.53 to 17.37 m (USDT, 2004).

[6]Greater London area covers 1572 km^2 with population of about 8,538,689 (mid 2014) (http://www.britannica.com/place/Greater-London).

Warehousing: For particular products/commodities to be distributed to retailers in a given urban area, warehousing is provided at ULDCs. These can be dedicated to the single or to a couple types or be shared by many products/commodities. In case that different commodities share the same ULDC, its warehousing space has to be appropriately designed and air-conditioned for maintaining the commodities in the consumable condition and shape before being distributed to the retailers (eg, air conditioning/refrigerating space is needed for generally warehousing food). In addition, ULDCs should have a certain warehousing capacity for particular types of products/commodities. This capacity depends on the dynamism of filling in and empting of inventories of particular products/commodities. Changing of these inventories primarily depends of time, number, and quantity of orders influenced by their demand/consumption rates at retailers. Due to an inherently uncertain demand, many retailers can also have some warehousing space in order to prevent shortage of products/commodities. In some cases, some type of products can be warehoused in the retailer shops' shelf.

Vehicle/truck capacity: Depending on the quantity and frequency of ordered products/commodities, trucks of different size (weight and payload capacity) can be used to distribute them from ULDC to particular retailers in a given urban area. The other influencing factors on the truck size are the accessibility of particular retailers regarding configuration of streets and local roads, prevailing traffic conditions, and regulation in the area. This latter relates to truck type (weight, size), access time, preferred routes, loading and unloading zones, and licenses (Huschebeck, 2001). For example, in the Greater London area, access of trucks heavier than 18 tons is restricted during the working days between 9 pm and 7 am, and during the weekend (www.londonlorrycontrol.com/routes). The gross weight of the conventional (rigid) and non-conventional (articulated) trucks can generally vary from 3.5 to 33 tons, and more than 33 tons (EC, 2004). Fig. 9.11 shows the relationship between the truck GVW (Gross Vehicle Weight) (tare + fuel + payload + driver) and its PL (Payload) (carrying capacity in terms of weight/mass).

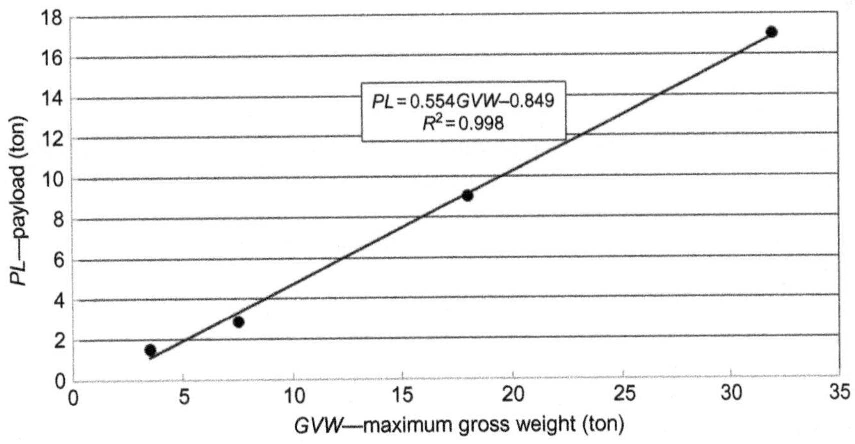

FIG. 9.11

Relationship between PL (Payload) and GVW (Gross Vehicle Weight) of trucks used in urban freight distribution (Browne et al., 2004).

As can be seen, this linear relationship indicates that the payload capacity shares about 55% of the truck gross weight. For example, in the case of the above-mentioned Greater London area, the largest trucks transported the largest proportion of entering products/commodities. They also achieved the highest load factor (about 60%), while at the same time carried out disproportionally lower amount of vehicle-km, ie, the offered transport work (about 14%). At the same time, the smallest trucks have carried out the greatest amount of vehicle-km, ie, offered transport work (55%) with a load factor of about 40%, despite carrying out the lowest proportion of products/commodities as shown in Fig. 9.12.

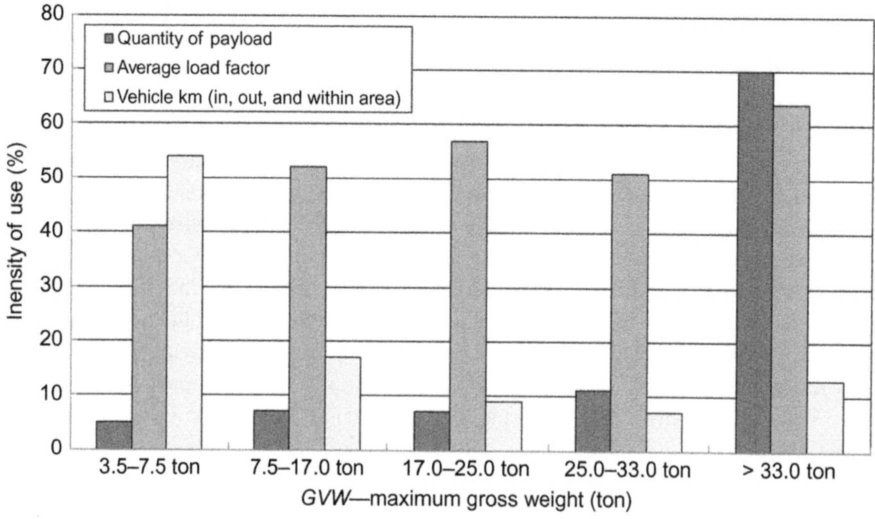

FIG. 9.12

Examples of use of different trucks for delivering products/commodities to Greater London area in 2002 (Browne et al., 2004).

In the given case, the daily volumes of products/commodities entering the area was about 1.8 million tons (6% from the rest of the UK and 94% from EU and the rest of the world) carried out by about 520 thousands trucks of all sizes (Browne et al., 2004; EC, 2004).

Vehicle/truck routing: The trucks deliver products/commodities from ULDC to retailers in a given urban area usually along the shortest routes under given conditions. These routes, depending on the vehicle routing strategy, consist of two or three segments. In the case of a route with two segments the truck runs full from ULDC to a single retailer (the products/commodities are only for it), leaves its load there, and returns empty. In the case of a route with three segments, the full truck runs from ULDC to the first and then succeeding retailers, drops the intended load there, and then after visiting the last one return back empty. In this case, the truck serves the same or different types of retailers dealing with the same or different types of products/commodities, respectively. In addition, the same trucks can take over some freight/goods to some other destinations on their return segments of the routes. Fig. 9.13 shows the simplified scheme of the mentioned cases.

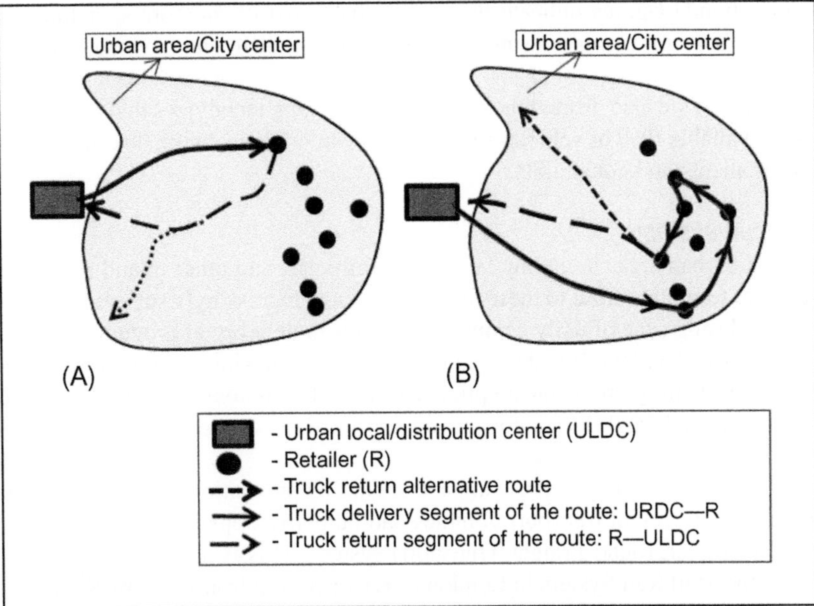

FIG. 9.13

Simplified scheme of a truck's routing in urban goods distribution. (A) Two-segment truck route and (B) Three-segment truck route.

In addition, Fig. 9.14 shows an illustration example of length of product/commodity delivery distance to the Greater London area in 2002. As can be seen, the average delivery distance was about 75 km. About a half of these lengths was from up to the distance of 100 km, and the other half from the distances between 100 and 300 km (Browne et al., 2004).

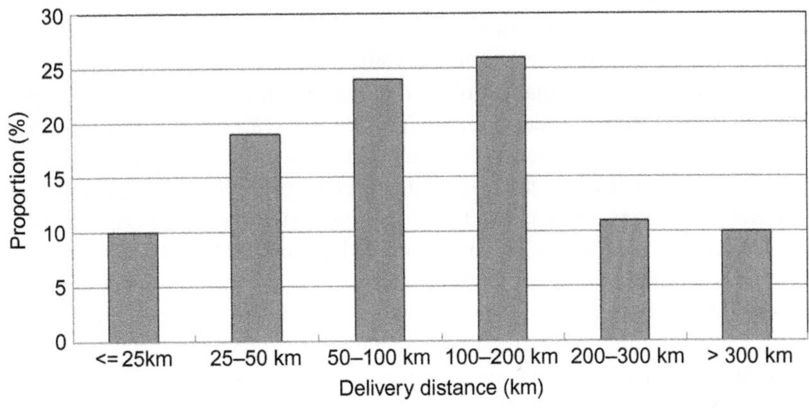

FIG. 9.14

Distribution of the length of delivering distances of products/commodities to Central London area in 2002 (Browne et al., 2004).

The above-mentioned figures indicate that last (urban) part of the corresponding supply chains is generally short, but not compromising use of trucks of different payload capacity relatively efficiently.

Later, in this chapter we also discuss the basic vehicle routing techniques that help decision-makers to efficiently use available fleet of vehicles to satisfy demand in the network, taking care about various operational requirements and constraints.

9.4.2.2 Advanced systems

Large urban and suburban areas including large ports, airports, and other inland road, rail, and intermodal freight/goods terminals close to these urban areas have increasingly suffered from severe traffic congestion causing losing time of daily commuters, delays in delivery of products/commodities to retailers; increased risk of incidents/accidents caused by trucks carrying out urban freight/goods distribution, and increases in energy/fuel consumption and related emissions of GHG (Green House Gases) by all vehicles (trucks and cars) due to congestion and local noise caused by such congested slowly moving traffic.

A prospective solution to mitigate these impacts including prevention of their future escalation due to prospective growth in freight transport volumes can be partial replacement of the ground moving trucks by the UFT (Underground Freight Transport) system. This is not new concept since the first such systems was the Mail Rail System in London (UK), operating from 1927 by Royal Mail for moving mail through the area. At present, some UFT systems have been successfully used in Japan to transport bulk material (Nippon/Daifuku and Sumitomo Electric Industries). In addition, two UFT systems have been operating in Georgia (Tbilisi) for the movement of crashed rock and in Russia (Petrograd) for moving garbage. The others are two automated capsule systems—one for containers and other for pallets—still being at the conceptual level with some pilot trails under the laboratory conditions (Rijsenbrij et al., 2006).

The concept is relatively simple: instead of moving products/commodities consolidated into pallets and/or containers by on the streets and roads by trucks, they are moved through the underground pipes/tunnels on the automatically controlled dedicated vehicles. The pipes/tunnels, which could be partially on the ground, would connect different ULDCs as origins and the large retailers as destinations of products/commodities in the given urban and/or suburban area(s). The vehicles/capsules, loaded by freight/goods shipments would move completely automatically though the pipes/tunnels. At the beginning these have been thought to be with a diameter/width of about 1 or 2 m mainly for transporting smaller shipments up to the size of pallets. Later on, the pipes/tunnels with much wider diameter/width have been considered for moving containers, swap-bodies and semi-trailers. This has resulted in emerging two fully automated distinguished concepts: (i) Automated Capsule System for Pallets such as CargoCap (Germany), Subtrans (Texas, U.S.), and MTM (US) and (ii) Automated Capsule System for Containers such as CargoCapContainer (Germany) and SAFE Freight Shuttle (TTI, Texas, US). Both concepts, but particularly the latter one(s), imply interoperability while connecting with the existing (conventional) freight/goods transport systems. All these systems require building up completely new underground infrastructure—tubes/tunnels, which developed as networks could be very costly, which is one of the main reasons why none of them has been fully commercialized (Liu, 2004; Rijsenbrij et al.,, 2006).

The main components of an UFT system can be considered to be the infrastructure network, vehicles/capsules and their power and guidance system, and transport service network.

Infrastructure network: The infrastructure network of UFT systems consists of the stations/terminals as the network nodes and the pipes/tunnels as the network links connecting them. In general, the infrastructure network can be designed as a single line or as the network of lines. In both cases, the lines connect the network stations/terminals, which act as the entries and exits of the vehicles/capsules to/from the network, respectively. In order to properly cover an urban or suburban area, the UFT network needs to have sufficient number of stations/terminals, each covering the specified "service area" (few hundred meters around the station at the street level. Some of these stations/terminals should also be close to locations where transfer between the UFT and other freight transport systems/modes, and vice versa, takes place. For example, these locations can be the port, airport, and inland container terminal(s), RLDCs, ULDCs, and large retailers in a given urban and suburban area. In any case, the network is expected to cover the short to medium products/commodities delivery distances, ie, at longest up to 350–400 km.

Stations/terminals of an UFT network are the underground structures designed as platforms with several tracks for handling of incoming and outgoing vehicles/capsules. The area of floor of a station depends on the number and length of tracks. The latter depends of the length of a single vehicle/capsule or the length of a "train" composed of few vehicles/capsules. The level of floor of a station is usually located above the level of pipes/tunnels thus enabling use of the gravity for acceleration of the incoming and deceleration of the outgoing vehicles/capsules, respectively. Each station can have a set of short tunnels at the same level, each ending with an elevator (if the station is deeply underground), which enables lifting vehicles/capsules to/from the ground/street level. From there, the vehicles/capsules are delivered to the final destinations, ie, retailers, and vice versa. Fig. 9.15 shows a simplified scheme of the layout of a station/terminal of an UFT system for pallets. The underground pipes/tunnels connect the given station/terminal to the rest of the network. Four elevators enable delivering of vehicles/capsules with or without pallets vertically to/from the neighboring streets (Liu, 2004).

Pipes/tunnels made of steel and concrete can be circular or rectangular/squared. The circle shape is recommended when the pipes/tunnels are built deeper (6–8 to 30–50 m or even deeper) underground in order to withstand higher internal pressure. Fig. 9.16 shows a simplified scheme of a circle-shaped design of the above-mentioned Automated Capsule System for Pallets.

The rectangular/squared pipes/tunnels appear to be more convenient respecting the shape of pallets and containers. Fig. 9.17 shows an example of the Automated Capsule System for Containers of 40 ft (2 TEU) (Liu, 2004).

As can be seen, the rectangular shape of pipe/tunnel fits better with the shape (profile) of the containers, swap-bodies, and semi-trailers. If setting up bi-directional tracks within a single pipe/tunnel, its diameter/width will have to be at least doubled. Otherwise, the single direction pipes/tunnel can be used for moving containers in the single direction independently.

Vehicles/capsules and their power and guidance system

Vehicles/capsules: These are mainly made from steel and can carry loading units such as standard pallets or containers, swap bodies, and/or semi-trailers. Their shape should fit with the shape of pipes/tunnels and that of loading units. They have wheels, which depending on the concept, can be steel-made or the rubber tires. In addition, the side wheels can be mounted at the vehicles/capsules in order to provide their stability and mitigate friction with the walls of pipes/tunnels. The vehicles/capsules can operate as single units or be coupled as a "train" of two or three units. At the UFT systems for

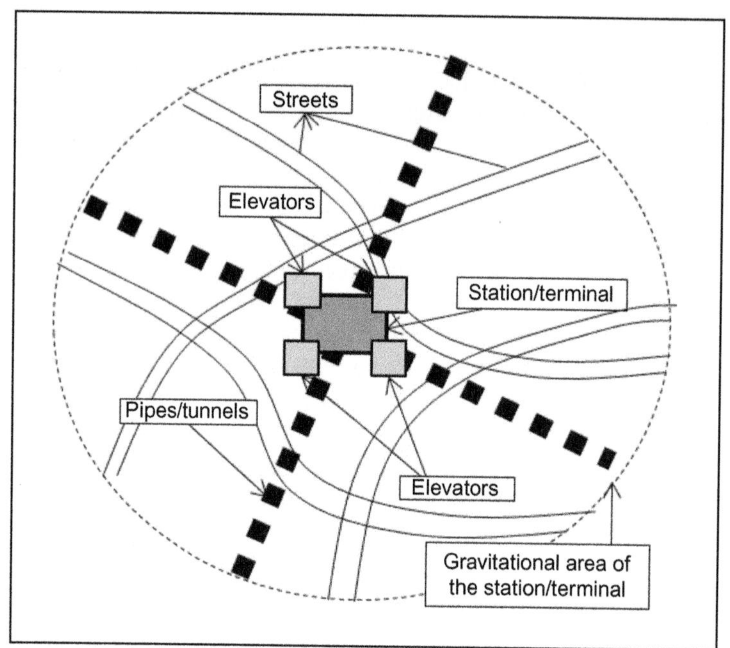

FIG. 9.15

A scheme of a station/terminal of an UFT system for pallets (Liu, 2004).

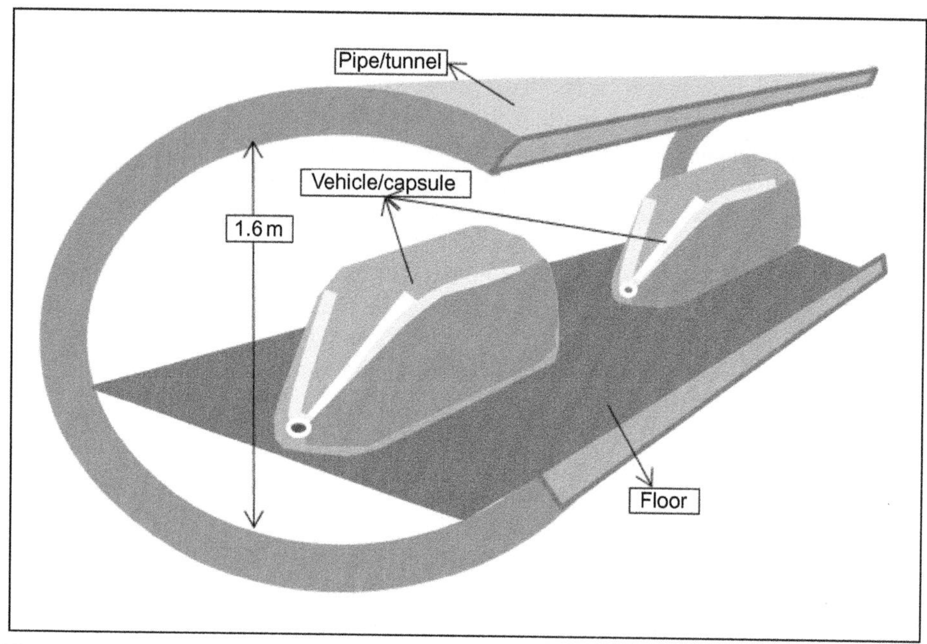

FIG. 9.16

The UFT Automated Capsule System for Pallets—CargoCap (CC, 1999).

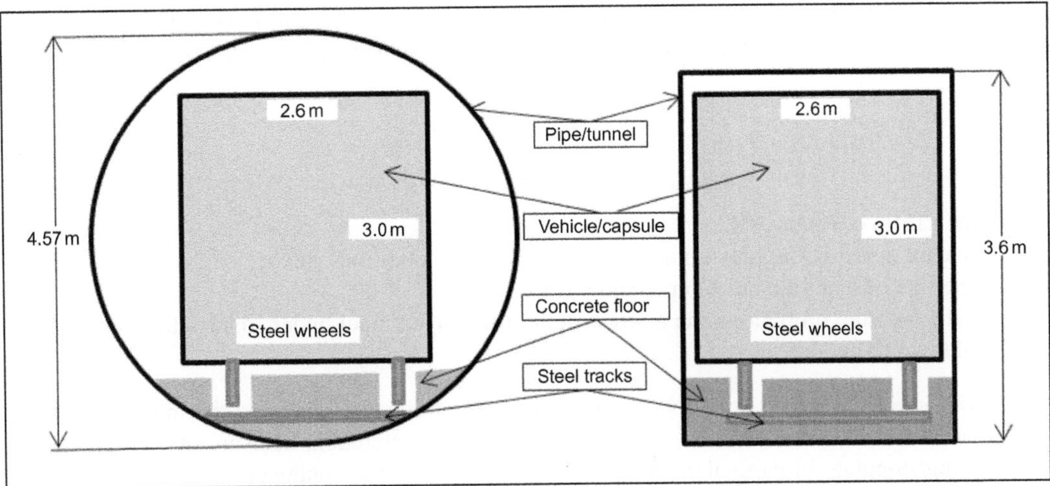

FIG. 9.17

Possible cross-sectional shapes of a pipe/tunnel of the UFT system for containers—single direction (Liu, 2004).

pallets, the typical capacity of a single vehicle/capsule usually corresponds to 2 Euro pallets ($800 \times 1200 \times 1500$ mm) or to about 2 tons of freight/goods. At the UFT systems for containers and swap bodies, the dimensions and carrying capacity of a single vehicle/capsule usually corresponds to one container or a swap body equivalent to 2 TEU. Table 9.1 gives some characteristics of the vehicle/capsule of the UFT system for pallets and containers.

Table 9.1 Characteristics of the UFT System Vehicle/Capsule for Pallets and Containers (CC, 1999; Liu, 2004)

Characteristic	UFT System	
	Pallets CargoCap (Germany)	*Containers* New York City (USA)
Length/width/height (m)	4/1.4/1.6	12.81/2.6/3.0
Empty weight (ton)	0.8	25
Payload (maximal) (ton)	1.5	64
Gross weight (ton)	2.3	89
Track gauge (mm)	800	1453

The vehicles/capsules can be loaded/unloaded at both underground and above the ground stations/terminals means by conveyor technique.

Power system: The vehicles/capsules are powered by LIMs (Linear Induction Motor(s)).[7] In general, the power, ie, thrust force, of a given LIM (MW—Megawatt)[8] can be estimated as follows (Liu, 2004):

$$P_{\text{in}} = e \times v_{\text{a}} \times A \times \Delta p \tag{9.12}$$

where:

- e is the LIM's efficiency (%);
- v_{a} is the air speed in the pipe/tunnel under steady-state conditions (m/s);
- A is the size of inner area of a pipe/tunnel (m^2); and
- Δp is the pressure drop along the entire length of the pipe/tunnel under steady-state conditions (Pa—Pascal (N/m^2)).

The thrust force (P_{in}) in Eq. (9.12) is most effectively controlled by changing the input current frequency, which enables the vehicles/capsules acceleration/deceleration, cruising at constant speed, and breaking/stopping. In particular, for the UFT systems handling containers and swap bodies, the required power of LIMs could reach a couple of hundreds MW and efficiency (e) should be about 70–80%.

Guidance system: In general, the guidance system of vehicles/capsules consists of the system for controlling their movement through the pipes/tunnels and the system for their vertical transport. The system for controlling movements identifies and sorts vehicles/capsules while moving through a given UFT network. At the entry into the network, the type of freight/goods is identified including shippers and receivers. At the exit from the network, the vehicle/capsule destination is identified in order to activate the appropriate switches. The sensors of the system are installed at the stations and inside the pipes/tunnels at each intersection/branching location. In general, the UFT systems for pallets and containers are recommended to use RFID (Radio Frequency Identification) system. The system for vertical transport of vehicles/capsules enables the vertical transfer of both loaded and unloaded vehicles/capsules between the street level and the UFT system. The simple straightforward solution is based on use of the common elevators, designed according to the purpose (UFT system case) and able to operate at a reasonable vertical speed of about 1–3 m/s (Liu, 2004).

Service network: The transport service network is operated at the above-described infrastructure network. The main performances of a given UFT service network are its capacity, transport work, productivity, the vehicle/capsule fleet size, and quality of service.

Capacity: The capacity of a given UFT service network can be expressed by the maximum number of pallets, containers and/or swap bodies processed during a given period of time under given conditions, ie, constant demand for service. For example, the "ultimate" transport capacity of the pipe/tunnel (i) of a given UFT network offered during the specified period of time (tons) can be estimated as follows:

$$C_{s/i}(\tau) = f_{s/i}(\tau) \times m_i \times PL = \frac{\tau}{h_{s/i}} \times m_i \times PL \tag{9.13}$$

[7]It can be considered as an electric machine, which converts electrical energy directly to mechanical energy in translational motion.

[8]1 MW = 1000 KW (kW—Kilowatt).

where:

τ is the time period under consideration (hour, day, week, year);
$h_{s/i}$ is the scheduled headway between two successive vehicle/capsule trains through pipe/tunnel (i) (s); this can also be the minimum headway ($h_{min/i}$);
$f_{s/i}(\tau)$ is the scheduled frequency of services through the pipe/tunnel (i) in a single direction during the time period (τ) counted at the given location;
m_i is the number of vehicles/capsules per train operating in the pipe/tunnel (i); and
PL is the payload capacity of a single vehicle/capsule common for all vehicles/capsules operating in the given UFT network (tons).

The headway ($h_{s/i}$) can be estimated as:

$$h_{s/i} = \delta_{s/i}/v_i \tag{9.14}$$

where:

δ_i is the scheduled distance between any two successive vehicle/capsule trains operating in the pipe/tunnel (i) (m); and
v_i is the average operating speed of a vehicle/capsule train in the pipe/tunnel (i) (km/h).

Typical average operating speed of the UFT systems for pallets and containers is expected to be about: $v = 30–40$ km/h and the maximum speed: $v = 80–90$ km/h. In addition, the capacity of a given UFT system's station in terms of the weight of products/commodities served during a given period of time can be estimated analogously as in relation (9.2). In such cases, the time interval ($h_{s/i}$) represents the average interval between injection of the successive vehicle/capsule trains at the entry station/terminal of the given line (i).

Transport work: The transport work offered in the given pipe/tunnel (i) during a given period of time (ton-km) can be estimated, based on relation (9.2), as follows:

$$TW_{s/i}(\tau) = C_{p/i}(\tau) \times l_i = f_{s/i}(\tau) \times m_i \times PL \times l_i = \frac{\tau}{h_{s/i}} \times m_i \times PL \times l_i \tag{9.15}$$

where l_i is the length of the pipe/tunnel (i) (m, km).

The other symbols are analogous to those in the previous expressions.

Productivity: The productivity of a pipe/tunnel of a given UFT system can be estimated from relation (9.2) as follows:

$$TP_{s/i}(\tau) = C_{s/i}(\tau) \times v_i \tag{9.16}$$

where all symbols are as in the previous equations. For example, let the given UFT system deals with containers and swap bodies. Its each vehicle/capsule has the payload capacity: $PL = 2$ TEU. It moves through the pipe/tunnel at an average speed of: $v_i = 35$ km/h. If the vehicles/capsules are injected at the entry station/terminal of a given pipe/tunnel every: $h_{s/i} = 60$ s, during the time period of: $\tau = 1$ h, the line/tunnel productivity will be: $TP_{s/i}(\tau) = (3600/60) \times 1 \times 2$ TEU $\times 35 = 4200$ TEU-km/h.

Fleet size: The size of fleet, ie, the number of vehicles/capsules required to operate through a given pipe/tunnel (i) of the UFT network in order to serve given volume of demand, can be estimated as follows:

$$M_i(\tau) = \left[f_{s/i}(\tau)\right] \times \tau_{tr/i} = \frac{Q_i(\tau)}{\theta_i \times m_i \times PL} \times \left[t_{l/i} + 2 \times (H_i/V + l_i/v_i) + t_{ul/i}\right] \quad (9.17)$$

where:

$\tau_{tr/i}$ is the turnaround time of a vehicle/capsule train through a given pipe/tunnel (i) (min, h);
$Q_i(\tau)$ is the demand of freight/goods to be transported during the time period (τ) through the pipe/tunnel (i) (ton);
θ_i is the average load factor of a vehicle/capsule train operating through the pipe/tunnel (i) (≤ 1.0);
$t_{l/i}$, $t_{ul/i}$ is the average loading and unloading time of a vehicle/capsule train at shippers and receivers at both ends of the pipe/tunnel (i), respectively (min, h);
H_i is the vertical distance between the level of the stations/terminals and the streets (m) of the pipe/tunnel (i) (m); and
V is the average vertical speed of movement of the vehicle/capsule between the street and terminal level common for all lines of the given UFT network (m/s).

The other symbols are as in the previous relations. For example, let the products/commodities weighting: $Q_i(\tau) = 1000$ tons need to be transported through a pipe/tunnel of a given UFT system during the time period: $\tau = 24$ h, ie, day. The length of line and its positioning underground: $l_i = 50$ km and $H_i = 70$ m. If the payload capacity of a single TEU is 21.7 tons, the number of fully loaded TEUs will be: 10,000/21.7 ≈ 460 TEU/24 h (http://www.interfreight.co.za/container_information.html). If the vehicles/capsules each with the payload capacity of: $PL = 2$ TEU are fully loaded in one and empty in the opposite direction, ie, $\theta_i = 0.5$, $m_i = 1$, their required scheduled service frequency will be: $f_{s/i}(\tau) = 460/(0.5 \times 1 \times 2) = 460$ dept/24 h ≈ 20 dept/h. Let their operating speed through the pipe/tunnel be: $v_i = 35$ km/h, the speed of their vertical movement: $V = 3$ m/s, and the loading and unloading time at the shippers and receivers, respectively: $t_l = t_{ul} = 0.25$ h (15 min). Then the turnaround time of a vehicle/capsule along the given pipe/tunnel will be: $\tau_{tr/i} = 0.25 + 2 \times [(70/3)/3600 + 50/35] + 0.25 = 3.37$ h/dept. As a result, the required fleet will be: $M_i(\tau) = 20$ dept/h × 3.37 h/vehicle/capsule dept ≈ 67 vehicles/capsules.

By summing up fleets from all pipes/tunnels, the required vehicle/capsule fleet for the entire UFT network operating under given conditions can be estimated.

Service quality: This is expressed by attributes such as accessibility and availability of services, delivery time, reliability, and punctuality of services.

Accessibility and availability of services depend on the spatial coverage of a given UFT system and its operating regime. The spatial accessibility is high if there is sufficient number of conveniently located stations/terminals providing access to customers from a given urban or sub urban area it serves. This implies that the preferred access distance should be about couple of hundreds of meters. Availability implies the system's accessibility in time. In general, the most UFT concepts are considered to operate, ie, to be available, over 24 h every day during the year.

The delivery time of products/commodities by a given UFT system depends on the length of pipes/tunnels, operating speed of the vehicle/capsule trains, and the time needed for their vertical transport between the system and the ground (street) level, and vice versa. This time is the part of the vehicle/capsule turnaround time in one direction based on relation (9.6) as follows: $t_{d/i} = H_i/V + l_i/v_i$, where all symbols are as in the previous relations.

The UFT systems are expected to provide reliable and punctual services by design. This can be possible thanks to their full automation based on monitoring and controlling the vehicle/capsule train movements by the centralized computer system.

9.5 BASICS OF LOCATION THEORY

The level of service in transportation, as well as the transportation system's total costs, highly depend on the location of various facilities in the network (production plants, warehouses, distribution centers, bus stops, metro stations, hub locations, vehicle depots, airport locations, etc). The position of a specific facility in the network is also highly influenced by the type of service that facilities provide to the clients. For example, it is clear that airports must be located, as far as possible, from the city center due to environmental reasons. On the other hand, they should not be too far away, since the airport that is at a considerable distance from the city down-town significantly decreases the level of air transportation service. The locations of transit bus stops are the consequence of an attempt to minimize users' walking distances, while the locations of the firefighting brigades, ambulance services and police stations are conditioned by the requirement to minimize the distance to the farthest user.

Location analysis (Weber, 1929; Mirchandani and Francis, 1990; Daskin, 1995; Hamacher and Nickel, 1998) tries to address key questions encountered in locating facilities, such as the following:

- How many facilities should be located?
- Where facilities should be located?
- How should demand for facilities service be allocated to facilities?

Production plants, distribution centers, warehouses, freight terminals, bus stops, airports, public garages, police stations, firefighter brigades, ambulances, hubs, schools, hospitals, restaurants, gasoline stations, swimming pools, and undesirable facilities are some of the facilities whose locations were analyzed and determined by various location analysis methods.

In years to come, location analysis methods and techniques will even be more important in planning and operating activities designed to minimize human pollution, improve transport and disposal of waste and hazardous materials, and to minimize the potential accidental risk (Teodorović et al., 1986; Rakas et al., 2004; Šelmić et al., 2010). Selecting good disposal sites for waste and hazardous materials and effectively transporting them with optimal routing, minimizes population exposure, and reduces further pollution of the environment.

Locating fire brigades, ambulance stations, and police stations are most often stipulated by the requirement to minimize the distance to the farthest user. This type of problem is usually called a *center problem* and sometimes it is called a *minimax problem.*

Many research projects and papers studied the problem of locating several facilities in order to minimize the average "distance" between facilities and service users. We would note that "distance" can also stand for travel time, travel costs, or some other quantity. This type of problem is known an a *median problem* and it is encountered when designing networks for distribution centers, locations for schools, post offices, shops, etc.

In some cases, facilities can be located at any point in the space (*continuous location problems*). In many other cases, facilities can be located only at pre-specified nodes in the network (*discrete location problems*). We study in this book exclusively discrete location problems, since majority of transportation terminals can be located only in pre-specified nodes due to various geographic, engineering, legal, organizational, and economic constraints (Teodorović et al., 2002).

The first paper devoted to location theory problems is related to Fermat's famous problem. Pierre de Fermat (1601–65) posited the following problem: "Given three points in a plane, find a fourth point such that the sum of its distances to the three given points is a minimum." The pioneer of modern location analysis is German economist *Alfred Weber* (1868–1958) who studied the locations of industrial facilities and warehouses.

9.5.1 LOCATION PROBLEMS CLASSIFICATION

Starting from the characteristic of the various location problems, as well as of the characteristic of already developed models and methods, we offer the following classification of the location problems:

A. The number of facilities in the network
One facility should be located.
More than one facility should be located (the number of facilities that should be located is pre-specified).
More than one facility should be located (the number of facilities that should be located is calculated as a part of optimization procedure).

B. Allowed location sites
Facilities can be located at any point in the space (continuous location problems).
Facilities can be located only at pre-specified nodes in the network (discrete location problems).

C. Desirable/undesirable facilities
Desirable facilities (the distance to, or from facilities should be minimized).
Undesirable facilities (waste disposal sites, for example) (there are few conflicting objectives).

D. Facility types
Medians (problem of locating several facilities in order to minimize the average "distance" between facilities and service users).
Centers (it is necessary to locate facilities in the network in such a way to minimize the distance to the farthest user).
Requirements location problems (the total number and the locations of facilities should be determined in order to achieve pre-specified standards/requirements).

E. Facility capacities
Facilities have unlimited capacity.
There are facility capacities constraints.

F. Type of demand
Deterministic demand.
Stochastic demand.

G. Type of service
Unique service is offered in the facilities.
Various types of services are offered in the facilities.

H. Client allocation
Clients are served in a nearest facility.
Demand is split among facilities.

It is absolutely possible to modify the proposed classification and to propose some different classification. The offered classification primarily shows various aspects that should be taken into account when solving complex location problems.

9.5.2 MEASURING DISTANCES BETWEEN FACILITIES AND DEMAND-GENERATING NODES

The distances between all pairs of nodes in the transportation network represent the basic input data for any location model. The distances could be measured in a various ways. In other words, there are various distance functions, or metrics. *Manhattan distance* and the *Euclidean distance* are the most common distance functions in a location analysis.

Fig. 9.18 shows Manhattan distance and the Euclidean distance between point $J(x_j, y_j)$ and point $I(x_i, y_i)$. Manhattan distance is characterized for the cities that have grid traffic network. Manhattan is typical example of grid traffic network. Manhattan has 12 avenues that run in parallel to the Hudson River. There are also 220 streets perpendicular to the river Hudson.

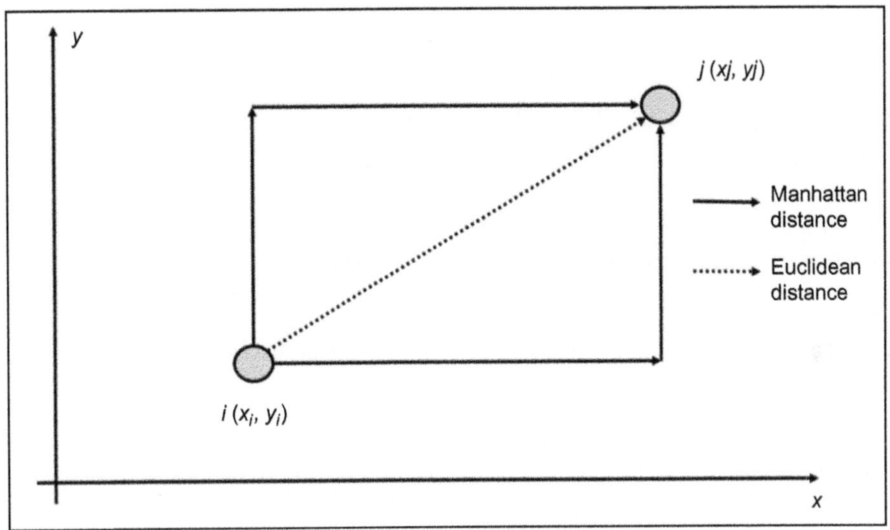

FIG. 9.18

Manhattan distance and Euclidean distance between two points.

Euclidean distance $e(I, J)$ between two points represents the length of the line segment that connect them (Fig. 9.19). *Manhattan distance* (city block distance) $m(I, J)$ between point $I(x_i, y_i)$ and point $J(x_j, y_j)$ represents sum of the absolute differences of their coordinates, ie:

$$m(I,J) = |x_i - x_j| + |y_i - y_j| \tag{9.18}$$

Euclidean distance $e(I, J)$ is given by the Pythagorean theorem, ie:

$$e(I,J) = \sqrt{(x_i - x_j)^2 + (y_i - y_j)^2} \tag{9.19}$$

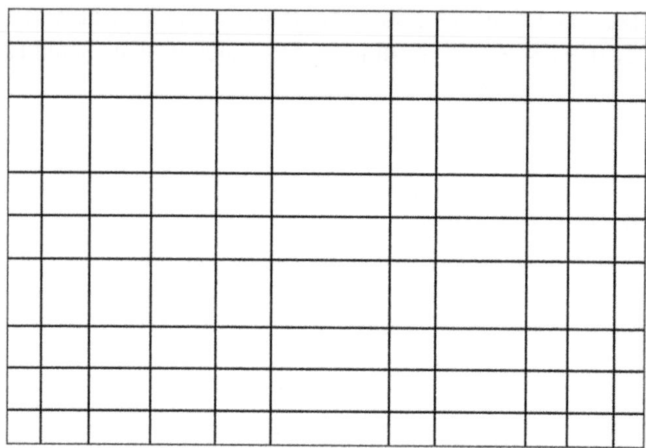

FIG. 9.19

Grid traffic network.

Both, Manhattan distance and Euclidean distance represent special cases of the l_p distance defined in the following way:

$$l_p(I,J) = \left[|x_i - x_j|^p + |y_i - y_j|^p\right]^{\frac{1}{p}} \tag{9.20}$$

In the case when $p=1$, l_p distance becomes Manhattan distance, ie:

$$m(I,J) = l_1(I,J) \tag{9.21}$$

Similarly, in the case when $p=2$, l_p distance becomes Euclidean distance, ie:

$$e(I,J) = l_2(I,J) \tag{9.22}$$

It has been shown that the road distances are usually 10–30% greater than the corresponding Euclidean distances. According to Love et al. (1988) the road distances between main American cities are 18% greater than the corresponding Euclidean distances. The distance to be used highly depends on the considered location problem context.

9.5.3 THE LOCATION SET COVERING PROBLEM

Some location problems are characterized by the existence of the maximal service distance D_s. This distance is the maximum distance that any client would have to travel to reach a facility. This distance defines so called *catchment area.* Airports, freight terminals, bus stops and other transportation facilities have corresponding catchment areas. Clients located within defined maximal service distance (clients located within catchment area) receive service. In other words, these clients are "covered" by the facility. The other clients do not receive service, and they are not "covered" (Fig. 9.20).

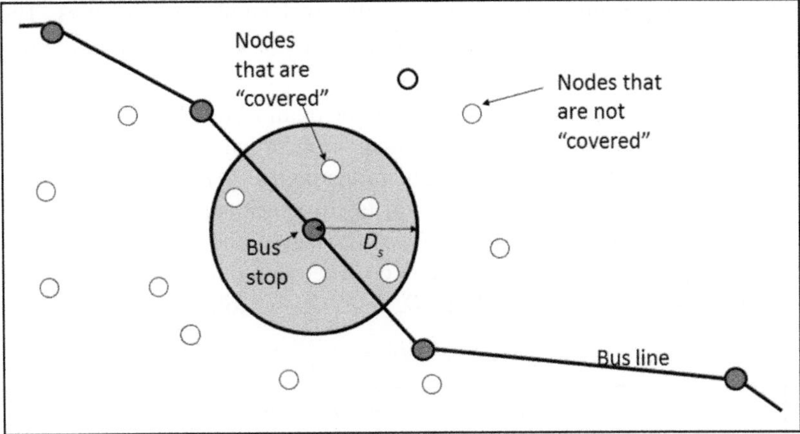

FIG. 9.20

Catchment area of the bus stop.

The location set covering problem could be formulated in the following way: Find minimal necessary number of facilities from the set of node-candidates for facility location such that all demand-generating nodes are covered (Toregas et al., 1971; Current et al., 2002). Let us introduce the following notation:

I	Set of demand nodes
i	Demand nodes index
J	Set of node-candidates for facility location
j	Facility nodes index
$d_{i,j}$	The shortest distance from node i to node j
D_p	Maximal service distance
$N_i = \{j \mid d_{ij} \leq D_p\}$	Set of node candidates that can cover demanding node i

$$x_j = \begin{cases} 1 & \text{if facility is located in node } j \\ 0 & \text{otherwise} \end{cases}$$

The location set-covering problem has the following mathematical formulation (Toregas et al., 1971; ReVelle et al., 1977; Current et al., 2002):

Minimize:

$$\sum_{j \in J} x_j \tag{9.23}$$

subject to:

$$\sum_{j \in N_i} x_j \geq 1 \quad \forall i \in I \tag{9.24}$$

$$x_j \in \{0, 1\} \quad \forall j \in J \tag{9.25}$$

The objective function to be minimized represents the total number of facilities. The constraints $\sum_{j \in N_i} x_j \geq 1 \ \forall i \in I$ request that every node must be covered by at least one facility. After solving the location set-covering problem, the analyst can discover the minimal number and the location of service facilities that enable that no demand point is further than the maximal service distance from the facility.

Let us consider the following eight demand-generating points: *A*, *B*, *C*, *D*, *E*, *F*, *G* and *H* (demand set). We also consider the following 6 candidates for facility locations: *I*, *J*, *K*, *L*, *M*, and *N* (facilities set). Specific facility "covers" specific demand-generating point if the distance between the object and the demand-generating point is less than or equal to 60. The matrix of minimum distances $[d(i,j)]$ between nodes *I*, *J*, *K*, *L*, *M*, *N* and nodes *A*, *B*, *C*, *D*, *E*, *F*, *G* is as follows:

$$[d(i,j)] =$$

	Demand nodes							
Objects	**A**	**B**	**C**	**D**	**E**	**F**	**G**	**H**
I	9	29	43	62	58	74	71	20
J	75	71	32	27	96	87	42	46
K	42	84	77	28	74	46	76	70
L	29	14	29	57	76	66	44	60
M	61	89	88	90	30	91	88	32
N	42	27	96	29	75	69	29	74

The location set covering problem is defined as follows: find the smallest number of candidates for facility locations necessary to cover the demand-generating nodes if it is required that the distance between the object and the demand-generating point is less than or equal to 60.

We first transform the matrix of minimum distances $[d(i,j)]$ in the so-called *coverage matrix* $[p(i,j)]$ in the following way:

$$p(i,j) = \begin{cases} 1, & \text{for} d(i,j) \leq 60 \\ 0, & \text{otherwise} \end{cases} \tag{9.26}$$

where $p(i,j)$ is an element of the coverage matrix $[p(i,j)]$.

The coverage matrix reads:

$$[p(i,j)] = \tag{9.27}$$

	Demand nodes							
Objects	**A**	**B**	**C**	**D**	**E**	**F**	**G**	**H**
I	1	1	1	0	1	0	0	1
J	0	0	1	1	0	0	1	1
K	1	0	0	1	0	1	0	0
L	1	1	1	1	0	0	1	1
M	0	0	0	0	1	0	0	1
N	1	1	0	1	0	0	1	0

The coverage matrix inform us which specific facility "covers" specific demand-generating point. For example, we conclude from the coverage matrix that node K covers nodes A, D and F. In order to solve the problem defined, we shall use simple matrix-reduction algorithm proposed by Larson and Odoni (1981). Their matrix-reduction algorithm consists of the following steps:

Step 1: Check if there is, as a minimum, one column in the coverage matrix that consists completely of zeroes. If the answer is yes, stop. In such a case no feasible solution exists. Required standards for coverage must be changed, or more nodes must be included in the facilities set. If the answer is no, go to step 2.
Step 2: If whichever columns have only one nonzero element (eg, in row i^*), then the facility must be located in the node corresponding to row i^*. Include that node in the list of facility sites and remove row i^* and all columns having a 1 in row i* from the matrix.
Step 3: Eliminate any row(s) i'' if all its elements are less than or equal to the corresponding elements of another row i'.
Step 4: Eliminate any column(s) j'' if all its elements are greater than or equal to the corresponding elements of another column j'.
Step 5: Reiterate steps 2–4 until:
(a) the coverage matrix becomes entirely empty, or
(b) no columns or rows are removed through an entire pass through steps 2–4.

EXAMPLE 9.4

Let us apply the matrix-reduction algorithm in the case of the following coverage matrix:

$$[p(i,j)] =$$

	Demand nodes							
Objects	A	B	C	D	E	F	G	H
I	1	1	1	0	1	0	0	1
J	0	0	1	1	0	0	1	1
K	1	0	0	1	0	1	0	0
L	1	1	1	1	0	0	1	1
M	0	0	0	0	1	0	0	1
N	1	1	0	1	0	0	1	0

The distances between all pairs of nodes are given in the previous example. We conclude that feasible solution exists, since there is no one column in the coverage matrix $[p(i, j)]$ that consists completely of zeroes. Step 1 of the algorithm is finished. We see in step 2 that column F has only one nonzero element at row K. We conclude that a facility must be located at K. This facility will serve at least demand-generating node F. We remove from further consideration row K and columns A, D and F (since $p(K, A)=p(K, D)=p(K, F)=1$). Reduced matrix reads:

(Continued)

EXAMPLE 9.4—cont'd

$$[p(i,j)] =$$

	Demand nodes				
Objects	B	C	E	G	H
I	1	1	1	0	1
J	0	1	0	1	1
L	1	1	0	1	1
M	0	0	1	0	1
N	1	0	0	1	0

In step 3, we remove rows J and M. All elements in row J are less than or equal to the corresponding elements of row L. This means that all nodes covered by the facility located in node J can be covered by the facility located in node L. The similar is valid respectively for rows M and I. The reduced matrix reads:

$$[p(i,j)] =$$

	Demand nodes				
Objects	B	C	E	G	H
I	1	1	1	0	1
L	1	1	0	1	1

In step 4, we remove columns B and C, since all elements in these columns are greater than or equal to the corresponding elements of the columns E and G respectively (The facility that covers node E will be able also to cover nodes B and C. After the fourth step, the reduced matrix reads:

$$[p(i,j)] =$$

	Demand nodes		
Objects	E	G	H
I	1	0	1
L	0	1	1

We also remove column H, since all elements in this column are greater than or equal to the corresponding elements in the columns E and G. The reduced matrix reads:

$$[p(i,j)] =$$

	Demand nodes	
Objects	E	G
I	1	0
L	0	1

EXAMPLE 9.4—cont'd

We return to step 2. Since both columns E and G have only one nonzero element, we conclude that facilities must be located in nodes I and L. By eliminating rows I and L, we create a coverage matrix that is empty. In this way, we finish with the algorithm.

We conclude that we need 3 facilities to cover points A, B, C, D, E, F, G, and H. These facilities should be located at nodes K, I, and L. The matrix of shortest distances between facilities located at nodes I, K, and L and demand-generating points reads:

	Demand nodes							
Objects	A	B	C	D	E	F	G	H
I	9	29	43	62	58	74	71	20
K	42	84	77	28	74	46	76	70
L	29	14	29	57	76	66	44	60

Every demand-generating point is covered from the closest facility. Facility I covers points A, E, and H. Facility K covers D and F, while facility L covers nodes B, C, and G. The distances between any facility and the nodes that facility serves are always less than 60.

9.5.4 MAXIMAL COVERING LOCATION PROBLEM

When considering location set covering problem, we try to discover the smallest number of candidates for facility locations necessary to cover *all* demand-generating nodes. On the other hand, in some cases (due to budget constraints) it is not possible to cover all demand-generating nodes. In such cases, it is logical to maximize coverage (population covered) within a prescribed maximal service distance by locating given number of service facilities. This problem is known as the Maximal Covering Location Problem (Church and ReVelle, 1974).

Let us introduce the following notation:

a_i—population to be served at demand-generating node i (demand at node i),

p—number of facilities to be located,

$$Z_i = \begin{cases} 1, & \text{if demand at the node } i \text{ is covered} \\ 0, & \text{otherwise} \end{cases} \tag{9.28}$$

Mathematical formulation of the maximal covering location problem reads:

Maximize

$$\sum_{i \in I} a_i z_i \tag{9.29}$$

subject to:

$$\sum_{j \in N_i} x_j - z_i \geq 0 \quad \forall i \in I \tag{9.30}$$

$$\sum_{j \in J} x_j = p \tag{9.31}$$

$$x_j = \{0, 1\} \quad \forall j \in J \tag{9.32}$$

$$z_i = \{0, 1\} \quad \forall i \in I \tag{9.33}$$

The objective function to be maximized represents the total number of people served within the predefined service distance. Constraint $\sum_{j \in N_i} x_j - z_i \geq 0 \;\; \forall i \in I$ treats node i as covered only when one or more facilities are located within maximal distance of demand node i. Constraint $\sum_{j \in J} x_j = p$ describes the fact that the total number of deployed facilities is equal to p.

9.5.5 MEDIANS

Median problems are particularly important for transportation activity since they appear when designing different distribution systems. In this case, we try to minimize the average "distance" between facilities where some service is provided and the user of that service. The p median problem was for a first time formulated by Hakimi (1964). We consider non-oriented network $G=(N, A)$ that has n nodes. Let us denote by a_i the number of service demands from node i.

We denote by d_{ij} the distance between node i and node j. We also denote by p the total number of facilities to be located in the network, and by a_i the population to be served at demand-generating node i. The facilities could be located at any of the network's nodes. We define binary variables x_{ij} in the following way:

$$x_{ij} = \begin{cases} 1, & \text{if users from the node } i \text{ are served in the node } j \\ 0, & \text{otherwise} \end{cases} \tag{9.34}$$

The mathematical formulation of the p-median problem reads:

Minimize

$$\min F = \sum_{i=1}^{n} \sum_{j=1}^{n} a_i d_{ij} x_{ij} \tag{9.35}$$

subject to

$$\sum_{j=1}^{n} x_{ij} = 1, \quad i = 1, 2, \ldots, n \tag{9.36}$$

$$\sum_{j=1}^{n} x_{jj} = p \tag{9.37}$$

$$x_{jj} \geq x_{ij}, \quad i, j = 1, 2, \ldots, n; i \neq j \tag{9.38}$$

$$x_{ij} \in \{0, 1\}, \quad i, j = 1, 2, \ldots, n \tag{9.39}$$

The defined objective function that we want to minimize represents the total distance traveled by the clients. The first constraint describes that every client is served by one facility. The second constraint requires that the total number of facilities in the network equals p. Every client from the node where facility is located receives service from that facility (constraint 9.27).

Hakimi (1964) showed that there is at least one set of p-medians in the nodes of network G, which means that p optimal locations for facilities in the network must be found exclusively in the network nodes. This fact significantly facilitates the procedure for finding p medians since only locations found in nodes must be examined.

The p-median problem was solved by various optimization and heuristic procedures. The algorithmic approach depends on the size of a problem considered, acceptable CPU time, as well as on analyst and decision-maker's readiness to accept non-optimal solutions.

9.5.5.1 Location of a single median

The algorithm (Hakimi, 1965) for determining one median for non-oriented network $G(N, A)$ is composed of the following algorithmic steps:

Step 1: Calculate the shortest distances d_{ij} between pairs of nodes (i, j) in network G and show them in a matrix of shortest distances.
Step 2: Multiply the jth column of the matrix by the number of service demands a_j from node j. Element $a_j \cdot d_{ij}$ of matrix $[a_j \cdot d_{ij}]$ is the distance traveled by users from node who are serviced in node i. Denote the matrix $[a_j \cdot d_{ij}]$ by D'.
Step 3: Add up the elements of every row i of matrix D'. Expression $\sum_{j=1}^{n} a_j \times d_{ij}$ is the total distance traveled by users when the facility is in node i.
Step 4: The node whose row corresponds to the least total distance traveled by users is the median location.

EXAMPLE 9.5

Let us study the transportation network shown in Fig. 9.21. Transportation network nodes are denoted respectively by $A, B, C, \ldots, H$. Daily service demands are given in parentheses next to each node. All link lengths are also denoted.

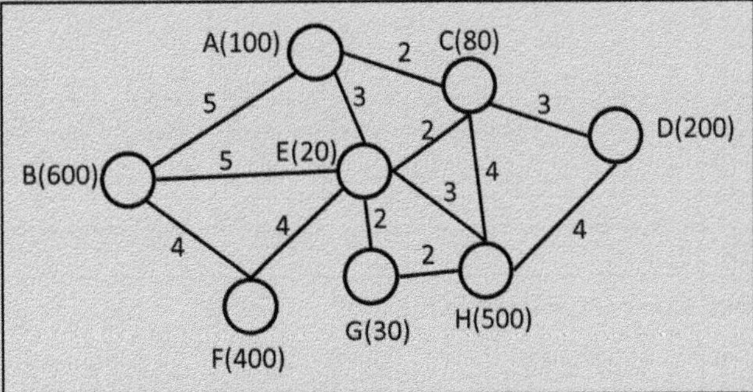

FIG. 9.21

Transportation network in which one median is to be found.

Our problem is as follows: Where should we locate a service facility so that the average distance covered by service users (located in the nodes) to the facility is minimized?

(Continued)

EXAMPLE 9.5—cont'd

Solution

Based on Hakimi's theorem, we can conclude that there are eight candidate spots in which to locate the facility. These are nodes $A, B, C, \ldots, H$. Using the shortest path algorithm we calculate shortest paths d_{ij} between all pairs of nodes (i, j) in the transportation network. These shortest distances are given in the following matrix:

$$
[d_{ij}] = \begin{array}{c} \\ A \\ B \\ C \\ D \\ E \\ F \\ G \\ H \end{array}
\begin{array}{c} \begin{array}{cccccccc} A & B & C & D & E & F & G & H \end{array} \\
\begin{bmatrix}
0 & 5 & 2 & 5 & 3 & 7 & 5 & 6 \\
5 & 0 & 7 & 10 & 5 & 4 & 7 & 8 \\
2 & 7 & 0 & 3 & 2 & 6 & 4 & 4 \\
5 & 10 & 3 & 0 & 5 & 9 & 6 & 4 \\
3 & 5 & 2 & 5 & 0 & 4 & 2 & 3 \\
7 & 4 & 6 & 9 & 4 & 0 & 6 & 7 \\
5 & 7 & 4 & 6 & 2 & 6 & 0 & 2 \\
6 & 8 & 4 & 4 & 3 & 7 & 2 & 0
\end{bmatrix} \end{array}.
$$

In the next step, we calculate expression $a_j \cdot d_{ij}$, by multiplying every column of the shortest paths matrix by the number of service demands in node j. Matrix $[a_j \cdot d_{ij}]$ is as follows:

$$
[a_j \times d_{ij}] = \begin{array}{c} \\ A \\ B \\ C \\ D \\ E \\ F \\ G \\ H \end{array}
\begin{array}{c} \begin{array}{cccccccc} A & B & C & D & E & F & G & H \end{array} \\
\begin{bmatrix}
0 & 3000 & 160 & 1000 & 60 & 2800 & 150 & 3000 \\
500 & 0 & 560 & 2000 & 100 & 1600 & 210 & 4000 \\
200 & 4200 & 0 & 600 & 40 & 2400 & 120 & 2000 \\
500 & 6000 & 240 & 0 & 100 & 3600 & 180 & 2000 \\
300 & 3000 & 160 & 1000 & 0 & 1600 & 60 & 1500 \\
700 & 2400 & 480 & 1800 & 80 & 0 & 180 & 3500 \\
500 & 4200 & 320 & 1200 & 40 & 2400 & 0 & 1000 \\
600 & 4800 & 320 & 800 & 60 & 2800 & 60 & 0
\end{bmatrix} \end{array}.
$$

By summing up the rows of matrix $[a_j \cdot d_{ij}]$, we get the number of kilometers traveled by users if the facility is located in the nods whose row is being summed up. It is clear that the facility should be located in the node whose summed up row is the smallest. By dividing the total sum by the total number of users in the entire transportation network (in this case, 1930), we get the average distance covered by one user in order to satisfy his or her demand in the facility. Table 9.2 gives the total number of kilometers traveled when the facility is located in a specific node in the network:

Table 9.2 Number of Kilometers Traveled

Facility Location	Number of Kilometers Traveled	Facility Location	Number of Kilometers Traveled
A	10,170	E	7620
B	8970	F	9140
C	9560	G	9660
D	12,620	H	9440

EXAMPLE 9.5—cont'd

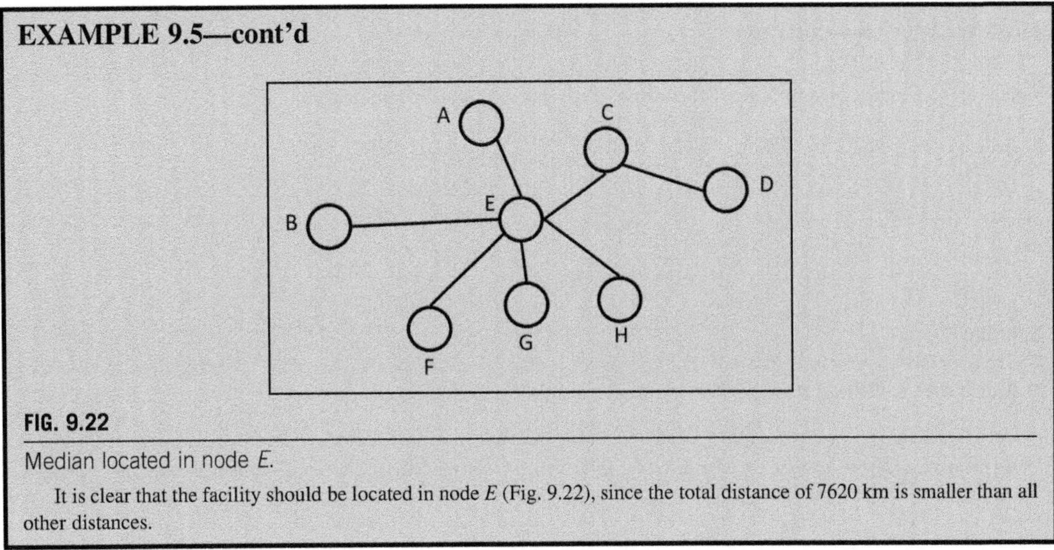

FIG. 9.22

Median located in node *E*.

It is clear that the facility should be located in node *E* (Fig. 9.22), since the total distance of 7620 km is smaller than all other distances.

The algorithm described could be also used in the case of oriented network. In the case of oriented network, we distinguish situations when clients travel from home to the facility, from the situations when service team travel from the facility to clients. The median in the first case is called the *inward median*, while the second type of median is called the *outward median*. The described algorithm for determining a single median location could be also used to determine locations of the inward and outward medians.

EXAMPLE 9.6

Determine the inward median location for the network shown in Fig. 9.23.

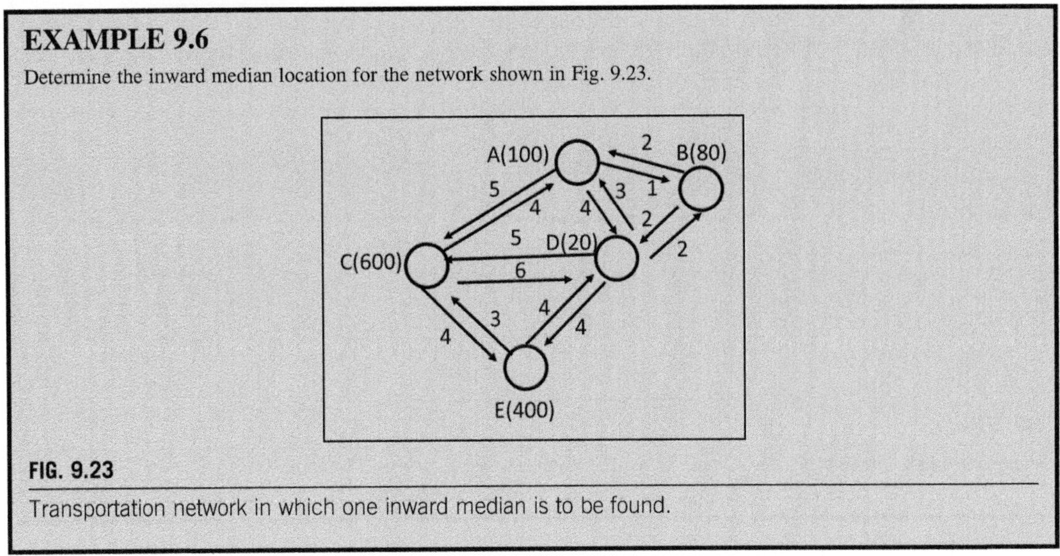

FIG. 9.23

Transportation network in which one inward median is to be found.

(Continued)

EXAMPLE 9.6—cont'd

Network nodes are denoted respectively by A, B, C, D, and E. Daily demands at nodes are given in parentheses. The link lengths are also indicated in the figure. The matrix of the shortest path distances reads:

$$[d_{ij}] = \begin{matrix} & A & B & C & D & E \\ A & 0 & 1 & 5 & 4 & 7 \\ B & 2 & 0 & 7 & 2 & 6 \\ C & 4 & 5 & 0 & 6 & 4 \\ D & 3 & 2 & 5 & 0 & 4 \\ E & 7 & 6 & 3 & 4 & 0 \end{matrix}$$

Solution

The matrix of the shortest path distances is not symmetric. Let us determine the location of the inward median. If we locate median in node A, the total travel distance of all clients equals:

$$100 \times 0 + 80 \times 2 + 600 \times 4 + 20 \times 3 + 400 \times 7 = 5420$$

In the same way, we calculate the total distances for the cases when median is located respectively in nodes B, C, D or E. These distances are given in Table 9.3.

Table 9.3 Inward Median Locations and the Total Traveled Distances

Inward Median Location	Total Number of Passenger Kilometers
A	5420
B	5540
C	2160
D	4240
E	3660

Clearly, the inward median should be located in node C (Fig. 9.24).

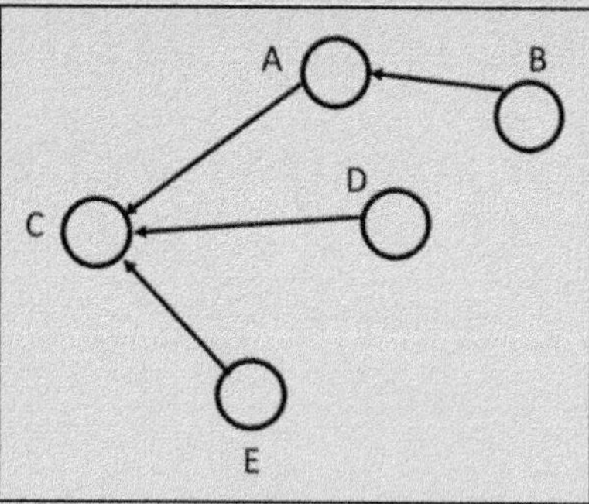

FIG. 9.24

Inward median located in node *C*.

EXAMPLE 9.7

Determine the locations of two medians for the network shown in Fig. 9.25.

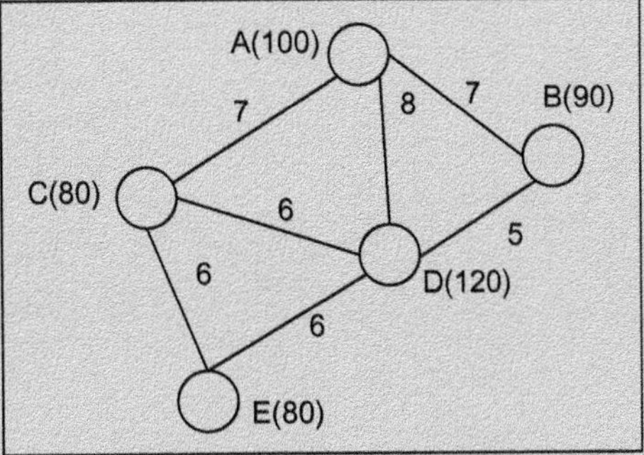

FIG. 9.25

Network for which locations for two medians should be determined.

The network nodes are denoted respectively by A, B, C, D, and E. Daily demands at nodes are given in parentheses. The link lengths are also indicated in the figure.

Solution

Given that there are five nodes in the network, the total number of combinations to locate two medians is equal to:

$$\binom{5}{2} = 10$$

The medians could be located in the following node pairs:

$$(A,B),(A,C),(A,D),(A,E),(B,C),(B,D),(B,E),(C,D),(C,E)i(D,E)$$

Let us calculate the total distance traveled by all users in the case when the medians are located in nodes A and B. The users from node A are served in node A. In the same way, users from node B are served in node B. The distance between C and A is smaller than the distance between C and B. As a consequence, the users from node C will be served in node A. The users from nodes D and E will be served in node B. The total distance traveled by all users equals:

$$0 \times 100 + 0 \times 90 + 7 \times 80 + 5 \times 120 + 11 \times 80 = 2040$$

The number of kilometers traveled for various median locations is shown in Table 9.4.

Table 9.4 The Number of Kilometers Traveled for Various Median Locations

Median Locations	Number of Kilometers Traveled	Median Locations	Number of Kilometers Traveled
(A, B)	2040	(B, D)	1660
(A, C)	2190	(B, E)	1780
(A, D)	1410	(C, D)	1630
(A, E)	1830	(C, E)	2410
(B, C)	1780	(D, E)	1730

(Continued)

EXAMPLE 9.7—cont'd

We conclude that the medians should be located in nodes *A* and *D* (Fig. 9.26).

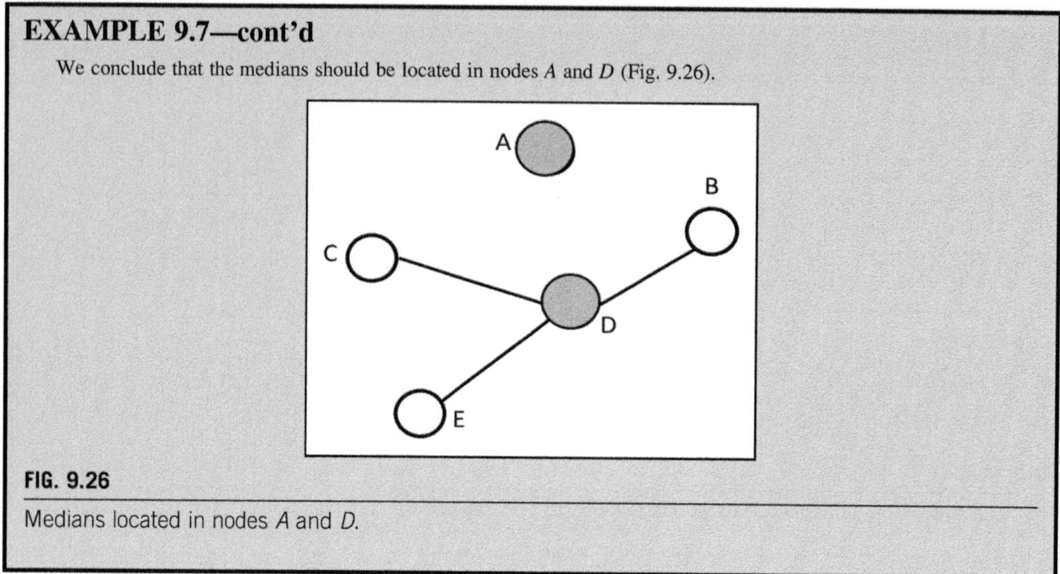

FIG. 9.26

Medians located in nodes *A* and *D*.

Problem of finding optimal locations of a few medians in a transportation network is combinatorial by its nature (as many other location problems). Most of the combinatorial optimization problems are difficult to solve either because of the large dimensionality or because it is very difficult to decompose then into smaller sub-problems. So, in many cases optimal solution cannot be discovered in a reasonable CPU time. Metaheuristic algorithms (Simulated Annealing, Genetic Algorithms, Taboo Search, Variable Neighborhood Search, Ant Colony Optimization, Particle Swarm Optimization, and Bee Colony Optimization (BCO)) are considered to be a general-purpose techniques capable of producing good solutions of a difficult discrete optimization problems in a reasonable computer time.

9.5.6 HUB LOCATION

The hub location problem is one of the most important problems in location theory. This problem appears in transportation systems that have large number of origin-destination pairs. Many transport and delivery companies organize hub networks, as flows between hubs are characterized by the economy of scale effect. Usually, it is cheaper for the carrier to avoid direct routing between all pairs of nodes between which flow occurs. Instead, hub objects perform the transfer of passengers (from one plane/bus line to another) or goods reloading (reloading letter mail and packages from small to larger aircraft, reloading of goods from trucks to ships etc.) with a higher frequency of operations and lower costs per unit. Passengers or goods travel longer and change means/modes of transport in the hubs (Fig. 9.27) but travel much more frequently to their destinations. This area of research was introduced to the literature by O'Kelly (1986, 1987) and Bryan and O'Kelly (1999). To organize goods delivery in a specific region, the delivery company must make appropriate decisions about the total number of hubs, their locations, and the allocation of demand for facilities' services to facilities.

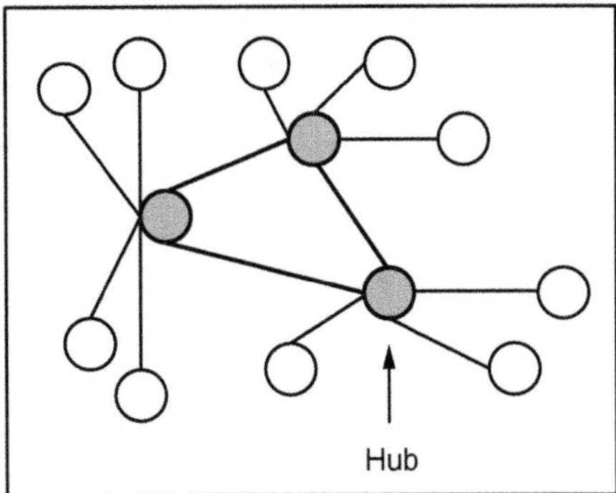

FIG. 9.27

Transportation network with hubs.

9.6 VEHICLE ROUTING AND SCHEDULING

Vehicle routing and scheduling problems appear in various activities of many companies and service providers. Distribution of goods from warehouses to shops and supermarkets, newspapers distribution, public health nursing, emergency services (ambulance, police, and fire), repair businesses, waste collection, street cleaning and sweeping, snow plowing, transporting students to schools, transporting elderly and handicapped people, and courier service in cities are some of the activities, than cannot be efficient, without proper vehicle routing and scheduling. Planners, engineers, and dispatchers in many companies make operational decisions on a daily basis. The fundamental question that they faced is how to efficiently use available fleet of vehicles to satisfy demand in the network, taking care about various operational requirements and constraints. The requirements could be related to roads in the network, traffic conditions, vehicle types, company procedures, drivers' work regulations, etc. Well-organized vehicle routing or a well-designed schedule can significantly contribute towards a decrease in transportation costs and increase the level of transportation services.

By solving vehicle routing and scheduling problems, an analyst designs routes and creates schedule for every vehicle in the fleet (Dantzig and Ramser, 1959). *Vehicle route* represents a sequence of clients that vehicle should visit, where vehicle starts and ends its journey at the depot. For example, the route of Vehicle #7 could be the route {depot, node 3, node 12, node 9, node 6, node 22, depot}. *Vehicle schedule* precisely defines planned vehicle activities along the route. For example, Vehicle #7 should depart from the depot at 6:00 am, arrive at node 3 at 6:15 am, depart from node 3 at 6:30, arrive at node 12 at 6:50 am, etc.

Vehicle *routing problems* (VRPs) do not have time constraints as to when services in different nodes should start or finish. Contrary to this, *scheduling problems* contain times, fixed in advance, within which service in each node must be completed. *Combined routing and scheduling problems* appear when service in certain nodes must be carried out within a specific time intervals (*time windows*). In the case of public transportation, postal service, courier services, and some trucking companies, service is performed only if delivery is timely.

The total service cost represents the usual objective function of most routing and scheduling problems in distribution systems, since the majority operators are private companies. The total costs are usually composed of vehicle capital costs, crew costs, and vehicle mileage.

The objectives in the public sector are related to the level-of-service offered to the clients. For example, when a school bus collects students from various locations, and transports them to the school, the natural objective is to create vehicle routes in such a way to minimize the total number of student-minutes on the bus. In the case of emergency services, the response time represents objective function that should be minimized.

9.6.1 VRPs TYPES

In some cases, vehicles have to traverse defined set of streets (cleaning and sweeping of streets, snow plowing, waste collection from houses, etc.). Problems of this type are called *edge-covering* problems (Fig. 9.28). In some other situations, vehicles must visit the defined set of nodes. These problems are known as *node-covering* problems (Fig. 9.29). Node-covering problems appeared primarily in the area of distribution systems.

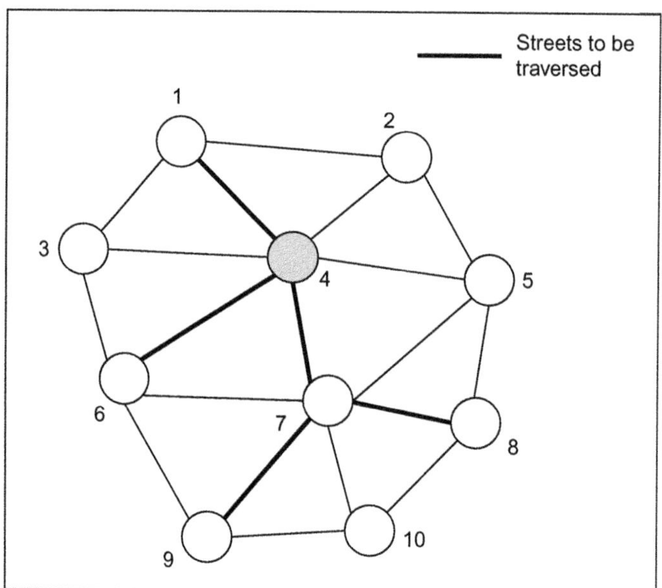

FIG. 9.28

Edge covering.

Starting with specific characteristics that describe certain types of routing or scheduling problems, Bodin and Golden (1981) proposed the following classification:

(a) Number of vehicle depots in the network:
- There is only one depot in the network.
- The network contains several depots.

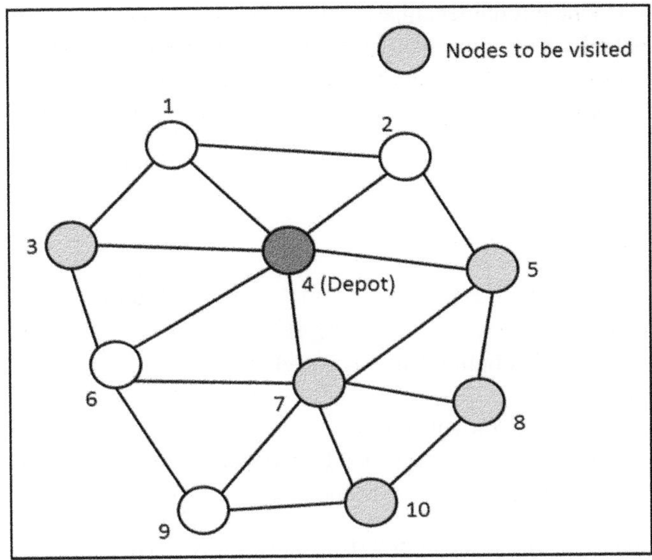

FIG. 9.29

Node covering.

(b) Size of vehicle fleet available:
- The fleet contains only one vehicle.
- The fleet contains several vehicles.

(c) Type of vehicles in the fleet:
- All vehicles in the fleet are the same.
- The fleet contains different types of vehicles.

(d) Nature of service demands:
- Deterministic demands appear in the network.
- Stochastic demands for service appear.

(e) Location of service demands:
- Service demands appear in the networks nodes.
- Service demands appear in the network's branches.
- Service demands appear in both nodes and branches.

(f) Type of transportation network:
- Oriented transportation network.
- Non-oriented transportation network.
- Mixed transportation network.

(g) Vehicle capacity constraints:
- All vehicles have the same regulated capacity constraints.
- There are differences between vehicles regarding regulated capacity constraints.
- There are no constraints regarding vehicle capacity.

(h) Maximum allowed vehicle route length:
  - All vehicles in the fleet have the same maximum allowed route length.
  - Some vehicles have different maximum allowed route lengths.
  - There are no constraints regarding the maximum allowed vehicle route length.

(i) Costs:
  - Variable.
  - Fixed.

(j) Operations carried out:
  - Picking up.
  - Delivering.
  - Picking up and delivering.

(k) Objective functions on which optimization is based:
  - Minimizing route costs.
  - Minimizing total fixed and variable costs.
  - Minimizing the number of vehicles needed to carry out transportation operations.

(l) Other constraints (depends on specific problem).

The above classification is very useful, particularly because it gives a better insight into specific transportation problems to be solved. It also helps when making analogies between problems in different branches of transportation and solving them by the same or similar methodological process.

9.6.2 VEHICLE ROUTING AND SCHEDULING PROBLEMS COMPLEXITY

Vehicle routing and scheduling problems are combinatorial by their nature. Many instances of these problems are difficult to solve, primarily of the large dimensionality. In many cases, optimal solution of the large VRPs cannot be discovered in a reasonable CPU time. For example, if we have in a problem considered 1000 binary variables that can take value 0, or 1, the total number of all possible solutions is equal to 2^{1000}.

The computer time needed to find optimal solution of various vehicle routing and scheduling problems increases very rapidly with the dimension of the problem (number of clients to be visited). In order to overcome this problem various *heuristic algorithms* were developed during the last 5 decades. Heuristic algorithms, could be described as a mixture of scientific methods, invention, experience, and intuition for problem solving. These algorithms are capable of producing good enough solution(s) in a reasonable amount of CPU time. On the other hand, heuristic algorithms that are based on experience and/or judgment cannot be guaranteed to generate the optimal solution.

9.6.3 TRAVELING SALESMAN PROBLEM

The Traveling Salesman Problem (TSP) is the most famous problem in the class of the node-covering problems (Lin, 1965). The traveling salesman leaves the home town and wants to visit a number of towns. We denote the number of towns to be visited by n. We assume that the distances between each pair of towns are known. The Traveling Salesman Problem (TSP) reads: what is the shortest possible route that visits every town just once and comes back to the home town? There are $(n-1)!$ different orderings of the towns to be visited. Since each generated traveling salesman tour can be run in any of two possible directions, there are $(n-1)!/2$ different solutions of the TSP. For example, when $n=15$, there are $14!=87,178,291,200$ various orderings, and $(15-1)!/2=43,589,145,600$ different solutions of the TSP. Real-life problem that appear in distribution systems can have hundreds and thousands of nodes to be visited. The computer time needed to discover an optimal solution of the TSP increases very rapidly with n. In order to prevail over this problem various *heuristic algorithms* were developed during the last 5 decades. Heuristic algorithms, could be described as a mixture of scientific methods, invention, experience, and intuition for problem solving. These algorithms are capable of producing good enough solution(s) in a reasonable amount of CPU time. On the other hand, heuristic algorithms that are based on experience and/or judgment cannot be guaranteed to generate the optimal solution.

The Nearest Neighbor algorithm is a very simple heuristic algorithm for solving the TSP. The algorithm consists of the following steps:

Step 1: Arbitrarily choose the initial node of the route.
Step 2: Find the node closest to the last node included in the route. The closest node included in the route.
Step 3: Repeat step 2 until all the nodes are not involved in the route. Connect the first and last node of the route.

EXAMPLE 9.8

Node 1 represents traveling salesman home town (Fig. 9.30). The traveling salesman should visit nodes 2, 3,..., 9, and return to node 1. Design the traveling salesman route by using the Nearest Neighbor algorithm.

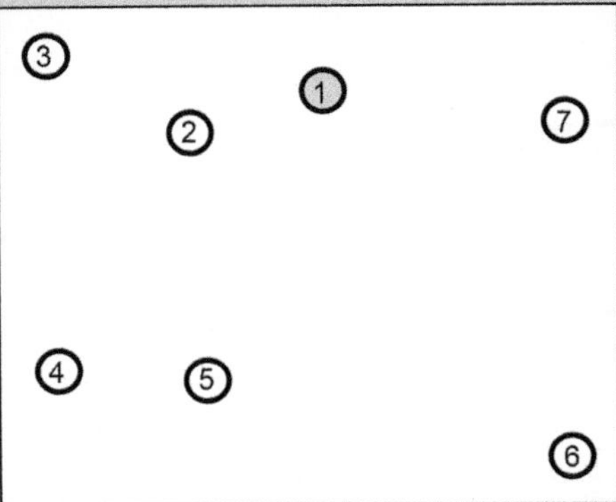

FIG. 9.30

Location of cities and traveling salesman home town.

The distances between all pairs of nodes are given in Table 9.5.

The route must start at node 1. The closest node to node 1 is node 2. Node 2 is included in the route. The traveling salesman route now reads: (1, 2). Node 3 is the closest node to node 2. Node 3 is included in the route. The route reads: (1, 2, 3). The next node to be included in the route is node 4, etc.

The final route is: (1, 2, 3, 4, 5, 6, 7, 1). The route is shown in Fig. 9.31.

(Continued)

EXAMPLE 9.8—cont'd

Table 9.5 Distances Between Nodes

	1	2	3	4	5	6	7
1	0	75	135	165	135	180	90
2	75	0	90	105	135	210	150
3	135	90	0	150	210	300	210
4	165	105	150	0	135	210	210
5	135	135	210	135	0	90	105
6	180	210	300	210	90	0	120
7	90	150	210	210	105	120	0

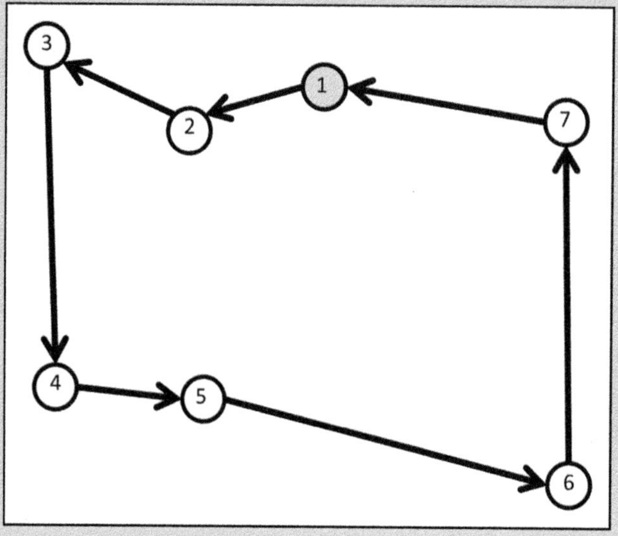

FIG. 9.31

Traveling salesman route obtained by the Nearest Neighbor algorithm.

9.6.4 VEHICLE ROUTING PROBLEM

Let there be n nodes that request service, each demanding v_i $(i=1, 2,\ldots, n)$ amount of some good. Vehicles are stationed at point B.

All vehicles have the same capacity V and when servicing nodes all vehicles must start and finish their trips at point B. We consider the case of homogenous fleet. In other words, all vehicles have the same capacity. Let capacity of the vehicles V be greater than any amount of goods v_i to be transported, so that each node is serviced by only one vehicle. Obviously, one vehicle can service one or more nodes. We also assume that we know distances between all pairs of nodes.

The VRP could be formulated in the following way: Determine the set of vehicle routes in such a way to minimize the total distance travelled by the entire fleet of vehicles, while vehicle capacity constraint is satisfied, each route starts and ends at the depot, and all clients are serviced (Fig. 9.32).

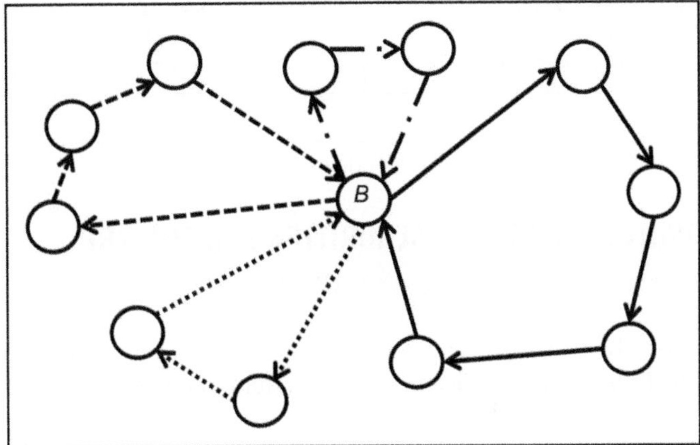

FIG. 9.32

Set of designed vehicle routes.

Various approaches have been developed for solving VRP during the last few decades (Gillett and Miller, 1974; Fisher and Jaikumar, 1981; Solomon, 1987; Teodorović and Radivojević, 2000; Cordeau et al., 2002; Laporte and Semet, 2002). The methods could be grouped into the following classes (Fig. 9.33):

- cluster-first route-second methods; and
- route-first cluster-second methods.

The *Cluster-first route-second* methods in the first step generate clusters of clients to be served. In the second step, these methods solve the TSP within each cluster (*f*).

The *Route-first cluster-second* methods (Beasley, 1983) in the first step relax vehicle capacity in order to generate a "giant tour" (TSP tour). Then, in the second step, the generated giant tour is split into feasible vehicle routes.

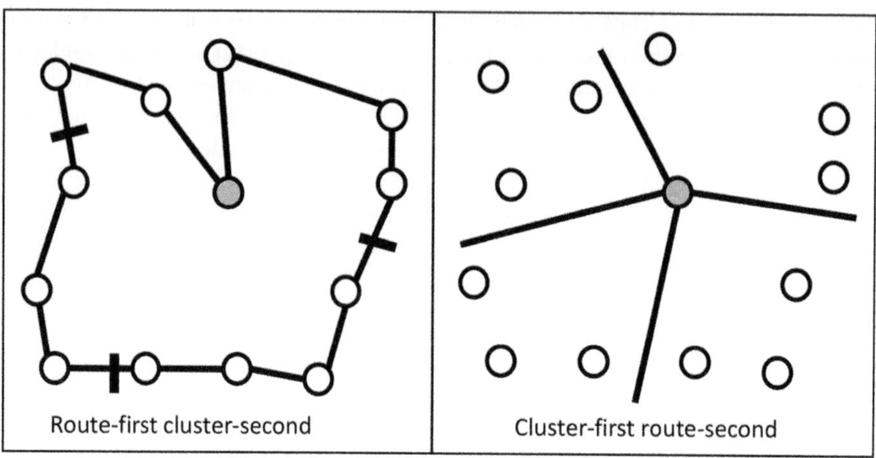

FIG. 9.33

Route-first cluster-second and Cluster-first route-second.

9.6.5 CLARK-WRIGHT'S "SAVINGS" ALGORITHM FOR THE VRP

One of the widely used algorithms for solving the VRP is Clark-Wright's "savings" algorithm (Clarke and Wright, 1964).

Let one vehicle service first point and return to point B. After that vehicle services second point and again return to point B. Then vehicle service third point, return to point B, etc. The total distance covered equals:

$$2\times b(B,1)+2\times d(B,2)+\cdots+2\times d(B,n)=2\times\sum_{i=1}^{n}d(B,i) \tag{9.40}$$

where $d(B,i),(i=1,2,\ldots,n)$ is the distance between point B and point i.

If one vehicle should service two points instead of one, for example, i and j, then there is a mileage saving $s(i,j)$ made which equals (Fig. 9.34):

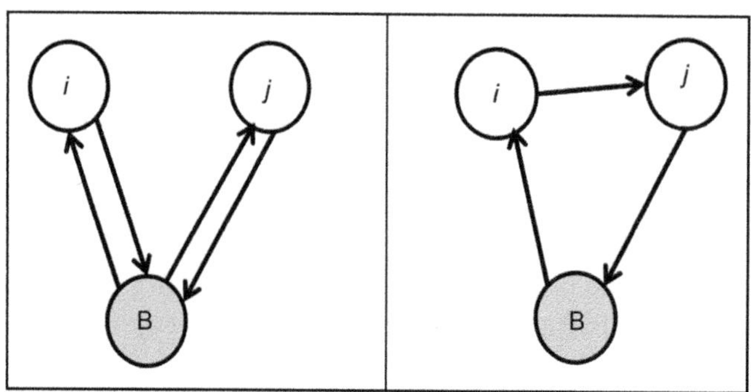

FIG. 9.34

Joining nodes i and j into one route.

$$s(i,j) = 2 \times d(B,i) + 2 \times d(B,j) - [d(B,i) + d(i,j) + D(b,j)] = b(B,i) + d(B,j) - d(i,j) \tag{9.41}$$

Quantity $s(i,j)$ is called the *saving* which is obtained by joining nodes i and j into one route. It is logical that the larger $s(i,j)$ becomes, the better it is to join i and j into one route. Nodes i and j cannot be joined into one trip if doing so violates one of the operational constraints (constraints regarding vehicle capacity, maximum allowed route length, number of nodes that can be visited, etc.).

Clark-Wright's "savings" algorithm is composed of the following algorithmic steps:

Step 1: Calculate saving $s(i,j) = d(B,i) + d(B,j) - d(i,j)$ for every pair (i,j) of nodes to be serviced.

Step 2: Make a list of savings in descending order.

Step 3: Deal with the saving list, starting with the highest saving. When considering saving $s(i,j)$, include the corresponding link (i,j) in the route, if doing so does not violate one of the given operational constraints and if:

(a) neither point i nor point j has been included in a route, in which case a new route is created that contains both nodes i and j;

(b) one point (either point i or point j) is previously included in a route and if that point is not an *internal point* on the route (an internal route point is not next in the route to starting point B) in which case the link (i,j) is included in that route; and

(c) both nodes i and j are included in different routes and neither one is an internal route point, in which case the routes can be merged.

Step 4: If the list of savings (after formation of the first route) is not completely used up, return to Step 3 and start from the beginning with the largest unused saving. Otherwise stop.

EXAMPLE 9.9

Clients that demand service are located in nodes 0, 1, 2, 3, 4,..., 9. The depot is located in node 0. Node coordinates and node demand are given in Table 9.6.

Table 9.6 Node Coordinates and Node Demand

Node	Coordinate x	Coordinate y	Demand
Node 0	82	76	0
Node 1	96	44	19
Node 2	50	5	21
Node 3	49	8	6
Node 4	13	7	19
Node 5	29	89	7
Node 6	58	30	12
Node 7	84	39	16
Node 8	14	24	6
Node 9	2	39	16

All vehicles that participate in the service have the same capacity. The vehicle capacity equals $V = 50$. The distances between all pairs of nodes are given in Table 9.7

(Continued)

EXAMPLE 9.9—cont'd

Table 9.7 The Distances Between all Pairs of Nodes

0	35	78	76	98	55	52	37	86	88
35	0	60	59	91	81	40	13	84	94
78	60	0	3	37	87	26	48	41	59
76	59	3	0	36	83	24	47	38	56
98	91	37	36	0	84	51	78	17	34
55	81	87	83	84	0	66	74	67	57
52	40	26	24	51	66	0	28	44	57
37	13	48	47	78	74	28	0	72	82
86	84	41	38	17	67	44	72	0	19
88	94	59	56	34	57	57	82	19	0

By using Clarke-Wright's savings algorithms, generate the set of vehicle routes.

Solution

The savings are respectively equal:

$S(1,2)=d(0,1)+d(0,2)-d(1,2)=35+78-60=53$
$S(1,3)=d(0,1)+d(0,3)-d(1,3)=35+76-59=52$
$S(1,4)=d(0,1)+d(0,4)-d(1,4)=35+98-91=42$
$S(1,5)=d(0,1)+d(0,5)-d(1,5)=35+55-81=9$
$S(1,6)=d(0,1)+d(0,6)-d(1,6)=35+52-40=47$
$S(1,7)=d(0,1)+d(0,7)-d(1,7)=35+37-13=59$
$S(1,8)=d(0,1)+d(0,8)-d(1,8)=35+86-84=37$
$S(1,9)=d(0,1)+d(0,9)-d(1,9)=35+88-94=29$
$S(2,3)=d(0,2)+d(0,3)-d(2,3)=78+76-3=151$
$S(2,4)=d(0,2)+d(0,4)-d(2,4)=78+98-37=139$
$S(2,5)=d(0,2)+d(0,5)-d(2,5)=78+55-87=46$
$S(2,6)=d(0,2)+d(0,6)-d(2,6)=78+52-26=104$
$S(2,7)=d(0,2)+d(0,7)-d(2,7)=78+37-48=67$
$S(2,8)=d(0,2)+d(0,8)-d(2,8)=78+86-41=123$
$S(2,9)=d(0,2)+d(0,9)-d(2,9)=78+88-59=107$
$S(3,4)=d(0,3)+d(0,4)-d(3,4)=76+98-36=138$
$S(3,5)=d(0,3)+d(0,5)-d(3,5)=76+55-83=48$
$S(3,6)=d(0,3)+d(0,6)-d(3,6)=76+52-24=104$
$S(3,7)=d(0,3)+d(0,7)-d(3,7)=76+37-47=66$
$S(3,8)=d(0,3)+d(0,8)-d(3,8)=76+86-38=124$
$S(3,9)=d(0,3)+d(0,9)-d(3,9)=76+88-56=108$
$S(4,5)=d(0,4)+d(0,5)-d(4,5)=98+55-84=69$
$S(4,6)=d(0,4)+d(0,6)-d(4,6)=98+52-51=99$
$S(4,7)=d(0,4)+d(0,7)-d(4,7)=98+37-78=57$
$S(4,8)=d(0,4)+d(0,8)-d(4,8)=98+86-17=167$
$S(4,9)=d(0,4)+d(0,9)-d(4,9)=98+88-34=152$
$S(5,6)=d(0,5)+d(0,6)-d(5,6)=55+52-66=41$
$S(5,7)=d(0,5)+d(0,7)-d(5,7)=55+37-74=18$
$S(5,8)=d(0,5)+d(0,8)-d(5,8)=55+86-67=74$
$S(5,9)=d(0,5)+d(0,9)-d(5,9)=55+88-57=86$
$S(6,7)=d(0,6)+d(0,7)-d(6,7)=52+37-28=61$
$S(6,8)=d(0,6)+d(0,8)-d(6,8)=52+86-44=94$
$S(6,9)=d(0,6)+d(0,9)-d(6,9)=52+88-57=83$

EXAMPLE 9.9—cont'd

$S(7,8) = d(0,7) + d(0,8) - d(7,8) = 37 + 86 - 72 = 51$
$S(7,9) = d(0,7) + d(0,9) - d(7,9) = 37 + 88 - 82 = 43$
$S(8,9) = d(0,8) + d(0,9) - d(8,9) = 86 + 88 - 19 = 155$

The list of savings in descending order is given in Table 9.8.

Table 9.8 The List of Savings in Descending Order

Pair of Nodes	Saving
(4,8)	167
(8,9)	155
(4,9)	152
(2,3)	151
(2,4)	139
(3,4)	138
(3,8)	124
(2,8)	123
(3,9)	108
(2,9)	107
(3,6)	104
(2,6)	104
(4,6)	99
(6,8)	94
(5,9)	86
(6,9)	83
(5,8)	74
(4,5)	69
(2,7)	67
(3,7)	66
(6,7)	61
(1,7)	59
(4,7)	57
(1,2)	53
(1,3)	52
(7,8)	51
(3,5)	48
(1,6)	47
(2,5)	46
(7,9)	43
(1,4)	42
(5,6)	41
(1,8)	37
(1,9)	29
(5,7)	18
(1,5)	9

(Continued)

EXAMPLE 9.9—cont'd

The biggest savings (167) is connected with the link (4, 8). We create the vehicle route {0, 4, 8, 0}. The total demand on this route is equal to 25. The second largest saving (155) is associated with the link (8, 9). Consequently, the first route is expanded to {0, 4, 8, 9, 0}, as the conditions under step of the algorithm are fulfilled by such an expansion. The total demand on the expanded route is equal to 41. Next in the savings list is the link (4, 9). Both nodes 4 and 9 are already included in the vehicle route. The fourth in the saving list is the link (2, 3) with a corresponding saving that is equal to 151. We create the second vehicle route {0, 2, 3, 0}. The total demand on the second route is equal to 27.

Next in the savings list is the link (2, 4). The expansion of the first vehicle route to {0, 2, 4, 8, 9, 0} is impossible because such a route would involve a load of 62 units. Such load would exceed the vehicle capacity ($V = 50$). Therefore, we reject the link (2, 4). We reject the link (3, 4), since the merging the route {0, 2, 3, 0} and the route {0, 4, 8, 9, 0} into the route {0, 2, 3, 4, 8, 9, 0} would violate vehicle capacity constraint. We reject the link (3, 8) since point 8 is the *interior* point in the vehicle route. For similar reasons we reject the links (2, 8), (3, 9), and (2, 9).

Next in the savings list is the link (3, 6) with the corresponding saving that is equal to 104. The second vehicle route is expanded to {0, 2, 3, 6, 0}}, as the conditions under step of the algorithm are fulfilled by such an expansion. The total demand on the expanded route is equal to 39. We reject the links (2, 6), (4, 6), and (6, 8).

Next in the savings list is the link (5, 9) with the corresponding saving that is equal to 86. The first route is expanded to {0, 4, 8, 9, 5, 0}. The total demand on the expanded route is equal to 48. We reject the links (6, 9), (5, 8), (4, 5), (2, 7), (3, 7), and (6, 7).

Next in the savings list is the link (1, 7) with the corresponding saving that is equal to 59. We create the third vehicle route {0, 1, 7, 0}. The total demand on the third route is equal to 35.

We reject the following links (4,7), (1,2), (1,3), (7,8), (3,5), (1,6), (2,5), (7,9), (1,4), (5,6), (1,8), (1,9), (5,7), and (1,5). We finish our algorithm, since all nodes are included in the vehicle routes. We created the following vehicle routes (Table 9.9):

Table 9.9 Routes Generated by Clarke-Wright's Savings Algorithm

Vehicle Route	Total Demand on Route
{0, 4, 8, 9, 5, 0}	48
{0, 2, 3, 6, 0}	39
{0, 1, 7, 0}	35

The total length of all generated vehicle routes is equal to 488. The vehicle routes are shown in Fig. 9.35.

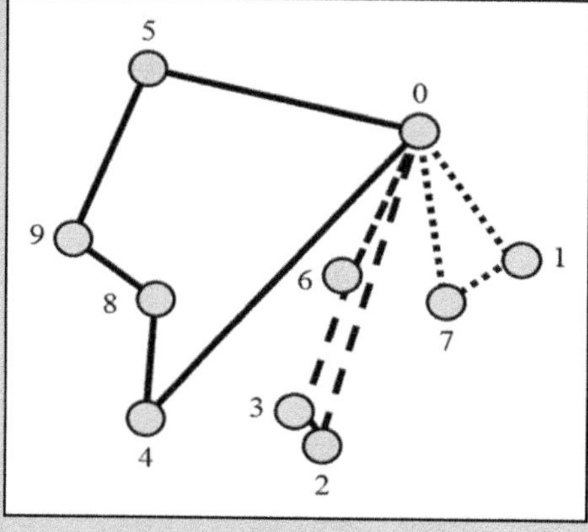

FIG. 9.35

The vehicle routes generated by Clarke-Wright's savings algorithm.

EXAMPLE 9.10

Clients that demand service are located in nodes 0, 1, 2, 3, 4,..., 9. The depot is located in node 0. Node coordinates and node demand are given in Table 9.10.

Table 9.10 Node Coordinates and Node Demand

Node	Coordinate *x*	Coordinate *y*	Demand
Node	42	68	0
Node1	77	97	5
Node 2	28	64	23
Node 3	77	39	14
Node 4	32	33	13
Node 5	32	8	8
Node 6	42	92	18
Node 7	8	3	19
Node 8	7	14	10
Node 9	82	17	18

All vehicles that participate in the service have the same capacity. The vehicle capacity equals $V=50$. The distances between all pairs of nodes are given in Table 9.11.

Table 9.11 The Distances Between all Pairs of Nodes

0	45	15	45	36	61	24	73	64	65
45	0	59	58	78	100	35	117	109	80
15	59	0	55	31	56	31	64	54	72
45	58	55	0	45	55	64	78	74	23
36	78	31	45	0	25	60	38	31	52
61	100	56	55	25	0	85	25	26	51
24	35	31	64	60	85	0	95	85	85
73	117	64	78	38	25	95	0	11	75
64	109	54	74	31	26	85	11	0	75
65	80	72	23	52	51	85	75	75	0

By using Clarke-Wright's savings algorithms, generate the set of vehicle routes.

Solution

Clarke-Wright's algorithms created the following vehicle routes:

Route I: {0 4 5 7 8 0} with the total demand equal to 50
Route II: {0 3 9 0} with the total demand equal to 32
Route III: {0 1 6 2 0} with the total demand 46

The total length of all routes is equal to 420. The generated vehicle routes are shown in Fig. 9.36.

(Continued)

EXAMPLE 9.10—cont'd

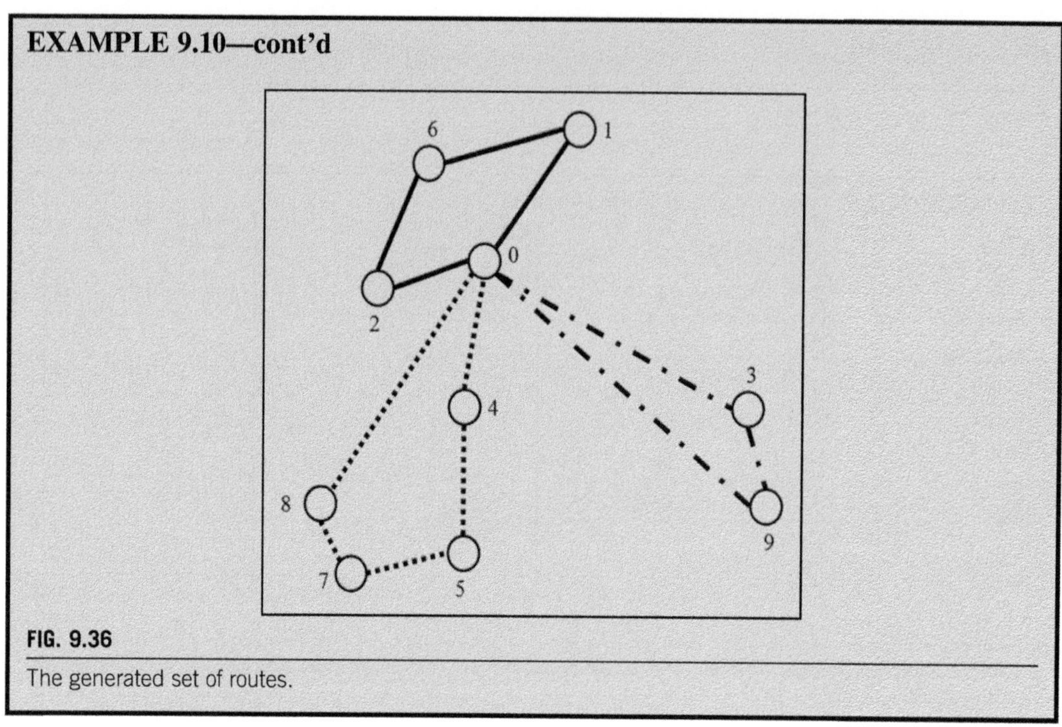

FIG. 9.36

The generated set of routes.

9.6.6 SWEEP ALGORITHM FOR THE VRP

The Sweep algorithm is applied to polar coordinates and the depot is considered to be the origin of the coordinate system. Then the depot is joined with an arbitrarily chosen point which is called the seed point (Fig. 9.37). All other points are joined to the depot and then aligned by increasing angles which are formed by the segments which connect the points to the depot and the segment which connects the depot to the seed point. The route starts with the seed point and then the points aligned by increasing angles are included, respecting given constraints all the while.

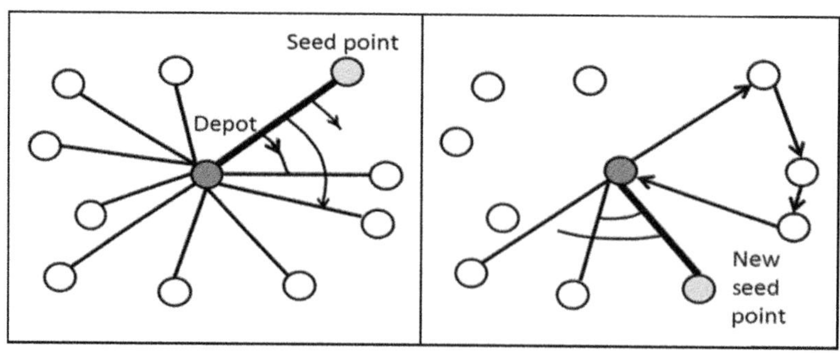

FIG. 9.37

Sweep algorithm.

When a point cannot be included in the route since this would violate a given constraint, this point becomes the seed point of a new route, etc. The process is completed when all points are included in a route.

EXAMPLE 9.11

Clients that demand service are located in nodes 2, 3, 4,..., 9. The depot is located in node 1 (Fig. 9.38).

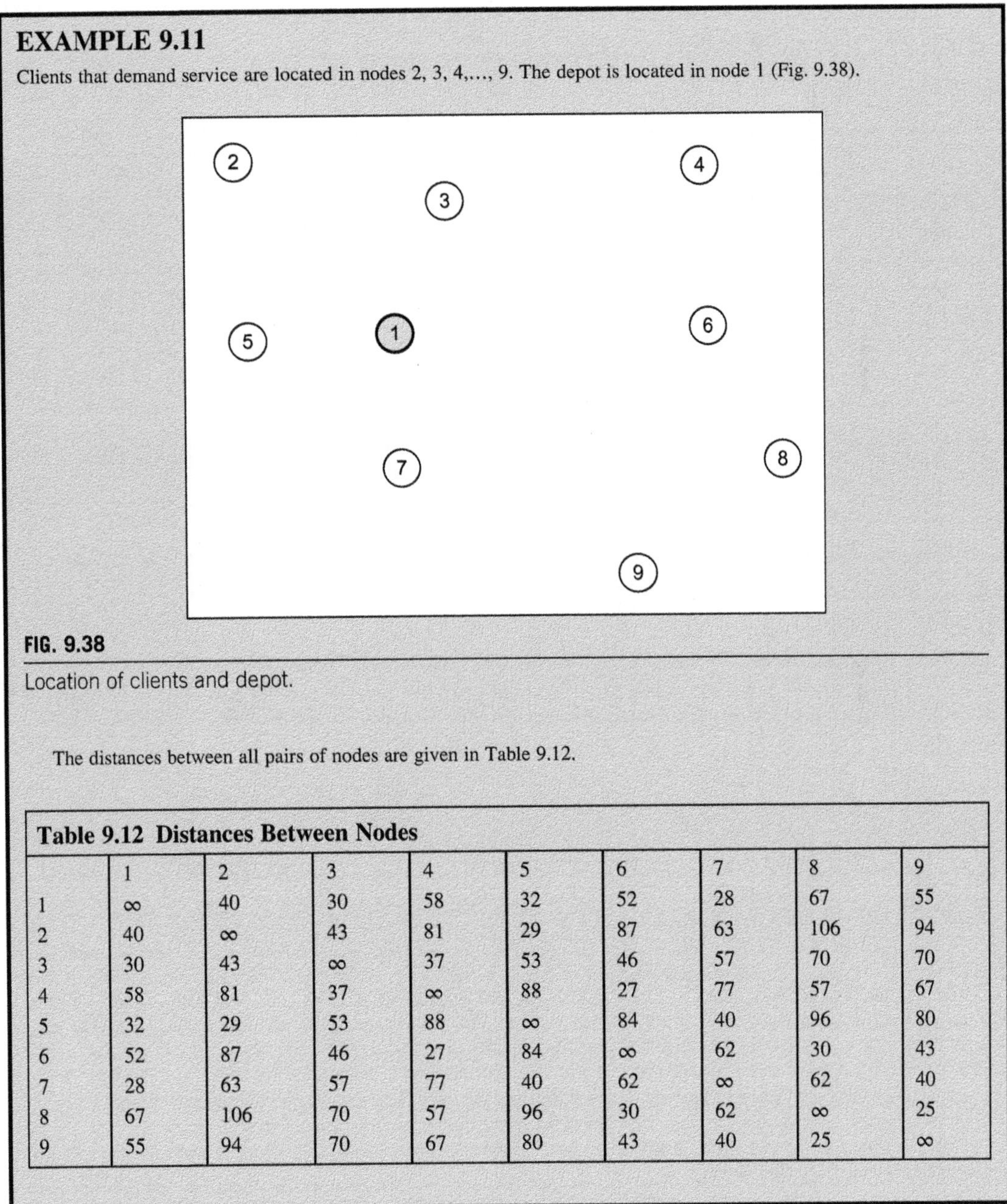

FIG. 9.38

Location of clients and depot.

The distances between all pairs of nodes are given in Table 9.12.

Table 9.12 Distances Between Nodes

	1	2	3	4	5	6	7	8	9
1	∞	40	30	58	32	52	28	67	55
2	40	∞	43	81	29	87	63	106	94
3	30	43	∞	37	53	46	57	70	70
4	58	81	37	∞	88	27	77	57	67
5	32	29	53	88	∞	84	40	96	80
6	52	87	46	27	84	∞	62	30	43
7	28	63	57	77	40	62	∞	62	40
8	67	106	70	57	96	30	62	∞	25
9	55	94	70	67	80	43	40	25	∞

(Continued)

EXAMPLE 9.11—cont'd

All vehicles that participate in the service have the same capacity. The vehicle capacity equals $V = 12$. The demand in nodes is given in Table 9.13.

Table 9.13 Demand at Nodes

Node i	2	3	4	5	6	7	8	9
Demand v_i	4	7	3	2	6	3	2	3

By using the Sweep algorithm generate the set of vehicle routes.

Solution

Node 1 (depot) is the system origin. Let us arbitrarily choose point 3 to be the seed point. In this case, we measure the angles in clockwise direction (the angle measurement direction is chosen arbitrarily). The origin of the system, seed point, other points and corresponding angles are shown on Fig. 9.39.

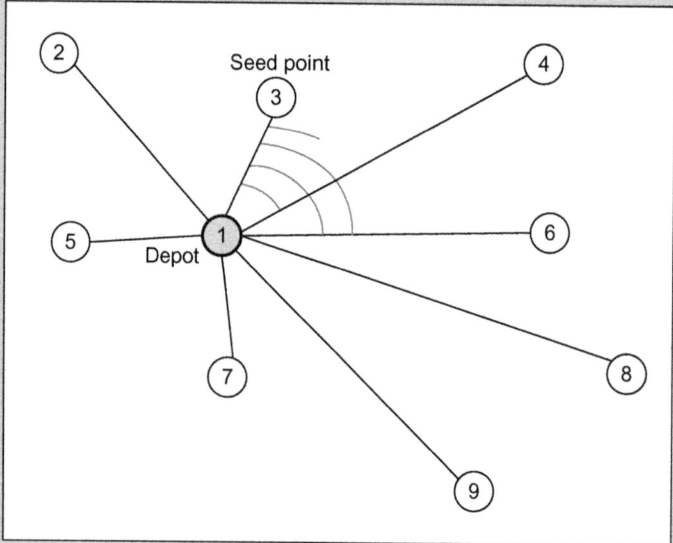

FIG. 9.39

Origin of the system (depot) and the seed point.

The route starts from the depot (point 1) and goes to the seed point (point 3). From point 3 should be taken 7 units of goods. The route now contains (1, 3). We now include the next point to the route from the set of points aligned in increasing angles. This is point 4 which has 3 units to be transported which makes a total of 10 units when added to those of point 3. Since this is less than the vehicle capacity of 12 units, point 4 can be included in the route. The new route now contains (1, 3, 4). Point 5 cannot be included in the route since the route (1, 3, 4, 6) would have to carry 16 units, which is in excess of the vehicle capacity. So we have finished with the first route structure which reads (1, 3, 4, 1).

Point 6 becomes the new seed point. Fig. 9.40 shows the first route (1, 3, 4, 1) and the new seed point—point 6, with the other points and corresponding angles.

EXAMPLE 9.11—cont'd

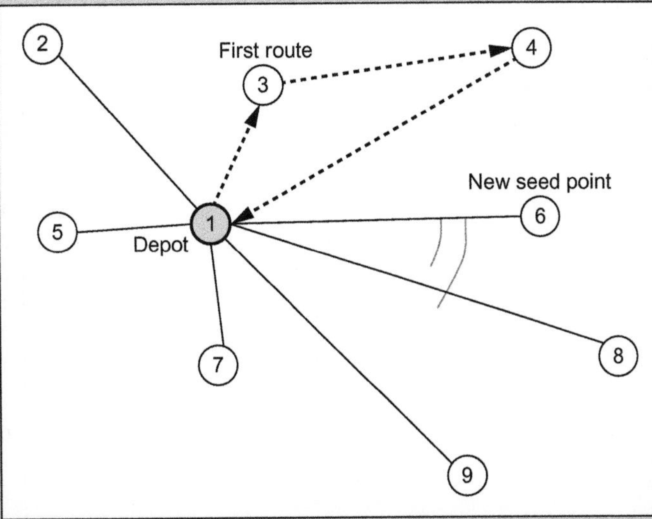

FIG. 9.40

Sweep algorithm after the second step (first route and new seed point).

By continuing with the described procedure, we get two more routes—route (1, 6, 8, 9, 1) and route (1, 7, 5, 2, 1). Fig. 9.41 shows the final set of vehicle routes obtained by the Sweep algorithm.

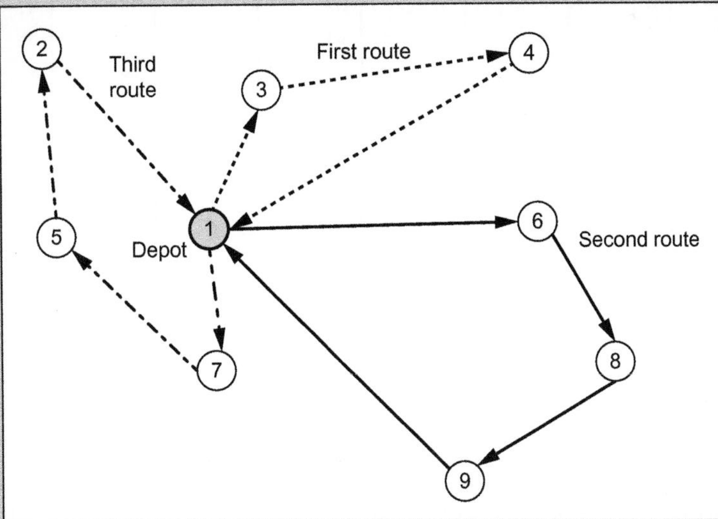

FIG. 9.41

The final set of routes obtained by the Sweep algorithm.

EXAMPLE 9.12

Clients that demand service are located in nodes 0, 1, 2, 3, 4,..., 9. The depot is located in node 0. Node coordinates and node demand are given in Table 9.14.

Table 9.14 Node Coordinates and Node Demand

Node	Coordinate x	Coordinate y	Demand
Node 0	82	76	0
Node 1	96	44	19
Node 2	50	5	21
Node 3	49	8	6
Node 4	13	7	19
Node 5	29	89	7
Node 6	58	30	12
Node 7	84	39	16
Node 8	14	24	6
Node 9	2	39	16

All vehicles that participate in the service have the same capacity. The vehicle capacity equals $V=50$. The distances between all pairs of nodes are given in Table 9.15.

Table 9.15 The Distances Between all Pairs of Nodes

0	35	78	76	98	55	52	37	86	88
35	0	60	59	91	81	40	13	84	94
78	60	0	3	37	87	26	48	41	59
76	59	3	0	36	83	24	47	38	56
98	91	37	36	0	84	51	78	17	34
55	81	87	83	84	0	66	74	67	57
52	40	26	24	51	66	0	28	44	57
37	13	48	47	78	74	28	0	72	82
86	84	41	38	17	67	44	72	0	19
88	94	59	56	34	57	57	82	19	0

By using combination of the Sweep algorithm and the Nearest Neighbor algorithm, generate the set of vehicle routes.

Solution

Nodes 5, 9, 8, and 4 belong to cluster 1 (Table 9.16). The total demand in cluster 1 is equal to 48. Nodes 6, 3, and 2 belong to cluster 39. The total demand of cluster 2 equals 39. Cluster 3 contains nodes 7, and 1. The total demand in cluster 3 equals 35. We start to create the traveling salesman route within cluster 1. The route starts from node 0. Since, min $\{55, 88, 86, 98\}=55$, we conclude that node 5 is the closest node to node 0. We include node 5 in the route. The partial route reads: $\{0, 5\}$. Since $\min\{57, 67, 84\}=57$, we conclude that node 9 is the closest to node 5 (that was the last node included in the route). We include node 9 in the route. The partial route reads: $\{0, 5, 9\}$. Proceeding in this way, we include in the route node 8, and then node 4. Since all nodes are included in the route, the route ends in node 0. The final traveling salesman route in the first cluster reads: $\{0, 5, 9, 8, 4, 0\}$. By using the described procedure we generate the following traveling salesman routes for cluster 2 and cluster 3:

Cluster 2: $\{0, 6, 3, 2, 0\}$
Cluster 3: $\{0, 1, 7, 0\}$

EXAMPLE 9.12—cont'd

Table 9.16 Node Characteristics

Node	Demand	Angle [°]
Node 5	7	166.22
Node 9	16	204.82
Node 8	6	217.41
Node 4	19	225
Node 6	12	242.45
Node 3	6	244.11
Node 2	21	245.74
Node 7	16	273.09
Node1	19	293.63

The final set of routes is shown in Table 9.17 and Fig. 9.42.

Table 9.17 The Set of Vehicle Route Generated by the Combination of the Sweep Algorithm and the Nearest Neighbor Algorithm

Route	Total Demand
{0, 5, 9, 8, 4, 0}	48
{0,6, 3, 2, 0}	39
{0, 1, 7, 0}	35

The total length of all generated vehicle routes is equal to 488.

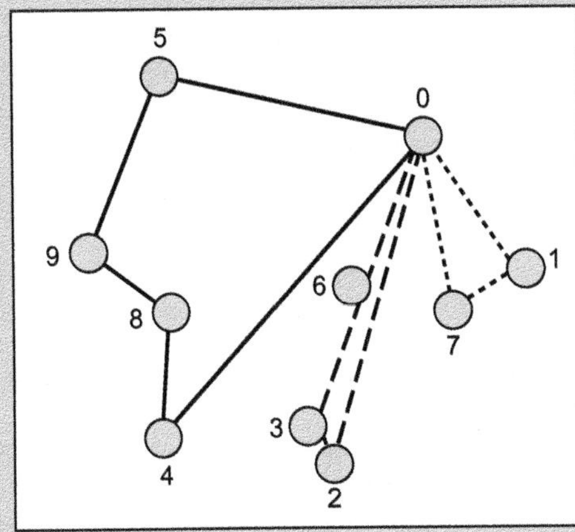

FIG. 9.42

The vehicle routes generated by the combination of the Sweep algorithm and the Nearest Neighbor algorithm.

9.7 PROBLEMS

1. Clients who demand service are located in nodes 1 to 6. The depot is located in node B. The distances between all pairs of nodes are given in the following table:

$$\begin{array}{c} B \\ 1 \\ 2 \\ 3 \\ 4 \\ 5 \\ 6 \end{array}\begin{bmatrix} - & 37 & 21 & 19 & 25 & 31 & 19 \\ 37 & - & 14 & 22 & 27 & 30 & 11 \\ 21 & 14 & - & 36 & 29 & 24 & 28 \\ 19 & 22 & 36 & - & 13 & 18 & 26 \\ 25 & 27 & 29 & 13 & - & 15 & 32 \\ 31 & 30 & 24 & 18 & 15 & - & 20 \\ 19 & 11 & 28 & 26 & 32 & 20 & - \end{bmatrix}$$

All vehicles that participate in the service have the same capacity. The vehicle capacity equals $V = 14$. The quantities of goods per node are:

Node i	1	2	3	4	5	6
Demand v_i	5	5	3	4	6	3

Using Clarke-Wright's savings algorithms, generate the set of vehicle routes.

2. Clients who demand service are located in nodes 2 to 6. The depot is located in node 1. The distances between all pairs of nodes are given in the following table:

$$\begin{array}{c} \\ 1 \\ 2 \\ 3 \\ 4 \\ 5 \\ 6 \end{array}\begin{array}{c} \begin{array}{cccccc} 1 & 2 & 3 & 4 & 5 & 6 \end{array} \\ \begin{bmatrix} - & 21 & 13 & 17 & 25 & 9 \\ 21 & - & 14 & 19 & 21 & 27 \\ 13 & 14 & - & 8 & 18 & 20 \\ 17 & 19 & 8 & - & 29 & 17 \\ 25 & 21 & 18 & 29 & - & 15 \\ 9 & 27 & 20 & 17 & 15 & - \end{bmatrix} \end{array}$$

All vehicles that participate in the service have the same capacity. The vehicle capacity equals $V = 15$. The quantities of goods per node are:

Node i	2	3	4	5	6
Demand v_i	4	2	6	6	4

Using Clarke-Wright's savings algorithms, generate the set of vehicle routes.

3. Solve the Traveling Salesman Problem for the case given in Problem 2. Use Clarke-Wright's algorithm.
4. Solve the Traveling Salesman Problem for the case given in Problem 2. Use the Nearest Neighbor algorithm.

5. Clients that demand service are located in nodes 2 to 6. The depot is located in node 1. The distances between all pairs of nodes are given in the following table:

$$
\begin{array}{c} 1\\2\\3\\4\\5\\6 \end{array}
\begin{bmatrix}
- & 5 & 7 & 2 & 4 & 1\\
5 & - & 3 & 6 & 2 & 8\\
7 & 3 & - & 4 & 5 & 5\\
2 & 6 & 4 & - & 2 & 7\\
4 & 2 & 5 & 2 & - & 9\\
1 & 8 & 5 & 7 & 9 & -
\end{bmatrix}
$$

Using Clarke-Wright's savings algorithms, solve the Traveling Salesman Problem.

6. Clients that demand service are located in nodes 2, 3, 4, 5, and 6. The depot is located in node 1 (Fig. 9.43).

The distances between all pairs of nodes are:

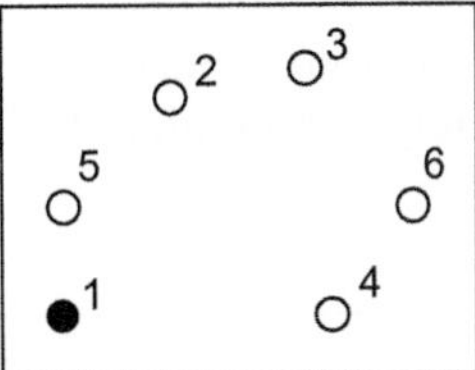

FIG. 9.43

Clients that demand service (nodes 2, 3, 4, 5, and 6).

$$
\begin{array}{c} 1\\2\\3\\4\\5\\6 \end{array}
\begin{bmatrix}
- & 8 & 12 & 6 & 3 & 11\\
8 & - & 9 & 21 & 7 & 10\\
10 & 9 & - & 11 & 8 & 6\\
6 & 19 & 11 & - & 14 & 20\\
4 & 7 & 12 & 13 & - & 9\\
11 & 10 & 3 & 20 & 9 & -
\end{bmatrix}
$$

All vehicles that participate in the service have the same capacity. The vehicle capacity equals $V=10$. The demand in nodes is:

Node i	2	3	4	5	6
Demand v_i	3	2	5	6	4

Using the Sweep algorithm, generate the set of vehicle routes.

7. Transportation network nodes are denoted respectively by *a, b, c,…, f.* Daily service demands are given in parentheses next to each node. All link lengths are also denoted (Fig. 9.44).

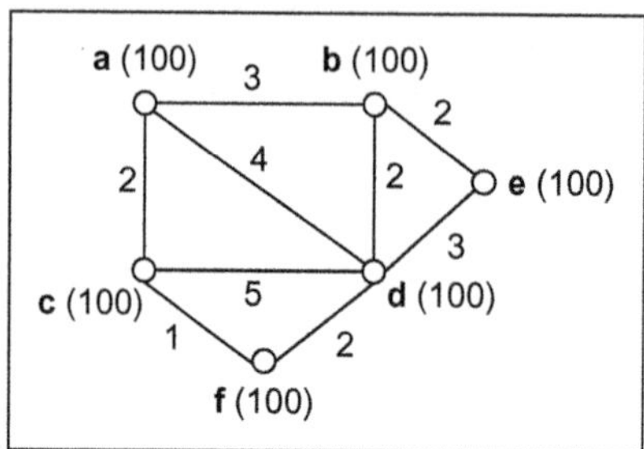

FIG. 9.44

Network in which optimal locations of service facilities should be determined.

(a) Where should we locate a service facility so that the average distance covered by service users (located in the nodes) to the facility is minimized?

(b) Where should we locate two retail objects so that the average distance covered by service users (located in the nodes) to the facility is minimized?

REFERENCES

ASECAP, 2010. Introduction of Longer & Heavier Vehicles: Impacts on Road Infrastructure. Association des Europeennes des Concessionaires d'Autoroutes et d'Ouvrages a Peage, Brussels, Belgium.

Bartholdi, J.J., Gue, K.R., 2004. The Best Shape for a Crossdock. Transp. Sci. 38 (2), 235–244.

Beasley, J.E., 1983. Route-first cluster-second methods for vehicle routing. Omega 11, 403–408.

Bodin, L., Golden, B., 1981. Classification in vehicle routing and scheduling. Networks 11, 97–108.

Browne, M., Allen, J., Christodoulou, G., 2004. Freight transport in London: a summary of current data and sources, Final Report. Transport Studies Group, University of Westminster for Transport for London, London, UK.

Bryan, D., O'Kelly, M., 1999. Hub-and-spoke networks in air transportation: an analytical review. J. Reg. Sci. 39, 275–295.

CC, 1999. Automated Underground Transportation of Cargo: The 5th Transportation Alternative for the Transport of Goods in Congested Urban Areas. Brochure, CargoCap GmbH, Bochum, Germany.

Church, R.L., ReVelle, C., 1974. The maximal covering location problem. Pap. Reg. Sci. Assoc. 32, 101–118.

Clarke, G., Wright, J.W., 1964. Scheduling of vehicles from a central depot to a number of delivery points. Oper. Res. 12, 568–581.

Cordeau, J.-F., Gendreau, M., Laporte, G., Potvin, J.-Y., Semet, F., 2002. A guide to vehicle routing heuristics. J. Oper. Res. Soc. 53, 512–522.

Current, J.R., Daskin, M.S., Schilling, D., 2002. Discrete network location models. In: Drezner, Z., Hamacher, H. (Eds.), Facility Location Theory: Applications and Methods. Springer-Verlag, Berlin, pp. 81–118 (Chapter 3).

Daganzo, C., 2005. Logistics Systems Analysis. Springer, Berlin.

Dantzig, G.B., Ramser, J.H., 1959. The truck dispatching problem. Manag. Sci. 6, 80–91.

Daskin, M.S., 1995. Network and Discrete Location: Models, Algorithms, and Applications. Wiley & Sons, New York.

DFDS, 2014. Annual report. DFDS A/S, Copenhagen, Denmark, www.dfds.com.

DHL, 2008. The Logistics Glossary, DHL Logbook, https://www.dhl-discoverlogistics.com/cms/en/glossary/.

Don Taylor, G. (Ed.), 2007. Logistics Engineering Handbook. CRC Press, Taylor & Francis Group.

EC, 2004. Quantification of Urban Freight Transport Effects I, Deliverable D 5.1, TREN/04/FP6TR/S07.31723/506384, BESTUFS II, Best Urban Freight Solutions II, European Commission, Brussels, Belgium.

EU, 2015. EU Transport in Figures: Statistical Pocketbook 2014. Publications Office of the European Union, Luxembourg.

Fisher, M.L., Jaikumar, R., 1981. A generalized assignment heuristic for the vehicle routing problem. Networks 11, 109–124.

Fraunhofer, 2009. Long-Term Climate Impacts of the Introduction of Mega-Trucks: Study for the Community of European Railway and Infrastructure Companies (CER). The Fraunhofer-Institute for Systems and Innovation Research, Karlsruhe, Germany.

Gillett, B.E., Miller, L.R., 1974. A heuristic algorithm for the vehicle dispatch problem. Oper. Res. 22, 340–349.

Hakimi, S.L., 1964. Optimal locations of switching centers and the absolute centers and medians of a graph. Oper. Res. 12, 450–459.

Hakimi, S.L., 1965. Optimum distribution of switching centers in a communications network and some related graph-theoretic problems. Oper. Res. 13, 462–475.

Hamacher, H.W., Nickel, S., 1998. Classification of location models. Locat. Sci. 6, 229–242.

Huschebeck, M., 2001. BEST urban freight solutions. Report of Deliverable 1.1 of Recommendations for Further Activities I, The EU 5th RTD Framework Programme, European Commission, Brussels, Belgium.

Laporte, G., Semet, F., 2002. Classical heuristics for the capacitated VRP. In: Toth, P., Vigo, D. (Eds.), The Vehicle Routing Problem. SIAM Monographs on Discrete Mathematics and Applications, Philadelphia, pp. 109–128.

Larson, R., Odoni, A., 1981. Urban Operations Research. Prentice Hall, Englewood Cliffs, New Jersey.

Lauber, V., 2001. The sustainability of freight transport across the Alps: European Union policy in controversies on transit traffic. In: Lenschow, A. (Ed.), Environmental Policy in Integration. Earthscan, London, pp. 153–174. Demokratiezentrum Wien.

Lin, S., 1965. Computer solution of the traveling salesman problem. Bell Syst. Tech. J. 44, 2245–2269.

Liu, H., 2004. Feasibility of underground pneumatic freight transport in New York, Final Report. Prepared for The New York State Energy Research and Development Authority (NYSERDA), Freight Pipeline Company, Columbia, Missouri, USA.

Love, R.F., Morris, J.G., Wesolowsky, G.O., 1988. Facilities Location: Models and Methods. North Holland, New York.

Mirchandani, P., Francis, R. (Eds.), 1990. Discrete Location Theory. John Wiley & Sons, New York.

O'Kelly, M., 1986. The location of interacting hub facilities. Transp. Sci. 20, 92–106.

O'Kelly, M., 1987. A quadratic integer program for the location of interacting hub facilities. Eur. J. Oper. Res. 32, 393–404.

Rakas, J., Teodorović, D., Kim, T., 2004. Multi-objective modeling for determining location of undesirable facilities. Transport. Res. D 9, 125–138.

ReVelle, C., Bigman, D., Schilling, D., Cohon, J., Church, R., 1977. Facility location: a review of context-free and EMS models. Health Serv. Res. 129–146.

Rijsenbrij, J.C., Pielage, B.A., Visser, J.G., 2006. State of the Art on Automated (Underground) Freight Transport Systems for the EU TREND Project. Delft University of Technology, Delft, The Netherlands.

Šelmić, M., Teodorović, D., Vukadinović, K., 2010. Locating inspection facilities in traffic networks: an artificial intelligence approach. Transp. Plan. Technol. 33, 481–493.

Simchi-Levi, D., Chen, X., Bramel, J., 2014. The Logic of Logistics: Theory, Algorithms, and Applications for Logistics Management. Springer, New York.

Solomon, M.M., 1987. Algorithms for the vehicle routing and scheduling problems with time windows. Oper. Res. 35, 254–265.

STC, 2015. Logistics Glossary, Stolt Tank Containers. http://www.logisticsglossary.com/.

Taniguchi, E., Thompson, R.G., Yamada, T., van Dui, J.H.R., 2001. City Logistics: Network Modelling and Intelligent Transport Systems. Pergamon, Amsterdam.

Taniguchi, E., 2014. Concepts of city logistics for sustainable and liveable cities. Procedia Soc. Behav. Sci. 151, 310–317.

Taniguchi, E., 2016. City logistics. In: Teodorović, D. (Ed.), Routledge Handbook of Transportation. Routledge, New York.

Teodorović, D., Radivojević, G., 2000. A fuzzy sets approach to the dynamic dial-a-ride problem. Fuzzy Sets Syst. 116, 23–33.

Teodorović, D., Radoš, J., Vukanović, S., Šarac, M., 1986. Optimal locations of emergency service depots for private cars in urban areas: case study of Belgrade. Transp. Plan. Technol. 11, 177–188.

Teodorović, D., Van Aerde, M., Zhu, F., Dion, F., 2002. Genetic algorithms approach to the problem of the automated vehicle identification equipment locations. J. Adv. Transp. 36, 1–21.

Toregas, C., Swain, R., Revelle, C., Bergman, L., 1971. The location of emergency service facilities. Oper. Res. 19, 1363–1373.

UIC, 2008. Mega Trucks Versus Rail Freight. International Union of Railways, Paris, France.

USDT, 2004. Federal Size Regulations for Commercial Motor Vehicles. U.S. Department of Transportation, Federal Highway Administration, Office of Freight Management and Operations, Washington DC, USA.

Weber, A., 1929. Theory of the Location of Industries. The University of Chicago Press, Chicago.

WEBSITE

http://www.alphaliner.com/top100/.

http://www.britannica.com/place/Greater-London.

https://www.dhl-discoverlogistics.com/cms/en/glossary/buchstabe_s.jsp.

http://www.eurolines.com/en/.

http://dot.gov/airconsumer/.

http://www.delta.com/content/www/en_US/traveling-with-us/where-we-fly/routes/downloadable-route-maps.html.

http://www.ect.nl/en/content/euromax-terminal-rotterdam.

https://www.google.nl/search?q=southwest+airlines+route+map).

http://www.interfreight.co.za/container_information.html.

www.londonlorrycontrol.com/routes/.

https://www.rta.ae/.

http://www.ryanair.com/en.

http://www.sea-distances.org/.

http://www.ship.gr/news6/hanjin28.htm.

http://en.wikipedia.org/wiki/Container_ship.

https://en.wikipedia.org/wiki/Instrument_landing_system.

http://en.wikipedia.org/wiki/List_of_largest_container_ships.

https://en.wikipedia.org/wiki/Low-cost_carrier.

https://en.wikipedia.org/wiki/UIC_classification_of_goods_wagons.

http://www.worldcontainerindex.com/.

http://www.worldslargestship.com/.

What are fixed and what are variable costs in transportation? What are economies of scale and what are economies of scope? What kind of relationships exist between the unit price and the demand and capacity supply curve? How do we evaluate transportation projects? When do we use a Cost-Benefit Analysis? What costs are related to the urban transit systems infrastructure construction? How big are operational costs and revenues of the High Speed Rail? What are the operating costs of air cargo transport?

TRANSPORT ECONOMICS 10

10.1 INTRODUCTION

All decisions related to planning, design, and improvement of transportation infrastructure have economic implications. Transport economics includes the issues such as transport location, movement of people and freight/goods, transport demand, transport planning and forecasting, direct and indirect cost of transport, pricing of transport services, investments in transport infrastructure and services, transport and social-economic development, and transport regulation (Button, 2010). In this chapter the transport economics is considered from the microeconomic perspective. We consider various aspects of the direct costs and pricing of transport infrastructure and services of different transportation modes.

Considered from this microeconomic perspective, it can be said that the transport sector consists of the demand and supply component. The transport demand is derived demand due to needs of people and freight/goods shipments to change the physical place. For example, many people live at one place but work and/or have a leisure on the others, which requires them to travel forward and backward. The location of companies providing raw materials is different than those of the users of these materials—manufacturers of the semifinal and final products, which requires transportation of these raw materials from the former to the latter. In addition, the manufacturers of the final products are often located far away from the retailers of these products, which again require transportation, this time of the final products. Thus, it can be said that the transport demand is derived demand. In many cases transportation demand is proportional to the volumes of peoples' activities and the quantities of final products they consume during a given period of time.

Transportation Engineering. http://dx.doi.org/10.1016/B978-0-12-803818-5.00010-X

The transport demand is handled by the transport supply/capacity provided by transport companies. The transport companies generally provide transport infrastructure with the supportive facilities and equipment, and rolling stock/vehicles carrying out transport services. In order to make them operational, the corresponding material, labor, ie, employed staff/personnel, and energy/fuel, are consumed. In terms of time, the transport infrastructure has particularly the long life-cycle, which is, for example, about 20, 30, 40, or even 60 years. That of rolling stock/vehicles is shorter (20–25 years) mainly due to its/their physical and also technological obsolescence, after they need to be replaced. In this context, two categories of transport companies can be distinguished: that providing transport infrastructure called "the infrastructure providers," and that providing transport services called "transport operators." They both constitute the transport systems within particular transport modes.

According to the economic jargon, the above-mentioned components of transport supply/capacity represent the main inputs to transport processes. The outputs of the transport process are the transport services produced in the given quantities and at the specified quality. They are consumed by users—passengers and/or freight shippers/receivers at the same time as they are produced. This implies that they are the short-lived similarly as transport demand without possibility to be stored/warehoused and left to be consumed sometimes latter on.

The most important economic categories of a given transport system are its costs, revenues, and their relationship.

The costs are expenses for maintaining the transport infrastructure and carrying out transport services. These costs are passed to users of these services (passengers and freight/goods shippers and receivers) in the form of prices/charges. These charges bring revenues to transport companies. In general, these prices/charges, ie, revenues, are set up to cover the company's costs and provide some profits. Nevertheless, the difference between revenues and costs can generally be positive or negative, thus representing the profits or losses. Profits or losses are calculated for the given period of time (usually for a quarter of year, year, or few years).

The prices charged to users of transport services represent for them direct costs. In addition, users are imposed the indirect costs, which are usually considered to be the cost of time during trip/travel/transportation. In such case, the sum of direct and indirect costs is called the users' generalized trip/travel/transportation costs for users.

We describe in this chapter the main elements of economics of transportation systems considered from the engineering perspective. The chapter analyzes direct costs and revenues of the providers of transport infrastructure and transport operators. Costs and revenues could be analyzed at the level of the individual component (infrastructure, rolling stock) and/or of the entire company. We start with the definitions of the main economic terms commonly used. We then focus on economics of particular transportation modes and system within them such as: urban transit systems (passenger road-and rail-based), interurban road, rail, and air passenger and freight/goods, and inland waterways, maritime, and intermodal freight/goods transport mode.

The shown cases of various transportation systems date from the period of last 20–25 years. This implies that the absolute values of particular economic variables have been changing, but the relationships between particular crucial variables have remained generic and as such useful in the given and eventually future considerations. The recently strongly emerging issues of the external costs (the costs of impacts of transportation sector on the environment and society), are considered in Chapter 11. When these costs are going to be fully internalized, they will certainly qualify to be fully considered in the scope of as a part of the overall transport economics issues.

10.2 DEFINITION OF THE MAIN TERMS

In order to understand transport economics fully, we start with the definition of the most important terms. These terms relate to direct costs, revenues, and their relationship (http://www.investopedia.com/terms).

10.2.1 TRANSPORT SECTOR/INDUSTRY

The transport sector of a specific region continent consists of transport companies providing transport infrastructure (infrastructure providers), and transport services (transport operators). They both could be fully public, fully private, and mixed public/private owned. The transport infrastructure providers obtain investments for building transport infrastructure usually from the national and international banks and monetary funds. Some of the well-known funds are International Monetary Fund, World Bank, and European Central Bank. The obtained investments are usually long-term credits, and they are spread over the life-cycle of the given infrastructure. The annuities on the loan, as well as the cost of current and capital maintenance, are covered by charging transport operators for using the infrastructure. On one hand, these charges represent the revenues of the infrastructure providers. On the other hand, they represent the part of the transport operators' operating costs. The charges enable entry of particular transport operators to the particular link/line/route, node, or a part of the infrastructure network. In general, more than one operator could be allowed to use the transportation infrastructure. This depends on the expected demand volumes, as well as on the capacity of the related infrastructure elements. Such policy of access enables competition between different operators, which in general reduces prices charged to their users (passengers and freight/goods shippers and receivers). For examples, road users—cars, buses, and trucks pay for accessing and use highways, the railway operators are charged for using the rail infrastructure after their separation, ships are charged for accessing ports, airlines pay fees for getting slots at airports, etc.

10.2.2 FIXED AND VARIABLE COSTS

Each transport system is characterized by internal costs and external costs. Internal costs are paid exclusively by the transportation system users. Internal costs are construction costs, maintenance costs, fuel costs, etc. The main external costs are air pollution, high level of noise, negative outcomes on wetlands, negative effects on wildlife habitat, and low water quality. These external costs are paid by the whole society.

The total costs of transport infrastructure providers, or transport operators can be divided into two main categories: fixed costs and variable costs (Fig. 10.1).

The fixed costs are not dependent on the number of passengers (or passenger-kilometers) and/or quantities of freight/goods (or ton-kilometers) transported and carried out. The fixed costs have to be paid even if no output is being generated. In the case of the infrastructure providers, they include the construction costs of a given infrastructure and supporting facilities and equipment, capital maintenance during its life-cycle, renting buildings, permanent labor, administration, etc. The variable costs refer to the cost of maintenance depending on the traffic volume, the cost of powering supporting facilities and equipment, the cost of collecting charges for using the infrastructure, some administration costs, etc. In the case of transport operators, the fixed costs usually include costs of acquiring the rolling stock/vehicle fleet and its capital maintenance during the life-cycle.

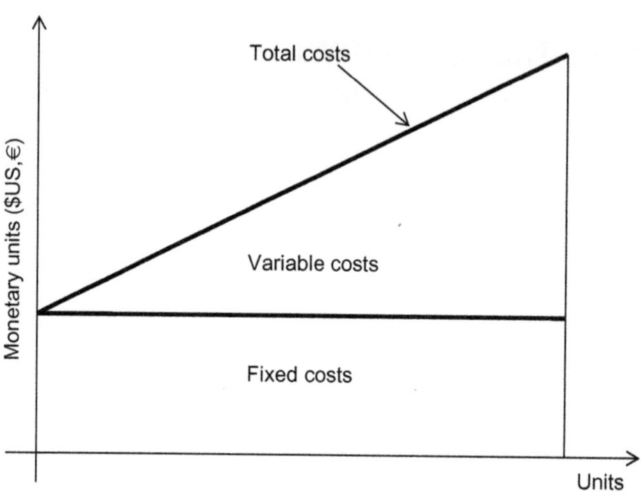

FIG. 10.1

Fixed costs and variable costs.

The variable costs, that depend on the level of utilization of a given rolling stock/vehicle fleet. They include costs of labor, material, and energy/fuel. These costs are not inevitable. They are lower with the lower utilization of a given rolling stock/vehicle fleet.

The total costs $TC(k)$ of any transport company represent the sum of fixed and variable costs (Fig. 10.1), ie:

$$TC(k) = FC(k) + VC(k) \tag{10.1}$$

where

k is the period of time (day, month, quarter, year)
$FC(k)$ is the fixed costs during the time period (k) (monetary units: $US, €)
$VC(k)$ is the variable costs during the time period (k) (monetary units: $US, €)

In addition, the fixed cost of the infrastructure providers and transport operators for construction of infrastructure and acquiring the rolling stock/vehicle fleet are covered by the investments. These, after uniformly spread over the given period of time, usually over the predicted life-cycle, are expected to bring benefits to the actors/stakeholders involved such as investors, transport infrastructure providers and transport operators, users/passengers and freight/goods shippers and receivers, and the local, regional, national, and global community, ie, the economy and society.

10.2.3 ECONOMIES OF SCALE AND ECONOMIES OF SCOPE

The average costs per unit of output are obtained by dividing the total costs (Eq. 10.1) by the volume of output. The output could be the number of vehicles, the number of passengers, quantity of freight/goods, the number of vehicle-kilometers, the number of passenger-kilometers, the number of freight-kilometers, etc. If this average cost decreases with increasing of the volume of given output, the economies of scale exist (Fig. 10.2). Otherwise, diseconomies of scale exist. In the former case, in general, the average cost decreases because of spreading the fixed costs over the larger number of units of output. Economies of scale could be interpreted as the cost advantages that companies achieve as a result of size, output, or scale of operations. For example, when the output increases from O_1 to O_2, the average cost of each unit decreases from AC_1 to AC_2 (Fig. 10.2).

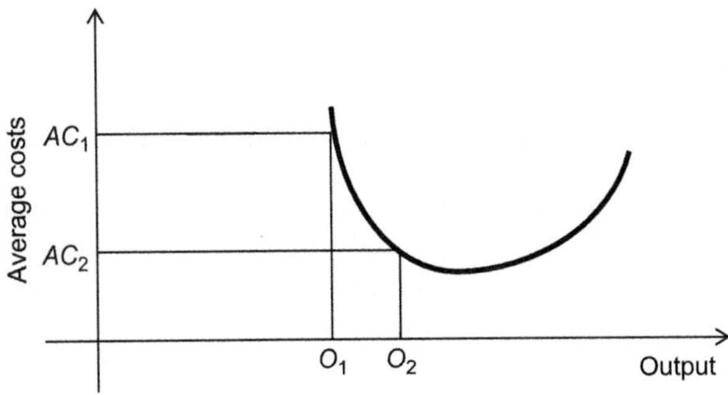

FIG. 10.2

Economies of scale.

Trucking, delivery companies, and airlines usually organize a hub-and-spoke networks as the flows between hubs are characterized by economies of scale. At hubs, goods or passengers are exchanged across vans, trucks, and planes.

Based on Eq. 10.1, the average cost per unit of output can be estimated in the following way:

$$AC(k) = TC(k)/VO(k) = [FC(k) + VC(k)]/VO(k) \tag{10.2}$$

where $VO(k)$ is the volume of output during the time period (k) (number, quantity).

The other symbols are the same like the symbols in Eq. (10.1).

Transportation projects and transportation operations are also characterized by marginal costs. Marginal costs are defined as the change in total cost caused by supplying an additional unit of transport capacity—transport service, seat at transport service, etc. Marginal costs have a tendency to decrease with transportation project size.

Economies of scope are defined as decreasing of the average total cost with increasing of the number of different services produced. In the case of a transport company, this could be offering transport services for both passengers and freight/goods. An airline operating both passenger and cargo aircraft fleet could be an example. Such airline can operate at lower cost than what it would be cost of two separate airlines (one operating passenger aircraft fleet, and the other operating cargo aircraft fleet).

10.2.4 THE COST FUNCTION AND REVENUES

The cost function can be determined for transport infrastructure and transport services. The total costs of a given transport infrastructure can be estimated as follows:

$$CI(k,d) = c_{\text{inf}}(k) \cdot d + \sum_{i=1}^{M} c_{\text{veh}/i}(k) \cdot V_i(k) \tag{10.3}$$

where

$c_{inf}(k)$ is the average unit cost of construction and capital maintenance of a given infrastructure in the (k)th year of its life-cycle (€/km or $US/km)
d is the length/size of a given infrastructure (km; km^2)
$c_{veh/i}(k)$ is the average costs of maintaining a given infrastructure due its tearing by the vehicle category (i) during the (k)th year of its life-cycle (€/veh km or $US/veh km)
$V_i(k)$ is the utilization of a given infrastructure by the vehicles of category (i) during (k)th year of its life cycle (veh km)
M is the number of categories of vehicles operating on a given infrastructure

The infrastructure can be a road used by different categories of vehicles (for example, in Europe, these are motor cycles, small and cars, buses/coaches, LDVs (light duty vans), and HDVs (high duty vehicle (s)) of the GVW (gross vehicle weight) of 5.5, 12, 24, and 40 ton, respectively), rail line handling both passenger (conventional and HS (high speed) trains, or conventional passenger and freight trains, the airport runway(s) accommodating different aircraft types, and the port quay handling ships of different size. Their length vary from several hundred kilometers (road and rail lines), 3–4 km (airport runway (s)), to few hundred meters (the port's quay(s)). The life cycle of infrastructure can be typically 30, 50, or even more years. Their average fixed costs per year of the life-cycle are usually expressed as the net present values (*NPVs*). The average maintenance costs due to tearing given infrastructure are usually determined empirically according proportion of its use by particular vehicle categories. In many cases, these costs are used as the basis for setting up the charges for accessing and use the infrastructure. From Eq. (10.3), the average cost per unit of output of a given infrastructure can be estimated as follows:

$$\overline{CI(k,d)} = \frac{CI(k,d)}{\sum_{i=1}^{M} V_i(k)} \tag{10.4}$$

where all symbols are as in the previous equations.

The average cost $\overline{CI(k,d)}$ in Eq. (10.4) are expressed in €/veh km or $US/veh km.

In addition, the volume of output $V_i(k)$ can be expressed also by the transport work in terms of seat km, pax km, TEU (twenty foot equivalent unit) km, or ton km, all carried out during the specified period of time (the (k)th year of the life-cycle of a given infrastructure).

The total cost of transport services have usually been estimated by using the empirical data. These have been mainly contained on the different levels of aggregation in the annual financial reports of the particular transport companies, or of the entire sector/industry. In order to estimate the cost on a given route carried out by the rolling stock/vehicle of a given capacity, three sets of data for the different routes (cross-sectional) or different time periods for the same route (time-sectional) are needed as follows: (i) the line/route cost per service; (ii) the route length; and (ii) the size/capacity of rolling stock/vehicles, which have carried out performed transport services on the given line/routes. By using usually the regression analysis,[1] the following casual relationships have been obtained:

[1]This is one of the statistical approaches for estimating the relationships between selected, one dependent and few independent variables, the latter usually used as some kind of "predictors." The least-square method is usually used.

(i) In the linear form

$$C_o(d,S) = a_0 + a_1 \cdot d + a_2 \cdot S \quad (10.5)$$

(ii) Log-linear form

$$C_o(d,S) = a_0 \cdot d^{a_1} \cdot S^{a_2} \quad (10.6)$$

where

a_0 is the fixed cost per service (€ or \$US)
a_1 is the average cost per unit distance (€/km or \$US/km)
a_2 is the average cost per unit of the offered capacity (€/space, €/ton, or €/TEU); or (\$US/space, \$US/ton, or \$US/TEU)
d is the length of route (km)
S is the payload capacity of rolling stock/vehicle (spaces, ton, TEU)

As could be intuitively expected, the route cost generally increases with increasing of the line/route length and the vehicle size/capacity. By dividing the cost functions in Eqs. (10.5) and (10.6) by either the route length (d), the capacity of rolling stock/vehicle (S), or by both, the average unit cost per given transport service can be obtained. In the latest case, the average unit cost will be

(i) From the linear equation

$$c_o(d,S) = \frac{a_0}{d \cdot S} + \frac{a_1}{S} + \frac{a_2}{d} \quad (10.7)$$

(ii) From the log-linear equation

$$c_o(d,S) = a_0 \cdot d^{a_1 - 1} \cdot S^{a_2 - 1} \quad (10.8)$$

where all symbols are as in the previous equations.

The average unit cost ($c_o(d,\cdot S)$) is expressed in the monetary terms per unit of output, ie, €/space km, €/ton km or €/TEU km, or \$US/space km, \$US/ton km or \$US/TEU km). In addition, by dividing Eqs. (10.5) and (10.6) by the load factor,[2] the corresponding average costs per unit of transported item—revenue passenger, ton, or TEU km can be obtained.

If the rolling stock/vehicle capacity is given, Eqs. (10.5)–(10.8) express the total and average cost, respectively, exclusively in dependence on the distance. In the case when the distance is given, they express the total and average cost, respectively, exclusively in dependence on the different capacity of rolling stocks/vehicles.

Similar analysis can be carried out for the prices of transport services, which usually reflect the above-mentioned cost under the equivalent conditions.

Depending on the values of particular coefficients (particularly those in the log-linear form) in Eqs. (10.6) and (10.8), the average cost can increase, be constant, or decrease with the increase of the vehicle payload capacity and/or route length, thus implying diseconomies, constant, and economies of scale, respectively.

[2]This is the ratio between the utilized and available (offered) capacity per given transport service.

Generalized costs are the monetary and nonmonetary expenses on the demand side of transport sector. Since transport is derived activity, it costs users (passengers and freight/goods shipments) to change the place. In general, these costs consist of the monetary and nonmonetary component. The former mainly relates to the price/charge paid for trip or transport of a given goods/freight shipment, respectively, between their origin and destination. The latter relates to some nondirect expenses, which in most cases include the value of users' time while in transportation. The sum of both components is called generalized travel or transport costs. Analytically, for a given transport mode, they can be expressed as follows:

$$C_i(d) = p_i(d) + \alpha_i \cdot [d/v_i(d)] \tag{10.9}$$

where

d is travel or transport distance, ie, the route length (km)
$p_i(d)$ is the price paid by user for travel/transport on the route (d) (€ or $US)
α_i is the user's value of time while traveling by transport mode (i) (€/h or $US/h)
$v_i(d)$ is the average travel/transport speed by mode (i) on the route (d)(km/h)

The value of time has shown to be particularly important for business passengers and the time sensitive freight/goods shipments (urgent parcels, fruit, vegetables, flowers, etc.). In most cases of estimating the probability of choice of a given transport mode, the above-mentioned form of the generalized cost function has been used (Logit model).

Revenues are amounts of money that a transport company receives during a specific period of time from charging its transport services to users. These are calculated by multiplying the price at which the given services are sold by their number carried out during the specified period of time.

Net income represents the difference between the revenues and costs, which a given transport company or transport sector has achieved during the specified period of time (usually a quarter or a year). It can be estimated as follows:

$$I(k) = R(k) - C(k) \tag{10.10}$$

where $R(k)$ is the revenues of a transport company achieved during the time period (k) (monetary units: $US, €).

This income can be positive (revenues are greater than costs) in which case, after being taxed, represents profits. Otherwise, the negative net income (revenues are lower than costs) represents loses in the given context.

Profit margin represents the ratio of net income and total revenues, as follows:

$$PM(k) = I(k)/R(k) = [R(k) - C(k)]/R(k) = 1 - C(k)/R(k) \tag{10.11}$$

where the other symbols are analogous to those in the previous equations. In effect, as defined above, the profit margin measures how much income a transport company has from each unit of the revenues earned. For example, a 10% profit margin implies that the company has a net income of €0.10 for each Euro of the total revenues.

10.2.5 RELATIONSHIP BETWEEN DEMAND AND SUPPLY

Transport companies supply transport capacity to satisfy the expected demand during a given period of time. This is carried out based on the unit price of output/transport service. The unit price is based on the revenues obtained by the given capacity supply and its related costs. In general, when the unit price decreases, the demand for transport services generally tends to increase, and vice versa. Under the same conditions, the total capacity supply and related costs tend to increase, and vice versa. Fig. 10.1 shows the simplified scheme of the relationship between demand and capacity supply depending of the unit price.

Quantity demanded and quantity supplied are the functions of the unit price. Fig. 10.1 does not follow the standard graphical representation of the functions. Traditionally, in economic literature, the unit price is on the vertical axis, and quantity on the horizontal axis.

As can be seen from Fig. 10.1, the demand and capacity supply curves intersect at the point A. At this point the quantity demanded (Q_A), at the current price $P_A(Q_A)$, is equal to the quantity supplied at the current price $P_A(Q_A)$, resulting in an economic equilibrium.

The relationship between the two curves indicates two phenomena: consumer and producer surplus. The former appears when the users—passengers or freight/goods sippers/receivers are willing to pay the price for service even above its market-balanced level (Triangle: $A-P(Q_A)-B$). The latter represents the difference between the amount that a transport company is willing and able to supply and the amount it actually receives by charging its supply at the market-balanced prices (Triangle: $A-P(Q_A)-C$). As such, this is considered as the benefit for the transport company operating in the market under given conditions. In addition, the sum of the consumer and producer surplus (Triangle: $A-B-C$ in Fig. 10.3) represents the economic welfare, ie, the utility, which could be gained by both users and producers, ie, suppliers of transport services under given (market) conditions.

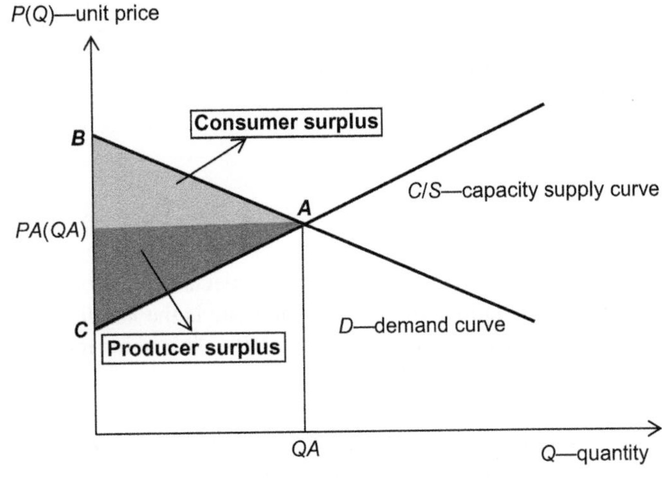

FIG. 10.3

A simplified scheme of the relationship between the unit price, demand, and capacity supply curve.

10.3 TRANSPORTATION PROJECTS EVALUATION

Airport extension, port extension, construction of a new freight center, waterway deepening, or construction of a new high speed rail (HSR) represent transportation projects faced by transportation experts. In the initial stage of these projects, it is necessary to perform their evaluation, that is, as precise as possible, analyze and review economic, environmental, equity, as well as other project impacts. The planners and engineers must properly answer the following questions: (a) Are the transportation project's benefits greater than the projects' costs?; (b) What is the best project's alternative in the case when project has few mutually exclusive alternatives?; (c) How to allocate available funds among competitive transportation projects?; (d) When to start the considered project?

Frequently, financial resources are scarce, and appropriate engineering economic analysis can significantly help planners and decision-makers to allocate available resources properly.

In the first step of any project evaluation, it is necessary to analyze the project's socio-economic context, and to clearly define the project's objectives. In the next steps, the analysts should clearly recognize the type of costs and benefits, compare them and make recommendations to the decision makers.

Transportation projects usually extend over many years. On the other hand, the purchasing power of money decreases over time. The main cause of this phenomenon is the inflation that exists in every society. A *discount rate* regulates the value of money for time. This rate is used to represent future monetary quantities in terms of their today's value. *Compounding* and *discounting* are techniques that enable us to compare money values at different points in time. Let us briefly explain compounding technique.

We assume that we want to invest \$100 these days, at an annual interest rate (r) of 5%. It will be worth \$100+\$5=\$105 in 1 year. After 2 years, it will be worth \$105+\$ 0.05 × 105=\$110.25. After 3 years we will have \$110.25+\$ 0.05 × 110.25=\$115.7625. Discounting represent reverse operation of compounding. The compounding technique helps us to find the answer to the following question: what is the present value (PV) of a known future amount of money?

In our example, the PV of \$105 next year, when $r=5\%$, is \$100. The PV equals

$$PV = V_t/(1+r)^t \quad t=0,1,2,\ldots,n \tag{10.12}$$

where

n is the project duration (in years)
r is the discount rate
V_t is the value in year t

There are two interest rates that are used in transportation projects evaluation. The first one is the *real interest rate* that is exclusive of inflation, while the second one is the *nominal interest rate* that is inclusive of inflation. Project value is usually expressed as a *NPV*. This value represents a project's value or cost for its whole life cycle in today's dollars.

When evaluating transportation projects many governments and funding agencies in the world (OECD, World Bank, etc.) require a cost-benefit analysis (CBA) to be performed. In this way, the CBA represents common evaluation language between the governments, funding agencies and the transportation project supporters.

10.4 COST-BENEFIT ANALYSIS

The CBA is a method that calculate and compare project's costs and benefits to society over period of time. The CBA monetizes all project's inputs and outputs. In other words, the CBA converts the inputs and the outputs into a monetary values. The CBA helps decision-makers to rank and prioritize various project's alternatives including also alternative "no action" ("no action," or "do nothing" case assumes continued operation of the existing facility, exclusive of any major investments). The specific transportation project should start only when the CBA clearly shows to the decision makers that the total benefits to society outweigh the total costs. When performing CBA, the analysts enumerate all project's costs and benefits to society. In the next step, they assign monetary values to costs and benefits, and discount them to a *NPV*. All costs, as well as all benefits are added into a single number. The transportation project is evaluated by using total costs, and the total benefits values.

In the first step of the CBA, it is necessary to identify transportation project's alternatives to be evaluated. Alternatives may represent "do nothing" case, rehabilitation of existing facility, construction of a new facility, etc. Transportation projects have consequences over time. The analysts should also define the time period over which the life cycle costs and benefits of all of the alternatives will be calculated.

The project's economic performances are measured by the following indicators:

NPV: net present value
IRR: internal rate of return
$\frac{B}{C}$: the benefit/cost ratio

The majority of experts consider the *NPV* as the most important CBA indicator. The *NPV* is defined in the following way:

$$NPV = \sum_{t=0}^{n} \frac{B_t - C_t}{(1+r)^t} \tag{10.13}$$

where

B_t: benefits in year t
C_t: costs in year t
n: project duration (in years)
r: interest rate
t: year index

Project's benefits and costs are forecast over the project duration. For example, benefits from road investment could be shorter traveling distance, shorter travel time, reduced number of traffic accidents, etc. The road improvement costs could be project design costs, labor costs equipment costs, material costs, etc.

The analysts use a *NPV* to express a project's worth for its complete life cycle in today's money value. We see that the *NPV* decreases if r (interest rate) increases. In the case when $NPV > 0$, the project may be accepted. In the opposite case, when < 0, the considered project should be rejected. Finally, when $NPV = 0$, we conclude that the considered project adds no monetary value. The final decision about such transportation project should be based on some additional criteria.

The internal rate of return (*IRR*) is the indicator that also measures the project's performances. The *IRR* is the discount rate/interest rate at which the $NPV = 0$. We calculate the *IRR* by solving the following equation:

$$\sum_{t=0}^{n} \frac{B_t - C_t}{(1+IRR)^t} = 0 \tag{10.14}$$

The average values of the observed *IRR*'s in a sample of investment projects sponsored by the European Union (EU) at the end of the 20th century are approximately equal to 15% in the cases of roads and highways, 10% in the cases of railways and underground, and 25% in the cases of ports and airports.

After calculation net benefits B and net costs C, the benefit/cost ratio $\left(\frac{B}{C}\right)$ should be also calculated. (Frequently, it is not easy to estimate future costs, and, especially project's benefits.) The benefit/cost ratio (B/C) informs us about the improvement in traffic operations (expressed in dollars) per dollar invested.

Analysts and engineers usually perform a *sensitivity analysis* to conclude how sensitive final results are to changes in hypothesis about the costs, benefits, and discount rate.

EXAMPLE 10.1

The comparison between four possible alternatives in improving transportation facility are given in Table 10.1 and Fig. 10.4.

Table 10.1 Benefits and Costs of the Transportation Project

	Alternative 1	Alternative 2	Alternative 3	Alternative 4
Benefits B	\$12,000,000	\$14,000,000	\$10,000,000	\$16,000,000
Costs C	\$16,000,000	\$18,000,000	\$8,000,000	\$12,000,000
B/C	0.75	0.78	1.25	1.33
Net benefits = benefits − costs	−\$4,000,000	−\$4,000,000	\$2,000,000	\$4,000,000

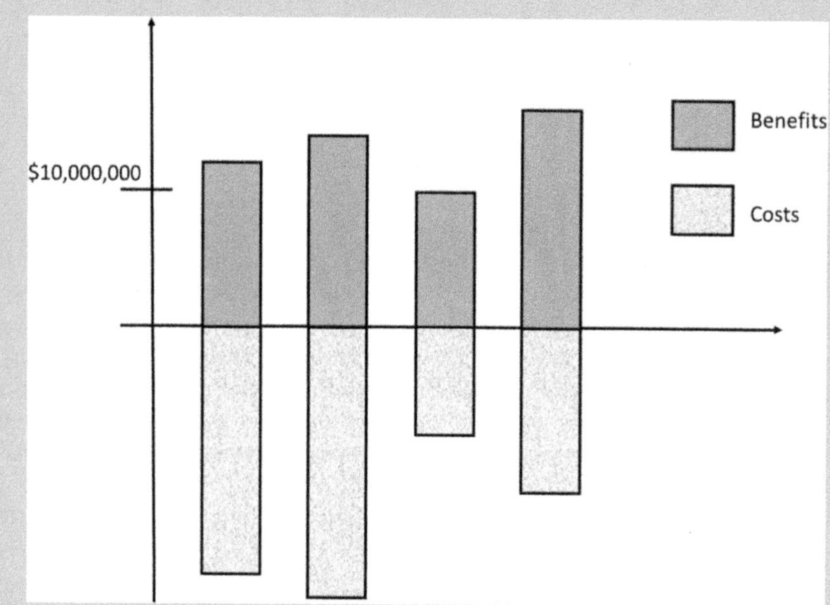

FIG. 10.4

Benefits and costs of the transportation project.

It is a common way to show benefits over the x-axis, and the costs below the x-axis. Obviously Alternative 4 should be chosen.

The main weakness of the CBA is that all transportation project benefits are evaluated only in monetary terms. It is very complicated to value all the costs and benefits of transportation projects in monetary terms. In other words, in many situations, social and environmental aspects of the considered transportation projects are not treated adequately. For example, many traffic safety programs, actions and projects involve the prevention of loss of life. The logical and ethical question is how should we value a life saved? A pure economic approach would suggest to us that the value of life is equal to the *PV* of lifetime earnings. Obviously, there are numerous opponents to such an oversimplified and ethically questioned approach.

The complexity of the decision-making processes has increased over time. The increase happened in the number of alternatives in the scope of particular solutions, the number of (usually conflicting) attributes/criteria per alternative, and the number of actors whose (very often diverse and conflicting) points of view needed to be taken into account. The MCDM (multicriteria decision making) and/or MADM (multiattribute decision making) methods have been recommended as more convenient tools for looking for the preferable among several alternatives of a given solutions. Some academic-research and professional-practical successful applications of the MCDM or MCA (multi criteria analysis) methods have included the SAW (simple additive weighting), AHP (analytical hierarchy process), TOPSIS (technique for order preference by similarity to ideal solution), ELECTRE [ELimination Et Choix Traduisant la REalité (elimination and choice expressing reality)], PROMETEE (preference ranking organization method for enrichment of evaluations), and many others methods with their modifications (Hwang and Yoon, 1981; Sauian,, 2010).

10.5 INFRASTRUCTURE COST

The infrastructure costs relate to the expenses to construct particular components of transport networks. The components could be links/lines of particular modes—urban transit systems, road highways or motorways, HSR lines, inland waterways, and the nodes such as airports and sea-ports.

In general, the infrastructure costs consist of the fixed and variable component. The fixed component generally relates to expenditures for building the new and/or expanding existing infrastructure components (links, nodes). The expenditures include infrastructure planning, land acquisition, carrying out earthworks (ground preparation, drains, etc.), building substructures (bases and frost protection layers), superstructures (binder and surface layers), and engineering works (bridges, tunnels, etc.), installing facilities and equipment (traffic signs, signals, power supply system), and carrying out capital maintenance in the specified time intervals of the life-cycle, not depending on the intensity of use, ie, the volumes of traffic. The variable component generally relates to the expenses for the current maintenance of both existing and new infrastructure that depend on the intensity of use, ie, the volume of traffic, collection of charges, and other temporal engagement of labor, material, and energy. Both total and average costs can be estimated, depending on the relevant data.

10.5.1 URBAN MASS TRANSIT SYSTEMS

Constructing the infrastructure of urban transit systems has generally required substantive capital (investment) costs. In this context, these costs are presented for the road-based BRT (bus rapid transit), and the rail-based streetcar (tramway), LRT (light rail transit), and subway/metro system, at different levels of details. The aim is to get evidence on the rank of order about the absolute and relative amounts of these costs.

10.5.1.1 General

The costs of urban mass public transit system consist of the cost of providing and maintaining infrastructure. These costs are usually considered as fixed during the given period of time (day, month, and/or, year). The experience up to date has shown that the average unit costs of this infrastructure increases with increasing of the capacity supply of particular urban transit systems, which implies moving from the road-based bus to the rail-based streetcar, LRT, heavy rail, and subway/metro system(s). Fig. 10.2 shows an example for the London urban transit systems such as conventional bus, two types of BRT systems, streetcar (tramway), LRT, and heavy rail, all considered as alternatives for passing over the river Thames (TfL, 2000; Fig. 10.5).

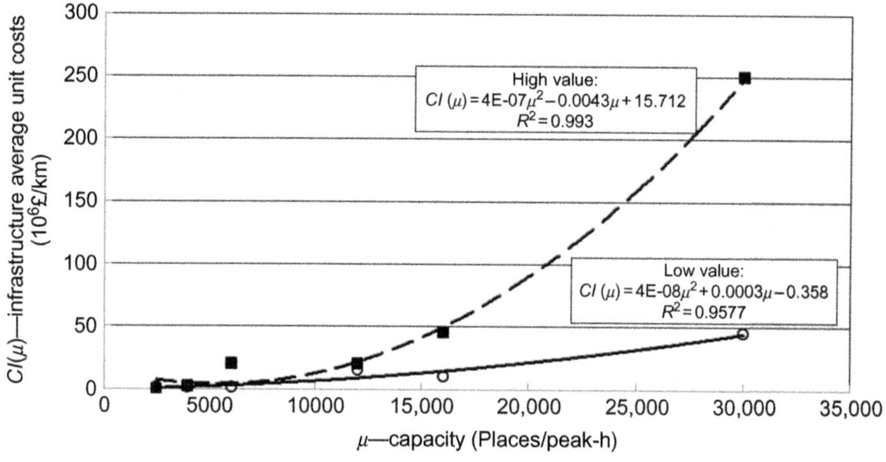

FIG. 10.5

Relationship between the infrastructure average unit cost and the capacity of urban transit modes (TfL, 2000).

As can be seen, in the given case, the average unit costs of infrastructure throughout considered systems for both high and low values have increased more than proportionally with increasing of the systems' capacity supply, thus implying existence of the corresponding diseconomies of scale, ie, increase in the capacity supply would require more than proportional increase in the average unit costs of required infrastructure. In addition, Fig. 10.6 shows the relationship between the operating speed throughout considered systems and the average unit cost of their infrastructure.

As can be seen, the average speed can be increased to the certain level with increasing of the investments and then decrease despite further increasing of investments. This practically means that choosing the rail-based streetcar (tramway) and/or LRT system at the higher infrastructure investment cost as compared to the road-based BRT system could contribute to increasing of the average travel speed, but at decreasing rate. Implementation of the heavy rail, despite higher construction costs, would not bring benefits in terms of increasing the average travel speed.

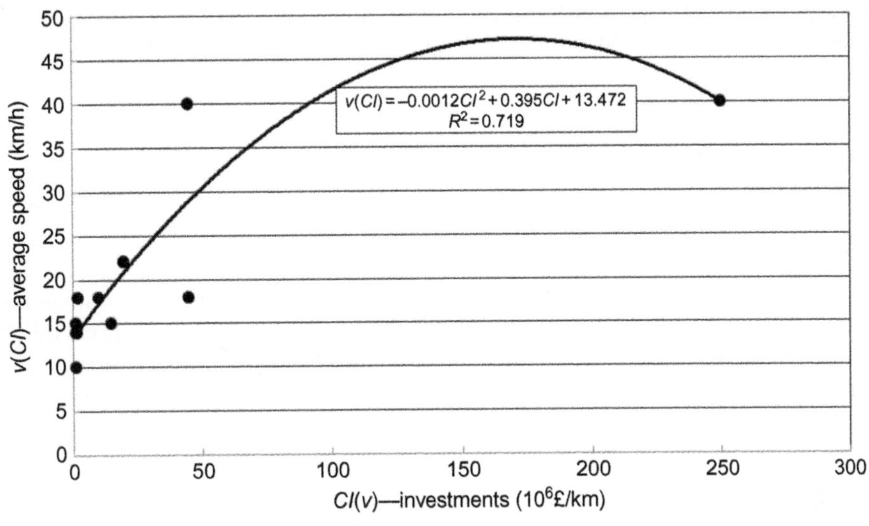

FIG. 10.6

Relationship between the average operating speed of urban transit systems on the infrastructure investment unit cost (TfL, 2000).

10.5.1.2 Streetcar (tramway)

In general, the costs of construction of the streetcar lines have substantively varied, mainly because their specific characteristics throughout different urban areas. In addition, in many cases it has been difficult to obtain their construction costs, because building these lines has often been part of larger projects. Nevertheless, Table 10.2 provides the information of the construction cost of five recent streetcar projects in the United States.

Table 10.2 Some Characteristics of the Streetcar (Tramway) Lines and Their Construction Costs in the Given Examples (United States) (APTA, 2015)

Place/City/Urban Area	Length of Line (Miles)	Cost (10^6 \$US/Mile)	Number of Streetcars Acquired[a]
San Francisco		30	
Kenosha, Wisconsin	2	2	5
Portland, Oregon	4.6	8.2	7
Tampa, Florida	2.3	8.10	8
Little Rock, Arkansas	2.1	5.3	3
San Pedro, California	1.5	2.2	3

[a]*Each streetcar (tramway) is assumed to cost: 0.6×10^6\$US. (The vehicle cost varies from 0.2 to 0.8×10^6\$US.)*
1 mile = 1.6010 km.

Nevertheless, the most recommendations have been to consider the construction cost of a streetcar line of about $\$10 \times 10^6$\$US/mile, at least for an initial assessment, ie, as an orientation (1 mile = 1.6010 km).

10.5.1.3 BRT and LRT

The construction costs of BRT and LRT system infrastructures have always been considered as a strong criterion when the decisions which system to further develop or implement in given urban areas have to be made. This is because both systems have been comparable regarding other (operational and economic) performances, while satisfying expected passenger demand under given conditions. In addition, these costs of both systems have shown to be inherently very diverse in various urban areas. This diversity is caused mainly by differences in the local physical conditions for building the infrastructure, as well as by the costs of inputs for undertaking the activities in terms of labor, material, and energy. Fig. 10.7 shows the total cost of constructing BRT and LRT infrastructure/lines in dependence on their length in the urban areas in the United States (GAO, 2001).

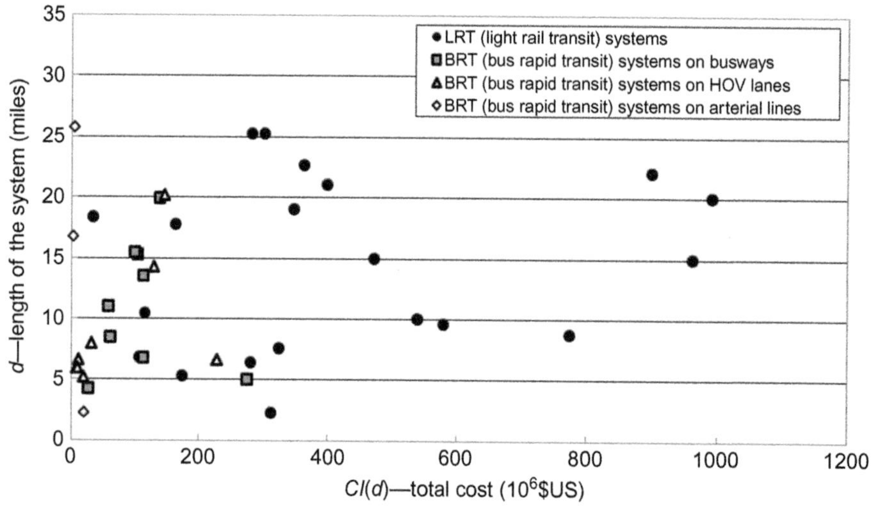

FIG. 10.7

Relationship between the total construction cost of LRT and BRT systems and their length (case of United States) (GAO, 2001).

As can be seen, at the LRT systems, these costs have generally increased with increasing of the length of lines. However, the differences have been considerable: for example, to build the line of length of 25 miles, the costs have varied from about 0.3–1.0 × 10^{10} \$US. At the BRT systems on busways and HOV (high-occupancy vehicle) lanes, these costs have mainly varied from 0.01 to 0.2 × 10^{10} \$US for the length of lines from about 5 to 20 miles. At the BRT systems on arterial lines these costs have been much lower for the same range of length of lines. Table 10.3 gives the aggregate figures on the total length, and total and average construction costs for the above-mentioned BRT and LRT systems.

As can be seen, the average unit construction costs of LRT system has been particularly for about three and four times greater than that of BRT systems on busways and HOV lanes, respectively.

Table 10.3 Examples of the Construction Costs of the LRT and BRT Systems in the United States (GAO, 2001)

System (No. of Cases)	Length (Miles)	Total Cost (10^6 $US)[b]	Average Cost (10^6 $US/Mile)
LRT systems (21)	3010.7	10,774.58	34.710
BRT system on busways (10)	1010.8	1346.62	13.410
BRT system on HOV[a] lanes (8)	72.10	653.82	8.107
BRT system on arterial streets (3)	44.7	30.36	0.68

[a]*High-occupancy vehicle lanes.*
[b]*2000 $US value.*
LRT, light rail transit; *BRT*, bus rapid transit.

10.5.1.4 Subway (metro)

The construction costs of subway/metro infrastructure have been inherently very different in particular urban areas mainly due to differences in the local physical conditions (proportion of being on and underground, length, deepness of the underground portion(s), necessity for bridges on the ground portion(s), etc.), as well as in the local costs of necessary inputs for construction such as labor, material, and energy. Fig. 10.8 shows the relationship between the total construction cost and length of subway/metro lines built, under construction, and planned to be built in 27 urban areas round the world (PO, 2013).

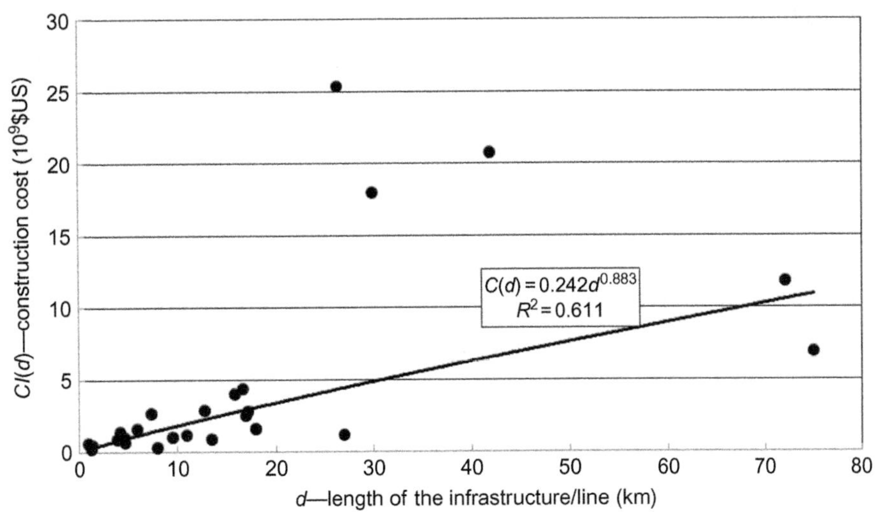

FIG. 10.8

Relationship between the construction costs and length of the subway/metro line in given examples (PO, 2013).

As can be seen, the total construction costs have increased more than proportionally with increasing of the length of line(s), but at decreasing rate. In given example, the average unit construction costs have been 241 × 10^6 $US/km. In should be mentioned that due to the nature of getting the sources of data, some errors of 10–20% could be expected, but anyway this could be used as an orientation of the rank of order of the construction costs of given infrastructure. Table 10.4 gives an additional information about the location, time of implementation, and the average unit construction costs of the above-mentioned subway/metro lines.

Table 10.4 Some Characteristics of the Subway/Metro Infrastructure Undertaken to be Built Over the Last 15–20 Years (PO, 2013)

Name of the System/Year of Implementation	Length (km)	Proportion Underground (%)	Cost (10^6 $US/km)
Singapore Thomson MRT Line (2013–21)	30	100	600
Singapore Downtown MRT Line (2017)	42	100	4103
Hong Kong Sha Tin to Central Link (2018)	1	100	586
Budapest Metro Line 4 (2014)	7.4	100	358
Fukuoka, Nanakuma Line/Hakata (2014–20)	1.4	100	321
Cairo Metro Line 3, Phase 1 (2006–12)	4.3	100	310
Kawasaki Subway (2018)	16.7	100	260
Stockholm City Line (2017)	6	81	2510
Sao Paulo Metro Line 4 (2004–14)	12.8		223
Sao Paulo Metro Line 6 (2014)	15.10	100	250
Dnipropetrovsk Metro Extension (2008–15)	4	100	214
Malmö City Tunnel (2005)	4.65	100	212
Bangalore Metro Phase 2 (2017)	72.1	110	164
San Juan Tren Urbano (1996–2004)	17.2	7.5	163
Lucern Zentralbahn (2008–13)	1.32	100	151
Sofia Metro Line 2 (2008–12)	17.0	82	148
Thessaloniki—I Phase (2016)	10.6	100	104
Thessaloniki—II Phase (2017)	4.8	100	135
Vancouver Evergreen Line (2012–16)	11	18	103
Dubai Metro (Lines 1 and 2) (2005–11)	75	17	102
Mexico City Metro Line 12 (2007–12)	26.4	410	100
Seoul Sin-Bundang Line (2005–11)	18	100	87
Seoul Subway Line 10 (2010)	27	100	43
Bangalore Metro, Phase 1 (2001–06)	18	100	87
Helsinki Westmetro (2010–15)	13.5	100	66
Barcelona Sants-La Sagrera (2008–11)	8	100	310

The very high diversity in the average unit construction costs is noticeable. It should be also mentioned that the time of building given line has varied between 3 and 10 years.

10.5.2 ROAD

The costs of highways and motorways relate to the expenses for their construction and operating. The former expenses generally include the capital costs for provision of the new capacity (ie, new lanes) and the fixed maintenance costs. The latter (operating) expenses are imposed by repairs and maintenance, policing, signaling, collecting charges, and administration. The *NPV* of the annual costs of a newly built road of the length (d) per year of its life cycle of (N) years (usually 50–65 years), can be estimated as follows:

$$NPV = [(c_{\mathrm{b}} + c_{\mathrm{r}} + c_{\mathrm{m}}) \cdot d] \cdot \left[\left(\frac{1-(1+i)^{-N}}{i}\right)\right] \tag{10.15}$$

where

c_{b} is the average unit costs for constructing a given road (€/km or \$US/km)
c_{r} is the average unit costs of renewal a given road (€/km or \$US/km)
c_{m} is the capital maintenance cost of a given road during its life-cycle (€/km or \$US/km)
i is the interest rate on the debt for implementing a given new road (%); (it is usually 4–5%).

The other symbols are analogous to those in the previous equations. The costs in Eq. (10.15) are expressed as the costs of amortization repaid through the annual annuities of the investments. Table 10.5 gives some figures of the construction costs of new highways in Europe.

Table 10.5 Capital Cost of New Highways in Europe (CE Delft, 2008)

Country	Cost (c_{b}) (10^6€/km)
Switzerland	1.100
Austria	0.780
Italy	0.700
France	0.520
Denmark	0.700
PT	0.052

As can be seen, they vary across the countries, conditioned by their specificity. The differences seem to come from the input prices for construction, but also they can be driven by the topography of the countries. However, there it is not obvious explanation why, for example, the costs in Austria are so much different from those in Switzerland. Table 10.6 gives an example of the average unit construction costs of highways in particular world's regions (continents).

As can be seen, these costs have been the highest in the United States and the lowest in Asia. The reasoning behind the differences can be the same as in the case of European counties: differences in the costs of inputs for construction—labor, material, and energy, and differences in the topography of the regions. The costs of construction of 4-lane highway are approximately twice higher than that of 2-lane highway.

Table 10.6 Example of the Average Construction Cost of the New 2-Lane Highway (WB, 2015; WSDT, 2004)

Region	Cost (10^6 2000 $US/km)	Cost (10^6 €/km)[a]
Asia	1.007	1.0610
Africa	1.023	1.086
Europe	1.234	1.310
Latin America and Caribbean	NA	NA
North America (United States)[b]	1.4210	1.1210
All regions	1.183	1.256

[a]*Exchange rate: 1.2 $US for 1 €=0.104206.*
[b]*WSDT (2004).*

The analysts could also assume that the new road would have an infinite lifetime. This could happen if the renewal and maintenance can prolong it. Then, the *NPVs* of a given investments in the (*n*)th year of the life time of a given new road can be estimated as follows:

$$NPV_n = \frac{[(c_b + c_r + c_m) \cdot d]}{(1+i)^n} \tag{10.16}$$

where all symbols are analogous to those in the previous equations.

In general, except the construction costs, all other costs vary with varying of the volumes and structure of expected traffic, the latter expressed in terms of the maximum GVW and corresponding axle load. These latest two represent the basis for allocation of the above-mentioned costs to the particular categories of vehicles as users. In Europe, the traffic structure generally consists of: motor cycles, small and cars, buses/coaches, LDVs, and HDVs of the GVW of 5.5, 12, 24, and 40 ton, respectively. In the United States, the vehicles are categorized in eight categories: Class 1—up to 2.75 ton, Class 2—2.75 to 4.60 ton, and Class 3—4.60–6.40 ton as LDVs; Class 4—6.4–7.3 ton, Class 5—7.3 to 10.0 ton, and Class 6—10 to 11.10 ton as medium duty vehicle(s), and Class 7—11.10 to 15.00 ton, and Class 8—greater than 15.00 ton, as HDVs (USDE, 2013). Under such conditions, the total cost are expressed as the average and marginal costs in €/veh km, which in turn represent the basis for charging the above-mentioned road users. For example, in EU210 countries, the average total costs of motorways independently on the vehicle category have estimated to be 18.2€ ct/v km. However, this cost has varied considerably with the vehicle weight, for example from 4€ ct/v km for a 5.5 ton lorry to 110€ ct/v km for a 40 ton truck and trailer combination. At the same time, the corresponding marginal cost in EU27 countries, has varied from 0.024€ ct/v km for a 5.5 ton lorry to 10.63€ ct/v km for a 40 ton truck and trailer combination (CE Delft, 2008).

10.5.3 RAIL

The construction and maintenance costs of the HSR lines are given to indicate the costs of rail infrastructure despite they carry on only passengers at HS. Specifically, in some countries such as Germany (Europe), these are also upgraded formerly conventional rail lines, but again carrying out only passengers). This implies that in this case, the same infrastructure is not shared by both passenger and freight/goods trains. Similarly, as at roads, these costs consist of the fixed expenses for construction and capital maintenance of a given line not depending on the volumes of traffic, and the operating expenses needed for current repair, maintenance of the line, power, and traffic signaling and managing system, and related administration depending on the volumes of traffic. Again, based on Eq. (10.16), the *NPV* of these costs in the *n*th year of the life cycle of the line of *N* years ($N=50$ years) can be estimated.

Similarly as in cases of the urban transit systems and roads, these costs have significantly varied across the countries where the HSR lines have been built. Table 10.7 gives an indication of the average unit costs of already built and planned HSR lines, which do not include the cost of planning, and land acquisition and preparation (Pourreza, 2011; de Rus and Nombela, 2007).

Table 10.7 Examples of the Construction Costs for the HSR Lines in Particular Countries (Pourreza, 2011; de Rus and Nombela, 2007)

	Cost (10^6 €/km)	
Country	**Built (In Service) Lines**	**Underconstruction Lines**
Austria	–	18.5–310.6
Belgium	16.1	15.0
France	4.7–18.8	10.0–23.0
Germany	15.0–28.8	21.0–33.0
Italy	25.0	14.0–65.8
The Netherlands	–	43.7
Spain	7.8–20.0	8.10–17.5
Japan	20.0–30.0	25.0–40.0
South Korea	–	34.2

As can be seen, the variation of the average unit costs for both already built and underconstruction HSR lines has been significant in both European and nonEuropean, ie, two Asian countries. In Europe, the lowest cost has been in France and Spain, and much higher in Italy, Germany, and Belgium. It can be shown that the average unit cost has been 18×10^6 €/km. In addition, the cost of building the new HSR lines in Asian countries (Japan, South Korea, except China) has been slightly higher than that in the European countries. In particular, in Japan these costs have been mainly caused by the substantive proportion of long viaducts and bridges along the particular lines (Pourreza, 2011; de Rus, 2010).

In addition, the average maintenance cost per unit of length of the HSR system infrastructure has also highly varied, mainly depending on the length of line(s) and local conditions. Some estimates indicate that this average cost in the European countries has amounted from about $13–72 \times 10^3$ €/year (Henn et al., 2013; Pourreza, 2011).

10.5.4 INLAND WATERWAYS

The infrastructure costs of inland waterways relate to expenses in constructing and maintaining the new canals and maintaining the existing ones including canalization of the existing rivers. The components under treatment are embankments, locks, and bridges, the latter two influenced by the size of towboats and the size and type of barges (motor vessels or nonpropelled) used. In most countries, these costs are mainly covered by the national and regional governments and their related bodies.

In general, the construction cost of a new channels have been estimated to be 5–20 × 10^6 €/km, which is comparable to the construction costs of railway lines (INA, 2005). Similarly as in the cases of road and rail, the total infrastructure costs in this case can be divided into the fixed and variable costs. The former do not and the latter do change with changes in the volume of vessel/barge traffic. The fixed costs of the already existing inland waterways comprise the costs of regular maintenance and renewal of a given waterway, which mainly relates to dredging. This has become of increasing importance with increasing of the size and draught of vessels/barges. The examples of the total maintenance costs of inland waterways in particular European countries and United States in the year 2000 are given in Table 10.8.

Table 10.8 The Average Maintenance Cost Per Unit of the Inland Waterway Infrastructure in Europe and United States (Year 2000) (INA, 2005)

Country	Length of the Network (km)	Average Cost (10^3€/km)
Germany	7656	66.607
France	5788	16.045
Finland	6000	2.107
Sweden	120	37.453
The Netherlands	5046	210.1210
United States	17,700	20.174

As can be seen, these costs have been quite diverse, which had been influenced mainly by the local (country) conditions. The variable costs have mainly included expenses for the traffic control, operation of locks and bridges, and operation of the patrol vessels and crews. For example, in The Netherlands, these total costs have amounted 210 × 10^6, 50 × 10^6, and 6 × 10^6 €/year, respectively. In this case, these total costs can be allocated to users—freight/cargo and recreational inland waterway rolling stock—the towboats, motor vessels/propelled and nonpropelled barges, and the motor passenger vessels, respectively, as charges for using the inland waterways. This can be carried out by dividing the total infrastructure costs with the total volume of kilometers (miles), carried out by particular type of rolling stock during the given period of time (usually 1 year) as follows:

$$c_k(\tau) = C_{\text{iww}}(\tau)/[p_k(\tau) \cdot V_{\text{iww}}(\tau)] \tag{10.17}$$

where

τ is the period of time (usually 1 year)
$c_k(\tau)$ is the average infrastructure costs allocated as a charge to the rolling stock (vessel/barge) of type (k) operating along the given inland waterway(s) during the time period (τ) (€/vehicle km)
$C_{\text{iww}}(\tau)$ is the total infrastructure costs of a given waterway(s) (a segment or the entire network) during the period of time (τ)
$V_{\text{iww}}(\tau)$ is the total volume of output (vessel/barge km) carried out along the given waterway(s) (a segment or the entire network) during the time period (τ)

$p_k(\tau)$ is the share of output (vessel/barge km) of the rolling stock (vessel/barge) of type (k) in the total volume of output (vessel/barge km) carried out during the period of time (τ)

One such allocation of these total fixed and variable infrastructure costs to the professional inland shipping in The Netherlands has resulted in the average marginal costs from about 1.105 €/vessel km (the motor vessels of size < 250 ton) to 8.64 €/vessel km (the motor vessels of size > 3000 ton) (van Donselaar and Carmigchelt, 2001; ECORYS and METTLE, 2005).

10.5.5 PORTS

In general, the infrastructure costs of ports include the expenses for construction and development of berths—quays with the supportive facilities and equipment such as fixed and mobile cranes, forklifts, tractors, trailers, and trucks operating in the scope of the dedicated terminals (UN, 1976). In particular, the quays represent the basic port infrastructure whose construction costs depend on many factors such as location, materials used (mainly concrete, steel), functional, and technical requirements. In this context, the costs of quay walls dominate in correlation with their importance. Some investigations have shown that about 75% of these costs depend on the retaining height[3] of the wall. Fig. 10.9 shows a simplified scheme.

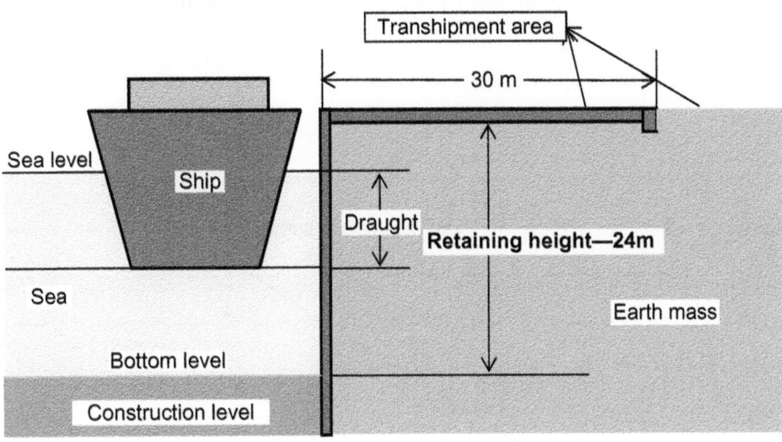

FIG. 10.9

Simplified scheme of a quay wall (de Gijt, 2010).

The data round the world show that the average unit cost depends on the retaining height at sea ports in the following way (de Gijt, 2010):

$$c(h) = 670.45 \cdot h^{1.273};\ R^2 = 0.571 \tag{10.18}$$

[3]This is defined as the height of the vertical structure supporting and separating the mass of earth from the sea, and vice versa, in the given context.

where

h is the retaining height (m)
$c(h)$ is the average unit cost of constructing the quay wall (€/m^2)

By multiplying the sum of retaining height and width by length of a given quay wall, its total area can be estimated. Then, multiplying this area by the average unit cost in relation (10.18) gives the total cost of constructing a given quay wall as follows:

$$C(h, w_q, L_q) = c(h) \cdot (h + w_q) \cdot L_q = (670.45 \cdot h^{1.273}) \cdot (h + w_q) \cdot L_q \tag{10.19}$$

where

w_q is the width of a quay wall as a part of the transhipment area (m)
L_q is the length of a given quay wall (m^2)

The other symbols area analogous to those in the previous relations.

For example, if on Fig. 10.9, $h = 24$ m, $w_q = 30$ m, and $L_q = 250$ m, the total construction cost of this quay wall will be equal to

$$C(24, 30, 250) = (670.45 \times 24^{1.273}) \times (24 + 30) \times 250 = 517.3 \times 10^6 \text{ €} (10^6 = \text{million})$$

In addition, Fig. 10.10 shows an example of the cost of construction of berths at the sea port in South China.

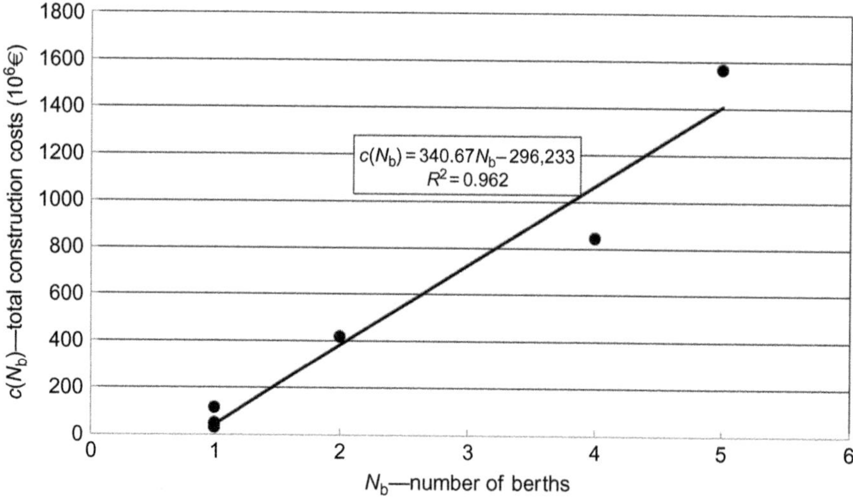

FIG. 10.10

An example of the relationship between the number of sea-port berths and the total construction costs of the selected cases in South China ("11th Five Year Plan" 2006–10 Guangdong) (Yue-man et al., 2006; GCD, 2006).

As can be seen in this example, the total construction (investment) costs increase linearly with increasing of the number of berths. The average unit cost has been about: 371×10^6 €/berth. In addition, these costs do not include the capital maintenance costs over the life-cycle of the constructions. As well, the average unit cost is much less that in the above-mentioned cost of quay wall, but, among other factors, this cost is also highly dependent on the local prices of inputs—material, energy, and labor.

10.5.6 AIRPORTS

The airport infrastructure costs generally relate to the expenses for construction and developments of their airside, passenger and cargo terminals, and landside infrastructure. An example from the United States has indicated that these costs have been planned as investments for five categories of commercial airports categorized according to the percentage of the annual boarding passengers as follows (ACI, 2010): large hubs (1%), medium hubs (0.25% but less than 1%), small hubs (0.05 but less than 0.25%), nonhub primary (10×10^3 but less than 0.05%), and nonprimary commercial service (2.5×10^3 but less than 10×10^3) airports. In the given case, the total planned investment costs during the period 2005–10 had been about 71.5 billion $US. During the period 2010–13 these costs have been 80.7 billion $US. The large hubs have accounted 55.3 (510%), medium hubs 13.3 (14%), and the small hubs 5.8 (6%) billion $US. Table 10.9 gives an indication of the total and unit planned investment cost for three among above-mentioned five categories of the US airports during the observed period.

Table 10.9 An Estimation of the Characteristics of Construction Cost at Selected US Airports During the Period 2010–13 (ACI, 2010)

Airport Category	Number	Total Costs (10^{10} $US)	Average Cost (10^6 $US)/Airport
Large hub	30	52.81010	1763.3
Medium hub	37	12.7410	344.7
Small hub	73	5.557	76.3
Nonhub	248	5.101	20.6

10^{10}—billion; 10^6—million.

As can be seen, the greatest amount of construction and development funds has been planned per a large hub airport and the lowest for the nonhub airport. In addition, Table 10.10 gives an indication of the planned distribution of these funds to particular airport components (ACI, 2010).

Table 10.10 An Example of Distribution of Construction and Development Funds to Particular Components at the US Airports (ACI, 2010)

Component/Airport Category	All (%)	Large Hubs (%)	Medium Hubs (%)	Small Hubs (%)
Airside	32.1	210.0	310.1	47.6
Terminal	46.6	50.7	36.3	22.8
Landside	21.3	20.3	24.6	210.5
Total	100	100	100	100

As can be seen, the highest proportion of the construction and development funds has generally been planned to be allocated to the airport terminals and the lowest to the airport landside area. The same have been at the large hubs. At the medium hubs, the allocated proportions to the airport airside area and terminals have been similar. However, at the small hubs, the largest proportion of construction and development funds has been planned to be allocated to the airport airside area.

Fig. 10.11 shows the relationship between the average unit capital cost (planned investments) and the realized number of enplaned passengers at top 50 US airports of the above-mentioned four categories (large, medium, and small hubs and nonhubs) during the period 2010–13.

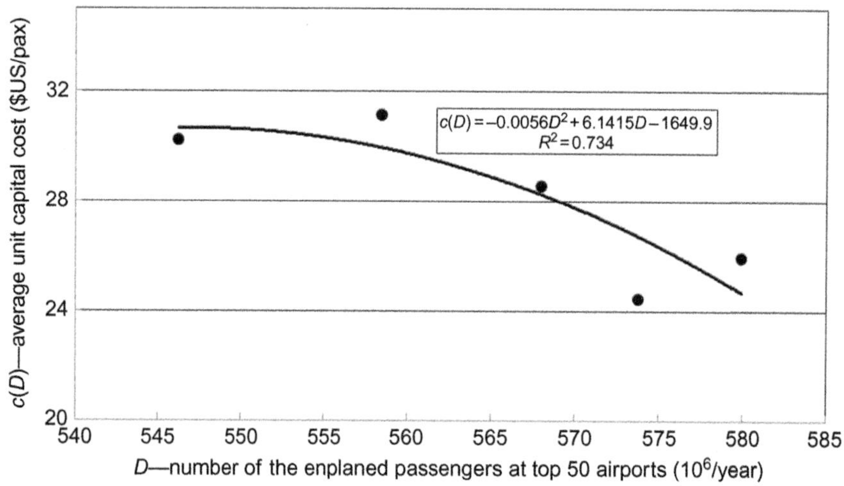

FIG. 10.11

Relationship between the average unit (planned) capital cost and the (realized) number of enplaned passengers at top 50 US airports during the period 2010–13 (ACI, 2010).

As can be seen, the average unit investment costs have decreased with increasing of the number of passengers from about 30 to 24 $US/pax during the observed period. As well the above-mentioned costs can be expressed per realized atm (air transport movement) (1 atm—1 landing or 1 take-off), as well as per unit of handled cargo. Anyway, they give an illustration of the possible scale and scope of the planned construction and development (investment) costs in both absolute (totals) and relative (per unit of given type of (realized) output) terms under given circumstances.

10.6 OPERATING COSTS AND REVENUES

10.6.1 INDIVIDUAL CARS

The total operating costs of individual cars consist of the fixed and variable component. The fixed component generally includes the costs for the car's registration, insurance, finance costs, and depreciation. As such, these costs do not depend on the intensity of car's use. The variable component changes with the intensity of car's use and generally includes the costs for fuel/energy, regular maintenance, repairs, and tires (AAA, 2013; BTS, 2014). Fig. 10.12A and B shows an example of evolution and structure of the total average unit cost of car use over time in the United Sates.

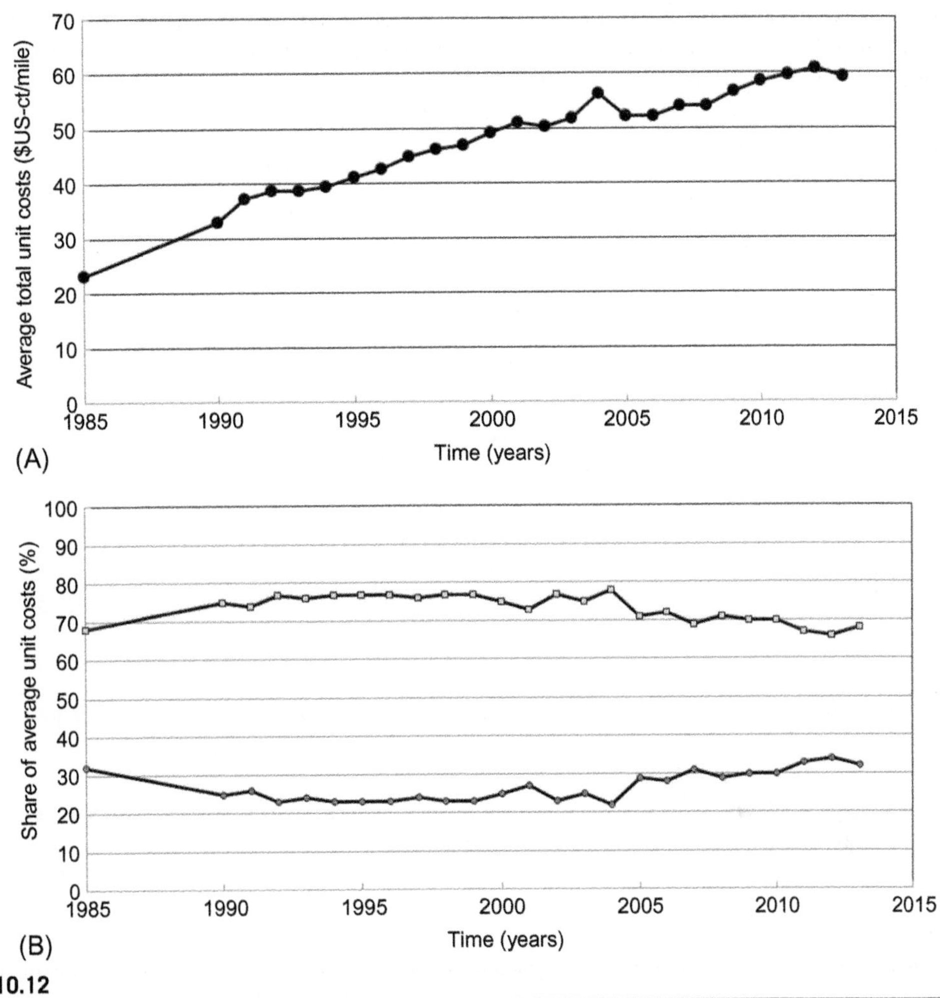

FIG. 10.12

Evolution of the average unit costs of car use and their structure over time in the United States (BTS, 2014): (A) amount and (B) structure.

The calculation has been carried out for a composite of three current American car models[4] with the utilization rate of 15×10^3 miles/year over the period of 5 years (ie, 75×10^3 miles/5 years) (1 mile = 1.6010 km). The current currency ($US) rate has been applied (BTS, 2014). Specifically, Fig. 10.12A shows that these average unit costs have generally and constantly increased during the observed period, from about 23 $US ct/mile in the year 1985 to about 60 $US ct/mile in 2013. At the same

[4]These are: *Small sedan* (Chevrolet Cruze, Ford Focus, Honda Civic, Hyundai Elantra, and Toyota Corolla); *Medium sedan* (Chevrolet Malibu, Ford Fusion, Honda Accord, Nissan Altima, and Toyota Camry); and *Large sedan* (Buick LaCrosse, Chrysler 300, Ford Taurus, Nissan Maxima, and Toyota Avalon) (http://exchange.aaa.com/automobiles-travel/automobiles/driving-costs).

time Fig. 10.12B shows that the share of fixed costs in the total unit costs has been for about two to three times higher that the share of the variable costs. In general, it has varied from 70% to 80% (fixed costs) compared to 30–20% (variable costs). In addition, Fig. 10.13 shows an example of the relationship between the average total unit costs and utilization of cars of three current American car models and their composite (AAA, 2013)

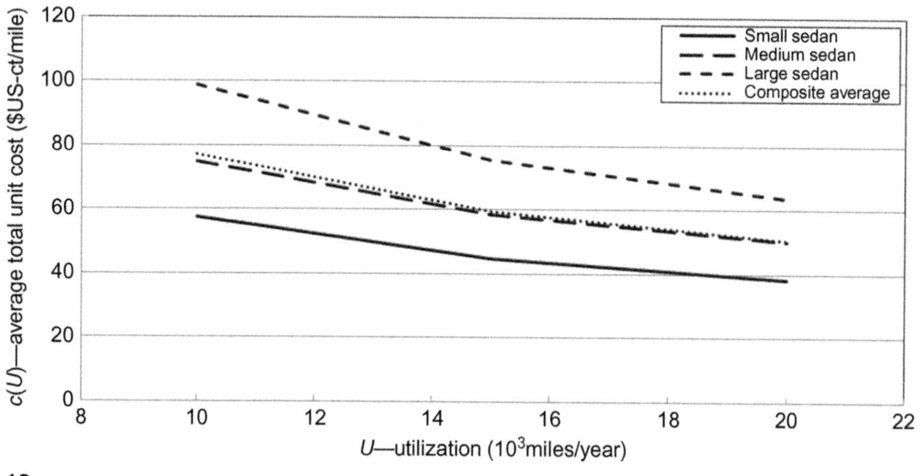

FIG. 10.13

Relationship between the average total unit costs and utilization of particular car models in the United States (AAA, 2013).

As can be seen, at all three models and their composite model the average unit costs have decreased more than proportionally with increasing of the annual utilization thus indicating existence of economies of scale. One of the reasons has been spreading the larger share of fixed costs over the greater number of miles during the given period of time (1 year). In addition, these unit costs have been higher for the larger than the smaller cars, as intuitively expected. The closeness of the average unit costs of a composite model and Medium sedan car indicates prevalence of the smaller and medium-sized cars in the considered mixture of different sized cars.

In addition, some estimates for the average unit costs of BEV (battery electric vehicles) have been provided just for the purpose of comparison with their ICE (internal combustion engine) diesel counterparts and shown in Fig. 10.14 (Crist, 2012). The life-cycle of both types of vehicles was assumed to be 15 years. The annual utilization of 4- and 5-door sedan car was assumed to be from 10 and 13×10^3 km/year. For the light commercial van, the annual utilization was assumed to be about 24×10^3 km/year.

In the given case, at the lower annual utilization the ICE diesel cars have shown to be more economical, ie, with the lower total unit operational costs, than their BEV counterparts, under given conditions. With increasing of their utilization, the BEVs could be more economical than their ICE diesel counterparts, again under given conditions. In the given example, the balance appears to be at the utilization of 20,000 km/year (Crist, 2012). This comparison between BEVs and their ICE diesel counterparts should be considered only as the case specific, ie, as an illustration in both absolute and

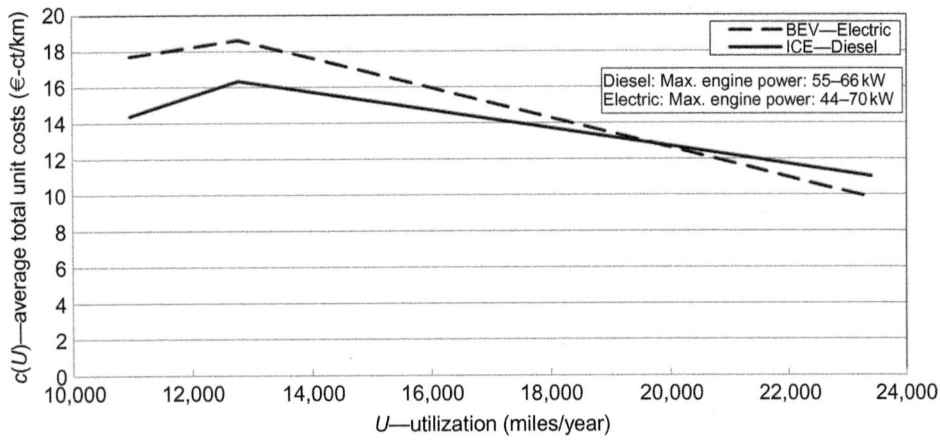

FIG. 10.14

Relationship between the average unit operational costs and the annual utilization of the diesel and electric cars as their counterparts in Europe (France) (Crist, 2012).

relative terms. More cases need to be considered before making judgement about the economic efficiency of both technologies, which will certainly be possible after increasing experience in using BEVs, of course after their more massive entering the car market worldwide.

10.6.2 URBAN MASS TRANSIT SYSTEMS

10.6.2.1 General

The operating costs of urban mass transit systems consist of the fixed and variable component. In general, the fixed component generally includes the expenses for acquiring, garaging, and capital maintenance of rolling stock/vehicles, which together with that for administration are fixed per period (year) of their life cycle. They do not change with changes of the intensity of their use. The variable component consists of the expenses for energy/fuel consumption, current repairs and maintenance, wages of staff, and information and ticketing systems. These expenses change with the intensity of use of rolling stock/vehicles, which in turn depends on the volumes of demand to be satisfied. When estimating these costs across particular urban mass transit systems operating under given scenario(s), it can be said that the average unit operating costs per unit of offered capacity generally decrease with increasing of the capacity supply under given conditions. Fig. 10.15 shows the example of the above-mentioned case for London (UK) (see Fig. 10.6).

As can be seen, the unit operating costs for both low and high value have decreased more than proportionally with increasing of the capacity supply of the given urban transit systems, thus indicating existence of economies of scale. The low and high values imply the "optimistic" and "pessimistic" estimate of costs, respectively, at the time of evaluation of the particular alternatives. These outcomes also suggest complexity in selecting the system: on the one hand road-bus based system has very low infrastructure average unit costs but much higher average unit operating costs. The case with heavy rail is quite opposite: very high average unit infrastructure and very low average unit operational costs.

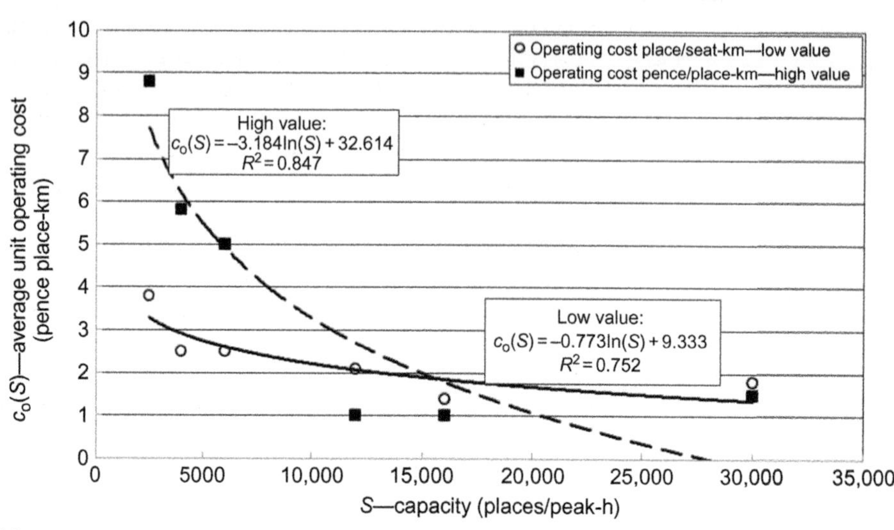

FIG. 10.15

Relationships between the unit operating cost and the space capacity of urban transit systems (TfL, 2000).

In addition, Fig. 10.16 shows the relationship between the unit fares and operating costs, and the annual output in terms of the volumes of pax-miles for the urban mass public systems—buses, streetcars, and light rails—operated in 82 urbanized areas in the United States (1 mile = 1.6010 km) (Harford, 2006).

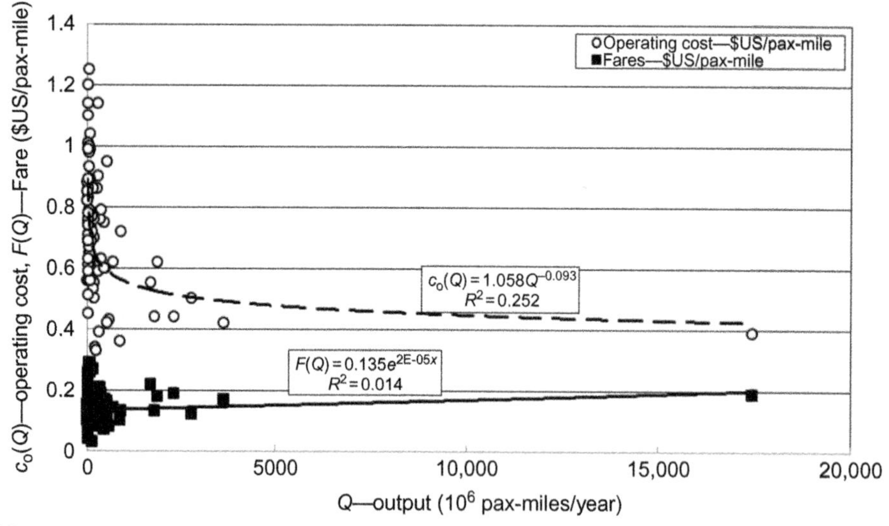

FIG. 10.16

Relationship between the average operating costs, fares, and the volumes of output of public mass transit systems in the US urbanized areas (Harford, 2006).

As can be seen, the annual volumes of output for all selected urban areas have been up to 5×10^6 pax miles/year, except the New York are where it has been about 17.5×10^6 pax miles/year. In addition, the average operating cost has been: $c_o(Q)=0.47$ \$US/pax mile and the average fare: $F(Q)=0.110$ \$US/pax mile. (The net income has been negative, ie, $0.110–0.47=-0.28$ \$US/pax mile; the profit margin has also been negative, ie $(0.110–0.47)/0.47=-1.474$.) In both cases, although represented by the rather weak regression equations, the economies of scale have generally existed, ie, the unit cost has decreased with increasing of the volumes of output across the selected urbanized areas. In addition, the unit fares have generally followed these costs. The obvious differences between the average unit operating costs and fares have been subsidized. The benefits have considered to be relieving congestion and its impacts in urban areas in terms of increased air pollution and vesting of time of individual car drivers and their accompanies by using the public transit systems (Harford, 2006). As well, Fig. 10.17 shows the relationship between the average fare and corresponding operating cost per passenger of the bus systems operating in 13 cities worldwide (Vancouver, Tokyo, Barcelona, Sidney, Singapore—2, Stockholm, Washington, District of Columbia, New York, London, Hong Kong, Dublin, Chicago) (References, 2011).

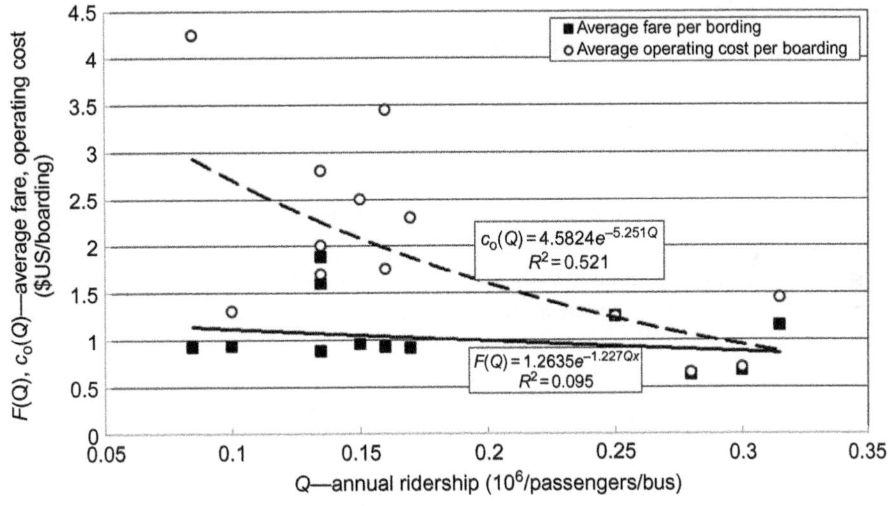

FIG. 10.17

Relationship between the average fares and costs, and the annual number of passengers per bus at the selected urban bus systems (References, 2011).

As can be seen, the average operating cost per boarding/passenger have decreased more with increasing of the number of passengers using the systems thus indicating existence of the economies of scale at the larger systems. At the same time the corresponding fares have been generally lower than the costs at most systems, which have implied subsidizing the corresponding systems. These subsidies have tended to be higher at the systems operating at the smaller than at the larger scale in terms of the number of passengers handled.

10.6.2.2 Streetcar (tramway)

The operating costs of the streetcar (tramway) include the expenses of acquiring, garaging, and capital maintenance of vehicles, which together with that for administration are fixed per period (year) of their life cycle, and those for energy consumption, current repairs and maintenance, wages of staff, information and ticketing system as dependent on the intensity of vehicles' use. This depends on the volumes of demand expected to be served by operations specified by the timetable during the day. In general, these costs are expressed in the absolute and relative terms. The first are the totals during given period of time. The latter the average unit costs per unit of the system output such as "€ or $US per passenger transported" or "€ or $US per passenger-mile or passenger-kilometer." Fig. 10.18A and B shows an example of these average unit costs, respectively, for the streetcar systems operating in seven US urban areas/cities—Little Rock, Memphis, New Orleans, Portland, Seattle, Tacoma, Tampa (Brown, 2013). As can be seen in both cases, these costs generally decrease more than proportionally with increasing of the volumes of the systems output, thus indicating existence of economies of scale. The corresponding information on the revenues from passenger fares and other commercial activities have not been available for comparison, but it may be also thinking that these systems belong to the wide range of the subsidized urban mass public systems in the United States, as shown in Fig. 10.18.

10.6.2.3 BRT and LRT

The economics of operation of BRT and LRT systems are presented by their operational costs and revenues.

10.6.2.4 Costs

The total operating costs of a BRT and LRT system generally include the expenses for acquiring, garaging, and capital maintenance of vehicles as amounts per period of time (year) of their life-cycle, current maintenance, fuel/energy consumption, wages of staff, information and ticketing system, and administration. Except those related to the life-cycle and administration, all other expenses depend on the intensity of vehicles' use, which in turn depends on the volumes of demand and ways of its satisfying during the day. Thus, the operating costs of either BRT or LRT system serving a given urban area during a given period of time can be estimated as the product of the average cost per unit of output and the volume of output carried out during the specified period of time (year) as follows (Janić, 2014):

$$C_o(V) = 365 \cdot V \cdot c_v(V) \tag{10.20}$$

where

V is the volume of the system's output per day (veh km, pax km, or pax per day)
$c_v(V)$ is the average unit cost per system's output (€ or $US/veh km, pax km, or pax)
365 is the number of days per year the systems are assumed to operate

The average unit cost $c_v(V)$ includes the annuities on bonds for acquiring the vehicles/buses, the vehicle/bus insurance costs, the wages of drivers and other supporting staff, the cost of vehicle/bus maintenance including the wages of personnel and spare parts, the energy/fuel cost, and the cost for using the infrastructure (taxes). In other way around, these include the cost of material, labor, and energy/fuel. The similar structure of this cost is at the LRT systems. Table 10.11 gives an example of the typical average costs for the selected US BRT and LRT systems.

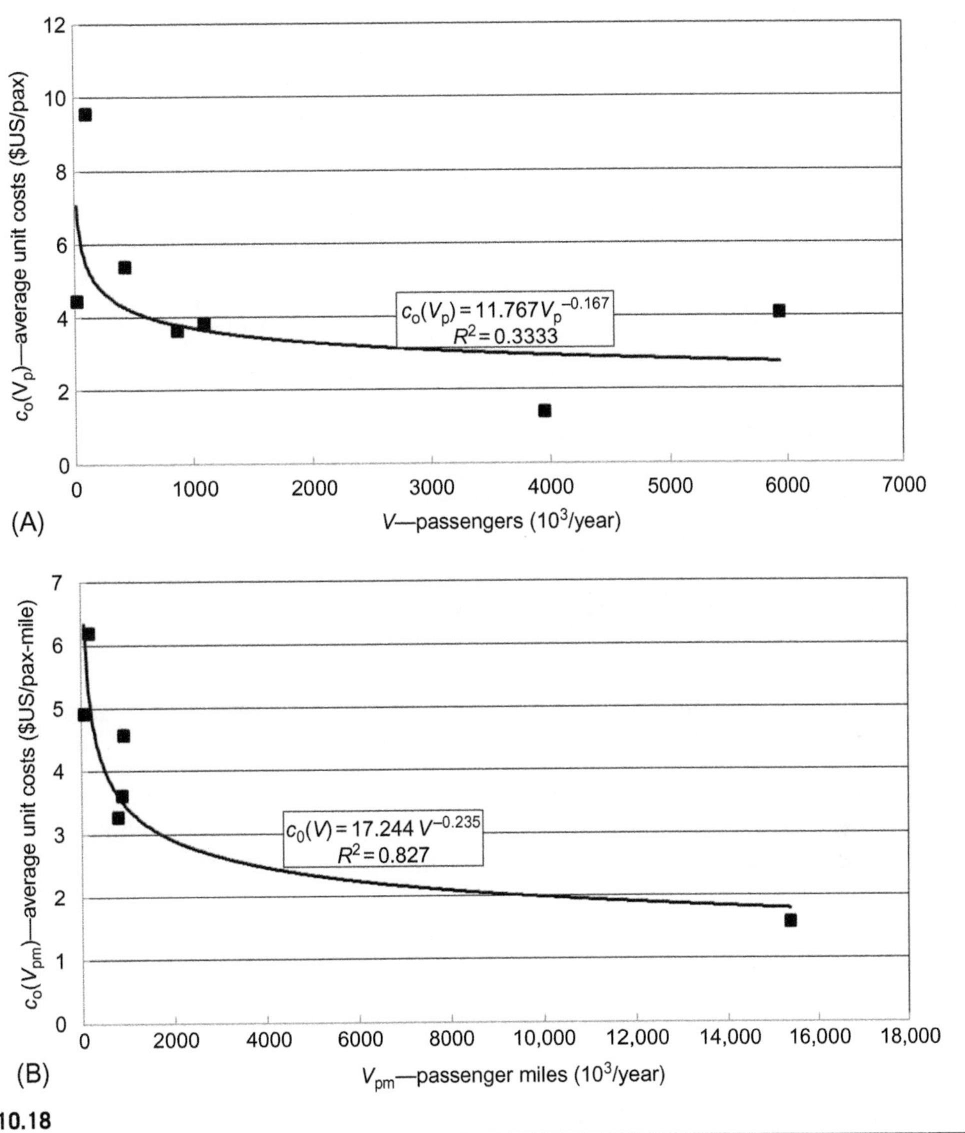

FIG. 10.18

Relationship between the average unit costs and the volume of output of the selected streetcar (tramway) systems in the United States (Brown, 2013): (A) costs per passenger and (B) costs per passenger-mile.

Table 10.11 shows that at the given systems, the average unit cost pert pax-mile and veh-km has been lower for BRT than for LRT. However, the cost per pax have been lower at LRT than at BRT systems. Fig. 10.19A and B shows an additional example of the unit costs of BRT and LRT systems serving six urban areas in the United States—Denver (Colorado), Huston (Texas), Minneapolis (Minnesota), Pittsburgh (Pennsylvania), Portland (Oregon), and San Diego (California) (HPW, 2010).

Table 10.11 Average Unit Costs of the Selected BRT and LRT Systems (Janić, 2014)		
System	**BRT**	**LRT**
Operating costs		
• $US/pax mile[a]	0.048	0.113
• $US/veh mile[b]	1.8106	5.531
• $US/pax[b]	3.20	2.57

[a]*Five BRT and fifteen LRT systems in the United States.*
[b]*Six BRT and LRT systems in the United States.*

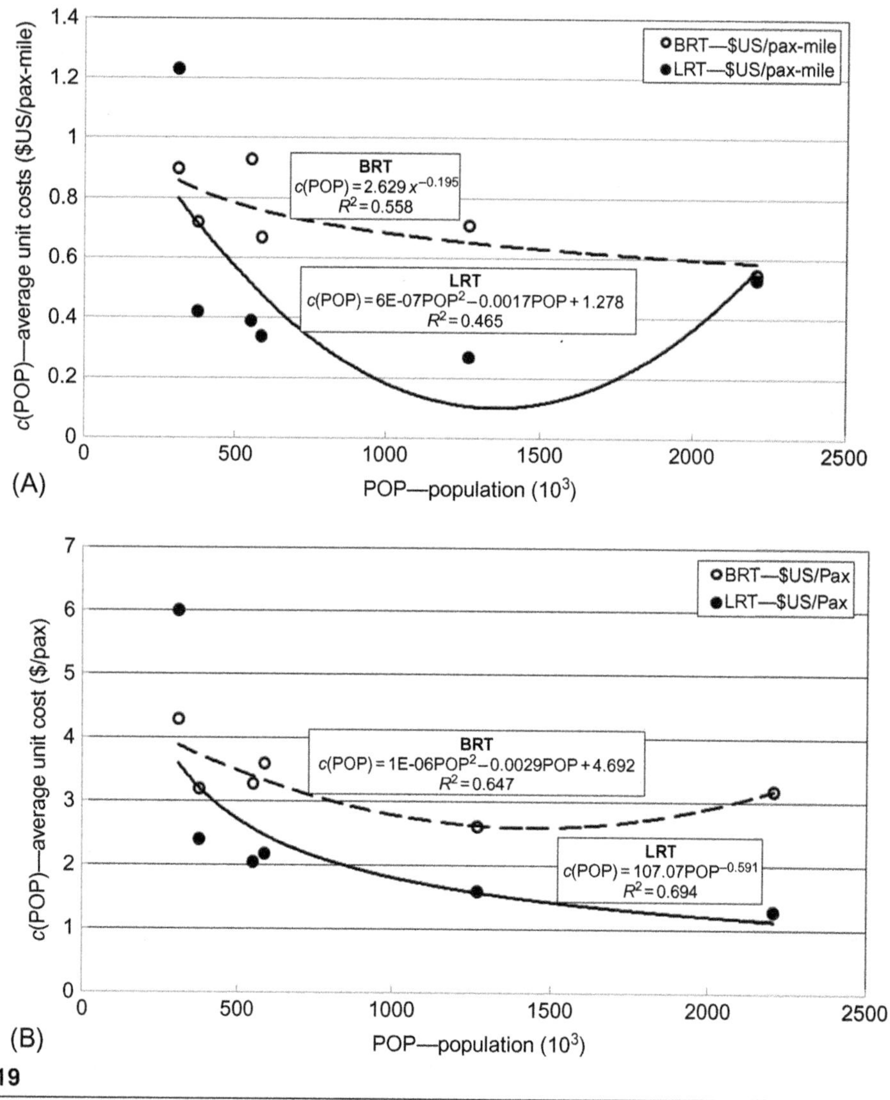

FIG. 10.19

Relationship between the average unit cost of BRT and LDR system in the selected urban areas of the United States (Denver (Colorado), Huston (Texas), Minneapolis (Minnesota), Pittsburgh (Pennsylvania), Portland (Oregon), San Diego (California)) (HPW, 2010): (A) average cost per pax-mile and (B) average cost per passenger.

As can be seen, the outcome is quite different than that in Table 10.11. The average unit cost per both types of output—pax/mile and pax, have been lower for the LRT than for BRT system. At LRT system, the unit cost per pax/mile has decreased and then increased more than proportionally with increasing of the population of urban area. At BRT system, this cost has continuously decreased more than proportionally (Fig. 10.19A). The unit cots per pax have had quite opposite behavior at both systems, LRT and BRT, respectively, with increasing of the population in urban area (Fig. 10.19B). If the size of population is assumed to reflect the potential demand, then this cost behavior can be considered as dependent on the potential or satisfied demand.

It should be pointed out that the BRT and LRT systems have usually placed in the urban areas and their parts with the substantive potential of passenger demand. Therefore, much better comparison could be made if both systems operating in the same urban (suburban) corridor(s) were evaluated. However, this cannot be carried out in practice since these corridors are usually served by one mode or the other. This suggest that the above-mentioned figures could be used only as an illustration of comparison of two systems, ie, just for such purpose, at the system level.

10.6.2.5 Revenues

The revenues from operating given BRT or LRT system are gained by collecting fares and different subsidies. For the period of 1 year, these revenues can be estimated as follows:

$$R = 365 \cdot q_p \cdot p + S_u \tag{10.21}$$

where

q_p is the daily number of users/passengers (users/passengers/day)
p is the average fare per user/passenger ($US/user/passenger)
S_u is the annual subsidies of a given BRT system

For example, the average fare of the above-mentioned 40 BRT systems operating round the world has been: 1.25 $US/pax. About 68% of these systems (27 of 40) have needed subsidies at an average level of 25–30%. Similarly, the LRT systems have also needed subsidies at a level of 20–25% (Janić, 2014). Such rather general need for subsidies has also been shown in Fig. 10.19 for the case of US urban transit systems.

10.6.2.6 Subway (metro)

The operating costs of subway (metro) system relate to the expenses of the similar type as those of the other above-mentioned urban mass transit systems. Their total per period of time can be estimated similarly as that of BRT and LRT (Eq. 10.22). The revenues can be estimated analogously (Eq. 10.23). In addition to the absolute and relative values of both operating costs and revenues, the question is always if the latter cover the former and to what extent. In it does not happen, the subsidies from the local authorities are needed. Some evidence about this can be obtained from the CoMET and Nova groups including 33 metros in 31 cities around the world in Europe, North and South America, and Asia (TfL, 2015; http://cometandnova.org/benchmarking/). Fig. 10.18 shows an example of the relative relationship between the operating costs and revenues for 27 of the above-mentioned 31 subways (metros) for the period of 2004–10 (Anderson, 2010). The operating costs have included that of transit operations providing services to users-passengers, maintenance, and administration. The revenues have consisted of that from the passenger fares and the nonfare commercial activities (Fig. 10.20).

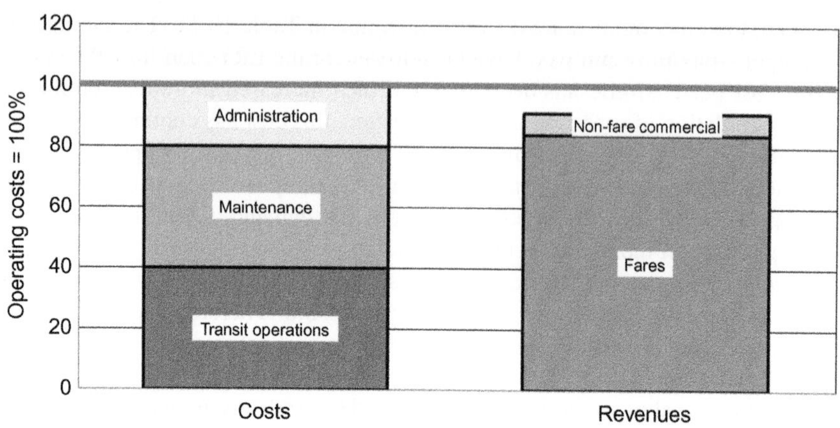

FIG. 10.20

Structure of the operating costs and revenues of an average among 27 CoMET and Nova subways (metros) (period: 2004–10) (Anderson, 2010).

As can be seen, at an average subway (metro), the structure of the total operating costs has been: 40% transit services, 40% maintenance, and 20% administration. The structure of total commercial revenues has been 84% from passenger fares and 7% from other nonfare commercial activities. The gap of 10% between the costs and revenues have existed and thus required public subsidies. In this context, a greater need for subsidizing the subways (metros) in Europe and North America that of those in Asia have also been noticed (Anderson, 2010; TfL, 2015). In addition, Fig. 10.21 shows the average operating costs and fares for eleven subways (metros) in ten cities around the world—Tokyo—2, Barcelona, Taipei, Singapore, Shanghai, Nexus Tyne &Wear, Hong Kong, New York, London, and Chicago.

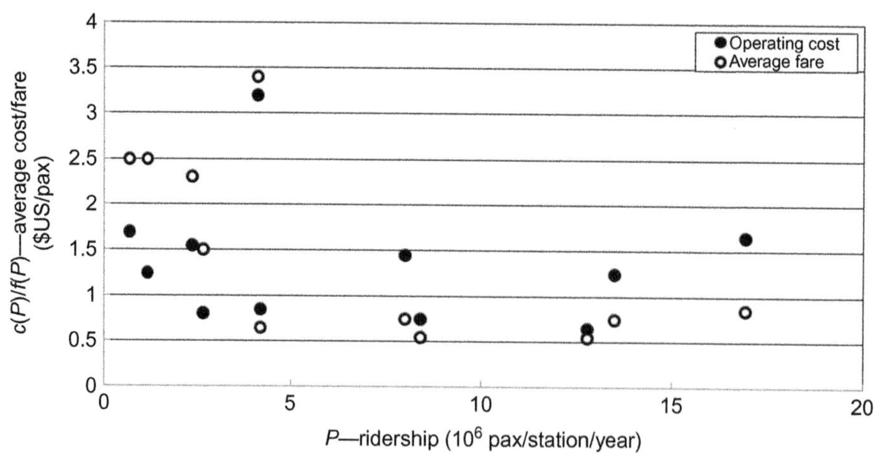

FIG. 10.21

Relationship between the average operating cost (excluding depreciation), fare, and the annual volume of passengers per station at the selected urban metro systems (References, 2011).

As can be seen, at the systems with the smaller annual number of passengers (up to 5 million per station), the average fares have mainly covered the corresponding operational costs. At those with the higher number of passengers per station, the fares have not covered the corresponding costs (References, 2011).

10.6.3 INTERURBAN MASS TRANSIT SYSTEMS

10.6.3.1 Road passenger transport

The operational costs of road interurban transport systems are illustrated for the bus systems. Two examples have been described. The first relates to estimating the costs of different bus technologies. The other relates to estimating the costs of operating the bus systems. Both cost calculations have been carried out respecting the given conditions.

In the first example, the total life-cycle costs of four different bus technologies have been estimated: diesel hybrid electric (DH), conventional diesel-fueled with ultra-low sulfur diesel (ULSD), biodiesel B20 (20%B100 Biodiesel and 80%ULSD), and compressed natural gas (CNG) (FTA, 2007). The conditions have been specifies as follows.

These buses of the length of 12 m and capacity of 40 seats have been assumed to operate at a national average speed of 20 km/h, and be used 60,000 km/year during the life-cycle of 12 years. The total costs have included the fixed/capital and the variable operation costs. The former have included the cost of acquiring the vehicle, the infrastructure costs (modification of refueling stations and depots), and the cost of emissions equipment. The latter included the cost of fuel, vehicle and facility maintenance, as the cost of replacing batteries. The costs of driver's wage(s) have not been taken into account. The total costs per unit of output ($US/seat km) calculated for the fleet of 100 vehicles are shown in Fig. 10.22.

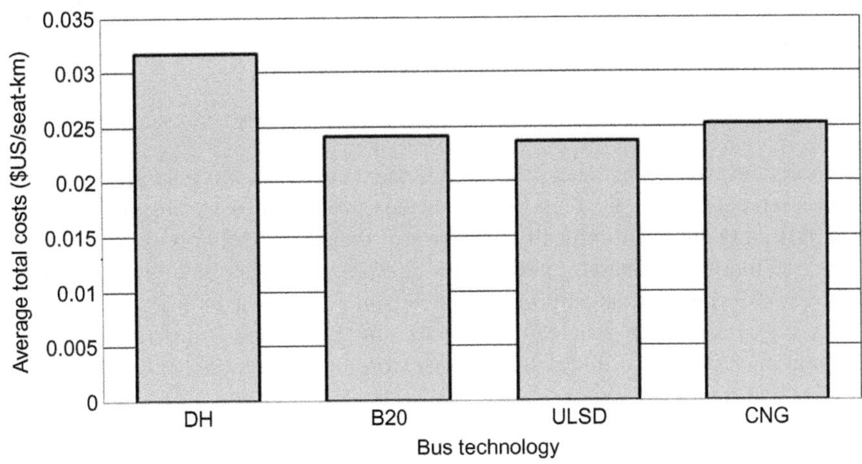

FIG. 10.22

The total life-cycle costs of the selected bus technologies (FTA, 2007).

As can be seen, the diesel hybrid electric buses (DH) have the highest and their conventional diesel-fueled with ULSD the lowest life-cycle unit costs calculated for the above-mentioned conditions. In addition, the bus capital cost has shared: 510% at DH, 47% at B20, 48% at ULSD, and 51% at CNG. The cost of fuel/energy has shared 25%, 41%, 42%, and 33%, respectively, in the above-mentioned total life cycle costs. These results indicate that the differences in technologies have influenced their costs, which has had to be fully respected in their overall evaluation and making choice.

In addition, Fig. 10.23 shows the relationship between average revenues and costs per passenger and the annual number of passengers transported by the intercity bus services operating in the United States.

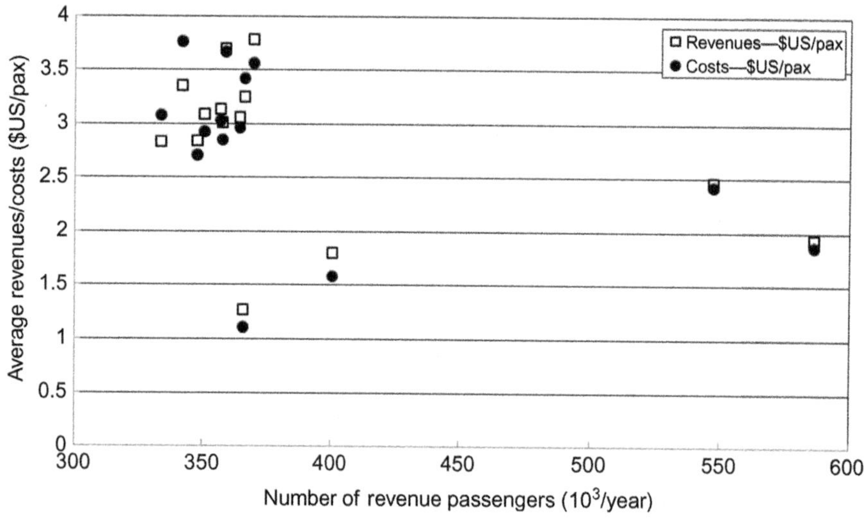

FIG. 10.23

Relationship between the annual number of passengers and the average revenues and costs for the intercity bus service in the United States (period: 1996–2003) (USDT, 2015).

As can be seen, the average revenues have been higher than their cost counterparts most of the time during the observed period with the tendency to decrease with increasing the annual number of passengers, thus indicating although indirectly existence of the economies of scale in the given context.

In addition, the following example again relates to estimating the costs of bus system operating in the United States. This time, for the purpose of comparison, the corresponding costs for the equivalent rail transport services have been provided. At present, the bus system consisting of about 4048 bus companies operating the fleet of about 16,000 buses provides the intercity services between 2766 stations in cities and towns throughout the country. About 20% of these companies offer the daily services. The rail system represented by the company AMTRAK rail currently operates 300 trains per day along 43 different routes connecting about 5000 cities and towns through United States. Fig. 10.24 shows the relationship between the average cost of providing a service per passenger and the length of route/trip distance for 20 selected routes served by both systems (both the capital/fixed variable/operating costs are included).

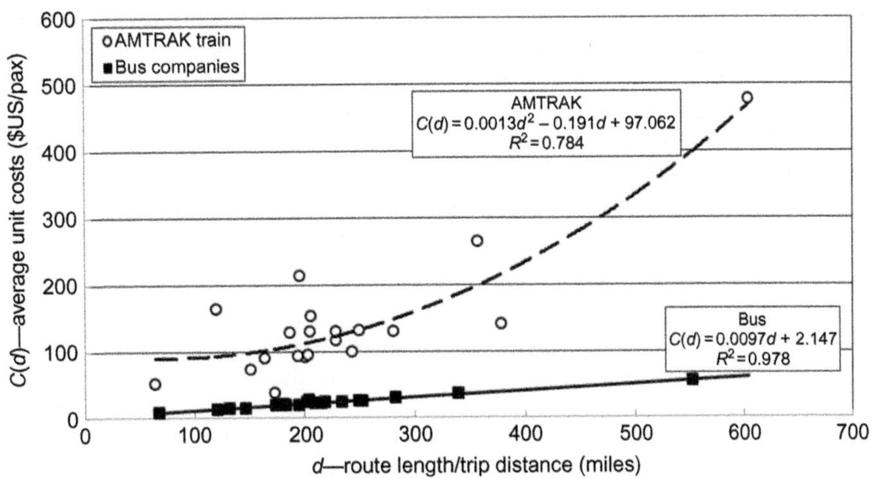

FIG. 10.24

Relationship between the average cost per passenger and the route length of the bus and the AMTRAK train on 20 selected routes in the United States (1 miles = 1.6010 km) (Lowell and Seamonds, 2013).

As can be seen, the average cost per passenger at the bus systems increases linearly with increasing of the route length/trip distance. The much higher corresponding average cost of the rail system increases more than proportionally with increasing of the route length/trip distance. In addition, the difference between the costs of two systems increases with increasing of the route length/trip distance. It is easy to show that if dividing the cost function $C(d)$ by the distance d, the average cost per unit of output of the bus system ($US/pax mile), will decrease more than proportionally with increasing of distance/route length, thus indicating existence of the economies of scale with respect to distance as shown in Fig. 10.25.

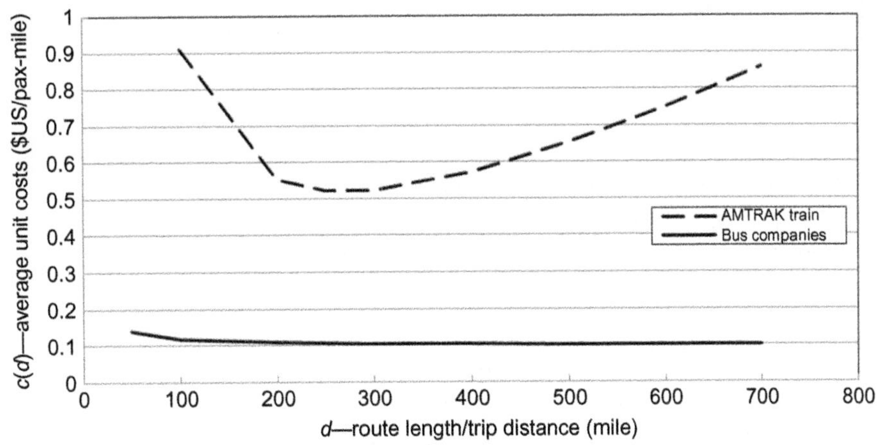

FIG. 10.25

Relationship between the average cost of output—passenger-mile and the route length/trip distance of the bus and the AMTRAK train on 20 selected routes in the United States (1 mile = 1.6010 km) (Lowell and Seamonds, 2013).

As can be seen, for the rail system, the economies of scale exist to the certain route length/trip distance (about 200 miles) and then disappear, ie, diseconomies of scale come up. In the given case, it should also be mentioned that the fares charged by the bus systems have mostly covered the above-mentioned costs because the considered services have been provided by the private profit-oriented bus companies. At AMTRAK rail system, the fares have been substantively lower than the corresponding costs, thus requiring their subsidizing by both the states and the federal authorities (Lowell and Seamonds, 2013).

10.6.3.2 Road freight transport

The operating costs of road freight transport are illustrated by analyzing the costs of heavy trucks. These costs consist of the fixed and variable component. The fixed component for a typical European 40 ton truck generally includes the expenses for acquiring the vehicles in terms of its annual depreciation and interest (12%), insurance (6%), wages of the staff (26%), and administration (18%). The variable component embraces expenses for the road tax (2%), tires (1%), and fuel (30%) (AEA, 2011). As expressed in per unit of output the averages in terms of monetary expenses ($US or €/vehicle km and $US or €/t km), the operating costs become the basis for setting up the prices of services. In general, these costs decrease with increasing of the utilization of these vehicles. Fig. 10.26 shows the example of the relationship between the total operating costs and annual utilization of trucks of different size, ie, GVW (LSB, 2006).

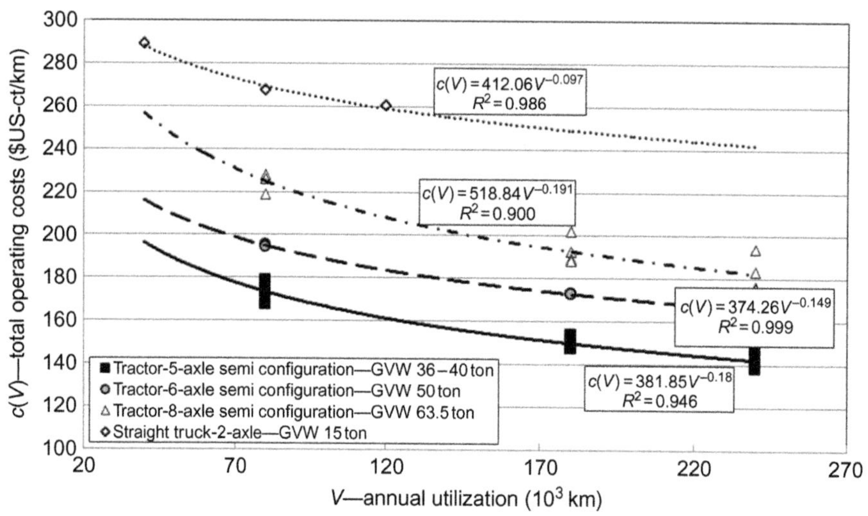

FIG. 10.26

An example of the relationship between the annual utilization and the total operating costs of the road trucks in Alberta (Canada) (year: 2005; profit margin: 2.5%) (LSB, 2006).

As can be seen, the operating costs decrease with increasing of the annual utilization more than proportionally at each category of trucks under given conditions. The costs of the lightest/smallest trucks have been the highest, also due to the lower annual utilization. At the heavier/larger trucks, these costs have increased with increasing of their GVWs for the entire range of possible annual utilization. In addition, Fig. 10.27 shows an example of the relationship between the total operating costs and one-way hauling distance in the road freight corridors in Canada (LSB, 2006).

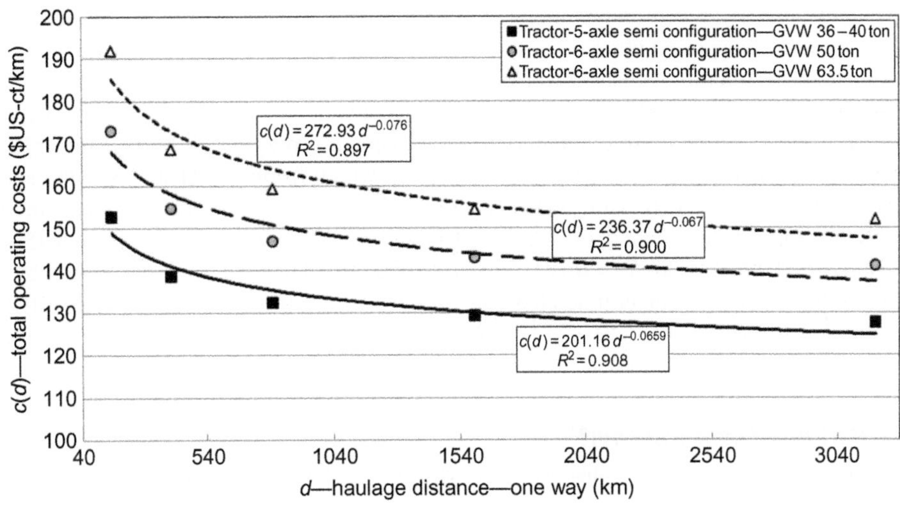

FIG. 10.27

An example of the relationship between the total cost and one-way hauling distance of the road trucks in Canada (year: 2005; profit margin: 2.5%) (LSB, 2006).

As can be seen, on the one hand these costs decease more than proportionally with increasing of the road haulage distance, while on the other increase with increasing of the truck size, ie, GVW. In both cases in Figs. 10.26 and 10.27, the economies of scale exists in the given context. In general, the cost of operating the road truck of a given size along a given haulage distance can be estimated by using the typical analytical (regression) relationship based on the empirical data as follows (Janić, 2007):

$$c(d) = a_0 \cdot d^{a1} \tag{10.22}$$

where

$c(d)$ is the average unit cost (€/vehicle km)
$a_{0,}$ a_1 are coefficients to be estimated
d is the distance (km)

By dividing the cost $c(d)$ in Eq. (10.22) by the vehicle payload capacity or the actual payload, the average cost per unit of payload per unit distance (€/t km) can be estimated as follows:

$$c(d, PL, \lambda) = \left(a_0 \cdot d^{a1}\right)/(\theta \cdot PL) \tag{10.23}$$

where

θ is the average load factor
P is the truck's payload capacity (ton, m^3)

The other symbols are analogous to those in Eq. (10.22).

The payload capacity and corresponding unit costs in Eq. (10.23) can also be expressed depending on the type of goods transported per a pallet and/or per unit of the available space on the vehicle (m^3).

An estimate of the operating costs of trucks by using Eq. (10.22) is carried out for the EU standard trucks of 40 ton GVW (Janić, 2007). The freight/goods is consolidated into TEU containers as the most common in Europe. Each of these containers has the average gross weight of 14.3 tons (12 tons of goods plus 2.3 tons of tare). The truck can carry 2 TEU, which implies that its payload capacity is: $PL=26$ ton. By applying the regression analysis based on Eq. (10.22) to the empirical data, the average costs for the standard truck loaded by $PL=2$ TEUs is estimated as follows: $c(d)=5.456\ d^{-0.277}$ (€/vehicle km) ($\lambda=0.85$; $N=26$; $R^2=0.781$; $25\leq d\leq 1600$ km). Then, dividing these costs by the corresponding payload capacity and assumed load factor, the average unit costs based on Eq. (10.23) have been obtained as: $c(d, 26, 0.85)=0.201d^{-0.277}$ (€/t km). Evidently, in both cases, economies of distance have existed.

10.6.3.3 Rail passenger transport

The economics of rail passenger transport is represented by the operational costs and revenues of the HSR.

Costs

The operational costs of HSR consist of the expenses for acquiring, operating and maintaining the rolling stock, wages of operating staff, selling services, and administration. The largest share is these costs are those of labor, material, and energy (UIC, 2005). Usually, these costs are expressed per unit of the HSR system output in terms of offered s km or realized p km (s km = seat kilometer; p km = passenger kilometer).

Analytically, the average operating cost of a given HSR system expressed per unit of its output (ie, p km) can be expressed as follows:

$$c_{\mathrm{av}}(T)=\frac{c_{\mathrm{o}}[T,f(\tau)]}{\rho(\tau)} \tag{10.24}$$

where

$f(\tau)$ is the service frequency of HSR services during the time (τ)
$\rho(\tau)$ is the average load factor per HSR service during the time (τ)
$c_{\mathrm{o}}[\tau, f(\tau)]$ is the average unit operating cost per HSR service at the service frequency $f(\tau)$ during the time (τ) (€ ct/s km)

In Eq. (10.24), the service frequency ($f(\tau)$) depends on the volume of demand during a given period of time (τ) and the "ultimate" or "practical" capacity of the given HS line/route and stations along it. The average unit operating cost during given period of time ($c_{\mathrm{o}}[\tau, f(\tau)]$) depends on the HS train's seat capacity, service frequency, and prices of inputs such as the train's acquisition cost,[5] labor, and (electrical) energy. In addition, the average unit maintenance cost of rolling stock is influenced by its utilization, ie, service frequency, and the prices and quantity of other necessary inputs such as material, energy, and labor. Consequently, the total average unit cost is mostly influenced by the rate of utilization of the capacity of infrastructure and rolling stock, which in turn depends on the volume of demand. Under conditions of complete absence of demand, this cost exclusively depends on the cost of capacity components.

The average unit operating cost of HSR services has differed throughout the European counties and the rest of the world. This cost has been mainly influenced by the local pricing of the particular above-mentioned inputs and type of HS trains. Some estimates indicate that this average unit cost for 12 types of HS trains operating in the corresponding European countries has been 0.14626€/seat km. The cost of maintenance of rolling stock has shared about 8.5% in this total. Under an assumption that the average load factor was 70%, the total average operating costs of HSR services throughout Europe would be 0.1105€/p km (Pourreza, 2011; de Rus, 2010). However, an example indicating the dependence of this cost on the train operating speed has shown much lower figures under given traffic scenario (Garcia, 2010). The analytical has been as follows: $c_{av}(v) = 5.527381 - 0.0192545 \cdot v + 0.0000427 \cdot v^2$ (€ ct/seat km), where (v) is the average train speed (km/h). Fig. 10.28 shows the relationship for the given train operating scenario.

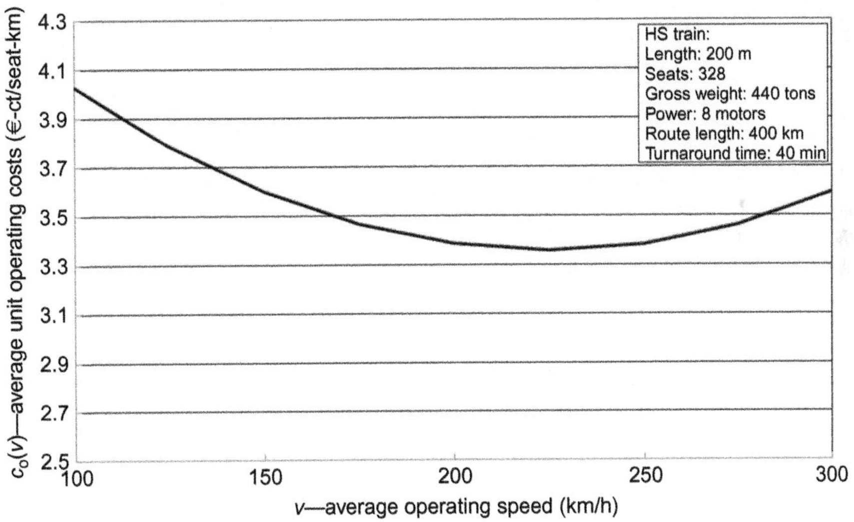

FIG. 10.28

Example of the relationship between the average direct operating cost of an HS train and its speed under given scenario (Garcia, 2010).

[5]The average construction cost for European-build HS train(s) is about 500,000 €/seat. For comparison, the corresponding cost of commercial aircraft (A318, A3110, B737-800) is about 300–3,500,000€/seat. In both cases, the amortization period is about 30–40 years (UIC, 2011).

As can be seen, the cost first decreases and then increases more than proportionally with increasing of the average speed. This can be explained as follows: in the former case, the cost components—train ownership, maintenance, and operating personnel—decrease more than the others increase. In the latter case, the cost components—energy consumption due to the higher speed and the infrastructure use—increase more than the others decrease.

Revenues

The HSR system has obtained revenues from different sources such as mainly transportation based on charging users/passengers, merchandise, and others (JR, 2014). In particular, the fares for users/passengers are set up to cover the total operating cost if subsidies are not provided as an element for enabling stronger competition with other transport modes such as conventional rail and particularly air transport. Fig. 10.29 shows some examples of the relationship between the fares of HSR in Europe, China, and Japan in dependence on the travel distance.

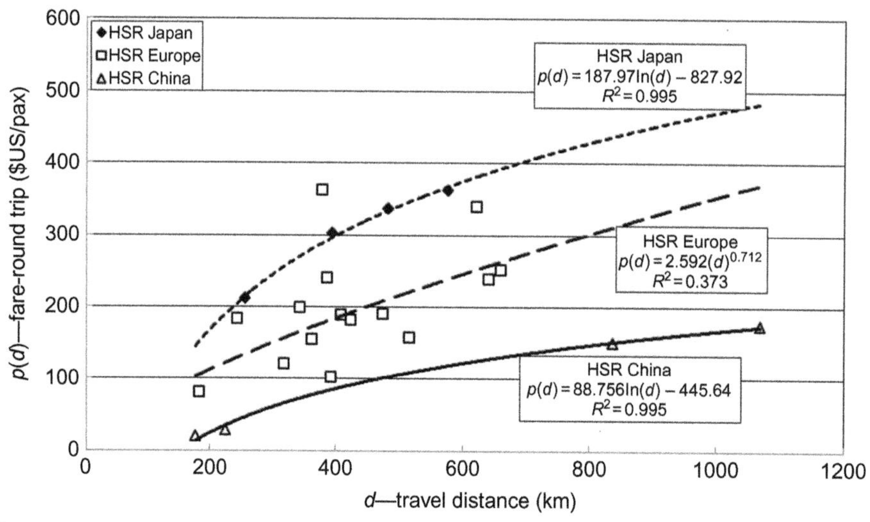

FIG. 10.29

Relationship between travel distance and the price for round trip charged 1 week in advance by the HSR and APT (air passenger transport) (Feigenbaum, 2013).

As can be seen, the fares of HSR services are the most dispersed in Europe and much less in China and Japan. In all three regions, they generally increase with increasing of the travel distance at decreasing rate. In addition, the fares in Japan are the highest and those in China the lowest. As well, the fares of HSR services have generally decreased more than proportionally with increasing of the volumes of passenger demand as shown in Fig. 10.30. If these fares are supposed to cover the operating cost, they indicate existence of economies of scale in the given context.

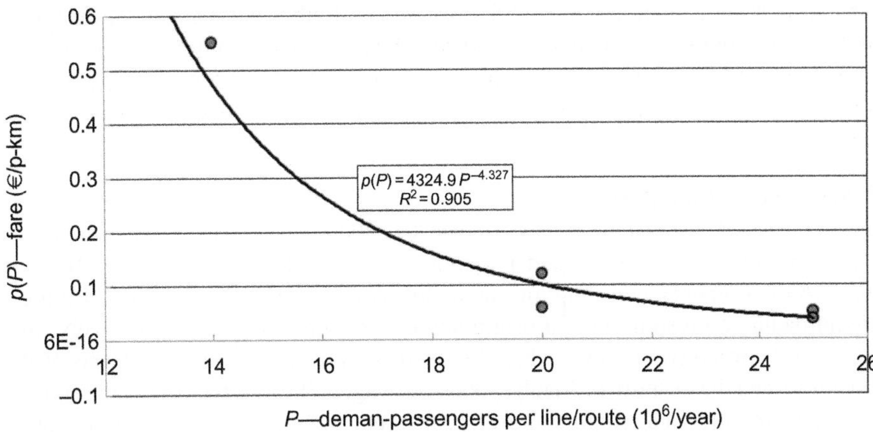

FIG. 10.30

Relationship between the unit fare of an HSR service and the volumes of passenger demand for five Chinese lines/routes (Wu, 2013).

Balancing revenues and costs

Similarly as the others, the HSR systems intend to operate in the profitable way, ie, to cover their total costs by the total revenues. Fig. 10.31 shows an example of profitability of one company operating both HSR and conventional rail services.

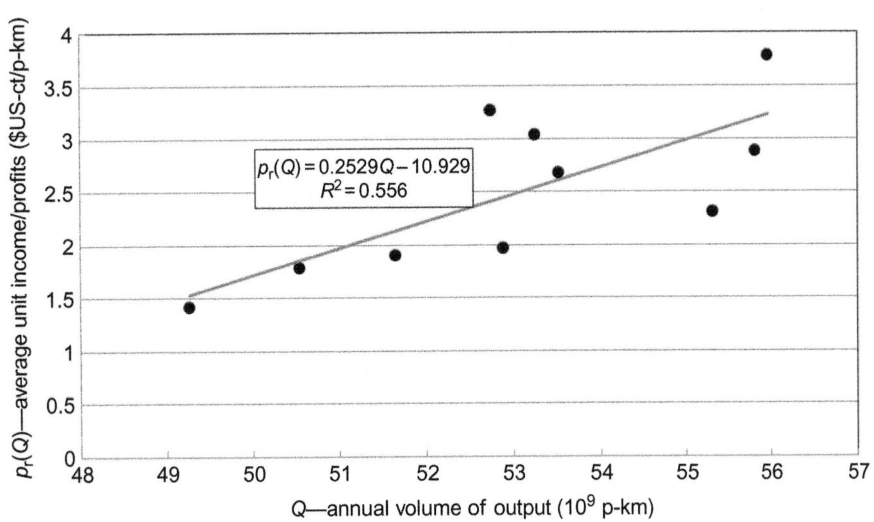

FIG. 10.31

Relationship between the annual volumes of output and the unit profits—Central Japan Company (period: 2004–13) (JR, 2014).

As can be seen, despite relatively high variations, the profitability has generally increased with increasing of the volume of the company's output during the given period of time. This case could be used as an example how the HSR system can be profitable in the medium- to long-term period of time—by careful balancing the revenues and costs while at the same time increasing the scale of operations.

10.6.3.4 Rail freight transport

The operational costs of rail freight transport consist of the expenditures on investment and capital maintenance of rolling stock-wagons and locomotive(s), use of the railway infrastructure, ie, the infrastructure charges, energy consumption, labor for assembling/decoupling and driving the train, and loading/unloading a given train(s) at origin and destination terminals, respectively. In this case, a train dispatched between two terminals has the fixed composition and load in terms of the number of wagons per train and the quantity of freight/goods per wagon. If this is the container train, the wagons are flatcars and the freight/goods shipments are TEUs (3 per car, each of the weight of about 14.3 ton), which fits the flatcar's capacity in terms of the maximum volume (3 TEU/car) and payload weight (50 ton). In the example presented here, the container train consists of 26 flatcars. The average load factor of such train is usually $\theta = 0.75$, which is the characteristic value for shuttle and direct trains operated in the important European long-distance corridors (Janić, 2007). Given the average weight of a TEU (14.3 ton), the number of TEUs per flatcar (3) and the number of flatcars per train (26), the average payload in estimated to be: $0.75 \times 14.3 \times 3 \times 26 \approx 837$ ton/train. The gross weight of such train is from 1100 to 1200 ton. For such trains, the total operating costs depending on the haulage distance are estimated means by the regression analysis as follows: $c(d) = 110.728d^{0.10805}$ (€/train) ($N = 52$; $R^2 = 0.848$; $100 \leq d \leq 2400$ km). This relationship shows an increase in the train operational costs at a slightly decreasing rate with increasing of haulage distance. Then, the average unit costs can be estimated as follows: $c_{av}(Q, d) = c(d)/(Q \cdot d) = 19.728d^{0.9805}/(837 \cdot d)$ (Janić, 2007). Fig. 10.32 shows the relationship between these average unit costs and the haulage distance.

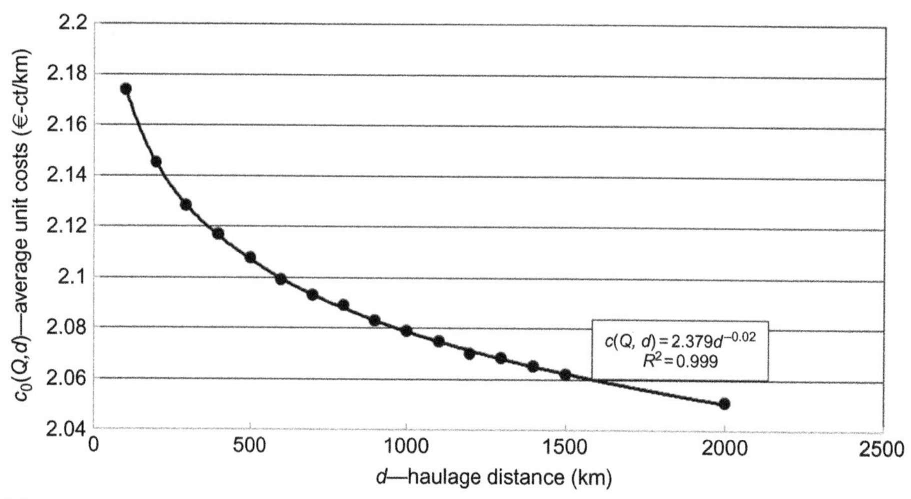

FIG. 10.32

Relationship between the average unit costs and haulage distance of the given container train (Janić, 2007).

As can be seen, these costs have decreased more than proportionally with increasing of the haulage distance, thus indicating existence of economies of distance in the given case.

10.6.4 INLAND WATERWAYS CARGO TRANSPORT

The freight/cargo and recreational vessels are used for transporting freight/cargo and passengers, respectively, along inland waterways (rivers and channels). Because the share of the freight/cargo operations and related outputs have generally been much higher than that of passengers (100–105% vs. 5–10%), only the operational costs of the former (freight/cargo) have been described. These costs of corresponding rolling stock/vehicles—towboat, motor vessels/self-propelled barges, and nonpropelled barges—operating along inland waterways are generally different. For towboats and motor vessels/self-propelled barges they generally include expenses for their replacement allocated to each year of their life-cycle, maintenance, insurance, administration, port taxes, fuel, crew, and others. These costs generally increase with increasing of the power (*HP*—horse power) of a towboat or motor vessel. For the nonpropelled barges the expenses relate to their acquiring and/or replacement again allocated to each year of their life cycle, operations including maintenance and repair, supplies, insurance, and other. In addition, the operating costs are generally different for different types of inland waterways rolling stock. For example, in Europe, these are the motor vessels/barges for transporting different solid bulk and containerized freight/cargo, and the motor vessels-tankers for transporting oil, liquids, and different kinds of chemicals. In the United States, these are towboats of different power (*HP*) and nonpropelled barges such as: open and covered hopper, deck and self-unloading, tank, and chemical tank barges. In this case, the common freight/cargo of hopper barges is coal, grain, and others. The cargo of deck and self-unloading barges is sand/gravel, cement, and stone. The cargo of thank barges includes different kinds of liquids—oil, asphalt, ammoniac, liquefied petroleum gas, chlorine, etc. The cargo of chemical tank barges includes different chemical in gas and liquid state such as urea, benzene, aqua ammoniac, alcohols, acids, and caustic soda (EGM, 2004).

The operating costs are calculated separately for towboats and nonpropelled barges, but also for a convoy consisting of a single towboat and several barges. These costs can be expressed as the averages per day, per ton of payload, and/or per unit of output—t km. In the latest case, for a convoy consisting of a towboat and several barges operating on a given route under given conditions the average unit operating costs can be estimated as follows:

$$c(HP, PL, d, n) = \frac{[c_{tb}(PW) + n \cdot c_b(PL)]}{\theta \cdot n \cdot PL \cdot d}\left(\frac{d}{24 \cdot v}\right) \tag{10.25}$$

where

$c_{tb}(PW)$ is the daily average unit costs of towboat of the power (PW) (\$US/day)
$c_b(PL)$ is the daily average unit costs of a barge of the payload capacity (PL) (\$US/day)
n is the number of barges in the given convoy
θ is the average load factor of a given convoy of (n) barges
d is the route length (km)
v is the average speed of a given convoy (km/h)
24 is the number of hour per day

Fig. 10.33 shows an example of estimating the average unit costs ($c_{tb}(HP)$) for Mississippi River System Towboats (United States) (CECW-CP, 2004).

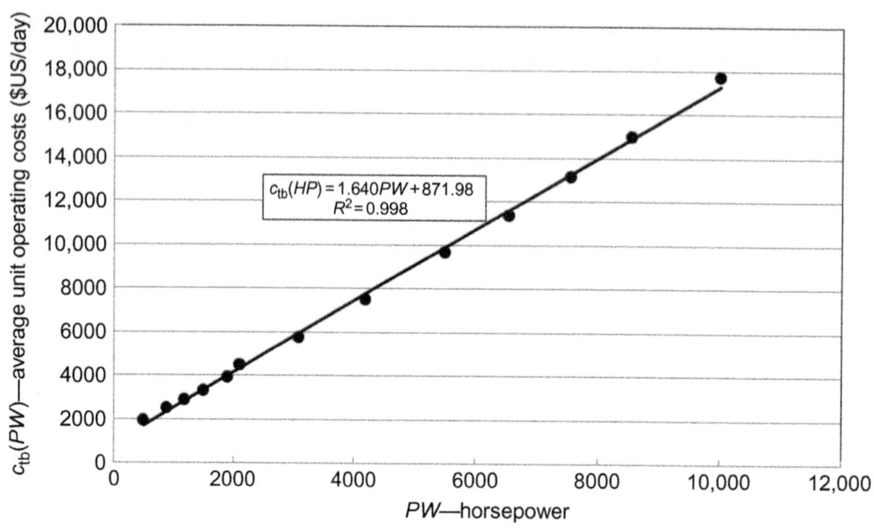

FIG. 10.33

Relationship between the daily average unit operating costs and power of the Mississippi River System Towboats (United States) ($US—2004) (CECW-CP, 2004).

As can be seen, these unit costs linearly increase with increasing of the power (*HP*) of towboats used continuously at the rate of 80% of their its maximum value. In this total daily costs, the administrative costs share about 12%. In addition, Fig. 10.34 shows an example of estimation of the average unit costs ($c_b(PL)$) for Mississippi River System (nonself-propelled) Barges (United States) (CECW-CP, 2004).

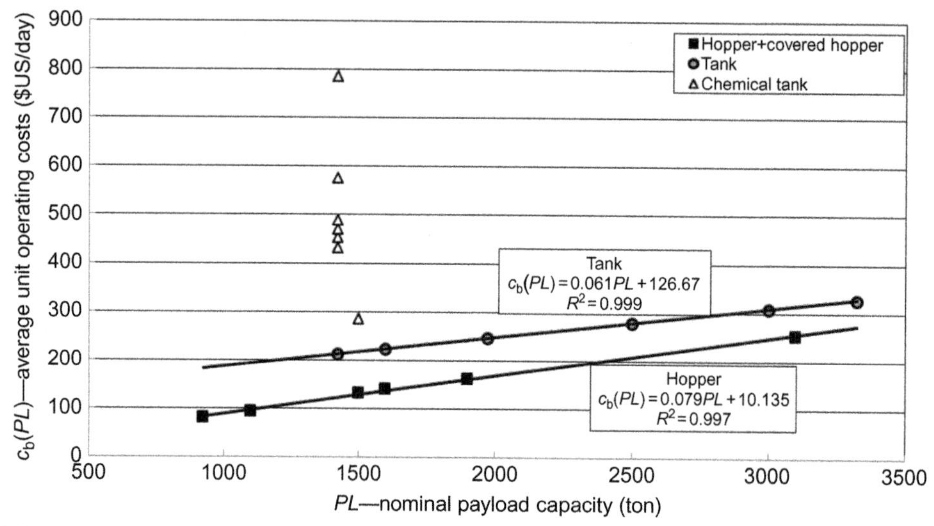

FIG. 10.34

Relationship between the daily average unit operating costs and the nominal payload capacity of the Mississippi River System Barges (United States) ($US—2004) (EGM, 2004).

As can be seen, the average daily unit operating costs for open and covered hopper barges were lower than that at their tank counterparts. At both types of barges, they increased linearly with increasing of the barge nominal payload capacity. However, at the chemical tank barges these costs significantly differed for barges of the same payload capacity, thus implicitly implied different complexity of dealing with their freight/cargo both on board and at the ports. In this total daily costs, the administrative costs shared from 15% to 30% at the hopper barges and decreased with increasing of the barge payload capacity. At the tank and chemical tank barges these costs have shared from 30% to 50%.

By using the relationships in Figs. 10.33 and 10.34 for the hopper barges and assuming that the average load factor of a convoy of (n) barges is: $\theta = 0.85$, Eq. (10.25) can be transformed as follows:

$$c(HP, PL, d, n) = \frac{[1.64PW + 871.98] + n \cdot [0.079PL + 10.135]}{0.85 \cdot n \cdot PL \cdot d} \left(\frac{d}{24 \cdot v}\right) \tag{10.26}$$

where all symbols area as in Eq. (10.25).

Then, the average costs per unit of output in terms of \$US/ton and \$US/t km can be calculated for the following scenario. One towboat of the power of: $PW = 3000$ HP pushes a convoy of: $n = 5$ to 20 open hopper barges, each with the payload capacity of $PL = 1500$ tons ($L \times W \times D = 1105 \times 35 \times 12$ ft or $510.5 \times 10.7 \times 3.7$ m; L—length; W—width; D—hull depth). The average load factor of the convoy of each size is assumed to be: $\theta = 0.85$. The convoy operating speed is adopted to be: $v = 6$–7 km/h along the Mississippi River (Gonzales et al., 2013). Respecting the inflation of \$US of 26% during the period 2004–15, the calculated costs are increased correspondingly. Fig. 10.35 shows the relationship of the average unit costs (\$US/ton) on the selected distances along the Mississippi river.

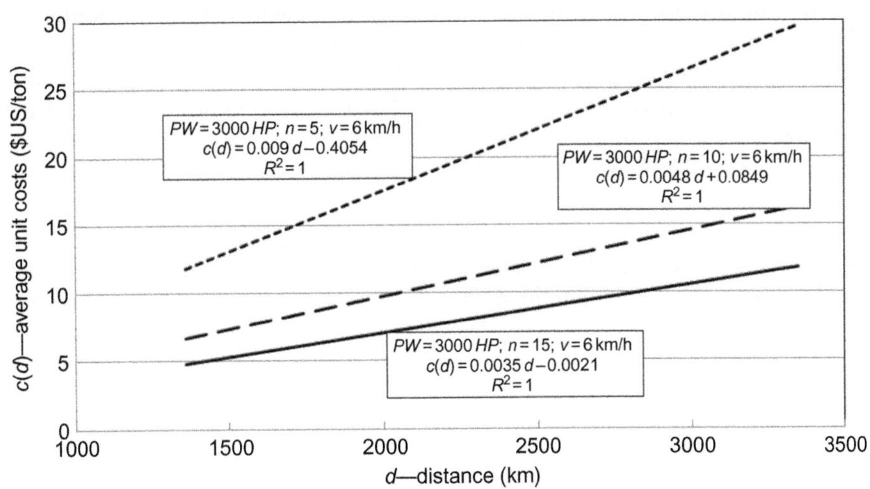

FIG. 10.35

Relationship between the average unit costs and distance along the Mississippi river (United States).

As can be seen, for a given composition of the convoy, these costs increase linearly with increasing of distances because the longer time is needed to cover them while sailing at the constant speed. In addition, for a given distance, these costs decrease with increasing of the size of convoy. Fig. 10.36 shows the relationship between the average unit costs (\$US ct/t km), the size of a convoy and its operating speed.

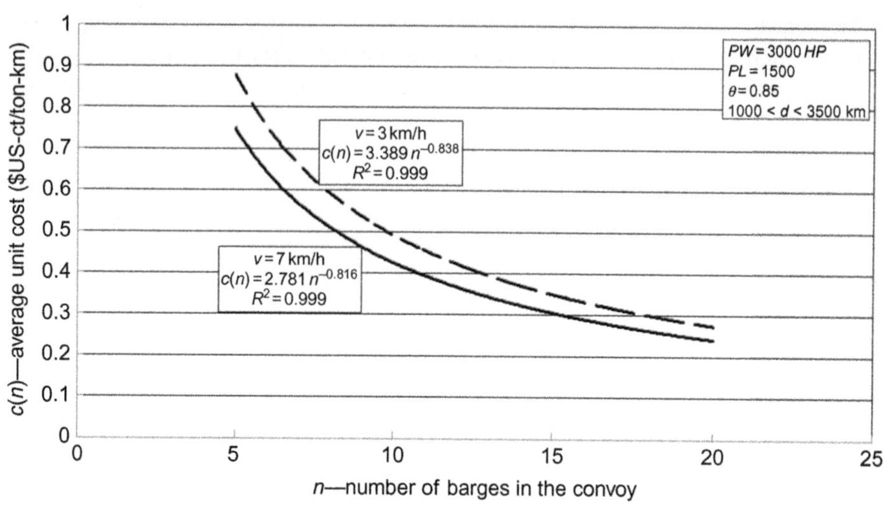

FIG. 10.36

Relationship between the average unit costs ($US ct/t km), and the size of convoy and its operating speed along the Mississippi river (United States).

As can be seen, the average unit costs decrease more than proportionally with increasing of the convoy size, thus indicating existence of economies of scale. This suggests that the larger convoys are always more profitable to operate, of course if there is sufficient demand. In addition, these costs decrease with increasing of the operating speed, which reduces time, ie, the number of days and related costs of spending on the given distance/route. In general, while sailing on the rivers, the upstream speed is lower than downstream speed, thus making the difference in operating time and related costs in different directions of a given route. In addition, Fig. 10.37 shows an example of the structure of average operating costs of different types of the motor vessels/barges in The Netherlands (Europe) (CE Delft, 2004).

As can be seen, these costs are categorized as that of labor, capital, fuel, and other. At all kind of vessels/barges except towboats the shares of labor and other costs have generally decreased and that of the capital and fuel costs increased with increasing of their size. The share of labor and capital costs have been absolutely the lowest at towboats. However they have had the highest shares of fuel and other costs.

10.6.5 MARITIME CARGO TRANSPORT

10.6.5.1 Ports

Costs

The port operational costs generally consist of the expenses for current maintenance of the port's infrastructure (quays/berths, terminals), material, energy, and labor providing handling ships and their payload/cargo under given conditions. These costs are covered by the revenues obtained from charging ships for using its infrastructure. The terminals in the ports specialized for handling specific type of

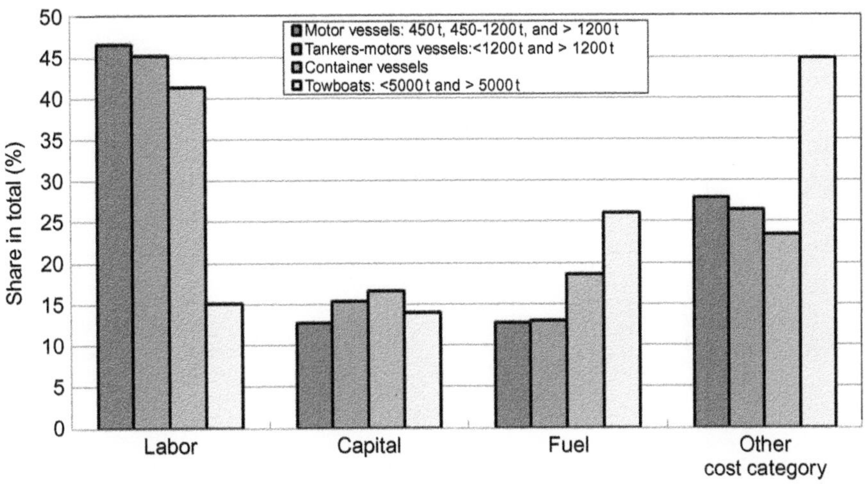

FIG. 10.37

Structure of the average operating costs of rolling stock—different types of vessels/barges—case of The Netherlands (CE Delft, 2004).

freight/cargo such as dry and liquid bulk, containers, Ro-Ro,[6] and others can be operated by the port authorities too. However, in most cases, these terminals are the private entities separately charging the shipping lines for providing services after their ships have been berthed. The terminal's charges are usually the matter of covering its costs, commercial negotiation, and the expected volumes of payload/cargo to be handled, usually over the multiyear period. Specifically, in the case of handling containers at the port's container terminals, the charges are called THCs (terminal handling charge(s)). These are actually collected from the freight/goods shippers and receivers by the shipping lines in order to cover their costs paid to the container terminals for handling—loading, unloading, storing—their containers including all other related costs, both at the origin and destination port(s) (EC, 2010).

In general, the total operating costs of ports are due to carrying out the activities such as port administration, handling cargo, pilotage, towage, mooring, and others. Fig. 10.38A and B shows an example of the characteristics of operational costs of the European ports estimated from the port tariffs (PwC, Pantela, 2013).

Specifically, as Fig. 10.38A shows the cost of handling ships and their payload/cargo dominates the structure of the port operational costs. At the same time, as shown on Fig. 10.35B, the average unit costs of handling containers is the highest and that of Ro-Ro payload/cargo the lowest. For example, if the container weights 15 ton, and if its average unit handling cost is 8.45 €/ton, the total cost of its handling will be: 126.75 €. (It should be mentioned that the actual rates of handling containers differ for 20 and 40 ft containers.) The main reasons for high differences between the average unit handling costs between containers and Ro-Ro shipments is that the former require unloading/loading, storage and

[6]This is wheeled payload/cargo, such as automobiles, trucks, semitrailer trucks, trailers, and railroad cars. The ships carrying them are called Ro-Ro ships.

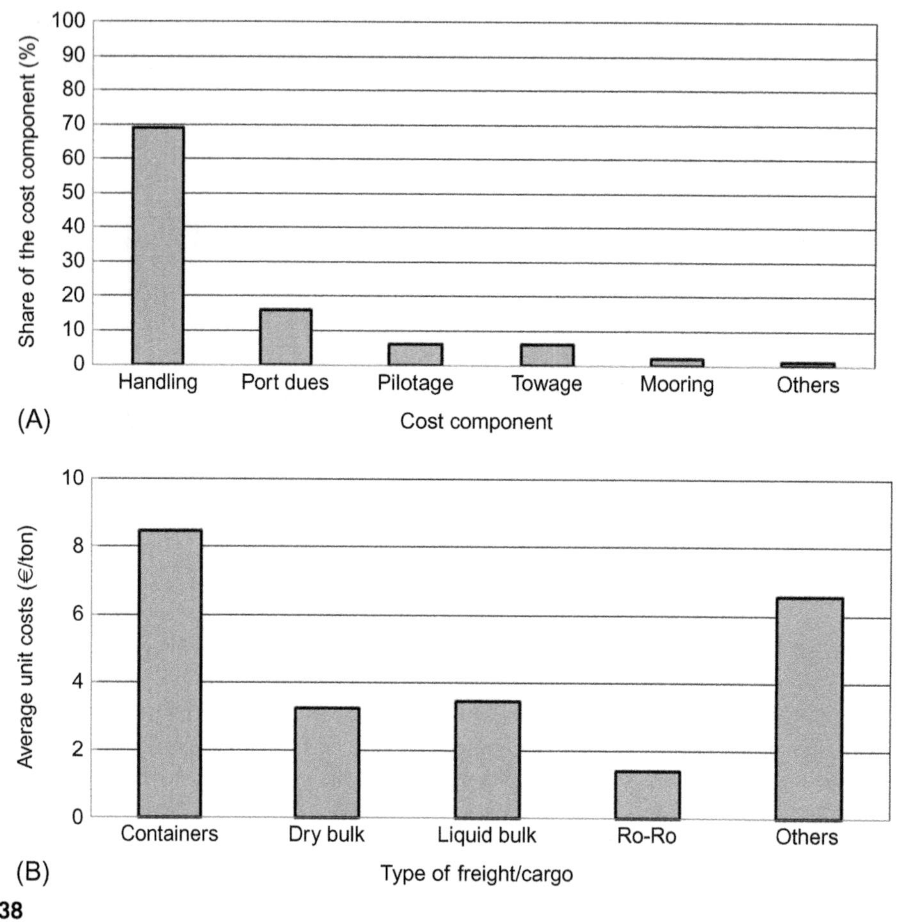

FIG. 10.38

An example of the port operating costs—the European ports (PwC, Pantela, 2013): (A) cost structure per activity/operation and (B) average unit costs per type of freight/cargo.

ground-based intermodal transhipment by using the container terminals' facilities and equipment, and the latter can be loaded/unloaded on the ships on "their own wheels."

Revenues

The above-mentioned costs are covered by charges as the sources of revenues for the port authorities. In general, the port charges can include the fairway/lighthouse dues, pilot fees, tonnage dues, towage, mooring/unmooring, waste disposal, and agency fees. These dues are generally based on the ship's GT (gross tonnage) based on its volume (ISL and amrie, 2006). Alternatively, some ports set up their charges based on the ship's GRT (gross registered tonnage) and NRT (net registered tonnage). In all cases, the weight of 1 ton corresponds to the volume of 2.83 m^3.

Specifically, the ship's GT is a nondimensional index related to the ship's overall internal volume. It can be calculated as follows:

$$GT = K \cdot V \tag{10.27}$$

where

V the ship's total volume embracing cargo, enclosed, and excluded spaces (length.beam.draught) (m^3))
K is a multiplier (coefficient) calculated as: $K = 0.2 + 0.02 \cdot \log_{10}(V)$; as such it can range from 0.22 to 0.32

The equation for the multiplier (K) indicates that it varies with the volume of ship, implying its lower values for the smaller and higher values for the larger ships.

Specifically, the fairway/lighthouse dues are charged to ships for using the navigational facilities and equipment while entering and leaving the ports. At many ports they are based on the ship's GT. The pilot fees are charged for the salaries of pilots guiding the ships during entering and leaving the port(s). They are calculated based on the ship's GT and the length of entering/leaving fairway/route. The tonnage dues are the basic charges to ships for their handling at the port(s). Based on the ship's GT, they are levied to each ship's entering and leaving the port including the time of its loading/unloading at the berth(s). The towage dues area also set up based on the ship's GT. The mooring/unmooring operation(s) gets an increasing difficulty with increasing of the ship size. Therefore the corresponding dues are based on the Ship's GT. As far as dues for the waste disposal are concerned, they are based on the volumes and type (liquid, solid) of the disposed waste, but most ports charge this based on the ship size. In additional, the overall port charges can include the environmental charges already in place at some ports. An example of the structure of port charges including the mentioned dues is given in Table 10.12.

Table 10.12 An Example of the Structure of Port Charges—Case of European Ports (ISL and amrie, 2006)

Type of Dues	Share (%)
Port	5–15
Pilotage, Towage, Berthing	2–5
Cargo handling	70–100
Agent fees	3–6

As can be seen, the cargo handling dues has dominated in the structure of the port charges, which coincides with the structure of the port costs shown in Fig. 10.38A. (One should bear in mind that this coincidence is not by chance: the costs on Fig. 10.38B are synthesized from the corresponding tariffs for particular operations/activities.)

In general, the port and terminal charges have been the matter of negotiations between ports and their users—shipping lines. The main criteria has been covering the operational costs, the volumes of payload/cargo to be handled from the shipping line over the given period of time, competitive position, trade routes, etc. Fig. 10.39 shows an example of the relationship between the total port charges and the number of the annual number of calls by the container ships at the world's 21 largest ports (PHK, 2006).

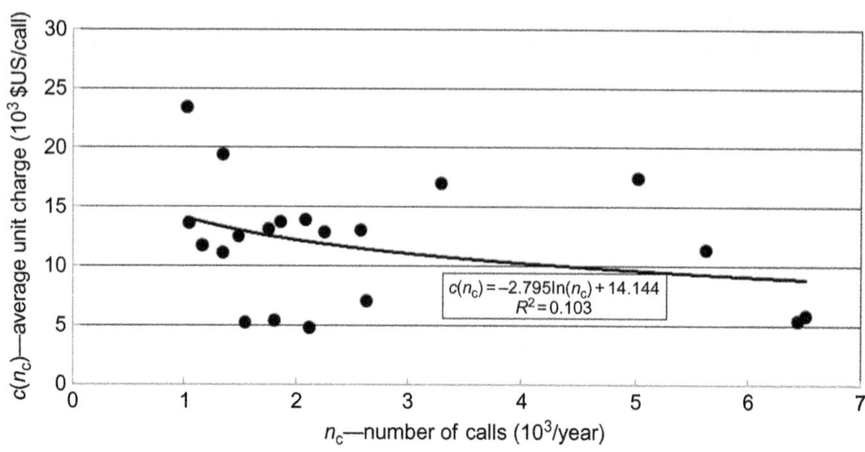

FIG. 10.39

Relationship between the average unit charge and the annual number of calls at the 21 largest ports handling containers (period: 2005) (PHK, 2006).

As can be seen, the average unit charges per call have been varying in the wide range from about 5000 to 20,000 $US/call. However, they have generally decreased with increasing of the annual number of calls thus indicating in some sense existing of economies of scale. This implies that the shipping lines calling the larger ports could expect the lower charges, for about four times in the extreme cases.

10.6.5.2 Shipping lines

Costs

The shipping lines operate the fleets of ships of different size and type respecting the category of freight/cargo—dry and bulk, tanker, container, and Ro-Ro. These ships have their operational costs, which generally consists of the fixed and variable component. The fixed costs component includes the ship's price converted into a series of equal annual payments over its life-cycle of about 25–30 years. The variable cost component generally consist of seven categories of expenses: crew, lubes and stores, maintenance and repair, insurance, administration, fuel, and dry docking. Table 10.13 gives an example of the structure of variable component of operating costs of two types of ships excluding the fuel and dry docking costs (Greiner, 2012).

As can be seen, the crew costs have been the highest and the insurance costs the lowest at both types of ships. Over time, these costs have been increasing. For the bulk cargo ships the average annual increase was 5.5%, tankers 5.10%, and containers 6.1% over the period 2000–11, a period of 11 years. In the absolute amounts these costs in 2012 were: 6606 $US/day for bulk (Panamax), 84,110 for tanker (Panamax), and 7730 $US/day for container (main liner: 2000–6000 TEU) ship (Greiner, 2012). In addition, Fig. 10.40 shows the structure of operating costs of the world's largest container shipping line—Maersk line over time (APMMG, 2014).

The costs have been categorized differently as those in Table 10.12 as the terminal, inland transportation, containers and equipment, ship, fuel, administrative and the other costs. The terminal costs

Table 10.13 Structure of the Variable Component of Operating Costs of Different Types of Ships (Greiner, 2012)

Cost Component	Type of Ship/Cost Share (%)	
	Dry and Bulk	Tanker
Crew	45	52
Lubes and stores	14	11
Maintenance and repair	14	14
Insurance	10	8
Administration	17	15
Total	100	100

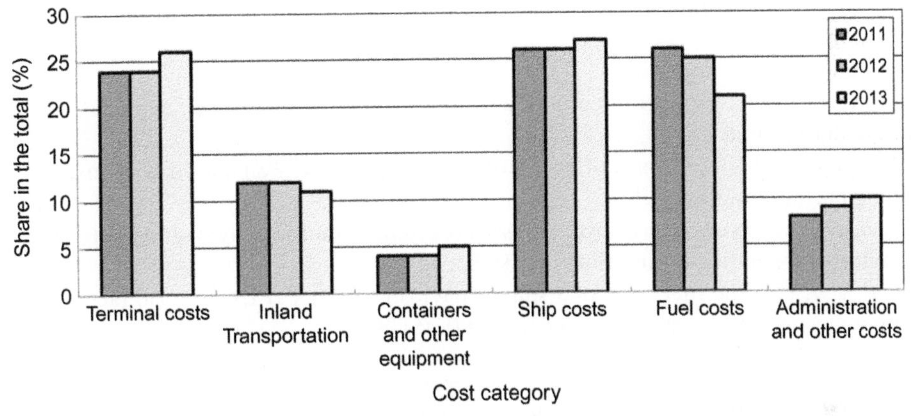

FIG. 10.40

An example of the structure of the operating costs of Maersk container shipping line (period: 2011–13) (APMMG, 2014).

have related mainly to loading/unloading of container content and storage of containers. The inland transportation costs have related to transporting containers inland by trucks and railways. The containers and other equipment costs have related to maintenance, repair, lease, and depreciation of containers. The ship costs have included expenses for port dues, Panama channel fees, crews, depreciation of owned and charges for hired vessels, lubrication, etc. The fuel costs have related to the fuel consumption of ships. The administration and the other costs have included expenses of the company's headquarters and service centers round the world, staff, offices, training consultancy, etc. (APMMG, 2014). As can be seen, the shares of particular categories of costs have been relatively stable over the given period of time. The highest variation (decrease) has been at the shares of fuel costs (mainly due to decreasing the prices of crude oil worldwide at that time). The shares of administrative and other costs (up to 10%), container and equipment (up to 5%), and inland transportation (up to 12–13%) have been much lower than the shares of terminal, ship, and fuel costs, all about 20–27%.

As far as the individual container ships are concerned, the experience up to date has shown that there is existence of economies of scale with increasing of the ship size, ie, the average unit costs per unit of payload/cargo have generally decreased with increasing of the ship's payload/cargo capacity (Sys et al., 2008). This can be investigated by estimating the average unit costs depending on the ship's payload capacity and its utilization while operating on a given route at the given speed during the specified period of time as follows:

$$c(PL, \theta, \tau, v) = \frac{C_c(PL, \tau, d) + C_o(PL, \tau, d) + C_f(PL, \tau, d, v)}{PL \cdot d \cdot \theta} \tag{10.28}$$

where

PL is the payload capacity of a ship (tons, m^3, TEU)
θ is the average load factor during a given period of time
τ is the period of time (day, month, year)
v is the average operating sped of a ship of payload capacity (PL) (kt; km/h) (1 kt (knot) = 1.852 km/h)
d is the route length (km, nm)
$c(PL, \theta, \tau, d)$ is the average unit costs of a ship of payload capacity (PL) and load factor (θ) operating during time (τ)
$C_c(PL, \theta, \tau, d)$ is the fixed costs of a ship of payload capacity (PL) and load factor (θ) operating during time (τ) ($US, €)
$C_o(PL, \theta, \tau, d)$ is the operating costs of a ship (excluding fuel) of payload capacity (PL) and load factor (θ) operating during time (τ) ($US, €)
$C_f(PL, \theta, \tau, d, v)$ is the fuel costs of a ship of payload capacity (PL) and load factor (θ) operating during time (τ) at the average speed (v)($US, €)

Fig. 10.41 shows an example of the average unit operational costs of container ships of the payload capacity of 4000 and 18,000 TEU operating at different speeds on the route shown in Fig. 5.53 (Chapter 5).

As can be seen, the average unit costs generally increase with increasing of the operating/cruising speed. This is because the increased speed requires greater fuel consumption, thus additionally raising the share of the already dominating fuel costs in the total ship's operating costs. In addition, in the given example, the unit costs of larger ship (in this case Triple E Maersk) are for about 5 and 3.5 times lower than that of the smaller ship, if operating at the speed of 20 and 15 kts, respectively. In addition, Fig. 10.42 shows an estimate of the relationship of average unit costs and size of container ships, when their annual load factor is assumed to be 85% (OECD/ITF, 2015).

As can be seen, the average annual unit costs per TEU decrease more than proportionally with increasing of the payload capacity of container ships thus indicating existence of economies of scale. For example these costs are for about four times lower for the ship of the payload capacity of 18,000 TEU (Triple E Maersk) than that of the ship of the payload capacity of 4000 TEU.

Revenues

The shipping lines obtain their revenues by charging their users-shippers and receivers by freight/cargo rates. These have usually been set to cover the lines' operating costs while respecting the highly

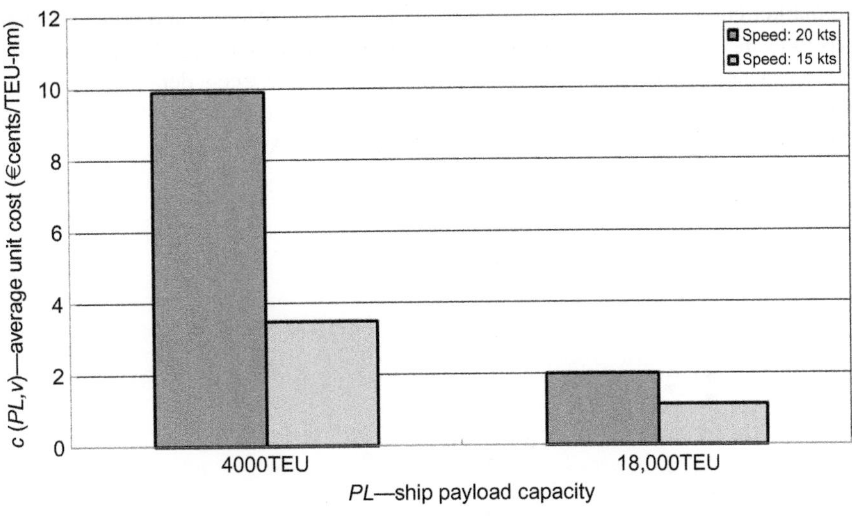

FIG. 10.41

The average unit cost of container ships of different size operating at different speeds.

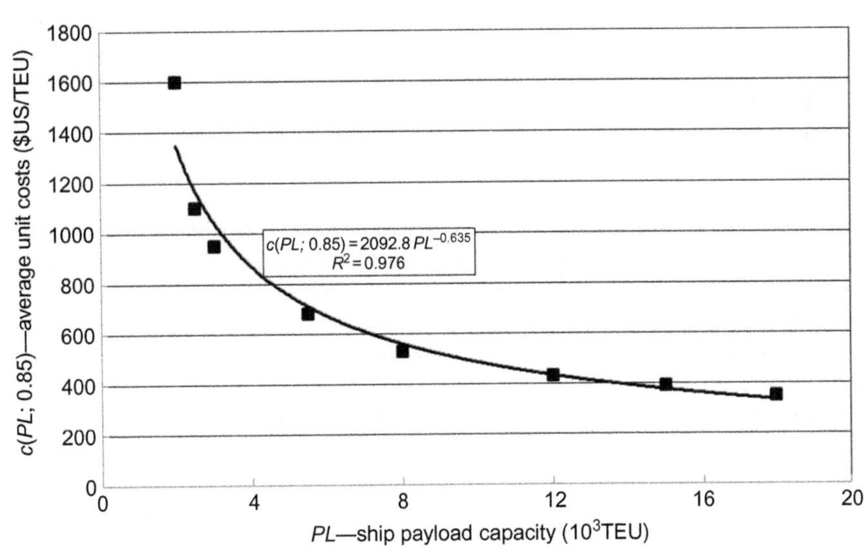

FIG. 10.42

Relationship between the average annual unit costs and the payload capacity of container ships (OECD/ITF, 2015).

competitive market conditions. Since these all have been quite different and changed over time, the freight/cargo rates have substantively varied across the shipping lines, trade markets, and prevailing micro- and macro-economic conditions. Specifically, they have been dependent on the ship size and transport distance. Fig. 10.43 shows an example of dependence of these rates for transporting containers on the ship type/size during the period 2002–12 (UNCTAD, 2015).

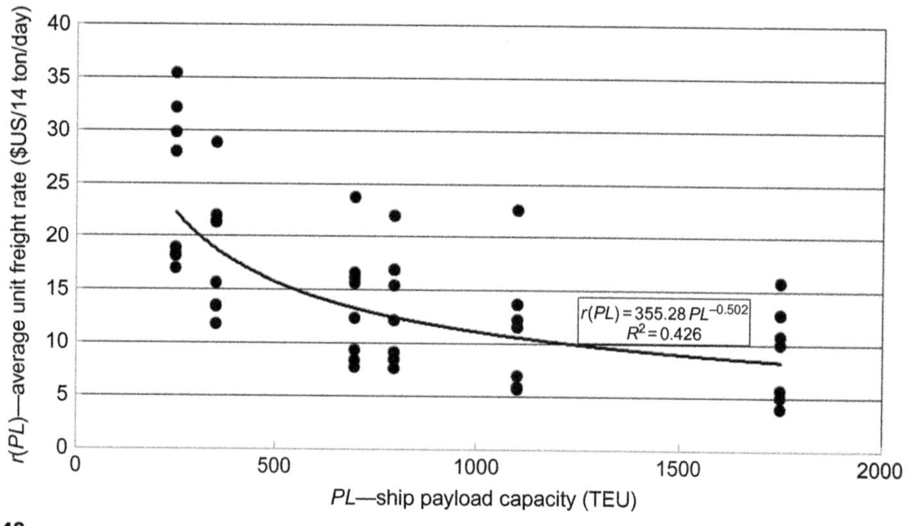

FIG. 10.43

Relationship between the average annual unit freight/cargo charter rates and the ship type/size ($US/14-ton slot/day) (period: 2002–12) (UNCTAD, 2015).

As can be seen these rates have generally decreased more than proportionally with increasing of the ship size, thus reflecting coverage of the average unit costs behaving according to the above-mentioned economies of scale. At the same time, the larger ships have operated at higher speeds, thus implying that the average unit rates have also decreased with increasing of the ships' operating speeds. The relationship between the ship speed and size/payload capacity in these cases has been: $v(PL) = 13.354 + 0.005PL, R^2 = 0.885; N = 66$. Fig. 10.44 shows the additional example of the relationship between average unit freight/cargo rates and distance, ie, route length between particular ports in different trade markets.

As can be seen, the average unit freight/cargo rates were very distinctive but generally increased with increasing of distance, ie, route length between ports in the same market(s). They increased linearly in Transatlantic, at decreasing rate in United States/Canada-Asia, and increasing rate in Europe-Asia market (DSC, 2010).

Fig. 10.45 shows the relationship between the average annual unit freight rates and costs for the container shipping line over the specified period of time.

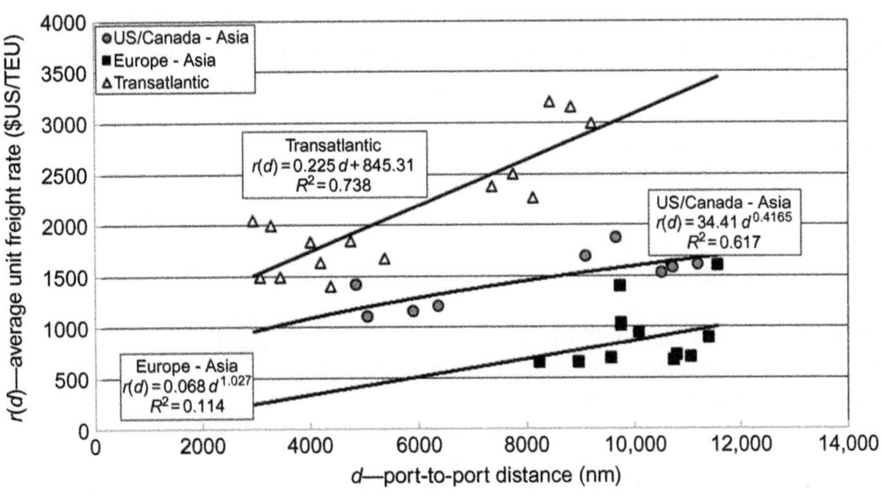

FIG. 10.44

Relationship between the freight/cargo rates and distance between ports in particular trade markets (period: March 2010) (DSC, 2010).

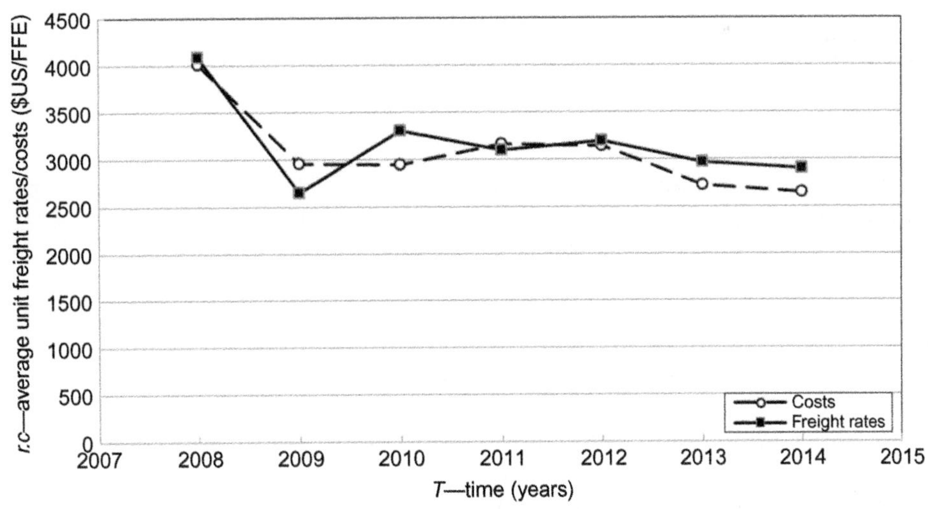

FIG. 10.45

An example of the relationships between the average unit freight rates and costs over time—case of Maersk container shipping line (period: 2008–14) (APMMG, 2014).

As can be seen, the freight rates (in \$US/FFE[7]) have been both higher and lower than the corresponding costs, thus providing the liner's profitability and loses in the corresponding years of the observed period. The question is how these rates have been set up. The main influencing factors have presumed to be the fuel costs and the volumes of expected demand (containers) to be served in the given intercontinental markets such as Asia-Europe, Africa, Transpacific, Latin America, Transatlantic, Oceania, and Intra-Asia. The multiple linear regressions have been obtained as follows:

$$r(Q,FP)=3720.167-281.160Q+2.522FP, R^2=0.515; N=42;$$
$$t\text{-stat}\ (4.248)\quad (-1.892)\quad (1.900)\quad F=2.127 \tag{10.29}$$

where

$r(Q, FP)$ is the average rate (\$US/FFE)
Q is the annual volumes of transported FFEs (million/year)
FP is the average fuel price (\$US/ton)
N is the number of elements in the sample

As can be seen, the shipping liner has adapted its fares accordingly: by reducing them with increasing of the volumes of handled demand (transported FFEs) reflecting economies of scale of its operating costs but at the same time by raising them in order to compensate increase in the fuel prices/costs, and vice versa. In addition, the freight rates have been adjusted subjects to the changes of shares of volumes of demand on the particular (intercontinental) routes/markets. Fig. 10.46 shows the example of such changes on the above-mentioned 7 routes/markets by the same container shipping line (Maersk) during the period 2008–11.

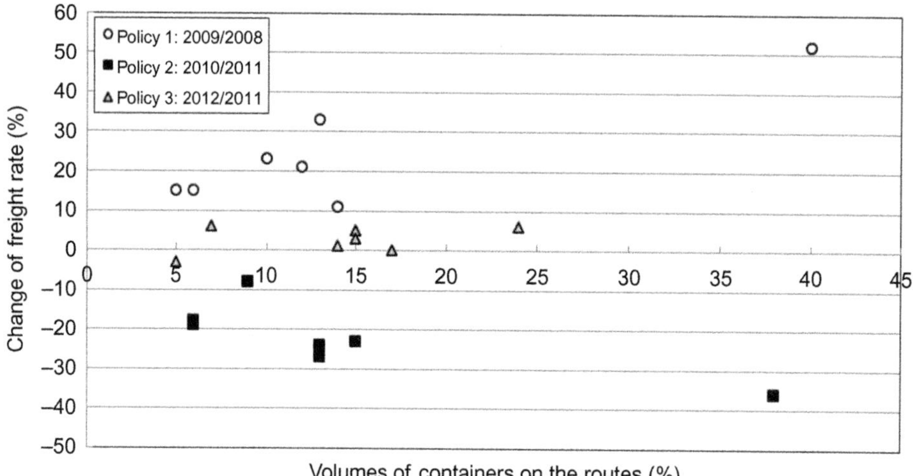

FIG. 10.46

Changing of the freight rates depending on the distribution of volumes of container demand across the routes/markets—case of Maersk container shipping line (APMMG, 2014).

[7]FFE—forty-foot equivalent unit defined as two TEU (twenty foot equivalent unit) (length: 40 ft (12.2 m); width: 8 ft (2.44 m); height: 8 ft 6 in (2.510 m); volume: 77 m^3 (OECD, 2007).

Three policies of adjusting the freight rates can be observed: Policy 1 with the higher relative increase in the freight rates on the routes/markets with the higher volumes of demand aiming at getting as much as possible revenues from there; Policy 2, quite opposite to the previous one (Policy 1), with higher decrease in the freight rates on the routes/markets with the higher volumes of demand aiming at attracting as much as possible of it; and Policy 3, with slight variation of the freight rates independently of the volumes of demand on the particular routes/markets.

10.6.6 AIR

Similarly as other transport modes, the operational costs of air transport system usually relates the costs, revenues, and profits of its components—airports, ATC (air traffic control), and airlines—all depending on the corresponding volumes of output (traffic).

The costs, revenues, and profits are characterized by their total amount and structure, mutual relationships and the relationship with the volumes of output(s), and the average and marginal values. The amount of operational costs is primarily dependent on the volume of resources (material, labor, energy/fuel) consumed for provision of the given volumes of services. The structure of operational cost can be expressed by the share of costs of particular physical inputs (labor, capital and energy/fuels) in the total costs. By division of these total costs by total volume of output(s), the average cost can be estimated. By deriving an analytical form of the total cost function subject to the volume of output being the independent variable, the marginal cost per unit of output can be obtained. In that case the cost function has to be continuous, eg, to have at least the first derivative at each point belonging to the range of possible outputs (Janić, 2000).

The economies of scale and economies of density can be used as the measures of operational efficiency of particular components. For example, if the analytical form of the cost function is known, the average and marginal cost per unit of output can be estimated. Let Q, $AC(Q)$ and $MC(Q)$ be the volume of output during a given period of time, the average, and the marginal cost, respectively, of a given component of the air transport system. Let: $\alpha(X) = AC(Q)/MC(Q)$, then, if $\alpha(Q) > 1$ there is economies of scale; if $\alpha(Q) = 1$ there is constant return to scale; and if $\alpha(Q) < 1$ there is diseconomies of scale. In addition to other quantitative characteristics some examples for the particular system's components have been provided as an illustration of existence of these concepts.

10.6.6.1 Airports

The operating costs are the expenses on the current maintenance of the components of airside area (runways, taxiways, apron-gate complex) and that of landside area (passenger and cargo terminals and supportive facilities and equipment), energy and labor for serving aircraft, passenger, and cargo, and administration. Some investigations have dealt considered simpler categorization of the airport operational costs based on the available data from the airport annual reports. For example, one of them carried out for 11 largest European airports identified four operational cost categories and their average share in the total airport operation costs as follows: material cost and services (210%), staff (40%), depreciation (20%), and others(11%) (EEC, 2010). These costs are covered by charges bringing the revenues to the airport, which can generally be categorized as aeronautical and nonaeronautical. The aeronautical charges are those for services and facilities directly related to serving aircraft, passengers, and air freight/cargo. These can for aircraft landing (and/or take-off), terminal-area air navigation, passenger and cargo services (at the corresponding terminals), the aircraft ground handling and parking, security, aircraft noise, emissions of green house gases, and en-route navigation. The nonaeronautical charges providing the revenues relate to the airport ancillary commercial services, facilities and amenities available at an airport such as: concession fees for aviation fuel, oil, and other commercial activities, car parking and renting space, land, and equipment, engineering services, etc. The additional revenues can be derived from activities such as real estate ventures, consulting, investments at other airports, etc. (ICAO, 2013; Janić, 2000). Fig. 10.47A and B shows the example of total costs and revenues for 21 largest European airports. Specifically, Fig. 10.47A shows the relationship between the mentioned averages and the total annual number of handled passengers.

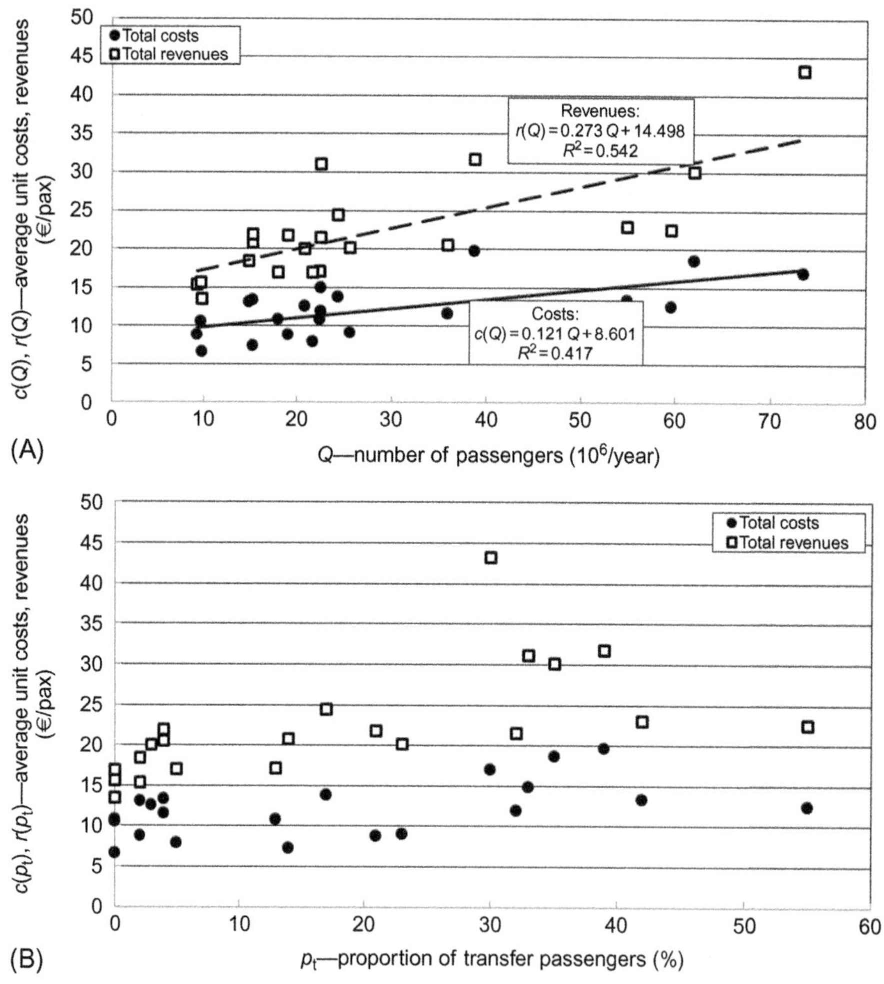

FIG. 10.47

Relationship between the average operating costs and revenues per passenger and the annual number and structure of passengers at 21 largest European airports (period: 2013–14) (Medeiros and Potter, 2015): (A) average cost and revenues vs. total number of handled passengers and (B) average costs and revenues vs. proportion of transfer passengers.

As can be seen, both the average costs and revenues have increased with increasing of the total annual number of passengers. At all airports these revenues have been greater than costs, thus indicating profitability of their operations under given conditions.

Fig. 10.47B shows that the average revenues and costs have also increased with increasing of the proportion of transfer passengers at these airports. Again, the revenues at all airport have been greater than the corresponding costs, thus indicating their profitability thanks to operating as the airline hubs characterized by substantive proportions of transfer passengers.

In order to control the costs of an airport handling both passengers and freight/cargo a unique measure of its landside area to be charged has been defined as WLU (workload unit) The WLU is an weight equivalent to one passenger or 100 kg of freight/cargo or mail (The weight of one passenger is assumed to be 80 kg, and the weight of his/her baggage 20 kg). By applying appropriate criteria, the aircraft movements served at the airport may also be easily converted into the WLUs (Doganis, 1992). In general, the costs of processing WLUs under given conditions depend on their volumes and prices of resources (inputs) consumed for provision of the appropriate services. Generally, if the volumes of WLU increase, the cost of their processing will also increase, and vice versa. Using the relevant data enables estimating relationship between these costs and the volumes of WLUs as outputs during a given period of time. Dividing these total costs by the total volume of the WLU processed during a given period of time, the average cost per WLU is obtained. Fig. 10.48 shows an example of such relationship.

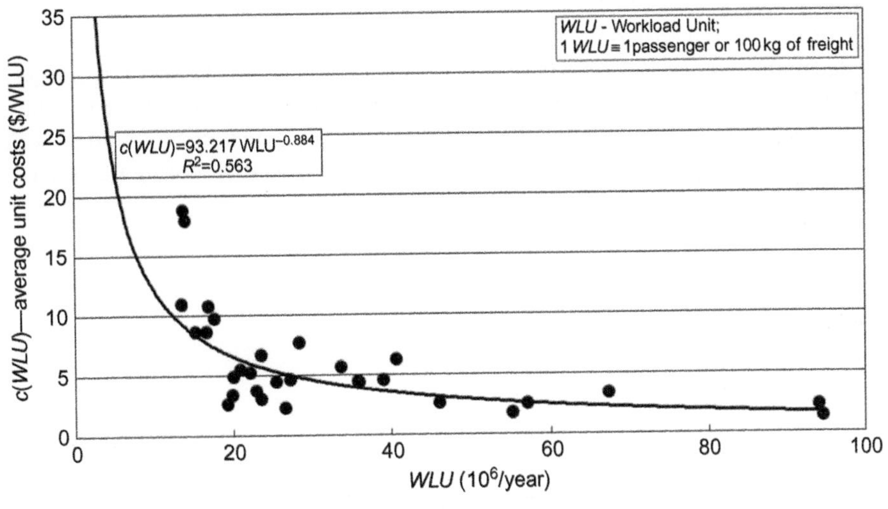

FIG. 10.48

Dependence of the average cost per WLU on the annual volume of WLU processed at an airport (Janić, 2000).

As can be seen, the average costs per WLU decrease more than proportionally with increasing of the volume of WLUs, thus indicating existence of economies of scale. This implies that the larger airports handling larger volumes of WLUs under given conditions have the higher total operating costs but lower average costs per unit of output, in this case WLU. This could be their driving force to tend to become even larger by benefiting from the lower average unit processing costs.

10.6.6.2 Air traffic control

The operation costs of ATC system consists of the expenses for maintaining the ground radio-navigation facilities and equipment, their depreciation, interest, operations and staff. For example, some investigation shew that at European centralized ATC system EUROCONTROL, the staff costs shared about 67%, depreciation of facilities and equipment 18%, operating costs 10%, and interest costs 5%. In addition, the structure of these costs differed among particular (38) state ATC systems, contracting members of EUROCONTROL. The operation costs of all ATC systems worldwide are covered by the en-route charges paid by users of ATC services—airline flights. The estimates, again for Europe, indicate that these charges have shared about 5–8% of the airline operational costs over the period 2000–14. The charge for each flight performed in the airspace of a contracting state is determined as follows (EEC, 2010; 2015):

$$c_i = \frac{d_i}{100} \cdot \sqrt{\frac{MTOW}{50}} \cdot R_i \tag{10.30}$$

where

i is the index charging zone, usually the entire or a part of the contracting country
d_i is the length of route, ie, flying distance in the charging zone (i) measured as the great circle (orthodrome) distance (km)
$MTOW$ is the aircraft maximum take-off weight (ton)
R_i is the unit rate of the charging zone (i) (€/flight)

The unit rate (R_i) is generally determined by dividing the forecasted cost of the given charging zone (i) by the expected number of service units (flights) to be handled in the zone during the specified period of time, usually 1 year. The distance (d_i 0 is usually reduced for about 20 km in order to take into account the aircraft maneuvering distances around airports during take-off and landing. during is Fig. 10.49 shows the average unit rate of 38 contracting states of EUROCONTROL during the observed period of time (2005–14) (EEC, 2015).

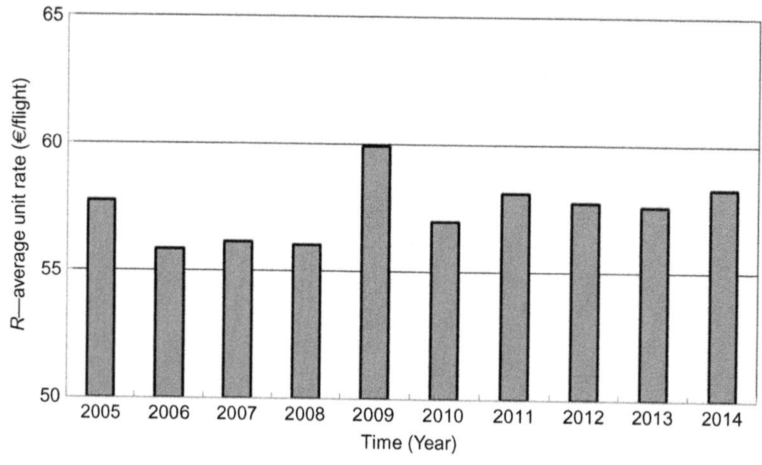

FIG. 10.49

The average state/national unit rate charged per flight in Europe over time (EEC, 2015).

As can be seen, in the given case, the average unit rate varied 56 and 60 €/flight during the observed period. Fig. 10.50 shows the example of relationships between the charge per flight, route length, and the aircraft *MTOW* calculated by Eq. (10.30).

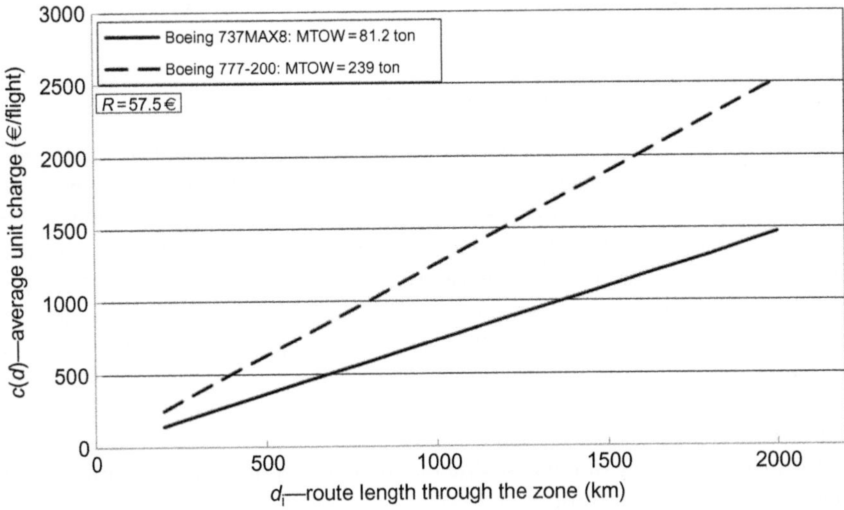

FIG. 10.50

Example of the relationship between the average unit charge per flight, route length in charging zone, and the aircraft *MTOW* in Europe.

The unit rate is specified as an average over the period of time as shown in Fig. 10.49. As can be seen, the charge linearly increases with increasing of the route length for the given aircraft *MTOW* as intuitively expected. For the given route length it increases with increasing of the aircraft *MTOW* but the difference becomes greater on the longer routes. The possible existence of economies of scale in the ATC system is illustrated by established causal relationship between the annual operating costs and the annual number of flights served by the ATC systems of 34 countries given in Table 10.14 (Janić, 2000).

Table 10.14 Relationship Between the ATC Annual Total Operating Costs and the Volume of Traffic (Janić, 2000)

Characteristic	Estimate
Sample size	34
Function	$C(n) = -20.108 + 0.000207117 \times n$
t-Values	(1.11010) (8.0210)
R^2	0.668
F-value	64.460
DW	1.718

C(n)—*the ATC annual operational cost(106 $US/year).*
n—*the number of flights served per year.*

As can be seen, generally, the ATC costs linearly increase with increasing of the volume of traffic, ie, the annual number of flights served. This implies that overall, the higher expenses were needed for operating the larger ATC systems serving the larger volumes of traffic. Due to linearity of the cost function, both average and marginal costs per flight were equal and constant (207.117 $US/flight), which implies that there was the constant return to scale at an average ATC system.

10.6.6.3 Airlines passenger transport

Costs

The operating costs of airlines can be considered for the airline industry as the whole, individual airlines, and particular aircraft types. The information and data on the airline costs have been standardized and classified into direct operating costs (DOC) and indirect operating costs (IOC) (expenses). The ICAO (International Civil Aviation Organization) specifies direct operating costs (expenses) as follows: the costs of flight operations (crew, fuel and oil, insurance and rental of flight equipment, flight crew training, and other flight expenses), the costs of maintenance and overhaul, and the costs of depreciation and amortization of the aircraft/flight and ground equipment. The IOC include user charges and station expenses (landing and associated airport charges, en-route facility charges, station expenses), passenger service, ticketing sales and promotion, general, administrative, and other operating expenses.

Fig. 10.51 shows an example of the cost structure of a conventional/legacy airline(s) (in this case British Midlands) and LCCs (low cost carrier(s)).

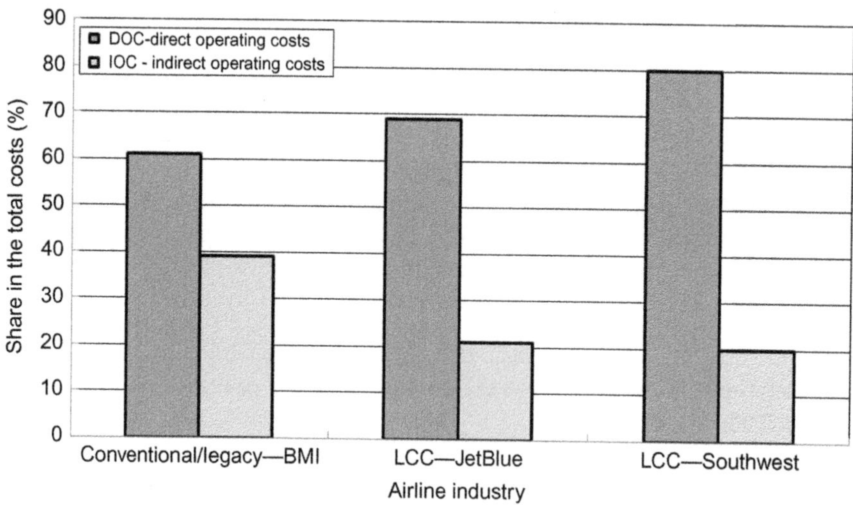

FIG. 10.51

An example of the structure of the operating costs of the conventional/legacy airlines and LCC (DLR, 2008; Southwest, 2015).

As can be seen, the structure of total costs of both types of airlines has been different. The conventional/legacy airline has had approximately 20% lower IOC than DOC. The LCCs have had for about 50–60% higher DOC that IOC.

As mentioned in Chapter 5, the conventional/legacy airlines generally operate either the "point-to-point" or "hub-and-spoke" networks. The LCCs operate exclusively "point-to-point" networks. The total cost of operating the "point-to-point" network with (N) airports as the network nodes can be estimated as follows (Janić, 2000):

$$C_1(N) = \sum_{i=1}^{N} \sum_{\substack{j=1 \\ i \neq j}}^{N} Q_{ij} \cdot d_{ij} \cdot AC_{ij} + c_f \cdot [N \cdot (N-1)]/2 \tag{10.31}$$

where

Q_{ij} is the volume of demand between airport (i) and (j) (pax)
d_{ij} is the length of direct route connecting airports (i) and (j) (km)
AC_{ij} is the average unit costs on the route (d_{ij}) ($US ct/pax km)
c_f is the airline average fixed cost is the fixed cost associated with maintaining each direct route of the network ($US/route)

Similarly, the total operating cost of a "hub-and-spoke" network with (N) airports as the network nodes can be estimated as follows (Janić, 2000):

$$C_2 = \sum_{i=1}^{N} \sum_{\substack{j=1 \\ i \neq j \neq h}}^{N} Q_{ij} \cdot \left(d_{ih} \cdot AC_{ih} + d_{hj} \cdot AC_{hj}\right) + c_f \cdot (N-1) \tag{10.32}$$

where

i, h, j is the origin, hub and sink airport, respectively
d_{ij}, d_{hj} is the length of routes connecting the airports (i) and (h), and the airports (h) and (j), respectively (km)
AC_{ih}, AC_{hj} is the average unit costs on the routes (d_{ih}) and (d_{hj}), respectively ($US ct/pax km)

The other symbols are analogous to those in Eq. (10.32). By simple manipulation with Eqs. (10.31) and (10.32), it can be shown that the "hub-and-spoke" network is preferable under the following conditions (Janić, 2000):

$$\sum_{i=1}^{N} \sum_{\substack{j=1 \\ i \neq j \neq h}}^{N} Q_{ij} \cdot \left[\left(d_{ih} \cdot AC_{ih} + d_{hj} \cdot AC_{hj}\right) - d_{ij} \cdot AC_{ij}\right] < (k/2)\left(N^2 - 3N\right) + 1 \tag{10.33}$$

It should be noted that an incremental cost of an indirect connection is always higher than the cost of equivalent direct connection, ie, $(d_{ih} \times AC_{ih} + d_{hj} \times AC_{hj}) > d_{ij} \times AC_{ij}$ for each (i, j, h). Thus, the total cost of "hub-and-spoke" networks will be higher than the cost of equivalent "point-to-point" network, which could deter the airlines to operate them. However, the total savings due to reducing the number of routes in the "hub-and-spoke" networks have been always greater than the total incremental cost, which has made them preferable with respect to the total operating cost.

In addition, the average costs per flight on the direct route (ij) of an airline "point-to-point" network in Eq. (10.34). Eqs. (10.31) and (10.32) can be estimated as follows (Janić, 2000):

$$c_{ij}\left(S_{ij}, d_{ij}\right) = \frac{Q_{ij}(T) \cdot AC_{ij} \cdot d_{ij}}{f_{ij}(T)} \tag{10.34}$$

where

$f_{ij}(T)$ is the flight frequency on the route (d_{ij}) during the time period (T)
$c_{ij}(S_{ij}, d_{ij})$ is the cost per flight carried out by the aircraft of the seat capacity (S_{ij}) on the route (d_{ij}) ($US/flight)
$Q_{ij}(T)$ is the number of passengers on the route (d_{ij}) during the time period (T)

The other symbols are analogous to those in the previous equations. The costs per flight on the indirect routes of the airline "hub-and-spoke" network (ij) and (hj) can be similarly estimated by Eq. (10.34). An example of the above-mentioned costs per flight has been estimation of the average cost per flight at the European and US airline industry by using regression analysis as given in Table 10.15 (Janić, 2000).

Table 10.15 The Average Cost Per Flight vs. the Aircraft Seat Capacity and Route Length at Conventional/Legacy European and US Airlines (Janić, 2000)

Airline Industry	Estimate
European airlines	
Sample size	21
Cost function	$c(S, d) = 7.1034 \times S^{0.603} \times d^{0.656}$
t-Statistics	(3.266) (4.3310) (4.773)
R^2	0.8106
F-statistic	77.477
DW-statistic	1.6102
US airlines	12
Cost function	$c\,(S, d) = -13240.377 + 6.663 \times S + 22.22 \times d$
t-Statistics	(−3.837) (0.201) (0.126)
R^2	0.1076
F-statistic	1710.867
DW-statistic[a]	2.323

[a]*Durbin-Watson; c(S, d)—the average cost ($US/flight); S—the aircraft capacity (seats); d—route length (km); $28 \leq S \leq 460$ (seats); $300 \leq d \leq 4800$ (km)*

The form of these relationships is the same as the generic ones.

When the average cost per flight is divided either by the aircraft seat capacity (S) (the route length (d) is fixed) or the route length (d) (the aircraft seat capacity (S) is fixed), the average cost per seat on a given route and the average cost per unit of flying distance, respectively, can be estimated. In both cases, the average costs decrease more than proportionally with increasing of (N) *or* (d). The first relationship shows existence of the economies of seat density of the flight, ie, the average cost per seat will decrease if a larger aircraft is engaged to carry out the flight. If the flight is performed on a longer nonstop route by the same aircraft, the average cost per unit of travel distance will decrease thus indication the economies of route length. If both the route length and the aircraft size increase, the cost per unit of output (expressed in $US ct/ASK—available seat kilometer) will decrease more than proportionally too (ASK). This illustrates the temporary endeavors of many conventional airlines to operate the large (hub-and-spoke) networks with a significant number of long-haul profitable flights carried out by the high-seat density (wide-body) aircraft.

Revenues

The airline operating revenues are obtained by charging passengers. These charges are called airfares. In addition to the size of aircraft engaged on the given route, the airfares are mostly dependent on the route length, ie, length of trip. Fig. 10.52 shows an example of the relationships between the airfare and route length for different trip purposes of the conventional/legacy European airline.

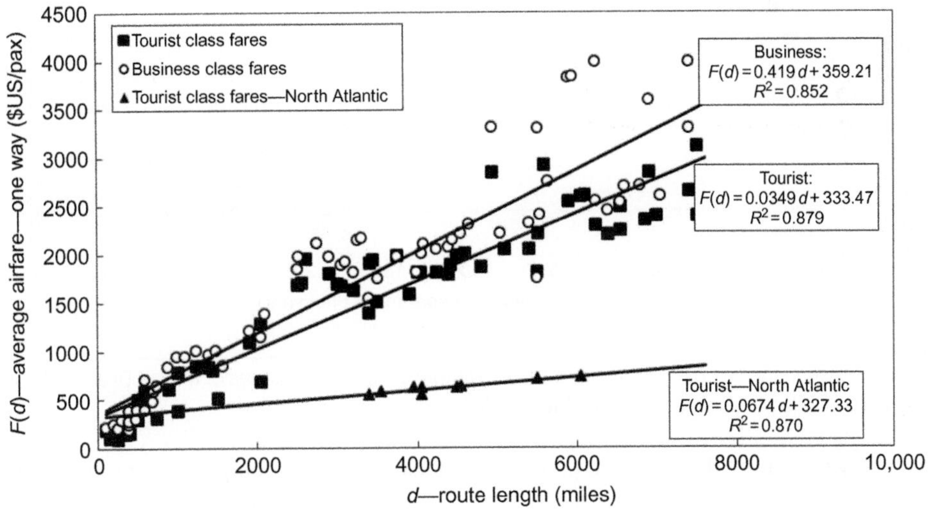

FIG. 10.52

An example of the relationship between the airfares, route length, and trip purpose—KLM routes scheduled from Airport Schiphol (Amsterdam, The Netherlands) (Janić, 2000).

As can be seen, in this example, three classes of airfares increased approximately linearly with increase in the route length. The business airfares were always higher than the tourist airfares. The other analyses also indicated the general linear relationship between the airfares and route length (Janić, 2000). The airfares can be used to measure the average revenue (ie, "yield") per unit of the airline output. For example, if one divides the airfare ($F(d)$) by the route length (d), the average yield per pax-mile can be obtained. The yield per passenger takes the form: $yi(d) = F(d)/d = a_o/d + a_1$. The average yield per passenger is higher on the shorter routes since the fare is set up to cover the higher corresponding operating costs. If the route length (d), at least theoretically, tends to infinity, the average yield approaches to constant marginal revenue (a_1).

Generally, it can be said that the airfares and operating costs are closely interrelated. Nevertheless, there have been some exceptions from this rule, which can be noticed on the specific highly competitive routes, in the cases of promotion of relatively new services, and in the cases when there are relatively high diversity of the structure of passenger demand. In addition, there is the difference between the revenues and cost and the conventional airlines and LCCs. Fig. 10.53 shows an example of the revenues and costs of the conventional/legacy US Delta and LCC Southwest airline operating the networks shown on Figs. 5.43 and 5.44 (Chapter 5).

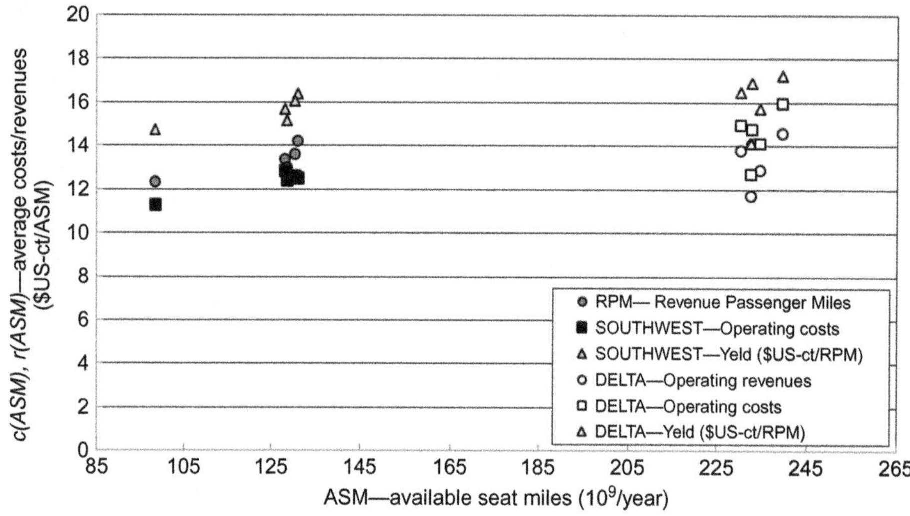

FIG. 10.53

Relationships between the volumes of output, and the average costs and revenues of the US Delta and Southwest airlines (period: 2010–14) (Delta, 2014; Southwest, 2015).

As can be seen, during the observed period, the annual volumes of output in terms of ASM (available seat miles) have been for about a twice higher at Delta that at Southwest Airlines. At the same time, Southwest has always covered its operating costs by revenues while increasing the annual volumes of output over time. Its average yield has varied between 14 and 16 $US ct/pax mile. Delta Airlines have kept its annual volumes of output relatively constant. The operating revenues have not always covered the operating costs, but the yield varying from about 14 to 17.5 $US ct/pax mile has always been positive, ie, greater than the operating costs, thus implying that also nonaeronautical activities have contributed to the overall airline profitability.

10.6.6.4 Airlines cargo transport

The operating costs of air freight/cargo transport are analyzed by taking into account the particular cargo aircraft types and their utilization when being operated by different air cargo operators, ie, integrators[8] of the door-to-door freight/cargo delivery services. The cargo aircraft usually designed to optimize the volume/payload ratio can be the new aircraft and the converted passenger aircraft

[8]These companies operate exclusively the cargo aircraft fleets. In addition, the air freight/cargo shipments are transported by the cargo fleets of passenger airlines and their passenger aircraft as well.

already used for about 15–20 years. After being converted and refurbished, these latter aircraft can be used for further 15–20 years. Table 10.16 gives an illustration of the costs of different versions of these aircraft (Vandenbossche, 2015; WB, 2010).

Table 10.16 Cost of Different Versions of Cargo Aircraft (Vandenbossche, 2015)

Version/Category	Small	Wide-Bosdy	Large
New (10^6 \$US)	35	75–100	140–80
Used for conversion (10^6 \$US)	8–12	7–20	35–45
Conversion (10^6 \$US)	4	13–14	22–8
Converted (10^6 \$US)	12–16	20–34	57–73

As can be seen, the cost of converted aircraft increases with their size. At the same time, the ratio between the costs of converted and the cost of new aircraft amounts between 30% and 50%. Some investigation has indicated that over the next 20 years, about 70% of cargo aircraft will be obtained by conversion of the passenger aircraft. In addition, the average structure of operating costs of these aircraft operated by particular logistics companies such as World, Evergreen, FedEx, Kalitta, UPS, Atlas, and DHL (Deutsche Post DHL Group) are estimated to be as given in Table 10.17.

Table 10.17 The Average Structure of the Operating Costs of Cargo Aircraft (Vandenbossche, 2015)

Cost Category/Share	Small	Wide-Body	Large
Flight crew (%)	24	8	11
Fuel (%)	36	55	61
Maintenance (%)	28	24	16
Depreciation (%)	5	5	6
Other (%)	7	8	6
Total (%)	100	100	100

The share of fuel costs is the highest at all aircraft categories followed by that of the maintenance and flight crew costs. As well, the average unit costs of the large, wide-body, and small cargo aircraft such as B747-400F, B747-100/200, MD-11F, DC-10-30F, B767-300F, A310F, A300F, B757-200F, B727-200F operated by the above-mentioned logistics companies (World, Evergreen, FedEx, Kalitta, UPS, Atlas, and DHL) have been estimated in dependence of the average stage length and the aircraft payload capacity means by the regression analysis as follows (Vandenbossche, 2015):

$$c_{av}(PL,d) = 671.278 - 0.031d - 3.312PL \qquad (10.35)$$

$$t = (10.504)(-1.114)(-2.104)$$

$$R^2 = 0.621;\; F = 16.583;\; N = 20$$

where

$C_{av}(d, PL)$ is the average unit costs ($US ct/ATK) (ATK—available ton kilometer)
d is the average stage length (km)
PL is the cargo aircraft payload capacity (ton)

As can be seen, the form of Eq. (10.35) indicates that the average unit costs decrease with increasing of the stage length and the aircraft payload capacity. This implies existence of economies of scale respecting both variables.

10.6.7 INTERMODAL—RAIL/ROAD FREIGHT TRANSPORT

10.6.7.1 General

The operational costs of intermodal transport systems are analyzed for the rail/road intermodal freight transport service network under conditions of operating in Europe as shown in Fig. 5.15 (see Chapter 5). The network delivers LU (load units)—containers from particular shippers to receivers in five steps: (i) collection in the origin "zone" and transportation to the "origin" rail-road intermodal terminal located in the "shipper" zone by road trucks; (ii) transhipment at the "origin" rail-road intermodal terminal 1 from trucks to CIFTs (conventional intermodal freight train(s)) or LIFTs (long intermodal freight train(s)); (iii) the rail haulage between the "origin" and "destination" intermodal terminal 1 and 2 (the CIFTs and/or LIFTs running between two terminals use the same technology (infrastructure, traffic control and signaling system, and rolling stock)) but have distinct technological/operational characteristics mainly in terms of the breaking system and operating speed; the former train has been operating for a long time; the later train similar to that described in Chapter 5 operating through Siberia is still the concept and under preparation for the pilot trials (in Europe) (Janić, 2007, 2008); (iv) transhipment at the "destination" intermodal terminal 2 in the "receiver" zone from the rail to the trucks; (v) distribution from the terminal 2 to the destination "zone" by trucks. The road haulage between particular shippers and receivers in the corresponding zones usually considered as the main competitor includes three steps: (i) collection in the origin/shipper "zone"; (ii) road haulage from the shopper to the receiver zone; and (ii) distribution in the destination/receiver zone (as shown in Fig. 5.15, Chapter 5) (Janić, 2007, 2008).

10.6.7.2 CIFTs

The CIFTs operating in many national and Trans-European corridors has a rather fixed composition of 25–30 four-axle rail flat wagons, each of the approximate length of 20 m. This gives typical length of CIFT of about 500–600 m. The empty weight of a flat wagon depends on type and varies from about 24 ton (for that carrying containers and swap bodies) to 32 ton (for that carrying the semitrailers). The carrying capacity of each flat wagon is up to 50 ton. In terms of space, the flat wagon can carry out an equivalent of three TEUs each of the length of 20 ft (6 m) (the most common in Europe) and the empty weight (tare) of 2.3 ton. Freight/goods in the TEUs and their tare represent payload for CIFT. The length of CIFT depends on the number of wagons. Its weight depends on the weight of empty flat cars and locomotive, and payload, ie, load factor of each car or the entire train. Fig. 10.54 shows the relationship between the length and typical and maximum gross weight of CIFT across particular European corridors. The prospective length and weight for LIFT are shown for the comparative purposes.

As can be seen, the length of CIFTs is within the current UIC (International Union of Railways) pneumatic braking regulation of 750 m while typical length is between 450 and 650 m. In addition, the CIFT has the fixed composition along corridors, which excludes additional shunting. Due to

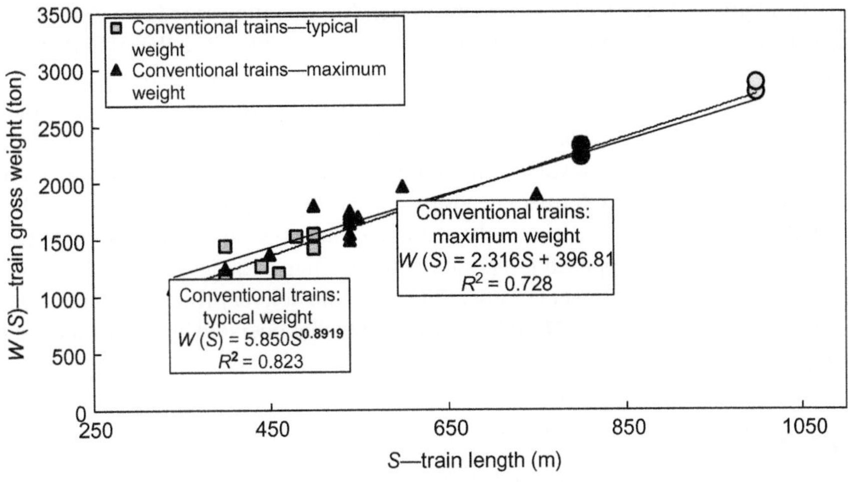

FIG. 10.54

Dependence of the gross weight of the length of CIFTs and LIFTs in Europe (Janić, 2007, 2008).

differences in the electric power supply in particular European countries (1.5 kV DC, 3 kV DC, 25 kV 50 Hz AC), a multisystem locomotive such as 6 MW ALSTOM PRIMA 6000B or SIEMENS CLASS 1810 weighting 86–810 tons is usually used to power the Trans-European CIFT enabling the necessary interoperability and eliminating the border crossing delays due to the locomotive-change. The CIFT can run at the maximum operating speed of about 120 km/h, but the average commercial speed tends to decrease with increasing in distance, which gives an average of 40–50 km/h between rather distant terminals. This is mainly due to frequent speed changes and intermediate stops. The frequency of dispatching CIFTs is usually once or twice per day during the weekdays depending on the volumes of demand. The frequency and speed make the CIFTs a rather competitive alternative to the road haulage in terms of delivery time on some longer distances (Janić, 2007, 2008).

10.6.7.3 LIFTs

The LIFT is considered to be an extended CIFT of either 800 or 1000 m composed of 38 or 48 rail flat wagons, respectively. The LIFT is also supposed to have the compact-fixed composition. Regarding the empty weight and load factor of 75%, the gross weight of LIFT could vary from about 2400 to 2800 tons, respectively (see Fig. 10.54). Due to such length and weight, the braking system for LIFT needs to be specifically designed and strengthened, which is actually the main technical/technological innovation on the one hand and the current barrier to implementation on the other as compared to CIFT[9]

[9]Freight trains longer than 750 m do not fulfill the current UIC breaking regulation. However, installing a remote controlled additional breaking device (ABD) at the rear end of the train in order to distribute pressure more symmetrically towards the middle of the train can shorten the pressure stabilization delay and thus enable shortening of the breaking distance(s) according to the prescribed standards, and consequently enable operating the trains of up to 1000 m. This ABD will be controlled by means of the radio-based communications system and related network, whose main features are autonomy without interference with those of other trains and interoperability across different European countries (EC, 2006).

(Janić, 2008). In addition, it in order to achieve the preferable operating speed two instead of one multi-system electric locomotive will frequently pull the LIFT. For example, the LIFT of 800 m pulled by one 6 MW electric locomotive is not able to run faster than 70 km/h; with two locomotives, it is able to run at the maximum speed of about 100 km/h. Similarly, the LIFT of 1000 m is able to run at the maximum speed of 65 and 100 km/h pulled by one and two 6 MW locomotives, respectively (Janić, 2008). Such operating performances will be achievable if the rail line follows a rather plain terrain. Consequently, regarding the speed changes and intermediate stops, the average commercial speed of LIFT between distant intermodal terminals will not seemingly be higher than 40 km/h. In order to maintain the competitive potential, the LIFT is supposed to operate at the similar frequency as CIFT—minimum once per weekday. Finally, despite the lower commercial speed, the delivery time by the intermodal system operating LIFT is not expected to be affected as compared to that operated by CIFTs and the road counterpart as well.

10.6.7.4 Costs

The operating costs of the above-mentioned rail/road network relate to the expenses of its operators performing particular operational steps of the door-to-door delivery of LUs. These costs generally depend on the transport frequencies, delivery distances, and the volume of transshipment activities between road and rail mode at the intermodal terminals. These all depend on the volumes of demand, ie, the number of LUs, and the number and spatial location of particular shippers and receivers in the corresponding areas. In addition, these expenses depend on the prices of the main inputs—material, labor, and energy.

The operational costs are constant in the short-term for the given volume of the network's activities. In addition to the other factors such as the inter- and intramodal competition, consumers' preferences and institutional constraints, they are mainly used as a basis for the cost-recovering charging of users, ie, pricing.

The operational costs of each step and component involved in delivering LUs between their shippers and receivers embraces the expenses for ownership (depreciation), insurance, repair and maintenance of transport vehicles and transshipment facilities and equipment, labor (drivers'-operators' salary packages), energy, taxes, and tolls/fees for using the transport and intermodal terminal infrastructure assumed to be already in place. The expenses for depreciation, maintenance, and repair of LUs are not considered since these tend to be mostly at the side of their owners—freight/goods shippers and receivers (Janić, 2007).

The generic models for estimating particular components of the operational costs of a given intermodal rail/road transport network can be as follows (Janić, 2007):

$$\begin{aligned}\text{Transport cost} &= (\text{frequency}) \times (\text{cost per frequency}) \\ &= [(\text{demand})/(\text{load factor} \times \text{vehicle capacity})] \times (\text{cost per frequency})\end{aligned} \quad (10.36)$$

$$\text{Handling cost} = (\text{demand}) \times (\text{cost per unit of demand}) \quad (10.37)$$

The generic variables in Eq. (10.36) are specific for the particular operational steps of the intermodal transport network: In the collection and distribution step, "Frequency" relates to the number of trucks' runs in collecting and/or distributing a given volume of LUs. In particular, in the "shipper" zone (k) "Frequency" (f_k) is proportional to the volume of demand-LUs (Q_k) (ton) and inversely proportional to the product of the truck payload capacity (PL_k) (ton) and load factor (λ_k) (≤ 1.0). The variable "Cost per

Frequency" relates to the operational costs of an individual truck and is usually expressed in relation to distance (ie, length of a tour) as ($c_{ok}(d_k)$ (€/veh km). The analogous reasoning for the truck service frequencies, distances, and related costs is applied to the distribution step in the "receiver" zone (l) (The number of zones in the "shipper" and "receiver" area is (K) and (L), respectively). In the line hauling step, "Frequency" f (departures/week) is proportional to the volume of LUs (Q) (ton) and inversely proportional to the product of the train payload capacity (PL_t) (ton) and load factor (λ_t) (≤ 1.0). The "Cost per Frequency" relates to the operational costs per train assumed to depend on the train's gross weight (W) (ton), payload (q) (ton), and the line haul distance (d) (km), $c_o(W, q, d)$ (€/train). Such expression of the costs enables making a clear distinction between the CIFTs and LIFTs.

The generic variables in Eq. (10.37) have the following meaning: the handling costs in the collection step in "zone" (k) are proportional to the volume of demand, ie, quantity of LUs (Q_k) (ton) and the unit handling costs (c_{hk}) (€/ton), respectively. These costs are analogous for the distribution step in zone (l). In the rail line-hauling step, the handling costs are proportional to the payload (q) (ton) and the unit handling costs at both intermodal terminals, (c_{h1}) and (c_{h2}) (€/ton), respectively. In many cases these costs have been considered as the cost of loading/unloading of a train. The more detailed equations for calculating particular cost components are as follows (Janić, 2007):

$$C_F = C_c + C_{lh} + C_d = \sum_{i=1}^{2}\sum_{k=1}^{K} C_{c/i/k} + C_{lh} + C_{ht} + \sum_{i=1}^{2}\sum_{l=1}^{L} C_{d/i/l} \tag{10.38}$$

where

Collection/distribution (transport costs) ($i=1$):

$$C_{c/1/k} = (Q_k/\lambda_k \cdot PL_k) \cdot c_{ok}(d_k) \text{ and } C_{d/1/l} = (Q_l/\lambda_l \times PL_l) \times c_{ol}(d_l) \tag{10.39}$$

Collection/distribution (handling costs) ($i=2$):

$$C_{c/2/k} = Q_k \cdot c_{hk} \text{ and } C_{c/2/l} = Q_l \cdot c_{hl} \tag{10.40}$$

Line-haul (train transport costs):

$$C_{lh} = f \cdot c_o(W, q, d) = (Q/\lambda_t \cdot PL_t) \cdot c_o(W, q, d) \tag{10.41}$$

Line-haul (handling—train loading/unloading costs):

$$C_{ht} = q \cdot (c_{h1} + c_{h2}) \tag{10.42}$$

In particular, the train's operational costs consist of five components: (i) investments in rolling stock—wagons and locomotive(s); (ii) the cost of maintenance of rolling stock-wagons and locomotives; (iii) the cost of using the railway infrastructure, ie, the infrastructure charge; (iv) the cost of energy consumption; and (v) labor cost for assembling/decoupling and driving the train (Janić, 2008). Dividing the total costs in Eq. (10.38) by the volume of demand and distance gives the average operational costs per unit of the network's output (€/t km), which can be used for comparisons of the intermodal transport networks operated by either CIFTs or LIFTs, or for comparisons between the intermodal and road transport networks. In the latter case, for the road transport system, the equations for the costs of collection/distribution can be modified using the door-to-door distance between "zone" (k) and (l).

The above-mentioned models for calculating the operational costs of given intermodal rail road transport network and its road counterpart have been applied by using the following input data (Janić, 2007, 2008):

The LUs are of the size of TEU, with the average gross weight of 14.3 ton (12 ton of payload and 2.3 ton own weight-tare). The characteristics of operating road trucks In each freight/goods collecting and distributing zone of the intermodal transport network have been given in Table 10.18:

Table 10.18 Characteristics of the Collection and Distribution of LUs in the Intermodal Transport Network in Given Example (Janić, 2007, 2008)

Parameter	Train Category		
	CIFT	LIFT—800 m	LIFT—1000 m
Collection/distribution distance by road ($d_{k/l}$) (km)	50	75	75
Truck's payload capacity ($PL_{k/l}$) (ton)	2 × 14.3	2 × 14.3	2 × 14.3
Load factor $\lambda_{k/l}$	0.60	0.60	0.60

The average collection/distribution distance is assumed to be longer for the LIFT than for the CIFT because more loading units are needed to fill in the latter. They are collected and distributed over the wider area implying the constant spatial concentration of shippers and receives and their freight/goods shipment generating and attracting potential.

The truck's operational costs during the collection/distribution based on the full load equivalent of 2 TEU have been estimated by the regression analysis as follows: $c_o(d) = 5.456\ d^{-0.277}$ (€/veh km) ($N = 26$; $R^2 = 0.78$; $25 \le d \le 1600$ km). The average load factor has assumed to be: $\lambda = 0.60$. These costs have already included the handling costs of LUs (Janić, 2007, 2008).

The characteristics of the intermodal CIFTs and LIFTs operating between two intermodal terminals of the given network are given in Table 10.19 (Janić, 2007, 2008).

Table 10.19 Characteristics of the CIFTS and LIFTs in the Given Example (Janić, 2007, 2008)

Parameter	Train Category		
	CIFT	LIFT—800 m	LIFT—1000 m
Train load factor (λ)	0.75	0.75	0.75
Payload (q) (ton)	14.3 × 3 × 26 × 0.75 = 837	14.3 × 3 × 38 × 0.75 = 1223	14.3 × 3 × 48 × 0.75 = 1544
Empty train weight (W_e)			
1 locomotive (ton)	24 × 26 + 810 = 713	24 × 38 + 810 = 1001	24 × 48 + 810 = 1241
2 locomotives (ton)	–	24 × 38 + 2 × 810 = 10,100	24 × 48 + 2 × 810 = 1330
Total train gross weight (W)			
1 locomotive (ton)	1550	2224	2785
2 locomotives) (ton)	–	2313	2874

Table 10.19 Characteristics of the CIFTS and LIFTs in the Given Example (Janić, 2007, 2008)—cont'd

Parameter	Train Category		
	CIFT	LIFT—800 m	LIFT—1000 m
Train operating speed (v)			
1 locomotive) (km/h)	110	70	65
2 locomotives) (km/h)	–	105	100
Train commercial speed (v)			
1 locomotive) (km/h)	60	50	45
2 locomotives) (km/h)	60	50	45
Anticipated delay (D) (h)	1	1	1

In addition, the operational costs of the CIFTs and LIFTs, $c_o(W, q, d)$ in Eq. (10.41), have been estimated as follows (Janić, 2008):

$$c_o(W,q,d) = (4.60n_l + 0.144n_w + 0.3)\cdot d + 12.98(n_l + n_w) + 0.0019\cdot W\cdot d + \\ + \sum_{n=1}^{N}[0.227\cdot 10^{-6}v_n^2/\ \ln d_n + 0.000774]\cdot W\cdot d + 33\cdot n_d\cdot\left(t_{dp} + d/v + D\right) \tag{10.43}$$

where

n_l is the number of locomotives per train
n_w is the number of locomotives per train
v_n is the train speed on the segment (n) of the route (km/h)
n_d is the number of drivers per train
t_{dp} is the driver's preparation and finishing time before and after the trip (h)

The particular cost rates have the following meaning: the first term represents the unit cost of depreciation and maintenance of the rolling stock (flat wagons and locomotives) and the monitoring cost of a train along the line (4.60, 0.144, 0.3, respectively) (€/km). The second term represents the unit cost of assembling/decomposing a train at both sides of the corridor (12.88) (€/train). The third term represents the unit cost of using the rail infrastructure (ie, the infrastructure charge) (0.000110) (€/t km). The fourth term represents the unit cost of the energy consumption along the line with (N) segments, ie, intermediate stops $\left[\sum_{n=1}^{N} 0.227\times 10^{-6}v_n^2/\ \ln d_n + 0.000774\right]$ €/t km. The last term represents the unit cost of the train's driver(s) (33) (€/h). For the road counterpart network, the operational costs of the road haulage between particular shippers and receivers are estimated by the same equation as for the collection/distribution step of LUs at intermodal transport network, but respecting the corresponding haulage distance and the average load factor of: $\lambda = 0.85$ (Janić, 2007, 2008).

Fig. 10.55 shows the relationships between average unit costs of the intermodal rail/road transport network operated by either CIFTs or LIFTs and its road truck counterpart, and the door-to-door delivery distance of LUs estimated by using the above-mentioned input data (Janić, 2007, 2008).

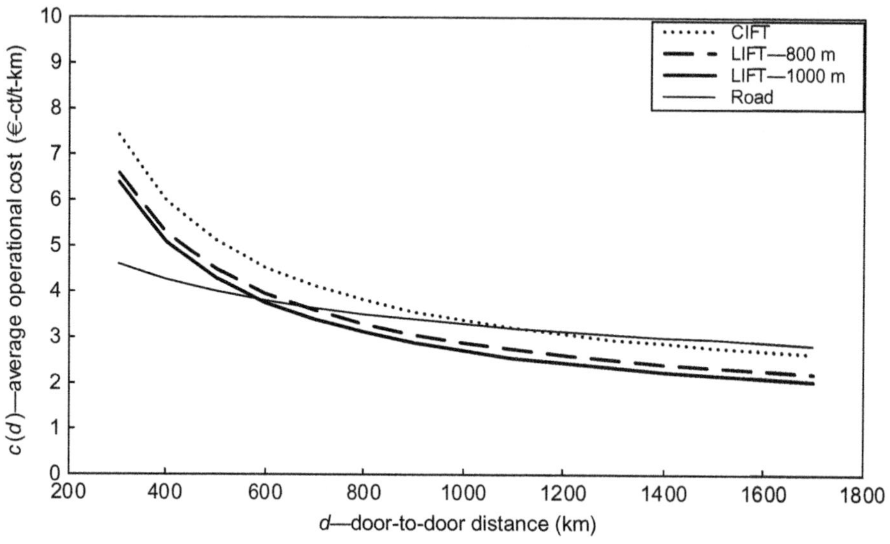

FIG. 10.55

Relationships between the average unit operating costs of different categories of intermodal freight trains (CIFT and LIFT) and the alternative road truck haulage, and door-to-door delivery distance of LUs.

As can be seen, the average unit costs decrease more than proportionally with increasing of the door-to-door distance at both (intermodal and road) network, thus indicating existence of the economies of distance. In general, these costs decrease at higher rate at intermodal than at road transport network. The costs of two networks equalize and become increasingly lower at the intermodal transport network beyond the distance of about 1000 km if the CIFTs and the distances of about 600 and 700 km if different categories of the LIFTs would be operated, respectively. This is because the average unit costs of the intermodal transport network operating the LIFTs of 800 m would be lower than that operating the CIFTs for about 12–18% for the range of door-to-door distances from 300 to 1300 km. If the LIFTs of 1000 m were operated these costs would be lower for about 18–27% on the same range of distances. Both figures indicate the existence of economies of scale of the LIFTs as compared to the CIFTs.

The relationships between the average operating unit costs of both systems indicate that the intermodal transport operating the CIFTs is currently a competitive alternative to the long-haul road transport beyond the above-mentioned "break-even" distances. Since the volumes of demand generally decrease with the distance, this might partially explain the current modal split between two modes in Europe. The average unit operating costs of road are lower than that of intermodal transport network over the short, medium and even some long distances—markets, which, in combination with the other market and regulatory factors has lead to the lower road prices. These have attracted more of the voluminous and price sensitive demand over short, medium and some long distances (about 100% up to 600 km). Therefore, introducing the LIFTs could seemingly raise competitiveness of the intermodal transport through decreasing of the cost-based prices over the range of shorter distance(s) where more voluminous demand has existed. Thus, deployment of LIFTs would generally contribute to improving efficiency of the intermodal (particularly rail) transport operators and consequently cause change of the current modal split (in Europe).

10.6.7.5 Revenues

The operating costs of the intermodal rail/road transport networks are covered by charging customers, in this case usually the shippers of containers. Fig. 10.56 shows an example of the relationships between these average unit charges/prices and the door-to-door delivery distances for the intermodal rail/road transport and its road counterpart in the United States. For the purpose of comparison with the costs in Fig. 10.55, the values are converted from the \$US into €.

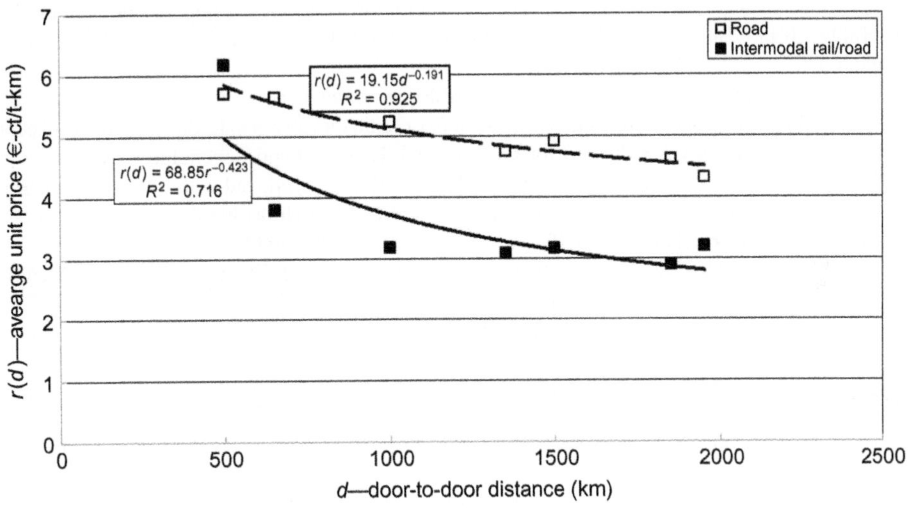

FIG. 10.56

Relationship between the average unit prices and distances for the intermodal rail/road and road transport in the United States (period: 2008) (http://www.ppiaf.org/sites/ppiaf.org/files/documents/toolkits/railways).

As can be seen, the average unit prices at both systems decrease more than proportionally with increasing of the door-to-door distances, thus indicating their following the similar behavior of operating costs. In addition, in given case, for distances longer than 500 km, the intermodal rail/road transport of containers have costed for about 20% less than its road counterpart. This gap has further increased with increasing of the door-to-door distance. As well, the absolute values of these prices/charges of intermodal transport system are in some way comparable to their cost counterparts shown in Fig. 10.55 (http://www.ppiaf.org/sites/ppiaf.org/files/documents/toolkits/railways).

REFERENCES

AAA, 2013. Your Driving Costs, Edition 2012. American Automobile Association, Heathrow, FL.

ACI, 2010. Airport Capital Development 2010–2013. Airport Council International-North America, Washington, DC.

AEA, 2011. Reduction and testing of greenhouse gas (GHG) emissions from heavy duty vehicles—lot 1: strategy. Final report to the European Commission, DG Climate Action Ref: DG ENV. 070307/20010/548572/SER/C, Didcot.

Anderson, R., 2010. Managing metro fares and funding. Metro Report International, 33–35.

APMMG, 2014. Annual Reports (2008–2014). A. P. Moller-Maersk A/S Group, Copenhagen.

APTA, 2015. APTA streetcar and heritage trolley site. Hosted by the Seashore Trolley Museum, APTA Streetcar Subcommittee, American Public Transportation Association.http://www.heritagetrolley.org/.

Brown, J., 2013. The modern streetcar in the US: an examination of its ridership, performance, and function as a public transportation mode. J. Public Transport. 16 (4), 43–61.

BTS, 2014. Average Cost of Owning and Operating an Automobile (a) (Assuming 15,000 Vehicle-Miles Per Year), Table 3–17. Bureau of Transport Statistics, United States Department of Transportation, Washington, DC.

Button, K., 2010. Transport Economics, third ed. Edward Elgar, Aldershot.

CE Delft, 2004. Charges for Barges? Preliminary Study of Economic Incentives to Reduce Engine Emissions From Inland Shipping in Europe. CE Solutions for Environment, Economy and Technology, Delft.

CE Delft, 2008. Road infrastructure cost and revenue in Europe. Report, produced within the study internalisation measures and policies for all external cost of transport (IMPACT)—deliverable 2, version 1, solutions for environment, economy and technology, Delft.

CECW-CP, 2004. Shallow draft vessels operating costs—fiscal year 2004. Economic Guidance Memorandum, 05-06, US Army Corps of Engineers Civil Works, Vicksburg, MS.

Crist, P., 2012. Electric vehicles revisited—costs, subsidies and prospects. Discussion Paper 2012-03, International Transport Forum at the OECD, Paris.

de Gijt, J.G., 2010. A history of quay walls: techniques, types, costs and future. PhD thesis, Delft University of Technology, Delft.

de Rus, G. (Ed.), 2010. Economic analysis of high speed rail in Europe: Informes 2010. Economía y Sociedad, Fundación BBVA, Bilbao.

de Rus, G., Nombela, G., 2007. Is investing in high-speed rail socially profitable? J. Transp. Econ. Policy 41 (1), 3–23.

Delta, 2014. Annual Report Pursuant to Section 13 or 15(d) of the Securities Exchange Act of 1934, Securities and Exchange Commission (Washington, D.C.,), FORM 10-K, Delta Airlines, Inc., Atlanta, Georgia, USA.

DLR, 2008. Analyses of the European Air Transport Market: Airline Business Models. German Aerospace Centre, Air Transport and Airport Research, Koln.

Doganis, R., 1992. The Airport Business. Routledge, London and New York, NY.

DSC, 2010. Container Freight Rate Insight: Bi-Monthly Pricing Benchmarks on the Container Market. Drewry Shipping Consultants Ltd, London.

EC, 2006. Long innovative intermodal interoperable freight trains. European Commission, INTERREG IIIB NWE, C041, Brussels, Belgium.

EC, 2010. Terminal handling charges during and after the liner conference. Competition reports, European Commission, Brussels.

ECORYS, METTLE, 2005. Charging and pricing in the area of inland waterways: practical guideline for realistic transport pricing. Final report, ECORYS Transport (NL), METTLE (F), Rotterdam.

EEC, 2010. Cost of the en-route air navigation services in Europe. EEC Note No. 8/1010, Project GEN-4-E2, EUROCONTROL Experimental Centre, Bretigny-sur-Orge CEDEX, France.

EEC, 2015. Report on the Operation of the Route Charges System in 2014. Central Route Charges Office (CRCO), EUROCONTROL, Brussels.

EGM, 2004. Shallow Draft Vessel Operating Costs, Economic Guidance Memorandum, FY 2004 05–06. hqplanning@usace.army.mil.

Feigenbaum, B., 2013. High-Speed Rail in Europe and Asia: Lessons for the United States. Reason Foundation, Los Angeles, USA. www.reason.org.

FTA, 2007. Transit bus life cycle cost and year 2007 emissions estimation. Report no. FTA-WV-26-7004.2007.1, Federal Transit Administration, US Department of Transportation, Washington, DC.

GAO, 2001. Mass transit: bus rapid transit show promise. Report to congressional requesters, United States General Accounting Office, Washington, DC.

Garcia, A., 2010. Relationship between rail service operating direct costs and speed. Study and Research Group for Economics and Transport Operation. International Union of Railways (UIC), Paris, France.

GCD, 2006. South China Port Development Industry. The Paper of the Guangdong Communication Department, Guangzhou.www.hollandinchina.org.

Gonzales, D., Searcy, M.E., Eksioglu, D.S., 2013. Cost analysis for high-volume and long-haul transportation of densified biomass feedstock. Transp. Res. A 410A, 48–61.

Greiner, R. (Ed.), 2012. Ship operating costs: current and future trends. Shipping. Moore Stephens, London.

Harford, D.J., 2006. Congestion, pollution, and benefit-to-cost ratios of us public transit systems. Transp. Res. D 11, 45–58.

Henn, L., Sloan, K., Douglas, N., 2013. European case study on the financing of high speed rail. In: Proceedings of Australasian Transport Research Forum, 2–4 October 2013, Brisbane.

HPW, 2010. Operating costs fact sheet. Rapid Transit, Moving Hamilton Forward. Hamilton Public Works.www.hamilton.ca/rapid-transit.

Hwang, L.C., Yoon, K. (Eds.), 1981. Multi attribute decision-making: a methods and applications. In: Lecture Series in Economics and Mathematical Systems, Springer-Verlag, Berlin.

ICAO, 2013. Airport Economic Manual, Doc 10562, third ed. International Civil Aviation Organization, Montreal.

INA, 2005. Economic aspects of inland waterways. International Navigation Association, Report of Working Group 21, PIANC General Secretariat, Brussels.

ISL and amrie, 2006. Tonnage measurement study. MTCP Work Package 2.1, Quality and Efficiency, Final report, Bremen/Brussels.

Janić, M., 2000. Air Transport System Analysis and Modelling: Capacity, Quality of Services, and Economics. Gordon and Breach Science Publishers, Amsterdam, The Netherlands.

Janić, M., 2007. Modelling the full costs of intermodal and road freight transport networks. Transp. Res. D 12 (1), 33–44.

Janić, M., 2008. An assessment of the performance of the European long intermodal freight trains (LIFTS). Transp. Res. Part A: Policy Pract. 42 (10), 1326–1339.

Janić, M., 2014. Advanced Transport Systems: Analysis, Modelling, and Evaluation Performances. Springer, London.

JR, 2014. Visitors Guide 2014. Central Japan Railway Company, Nagoya.

Lowell, D., Seamonds, D., 2013. Supporting Passenger Mobility and Choice by Breaking Modal Stovepipes: Comparing Amtrak and Motorcoach Service. M.J. Bradley & Associates LLC, Manchester, NH.

LSB, 2006. Operating Costs of Trucks in Canada 2005. Logistics Solutions Builders Inc., Calgary. p. 73.

Medeiros, A., Potter, J., 2015. European Airport Operating Cost Benchmarking, 2010–2014, Quick Reference Guide. PWC, London, UK. www.strategyand.pwc.com.

OECD, 2007. Twenty foot equivalent unit (TEU). Glossary of Statistical Terms, Organization for Economic Co-operation and Development, Paris.

OECD/ITF, 2015. The Impact of Mega-Ships: Case-Specific Policy Analysis. International Transport Forum, Organization for Economic Co-operation and Development, Paris.

PHK, 2006. Port Benchmarking for Assessing Hong Kong's Maritime Services and Associated Costs With Other Major International Ports. Maritime Department, Planning Development and Port Security Branch, Port of Hong Kong, Hong Kong.

PO, 2013. Comparative subway construction costs, revised. Pedestrian Observations. https://pedestrianobservations.wordpress.com/2013/06/03/comparative-subway-construction-costs-revised/.

Pourreza, S., 2011. Economic Analysis of High Speed Rail. NTNU (Norwegian University of Science and Technology), Trondheim.

PwC, Pantela, 2013. Impact Assessment on Measures to Enhance Efficiency and Quality of Port Services in the EU. PWC, London, UK.

References, 2011. Comparisons of public transport operations. Journeys (November) 71–78.

Sauian, S.M., 2010. MCDM: a practical approach in making meaningful decisions. In: Proceedings of the Regional Conference on Statistical Sciences 2010 (RCSS'10), June, pp. 139–146.

Southwest, 2015. 2014 Annual Report to Shareholders. Southwest Airlines Co., General Offices, Dallas, TX.

Sys, C., Blauwens, G., Omey, E., Van de Voorde, E., Vitlox, F., 2008. In search of the link between ship size and operations. Transp. Plan. Technol. 31 (4), 435–463.

TfL, 2000. Cross River Transit: Summary Report. Transport for London, London.

TfL, 2015. Rail and underground international benchmarking report. Rail and Underground Panel, Transport for London, London.

UIC, 2005. High Speed Rail's Leading Asset for Customers and Society. UIC Publications, International Union of Railways, Paris.

UN, 1976. Port Performance Indicators, United Nations Conference of Trade and Development, Geneva, Switzerland.

UNCTAD, 2015. Review of maritime transport. In: United Nations Conference of Trade and Development. United Nations, New York, NY.

USDE, 2013. Trucks and heavy-duty vehicles technical requirements and gaps for lightweight and propulsion materials. Workshop Report, DOE/EE-0867, Vehicle Technologies Office, US Department of Energy, Washington, DC.

USDT, 2015. Bus Profile. US Department of Transportation, Federal Highway Administration, Highway Statistics, Washington, DC.

van Donselaar, P., Carmigchelt, H., 2001. Container transport on the Rhine marginal cost: case study infrastructure, environmental—and accident costs for Rhine container shipping. Competitive and Sustainable Growth Program (European Commission), Workpackage 5/8/10, Version 2, NEI B. V., Amsterdam.

Vandenbossche, P.K.J.W., 2015. Binary mixed-integer model of fleet disposal and acquisition strategy for freight airlines. MSc thesis, Imperial College London, London, Delft University of Technology, Delft.

WB, 2010. Air freight: a market study with implications for landlocked countries. The World Bank Group, Transport Papers, TP-26, World Bank, Washington, DC.

WB, 2015. World Development Indicators Database. World Bank, 8 July 2014, Washington, DC, USA.

WSDT, 2004. Highway construction costs: are WSDOT's highway construction costs in line with national experience? In: Presentation. Washington State Department of Transportation, Washington, DC.

Wu, J., 2013. The financial and economic assessment of China's high speed rail investments: a preliminary analysis. Discussion Paper No. 2013/28 Prepared for the Roundtable on The Economics of Investment in High Speed Rail, 18-110 December 2013, New Delhi InternationalTransport Forum, Paris.

Yue-man, Y., Jianfa, S., Li, Z., 2006. China's 11th five-year plan: opportunities and challenges for Hong Kong. Final report, Hong Kong Institute of Asia-Pacific Studies, The Chinese University of Hong Kong, Hong Kong.

WEBSITES

http://cometandnova.org/benchmarking.
http://exchange.aaa.com/automobiles-travel/automobiles/driving-costs.
http://www.investopedia.com/terms.
http://www.ppiaf.org/sites/ppiaf.org/files/documents/toolkits/railways.

When does traffic congestion occur? When does sound become noise? What is the "noise footprint" around an airport? What is the target for reduction of total CO_2 emissions? What factors influence energy consumption of transport vehicles? How do we estimate emissions of GHG and particularly CO_2 from the transport sector? What is an acceptable proportion of urban land to be devoted to transport infrastructure?

TRANSPORTATION, ENVIRONMENT, AND SOCIETY 11

11.1 INTRODUCTION

Transportation systems are among the most important energy consumers. At the same time, transportation is the biggest polluter in the modern world. Transportation has a great damaging effect on the natural environment. On a daily basis, transportation makes happen atmospheric and noise pollution. The social-economic development during the 20th century and at the beginning of the 21st century has shown that modern societies have seemingly been unable to provide sustainable transportation for the human population. Sustainable transportation assumes sustainability in the senses of social, environmental and climate impacts. Sustainable transportation also assumes capability to provide the source energy ad infinitum.

Generally, the concept of "sustainable" development has been defined as the "development that meets the present needs without compromising the ability of the future generations to meet their needs" (Brundtland, 1987; Kelly, 1998; OECD, 1998).

Transportation systems have provided numerous socio-economic benefits to the society. On the other hand, they have generated a series of social and environmental impacts, particularly by taking land for transportation infrastructure, causing congestion, local noise, increased fuel/energy consumption, emissions of Green House Gases (GHG), and traffic accidents/incidents. Since these effects have been continuously growing, the calls for the transportation systems sustainable development have become more frequent and louder. The definitions of the sustainable transportation have emerged by both academic and practitioners communities. "Sustainable transportation is that, which does not endanger public health or ecosystems and that meets needs for access consistent with (a) use of renewable

Transportation Engineering. http://dx.doi.org/10.1016/B978-0-12-803818-5.00011-1

resources that are below their rates of regeneration, and (b) use of nonrenewable resources below the rates of development of renewable substitutes" (OECD, 2001).

An example of setting up the quantitative targets on the transportation systems' impacts on the environment and society was that by the European Union (EU), as given in Table 11.1 (CEC, 2000):

Table 11.1 The Targets for Reduction of Some Social and Environmental Impacts of Transportation Systems in the EU Until the Year 2020 (CEC, 2000)

Impact	Long-Term Targets
CO_2	Should not exceed 20% of the total CO_2 emissions in 1990
VOCs[a]	Should not exceed 10% of the total VOCs emissions in 1990
NO_x	Should not exceed 10% of the total transport-related NO_x emissions in 1990
Particulates	Reduction of fine particulate (PM_{10}) emissions from transport for 55–99%
Noise	Should not exceed a maximum of 55–70 dB during the day and 45 dB at night and indoors
Land-use	Compared to 1990, a smaller proportion of urban land devoted to transport infrastructure

[a]*Volatile organic compound(s).*

These targets have been one of the best-known examples of the international policies addressing sustainability of the transport systems. Consequently, citizens should be provided with "safe, environmentally and consumer friendly, and quality driven transport systems and services." In other words, it appears that the long-term sustainable development of a given transportation system could be achieved if its overall positive contribution to the economic and social welfare would continuously increase and the total negative impacts on people's health and environment decrease to or below given absolute targets (Janić, 2003).

This chapter analyses the direct impacts of transportation systems on the society and environment, and their costs/externalities. Only the impacts and related costs/externalities from providing transport services are considered. The main impacts on the society considered include congestion, noise, and traffic incidents/accidents (ie, safety). The main impacts on the environment considered include the energy/fuel consumption and related emissions of GHG, land use, and waste.

If these impacts are charged usually in the form of different environmental taxes/fees, they represent the external costs/externalities imposed on those causing them, that is, transport operators as providers of transport services and their users. If actually being paid, these costs/externalities can also be considered as components of the total costs of particular transport systems.

The direct impacts of transportation systems on the society and environment and their costs/externalities are described for road, rail, water, and air-based transport systems. Some of them are considered transporting both passengers and freight/cargo shipments, the others transporting exclusively freight/cargo shipments, but in both cases at different levels of details and spatial/geographical scale.

11.2 CATEGORIZATION AND MODELING IMPACTS

11.2.1 CONGESTION

Congestion occurs whenever demand exceeds the available capacity of the given, usually the infrastructure, component of transport system(s)—urban street(s)/intersection(s), interurban road lane(s), rail tracks/stations, inland waterway segment(s), port quay/berth area(s), airport runway(s), etc. The ultimate causes lies on both capacity and demand side of the systems' components. The capacity of each component is limited by its design and constructive characteristics, and operational rules and procedures of its use (safely, efficiently, and effectively). The demand causes congestion because its units intend to use such limited capacity at the same time, thus interfering among each other while waiting to be served. Such demand/capacity interaction results in the shortage of given component's capacity during that relatively short times called "peak periods" and plenty of it otherwise.

Congestion causes delays of units of demand, generally vehicles, and consequently prolongs their door-to-door travel times. Therefore, congestion is considered as an impact due to delays it causes, which are assumed to be lost times for those affected. These typically can be the morning and late afternoon/evening commuters in urban areas and the time sensitive freight/good shipments waiting at the congested road at trucks, rail lines at trains, and ports at ships waiting to be docked and unloaded. In this context, congestion does not include the time loses during irregular system's/component's operations caused disruptions due to the system-internal and external causes. The former are the failures of particular systems components due to internal reasons. The latter are bad weather affecting regular system's/component's operations and natural disasters (volcano eruptions, earthquakes, flooding) damaging the system's components and consequently affecting their regular operations.

In addition to the traffic flow theory, elaborated in Chapter 4, the most common way for expressing the congestion delay has been the well-known simple formula from *M*/*D*/1 queuing theory (elaborated in Chapter 3) as follows *M*—Poisson flow of arrivals; *D*—deterministic service time; 1—single server:

$$D = \frac{\rho}{2\mu(1-\rho)} \tag{11.1}$$

In this case, $\rho = \lambda/\mu < 1.0$, where λ and μ is the intensity of demand and the service rate, respectively (units of demand/unit time).

11.2.2 NOISE

Sound becomes noise when it affects the common quality of life. It is caused by a fluctuation in pressure transmitted as a wave through the air (or any other elastic medium, including water). Noise is usually expressed by the so-called pressure level L_p as follows:

$$L_p = 10\log_{10}\frac{p_{rms}^2}{p_{ref}^2} = 20\log_{10}\frac{p_{rms}}{p_{ref}} \tag{11.2}$$

where

p_{rms} is the pressure mean square value; and
p_{ref} is the reference pressure.

The pressure mean square value p_{rms} in Eq. (11.2) may vary in practice from between 10^{-5} and 10^3 Pa (Pa—Pascal). The reference pressure is usually taken to be: $p_{ref} = 2 \times 10^{-5}$ Pa, which is the pressure just audible to the human ear at a frequency of 1000 Hz (Hz—Hertz). The unit of sound pressure is decibel—dB. Therefore, the threshold of hearing at the frequency of 1 kHz corresponds to $p_{rms} = p_{ref} = 1$, thus giving $L_p = 0$. According to this definition, doubling p_{rms}^2 implies an increase in L_p of about 3 dB. However, these relationships have not always appeared to be particularly appreciated or clearly understandable. Another important aspect of dealing with noise is the distance from the source. In order to assess the influence of distance, the noise power is expressed in W (watts). Under such circumstances the noise power can vary between 10^{-9} and 10^4 W, a scale similar to the one for

(p_{rms}^2) and twice as wide as the one for p_{rms}. Taking that into account the sound power level can be expressed as:

$$L_W = 10\log_{10}\frac{W}{W_{ref}} \tag{11.3}$$

The value of $W_{ref} = 10^{-12}$ W (Watts). Consequently the sound pressure level $L_p(R)$ at distance d from a noise source can be approximated as (Boeker and Grondelle, 1999):

$$L_p(d) = L_W - 10\log_{10}\left(4\pi d^2\right) + 0.14 \tag{11.4}$$

where L_W is determined from Eq. (11.3). In general, Eq. (11.4) implies that the distance from the noise source is important and that the sound power decreases with increasing of the distance from the source. For example, by doubling the distance from d to $2d$, the sound power L_p decreases by about 6 dB. For example, the human ear can register the minimum sound of 0 dB. The car and truck passing by create noise of about 70 dB and 90 dB, respectively.

Regarding everyday life and activities, the commonly used measure for noise is L_{dn} (ie, average day/night Average Sound Level). This is the measure of the cumulative noise the individuals or group of people are exposed during 24 h of the day. It weights the noise during the night by adding a decibel "penalty" as follows:

$$L_{den} = 10\log\left[\frac{12}{24}\cdot 10^{L_{eq/day}/10} + \frac{4}{24}\cdot 10^{\left(L_{eq/evening}+5\right)/10} + \frac{8}{24}\cdot 10^{\left(L_{eq/night}+10\right)/10}\right] \tag{11.5}$$

where $L_{eq/day}$, $L_{eq/evening}$, $L_{eq/night}$ are the annual average equivalent sound pressure levels during the day, evening, and night, respectively.

In addition, the L_{eq} (Equivalent Continuous Sound) is expressed in A-weighted dBA scale as an average of the acoustical energy measured also over a shorter period of time (min, h) (Janić, 2007). It has shown to be particularly convenient for estimating noise generated by the successive vehicles passing near the affected population as follows:

$$L_{eq} = \overline{L_{AE}} + 10\log_{10}N - 10\log_{10}\tau \tag{11.6}$$

where

$\overline{L_{AE}}$ is the average of the maximum noise levels L_{amax} (dBA) generated by N successive noise events—vehicles passing by an affected observer—during the period τ (dBA);
N is the number of vehicles passing by affected population during time period τ; and
τ is the period of time (min, h).

In addition, in any case, while dealing with the excessive noise, the distance between the moving vehicles as the sources of noise and the potentially affected observer(s)/population needs to be taken into account. In general, the level of noise attenuates with distance as follows:

$$L_{AE}(d_2) = L_{AE}(d_1) - 20\log_{10}\left(\frac{d_2}{d_1}\right) \tag{11.7}$$

where

$L_{AE}(d_1), L_{AE}(d_2)$ is the noise level from the source at the distance d_1 and d_2, respectively (dBA); and
d_1, d_2 is the distance at which the noise is measured/registered ($d_1 < d_2$) (m).

For example, if the noise level measured from the source at the distance $d_1 = 25$ m is $L_{AE}(d_1) = 90$ dBA, the noise from the same source at the distance $d_2 = 50$ m will be: 90 dBA − 20 * $\log_{10}(50/25)$ = 90 dBA − 6.02 = 83.98 dBA.

Last but not least, some standards on the maximum acceptable levels of noise L_{DEN} and L_{eq} the affected population can be exposed to have been set up. One of the cases is that by the U.S. FHWA (Federal High-Way Administration): the outside noise levels in the residential areas should be up to about: $L_{DEN} = 55$ dBA; the perceived noise inside the home should be for 12–17 dBA and 20 dBA lower if the windows are opened and closed, respectively. Otherwise, the noise barriers need to be used in order to protect the affected population from the excessive noise. For example, the walls constructed along the highways and railway tracks/lines can reduce noise for about 10–15 dBA and in some cases even for about 20 dBA.

11.2.3 TRAFFIC INCIDENTS/ACCIDENTS (SAFETY)

Road traffic injuries are the main reason of death for young people. The average world road traffic fatality rate is 18 per 100,000 population. The transportation systems are examples of the systems where various types of risks are inherently present. The risk can be defined as the probability of hazardous events, specific activities or actions occurring in a more or less random manner, whose outcome may be the individual's (or group's) death or injury. Frequently, the risk is related to the statistically expected value of loss. In other words, it represents the statistical likelihood of random exposure of an individual to some hazardous event, in which case a measure of the probability of severity of the impacts is involved. In such a context, safety represents the acceptable level of risk.

Up to date, the following four types of societal risk have been identified (Sage and White, 1980):

- real risk, which may be determined based on the circumstances after they have fully developed;
- statistical risk, which may be determined by the available statistics about a particular type of accidents;
- predicted risk, which may be predicted analytically from the dedicated models; and
- perceived risk, which may intuitively be felt by individuals.

The transportation system is an example of the systems where all the above-mentioned types of risks are inherently present and related to occurrence of traffic accidents. These accidents are generally characterized by randomness in terms of time of occurrence, the number of people affected, consequences in terms of fatalities and injuries of people involved and those of the third party including damaging properties corresponding properties, and permanency of threat. These characteristics can vary at transport systems operated by the same and/or different transport modes.

In general, most traffic accidents are caused by a complex of mutually dependent causes happening in a sequential order, which can be classified according to two criteria. First, according to the current state-of-knowledge they can be "known and avoidable" and "unknown and unavoidable." The term "unknown and unavoidable" should be considered only conditionally, since just after an accident the real causes might not be known. As the accident investigation progresses, causes often become "known and avoidable" except in some rare cases. Second, according to the type, causes of traffic accidents can be classified into "human errors," "mechanical failures," "hazardous weathers," "sabotage/terrorist attacks," and "military operations" (Janić, 2007). Consequently, if the traffic accidents in a given system occur due to the already known and avoidable causes, the system should be considered as unsafe. Otherwise, it should be considered as safe (Kanafani, 1984).

The safety of a given transport system operating by a given transport mode can be analyzed in two ways: (i) absolute, when the number of events (vehicle accidents/crashes, fatalities and injuries of both users/passengers and freight shipments on board, the system employees, and damages to the third party of any kind) is considered over time; and (ii) relative, the number of events is considered as dependent of the given transport system's output, realized during a given period of time. The former approach enables monitoring the safety trend over time just based on the number of events. The latter approach enables monitoring the system's safety depending on the scale of its operations but also implicitly over time. In this later case, the safety, that is, the perceived risk of an accident and/or related fatalities and/or injuries in a given transport system can be estimated as follows:

$$AC_r = N_{ac}(\tau)/Q(\tau) \tag{11.8}$$

where

AC_r is the rate of traffic accidents, fatalities, injuries, etc. (number/v-km; number/p-km; number/t-km; number/TEU-km);
$N_{ac}(\tau)$ is the number of events—traffic accidents, fatalities, and/or injuries occurred during the time period τ; and
$Q(\tau)$ is the volume of output of a given transport system during the time period τ (v-km/year; p-km/year; t-km/year; TEU-km/year) (v—vehicle).

Usually, the obtained values in (11.8) have been compared to some targets set up in advance in order to assess and monitor the systems' progress towards them. It has usually been expected that the accident rate expressed in any terms diminishes (often proportionally or even more than proportionally) over time and with increasing of the (annual) volumes of output(s), thus indicating that the given system becomes less risky, that is, safer.

In addition, the accident rate at particular transport systems can include, in more details, other indicators such as the loss of vehicles and damages to the third parties. In addition, this rate can be analyzed depending on the most important causes of given accidents in which case the regression analysis has often shown to be a useful analytical tool.

11.2.4 ENERGY/FUEL CONSUMPTION AND EMISSIONS OF GHG

11.2.4.1 Energy/fuel consumption

Energy consumption of transport vehicles generally depends on the many factors such as characteristics of the vehicle(s), infrastructure (roads, rail tracks), type of transport services characterized by the stops, acceleration/deceleration along the given route(s), and driving behaviors. In order to estimate this energy consumption, it is needed to analyze the forces acting on vehicles during their motion. These are the resistance force and the tractive force with the following relationship:

$$F = TF - R \tag{11.9}$$

where

F is the resulting acting force on the vehicle in motion (N or Newton);
TF is the tractive force (N); and
R is the resistance force (N).

In general, if the resulting force $F=0$, the vehicle moves at the constant speed; if $F<0$, the vehicle decelerates, and if $F>0$, the vehicle accelerates. The tractive force TF depends on the type and characteristics of the vehicles. The resistance force R depends on the complex of factors, which can be generally classified into the mechanical and air resistance. The mechanical resistance includes the rolling resistance influenced by the vehicles' mass/weight and the way resistance depending on the characteristics of the infrastructure (roads and rail tracks). The air resistance depends on the aerodynamic characteristics of the vehicles. Both resistances increase with increasing of the vehicles' speeds, which can be expressed as follows (Vuchic, 2007):

(a) For the road vehicle(s):

$$R_{\mathrm{r}}=(c_{1\mathrm{r}}+c_{2\mathrm{r}}\cdot v_{\mathrm{r}})\cdot M_{\mathrm{r}}+c_{\mathrm{ar}}\cdot A_{\mathrm{r}}\cdot v_{\mathrm{r}}^{2} \tag{11.10}$$

(b) For the rail vehicle(s):

$$R_{\mathrm{t}}=\left(c_{1\mathrm{t}}+\frac{c_{3\mathrm{t}}}{p_{\mathrm{t}}}+c_{2\mathrm{t}}\cdot v_{\mathrm{r}}\right)\cdot M_{\mathrm{t}}+c_{\mathrm{at}}\cdot A_{\mathrm{t}}\cdot v_{\mathrm{t}}^{2} \tag{11.11}$$

where

$c_{1\mathrm{r}}$, $c_{2\mathrm{r}}$, $c_{1\mathrm{t}}$, $c_{2\mathrm{t}}$, $c_{3\mathrm{t}}$ are the empirically determined coefficients of the mechanical resistance to the motion of road and rail vehicle(s);
c_{ar}, c_{at} is the empirically determined coefficient of the aerodynamic resistance to the motion of road and rail vehicle(s), respectively;
M_{r}, M_{t} is the mass of the road and rail vehicle(s), respectively (kg);
p_{t} is the number of axes of the rail vehicle;
v_{r}, v_{t} is the speed of the road and rail vehicle, respectively (m/s); and
A_{r}, A_{t} is the area of profile of the road and rail vehicle(s), respectively (m^2).

The particular coefficients to the mechanical and air resistance to motion have been experimentally determined for different types of vehicles operated by different transport modes. This also relates to their area of profile (Vuchic, 2007).

As simplified, relations (11.10) and (11.11) known as Davis's equation, specifically used for railways, can be expressed as follows:

$$R=A+B\cdot v+C\cdot v^{2} \tag{11.12}$$

where

R is the vehicle resistance (N);
v is the vehicle speed (m/s); and
A, B, C are the empirically estimated coefficients (A (N); B (N s/m); C (N s^2/m^2)).

In Eq. (11.12), the coefficients A and B account for mass and mechanical resistance and the coefficient C accounts for air resistance. In this case the grade and rail track curvature are not taken into account.

In general, the energy consumption of a vehicle is approximately equivalent to the transport work E in (N-m) carried out along the given distance as follows (N-m = 0.000000278 kWh):

$$EC=TF\cdot d \tag{11.13}$$

where d is the given distance (m).

The other symbols are as in the previous equations. As mentioned above, when the vehicle moves at constant speed, the tractive force TF_c in (N) acting in the opposite direction to the resistance force R is equal to it as follows:

$$TF_c = R = 2650 \cdot \frac{P \cdot \eta}{v} \tag{11.14}$$

where

P is the power output delivered by the engine (HP—Horse Power);
η is the efficiency in converting the vehicle's power output into the tractive force (usually $\eta = 0.70$–0.85); and
v is the vehicle speed (km/h).

Based on Eqs. (11.13), (11.14), the energy consumption EC_c in (N-m) of the vehicle (train) with the engine power output P in (HP) and its utilization efficiency η moving along the flat, straight route d in (m) at the cruising speed v in (m/s) can be estimated as follows:

$$EC_c = (1+\varepsilon_c) \cdot TF_c \cdot d = (1+\varepsilon) \cdot R_c \cdot d = (1+\varepsilon) \cdot 2650 \cdot \frac{P \cdot \eta}{v} \cdot d \tag{11.15}$$

where ε_c is the share of additional energy spend for providing the comfort on board the vehicle such as for example heating, lighting, etc. during the cruising phase of a trip (usually $\varepsilon = 0.10$–0.20).

The other symbols are analogous to those in the previous equations. When the vehicle accelerates at the constant rate, the required tractive force TF_a in (N), based on the Newton's second law, can be estimated as follows:

$$TF_a = M \cdot a \tag{11.16}$$

where

M is the vehicle mass/weight (kg); and
a is the vehicle acceleration (m/s^2).

The energy consumption EC_a in (N) during the acceleration at the constant rate a in (m/s^2) in order to increase the current speed for an increment Δv in (m/s) can then be calculated as follows:

$$EC_a = (1+\varepsilon_a) \cdot TF \cdot d_a = (1+\varepsilon_a) \cdot \frac{1}{2} \cdot M \cdot \frac{(\Delta v)^2}{a} \tag{11.17}$$

where ε_a is the share of additional energy spend for providing the comfort on board the vehicle such as for example heating, lighting, etc. during the acceleration/deceleration phase of a trip (usually: $\varepsilon = 0.1$–0.2).

The other symbols are analogous to those in the previous equations. When the vehicle decelerates the tractive force TF is not applied and consequently is equal to zero. Therefore, the total energy consumed by a given vehicle moving along the route d is equal to the sum of the energy consumed for all accelerations and the cruising phases of a trip performed at the specified speed(s), based on Eqs. (11.16), (11.17), as follows:

$$EC_T = n \cdot EC_a + m \cdot EC_c \tag{11.18}$$

where n, m is the number of equal or equivalent acceleration and cruising phases during a given trip.

The other symbols are analogous to those in the previous equations.

By dividing the total energy consumed EC_T in Eq. (11.18) by the actual route length and by the number of passengers on board, the average unit energy consumption or the specific energy consumption (kWh/p-km) can be obtained as follows:

$$SEC = \frac{E_T}{\theta \cdot s \cdot d} \tag{11.19}$$

where

θ is the load factor of the vehicle;
s is the train's seat capacity (seats); and
d is the route length (km).

The other symbols are analogous to those in the previous equations.

11.2.4.2 Emissions of GHG

Air pollutants

Transport vehicles consume energy/fuel, which, at present, is derived from the primary sources such as crude oil, natural gas, biomasses, coal, nuclear, and water energy. The first three are used for manufacturing petrol/gasoline, diesel, jet fuel, and gas. The last three are used for manufacturing gas and electric energy. Except the water, all other primary sources are nonrenewable. In addition, except the nuclear and water, all other primary sources emit air pollutants during manufacturing the final products—energy/fuels—for the final consumption by transport services of which some are GHG.[1] Then, the fuels as the final products emit air pollutants (ie, GHG) while being consumed for carrying out transport services. In the given context, emissions from the manufacturing energy/fuels are called the secondary emissions. Those from carrying out transport services and going to be under consideration are called the primary air pollutants/emissions as follows:

Nitrogen Oxides (NO_x): The symbol NO_x implies the nitrogen oxides NO and NO_2 ($NO_x = NO + NO_2$). They are produced by any combustion in which air in the form of N_2 and O_2 is brought to a high temperature by burning fuel. NO_x is formed in the flame at a very high temperature and generally its formation increases as the burning temperature increases. The main impact of NO_x is that they affect the ozone layer as described below.

Sulfur Oxides (SO_x): Crude oil and its derivative—petrol, diesel, and jet aviation fuel—may have considerable amounts of Sulfur. In a chemical reaction with the water vapor in the atmosphere it creates acid rain, damaging trees other dependent natural habitats. In order to diminish its presence the different catalysts are added.

Carbon Oxides—CO and CO_2: Carbon monoxide (CO) is always produced during the burning of fossil (crude-oil derived) fuels. It reacts with the oxygen (O_2) in the atmosphere and forms carbon dioxide (CO_2), which appears to be one of the most important GHG. The emitted CO_2 has a long lifetime in the atmosphere (about 100 years). There is no remedy for reducing the quantity of emissions of CO_2 by improving the fuel burning process in the transport vehicles' engines simply because of the chemistry of fuel. The only option remains reducing the amount of fuel burnt and development of new technologies using some other type of fuel. The latter option—changing the fuel type—is a

[1]GHG are the air pollutants identified to contribute to global warming.

technological breakthrough for the future, probably after developing engines using, for example LH_2 (Liquid Hydrogen) as the main energy component. This fuel would certainly eliminate emissions of CO_2 but would also increase emissions of the water vapor (H_2O). Therefore, its net contribution still needs to be investigated.

Non-Methane Hydrocarbons—(NM)HCs: The hydrocarbons (HCs) contribute to the formation of smog and consequently to global warming. This has been identified through: (i) production of ozone (O_3), (ii) extending the lifetime of methane (CH_4), and (iii) their conversion into Carbon dioxide (CO_2) and water vapor (H_2O), as the most important GHG (Archer, 1993).

Water Vapor (H_2O): Water vapor (H_2O) emitted after burning the fossil fuels influences climate change through contribution to more intensive formation of clouds both near the earth's surface and in the troposphere (10–12 km above the earth's surface). In particular, clouds near the earth's surface affect the atmosphere by reducing the amount of solar radiation returning to space and by increasing the amount of solar radiation reflected from the atmosphere. Consequently, the surface becomes warmer in order to keep the radiative forces in balance. Clouds in the troposphere called contrails are formed by the jet fuel burned by commercial aircraft. They have been identified as one of the causes for diminishing the intensity of sunlight reaching the Earth's surface, thus causing the cooling of temperatures during the daytime and the warming of temperatures during the night. Consequently, contrails have proved to contribute to global warming directly or indirectly (Janić, 2007).

Contribution to global warming/climate change

Generally, life on earth is related to some physical properties of the sun-earth system. The surface temperature of the sun is about 5800 K (K—Kelvin degree), which results in an emission spectrum with a maximum wavelength of 500 nm* (nm*—nanometer; 1 nm* $= 10^{-9}$ m). This makes the solar temperature of the exact magnitude to induce photochemical reactions. Depending on the radius of the sun and the earth, their distance and the above-mentioned surface solar temperature, one can estimate that the earth receives energy of 1379 W/m^2 (watts per square meter), although the solar constant is always taken a bit lower, that is, $S \approx 1370$ W/m^2. With the support of some gases in the earth's atmosphere, this energy appears sufficient to maintain an average temperature on the earth's surface of $T = 288$ K (+15 °C). A part of the received energy is reflected from the earth's surface back into the space. This is called albedo a from the Latin term "albus" meaning "white." Astronomers usually use albedo to express the brightness of the earth as seen from space. Consequently, the energy equation can be set up as follows: $(1-a)\pi R^2 S = 4\pi R^2 \sigma T^4$, where R is the earth's radius (6400 km) and σ is the Stefan-Boltzmann constant ($\sigma = 5.672 \times 10^{-8}$ W/m^2/K^4). With an estimated value of albedo $a = 0.34$, one can obtain a temperature of the earth's atmosphere of $T = 250$ K (Usually it is taken to be $T = 255$ K). This is lower than the earth's surface temperature (288 K), which is mainly due to the presence of gases such as Carbon dioxide (CO_2), Ozone (O_3), Nitrogen dioxide (NO_2), Methane (CH_4), and water vapor (H_2O). Otherwise, this temperature would be lower by about 30 K. In general, these gases absorb most of the heat radiation from the earth and reemit it back towards the earth surface, which is the process called the "greenhouse effect." The mentioned gases are therefore called GHG. Currently, the concentration of GHG is continuously increasing due to both natural and human causes. For example, the concentration of CO_2 has increased by about 25% over the past 200 years; the level of CH_4 has doubled during the last 100 years, while the concentration of NO_x has been increasing by about 0.25% per year. Increasing the concentration of GHG might strongly influence the climate by increasing the average temperature on the earth surface. During the past 130 years, the average global temperature has increased by about 0.6 K. The speed and scope of the process is still not precisely known. The carbon dioxide (CO_2) has a very long residence time in the atmosphere where it mixes well with other gases. Some simple estimates can show that, for example, an instant doubling of the concentration of CO_2 relative to the present concentration would increase the average temperature on the earth's surface by about 1.4 K. This phenomenon can be explained as follows: increasing the concentration of CO_2 will reduce the earth's long wavelength radiation at the top of the atmosphere by a certain amount and consequently reduce the inward flux there by the same amount. The energy balance at the top of the atmosphere requires a constant flux. Therefore the earth's surface temperature should rise in order to compensate such an imbalance. This effect is called the radiative forcing. Some estimates have shown that the air transport system might contribute to increasing of the radiative forcing by about 0.02 W/m^2 (watts per square meter). An example of contributions of the radiative forcing to the increase of the earth's surface temperature is presented below. Such or any increase in the global temperature might cause additional effects—increasing or mitigation of the concentration of CO_2 as the reversible process. Some estimates suggest that the current concentration of CO_2 in the Earth's atmosphere is around 382 ppm and the tendency is at it will increase by an annual rate of about 1.2 ppm over the next 40 years (until the year 2050) (ppm—parts per million). Some other estimates indicate that when the total known reserves of crude oil of about 1650 billion (10^{12}) US barrels are exhausted by the end of the 21st century, the concentration of CO_2 will contribute to the increasing of the average global temperature by about 2.5 K (1 US barrel = 158.987 L) (Boeker and Grondelle, 1999). Equally important gas in the earth's atmosphere is ozone (O_3). Its presence protects the earth from the harmful solar UV radiation by absorbing all light with a wavelength less than 295 nm* (nanometer). The layer of O_3 in the earth's atmospheres is relatively thin—about 0.3–0.4 cm— under constant temperature and atmospheric pressure. The gas is present throughout the atmosphere but it is maximally concentrated in the stratosphere at the altitudes of about 20–26 km from the earth's surface. It is permanently formed through the reaction of the molecular oxygen (O_2) and the atomic oxygen (O) influenced by the solar UV radiation. Most of the ozone is formed above the equator where the amount of UV solar radiation is maximal. From there, it moves towards the poles where it is "accumulated" up to a thickness of about 0.4 cm during the winter period. However, ozone is sensitive to the free radicals such as the atomic chlorine (CI), nitric oxide (NO), and hydroxyl radicals (OH). They are formed from the water vapor (H_2O) and chlorofluorocarbons (CFCs), products of burning aviation fuel, which escape from the troposphere (10–12 km from the earth surface) where most commercial flights take place to the stratosphere where the ozone layer is formed. At these altitudes free radicals including NO_x lead to depletion of the ozone layer. Those that do not escape remain extremely stable in the troposphere where they, together with NO_x, c contribute to thickening of the ozone layer. The residence time of NO_x in these regions increases with altitude. Therefore, NO_x affects the ozone layer regionally if injected into the troposphere and globally if injected into the stratosphere (IPCC, 1999). In any case, the increased concentration of the above-mentioned pollutants might generally cause depletion of the ozone layer with inevitable impacts. For example, depletion of this layer by about 10% may cause an increase in the UV radiation by about 45%, which certainly inflicts damages to almost all biological cells and in particular causes skin cancer in people who expose their skins.

Some estimation of emissions of GHG

In general, two methods have been used for estimating emissions of GHG and particularly CO_2 from the transport sector: activity-based method and energy-based method. The activity-based method is as follows:

$$EM_{CO_2} = Q \cdot d \cdot e_{r/CO_2} \tag{11.20}$$

where

Q is the volume of transportation during a given period of time (passengers, ton, TEU per h, day, year);
d is the average transport distance (km); and
e_{r/CO_2} is the emission rate of vehicles operated by given transport mode and its systems (gCO_2/p-km or gCO_2/t-km).

The energy-based method is as follows:

$$EM_{CO_2} = FC \cdot e_{f/CO_2} \tag{11.21}$$

where

EM_{CO_2} is the emission of CO_2 by the fuel-powered vehicle(s) ($kgCO_2$);
FC is the vehicle fuel consumption (liters of fuel); and
e_{f/CO_2} is the emission factor ($kgCO_2$/L of fuel).

The examples of emission factors and rates e_{CO_2} in Eq. (11.21) from burning transport fuels are given in Tables 11.2 and 11.3 (EIA, 2015a,b).

Table 11.2 Typical Emission Factor of CO_2 From Burning Transport Fuels (EIA, 2015a,b)

Transportation Fuel Type	Emission Factor e_{f/CO_2} ($kgCO_2$/L of Fuel)
Aviation gasoline	2.2
Biodiesel	
B100	0.00
B20	2.15
B10	2.41
B5	2.55
B2	2.55
Diesel fuel (no. 1 and no. 2)	2.68
Ethanol/ethanol blends	
E100	0.00
E85	0.35
E10	2.12
Methanol/methanol blends	
M100	1.09
M85	1.28
Motor gasoline	2.35

Table 11.2 Typical Emission Factor of CO_2 From Burning Transport Fuels (EIA, 2015a,b)—cont'd

Transportation Fuel Type	Emission Factor e_{f/CO_2} (kgCO_2/L of Fuel)
Jet fuel, kerosene	2.53
Natural gas	14.42
Propane	1.51
Residual fuel (Np. 5 and no. 6 fuel oil)	3.11

Table 11.3 Typical Emission Rates and Factors of NO_2 (Nitrous Oxide) and CH_4 (Methane) for Different Types of Vehicles in the United States (EIA, 2015a,b)

Vehicle Type/Highway Conventional Fuels	Emission Rate e_r (g/km)	
	NO_2	CH_4
Gasoline passenger cars	0.0022–0.0122	0.0108–0.1106
Gasoline light-duty trucks	0.0041–0.0137	0.0101–0.1258
Gasoline heavy-duty vehicles	0.0083–0.0309	0.0207–00.2861
Diesel passenger cars	0.0006–0.0008	0.0003–0.0004
Diesel light-duty trucks	0.0009–0.0011	0.0006–0.0007
Diesel heavy-duty vehicles	0.030–0.030	0.0032–0.0032
Vehicle type/highway alternative fuels[a]	**Emission rate e_r (g/km)**	
	NO_2	CH_4
Light-duty trucks	0.0311–0.0416	0.0112–0.458
Heavy-duty vehicles	0.1087–0.1087	0.041–1.2216
Buses	0.1087–0.1087	0.041–1.2216
Vehicle type/non-highway	**Emission factor e_f (g/L of fuel)**	
	NO_2	CH_4
Ships and boats (diesel)	0.058–0.079	0.172–0.227860
Aircraft—jet fuel	0.082	0.0713
Aviation gasoline	0.029	1.86

[a]*Methanol, ethanol, CNG (compressed natural gas), LPG (liquefied petroleum gas), LNG (liquefied natural gas).*

As can be seen, the emission factors of CO_2 are generally much higher than that of the other considered air pollutants. Therefore, in most analyses, either exclusively CO_2 or the so-called CO_{2e} (Carbon-Dioxide Equivalents) embracing all air pollutants from burning transport fuels have been usually considered.

In addition to the liquid and gaseous fuels, the electric energy is consumed for performing transport services mainly by railways performing transport services in urban, sub-urban, and interurban areas. In the future, it is likely that electric energy be also consumed by the vehicles operated by other transport modes (road, inland waterways, maritime). A mentioned above, at present, the electric energy can be obtained from one or a combination of different primary nonrenewable sources, which have the emission factor of CO_2 as give in Table 11.4.

Table 11.4 The CO_2 Emission Factors From the Primary Nonrenewable Sources for Obtaining the Electric Energy in the United States (EIA, 2015b)

Fuel Type	Emission Factor e_f ($kgCO_2/kWh$)
Bituminous coal	0.318
Distillate fuel oil	0.249
Geothermal	0.026
Jet fuel	0.242
Kerosene	0.246
Lignite coal	0.333
Municipal solid waste	0.142
Natural gas	0.181
Petroleum coke	0.348
Propane gas	0.215
Residual fuel oil	0.114
Coal-derived synthesis gas	0.181
Synthesis gas from petroleum coke	0.181
Subbituminous coal	0.331
Tire-derived fuel	0.293
Waste coal	0.318
Waste oil	0.325

According to the above-mentioned energy-based method, the CO_2 emissions from the vehicle's consumption of electric energy can be estimated as follows:

$$EM_{CO_2} = E \cdot e_{f/CO_2} \tag{11.22}$$

where

EM_{CO_2} is the emission of CO_2 by the electric-powered vehicle ($kgCO_2$);
E is the electric energy consumption (kWh); and
e_{f/CO_2} is the emission factor from the electricity production from a given combination of the primary sources ($kgCO_2/kWh$).

11.2.5 LAND USE

The transportation systems use land for settling their infrastructure, whose area can be substantive in an absolute sense—hundreds and thousands of hectares (ha) or km^2. An important indicator is the total area of land used by infrastructure of the particular transport modes and systems. It is usually expressed by the ratio between the area of land taken by a given transport mode/system's infrastructure and the area of a given urban area, region, country, or continent. This ratio can be approximated as follows:

$$r_{lu} = A_l / A_c \tag{11.23}$$

where

A_l is the area of land occupied by the given transport system infrastructure or its component (ha, km^2); and
A_c is the area of a given urban area, region, country, or continent (km^2).

The another attribute is, after the land is taken, the intensity of its use by the volumes of transport services carried out during the specified period of time under given conditions. For example, this intensity can be expressed as follows:

$$\rho_{lu}(\tau) = Q(\tau)/A_l \tag{11.24}$$

where

$\rho_{lu}(\tau)$ is the intensity of land use during the time period τ (p-km/, t-km/, or TEU-km/ha or km^2); and
$Q(\tau)$ is the volume of output during the time period τ (p-km, t-km, TEU-km).

The other symbols are as in the previous equations. As indicated in Eq. (11.24), for the fixed size of area of the occupied land, the intensity of its use increases with the growth of traffic. When the existing infrastructure reaches saturation, its expansion is made. This takes additional land and consequently, at least temporarily, decreases the intensity of the overall occupied land. After that, this intensity continues to increase, of course, if traffic continues to grow.

11.2.6 WASTE

Waste is generated before, during and after providing transport services. In general, this waste can be segregated into nonindustrial and industrial waste. The nonindustrial waste originates from the passenger service on board and consumption of the transport systems' employees and visitors (food, newspapers, cans, paper). The industrial waste originates from daily activities such as washing and cleaning the vehicles, and their maintenance and repair including painting and metal work, engine testing, etc. The industrial waste is further categorized into hazardous and nonhazardous waste. The former is managed according to the strict national and airport regulations governing collection, treatment, storage and disposal. The aim is to prevent contamination of the soil and drinking water at and around the given transport system and its particular components.

It is easy to conclude that the larger transport systems carrying out the larger volumes of transport services produce the larger quantities of waste in the absolute terms (kg, ton). However, they can generate lower quantities of waste per unit of output (kg/p-km, t-km, or TEU-km).

11.3 ROAD-BASED SYSTEMS

11.3.1 CONGESTION

11.3.1.1 Cars

Impact

The traffic flow theory that was elaborated in Chapter 4 has shown that, the higher the traffic density, the lower the average speed. Such reduction of speed causes increase in the travel time as compared to the time of traveling at the free flow speed in urban areas of 88 km/h (55 mph) (mph—miles per hour). This usually happens in the urban areas during the morning and afternoon peak(s) causing, as mentioned above, two impacts: the extra travel times (delays) of commuters and the related extra fuel consumption just due to this extra travel times (delays). This latter also causes the extra emissions of GHG. In general, the most common functional form for expressing the travel time per mile (km) as an inverse of the speed to traffic flow(s) has been as follows (Parry, 2008):

$$\tau = \tau_f \cdot \left(1 + \alpha \cdot q^{\beta}\right) \tag{11.25}$$

where

τ_f is travel time per mile (km) under conditions of free traffic flowing (h/mile; h/km);
q is the volume of traffic flow (v/h) (v—vehicle); and
α, β are the coefficients (typical values are: $\beta=2.5$–5.0; $\alpha=0.15$).

Then, the positive difference $(\tau - \tau_f)$ represents the extra travel time (min or h/mile or km), which multiplied by the length of travelled route(s) gives the total extra (delay) travel time(s). The ratio between this total travel time including delay(s) (commonly during the peak period(s)) and the travel time under the free-flow travel conditions represents the "Freeway Travel Time Index," which usually appears to be greater than one. During the given period of time the extra travel time during commuting accumulate at those affected (daily commuters), who in many cases are not aware of its scale and also not about the related costs—of their time and extra fuel consumed. Fig. 11.1 shows an example of the average annual delays of a commuter in 370 urban areas in the United States (Schrank et al., 2015).

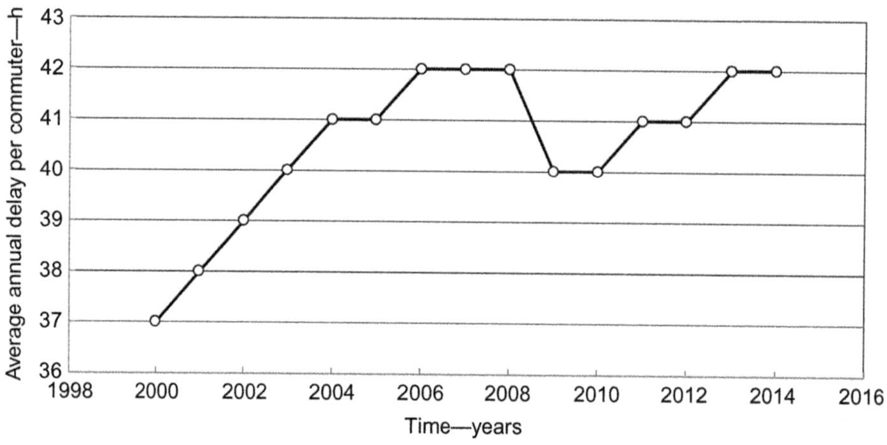

FIG. 11.1

The average annual delay per commuter due to congestion is 370 urban areas in the United States (period: 2000–14) (Schrank et al., 2015).

As can be seen, this annual delay has generally increased during the period 2000–08, then dropped significantly over the next 2 years, and then continued to raise again reaching about 42 h/year per commuter. For comparison, in the year 2013 the average delays in urban areas in the United Kingdom, France, and Germany have been estimated to be 40.1, 43.9, and 38.2 h/year per commuter, respectively (Cebr, 2014). One of the mitigating measures for reducing congestion in many urban areas round the world has been introducing congestion charges.

Table 11.5 gives an example for London Central area (TfL, 2008).

As can be seen, the charge has been effective in reducing traffic entering the central area of the city but one has to bear in mind that these are the average figures related to both chargeable and non-chargeable vehicles. The net impact on the chargeable trips by cars and minicabs has been much higher—reduction of 53% as the response to the charge of 8 £/day. The data from TfL's (Transport for London) annual report (2006–07) show that that the revenues from congestion charges were

Table 11.5 Some Effects of Charging Congestion in Central London (UK) (TfL, 2008).

Year/Time	Charge (£/Day)[a]	Year-on-Year	Changes in Traffic (%)
February 2003 Time: 7:00–18:30 h	5	2003 vs 2002	−14
July 2005 Time: 7:00–18:00 h	8	2005 vs 2004 2006 vs 2002	−2 −16

[a]*Central charging zone (CCZ).*

£252.4 million and the net profit of £89.1 million for TfL (TfL, 2007). However, the initial operating revenues from the congestion charge did not reach the levels that were originally expected and that the target was not achieved. Consequently, the charge has further increased to £10 in Jan. 2011 (25% increase) and again to £11.5 (15% increase). As a result, the net income has increased to £131 million in the year 2013 (TfL, 2013).

11.3.1.2 Buses

Buses usually operate at lower speeds than the passenger cars/vehicles either in urban areas or on the open roads and highways. These speeds become different in two cases. First, when the buses share the same lanes with other traffic in urban and sub-urban areas (ie, these are mainly buses of the public transit systems), their speeds drop to the speeds of passenger cars when the traffic congestion occurs. In such case, both categories of vehicles impose delays on each other, but buses do that more due to their size and lower maneuverability. On the open two-lane roads in both directions, they slow down passenger cars/vehicles queuing behind and awaiting to overtake them. On the two-lane motorways/highways in each direction, this queuing also occurs but much rarely, except in some specific circumstances causing closing the lane(s) (eg, traffic incidents/accidents, weather-related impacts, etc.). However, in any case, but particularly in urban and sub-urban areas, buses can replace many individual cars due to their much higher carrying capacity, and consequently contribute to reducing traffic congestion and related delays of users-passengers there (see Chapter 7). Buses are the only road-based public transit system in many cities round the world. They are also the key-support to the urban and sub-urban rail-based transit systems. The dedicated bus corridors in the form of "congestion-free bus networks" covering the entire cities and metropolitan areas have been implemented in many cities. These networks usually include the main trunk roads and expressways served by buses operating the radial routes penetrating into the metropolitan center. One of these has been Bus Rapid Transit (BRT) system. Its congestion and related delays of services can be considered from three aspects. The first implies congestion caused by interference between the BRT vehicles/buses and other traffic, and vice versa, while operating along the mixed traffic bus lanes. However, thanks to mostly dedicated routes of the "congestion-free bus networks" free of other traffic, this congestion has happened rarely, rather exceptionally. The second implies congestion due to clustering of the BRT buses operating along the dedicated routes of the network particularly those with single lanes in each direction and without passing lanes at the terminals/stations. This can happen in corridors with several BRT routes/lines operating relatively frequent services. The trunk part of the feeder-trunk network can particularly suffer from this kind of induced congestion causing delays of the affected services. The last implies contribution of the

BRT systems to savings of own congestion and that of the other traffic, which both contribute to savings in the overall user/passenger travel time. For example, the savings in travel time compared to previously used transit services vary from 5% to 35% at 16 US BRT systems. In addition, as compared to individual traffic, savings of 32% at TransMilenio (Bogotá, Colombia), 35% at Metrobús (Mexico City, Mexico), and 45% at Metrobüs (Istanbul, Turkey) have been reported (Janić, 2014a,b).

11.3.1.3 Trucks

The growth in freight transport volumes in many countries worldwide has been one of the major contributors to congestion and related delays of road/truck freight transport services compromising their punctuality and reliability in urban, sub-urban, and interurban areas. In particular, the long-haul freight transportation carried out by heavy trucks has shown to be a frequent significant contributor to local (urban and sub-urban) congestion and delays. Congestion and delays of trucks have usually happened close to the large national and international freight hubs such as freight terminals, rail yards (in the case of rail/road intermodality), ports, airports, and border crossings. Different bottlenecks in road infrastructure connecting particular large freight hubs force converging both freight and passenger traffic at road/motorway/highway intersections, steep grades, lane reductions. In particular, the intersections and road/street lanes where the flows of passenger cars/vehicles and trucks have mixed have acted as severe bottlenecks. The other causes of congestion of trucks have restrictions on freight movement due to the lack of sufficient space in dense urban areas and limited delivery and pick-up times at terminals, ports, airports, and freight shipper/receiver doors. Consequently, this congestion and related delays of the affected parties, trucks and passenger cars/vehicles converge and diverge to/from the large freight hubs and urban and sub-urban areas where the freight shippers and receivers are mostly spatially concentrated.

In general, presence of trucks of different size (light, medium, heavy) in traffic flows causes congestion and delays imposed on themselves and other affected passenger car/vehicles. Under given conditions of road infrastructure and supporting facilities and equipment, the increasing intensity of these flows changing during the peak and non-peak periods reduces the operating speed of all vehicles including trucks. Such reduction of speed causes congestion and delays of the affected vehicles including trucks involved, which in turn compromises punctuality and reliability of their (freight) transport services (see also Chapter 9). In particular, the truck operating speed during the traffic peak has usually been lower than that during the non-peak periods. Fig. 11.2 shows an example of the relationship between the peak—and non-peak period operating speeds at the selected locations in the United States (USDT, 2013c).

As can be seen, the peak-operating speed of trucks has linearly increased with increasing of the non-peak operating speed. At all 25 locations both speeds have been lower than the speed of free traffic flow of 55 mph, which clearly indicates congestion conditions. In addition, Fig. 11.3 shows the effects of improvements during the period 2011–12 in removing bottlenecks causing congestion and delays and consequent decreasing of operating speeds during the peak—and non-peak periods at some of the above-mentioned locations in the United States (USDT, 2013c).

As can be seen, the peak-period operating speed has again increased almost linearly with increasing of the non-peak operating speed. The trend line has been upward steeper for speeds in the year 2012 that for those in the year 2011, thus indicating clearly benefits of improvements in mitigating bottlenecks at the given locations. However, again, in both cases, before and after improvements, the truck operating speeds have, except in one case, remain below the free-flow speed of 55 mph (mph—miles per hour; 1 mile = 1.609 km) (USDT, 2013c).

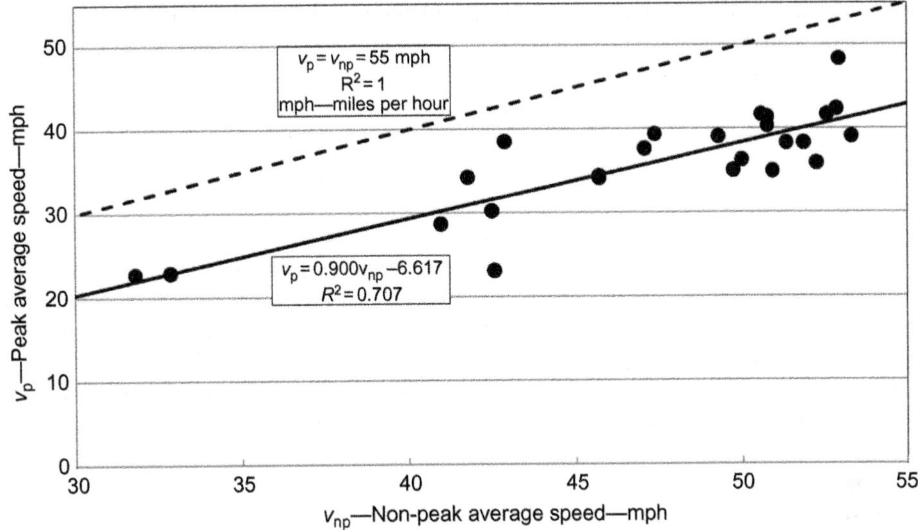

FIG. 11.2

Relationship between the peak and non-peak speeds of trucks—top 25 (of 250 identified) Congested Freight-Significant Locations (period: 2011–12) (USDT, 2013b).

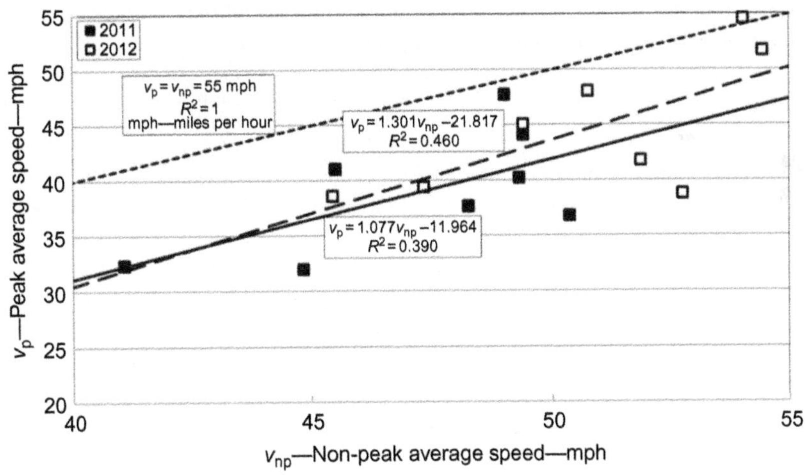

FIG. 11.3

Relationship between the peak and non-peak speeds of trucks—9 of top 25 (of 250 identified) Congested Freight-Significant Locations (period: 2011–12) (USDT, 2013b).

11.3.2 NOISE

11.3.2.1 Cars

Character of impact

Regarding of the type of fuel/energy used, passenger cars are generally categorized into three categories: Conventional ICEVs (Internal Combustion Engine Vehicle(s)) using gasoline and diesel fuel, HYVs (Hybrid Vehicle(s)) powered by conventional gasoline or diesel ICEs (Internal Combustion

Engine(s)) and an electromotor using energy from the batteries on board, which is filled in by ICEs; and BEVs (Battery Electric Vehicle(s)) propelled by electro-motors using the electric energy stored in batteries on-board the vehicle, which are recharged from the power grid (at home or at street/shop charging stations) (This time the forthcoming driverless passenger cars and their noise are not particularly considered). The main sources of noise from all above-mentioned passenger car categories, similarly as at the other road vehicles, are the rolling noise due to tire/road interaction and the propulsion noise produced by the driveline (engine, exhaust, etc.) of a vehicle. Due to the scale and scope of potential impact on the human health, modeling and real-life measurements of noise from passenger cars have been substantive and permanent aiming at reducing it as much as possible under given conditions, which have usually been specified by the car engine technology and the state of road pavement. For such purposes the passenger cars have also been treated separately from the other road vehicles such as the articulated trucks, non-articulated trucks, buses, and motorcycles (Brown and Tomerini, 2011). In some other cases, the passenger cars have been categorized as light vehicles (capacity up to 10 passengers) together with small-sized vehicles (the overall length of 4.7 m or less). The rest of the road vehicles have been the medium-sized (longer than 4.0 m and the capacity of 11–29 passengers) and large-sized vehicles (heavier than 8 ton with payload greater than 5 ton, and the capacity greater than 630 passengers), and motorcycles (Okada et al., 2014).

Both modeling and measurements have indicated that, under given other conditions, the noise generated by passing by passenger cars mainly depends on their speed and the distance from the measurement location (ie, observer). Fig. 11.4 shows an example of the relationships of this noise and speed of passing by passenger cars measured in the Brisbane urban area (Australia), Japan, and estimated by the EU Harmonise/IMAGINE model. The values have been recalculated for the distance of 25 m from the center of the road (Brown and Tomerini, 2011; Okada et al., 2014).

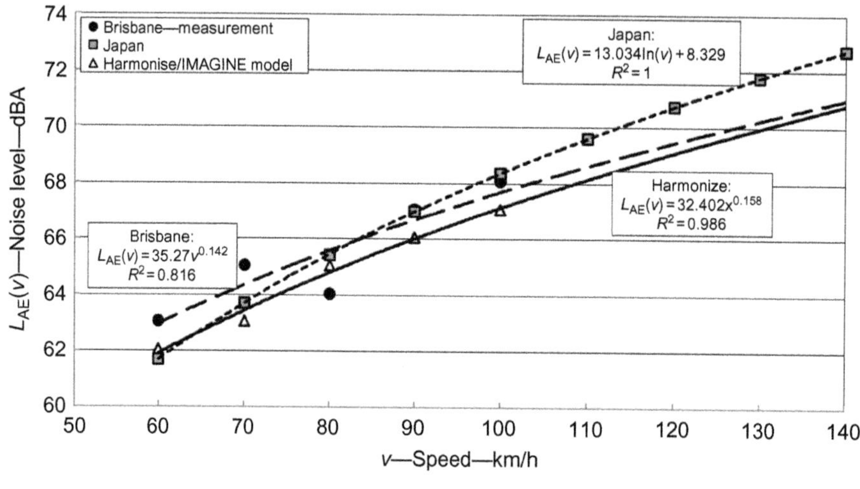

FIG. 11.4

Relationship between the noise sound power level from the passing by passenger cars as sources in Brisbane urban area (Australia) (Brown and Tomerini, 2011).

As can be seen, the noise level has increased with increasing of the speed of passing by cars (at decreasing rate) as intuitively expected because both the propulsion noise and noise from tire/pavement interaction increases with increasing of the vehicle speed (Jonasson, 2007). In addition, Fig. 11.5 shows the relationship between the noise and speed of ICEV, HYV, and BV passenger cars.

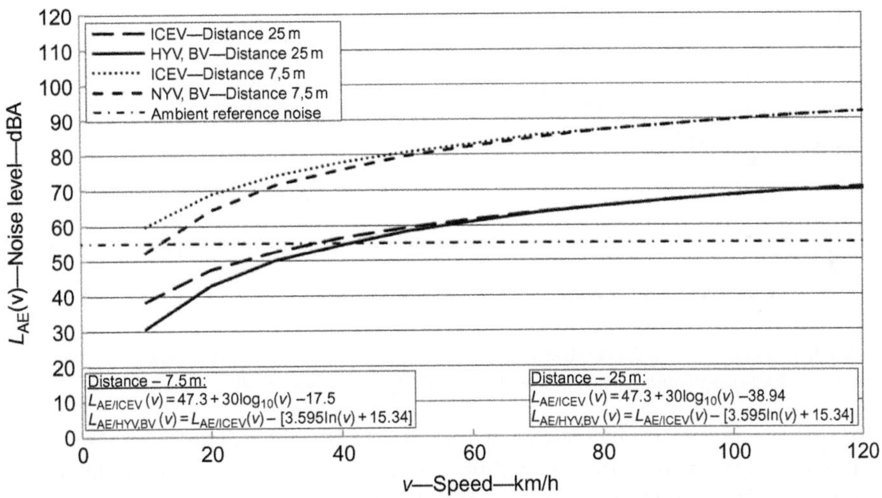

FIG. 11.5

Relationship between the level of noise and speed of different passenger car/vehicle technologies.

As can be seen, at the lower speeds (up to 50–55 km/h), the noise from HYVs and BVs is lower than that of their ICEV counterparts. The propulsion noise from (electric) engines of these cars at these speeds is negligible, but the rolling noise increases with increasing of their speeds. At the speeds of about 55–60 km/h, the noise equalizes at all types of cars. At HYVs, the ICE starts to power the car. At BVs, the rolling resistance noise starts to dominate and becomes equal to that of both ICEVs and HYVs. In addition, the noise from all cars decreases with increasing of the distance from the measurement location (ie, observer). It can be noticed that the noise of all types of cars at speeds below 40–50 km/h is below the ambient noise set up to be 55 dBA. As can be seen latter, this is important when considering safety of operating HYVs and BEVs at low speeds in urban streets because, for most visually impaired people who register closeness of cars based on their noise, it would be difficult to register their approach except when they come very close. This is because these cars use electric engines at these (low) speeds, which emit the noise at very low frequencies difficult to register as mixed with the ambient noise.

In addition, Fig. 11.6 shows the calculated relationship between the continuous noise level, the intensity of flow and speed of passenger cars (ICEVs) during the day (07:00–22:00 h). The reference location for the noise calculation has been at the distance of 25 m and at the height of 5 m from the center of the road (http://rigolett.home.xs4all.nl/ENGELS/vlgcalc.htm/).

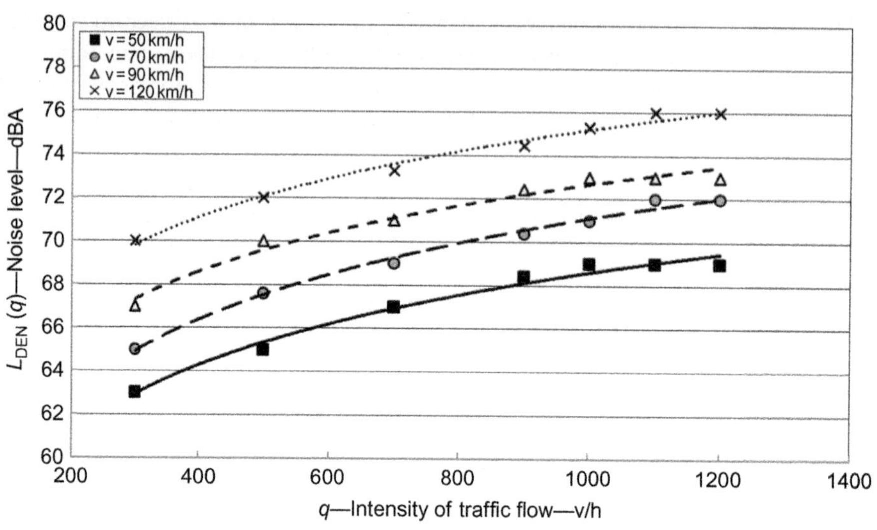

FIG. 11.6

The examples of the relationships between the daily noise level, the intensity of traffic flow and speed of passenger cars/vehicles (v—vehicle) (http://rigolett.home.xs4all.nl/ENGELS/vlgcalc.htm/).

As can be seen, the noise level increases with increasing of the intensity of flow of passenger cars at decreasing rate under given conditions. In addition, it increases with increasing of this flow independently on its intensity. It should be mentioned that the level of noise should always be considered under the specified conditions.

Some mitigation measures

The noise from passenger cars has been recognized as a cause of healthy problems at exposed people by the World Health Organization (WHO). Regarding such damaging impact, the endeavors to mitigate this noise from passenger cars have been persistent over time as follows:

- Reducing of the intensity of using passenger cars, particularly in urban areas, by introducing the noise charging system, which would, similarly, as the congestion charging system, stimulate people (usually daily commuters) to switch to other mass urban transit modes (bus, streetcar, Light Rail Transit (LRT), subway/metro);
- Noise regulation requiring the passenger car manufacturers to reduce noise at source, that is, already during the car design and construction phase;
- Improvement of road pavements in order to reduce the rolling resistance noise;
- Setting up the noise barriers (walls) at the particularly exposed locations (usually along the highways); and
- Locating the noise-sensitive population activities at the reasonable distances from noisy roads and highways.

An example of simultaneous mitigating noise at the same time as congestion has been, as mentioned above, introducing congestion charging system in London urban area. The noise emitted by passenger cars has been under regulation for a long time—since 1929 when the Motor Cars (Excessive Noise) regulations were introduced. Over time, this regulation has become increasingly stricter in terms of the allowed maximum level(s) of emitted noise. The most recent been the noise regulation act by the EU (EC, 2014). According to this act, the passenger cars have been categorized according to the engine output power/mass ratio and mass/weight with the specified maximum permissible noise level as given in Table 11.6.

Table 11.6 The European Union (EU) Limits on the Noise by Passenger Cars (EC, 2014)

Car Category	Description of Car Category	Noise Limits (dBA)
	P/M (power/mass) ratio; M—mass	a, b, c
M1	≤120 kW/1000 kg	72/70/68
M1	120 kW/1000 kg ≤ P/M ≤ 160 kW/1000 kg	73/71/69
M1	PM < 160 kW/1000 kg	75/73/71
M1	P/M > 200 kW/1000 kg; no of seats ≤ 5	75/74/72
M2	M ≤ 2500 kg	72/70/69
M2	2500 ≤ M ≤ 3500 kg	74/72/71
M2	3500 ≤ M ≤ 5000 kg and P ≤ 135 kW	75/73/72
M2	3500 M ≤ 5000 kg and P > 135 kW	75/74/72
M3	P ≤ 150 kW	76/74/73
M3	150 kW ≤ P ≤ 250 kW	78/77/77
M3	P > 250 kW	80/78/77

[a]*Phase 1: Applicable to new vehicle types from July 1, 2016.*
[b]*Phase 2: Applicable to new vehicle types from July 1, 2020 and for first registration from July 1, 2022.*
[c]*Phase 3: Applicable to new vehicle types from July 1, 2024 and for first registration from July 1, 2026.*

These values of noise have been determined for the specified conditions of measurement: in general, at the distance of 7.5 m and car speed of 50 km/h. As can be seen for each category of cars, the noise is supposed to be reduced in three phases. As a result, the noise of the cars with P/M ratio lower than 120 kW/1000 kg is to be reduced from 72 dBA in Phase 1–68 dBA in Phase 3. At the same time, the noise from cars with $P > 250$ kW is to be reduced from 80 dBA in Phase 1–77 dBA in Phase 3. In particular, the "silent" electric and hybrid cars will have to become noisier, which will be achieved by equipping them with sound generating devices such as for example Acoustic Vehicle Alerting Systems (AVAS). This would make these them safer for pedestrians/visually impaired persons.

Improvements of road pavements have been permanently carried out during maintenance. This has also included setting up the noise barriers along the nose sensitive locations.

Location of noise-sensitive population activities has also been taking place through urban re-planning processes including relocation of some of them (housing, schools, hospitals), and improving insulation of those remained at the same place.

11.3.2.2 Buses

Buses operating in the urban and sub-urban areas and on the open roads and motorways/highways generate noise, which could be disturbing to individuals/persons causing different health problems. In general, the noise has shown to increase with increasing of the bus operating speed independently on the power technology such as diesel, hybrid, or electric-trolleybuses. Some reasoning results in judging that noise levels of all technology buses are expected to be relatively similar since the noise from the tire/pavement-interaction begins to dominate. Some measurements have shown that the average noise level generated by passing-by urban buses at the distance of 5 m from an observer at the speed of about 60 km/h is about 80–85 dBA. The former (80 dBA) has been considered as the maximum tolerable noise level in urban environment.

In addition, the above-mentioned vehicles/buses of BRT systems also generate noise during performing their transit services. As in the case of other transport systems, this noise generally depends on their constructive-technical/technological characteristics and the pass-by speed. It has already been mentioned that BRT systems operate vehicles/buses powered by different engine technologies, which crucially influence levels of their noise. For example, the noise of BRT diesel and Compressed Natural Gas (CNG) buses comes from their exhaust system, engine block, cooling system, air intake components, and tire/pavement-interaction. The noise from BRT hybrid (diesel-electric) vehicles/buses comes from both diesel and electric motors. The main noise sources of BRT trolley buses are interaction between the catenary wire and the pantograph, electric motor, auxiliary equipment, and tire-pavement interaction. Important factors influencing received noise from BRT systems are: (i) the distance from the noise source, that is, passing-by vehicle(s), and (ii) the existence of noise barriers along the lanes. Fig. 11.7 shows an example of the dependency of the noise on the operating speed of BRT vehicles. These are 12–18 m long, weighting 13–17 tons empty and 32 tons full (vehicle + driver + passengers) with a capacity of 75–100 spaces (Janić, 2014a,b).

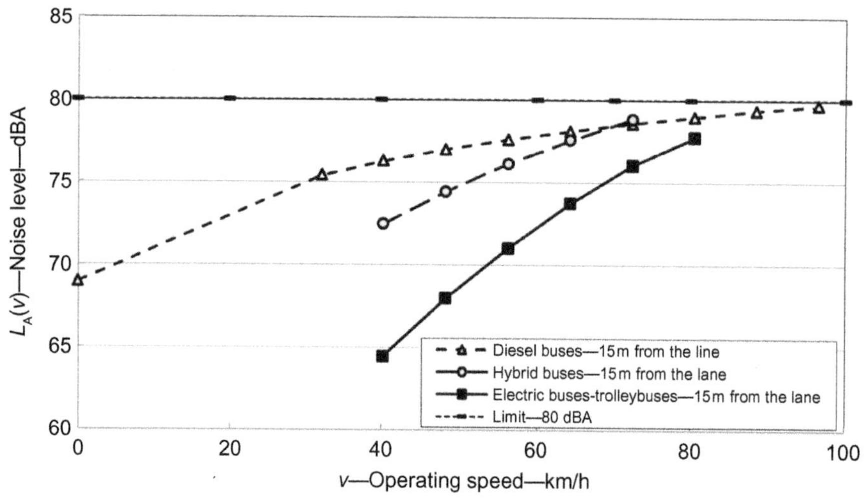

FIG. 11.7

Relationships between the noise and speed of BRT systems (Janić, 2014).

As can be seen, the noise of BRT buses/vehicles increases in line with the operating speed at a decreasing rate. As intuitively expected, the noise level is the highest at diesel-powered and the lowest at electric-powered buses-trolleybuses. The noise level of hybrid buses is somewhere in between. In addition, it is shown that the noise levels become very similar at diesel and hybrid-powered buses operating at speeds at about 60 km/h, while the electric-powered buses—trolleybuses remain are 4–5 dBA quieter. With further increase in speed, the difference in noise levels between all bus technologies continue to decrease thus confirming dominance of the tire/pavement-interaction in generating noise above the certain speed. In addition, the noise level of all three bus technologies for the given range of speeds up to 100 km/h is lower than the prescribed maximum level of 80 dBA. Finally, the noise barriers of a sufficient height built of brick or concrete along BRT routes contribute to decreasing noise to and below the sustainable level of about 55 dBA (Janić, 2014a,b).

11.3.2.3 Trucks

Noise from trucks and particularly the heavy ones originates from a variety of sources but mostly from engines, exhaust pipes, tires-pavement interaction, and aerodynamics. The experienced aggregate noise level by an observer depends mainly on the truck passing-by speed and the distance from the observer. The noise by an individual truck is commonly measured as the standard at the right-angle distance of 25 m (m—meter) from the source (ie, passing-by trucks) and at height of 3 m. Typical values depending of the operating speed of the standard truck in Europe have been as follows (Janić, 2014a,b):

$$L_{eq}(25, v_j) = 5.509 \ln v_j + 25.36\text{dB(A)}; \quad R^2 = 0.988; \quad 10 < v_j < 90(\text{km/h}) \tag{11.26}$$

After discussion of introducing mega trucks in Europe, their noise is expected to be higher mainly because of their stronger engines and increased rolling resistance due to the greater number of axles. In such case the above-mentioned expression will need to be slightly modified. In addition, the exposure to noise of a single mega truck will last longer, likely in proportion to the difference in its length and the length of a standard truck (25.25/18.75, which is about 35%). These estimates can be further generalized by including the actual frequency of noise events and the number of population located close and consequently exposed to passing-by trucks of both categories (Janić, 2014a,b). In addition, Fig. 11.8 shows examples of the relationship between the truck noise, its operating (passing-by) speed, and the distance from the observer, that is, measurement location, in the United States and Europe (Donavan et al., 2009; M+P, 2006).

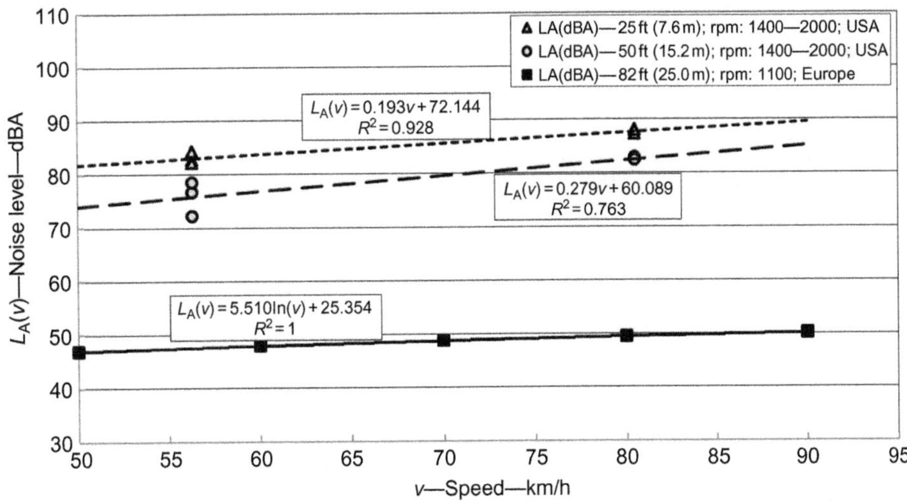

FIG. 11.8

Relationship between the truck noise, speed, and distance in the United States and Europe (rpm—(engine) rate per minute) (Donavan et al., 2009; M+P, 2006).

As can be seen, the truck noise has linearly increased with increasing of its operating speed at the given distance. In addition, for a given range of operating speeds this noise has increased with decreasing distance between the source-passing by truck and the observer-measurement location, and vice versa.

In considering the actual exposure of population close to roads/highways/motorways to noise by passing-by trucks, it is necessary to take into account the influence of noise protective barriers. As at buses, they are usually set up to protect particularly noise-sensitive areas by absorbing the maximum level of noise of about 20 dBA (single barrier) and 25 dBA (double barrier) (Janić, 2014a,b).

11.3.3 TRAFFIC ACCIDENTS/INCIDENTS (SAFETY)

11.3.3.1 Cars

Character of impact

The safety of passenger cars have usually been expressed by the rate of the number of traffic accidents/incidents happened during the specified period of time (usually 1 year) in terms of the number of vehicle crashes, and/or fatalities and injuries at both vehicles involved and third parties, and the volume of the system output, in this case, the given volume of passenger car kilometers (miles). Fig. 11.9 shows the rate of passenger car crashes and fatalities in the United States over time (USDT, 2011).

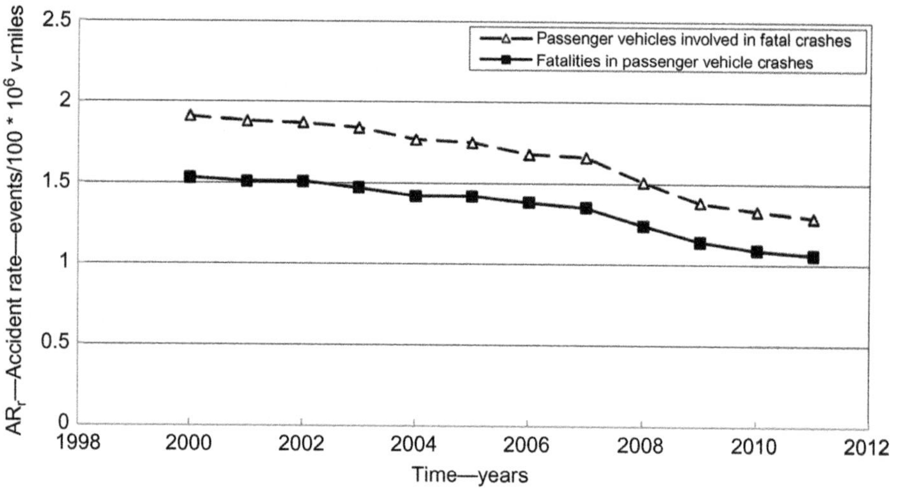

FIG. 11.9

The rate of passenger car crashes and related fatalities in the United States over time (period: 2000–11) (v—vehicle) (USDT, 2011).

As can be seen both the rate of car crashes and fatalities have been permanently decreasing during the observed period—for about 50% in the former and 75% in the latter case. Such development has indicated that driving passenger cars in the country has become, at least statically, less risky. In addition, Fig. 11.10 shows that the statistical risk of losing life in a car crash in the United States and EU-27 Member States has also decreased over the past almost two decades despite increasing of the volume of car use. This again indicates that both systems have been becoming safer, that is, less risky, according to the given statistics.

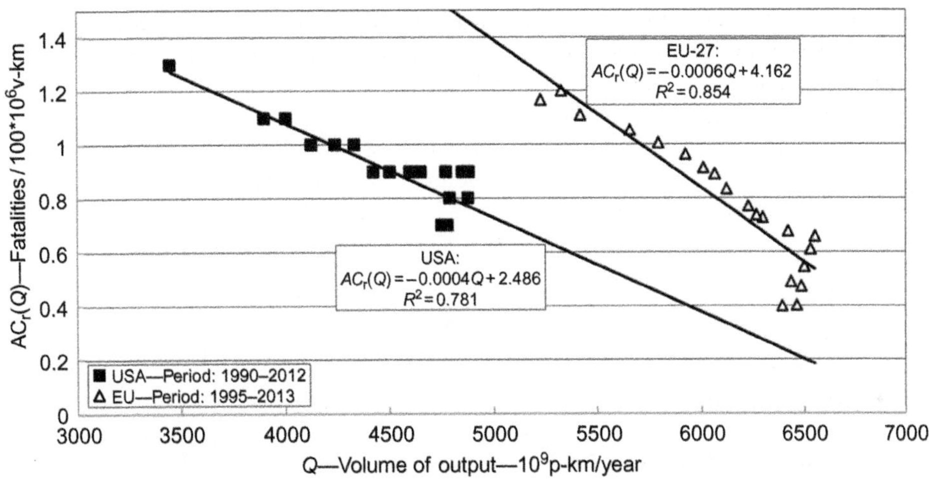

FIG. 11.10

Relationship between the annual volume of output and the fatality rate at the EU and US motor vehicles (all kinds) (v—vehicle) (EU, 2015; USDT, 2013a,b,c).

As can be seen, the passenger car driving in the EU has been approximately equally and even less risky (ie, safer) despite its higher volumes. For example, $AC_r = 0.8$ fatalities/100 * 10^6 v-km at the volume of about 5000 billion p-km in the United States and $AC_r = 0.4$ fatalities/100 * 10^6 v-km at the volume of about 6000 billion p-km in the EU.

Some mitigating measures

Improvements of the road traffic safety regarding the passenger cars are expected to be achieved through mitigating measures such as additional improvements of safety of cars at the manufacturers, proper maintenance of roads and highways, strengthening traffic surveillance, management, and control, and improvement of driving skills and education of drives, the latest in order to minimize their errors as the most frequent cause of accidents/incidents.

11.3.3.2 Buses

The traffic accidents/incidents and related causalities—fatalities and injuries of all categories (light, moderate, severe) reflect the safety and security of a given bus transit system. These accidents/incidents have been usually caused by collisions of buses with other buses and with other vehicles, bikers, and pedestrians, all often resulting fatalities and injuries, as well as damages to property. The number of events (fatalities, injuries, vehicle/bus crashes) and the number of these events per unit of the system's output expressed in the number of passenger-km (miles) carried out during the specified period of time as the accident/incident rate have shown to be convenient measures for monitoring the level of safety of bus systems at different geographical/spatial scales (1 mile = 1.609 km). Fig. 11.11 shows an example of the total annual events (fatalities) at the bus systems operating in the United States (USDT, 2010, 2013a,b,c).

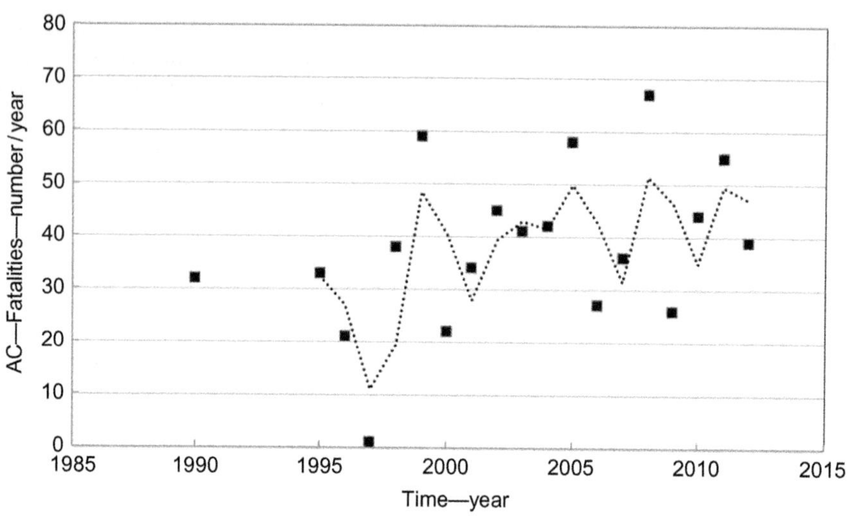

FIG. 11.11

Total annual fatalities at the US (school, local transit, intercity) bus systems over time (period: 1990–2012) (USDT, 2010, 2013a,b,c).

As can be seen, the number of fatalities has generally increased during the observed period, which has risen the concern if the system has been becoming less safe over time. This can be confirmed if the above-mentioned figures are considered in the absolute terms. In addition, some other measures indicate the opposite development, that is, that the system has become safer over time. Fig. 11.12 shows example of the accident rate in terms of the number of events (vehicle crashes, fatalities) per unit of output of the bus systems operating in the United States (USDT, 2013b).

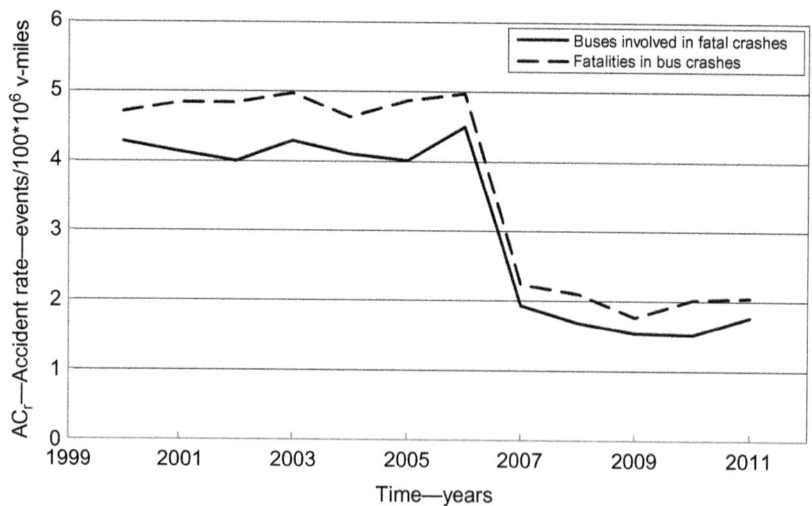

FIG. 11.12

The accident rate over time at the bus system in the United States (period: 2000–11) (USDT, 2010, 2013a).

As can be seen, during the entire observed period, the fatality rate has been higher than that of the vehicle/bus crash rate, which has been intuitively expected regarding the bus size, capacity, and the number of passengers/occupants on board. In addition, during the first part of the observed period (2000–06) these rates have been much higher (4–5 events/100 * 10^6 v-mile) than that during the second part (2007–11) (about 1.5–2 events/100 * 10^6 v-mile). This could be an indication of substantive improvement of the overall safety of the US bus system in the relative terms as considered in the given context.

Fig. 11.13 shows an example of the accident rate of urban BRT system operating in Bogota (Columbia).

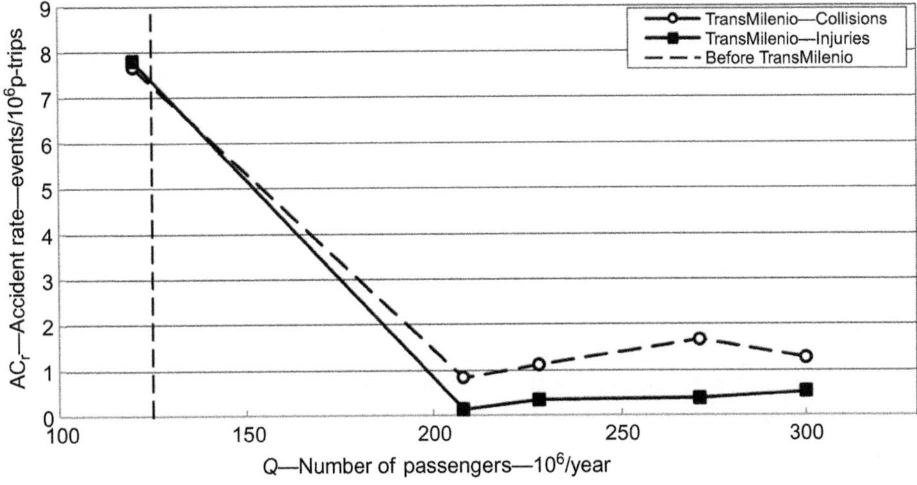

FIG. 11.13

Dependence on the number of accidents of the annual volume of traffic at the selected BRT systems (Janić, 2014a,b).

As can be seen, the accident rate of the conventional urban bus transit system before implementation of the BRT system has been about 7.7/10^6 p-trips. Over the period 2000–05, thanks to the BRT TransMilenio, despite increasing of the number of trips, this rate dropped to about 1–2/10^6 p-trips. That said, in the given and similar cases the BRT systems could make urban transit overall safer than the conventional ones mainly thanks to operating on dedicated routes free of other traffic and thus being exposed to the lower risk of collisions with it and related consequences. In addition, the systems themselves need to be designed and operated in a way that accidents/incidents due to the already known reasons do not occur (Janić, 2014a,b).

11.3.3.3 Trucks

Similarly as at buses, the traffic accidents/incidents and related causalities—fatalities and injuries of all categories (light, moderate, severe)—reflect the safety and security of trucks. These accidents/incidents have been usually caused by collisions of trucks with other trucks and with other vehicles, bikers, and pedestrians, all often resulting fatalities and injuries, as well as damages to property. The number of events (fatalities, injuries, truck crashes) per unit of the system's output expressed by the number of truck-km (miles) carried out during the specified period of time has shown to be a convenient measure for monitoring the level of safety of truck operations at different geographical/spatial scales (1 mile = 1.609 km). Fig. 11.14 shows the relationship between the large truck accident rate (crashes and fatalities) and the volume of output in the United States during the specified period of time (USDT, 2013b).

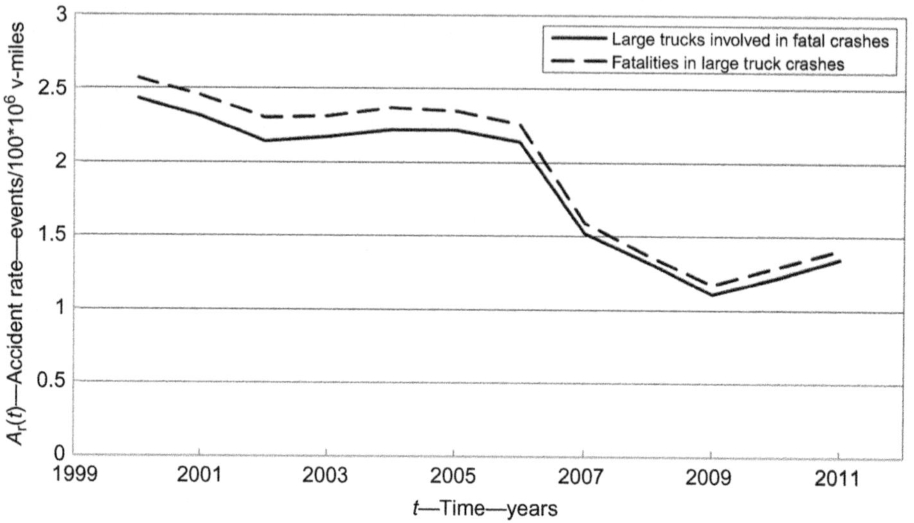

FIG. 11.14

The accident rate (crashes and related fatalities) at the large trucks in the United States over time (period: 2000–11) (USDT, 2013a,b).

As can be seen, the accident rates in terms of the number of trucks involved in fatal crashes and the number of fatalities per unit of output (v-miles per year) has been very similar. Both have been around 2.25–2.50/100 * 10^6/v-miles during the sub-period 2000–06, and then dropped to about 1–1.5/ 100 * 10^6/v-miles during the period 2007–11. These developments indicate that significant improvements in the safety of large truck operations were achieved, particularly during the latter part of the observed period. Again, due to the lack of specific safety targets in the given case, monitoring of developments as presented has remained to be useful for assessing the safety levels.

11.3.4 ENERGY/FUEL CONSUMPTION AND EMISSIONS OF GHG

11.3.4.1 Cars

Fuel consumption and related emissions of GHG of passenger cars are, in addition to other factors specified by driving conditions, mainly influenced by their technologies. At present five types of these technologies can be considered: conventional ICEVs, HYVs, BEVs, HVs (Hydrogen Vehicle(s)), and HFCVs (Hydrogen Fuel Cell Vehicle(s)).

Categories

- *Conventional ICEV* (*Internal Combustion Engine Vehicles*): These are considered as relatively low energy/fuel efficient due to the fact that as a result of converting fuel into propulsion, most of the energy is emitted as heat. Typical petrol ICEVs engines effectively use only 21% of the fuel energy content to move the vehicle and their diesel ICE counterparts are efficient up to 25%. This WTW (Well-to-Wheel) efficiency includes the energy consumed to produce and deliver fuel to the station (WTT (Well-to-Tank)) and the energy used to fill and consume it in the car (TTW (Tank-to-Wheel)) (Bodek and Heywood, 2008; Janić, 2014b).

For example, currently, in the EU-27 Member States, conventional ICEVs are categorized into three categories depending on the engine volume: Small <1.4 L, Medium >1.4 and ≤2.0 L, and Big >2.0 L (L—liter). Regardless of the fuel used, Small cars are the most and their Big counterparts the least numerous. The typical engine power of these cars is about 60–80 kW. The engine volume is correlated to the car weight, which is related to the average unit fuel consumption, that is, efficiency, as follows: $r = 0.004 + 5.249\ W$ ($R^2 = 0.839$) (r is fuel efficiency, ie, the average fuel consumption (L/100 km)); W is the car weight (kg). In addition, the average unity fuel consumption of an average car using petrol, diesel and/or gas amounts to: $r = 6.7$ L/100 km (0.683 kWh/km). Specifically, the average fuel consumption of an average petrol car is: $r = 7.3$ L/100 km (0.706 kWh/km) (This is expected to decrease to: $r = 5.8$ L/100 km (0.561 kWh/km) by the year 2020), and that of an average diesel/gas car: $r = 5.8$ L/100 km (0.594 kWh/km), which is expected to decrease to 4.6 L/100 km (0.493 kWh/km) by 2020. The average age of a passenger car in the EU-27 is 7.5 years (This is expected to increase to about 11–13 years by the year 2020) (IPTS, 2008; Janić, 2014b). In addition to the above-mentioned free-traffic-flow operation, the car fuel consumption additionally increases while operating in the congested urban and sub-urban areas. Fig. 11.15 shows an example of the relationship between the annual delays due to congestion and related additional fuel consumption of a commuter in 15 Very Large Urban Areas in the United States (Schrank et al., 2015).

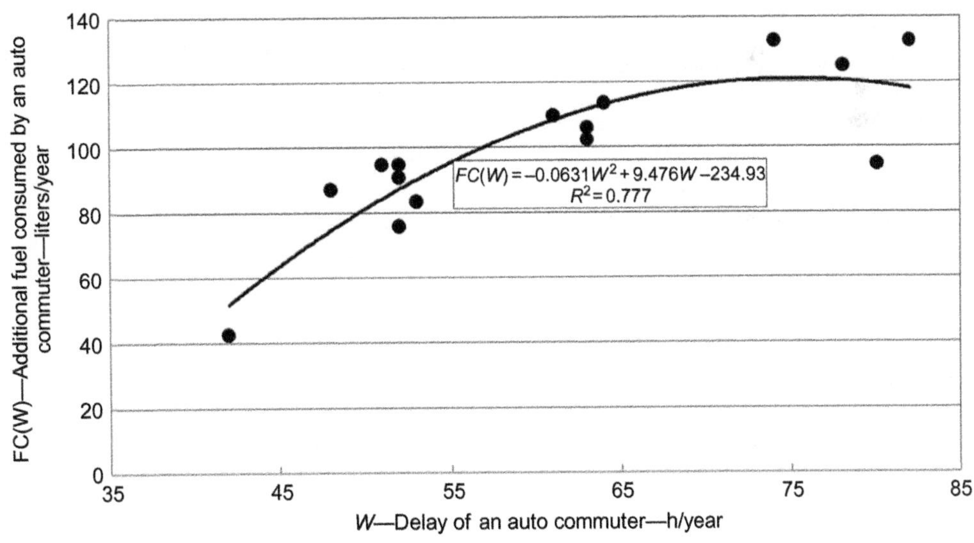

FIG. 11.15

Relationship between the additional fuel consumption and delays due congestion of an auto commuter in 15 Very Large Urban Areas in the United States (Schrank et al., 2015).

As can be seen, in this case, the annual fuel consumption has increased at a rather decreasing rate with increasing of delays due to urban congestion under given conditions.

The emissions of GHG by conventional ICEVs are usually considered as closely related to their WTW energy/fuel efficiency. In many cases, both can be standardized and as such become a country or region specific. For example, the standards set up for the EU-27 Member States in 2007–08 were: $r=6$–8 L/100 km (0.612–0.760 kWh/km) of the fuel consumption and $e_{CO_2} = 165 - 200\,gCO_2/km$. The newly proposed standards are around: $r=6.2$ L/100 km (0.632 kWh/km) and $e_{CO_2} = 140\,gCO_2/km$. The targets to be achieved by the year 2030 are: $r=3.5$ L/100 km (0.357 kWh/km) for the fuel consumption and $e_{CO_2} = 82 - 8 - 84\,gCO_2/km$ for emissions of CO_2 (carbon dioxide) (IPTS, 2008; Janić, 2014b).

- *HYV* (*Hybrid Vehicles*): These can be considered as an advanced passenger car technology. They are powered by conventional petrol or diesel ICE and an electromotor. While the former uses petrol or diesel fuel, the latter uses electric energy stored in on-board batteries, which are charged by the energy from the ICE engine. This means that recharging batteries by plugging in at street stations and/or at home is not possible. In general, the electromotor is used for driving at low speeds predominantly in urban areas, while the power switches to ICE when driving at higher speeds requiring greater engine power. The WTW energy/fuel efficiency of these cars is about 40% (Toyota Prius) and is expected to improve to about 55% in the mid-term future. For example, the most efficient hybrid car in 2005 was the Honda Insight whose WTW energy/fuel efficiency was 0.64 km/MJ (0.391 kWh/km) followed by the Toyota Prius with 0.56 km/MJ (0.491 kWh/km), and the petrol ICE Honda Civic VX with 0.52 km/MJ (0.534 kWh/km) (MJ—Mega Joule; kWh—kilowatt hour). In general, for example, in 2010, the fuel consumption of an average hybrid electric-petrol car amounted to about 5.4 L/100 km (0.799 kWh/km) and that of an average hybrid electric-diesel car to about 4.51 L/100 km (0.483 kWh/km). The corresponding emissions of GHG were 125 gCO_2/km and 90 gCO_2/km, respectively. Fig. 11.16 shows

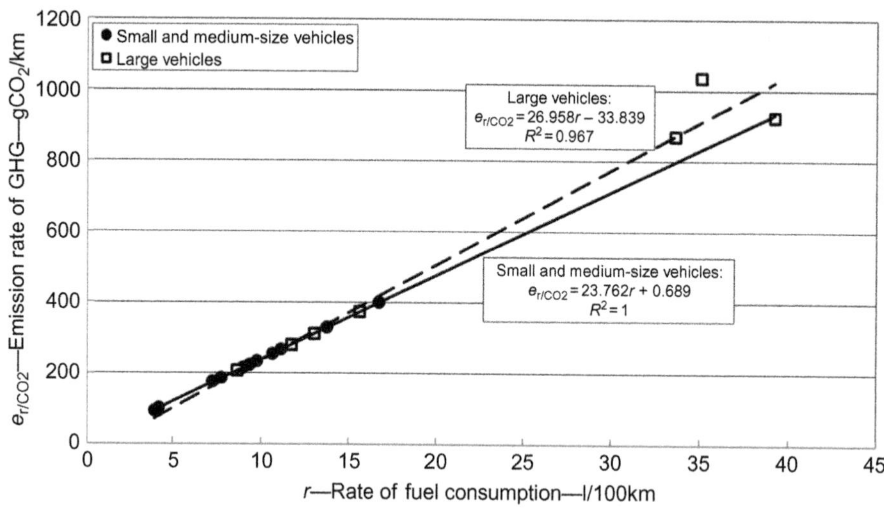

FIG. 11.16

Relationship between the average rate of emissions of GHG (CO_2) and the corresponding rate of fuel consumption of different kinds of passenger cars/vehicles and fuels (gasoline, diesel, gas, hybrid) in the United States (www.epa.gov/autoemissions).

the relationship between the average rate of emissions of GHG (CO_2) and the corresponding rate of fuel consumption of the passenger car/vehicles of different size and type in the United States (www.epa.gov/autoemissions).

As can be seen, the average emissions of GHG (CO_2) linearly increases with increasing of the average fuel consumption for cars (vehicles) of considered technologies and size. In addition, some improvements particularly to the fuel supply systems in HYVs lead to expectations that their consumption will decrease by the year 2035 to about: $r = 3.4$ L/100 km (0.329 kWh/km) at the hybrid electric-petrol and to about: $r = 2.45$ L/100 km (0.251 kWh/km) at the hybrid electric-diesel car. The corresponding emissions of GHG are expected to be: $e_{CO_2} = 52\,gCO_2/km$ and $e_{CO_2} = 47\,gCO_2/km$, respectively. This implies that in terms of energy/fuel efficiency and related emissions of GHG, electric/petrol and electric/diesel HYVs are more efficient than their conventional ICE counterparts by about 25% and 30%, respectively (Bodek and Heywood, 2008; Janić, 2014a,b).

The average rates of energy/fuel consumption and emissions of GHG of different passenger car/vehicle technologies expected to be used in the EU-27 during the period 2020–35 are given in the self-explaining Table 11.7.

Table 11.7 Average Rates of Energy/Fuel Consumption and Emissions of GHG of the Different Passenger Car/Vehicle Technologies (Janić, 2014b)

Technology	Basic Energy/ Fuel	Efficiency (%)	WTW Energy Efficiency (kWh/km)	WTW Emissions of GHG (gCO_{2e}/km)
Conventional ICEVs	Petrol/diesel/	20–21	0.612–0.760	165–200
	Auto gas (e)	25	0.955	-
	Petrol/diesel/ auto gas(n)	20–21	0.632	140
	Petrol/diesel/ auto gas (n)	20–25	0.357	82–84
HYVs	Petrol (e)	40	0.799	125
	Diesel (e)	55	0.483	90
	Petrol (e)	50	0.329	52
	Diesel (e)	65	0.251	47

Note: *e—Existing standards; n—new proposed standards to be in place over the period 2020–2035.*

In order for conventional ICEVs and their hybrid versions to fulfill the above-mentioned targets, improvements in both WTT and particularly TTW efficiency will be needed. In the former case, such improvements will be rather difficult to achieve, while in the latter, improvements will mainly stem from advanced car design including increased use of generally lighter (composite) materials. In any case, the average annual rate of improvements of the WTW efficiency and related emissions of GHG by 2020/2035 will need to be about 4–5% for conventional petrol ICEVs, 4.5–5.5% for conventional diesel ICEVs and 4.7–5.7% for HYVs. The WTW energy efficiency of BEVs and HFCVs will also improve while their emissions of GHG will strongly depend on the primary sources for the production of electric energy (EU-27—2010/15–65).

- *BEV* (*Battery Electric Vehicle*(*s*)): These can be considered as an advanced passenger car technology. They are propelled by electro-motors using the electric energy stored in batteries on-board the vehicle. The batteries are recharged from the power grid (at home or at street/shop charging stations). The WTW energy efficiency of electric cars is expected to reach up to about 80%. This can be achieved, among other factors, also thanks to converting the stored energy into propelling the car, not consuming energy while stopping, and regenerating some (about 20%) through regenerative braking. For example, the Tesla Roadster BEV has a WTW energy efficiency of about 1.14 km/MJ (0.235 kWh/km). Other typical electric cars are expected to have a WTW energy efficiency of about 1.125 km/MJ (0.247 kWh/km) and 1.583 km/MJ (0.175 kWh/km) (Toyota Rav4EV). It should be mentioned that about 20% of this energy consumption is due to inefficiencies in recharging the on-board batteries. These are the most sensitive parts of electric cars in terms of their specific energy capacity versus the weight, replacement, durability, and the short and full charging time. With a single charge they need to provide sufficient energy for the car to cover a reasonable distance at a reasonable speed as compared to conventional ICE petrol/diesel cars. Contemporary lithium batteries usually have a specific energy capacity of about 130 W-h/kg, which is one of the reasons for their frequent use despite their rather limited lifespan. Modified Lithium iron phosphate and Lithium-titan batteries have an extended lifespan of up to several thousand cycles and are relatively easily replaced. Their recharging time also needs to be reasonable. This is not particularly important if recharging takes place at home during off-peak hours, however it becomes very important if recharging takes place at street stations. Depending on the car's charger and battery technology, the recharging time can be 10–30 min to fill the batteries to about 70% of their capacity. For example, the models in the EU-27 market in 2011 such as Nissan Leaf, Renault Fluence Z.E. and Hyundai Blue have ranges between 140 and 170 km, top speeds between 130 and 145 km/h, full charging times of 6–8 h, and rapid charging times (up to 80%) of about 25–30 min. The above-mentioned-characteristics make these cars particularly convenient for use in urban and suburban areas with rather short daily driving distances. Electricity for BEVs can be obtained from different primary nonrenewable and renewable primary sources. The former include coal, crude oil, natural gas, biomass and nuclear energy, and the latter solar, wind, and hydro energy. The shares of the above-mentioned primary sources (usually country or region specific) make GHG emissions by BEVs exclusively dependent on their WTT energy/fuel efficiency (Janić, 2014b).

- *HV (Hydrogen Vehicle(s)) and HFCV (Hydrogen Fuel Cell Vehicle(s))*: These are powered by hydrogen fuel. Two categories of these vehicles can be distinguished. The first are slightly modified conventional ICEs that use hydrogen instead of petrol/diesel/gas as fuel—HV. In order to cover a reasonable distance, hydrogen is highly compressed in the fuel storage tanks of these vehicles, mainly thanks to its low density. HFCVs represent an advanced technology in passenger cars. They consist of five components which distinguish them from their HV counterparts: their fuel cell stack, electric motor, power control unit, hydrogen storage tank, and high output batteries. Specifically, the fuel cell stack consists of individual fuel cells whose number depends on their size and the required electric energy. Each fuel cell uses either pure hydrogen from hydrogen-rich sources, or oxygen to generate electric energy. Fuel is used to feed the electric motor that actually propels the car. The intensity of electric energy delivered from the fuel cells to the electric motor is regulated by the power control unit. Hydrogen as the source of electricity is stored in the hydrogen storage tank either as a liquid or as a highly compressed gas. In addition, high-output batteries are installed to accumulate the electric energy from the regenerative braking, thus providing additional power to the electric motor.

Hydrogen as a prospective fuel exists in nature as a component of natural gas (CH_4) and water (H_2O). This means that in order to provide hydrogen as fuel for hydrogen fuel cell cars, it needs to be extracted from the above-mentioned sources. This can be carried out by reforming natural gas or through the water electrolysis either at large plants or at local fuel supply stations. In the former case, distribution from the producing plants to local supply stations needs to be provided either by truck or an underground pipeline network. Hydrogen has more energy per unit of mass than other crude-oil-based fuels including natural gas. On the other hand, it is much less dense. The design of the fuel tanks of HFCVs will have to take the above-mentioned facts into consideration. Nevertheless, the volume of these tanks should not be much greater than that of conventional ICEVs as more energy per unit of mass of hydrogen is expected to compensate its lower density to a large extent. In addition, this will enable a similar pattern of utilization of HFCVs compared to their modern conventional ICEV counterparts.

The primary sources for obtaining hydrogen heavily influence the energy/fuel efficiency of HFCVs. At present, in practice, the WTW energy efficiency of HFCVs reaches about 50–60% (ie, 0.85 km/MJ or 0.327 kWh/km) if hydrogen is obtained from reforming natural gas and to only about 22% (ie, 0.30 km/MJ or 0.926 kWh/km) if it is obtained through water electrolysis. However, the theoretical overall efficiency of HFCVs can be nearly 100% (ie, 1.39 km/MJ or 0.198 kWh/km and 2.78 km/MJ or 0.102 kWh/km, respectively) (Janić, 2014b; http://www.fueleconomy.gov/FEG/fuelcell.shtml).

If hydrogen is derived from water electrolysis, the emissions of GHG by HFCVs will mainly depend on the primary sources of the electric energy used for this electrolysis. This can be from both nonrenewable and renewable sources, which influences the total WTW emissions of GHG. In the WTT segment, these will be zero if electricity is obtained exclusively from renewable sources and much higher otherwise. In the TTW segment, the emissions will be zero except for those of water vapor (H_2O), which will increase by about three times as compared to those from conventional ICEV and HYV fuels. Table 11.8 gives a summary of the average rates of energy/fuel consumption and related emissions of GHG (CO_{2e}) of BEVs, HVs, and HFCVs (Janić, 2014a,b).

Table 11.8 Average Rates of Energy/Fuel Consumption and Emissions of GHG of Different Passenger Car Technologies (Janić, 2014b)

Technology	Basic Energy/Fuel	Efficiency (%)	WTW Energy Efficiency (kWh/km)	WTW Emissions of GHG (gCO_{2e}/km)
BEVs	Electricity	80	0.175–0.235	[a]
HVs	Hydrogen	50	0.926	[a]
HFCVs	Hydrogen	95–100	0.010–0.327	[a]

[a]*Depending of the primary source for obtaining electricity and/or hydrogen.*

As can be seen, it is not certain what the emission rates of GHG of these vehicles will be because they will be mainly dependent on the primary sources for obtaining electricity and/or hydrogen.

Some effects of different passenger car technologies

Figs. 11.17–11.20 show some possible effects of introducing the particular passenger car technologies on the emissions of GHG at the global scale in Europe. In particular, Fig. 11.18 shows one of the possible scenarios of market penetration of these cars in Europe (Janić, 2014b).

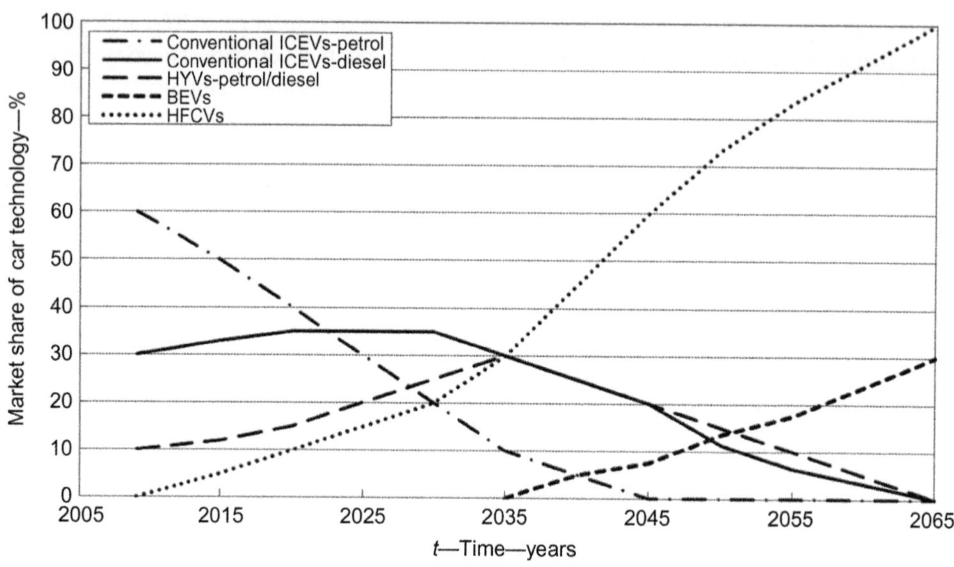

FIG. 11.17

Possible scenario of the market penetration of different passenger car technologies in Europe (EU-27) (period: 2010/15–65) (Janić, 2014b).

As can be seen, in all scenarios, the proportion of conventional petrol ICEVs is expected to decrease to 0%. The proportion of conventional diesel ICVs is expected first to increase and then continuously decrease to 0% until the end of the observed period. The proportion of HYVs is expected to increase to 35% during the first half and then decrease to 0% during the second half of the observed period. From the time of entering the market, the proportion of BEVs and HFCVs is expected to increase in and reach about 70% and 30%, respectively, by the end of the observed period.

Fig. 11.18 shows the possible scenario of developing share of renewable primary sources in the energy/fuel supply, energy consumption by transport sector, and particularly by new passenger car technologies is shown in Janić (2014b).

As can be seen, the proportion of renewable primary sources in the energy supply is expected to continuously grow during the given period of time according to the "S-curve power law." This implies relatively modest growth rates at the beginning, higher in the middle, and again lower growth rates at the end of the observed period. Such dynamism is reasonable if EU-27 states intend to mitigate their currently increasing dependency from imported and volatile priced depleting crude oil reserves. Since these reserves are expected to be exhausted by the end of the observed period, renewable sources will remain the exclusive primary energy supply sources in the region (EU-27). A substantial proportion of such obtained energy will be consumed by the transport sector—in proportion to that of the overall supply. Since this is actually electricity, this energy will be mainly consumed by new passenger car technologies dominating the transport sector during the observed period according to Scenario shown on Fig. 11.17.

Fig. 11.19 shows the annual energy/fuel consumption for the above-mentioned Scenario of introducing advanced passenger car/vehicle technologies during the observed period and for Scenario when the conventional petrol and diesel ICEVs would exclusively remain in the market (Janić, 2014b).

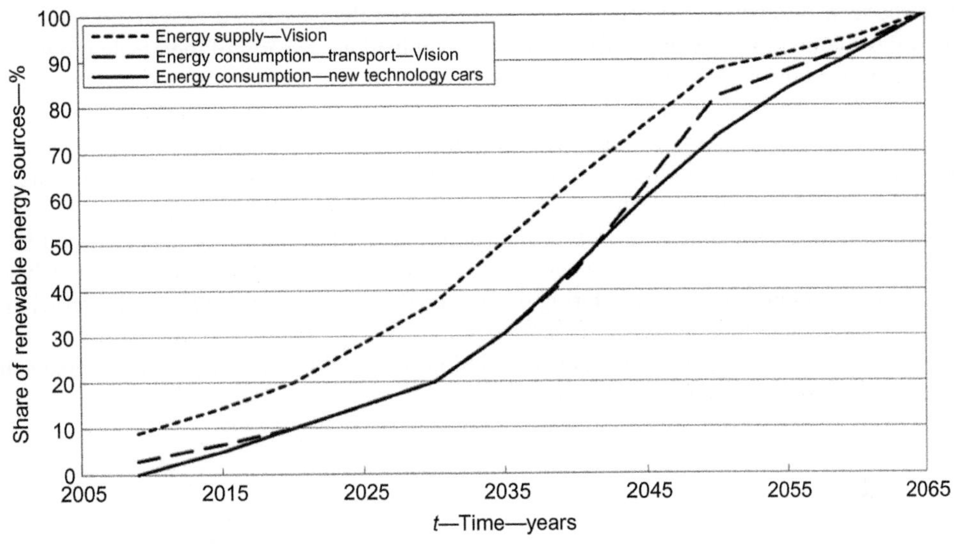

FIG. 11.18

Possible scenario of developing the share of renewable primary sources for the energy supply, consumption by transport sector, and new passenger car technologies (EU-27) (period: 2010/15–65) (Janić, 2014b).

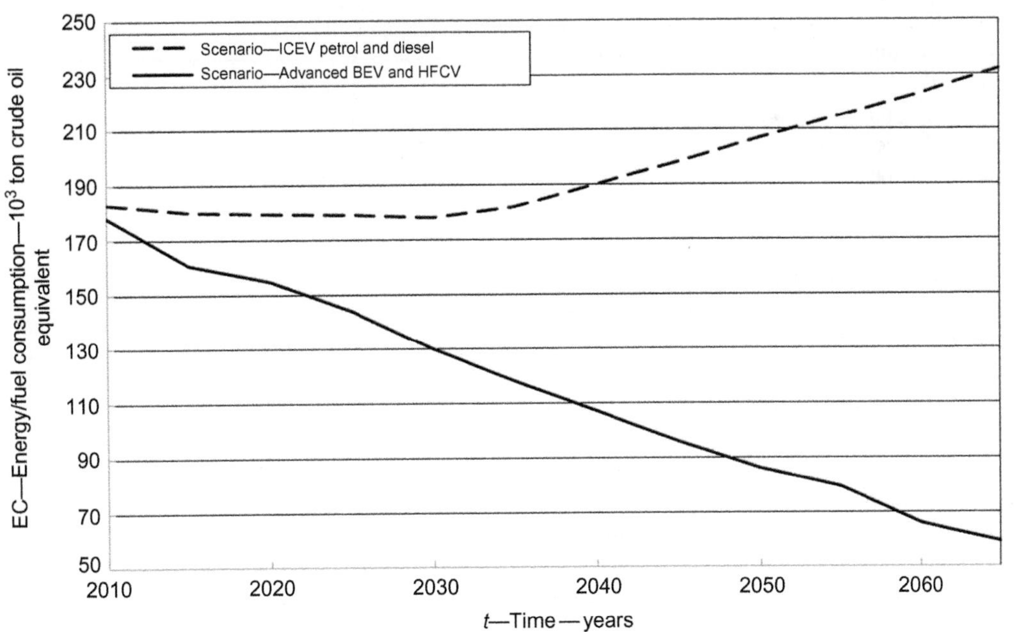

FIG. 11.19

Energy/fuel consumption of different passenger car/vehicle technologies in terms of crude oil equivalents in the given example (EU-27) (period: 2010/15–65) (Janić, 2014b).

As can be seen, the annual energy consumption differs and changes over the observed period driven mainly by the volumes of passenger car/vehicle use, combination of shares of their technologies in the market, and improvements of their WTW energy efficiency. In particular, in Scenario when only conventional ICEVs were used the annual energy consumption would decrease at the beginning of the observed period despite growing volumes of passenger car use mainly thanks to improvements in their energy efficiency (Tables 11.7 and 11.8). When these improvements were exhausted, the annual energy consumption would begin and continue to increase until the end of the observed period mainly driven by growing volumes of passenger car/vehicle use. In Scenario when the advanced BEVs and HFCVs were increasingly introduced, the annual energy consumption, permanently lower than that in the other scenario, would continuously decrease during the observed period (2010/15–65) despite growing volumes of passenger car/vehicle use (Fig. 11.17).

Fig. 11.20 shows the corresponding annual emissions of GHG (CO_{2e}) by passenger cars/vehicles according to Scenarios shown in Fig. 11.19.

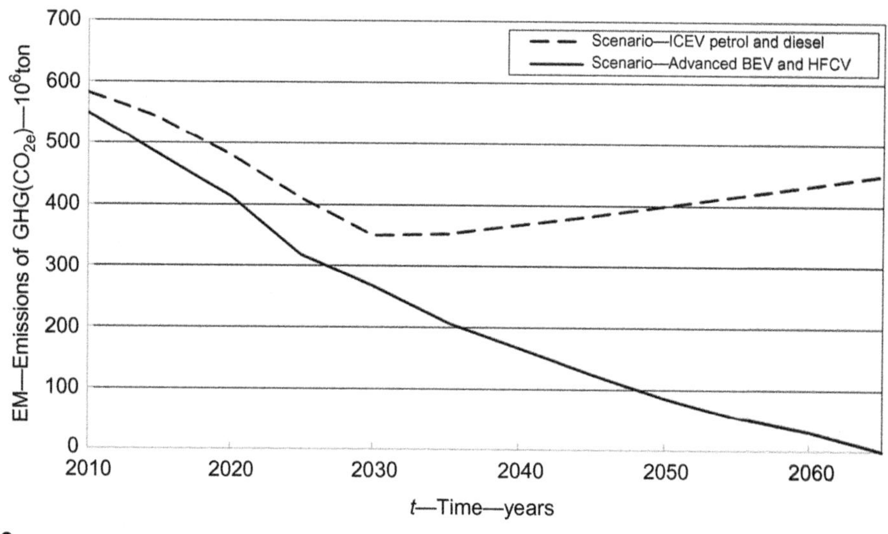

FIG. 11.20

Emissions of GHG (CO_{2e}) in the EU-27 under given scenario (period: 2010/15–65) (Janić, 2014b).

As can be seen, the general trends are similar to those of the energy consumption. In Scenario when the conventional ICEVs were exclusively remained in the market, the annual emissions of GHG (CO_{2e}) would decrease during the first part and then increases until the end of the observed period. In Scenario of gradually increasing introduction of the advanced BEVs and HFCVs, the annual emissions of GHG (CO_{2e}) would stay lower than in the previous Scenario and decrease over the entire observed period. At the end of the observed period (in 2065), the annual emissions may come close to zero, implying that the energy/electricity for advanced passenger cars/vehicles dominating the market is obtained exclusively from the renewable primary sources (Janić, 2014b).

11.3.4.2 Buses

As mentioned above, different bus technologies in terms of power and consequently energy/fuel use have been operated in urban, sub-urban, and inter-urban transit in different countries. The most common have been diesel bus vehicle types, which can use: diesel, compressed natural gas, diesel-electric energy (hybrid), and electric energy (trolleybus with overhead catenary). At present the conventional diesel buses have been the most prevalent in many counties. For example, in the North American fleet they shared 80%, followed by the shares of CNG (Compressed Natural Gas) and LNG (Liquefied Natural Gas) (15%). In the year 2006, only four transit agencies utilized trolleybuses, which share a small proportion of the total fleet.

The bus technologies influence their energy/fuel consumption and related emissions of GHG. In general, these can be considered as direct on the one hand, and indirect as expressed in terms of savings in these both thanks to the modal shift from other urban, sub-urban and interurban transit systems on the other.

Direct energy/fuel consumption and related emissions of GHG

The direct energy/fuel consumption and related emissions of GHG are usually expressed in relative terms, that is, as the average quantities per unit of the system's output, that is, g/p-km or g/s-km (grams/passenger-kilometer or grams/space-kilometer). These are usually estimated for the specified vehicle size and occupancy rate (load factor) while always bearing in mind the specific conditions in which a given bus system operates. Then, the absolute values have been easily obtained by multiplying these relative values by the corresponding volumes of output over the specified period of time, or vice versa.

In particular, the relative values of the energy/fuel consumption and related emissions of GHG are convenient for comparison of different bus systems between themselves and with other transit systems. Table 11.9 gives an example of the energy/fuel consumption and related emissions of GHG (CO_2) of different urban mass (BRT, LRT) and individual (car) transit systems (Janić, 2014a,b).

Table 11.9 Energy/Fuel Consumption and Related Emission of GHG (CO_2) Rates of the Selected Urban Transport Modes (Averages) (Janić, 2014a,b)

Impact/System-Mode	BRT[b]	BRT[c]	LRT[b]	Car[b]
Vehicle (length/no. of units)	12–18 m	18 m	2 units	1 unit
Energy/fuel consumption rate (f_c) (g/p-km)[a]	8.70–11.69	8.09	5.73	40.79
Emission rate of GHG (CO_2) (e_{CO_2}) (g/p-km)	27.85–37.41	25.9	18.37	130.53

[a]*Diesel fuel.*
[b]*US system(s).*
[c]*BRT TransMilenio (Bogota, Columbia) (based on 75 passengers per BRT and/or LRT vehicle, and 2 passengers per car).*

As can be seen, at both BRT systems, conventional buses (12 m long) mainly used in small and medium-sized direct or convoy systems/networks and for feeder services in larger trunk-feeder and hybrid networks consume and generate less energy/fuel and related emissions of GHG (CO_2), respectively, than their larger articulated counterparts (18 m long). In addition, in this case, the BRT systems have been inferior compared to the LRT systems on the one hand, but superior compared to individual passenger cars on the other. In addition, Table 11.10 gives the emissions of the other than CO_2

GHG—VOC (Organic Compounds), NO_x (Nitrogen Oxide), and CO (Carbon Monoxide)—generated by the different technologies of US BRT systems.

Table 11.10 Emission Rate of Other GHG by the US BRT Systems (Averages) (Janić, 2014a,b)

Emission Rate	BRT Technology		
GHG	Diesel	Hybrid	CNG[a]
NO_x (g/p-km; g/s-km)	0.7150	0.439; 0.336	0.2300; 0.1590
VOC (g/p-km; g/s-km)	0.0063	0.003159; 0.002418	0.0112; 0.0074
CO (g/p-km; g/s-km)	0.0713	0.00238; 0.00182	0.2570; 0.1770

[a]*CNG—compressed natural gas.*

As can be seen, regarding the type of GHG, the particular bus technologies are differently convenient in the relative terms: CNG regarding emissions of NO_x, Hybrid regarding emissions of VOCs, and Hybrid regarding emissions of CO.

Indirect energy/fuel consumption and related emissions of GHG—savings by modals shift

Given the vehicle/bus technologies and efficiency of their operations in urban and sub-urban areas, savings in the energy/fuel consumption and related emissions of GHG can be achieved as follows: (i) keeping existing passengers on-board; (ii) attracting those passengers currently using individual car as the system (j) to shift to the bus as the system (i); and (iii) attracting the new users of public transit systems. For example, the potential savings on a given route can be estimated as follows (Janić, 2014a,b):

$$s_{ji} = \left[(\lambda_i N_i)/\lambda_j N_j\right) \cdot f_{c/j} - f_{c/i}\right] \cdot d \tag{11.27}$$

where

N_i, N_j is the average vehicle capacity of transit systems i and j, respectively (spaces);
λ_i, λ_j is the load factor of vehicles of transit systems i and j, respectively;
$f_{c/i}$, $f_{c/j}$ is the average rate of energy/fuel consumption and/or emissions of GHG by transit systems i and j, respectively (g/p-km); and
d is the travel distance (km).

For example, a BRT bus carrying 75 passengers can replace 37 individual cars, each with 2 passengers (Table 11.5). If the average commuting distance is $d = 19$ km, the savings in the energy/fuel consumption by such car/BRT modal shift would be about $s_{ji} = 28.5$ (bus—12 m long) and 28.7 (bus—18 m long) kg of diesel fuel. The corresponding savings in the emissions of GHG would be about: $s_{ji} = 91.2$ and 91 $kgCO_2$, respectively (Janić, 2014a,b).

11.3.4.3 Trucks

Trucks consume diesel fuel as a derivative of crude oil. The quantity of fuel consumed on a given distance/route can be estimated as follows:

$$FC = f_c \cdot (100 \cdot d) \tag{11.28}$$

where

f_c is the fuel consumption rate (L/100 km); and
d is the travel distance (km).

The main operational factors influencing the truck fuel consumption are the operating speed, and the aerodynamic and the rolling resistance. With increasing speed, the total resistance linearly increases mainly thanks to an increase in the aerodynamic resistance on the account of the rolling resistance (AEA, 2011). Fig. 11.21 shows an example of the relationship between the average unit fuel consumption rate and (operating) speed of a European standard truck (GVW 40 ton and payload 25 ton—100% load factor).

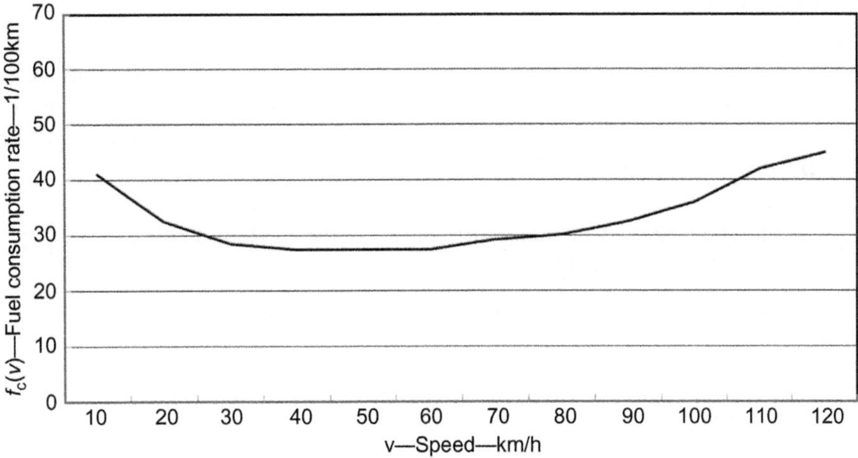

FIG. 11.21

Relationship between the road truck average fuel consumption rate and (operating) speed (AEA, 2011; Janić, 2014a,b).

As can be seen, the average fuel consumption rate first decreases with increasing of the operating speed, then remains relatively constant for speeds between 40 and 60 km/h, and then increases again. In most European countries, the maximum speed of these trucks is limited to 90 km/h on the highways/motorways and 70 km/h on other roads, which implies that the corresponding fuel consumption rate is about 29–32 L/100 km.

Emissions of GHG by trucks consuming diesel fuel can be considered in terms of CO_{2e} (Carbon-Dioxide equivalents). For distance d, these emissions can be estimated, based as follows (Janić, 2014a,b):

$$EM = FC \cdot e_{CO_{2e}} = f_c \cdot (100 \cdot d) \cdot e_{CO_{2e}} \tag{11.29}$$

where $e_{CO_{2e}}$ is the emission factor ($kgCO_{2e}$/L of fuel).

The other symbols are as in the previous equations. The emission factor ($e_{CO_{2e}}$) usually relates to the on-wheel emissions including the emissions from manufacturing fuel and those from the direct burning of fuel. In particular, as far as the direct burning of fuel is concerned, if the above-mentioned standard trucks operate at an average speed of: $v = 70$ km/h, its fuel consumption rate will be $f_c = 29$ L/100 km (see Fig. 11.21). The specific gravity of diesel fuel is $SG = 0.82–0.95$ kg/L, and its calorific value: $CV = 12.777$ kWh/kg, which produces an average rate of energy consumption of about: $SEC = 3.28$ kWh/km. The emission rate of GHG is $e_{CO_{2e}} = 0.324 kgCO_{2e}/kWh$ (CO_{2e} includes CO (Carbon Oxide), HC (Hydro Carbons), NO_x (Nitrogen Oxides), and PM (Particulate Matters)), which gives the emission rate of $em_r = 3.28 * 0.324 = 1.062$ $kgCO_{2e}$/v-km. By dividing this amount with the full payload of the standard truck of 25 ton, the average emission rate of GHG can be obtained as $em_r = 1.062/25 = 0.0424$ $kgCO_{2e}$/t-km. In addition, by multiplying this rate with the total distance between the freight shipper and receiver doors and dividing it by the actual load factor ($\leq 100\%$), the total emissions of GHG per transport service carried out by the given truck service can be obtained.

11.3.5 LAND USE

11.3.5.1 Cars

The road transport infrastructure used by passenger cars/vehicles occupies land in urban, sub-urban, and inter-urban area(s). The area of land occupied by roads depends on their length and width. The road width depends on its category—motorway or non-motorway—and the number of lanes per direction. Table 11.11 gives an example for the US roads (AASHTO, 2001).

Table 11.11 Some Characteristics of the Width of Road Profile in the United States (AASHTO, 2001)

Road Category	Total Width (m)
Motorway	
Dual 2 lane	40.2
Dual 3 lane	47.6
Dual 4 lane	54.8
Non-motorway	
All-purpose rural	7.3

In particular, the width of the standard road lane in the United States is specified to be 3.7 m for the interstate highway systems, while the narrower lanes are used on lower classification roads. In Europe, the road and lane width vary by country, but the minimum width of lane is generally from 2.5 to 3.25 m. In addition, the federal interurban network in Germany specifies the minimum width of lane of 3.5 m for each the smallest two-lane roads with an additional 0.25 m on the outer sides and shoulders of at least 1.5 m on each side. A modern highway will have two lanes per direction which are 3.75 m wide with an additional clearance of 0.50 m on each side. Three lanes per direction are set at 3.75 m for the rightmost lane and 3.5 m for the other lanes. Urban access roads and roads in low-density areas may have the minimum width of lanes of 2.75 m with shoulders of at least 1.0 m. In principle, the lane will be wider if the persistent and more intensive truck traffic is expected. In addition, if the lane width decreases, the traffic capacity decreases too. Consequently, a full-width freeway lane typically has the capacity of 2000 cars/h (see Chapter 5). In addition to roads, the land is occupied due to providing parking for passenger cars/vehicles. For example, in Europe, the standard parking place is of the net rectangular shape 4.8 m × 2.4 m (11.52 m^2). However, the architectural norm is 24 m^2/parking place. In addition, considerable areas of land are used for petrol/gas stations, motorway service, and all other service and ancillary activities related to the passenger car/vehicle use and maintenance (Whitelegg, 1994).

For example, the total area of road is Europe is about $5 * 10^6$ km of which 1.302% are two—direction two and three lanes/direction motorways and the rest (96.98%) are other non-motorway paved roads (ERF, 2011). If the average total width of a profile of a motorway including margins on both sides is assumed to be 38.0 m and of that of non-motorway 11.0 m, the total area taken by roads in Europe is approximately equal to $A = 0.01302 * (0.038 * 5{,}000{,}000) + 0.9698 * (0.011 * 5{,}000{,}000) = 55{,}617.5\ \text{km}^2$. If the area of EU is 4,381,400 km^2, the proportion occupied by roads is equal to 55,617.5/4,381,400 = 1.269%. If the passenger cars were carried out $4672 * 10^9$ p-km, the intensity of use of road network would be: $ILU = 4672 * 10^9 / 5.56175 * 10^6 \approx 840 * 10^3$ p-km/ha (EU, 2015). Since there are no specific guidelines and criteria for judgment on the intensity of land use in the above-mentioned context, the above-mentioned estimations can be used for the illustrative purposes and as an initial step for further more detailed studies.

11.3.5.2 Buses

The buses operate along the urban, suburban, and interurban streets, roads, and motorways/highways, respectively. Under such conditions they share the same road infrastructure with other road vehicles—passenger cars and trucks. In some specific cases, such as the above-mentioned BRT systems, the land is exclusively taken and used by them. Their infrastructure networks consisting of the corridors/routes with dedicated bus-ways and terminals/stations spread over, pass by and/or through densely populated/demand attractive areas of the given urban agglomerations—the city center(s) or CBDs (Central Business District(s)). These bus-ways passing through the high density areas continue outside it as right-of-way bus lanes. In some cases, the BRT dedicated bus-ways or bus-only roadways are built along old rail corridors/lines. In addition, the dedicated bus-ways are usually designed as two-way lanes in different directions in mixed traffic, two-way lines on the same side or in the middle, or as a single line in each direction on different sides of the given corridor/route. In some cases, the bus-way is split into two one-way lanes/segments. Given the length of a given BRT corridor usually defined as the distance between the initial and the end terminal/station, width, and the number and area of the terminals/stations along it, the total area of land directly taken for building this infrastructure can be estimated as follows (Janić, 2014a,b; Vuchic, 2007):

$$A = D \cdot L + n \cdot (l \cdot d) \tag{11.30}$$

where

- D is the width of the corridor (m);
- L is the length of corridor (km);
- n is the number of stations/platforms along the corridor; and
- l, d is the length and width of the plot of land occupied by the terminal/station (m, m), respectively.

For example, the width D of an exclusive bus-way (both directions) within a BRT corridor varies from 10.4 to 11.6 m (for moderate speeds $\leq$70 km/h) to 14.60 m (for speeds up to 90 km/h). The typical length of the bus stops varies from $l = 18$–26 m depending of the bus length (for a single bus). The minimum width of the bus stop at the terminal/station is about $d = 3.0$–3.5 m. However, the width of the area occupied by the terminal/station itself with the supporting facilities and equipment could be up to 9.0 m (Janić, 2014a,b; Vuchic, 2007). Consequently, the BRT corridor of length of: $L = 20$ km, width $D=14.60$ m, and 20 stations/stops on both sides, each of the area of $l*d=26.0*9.0$ m $=234$ m^2 takes the land of $A=0.0146*20+20*(0.026*0.009)=0.29668$ km^2 or 29.668 ha. If BRT vehicles operate at the frequency of 2 departures/h pre direction, each carrying out about 75 passengers during 12 h per day, the daily volume of output will be $Q=2*2*75*20*12=72{,}000$ p-km/day. The intensity of land use per day will be $ILU=Q/A=72{,}000/29.688=2697.84$ p-km/ha per day. The annual utilization can be also estimated straightforward respecting the daily and monthly volumes of operations/traffic/transport.

11.3.5.3 Trucks

Trucks carrying out freight/cargo shipments between doors of shippers and receivers share the road and motorway/highway infrastructure with buses and passenger cars/vehicles. Therefore, it can be said that the additional land for their operations is not taken. As far as the intensity of land use is concerned, the area of land taken by roads in Europe has been estimated to be $A=55617.5$ km^2. For example, if the annual volume of output of road freight trucks in the year 2013 was: $Q=1719.4*10^9$ t-km, the intensity of land use was $ILU=A/Q=1719.4*10^9/55617.5=30.915*106$ t-km/km^2 (EU, 2015). This should be considered again only as an illustration because the specific targets about low, medium, or high land take and use do not exist in the given context. In addition, the estimated land used and taken by roads in Europe and elsewhere do not take into account the required parking spaces, in this case for trucks at least at four locations: doors of the freight shippers and receivers (loading, unloading, respectively), hub terminals (loading, unloading, transit), the drivers' resting places and fuel-supply stations along the long routes (NTSB, 2000). As an example, a single parking place for a heavy truck including the maneuvering space takes the area of land of 45.1 m $\times$ 4.3 m $=192.74$ m^2 when entry is at the angle of 90 degrees, and less if it is at the lower angles (30 degrees, 45 degrees). In the case of parking in the covered space (garage), the minimum headroom is 4.5–4.75 m (This is the vertical distance between the floor and the lowest barrier in the garage) (Waco, 2010).

11.4 RAIL-BASED SYSTEMS

The social and environmental impacts of the rail-based urban streetcar (tramway), LRT, and subway (metro), and inter-urban passenger and freight transport systems are analyzed. The social impacts include traffic congestion and related delays, noise, and traffic incidents/accidents. The environmental impacts embrace energy consumption and related emissions of GHG and land use.

11.4.1 CONGESTION

11.4.1.1 Streetcar (tramway)

The streetcars (tramways) are free of traffic congestion and related delays while operating under regular operating conditions. Therefore their congestion and delays are not particularly considered in the given context. However, in some cases the streetcars (tramways) can impose additional congestion and delays on other vehicles—passenger cars and buses—sharing the same lanes, streets, and corridors in the particular urban (and sometimes sub-urban) areas. These (additional) congestion and delays are imposed on cars and buses due to assigning higher priorities to the streetcars (tramways) while sharing the same infrastructure at the same time.

11.4.1.2 LRT (Light Rail Transit)

Similarly as the streetcar (tramway) system, the LRT system(s) is free of traffic congestion and related delays while operating under regular operating conditions due to three reasons. The first is the nature of operations implying the space-time separation of the successive LRT vehicles moving along the same track in the same direction. The second is the double track lines enabling simultaneous operation of the LRT vehicles in both directions without interfering with each other. The last is positioning of the lines/tracks, usually along the isolated corridors preventing interference of the LRT vehicles with other urban and sub-urban vehicles-traffic, and vice versa. Therefore, congestion of the LRT systems is not particularly elaborated in the given context.

11.4.1.3 Subway (metro)

The subway (metro) system(s) is also free of traffic congestion and related delays thanks to operating vehicles on different tracks while moving in different directions, and the time-space separation between vehicles moving along the same track(s) in the same direction—all under regular operating conditions. Therefore, traffic congestion and related delays are not particularly elaborated in the given context.

11.4.1.4 Passenger inter-urban trains

Passenger inter-urban trains are free of traffic congestion and related delays similarly as the other rail-based systems, mainly due to the planned operations in the form of timetable, which enables matching the train frequency with the capacity of rail lines/tracks and stations under regular operating conditions. Nevertheless, due to deviations of the actual from the planned realization of the timetable, congestion and delays of trains happen along the lines and at the stations.

11.4.1.5 Freight trains

Similarly as the passenger, the freight trains are free of traffic congestion and related delays thanks to planning their operations according to the timetable either along the exclusively used rail lines/tracks or that shared with the passenger trains. Therefore, congestion and delays as the social impacts and related externalities are not particularly considered.

11.4.2 NOISE

11.4.2.1 Streetcar (tramway)

The streetcars (tramways) generate direct and indirect noise. The direct noise is generated during their operations (the noise during construction of the streetcar lines is not considered). Similarly as at the other rail-based systems, the level of this direct/operational noise depends on the vehicle's speed. At lower speeds, the wheel/rail noise, the noise from vehicle's traction motors, and the noise from vehicle's auxiliary equipment (ie, air conditioning, compressors, and motor controllers) influence the overall operational noise levels. The wheel/rail noise strongly depends on the condition of the wheels and the rails, which implies that they both should be maintained in good conditions. At the higher speeds, the operational noise is mainly the wheel/rail noise. The streetcars (tramways) also use bells as audible warning devices regularly during operations including stops and starts due to any reasons, and for alerting pedestrians and other vehicle drivers of a potential safety risk. In addition to speed, the direct noise from operating streetcars (tramways) has been depending on the distance of the observer from the source of noise—moving streetcar. The measurements of noise from streetcars (tramways) have shown that below and up to the speed of 32 km/h, the noise from traction motors, air conditioning, and other auxiliary equipment on the vehicles dominates. Above 40 km/h, the rolling noise from the wheel-rail interface dominates. As a result, the noise from streetcars (tramways) expressed in dBA varies with operating speed as follows (DDOT, 2013):

$$L_{max}(v) = \begin{bmatrix} \text{Speed} - \text{independent};\ v < 24\text{km/h} \\ 12 \cdot \log_{10} v;\ 24 < v < 40\text{km/h} \\ 30 \cdot \log_{10} v;\ > 40\text{km/h} \end{bmatrix} \tag{11.31}$$

where v is the operating speed of a streetcar (tramway) (km/h).

At the reference distance from the centerline of the track, the maximum noise level of the streetcar in dependence of the speed has been as shown on Fig. 11.22.

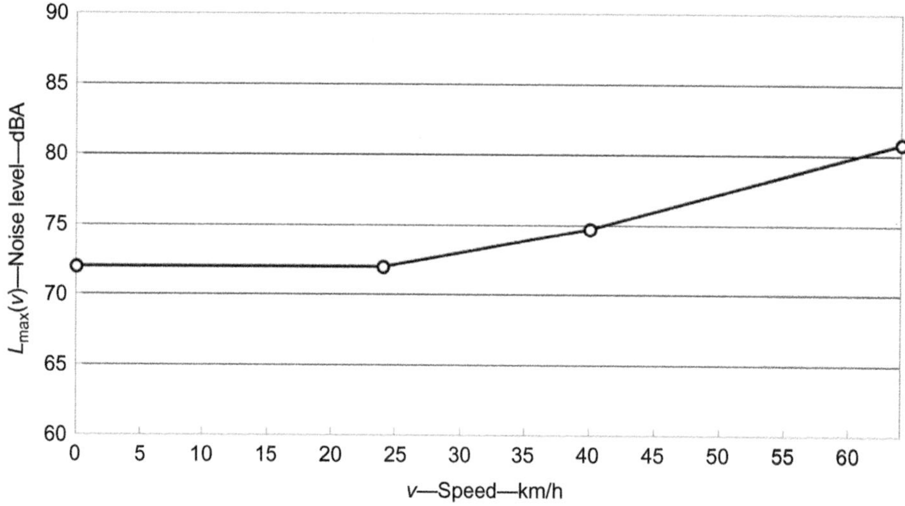

FIG. 11.22

Example of the relationship between the noise level and (operating) speed of streetcar (tramways) measured at the distance of $y = 15$ m (DDOT, 2013).

As can be seen, the noise remains at the constant level of about 72 dBA up to the speed of 24 km/h, then slightly increase to about 75 dBA with increasing of the speed from 24 to 40 km/h, and then continues to sharper increase to about 80 dBA at the speed of 64 km/h. The length of the streetcar (tramway) vehicle has been 20 m (DDOT, 2013). In addition, Fig. 11.23 shows the relationship of continuous daily noise from the streetcar (tramways) depending on the distance of an observer (ie, the noise measurement location) from the source (DDOT, 2013).

As can be seen, the noise level has decreased more than proportionally with increasing of the distance between the observer and its source (operating streetcars (tramways)).

The indirect noise generated by the streetcars (tramways) can happen after introducing the new lines, which sometimes can cause substantial changes in the traffic patterns and volumes. For example, this traffic may shift from the streetcar route to other roads/streets, which would reduce the overall traffic noise along the streetcar (tramway) route(s) and increase the noise at these other routes/streets.

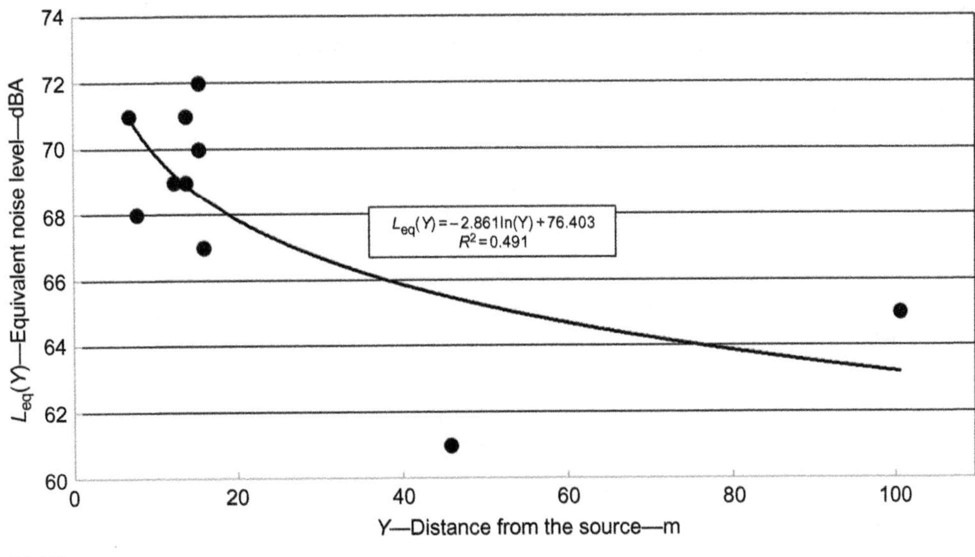

FIG. 11.23

Relationship between the equivalent continuous noise level by streetcar(s) (tramway(s)) and distance of the observer from the source (DDOT, 2013).

11.4.2.2 LRT (Light Rail Transit)

The LRT vehicles usually consist of 2–3 cars operating at an average speed of 55–60 km/h on the lines with more dense stops/stations and 65–70 km/h along the lines with less dense stations. The LRT service frequency is usually adjusted to demand and differs during peak and off-peak hours of the day—maximally from 12 dep/h (every 5 min) to 4 dep/h (every 15 min) during the day and 1–2 dep/h (every half or an hour) during the night.

Such operational patterns generate noise by each individual vehicle and the successive vehicles from the sources similar as the streetcar (tramways) such as the wheel/rail noise, the noise from the vehicle traction motors, the noise from the vehicle's auxiliary equipment (ie, air conditioning, compressors, and motor controllers), and warning devices.

For example, some measurements have shown that the LRT vehicle operating at the speed of 80 km/h creates the maximum level of noise $L_{max} = 80$ dBA at the distance of 15 m from the tracks. In addition, Fig. 11.24 shows relationship the level of noise and speed for the LRT vehicles. The noise from other vehicles has been given for the comparative purposes (Janić, 2014a,b).

As can be seen, the noise from LRT increases at decreasing rate with increasing of the vehicle's operating speed. At the same time, the noise level of LRT vehicle is greater than the noise level generated by other urban and sub-urban transit vehicles/systems/modes at any speed mainly due to the shorter distance between the vehicle and the observer. In addition, at the speed of 80 km/h, the LRT vehicle noise has increased for about 3 dBA with decreasing distance from 15 to 3 m. The LRT vehicles also generate noise from on-board warning devices including gong, bells, and horn. The gong or the bells are used for alerting when the LRT vehicles enter a station to alert passengers on the platforms of oncoming vehicles. The louder horns are used at grade crossings. The maximum

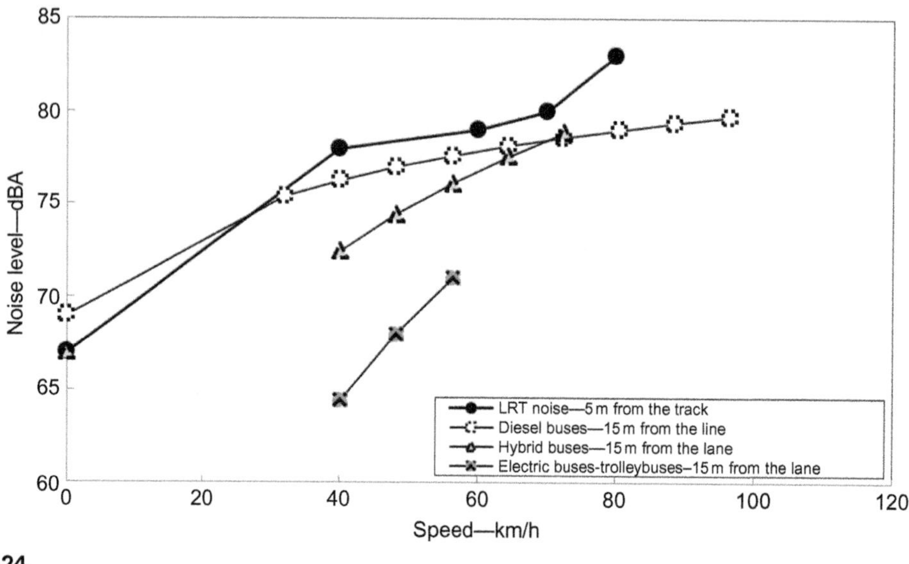

FIG. 11.24

Relationships between the noise and speed of LRT and BRT (Bus Rapid Transit) systems (Janić, 2014a,b).

sound levels obtained at the distance of 3 m has measured to be 75 dBA by the gong, 95 dBA by the bells, and 112 dBA by the horn. In addition, by using the LRT source noise levels, the average hourly daytime and night time operations, and speed, the LRT pass-by hourly and daily noise levels can be estimated (KMCEI, 2007).

11.4.2.3 Subway (metro)

The subway (metro) systems mostly operate though the underground tunnels, thus preventing spreading noise from their passing-by vehicles along wider along the lines. However, parts of lines of some subway (metro) systems are also elevated (on the ground surface), thus making considering their noise relevant. In general, given the track conditions, line and tunnel design, construction quality, and geology of the terrain, the main source of noise during operation of the subway (metro) vehicles has shown to be the wheel/track noise, which generally increases with increasing of their operating speed. The other sources have been the curve and brake squeal noise, which however have been lower than the straight line segment operating noise primarily due to lower operating speed(s). Some recent measurements have shown that the noise of subway (metro) vehicles primarily increases with speed, but also differs if they move through the underground tunnels or elevated tracks. Fig. 11.25 shows an example of the measured noise just outside and inside the vehicles operating at different speeds on the newly built subway (metro) line in Beijing (China) (Lu et al., 2014).

As can be seen, the equivalent noise level L_{eq}(dBA) has increased with increasing of the subway (metro) operating speed. The outside noise has been substantively higher than the inside noise. In addition, both inside and outside noise along the elevated tracks has been a bit higher than the noise along the tunnel tracks.

As far as the external observers-passengers and accompanies are concerned, they experience the noise from the subway (metro) vehicles while they are approaching, standing, and departing from

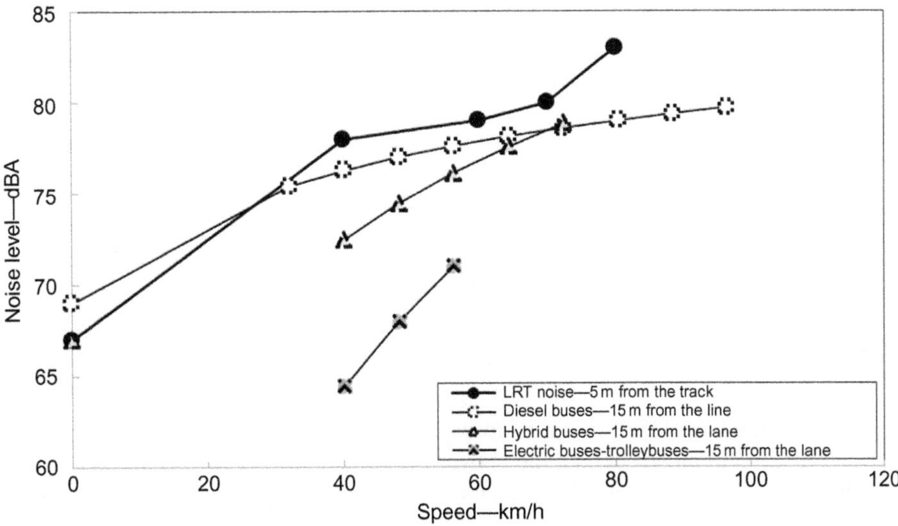

FIG. 11.25

Relationship between the equivalent noise level and (operating) speed of the subway (metro) vehicles—the new line in Beijing (China) (Lu et al., 2014).

the subway (metro) stations and inside the vehicles. The research including the measurement of the subway (metro) noise at the scene has indicated that both outside and inside noise could be substantive. For example, the average and maximum noise levels on the platforms of subway stations were measured inside the subway (metro) vehicles. The average noise level measured on platforms of the stations of the New York subway (metro) system has amounted 86 $\pm$ 4 dBA, while the maximum level has been 106 dBA. The noise level inside the subway (metro) vehicles has ranged from 84 to 112 dBA. These values indicate that this subway (metro) noise can exceed the recommended noise exposure guidance specified by WHO and other national regulatory bodies. Some agencies have recommended the 8-h continuous exposure noise level of 85–90 dBA. The WHO have established guidelines for community noise exposure of 75 dBA 8-h or 70 dBA 24-h daily average noise exposure level, the later during the period of 40 years. In practice, the approximate noise levels are 45–60 dBA for normal conversation, 100 dBA for a chainsaw, and 140 dBA for a gun blast. Due to the logarithmic nature of decibels, an increase of sound for 10 dB implies a 10-fold increase in the noise intensity; therefore, a 90 dB sound is 10 times as intense as an 80 dB sound, 100 times as intense as a 70 dB sound, and 1000 times as intense as a 60 dB sound (Gershon et al., 2006).

11.4.2.4 Passenger inter-urban trains

Both conventional and High Speed (HS) trains generate noise, which comprises rolling, aerodynamic, equipment, and propulsion sound. This noise mainly depends on its level generated by the source, that is, moving train(s), and its distance from an exposed observer(s). Fig. 11.26 shows a scheme of changing the distance and time of exposure to noise of an observer by a passing-by train.

The shadow polygon represents a train of length L passing-by an observer (a small triangle at the bottom) at the speed v. He/she starts to consider noise of an approaching train when it is at distance β

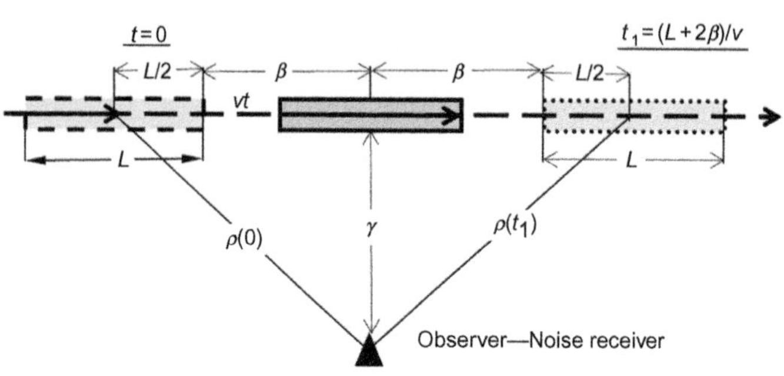

FIG. 11.26

Scheme for determining the noise exposure of an observer to noise of passing-by train (Janić and Vleugel, 2012).

from the point along the line, which is at the closest right angle distance γ from him/her. The consideration stops after the train moves behind the above-mentioned closest point again for the distance β. Under such circumstances, the distance between the observer and the passing-by HS train changes over time as follows:

$$\rho^2(t) = (L/2 + \beta - v \cdot t)^2 + \gamma^2 \quad \text{for } 0 < t \le (L + 2 \cdot \beta)/v \tag{11.32}$$

where the last term represents duration of the noise event, that is, the time needed for a train to pass by the observer (the length of passenger conventional and HS trains is usually 200–450/500 m). If the level of noise received from the train passing-by an observer with the speed (v) at the shortest distance (γ) is $L_{eq}(\gamma, v)$, the level of noise at any time (t) can be estimated as follows:

$$L_e[\rho(t), v] = L_e(\gamma, v) - 8.6562 \ \ln[\rho(t)/\gamma] \tag{11.33}$$

The second term in Eq. (11.33) represents the noise attenuation with distance over the area free of barriers. The total noise exposure of the observer from $f(\tau)$ successive trains passing by during the period (τ) can be estimated as follows:

$$L_{eq}[f(\tau)] = 10\log\sum_{k=1}^{f(\tau)} 10^{\frac{L_e[k, \rho(t), v]}{10}} \tag{11.34}$$

In particular, as a standard approach, the noise from conventional and HS passenger trains is measured at the right angle distance of $\gamma = 25$ m from the center of the line/track(s). Fig. 11.27 shows the results of some such noise measurements of the HS trains operating in Europe in dependence on their maximum operating speed.

As can be seen, the noise has generally linearly increased with increasing of the train's operating speed: at the lower rate for the speeds up to about $v = 300$ km/h, and at the higher rate for the speeds above $v = 300$ km/h. The variation of noise level at the given speed has been about 3–4 dBA. This noise has included the train's rolling (wheel), pantograph/overhead, and aerodynamic noise. Some additional measurements have shown that the rolling and pantograph/overhead noise has predominated and increased with increasing of the HS train's speed approximately at the rate of 30 $\log_{10} v$ up to the speed(s) of about $v = 300$ km/h (some data have shown that this is $v = 370$ km/h). The aerodynamic noise depending on the HS train's (aerodynamic) design has also increased, equalized with the rolling noise at the above-mentioned (transition) speed(s), started predominating and further increasing at an

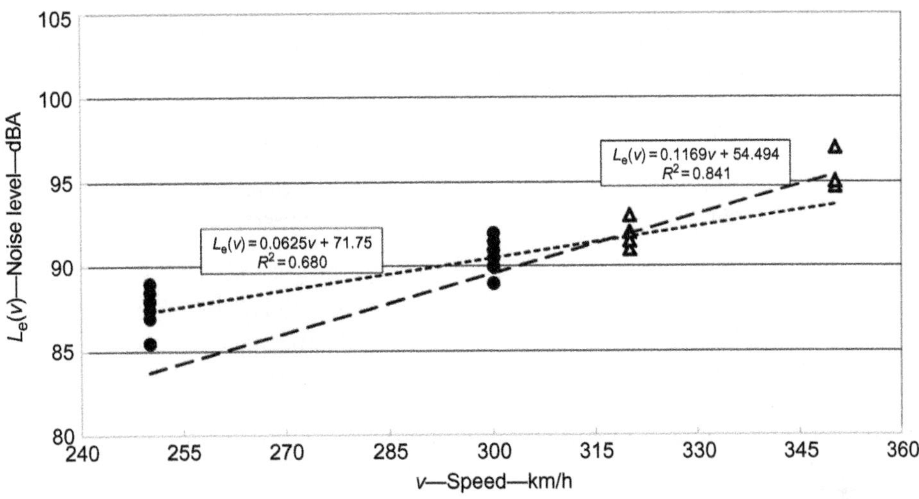

FIG. 11.27

Relationship between the noise and the maximum operating speed of the passing-by HS train(s) measured at the right-angle distance of $\gamma = 25$ m (Belgium, France, Germany, Spain, Italy) (Gautier and Letourneaux, 2010; Janić, 2016a,b).

approximate rate of 80 $\log_{10}v$ (Thompson et al., 2015). In addition, in cases when the frequent High Speed Rail (HSR) services are carried out along the particular lines/routes, their noise becomes persistent over time and can be estimated from Eq. (11.34). In addition, the time of exposure of an observer to noise by a passing by HS train can be estimated from Eq. (11.32). If $\beta = 0$ m, $L = 200$ m, and $v = 250$ km/h, this exposure time to the maximum noise will be about $t_1 = 3$ s; if $v = 350$ km/h, this time will be about $t_1 = 2$ s. The noise from trains has become a sensitive issue mainly because many existing rail lines have been passing through or close to densely populated areas, or the newly built particularly HS rail lines have been built in areas where the pre-existing noise was very low. Since it has been shown that this noise has mainly increased with increasing of the trains' operating speed, the initiatives for controlling/limiting noise by influencing speed have emerged worldwide. One of these has been that of UIC (International Union of Railways), which specified the noise limits for the rail passenger vehicles of 75 dBA at the speed of $v = 80$ km/h, 84 dBA at the speed of $v = 160$ km/h, and 87 dBA at the speed of $v = 200$ km/h (UIC, 2002). In addition, specifically in Europe, the large scale measurements of HSR noise were carried out and consequently the noise limits for HS trains depending on their maximum operating speed have been set up by the EC (European Commission) according to the ISO 3095 as follows: after the year 2004, these limits for the existing rolling sock have been 88 ± 2 dBA for the speed of $v = 250$ km/h, 92 ± 2 dBA for the speed of $v = 300$ km/h, and 92 ± 2 dBA for the speed of $v = 320$ km/h. The corresponding limits for the future rolling stock have been 88, 91, and 92 dBA, for the speeds $v = 250$, 300, and 320 km/h, respectively (EC, 2002). Last but not least, while considering the actual exposure of the population located close to passing-by trains, it is necessary to take into account the noise-mitigating barriers protecting the particular land use activities, that is, a quiet land with intended outdoor use, a land with the residence buildings objects, and a land with the daytime activities (businesses, schools, libraries, etc.), all by absorbing the maximum noise levels for about 20 dB(A) (single barrier) and 25 dB(A) (double barrier) (Janić, 2016b).

11.4.2.5 Freight trains

The current practice in many countries has considered the noise level of freight trains at the distance of $\gamma = 25$ m from the track's axis in dependence on their operating—passing-by speed. It has been 85 dBA at the speed of $v = 80$ km/h, 88 dBA at the speed of $v = 100$ km/h and 90 dBA at the speed of $v = 120$ km/h. In addition, UIC has proposed the noise limits for existing freight trains operating at the speed of 80 km/h along the tracks in good conditions. It has been 78 dBA and 77 dBA for diesel and electric locomotives, respectively, at the distance of $\gamma = 25$ m from the track's axle, and 85 dBA and 84 dBA, respectively, at the distance of $\gamma = 7.5$ m from the track's axle. The noise level of freight wagons under the same conditions has been 80 and 87 dBA, respectively (Hemsworth, 2008; UIC, 2002). Similarly as at the passenger trains, the noise barriers along the tracks can be set up to mitigate noise from the freight trains passing-by near the noise sensitive areas.

11.4.3 TRAFFIC ACCIDENTS/INCIDENTS (SAFETY)

11.4.3.1 Streetcar (tramway)

Safety of system operating in a given urban are can be expressed as the perceived risk that a person on board loses life or be seriously or lightly injured due to incidents and/or accidents including the vehicles' derailment, collision with any other persons/pedestrians or other vehicles, infrastructure, obstructions or objects, which result in significant property damage. This perceived risk is measured by different indicators, but most frequently by the number of events and the number of events per unit of the system output (veh-km or pax-km) specified period of time. For example, Table 11.12 gives an indication of the number of different accidental/incidental events happened during the specified period of time at Victorian Tram Operators (Melbourne, Victoria, Australia) (VTO, 2014).

Table 11.12 The Number of Victims (Fatalities and Serious Injuries) and Accidents/Incidents at Victorian Tram Operators Over Time (Period: 2010–14) (VTO, 2014)

Event	Year				
	2010	2011	2012	2013	2014
Fatalities	1	2	1	3	1
Serious injuries	23	25	15	31	31
Line derailment	43	31	24	20	21
Collisions between trams	21	17	15	16	15
Collisions with persons	47	46	37	28	35
Collisions with infrastructure	8	9	4	4	5
Collisions with obstructions	0	3	1	0	0
Collisions with road vehicles	819	956	877	834	817
Fall on tram, slip, trip	172	184	157	188	180
Fall on tram platform, slip, trip	11	9	11	10	10
Broken rails	2	1	1	1	2
Fire on trams	1	1	2	4	7
Total	1148	1284	1145	1139	1124

As can be seen, during the observed period of time, the most numerous accidents/incidents have been collisions with road vehicles and falls on tram. The least frequent have been fatalities on board the trams, collisions with obstructions and infrastructure, broken rails and fire on the trams. However, their total number has been greater than 1000/year during the observed period. For some comparison, at the France's 41 streetcar (tram) networks, the total number of accidents/incidents was rising during the period 2003–12, from about 800 in the year 2003 to 1700 in the year 2012 (De Labonnefon and Passelaigue, 2014). However, at the same time, the rate of accidents/incidents per unit of the system output, which is in this case the given volume of streetcar (tram) kilometers carried out, was mainly decreasing during the observed period as shown on Fig. 11.28—from 0.45 in the year 2003 to 0.39 in the year 2012 for the French systems, and from 0.30 in the year 2003 to 0.173 in the year 2012 at the Croatian system, all per 10^5 veh-km/year.

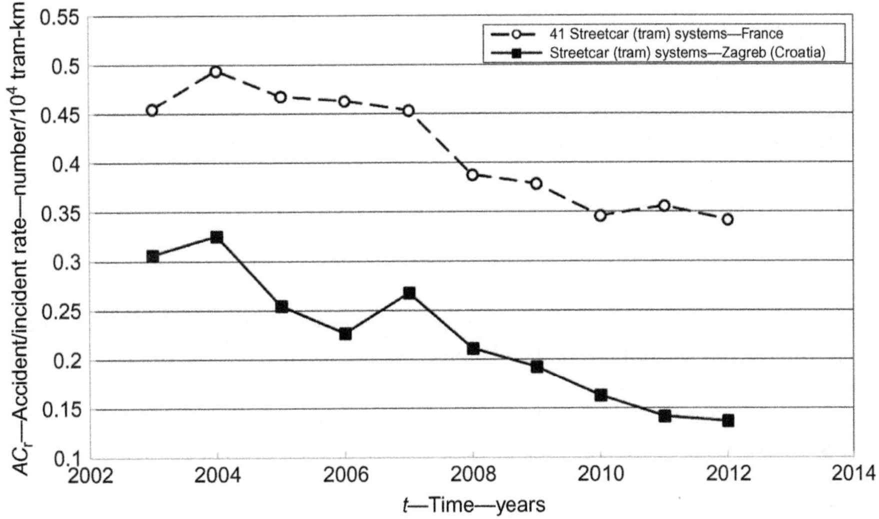

FIG. 11.28

The accident/incident rate at the streetcar (tram) systems in France and Croatia (Zagreb) over time (period: 2003–12) (Brčić et al., 2013; De Labonnefon and Passelaigue, 2014).

At the system operating in Zagreb (Croatia) also shown in Fig. 11.28, the rate of accidents/incidents embracing traffic accidents, loses of life, and injuries of passengers on board the vehicles has permanently decreased during the observed period (Brčić et al., 2013). Therefore, it can be said that the perceived risk of accidents/incidents at both systems has been decreasing and the safety as the risk's counterpart increased.

11.4.3.2 LRT (Light Rail Transit)

The safety of LRT systems operating in given urban and sub-urban areas can be expressed as the perceived risk that a person on board loses his/her life or be seriously or lightly injured in an incident and/or accident including the LRT vehicle derailment, collision with any other persons/pedestrians or other vehicles, infrastructure, obstructions or objects all resulting in substantive property damage. Again, this perceived risk is measured by different indicators, but most frequently by the number of events and the number of events per unit of the system output (veh-km or p-km) specified period of time. Fig. 11.29 shows an example of the number of accidents/incidents, fatalities and injuries for the LRT systems in the United States during the observed period (USDT, 2015a,b).

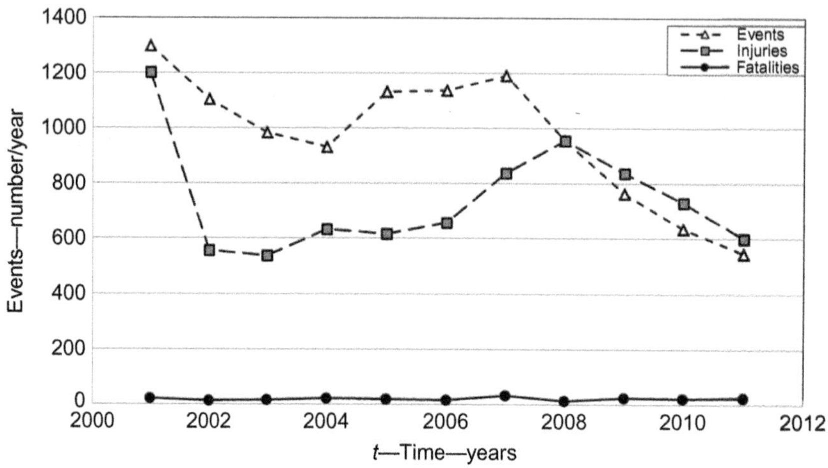

FIG. 11.29

The safety statistics of the US LRT systems over time (period: 2001–11) (USDT, 2015a,b).

As can be seen, the number of fatalities has been negligible compared to the total number of events and injuries during the observed period. The latest two have significantly changed over time and decreased towards the end of the observed period. The high difference between the number of injuries and fatalities also indicates that the accidents/incidents have caused mainly the former and much less the latter damages. In addition, Fig. 11.30 shows an example of the relationship between the accidents/incident rate (events/veh-km) as the perceived risk and the volume of the US LRT systems' annual output.

As can be seen, the accident/incident rate as the perceived risk has been decreasing with increasing of the annual output of the LRT systems, thus indicating improvements of their level of safety in the given case.

11.4.3.3 Subway (metro)

Similarly as at its other rail-based counterparts, the safety of subway (metro) systems can be expressed as the perceived risk that a person on board loses his/her life or be seriously or lightly injured in an incident and/or accident caused by the vehicle's derailment and/or collision with other vehicle(s), objects, people, and fires, all resulting also in the own and third-party property damage. Fig. 11.31 shows the number of incidents and injuries happened during the specified period of time at the subway (metro) systems in the United States (USDT, 2015a,b).

As can be seen, the annual number of incidents has been varied between 6000 and 9000 with substantive increase in the last year (2011) of the observed period. At the same time, the total number of injuries has changed approximately in line with the number of incidents except in the last year of the observed period, when it decreased. In addition, the number of fatalities during the observed period has

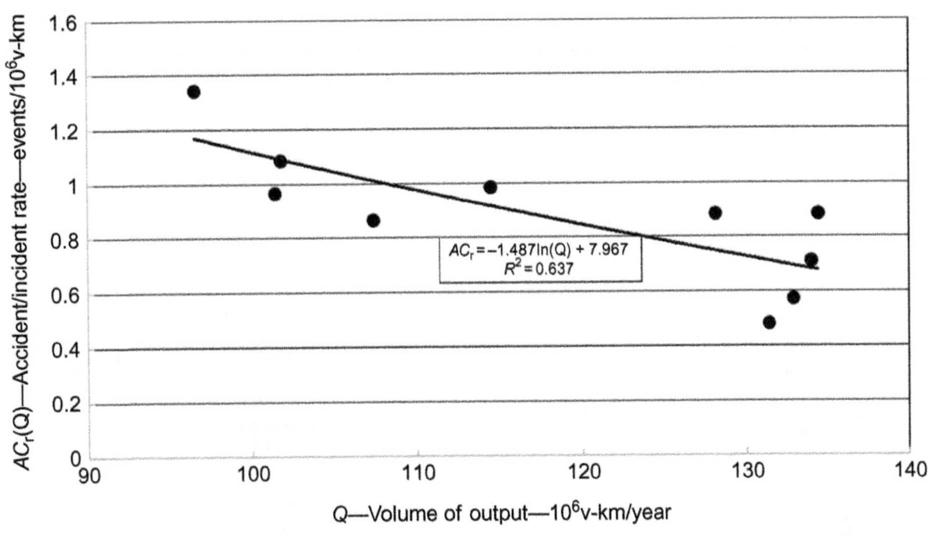

FIG. 11.30

Relationship of the accident/incident rate and the volume of output of the US LRT systems (period: 2001–10) (USDT, 2015a,b).

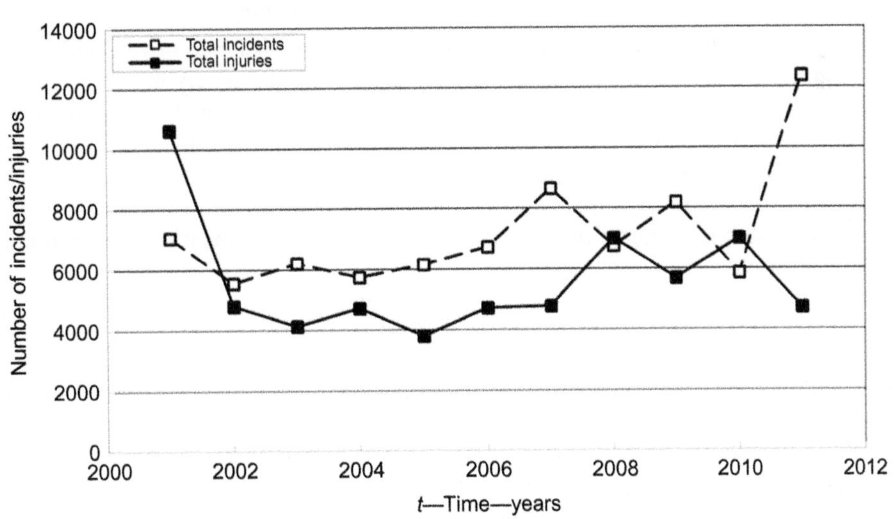

FIG. 11.31

The number of incidents at the subway (metro) systems in the United States over time (period: 2001–11) (USDT, 2015a,b).

varied between 23 and 87. However, these developments still do not indicate what has can be the level of safety at these systems. Fig. 11.32 provides an indication by showing the relationship between the annual rate of incidents in terms of events/v-km and the volume of vehicle-km carried out by the same subway (metro) systems in the United States during the specified period of time (USDT, 2015a,b).

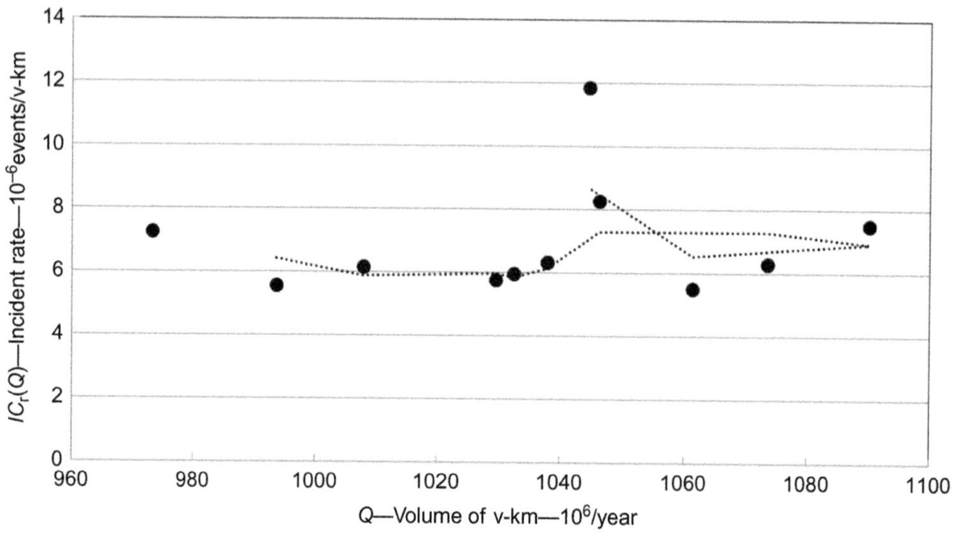

FIG. 11.32

Relationship between the incident rate and the annual volume of v-km carried out by US subway (metro) systems (period: 2001–11) (USDT, 2015a,b).

As can be seen, the incident rate has changed with changes of the volumes of vehicle-km during the observed period. During the entire period (except in 2 years) it has been very low indicating the very low perceived risk of having an incident of a vehicle of the above-mentioned subway (metro) system (about $5–8 * 10^{-6}$/v-km). In addition, Fig. 11.33 shows relationship between the fatality rate in terms of the number of fatalities/p-km and the annual volumes of p-km for the same subway (metro) systems.

As can be seen, the fatality rate, that is, the perceived risk of losing life while using the US metro systems has been varying and slightly increasing with increasing of the annual volumes of p-km carried out, but remained all the time very low ($1–3.5 * 10^{-9}$/p-km).

11.4.3.4 Passenger inter-urban trains

Traffic accidents/incidents have occurred at the passenger trains operating worldwide. They have included events such as collisions and derailments of trains, level crossing and accidents to persons caused by rolling stock in operation, fires in rolling stocks, and other accidents. The victims have been rail passengers, employees, level crossing users, unauthorized persons, and other persons. Fig. 11.34 shows an example of developments in the EU-27 Member States (EU, 2015).

As can be seen, more than 2000 significant accidents occurred occur during each year of the observed period. More than 75% of the accidents resulting in about 1200 fatalities and the same number of serious injuries were caused by the rolling stock in motion and the level-crossing accidents. Over the

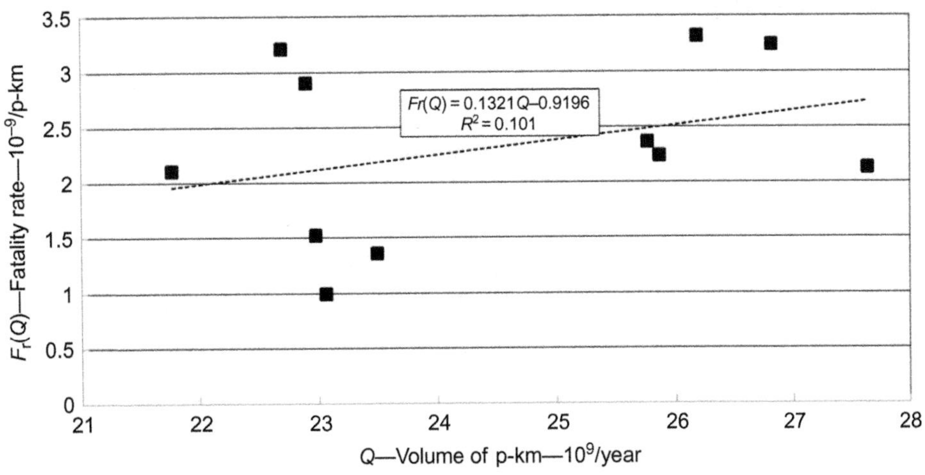

FIG. 11.33

Relationship between the fatality rate and the annual volume of p-km carried out by US subway (metro) systems (period: 2001–11) (USDT, 2015a,b).

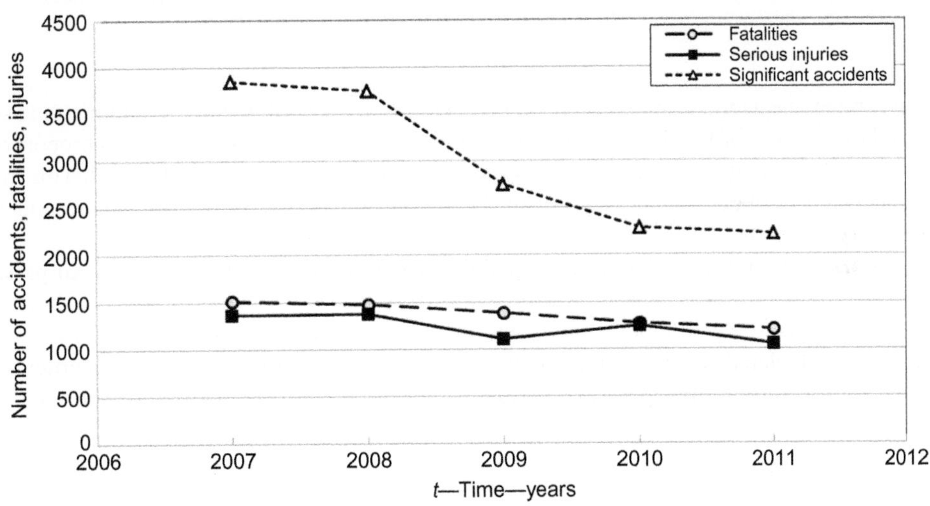

FIG. 11.34

The number of significant accidents, fatalities, and serious injuries at the railways in the EU-27 Member States.

last 3 years of the observed period there was approximately a single fatality or serious injury per an accident. However, in general, these three sets of numbers have generally decreased during the observed period, thus indicating raising of the railway system safety in the given area (EU-27) during the given period of time. In addition, Fig. 11.35 shows the accident rate in dependence of the volume of output of the passenger rail system in EU-27 Member States (EU, 2015).

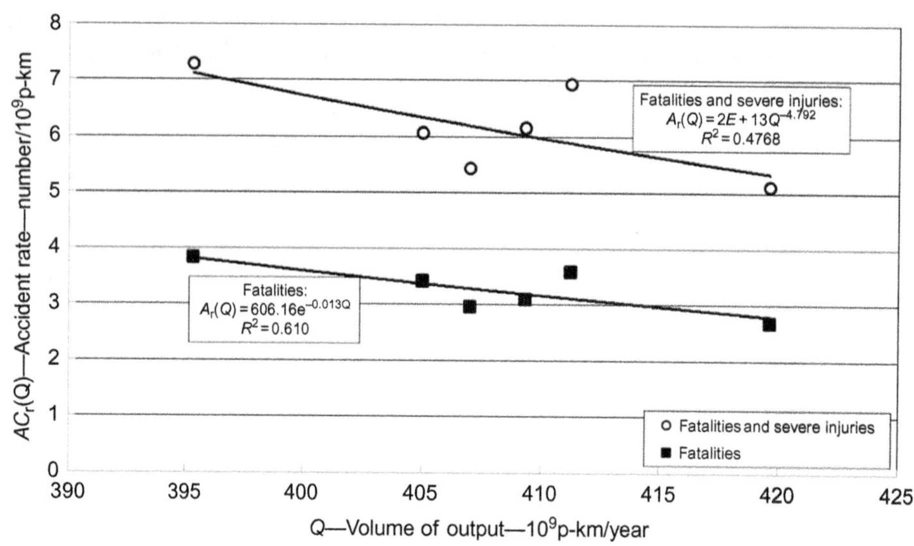

FIG. 11.35

Relationship between the accident rate (the number of fatalities and severe injuries) and the annual volume of output of passenger railways in EU-27 Member States (period: 2007–12) (EU, 2015).

As can be seen, the average annual accident rate in terms of the total number of fatalities and severe injuries, and the number of fatalities per unit of the system output decreased more than proportionally with increasing of the annual volumes of the system output during the observed period. This implies that the perceived risk of the person's fatality and/or injury decreased, thus indicating increasing of the level of safety.

In addition, the experience so far has indicated that the HSR has been (together with Air passenger Transport (APT)) the safest transport system/mode in which the traffic incidents/accidents have rarely occurred, usually due to the previously unknown reasons. This means that the number of traffic incidents/accidents and related person injuries, deaths, and the scale and cost of damaged properties both of the systems and the third parties per, for example, 10^9 s-km and/or p-km carried out during a given period of time have been extremely low. Such high safety of the HSR systems has been achieved also a prior by designing the completely new grade-separated lines and the other supportive built-in safety features at both infrastructure and rolling stock. This implies that the safety has been achieved on the account of increased investments and maintenance cost. In addition, the HSR operators and infrastructure managers have continuously practiced the risk management and training approach aiming at maintaining the high level of safety and particularly with increasing of the maximum speeds. Nevertheless, the HSR systems in different countries have not been completely free from traffic incidents/accidents. For example, some relevant statistics for the TGV (Train á Grande Vitesse) system in France indicate that there have not been accidents with fatalities (deaths) and severe injuries of users/passengers, staff, and/or third parties since starting the services started in the year 1981 despite the trains have been carrying out annually about $10 * 10^6$ p-km. Some incidents happened at the HSR lines/routes such as broken windows, opening of the passenger doors during operating at the cruising speed, couple of fires on board, collision with animals and concrete block on the tracks, and the terrorist attempts to bomb the tracks. The incidents and accidents of TGV trains operated on the conventional tracks have been more frequent with the fatalities, injuries, and damages of properties but all at the relatively low scale. In these cases, the HS trains have been exposed to the external risk similarly as their conventional counterparts (http://www.railfaneurope.net/tgv/wrecks.html). Similarly, since started in the 1960s, the Japan's Tokaido Shinkansen HS services[2] have also been free from accidents causing the user/passenger and staff fatalities and injuries due to derailments and collisions of trains. This has been achieved despite the services have been exposed to the permanent threat of the relatively frequent (and sometimes strong) earthquakes.

Nevertheless, the fatal accidents resulting in fatalities and injuries of users/passengers and staff happened at the HSR systems in Germany, Spain, and China (one in each country). Table 11.13 gives the main characteristics of these three accidents.

Table 11.13 Characteristics of the HSR Fatal Accidents (NDTnet, 2000; Qiao, 2012; Puente, 2014)

Country/System/ Number of Trains	Date	Cause	Passengers on Board	Fatalities	Injuries
Germany/ICE/1	03/06/1998	Wheel disintegration	287	101	88
China/2	23/07/2011	Railway signal failure	1630	40	>210
Spain/Alvia/1	24/07/2013	Derailment due to excessive speed	222	>79	139

The accident of German ICE HS train is known as the "Eschede train disaster" called according to the place it occurred—near the village of Eschede in the Celle district of Lower Saxony (Germany). The HS train derailed and crashed into a road bridge, which collapsed latter on with the impact mentioned in Table 11.13. This has been the worst rail accident in the railway history of the Federal Republic of Germany and the worst of the HSR in the world. The wheel disintegration as the direct cause was due to a single crack in a wheel. When it failed, the train derailed at a switch.

The accident in China included collision of two HS trains operating on the Yongtaiwen line on a viaduct in the suburbs of Wenzhou, Zhejiang province (China) with the impact given in Table 11.13. The collision caused derailment of both trains, of which four cars fell off the viaduct. The main cause of collision was the failure of the signaling system.

The third accident occurred in Spain known as the "Santiago de Compostela rail disaster." It occurred when a HS train operating on the route Madrid—Ferrol derailed due to entering a bend at the excessive high speed (twice higher than the allowed of 80 km/h). The derailment took place about 4 km outside the railway station at Santiago de Compostela (the north west of Spain) with the impacts given in Table 11.13. Similarly as in Germany, this accident was the worst at Spanish railways in the past 40 years.

[2]Tokaido Shinkansen line/route of the length of 552.6 km connects Tokyo and Shin Osaka station is free of the level crossings. The trains operate at the maximum speed of 270 km/h covering the line/route in 2 h and 25 min. The route/line capacity is $\mu_l = 13$ trains/h/direction. The number of passengers carried is about 386,000/day and 141 million/year (2011) (JR Central, 2012).

11.4.3.5 Freight trains

The traffic accidents/incidents have also occurred at the freight trains. They have included collisions and derailments of trains, level crossing and accidents to persons caused by rolling stock in operation, fires in rolling stocks, and other accidents. The victims have been rail employees, level crossing users, unauthorized persons, other persons. In addition, these have been destroyed freight/cargo, rolling stock, and other damages to the third parties.

Fig. 11.36 shows an example of the total number of accidents and incidents and particularly the number of accidents happened in the US freight rail network[3] during the specified period of time (2006–14) (USDT, 2015b).

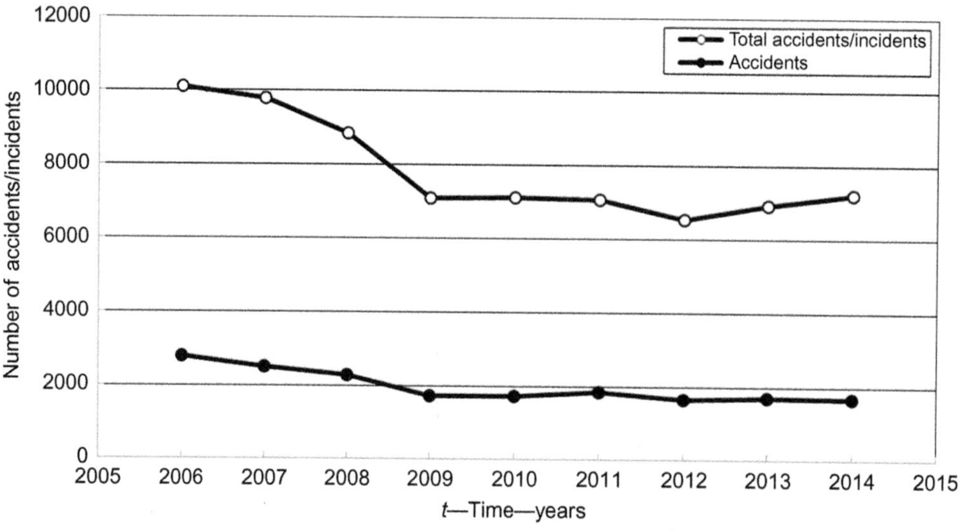

FIG. 11.36

The freight train accidents/incidents in the United States (period: 2006–14) (USDT, 2015a).

As can be seen the total number of accidents/incidents have been decreasing during the first and remained relatively stable during the second half of the observed period. At the same time, the number of accidents has generally been decreasing, thus indicating in some way improvements of the system safety. In addition, Fig. 11.37 shows the relationship between the accident/incident and the accident rate in dependence on the volume of output of the freight trains in the United States during the same period of time (USDT, 2015b).

As can be seen, both the total accident/incident rate and the accident rate were very low and relatively slightly variating with changing/increasing of the volume of output. The former was for almost four times higher than the latter mainly due to the higher number of counted events. Consequently, regarding this indicator(s) it can be said that the system's safety was relatively stable over

[3]In the year 2014, the total length of the US rail network was over 250,000 km, of which 80% were freight lines (http://data.worldbank.org/indicator/IS.RRS.TOTL.KM). For comparison, the length of the European rail network in the year 2013 was 215,258 km, of which 53.8% was electrified (EU, 2015).

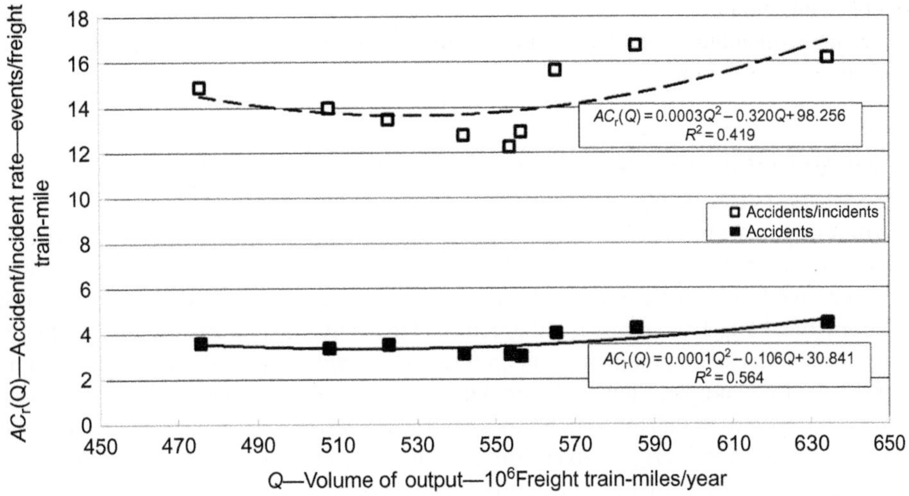

FIG. 11.37

Relationship between the accident/incident rate and the volume of output of freight trains in the United States (period: 2006–14) (USDT, 2015a).

the given period of time. In addition, Fig. 11.38 shows the relationship between the number of fatalities and the total number of accidents/incidents during the same period of time (USDT, 2015b).

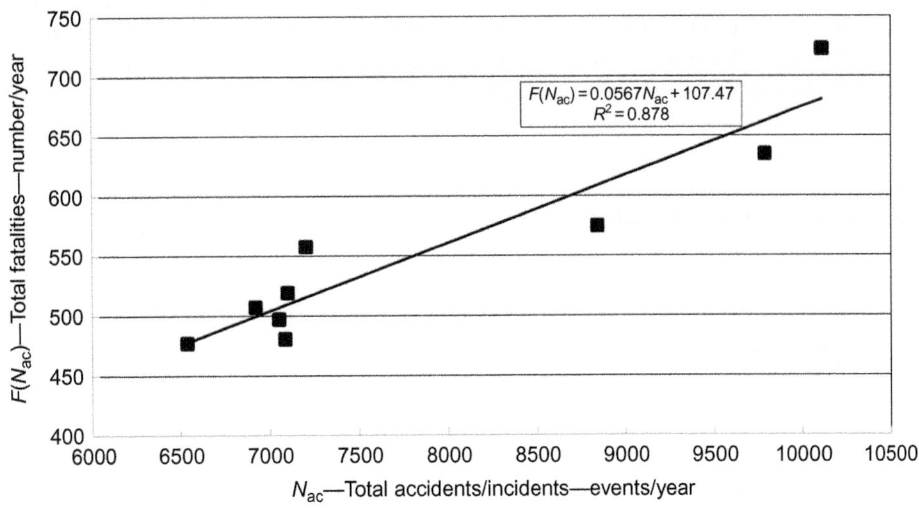

FIG. 11.38

Relationship between the total number of fatalities and the total number of accidents/incidents at the US freight railways (period: 2006–14) (USDT, 2015a).

As can be seen, the number of fatalities approximately linearly increases with increasing of the number of accidents/incidents, which could be intuitively expected. The average rate in the given case has been about $5.67 * 10^{-2}$ fatalities per an accident/incident.

11.4.4 ENERGY/FUEL CONSUMPTION AND EMISSIONS OF GHG

11.4.4.1 Streetcar (tramway)

Streetcars (tramways) are powered by electric energy, which can be obtained from different primary sources. They consume energy but also save it thanks to regenerative braking technology, which enables them to convert the kinetic energy of the moving vehicle into electrical energy when it starts decelerating and breaking. This energy is either returned to the overhead wires for use by other vehicles or for powering auxiliary equipment such as on board heating/cooling systems. Some contemporary streetcars such as, for example, Siemens' Combino Plus, can recover about 30% of the energy used to power the vehicle through this process. Some additional investigations have shown that at the lower speeds (eg, 19–20 km/h) the energy recovery from regenerative breaking can be more than 42%. Under conditions when the number of passengers on board has been 25 (the vehicle capacity has been 155 spaces), the average unit energy consumption of a contemporary streetcar (tramway) has been about: $SEC = 0.07$ kWh/p-km, which is lower than that of LRT (0.08), trolleybus (0.225), and the hybrid car Toyota Prius (0.40) (Condon and Dow, 2009).

The electricity consumed by streetcars (tramways) can be obtained from different primary sources, which each has emissions factor (e_f) of GHG: Coal 206 gCO_2/kWh, Natural Gas 106 gCO_2/kWh, Hydro 4.4 gCO_2/kWh, and Nuclear 0.0 gCO_2/kWh. These values can vary across particular countries depending on the specificity of technologies used for obtaining electric energy from particular primary source (Spadaro et al., 2000). Consequently, the average emission rate of GHG in terms of CO_{2e} by a contemporary streetcar (tramway) can be estimated as follows: $e_r = SEC * e_f = 0.07 * 206 = 14.420$ gCO_{2e}/kWh if coal, $0.07 * 106 = 7.420$ gCO_{2e}/kWh if natural gas, $0.07 * 4.4 = 0.308$ gCO_{2e}/kWh if hydro, and $0.07 * 0.0 = 0.000$ gCO_{2e}/kWh if nuclear source for obtaining electric energy is primary and exclusively used (Condon and Dow, 2009). In most countries, the electric energy is obtained from the mixture of different sources. Therefore, the average unit emissions in the given context will be between the two extremes, when it is obtained exclusively by carbon and exclusively by nuclear-based primary source.

11.4.4.2 LRT (Light Rail Transit)

The LRT systems are powered by electric energy, which, as in the case of the streetcar (tramway) systems, can be obtained from different primary sources. The use of these sources exclusively or in different combinations depends on the local urban, regional, and country's conditions, which in turn influence the emissions of GHG from the LRT systems. Fig. 11.39 shows the relationship the annual energy consumption and the volume of output of the LRT system operated in the United States (Henry et al., 2009).

As can be seen, the total annual energy consumption of the selected 18 LRT systems has linearly increased with increasing of their annual output in terms of p-km. The average rate of the energy consumptions has been $SEC = 5.367$ kWh/p-km. In addition, the average emission rate of GHG (CO_2) has been $e_r = 58.7$ gCO_2/p-km. This implies that the average emission factor of CO_2 from combination of the primary sources for obtaining the electric energy for powering these LRT systems has been $e_f = e_r / SEC = 58.7/5.367 \approx 10.94$ gCO_2/kWh.

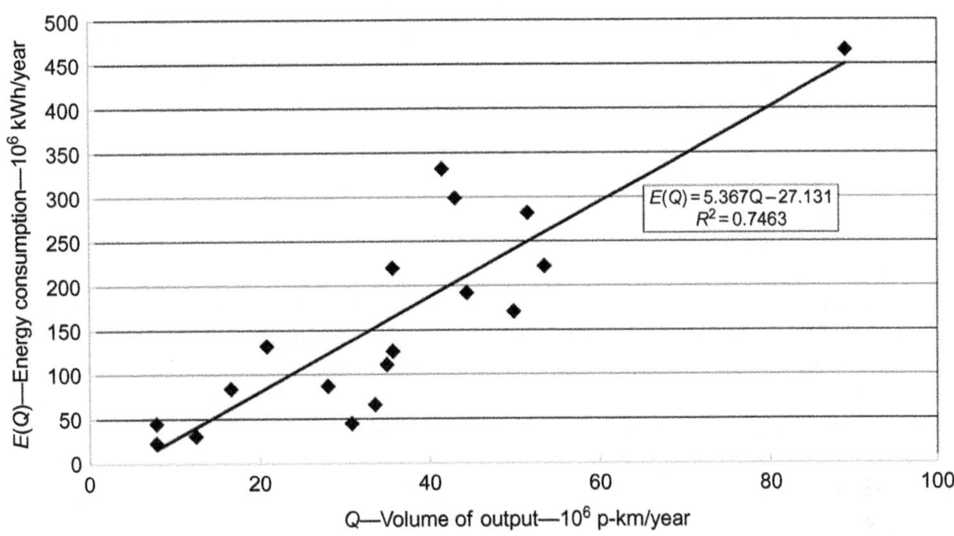

FIG. 11.39

Relationship between the total annual energy consumption and the volume of output of the selected US LRT systems (period: 2007) (Henry et al., 2009).

11.4.4.3 Subway (metro)

The subway (metro) vehicles are powered by the electric energy, which they consume to overcome the rolling (and partially air) resistance during acceleration and cruising phase of their movement between any two stations along a given line. During deceleration phase of the movement before stopping at a station, they apply the regenerative braking returning the energy into the power network. The energy efficiency of the subway (metro) systems has usually been expressed in the relative terms, that is, by the average rate of energy consumption in kWh/veh-km, kWh/space-km, and/or kWh/p-km. Some characteristic values for the contemporary systems with the vehicle capacity of 150–200 spaces/veh have been $SEC = 3.5–5$ kWh/veh-km and $SEC = 0.0175–0.200$ kWh/p-km (Vuchic, 2007). The more recent analysis has shown that, for example, London Underground (LU) system consumes electric energy at an average rate of $SEC = 0.160$ kWh/p-km. If the average emission factor of GHG (CO_{2e}) from the electricity production in the United Kingdom is $e_r = 463$ gCO_2/kWh, the emission rate of LU system will be $e_r = SEC * e_f = 0.160 * 463 = 74$ gCO_2/p-km (Palacin et al., 2014; TfL, 2008; http://www.carbon-calculator.org.uk/). Table 11.14 summarizes the average emission rates from different subway (metro) systems in the United States in the year 2010 (FTA, 2010).

As can be seen, the lowest CO_2 emission rate has been at San Francisco and the highest at Baltimore subway (metro) system mainly due to the very high utilization of the former and the very low utilization of the latter. But all these values are comparable to that of the above-mentioned LU system. For comparison, the CO_2 emission rate of Melbourne subway system has been $e_r = 145$ gCO_2/p-km primarily due to obtaining electricity primarily from coal. For comparison, the average rate of energy consumption of the Rio de Janeiro (Brazil) subway (metro) system in the year 2012 was $SEC = 0.0981$ kWh/p-km and the corresponding emission rate of CO_2, $e_r = 6.4$ gCO_2/p-km (De Andrade et al., 2014).

Table 11.14 The Average Emission Rates of GHG (CO_2) of the Selected US Subway (metro) Systems (Period: 2010) (FTA, 2010)

System/City	Emission Rate e_r (gCO_2/p-km)
San Francisco	24
Atlanta	69
Los Angeles	79
Boston	95
Washington, DC	98
Baltimore	259

11.4.4.4 Passenger inter-urban trains

The passenger trains are powered by diesel and electric locomotives, which consume the corresponding fuel and energy, respectively. In most countries and larger continental areas, the rail networks have been partially electrified, thus enabling use of the electric locomotives For example, in Europe, in 30% of the rail network has been electrified in the year 1990 and 50% in the year 2009. Fig. 11.40A–D shows characteristics of the energy consumption and related emissions of GHG of the passenger trains operating in the EU-27 Member States during the period 1990–2012 (EU, 2015; UIC, 2014).

As can be seen, the specific energy consumption over time and per unit of output (p-km) has decreased thus indicating the improvement of the energy efficiency of passenger trains operating in the EU-27 Member States during the observed period of time. The corresponding rates of emissions of GHG (CO_2) have also decreased in proportion with decreasing of the energy consumption. It is easy to show that the average emission factor during the observed period of time was about 48 gCO_2/kWh.

Specifically, the energy consumption and related emissions of GHG can be considered exclusively from operations of the HSR systems (UIC, 2010a). In general, similar as their conventional counterparts, the HS trains consume electric energy for accelerating up to the operating/cruising speed and then for overcoming the rolling/mechanical and aerodynamic resistance to motion at that speed. The additional energy is consumed for overcoming resistance of grades and curvatures of tracks along the given line/route. In addition, the energy is consumed for powering the equipment onboard the trains. In particular, during the acceleration phase of a trip the electric energy is converted into the kinetic energy at an amount proportional to the product of the train's mass and the square of its speed(s). A part of this energy recovers during deceleration phase before the train's stop means by the regenerative breaking. During cruising phase of a trip, the energy is mainly consumed to overcome the rolling/mechanical and the aerodynamic resistance, which for a given type of HS train can be expressed as follows (Rochard and Schmidt, 2000):

$$R = R_M + R_A = (a + b \cdot V) \cdot W + c \cdot V^2 \tag{11.35}$$

where

R_M, R_A are the rolling/mechanical and aerodynamic resistance respectively (N) (N—Newton);
W is the weight of a train (tons);

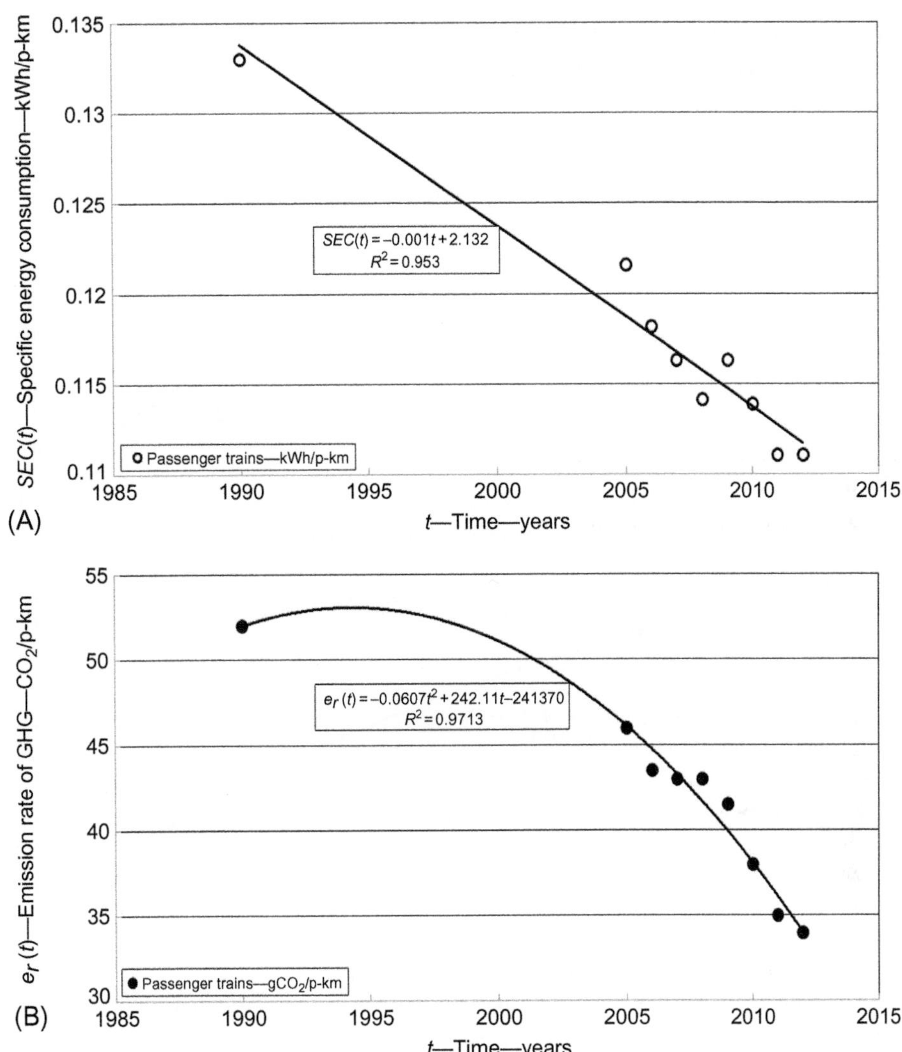

FIG. 11.40

Characteristics of the specific energy consumption and related emission rate of GHG (CO_2) of the passenger trains operating in the EU-27 Member States (period: 1990–2012) (EU, 2015; UIC, 2014). (A) Specific energy consumption over time. (B) Emission rate of GHG (CO_2) over time.

(Continued)

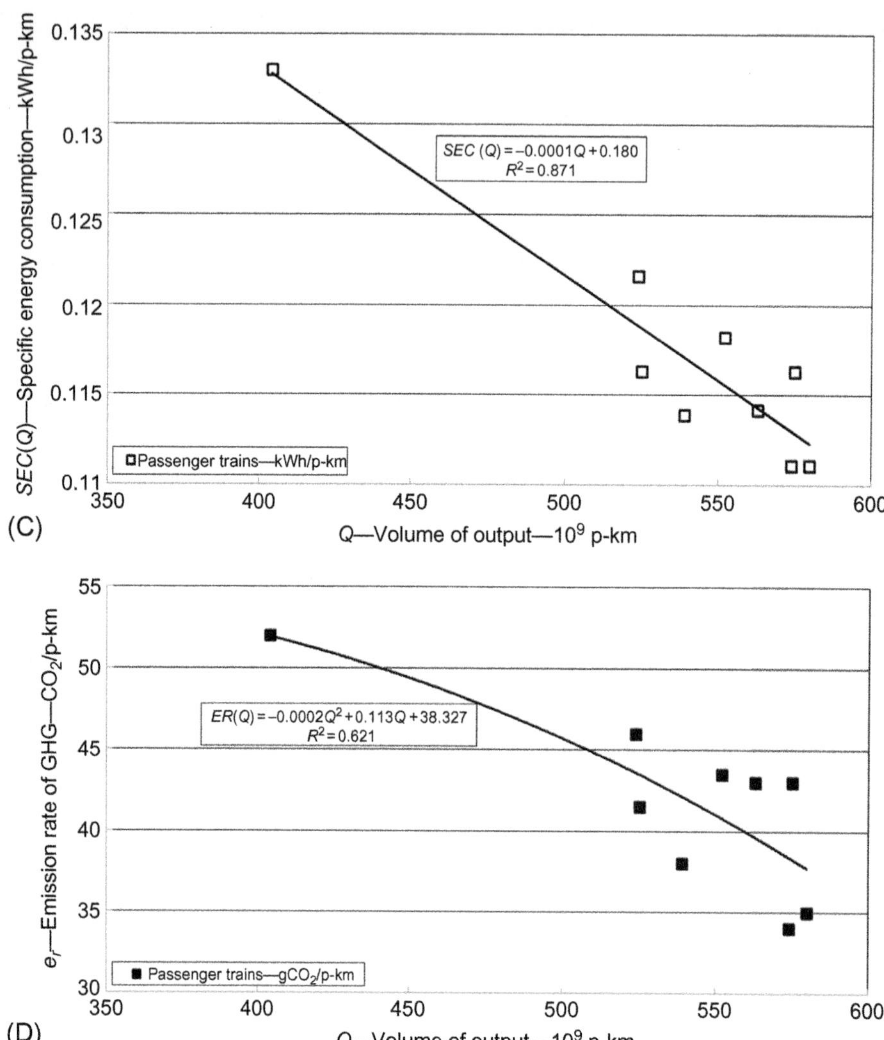

FIG. 11.40, CONT'D

(C) Relationship between the specific energy consumption and the annual volume of output. (D) Relationship between the emission rate of GHG (CO_2) and the annual volume of output.

V is the operating/cruising speed of a train (km/h); and
a, b, c are the experimentally estimated coefficients.

Eq. (11.35) essentially reflects Davis's equation with the corresponding coefficients. It indicates that the aerodynamic resistance generally increases with the square of operating/cruising speed. The rolling mechanical resistance increases linearly with increasing of this speed and weight of the HS train. As an example, the above-mentioned Davis's equation has been estimated for three types of HS trains (one

Japan-built and two French-built) operating at the constant speed (v) along the route of length of $d = 100$ km as follows:

i. *Shinkansen Series 200* (Richard and Schmidt, 2000):

$$EC_c(v,d) = 2.7778 \cdot 10^{-3} \cdot \left[8.202 + 0.106568 \cdot (v/3.6) + 0.01193 \cdot (v/3.6)^2\right] \cdot d \tag{11.36}$$

ii. *TGV-R* (SYSTRA, 2011):

$$EC_c(v,d) = 2.7778 \cdot 10^{-3} \cdot \left(270 + 3.3 \cdot v + 0.051 \cdot v^2\right) \cdot d \tag{11.37}$$

iii. *AGV-11* (SYSTRA, 2011):

$$EC_c(v,d) = 2.7778 \cdot 10^{-3} \cdot \left(250 + 2.9 \cdot v + 0.045 \cdot v^2\right) \cdot d \tag{11.38}$$

In Eqs. (11.36)–(11.38), the speed v is expressed in (km/h), the distance d in (km) and the energy consumption EC_c in (kWh). These equations have not included the energy consumed for acceleration and deceleration of trains and that for providing the passenger comfort on board. In addition, efficiency in converting the vehicle's power output into the tractive force has assumed to be 100% (ie, $\eta = 1.0$). Fig. 11.41 shows the obtained relationships.

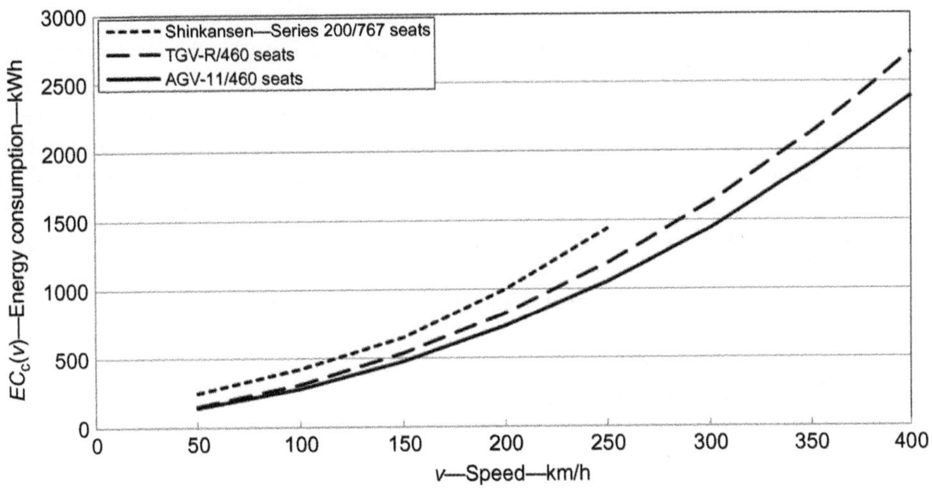

FIG. 11.41

Relationship between the energy consumption and operating speed of the selected types of HS trains (SYSTRA, 2011; Richard and Schmidt, 2000).

As can be seen, the energy consumption increases more than proportionally with increasing of the HS train speed under given conditions. This is because disproportionally more energy has been spent to overcome the air resistance due to its increasing with the square of operating speed. The above-mentioned relationship emphasizes importance of reducing both the weight of train and its aerodynamic resistance in order to achieve savings in the energy consumption during the longest phase of trip—cruising at high speed.

Estimates of the energy consumption by different types of HS trains including acceleration/deceleration/cruising phase of a trip have differed and changed over time, just thanks to the above-mentioned permanent improvements of their both characteristics (aerodynamic, weight) and operations. Table 11.15 gives some recent estimates of this energy efficiency for different types of the HS trains (ATOC, 2009; Siemens, 2014).

Table 11.15 Energy Efficiency of Different HS Trains (ATOC, 2009; Siemens, 2014)

Train Type	Operating Speed (km/h)	Seat Capacity (s-seats)	SEC (Energy Efficiency) (kWh/s-km)
Shinkansen Series 700	300	1323	0.029
AVG	300	650	0.033
TGV Reseau	300	377	0.031
TGV Duplex	300	545	0.032
Pendolino Class 300	300	439	0.033
Eurostar Class 323	300	750	0.041
Velaro D	320	601	0.030

As can be seen, the Japanese Shinkansen are the most and the Eurostar the least energy efficient trains. One of the reasons is the relative large difference in the seat capacity between them. As an indication, at present, the average energy efficiency of a HS train is assumed to be about $SEC = 0.033$ kWh/s-km (s-seat). Respecting this and taking into account the emission rates of the primary sources for producing electricity in Japan, the average emission rate of GHG by Shinkansen trains has been $e_r = 42$ gCO_2/s-km (JR Central, 2012). Under the analogous conditions, in Europe, this rate has been $e_r = 21$ gCO_2/s-km with an ambition to be reduced to 5.9 gCO_2/s-km by the year 2025, 1.5 gCO_2/s-km by the year 2040, and 0.9 gCO_2/s-km by the year 2055. This reduction is expected to be achieved through further improvement of the energy efficiency of HS trains and their operations on the one side and by changing type and composition of the primary sources for producing electric energy on the other. In the latter case, the aim is to produce as much as possible electric energy from the renewable decarbonized primary sources (UIC, 2010a, 2011).

11.4.4.5 Freight trains

Freight trains are powered by both diesel fuel and electric energy, depending on the level of electrification of the rail network in particular countries. In such case, the energy consumption and related emissions of GHG (CO_2) can be separately estimated for diesel and electric powered train operations. Fig. 11.42A and B shows an example of the (specific) energy consumption and related emissions of GHG (CO_2) per unit of output (kWh/t-km) of freight trains operating in EU-27 Member States during the specified period of time (2005–12) (UIC, 2014).

As can be seen, the specific energy consumption decreased more than proportionally over time while varying between $SEC = 0.05$ and 0.06 kWh/t-km. The related rates of emissions of GHG (CO_2) also decreased more than proportionally during the specified period of time while varying between about $e_r = 18$ and 23 gCO_2/t-km. This is an indication that the energy efficiency of freight train operations in the given case has been improving over time. In addition, Fig. 11.43A and B shows the

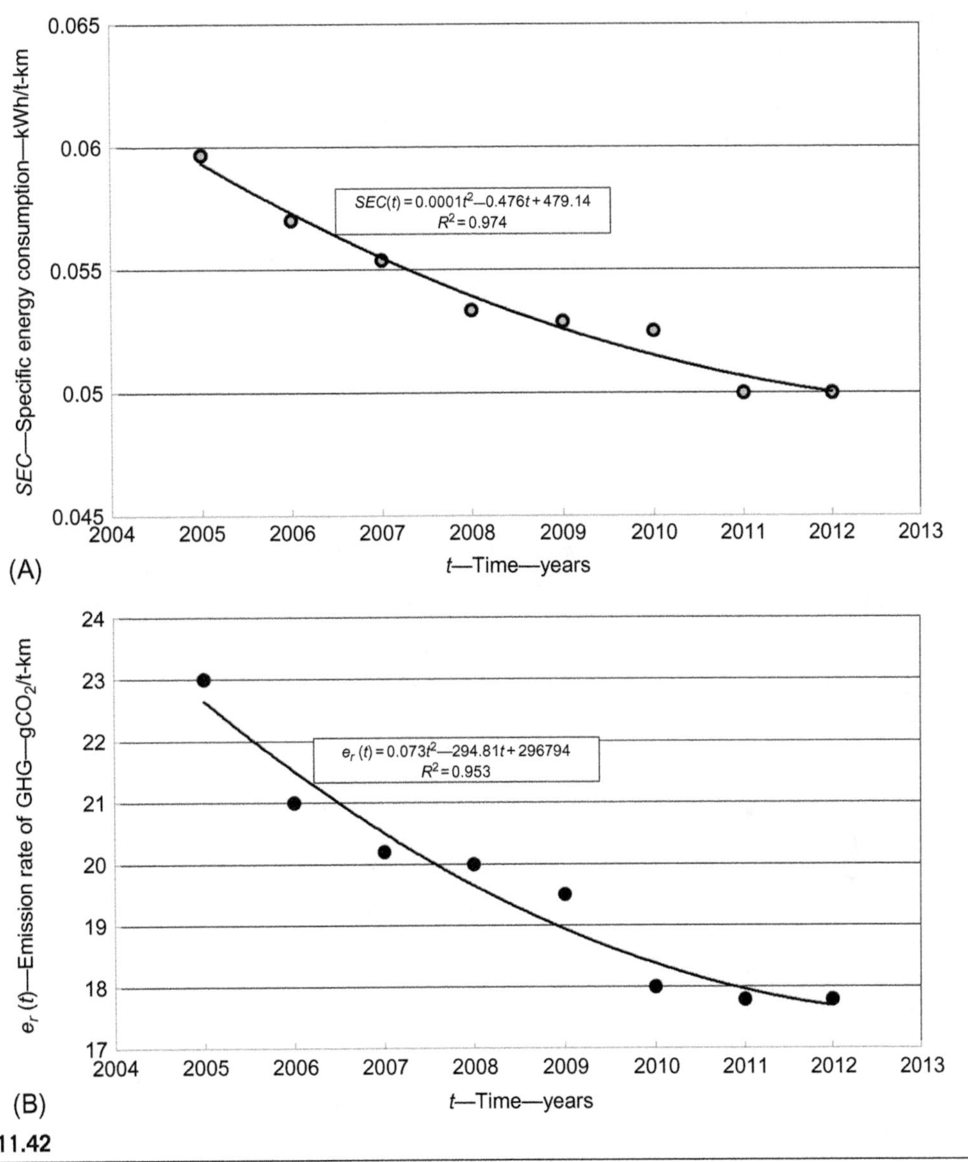

FIG. 11.42

Specific energy consumption and related emissions of GHG (gCO_2) over time by freight trains operating in EU-27 Member States (period: 2005–12) (UIC, 2014). (A) Specific energy consumption. (B) Emission rate of GHG (CO_2).

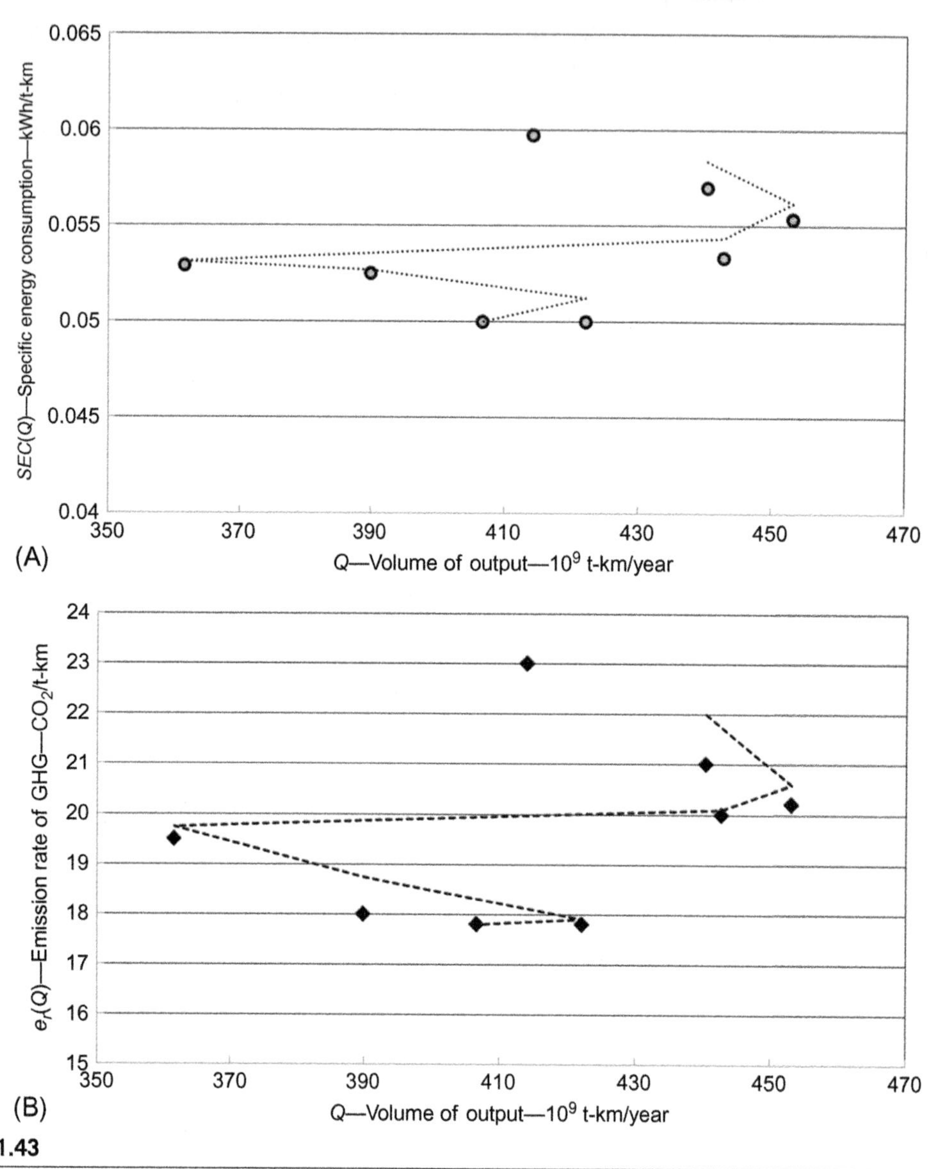

FIG. 11.43

Relationship between the specific energy consumption and emission rate of GHG(CO_2) and the annual volume of output of freight trains operating in EU-27 Member States (period: 1990–2007) (EU, 2015; UIC, 2014). (A) Specific energy consumption. (B) Emission rate of GHG.

(specific) energy consumption and related emissions of GHG (CO_2) of freight trains for the same case depending on the annual volumes of their output (t-km) (UIC, 2014).

As can be seen, both energy consumption and related emissions of GHG (CO_2) have changed with changing/increasing of the volumes of output of freight trains in the given case, but without indicating a rather stronger causal relationship(s).

11.4.5 LAND USE

11.4.5.1 Streetcar (tramway)

Streetcars (tramway) lines with two or more tracks enabling operations of the vehicles in both directions independently and stops/stations occupy land, which could be used exclusively by the system or partially shared with other urban and somewhere sub-urban road transport systems—passenger cars, buses, and light trucks. The typical distance between the track centers is 6.4 m, width of the vehicle 2.4 m, width of the shoulders along the tracks 1.2 m, which give the total width of profile of $w = 11.2$ m, excluding land for the stations along the line and at its both ends (Vuchic, 2007). Some statistics have shown that 176 streetcar (tramway) systems with 1149 lines of the total length of $d = 14{,}676$ km operate in Europe, which make the average of 6.5 lines/system, 12.8 km/line, and 84 km of network (ERRAC, 2012). By multiplying the total length of lines by the above-mentioned profile of a two-way line, the total area of occupied land by the streetcar (tramway) infrastructure can be approximately estimated as $A = d * w = (14{,}676 * 10^3/2 * 11.2)/10^5 = 8218.6$ ha (ha—hectare). Again this area does not include the land for stations and other supportive facilities and equipment. However, these numbers should be considered only for the illustrative purposes because they contain information on both streetcar (tramway) and LRT networks/system, although the former actually predominates.

The intensity of use of taken land by the streetcar (tramway) systems can be estimated as the ratio between the total systems' output in terms of the number of passengers and/or p-km carried out during given period of time and the total area of taken/occupied land. As an example, in the above-mentioned case, the total number of passengers carried out by 176 systems in the year 2006 has been about $Q = 9.7 * 10^9$. This gives the average intensity of land use as follows: $ILU = Q/A = 10.4 * 10^9/8218.6 * 10^5 = 126.54$ p/m^2. At the same time the average number of trips per an inhabitant has been $10.4 * 10^9/586 * 10^6 \approx 17.75$ trips/inhabitant (p—passenger) (ERRAC, 2012).

11.4.5.2 LRT (Light Rail Transit)

The land occupied by the LRT system(s) infrastructure can be determined similarly as that occupied by the streetcar (tramway) system(s). The typical distance between the centers of two closest parallel tracks is 6.4 m, the vehicle width 2.4 m, and the shoulder area along the tracks 1.2 m, which gives the total width of profile of the two-track LRT system line of 11.2 m (Vuchic, 2007). Then, by multiplying the length of lines by the profile, the total area of land can be estimated, again not including the length occupied by stations and other supportive facilities and equipment. In the case of the above-mentioned 18 LRT systems in the United States, their total length in the year 2009 has been $d = 936.3$ km (one-way line) (Schumann, 2009). Respecting the width of profile of $w = 11.2$ m, the total area of occupied land can be estimated as: $A = d * w = (936.3 * 10^3 * 11.2)/10^5 = 1048.656$ ha. The total annual volumes of p-km carried out by these systems in the year 2007 have been $Q = 2932.84 * 10^6$ (Henry et al., 2009). Then, the annual intensity of land use is estimated as follows: $ILU = Q/A = 2932.84 * 10^6/1408.56 = 2.797 * 10^6$ p-km/ha.

11.4.5.3 Subway (metro)

The subway (metro) systems use land if their lines/tracks are elevated. Otherwise, when they spread through the underground tunnels they are not considered to use land in the given context. For example, some standards (Germany) specify the width and height of the tunnel with double track lines of 7.30 m and 4.30 m, respectively (Vuchic, 2007). If this is applied to the elevated lines/tracks, the minimum profile would be 4.30 m plus 1.20 m at each side of the line(s), which will amount $w=6.7$ m. By multiplying this profile by the length of the elevated lines, the total area of land taken by a given subway (metro) line can be estimated. This however does not take into account the area of land taken for the stations and supporting facilities and equipment of the line. Some investigation has shown that 45 subway (metro) systems in Europe operate 169 lines with the total length of $d=2675$ km. These give the average length of the network of 59.4 km, the average number lines/network of 3.7, and the average line length of 15.8 km. The evidence has not been provided about the elevated portions of these networks. Nevertheless, as an illustration, regarding that about $Q=9.9*10^9$ passengers per year use these networks, the average intensity of land use under given conditions is equal to $ILU=Q/A=9.9*10^9/(2675*10^3*4.3)*10^5=8.606*10^6$ p-km/ha/year (ERRAC, 2012).

11.4.5.4 Passenger inter-urban trains

The HSR infrastructure directly occupies the area of land, which is much smaller than that of its road-highway counterpart. For example, if the width of a profile of land taken by a HSR line is (w) and the length (d), the total occupied land can be estimated as follows:

$$A=w*d \tag{11.39}$$

For example, if: $w=32$ and $d=1$ km line, the total area of directly taken land will be $A=32*1000=32{,}000\ \text{m}^2=3.5$ ha (ha—hectare) (the average gross area of taken land is $A=3.2$ ha). For example, the total length of HSR network in Europe in the year 2013 has been $d=7298$ km. If the width of taken land is assumed to be $w=32$ m, the area of total occupied land will be $A=7{,}298{,}000*25=233{,}536{,}000\ \text{m}^2=23{,}353.6$ ha. At the same time, the annual volumes of p-km carried out have been $Q=111.67*10^9$ p-km. Consequently, the average utilization of the occupied land has been $ILU=Q/A=111.67*10^9/23{,}353.6=4.7817*10^6$ p-km/ha (EU, 2015). In addition, if the capacity of a given HSR line/route in both directions is $2*\mu_1=12$–14 trains/h, that is, 24–28 trains/h and if each train carries about 600 passengers, the intensity of land use of line of the length $d=1$ km will be $ILU=Q/A=24$–$28*600/3.2=14{,}400/3.2$–$16{,}800/3.2=4500$–5250 p/h/ha (p—passenger) (UIC, 2010b).

11.4.5.5 Freight trains

The rail infrastructure used by freight trains consists of lines/tracks, shunting yards, freight terminals, and other supportive facilities and equipment for handling rolling stock and freight/cargo shipments. Specifically, the area of land occupied by the rail network used by freight trains can be estimated as the product of its length (d) and the width of profile (w) similarly as in Eq. (11.39). For example, if the profile $w=25$ m, then the area of land occupied by the US network used exclusively by freight trains in the year 2014 is $A=(228{,}218*10^3*0.8*25)/10^5=456{,}436$ ha (0.8 is the proportion of the total length of the US railway network used exclusively by the freight trains). The volume of output in the same year was $Q=890.5*10^6$ train-km. Thus, in that year, the average intensity of land use by freight trains was $ILU=Q/A=890.5*10^6/456{,}436=1950.9$ train-km/ha (USDT, 2015b).

11.5 WATER-BASED SYSTEMS

Congestion, noise, traffic accidents/incidents, that is, safety, as the social, and the energy/fuel consumption and related emissions of GHG, land use, and waste as the environmental impacts are considered for the water-based transport systems including inland waterways and maritime systems and their infrastructure components—ports and lines/routes, and corresponding transport services. In particular, the impacts of freight/cargo transport are under focus, which however does not intend to compromise in any way the importance of its passenger transport counterpart.

11.5.1 CONGESTION

In particular, the important components of ground access systems of some important seaports in Europe are rivers and canals. For a long time, they have operated less efficiently and effectively due to the capacity bottlenecks, for example, at the main European seaports Rotterdam and Antwerp. Because of the capacity shortage at the container terminals of these seaports compromising efficient and effective loading and unloading of containers the inland container barges and vessels have experienced the long waiting times imposing substantive corresponding costs on the inland waterways operators.

The other factors causing congestion and related delays at seaports can be bad weather affecting unloading/loading of ships and other port-related freight/cargo operations, accidents/incidents causing destroys and damages of the port equipment, ship, and/or freight/cargo, industrial action of the staff causing partial or complete stoppage of the port operations, and unplanned/unexpected sudden increase of the intensity and volume of freight/cargo demand due to the diverted ship traffic from other ports or due to increase in the ship size. The most seaports worldwide have experienced congestion and delays of ships and their freight/cargo shipments due to a single or a combination of the above-mentioned factors/causes. The examples have been the ports in China and Western Europe faced with the capacity bottlenecks at both water—and land-side area. In the short-term, the remedies have shown to be improving efficiency and effectiveness of utilization of the existing/available capacity of particular overloaded seaport's components. In order to achieve this, at least three main actors have to coordinate their activities: seaport/terminal operators, shipping lines, and operators of the seaport ground access systems—rail, road, and inland waterways. In particular, in order to efficiently and effectively handle larger ships, the seaports expect that shipping lines will provide reasonably reliable forecasts of freight/cargo volumes and keep the ships' arrivals as scheduled, this enabling efficient and effective management of the berths' availability, that is, in the predictable manner.

The shipping lines expect that the seaports/terminals will be able to provide sufficient capacity efficiently and effectively to handle the forecasted freight/cargo volumes. The experience so far has shown that handling larger vessels and their inherently more concentrated freight/cargo has been more efficient and effective, and as such preferred by many seaports (Meersman et al., 2012; WSC, 2015). The seaport ground access system operators expect that operations of the former two actors do not affect their schedules and consequently efficiency and effectiveness of their own operations. In addition, they are also expected to provide sufficient capacity in order not to compromise the expected efficiency and effectiveness of other two actors, that is, not to act as the capacity bottlenecks in anyway. The similar factors as above-mentioned at the seaports can cause congestion and delays of inland barges and vessels and their freight/cargo at the inland waterways ports, but acting specifically under the specific given conditions. At both sea—and inland waterways ports, congestion is considered as externality due to the additional costs it imposes on the freight/cargo shippers/receivers and operators of the corresponding supply chains.

11.5.2 NOISE

11.5.2.1 Seaports

Sea and inland waterways ports, and their related activities can generate high noise levels originating from the different primary sources. These can be broadly divided in to those within the given port's area and those outside this area but closely related to the operation of the port and its particular components. The former primary sources mainly include shipping, cargo handling, and maintenance of facilities and equipment if carried out within the given port's area, that is, the industrial noise. The latter sources are mainly the port's ground access systems, usually rail, road, and inland waterways, and other related traffic. At some ports, the noise levels can be clearly distinguished between the ground access systems such as rail and road and the industrial noise prominently present in the port areas. The excessive noise by ports generally negatively impacts on their surroundings, particularly to the close population under an assumption that the ports' employees are appropriately protected. In some cases, such as the Rijnmond area around the port of Rotterdam (the Netherlands), the level of noise has been higher than 75 dBA. This has initiated permanent noise monitoring and control. This has included division of this (Rijnmond) area into several (noise) zones. Each zone has been granted an average noise quota for the industrial noise as an environmental permit expressed by the maximum noise level per square meter. Then, this quota has been distributed to the particular actors-noise generating sources-operating in the given zone(s) while maintaining flexibility of allocation as long as the noise levels have been maintained within the quota. In addition, the quotas have been stricter during the night, thus preventing operations at the particular locations during the entire day-24 h (www.portofrotterdam.com/nl/nieuws/persberichten/2009/20090406_02.jspww).

Consequently, expansion of the given port has been temporary constrained. However, simultaneously some noise mitigating measures including application of existing sound reducing techniques and advanced technologies have been deployed contributing to decreasing the overall noise levels and consequently opening the space for further expansion of the port, while maintaining these noise levels within the prescribed quotas. One of the options has been implementation of the installation of shoreside electricity for the inland waterways vessels and barges. As they switch from their auxiliary engines to the electricity grid, the noise generated by the auxiliary engine vanishes. For these vessels and barges, the use of their auxiliary engines has been banned at several locations where these power outlets are available (Den Boer and Verbraak, 2010).

11.5.2.2 Shipping lines and inland vessels/barges

The noise of sea ships and inland vessels/barges is not particularly elaborated. In the former case, the sea ships operate within the seaports at the low speed and low engine rate(s), not creating substantive (excessive and disturbing) noise. In the latter case, the noise from inland vessels and barges becomes relevant for consideration if the populate areas are located close to the inland waterways. The mitigating measures are similar as at the other above-mentioned inland transport modes—rail and road—and their systems.

11.5.3 TRAFFIC ACCIDENTS/INCIDENTS (SAFETY)

Similarly as at the other transport modes and their systems, the traffic accidents/incidents also happen at sea and inland waterway ports and corresponding shipping lines. They are usually statistically recorded and then used to estimate safety in the given context expressed as the number of accidents/incidents per unit of output carried out during the specified period of time. This time the accidents/incidents, that is, safety, of seaports and maritime shipping lines are considered as sufficiently illustrative for dealing with this kind of social impacts of the water-based systems. The impacts imply people fatalities and injuries, destroying and damaging facilities and equipment at seaport terminals, and damages and losses of ships of shipping lines.

11.5.3.1 Seaports

For example, the main most frequent locations of accidents (79% of the overall total) at the selected UK seaports during the period 2010–14 have been ship and craft, handling equipment, quayside, sheds and warehouses, and container berths. The total number of fatal accidents has been 1–2 per year during the same period of time. In addition, Fig. 11.44 shows the number of different types of accidents and dangerous occurrences a these seaports over the same period of time (PSS, 2014).

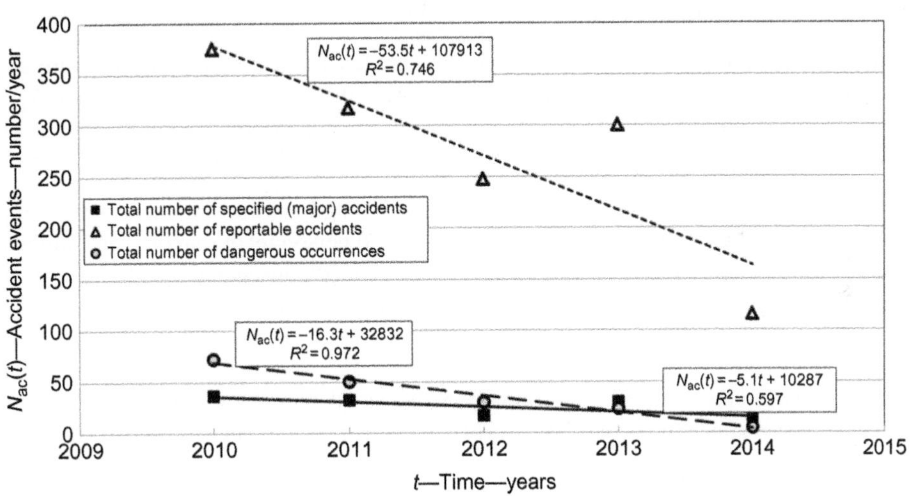

FIG. 11.44

The number of different accidental events at the selected UK seaports (period: 2010–14) (PSS, 2014).

As can be seen, with some variations, the number of all three types of considered accidental events has generally decreased during the observed period of time, thus indicating improvements of safety of the seaports expressed in the absolute terms—the number of events. In addition, Fig. 11.45 shows the relationship between the accident rate and the number of employees at the selected UK ports during the same period of time (PSS, 2014).

As can be seen, the accident rate during the observed period has varied between $AC_r = 1.4$ and 1.8 accidents/100 employees for the number of employees varying between 16,500 and 19,500. This

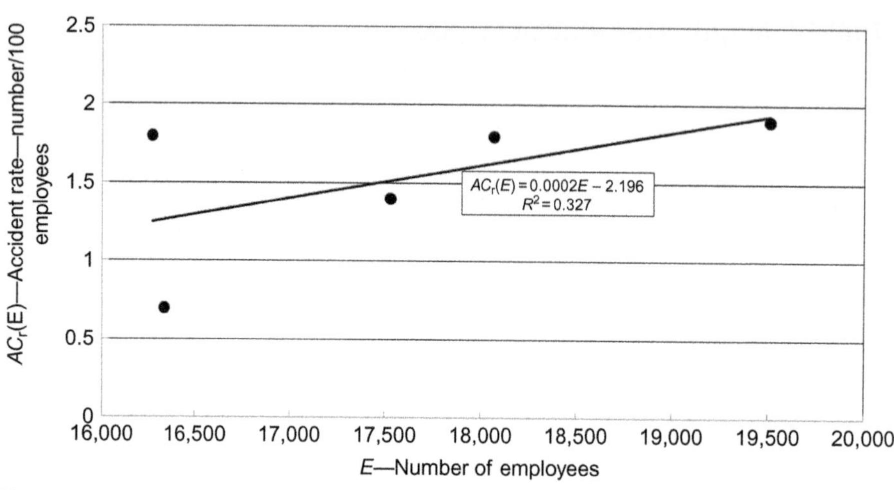

FIG. 11.45

Relationship between the accident rate and the number of employees at the selected UK seaports (period: 2010–14) (PSS, 2014).

implies that it could be said that in the given case the greater number of employees were generally exposed to the higher risk of experiencing an accident during the observed period of time (PSS, 2014).

11.5.3.2 Shipping lines

The accidents/incidents have happened to the maritime shipping lines causing the losses of ships, that is, their capacities, fatalities and injuries of crew onboard and personnel at seaports, and spillage of oil and hazardous freight/cargo. The accidents/incidents and related consequences expressed in both absolute (the number per period of time) and relative (the number per unit of output) terms can be used, similarly as at the other transport modes and their systems, to assess the perceived risk of accidents/incidents and related causalities, that is, safety under given conditions. Fig. 11.46 shows an example of the relationship between losses of the total annual capacity and the number of different types of freight/cargo ships with 500 gt and greater round the world during the period 1996–2013 (gt = gross registered tonnage) (EU, 2015).

As can be seen, the total lost capacity has generally increased with increasing of the number of lost ships, which has been intuitively expected. The capacity has been lost at the highest rate at bulkers and combined carriers and at the lowest rate at the other ships indicating the difference in their capacity, that is, loss of the smaller number of larger ships resulted in the substantive losses of their carrying capacity, and vice versa. In addition, the correlation between the lost capacity and the number of lost ships is rather low/weak indicating the actual randomness of accidents at particular ships categories and size during the observed period.

The fatalities from accidents/incidents of shipping lines usually happen during particular stages of operations such as loading and unloading at the port terminals, operations in ports, restricted and coastal waters, and open sea transit. Fig. 11.47 shows an example of the total number of fatalities at different categories of ships of the world's merchandise fleet during the specified period of time (IMO, 2012).

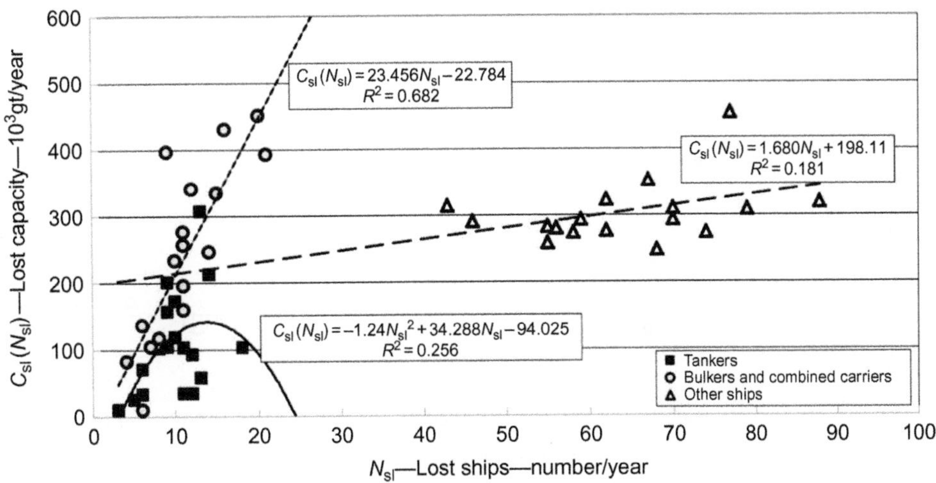

FIG. 11.46

Relationship between the total lost annual ship capacity and the number of lost ships of the world merchant fleet (period: 1996–2013) (EU, 2015).

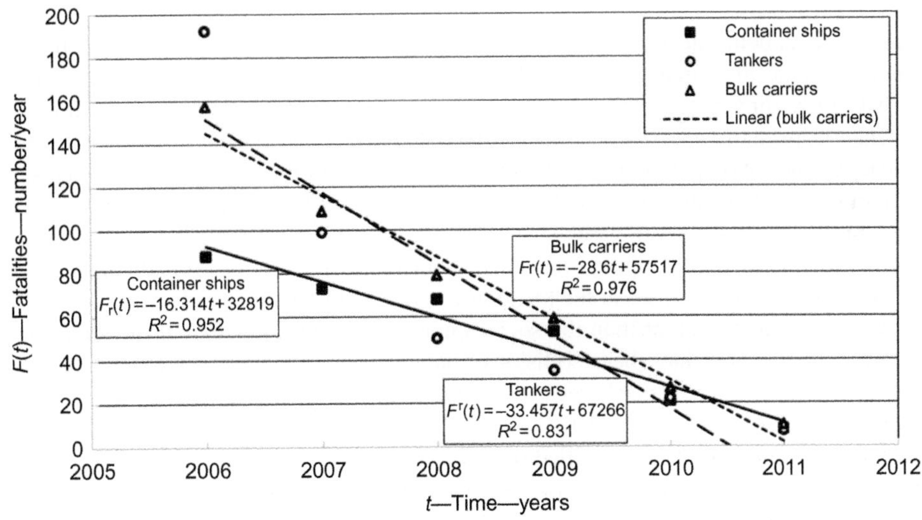

FIG. 11.47

The number of fatalities at the world merchandise fleet over time (period: 2006–11) (IMO, 2012).

As can be seen, the annual number of fatalities has generally decreased proportionally and more than proportionally at particular ship categories during the observed period of time thus indicating improvements in safety in the absolute terms. In addition, Fig. 11.48 shows the relationship between the fatality rate and the volume of output of the world's container and tanker ship fleet during the period 2006–11 (IMO, 2012; UNCTAD, 2015).

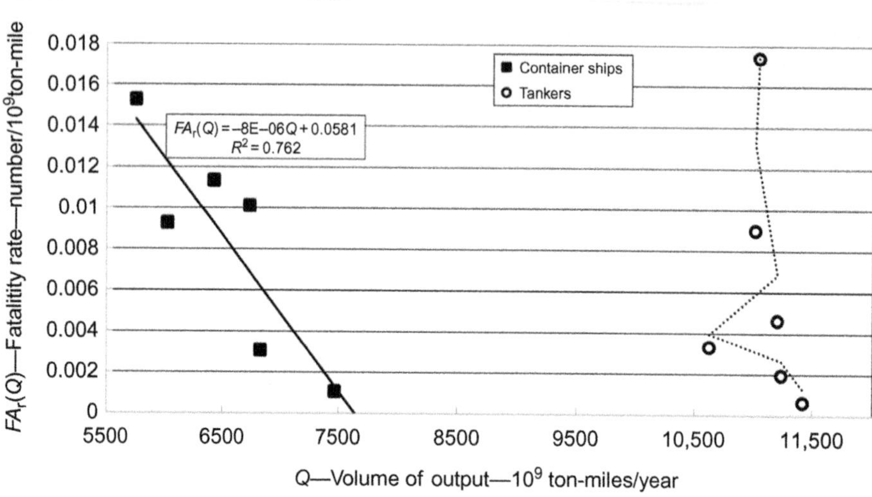

FIG. 11.48

Relationship between the fatality rate and the annual volume of output of the world container and tanker ship fleet (period: 2006–11) (IMO, 2012; UNCTAD, 2015).

As can be seen, at both categories of ships, the fatality rate has generally decreased with increasing of the annual volume of output. At the container ships it decreased from 0.016 to 0.001 per 10^9 ton-miles with increasing of the annual volume of output from 5600 to 7500 * 10^9 ton-miles, respectively. At tankers the fatality rate has decreased from 0.018 to 0.0005 per 10^9 ton-miles at the relatively stable annual volumes of output of about 10,500–11,500 * 10^9 ton-miles. These developments indicate that the risk of losing life while operating these categories of ships has been decreasing, thus reflecting improvement of safety in the given cases.

The accidents/incidents of shipping lines have also caused spillages of oil and hazardous freight/cargo, which have contaminated sea at the locations. In cases of no fatalities, these could also be classified as the environmental rather than the social impacts. The typical accidents/incidents at the oil tankers causing spillage of oil have been collisions, contacts, fire/explosions, war losses, structural failures, transfer spills, unauthorized discharges, and groundings. For example, during the period 1992–94, the frequency of events has been $6.9 * 10^{-3}$ spills per ship year, the oil spilled quantity 17.43 ton per ship year, and the average quantity of spilled oil 2522 ton (OGP, 2010). In addition, Fig. 11.49 shows the relationship between the annual quantities of oil discharged into the sea by cargo and bunker ships and the annual quantity of seaborne trade of crude oil round the world during the period 2002–09 (IMO, 2010).

As can be seen, the annual quantities of discharged oil by both cargo and bunker ships decreased more than proportionally with increasing of the annual quantities of seaborne traded crude oil, which indicates reducing the risk of such events. In addition, the oil spills from ships performing the offshore activities have happened during loading/unloading, accidental discharges overboard, and/or ruptured tanks. However, the reporting on these accidents/incidents and related consequences has been relatively scarce (www.ogp.org.uk).

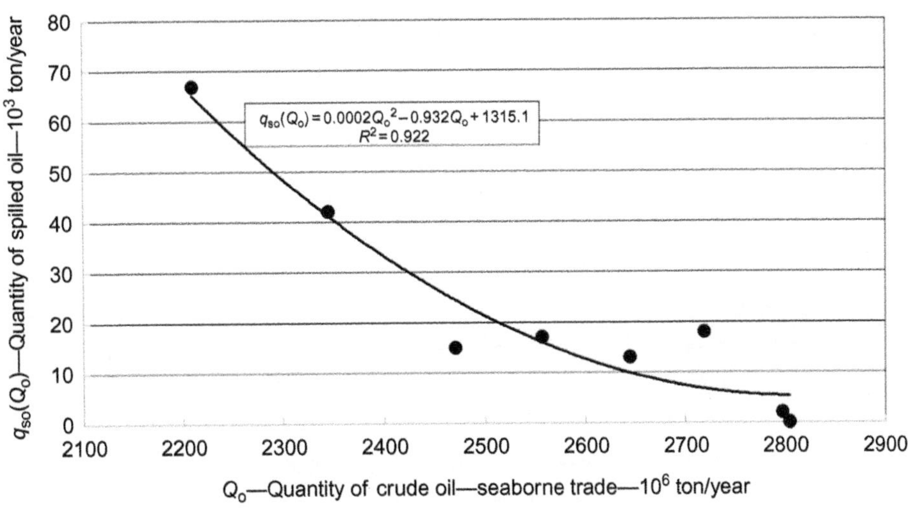

FIG. 11.49

Relationship between the annual quantity of oil discharged into the sea by cargo and bunker ships and the annual quantity of seaborne traded crude oil round the world (period: 2002–09) (IMO, 2010).

11.5.4 ENERGY/FUEL CONSUMPTION AND EMISSIONS OF GHG

The energy/fuel consumption and related emissions of GHG is analyzed for inland waterways vessels/barges, and seaports and maritime shipping lines.

11.5.4.1 Inland waterways

The inland waterways vessels/barges generally consume liquid fuels including diesel fuel, Bunker C residual fuel oil, other Fuels, and/or gasoline. The rate of fuel consumption mainly depends of the size of vessels/barges and conditions of their operations. The former relates to the class of vessel/barge expressed by the load capacity. The latter depends on the rate of engine use, operating speed, and other conditions prevailing at the rivers as the inland waterways (upstream/downstream sailing). For example, the vessels/barges operating along the European inland waterways are classified regarding the load capacity into nine classes: A (251–450 ton), B (451–650 ton), C (651–850 ton), D (851–1050 ton), E (1051–1250 ton), F (1251–1800 ton), G > 1800 ton, H (Push tug (2)—5800 ton), I (Push tug (4)—10,800 ton) The barges used on the River Mississippi (USA) have the capacity of 925, 1100, 1500, 1990, and 3100 ton. The tanks for chemical are of the capacity of 1425 ton (CECW-CP, 2004). Fig. 11.50 shows an example of the relationship between the Specific Fuel Consumption (SFC) and the average capacity of vessels/barges operating along in the European inland waterways (Donselaar van and Carmigchelt, 2001).

As can be seen, the SFC expressed in kg/103 t-km generally decreases more than proportionally with increasing of the average capacity of vessel/barge independently on the load factor. This indicates existence of scale in terms of fuel consumption. In addition, this SFC is higher if the load factor is lower, and vice versa, independently on the vessel/barge capacity.

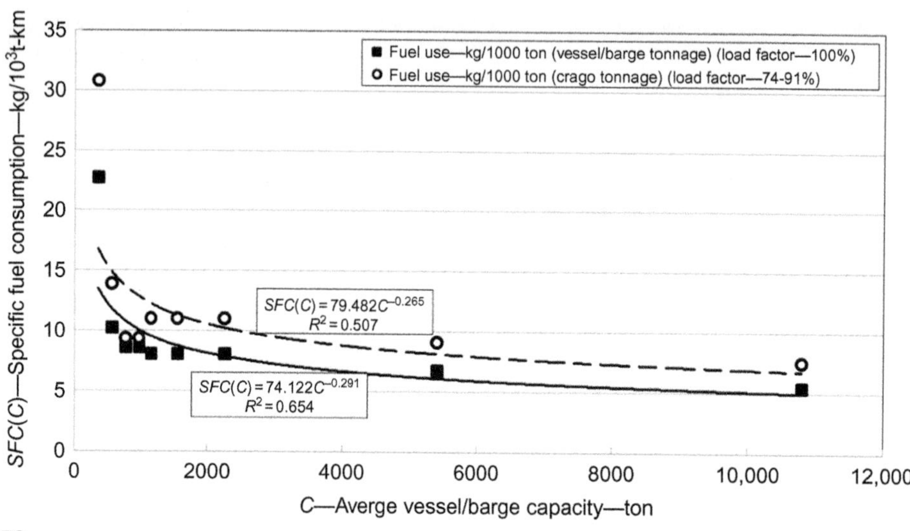

FIG. 11.50

Relationship between the average fuel consumption per unit of output (t-km) and the capacity of vessels/barges operating along the European inland waterways (steady-state operations at constant speed) (Donselaar van and Carmigchelt, 2001).

In addition, the average daily fuel consumption of towboats operating along the River Mississippi (USA) has been estimated in dependence of their maximum engine power and average rate of their use (80%) as follows (CECW-CP, 2004):

$$FC = 0.07743 \cdot HP^{1.24127};\ R^2 = 0.87;\ N = 13 \tag{11.40}$$

where

FC is the fuel consumption (gallons/day) (1 US gallon = 3.7854 L); and
HP is the engine power of towboat (HP—Horse Power) ($400 < HP < 10{,}000$).

The towboats have been of the above-mentioned capacity of 925–3100 ton. Eq. (11.40) indicates that 1% increase in the engine power operating under given conditions results in 1.24% increase in the fuel consumption of given towboats.

Burning fuels by the vessels/barges operating along inland waterways generate emissions of GHG, which in general contribute to local air pollution and global warming. The main emitted GHG are CO_2 (Carbon Dioxide), CO (Carbon Monoxide), NO_x (Nitrogen Oxides), SO_2 (Sulfur Dioxide), PM_{10} (Respirable Fraction Particles), HC (Hydro Carbons), and NMHC (Non Methane Hydro Carbons). These emissions are closely related to the average rate of fuel consumption, which, as mentioned above, depends of the capacity of vessel/barge size and its use, that is, load factor. For example, as an illustration, a container vessel/barge of the capacity of 3300 ton operating along the River Rhine at the engine power rate of 95% of its maximum and load factor of 74% emits about $e_r = 29.54$ kgCO_2/10^3 t-km, 25.5 gCO/10^3 t-km, 472 gNO_x/10^3 t-km, 32 gSO_2/10^3 t-km, 9.5 gPM_{10}/10^3 t-km, 9.5 gHC/10^3 t-km, and 9.0 gNMHC/10^3 t-km. In addition, Fig. 11.51 shows an example of the relationship between the average emissions rate of CO_2 and the capacity of container vessels/barges operating along the River Rhine (steady-state operations at constant speed) (Donselaar van and Carmigchelt, 2001).

The above-mentioned values have been calculated operating at constant speed downstream, engine power rate of 95%, 50%, and 25% of its maximum, and an average load factor of 74%. It should be mentioned that in order to maintain the same constant speed upstream, the engine power rate needs to be higher, which causes higher fuel consumption and corresponding emissions of GHG. Anyway, as can be seen, the emission rate, similarly as the average unit fuel consumption, decreases more than proportionally with increasing of the vessel/barge capacity.

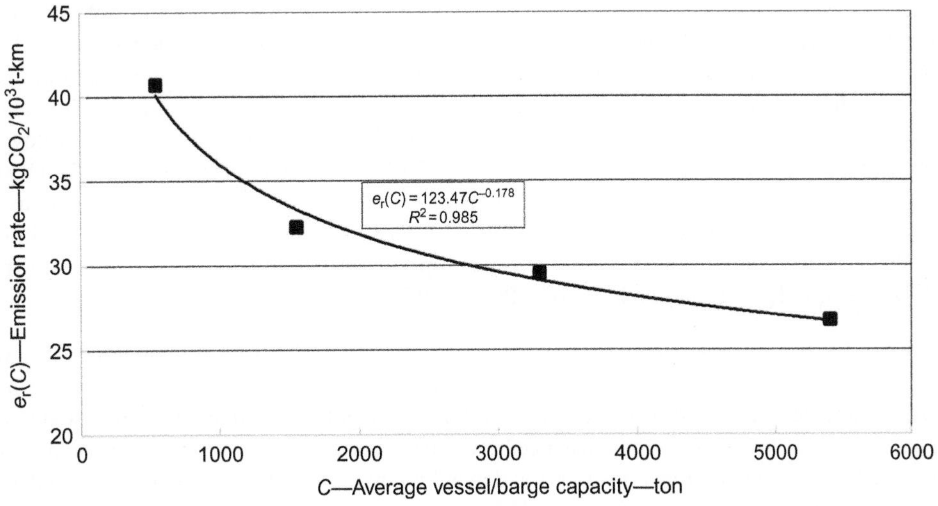

FIG. 11.51

Relationship between the average emission rate of CO_2 and the capacity of container vessels/barges operating along the River Rhine (Donselaar van and Carmigchelt, 2001).

11.5.4.2 Seaports

Energy/fuel consumption

The seaport-related operations require consumption of energy/fuel, which in turn also result in emissions of GHG. In general, the energy/fuel is consumed for handling the incoming and outgoing freight/cargo shipments y. The main energy consumers for bringing freight/cargo shipments to and from the port are sea ships maneuvering within the port area, the port ground access systems—freight rail, road, inland waterways vessels/barges, and pipelines. The main consumers of energy/fuel for carrying out industrial and freight/cargo shipment handling operations and within the port area itself are loading/unloading and transport facilities and equipment. Many of these consumers a relatively large amounts of mostly fossil fuels and thereby result in the emissions of GHG. Alternatively, the consumers of energy/fuel consumption as sources of related emissions of GHG can be broadly classified into mobile and stationary ones. The former (mobile) consumers/sources generally include cargo handling equipment not designed to operate on outside the port area, transport vehicles belonging to the port's ground access systems (road trucks, freight trains, inland waterways vessel/barges, pipelines, etc.), and smaller usually road vehicles transporting people, such as cars and vans, railroad locomotives, and ships. The latter (stationary) sources include fuel-fired heating units, portable or emergency generators, electricity consuming equipment and buildings, and refrigeration/cooling equipment. The above-mentioned categorization is relative since there may be some overlapping in exclusively mobile or stationary sources. For example, the fixed fuel-burning cranes belong to the freight/cargo handling facilities and equipment. In addition, the mobile electrically powered forklifts also belong to the same category of the facilities and equipment.

Emissions of GHG

In assessments and evaluations of the energy/fuel consumption and related emissions of GHG by seaports, the usual categorization of the energy/fuel consumers as sources of emission of GHG has been as follows: Shipping, that is, Ocean-Going Vessels (OGV); Harbor Craft (HC); Cargo Handling Equipment (CHE); Heavy-Duty Road Trucks (HDRT); Rail Locomotives (RL); and other equipment used within the port landside area. The considered GHG (air pollutants) have usually been as follows: CO (Carbon Monoxide), CO_2 (Carbon Dioxide), CO_{2e} (Carbon Dioxide Equivalent), NO_x (Nitrogen Oxides), N_2O (Nitrous Oxide), SO_x (Sulfur Oxides), CH_4 (Methane), HC (Hydrocarbons), and PM_s (Particulate Matter(s)) (Cannon, 2014; ENVIRON, 2013; PAEH, 2011; PLB, 2014; Shin and Cheong, 2011). Specifically, CO_{2e} is obtained by multiplying the mass and the global warming potential of $CO_2=1.0$, $CH_4=21$, and $NO_2=310$ (IPCC, 1995).

- *Shipping, that is, OGV (Ocean-Going Vessels)*: Emissions of GHG from shipping operating within the given port can be estimated for the specified period of time (usually 1 year) based on the number of particular categories of ships respecting their Gross Weight (GW), maneuvering distance within the port seaside area (the round trip distance between the berth terminal(s) and the boundaries of the port), and the average unit fuel consumption depending on the engine operating regime under given conditions as follows (Shin and Cheong, 2011):

$$EM_{OGV} = \sum_{i=1}^{N}\sum_{j=1}^{M} \frac{l_i}{FC_i} \cdot e_j \tag{11.41}$$

where

N is the total number of ships handled at the given port during the specified period of time (eg, 1 year);

M is the number of different GHG (air pollutants) considered;

l_i is the maneuvering distance of the ship (i) within the given port (km);

FC_i is the average unit fuel consumption of the ship (i) while maneuvering within the port (L/km or ton/km) (L—liter);

e_j is the emission factor of GHG (air pollutant) of type (j) per unit of fuel consumed by a ship (kg of pollutant/L or kg of fuel).

The energy consumption and related emissions of GHG from hoteling the ships at berths also need to be added to the amounts of GHG estimated by Eq. (11.41). These are dependent on the fuel consumption (about 20% of the consumption with the full engine power) and the hoteling (berth occupancy) time.

- *CHE (Cargo Handling Equipment)*: In seaports, this can be different regarding type, the number of units of a given type, and type of the energy/fuel use. This latest can be electricity and diesel/petrol fuels. Some typical types of CHE are container and transfer cranes, reach stackers, and yard tractors, all used for loading/unloading of the freight/cargo shipments at terminals. Under such conditions, the total emissions of GHG (air pollutants) during the specified period of time (usually 1 year) by a given port's terminal can be estimated as follows (Shin and Cheong, 2011):

$$EM_{CHE} = \sum_{k=1}^{K}\sum_{j=1}^{M} n_k \cdot EC_k \cdot Q \cdot e_{jk} \tag{11.42}$$

where

K is the number of different types of CHEs regarding the type of energy consumed;

n_k is the number of CHEs of type (k);

Q is the volume of handled freight/cargo shipments (ton);

EC_k is the average unit energy/fuel consumption of the CHE of type (k) (L or kg of fuel and/or kWh of electricity/ton of freight/cargo handled); and

e_{jk} is the emission factor of GHG (air pollutant) of type (j) per unit of energy/fuel consumed by CHE of type (k) (kg of pollutant/L or kg of fuel and/or kWh of electricity consumed).

The other symbols are analogous to those in the previous equations.

- *HDRT* (*Heavy-Duty Road Trucks*): These are the component of ground access system transport the incoming and outgoing freight/cargo shipments to and from a given port, respectively. Under such conditions, the emissions of GHG by HDRTs operating within the port landside area during the specified period of time (usually 1 year) can be estimated as follows (Shin and Cheong, 2011):

$$EM_{HDRT} = \sum_{l=1}^{L}\sum_{j=1}^{M} NT_l * FC_l * d_l * e_j \tag{11.43}$$

where

L is the number of road segments within the port landside area;

NT_l is the number of HDRTs operating along the segment (l) of the port's road infrastructure per period of time (vehicles/year);

d_l is the length of segment (l) of the port's road infrastructure (km);

FC_l is the average unit fuel consumption of a HDRT (L or kg of fuel /km); and

e_j is the emission factor of GHG (air pollutant) of type (j) per unit of fuel consumed by a HDRT (kg of pollutant/L or kg of fuel).

The other symbols area analogous as in the previous equations.

- *RL* (*Rail Locomotives*): Similarly as at HDRTs, the emissions of GHG by RL pulling and pushing trains composed of the number of cars loaded with freight/cargo shipments during the specified period of time (usually 1 year) can be estimated for activities and operations within the port's landside area as follows (Shin and Cheong, 2011):

$$EM_{LR} = \sum_{l=1}^{L}\sum_{j=1}^{M} N_{tr} \cdot n_{c/tr} \cdot d_{tr} \cdot FC_{c/tr} \cdot e_j \tag{11.44}$$

where

N_{tr} is the number of trains per period of time (year);

$n_{c/tr}$ is the average train composition (cars/train);

$FC_{c/tr}$ is the average unit fuel consumption per car of a train (L or kg of fuel, or kWh of electricity/km); and

e_j is the emission factor of GHG (air pollutant) of type (j) from the energy/fuel consumed by a train car (kg of pollutant/L or kg of fuel or kWh of electricity).

The other symbols are analogous to those in the previous equations.

The energy/fuel consumption of HC can be estimated, as well as the other equipment used within the port respecting the volumes of their operations and intensity and type of energy/fuel used.

Estimation of GHG means by the above-mentioned equations is illustrated by the case of seaport of Busan (South Korea). In the years 2001 and 2002, it had been the third busiest container port in the world in terms of the total annual number of Twenty Foot Equivalent Unit (TEU) handled. During the period 2003–13, it was at the fifth and in the year 2014 at the sixth place. Fig. 11.52 shows developments of the annual volumes of freight/cargo shipments and related emissions of GHG by four above-mentioned main sources at the seaport of Busan during the period 2000–07 (OECD, 2010; Shin and Cheong, 2011).

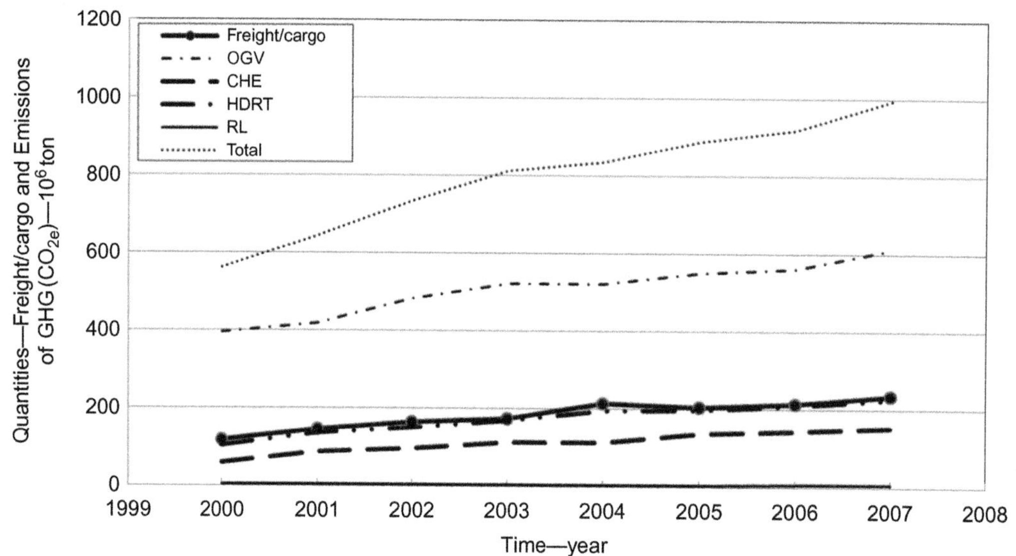

FIG. 11.52

Developments of the annual volumes of freight/cargo shipments and related emissions of GHG at the seaport of Busan (South Korea) (period: 2000–07) (OECD, 2010; Shin and Cheong, 2011).

As can be seen, the volumes of the freight/cargo, the total and particular emissions of GHG handled were continuously increased during the observed period. The emissions by OGV were the highest (share of about 63%), followed by that of HDRT (share about 22%), CHE (share about 14%), and LR (share about 1%) as the lowest. In addition, Fig. 11.53 shows the relationship between the total emission rate of GHG and the annual volumes of freight/cargo shipments handled by the given seaport, again during the period 2000–07.

As can be seen, the average emission rate of GHG decreased with increasing of the annual volumes of freight/cargo handled, thus indicating improvement of the seaport environmental performances in the given context. In addition, Fig. 11.54 shows an example of the share of the primary sources in the emissions of GHG (CO_{2e}) at the seaport of Los Angeles (USA).

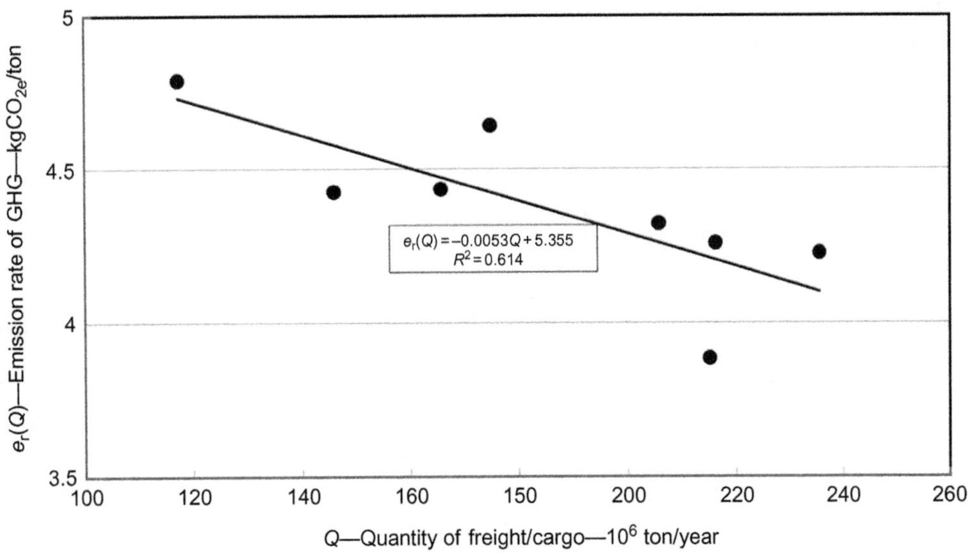

FIG. 11.53

Relationships between the average total unit emissions of GHG and the annual volumes of freight/cargo shipments at the seaport of Busan (South Korea) (period: 2000–07) (OECD, 2010; Shin and Cheong, 2011).

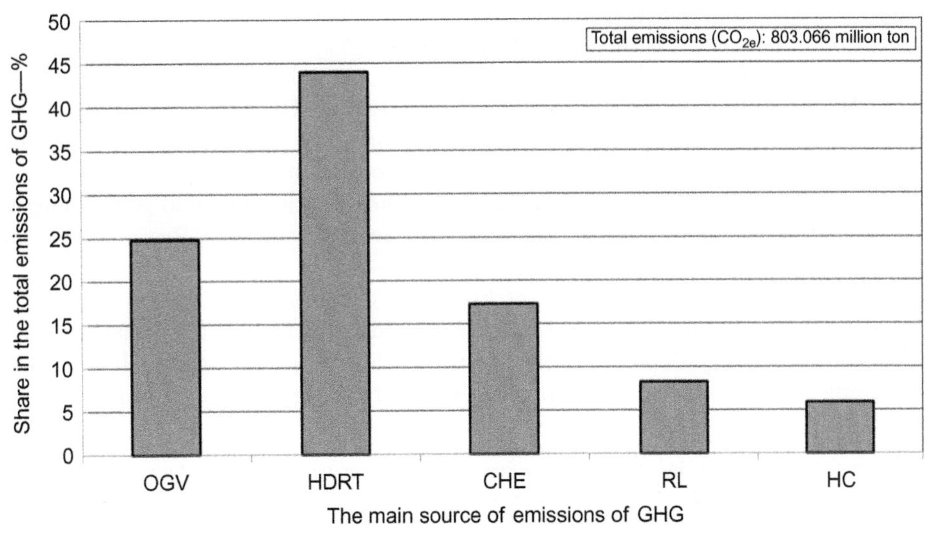

FIG. 11.54

Structure of emissions of GHG (CO_{2e}) by the primary sources at the seaport of Los Angeles (USA) (period: 2013) (Cannon, 2014).

As can be seen, over 90% of the total emissions of GHG (CO_{2e}) were from the sources operated by parties out of control by the given seaport authorities.

Some mitigating measures

This raises the question how much the seaports themselves can contribute to reducing their energy/fuel consumption and related emissions of GHG, that is, to improve "carbon footprint" and consequently mitigate impacts on the environment. At the global scale the mitigating measures has included eight (8) projects to be coordinated by The International Association of Ports & Harbors Initiatives (IAPH) as follows (Narusu, 2015):

- *Carbon Footprinting for Seaports* as the web-based manual is for calculating the current GHG emissions by seaports according to more and less (surrogate) detailed approach;
- *IAPH Tool Box for Port Clean Programs* as the web-based guidance documents aims at providing relevant information on air and climate issues of seaports and enable its users how to prepare a "Port Clean Program" for their ports respecting all relevant GHG (CO_{2e}). The seaports of Los Angeles and Long Beach (USA) have started implementation of the Clean Air Programs in practice resulted in substantive reducing of emissions of GHG 82% decrease in PMs, 54% in NO_x, and 90% in SO_x, during the period 2006–15.
- *On-shore Power Supply web-based manual* provides guidelines for the on-shore power supply of electricity from the local grid to ships in order to meet their power demand. The implementation has already taking place at the seaports worldwide. For example, currently 120 berths are fitted with the on-shore power connections in the world: 60 in North America; 50 in Europe; and 10 in the Middle East/Asia-Pacific. In China, all new seaport terminals should have the on-shore power in their plans.
- *Sustainable Lease Agreement Template* is still under development.
- *Cargo-Handling Equipment* is still under development.
- *Environmental Ship Index (ESI)* already been joined by many ships and seaports identifies ships performing better than average in terms of emission of GHG and consequently get some incentives from seaports in terms of reducing seaport dues. The ESI is composed of credits (0–100) for GHG such as CO_2, NO_x, SO_x, and PMs allocated to each ship by seaports. Starting from Jul. 1, 2015, about 3800 ships and 35 incentive providers including 30 ports among which are Amsterdam, Rotterdam, Hamburg, Antwerp, Le Havre, Los Angeles, Busan, Tokyo, and others have taken part in the system.
- *LNG-fueled Vessels and Ports (safety check lists for LNG bunkering)*

  The LNG (Liquid Natural Gas)[4] has been considered as the fuel for reducing emissions of GHG by sea ships during all phases of their trips including in seaports. Currently about 30 sea ships in North Europe are fueled by LNG. The new LNG-fueled engines are also under development

[4]LNG is an alternative fuel whose main component is methane. It has come under focus mainly due to its CO_2 content which is about 20–25% lower than that of HFO. Also, it can reduce emissions of SO_x (Sulfur Oxide) by about 90–95% and NO_x to the level complying with IMO Tier III limits to be in effect from 2016. In addition, the price of LNG would be comparable to that of HFO. However, the tanks for storing LNG are much larger than those for storing HFO, thus requiring more space, which can compromise the ship's loading capacity. Nevertheless, this could be compensated by improved energy efficiency. The engines powering the ship would be the dual-fuel hybrid constructions enabling operation in both HFO and LNG mode. Some designs such as Quantum (DNV—Det Norske Veritas) indicate that the EEDIs of large container ships using a mixture of LNG/HFO could be significantly lower than that required beyond 2025 (by about 30%) (GL, 2012).

by the manufacturers. A prospective increased use of LNG has initiated IAPH to set up the web-based safety guidelines and checklists regarding its logistics within the seaports (storage at thanks, transfers between ships, trucks and ships, tanks to ships, and vice versa, etc.).

- *Terminal Automation* at seaports has taken place firstly at container terminals. It has started in Europe in the year 2002 at the seaports of Hamburg (Germany) and Rotterdam (the Netherlands) aiming at improving efficiency of the seaport operations, reduce labor costs, and improve overall safety. Currently, it has become the norm for large scale container terminals means by electric or hybrid CHE, contributing to reducing emissions of GHG.

11.5.4.3 Shipping lines

Energy/fuel consumption and emissions of GHG

The sea freight/cargo container, bulker, and tanker ships consume MDO (Marine Diesel Oil), sometimes also known as No. 6 Diesel or Heavy Fuel Oil (HFO) or Bunker C fuel adapted to the 2005 standards as Marine Distillate Fuels (MDF) (http://en.wikipedia.org/wiki/Heavy_fuel_oil). These are largely unrefined very thick crude oil derivatives, often needed to be heated by steam in order to reduce their viscosity and thus enable them to flow. In general, the fuel consumption of larger freight/cargo ships is substantive. Fig. 11.55 shows an example of the relationship between daily fuel consumption and operating/cruising speed of container ships of different size, that is, payload capacity (AECOM/URS, 2012; Churchill and Johnson, 2012; Janić, 2014a,b; Notteboom and Carriou, 2009).

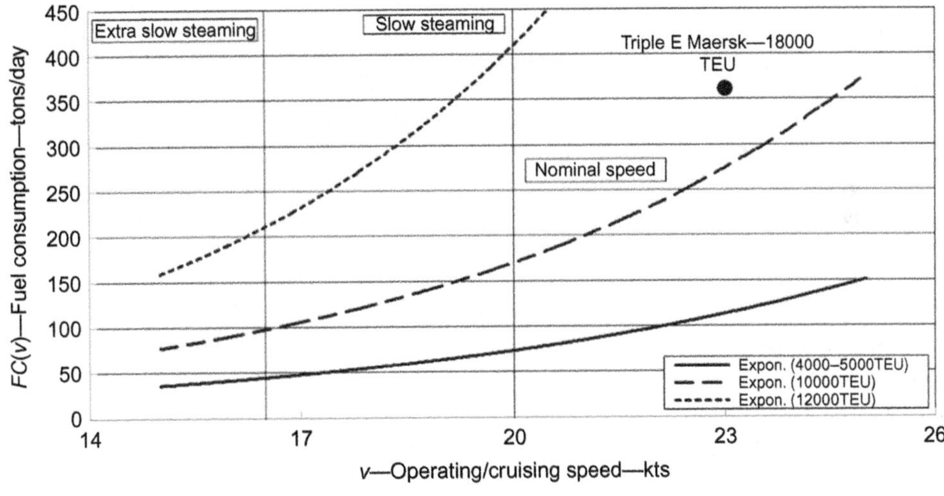

FIG. 11.55

Relationship between the fuel consumption and the operating/cruising speed of container ships of different size, that is, payload capacity (AECOM/URS, 2012; Churchill and Johnson, 2012; Janić, 2014a,b; Notteboom and Carriou, 2009).

As can be seen, the fuel consumption of a ship of a given size increases more than proportionally with increasing of the operating/cruising speed. For example, the ships of a capacity of 12,000 TEU such as Emma Maersk consume about 400 tons of fuel per day while cruising at a speed of about 20 kts (the length of route is about 15,000 nm) (AECOM/URS, 2012). Ships with a capacity of 10,000 TEU consume about 375 and 200 tons of fuel per day while cruising at the (designed) speed of 25 kts and reduced speed of 21 kts, respectively. For container ships of 4000–5000 TEU the corresponding fuel consumption is 150 and 85 tons per day, respectively. The largest Triple E Maersk ship of a capacity of 18,000 TEU consumes about 360 tons of fuel per day while cruising at the speed of 25 kts. These figures indicate the very high sensitivity of fuel consumption to the ship's operating/cruising speed independently on their size (TEU). In addition, fuel consumption can be expressed differently. For example, the world's largest single diesel Wärtsilä-Sulzer 14RTFLEX96-C engine powers the Emma Maersk container ship. At the maximum rate of 80–81 MW enabling the designed operating/cruising speed of 25 kts it consumes about $19*10^3$ L or 16.7 tons of HFO/h or 198 g of HFO/KWh. The larger Triple E Maersk container ship with two MAN diesel engines of total power of 64 MW enabling an operating/cruising speed of 23 kts consumes 15.04 tons of HFO/h or about 231 g of HFO/kWh. At the designed operating/cruising speeds, these give 1.39 kg of HFO/TEU/h for Emma Maersk and 0.84 kg of HFO/TEU/h for Triple E Maersk ship, which is a reduction of about 40% (Janić, 2014a,b).

Burning HFO produces GHG such as SO_x (Sulfur Oxides), NO_x (Nitrogen Oxides), PM (Particulate Matters), and CO_2 (Carbon Dioxide). Fig. 11.56 shows an illustration of the relationship between the average unit emissions of GHG (CO_2) and the annual volume of output of the world's merchant fleet during the period 1990–2008 (IMO, 2010)

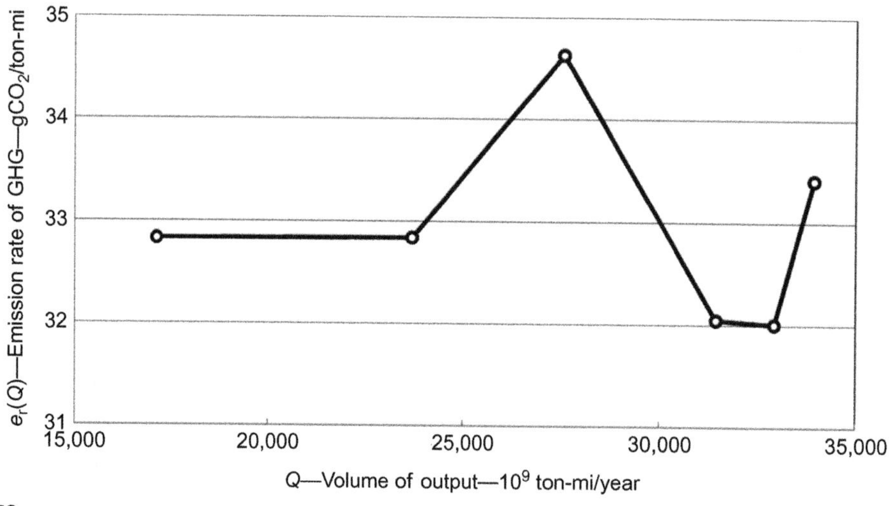

FIG. 11.56

Relationship between the emission rate of GHG (CO_2) and the annual volume of output of the world's merchandise ship fleet (period: 1996–2008) (mi—mile; 1 mile = 1.609 km) (IMO, 2010).

As can be see, there was observable variation of the emission rate of GHG (CO_2) (32–34.6 g/t-mile) with increasing of the annual volumes of output (t-miles) in the given case.

Some mitigating measures

The increased quantities of emitted CO_2 contribute to global warming, thus requiring efforts of the maritime industry and its national and international organizations to at least control them. For example, the World Shipping Council (WSC) and its members have been engaged through the International Maritime Organization (IMO) in the numerous initiatives aiming at improving the energy/fuel efficiency of the maritime sector through reducing the ships' fuel consumption and related emissions of GHG. At the level of individual shipping lines, this has been carried out through the medium-to long-term sustainability plans. Fig. 11.57 shows an example of achievements of a large Maersk Line during the period 2006–11 (Janić, 2014a,b; ML, 2011).

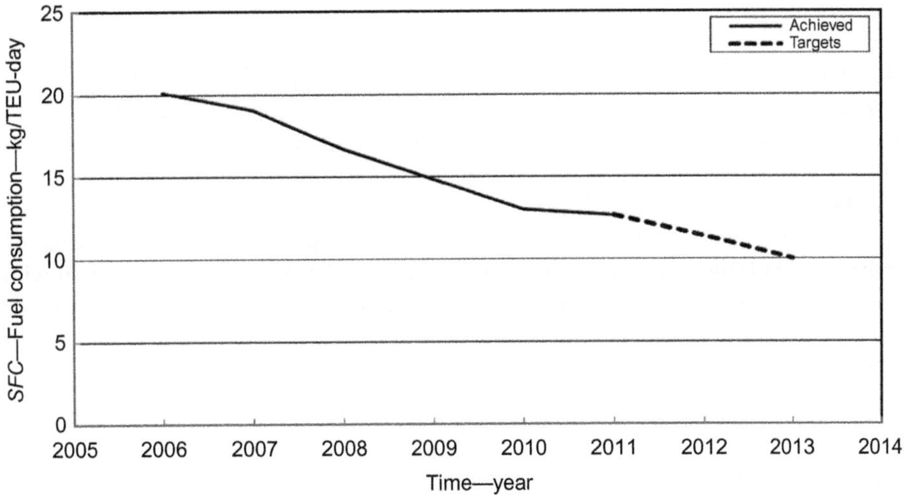

FIG. 11.57

Changes of the SFC (Specific Fuel Consumption) over time—Maersk Line (period: 2006–11) (Janić, 2014a,b; ML, 2011).

As can be seen, the company has followed a downward path towards the established target of an average *SFC* of about 10 kg/TEU/day achieved in the year 2013. This also implies reduction of the corresponding emission rates of GHG.

The other efforts have been institutionalized through Annex VI of MARPOL, an international treaty developed through IMO, which has established legally-binding international standards for regulating the energy/fuel efficiency of existing and future freight/cargo ships. Consequently, the main environmental performances of particular categories of ships are contained in these standards specified by the Marine Environment Protection Committee (MEPC) of the IMO aimed at reducing the energy consumption and related emissions of GHG over the 2013/14–2025 period and beyond. For example, the main characteristics of these standards for container ships with a deadweight of over 15,000 tons imply the following targets for reducing GHG (CO_2) emissions: Phase 0—0% over 2013–14; Phase 1—10% over 2015–19; Phase 2—20% over 2020–24; and Phase 3—30% beyond 2025 (MEPC, 2012). This is expected to be achieved individually and/or in combination of technical/technological, operational, and economic measures.

- *Technical measures*: Some of the technical/technological measures to be applied to the existing and new ships have been as follows (IMO, 2012): reducing propulsion resistance by modifying the ship's hull form; ensuring enhanced propulsion efficiency by modified propeller(s); increasing the ship's hull size in order to increase the deadweight (capacity); using energy from exhaust heat recovery; and using renewable energy (wind, solar power, etc.). In order to evaluate effects of the above-mentioned particular measures the IMO has proposed Energy Efficiency Design Index (EEDI), whose rather complex expression has been simplified as follows (IMO, 2011; Janić, 2014a,b):

$$EEDI_{ref} = \frac{P \cdot SFC \cdot C_f}{DWT \cdot v} (gCO_2/\text{ton-mile}) \tag{11.45}$$

where

P is the engine power including the main engine and auxiliary engines (kW);
SFC is the specific fuel consumption (the recommended value is 190 g/kWh);
C_f is the carbon emissions factor (3.1144 gCO_2/g of HFO);
DWT is the ship's deadweight (tons); and
v is the speed that can be achieved at 75% of (P) of the main engine.

Eq. (11.45) indicates that EEDI decreases more than proportionally with increasing of the ship's deadweight (DWT) and operating/cruising speed (v). It increases in proportion with increasing of the ship's engine power (P) and fuel efficiency (SFC). For example, using the data for container ships built over the period 1999–2008, the average EEDI as the reference value has been interrelated with the deadweight (DWT) as follows (Janić, 2014a,b):

$$EEDI_{ref} = 174.22 \cdot DWT^{-0.201} \tag{11.46}$$

where all symbols are as in the previous equations.

If the attained EEDI value of a given new-built ship is above reference value in Eq. (11.46), the ship is considered energy inefficient, and vice versa. Consequently, the required EEDI can be defined as the allowable maximum attained EEDI for a given container ship, which is below and/or at most at the reference value. Regarding the above-mentioned policy targets for improving the energy efficiency of container ships over the forthcoming period (ie, by 2025 and beyond), the required EEDI can be estimated as follows (IMO, 2011; Janić, 2014a,b):

$$EEDI_{req} = (1 - X/100) \cdot EEDI_{ref} \tag{11.47}$$

where X is the target for improving the energy efficiency of container ships during the specified period of time (%).

Fig. 11.58 shows an example of the attained and required EEDI, the latter respecting the above-mentioned energy/fuel efficiency improvement targets. The attained EEDI has been calculated for existing Post-Panamax large container ships, which entered service during the period 2003–08. For these ships the capacity utilization (DWT) has been assumed to be 70%, the engine power 75% of its maximum, and the speed 1 kt below the maximum designed speed (Janić, 2014a,b).

As can be seen, all considered container ships have fulfilled the required 2013–14 EEDI. Ships larger than 85,000 DWT will be able to satisfy the required 2015–19 EEDI. None of these ships will be able to satisfy the required 2020–24 EEDI, or those set for 2025 and beyond.

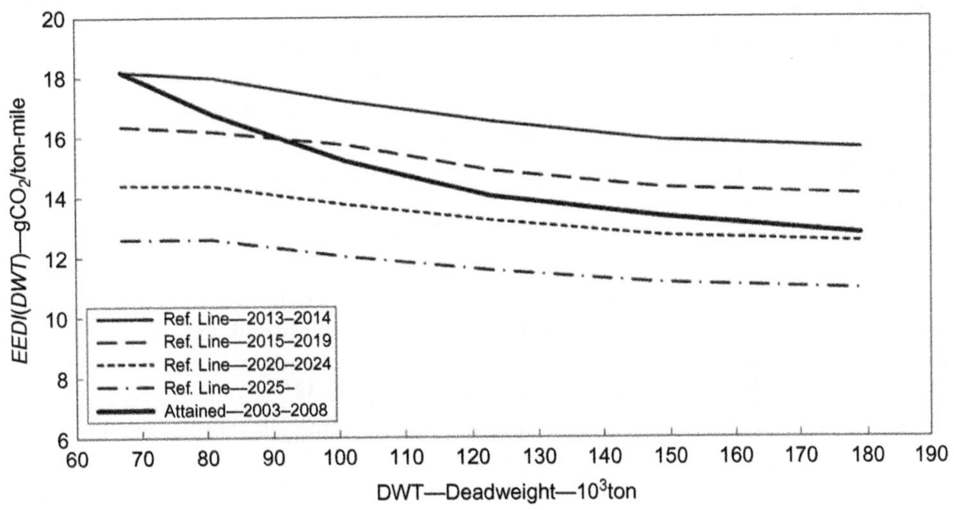

FIG. 11.58

Relationship between the existing/attained and required/target EEDI, and the capacity of large container ships (Janić, 2014a,b).

- *Operational measures*: These are aimed to improve the energy/fuel efficiency of freight/cargo ships through innovative operations. They are supposed to be applied by shipping companies to existing ships in the scope of their efforts to improve energy and consequently economic efficiency. Some of these measures are as follows (Janić, 2014a,b): optimizing operations of individual ships and fleets; operating/cruising at reduced speed, that is, slow steaming; entering and leaving ports on time; maintaining the hull clean in order to reduce propulsion resistance; and ensuring regular maintenance of the ship's overall machinery. In order to make some and/or all above-mentioned operational measures applicable to existing freight/cargo ships, IMO has proposed two indicators/tools: Energy Efficiency Operational Indicator (EEOI) and Ship Energy Efficiency Management Plan (SEEMP). The EEOI was introduced voluntarily in the year 2005 as an indicator to express the energy/fuel efficiency of a given ship in operation as follows (IMO, 2012; Janić, 2014a,b):

$$EEOI = \frac{FC \cdot C_f}{W_c \cdot d} (\text{gCO}_2/\text{ton-mile}) \tag{11.48}$$

where

FC is the fuel consumption during a trip (ton);
W_c is the actual weight of freight/cargo (ton); and
d is the length of route, that is, the actual trip distance (nm).

The other symbols are analogous to those in the previous equations.

Eq. (11.48) indicates that EEOI is proportional to the fuel consumed during a given trip and is inversely proportional to the actual weight of freight/cargo on-board and length of route. Thus, the

EEOI can be improved by decreasing the fuel consumption, as mentioned above, through reducing the operating/cruising speed, that is, slow steaming, while transporting larger quantities of freight/cargo on longer routes.

Despite being expressed in the same units as EEDI, the EEOI is estimated from the values of particular variables measured during or just after a given trip. Therefore, it can be used for measuring changes in the energy efficiency of the same ship operating along different routes/markets under different conditions.

The SEEMP as a tool has been aimed for improvements in the ship and fleet's energy/fuel efficiency means by operational measures. These measures include planning the trip in terms of weather routing, arrivals and departures from seaports on time, optimization of operating/cruising speed, etc., optimizing the handling and maintenance of the ship's hull, use of engines and waste heat recovery, and energy management and reporting. This voluntary tool can be implemented by shipping lines through five procedures contributing to improving the ship/trip energy/fuel efficiency cycle as follows (Janić, 2014a,b): planning, implementation, monitoring, self-evaluation, and publication of achieved results. In addition, due to the need for collecting a relatively large quantity of information even for a single ship/trip, different support systems have been developed for calculating, analyzing, and preparing reports on the ship/trip energy/fuel efficiency cycle(s).

- *Economic measures*: These measures aim at promoting and implementing the above-mentioned technical/technological and operational measures. For such a purpose, IMO has proposed several measures classified into two broad categories as follows: the fuel pricing system (proposed by Denmark); and the emissions trading system (proposed by Norway, Germany, and France). The former measure implies automatically charging an amount at the moment of purchasing fuel. A part of the collected charges would be used for the projects aiming at reducing emissions of GHG, particularly those in developing countries. In addition, other part of the collected charges would be returned back to ships achieved substantial improvements in energy/fuel efficiency. The latter measure implies that the total amount of GHG generated by the shipping industry would be regulated by the emission trading scheme. In such a case, each ship would be assigned an annual allowable quantity, that is, quota, of emitted GHG (CO_2). The positive difference between the actually emitted and assigned quantity of CO_2 would be traded with other ships and/or with the rest of transport and other non-transport sectors (Janić, 2014a,b).
- *Potential effects of proposed measures on global emissions of GHG*: The above-mentioned EEDI and SEEMP measures have been expected to significantly reduce energy/fuel consumption and related emissions of GHG from the world's freight/cargo ships over the long-term period 2010–50. Fig. 11.59 shows one such scenario (IMO, 2011; Janić, 2014a,b; MEPC, 2012).

As can be seen, both EEDI and SEEMP measure will not be able to prevent a further increase in the total cumulative emissions of GHG according to an upward trend starting from the emission levels in the year 2010 mainly driven by the expected growth in global trade. Nevertheless, this increase will be at reduced rates causing the absolute reduction in the emitted quantities of GHG of about 13%–23% over the period 2020–30 compared to that at the Business As Usual (BAU) scenario (Janić, 2014a,b).

Future technologies

The main drivers of design of future large advanced container ships will be conditioned by the strategic plans of shipping companies and environmental constraints aiming at the following (Janić, 2014a,b):

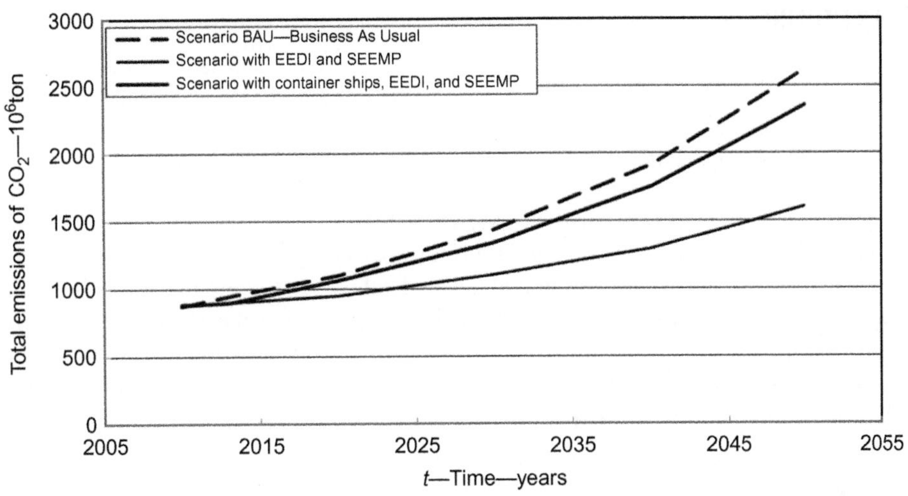

FIG. 11.59

Development of cumulative emissions of GHG (CO_2) over time by the global freight/cargo ship fleet (IMO, 2011; Janić, 2014a,b; MEPC, 2012).

- Improving economics by reducing the staff and thus increasing productivity;
- Increasing flexibility of services by modifying routes and networks, and ship deployment;
- Optimizing utilization of containers, that is, securing return freight/cargo volumes;
- Minimizing delays at seaports; and
- Minimizing fuel consumption and related emissions of GHG by meeting the current and prospective energy efficiency regulatory requirements, that is, required EEDI.

In particular, the options for minimizing fuel consumption and related emissions of GHG through ship design include: (i) Reduction of power; (ii) New technology for power generation; and (iii) Renewable fuel/energy primary sources.

(i) Reduction of power can generally be achieved by designing/developing the hull form, reducing the weight and power for the ship's own use, frictional and wind resistance, and improving the engine efficiency;

(ii) New technologies for power generation include use of alternative fuels such as biofuels, LNG and LH_2 and fuel cells; and

(iii) Renewable fuel/energy primary sources include solar and wind energy.

In addition, an option for minimizing the fuel consumption and related emissions of GHG includes forthcoming trip support systems, one of which is "Sea-Navi." These are designed to support on-line optimization of the ship's routing respecting the shortest distance, weather, characteristics, and regime of engine operation, thus contributing to improving EEOI and SEEMP (Janić, 2014a,b).

11.5.5 LAND USE

The sea and inland waterways large ports usually occupy relatively substantive area of land, thus in many cases compromising its use for housing, agriculture, or natural habitat. In such way, these ports are considered to make an impact on the environment. In particular, the large area of land is taken by large seaports handling large ships of different categories—container, tanker and bulker. In general, the area of land taken by a given seaport is mainly influenced by the ship and freight/cargo characteristics. In general, the length of ship influences the length of berths, which has shown to be increasingly important characteristic of the seaports' flexibility in responding to the current trend of increasing of size of ships, that is, their length and draught. The depth of land along the berths depends on the payload capacity and dynamism of unloading/loading of ships, and the stacking characteristics of freight/cargo shipments. In general, the area of land in a seaport used for handling the given type of freight/cargo shipments transported by ships of the similar payload capacity and load factor can be roughly estimated as follows (UN, 1983):

$$A = A_0 + \frac{n \cdot \max[0; (R - r)] \cdot (\theta \cdot C/R)}{q_s} \tag{11.49}$$

where

A_0 is the area of land for the long-storage of the given freight/cargo shipments (ha);
n is the number of ships simultaneously being at the berths;
R is the unloading/loading rate of a ship with the given freight/cargo shipments (ton or TEU/day);
r is the rate of departing/arrival of the given freight/cargo shipments from/to the port, respectively, by the ground transport modes/systems (ton or TEU/day);
θ is the average load factor of a ship carrying the given freight/cargo shipments (ton or TEU/ship);
C is the average payload capacity of a ship carrying the given freight/cargo shipments (ton or TEU/ship); and
q_s is the stacking characteristics of the given freight/cargo shipments (ton or TEU/ha).

For example, let's consider handling container ships at the seaport's container terminals. Each terminal consists of berths at the port's water-side for ship docking, a coastal area of land for the storage of containers, CHE including the specialized berth and yard cranes for container loading/unloading to/from the ships and within the storage area and between them, respectively, gates for HDRT and in many cases RL yards, and various maintenance and administrative buildings. Fig. 11.60 shows an example of the relationship between the number of seaport's berths and area of taken/used land (CGI, 2007).

As can be seen, the total area of land taken by the container terminals increases linearly with increasing of the number of berths. In this case, the average area of land taken per single berth is about 28 ha.

In addition, most seaports set aside, that is, reserve, the "land banks" for future expansion, which can range from a few hundreds to a few thousand percent of the land occupied by existing terminals (Janić, 2014a,b). The above-mentioned area of land taken is also dependent on the stacking characteristics of the freight/cargo, which in this case are containers.

The particular indicator of efficiency and effectiveness of operations of seaports in the given context is the intensity of land use. This can be expressed by the volumes of freight/cargo shipments handled per unit of land occupied by a given port. Fig. 11.61 shows an example of the intensity of land use for the seaport of Rotterdam (Europe) during the specified period of time (PRA, 2015).

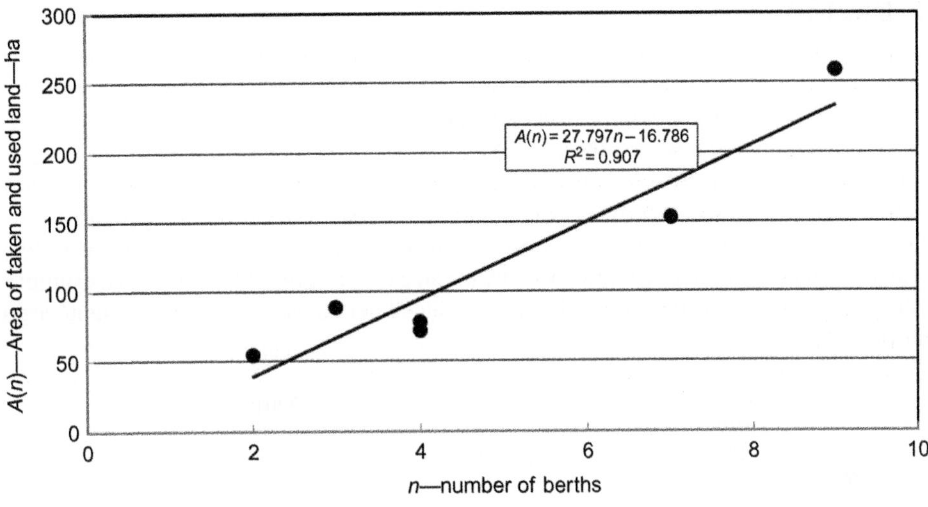

FIG. 11.60

Relationship between the used land by the container terminals and the number of berths at the selected US ports (CGI, 2007).

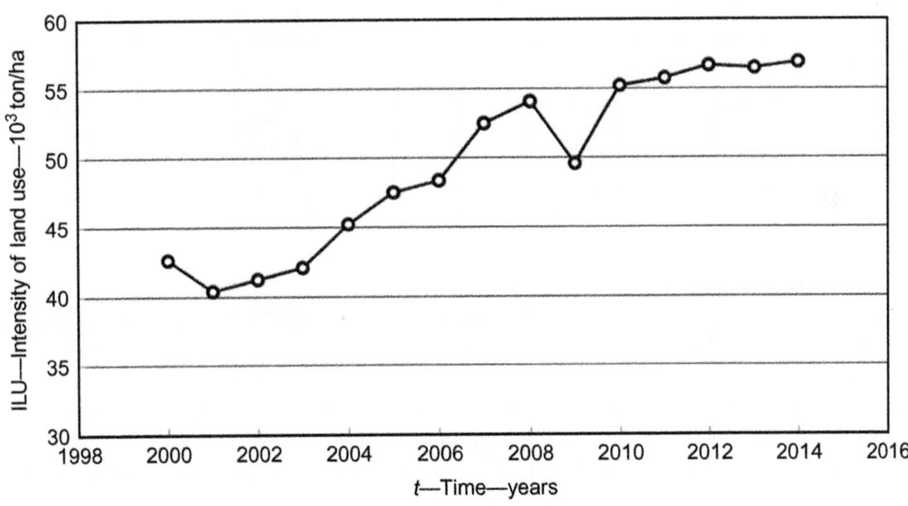

FIG. 11.61

The intensity of land use at the seaport of Rotterdam (the Netherlands) (period: 2000–14) (PRA, 2015).

The total area occupied by the port is 12,603 ha. The area of the waterside is 4810 ha and of the landside 7793 ha. As can be seen, as the landside area is fixed, the intensity of land use increases with changes with changing of the annual volumes of handled freight/goods shipments. \in this case, the intensity of land use has generally increased during the observed period. This implies that once the area of land has been taken, it needs to be used intensively for the purpose.

11.5.6 WASTE

Sea ships generate waste, which, if not handled properly, can affect the water-side of the seaports and consequently damage the environment and existing ecosystems there. Some waste can also affect the land side of the ports threatening to the health of their employees. In general, the waste generated by sea ships and considered by seaports as potentially harmful has been categorized as follows: solid waste (garbage), sewage, ballast water, oil, anti-fouling paint scraps, and hazardous freight/cargo shipments and other maintenance material (http://www.ukmarinesac.org.uk/activities/ports/ph6.htm).

As far as the solid waste (garbage) is concerned, the quantity produced has been generally dependent on the number of crew on board influenced by the size of ships. Fig. 11.62 shows an example of the relationships between the quantity of solid waste produced per day and the size of freight/cargo ships (Mohammad, 2000).

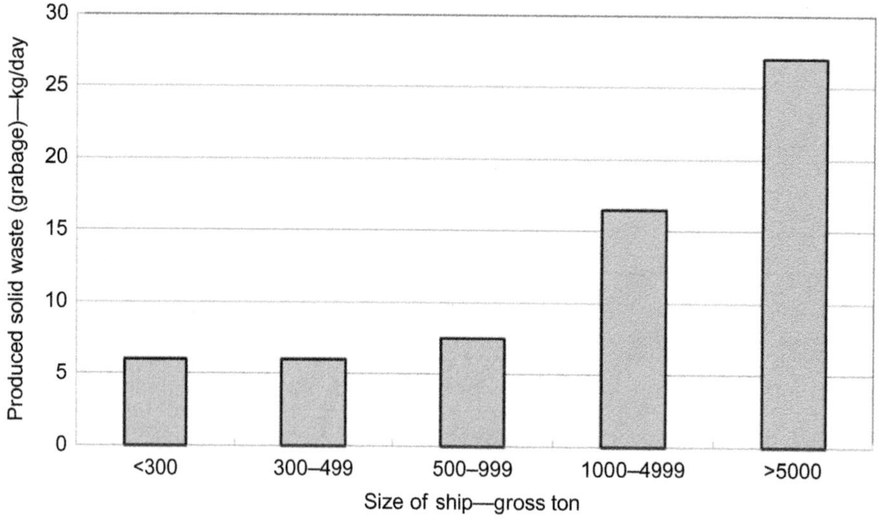

FIG. 11.62

The quantity of solid waste produced on board freight/cargo ships of different size categories of ship—mixed cargo, bulk, container, oil and gas tanker (Mohammad, 2000).

As can be seen, the quantity of solid wastes increases more than proportionally with increasing of the size of ships influencing the number of crew on board. This is 4 for ships <499, 5 for ships from 500 to 999, 11 for ships from 1000 to 4999, and 18 for ships >5000 gross tons (Mohammad, 2000).

Disposing waste from ships has been the matter of international regulation and conventions. At the global scale, the most know is the MARPOL Convention 73/78 (International Convention for the Prevention of Pollution from Ships) 1973, as modified by the Protocol of 1978, as amended (IMO, 1997). At the narrower scale, it has been EC directive 2000/59/EC on the port reception facilities for ship-generated waste and cargo residues (EC, 2000). According to this regulation some solid waste (garbage) and sewage except plastic can be disposed of at sea legally. The rest has to be deposited at the waste reception facilities at ports, for which the ships have been charged. One element of this regulation specifies that the sea ships are charged for waste disposal independently if they use or not the waste reception facilities. The charge/fee is set up as the product of the mandatory charge/fee and the factor depending on the ship size shown in Fig. 11.63 (Van den Dries, 2013).

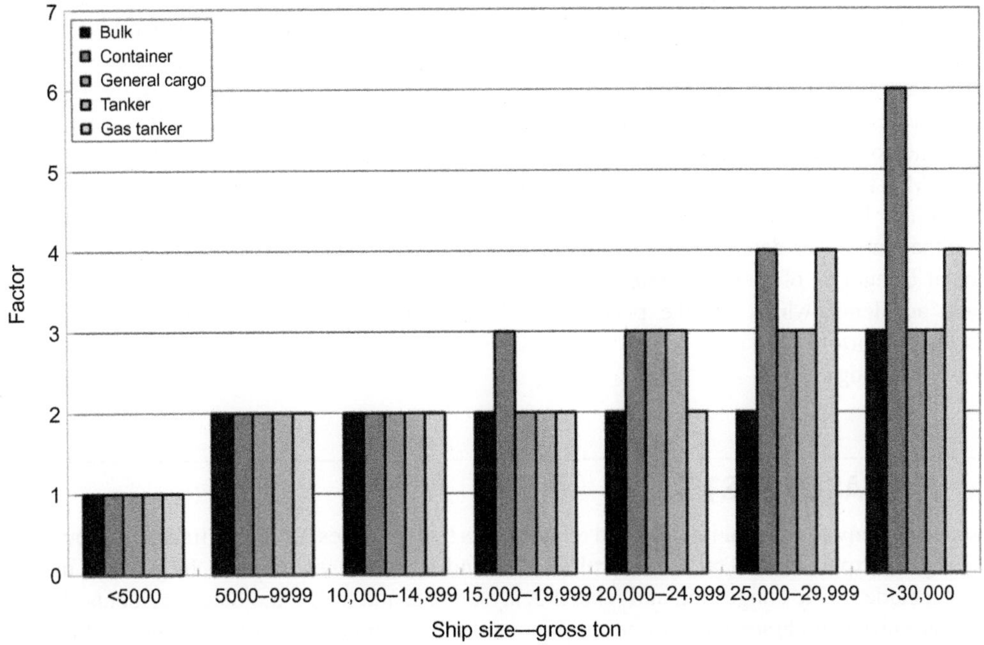

FIG. 11.63

The factor for charging waste from different categories of ships by the port of Antwerp (Belgium) (Van den Dries, 2013).

As can be seen, the factor for charging waste is the least at the bulk and the most different at the container ships with increasing of the ship size. In general, it tends to increases with increasing of the ship size at all categories of ships. In this case the mandatory fee as the basis has been 85€. In addition, in order to stimulate ships to deliver the waste at the reception facilities, the port authority makes deduction of the total charge/fee for 20–20€/m^3 (Van den Dries, 2013).

The oil-based wastes can be roughly divided into that, which can be incinerated on board ships and that, which has to be to be deposited at the dedicated waste collection facilities at ports for further recycling. The latter quantities are the also subject to charging system, which again stimulates their disposal at the waste collection facilities. However, unpredictable oil spills due to a range or reasons have frequently happened at many ports. In order to prevent them or mitigate their impacts, the specified checking procedures have been introduced and the specialized ships to deal with them deployed, respectively, all aiming at minimizing the possible negative impacts on the environment.

Additional impacts on the environment in the port's waterside area come from sewage and ballast water used by sea ships for controlling their balance and buoyancy. In particular, by intake and release of the ballast water, species having a severe impact on the ecosystem can be released. In particular, the North Sea is subject to a negative impact of such invasive species. The mitigating measures have included filtering or treating the ballast water with chemical substances.

In addition, the hulls of sea ships under the waterline are prone to fouling of organisms, which increases the drag and weight of the hulls and consequently the water resistance and related energy consumption and emissions of GHG. Therefore, the special paints have been applied to prevent this fouling of organisms. However, if these paints frequently leach, they can expose the close water/environment at risk of contamination due to high toxicity. In addition to leaching process, these paints can also be released into the environment through the ship maintenance (sanding and grinding). In order to mitigate the possible impacts, use of some of these antifouling substances has been banned in the year 2003 by IMO. Last but not least, the hazardous freight/cargo shipments and maintenance material need to be carefully handled in order to minimize the risk of their impact generally on the environment and also on the employees dealing with. Consequently, most ports have prescribed the rules of handling particular categories of these freight/cargo shipments aiming at maintaining a standard for the risk on fatal accidents, which for the port of Rotterdam (the Netherlands) amounts $1/10^6$ per year (https://www.portofrotterdam.com/en/shipping/sea-shipping/ships%E2%80%99-waste-from-seagoing-shipping).

11.6 AIR-BASED SYSTEMS

The social and environmental impacts and related costs (externalities) of the air transport sport systems consisting of airports, Air Traffic Control (ATC), and airlines are analyzed. In general, the social impacts include traffic congestion and related delays, noise, and traffic incidents/accidents. The environmental impacts embrace energy consumption and related emissions of GHG, and land use.

11.6.1 CONGESTION

Traffic congestion and related delays of aircraft/flights and their users-air passengers and air freight/cargo shipments occur whenever demand for service exceeds the available capacity of the service facility, that is, component of the system, independently on the cause(s). They have been considered as the social impacts mainly due to additional time and its costs imposed on airlines and their users/air passengers and air freight/cargo shipments under given conditions. The evidence so far has shown that the most common causes of aircraft/flight delays in the well-developed matured air transport systems such as those in Europe and United States have been the general imbalance between the aircraft/flight demand and the airport and ATC capacity under regular operating conditions, bad weather, the ATC staffing, failures of the ATC facilities and equipment, and others. They all generally restrict the airport and ATC capacity causing slowing down of the affected aircraft/flights and consequently create congestion and related delays.

11.6.1.1 Shortage of the airport and ATC capacity

In general, the shortage of airport and/or ATC capacity under any operating conditions causes their imbalance with demand causing the aircraft/flight congestion and delays. Three levels of such imbalance can happen as follows:

- The demand temporarily exceeds the capacity for relatively short period of time (several minutes). In such case, the arriving aircraft/flights are airborne waiting in the holding pattern for landing in the vicinity of destination airport(s). Departing aircraft/flights wait in the departure queue at given airport. Different reasons may cause these delays such the imprecision in keeping the arrival and departure schedule, change of the runway in use and/or temporal bad weather. These are General Arrival and Departure Delays.
- The prospective demand at given destination airport is expected to exceed the capacity substantively for the longer period of time (several hours during the day). In this case the airport arrival and/or departure capacity significantly deteriorates below the scheduled demand thus limiting the number of aircraft/flights that can be accommodated. This causes rather long delays carried out according to the Ground Holding Program (GHP) usually at the aircraft/flight origin airports in order to reduce the increased cost of delays if they would be airborne. Different causes can significantly deteriorate the capacity at destination airport. The most influential is bad weather usually requiring increasing of the ATC separation rules between the arriving/departing aircraft and consequently decreasing of the capacity on the one hand and even closure parts of the airport (runways) on the other. These are Ground Holding Delays.
- The temporal (short time) deterioration of the arrival capacity at the destination airport(s) requires imposing delays on departing aircraft/flights. In such cases, the affected departures are temporarily held until the capacity at the destination airport fully recovers. The most frequent reason for these delays is the short lasting severe bad weather at destination airport. These are Ground Stop Delays.

The congestion and delays, depending on the purpose, have been recorded at different "reference locations" within the airport airside area for both arriving and departing aircraft/flights. In all cases, delays have been measured as the difference between the actual and the scheduled (in Flight Plan) time of passing through a given "reference location." Consequently the following categories of delays have been distinguished (https://aspm.faa.gov/).

(i) *Arriving aircraft/flights*
- *Airborne Delay* is the difference between the actual airborne time and estimated flight time. This delay includes the airborne delay en-route and that just before landing.
- *Taxi-in Delay* is the difference between the actual taxi-in time and the unimpeded taxi-in time. This delay occurs when a gate for an arriving aircraft is not available and/or if there is the interference between given aircraft/flight and other incoming and outgoing traffic using taxing within the same taxiway system.
- *Gate Arrival Delay* is the difference between the actual gate in time and the scheduled gate in time. This delay is an aggregate of the airborne delay before landing and taxi-in delay.

(ii) *Departing aircraft/flights*
- *Gate Departure Delay* is the difference between the actual gate time out and the scheduled gate time out. This delay can be caused by many causes such as Gate Arrival Delay and delays in the processing passengers and freight in the airport landside area (passenger and freight terminal buildings).
- *Taxi-Out Delay* is the difference between the actual taxi-out time and the unimpeded taxi-out time. This may be caused by the interference with other outgoing and incoming aircraft/flights using the same taxiway system.

- *Departure Delay* is the difference between the actual take-off time and the scheduled gate departure and unimpeded taxi-out time. Consequently this delay embraces the average gate departure, taxi-out, and delays due to waiting in the departure queue just before taking off.

The above-mention definition of Gate Arrival and Departure Delays implies that they comprise all causes of slowing down the affected aircraft/flights in other parts of the system (airports and airspace), the given airport itself, and their mutual interdependency. In particular, the mutual interdependency at the given airport includes delaying of the departure aircraft/flights due to the late arrivals of their aircraft, which could not be neutralized during the aircraft turnaround time. In such case, the arrival delays pass to the departure delays and consequently further throughout the aircraft daily itinerary. Sometimes, the terms "primary" of the initial and the "reactionary" for delays passed to forthcoming flight(s) by the same aircraft are used (ITA, 2000). Fig. 11.64 shows an example of such interdependency between Gate Arrival and Departure delays at six US airports—Atlanta Hartsfield, Dallas/Fort Worth, NY LaGuardia, Chicago O'Hare, San Francisco and Los Angeles International airport for the period 1999–2006 (http://www.apo.data.faa.gov/).

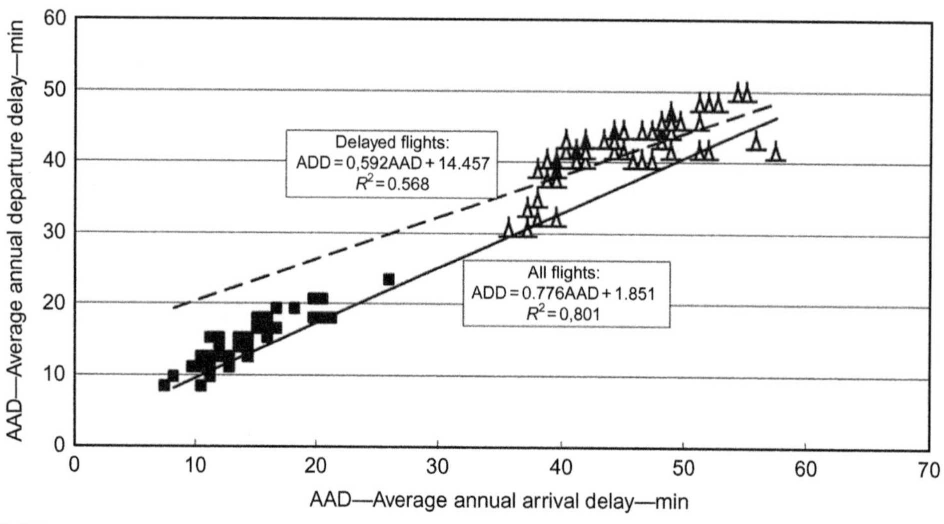

FIG. 11.64

Relationship between the departure and the arrival delays at six busiest US airports (period: 1999–2006) (http://www.apo.data.faa.gov/).

As can be seen, for all flights, the average arrival delays have ranged between 8 and 28 min causing the average departure delays to range between 8 and 23 min. Both are in the rather strong linear relationship. In addition to the unaffected departure delay of about 2 min/flight, each minute of the arrival has generated about 0.8 min of the departure delays. The average delays of delayed arrival flights have ranged between 35 and 60 min causing delays of departure flights from 30 to 50 min. In this case, the average unaffected departure delay has been about 14.5 min. Each minute of the arrival delay has generated about 0.6 min of the departure delay. Both cases indicate tendency of the affected airlines to neutralize propagation of delays from the arrival to the forthcoming departure flights, in the case of delayed flights more strongly.

In addition, the interdependency between Gate Arrival and Departure delays can be an inherent feature of the airline hub-and-spoke network operations when several outgoing flights may wait for a single (or several) delayed incoming flight feeding them, or vice versa. For example, at large European airlines the former are the long-haul intercontinental flight(s) and later the short—and medium-haul national and continental (European) flights (Janić, 2005).

11.6.1.2 Demand/capacity relationship at airports

As mentioned above, under regular conditions the aircraft/flight congestion and delays at airports happen as soon as the instant (scheduled) demand exceeds the airport available capacity. Such relationship (s) between demand and capacity has been commonly expressed by the demand/capacity or the volume/capacity ratio. As mentioned in Chapter 5, this ratio can take the values lower, equal, or greater than 1.0, that is, $\rho = D/C <, =,$ or >1.0 (D—Demand; C—Capacity), Fig. 11.65 shows some examples of changing of the demand/capacity ratio during the day at the selected US and European busy airports (Janić, 2007).

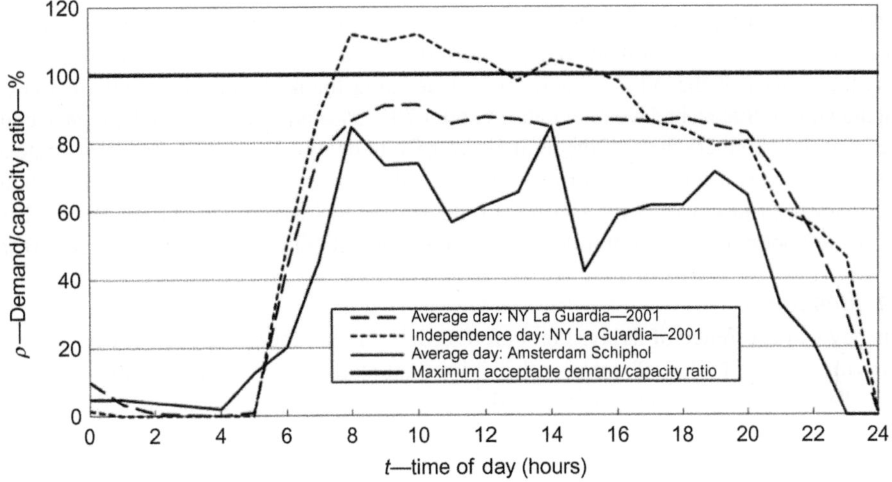

FIG. 11.65

The daily variations of demand/capacity ratio in the given examples (Janić, 2007).

As can be seen, the demand at Amsterdam Schiphol and New York LaGuardia airport was always below the capacity during an average day in 2001 ($\rho < 1$). However, at NY LaGuardia airport the scheduled demand exceeded the capacity on the Independence Day in 2001 staring from the early morning and lasting until the middle afternoon ($\rho > 1$). The first two cases suggest occurrence of the general arrival and departure delays. The last case reflects occurrence of Ground-Holding (usually long) delays.

11.6.1.3 Some other causes

In addition to imbalance between demand and capacity occurring under regular operating conditions, the other above-mentioned causes also cause aircraft/flight congestion and delays. Without compromising generosity, Table 11.16 gives some past but illustrative statistics of these causes in the European airspace during the period 2000–06 (EEC, 2006).

Table 11.16 Causes of the Aircraft/Flight Delays in European Airspace (EEC, 2006)

	Cause (%)			
Year	Airport Capacity	ATC Capacity	Weather	Other/ATC Staffing ATC Equipment
2000	6	72	11	11
2001	7	62	13	18
2003	8	46	20	18
2003	14	35	24	27
2004	15	42	25	18
2005	17	18	27	34
2006	20	19	28	33

As can be seen, during the observed period, the airport capacity and bad weather were increasing causes of aircraft/flight congestion and delays, from 6% to 20% and 11% to 28%, respectively. At the same time, the impact of the ATC capacity was decreasing thanks to developments in EUROCONTROL. In the United States, bad weather caused about 70–75% and the airport and airspace congestion (ie, traffic volume) about 20–30% of the total aircraft/flight delays. In general, in the en-route airspace bad weather initiated restrictions and/or rerouting of traffic flows. At airports, it required changing of the flight operating rules from VFR (Visual Flight Rules) to Instrumental Flight Rules (IFR), thus diminishing the "optimal" arrival and departure rates (capacities) for about 20% (Janić, 2007).

11.6.1.4 Frequency

The frequency of occurrence of delays compared to the total number of flights carried out in a given airspace during a given period of time is other important statistic related to aircraft/flight congestion and delays. Table 11.17 gives an example of this frequency at the selected busiest European and US airports (Janić, 2007).

As can be seen, in both regions, the proportions of delayed arrival and departure aircraft/flights were different at different airports. In Europe, they ranged from 17% to about 30% for arriving and 8% to 24% for departing aircraft/flights. In the United States, they ranged from about 22% to 40% for arriving and 19% to 38% for departing aircraft/flights. This implies that the delayed aircraft/flights were more frequent at the US than at the European airports (Janić, 2007).

11.6.2 NOISE

Noise by operating air transport system primarily comes from the aircraft engines while flying near the ground, that is, around airports during approach and landings, flyovers, and taking-offs.

11.6.2.1 Aircraft noise

This noise spreads in front of and behind the aircraft engine(s). The front noise-spreading generators are the engine(s) compressor and fan. The back noise-spreading generators are the turbine, fan, and jet-afflux. The aircraft noise is considered as the impact of air transport on the society due to disturbing

Table 11.17 Frequency of Occurrence of Delays at the Selected European and US Airports (Janić, 2007)

European Airports (2001)	Delayed Flights		US Airports (1999)	Delayed Flights	
	Arrivals (%)	Departures (%)		Arrivals (%)	Departures (%)
Paris CDG	24.6	21.8	Chicago-O'Hare	33.6	29.9
London Heathrow	17.4	21.0	Newark	38.4	31.0
Frankfurt	30.8	18.9	Atlanta	30.9	26.8
Amsterdam	25.7	23.2	NY-La Guardia	40.1	28.9
Madrid/Barajas	19.6	20.0	San Francisco	32.1	21.5
Munich	19.0	19.0	Dallas-Ft. Worth	21.7	23.7
Brussels	29.8	27.7	Boston Logan	37.7	29.3
Zurich	23.2	23.8	Philadelphia	40.4	37.9
Rome/Fiumicino	-	12.5	NY-Kennedy	28.0	19.0
Copenhagen/K	17.8	10.3	Phoenix	29.6	30.8
Stockholm/Arlanda	-	8.0	Detroit	24.6	26.3
London/Gatwick	19.6	24.3	Los Angeles	26.1	20.8

population living near the airports. As mentioned above, exposure to the persistence excessive noise causes damages of the air system and other health problem, and as such also requires the cost of protective measures on the ground such as, for example, extra insulation of houses. Therefore, under conditions of continuously growing air traffic and associated impacts, this noise has also continuously been the subject of regulation. For the first time, it had been regulated in the year 1959 by setting the acceptable noise limit of 112 PNdB to the sound generated by aircraft operating at particular airports. Latter, the International Civil Aviation Organization (ICAO) established the international certification standards for commercial jet aircraft in 1971. In the late 1970s the new noise restrictive standards were included in Chapters 2 and 3 of the ICAO Annex 16, Vol. 1 (Environmental Protection) (Walder, 1993). These standards have been applied to all jet aircraft that have entered service since Oct. 1977. More recently, they have been reconfirmed in Chapter 4. According to this chapter, the maximum noise at these locations must not exceed 108 EPNLdB (EPNLdB—Effective Perceived Noise Level in decibels). This noise is equivalent to about 96 dBA (dBA—A noise weighted scale) (Janić, 2007).

In general, the noise level from aircraft operating around and at airports depends on their size, that is, Take-Off Weight (TOW). Fig. 11.66 shows an example of the relationship between the aircraft Specific Noise (SN) (ie, EPNLdB/ton of TOW) and TOW (FAA, 1997; EASA, 2011).

As can be seen, the specific noise level decreases more than proportionally with increasing of the aircraft TOW, thus indicating in some sense "economies of scale" of larger aircraft in terms of noise. In addition, Table 11.18 gives the noise characteristics of selected commercial passenger aircraft at particular above-mentioned noise-certification locations (EASA, 2011).

As can be seen, the most recent B787-8 aircraft generates about 5–7, 7, and 0.6–6 dB lower certificated noise than its counterparts while taking-off, flying over, and approaching, respectively. In addition, making a broader judgment concerning mitigation of noise by introducing B787-8 can be

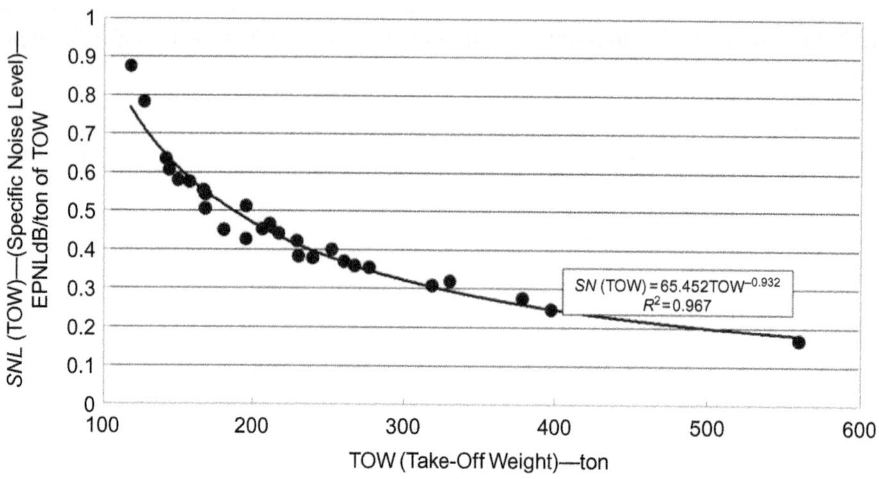

FIG. 11.66

Relationship between the SNL (Specific Noise Level) and TOW (Take-Off-Weight)—Heavy aircraft (FAA, 1997; EASA, 2011).

Table 11.18 Noise Characteristics of the Selected Commercial Passenger Aircraft (EASA, 2011)

Aircraft Type	Noise Level (EPNLdB)[c]			
	TOW[a]	Lateral	Flyover	Approach
B767-200	144	95.7	91.5	102.1
B767-200ER	168	97.8	91.1	98.6
B767-300	158	96.0	91.3	98.5
B767-30ER	180	95.7	91.5	99.7
A330-200	230	97.0	94.4	96.8
A330-300	217	97.6	91.6	98.9
B787-8	220	90.5	83.0	96.2
A350-800[b]	259	89.0	83.0	95.0

[a]*Typical TOW (Take-Off-Weight).*
[b]*Preliminary data.*
[c]*EPNLdB—Effective Perceived Noise.*
Note: *Level in decibels (typical engines).*

made by assuming that it replaces B767-200/200ER aircraft. This implies the gradual increase in the number of replacing aircraft (B787-8) on the account of gradually replaced aircraft (B767-200/200ER). The total number of B767-200/200ER aircraft to be replaced is assumed to be 800 (based on the current orders of B787-8). Regarding operating long-haul flights, each aircraft of both fleets is assumed to perform the same number of flights (2/day). The example is shown in Fig. 11.67 (Janić, 2014a,b).

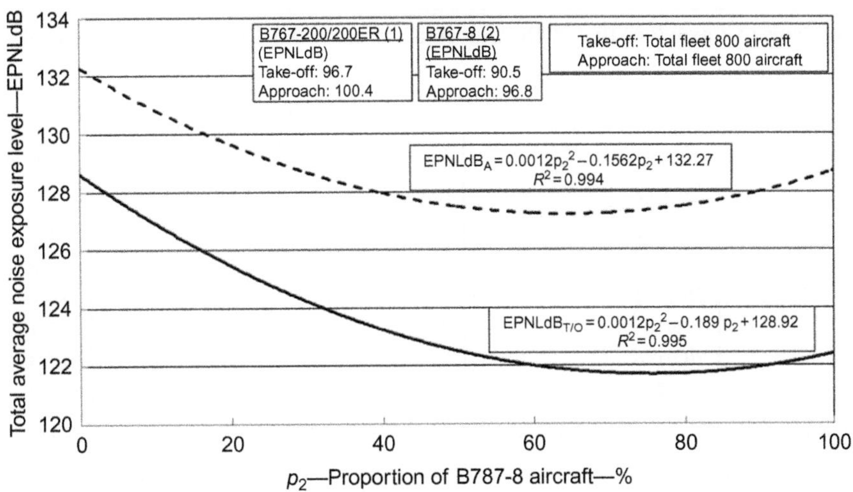

FIG. 11.67

Relationship between the average noise exposure and the proportion of B787-8 aircraft in the airline fleet (Janić, 2014a,b).

As can be seen, the total potential noise exposure by the entire replacing and replaced fleet decreases more than proportionally with increasing of the proportion of advanced B787-8 aircraft. By full replacement, this exposure could be reduced by about 3 and 6 dB during approach and take-off, respectively. Consequently, the "noise contour" or "noise footprint" as the area of constant noise generated by B787-8 aircraft around an airport can be by about 60% smaller than those of its counterparts, thus ensuring that noise level above 85 dB certainly does not spread outside the airport boundaries. This is achieved mainly thanks to the improved aerodynamics design on the one hand, and the lower fan speed and low jet velocity of the RR Trent 1000 engines of B787-8, on the other. The forthcoming A350-800 aircraft is expected to be even quieter (Janić, 2007, 2014a,b). In addition, Table 11.19 gives the noise level at noise-certified locations for the selected commercial freight/cargo aircraft (EASA, 2011).

As can be seen, depending on the noise certification location both B747-8F and A380-800F aircraft are much quieter than the current long-haul freight/cargo aircraft including their closest counterpart—B747-400F, by about 2.9–4.6 dB. Fig. 11.68 shows that the larger aircraft are also superior regarding their SN during arrivals and departures.

Consequently, replacing the current aircraft with either B747-8F and/or A380-800F will significantly contribute to mitigating the aircraft noise around airports.

11.6.2.2 Airport noise

The above-mentioned aircraft noise materializes around airports during aircraft approach, landing, fly-over, and taking-off. This noise disturbs too close population by spreading over relatively wide area of land around many airports. This area of land called "noise footprint" is influenced, in addition to the aircraft noise at source, by the intensity, that is, volume of traffic carried out during the specified period

Table 11.19 Noise Characteristics of the Selected Commercial Freight/Cargo Aircraft (EASA, 2011)

		Noise Level (EPNLdB)		
Aircraft Type	**MTOW/MLW**	**Lateral**	**Flyover**	**Approach**
B787-4F	448/346	94.0	94.0	100.9
A380-800 F	590/427	94.2	95.6	98.0
B747-400F	386/296	98.3	98.6	103.8
MD-11F	286/223	96.1	95.8	104.4
A330-200F	233/187	97.4	90.7	97.1
MD10-30F	263/198	97.9	97.4	106.3
B777F	287/221	98.7	87.0	99.7

Note: *MTOW (Maximum Take-Off-Weight); MLW—Maximum Landing Weight; EPNLdB—Effective Perceived Noise Level in decibels (typical engines).*

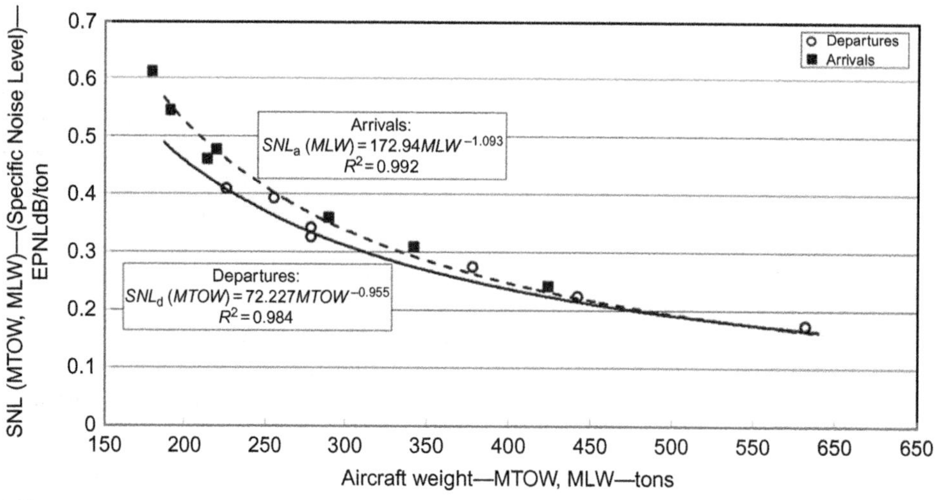

FIG. 11.68

Relationship between the SN (Specific Noise) and weight of the commercial freight aircraft (MTOW—Maximum Take-Off Weight; MLW—Maximum Landing Weight) (EASA, 2011; FAA, 1997).

of time. Fig. 11.69 shows an example of the noise footprints with different continuous noise levels in dependence on the volume of annual traffic at Frankfurt airport (Germany) (Janić, 2007).

As can be seen, the footprint for each continuous noise level has decreased more than proportionally with increasing of the annual volume of airport traffic. In addition, the areas with lower level of continuous noise were much larger than that with the higher continuous noise (L_{eq}). This confirms the above-mentioned statements that the aircraft have become less noisy or "more silent." In addition,

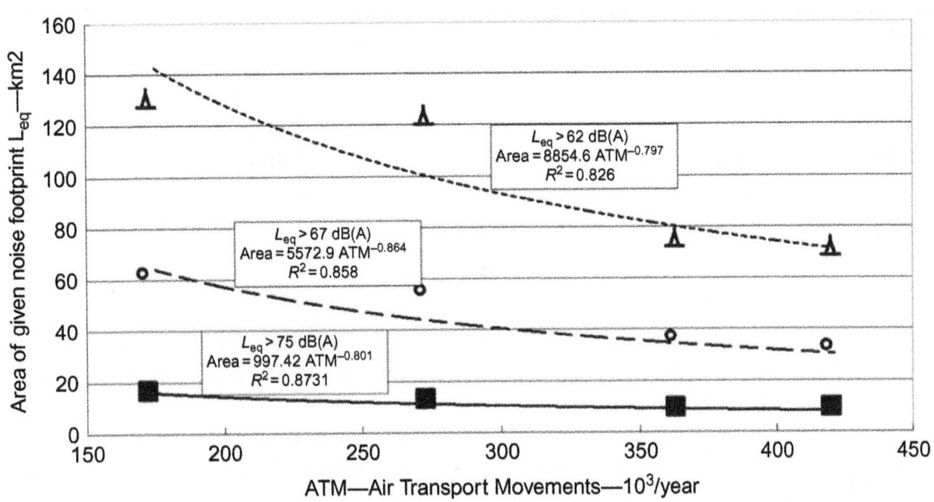

FIG. 11.69

Relationship between the area of noise footprint and the volume of traffic at Frankfurt Airport (Germany) (period: 1987–99) (Janić, 2007).

the innovative operational procedure particularly the continuous descent approach and landing have contributed to reducing the areas exposed to the excessive noise. In general, the noise exposure can be expressed by the noise efficiency, which relates to the noise energy in terms of the equivalent long-term noise level L_{eq} (in decibels—dB(A)) generated by air transport movements (atm) over the populated area of land close to an airport. Table 11.20 gives an example of reducing noise, the exposed area, related population, and the number of exposed households around London Heathrow airport (UK) (CAA, 2012; Janić, 2007).

Table 11.20 An Example of Estimating the Area, Population, and the Number of Households Within the Noise Contours L_{den} at London Heathrow Airport—LHR (UK) (CAA, 2012; Janić, 2007)

Contour Level (dB(A))	Area (km^2)	Population (10^3)	Household (10^3)
>55	302.3	782.9	344.9
>60	114.3	260.5	109.8
>65	47.7	74.5	29.9
>70	20.8	16.6	6.5
>75	7.5	1.7	0.7

As can be seen, the area, population, and the number of households within particular footprints have decreased with increasing of the level of continuous noise. This practically means that fewer and fewer individuals and their households closer to the airport have been exposed to higher level of noise, in this case 70–75 dBA.

11.6.3 TRAFFIC ACCIDENTS AND INCIDENTS (SAFETY)

11.6.3.1 Risk

Civil aviation has emerged as the example where the real risk, statistical risk, predicted risk, and perceived risk have been inherently present.[5] They are related to the occurrence of aircraft accidents/incidents. These are considered as the impacts on society due to resulting in fatalities and injuries of users-air passengers and aircraft crew(s) on board, eventual causalities on the ground and damages to properties to the third parties, all imposing the corresponding costs (externalities).

The aircraft accidents/incidents possess some specific characteristics in comparison to accidents happened at other transport modes, which can be summarized as follows (Janić, 2000):

- since the aviation operations (ie, flying) may take place over large area(s), the accidents can occur at any point of time and/or space, thus making individuals over large space n exposed to both individual and global hazard;
- the air passengers and aircraft crews, who directly take part in transport operations, have primarily been the target groups exposed to the risk of accidents/incidents. In addition, the individuals on the ground have been exposed to these accidents but to a much lesser extent;
- although being rare events with very low probability of occurrence compared to the volumes of operations during the specified period of time, any aircraft accident/incident has always caused high consequences in terms of fatalities and/or severe injuries of on board passengers and crew and affected population on the ground, including the loss and damages of the corresponding properties;
- if any aircraft movement has been considered as an inherently risky event, then, according to probability theory, the aircraft accidents/incidents could be classified into class of highly unlikely (although possible) events; and
- respecting the time dependency, the risk of aircraft accident/incidents has permanently been existed over a given time and space horizon, that is, whenever and wherever the flying has taken place.

11.6.3.2 Causes

The evidence so far has indicated that most of the air transport accidents/incidents have been happened due to the occurrence of a complex system of the mutually dependent sequential causes. These causes can be classified according to different criteria. First, according to the current state-of-knowledge they can be the group of "known and avoidable" and the group of "unknown and unavoidable" causes. The term "unknown and unavoidable" causes should be considered only conditionally, since just after the accident(s) happened the real causes have not been known. As the investigation has been progressed and finished, the causes have been uncovered, and thus become "known and avoidable." Of course, the causes of some accidents have never been uncovered. Second, with respect to the type, the main causes of air accidents can be conditionally classified into five groups. These are: the "human errors," the mechanical failures, the hazardous weather, and the sabotages and military operations (Janić, 2000).

The "human errors" have shown to be able to be managed and reduced by proper training the aviation staff (in the aircraft and on the ground) and proper organizing the traffic pattern(s) to avoid

[5]Real risk is the risk to an individual, which may be determined based on future circumstances when they fully developed; Statistical risk may be determined by available data on the accidents/incidents in question; predicted risk may be predicted analytically from the system models structured from relevant research; perceived risk, may intuitively be seen by individuals (Sage and White, 1980).

stressful and strain situations. The other factors like hazardous weather, "hidden" mechanical errors causing the failures of vital airborne and ground equipment, sabotages and military operations have been often uncertain, and thus much less unavoidable and uncontrollable. Since these factors have been inherently and permanently present at the aviation activities, apparently we can never be certain that accidents will not occur. Does it mean that the system will never be safe? The answer is negative since the accidents/incidents do not necessarily mean that the system is unsafe. In order to properly judge with this matter, the safety should be considered with respect to basic causes of accidents. Namely, if accidents/incidents occur due to the already *known* and *avoidable* reasons, the system should be considered *unsafe*. Otherwise, if accidents/incidents occur due to the *unknown* and *unavoidable* reasons, the system should be considered *safe* (Kanafani, 1984).

11.6.3.3 Assessment

In order to assess the risk and safety trends by using the available statistics, few measures can million p-km (ie, the "fatality rate"). The other one has been the number of air accidents per $100 * 10^6$ ac-km flown during the specified period of time (given year) (ie, the "accident rate") (ac—aircraft). In both cases, the "fatal accident" has been defined as the event where one or more people have died during the affected flight. In addition, the number of accidents occurred per given number of flights (departures) carried out either by an airline or an aircraft type have shown to be a sufficiently "good" measure of the risk and safety (again this measure can be titled as the "accident rate").

At the global scale, the "fatality rate" (dependent variable F_R—$100 * 10^6$/p-km) can be estimated in dependence of the total number of fatalities per aircraft accident (crash) (dependent variable N_D—number/event) and annual volume of passenger kilometers (dependent variable PKM—$100 * 10^6$ p-km). An example of such estimation carried out by regression analysis been as follows (Janić, 2000):

$$F_R = \underset{(2.983)}{3.801 \cdot 10^{-10}} + \underset{(10.674)}{4.196 \cdot 10^{-11}} \cdot N_D - \underset{(3.446)}{2.095 \cdot 10^{-16}} \cdot PKM \\ R^2 = 0.901;\ F = 69.296;\ N = 16 \tag{11.50}$$

The regression equation has been significant at both 1% and 5% level (F-value). Particular coefficients of the independent variables have also been significant (t-statistics given in parenthesis below them). They also possess a relatively high explanatory power (R^2) (Janić, 2000). As it can be seen, the "fatality rate" has increased with increase in the number of fatalities per single aircraft crash and decreased with increase in the volume of air transport output. This has explained two facts: First, an accident (crash) of larger aircraft with the larger number of passengers on board increases risk of greater number of fatalities. Second, introducing faster and larger aircraft over time has increased the air transport output and at the same time made flying more reliable, thus significantly reducing the risk of accidents (crashes). In addition, more recent figures of the relationship between the "fatality rate" and the volume of output of the US and EU airline industry are shown in Fig. 11.70.

As can be seen, the fatality rate has been lower at the European airline industry, which also carried out the lower annual volumes of p-miles. At both industries, this actually the very low rate has decreased with increasing of the annual volumes of p-miles, thus indicating the lower individual's perceived risk of losing life if undertaking flying.

The "accident rate" has been estimated by regressing the number of accidents per million of flights (dependent variable A_R) and the total (cumulative) number of flights carried out by an airline during given period of time (independent variable F) (Janić, 2000). Fig. 11.71 shows the observed trend.

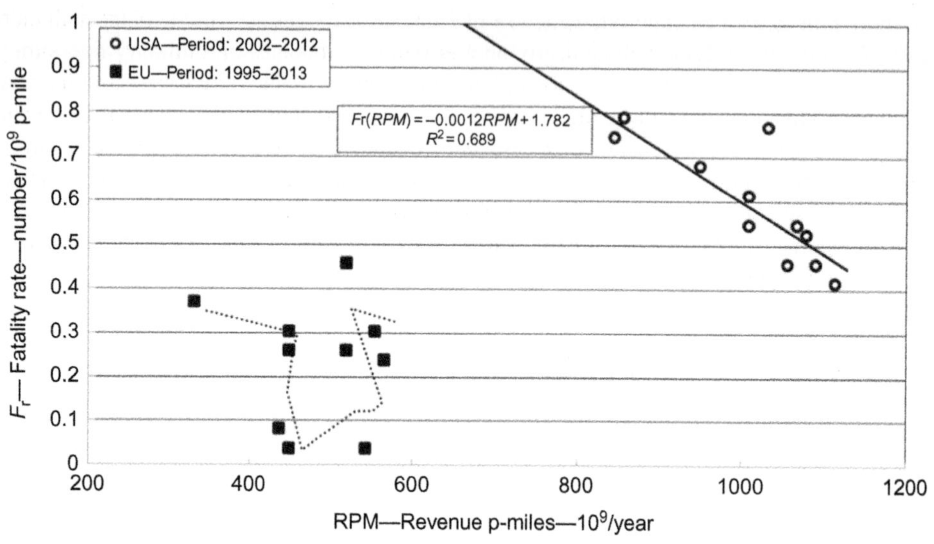

FIG. 11.70

Relationship between the average annual fatality rate and the volume of output of US and EU commercial airline industry (period: 2002–12; 1995–2013) (EU, 2015; USDT, 2013a,b,c).

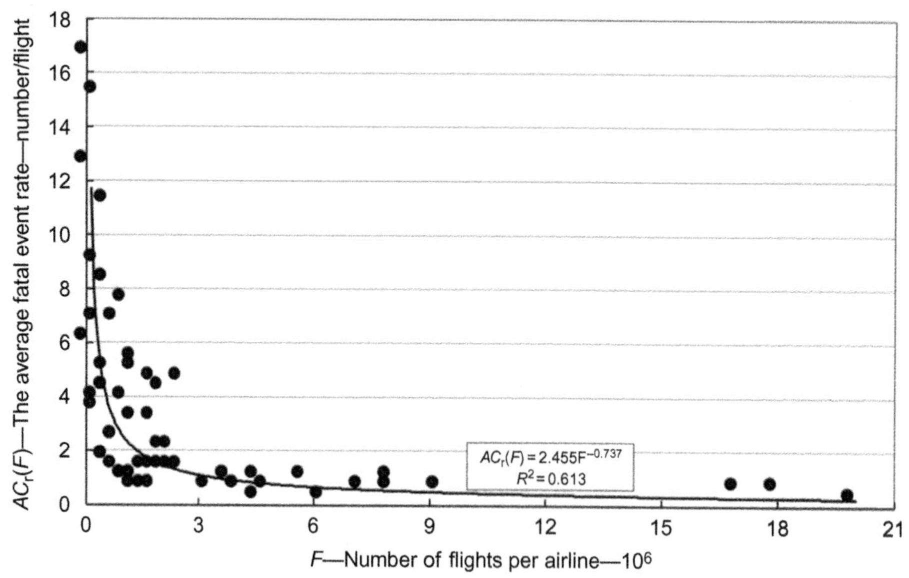

FIG. 11.71

Relationship between the "accident rate" and the volume of output—the number of airline flights (period: 1970–97) (Janić, 2000).

As it can be seen, the "accident rate" per airline decreased more than proportionally with increase in its cumulative number of flights. Since large airlines performed a larger number of flights they could be, at least according to this figure, considered as lower risky (ie, more safe) than the smaller ones. In addition, during the observed period, the US and European airlines experienced much lower "accident rate" than the airlines from other world regions under given conditions. In addition, the "accident rate" can be estimated for particular aircraft types by regressing the number of accidents per aircraft type (dependent variable A_R), the number of flights per aircraft type carried out during specified period of time (variable FL—10^6 flights from entering the service to the year 1992) and the average age of particular aircraft types (the variable AGE—years) (Fokker F28, Fokker F70/F100; Airbus A300, A310, A320; Lockheed L1011; British Aerospace BAe146; Boeing B727, B737-1/200, B737-3/4/500, B747, B757, B767; McDonnell Douglas DC9, MD80). The regression equation has taken the following form (Janić, 2000):

$$A_R = \underset{(0.692)}{1.206} + \underset{(6.355)}{1.743} \cdot FL + \underset{(3.887)}{0.900} \cdot AGE \quad (11.51)$$
$$R^2 = 0.929;\; F = 84.640;\; N = 16$$

As can be seen, the independent variables possess a high explanatory power (R^2). The regression equation and its particular coefficients are significant at both 1% and 5% level as indicated by the corresponding (F) and (t) statistics, respectively.

This regression equation may explain what happened in the past, that is, that the most air accidents happened at those aircraft, which were "older" and which, at the same time, carried out a greater cumulative number of flights. Of course, both the airlines and passengers had not known a priory these facts, since they had not been able to observe them, as it had been possible after this a posterior analysis. In some sense, this regression indicates that the cause of air accidents might have generally been existence of the so-called "geriatric problem" that escalated much faster at more utilized and older aircraft. In the absence of careful and preventive maintenance actions, uncovered and unpredictable metal fatigues caused the failures of vital aircraft systems, which led to accidents/incidents. Since the hypothesis has been that the aircraft accidents have always happened as the random events, this equation does not imply that more utilized and "older" aircraft have been "less safe." It only indicates the fact that the perceived "risk" of travelling by these aircraft has been higher.

11.6.4 ENERGY/FUEL CONSUMPTION AND EMISSIONS OF GHG

Three main components of air transport systems—airline aircraft, airport and ATC—consume energy/fuel for their operations and functioning. The direct energy/fuel consumption of airline aircraft/flights has appeared to the most substantive and therefore is under focus in the given context.

11.6.4.1 Aircraft energy/fuel consumption

The burning of aviation fuel as the derivative of crude oil, in addition to other man-made emissions, contributes to the increase in concentration of GHG in the atmosphere and consequently to the climate change called "global warming." In addition, it depletes the reserves of fossil fuels (crude oil) as non-renewable energy resources. These and many other airline economic reasons have driven the contemporary air transport system worldwide, and particularly the aircraft manufacturers, to reduce fuel consumption. At the beginning, the main reason was the economic efficiency of airline operations, which in turn has enabled lower prices and consequently cheaper air travel. The reduction of travel prices has been an important driving force in the significant and constant growth of air transport demand in terms of the volumes of passengers, passenger and freight kilometers, and aircraft operations. More recently, with the increasing awareness of the depletion of the reserves of crude oil and the consequently potentially limited availability of jet fuel as its derivative at given prices as well as on the harmful impacts of the products of burning this fuel on people's health, natural habitats, and the earth's atmosphere, aerospace manufacturers have made a lot efforts to improve jet engine fuel efficiency, and consequently reduce the emissions of GHG. The design of such engines has required solving a range of complex problems, of which the most complex have included balancing the engines' propulsion and thermal efficiency. Better propulsive efficiency has provided a greater propulsive power from the combustion process while the improved thermal efficiency has generated a higher overall engine pressure ratio and turbine temperature using the same amount of fuel (energy). Other problems have related to proper balancing between the engine weight, drag, noise, and emissions of GHG.

In order to obtain a higher propulsive efficiency it has been necessary to reduce the waste energy in the engine exhaust stream, which has decreased the jet velocity. Since the engine thrust is the product of exhaustive mass flow and its velocity, if this velocity was reduced the mass of flow would be increased to retain the desired level of thrust. This has implied an increase in the bypass ratio (BR) defined as the rate between the amount of air flowing round the engine core and the amount of air passing through the engine itself. The engines with the higher BR usually have lower SFC, defined as the ratio of the fuel burned per hour per ton of the net thrust (Janić, 1999, 2007). The SFC of the most contemporary jet aircraft engines amounts to about 0.25–0.30 kg of fuel/kg of thrust/h, which has been diminishing until nowadays to less than 0.184 kg of fuel per/kg of thrust/h (h—hour). The SFC relates to the jet engine BR whose generic nature has been illustrated by using data for 20 engine types produced by the different airspace manufacturers. The regression relationships in which BR is considered as the independent and the SFC as the dependent variable is shown in Fig. 11.72.

As can be seen, *SFC* has, independently of the phase of the flight (take-off or cruising), decreased more than proportionally with the increasing engine BR, which might be useful information for estimating the trend of development of jet engines for commercial aircraft (Janić, 2007).

Improvements in the aircraft aerodynamic performances have also played an important role in the improvement of their fuel efficiency. The case of the development of the most recent Boeing B787 and A350 aircraft are illustrative. Fig. 11.73 shows that the most recent B787-8 aircraft has been more fuel efficient by about 8%, 9%, and 18% than its B767-200ER, B777-200ER, and 777-300ER counterparts, respectively. The newest A350-800 is expected to be even more fuel efficient.

The most recent figures obtained by ANA (Air Nippon Airlines, Japan) show that the fuel savings by operating B787-8 powered by RR Trent 1000 engines on short-haul routes are 15–20% and up to 21% on long-haul (international) routes as compared to the B767-200/300ER aircraft. Some additional savings of up to about 3% have been reported by JAL (Japan Airlines) using the B787-8 aircraft powered by GEnx 1B engines (Janić, 2014a,b). In addition, Airbus expects the fuel consumption of the A350-800 to be by about 6% lower than that of B787-8 aircraft.

11.6.4.2 Airline energy/fuel consumption

Parallel to improving the engine/aircraft fuel efficiency, two groups of air traffic and transport operational measures have been developed and implemented in the commercial airline industry. The first group has aimed at cutting the overall fuel consumption by improvements of the ATC system, that is, by

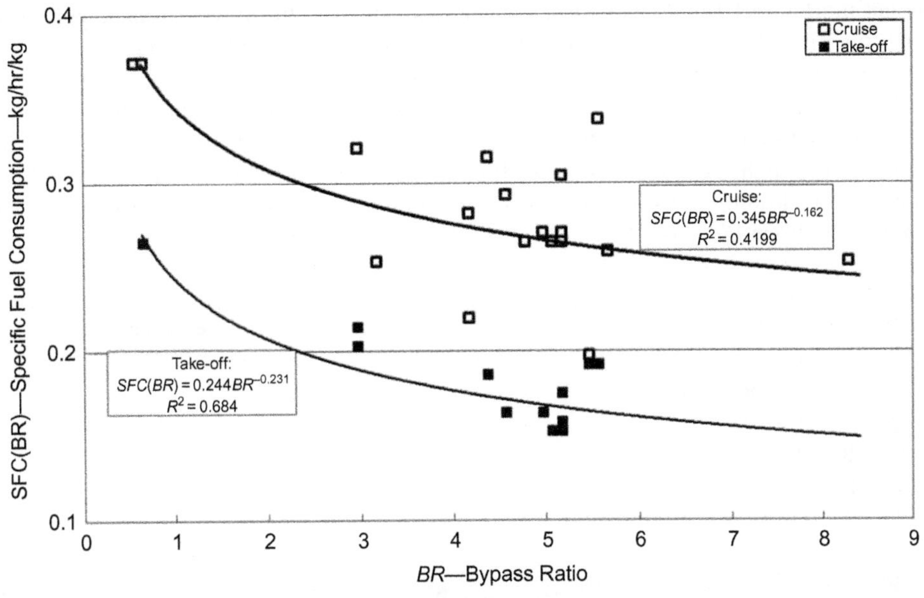

FIG. 11.72

Specific Fuel Consumption (SCF) vs Bypass Ratio (BR) for jet engines (Janić, 2007).

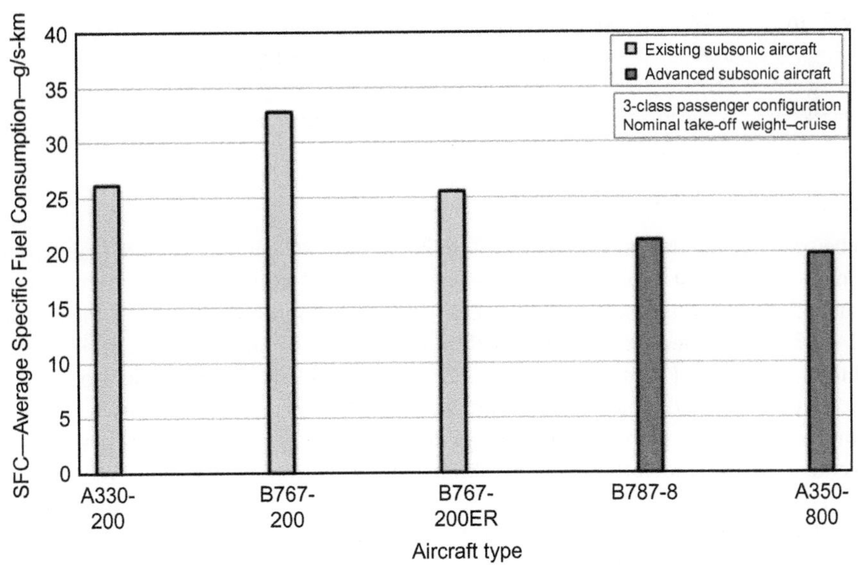

FIG. 11.73

Fuel efficiency of the selected commercial aircraft (Janić, 2014a,b).

carrying out more direct flights and consequently reducing the total travel distance per passenger, optimizing the aircraft climb/descent profiles in terms of the fuel consumption, reducing cruising speed(s), harmonizing fuel prices, improving load factor, and reducing the long taxiing and towering of aircraft at airports. Another group of measures has included optimizing the cruising altitudes and removing restrictions on the flight routings, reducing the number of flights during daylight, widening (or narrowing) the flight corridors and repositioning the flight corridors in order to avoid extreme weather conditions leading to excessive fuel consumption. Fig. 11.74 shows an example of achievements of the US airline industry.

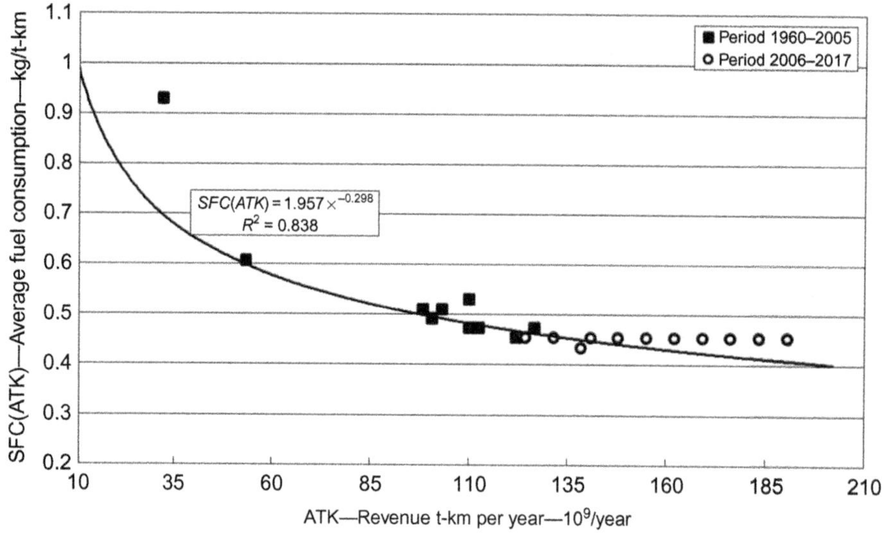

FIG. 11.74

Relationship between fuel efficiency (SFC—Specific Fuel Consumption) and the volume of output of the US airline industry (period: 1960–2017) (Janić, 2007).

As can be seen, during the period from 1960 to 2005, the SFC decreased more than two times with increasing of the volumes of industry's output—from about 0.92 to about 0.43 kg/t-km. Some forecasts for the decade (2006–17) have expected the average fuel consumption to remain at the present level of about 0.42–0.43 kg/t-km despite further growth in the volume of output (Janić, 2007).

In addition, the general trend in terms of the aircraft fuel consumption and energy efficiency during the period 1990–2015 has been illustrated using the regression analysis where the average *SFC* (g/t-km) is considered as the dependent variable whose values have been estimated by assuming an improvement in the fuel efficiency of about 2.5%/year during the observed period (t-km—ton-kilometer). Two independent variables have been time (*T*) (years of the period 1990–2015) and the annual volume of output (*TKM*) (10^6 ton-kilometers). This variable has been estimated by taking into account the prospective renewal of the aircraft fleet and relatively stable growth of the air passenger and freight/cargo transport demand at an average annual rate of 5% and 6.5%, respectively (in this case: 1 t-km = 10 p-km). The average load factor has assumed to be 0.55. The resultant equation has been as follows (Janić, 2007):

$$SFC = \underset{(143.67)}{30,690.22} \cdot TKM^{\underset{(-7.234)}{-0.355}} \cdot T^{\underset{(-62.390)}{-0.017}}$$
$$R^2 = 0.993;\ F = 10,023.563;\ N = 25 \quad (11.52)$$

The values of t-statistics in parenthesis below the coefficients and F-statistic indicate that the independent variables and the equation are significant at the level of 1% and 5%, respectively. R_2 statistics confirm the strong explanatory power of the selected independent variables. Based on this relationship, it can be said that the average fuel consumption has decreased by about 18% during the period 1990–2000, and will continue to decrease further by about 65% over the period 1990–2015 (Janić, 1999, 2007). This estimate coincides with the estimates of the Intergovernmental Panel of Climate Change (IPCC), which have stated 70% reduction in the average SFC during the period 1960–2000. Nevertheless, there have been some doubts about the further annual improvements of fuel efficiency of about 2.5–3% until the year 2020 (Peeters et al., 2005). It seems more reasonably that it will be about 1.5% until the year 2020 (IATA—International Air Transport Association) or 2% until the year 2050 (ICAO) (Chandra et al, 2014). The reciprocal of the average unit fuel consumption (g/p-km, g/t-km, or g/ASK) is called Energy Intensity (EI) (ASK—Available Seat Kilometers). Fig. 11.75 shows an example of such developments at the US airline industry (USDT, 2012).

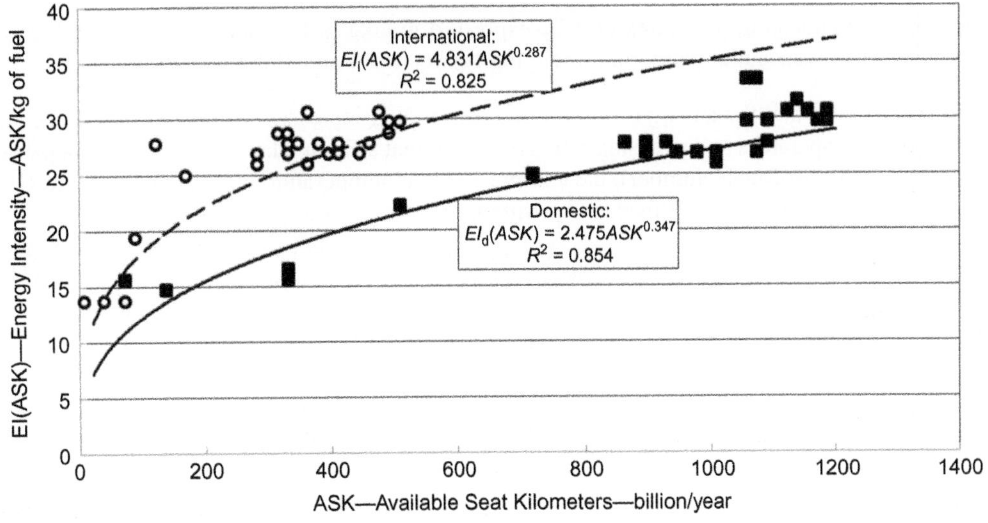

FIG. 11.75

Relationships between Energy Intensity (EI) and the available capacity of the US certificated airlines (period: 1960–2010) (USDT, 2012).

As can be seen, the EI has increased with increasing of the volumes of available industry's annual output at decreasing rate at both domestic and international operations. As have been expected, due to the nature of operations (Short—and medium-haul flights carried out by the corresponding aircraft types), the EI of domestic operations has been lower than that of the international operations (long-haul flights carried out by the corresponding flights). Such developments also reflect decreasing of AFC more than proportionally with increasing of the industry's available output thus indicating continuous improvements during the observed period of time.

11.6.4.3 Aircraft and airline emissions of GHG

The main GHG from the burning jet-A aviation fuel (kerosene) used by the commercial aircraft worldwide are: CO_2 (Carbon Dioxide), H_2O (water vapor), NO (Nitric Oxide) and NO_2 (Nitrogen Dioxide), which together form NO_x (Nitrogen Oxides), SO_x (Sulfur Oxides) and smoke. The emission rates of CO_2, H_2O, SO_2 are relatively constant—3.18 kg/kg of fuel, 1.23 kg/kg of fuel, and up to 0.84 g/kg of fuel, respectively (Janić, 2007). According to the currently available evidence, the emission rate of NO_x changes, that is, increases with increasing of the jet engine pressure ratio, which in turn increases the jet engine thermal efficiency. The engine pressure ratio is defined as the ratio of the total pressure at the compressor discharge and at the compressor entry. The other jet engine performances such as thrust, fuel consumption, and efficiency also depend on this ratio. For the contemporary turbofan engines, this ratio amounts to about 10–50. This originates from the typical design of combustion chambers of these engines (Huenecke, 1997). Experiments to investigate the relationship between the engine emission index of NO_x, compressor outlet temperature, and pressure ratio have resulted in the regression equation as follows (RAS, 2003):

$$EI_{NO_x} = 0.17282 \cdot e^{0.00676593 \cdot T_s} \tag{11.53}$$

where

EI_{NO_x} is the engine emission index of NO_x expressed in g/kg of fuel; and
T_s is the compressor outlet temperature ranging between 280 and 1080 K (Kelvin degree).

Eq. (11.53) indicates that the jet engine emission index of NO_x increases with increasing of the compressor's outlet temperature at decreasing rate. Assuming that a flight takes place in the tropopause at the speed of M0.85 (M-Mach Number), the compressor inlet temperature will be about 250 K. Given the compressor efficiency of 0.9, this can be written:

$$T_s = 250r^{(2/(7*0.9))} = 250r^{(1/3.15)} \tag{11.54}$$

where r is the engine overall pressure ratio ranging from 10 to 50.

Combining Eqs. (11.53), (11.54), the emission index of NO_x becomes:

$$EI_{NO_x} = 0.17282e^{1.69158 \cdot r(1/3.15)} \tag{11.55}$$

Eq. (11.55) confirms that the emission index of NO_x increases with increasing of the engine pressure ratio at an increasing rate, that is, more than proportionally. Therefore, a trend towards increasing the engine pressure ratio of larger aircraft might compromise and even diminish other effects obtained by reduction of the fuel consumptions and related emissions of other GHG—CO_2 and H_2O.

Table 11.21 gives an example of the average unit fuel consumption and related emissions of GHG for the largest commercial freight/cargo aircraft (Janić, 2014a,b).

As can be seen, the *SFC* and related emission rate of GHG of the aircraft B747-8F are lower than those of the aircraft B747-400F by about 15–22%. The SFC of the aircraft A380-800F derived from its passenger version appears to be for about 5–6% higher than the lowest one of aircraft B747-8F. Nevertheless, despite such differences and the inherent uncertainty of the figures for the A380-800F aircraft, the SFC n and related emission rates f GHG (CO_{2e}) of both aircraft appear to be quite comparable.

Table 11.21 The Specific Fuel Consumption (SFC) and Related Emission Rates of GHG by Large Commercial Freight/Cargo Aircraft (Janić, 2014a,b)

	Aircraft Type/Engine		
Indicator/Measure	**B747-400F/PW, GE, RR**	**B747-8 F/GP7200**	**A380-800 F/RR Trent 900**
SFC (g of fuel/ATK)	117	91–101	88–103/113
Emission rate of GHG (gCO_{2e}/ATK)	510	399–438	383–449/492

Note: *CO_{2e}—Carbon-Dioxide equivalents; PW—Pratt and Whitney; GE—General Electric; GP—Engine Alliance; RR—Rolls Royce; ATK—Available Ton Kilometer.*

For comparison, the fuel consumption of the largest An-225 aircraft amounts about 23.5 ton/h while flying at the average cruising speed of 800 km/h with a payload of 200 ton. These give an average *SFC* of about 147 g/ATK and emission rate of GHG of about 639 gCO_{2e}/ATK (http://www.antonov.com/aircraft/transport-aircraft/an-225-mriya).

11.6.4.4 Airport energy/fuel consumption and emissions of GHG

At the local scale, the aircraft consume fuel and emit GHG during LTO[6] cycles carried out at airports. The GHG emitted during the LTO cycles have shown to be the greatest relative contributor to the total emissions of GHG by a given airport (about 60%), followed by the ground aircraft servicing at the apron/gate complex (about 20%), the airport ground access systems/modes (15%), and electricity consumption in the airport buildings (about 5%). The intensity of fuel consumption in the airport airside area by the aircraft LTO cycles can be estimated as follows (Janić, 2007, 2016a,b):

$$FC(A,\tau) = \left[\sum_{i=1}^{N(\tau)} FC_{i/\mathrm{LTO}}(\tau)\right] / A(\tau) \tag{11.56}$$

where

$N(\tau)$ is the number of LTO cycles carried out during time (τ);
$FC_{i/\mathrm{LTO}}(\tau)$ is the fuel consumption per LTO cycle carried out by the aircraft of category (i) during the time (τ) (tons/LTO cycle/year); and
$A(T)$ is the area of land occupied by a given airport during the time (τ) (ha or km^2).

The fuel consumption $FC_{i/\mathrm{LTO}}(\tau)$ in Eq. (11.56) (ton of fuel/ha) (ha—hectare) depends of the aircraft size and the actual length of the LTO cycle. Experience so far has indicated that the energy/fuel consumed by other operations and activities in the airport airside and landside area can be easily allocated

[6]Despite the actual differences at particular airports, a LTO cycle has been standardized by ICAO in terms of four time-based components as follows: 0.7 min for take-off, 2.2 min for climbing, 4.0 min for approach and landing, and 26.0 min for taxiing/idle phase of operation(s). This implies that one LTO cycle contains two atms (air transport movements) (1 atm = 1 landing or 1 taking-off) (ICAO, 2002).

to that of LTO cycles in order to get the total energy/fuel consumption under the given conditions. In addition, the intensity of related emissions of GHG by LTO cycles at a given airport can be estimated as follows (Janić, 2016a,b):

$$EM_{GHG}(A,\tau) = \left[\sum_{i=1}^{N(T)} FC_{i/LTO}(\tau) * e_{i/LTO}(CO_{2e})\right] / A(\tau) \qquad (11.57)$$

where $e_{i/LTO}(CO_{2e})$ is the emission rate of CO_{2e} (carbon dioxide equivalents) per LTO cycle carried out by the aircraft category (i) (kgCO_{2e}/kg of fuel).

The other symbols are analogous to those in Eq. (11.56). The intensity of emissions of GHG in Eq. (11.57), $EM_{GHG}(A, \tau)$ is expressed in the quantities of CO_{2e}/ha. The emission rate $e_{i/LTO}(CO_{2e})$ is based on the above-mentioned emission rates and relative proportions of particular in the total above-mentioned emitted GHG. Expressed in tons of, and thus expressing GHG concentration over the area of a given airport, this measure is preferred to be as low as possible. Fig. 11.76 shows an example of the intensity of energy/fuel consumption and related emissions of GHG at London Heathrow airport (UK) (Janić, 2016a,b).

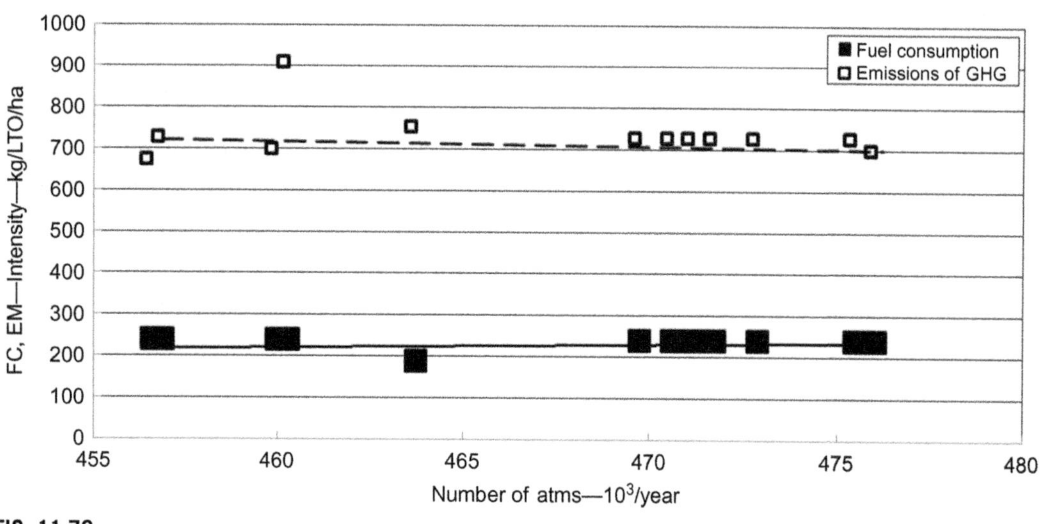

FIG. 11.76

Relationship between the intensity of fuel consumption and related emissions of GHG and the annual number of atms at London Heathrow airport—LHR (UK) (period: 2001–12) (Janić, 2016a,b; http://www.acl-uk.org/).

As can be seen, with increasing of the annual number of atms, the intensity of fuel consumption and related emissions of GHG has stabilized at the relatively constant level of about 250 and 700 ton/LTOcycle/ha, respectively, despite increasing of the volumes of accommodated demand on a constant area of land (1227 ha) during the observed period. This indicates an improvement of carrying out the LTO cycles, on the one hand, and at the same time increased use of more fuel-efficient aircraft, on the other.

11.6.5 LAND USE

The component of air transport system taking and using land are primarily airports. In general, airports occupy land directly and indirectly. Their direct occupation includes land used for aeronautical activities—accommodation of aircraft, passengers, and air cargo at the airport airside and landside area. The airside area includes runways, taxiways, and the apron/gate complex handling the aircraft. The landside area consists of passenger and cargo terminal complexes and the airport ground access transport modes (usually road and different categories of rail-based systems) (Janić, 2014a,b). Indirect occupation includes land occupied by non-aeronautical activities around airports, which otherwise would not be there. For example, these can be financial, retailing, and logistics businesses, shopping malls, sport centers, etc. In some cases, particularly around large airports, these activities, together with the airports themselves, form urbanized entities called "airport cities" (Stevens and Baker, 2013; Reiss, 2007; http://www.globalairportcities.com/airports-and-partners/what-is-an-airport-city).

The area of taken and used land by airports depends on layout specified by their category (ie, reference code). Fig. 11.77 shows the simplified scheme of six typical (theoretical) airport land layouts, that is, "footprints" with their main geometrical parameters. These are (i) a single runway used for both landings and take-offs; (ii) two parallel runways used both for landings and take-offs; (iii) two parallel runways of which one is used for landings and the other for take-offs; (iv) two converging runways each used for both landings and take-offs depending on the prevailing wind; (v) two parallel plus one crossing runway each used for landings and take-offs; and (vi) two pairs of parallel runways of which two outer runways are used for landings and two inner runways for take-offs (Horonjeff and McKelvey, 1994; Janić, 2016a,b).

The minimal (standard) values of particular parameters have been recommended by the ICAO (1987). For example, for airports accommodating the largest aircraft (Category D and E) these are as follows: $d=300$ m; $h=500$ m; $l=500$ m; $L=4500$ m; $d_0=2000$ m; $d_{01}=d_{02}=1050$ m. Consequently, the minimum area of occupied land can be $A=260$ ha for configuration (a), 1035 ha for configuration (b) and (c), 878 ha for configuration (d), 1179 ha for configuration (e), and 1980 ha for configuration (f) (Horonjeff and McKelvey, 1994; Janić, 2016a,b). However, in practice, most airports actually occupy a wider area of land than the above-mentioned (theoretical) one. Table 11.22 shows an example of the actually taken land by the selected airports in the United Kingdom (Whitelegg, 1994).

As can be seen, Heathrow airport operating two parallel runways occupied the largest area of land, which was for about 15.7% larger than the above-mentioned theoretical ones (Configurations (b) or (c)). The other single-runway airports occupied for about 1.05–3.75 times larger area of land than the above-mentioned theoretical one (Case (a)). In addition, the land taken in the airport airside area for car parking shared about 4.9–9% of the total area of occupied land. An additional evidence for 30 largest airports worldwide in terms of the annual number of passengers (enplaned, deplaned, transit/transfer) indicates that they all have occupied the land of about 58,150 ha, which is more or less equivalent to the land occupied by Singapore city. The average area of land taken by one of these airports is 1938 ha (ACI, 2012; http://en.wikipedia.org/wiki/World's_busiest_airports_by_passenger_traffic).

Once occupied by airports, the land is generally characterized by intensity of its use, which is mainly dependent on the volumes of air transport demand accommodated during given period of time. However, this intensity is always limited by the capacity of airport airside and landside infrastructure.

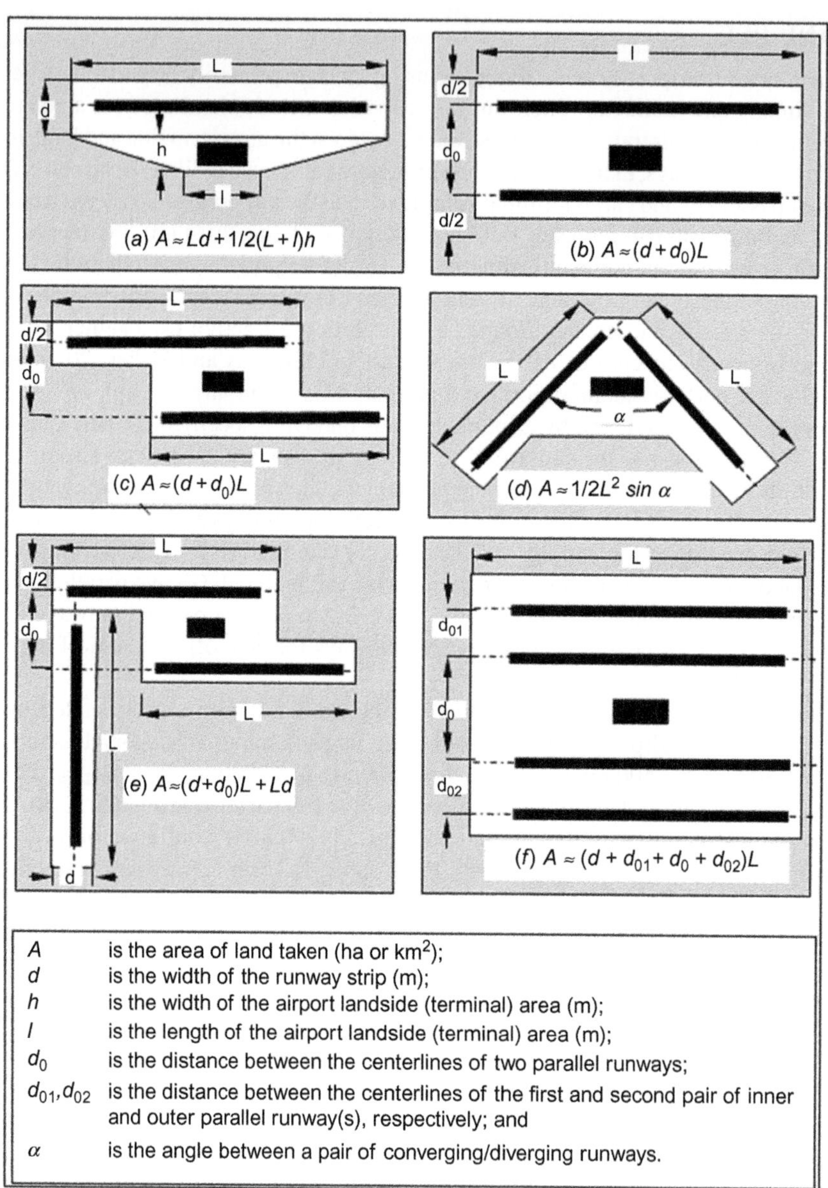

FIG. 11.77

Simplified scheme of different airport layouts, that is, "footprints" (Horonjeff and McKelvey, 1994; Janić, 2016a,b).

Table 11.22 The Area of Land Occupied by the Selected UK Airports (Whitelegg, 1994)

Airport	Total Area of Land Taken (ha)	Number of Car-Parking Place	Area of Land Occupied by Car-Parking (ha)
Heathrow[a]	1197	24,249	58.19
Gatwick[a]	759	28,571	68.57
Stanstead[a]	975	24,000	57.60
Luton	274	n.a.	n.a.
Manchester	607	14,280	34.27
Total	3812	91,100	218.63

[a]*London airports; n.a.—not available.*

Under such condition, the intensity of land use can be expressed by the volume of Workload Units (WLU) accommodated during a given period of time (year) per unit of occupied land (1 WLU = 1 pax or 100 kg of freight/cargo) as follows (Janić, 2016a,b):

$$ILU(A,\tau) = WLU(\tau)/A(\tau) \tag{11.58}$$

where

$WLU(\tau)$ is the number of WLUs accommodated at given airport during the period of time (τ); and
$A(\tau)$ is the area of land occupied by a given airport during the period of time (τ) (ha or km^2).

Fig. 11.78 shows an example of the land intensity of land use at Amsterdam Schiphol airport (the Netherlands) during the period before and after building the new (sixth) runway in the year 2002.

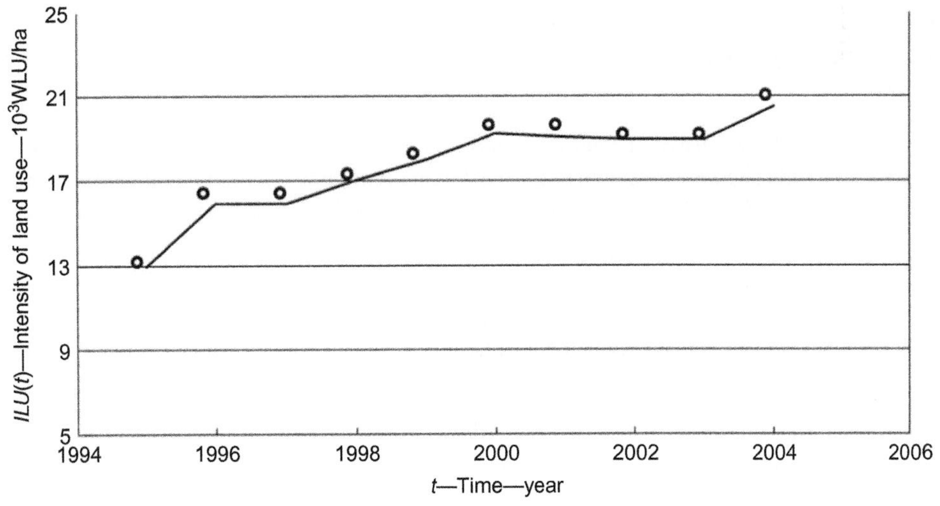

FIG. 11.78

An example of the efficiency of land use at Amsterdam Schiphol airport (Schiphol Airport 2004; Janić, 2007).

As can be seen, this intensity of land use had been increasing due to increasing of the volumes of air transport demand before the year 2001 (the year of crisis caused by the Sept. 11 terrorist attack on the US). Over the next 3 years it stagnated due to a combination of factors such as the stagnation of growth of air transport demand and implementation of the new runway (2002), which actually increased the area of land taken by the airport. Later on, starting from the year 2004, the intensity of land use recovered again thanks to recovering and continuation of the growth of air transport demand. In addition, Fig. 11.79 shows the relationship between the intensity of land use in terms of the number of atms at the level of airside capacity and the area of land occupied by selected different airports worldwide (1 atm = 1 landing or 1 taking-off) (Horonjeff and McKelvey, 1994; Janić, 2007, 2016a,b).

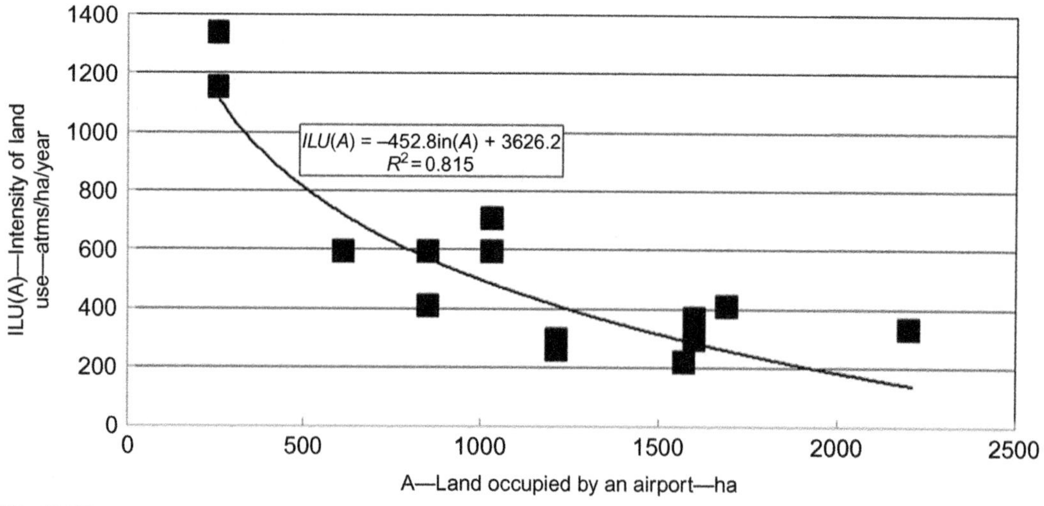

FIG. 11.79

The intensity of land use at the selected airports worldwide (Horonjeff and McKelvey, 1994; Janić, 2007, 2016a,b).

As can be seen, the intensity of land use under given conditions (the maximum number of atms carried out under IFR) decreases more than proportionally with increasing of the area of land occupied by airport(s). In some sense, this indicates diseconomies of land use respecting the airport operational capacity.

11.7 COSTS OF IMPACTS—EXTERNALITIES

11.7.1 DEFINITION

As mentioned above, operations of particular systems of different transport modes make direct physical impacts on the society and environment. In contrast to the obvious benefits, the costs of these impacts called the external costs are not fully covered by providers and users of transport services, the latter while making choice of transport mode and its system(s) usually based on the direct internal (out-off pocket) costs of their use.

Under such conditions, two problems have arose: on the one hand, the users of transport services have been faced with the wrong incentives compromising individual and overall social welfare; on the other, the impacts of transport systems on the society and environment have been growing in the absolute terms (quantities) driven by the overall social-economic growth (Gross Domestic Product (GDP)) and the above-mentioned wrong incentives of choosing transport mode and system based only on the direct costs of their use.

Consequently, recognizing these emerging and increasingly serious problems, the academic and policy endeavors and efforts in Europe and worldwide have been made to internalize the costs of impacts of transport systems on the society and the environment. The internalization in the given context has implied estimation of the average and marginal costs of particular impacts (ie, external costs) and including them into the decision-making processes of choice of transport mode and its system(s) under given conditions. According to the welfare theory approach, such internalization of external costs of particular impacts through the market-based instruments may provide more efficient use of transport infrastructure and means, reduce the negative impacts, and improve the fairness between transport users. Therefore, if being fully internalized the external costs also called "externalities"[7] could become an integral part of the overall economic characteristics of both demand and supply/capacity component of particular transport modes and their systems in both short—and long-term. The former implies imposing charges on users as fees/taxes. When these fees/taxes are equal to the costs imposed on the society and environment, transport users will take all them into account in their decision making. Then, they are likely to change their behavior, resulting in changing vehicle, system, and transport mode, thus causing changing of their utilization and consequently their overall transport volumes. The latter implies including these externalities in evaluation of feasibility of particular transport infrastructure projects, for example, in Cost Benefit Analysis (CBA) and/or for general policy development.

The taxes/fees are supposed to be based on the average and/or marginal costs of particular impacts on the society and environment. Therefore, they are expressed in the monetary terms per unit of the given system output ($ or € per v-km, p-km, t-km, TEU-km, etc.). These average and marginal costs have been estimated using different methods of which the most common and simplest in the given context have been "Damage"—and "Protection"-based methods. The "Damage" based methods presume that an impact as externality causes an amount of damage by, for example, lowering the property values, deteriorating the quality of life, and compromising health. The "Protection"-based methods estimates the cost to protect against a certain amount of an impact as externality through abatement, defense, or mitigation. Both methods use different techniques for costing, which can be broadly classified into three categories: revealed preference, stated preference, and implied preference. The revealed preference takes into account observed conditions and behavior of individuals subject to the externality. The stated preference is obtained from surveys of individuals in the rather "what-if" hypothetical situations. The implied preference considers the costs based on legislation.

A substantial number of research projects, including projects supported by the EC have suggested that internalizing the costs of particular impacts based on the average and marginal social cost pricing as the market-based instruments could produce substantive overall social benefits. In addition, "fair and

[7]In general, "externalities are costs or benefits generated by a given transport system, including infrastructure and vehicle/carrier operations, and borne in part or in whole by parties outside the system" (https://en.wikibooks.org/wiki/Transportation_Economics/Negative_externalities). In the given context, the impacts of transport systems on the society and environment are considered as the negative externalities (Button, 1994).

efficient transport pricing" has also been proposed by the EU in the numerous transport policy documents, of which the most known was the 2011 White Paper on Transport (EU, 2011). These projects have been dealing with estimation of the average and marginal costs of particular impacts using the above-mentioned methods, thus providing the basis for their internalizing. In addition, when combining the average costs with the volumes of output of particular transport systems, the total external costs can be calculated. Under such conditions, the social and environmental performances of the particular transport modes and their systems can also be compared (CE Delft, Infras, Fraunhofer ISI, 2011; Delucchi and McCubbin, 2010; Ricardo-AEA, 2014).

11.7.2 SOME MODELING

Modeling the costs of impacts of transport systems on the environment mainly relate to the average costs. This is because the marginal costs as the costs, which an additional unit of transport demand imposes on all other affected and unaffected units are very complex to estimate and particularly to implement in the real systems in the form of charges—taxes/fees. Under these conditions, the above-mentioned charges of road congestion could be considered as an exception.

11.7.2.1 Congestion

Congestion causes delays considered as losses of time of vehicles, passengers, and/or freight/cargo shipments during the trip between their origins and destinations. In this context, congestion and related delays happen either because many users tend to use the same part of transport infrastructure at the same time or due to actual scarcity of its capacity. As expressed in terms of the cost of lost time, congestion and related delays can be categorized as the social externality. This average cost of a given unit of demand using the given transport infrastructure (a vehicle, a passenger, a time-sensitive freight shipment) user can be estimated as follows:

$$c_e(W) = (c_e W)/d \tag{11.59}$$

where

c_e is the average unit costs of time of given unit of demand ($US or €/v-min or h, p-min or h, t-minor h) (v-min or h—vehicle minute or hour; p-min or h—passenger minute or hour; t-min or h—ton minute or hour); and
W is the average delay of a vehicle, passenger, or freight/cargo shipment (min, h).
d is the length of trip/route (km).

The average cost in Eq. (11.59) is expressed in $US or €/v-, p-, or t-km. The alternative approach to estimate the congestion cost as an externality particularly for the road users is by the method of charging congestion.

A road user, for example a commuter in an urban area, travelling during peak-period(s) experience congestion and delays, the latter in terms of the extra travel time and related costs of time. At the same time, he/she imposes delays and related costs on other commuters, just by its presence in the congested traffic flow(s). Under such circumstances, the measures need to be undertaken to mitigate this congestion and related total costs of delays of all affected commuters. The congestion charge as one of such measures aims at increasing the total commuter's cost to the level to give up from commuting by car and pass to one of available urban transport modes (bus, streetcar, LRT (Light Rail), or subway/metro).

The presumption is that the commuters represent the elastic "demand," which is sensitive to the imposed congestion charge as shown in Fig. 11.80.

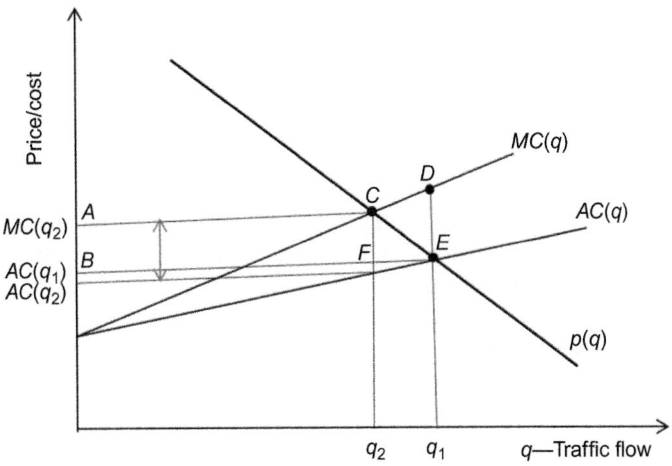

FIG. 11.80

A simplified scheme to determine the optimal congestion charge (Verhoef, 1999).

The decreasing curve ($p(q)$) expresses the individual and the total social benefits of commuting, that is, if the commuting flow (q) increases the congestion also increases and the individual and the overall social benefits consequently deteriorate, and vice versa. Such behavior of the "commuting demand" curve ($p(q)$) indicates its inherent elasticity. At the same time, the average individual trip costs ($AC(q)$) consisting of the cost of extra (delay) time and fuel consumed just due to this extra time increases approximately linearly with increasing of the of traffic flow (q). Under such conditions, the total costs of (q) commuting trips will be $TC(q) = q * AC(q)$. The marginal costs of an additional trip will be equal to $MC(q) = \partial TC(q) = AC(q) + q * \partial AC(q)/\partial q$ (Verhoef, 1999). As can be seen, at the point E where the curves $AC(q)$ and $p(q)$ intersect, the balanced traffic (commuter) flow will be (q_1). This implies that the commuters pay the average price $p(q)$, which does not include the costs they impose on the others in the flow (q). However, if these costs are considered, each commuter will be imposed the costs according to the curve $MC(q)$, which are higher that their average costs. Under such conditions, the commuters ($q_1 - q_2$) will give up of using the car and pass to some other available urban mass transit system. The new balance, that is, equality between the demand curve $p(q)$ and $MC(q)$ will be established at the level of flow (q_2) remaining there under given conditions. The average travel cost will be $AC(q_2)$ but all commuters will be forced to pay the total price equal to the marginal costs $MC(q_2)$ aiming at achieving the socially optimal use of urban roads under given conditions. The difference $\Delta C = MC(q_2) - AC(q_2)$ represents the congestion charge imposed under given conditions. From the microeconomic perspective, it increases the social surplus equivalent to the area CDE in Fig. 11.80. But it also generally rates losses for commuters if it is not intended to improve for their benefits. Each remaining commuter from the flow (q_2) suffers losses $MC(q_2) - AC(q_1)$. The total loses are equal to the area $ABCF$ in Fig. 11.80. The given up commuters ($q_1 - q_2$) suffer the total loses equivalent to the area CFE in Fig. 11.80. In addition, when the vehicles trapped in the congestion have their engines on, the fuel/energy will be consumed and consequently GHG emitted. In such case, congestion and delays can also be considered as a contributor to the environmental externality.

The similar concept of charging congestion could be applied to the capacity scarce airports (Janić, 2005). In addition, in general, for the scheduled transport modes and their systems such as, for example, rail and air, the external costs in this context, can be expressed as the difference between the charges/prices willing to pay for the scarce access slot(s) and the existing access slot charges (Ricardo-AEA, 2014).

11.7.2.2 Noise

The most well-known techniques for estimating the costs of noise damage are the hedonic price and contingent valuation methods. Both belong to the direct valuation methods. Specifically, the former method is based on the revealed and the latter method on the stated behavior (Janić, 2007). Specifically, the hedonic valuation method usually uses the property values determined by location, attributes of the neighborhood and community, and the environmental quality in the form of the property value function: $P = f(S, N, Q)$ (S is the vector of location characteristics, N is the vector of the neighborhood characteristics, and Q is the vector of environmental characteristics) (Morrell and Lu, 2000). The noise is considered as an attribute (i) of the vector of environmental characteristics and $\partial P/\partial Q_i$ as the hedonic price considered as the marginal price of the noise social cost. Subsequently the so-called Noise Degradation Index (NDI) expressing the percentage of the reduction in the house price per each additional A-weighted decibel (dBA) of noise above the common background noise has been empirically derived by the regression analysis. Dependent on the studies and their time of being carried out, the values of NDI have varied from 0.3% to 0.7% per an additional (dBA). These values have also been dependent on the noise scale in use (Levison et al., 1996). According to the hedonic price method, the total annual social cost of noise burden (C_n) can be estimated as follows:

$$C_n = \sum_{k=1}^{M} (I_{\text{NDI}} P_{\text{v}}) \cdot (N_{ak} - N_0) \cdot H_k \tag{11.60}$$

where

M is the number of zones per noise contour;
I_{NDI} is the NDI expressed as a percentage of the property value (%);
P_{v} is the annual average house rent in the vicinity of a given airport ($US or €/property unit);
N_{ak} is the average noise level for the (k)-th noise zone (dB(A));
N_0 is the background noise level (dBA); and
H_{k} is the number of residents within the i-th noise zone (population).

In Eq. (11.60), the product ($I_{\text{NDI}} * P_{\text{v}}$) represents the annual noise social cost per resident per noise unit (dB(A)). In about 30 studies the average value of (I_{NDI}) has been estimated to be 0.81 with a standard deviation of 0.72 (Schipper, 2004). In many other studies the value of (I_{NDI}) has been assumed to be about 0.48. The factor ($I_{\text{NDI}} * P_{\text{v}}$) should take into account that the annoyance caused by noise increases more than proportionally with increasing of the noise level above the ambient noise. The annual rent of the house located close to the noise source can be calculated as the product of its average price/value (P) and the capital recovery factor based on the mortgage interest rate (r) over the time period of (n) years as follows (Levison et al., 1996):

$$P_v = P\left(\frac{r(1+r)^n}{(1+r)^n - 1}\right) \tag{11.61}$$

The total cost of noise burden (C_n) in Eq. (11.60) can be used for setting up the noise charging policies at the affected locations. In such a context the revenues collected from charging passing by vehicles should cover these costs. Under such conditions, the noise charge for a specific vehicle (k) can be set up as follows (Morrell and Lu, 2000):

$$c_k = \frac{C_n}{\sum_{k=1}^{N} L_k p_k} L_k \tag{11.62}$$

where

C_n is the total costs of noise covered by the revenues collected by charging for noise at given location ($US or €);
p_k is the proportion of vehicles with the engine characteristics (k) (cars, trucks of different size, different trains—passenger, freight, aircraft);
L_k is the noise impact index of the vehicle with engine characteristics (k); and
N is the number of vehicle engines with different characteristics.

The costs in Eq. (11.62) can also be determined using some other criteria. One of them is the cost of investments in the noise mitigation measures such as for example the cost of isolating houses or building up the noise protection barriers around the affected location(s).

11.7.2.3 Traffic accidents/incidents (safety)

The principal approach in dealing with traffic accidents/incidents as an externality involves estimating the costs of the damages and losses. In general, this externality should include the cost of damaged and/or lost property (in this case this vehicle and the third party properties), the cost of the loss of life, and the cost of the time needed for recovery from the injuries of those who survived. Numerous studies have dealt with estimations of the value of life. In most cases this value has been determined for different countries in relation to the individual's contribution to the national GDP during their working age. Obviously the differences in the values of life have emerged due to the differences in the national GDP. For example, the value of life has been estimated to be between $US 1.6 and 4.7 million (Levison et al., 1996). The costs of injuries have also been expressed as an externality by converting the duration of injuries (time of nonworking) into the equivalent years of life and then using the concept of the value of life. Research has indicated that the functional years lost due to injuries are mainly dependent on the degree of injury. Some examples have shown that they vary from about 0.07 years in cases of minor injury to about 42.7 years in cases of fatal injury. In addition, on top of the cost of losing functional years, the costs for hospitalization, rehabilitation, and provision of the emergency services to the injured individuals need to be taken into account. Using the concept of the value of life and injuries, the average unit cost of a traffic accident ($US or €/p-km) can be estimated as:

$$C_{ac} = AC_r * n * [p_d * \beta_d + (1 - p_d) * \beta_i] \tag{11.63}$$

where

a_r is the traffic accident rate (the number/p-km);
n is the number of people on board the vehicle (pax/car; pax/bus; pax/train; pax/aircraft);
p_d is the probability of death during the accident;
β_d is the average value of life of a passenger killed in the accident ($US or €/pax); and
β_i is the cost of recovering from injuries ($US or €/pax).

Eq. (11.63) indicates that the social unit cost will increase in proportion with the overall accident rate, the number of passengers on board, that is, the vehicle size, probability of death/survival, and the average value of life and cost of recovery from injuries.

11.7.2.4 Energy consumption and emissions of GHG

Dealing with charging air pollution from the transport systems has been a very complex task. On the one hand, quantification of the fuel consumption and the associated emissions of GHG are shown to be a straightforward task given the volumes of output of particular systems and modes. On the other hand, while the contribution that fuel consumption makes to the depletion of the natural fuel stocks is quite clear, estimation of the marginal damages caused by the associated emissions of GHG seems to be rather complex. What is certainly known is that GHG affects society and the environment at both local near the earth's surface and global troposphere and stratosphere scale. Therefore, the polluters, in this case the transport systems, should be charged for such damages. Such a charge would generally cover four categories of air pollution costs: Photo-Chemical Smog, Acid Deposition, Ozone Depletion, and Global Warming (Levison et al., 1996). However, if one really intends to implement such a charging system, controversy emerges due to the lack of direct translation between the quantity of emitted pollutants and the damage they have already caused and/or are expected to cause in the future. Therefore, substantive research has been carried out aiming to estimate the cost of such short—and long-term damages. The first general research has dealt with the estimation of the physical damage to human health and the contribution of GHG such as CO_2 and NO_x to global warming. The following step has been expressing these damages in monetary values. These figures have been applied to set up different taxes on the emissions of pollutants such as CO_2 and carbon C. The second research has related to the estimation of the costs of protective, mitigating and defense measures from the impacts of GHG. One example of mitigating/protective measures has consisted of planting trees in order to soak up the emitted CO_2. Finally, research on how much people would be willing to pay to avoid the impact of GHG up to a certain level has been carried out, mainly using the stated preference surveys combined with analysis of the costs of preventing air pollution.

Some figures on the quantification of damage to people's health and the environment using a macro-economic/global model has been shown to be relatively convenient. In this way the "carbon-tax," reflecting the cost of damages caused by the transport system has been estimated. This tax has been supposed to be applied at a given point in time to optimize the amount of GHG on the one hand and trade-off the economic costs of damages by the GHG on the other. In addition, some other proposals for Carbon tax have been made. For example in Europe suggestions have been that it should be from about 53 to 123 $US/ton of CO_2. In the United States, it has been proposed to be between 83 and 179 $US/ton of CO_2 (Levison et al., 1996). In some other cases, these taxes have been proposed to be equivalent to the costs of perceived damages—from about 31 to 171 or 30 to 50€/ton of CO_2 depending on the social discount rate as the main governing factor. The damages from the total human-made emissions of GHG have also been estimated in rather global terms. The figures have ranged from about 1.0% to 2.7% of GDP, depending on the rate of increase in the average global temperature—from about 2.5°C to 3.0°C respectively (Tol, 1997).

As far as the transport systems operated by different transport modes are concerned, the average cost ($US or €/p-km, t-km, TEU-km) of perceived damage to people's health and the environment by the emissions of GHG of vehicle performing transport service on a given route can be estimated as follows

$$C_{ap} = (\tau \cdot FC_r \cdot e_f \cdot p)/(PL \cdot \theta \cdot d) \tag{11.64}$$

where

τ is the duration of trip along the given route with the vehicle's engines on (h);
FC_r is the rate of fuel consumption of a given vehicle type (kg or ton/min, h, day);
e_f is the emission factor of GHG of a given vehicle type from a given type of fuel (kg GHG/kg of fuel);
p is the charge (ie, tax) to be paid for the inflicted or perceived damage ($US or €/kg of GHG);
PL is the payload (seat) capacity of a vehicle (seats, ton, TEU/veh);
θ is the load factor ≤ 1.00; and
d is the length of route (km).

By multiplying Eq. (11.64) by the total volume of output (p-km, t-km, or TEU-km) carried out by a given transport system during a given period of time, the total costs of inflicted and prospective damages can be estimated.

11.7.2.5 Land use

Considering the land used for building and/or expanding transport systems' infrastructure as an externality is generally ambiguous. The question is whether it is more socially feasible to take land for such infrastructure or to use it for some other economical or noneconomical purposes such as housing, agriculture, recreation, and/or the natural environment (green area with intact flora and fauna). In all these mutually exclusive cases, the land taken has a certain value, which in general can be the economic, noneconomic, and market-based value. While assessing the social cost of such land, the economic value of land is relevant under the specific conditions. For example, let R_t and C_t be the total social revenues and costs, respectively, from operating the given transport system whose infrastructure occupies the area of land A_l. In addition, let R_o and C_o be the total social revenues and costs, respectively, from some other economic or noneconomic activity (j) carried on the same area of land S. The average unit value (ie, cost) of such alternatively (but exclusively) occupied land ($US or €/ha or km^2) can be estimated as follows (Janić, 2007):

$$C_l = \frac{[(R_t - R_o) - (C_t - C_o)]}{A_l * r} \tag{11.65}$$

where r is the rate converting the future into the present monetary values.

The nominator of Eq. (11.65) is often called the annual return to land. In addition, Eq. (11.65) also reflects the intensity of land use in the monetary terms.

11.7.2.6 Waste

The waste generated by operation of transport systems can be considered as externality if it is expected to inflict damages to the people and the environment. In the case of the industrial waste, these are usually the costs of its recycling. In the case of industrial both nonhazardous and hazardous waste, these are usually the costs of its neutralizing (recycling, storage). Both cost are usually expressed in the monetary terms per unit of waste weight ($US or €/kg or ton).

11.7.3 SOME ESTIMATION/QUANTIFICATION

As mentioned above, many projects and studies carried out by the academic and practitioner communities round in Europe and round the world have been dealing with estimation/quantification of the average and marginal costs—externalities—of above-mentioned impacts of different transport modes and their systems on the society and environment. It can be said that the most comprehensive and detailed have been those carried out for the EU and United States. It should also be mentioned that these estimates have been regularly updated, thus maintaining their above-mentioned usefulness mentioned above. Table 11.23 gives an example of estimation of the average cost of particular impacts on the society and environment by particular modes carrying out passenger transport in the EU-27 Member States in the year 2008 (CE Delft, Infras, Fraunhofer ISI, 2011).

Table 11.23 Average External Costs of Particular Transport Modes for Passengers by Category of Impact Excluding Congestion—EU-27 Member States (Period: 2008) (CE Delft, Infras, Fraunhofer ISI, 2011)

	Passenger Transport			
	Road			
	Cars	Buses	Rail	Air
Cost/impact category	¢/p-km	¢/p-km	¢/p-km	¢/p-km
Congestion[a]	n.a.	n.a.	n.a.	n.a.
Noise	0.75–1.09	0.46–0.65	0.75–1.37	0.72–1.19
Accidents	4.748	1.808	0.088	0.074
Air pollution/climate change	¢1.25–3.35	1.12–2.22	0.43–0.60	1.30–7.03
Land use[b]	0.031	0.294	0.0191	0.103
Total	6.779–9.219	3.682–4.972	1.287–2.077	2.197–8.394

[a]*Not included.*
[b]*Embraces: nature and landscape, biodiversity losses, soil and water pollution, and urban effects (€/$US = 1.471 (2008); ¢—$US cent).*

As can be seen, the highest total average costs/externalities were estimated at the road and the lowest at the rail passenger transport under given conditions. The lower and higher values of particular impacts came from considering the low and high scenario of particular impacts.

Table 11.24 gives an analogous example of estimation of the average cost of particular impacts on the society and environment by particular transport modes carrying out freight/cargo transport in the EU-27 Member States in the year 2008 (CE Delft, Infras, Fraunhofer ISI, 2011).

Table 11.24 Average External Costs of Particular Transport Modes for Freight/Cargo by Category of Impact Excluding Congestion—EU-27 Member States (Period: 2008) (CE Delft, Infras, Fraunhofer ISI, 2011)

	Freight/Cargo Transport		
Cost/Impact Category	**Road (¢/t-km)**	**Rail (¢/t-km)**	**Waterborne (¢/t-km)**
Congestion[a]	n.a.	n.a.	n.a.
Noise	0.764–1.058	0.499–0.764	0.118–0.191
Accidents	2.499	0.0294	0.000
Air pollution/climate change	1.617–3.425	0.191–0.294	0.882–1.323
Land use[b]	0.456	0.0735	0.132
Total	5.336–7.438	0.793–1.161	1.132–1.646

[a]*Not included.*
[b]*Embraces: nature and landscape, biodiversity losses, soil and water pollution, and urban effects; n.a., not available; rate: €/$ US = 1.471 (the year 2008).*

As can be seen, the road had again the highest and the real freight/cargo mode the lowest average costs/externalities for two low and high scenarios of particular impacts. In addition, Table 11.25 gives the example of estimated average costs/externalities of particular social and environmental impacts for particular transport modes carrying both passengers and freight/cargo shipments (Delucchi and McCubbin, 2010).

Table 11.25 Average External Costs of Particular Transport Modes for Passenger and Freight/Cargo by Category of Impact—United States (Period: 2006 Recalculated for the Year 2008) (Delucchi and McCubbin, 2010)

	Passenger Transport				Freight/Cargo Transport			
Cost/Impact Category	**Road**	**Rail**	**Air**	**Water**	**Road**	**Rail**	**Air**	**Water**
	¢/p-km				¢/t-km			
Congestion/delay	0.94–0.03	n.a.[a]	0.375	n.a.	0.578	0.032	n.a.	n.a
Accidents	1.498–15.408	n.a	n.a	n.a	0.118–2.140	0.235	n.a.	n.a.
Air pollution/climate change	0.161–12.305	0.546–1.819	0.096–0.417	1.194	0.128–26.322	0.021–0.877	2.515	0.086–2.065
Noise	3.745	0.556–0.952	0.942	n.a.	5.671	0.054	n.a.	n.a.
Water pollution	0.011–0.054	n.a.	n.a.	n.a.	0.003–0.054	n.a.	n.a.	n.a.

[a]*n.a.—not available.*

As can be seen, the particular impacts were slightly differently categorized than in the case of the EU-27 Member States. Nevertheless, differences in particular average costs/externalities in both regions (EU-27 and USA) are noticeable. This clearly indicates that the nature of physical acting of particular impacts and assessment of their costs have been and should continue to be the region/country specific and as such carefully considered—not to be overestimated but also not underestimated.

Multiplying the above-mentioned average estimated costs/externalities by the volumes of output carried out by particular transport modes and their systems during the specified period of time (usually 1 year) can provide the total costs of impacts of these systems individually and together on the society and the environment. Fig. 11.81 shows the example of assessed total costs of the social and environmental impacts of particular transport modes operated in the EU-27 during the year 2008.

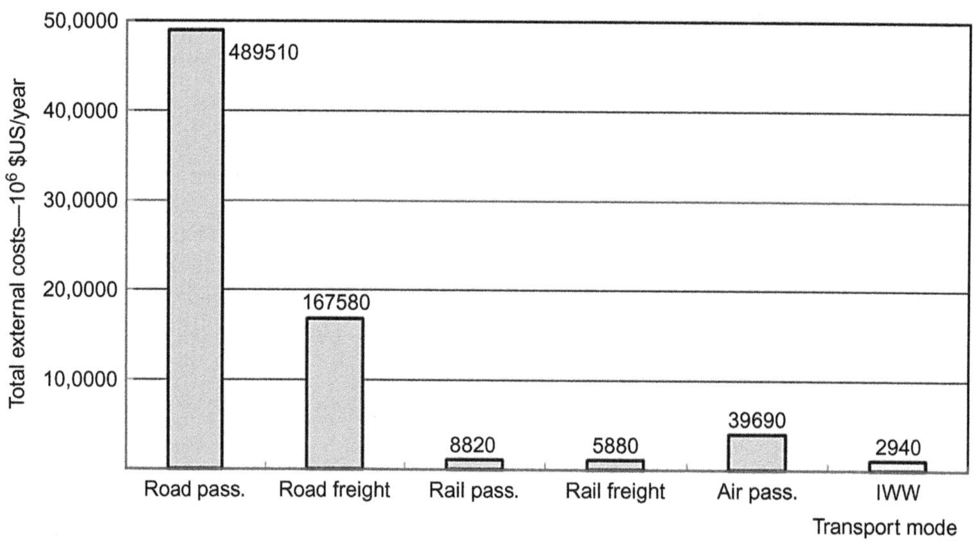

FIG. 11.81

Total costs of impacts on the society and environment by transport sector in the EU-27 Member States in the year 2008 (CE Delft, Infras, Fraunhofer ISI, 2011).

As can be seen, the road transport mode had the highest and inland waterways (IWW) the lowest external costs. The external costs of the entire transport sector were 714,420 million \$US (486,000 million €), which amounted almost 4% of the GDP in the given case (EU-27 Member States). For comparison, the fuel, environmental and other transport taxes were decreasing from 5.3% in the year 2005 to 4.8% of GDP in the year 2012 (EU, 2015). In addition, the external costs of waste were not calculated, but it is just the matter of time if they will start to be a part of the overall cost figure.

In general, these total costs of impacts of transport sector on the society and environment are dependent on the two key variables—the average cost per unit of output and the volume of output during a given period of time (usually 1 year). As mentioned above, the particular impacts in the relative terms—per unit of output—have been generally decreasing, which gives an indication that the average cost of particular impacts will also decrease. This has been achieved by development of innovative and advanced technologies and improvements of efficiency and effectiveness of operations, but also thanks to stricter corresponding regulation at different local, national, and international level. At the same time the volumes of output of particular transport modes have been increasing mainly driven by the overall economic growth, globalization of economy and trade, and vice versa. As a result, at least three scenarios regarding the future impacts and their costs/externalities in the absolute terms can be expected: (i) further increase at increasing rate if technological progress and supporting regulatory measures will not be able to neutralize the rate of growth of impacts from growing volumes of transport output (unsustainable development); (ii) further increase at decreasing rate if technological progress and supporting regulatory measures will be able to partially neutralize the rate of growth of impacts from growing volumes of transport output (balanced but still unsustainable development); and (iii) further stagnation and decrease at constant and/or decreasing rate if technological progress and supporting regulatory measures will be able to completely neutralize and even decrease the rate of growth of impacts from growing volumes of transport output (unsustainable development).

REFERENCES

AASHTO, 2001. AASHTO Green: A Policy on Geometric Design of Highways and Streets, fourth ed. 23 CFR 625.4, American Association of State Highway and Transportation Officials, Official Incorporator: the Executive Director Office of the Federal Register, Washington, DC.

ACI, 2012. Annual Traffic Data. Airport Statistics and Data Centre, Airport Council International, Montreal, Canada. http://www.aci.aero/Data-Centre.

AEA, 2011. Reduction and testing of greenhouse gas (GHG) emissions from heavy duty vehicles—Lot 1: strategy, Final Report to the European Commission, DG Climate Action Ref: DG ENV. 070307/2009/548572/SER/C, Didcot, Oxfordshire.

AECOM/URS, 2012. NC Maritime Strategy: Vessel Size vs Cost. Prepared for the North Carolina Department of Transportation, Architecture, Engineering, Consulting, Operations and Maintenance, Los Angeles, CA.

Archer, L.J., 1993. Aircraft Emissions and the Environment, EV 17. Oxford Institute for Energy Studies, Oxford.

ATOC, 2009. Energy Consumption and CO_2 Impacts of High Speed Rail. Association of Train Operating Companies Ltd, London.

Bodek, K., Heywood, J., 2008. Europe's Evolving Passenger Vehicle Fleet: Fuel Use and GHG Emissions Scenarios Through 2035. Laboratory for Energy and Environment, Massachusetts Institute of Technology (MIT), Cambridge, MA.

Boeker, E., Grondelle, R., 1999. Environmental Physics, second ed. John Wiley and Sons, Ltd., New York.

Brčić, D., Ćosić, M., Tepeš, K., 2013. An Overview of Tram Safety in the City of Zagreb. In: Proceedings of International Scientific Conference Planning and Development of Sustainable Transport System (ZIRP 2013), Zagreb, Croatia, pp. 68–76.

Brown, L.A., Tomerini, D., 2011. Distribution of the noise level maxima from the pass-by of vehicles in urban road traffic streams. Road Transp. Res. 20 (3), 50–63.

Brundtland, G.H., 1987. Our Common Future. Oxford University Press, Oxford.

Button, K., 1994. Alternative approaches toward containing transport externalities: an international comparison. Transp. Res. A 28A (4), 289–305.

CAA, 2012. Noise Exposure Contours for Heathrow Airport 2011. ERCD Report 1201, Environmental Research and Consultancy Department, Directorate of Airspace Policy, Civil Aviation Authority, UK.

Cannon, C., 2014. Actions to Reduce Greenhouse Gas Emissions by 2050. City of Los Angeles Harbour Department, Environmental Management Division & Starcrest Consulting Group, LLC, San Pedro, CA.

CE Delft, Infras, Fraunhofer ISI, 2011. External Costs of Transport in Europe: Update Study for 2008. Report commissioned by International Union of Railways UIC, CE Delft, Delft.www.cedelft.eu.

Cebr,, 2014. The Future Economic and Environmental Costs of Gridlock in 2030: An Assessment of the Direct and Indirect Economic and Environmental Costs of Idling in Road Traffic Congestion to Households in the UK, France, Germany and the USA. Centre for Economic and Business Research, London.

CEC, 2000. Defining an Environmentally Sustainable Transport System. Commission Expert Group on Transport and Environment, Report, Commission of the European Communities, Brussels.

CECW-CP, 2004. Shallow Draft Vessels Operating Costs = Fiscal Year 2004. Economic Guidance Memorandum, 05-06, U.S. Army Corps of Engineers Civil Works, Vicksburg, MS.

CGI, 2007. Container Terminal Parameters: A White Paper. Prepared for Marine Department of Transportation, The Cornell Group, Inc., Fairfax, VA.

Chandra, S., Chitgopeker, K.C., Crawford, B., Dwyer, J., Gao, Y., 2014. Establishing a benchmark of fuel efficiency for commercial airlines. J. Aviat. Tech. Eng. 4 (1), 32–39.

Churchill, J., Johnson, B., 2012. Saving billions on bunkers. Maersk Post, 9–12 (May 2012).

Condon, M.P., Dow, K., 2009. A Cost Comparison of Transportation Modes. Sustainability by Design, Foundational Research Bulletin, No. 7, November, Design Centre for Sustainability, School of Architecture and Landscape Architecture, Vancouver, BC, Canada.

DDOT, 2013. Noise and Vibration Technical Report for H Street/Benning Road Streetcar Project. The District Department of Transportation, DC Streetcar, Washington, DC. http://www.dcstreetcar.com/about/.

De Andrade, S.E.C., D'Agosto, de A.M., Junior, L.C.I., Guimarães, de A.V., 2014. CO2 emissions per passenger-kilometre from subway systems: application in the Rio de Janeiro subway. In: Proceedings of the Second International Conference on Advances in Civil, Structural and Environmental Engineering—ACSEE 2014, Zurich, Switzerland, pp. 223–227.

De Labonnefon, V., Passelaigue, J.-M., 2014. Accidentology of Tramways: Analysis of Reported Events—Year 2012: Evolution 2004–2012. Reports, Ministry of Ecology, Sustainable Development and Energy, MEDDE—DGITM, Technical Office for Mechanical Lifts and Guided Transport Systems (STRMTG), Saint Martin d'Hères.

Delucchi, M., McCubbin, D., 2010. External costs of transport in the U.S. In: de Palma, A., Lindsey, R., Quinet, E., Vickerman, R. (Eds.), Handbook of Transport Economics. Edward Elgar Publishing Ltd, Aldershot.

Den Boer, E., Verbraak, G., 2010. Environmental Impacts of International Shipping: A Case Study of the Port of Rotterdam (CE Delft). Part of the project "Environmental Impacts of International Shipping: The Role of Ports" of the Working Group on Transport under OECD's Environment Policy Committee, Organisation for Economic Co-operation and Development, Paris.

Donavan, R.P., Gurovich, A.Y., Plotkin, J.K., Robinson, H.D., Blake, K.W., 2009. Acoustic Beamforming: Mapping Sources of Truck Noise. National Cooperative Highway Research Program, NCHRP Report 635, Transportation Research Board, Washington DC. www.TRB.org.

Donselaar van, P., Carmigchelt, H., 2001. Infrastructure, Environmental and Accident Costs for Rhine Container Shipping. UNIfication of Accounts and Marginal Costs for Transport Efficiency Container Transport on the Rhine—UNITE, Version 2, EC Competitive and Sustainable Growth Programme, WP 5/8/9, NEI B.V., Amsterdam.

EASA, 2011. Type-Certificate Data Sheet for Noise A380. TCDSN EASA, Issue 6, European Aviation Safety Agency, Koln.

EC, 2000. Directive 2000/59/EC. European Commission, Off. J. Eur. Communities L332 (43), 81–89.

EC, 2002. WG Railway Noise of the European Commission. Position Paper on the European Strategies and Priorities for Railway Noise Abatement, European Commission, Directorate-General for Energy and Transport, Terms of Reference for the Working Group on Railway Noise, Brussels.

EEC, 2006. Report on Punctuality Drivers at Major European Airports. EUROCONTROL, Brussels.

EIA, 2015a. Voluntary Reporting of Greenhouse Gases Program: Fuel Emissions Coefficients. U.S. Energy Information Administration, U.S. Department of Energy, Washington, DC. http://www.eia.gov/oiaf.

EIA, 2015b. Carbon Dioxide Uncontrolled Emission Factors (Table A-3). U.S. Energy Information Administration, U.S. Department of Energy, Washington, DC. http://www.eia.gov/oiaf.

ENVIRON, 2013. Port of Oakland 2012 Seaport Air Emissions Inventory. ENVIRON International Corporation, Novato, CA.

ERF, 2011. ERF 2011 European Road Statistics. European Union Road Federation, Brussels. www.erf.be.

ERRAC, 2012. Metro, Light Rail and Tram Systems in Europe. The FP7 project of ERRAC, European Rail Research Advisory Council, Brussels. http://www.errac.org.

EU, 2011. White Paper for Transport: Roadmap to a Single European Transport Area—Towards a Competitive and Resource-Efficient Transport System, European Union. Publications Office of the European Union, Luxembourg.

EU, 2014. Regulation (EU) No. 540/2014 of the European Parliament and of the Council of 16 April 2014 on the Sound Level of Motor Vehicles and of Replacement Silencing Systems, and Amending Directive 2007/46/EC and Repealing Directive 70/157/EEC, Official Journal of the European Union, Brussels, Belgium, pp. 131–195.

EU, 2015. EU Transport in Figures: Statistical Pocketbook 2014. Publications Office of the European Union, 2014, Luxembourg.

FAA, 1997. Noise Level for U.S. Certified and Foreign Aircraft. AEE-110 Federal Aviation Administration, U.S. Department of Transportation, Washington, DC.

FTA, 2010. Public Transportation's Role in Responding to Climate Change. Federal Transit Administration, Washington, DC. http://www.fta.dot.gov/documents/.

Gautier, P.-E., Letourneaux, F.P., 2010. High speed trains external noise: a review of measurements and source models for the TGV case up to 360 km/h. http://pdf-ebooks.org/ebooks/high-speed-trains-pdf.html.

Gershon, R.M.R., Neitzel, R., Barrera, A.M., Akram, M., 2006. Pilot survey of subway and bus stop noise levels. J. Urban Health 85 (5), 802–812.

GL, 2012. Guidelines for Determination of the Energy Efficiency Design Index, Germanischer Lloyd SE, Hamburg, Germany.

Hemsworth, B., 2008. Environmental Noise Directive Development of Action Plans for Railways, Development of Action Plans for Railways. International Union of Railways, Paris.

Henry, L., Dobbs, D., Drake, A., 2009. Energy efficiency of light rail versus motor vehicles. In: Joint International Light Rail Conference: Growth and Renewal, Transportation Research Circular No. E-C145, Los Angeles, CA, pp. 78–89.

Horonjeff, R., McKelvey, F.X., 1994. Planning & Design of Airports, fourth ed. McGraw Hill, Inc., New York.

Huenecke, K., 1997. Jet Engines: Fundamentals of Theory, Design and Operations. Airlife Publishing Ltd., Shrewsbury.

ICAO, 1987. Airport Planning Manual, Part 1, Master Planning, second ed. International Civil Aviation Organization, Montreal, QC, Canada.

ICAO, 2002. Airport Planning Manual—Part 2: Land Use and Environmental Control, third ed. International Civil Aviation Organization, Montreal, QC, Canada.

IMO, 1997. MARPOL 73/78 Consolidated Edition. IMO, London.

IMO, 2010. Monitoring Performances, Strategy and Planning. 105th Session Council, International Maritime Organization, London.

IMO, 2011. Main Events in IMO's Work on Limitation and Reduction of Greenhouse Gas Emissions From International Shipping. International Maritime Organization, London.

IMO, 2012. Causality Statistics and Investigations. Loss of Life From 2006 to Date, Sub-Committee on Flag State, 20th Session, International Maritime Organization, London.

IPCC, 1995. IPCC Second Assessment Climate Change 1995. Intergovernmental Panel on Climate Change. Cambridge University Press, Cambridge.

IPCC, 1999. Aviation and the Global Atmosphere, Intergovernmental Panel on Climate Change. Cambridge University Press, Cambridge.

IPTS, 2008. Environmental Improvement of Passenger Cars (IMPRO-Car). IPTS, Seville.

ITA, 2000. Cost of Air Transport Delay in Europe. Final Report, Institut de Transport Aèrien, Paris.

Janić, M., 1999. Aviation and externalities: the accomplishments and problems. Transp. Res. D 4 (3), 159–180.

Janić, M., 2000. An assessment of risk and safety in civil aviation. J. Air Transp. Manag. 6 (1), 43–50.

Janić, M., 2003. An assessment of the sustainability of air transport system: quantification of indicators. In: 2003 ATRS (Air Transport Research Society) Conference, 10–12 July, Toulouse, France, p. 35.

Janić, M., 2005. Modelling airport congestion charges. Transp. Plan. Technol. 28 (1), 1–26.

Janić, M., 2007. The Sustainability of Air Transportation: A Quantitative Analysis and Assessment. Ashgate Publishing Limited, Aldershot, England.

Janić, M., 2014a. Advanced Transport Systems: Analysis, Modelling and Evaluation of Performances. Springer-Verlag, London.

Janić, M., 2014b. Estimating the Long-Term Effects of Different Passenger Car Technologies on Energy/Fuel Consumption and Emissions of Greenhouse Gases in Europe. Transp. Plan. Technol. 37 (5), 409–429.

Janić, M., 2016a. Analyzing, modelling, and assessing the performances of land use by airports. Int. J. Sustain. Transp. http://dx.doi.org/10.1080/15568318.2015. 1104566.

Janić, M., 2016b. A multidimensional examination of the performances of HSR (high speed rail) systems. J. Modern Transp. 24 (1), 1–24

Janić, M., Vleugel, J., 2012. Estimating potential reductions in externalities from Rail–Road substitution in Trans-European freight transport corridors. Transp. Res. D 17, 154–160.

Jonasson, H.G., 2007. Acoustical source modelling of road vehicles. Acta Acust. United Acust. 93, 173–184.

JR Central, 2012. Data Book 2011. Central Japan Railway Company, Nagoya.

Kanafani, A., 1984. The Analysis of Hazards and the Hazards of Analysis: Reflections on Air Traffic Safety Management. Working paper, UCB-ITS-WP-84-1, Institute of Transportation Studies, University of California, Berkeley, CA.

Kelly, K.L., 1998. A system approach to identifying decisive information for sustainable development. Eur. J. Oper. Res. 109, 452–464.

KMCEI, 2007. Noise and Vibration Analysis. Technical Report (Final) for the Denver-West Corridor Light Rail Transit Project Final Design Assessment, KM Chng Environmental, Inc., Denver, CO.

Levison, D., Gillen, D., Kanafani, A., Mathieu, J.M., 1996. The Full Cost of Intercity Transportation—A Comparison of High-Speed Rail, Air and Highway Transportation in California. Research Report, UCB-ITS-RR-96-3, Institute of Transportation, University of California, Berkeley, CA.

Lu, F., Ouyang, S., Zhan, P., Sui, F., 2014. A noise journey of a new subway line in Beijing. In: Proceedings of the 21st International Congress on Sound and Vibration—ICSV 21, 13–17 July, Beijing, China.

M+P, 2006. IMAGINE: improved methods for the assessment of the generic impact of noise in the environment-assessment programme for parameters of the "General" European Vehicle Fleet. In: Deliverable D3 of the IMAGINE project, EU 6th Framework Programme, Contract Number: SSPI-CT-2003-503549-IMAGINE, M+P Raadgevende ingenieurs by Noise and Vibration Consultancy, Vught, The Netherlands.

Meersman, H., Van de Voorde, E., Vaneslander, T., 2012. Port congestion and implications to maritime logistics. In: Maritime Logistics: Contemporary Issues. Emerald Group Publishing Limited, Bingley, pp. 49–68 (Chapter 4).

MEPC, 2012. 2012 Guidelines on the Method of Calculation of the Attained Energy Efficiency Design Index (EEDI) for New Ships. Annex 8, Resolution MEPC 212(63), The Marine Environment Protection Committee, London.

ML, 2011. Sustainability Progress Report 2011-Route 2. Maersk Line, Copenhagen.

Mohammad, A.H.M., 2000. Feasibility Study for the Establishment of Port Waste Reception Facility in Context of Ports in South Asian Countries (Dissertations). World Maritime University. Paper 304.

Morrell, P., Lu, C.H.-Y., 2000. Aircraft noise social cost and charge mechanisms—a case study of Amsterdam Airport Schiphol. Trans. Res. D 5, 305–320.

Narusu, S., 2015. IAPH'S (The International Association of Ports & Harbors Initiatives) to reduce emissions from ports. In: Multi-Year Expert Meeting on Transport, Trade Logistics and Trade Facilitation Sustainable Freight Transport Systems: Opportunities for Developing Countries. UNCTAD (United Nations Conference on Trade and Development), Geneva.

NDTnet, 2000. ICE train accident in Eschede—recent news summary. J. Nondestruct. Testing Ultrason. 5(2). http://www.ndt.net/news/2000/eschedec.htm/.

Notteboom, T., Carriou, P., 2009. Fuel surcharge practices of container shipping lines: is it about cost recovery or revenue making? In: Proceedings of the 2009 International Association of Maritime Economists (IAME) Conference, June, Copenhagen, Denmark.

NTSB, 2000. Highway Special Investigation Report: Truck Parking Areas. PB2000–917001, NTSB/SIR-00/01, National Transportation Safety Board, Washington, DC.

OECD, 1998. Sustainable Development: A Renewed Effort by the OECD. OECD Policy Brief No. 8, Organization for Economic Co-operation and Development, Paris. www.oecd.org/.

OECD, 2001. Policy Instruments for Achieving Project Environmentally Sustainable Transport. Organization for Economic Co-operation and Development-OECD, Paris. www.oecd.org/.

OECD, 2010. Environmental Impacts of International Shipping: A Case Study of the Port of Busan. ENV/EPOC/WPNEP/T(2010)2/FINAL, Environment Directorate Environment Policy Committee, Working Party on National Environmental Policies Working Group on Transport, Organisation for Economic Co-operation and Development, Paris.

OGP, 2010. Water Transport Accident Statistics. Report No. 434-10, Risk Assessment Data Directory, International Association of Oil & Gas Producers, London; Brussels.

Okada, Y., Tajika, T., Sakamoto, S., 2014. Road traffic noise prediction model "ASJ RTN-Model 2013". Proposed by the Acoustical Society of Japan—Part 2: Study on Sound Emission of Road Vehicles, In: Proceedings of Inter-Noise Conference, Melbourne, VIC, Australia, p. 8.

PAEH, 2011. Potential Measures for Air Emissions from NSW Ports, Preliminary Study. Prepared for New Office for Environment & Heritage, PAE Holmes, Environmental Consultants and Services, Glenvale, QLD, Australia.

Palacin, R., Correira, J., Zdziech, M., Cassese, T., Chitakova, T., 2014. Rail environmental impact: energy consumption and noise pollution assessment of different transport modes connecting Big Ben (London, UK) and Eiffel Tower (Paris, Fr). Transp. Probl. 9, 9–27.

Parry, W.H.I., 2008. Pricing Urban Congestion. Discussion Paper, RFF DP 08-35, Resources for the Future, Washington, DC.

Peeters, P.M., Middel, J., Hoolhorst, A., 2005. Fuel Efficiency of Commercial Aircraft: An Overview of Historical and Future Trends. National Aerospace Laboratory, Amsterdam. NLR, NLR-CR-2005-669.

PLB, 2014. 2013 Air Emissions Inventory. Port of Long Beach, Long Beach, CA.

PRA, 2015. Incoming and Outgoing Goods by Commodity. Port of Rotterdam Authority, Rotterdam.

PSS, 2014. Port Industry Accident Statistics 2014 Half Year Collated by Port Skills and Safety. Port Skills and Safety, London.

Puente, F., 2014. Driver error "Only Cause" of Santiago accident, says report. Int. J. Railw. http://www.railjournal.com/index.php/high-speed/wenzhou-crash-report-blames-design-flaws-and-poor-management.html/. Falmouth, Cornwall, UK.

Qiao, H., 2012. Wenzhou crash report blames design flaws and poor management. Int. J. Railw. http://www.railjournal.com/index.php/high-speed/wenzhou-crash-report-blames-design-flaws-and-poor-management.html/. Falmouth, Cornwall, UK.

RAS, 2003. Air Travel-Greener by Design. Royal Aeronautical Society, Environmental Group, Report of the Technology Sub-Group, London.

Reiss, B., 2007. Maximizing non-aviation revenue for airports: developing airport cities to optimize real estate and capitalize on land development opportunities. J. Airport Manag. 1 (3), 284–293.

Ricardo-AEA, 2014. Update of the Handbook on External Costs of Transport. Final Report for EC (European Commission)—DG Mobility and Transport, Ricardo-AEA, London.

Richard, P.B., Schmidt, F., 2000. A review of methods to measure and calculate train resistance. Proc. Inst. Mech. Eng. F: J. Rail Rapid Transit 214 (4), 185–199.

Rochard, P.B., Schmidt, F., 2000. A review of methods to measure and calculate train resistances. Proc. Inst. Mech. Eng. F 214, 185–199.

Sage, A.P., White, E.B., 1980. Methodologies for risk and hazard assessment: a survey and status report. IEEE Trans. Syst. Man Cybern. SMC-10, 425–441.

Schiphol Airport, 2004. Amsterdam Airport Schiphol. Schiphol Group Corporate Communications, Schiphol, The Netherlands.

Schipper, J., 2004. Environmental costs in European Aviation. Transp. Policy 11, 141–154.

Schrank, D., Eisele, B., Bak, J., 2015. 2015 Urban Mobility Scorecard. Texas A&M Transportation Institute, The Texas A&M University System & INRIX, Huston, TX.

Schumann, W.J., 2009. Status of North American light rail transit systems: year 2009 update. In: Joint International Light Rail Conference: Growth and Renewal, Los Angeles, CA, pp. 3–14. Transportation Research Circular No. E-C145.

Shin, K., Cheong, J.-P., 2011. Estimating transportation-related greenhouse gas emissions in the Port of Busan, S. Korea. Asian J. Atmos. Environ. 5 (1), 41–46.

Siemens, 2014. Valero CN high speed trains for China Railways. Mobility Division, Siemens AG, Berlin.

Spadaro, J.V., Langlois, L., Hamilton, B., 2000. Greenhouse Gas Emissions of Electricity Generation Chains: Assessing the Difference. IAEA Bulletin 42/2/2000. IAEA Planning and Economic Studies Section, Department of Nuclear Energy, Vienna.

Stevens, N., Baker, D., 2013. Land use conflict across the airport fence: competing urban policy, planning and priority in Australia. Urban Policy Res. 31 (3), 301–324.

SYSTRA, 2011. Carbon impacts of HS2: factors affecting carbon impacts of HSR. Version 3.1, 28 November 2011, www.greengauge21.net/.

TfL, 2007. Annual Report and Statement of Accounts 2006/07. Transport for London, TfL Group Publishing, London.

TfL, 2008. Demand Elasticities for Car Trips to Central London as Revealed by the Central London Congestion Charge. Prepared by Reg Evans for the Modelling and Evaluation Team, Transport for London Policy Analysis Division, London.

TfL, 2013. Annual Report and Statement of Accounts 2012/13. Transport for London, London.

Thompson, J.D., Eduardo, L.E., Liu, X., Zhu, J., Hu, Z., 2015. Recent developments in the prediction and control of aerodynamic noise from high-speed trains. Int. J. Rail Transp. 3 (3), 119–150.

Tol, R.S.J., 1997. A decision analytic treaties of the enhanced greenhouse effect (PhD Thesis). Vrije Universiteit, Amsterdam.

UIC, 2002. Noise Creation Limits for Railways. Main Report on the Railway's Position, International Union of Railways, Paris.

UIC, 2010a. High Speed, Energy Consumption and Emissions. UIC Publications, International Union of Railways, Paris.

UIC, 2010b. High Speed Rail: Fast Track to Sustainable Mobility. International Union of Railways, Paris.

UIC, 2011. High Speed Rail and Sustainability. UIC Publications, International Union of Railways, Paris.

UIC, 2014. Zero Carbon Railways. Final Report, International Union of Railways, Paris.

UN, 1983. Planning Land Use in Port Areas: Getting the Most Out of Port Infrastructure. UNCTAD Monographs on Port Management Monograph No. 2, United Nations, New York.

UNCTAD, 2015. Review of maritime transport. In: United Nations Conference on Trade and Development. United Nations Publication, New York and Geneva. http://unctad.org/rmt.
USDT, 2010. Bus Profile. U.S. Department of Transportation, National Highway Traffic Safety Administration, Washington, DC.
USDT, 2011. Highway Statistics 2011: Fatal Crashes, Vehicles Involved, and Fatalities. U.S. Department of Transport, National Highway Traffic Administration, Washington, DC.
USDT, 2012. National Transport Statistics. U.S. Department of Transportation Research and Innovative Technology Administration Bureau of Transportation Statistics, Washington, DC.
USDT, 2013a. Transportation Fatalities by Mode (Number of people), Table 3-1: Section 3: Transportation Safety. U.S. Department of Transportation, Research and Innovative Technology Administration (RITA), Bureau of Transportation Statistics, National Transportation Statistics, Washington, DC.
USDT, 2013b. Large Truck and Bus Crash Facts 2011. Federal Motor Carrier Safety Administration Analysis Division, U.S. Department of Transportation, Washington, DC.
USDT, 2013c. Freight Facts and Figures 2013. U.S. Department of Transportation, Federal Highway Administration, Bureau of Transport Statistics, Washington, DC.
USDT, 2015a. 1.13—Freight/Passenger Operations: Ten Year Overview. U.S. Department of Transportation, Federal Railroad Administration, Office of Safety Analysis, Washington, DC.
USDT, 2015b. Safety and Security Statistics. Federal Transit Administration, U.S. Department of Transportation, Washington, DC. http://transit-safety.volpe.dot.gov/Data/samis/default.aspx.
Van den Dries, P., 2013. Management of ship's waste: the Antwerp Port authority model. In: Port Waste Reception and Responsible Cargo Management, Sustainable Ocean Summit—World Ocean Council, 23–24 April, Washington, DC.
Verhoef, E.T., 1999. Time, speeds flows and densities in static models of road traffic congestion and congestion pricing. Reg. Sci. Urban Econ. 29, 341–369.
VTO, 2014. 2014 Annual Incident Statistics. Victorian Tram Operators, Melbourne, VIC, Australia.
Vuchic, V., 2007. Urban Transit Systems and Technology. John Wiley & Sons, Inc., Hoboken, NJ.
Waco, 2010. Parking and Access Design Standards for Site Development from the Waco, Development Guide. Waco, City of Waco, TX.
Walder, R., 1993. Ageing aircraft programme entails major effort and expense. ICAO J. 48 (November), 6–8.
Whitelegg, J., 1994. Transport and Land Take: A Report for CPRE. Eco-Logica Ltd White Cross, Lancaster.
WSC, 2015. Some Observations on Port Congestion, Vessel Size and Vessel Sharing Agreements. World Shipping Council, Partners in Trade, Washington, DC.

WEBSITE

http://www.antonov.com/aircraft/transport-aircraft/an-225-mriya.
http://www.apo.data.faa.gov/.
https://aspm.faa.gov/.
http://www.bts.gov/publications/national_transportation_statistics/.
http://www.carbon-calculator.org.uk/.
http://data.worldbank.org/indicator/IS.RRS.TOTL.KM.
www.epa.gov/autoemissions.
http://www.fueleconomy.gov/FEG/fuelcell.shtml.
http://www.globalairportcities.com/airports-and-partners/what-is-an-airport-city.

www.ogp.org.uk.
www.portofrotterdam.com/nl/nieuws/persberichten/2009/20090406_02.jspww.
https://www.portofrotterdam.com/en/shipping/sea-shipping/ships%E2%80%99-waste-from-seagoing-shipping.
http://www.railfaneurope.net/tgv/wrecks.html.
http://rigolett.home.xs4all.nl/ENGELS/vlgcalc.htm/.
http://www.ukmarinesac.org.uk/activities/ports/ph6.htm.
http://en.wikipedia.org/wiki/Heavy_fuel_oil.
http://en.wikipedia.org/wiki/World's_busiest_airports_by_passenger_traffic.
https://en.wikibooks.org/wiki/Transportation_Economics/Negative_externalities.

Index

Note: Page numbers followed by *f* indicate figures, *t* indicate tables, and *b* indicate boxes.

A

Acceleration, 65*f*
Acoustic Vehicle Alerting Systems (AVAS), 741
Activity-based travel demand models, 559–563, 561–562*f*
Actuated signal control, 315–316
Adaptive traffic control systems, 319, 319*t*
Addition law, 95
Aeronautical Fixed Telecommunication Network (AFTN), 371–372
Aeronautical information services (AIS), 330
Air-based transport systems, 816
- congestion, 816–820
- energy/fuel consumption, 829–836
- GHG emissions, 829–836
- land use, 837–840
- noise, 820–825, 822*t*
- traffic accidents/incidents, 826–829

Airborne delay, 817
Aircraft
- energy/fuel consumption, 829–830, 835–836, 836*f*
- fuel efficiency, 831*f*
- GHG emissions, 834–836, 835*t*, 836*f*
- noise, 820–823, 822*t*, 823*f*, 824*t*
- type, 824*t*

Airlines
- energy/fuel consumption, 830–833
- GHG emissions, 834–835, 835*t*
- operational costs
  - cargo transport, 704–706, 705*t*
  - passenger transport, 700–703, 700*f*, 702*t*
- revenue management, 486–489, 488*f*, 703–704, 703–704*f*

Airline scheduling, 68*f*
- aircraft rotations, 485*f*
- aircraft route, 484–486
- crew scheduling problem, 486
- departure times, 483, 483*t*
- fleet assignment, 485*f*
- long-haul traffic, 482
- meteorological conditions, 482
- passengers, 481–482
- planning process, 483–486, 484*f*

Air pollutants, 727–728
Airports, 52–53
- apron/gate complex, 269–270, 278–279
- ATC infrastructure, 367–368, 368*f*
- capacity for mixed operations, 274–277
- components, 270*f*
- demand/capacity relationship at, 819, 819*f*
- European and US, 821*t*
- infrastructure cost, 659–660, 659*t*
- landing capacity, 273–274
- layouts, 838*f*
- Ljubljana, 366*f*, 372–373
- noise, 823–825, 825*f*
- operational costs, 695–697, 696–697*f*
- practical capacity, 279–282
- reference location, 817–818
- service level, 279–282
- shortage of, 816–819
- take-off capacity, 274
- taxiways, 269–270, 277
- in UK, 839*t*
- ultimate capacity, 270–279

Airspace, 363–367, 364*f*
- airport zone, 364, 365*f*
- high altitude, 367, 368*f*
- low altitude, 366, 367*f*, 371
- terminal, 364, 366*f*, 370
- ultimate and practical capacity, 287*f*
- in vertical plane, 365*f*

Air traffic control (ATC), 294, 362–378
- air navigation services system, 369, 369*f*
- airports, 367–368, 368*f*
- airspace, 363–367, 364*f*
- automation system, 373–376
- communication equipment, 369
- components, 374, 376
- controller and pilots, 371–373
- infrastructure, 363–368
- level of service, 286–287
- navigational facilities, 369
- operational costs, 698–700, 698*f*, 699*t*
- practical capacity, 286–287
- radar screen, 371–372, 372*f*
- sector capacity, 377–378, 379*f*
- separation rules and procedures, 370–371
- shortage capacity, 816–819
- staff, 371–373
- surveillance equipment, 369
- ultimate capacity, 283–286
- view point, 496–497
- workload and capacity of, 376–378, 378*f*

Air transportation
airports, 52–53
apron/gate complex, 269–270, 278–279
capacity for mixed operations, 274–277
components, 270*f*
landing capacity, 273–274
practical capacity and service level, 279–282
take-off capacity, 274
taxiways, 269–270, 277
ultimate capacity, 270–279
air traffic control, 51
practical capacity and service level, 286–287
ultimate capacity, 283–286
Boeing B737-800 aircraft, 460*f*
commercial aviation, 456, 457*f*
demand, 457–459
fleet size, 478–479
infrastructure of, 16
level of service, 479–480
market selection, 457
networks, 463–469
operating modes, 16–17
percentage of passengers, 458*f*
planning process, 483–486
ranking of U.S. airlines, 481*f*
revenue management, 486–489
supply and capacity, 459–463
traffic volumes, 52*t*
transport service networks, 53–56
work and productivity, 475–477, 477–478*f*
AIS. *See* Aeronautical information services (AIS)
Aldrin, Buzz, 13
ALINEA ramp metering strategy, 335, 335*f*
Allsop's formula, 305–306
Amber Routes, 7
Amsterdam Schiphol airport, 372–373
land use, 839, 839*f*
operating costs and revenues, 703*f*
Annual average daily traffic (AADT), 165
Apollo 11, 13
Area-wide traffic control systems, 320–324
Armstrong, Neil A., 13
Arterial streets, traffic control for, 317–319, 317–318*f*
Artificial neural networks, 149–150
artificial neurons, 151–152, 151*f*
biological neurons, 150–151, 150*f*
characteristics of, 152–153
DVMT forecasting model, 154–155, 155*f*
multilayered feedforward neural network, 153, 153*f*
training, 154
validation, 154
Artificial neurons, 151–152, 151*f*
ATC. *See* Air traffic control (ATC)
ATO. *See* Automatic train operation (ATO)
Attribute weights, 125
Auctions, 345
Automated Capsule System for Pallets, 588–589, 590*f*
Automated Light Rail Transit (ALRT), 395
Automatic train control (ATC) metro system, 351, 358–359
Automatic train operation (ATO), 357
Automatic train protecting (ATP) system, 347, 348*f*, 357
Automatic train supervision (ATS), 357
Automation
air traffic control, 373–376
intelligent transportation systems, 329–330
Autonomous intersection management (AIM), 330–332
Available seat miles (ASM), 463
Average case analysis, 93
Average load factor (ALF), 463

B

Bangkok mass transit system, 392*f*
Base conditions, 202
Battery electric vehicles (BEV), 662–663, 737–739, 751
Belgrade Airport, passenger traffic at, 501–502*b*, 502–503, 502*t*, 504*t*, 507, 508*t*
Benz, Karl Friedrich, 12
Biological neurons, 150–151, 150*f*
Bottleneck segment, 216
Braess' paradox, 337, 543–546
Brenner Motorway Tunnel, 574
Buses
congestion, 735–736
energy/fuel consumption, 757–758
GHG emissions, 757–758, 758*t*
land use, 761–762
noise, 741–743
traffic accidents/incidents, 745–747
urban and sub/urban system, 22–23
Bus rapid transit (BRT) system, 735–736, 742–743, 742*f*
BRT TransMilenio, 23, 25*t*
infrastructure cost, 650, 650*f*, 651*t*
operating costs, 666–669, 668*f*, 668*t*
Bypass ratio (BR), 830, 831*f*

C

Caesar, Julius, 7
Capacity, 197
freight train, 234
highway capacity, 198–211, 199*f*
freeways, 201–205
methodology, 205–208
number of lanes requirement, 209–211

speed-flow curves, 204*t*
and traffic demand variations, 199–200
practical capacity, 197, 212–215
seaside and landside capacity, 262–263
shipping lines, 263–265, 267–268
single-track line, 215–219, 216*f*
transportation expansions, 543–546
ultimate capacity, 197, 212–215
vehicle fleet, 230–235
water side area, 240–243
Capacity restraint algorithm, 534–535, 534–535*b*, 535*t*
Capital cost, new highways in Europe, 653*t*
Carbon dioxide (CO_2), 728–729, 730–732*t*
Carbon footprinting, for seaports, 804
Carbon oxides, 727–728
Car-following model, 187–189, 189*f*
Cargo handling equipment (CHE), 799–800
Carpooling, 338–339
Cars
congestion impact, 733–735
energy/fuel consumption, 748–756, 755*f*
land use, 760–762
noise, 737–741
operating costs, 660–663, 663*f*
traffic accidents/incidents, 744–745
Center problem, 595
Centralized traffic control (CTC), 354–356
Change interval, calculation, 311–312
Chinese Rail High (CRH) speed network, 38*t*
Chiquita Brands International Inc., 569
City logistics, 571, 581–595
concept, 584
urban freight distribution, 582–595
Clark-Wright's "savings" algorithm, 618–624, 619–624*b*, 622*f*, 622*t*
Clipper ships, 11
Cluster-first route-second methods, 617, 618*f*
Collins, Michael, 13
Columbus, Christopher, 7–8, 8*f*
Comet, 10
Computational intelligence techniques, 135–136
artificial neural networks, 149–154
fuzzy sets, 136–148
Computer-aided traffic control (COMTRAC), 355
Conditional probability law, 95
Congestion
buses, 735–736
cars, 733–735
causes, 842–844
in Central London, 735*t*
freight trains, 763
light rail transit, 763
passenger inter-urban trains, 763
streetcar, 762
subway/metro system, 763
transportation systems, 720–721
air, 816–820
rail, 762–763
road, 733–736
water, 791–792
trucks, 736
Congestion charges, 341–343
Congestion pricing, 339–341, 343*t*
Constant speed, 65*f*
Construction cost
airports, 659*t*
BRT and LRT system, 650, 650*f*, 651*t*
high speed rail, 655*t*
streetcar, 649, 649*t*
subway/metro infrastructure, 651–652, 651*f*
2-lane highway, 654*t*
US airports, 659*t*
Contrails, 728
Control systems
adaptive traffic, 319, 319*t*
area-wide traffic, 320–324
interurban rail transport systems, 40
in transportation, 2
TRM system, 40
Conventional intermodal freight train(s) (CIFTs), 706–707, 707*f*, 710–711*t*
Cook, James, 9–10
Corvée, 10
Cost-benefit analysis (CBA), 644–647
Cost function, 639–642
Crew scheduling
problem, 486
in public transportation, 444–445
run-cutting, 444*f*
Critical lane volumes, 310
Cross-docking, 572, 573*f*
Cumulative density function, 99
Cycle length, 295–296
calculation, 312–313
at signalized intersection, 302–303*b*
two phases, 297*f*
Cycle time, 295

D

Da Gama, Vasco, 8
Daily vehicle miles of travel (DVMT) forecasting model, 154–155, 155*f*
DARPA Grand Challenge, 329

Data envelopment analysis (DEA)
decision-making units, 132–133
efficiency, 133, 134–135*b*
ratios, 130–132, 130*t*
relative efficiency, 133
da Vinci, Leonardo, 8–9
Davis's equation, 725
Deceleration, 65*f*
Decision-making units (DMU), 132–133
Delphi method, 500
Delta Airlines, 469, 469*f*
Demand and supply, 642–643
Demand-generating nodes, 597–598
Demand response transit operations, 328
Demand-responsive transportation (DRT) systems
dial-a-ride, 450–453
routing and scheduling, 448–449
service complements, 448
Departure delay, 818
Dial-a-ride system, 390, 390*f*, 450–453, 450–451*f*
Dijkstra's algorithm, 73–76
Dilemma zone, 311, 311*f*
Dimensionality
small dimensionality, 91–92
in traffic and transportation, 91–92
Direct shipment, 571
Discount rate, 644
Discrete choice models, transportation demand, 547–549
Disjunctive system of rules, 144, 144*f*
Disruption management
bus bunching, 445–446, 445*f*
modification of transit network, 447*f*
in public transit, 445–446
service frequencies changes, 446*f*
Distribution system
configuration, 570–571
urban freight, 582–595
Door-to-door travel time, 427
Driver agents, 330
DVMT forecasting model. *See* Daily vehicle miles of travel (DVMT) forecasting model
Dynamic ridesharing, 329
Dynamic route guidance, 328–329
Dynamic traffic assignment (DTA), 546

E

Earhart, Amelia, 12–13
Economics of transportation, 635–636
cost-benefit analysis, 644–647
cost function and revenues, 639–642
fixed costs, 637–638, 638*f*
HSR evaluation, 643–644
infrastructure cost, 647–660
scale/scope, 638–639, 639*f*
transport sector/industry, 637
variable costs, 637–638, 638*f*
Edge-covering problems, 612, 612*f*
Energy consumption
aircraft, 829–830, 835–836, 836*f*
airline, 830–833
buses, 757–758
cars, 748–756, 755*f*
inland vessels/barges, 797–799, 798*f*
seaports, 799–800
shipping lines, 805–806, 805*f*
transportation systems, 724–727, 846–847
air, 829–836
rail, 780–789
road, 748–760
water, 797–811
trucks, 758–760
Energy Efficiency Design Index (EEDI), 808, 809*f*, 810
Environmental Ship Index (ESI), 804
Equation of motion, 63–64
Erie Canal, 11
Erlang, Agner Krarup, 106
Eschede train disaster, 777
Euclidean distance, 597–598, 597*f*
European rail traffic management system (ERTMS), 350–352, 352*f*
European train control and command system (ETCS), 350
E-waste, 572–573
Expected value, 100
Exponential algorithms, 93
Exponential distributions, 101–102, 102*f*
of headways, 166–168, 167*f*
Exponential smoothing, 503, 503–504*b*

F

Federal-Aid Highway Act of 1956, 12
Federal Aid Road Act of 1916, 12
Federal High-Way Administration (FHWA) algorithm, 536, 536–537*b*, 536–537*t*
First Transcontinental Railroad, 11
Fixed block systems, 347–350
Fixed costs, 637–638, 638*f*
Fixed-time control strategies, 296
determination of, 306
at isolated intersection, 296–300
Fleet vehicles
capacity, 230–235

freight trains, 233–235
interurban rail transportation, 230–235
level of service, 230–235
passenger trains, 231–233
railways rolling stock, 231*f*
Flight arc, 67–68, 464
Flight frequency
airline dynamic capacity, 469
gaining market share, 470–472
minimizing total route cost, 472–474, 474*f*
production plan, 469–470
satisfying demand, 470
Flight leg, 462–463, 462*f*
Flight number, 463
Flight segment, 462
Flow-density relationship, 173–175, 174*f*
Flow rate, 199–200, 200*f*
Ford, Henry, 6, 12
Forecasting techniques
qualitative and quantitative, 500–501
time series, 501–503, 501–502*b*
transportation demand, 499–509, 500*f*
trend projection, 504–509, 505*f*, 505*t*, 507*f*
Frankfurt Main airport, 275, 275*f*, 276*t*
Free-flow speed (FFS), 145, 171, 202, 202*t*, 206*t*
Freeways, 201–205
FFS values, 202*t*
highway capacity, 201–205
merge and diverge influence areas, 201, 201*f*
traffic control, 332–337, 332*f*
driver information and guidance systems, 336–337
measurement, 333–337
ramp metering, 333–336, 333*f*
weaving influence area, 201, 201*f*
Freeway Travel Time Index, 734
Freight shipments
hub-and-spokes network, 43*f*, 44
interurban
rail transport systems, 42–44, 43*f*
road transport systems, 31–33
line/ring network, 43*f*, 44
point-to-point network, 42–43, 43*f*
trunk line, 43–44, 43*f*
urban and sub/urban, 29–30, 29*f*, 30*t*
Freight terminals, 575–577
ultimate/practical capacity, 224–225
Freight trains
capacity, 234
congestion, 763
land use, 790
noise level, 770
traffic accidents/incidents, 778–780, 778–779*f*
Freight transportation
congestion, 736
infrastructure network, 573–577
load factors, 582, 586
service networks, 577–581
urban, 582–595
Frontier analysis. *See* Data envelopment analysis (DEA)
Fuel consumption
aircraft, 829–830, 835–836, 836*f*
airline, 830–833
buses, 757–758
cars, 748–756, 755*f*
inland vessels/barges, 797–799, 798*f*
seaports, 799–800
shipping lines, 805–806, 805*f*
transportation systems, 724–727
air, 829–836
rail, 780–789
road, 748–760
water, 797–811
trucks, 758–760
Fully actuated signal timing, 296
Fully automated rail traffic control, 356–359
Functional airspace blocks (FABs), 363
Fuzzy inference rules, 188–189, 189*f*
Fuzzy reasoning, 140, 142
Fuzzy sets
approximate reasoning algorithm, 142–143, 143*f*
basics, 138–140
concept of, 136–137
disjunctive system of rules, 144, 144*f*
elements, 140–148, 141*f*
intersection, 139
membership functions, 139–140*f*, 142*f*
union, 139

G

Gagarin, Yuri, 13
Galilei, Galileo, 10
Gate arrival delay, 817
Gate departure delay, 817
Generalized cost, 642
GlaxoSmithKline (GSK), 569
Global positioning system (GPS), 325
Global system for mobile communications—railway (GSM-R), 350–351
Global warming, 728–729
Gompertz curve, 506–507
Grand Canal, 6–7
Gravity model, transportation demand, 514–517

Green house gases (GHG) emissions
  aircraft, 835–836, 836*f*
  buses, 757–758, 758*t*
  global freight/cargo ship fleet, 811*f*
  inland vessels/barges, 797–799, 799*f*
  seaports, 800–804, 802–803*f*
  shipping lines, 805–806, 806*f*
  transportation systems, 727–732, 846–847
    activity-based method, 730
    air, 829–836
    energy-based method, 730
    rail, 780–789
    road, 748–760
    water, 797–811
Greenshields model, 171
Green time, 294–295, 299
  allocation, 314
  effective, 299
  minor-street, 315–316
  pedestrian, 314–315
  timing of actuated, 316*f*
Grid transit network, 430
Ground holding arc, 67–68, 68*f*, 464, 465*f*
Ground Holding Program (GHP), 817
Gumbel's probability density function, 549

H

Hakimi's theorem, 606
Harrison, John, 10
Headways, 66–67
Heavy-duty road trucks (HDRT), 801
Heuristic algorithms, 92, 614–615
High-occupancy toll (HOT) lanes, 344
High-occupancy vehicle (HOV)
  facilities, 344
  lane management, 329
  traffic sign, 344*f*
High speed rail (HSR)
  infrastructure cost, 654–655, 655*t*
  networks, 20, 21*f*
    development, 37*t*
    location of stations, 20
    in Spain, 37*f*
    spatial topology, 36, 36*f*
  projects evaluation, 643–644
  services, 768–769, 777, 777*t*, 790
  traffic accidents/incidents, 776–777, 777*t*
Highway capacity, 173, 198–211, 199*f*
  base condition for lane width, 206
  defining level of service, 202
  freeways, 201–205
  methodology, 205–208
  number of lanes requirement, 209–211
  speed-flow curves, 204*t*
  and traffic demand variations, 199–200
Highway space inventory control system, 344–345
Horse-drawn omnibus, 391, 391*f*
Horse drawn vehicles, 14
HOV. *See* High-occupancy vehicle (HOV)
HSR. *See* High speed rail (HSR)
Hub-and-spoke networks, 48, 54, 429*f*, 577
Hub location problem, 610
Hybrid Vehicles (HYVs), 737–739, 750
Hydrogen Fuel Cell Vehicle (HFCV), 752
Hydrogen Vehicle (HV), 752

I

Incremental assignment algorithm, 537–539, 538–540*b*, 538–540*t*
Independence of irrelevant alternatives (IIA) property, 553
Infrastructure cost, 647
  airports, 659–660, 659*t*
  BRT system, 650, 650*f*, 651*t*
  high speed rail, 654–655, 655*t*
  inland waterways, 655–657, 656*t*
  LRT system, 650, 650*f*, 651*t*
  ports, 657–659, 658*f*
  road, 653–654
  streetcar, 649–650, 649*t*
  subway/metro, 651–652, 651*f*, 652*t*
  urban mass transit systems, 647–652, 648*f*
Infrastructure providers, 636
Inland waterways, 236–237
  for cargo shipments, 45–48
  classification, 45, 46*t*, 237–239, 238–239*t*
  energy/fuel consumption, 797–799, 798*f*
  European network, 46*f*
  freight transportation, 47, 48*f*
  GHG emissions, 797–799, 799*f*
  infrastructure, 16
    cost, 655–657, 656*t*
    network, 45, 239–249
  noise, 792
  operating costs, 681–684, 682–685*f*
  operating modes, 16–17
  transport service network
    rolling stock/vehicles, 250
    route and network, 251–252
Instrument meteorological conditions (IMC), 270
Integer programming
  gate assignment problem, 89–91*b*
  in traffic and transportation, 88–90
Intelligent Transportation Society of America, 325
Intelligent transportation systems (ITS), 325–332
  architecture, 326, 326*f*
  autonomous intersection management, 330–332

autonomous vehicles, 329–330
concepts, 325
technologies, 325–326
user services, 327–329
Interest rate, 644
Interior point method, 82
inventory control problem, 83, 84–85*f*
Intermodal transport systems
CIFTs, 706–707, 707*f*, 710–711*t*
LIFTs, 707–708, 707*f*, 710–711*t*
operational costs, 706–714
revenues, 713–714, 713*f*
Internal combustion engine (ICE), 662–663, 737–738
Internal Combustion Engine Vehicles (ICEVs), 737–739, 748
Internal rate of return (IRR), 645
International Air Transport Association (IATA), 463
The International Association of Ports & Harbors Initiatives (IAPH), 804–805
International Civil Aviation Organization (ICAO), 363, 820–821
International Space Station, 13
Intersection manager, 330–332
Interurban
mass transit systems, operating costs, 671–681
road freight transport, 674–676, 674*f*
road passenger transport, 671–674
rail transport systems, 34–44 (*see also* High speed rail (HSR), networks)
commercial use, 41–42
control systems, 40
European railway core network, 35*f*
freight shipments, 42–44
infrastructure, 38–39
levitation and propulsion, 39–40
longest rail infrastructure networks, 34*t*
passengers, 36–42
rolling stock and operating speed, 39
weight and energy consumption, 40–41
road transportation, 30–33
service networks, 453–455
ultimate and practical capacity, 453
Inventory control problem, 83, 84–85*f*
Inversion method, 123
Inward median, 607–610, 607–608*b*
Isolated intersection, 296–300
ITS. *See* Intelligent transportation systems (ITS)
IWT motor container vessel, 250*f*

J

Jam density, 171
Jet engines, 830, 831*f*

K

The Kaleidoscope, 10

L

Landside capacity, 262–263, 273–274
Land use
air transport system, 837–840, 838*f*, 839*t*, 840*f*
Amsterdam Schiphol airport, 839, 839*f*
buses, 761–762
cars, 760–762
freight trains, 790
light rail transit, 789
passenger inter-urban trains, 790
seaport, 813*f*
streetcar, 789
subway/metro systems, 790
transportation systems, 732–733, 847
air, 837–840
rail, 789–790
road, 760–762
water, 812–813, 813*f*
trucks, 762
Level of service
airports, 279–282
air traffic control, 286–287
freeways, 201–205, 204*f*, 205*t*
highway capacity and, 198–211, 199*f*
by linguistic expressions, 203*f*
maximum service flow rates, 210*t*
methodology, 205–208, 205*f*
number of lanes requirement, 209–211
productivity, 235
rating scale for, 203*t*
road freight terminals, 575–577, 575*f*
road truck roads, 574
shipping lines, 265–266, 268
size of rolling stock, 235
speed-flow curves, 204*t*
and traffic demand variations, 199–200
transportation demand/supply, 497
at U.S. railways, 222*t*
water side area, 240–243
Level terrain, 207
Lex Julia Municipalis, 15
Light Rail Rapid Transit (LRRT), 395
Light rail transit (LRT), 395
congestion, 763
infrastructure cost, 650, 650*f*, 651*t*
land use, 789
noise, 765–766, 766*f*
operating costs, 666–669, 668*f*, 668*t*
safety, 771–772, 772–773*f*
urban and sub/urban, 25–27, 26–27*f*, 27*t*

Lindbergh, Charles, 12–13
Linear Induction Motor(s) (LIMs), 592
Linear programming (LP)
  on-ramps on freeway, 86*f*
  ramp demands, 87, 87*t*
  resource allocation problem, 85–86
  section capacities, 86, 86*t*
  in traffic and transportation, 82–87
Line networks, 48
Lineside equipment unit (LEU), 350
Link-based measurements, 189–190
Link lengths, 72
Liquid Natural Gas (LNG), 804
Little's formulae, 575–576
Little's Law, 109–113, 342
Ljubljana airport, 366*f*, 372–373
Location problem
  classification, 596–597
  Euclidean distance, 597–598, 597*f*
  hub, 610
  Manhattan distance, 597–598, 597*f*
  maximal covering, 603–604
  set covering problem, 598–603
Locomotion No. 1, 11
Logistics, 569–573
  city, 581–595
  costs *vs.* number of warehouses, 571*f*
  cross-docking, 572, 573*f*
  direct shipment, 571
  flow of, 574*f*
  reverse, 572–573
Logit model
  estimation, 554–556, 556–557*b*
  independence of irrelevant alternatives property, 553
  maximum likelihood method, 549, 554
  multinomial, 549
  parameters of, 554
  transportation demand, 549–556, 550*f*
London Heathrow airport, 280, 280*f*
  GHG emissions, 836, 836*f*
  noise, 824–825, 825*t*
Long intermodal freight train(s) (LIFTs), 707–708, 707*f*, 710–711*t*
Longitude Act in 1714, 10
LRT. *See* Light rail transit (LRT)

M

Macro-simulation traffic models, 186
Magellan, Ferdinand, 9, 9*f*
Mandl's road network, 435*f*, 438–441*b*
Manhattan distance, 597–598, 597*f*
Marine Distillate Fuels (MDF), 805
Maritime freight/cargo transport system
  networks, 69
  operating costs, 684–695
  ports, 684–688, 686*f*, 687*t*
    capacity—access modes, 260–263
    capacity—landside area, 257–260
    capacity—seaside area, 254–257, 256*f*
    configuration—layout, 253–254, 254*f*
    freight/cargo handling equipment, 259–260
    service level—seaside area, 257
    terminal yard/area, 258–259
  shipping lines, 688–695
    network, 267–268
    route, 263–266
Mathematical programming applications
  algorithm, 82
  complexity of algorithms, 92–93
  components, 81–82
  constraints, 81–82
  dimensionality, 91–92
  elementary operations, 92
  integer programming, 88–90
  linear programming, 82–87
  optimal solution, 82
  optimization, 82
  unknowns, 81–82
  worst-case conditions, 92
Matrix-reduction algorithm, 601–603, 601–603*b*
Maximal covering location problem, 603–604
Maximax method, 127
Maximum likelihood method, 549, 554
Maximum load method, 415–416
Maximum service frequency, 412–413
Maximum take-off weight (MTOW), 459
McAdam, John Loudon, 10
Mean Absolute Deviation (MAD), 509
Mean Forecast Error (MFE), 508
Median problems, 595, 604–610
Metaheuristic algorithms, 92
Metro automation, 356, 357*f*
  components, 357–358
  concept of, 356
Metropolitan Railway, 11
Metro system, 396–397, 397*f*
  congestion, 763
  infrastructure cost, 651–652, 651*f*, 652*t*
  land use, 790
  noise, 766–767, 767*f*
  operating costs, 669–671
  safety, 772–774, 773–775*f*
  urban and sub/urban, 28

Micro-simulation traffic models, 186–187, 186*f*
Migration, 1–2
Minimax method, 126–127
Minimax problem, 595
Mississippi River System, operating costs, 681–682, 682*f*
Mixed-integer programs, 88
Mixed network, 69, 70*f*
Modal split analysis, 509, 510*f*, 518–521, 550
Monte Carlo simulation method, 121–123
Motion, 63–64
  acceleration, 65*f*
  constant speed, 65*f*
  deceleration, 65*f*
  macroscopic approach, 64–65
  object's position, 65*f*
  time-space diagram, 64, 64*f*
  traffic signal coordination, 67*f*
Mountainous terrain, 207
Multiattribute decision making (MADM), 124–125, 647
  attribute weights, 125
  Maximax method, 127
  Minimax method, 126–127
  negative-ideal solution, 128
  positive-ideal solution, 128
  SAW method, 127–128
  TOPSIS, 128–130
Multicriteria decision making (MCDM), 647
Multi-hub-and-spoke networks, 54, 465–469, 466*f*
Multilayered feedforward neural network, 153, 153*f*

N

National Electrical Manufacturers Association (NEMA), 295
Nearest Neighbor algorithm, 628–629*b*
  Traveling Salesman Problem, 615–616, 616*f*
Nested reservation system, 83, 487
Net income, 642
Net present value (NPV), 645
Network design, public transportation, 431–437
Network flow diagram
  generalized traffic flow variables, 190–191, 191*f*
  link-based measurements, 189–190
  trajectory-based measurements, 189, 192–193
Newton, Isaac, 10
Nitrogen oxides, 727
Node-covering problems, 612, 613*f*
Nodes, 69
  degree of, 69–71
  indegree of, 69–71
  outdegree of, 69–71
  shortest paths between all pairs, 77–80
Noise, 721–723
  aircraft, 820–823, 822*t*, 823*f*, 824*t*
  airport, 823–825, 825*f*
  buses, 741–743
  cars, 737–741
  freight trains, 770
  light rail transit, 765–766, 766*f*
  passenger inter-urban trains, 767–769, 768–769*f*
  seaports, 792
  shipping lines and inland vessels/barges, 792
  from streetcar, 763–764, 764–765*f*
  subway/metro system, 766–767, 767*f*
  transportation systems, 844–845
    air, 820–825, 822*t*
    rail, 763–770
    road, 737–744
    water, 792
  from trucks, 743–744
Non-methane hydrocarbons, 728
Nonoriented network, 69–71, 70–71*f*
Nonpolynomial algorithms, 93
Normal distribution, 102–105, 102*f*
  of headway, 168–170, 168*f*

O

Object motion, 63–68, 64*f*
Ocean-going vessels (OGV), 800
Operating costs
  air, 695–706
  airlines cargo transport, 704–706, 705*t*
  airlines passenger transport, 700–703, 700*f*, 702*t*
  airports, 695–697, 696–697*f*
  air traffic control, 698–700, 698*f*, 699*t*
  BRT, 666–669, 668*f*, 668*t*
  cars, 660–663, 663*f*
  CIFTs, 706–707, 707*f*, 710–711*t*
  inland waterways, 681–684, 682–685*f*
  intermodal, 706–714
  interurban mass transit systems, 671–681
  LIFTs, 707–708, 707*f*, 710–711*t*
  LRT, 666–669, 668*f*, 668*t*
  maritime cargo transport, 684–695
  Mississippi River System, 681–682, 682*f*
  ports, 684–686, 686*f*
  rail freight transport, 680–681
  rail passenger transport, 676–680, 677*f*
  road freight transport, 674–676, 674*f*
  road passenger transport, 671–674
  shipping lines, 688–690, 689*t*, 689*f*
  streetcar, 666, 667*f*
  subway/metro system, 669–671
  urban mass transit systems, 663–671, 664*f*

Operating Empty Weight (OEW), 459
Operations control centre (OCC), 359
Oriented network, 69–71, 70–71*f*, 607–610
Origin-Destination matrix (O-D matrix), 498–499, 546
Outward median, 607–610
Overnight arc, 464, 465*f*
Oversaturated traffic conditions, 295

P

Panama Canal, 11–12, 12*f*
Papin, Denis, 10
Paris Metro network, 430, 431*f*
Passenger-car equivalents (PCEs), 202, 207
Passenger flows
 along transit line, 408–410, 409*f*
 daily variations, 407, 407*f*
 hourly variations, 407–408, 407*f*
 maximum load section, 409–410
 maximum passenger volume, 409–410, 410*f*
 number of vehicle departures, 408, 408*f*
Passenger inter-urban trains
 congestion, 763
 land use, 790
 noise, 767–769, 768–769*f*
 traffic accidents/incidents, 774–777
Passenger paths, 463
Passenger waiting time, 413
Pavage tolls, 7
Peak hour factor (PHF), 200
Peak hour volume, 165
Pearson Type III distribution, 168–170
Pedestrian crossing, 314–315
Pedestrian green time (PGT), 314–315
Persian Royal Road, 6
Peter the Great, 10
Pick-up-and-delivery networks, 48
Pipes/tunnels, 589–591, 593
Platform door interface unit (PDIU), 359
Platoon, 317–318
p-median problem, 604–605
Point-to-point networks, 54, 465–468, 466*f*, 577
Poisson distribution, 101, 166–168
Polynomial algorithms, 93
Ports
 capacity, 239–245
 components, 239–245
 infrastructure cost, 657–659, 658*f*
 maritime freight/cargo transport system
  capacity—access modes, 260–263
  capacity—landside area, 257–260
  capacity—seaside area, 254–257, 256*f*
  configuration—layout, 253–254, 254*f*
  freight/cargo handling equipment, 259–260
  service level—seaside area, 257
  terminal yard/area, 258–259
 operational costs, 684–686, 686*f*
 operations, 239–245
 revenues, 686–688, 687*t*, 688*f*
 simplified layout, 240*f*
Practical capacity, 197, 212–215
 air traffic control, 286–287
 rail inter-urban transport systems, 224
 rail station, 222–230
 rivers and man-built channels, 248–249
 road freight terminals, 575–577, 575*f*
 road truck roads, 574
 single-track line, 220–221, 221*f*
 water side area, 243, 245
Price of anarchy, 542–543
Probability density function (PDF), 96–98, 98*f*
 and cumulative density function, 99, 99*f*
 example, 96–97*b*
 of normal distribution, 102, 102*f*
 in traffic and transportation, 100*t*
Probability theory, and traffic
 addition law, 95
 conditional probability law, 95
 conditioning events, 96
 definition, 94
 exponential distributions, 101–102, 102*f*
 intersection, 95, 95*f*
 normal distribution, 102–105
 poisson distribution, 101
 random variables, 93, 96–105
 sample space, 94
Public transportation
 air transportation
  demand, 457–459
  level of service, 479–480
  networks, 463–469
  planning process, 483–486
  scheduling, 481–483
  supply and capacity, 459–463
  work and productivity, 475–477
 amorphous transit network, 429*f*
 availability, 405–406
 crew scheduling, 444–445
 disruption management, 445–446
 DRT
  dial-a-ride, 450–453
  routing and scheduling, 448–449
 ferries, 387

flight frequency
  gaining market share, 470–472
  minimizing total route cost, 472–474
  satisfying demand, 470
freeway lane, 388*f*
grid transit network, 429*f*
hub-and-spoke transit network, 429*f*
interurban road transportation, 453–455
Mandl's road network, 435*f*, 438–441*b*
network design, 431–437
number of transported passengers *vs.* served vehicles, 388–389, 388–389*f*, 389*t*
passenger flows
  along transit line, 408–410
  daily variations, 407, 407*f*
  hourly variations, 407–408, 407*f*
planning process, 447, 447*f*
problems, 388
radial/circumferential transit network, 430*f*
radial network, 429*f*
rapid transit rail systems, 387
service frequency and headways
  bus headways, 411, 412*f*
  determination, 438–440
  maximum load method, 415–416
  maximum service frequency, 412–413
  passenger waiting time, 413
  public transit line, 411, 411*f*
  square root formula, 413–415
  and vehicle departure times, 410–411
timetable, 416–417, 417*f*
transit line capacity
  approximate capacity values, 420*t*
  line/route ultimate capacity, 418
  practical capacity, 418–419
  transit route lengths, 418
  utilization, 420–422, 421*f*
transportation modes, 387
urban public transit
  dial-a-ride system, 390, 390*f*
  horse-drawn omnibus, 391
  infrastructure, 398–404
  links and indicators, 401–404
  performances, 423–428
  rail-based urban transit systems, 395–397
  road-based urban transit systems, 393–394
  simple greedy algorithm, 434–437
  stops/stations in, 399–401
  taxi system, 390
  topology and relationship, 403–404
  types, 429–431
vehicle scheduling, 441–444

Q

Quandt-Baumol model, 520
Queueing, 108*f*
  arrival process, 107
  average waiting time, 106–107
  D/*D*/1 queueing, 109
  and investments, 119–121, 120*f*
  length and waiting time, 110*f*, 111
  Little's Law, 109–113
  M/*M*/1 queueing, 114
  M/*M*/*s* queueing, 115–118
  number of servers, 107
  queue capacity, 107
  queue discipline, 107
  service process, 107
  toll collection, 109*f*

R

Radiative forcing effect, 728–729
Radio Frequency Identification (RFID) system, 592
Rail freight transport, operating costs, 680–681
Rail locomotives (RL), 801
Rail shunting yard, 226–230, 226*f*
  classification yard, 228, 229*f*
  departing yard, 230
  entire shunting yard, 230
  hump, 227
  receiving yard, 227
Rail transportation systems, 762
  balancing revenues and costs, 679–680, 679*f*
  capacity—access modes, 260–261
  congestion, 762–763
  energy/fuel consumption, 780–789
  European railway core network, 35*f*
  GHG emissions, 780–789
  infrastructure of, 16
  interurban, 34–44
    commercial use, 41–42
    control systems, 40
    European railway core network, 35*f*
    fixed block sections, 216
    freight shipments, 42–44
    infrastructure, 38–39
    levitation and propulsion, 39–40
    longest rail infrastructure networks, 34*t*
    passengers, 36–42
    rolling stock and operating speed, 39
    time-distance diagram, 218–219*f*
    ultimate and practical capacity, 215–235
    weight and energy consumption, 40–41

Rail transportation systems *(Continued)*
in Japan, 13
land use, 789–790
longest rail infrastructure networks, 34*t*
network modes, 69
noise, 763–770
operating costs, 676–678, 677*f*
operating modes, 16–17
revenues, 678, 678–679*f*
3-aspect signaling system, 347–348
traffic accidents/incidents, 770–780
traffic control, 345–362
European rail traffic management system, 350–352, 352*f*
fixed block systems, 347–350
infrastructure, 346–347
management system, 354–359
moving block system, 353–354
supportive facilities and equipment, 347–359
workload and capacity of train dispatcher, 359–362
urban transit systems
complementarity of systems, 397–398
light rail transit, 395
regional rail, 397
streetcars/tramways, 395, 396*f*
subway/metro, 396–397, 397*f*
Ramp metering, 333–336, 333*f*
ALINEA, 335, 335*f*
categories, 333–334
demand capacity strategy, 334*f*, 335
system wide traffic responsive, 336, 336*f*
Random variables, 96–105
Ratios, 130–132, 130*t*
efficiency of airport, calculation, 132*f*
efficient frontier, 131*f*
Red light, 295
Reference location, 271, 739, 817–818
Regional Logistics/Distribution Center (RLDC), 582–583
Regional rail, 397, 398*f*
Resource allocation problem, 85–86
Revenue Passenger Mile (RPM), 458
Revenues, 639–642
airlines passenger transport, 703–704, 703–704*f*
intermodal rail/road transport network, 713–714
ports, 686–688, 687*t*, 688*f*
rail passenger transport, 678–680, 678–679*f*
shipping lines, 690–695, 692–694*f*
urban mass transit systems, 669
Reverse logistics, 572–573, 574*f*
Ride sharing, 338–339, 338*f*
Right-of-way (ROW), 392
Rivers and man-built channels
practical capacity, 248–249
ultimate capacity, 245–247
Road-based urban transit systems
regular buses, 393, 393*f*
semi-rapid buses, 394, 395*f*
trolleybuses, 393–394, 394*f*
Road freight transport systems, 31–33
operating costs, 674–676, 674*f*
Road infrastructure cost, 653–654
Road network, 431, 432*f*
capacities, 332
urban, 293
Road traffic injuries, 58
Road transportation
capacity—access modes, 261–262
congestion, 733–736
energy/fuel consumption, 748–760
GHG emissions, 748–760
infrastructure of, 16
interurban, 30–33
service networks, 453–455
ultimate and practical capacity, 453
land use, 760–762
network modes, 69
noise, 737–744
operating costs, 671–674
operating modes, 16–17
traffic accidents/incidents, 744–748
Rolling stock/vehicles, 250
Rolling terrain, 207
Route and network, 251–252
Route choice model
system optimum, 540–541, 541–542*b*
transportation demand, 510, 511*f*, 521–523
Route deviation system, 448–449, 449*f*
Route-first cluster-second methods, 617, 618*f*
Runway system
arriving and departing flight paths, 272*f*
at Frankfurt Main airport, 277*f*
time-space diagram, 273*f*
ultimate capacity, 271

S

Santiago de Compostela rail disaster, 777
Saturation flow, 295
S-curve power law, 754
Sea lane network, 45*f*
Seaports
carbon footprinting for, 804
energy/fuel consumption, 799–800
GHG emissions, 800–804, 802–803*f*
land use, 813*f*
noise, 792
safety, 793–794, 793–794*f*

Sea shipping transport systems
  average ship size deployed per country, 50–51, 51*f*
  for cargo shipments, 48–51
  container transport development, 49*f*
  infrastructure of, 16
  largest container shipping lines, 50*t*
  operating modes, 16–17
  volumes of traded containers, 50*t*
Seaside capacity, 262–263
SEEMP. *See* Ship Energy Efficiency Management Plan (SEEMP)
Self-driving cars, 329–330
Semiactuated signal, 296
SemiAIM, 332
Semi-rapid buses, 394, 395*f*
Service frequency
  maximum, 412–413
  public transportation
    bus headways, 411, 412*f*
    determination, 438–440
    maximum load method, 415–416
    maximum service frequency, 412–413
    passenger waiting time, 413
    public transit line, 411, 411*f*
    square root formula, 413–415
    and vehicle departure times, 410–411
  transportation networks, 251–252
Service level
  airports, 279–282
  air traffic control, 286–287
  freeways, 201–205, 204*f*, 205*t*
  highway capacity and, 198–211, 199*f*
  by linguistic expressions, 203*f*
  maximum service flow rates, 210*t*
  methodology, 205–208, 205*f*
  number of lanes requirement, 209–211
  productivity, 235
  rating scale for, 203*t*
  road freight terminals, 575–577, 575*f*
  road truck roads, 574
  shipping lines, 265–266, 268
  size of rolling stock, 235
  speed-flow curves, 204*t*
  and traffic demand variations, 199–200
  transportation demand/supply, 497
  at U.S. railways, 222*t*
  water side area, 240–243
Service networks
  air transportation, 53–56
  freight transportation, 577–581
  inland waterways
    rolling stock/vehicles, 250
    route and network, 251–252
  in interurban road transportation, 453–455
  UFT system, 592
Set covering problem, 598–603
SFC. *See* Specific Fuel Consumption (SFC)
Shanghai Maglev Train, 41
Shinkansen trains, 3, 13
Ship Energy Efficiency Management Plan (SEEMP), 809–810
Shipping lines
  capacity, 263–265, 267–268
  economic measures, 810
  energy/fuel consumption, 805–806, 805*f*
  future technologies, 810–811
  GHG emissions, 805–806, 806*f*
  mitigating measures, 807–810
  network, 267–268
  noise, 792
  operational costs, 688–690, 689*t*, 689*f*
  operational measures, 809
  revenues, 690–695, 692–694*f*
  route, 263–266
  technical measures, 808
  traffic accidents/incidents, 794–796, 795–797*f*
Shock waves, 182–185, 182*f*
Shortest paths
  in probabilistic network, 72
  problem, 72
Signalized intersections
  change interval calculation, 311–312
  critical lane volumes, 310
  cumulative vehicle arrivals/departures, 300*f*
  fixed-time control strategies, 296–300, 306
  pedestrian crossing time check, 314–315
  signal phasing selection, 307–308
  traffic control at, 294–316
  vehicle delays at, 300–305
Signal timing, 294–295
Simple additive weighting (SAW) method, 127–128
Simple greedy algorithm, 434–437, 434*f*
Simulation
  Monte Carlo simulation method, 121–123
  statistical experiments, 121
Single hub-and-spoke network, 465–469, 466*f*
Single-track line
  capacity, 215–219, 216*f*
  practical capacity, 220–221, 221*f*
  ultimate capacity, 215–219, 215*f*
Slot auction, 345
Societal risk, 723
Space-time networks, 464, 464–465*f*
Spatial network, 464, 464*f*

Specific Fuel Consumption (SFC), 41, 830
  aircraft type/engine, 835*t*
  inland waterways, 797
  jet engines, 831*f*
  shipping lines, 807, 807*f*
Speed-density relationship, 171–172, 172–173*f*
Speed-flow relationship, 175–181, 176–177*f*
Speeds, 169–170, 169*f*
Split, cycle, offset optimization technique (SCOOT), 324
Square root formula, 413–415
Station controller (STC), 359
Steamboats, 10
Stochastic user equilibrium, 526
Streetcars, 395
  congestion, 762
  infrastructure cost, 649–650, 649*t*
  land use, 789
  noise from, 763–764, 764–765*f*
  operating costs, 666, 667*f*
  safety, 770–771, 771*f*
  urban and sub/urban system, 24
Subway/metro systems, 396–397
  congestion, 763
  infrastructure cost, 651–652, 651*f*, 652*t*
  land use, 790
  noise, 766–767, 767*f*
  operating costs, 669–671
  safety, 772–774, 773–775*f*
  urban and sub/urban, 28
Suez Canal, 11
Sulfur oxides, 727
Supply chains, urban freight/goods distribution, 582–584, 583*f*
Sustainable transportation, 719–720
Sweep algorithm, 624–629, 624*f*, 625–629*b*, 627*f*, 629*t*, 629*f*
System optimum
  route choice, 540–541, 541–542*b*
  user equilibrium, 523–533, 531–533*b*

T

Take-off capacity, 274
Take-Off Weight (TOW), 821, 822*f*
Taxi system, 390
  taxi-in delay, 817
  taxi-out delay, 817
Taxiways, ultimate capacity, 269–270, 277
TDM. *See* Transportation demand management (TDM)
Technique for Order Preference by Similarity to Ideal Solution (TOPSIS), 128–130
Telework, 339
Terrain, 207
Time series forecasting, 501–503, 501–504*b*
Time-space diagrams, 63–68, 64*f*
  movement of cars, 66*f*
  traffic signal coordination, 67*f*
TOW. *See* Take-Off Weight (TOW)
Toyota, 569
Toyota Prius, 14, 14*f*
Traffic
  congestion, 2, 145
  demand variations and highway capacity, 199–200
  engineers and planners, 2
  linear programming in, 82–87
  mathematical programming applications in, 81–93
  probability theory and, 93–105
  transportation and, 2–3
Traffic accidents/incidents, 723–724
  buses, 745–747
  cars, 744–745
  freight trains, 778–780, 778–779*f*
  light rail transit, 771–772, 772–773*f*
  passenger inter-urban trains, 774–777
  seaports, 793–794, 793–794*f*
  shipping lines, 794–796, 795–797*f*
  streetcar, 770–771, 771*f*
  subway/metro system, 772–774, 773–775*f*
  transportation systems, 845–846
    air, 826–829
    rail, 770–780
    road, 744–748
    water, 793–796
  trucks, 747–748
Traffic assignment
  dynamic, 546
  scheme, 523*f*
  technique, 524–525
  transportation demand, 510, 521–523
Traffic control, 5
  actuated signal control, 315–316
  air, 294, 362–378
  area-wide control systems, 320–324
  for arterial streets, 317–319, 317–318*f*
  auctions, 345
  change interval calculation, 311–312
  critical lane volumes selection, 310
  fixed-time strategies, 296–300, 306
  freeway, 332–337
  highway space inventory control system, 344–345
  HOV facilities, 344
  intelligent transportation systems, 325–332
  pedestrian crossing time check, 314–315
  rail, 294, 345–362
  at signalized intersections, 294–316

signal warrants, 324–325
strategies, 295
study location, 324–325
techniques, 293
transportation demand management, 337–343
Traffic engineers, 295, 310, 315–316
delay formulas, 305–306
and planners, 293
Traffic flow theory, 163–164, 721
car following models, 187–189
changes over hours during day, 496*f*
exponential distribution, 166–168
flow-density relationship, 173–175
flow measurement, 163, 165*f*
fundamental diagram, 182, 182*f*
loop inductance, 164, 164*f*
macroscopic approach, 163–164
micro-simulation traffic models, 186–187
network flow diagram, 189–193
normal distribution, 168–170
Pearson Type III distribution, 168–170
poisson distribution, 166–168
shock waves, 182–185
speed-density relationship, 171–172
speed estimation, 164, 164*f*
speed-flow relationship, 175–181
speeds, 169–170
variables measurement, 164–165
vehicle headways and flow, 165–166
Tragedy of commons, 339
Train dispatcher(s)
monitoring capacity, 361*f*
quantification of, 359–360
workload and capacity of, 359–362, 361*f*
Trajectory-based measurements, 189, 192–193, 192*f*
Tramways, 395, 396*f*
congestion, 762
infrastructure cost, 649–650, 649*t*
land use, 789
noise, 763–764, 764–765*f*
operating costs, 666, 667*f*
safety, 770–771, 771*f*
traffic accidents/incidents, 770–771, 771*f*
urban and sub/urban system, 24
Transit Management Subsystem, 328
Transit signal priority, 328
Transit vehicle tracking, 328
Transmission vole machine (TVM) system, 353–354, 354*f*
Transport and Road Research Laboratory (TRRL), 320
Transportation demand management (TDM), 337–343
congestion charges, 341–343
congestion pricing, 339–341
improved walkability, 339
park and ride, 339
ride sharing, 338–339, 338*f*
telework, 339
Transportation modes, 15
air transport system, 51–56
bus system, 22–23
components of, 16–17
freight shipments, urban and sub/urban, 29–30, 29*f*, 30*t*
inland waterways, 44–51
interurban rail transport systems, 34–44
interurban road transport systems, 30–33
LRT system, 25–27, 26–27*f*, 27*t*
relationships between, 20–22
sea shipping systems, 44–51
streetcar/tramway system, 24
structure of, 17, 18*f*
subway/metro systems, 28
technologies
guidance, 19
propulsion, 19–20
support, 17–18
Transportation networks
degree of node, 69–71
Dijkstra's algorithm, 73–76
graphs, 69
indegree of node, 69–71
maritime transport mode, 69
mixed network, 69, 70*f*
nodes, 69
nonoriented network, 69–71, 70–71*f*
oriented network, 69–71, 70–71*f*
outdegree of node, 69–71
path, 69–71
queueing in, 106–121
rail transport mode, 69
road transport mode, 69
service frequency, 251–252
service networks, 53–56, 53*f*
shortest path
between all pairs of nodes, 77–80
problem, 72
strongly connected oriented network, 69–71, 72*f*
Transportation systems, 719–720
air-based, 816–840
analysis and design, 2
congestion, 56–59
control, 2, 56–59
costs of impacts, 840–851
elements of, 2
environment protection, 56–59
external costs, 637

Transportation systems *(Continued)*
history of, 6–15
intelligent, 325–332
internal costs, 637
issues, 56–59
linear programming in, 82–87
mathematical programming applications, 81–93
necessity for, 2
planning, 56–59
rail (*see* Rail transportation systems)
road-based, 733–762
safety, 56–59
science, 3
sector, 15, 15*f*, 637
social and environmental impacts, 720*t*
and traffic systems, 2–3
water-based, 791–816
Transport economics. *See* Economics of transportation
Transport work
freight train, 234–235
route and network, 252
TransRapid MAGLEV (TRM) system
components, 40*f*
control systems, 40
cross-section profile, 39*f*
developing, 38*t*
empty weight of, 40
operational characteristics, 41*t*
Trans-Siberian Railway, 236*b*, 236*f*
TRANSYT model, 320
Travel demand modeling, 495, 635–636
activity-based models, 559–563, 561–562*f*
Braess' paradox, 543–546
components of, 500*f*
computational intelligence techniques, 557–558
on discrete choice models, 547–549
dynamic traffic assignment, 546
forecasting techniques, 499–509, 500*f*
four-step planning procedure, 509–523
gravity model, 514–517
logit model, 549–556, 550*f*
macroscopic/microscopic models, 499
modal split analysis, 509, 510*f*, 518–521
modeling, 497–499
price of anarchy, 542–543
route choice, 510, 511*f*, 521–523
and supply, 496–497
system optimal route choice, 540–541, 541–542*b*
traffic assignment, 521–523
transportation capacity expansions, 543–546
trend projection, 504–509, 505*f*, 505*t*, 507*f*
trip distribution, 510*f*, 513–514
trip generation, 509*f*, 511–512
user equilibrium
capacity restraint algorithm, 534–535, 534–535*b*, 535*t*
FHWA algorithm, 536, 536–537*b*, 536–537*t*
incremental assignment algorithm, 537–539, 538–540*b*, 538–540*t*
problem formulation, 527–533, 527*f*, 528*t*
stochastic model, 526
and system optimum, 523–533, 531–533*b*
weighted moving average method, 502–503
Traveling Salesman Problem (TSP), 614–616, 615–616*b*
Trip-based model, 559–560
Trip distribution, 510*f*, 513–514
Trip generation, 509*f*, 511–512
Trip interchange matrix, 498, 498*t*, 515, 515*t*
Trolleybuses, 393–394, 394*f*
Trucks, 573–574
capacity, 585
congestion, 736
energy/fuel consumption, 758–760
GHG emissions, 758–760
land use, 762
level of service, 574
noise from, 743–744
passenger-car equivalents, 207
peak and non-peak speeds, 737*f*
practical capacity, 574
routing, 586, 587*f*
service level of, 574
traffic accidents/incidents, 747–748
ultimate capacity, 574

U

UE. *See* User equilibrium (UE)
ULDC. *See* Urban Local Distribution Center (ULDC)
Ultimate capacity, 197, 212–215
airports, 270–279
air traffic control, 283–286
rail inter-urban transport systems
freight handling station, 224
freight terminals, 224–225
passenger stations, 222–223
rail shunting yard, 226–230
single-track line, 215–221
rail station, 222–230
rivers and man-built channels, 245–247
road freight terminals, 575–577, 575*f*
road truck roads, 574
single-track line, 215–219, 215*f*
water side area, 241–244

Underground Freight Transport (UFT) system, 588
capacity, 592
components of, 588
fleet size, 593–594
infrastructure network, 589
for pallets and containers, 589, 590–591*f*, 591*t*
pipe/tunnel productivity, 593
service network, 592
service quality, 594
stations/terminals of, 589
Underground pipes/tunnels. *See* Pipes/tunnels
Undersaturated traffic conditions, 295
United States Department of Transportation, 308
Urban and sub/urban transport systems
bus system, 22–23
for freight shipments, 29–30
LRT system, 25–27, 26–27*f*, 27*t*
streetcar/tramway system, 24
subway/metro systems, 28
Urban freight distribution, 582–595
advanced systems, 588–595
conventional systems, 584–588
Urban Local Distribution Center (ULDC), 583
vehicle/truck routing, 586
warehousing, 585
Urban MAGLEV, 42
Urban mass transit systems
infrastructure cost, 647–652, 648*f*, 649*t*, 650–651*f*, 651–652*t*
operating costs, 663–671, 664*f*
revenues, 669
Urban public transit
bicycle parking, 406*f*
dial-a-ride system, 390, 390*f*
horse-drawn omnibus, 391
infrastructure
Dubai RRT-metro infrastructure network, 404–405*b*
fixed-route system, 398–399, 399*f*
indicators, 402*t*
links and indicators, 401–404
stops/stations, 398–401, 400*f*, 402*f*
topology and relationship, 403–404, 403*t*
performances, 423–428
rail-based urban transit systems
complementarity of systems, 397–398
light rail transit, 395
regional rail, 397
streetcars/tramways, 395, 396*f*
subway/metro, 396–397, 397*f*
reliability of service, 426–427
road-based urban transit systems
regular buses, 393
semi-rapid buses, 394
trolleybuses, 393–394
service frequencies determination in, 438–440
simple greedy algorithm, 434–437
taxi system, 390
time-space diagram, 425*f*
topology and relationship, 403–404
transfer time, 426
travel speed, 426
types, 429–431
Urban rail-based transit systems
complementarity of systems, 397–398
light rail transit, 395
regional rail, 397
streetcars/tramways, 395, 396*f*
subway/metro, 396–397, 397*f*
User equilibrium (UE), 531*f*
capacity restraint algorithm, 534–535, 534–535*b*, 535*t*
FHWA algorithm, 536, 536–537*b*, 536–537*t*
incremental assignment algorithm, 537–539, 538–540*b*, 538–540*t*
problem formulation, 527–533, 527*f*, 528*t*
stochastic model, 526
and system optimum, 523–533, 531–533*b*
User services, ITS, 327–329

V

Vanpool, 338–339
Variable costs, 637–638, 638*f*
Variable message signs (VMS), 333, 337
Vehicle on-board controller (VOBC), 359
Vehicle routing problems (VRPs), 611
Clark-Wright's "savings" algorithm, 618–624, 619–622*b*, 622*t*
methods, 617, 618*f*
and scheduling problems, 611, 614
Sweep algorithm, 624–629, 624*f*, 625–627*b*, 627*f*, 629*t*, 629*f*
Traveling Salesman Problem, 614–616, 615–616*b*
types, 612–614
Vehicles, 2
delay at signalized intersections, 300–305
detector actuation, 316*f*
fleet (*see* Fleet vehicles)
headways and flow, 165–166
scheduling in public transit, 441–444, 442–443*f*
Vespucci, Amerigo, 8
Via Appia, 7
Vigiles, 7
VMS. *See* Variable message signs (VMS)
Volume, 199

Volume-to-capacity ratio, 299–300
Vostok, 13
VRPs. *See* Vehicle routing problems (VRPs)

W

Wardrop's first principle, 525–526
Warehousing, 585
Waste
  transportation systems, 733, 848
  of water-based transport systems, 814–816, 814–815*f*
Water-based transport systems, 791
  congestion, 791–792
  energy/fuel consumption, 797–811
  GHG emissions, 797–811
  land use, 812–813
  noise, 792
  traffic accidents/incidents, 793–796
  waste of, 814–816, 814–815*f*
Water side area, 240–243
Water vapor, 728
Webster's formula, 305, 305–306*b*, 312
Weighted moving average method, 502–503
Wheeled vehicles, 6
Worst case analysis, 93
Wright, Orville, 12, 13*f*
Wright, Wilbur, 12

Y

Yellow light, 295

CPI Antony Rowe
Eastbourne, UK
May 10, 2022